Human Anatomy and Physiology

fourth edition

John W. Hole, Jr.

wcb

Wm. C. Brown Publishers
Dubuque, Iowa

Book Team

Executive Editor
Edward G. Jaffe
Developmental Editor
Lynne M. Meyers
Designer
Anne Scheid
Production Editor
Vickie Putman Caughron
Production Editor
David Welsh
Photo Research Editor
Carol M. Smith
Permissions Editor
Mavis M. Oeth
Product Manager
Matt Shaughnessy

Cover: Dick Oden

The credits section for this book begins on
page 949, and is considered an extension
of the copyright page.

wcb

Wm. C. Brown Publishers, College Division

G. Franklin Lewis
Executive Vice-President, General Manager
E. F. Jogerst
Vice-President, Cost Analyst
George Wm. Bergquist
Editor-in-Chief
Beverly Kolz
Director of Production
Chris C. Guzzardo
Vice-President, Director of Marketing
Bob McLaughlin
National Sales Manager
Craig S. Marty
Manager, Marketing Research
Marilyn A. Phelps
Manager of Design
Colleen A. Yonda
Production Editorial Manager
Faye M. Schilling
Photo Research Manager

wcb group

Wm. C. Brown
Chairman of the Board
Mark C. Falb
President and Chief Executive Officer

For Shirley

Brief
Contents

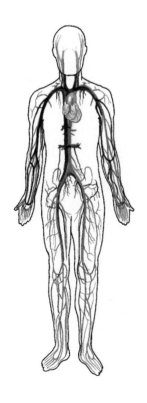

Unit 1
Levels of Organization 3

Unit 2
Support and Movement 177

Unit 3
Integration and Coordination 315

Unit 4
Processing and Transporting 487

Unit 5
The Human Life Cycle 803

Expanded Contents

Unit 1

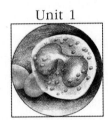

Levels of Organization 3

Unit 2

Support and Movement 177

Unit 3

Integration and Coordination 315

10 The Nervous System I
Basic Structure and Function 317

Processing and Transporting 487

Unit 5

The Human Life Cycle 803

23 Human Growth and Development 853

24 Human Genetics 885

Clinical, Practical, and Laboratory Applications

Plates

Charts

Preface

The fourth edition of **Human Anatomy and Physiology** is designed to provide accurate, current information about the structure and function of the human body in an interesting and readable manner. It is especially planned for students pursuing careers in allied health fields, who have minimal backgrounds in physical and biological sciences.

Organization

The text is organized in units, each containing several chapters. These chapters are arranged traditionally beginning with a discussion of the physical basis of life and proceeding through levels of increasing complexity.

Unit 1 introduces the human body and its major parts. It is concerned with the structure and function of cells and tissues; it introduces membranes as organs and the integumentary system as an organ system.

Unit 2 deals with the skeletal and muscular systems that support and protect body parts and make movements possible.

Unit 3 concerns the nervous and endocrine systems that integrate and coordinate body functions.

Unit 4 discusses the digestive, respiratory, circulatory, lymphatic, and urinary systems. These systems obtain nutrients and oxygen from the external environment; transport these substances internally; use them as energy sources, structural materials, and essential components in metabolic reactions; and excrete the resulting wastes.

Unit 5 describes the male and female reproductive systems, their functions in producing offspring, and the growth and development of this offspring. The final chapter concerns genetics and explains the determination of individual traits and their passage from parents to offspring.

Special Features of the Fourth Edition

The fourth edition of **Human Anatomy and Physiology** has been updated throughout and contains a number of new features, including the following:

1. The discussion of body organization and terms used to describe body parts have been incorporated into introductory chapter 1;

2. A new chapter dealing with joints of the skeletal system has been added;

3. The illustration program has been reviewed carefully and improved. Color has been added to many of the figures to emphasize particular parts, and several new full-color figures have been included;

4. Certain major topics have been revised or expanded. Among these are cellular respiration, membrane structure, neurotransmitters, autonomic nervous system, respiratory center, immune responses, allergic reactions, and menstrual cycle;

5. More than 40 new boxed asides have been written. Some of these deal with sports injuries and exercise physiology, while others present information of general or clinical interest;

6. Twelve longer asides—clinical, practical, and laboratory applications—have been added. These present such topics as ionizing radiation, PET and NMR imaging, skin cancer, osteoporosis, acquired immune deficiency syndrome, and in vitro fertilization.

7. Phonetic pronunciation of anatomic terms has been included within the narrative of chapters dealing with the skeletal, muscular, and circulatory systems.

Readability

Readability is an important asset of this text. The writing style is intentionally informal and easy to read. Technical vocabulary has been minimized, and summary paragraphs and review questions occur frequently within the narrative. Numerous illustrations, summary charts, and flow diagrams are carefully positioned near the discussions they complement.

Pedagogical Devices

The text includes an unusually large number of pedagogical devices intended to increase readability and to involve students in the learning process, to stimulate their interests in the subject matter, and to help them relate their classroom knowledge to their future clinical experiences. For an annotated listing of these devices, see Aids to the Reader, which follows this Preface.

Supplementary Materials

Supplementary materials designed to help the instructor plan class work and presentations and to aid students in their learning activities are also available:

Instructor's Resource Manual and Test Item File by John W. Hole, Jr., and Karen A. Koos, which contains chapter overviews, instructional techniques, suggested schedules, discussions of chapter elements, lists of related films, and directories of suppliers of audio-visual and laboratory materials. It also contains approximately fifty test items for each chapter of the text, designed to evaluate student understanding.

A Student Study Guide to Accompany Human Anatomy and Physiology by Nancy Corbett, St. Agnes Medical Center, Philadelphia, which contains chapter overviews, chapter objectives, focus questions, mastery tests, study activities, and answer keys corresponding to the chapters of the text.

Transparencies include a set of 90 acetate transparencies designed to complement classroom lectures or to be used for short quizzes.

Color Slides include a set of 72 light micrographs of tissues, organs, and other body features described in the textbook to complement classroom instruction.

Laboratory Manual to Accompany Human Anatomy and Physiology by John W. Hole, Jr. is designed specifically to accompany the fourth edition of *Human Anatomy and Physiology*.

Aids
to the Reader

This textbook includes a variety of aids to the reader that should make your study of human anatomy and physiology more effective and enjoyable. These aids are included to help you master the basic concepts of human anatomy and physiology that are needed before progressing to more difficult material.

Unit Introductions

Each unit opens with a brief description of the general content of the unit, and a list of chapters included within the unit (see page 3 for an example). This introduction provides an overview of chapters that make up a unit and tells how the unit relates to the other aspects of human anatomy and physiology.

Chapter Introductions

Each chapter introduction previews the chapter's contents and relates that chapter to the others within the unit (see page 5).

After reading an introduction, browse through the chapter, paying particular attention to topic headings and illustrations so that you get a feeling for the kinds of ideas that are included in the chapter.

Chapter Outlines

The chapter outline includes all the major topic headings and subheadings within the body of the chapter (see page 6). It provides an overview of the chapter's contents and helps you locate sections dealing with particular topics.

Chapter Objectives

Before you begin to study a chapter, carefully read the chapter objectives (see page 6). These indicate what you should be able to do after mastering the information within the narrative. The review activities at the end of each chapter (see page 31) are phrased like detailed objectives, and it is helpful to read them also before beginning your study. Both sets of objectives are guides that indicate important sections of the narrative.

Key Terms

The list of terms and their phonetic pronunciations, given at the beginning of each chapter, helps build your science vocabulary. The words included in these lists are used within the chapter and are likely to be found in subsequent chapters as well (see page 7). There is an explanation of phonetic pronunciation on page 931 of the Glossary.

Aids to Understanding Words

Aids to understanding words, at the beginning of each chapter, also helps build your vocabulary. This section includes a list of word roots, stems, prefixes, and suffixes that help you discover word meanings. Each root and an example word, which uses that root, are defined (see page 7).

Knowing the roots from these lists will help you discover and remember scientific word meanings.

Review Questions within the Narrative

Review questions occur at the ends of major sections within each chapter (see page 8). When you reach such questions, try to answer them. If you succeed, then you probably understand the previous discussion and are ready to proceed. If you have difficulty, reread that section before proceeding.

Illustrations and Charts

Numerous illustrations and charts occur in each chapter and are placed near their related textual discussion. They are designed to help you visualize structures and processes, to clarify complex ideas, to summarize sections of the narrative, or to present pertinent data.

As mentioned earlier, it is a good idea to skim through the chapter before you begin reading it, paying particular attention to these figures. Then, as you read for detail, carefully look at each figure and use it to gain a better understanding of the material presented.

Sometimes the figure legends contain questions that will help you apply your knowledge to the object or process the figure illustrates. The ability to apply information to new situations is of prime importance. These questions concerning the illustrations will provide practice in this skill.

There are also two sets of special reference figures that you may want to refer to from time to time. One set (see page 33) is designed to illustrate the structure and location of the major internal organs of the body. The other set (see page 309) will help you locate major features on the body surface.

Boxed Information

Short paragraphs set off in boxes of colored type occur throughout each chapter (see page 10). These boxed asides often contain information that will help you apply the ideas presented in the narrative to clinical situations. Some of them contain information about changes that occur in the body's structure and function as a person passes through the various phases of the human life cycle. These will help you understand how certain body conditions change as a person grows older.

Clinical, Practical, and Laboratory Applications

Other, longer asides are entitled Clinical, Practical, or Laboratory Applications. They discuss pathological disorders, laboratory techniques, and pertinent information of more general interest (see page xviii for a list of these topics).

Clinical Terms

At the ends of certain chapters are lists of related terms and phonetic pronunciations that are sometimes used in clinical situations (see page 27). Although these lists and the word definitions are often brief, they will be a useful addition to your understanding of medical terminology.

Chapter Summaries

A summary in outline form at the end of each chapter will help you review the major ideas presented in the narrative (see page 28). Scan this section a few days after you have read the chapter. If you find portions that seem unfamiliar, reread the related sections of the narrative.

Application of Knowledge

These questions at the end of each chapter (see page 31) will help you gain experience in applying information to a few clinical situations. Discuss your answers with other students or with an instructor.

Review Activities

The review activities at the end of each chapter (see page 31) will check your understanding of the major ideas presented in the narrative. After studying the chapter, read the review activities; if you can perform the tasks suggested, you have accomplished the goals of the chapter. If not, reread the sections of the narrative that need clarification.

Appendixes, Glossary, and Index

The Appendixes, following chapter 24, contain a variety of useful information. They include the following:

1. Table of International Atomic Weights and a Periodic Table of Elements (see page 914).
2. Lists of various units of measurements and their equivalents together with a description of how to convert one unit into another (see page 916).
3. Lists of clinical laboratory tests commonly performed on human blood and urine. These lists include test names, normal adult values, and brief descriptions of their clinical significance (see page 918).

The Suggestions for Additional Reading (see page 922) will help you locate library materials that can extend your understanding of topics discussed within the chapters. If a particular idea interests you, check the list of readings for items related to it.

The Glossary defines the more important textual terms and provides their phonetic pronunciations (see page 931). It also contains an explanation of phonetic pronunciation on page 931.

The Index is complete and comprehensive.

Acknowledgments

Once again, I want to express my gratitude to the users of earlier editions of **Human Anatomy and Physiology,** who supplied many helpful suggestions for improving this textbook. The changes appearing in the fourth edition are largely responses to their input. I also want to acknowledge the valuable contributions of the reviewers for the fourth edition who provided detailed criticisms and ideas for improving the manuscript as it was being prepared. They include:

Thomas S. Kaufman
Montgomery College

Robert E. Nabors
Tarrant County Junior College

James W. Russell
Georgia Southwestern College

Louise Squitieri
Bronx Community College

Howard M. Fuld
Bronx Community College

Gerald R. Dotson
Front Range Community College

Robert D. Morden
University of Wisconsin-Superior

Ahmad Kamal
Olive-Harvey College

In addition, I wish to thank my friend and former colleague, Professor Karen A. Koos, Rio Hondo College, who read the entire manuscript and provided assistance and suggestions as it was put into final form.

Special appreciation is extended to the members of the editorial and production staffs of Wm. C. Brown Publishers for their help. I am especially grateful to Ed Jaffe, Executive Editor, who encouraged me along the way, and to Lynne Meyers, Senior Developmental Editor, who took care of the details involved with planning and developing this fourth edition. It has been a great pleasure to work with all of these very talented people.

Finally, I want to express my heartfelt thanks to my wife Shirley and our children—Michael, Karen, and Larry—and daughter-in-law Andrea, who provided the loving support needed to complete this project.

John W. Hole, Jr.

Human Anatomy and Physiology

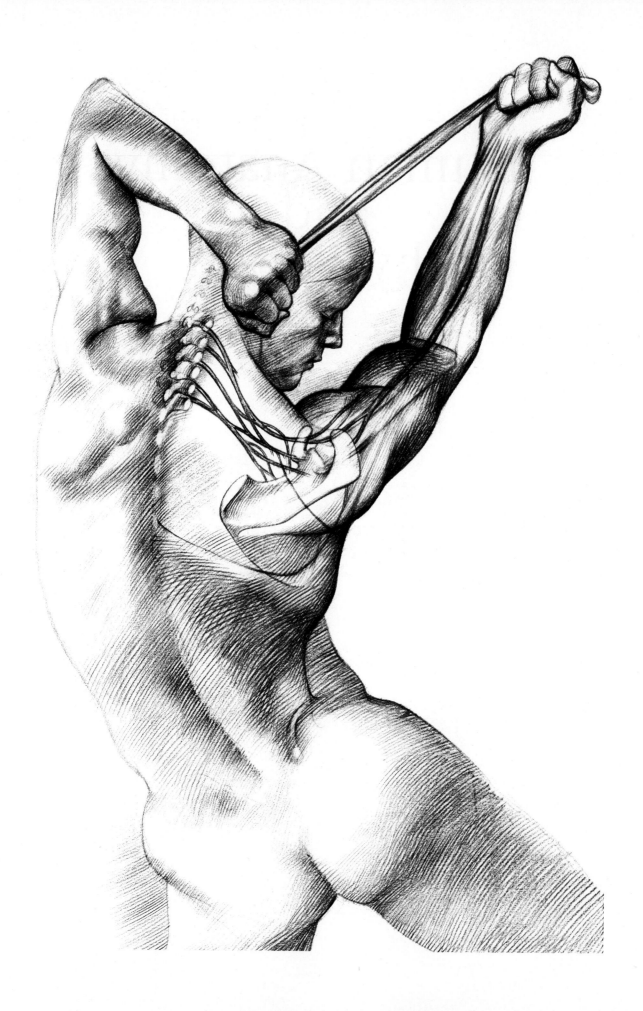

Unit 1

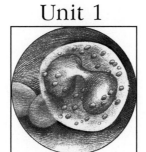

Levels of Organization

The chapters of unit 1 introduce the study of human anatomy and physiology. They are concerned with the ways the human is similar to other living things, the way the body is constructed of parts with increasing degrees of complexity, and the ways these parts interact with each other and with environmental factors to help ensure survival of the whole structure.

They describe each of the levels of organization within the body—chemical substances, cellular organelles, cells, tissues, organs, organ systems, and the human organism—and prepare the reader for the more detailed study of the organ systems presented in units 2–5.

1

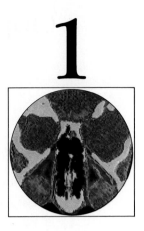

An Introduction to Human Anatomy and Physiology

Since humans represent a particular kind of living organism, they have traits that are shared by other organisms. For example, like all organisms, humans carry on life processes and demonstrate the characteristics of life. They also have needs that must be met if they are to survive, and their lives depend upon maintaining a stable internal environment.

A discussion of traits that humans have in common with other organisms and the ways their complex bodies are organized provides a beginning for the study of human anatomy and physiology.

Chapter 1 also introduces the organization of the human body and a special set of terms used to describe its major parts.

Chapter Outline

Chapter Objectives

After you have studied this chapter, you should be able to

1. Define *anatomy* and *physiology* and explain how they are related.

2. List and describe the major characteristics of life.

3. List and describe the major needs of organisms.

4. Define *homeostasis* and explain its importance to survival.

5. Describe a homeostatic mechanism.

6. Explain what is meant by levels of organization.

7. Describe the location of the major body cavities.

8. List the organs located in each of the body cavities.

9. Name the membranes associated with the thoracic and abdominopelvic cavities.

10. Name the major organ systems of the body and list the organs associated with each.

11. Describe the general functions of each organ system.

12. Properly use the terms that describe relative positions, body sections, and body regions.

13. Complete the review activities at the end of this chapter. Note that the items are worded in the form of specific learning objectives. You may want to refer to them before reading the chapter.

Key Terms

absorption   (ab-sorp′shun)

appendicular   (ap″en-dik′u-lar)

assimilation   (ah-sim″ĭ-la′shun)

axial   (ak′se-al)

circulation   (ser-ku-la′shun)

digestion   (di-jest′yun)

excretion   (ek-skre′shun)

homeostasis   (ho″me-ō-sta′sis)

metabolism   (mĕ-tab′o-lizm)

negative feedback   (neg′ah-tiv fēd′bak)

organism   (or′gah-nizm)

parietal   (pah-ri′ĕ-tal)

pericardial   (per″ĭ-kar′de-al)

peritoneal   (per″ĭ-to-ne′al)

pleural   (ploo′ral)

reproduction   (re″pro-duk′shun)

respiration   (res″pĭ-ra′shun)

thoracic   (tho-ras′ik)

visceral   (vis′er-al)

The accent marks used in the pronunciation guides are derived from a simplified system of phonetics standard in medical usage. The single accent (′) denotes the major stress. Emphasis is placed on the most heavily pronounced syllable in the word. The double accent (″) indicates secondary stress. A syllable marked with a double accent receives less emphasis than the syllable that carries the main stress, but more emphasis than neighboring unstressed syllables.

Aids to Understanding Words

append-, to hang something: *append*icular—pertaining to the arms and legs.

cardi-, heart: peri*cardi*um—membrane that surrounds the heart.

cran-, helmet: *cran*ial—pertaining to the portion of the skull that surrounds the brain.

dors-, back: *dors*al—a position toward the back of the body.

homeo-, the same: *homeo*stasis—the maintenance of a stable internal environment.

-logy, the study of: physio*logy*—the study of body functions.

meta-, a change: *meta*bolism—the chemical changes that occur within the body.

nas-, nose: *nas*al—pertaining to the nose.

orb-, circle: *orb*ital—pertaining to the portion of skull that encircles the eye.

pariet-, wall: *pariet*al membrane—membrane that lines the wall of a cavity.

pelv-, basin: *pelv*ic cavity—basin-shaped cavity enclosed by the pelvic bones.

peri-, around: *peri*cardial membrane—membrane that surrounds the heart.

pleur-, rib: *pleur*al membrane—membrane that encloses the lungs within the rib cage.

-stasis, standing still: homeo*stasis*—the maintenance of a stable internal environment.

-tomy, cutting: ana*tomy*—the study of structure, which often involves cutting or removing body parts.

THE STUDY OF the human body has a long and interesting history. Its beginnings stretch back to our earliest ancestors, for they must have been curious about their body parts and functions as we are today. At first their interests most likely concerned injuries and illnesses, because healthy bodies demand little attention from their owners. Even primitive people suffered from occasional aches and pains, injured themselves, bled, broke bones, and developed diseases. When they were sick or felt pain, these people probably sought relief by visiting shamans. The treatment they received, however, may have been less than satisfactory because primitive doctors relied heavily on superstitions and notions about magic and religion to help their patients. Still, these early medical workers began to discover useful ways of examining and treating the sick. They observed the effects of injuries, noticed how wounds healed, and attempted to determine causes of deaths by examining dead bodies. They also began to learn how certain herbs and potions affected body functions and how some could be used to treat coughs, headaches, and other common problems.

However, in ancient times it was generally believed that most natural processes were controlled by spirits and supernatural forces that humans could not understand. About 2500 years ago, attitudes began to change. The belief that natural processes were caused by forces that humans could understand grew in popularity.

This new idea stimulated people to look more closely at the world around them. They began to ask questions and seek answers. In this way the stage was set for the development of modern science. As techniques were developed for making accurate observations and performing careful experiments, knowledge of the human body expanded rapidly.

At the same time, many new terms were devised to name body parts, to describe their locations, and to explain their functions. These early terms, most of which originated from Greek and Latin, formed the basis for the language of anatomy and physiology. (See figure 1.1.)

There is a list of some modern medical and applied sciences on pages 27 and 28.

---

1. What factors probably stimulated early interest in the human body?
2. What idea helped people to begin studying the natural world?
3. What kinds of activities helped promote the development of modern science?

Figure 1.1

The study of the human body has a long history, as indicated by these illustrations (*a* and *b*) from the second book of *De Humani Corporis Fabrica* by Andreas Vesalius, issued in 1543.

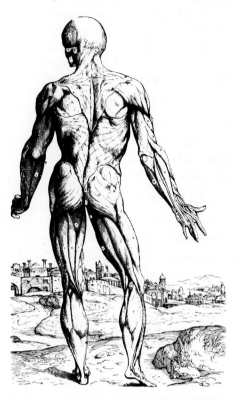

(a)

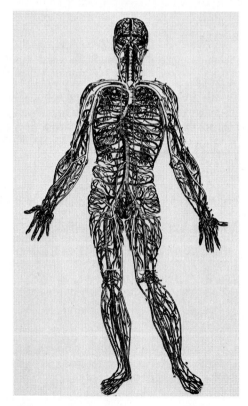

(b)

Anatomy and Physiology

Anatomy is the branch of science that deals with the structure of body parts—their forms and arrangements. Anatomists observe body parts grossly and microscopically and describe them as accurately and in as much detail as possible. **Physiology,** on the other hand, is concerned with the functions of body parts—what they do and how they do it. Physiologists are interested in finding out how such parts carry on life processes. In addition to using the same observational techniques as the anatomists, physiologists are likely to conduct experiments and make use of complex laboratory equipment.

Actually, it is difficult to separate the topics of anatomy and physiology because the structures of body parts are so closely associated with their functions. These parts are arranged to form a well-organized unit—the **human organism**—and each part plays a role in the operation of the unit as a whole. This role, which is the part's function, depends upon the way the part is constructed—that is, the way its subparts are organized. For example, the arrangement of parts in the human hand with its long, jointed fingers is related to the function of grasping objects. The tubular blood vessels are designed to transport blood to and from the hollow chambers of the heart; the heart's powerful muscular walls are structured to contract and cause the blood to move out of the chambers; and the valves associated with the vessels and chambers ensure that the blood will move in the correct direction. The shape of the mouth is related to the function of receiving food; the teeth are designed to break solids into smaller pieces; and the muscular tongue and cheeks are constructed to help mix food particles with saliva and prepare it for swallowing.

While reading about the human body, keep the connection between form and function of body parts in mind. Although the relationship between the anatomy and the physiology of a particular part is not always obvious, such a relationship is sure to exist.

---

1. What are the differences between an anatomist and a physiologist?
2. Why is it difficult to separate the topics of anatomy and physiology?
3. List several examples to illustrate the idea that the structure of a body part is closely related to its function.

The Characteristics of Life

Before beginning a more detailed study of anatomy and physiology, it is helpful to consider some of the traits humans share with other organisms. These traits, which are called *characteristics of life,* include the following:

1. **Movement.** Movement usually refers to a self-initiated change in an organism's position or to its traveling from one place to another. However, the term also applies to the motion of internal parts, such as the heart.

2. **Responsiveness.** Responsiveness (or irritability) is the ability of an organism to sense changes taking place inside or outside its body and to react to these changes. Drinking water to quench thirst is a response to loss of water in body tissues. Another example of responsiveness is pulling away from a hot stove.

3. **Growth.** Growth is an increase in body size, usually without any important change in shape. It occurs whenever an organism produces new body materials faster than the old ones are worn out or used up.

4. **Reproduction.** Reproduction is the process of making a new individual, as when parents produce an offspring. It also indicates the process by which microscopic cells produce others like themselves, as they do when body parts are repaired or replaced following an injury.

5. **Respiration.** Respiration is the process of obtaining oxygen, using oxygen in releasing energy from food substances, and transporting the resulting gaseous wastes (carbon dioxide) to the outside.

6. **Digestion.** Digestion is the process by which various food substances are chemically changed into simpler forms that can be absorbed and used by body parts.

7. **Absorption.** Absorption is the passage of the digestive products through the membranes that line the digestive organs and into the body fluids.

8. **Circulation.** Circulation is the movement of substances from place to place within the body by means of the body fluids.

9. **Assimilation.** Assimilation is the changing of absorbed substances into forms that are chemically different from those that entered the body fluids.

10. **Excretion.** Excretion is the removal of wastes that are produced by body parts as a result of their activities.

Chart 1.1 Characteristics of life			
Process	Examples	Process	Examples
Movement	Change in position of the body or a body part; motion of an internal organ.	Digestion	Breakdown of food substances into simpler forms.
Responsiveness	Reaction to a change taking place inside or outside the body.	Absorption	Passage of digestive products through membranes and into body fluids.
Growth	Increase in body size without change in shape.	Circulation	Movement of substances from place to place in body fluids.
Reproduction	Production of new organisms.	Assimilation	Changing of absorbed substances into chemically different forms.
Respiration	Obtaining oxygen, using oxygen in releasing energy from foods, and transporting carbon dioxide to the outside of the body.	Excretion	Removal of wastes produced by metabolic reactions.

Each of these characteristics of life—in fact, everything an organism does—depends upon physical and chemical changes that occur within body parts. Taken together, these physical and chemical changes or reactions are referred to as **metabolism.**

The characteristics shared by all living organisms are summarized in chart 1.1.

Vital signs are among the common observations made by physicians and nurses working with patients. They include measuring body temperature and blood pressure, and observing rates and types of pulse and breathing movements. There is a close relationship between these signs and the characteristics of life, since vital signs are the results of *metabolic activities.* In fact, death is recognized by the absence of such signs. More specifically, a person who has died will lack spontaneous muscular movements (including those of breathing muscles), responses to stimuli (even the most painful that can be ethically applied), reflexes (such as the knee-jerk reflex and pupillary reflexes of the eye), and brain waves (demonstrated by a flat encephalogram, which reflects a lack of metabolic activity in the brain).

---

1. What are the characteristics of life?
2. How are the characteristics of life related to metabolism?

The Maintenance of Life

With the exception of an organism's reproductive structures, which function to ensure that its particular form of life will continue into the future, the structures and functions of all body parts are directed toward achieving one goal—the maintenance of the life of the organism.

Needs of Organisms

Life is fragile, and it depends upon the presence of certain environmental factors for its existence. These factors include the following:

1. **Water.** Water is the most abundant chemical substance within the body. It is required for a variety of metabolic processes, and it provides the environment in which most of these reactions take place. Water also transports substances within organisms and is important for the regulation of body temperature.

2. **Food.** Food is a general term for substances that provide organisms with necessary chemicals in addition to water. Some of these chemicals are used as energy sources, others as raw materials for building new living matter, and still others enter into vital chemical reactions.

3. **Oxygen.** Oxygen is a gas that makes up about one-fifth of the air. It is used in the process of releasing energy from food substances. The energy, in turn, is needed to drive metabolic processes.

4. **Heat.** Heat is a form of energy. It is a by-product of metabolic reactions, and the rate at which these reactions occur is partly governed by the amount of heat present. Generally, the greater the amount of heat, the more rapidly the reactions take place. The measurement of heat intensity is called *temperature.*

5. **Pressure.** Pressure is a force that creates a pressing (or compressing) action. For example, the force acting on the outside of a land organism due to the weight of air is called *atmospheric pressure.* In humans, this pressure plays an important role in breathing. Similarly, organisms living under water are subjected to *hydrostatic pressure*—a pressure exerted by a liquid—due to the weight of water above them.

Chart 1.2 Needs of organisms

Factor	Characteristic	Use	Factor	Characteristic	Use
Water	A chemical substance	Needed for metabolic processes, as a medium for metabolic reactions, and to transport substances	Oxygen	A chemical substance	Needed to release energy from food substances
			Heat	A form of energy	Needed to help regulate the rates of metabolic reactions
Food	Various chemical substances	Needed to supply energy and raw materials for the production of living matter and for the regulation of vital reactions	Pressure	A force	Atmospheric pressure needed for breathing; hydrostatic pressure needed to help move blood

In complex organisms, such as humans, heart action creates blood pressure (another form of hydrostatic pressure), which forces blood through the blood vessels.

Although organisms need water, food, oxygen, heat, and pressure, the presence of these factors alone is not enough to ensure their survival. Both the quantities and the qualities of such factors are important. For example, the amount of water entering and leaving an organism must be regulated as must the concentration of oxygen in the body fluids. Similarly, survival depends on the quality as well as the quantity of food available—that is, it must supply the correct chemicals in adequate amounts. Chart 1.2 summarizes the major needs of organisms.

Homeostasis

As an organism moves from place to place, factors in its external environment change. However, if the organism is to survive, conditions within the fluids surrounding its body cells must remain relatively stable. In other words, body parts function efficiently only when the concentrations of water, food substances, and oxygen and the conditions of heat and pressure remain within certain limited ranges. Thus, it is not surprising that the metabolic activities of an organism are directed toward maintaining such steady conditions. This tendency to maintain a stable internal environment is called **homeostasis.**

To better understand the idea of maintaining a stable internal environment, imagine a room equipped with a furnace and an air conditioner. Suppose the room temperature is to remain near 20°C (68°F), so the thermostat is adjusted to a set point of 20°C. Because a thermostat is sensitive to temperature changes, it will signal the furnace to start whenever the room temperature drops below the set point. If the temperature rises

above the set point, the thermostat will cause the furnace to stop and the air conditioner to start. As a result, a relatively constant temperature will be maintained in the room (figure 1.2).

A *homeostatic mechanism* in the human body achieves similar results in regulating body temperature. The thermostat is a temperature-sensitive region of the brain. In healthy persons the set point of the brain's thermostat is at or near 37°C (98.6°F).

If a person is exposed to a cold environment and the body temperature begins to drop, the brain senses this change and triggers heat-generating and heat-conserving activities. For example, small groups of muscles may be stimulated to contract involuntarily, which causes shivering. Such muscular contractions produce heat as a by-product, which helps to warm the body. At the same time, blood vessels in the skin may be signaled to constrict so that less warm blood reaches the skin, and heat that might otherwise be lost is held in deeper tissues.

If a person is becoming overheated, the brain may trigger a series of changes that leads to increased loss of body heat. For example, it may stimulate the sweat glands of the skin to secrete watery perspiration and as the water evaporates from the surface, some heat is carried away and the skin is cooled. At the same time, it causes blood vessels in the skin to dilate. This allows blood that carries heat from deeper tissues to reach the surface where it loses some of this heat to the outside. Also, the heart rate is increased, causing a greater volume of blood to move into the surface vessels, and the breathing rate is increased, allowing more heat-carrying air to be expelled from the lungs.

Another homeostatic mechanism functions to regulate blood pressure in vessels leading away from the heart. In this instance, pressure-sensitive parts in the walls of these vessels are stimulated if the blood

Figure 1.2

A thermostat that can signal an air conditioner and a furnace to turn on or off helps to maintain a relatively stable room temperature. This system represents a homeostatic mechanism.

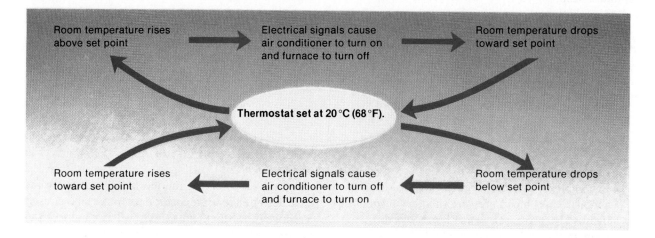

Figure 1.3

Homeostatic mechanism that regulates blood sugar concentration.

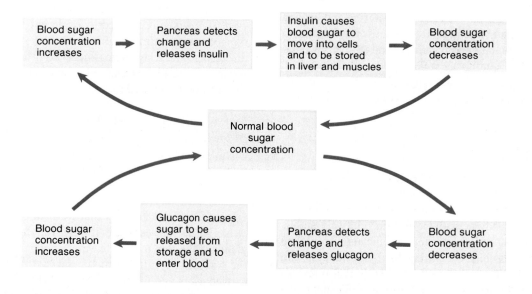

pressure increases above normal. When this happens, the sensors signal the brain, which then signals the heart to contract more slowly and with less force. Because of decreased heart action, less blood enters the blood vessels, and the pressure inside the vessels decreases. If the blood pressure is dropping below normal, the brain signals the heart to contract more rapidly and with greater force so that the pressure in the vessels increases.

Similarly, a homeostatic mechanism regulates the concentration of sugar (glucose) in the blood. If, for example, the amount of blood sugar increases following a meal, an organ called the pancreas detects this change and releases a chemical (insulin) into the blood.

This substance causes sugar to move from the blood into various body cells and to be stored in the liver and muscles. As this occurs, the concentration of blood sugar decreases, and when it reaches the normal set point, the pancreas ceases its release of insulin. (See figure 1.3.) If, on the other hand, the blood sugar concentration becomes abnormally low, the pancreas detects this change and releases a different chemical (glucagon) that causes sugar to be released from storage into the blood.

In these examples, homeostasis is maintained by a self-regulating control mechanism that can sense changes away from the normal set point and can cause reactions that tend to return conditions to normal. Since the changes away from the normal state stimulate changes to occur in the opposite direction, the responses are called *negative*, and the homeostatic control mechanism is said to act by a process of **negative feedback,** which is discussed in chapter 4.

Sometimes changes occur that stimulate still other similar changes. Such a process that causes movement away from the normal state is called a *positive feedback mechanism* (or vicious circle).

Although most feedback mechanisms in the body are negative, a positive system operates for a short time when a blood clot forms, because the chemicals present in a clot promote still more clotting. (See chapter 17.)

Since positive feedback mechanisms often produce unstable conditions they are most commonly associated with diseases and may lead to death.

Details about these and other homeostatic mechanisms are presented in later chapters. The maintenance of homeostasis is of primary importance to survival, and it is not surprising that metabolic activities are largely directed toward maintaining stable internal conditions.

---

1. What needs of organisms are provided from the external environment?
2. What is the relationship between the use of oxygen and the production of heat?
3. Why is homeostasis so important to survival?
4. Describe three homeostatic mechanisms.

Levels of Organization

Early investigators focused their attention on the larger body structures, for they were limited in their ability to observe small parts. Studies of small parts had to wait for the invention of magnifying lenses and microscopes, which came into use about 400 years ago. Once these tools were available, it was discovered that larger body structures were made up of smaller parts, which, in turn, were composed of still smaller parts.

Today, scientists recognize that all materials, including those that comprise the human body, are composed of chemical substances. These substances are made up of tiny, invisible particles called **atoms,** which are commonly bound together to form larger particles called **molecules;** small molecules may be combined in complex ways to form larger molecules (macromolecules).

Within the human organism the basic unit of structure and function is a microscopic part called a **cell.** Although individual cells vary in size, shape, and specialized functions, all have certain traits in common.

For instance, all cells contain tiny parts called **organelles** that carry on specific activities. These organelles are composed of aggregates of large molecules, including those of such substances as proteins, carbohydrates, lipids, and nucleic acids.

Cells are organized into layers or masses that have common functions. Such a group of cells constitutes a **tissue.** Groups of different tissues form **organs**—complex structures with special functions—while groups of organs are arranged into **organ systems.** Organ systems make up an **organism.** Thus, various body parts occupy different levels of organization. Furthermore, these parts vary in complexity from one level to the next. That is, atoms are less complex than molecules, molecules are less complex than organelles, organelles are less complex than cells, cells are less complex than tissues, tissues are less complex than organs, organs are less complex than organ systems, and organ systems are less complex than the whole organism (figure 1.4).

Chapters 2–6 discuss these levels of organization in more detail. Chapter 2, for example, describes the atomic and molecular levels; chapter 3 deals with organelles and cellular structures and functions; chapter 4 explores cellular metabolism; chapter 5 describes tissues; and chapter 6 presents membranes as examples of organs, and the skin and its accessory organs as an example of an organ system. Beginning with chapter 7, the structure and functions of each of the organ systems are described in detail.

---

1. How can the human body be used to illustrate the idea of levels of organization?
2. What is an organism?
3. How do body parts that occupy different levels of organization vary in complexity?

Organization of the Human Body

The human organism is a complex structure composed of many parts. The major features of the human body include various cavities, a set of membranes, and a group of organ systems.

Body Cavities

The human organism can be divided into an **axial portion,** which includes the head, neck, and trunk, and an **appendicular portion,** which includes the arms and legs. Within the axial portion of the body there are two major cavities—a **dorsal cavity** and a larger **ventral cavity.**

Figure 1.4

A human body is composed of parts within parts, which vary in complexity.

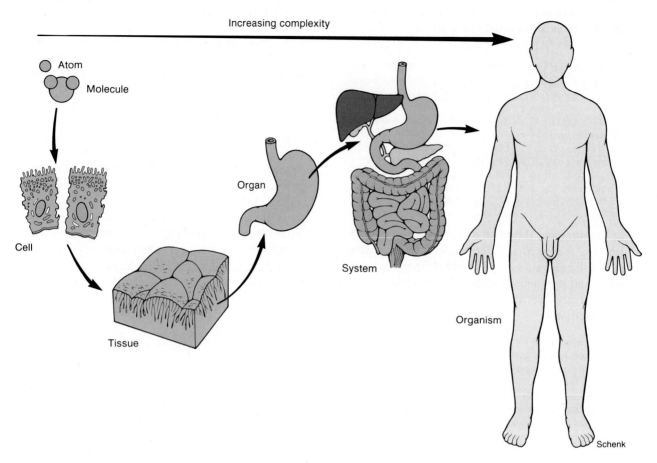

Increasing complexity

Atom

Molecule

Cell

Tissue

Organ

System

Organism

Schenk

These cavities are filled with organs; as a group these organs are called *visceral organs.* The dorsal cavity can be subdivided into two parts—the **cranial cavity,** which houses the brain; and the **spinal cavity,** which contains the spinal cord and is surrounded by sections of the backbone (vertebrae). The ventral cavity consists of a **thoracic cavity** and an **abdominopelvic cavity.** These major body cavities are shown in figure 1.5.

The thoracic cavity is separated from the lower abdominopelvic cavity by a broad, thin muscle called the **diaphragm.** When it is at rest, this muscle curves upward into the thorax like a dome. When it contracts during inhalation, it presses down on the abdominal organs. The wall of the thoracic cavity is composed of skin, skeletal muscles, and various bones. The viscera within it include the lungs and a region between the lungs called the *mediastinum.* The mediastinum separates the thorax into two compartments that contain the right and left lungs. The rest of the thoracic viscera—heart, esophagus, trachea, and thymus gland—are located within the mediastinum.

The abdominopelvic cavity, which includes an upper abdominal portion and a lower pelvic portion, extends from the diaphragm to the floor of the pelvis. Its wall consists primarily of skin, skeletal muscles, and bones. The visceral organs within the abdominal cavity include the stomach, liver, spleen, gallbladder, and most of the small and large intestines.

The **pelvic cavity** is the portion of the abdominopelvic cavity enclosed by the pelvic bones. It contains the terminal end of the large intestine, the rectum, the urinary bladder, and the internal reproductive organs.

In addition to the relatively large dorsal and ventral body cavities, there are several smaller cavities within the head. They include the following:

1. *Oral cavity,* containing the teeth and tongue.
2. *Nasal cavity,* located within the nose and divided into right and left portions by a nasal septum. Several air-filled *sinuses* are connected to the nasal cavity.
3. *Orbital cavities,* containing the eyes and associated skeletal muscles and nerves.
4. *Middle ear cavities,* containing the middle ear bones (figure 1.6).

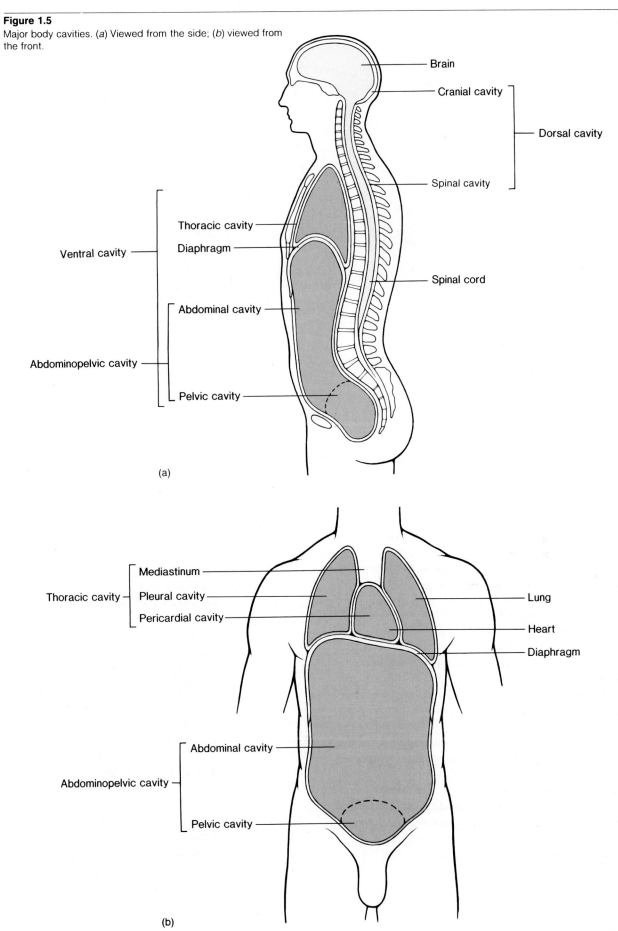

Figure 1.5
Major body cavities. (*a*) Viewed from the side; (*b*) viewed from the front.

Brain
Cranial cavity
Dorsal cavity
Spinal cavity
Spinal cord

Ventral cavity
Thoracic cavity
Diaphragm
Abdominal cavity
Abdominopelvic cavity
Pelvic cavity

(a)

Thoracic cavity
Mediastinum
Pleural cavity
Pericardial cavity
Lung
Heart
Diaphragm

Abdominal cavity
Abdominopelvic cavity
Pelvic cavity

(b)

Figure 1.6
The cavities within the head include the oral, nasal, orbital, and
middle ear cavities, and several sinuses.

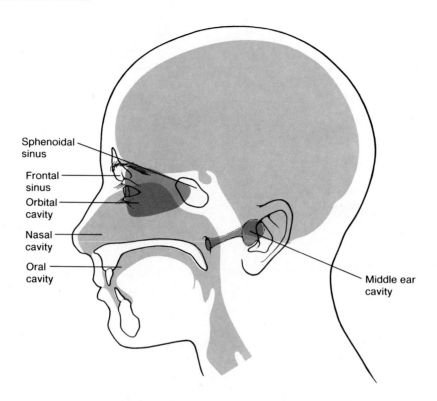

Sphenoidal
sinus

Frontal
sinus

Orbital
cavity

Nasal
cavity

Oral
cavity

Middle ear
cavity

Thoracic and Abdominopelvic Membranes

The walls of the right and left thoracic compartments,
which contain the lungs, are lined with a serous membrane called the *parietal pleura*. The lungs themselves
are covered by a similar membrane called the *visceral
pleura*.

    The parietal and visceral **pleural membranes** are
separated by a thin film of watery serous fluid that they
secrete. Although there is normally no actual space between these membranes, a possible or potential space
called the *pleural cavity* exists between them.

    The heart, which is located in the broadest por-
tion of the mediastinum, is surrounded by serous membranes called **pericardial membranes**. A thin *visceral
pericardium* (epicardium) covers the heart's surface
and is separated from the much thicker, fibrous *parietal pericardium* by a small amount of serous fluid. The
potential space between these membranes is called the
pericardial cavity. Figure 1.7 shows the membranes
associated with the heart and lungs.

    In the abdominopelvic cavity, the serous mem-
branes are called **peritoneal membranes**. A *parietal
peritoneum* lines the wall, and a *visceral peritoneum*

covers each organ in the abdominal cavity. The potential space between these membranes is called the *peritoneal cavity* (figure 1.8).

---

1. What is meant by visceral organ?
2. What organs occupy the dorsal cavity? The ventral
   cavity?
3. Name the cavities of the head.
4. Describe the location of the membranes associated
   with the thoracic and abdominopelvic cavities.

Organ Systems

The human organism is composed of several organ systems. Each of these systems includes a set of interrelated organs that work together to provide specialized
functions vital to the survival of the organism. As you
read about each organ system, you may want to consult
the illustrations of the human torso and locate some of
the features listed in the descriptions. (See pages 19,
21, and 22.)

Body Covering

The organs of the **integumentary system** (discussed in
chapter 6) include the skin and accessory organs such
as the hair, nails, sweat glands, and sebaceous glands.

Figure 1.7
(*a*) A transverse section through the thorax reveals the serous membranes associated with the heart and lungs. (*b*) What features can you identify in this cadaveric section through the thorax?

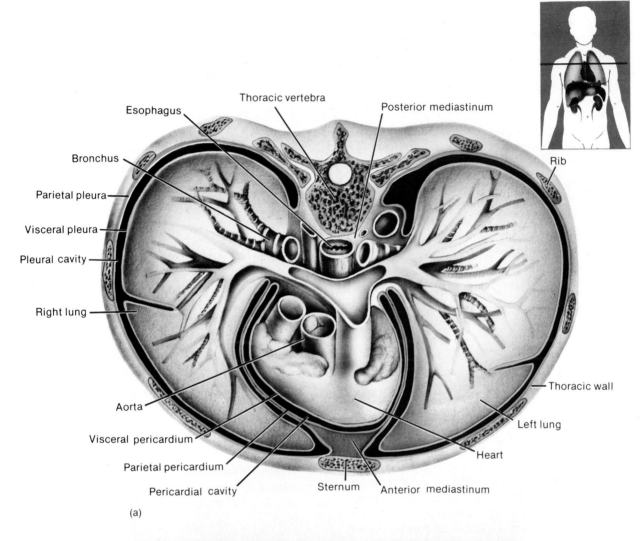

Esophagus

Thoracic vertebra

Posterior mediastinum

Bronchus

Rib

Parietal pleura

Visceral pleura

Pleural cavity

Right lung

Aorta

Visceral pericardium

Parietal pericardium

Pericardial cavity

Sternum

Anterior mediastinum

Heart

Left lung

Thoracic wall

(a)

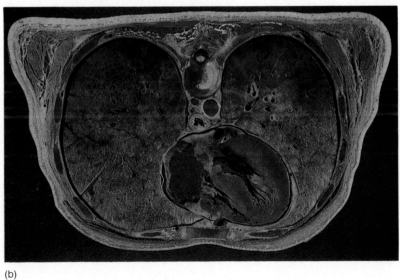

(b)

Figure 1.8
(*a*) A transverse section through the abdomen at the level of the stomach. How would you distinguish between the parietal and the visceral peritoneum? (*b*) What features can you identify in this cadaveric section through the abdomen?

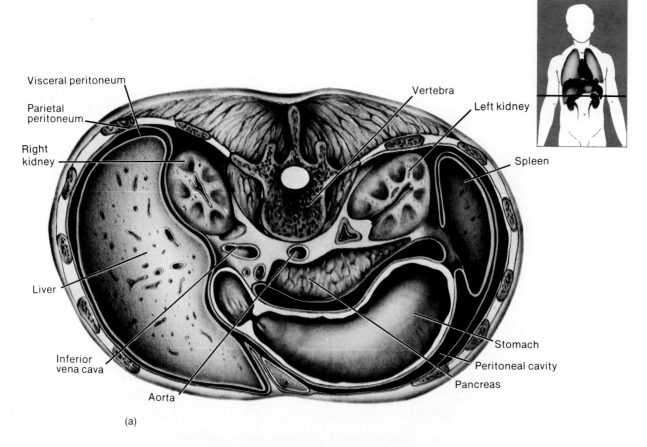

Visceral peritoneum

Parietal peritoneum

Right kidney

Liver

Inferior vena cava

Aorta

Vertebra

Left kidney

Spleen

Stomach

Peritoneal cavity

Pancreas

(a)

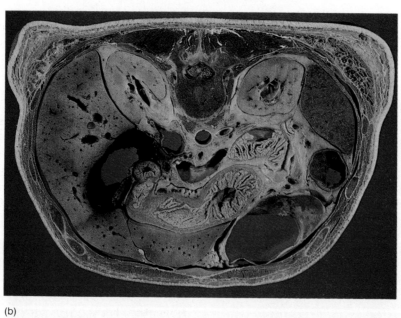

(b)

Figure 1.9
The integumentary system forms the body covering.

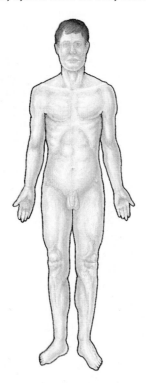

Integumentary system

Figure 1.10
Organ systems associated with support and movement.

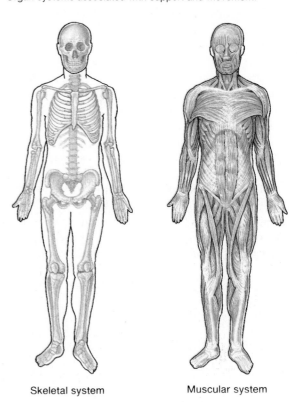

Skeletal system   Muscular system

Figure 1.11
Organ systems associated with integration and coordination.

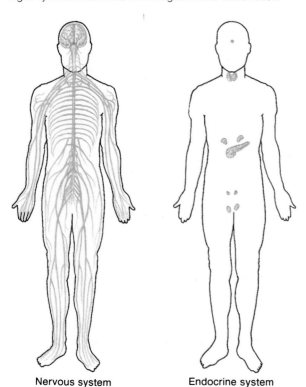

Nervous system   Endocrine system

These parts protect underlying tissues, help regulate the body temperature, house a variety of sensory receptors, and synthesize certain cellular products (figure 1.9).

Support and Movement
The organs of the skeletal and muscular systems (discussed in chapters 7, 8, and 9) function to support and move body parts (figure 1.10).

The **skeletal system** consists of the bones as well as the ligaments and cartilages that bind the bones together at joints. These parts provide frameworks and protective shields for softer tissues, serve as attachments for muscles, and act together with muscles when body parts move. Tissues within bones also produce blood cells and store inorganic salts.

The muscles are the organs of the **muscular system.** By contracting and pulling their ends closer together, they furnish the forces that cause body movements. They also function in maintaining posture and are largely responsible for the production of body heat.

Integration and Coordination
To act as a unit, body parts must be integrated and coordinated. That is, their activities must be controlled and adjusted, from time to time, so that homeostasis is maintained. This is the general function of the nervous and endocrine systems (figure 1.11).

The **nervous system** (discussed in chapters 10 and 11) consists of the brain, spinal cord, and associated nerves. Nerve cells within these organs use electrochemical signals called nerve impulses (action potentials) to communicate with one another and with other structures. Some nerve cells act as specialized sensory receptors that can detect changes occurring inside and outside the body. Others receive the impulses transmitted from these sensory units and interpret and act on the information received. Still others carry impulses from the brain or spinal cord to muscles or glands and stimulate these parts to contract or to secrete various products.

The **endocrine system** (discussed in chapter 13) includes all the glands that secrete **hormones.** The hormones, in turn, travel away from the glands in body fluids such as blood. Usually a particular hormone only affects a particular group of cells, which is called its target tissue. The effect of a hormone is to alter the function of the target tissue in some way and thus aid in controlling metabolism.

The organs of the endocrine system include the pituitary, thyroid, parathyroid, and adrenal glands, as well as the pancreas, ovaries, testes, pineal gland, and thymus gland.

Processing and Transporting

The organs of several systems are involved with processing and transporting nutrients, oxygen, and various wastes. The organs of the **digestive system** (discussed in chapter 14) receive foods from the outside. Then, they convert some of the food molecules into forms that can pass through cell membranes and, thus, be absorbed. Materials that fail to be absorbed are eliminated by being transported to the outside of the body. These organs also produce a variety of hormones, described in chapter 13.

The digestive system includes the mouth, tongue, teeth, salivary glands, pharynx, esophagus, stomach, liver, gallbladder, pancreas, small intestine, and large intestine.

The organs of the **respiratory system** (discussed in chapter 16) provide for the intake and output of air and for the exchange of gases between the blood and the air. More specifically, oxygen passes from air in the lungs into the blood, and carbon dioxide leaves the blood and enters the air. The nasal cavity, pharynx, larynx, trachea, bronchi, and lungs are parts of this system.

The **circulatory system** (discussed in chapters 17 and 18) includes the heart, arteries, veins, capillaries, and blood. The heart functions as a muscular pump that forces blood through the blood vessels. The blood serves as a fluid for transporting oxygen, nutrients, hormones, and wastes. It carries oxygen from the lungs and nutrients from the digestive organs to all body cells, where these substances are used in metabolic processes. It also transports hormones from various endocrine glands to their target tissues and wastes from body cells to the excretory organs, where they are removed from the blood and released to the outside.

The **lymphatic system** (discussed in chapter 19) is sometimes considered to be part of the circulatory system. It is composed of the lymphatic vessels, lymph fluid, lymph nodes, thymus gland, and spleen. This system transports some of the fluid from the spaces within tissues back to the bloodstream and carries certain lipids away from the digestive organs. It also aids in defending the body against infections by removing particles, such as microorganisms, from the tissue fluid and by supporting the activities of the cells (lymphocytes) responsible for the immune reactions against specific disease-causing agents.

The **urinary system** (discussed in chapter 20) consists of the kidneys, ureters, urinary bladder, and urethra. The kidneys remove various wastes from the blood, and assist in maintaining fluid and electrolyte balance. The product of these activities is urine. Other portions of the urinary system are concerned with storing urine and transporting it to the outside of the body.

Sometimes the urinary system is called the *excretory system*. However, excretion or waste removal is also a function of the respiratory, digestive, and integumentary systems.

The systems associated with transporting and processing are shown in figure 1.12.

Reproduction

Reproduction is the process of producing offspring. Cells reproduce when they undergo mitosis and give rise to daughter cells. The **reproductive system** of an organism, however, is concerned with the production of new organisms. (See chapter 22.)

The male reproductive system includes the scrotum, testes, epididymides, vasa deferentia, seminal vesicles, prostate gland, bulbourethral glands, penis, and urethra. These parts are concerned with the production and maintenance of the male sex cells, or sperm. They also function to transfer these cells into the female reproductive tract.

The female reproductive system consists of the ovaries, uterine tubes, uterus, vagina, and vulva. These organs produce and maintain the female sex cells, or

Figure 1.12
Organ systems associated with processing and transporting.

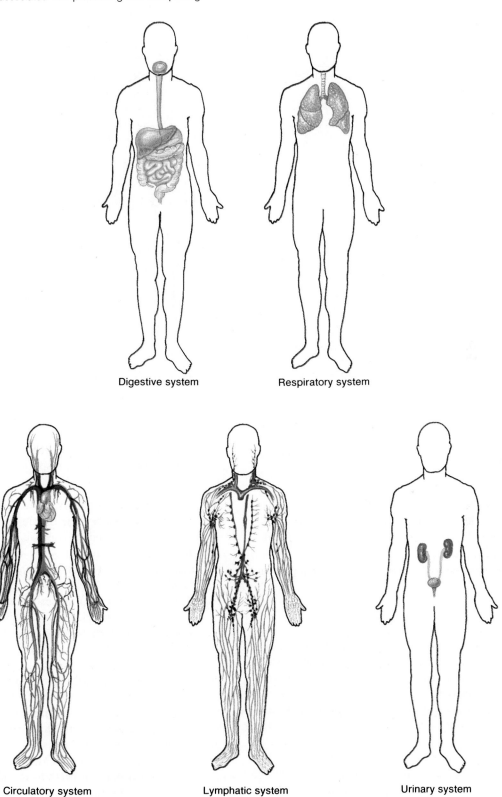

Digestive system

Respiratory system

Circulatory system

Lymphatic system

Urinary system

Figure 1.13
Organ systems associated with reproduction.

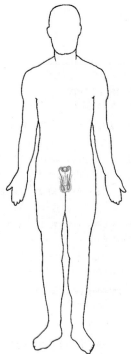

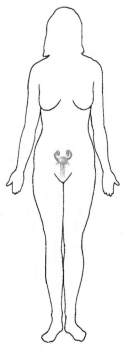

Male reproductive system   Female reproductive system

eggs, receive the male cells, and transport these cells within the female system. The female reproductive system also provides for the support and development of embryos and functions in the birth process.

Figure 1.13 illustrates female and male reproductive systems.

---

1. Name each of the major organ systems and list the organs of each system.
2. Describe the general functions of each organ system.

Anatomical Terminology

To communicate with each other more effectively, investigators over the ages have developed a set of terms with precise meanings. Some of these terms concern the relative positions of body parts, others relate to imaginary planes along which cuts may be made, and still others are used to describe various regions of the body.

When such terms are used, it is assumed that the body is in the **anatomical position;** that is, it is standing erect, the face is forward, and the arms are at the sides, with the palms forward.

Relative Position

Terms of relative position are used to describe the location of one body part with respect to another. They include the following:

1. **Superior** means a part is above another part, or closer to the head. (The thoracic cavity is superior to the abdominopelvic cavity.)
2. **Inferior** means situated below something, or toward the feet. (The neck is inferior to the head.)
3. **Anterior** (or *ventral*) means toward the front. (The eyes are anterior to the brain.)
4. **Posterior** (or *dorsal*) is the opposite of anterior; it means toward the back. (The pharynx is posterior to the oral cavity.)
5. **Medial** relates to an imaginary midline dividing the body into equal right and left halves. A part is medial if it is closer to this line than another part. (The nose is medial to the eyes.)
6. **Lateral** means toward the side with respect to the imaginary midline. (The ears are lateral to the eyes.) **Ipsilateral** pertains to the same side (the spleen and the descending colon are

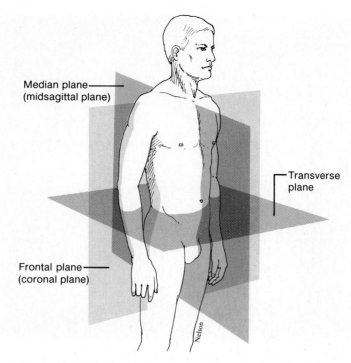

Median plane
(midsagittal plane)

Transverse plane

Frontal plane
(coronal plane)

Nelson

ipsilateral), while **contralateral** refers to the opposite side (the spleen and the gallbladder are contralateral).

7. **Proximal** is used to describe a part that is closer to a point of attachment or to the trunk of the body than something else. (The elbow is proximal to the wrist.)

8. **Distal** is the opposite of proximal. It means something is further from the point of attachment or the trunk than something else. (The fingers are distal to the wrist.)

9. **Superficial** means situated near the surface. (The epidermis is the superficial layer of the skin.) **Peripheral** also means outward or near the surface. It is used to describe the location of certain blood vessels and nerves. (Nerves that branch from the brain and spinal cord are peripheral nerves.)

10. **Deep** is used to describe more internal parts. (The dermis is the deep layer of the skin.)

Body Sections

To observe the relative locations and arrangements of internal parts, it is necessary to cut or section the body along various planes (figures 1.14 and 1.15). These terms are used to describe such planes and sections:

1. **Sagittal.** This refers to a lengthwise cut that divides the body into right and left portions. If the section passes along the midline and divides the body into equal halves, it is called *median* (mid-sagittal).

2. **Transverse.** A transverse cut is one that divides the body into superior and inferior portions.

3. **Frontal** (or *coronal*). A frontal section divides the body into anterior and posterior portions.

Sometimes an elongated organ such as a blood vessel is sectioned. In this case, a cut across the structure is called a *cross section,* a lengthwise cut is called a *longitudinal section,* and an angular cut is called an *oblique section* (figure 1.16).

Figure 1.15
A human brain sectioned along (*a*) the sagittal plane,
(*b*) transverse plane, and (*c*) the frontal plane.

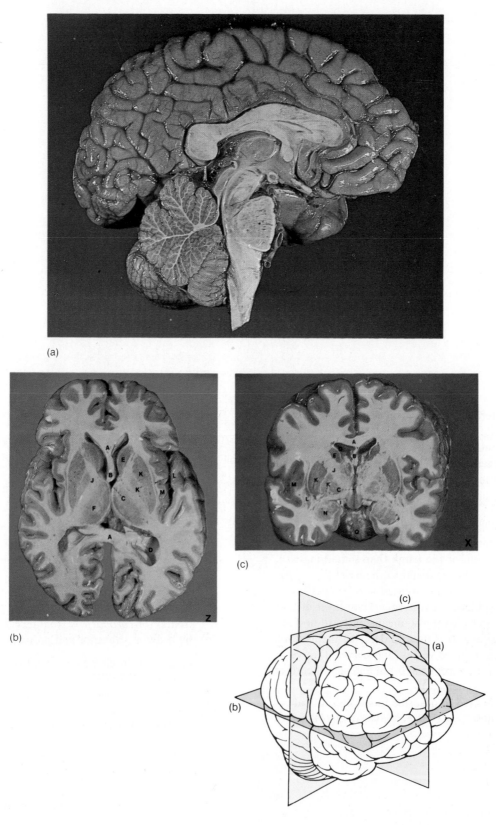

(a)

(b)

(c)

Figure 1.16
Tubular parts, such as blood vessels, may be cut in (*a*) cross section, (*b*) oblique section, or (*c*) longitudinal section.

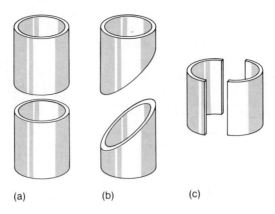

(a)   (b)   (c)

Figure 1.17
The abdominal area is subdivided into nine regions. How do the names of these regions describe their locations?

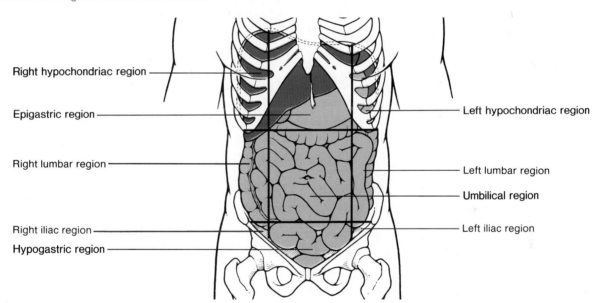

Right hypochondriac region

Epigastric region

Right lumbar region

Right iliac region

Hypogastric region

Left hypochondriac region

Left lumbar region

Umbilical region

Left iliac region

Body Regions

A number of terms are used to designate various body regions. The abdominal area, for example, is subdivided into the following nine regions, as shown in figure 1.17:

1. **Epigastric region**   the upper middle portion.
2. **Left** and **right hypochondriac regions**   on each side of the epigastric region.
3. **Umbilical region**   the central portion.
4. **Left** and **right lumbar regions**   on each side of the umbilical region.
5. **Hypogastric region**   the lower middle portion.
6. **Left** and **right iliac regions**   on each side of the hypogastric region.

Some other terms commonly used to indicate regions of the body include the following (figure 1.18 illustrates some of these regions):

Abdominal (ab-dom′ĭ-nal)   pertaining to the region between the thorax and pelvis.

Acromial (ah-kro′me-al)   pertaining to the point of the shoulder.

Antebrachium (an″te-bra′ke-um)   pertaining to the forearm.

Antecubital (an″te-ku′bĭ-tal)   pertaining to the space in front of the elbow.

Axillary (ak′sĭ-ler″e)   pertaining to the armpit.

Brachial (bra′ke-al)   pertaining to the upper arm.

Figure 1.18

Some terms used to describe body regions. (*a*) Anterior regions;
(*b*) posterior regions.

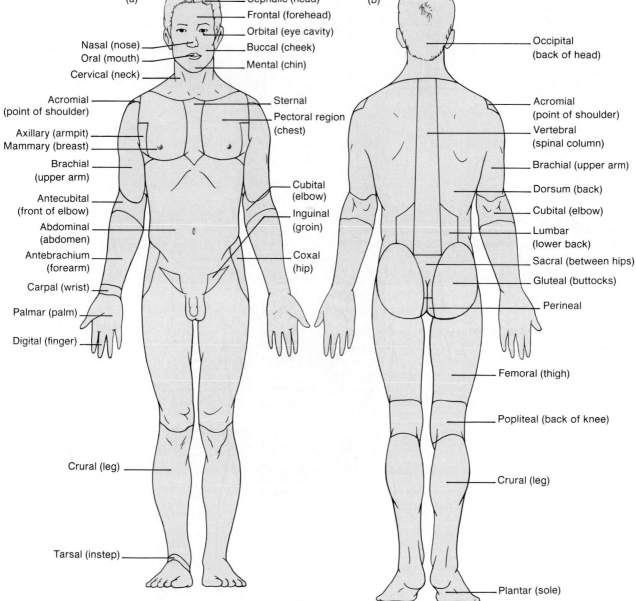

(a)

Cephalic (head)
Frontal (forehead)
Orbital (eye cavity)
Nasal (nose)
Buccal (cheek)
Oral (mouth)
Mental (chin)
Cervical (neck)

Acromial
(point of shoulder)
Sternal
Pectoral region
(chest)
Axillary (armpit)
Mammary (breast)
Brachial
(upper arm)
Cubital
(elbow)
Antecubital
(front of elbow)
Inguinal
(groin)
Abdominal
(abdomen)
Coxal
(hip)
Antebrachium
(forearm)
Carpal (wrist)
Palmar (palm)
Digital (finger)
Crural (leg)
Tarsal (instep)

(b)

Occipital
(back of head)
Acromial
(point of shoulder)
Vertebral
(spinal column)
Brachial (upper arm)
Dorsum (back)
Cubital (elbow)
Lumbar
(lower back)
Sacral (between hips)
Gluteal (buttocks)
Perineal
Femoral (thigh)
Popliteal (back of knee)
Crural (leg)
Plantar (sole)

Buccal (buk'al) pertaining to the cheek.
Carpal (kar'pal) pertaining to the wrist.
Celiac (se'le-ak) pertaining to the abdomen.
Cephalic (sĕ-fal'ik) pertaining to the head.
Cervical (ser'vĭ-kal) pertaining to the neck.
Costal (kos'tal) pertaining to the ribs.
Coxal (kok'sal) pertaining to the hip.
Crural (krōōr'al) pertaining to the leg.
Cubital (ku'bĭ-tal) pertaining to the elbow.
Digital (dij'ĭ-tal) pertaining to the finger.
Dorsum (dor'sum) pertaining to the back.
Femoral (fem'or-al) pertaining to the thigh.
Frontal (frun'tal) pertaining to the forehead.
Gluteal (gloo'te-al) pertaining to the buttocks.
Inguinal (ing'gwi-nal) pertaining to the depressed area of the abdominal wall near the thigh (groin).
Lumbar (lum'bar) pertaining to the region of the lower back between the ribs and the pelvis (loin).
Mammary (mam'er-e) pertaining to the breast.
Mental (men'tal) pertaining to the chin.
Nasal (na'zal) pertaining to the nose.
Occipital (ok-sip'ĭ-tal) pertaining to the lower back region of the head.
Oral (o'ral) pertaining to the mouth.
Orbital (or'bi-tal) pertaining to the eye cavity.
Palmar (pahl'mar) pertaining to the palm of the hand.
Pectoral (pek'tor-al) pertaining to the chest.
Pelvic (pel'vik) pertaining to the pelvis.
Perineal (per''ĭ-ne'al) pertaining to the region between the anus and the external reproductive organs (perineum).
Plantar (plan'tar) pertaining to the sole of the foot.
Popliteal (pop''lĭ-te'al) pertaining to the area behind the knee.
Sacral (sa'kral) pertaining to the posterior region between the hip bones.
Sternal (ster'nal) pertaining to the middle of the thorax, anteriorly.
Tarsal (tahr'sal) pertaining to the instep of the foot.
Vertebral (ver'te-bral) pertaining to the spinal column.

1. Describe the anatomical position.
2. Using the appropriate terms, describe the relative positions of several body parts.
3. Describe three types of sections.
4. Describe the nine regions of the abdomen.

Some of the Medical and Applied Sciences

cardiology (kar''de-ol'o-je) branch of medical science dealing with the heart and heart diseases.
cytology (si-tol'o-je) study of the structure, function, and diseases of cells.
dermatology (der''mah-tol'o-je) study of skin and its diseases.
endocrinology (en''do-krĭ-nol'o-je) study of hormone-secreting glands and the diseases involving them.
epidemiology (ep''ĭ-de''me-ol'o-je) study of infectious diseases, their distribution and control.
gastroenterology (gas''tro-en''ter-ol'o-je) study of the stomach and intestines and diseases involving these organs.
geriatrics (jer''e-at'riks) branch of medicine dealing with aged persons and their medical problems.
gerontology (jer''on-tol'o-je) study of the process of aging and the problems of elderly persons.
gynecology (gi''nĕ-kol'o-je) study of the female reproductive system and its diseases.
hematology (hem''ah-tol'o-je) study of blood and blood diseases.
histology (his-tol'o-je) study of the structure, function, and diseases of tissues.
immunology (im''u-nol'o-je) study of body resistance to disease.
nephrology (nĕ-frol'o-je) study of the structure, function, and diseases of the kidneys.
neurology (nu-rol'o-je) study of the nervous system in health and disease.
obstetrics (ob-stet'riks) branch of medicine dealing with pregnancy and childbirth.
oncology (ong-kol'o-je) study of tumors.
ophthalmology (of''thal-mol'o-je) study of the eye and eye diseases.
orthopedics (or''tho-pe'diks) branch of medicine dealing with the muscular and skeletal systems and problems of these systems.

Ultrasonography and CAT Scanning
A Clinical Application

Often noninvasive procedures that allow internal organs to be visualized are employed to help in diagnosing abnormal body conditions. Two such procedures are ultrasonography and CAT (computerized axial tomography) scanning.

Ultrasonography makes use of high frequency sound waves—those that are beyond the range of human hearing. In this procedure, a transducer that emits sound waves is pressed firmly against the skin and moved over the surface of the region being examined.

The sound waves travel into the body, and when they reach a border (interface) between structures of slightly different densities, some of the waves are reflected back to the transducer. Other sound waves continue into deeper tissues, and some of them are reflected back by still other interfaces.

As the reflected sound waves reach the transducer, they are converted into electrical impulses that are amplified and used to create a sectional image of the body's internal structure on a viewing screen.

Ultrasonography usually is not used to examine very compact organs such as bones or those containing air-filled spaces such as lungs, because the resulting images are of poor quality. However, useful images of medium density organs often can be obtained, making it possible to locate a fetus in the uterus, abnormal masses, accumulations of fluids, or stones (figure 1.19).

CAT scanning involves the use of X rays. In this procedure, an X-ray-emitting part is moved around the region of the body being examined. At the same time, an X-ray

Figure 1.19
This image resulting from the ultrasonographic procedure reveals the presence of gallstones (arrow) in a gallbladder.

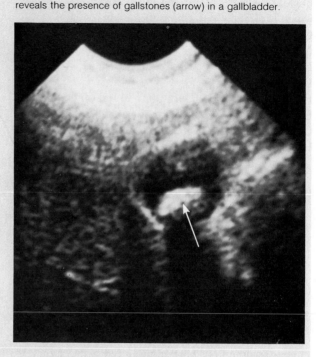

otolaryngology (o″to-lar″in-gol′o-je)   study of the ear, throat, larynx, and diseases of these parts.

pathology (pah-thol′o-je)   study of body changes produced by diseases.

pediatrics (pe″de-at′riks)   branch of medicine dealing with children and their diseases.

pharmacology (fahr″mah-kol′o-je)   study of drugs and their uses in the treatment of diseases.

podiatry (po-di′ah-tre)   study of the care and treatment of the feet.

psychiatry (si-ki′ah-tre)   branch of medicine dealing with the mind and its disorders.

radiology (ra″de-ol′o-je)   study of X rays and radioactive substances and their uses in the diagnosis and treatment of diseases.

toxicology (tok″sǐ-kol′o-je)   study of poisonous substances and their effects upon body parts.

urology (u-rol′o-je)   branch of medicine dealing with the kidneys and urinary system and their diseases.

Chapter Summary

Introduction

1. Early interest in the human body probably developed as people became concerned about injuries and illnesses.
2. Primitive doctors began to learn how certain herbs and potions affected body functions.
3. In ancient times it was believed that natural processes were caused by spirits and supernatural forces.
4. The belief that humans could understand forces that caused natural events led to the development of modern science.
5. A set of new terms, originating from Greek and Latin, formed the basis for the language of anatomy and physiology.

Anatomy and Physiology

1. Anatomy deals with the form and arrangement of body parts.
2. Physiology deals with the functions of these parts.
3. The function of a part depends upon the way it is constructed.

Figure 1.20

CAT scans of (a) the head and (b) the abdomen. What organs can you identify?

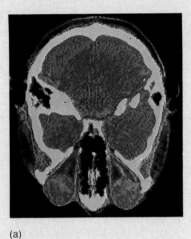

(a)

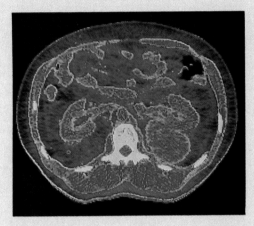

(b)

detector is moved in the opposite direction on the other side of the body. As the parts move, an X-ray beam is passed through the body from hundreds of different angles. Since tissues and organs of varying densities within the body absorb X rays differently, the amount of X ray reaching the detector varies from position to position.

The measurements made by the detector are recorded in the memory of a computer, which later combines them mathematically and creates a sectional image of the internal body parts on a viewing screen.

While ordinary X-ray techniques produce two-dimensional pictures, the CAT scan provides three-dimensional information. The CAT scan also makes it possible to differentiate clearly between soft tissues of slightly different densities, such as liver and kidneys, which cannot be seen in a conventional X ray. Thus, abnormal tissue growths often can be located in a scan (figure 1.20).

Characteristics of Life

Characteristics of life are traits shared by all organisms.

1. These characteristics include:
   a. Movement—changes in body position or motion of internal parts.
   b. Responsiveness—sensing and reacting to internal or external changes.
   c. Growth—increase in size without change in shape.
   d. Reproduction—production of offspring.
   e. Respiration—obtaining oxygen, using oxygen to release energy from foods, and transporting gaseous wastes to the outside.
   f. Digestion—changing food substances into forms that can be absorbed and used by cells.
   g. Absorption—passage of digestive products into body fluids.
   h. Circulation—movement of substances in body fluids.
   i. Assimilation—changing of substances into chemically different forms.
   j. Excretion—removal of body wastes.
2. Together these activities constitute metabolism.

Maintenance of Life

The structures and functions of body parts are directed toward maintaining the life of the organism.

1. Needs of organisms
   a. Water is used in a variety of metabolic processes, provides the environment for these reactions, and transports substances.
   b. Food is needed to supply energy, to provide raw materials for building new living matter, and to supply chemicals necessary in vital reactions.
   c. Oxygen is used in releasing energy from food materials; this energy drives metabolic reactions.
   d. Heat is a by-product of metabolic reactions and helps govern the rates of these reactions.
   e. Pressure is a force; in humans, atmospheric and hydrostatic pressures help breathing and fluid movements respectively.
   f. Survival of an organism depends upon the quantities and qualities of these factors.

2. Homeostasis
   a. If an organism is to survive, the conditions within its body fluids must remain relatively stable.
   b. The tendency to maintain a stable internal environment is called homeostasis.
   c. Homeostatic mechanisms regulate body temperature, blood pressure, and blood sugar concentration.
   d. Homeostatic mechanisms employ negative feedback.

Levels of Organization

The body is composed of parts that occupy different levels of organization.

1. Material substances are composed of atoms.
2. Atoms bond together to form molecules.
3. Organelles contain aggregates of large molecules.
4. Cells, which are composed of organelles, are the basic units of structure and function within the body.
5. Cells are organized into layers or masses called tissues.
6. Tissues are organized into organs.
7. Organs are arranged into organ systems.
8. Organ systems constitute the organism.
9. These parts vary in complexity progressively from one level to the next.

Organization of the Human Body

1. Body cavities
   a. The axial portion of the body contains the dorsal and ventral cavities.
      (1) The dorsal cavity includes the cranial and spinal cavities.
      (2) The ventral cavity includes the thoracic and abdominopelvic cavities, which are separated by the diaphragm.
   b. Body cavities are filled with visceral organs.
   c. Other body cavities include the oral, nasal, orbital, and middle ear cavities.
2. Thoracic and abdominopelvic membranes
   a. Thoracic membranes
      (1) Pleural membranes line the thoracic cavity and cover the lungs.
      (2) Pericardial membranes surround the heart and cover its surface.
      (3) The pleural and pericardial cavities are potential spaces between these membranes.
   b. Abdominopelvic membranes
      (1) Peritoneal membranes line the abdominopelvic cavity and cover the organs inside.
      (2) The peritoneal cavity is a potential space between these membranes.

3. Organ Systems
The human organism is composed of several organ systems. Each system consists of a set of interrelated organs that carry on vital functions.
   a. Integumentary system
      (1) The integumentary system provides the body covering.
      (2) It includes the skin, hair, nails, sweat glands, and sebaceous glands.
      (3) It functions to protect underlying tissues, regulate body temperature, house sensory receptors, and synthesize various cellular products.
   b. Skeletal system
      (1) The skeletal system is composed of bones and parts that bind bones together.
      (2) It provides framework, protective shields, and attachments for muscles; it produces blood cells and stores inorganic salts.
   c. Muscular system
      (1) The muscular system includes the muscles of the body.
      (2) It is responsible for body movements, maintenance of posture, and production of body heat.
   d. Nervous system
      (1) The nervous system consists of the brain, spinal cord, and nerves.
      (2) It functions to receive impulses from sensory parts, interpret these impulses, and act on them by causing muscles or glands to respond.
   e. Endocrine system
      (1) The endocrine system consists of glands that secrete hormones.
      (2) Hormones help regulate metabolism by stimulating target tissues.
      (3) It includes the pituitary, thyroid, parathyroid, and adrenal glands; the pancreas, ovaries, testes, pineal gland, and thymus gland.
   f. Digestive system
      (1) The digestive system receives foods, converts their molecules into forms that can pass through cell membranes and eliminates the materials that are not absorbed.
      (2) Certain portions produce hormones.
      (3) It includes the mouth, tongue, teeth, salivary glands, pharynx, esophagus, stomach, liver, gallbladder, pancreas, small intestine, and large intestine.
   g. Respiratory system
      (1) The respiratory system provides for the intake and output of air and for the exchange of gases between the blood and the air.
      (2) It includes the nasal cavity, pharynx, larynx, trachea, bronchi, and lungs.

h. Circulatory system
   (1) The circulatory system includes the heart, which pumps blood, and the blood vessels, which carry blood to and from body parts.
   (2) Blood transports oxygen, nutrients, hormones, and wastes.
i. Lymphatic system
   (1) The lymphatic system is composed of lymphatic vessels, lymph nodes, thymus, and spleen.
   (2) It transports lymph from tissue spaces to the bloodstream, carries lipids from the digestive organs, and aids in defending the body against infections.
j. Urinary system
   (1) The urinary system includes the kidneys, ureters, urinary bladder, and urethra.
   (2) It filters wastes from the blood and helps maintain fluid and electrolyte balance.
k. Reproductive system
   (1) The reproductive systems are concerned with the production of new organisms.
   (2) The male reproductive system includes the scrotum, testes, epididymides, vasa deferentia, seminal vesicles, prostate gland, bulbourethral glands, penis, and urethra, which produce, maintain, and transport male sex cells.
   (3) The female reproductive system includes the ovaries, uterine tubes, uterus, vagina, and vulva, which produce, maintain, and transport female sex cells.

Anatomical Terminology

1. Relative position
   These terms are used to describe the location of one part with respect to another part.
2. Body sections
   Body sections are planes along which the body may be cut to observe the relative locations and arrangements of internal parts.
3. Body regions
   Various body regions are designated by special terms.

Application of Knowledge

1. In many states, death is defined as "irreversible cessation of total brain function." In others, it is equated with failure of heart and breathing activities. How is death defined in your state? How is this definition related to the characteristics of life?
2. In health, the body parts function together effectively to maintain homeostasis. In illness, the maintenance of homeostasis may be threatened and various treatments may be needed to reduce this threat. What treatments might be used to help control (a) a patient's body temperature, (b) blood oxygen concentration, and (c) water content?
3. Suppose two individuals are afflicted with benign (noncancerous) tumors that produce symptoms because they occupy space and crowd adjacent organs. If one of these persons has the tumor in the ventral cavity and the other has it in the dorsal cavity, which would be likely to develop symptoms first? Why?

Review Activities

Part A

1. Briefly describe the early development of knowledge about the human body.
2. Distinguish between the activities of anatomists and physiologists.
3. Explain the relationship between the form and function of human body parts.
4. List and describe ten characteristics of life.
5. Define *metabolism*.
6. List and describe five needs of organisms.
7. Explain how the idea of homeostasis is related to these five needs.
8. Distinguish between heat and temperature.
9. Define two types of pressures that may act upon the outsides of organisms.
10. Explain how body temperature, blood pressure, and blood sugar concentration are controlled.
11. Explain why homeostatic mechanisms are said to act by negative feedback.
12. Explain what is meant by levels of organization.
13. List the levels of organization within the human.
14. Distinguish between the axial and the appendicular portions of the body.
15. Distinguish between the dorsal and ventral body cavities and name the smaller cavities that occur within each.
16. Explain what is meant by a visceral organ.
17. Describe the mediastinum.
18. Describe the location of the oral, nasal, orbital, and middle ear cavities.
19. Distinguish between a parietal and a visceral membrane.
20. Name the major organ systems and describe the general functions of each.
21. List the major organs that comprise each organ system.

Part B

1. Name the body cavity in which each of the following organs is located:

   a. Stomach   f. Rectum
   b. Heart   g. Spinal cord
   c. Brain   h. Esophagus
   d. Liver   i. Spleen
   e. Trachea   j. Urinary bladder

2. Write complete sentences using each of the following terms correctly:

   a. Superior   h. Contralateral
   b. Inferior   i. Proximal
   c. Anterior   j. Distal
   d. Posterior   k. Superficial
   e. Medial   l. Peripheral
   f. Lateral   m. Deep
   g. Ipsilateral

3. Prepare a sketch of a human body and use lines to indicate each of the following sections:

   a. Sagittal   c. Transverse
   b. Median   d. Frontal

4. Prepare a sketch of the abdominal area and indicate the location of each of the following regions:

   a. Epigastric   d. Hypochondriac
   b. Umbilical   e. Lumbar
   c. Hypogastric   f. Iliac

5. Name the region to which each of the following terms pertains:

   a. Acromial   l. Occipital
   b. Antebrachium   m. Orbital
   c. Axillary   n. Palmar
   d. Buccal   o. Pectoral
   e. Celiac   p. Perineal
   f. Coxal   q. Plantar
   g. Crural   r. Popliteal
   h. Femoral   s. Sacral
   i. Gluteal   t. Sternal
   j. Inguinal   u. Tarsal
   k. Mental   v. Vertebral

Reference Plates
The Human Organism

The following series of illustrations shows the major organs of the human torso. The first plate illustrates the anterior surface and reveals the superficial muscles on one side. Each subsequent plate exposes some deeper organs, including those in the thoracic, abdominal, and pelvic cavities.

The purpose of chapters 6–22 of this textbook is to describe the organ systems of the human organism in some detail. As you read them, you may want to refer to these plates to help yourself visualize the locations of various organs and the three-dimensional relationships that exist between them.

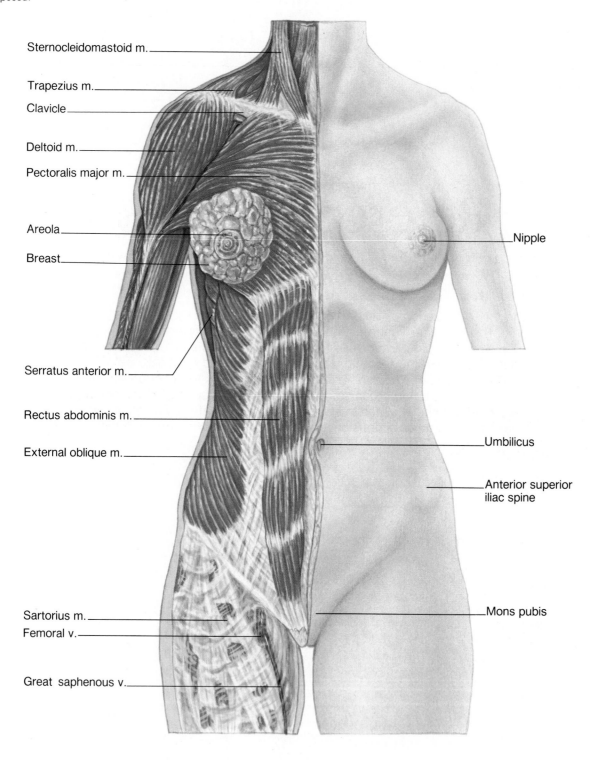

Sternocleidomastoid m.

Trapezius m.

Clavicle

Deltoid m.

Pectoralis major m.

Areola

Breast

Nipple

Serratus anterior m.

Rectus abdominis m.

External oblique m.

Umbilicus

Anterior superior
iliac spine

Sartorius m.

Femoral v.

Mons pubis

Great saphenous v.

Plate 2
Human torso, with deeper muscle layers exposed.

Levator scapulae m.

Subscapularis m.

Coracobrachialis m.

Pectoralis major m.
(cut head)

Long head biceps
brachii m.

Short head biceps
brachii m.

Serratus anterior m.

Ext. intercostal m.

Latissimus dorsi m.

Rectus abdominis m.

Transversus abdominis m.

Internal oblique m.

Anterior superior
iliac spine

Femoral n.

Femoral v.

Sartorius m.

Rectus femoris m.

Sternocleidomastoid m.

Trapezius m.

External intercostal m.

Deltoid m.

Teres major m.

Pectoralis minor m.

Pectoralis major m.

External oblique m.

Linea alba

Gluteus medius m.

Tensor fasciae latae m.

Inguinal canal

Penis

Great saphenous v.

Plate 3
Human torso, with abdominal viscera exposed.

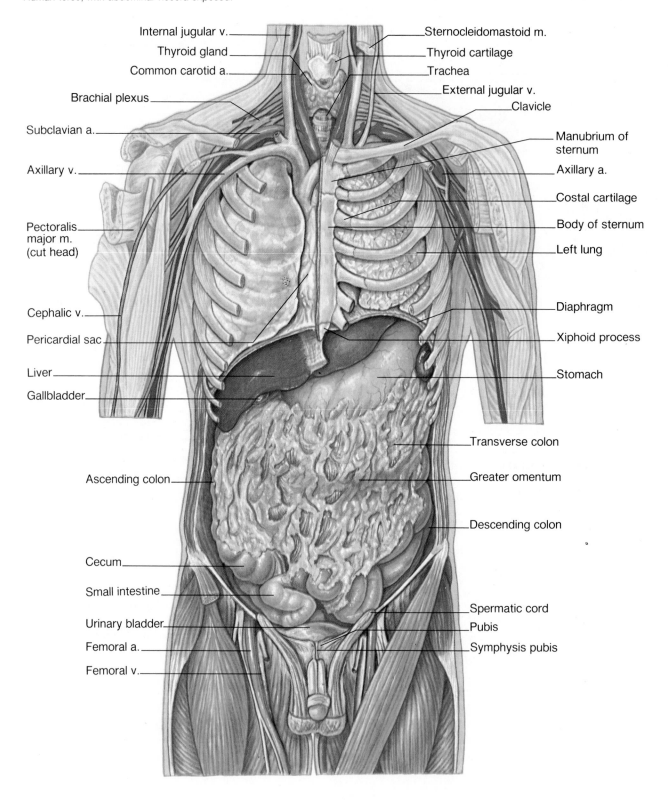

Internal jugular v.

Thyroid gland

Common carotid a.

Brachial plexus

Subclavian a.

Axillary v.

Pectoralis
major m.
(cut head)

Cephalic v.

Pericardial sac

Liver

Gallbladder

Ascending colon

Cecum

Small intestine

Urinary bladder

Femoral a.

Femoral v.

Sternocleidomastoid m.

Thyroid cartilage

Trachea

External jugular v.

Clavicle

Manubrium of
sternum

Axillary a.

Costal cartilage

Body of sternum

Left lung

Diaphragm

Xiphoid process

Stomach

Transverse colon

Greater omentum

Descending colon

Spermatic cord

Pubis

Symphysis pubis

Plate 4
Human torso, with thoracic viscera exposed.

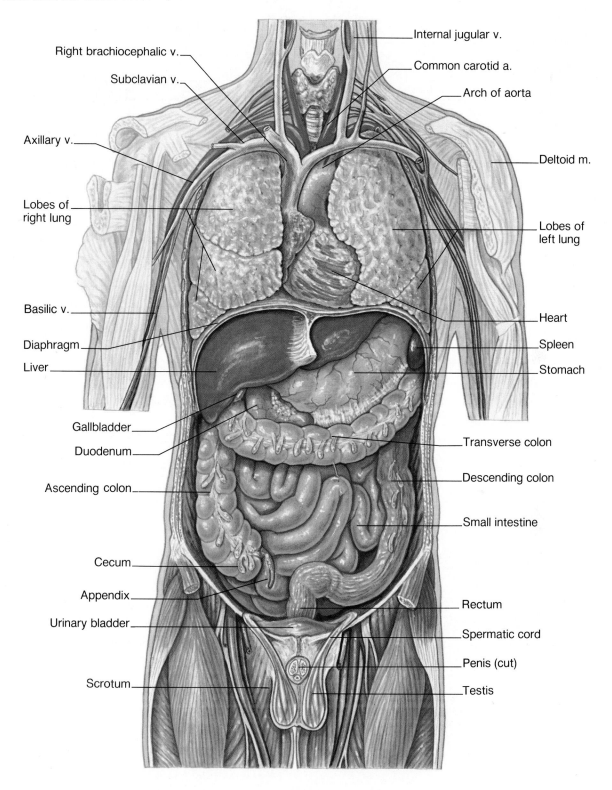

Internal jugular v.

Right brachiocephalic v.

Common carotid a.

Subclavian v.

Arch of aorta

Axillary v.

Deltoid m.

Lobes of
right lung

Lobes of
left lung

Basilic v.

Heart

Diaphragm

Spleen

Liver

Stomach

Gallbladder

Transverse colon

Duodenum

Ascending colon

Descending colon

Small intestine

Cecum

Appendix

Rectum

Urinary bladder

Spermatic cord

Penis (cut)

Scrotum

Testis

Plate 5

Human torso, with lungs, heart, and small intestine sectioned.

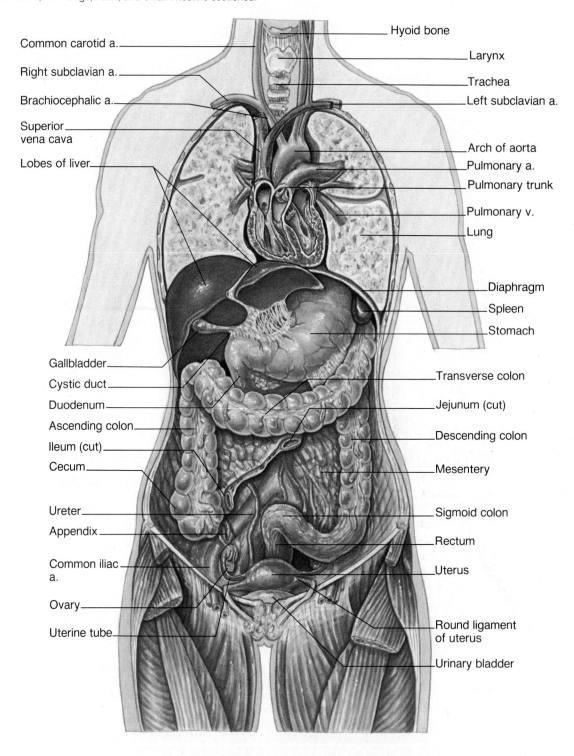

Common carotid a.

Right subclavian a.

Brachiocephalic a.

Superior vena cava

Lobes of liver

Gallbladder

Cystic duct

Duodenum

Ascending colon

Ileum (cut)

Cecum

Ureter

Appendix

Common iliac a.

Ovary

Uterine tube

Hyoid bone

Larynx

Trachea

Left subclavian a.

Arch of aorta

Pulmonary a.

Pulmonary trunk

Pulmonary v.

Lung

Diaphragm

Spleen

Stomach

Transverse colon

Jejunum (cut)

Descending colon

Mesentery

Sigmoid colon

Rectum

Uterus

Round ligament of uterus

Urinary bladder

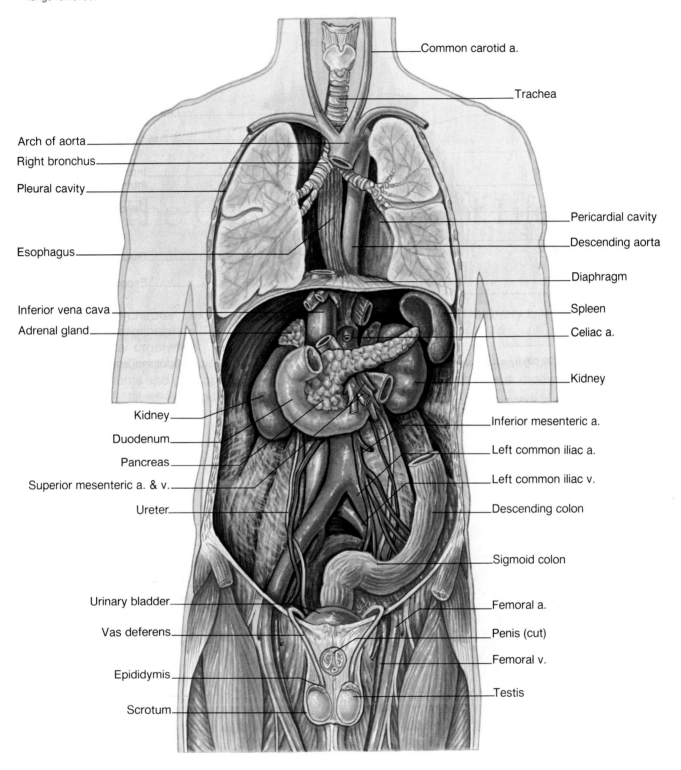

Common carotid a.

Trachea

Arch of aorta

Right bronchus

Pleural cavity

Pericardial cavity

Descending aorta

Esophagus

Diaphragm

Inferior vena cava

Spleen

Adrenal gland

Celiac a.

Kidney

Kidney

Inferior mesenteric a.

Duodenum

Left common iliac a.

Pancreas

Left common iliac v.

Superior mesenteric a. & v.

Descending colon

Ureter

Sigmoid colon

Urinary bladder

Femoral a.

Vas deferens

Penis (cut)

Femoral v.

Epididymis

Testis

Scrotum

CHEMISTRY IS THE branch of science dealing with the composition of substances and changes that take place in their composition. Although it is possible to study anatomy without much reference to this subject, knowledge of chemistry is essential for understanding physiology, because body functions involve chemical changes that occur within cells.

As interest in the chemistry of living organisms grew, and knowledge in this area expanded, a new subdivision of science called biological chemistry or *biochemistry* emerged.

Biochemical studies have been important not only in helping to explain physiological processes, but also in making possible the development of many new drugs and methods for treating diseases.

---

1. Why is a knowledge of chemistry essential to an understanding of physiology?
2. What is biochemistry?

Structure of Matter

Matter is anything that has weight and takes up space. This includes all the solids, liquids, and gases in our surroundings as well as in our bodies.

Elements and Atoms

Studies of matter have revealed that all things are composed of basic substances called **elements.** At present, 106 such elements are known, although most naturally occurring matter on earth includes only 91 of them. Among these elements are such common materials as iron, copper, silver, gold, aluminum, carbon, hydrogen, and oxygen. Although some elements exist in a pure form, they occur more frequently in mixtures or in chemical combinations of 2 or more elements known as **compounds.**

About 20 elements are needed by living things, and of these oxygen, carbon, hydrogen, and nitrogen make up more than 95% of the human body. A list of the more abundant elements in the body is shown in chart 2.1. Notice that each element is represented by a one- or two-letter symbol.

Elements are composed of tiny, invisible particles called **atoms.** Although the atoms that make up each element are very similar to each other, they differ from the atoms that make up other elements. Atoms vary in size, weight, and the ways they interact with each other. Some, for instance, are capable of combining with atoms like themselves or with other kinds of atoms, while others lack this ability.

Chart 2.1	Most abundant elements in the human body	
Major elements	**Symbol**	**Approximate percentage of the human body**
Oxygen	O	65.0
Carbon	C	18.5
Hydrogen	H	9.5
Nitrogen	N	3.2
Calcium	Ca	1.5
Phosphorus	P	1.0
Potassium	K	0.4
Sulfur	S	0.3
Chlorine	Cl	0.2
Sodium	Na	0.2
Magnesium	Mg	0.1
Trace elements		
Cobalt	Co	
Copper	Cu	
Fluorine	F	
Iodine	I	Less than 0.1
Iron	Fe	
Manganese	Mn	
Zinc	Zn	

Figure 2.1

An atom of lithium includes 3 electrons in motion around a nucleus, which contains 3 protons and 4 neutrons.

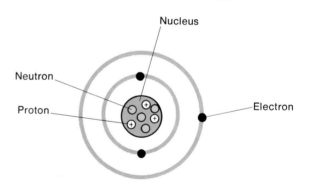

Lithium atom

Atomic Structure

An atom consists of a central core called the **nucleus** and one or more **electrons** that are in constant motion around the nucleus. The nucleus itself contains some relatively large particles called **protons** and **neutrons,** whose weights are about equal (figure 2.1).

Electrons, which are extremely small and have almost no weight, carry a single, negative electrical charge (e^-), while protons each carry a single, positive electrical charge (p^+). Neutrons are uncharged and thus are electrically neutral (n^0).

CHEMISTRY IS THE branch of science dealing with the composition of substances and changes that take place in their composition. Although it is possible to study anatomy without much reference to this subject, knowledge of chemistry is essential for understanding physiology, because body functions involve chemical changes that occur within cells.

As interest in the chemistry of living organisms grew, and knowledge in this area expanded, a new subdivision of science called biological chemistry or *biochemistry* emerged.

Biochemical studies have been important not only in helping to explain physiological processes, but also in making possible the development of many new drugs and methods for treating diseases.

---

1. Why is a knowledge of chemistry essential to an understanding of physiology?
2. What is biochemistry?

Structure of Matter

Matter is anything that has weight and takes up space. This includes all the solids, liquids, and gases in our surroundings as well as in our bodies.

Elements and Atoms

Studies of matter have revealed that all things are composed of basic substances called **elements.** At present, 106 such elements are known, although most naturally occurring matter on earth includes only 91 of them. Among these elements are such common materials as iron, copper, silver, gold, aluminum, carbon, hydrogen, and oxygen. Although some elements exist in a pure form, they occur more frequently in mixtures or in chemical combinations of 2 or more elements known as **compounds.**

About 20 elements are needed by living things, and of these oxygen, carbon, hydrogen, and nitrogen make up more than 95% of the human body. A list of the more abundant elements in the body is shown in chart 2.1. Notice that each element is represented by a one- or two-letter symbol.

Elements are composed of tiny, invisible particles called **atoms.** Although the atoms that make up each element are very similar to each other, they differ from the atoms that make up other elements. Atoms vary in size, weight, and the ways they interact with each other. Some, for instance, are capable of combining with atoms like themselves or with other kinds of atoms, while others lack this ability.

Chart 2.1	Most abundant elements in the human body	
Major elements	Symbol	Approximate percentage of the human body
Oxygen	O	65.0
Carbon	C	18.5
Hydrogen	H	9.5
Nitrogen	N	3.2
Calcium	Ca	1.5
Phosphorus	P	1.0
Potassium	K	0.4
Sulfur	S	0.3
Chlorine	Cl	0.2
Sodium	Na	0.2
Magnesium	Mg	0.1
Trace elements		
Cobalt	Co	
Copper	Cu	
Fluorine	F	
Iodine	I	Less than 0.1
Iron	Fe	
Manganese	Mn	
Zinc	Zn	

Figure 2.1

An atom of lithium includes 3 electrons in motion around a nucleus, which contains 3 protons and 4 neutrons.

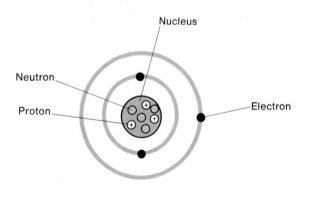

Lithium atom

Atomic Structure

An atom consists of a central core called the **nucleus** and one or more **electrons** that are in constant motion around the nucleus. The nucleus itself contains some relatively large particles called **protons** and **neutrons,** whose weights are about equal (figure 2.1).

Electrons, which are extremely small and have almost no weight, carry a single, negative electrical charge (e^-), while protons each carry a single, positive electrical charge (p^+). Neutrons are uncharged and thus are electrically neutral (n^0).

Key Terms

atom   (at′om)

carbohydrate   (kar″bo-hi′drāt)

decomposition   (de″kom-po-zish′un)

electrolyte   (e-lek′tro-līt)

formula   (fōr′mu-lah)

inorganic   (in″or-gan′ik)

ion   (i′on)

isotope   (i′so-tōp)

lipid   (lip′id)

molecule   (mol′ĕ-kūl)

nucleic acid   (nu-kle′ik as′id)

organic   (or-gan′ik)

protein   (pro′te-in)

radiation   (ra″de-a′shun)

synthesis   (sin′thĕ-sis)

Aids to Understanding Words

bio-, life: *bio*chemistry—branch of science dealing with the chemistry of life forms.

di-, two: *di*saccharide—a compound whose molecules are composed of two saccharide units bound together.

glyc-, sweet: *glyc*ogen—a complex carbohydrate composed of sugar molecules bound together.

iso-, equal: *iso*tope—atom that has the same atomic number as another atom but a different atomic weight.

lip-, fat: *lip*ids—group of organic compounds that includes fats.

-lyt, dissolvable: electro*lyte*—substance that dissolves in water and releases ions.

mono-, one: *mono*saccharide—a compound whose molecule consists of a single saccharide unit.

nucle-, kernel: *nucle*us—central core of an atom.

poly-, many: *poly*unsaturated—molecule that has many double bonds between its carbon atoms.

facchar-, sugar: mono*facchar*ide—sugar molecule composed of a single saccharide unit.

syn-, together: *syn*thesis—process by which substances are united to form a new type of substance.

-valent, having power: electro*valent* bond—chemical bond created by the attraction between two ions with opposite electrical charges.

Chapter Outline

Chapter Objectives

After you have studied this chapter, you should be able to

1. Explain how the study of living material is dependent on the study of chemistry.

2. Describe the relationships between matter, atoms, and molecules.

3. Discuss how atomic structure is related to the ways in which atoms interact.

4. Explain how molecular and structural formulas are used to symbolize the composition of compounds.

5. Describe three types of chemical reactions.

6. Discuss the concept of pH.

7. List the major groups of inorganic substances that are common in cells.

8. Describe the general roles played in cells by various types of organic substances.

9. Complete the review activities at the end of this chapter. Note that the items are worded in the form of specific learning objectives. You may want to refer to them before reading the chapter.

2

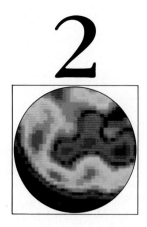

The Chemical Basis of Life

As chapter 1 explained, an organism is composed of parts that vary in level of organization. These include organ systems, organs, tissues, cells, and cellular organelles.

Organelles are in turn composed of chemical substances made up of atoms and groups of atoms called molecules. Similarly, atoms are composed of still smaller units called protons, neutrons, and electrons.

A study of living material at its lower levels of organization must be concerned with the chemical substances that form the structural basis of all matter and that interact in the metabolic processes carried on within organisms. This is the theme of chapter 2.

Plate 7
Human torso, with posterior walls of thorax, and abdominal and
pelvic cavities exposed.

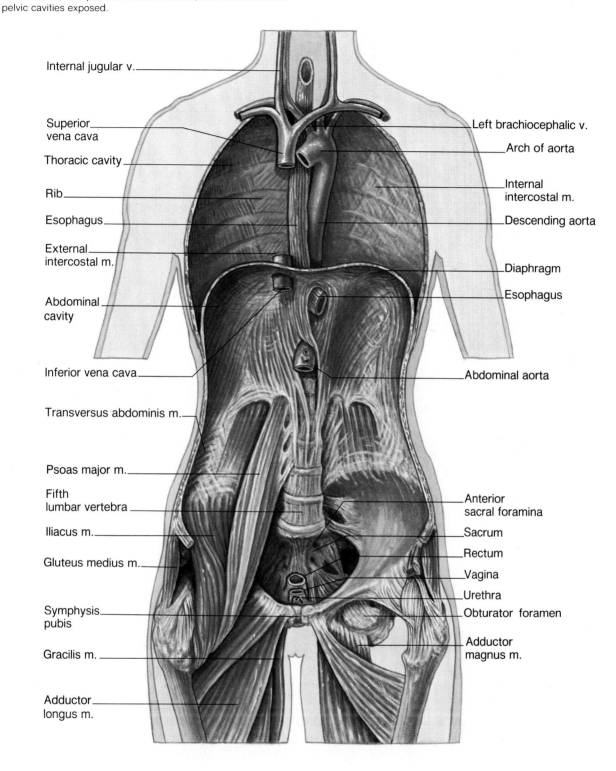

Internal jugular v.

Superior
vena cava

Thoracic cavity

Rib

Esophagus

External
intercostal m.

Abdominal
cavity

Inferior vena cava

Transversus abdominis m.

Psoas major m.

Fifth
lumbar vertebra

Iliacus m.

Gluteus medius m.

Symphysis
pubis

Gracilis m.

Adductor
longus m.

Left brachiocephalic v.

Arch of aorta

Internal
intercostal m.

Descending aorta

Diaphragm

Esophagus

Abdominal aorta

Anterior
sacral foramina

Sacrum

Rectum

Vagina

Urethra

Obturator foramen

Adductor
magnus m.

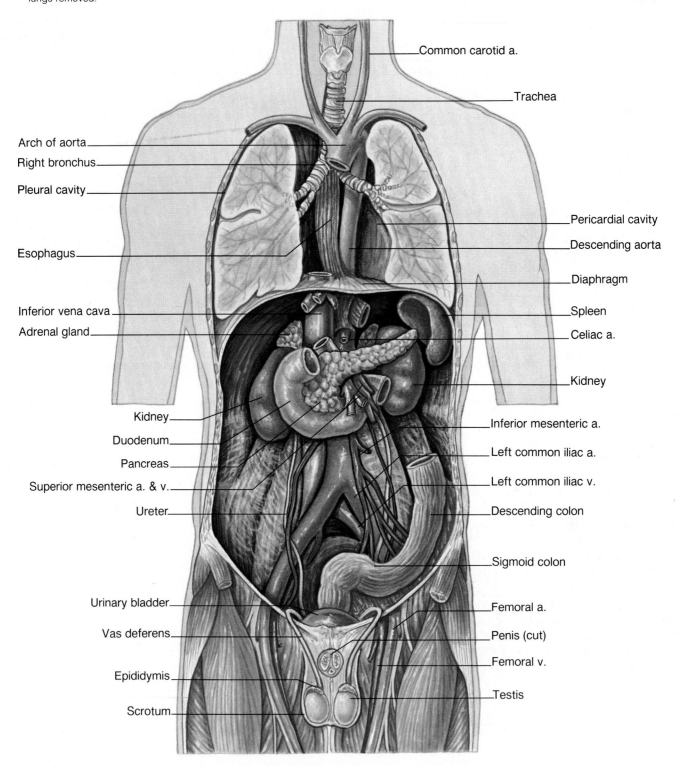

Common carotid a.

Trachea

Arch of aorta

Right bronchus

Pleural cavity

Pericardial cavity

Descending aorta

Esophagus

Diaphragm

Inferior vena cava

Spleen

Adrenal gland

Celiac a.

Kidney

Kidney

Inferior mesenteric a.

Duodenum

Left common iliac a.

Pancreas

Left common iliac v.

Superior mesenteric a. & v.

Descending colon

Ureter

Sigmoid colon

Urinary bladder

Femoral a.

Vas deferens

Penis (cut)

Femoral v.

Epididymis

Testis

Scrotum

Chart 2.2 Atomic structure of elements 1 through 12

Element	Symbol	Atomic number	Atomic weight	Protons	Neutrons	Electrons in shells		
						First	Second	Third
Hydrogen	H	1	1	1	0	1		
Helium	He	2	4	2	2	2	(inert)	
Lithium	Li	3	7	3	4	2	1	
Beryllium	Be	4	9	4	5	2	2	
Boron	B	5	11	5	6	2	3	
Carbon	C	6	12	6	6	2	4	
Nitrogen	N	7	14	7	7	2	5	
Oxygen	O	8	16	8	8	2	6	
Fluorine	F	9	19	9	10	2	7	
Neon	Ne	10	20	10	10	2	8	(inert)
Sodium	Na	11	23	11	12	2	8	1
Magnesium	Mg	12	24	12	12	2	8	2

Chart 2.3 Isotopes of oxygen

Isotope	Protons	Neutrons	Atomic number	Atomic weight	Type of isotope
Oxygen-13	8	5	8	13	Radioactive
Oxygen-14	8	6	8	14	Radioactive
Oxygen-15	8	7	8	15	Radioactive
Oxygen-16*	8	8	8	16	Stable
Oxygen-17	8	9	8	17	Stable
Oxygen-18	8	10	8	18	Stable
Oxygen-19	8	11	8	19	Radioactive
Oxygen-20	8	12	8	20	Radioactive

*Most abundant isotope.

Since the atomic nucleus contains the protons, this part of an atom is always positively charged. However, the number of electrons outside the nucleus is equal to the number of protons, so a complete atom is electrically uncharged or *neutral*.

The atoms of different elements contain different numbers of protons. The number of protons in the atoms of a particular element is called its *atomic number*. Hydrogen, for example, whose atoms contain one proton, has the atomic number 1; and carbon, whose atoms have six protons, has the atomic number 6.

The weight of an atom of an element is due primarily to the protons and neutrons in its nucleus, because the electrons have so little weight. For this reason, an atom of carbon with six protons and six neutrons weighs about twelve times as much as an atom of hydrogen, which has only one proton and no neutrons.

The number of protons plus the number of neutrons in each of its atoms is approximately equal to the *atomic weight* of an element. Thus, the atomic weight of hydrogen is 1, and the atomic weight of carbon is 12 (chart 2.2).

Isotopes

All the atoms of a particular element have the same atomic number because they have the same number of protons and electrons. However, the atoms of an element may vary in the number of neutrons in their nuclei; thus, they vary in atomic weight. For example, all oxygen atoms have 8 protons in their nuclei. Some, however, have 8 neutrons (atomic weight 16), others have 9 neutrons (atomic weight 17), and still others have 10 neutrons (atomic weight 18) (see chart 2.3). Atoms that have the same atomic numbers but different atomic weights are called **isotopes** of an element. Since a sample of an element is likely to include more than one isotope, the atomic weight of the element is the average weight of the isotopes present.

The ways atoms interact with one another are due largely to the number of electrons they possess. Because the number of electrons in an atom is equal to its number of protons, all the isotopes of a particular element have the same number of electrons and react

Uses of Radioactive Isotopes
A Clinical Application

Atomic radiation includes three common forms called alpha (α), beta (β), and gamma (γ). Alpha radiation consists of particles from atomic nuclei that travel relatively slowly and have weak ability to penetrate matter. Beta radiation consists of much smaller particles (electrons) that travel more rapidly and penetrate matter more deeply. Gamma radiation is similar to x ray and is the most penetrating of these forms.

Each kind of radioactive isotope produces one or more of these forms of radiation with particular energy levels. Furthermore, each of these isotopes tends to lose its radioactivity at a particular rate. The time required for an isotope to lose one-half of its activity is called its *half-life*. Thus, iodine-131 which emits one-half of its radiation in 8.1 days has a half-life of 8.1 days. Similarly, the half-life of iron-59 is 45.1 days, that of phosphorus-32 is 14.3 days, that of cobalt-60 is 5.26 years, and that of radium-226 is 1,620 years.

Since it is possible to detect the presence of atomic radiation by using special equipment, such as Geiger-Müller counters and scintillation counters, radioactive substances are useful in studying life processes. A radioactive isotope, for example, can be introduced into an organism and then traced with detectors as it enters into metabolic activities. Since the human thyroid gland makes special use of iodine in its metabolic processes, radioactive iodine-131 has been used to study its functions and to evaluate patients with thyroid disease (figures 2.2 and 2.3). Likewise, a form of technetium-99 with a half-life of 6.0 hours is sometimes used to assess the size and functions of heart chambers. It also may be used to locate regions of damaged heart tissue following a heart attack or to locate abnormal growths or injured tissues in the brain. Such procedures involve injecting the isotope into the blood and following its path using detectors that record images on paper or film.

Other clinical uses of radioactive isotopes include measuring kidney functions, estimating the concentration of hormones in body fluids, measuring blood volume and red blood cell mass, and studying changes in skeletal structures.

Atomic radiations also can cause changes in the structures of various chemical substances and in this way can alter vital processes within cells. This is the reason radioactive isotopes, such as cobalt-60, are sometimes used to treat cancers. The radiation coming from the cobalt can affect the chemicals within the cancerous cells and cause their deaths. Since cancer cells are more susceptible to such damage than normal cells, the cancer cells are destroyed more rapidly than the normal ones.

Figure 2.2
Scintillation counters, such as this, are used to detect the presence of radioactive isotopes.

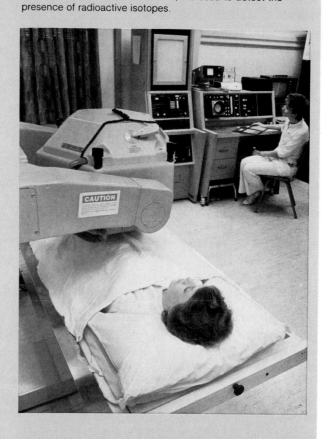

Figure 2.3
A scan of the thyroid gland 24 hours after the patient was administered radioactive iodine (see figure 13.17).

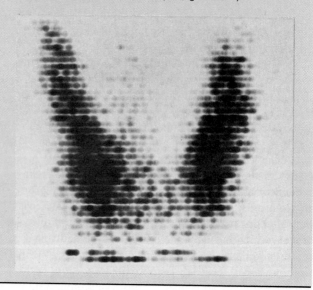

Figure 2.4
The single electron of a hydrogen atom is located in its first shell; the 2 electrons of a helium atom fill its first shell; and the 3 electrons of a lithium atom are arranged with 2 in the first shell and 1 in the second shell.

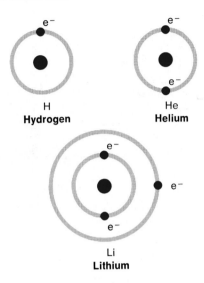

Figure 2.5
A diagram of a sodium atom.

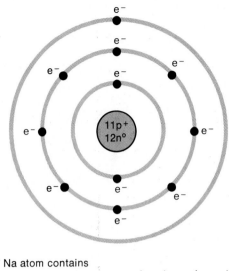

Na atom contains
11 electrons (e^-)   Atomic number = 11
11 protons (p^+)   Atomic weight = 23
12 neutrons (n^o)

chemically in the same manner. Therefore, any of the isotopes of oxygen might play the same role in the metabolic reactions of an organism.

Although some of the isotopes of an element may be stable, others may have unstable atomic nuclei that tend to decompose, releasing energy or pieces of themselves. Such unstable isotopes are called *radioactive*, and the energy or atomic fragments they give off are called *atomic radiations*. Examples of elements that have radioactive isotopes include oxygen, iodine, iron, phosphorus, and cobalt.

1. What is the relationship between matter and elements?
2. What elements are most common in the human body?
3. How are electrons, protons, and neutrons positioned within an atom?
4. What is an isotope?
5. What is meant by atomic radiation?

Bonding of Atoms

When atoms combine with other atoms, they gain or lose electrons, or share electrons with other atoms.

The electrons of an atom are arranged in one or more *shells* around the nucleus. The maximum number of electrons that each of the first three inner shells can hold is as follows:

First shell (closest to the nucleus)   2 electrons
Second shell   8 electrons
Third shell   8 electrons

(The third shell can hold 8 electrons for elements up to atomic number 20. More complex atoms may have as many as 18 electrons in the third shell.)

The arrangements of electrons within the shells of atoms can be represented by simplified diagrams such as those in figure 2.4. Notice that the single electron of a hydrogen atom is located in the first shell, the 2 electrons of a helium atom fill its first shell, and the 3 electrons of a lithium atom are arranged with 2 in the first shell and 1 in the second shell.

Atoms such as helium, whose outermost electron shells are filled, have stable structure and are chemically inactive (inert). Atoms with incompletely filled outer shells, such as those of hydrogen or lithium, tend to gain, lose, or share electrons in ways that empty or fill their outer shells. In this way they achieve stable structures.

An atom of sodium, for example, has 11 electrons, arranged as shown in figure 2.5: 2 in the first shell, 8 in the second shell, and 1 in the third shell. This atom tends to lose the single electron from its outer shell, which leaves the second shell filled and the form stable.

A chlorine atom has 17 electrons arranged with 2 in the first shell, 8 in the second shell, and 7 in the third shell. An atom of this type will tend to accept a single electron, thus filling its outer shell and achieving a stable form.

Figure 2.6

Figure 2.6
(*a*) If a sodium atom loses an electron to a chlorine atom, the sodium atom becomes a sodium ion and the chlorine atom becomes a chloride ion. (*b*) These oppositely charged particles are attracted electrically to one another and become united by an electrovalent bond.

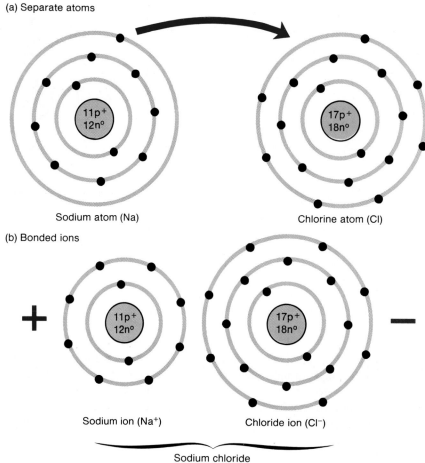

(a) Separate atoms

Sodium atom (Na) Chlorine atom (Cl)

(b) Bonded ions

Sodium ion (Na$^+$) Chloride ion (Cl$^-$)

Sodium chloride

Since each sodium atom tends to lose a single electron and each chlorine atom tends to accept a single electron, sodium and chlorine atoms will react together. During this reaction, a sodium atom loses an electron and is left with 11 protons (11+) in its nucleus and only 10 electrons (10−). As a result, the atom develops a net electrical charge of 1+ and is symbolized Na$^+$. At the same time, a chlorine atom gains an electron, which leaves it with 17 protons (17+) in its nucleus and 18 electrons (18−). Thus it develops a net electrical charge of 1−. Such an atom is symbolized Cl$^-$.

Atoms that have become electrically charged by gaining or losing electrons are called **ions,** and two ions with opposite charges are electrically attracted to one another. When this happens, a chemical bond called an *electrovalent bond* (ionic bond) is created between them, as shown in figure 2.6. When sodium ions (Na$^+$) and chloride ions (Cl$^-$) unite in this manner, the compound sodium chloride (NaCl, common salt) is formed.

Similarly, a hydrogen atom may lose its single electron and become a hydrogen ion (H$^+$). Such an ion can bond with a chloride ion (Cl$^-$) to form hydrogen chloride (HCl, hydrochloric acid).

Atoms can also bond together by sharing electrons rather than by gaining or losing them. A hydrogen atom, for example, has 1 electron in its first shell, but needs 2 to achieve a stable structure. It may fill this shell by combining with another hydrogen atom in such a way that the 2 atoms share a pair of electrons. As figure 2.7 shows, the 2 electrons then encircle the nuclei of both atoms so that each achieves a stable form. In this case, the atoms are held together by a *covalent bond*.

Carbon atoms, with 2 electrons in their first shells and 4 electrons in their second shells, always form covalent bonds when they unite with other atoms. In fact, carbon atoms may bond to other carbon atoms in such a way that 2 atoms share 1 or more pairs of electrons. If 1 pair of electrons is shared, the resulting bond is called a *single covalent bond*; if 2 pairs of electrons are shared, the bond is called a *double covalent bond*.

Figure 2.7
A hydrogen molecule is formed when 2 hydrogen atoms share a pair of electrons and are united by a covalent bond.

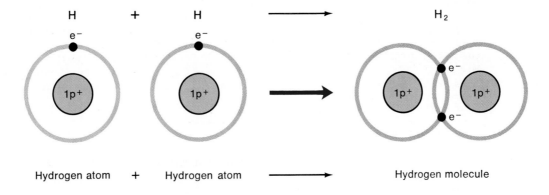

Figure 2.8
Hydrogen molecules (H₂) combine with oxygen molecules (O₂) to form water molecules (H₂O).

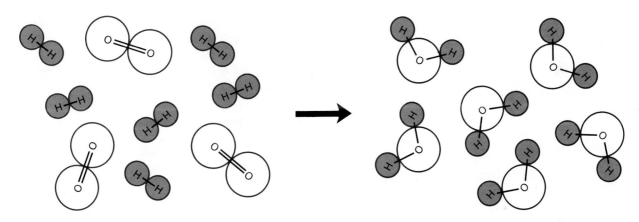

Molecules and Compounds

When 2 or more atoms bond together, they form a new kind of particle called a **molecule.** When atoms of the same element combine, they produce molecules of that element. The gases of hydrogen (H_2), oxygen (O_2), and nitrogen (N_2) contain such molecules.

When atoms of different elements combine, molecules of substances called **compounds** form. Two atoms of hydrogen, for example, can combine with 1 atom of oxygen to produce a molecule of the compound water (H_2O), as shown in figure 2.8. Sugar, table salt, natural gas, alcohol, and most drugs are examples of compounds.

A molecule of a compound always contains definite kinds and numbers of atoms. A molecule of water (H_2O), for instance, always contains 2 hydrogen atoms and 1 oxygen atom. If 2 hydrogen atoms combine with 2 oxygen atoms, the compound formed is not water, but hydrogen peroxide (H_2O_2).

Chart 2.4 lists some particles of matter and their characteristics.

1. What is an ion?
2. Describe two ways in which atoms may combine with other atoms.
3. Distinguish between a molecule and a compound.

Formulas

The numbers and kinds of atoms in a molecule can be represented by a *molecular formula.* Such a formula consists of the symbols of the elements in the molecule together with numbers to indicate how many atoms of each element are present. For example, the molecular formula for water is H_2O, which means there are 2 atoms of hydrogen and 1 atom of oxygen in each molecule. The molecular formula for the sugar called glucose is $C_6H_{12}O_6$, which means there are 6 atoms of carbon, 12 atoms of hydrogen, and 6 atoms of oxygen in a molecule.

Usually the atoms of each element will form a specific number of bonds—hydrogen atoms form single bonds, oxygen atoms form 2 bonds, nitrogen atoms form

Chart 2.4 Some particles of matter

Name	Characteristic	Name	Characteristic
Atom	Smallest particle of an element that has the properties of that element.	Neutron (n^0)	Particle with about the same weight as a proton; uncharged and thus electrically neutral; found within the nucleus of an atom.
Electron (e^-)	Extremely small particle with almost no weight; carries a negative electrical charge and is in constant motion around an atomic nucleus.	Ion	Atom that is electrically charged because it has gained or lost 1 or more electrons.
Proton (p^+)	Relatively large atomic particle; carries a positive electrical charge and is found within the nucleus of an atom.	Molecule	Particle formed by the chemical union of 2 or more atoms.

Figure 2.9

Structural formulas of molecules of hydrogen (H_2), oxygen (O_2), water (H_2O), and carbon dioxide (CO_2).

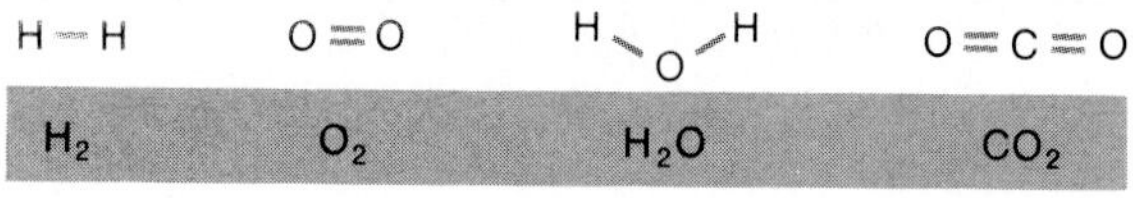

H—H $\qquad$ O=O $\qquad$ H$\diagdown_O\diagup$H $\qquad$ O=C=O

H_2 $\qquad\qquad$ O_2 $\qquad\qquad$ H_2O $\qquad\qquad$ CO_2

Figure 2.10

A water molecule (H_2O) can be represented by a three-dimensional, space-filling model. (a) In this model, the white parts represent hydrogen atoms, and the red part represents oxygen; (b) in this model of a carbon dioxide molecule (CO_2), the black part represents carbon.

(a)

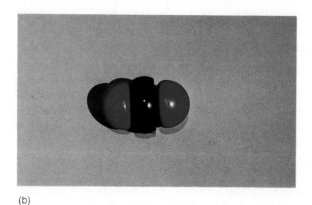

(b)

3 bonds, and carbon atoms form 4 bonds. The bonding capacity of these atoms can be represented by using symbols and lines as follows:

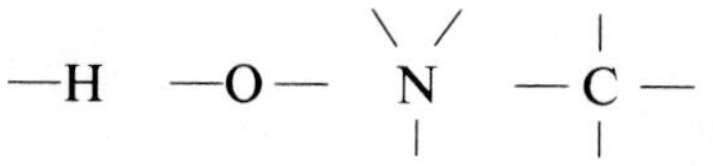

$$-H \qquad -O- \qquad \overset{\diagup}{\underset{|}{N}} \qquad -\overset{|}{\underset{|}{C}}-$$

These representations can be used to show how atoms are bound and arranged in various molecules. Illustrations of this type are called *structural formulas* (figure 2.9). Three-dimensional models that reflect these formulas can be built of colored parts representing different kinds of atoms. (See figure 2.10.)

Chemical Reactions

When atoms, ions, or molecules react together, bonds between atoms are formed or broken. As a result, new combinations of atoms are created.

Two kinds of chemical reactions are **synthesis** and **decomposition.** For example, when 2 or more atoms (reactants) bond together to form a more complex structure (end product), as when atoms of hydrogen and oxygen bond together to form molecules of water, the reaction is called *synthesis.* Such a reaction is symbolized in this way:

$$A + B \rightarrow AB$$

In a *decomposition* reaction the bonds within a reactant molecule break so that simpler molecules, atoms, or ions form. Thus, molecules of water can decompose to yield the end products hydrogen and oxygen. Decomposition is symbolized as follows:

$$AB \rightarrow A + B$$

Synthetic reactions are particularly important in the growth of body parts and the repair of worn or damaged tissues, which involve the buildup of larger

Ionizing Radiation
A Practical Application

The various forms of atomic radiation described previously are said to be **ionizing** because they are capable of causing the formation of ions. Such ions are produced when radiation is absorbed by matter. At the same time, the radiation imparts enough energy to add or remove electrons from electrically neutral atoms or molecules. (See figure 2.11.) Other forms of *ionizing radiation* include X rays from X-ray tubes, cosmic rays that arrive continually from outer space, and neutrons that sometimes are forced out of atomic nuclei.

The ions produced or the electrons dislodged from atoms by ionizing radiation may, in turn, cause changes in other nearby substances. Such changes may lead to a variety of effects in living things, including clouding of the eye lens, the formation of cancers, and interference with normal growth and development. Consequently, ionizing radiation can constitute a health hazard, and any unnecessary exposure to it should be limited whenever possible to minimize health risks.

The quantity of ionizing radiation to which specific individuals are exposed varies greatly; however, in the United States, members of the general population are exposed to very low levels. About 50% of this exposure comes from *background radiation,* which originates from natural environmental sources. These sources include cosmic rays from outer space, radioactive elements that occur in the earth's surface, rocks and clay used in building materials that contain small amounts of radioactive elements, and radioactive substances such as phosphorus-40 and carbon-14 that are found normally in the body. Since background radiation is part of the natural environment, little if anything can be done to reduce exposure to it.

The next largest source of exposure to ionizing radiation (about 40% of the total) comes from X rays and radioactive substances used in medical and dental procedures to diagnose and treat illness and disease. Although

Figure 2.11

(*a*) Ionizing radiation may dislodge an electron from an electrically neutral hydrogen atom; (*b*) without its electron, the hydrogen atom becomes a positively charged hydrogen ion (H^+).

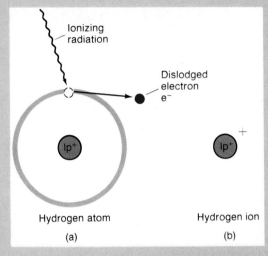

Ionizing radiation

Dislodged electron
e^-

$1p^+$

$1p^+$

Hydrogen atom

Hydrogen ion

(a)

(b)

this source of exposure can be controlled, the benefits derived from the medical and dental uses of ionizing radiation are thought to outweigh any health risks.

The remaining proportion of the total exposure to ionizing radiation comes from a variety of sources. They include (a) the development, production, and testing of nuclear weapons; (b) activities involved with the mining and processing of radioactive minerals; (c) the use of radioactive fuels in nuclear power plants to generate electricity; and (d) the inclusion of radioactive substances in consumer products, such as luminescent dials, certain smoke detectors, and some components of color television sets.

---

molecules from smaller ones. When food substances are digested or energy is released from food substances through cellular respiration, the processes are largely decomposition reactions.

A third type of chemical reaction is an **exchange reaction.** In an exchange reaction, parts of 2 different kinds of molecules trade positions with one another. The reaction is symbolized as follows:

$$AB + CD \rightarrow AD + CB$$

An exchange reaction occurs when an acid reacts with a base to produce water and a salt.

Many chemical reactions are reversible. This means the end product or products of the reaction can change back to the reactant or reactants that originally underwent the reaction. A **reversible reaction** can be symbolized using a double arrow, as follows:

$$A + B \rightleftharpoons AB$$

Whether a reversible reaction proceeds in one direction or another depends on such factors as the relative proportions of reactants and end products, the amount of energy available to the reaction, and the presence or absence of *catalysts*—particular atoms or molecules that can change the rate of a reaction without themselves being consumed by the reaction.

Figure 2.12

When crystals of table salt (NaCl) dissolve in water, ions of sodium (Na⁺) and ions of chlorine (Cl⁻) are released.

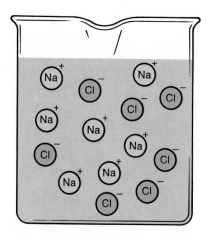

Chart 2.5	Types of electrolytes	
	Characteristic	Examples
Acid	Ionizes to release hydrogen ions (H^+).	Carbonic acid, hydrochloric acid, acetic acid, phosphoric acid
Base	Ionizes to release ions that can combine with hydrogen ions.	Sodium hydroxide, potassium hydroxide, magnesium hydroxide, aluminum hydroxide
Salt	Substance formed by the reaction between an acid and a base.	Sodium chloride, aluminum chloride, magnesium sulfate

Chart 2.5 summarizes the 3 types of electrolytes—acids, bases, and salts.

Acids, Bases, and Salts

Some compounds release ions (ionize) when they dissolve in water or react with water molecules. Sodium chloride (NaCl), for example, releases sodium ions (Na^+) and chloride ions (Cl^-) when it dissolves, as is represented by the following (figure 2.12):

$$NaCl \rightarrow Na^+ + Cl^-$$

Since the resulting solution contains electrically charged particles (ions), it will conduct an electric current. Substances that ionize in water are, therefore, known as **electrolytes.** Electrolytes that release hydrogen ions (H^+) in water are called **acids.** For example, the compound, carbonic acid (H_2CO_3), ionizes in water to release hydrogen ions and bicarbonate ions (HCO_3^-).[1]

$$H_2CO_3 \rightarrow H^+ + HCO_3^-$$

Electrolytes that release ions that combine with hydrogen ions are called **bases.** The compound, sodium hydroxide (NaOH), for example, ionizes in water to release hydroxyl ions (OH^-).

$$NaOH \rightarrow Na^+ + OH^-$$

The hydroxyl ions, in turn, can combine with hydrogen ions to form water; thus, sodium hydroxide is a base.

Acids that release hydrogen ions and bases that release hydroxyl ions can react to form water and electrolytes called **salts.** For example, hydrochloric acid and sodium hydroxide react to form water and sodium chloride. This reaction is symbolized as follows:

$$HCl + NaOH \rightarrow H_2O + NaCl$$

Acid and Base Concentrations

The chemical reactions involved with life processes are often affected by the presence of hydrogen and hydroxyl ions; therefore the concentrations of these ions in body fluids are important. Such concentrations are measured in units of **pH.** These units indicate the concentration of *hydrogen ions* in a solution expressed in grams of ions per liter. Actually, the pH value is equal to the negative logarithm of the hydrogen ion concentration. For example, a solution with a hydrogen ion concentration of 0.1 grams per liter has a pH value of 1.0; a concentration of 0.01 g H^+/l has pH 2.0; 0.001 g H^+/l has pH 3.0, and so forth. Thus, between each whole number on the pH scale (which extends from pH 0 to pH 14.0), there is a tenfold difference in the hydrogen ion concentration, which decreases as pH rises.

In pure water, which ionizes only slightly, the hydrogen ion concentration is 0.0000001 g/l, and the pH is 7.0. Since water ionizes to release equal numbers of acidic hydrogen ions (H^+) and basic hydroxyl ions (OH^-), it is said to be *neutral.*

$$H_2O \rightarrow H^+ + OH^-$$

Therefore, solutions with more hydrogen ions than hydroxyl ions are said to be *acidic;* they have pH values of less than 7.0. (See figure 2.13.) Solutions with fewer hydrogen ions than hydroxyl ions are said to be *basic* (alkaline); they have pH values of more than 7.0. Chart 2.6 illustrates the relationship between hydrogen ion concentration and pH. Figure 2.14 indicates the pH values of some common substances.

1. Some ions, such as HCO_3^- ions, contain two or more atoms. However, such a group behaves like a single atom and usually remains unchanged in a chemical reaction.

Figure 2.13

As the concentration of hydrogen ions (H⁺) increases, a solution becomes more acidic, and the pH value decreases. As the concentration of hydrogen acceptors (OH⁻, HCO₃⁻, etc.) increases, a solution becomes more basic, and the pH value increases.

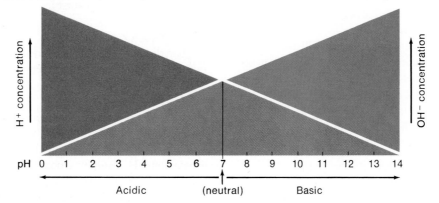

Chart 2.6	Hydrogen ion concentrations and pH	
Grams of H⁺ per liter	**pH**	
1.0	0	
0.1	1	
0.01	2	
0.001	3	Increasingly acidic
0.0001	4	
0.00001	5	
0.000001	6	
0.0000001	7	Neutral—neither acidic nor basic
0.00000001	8	
0.000000001	9	
0.0000000001	10	
0.00000000001	11	Increasingly basic
0.000000000001	12	
0.0000000000001	13	
0.00000000000001	14	

Note in figure 2.14 that the pH of human blood is normally about 7.4. If this pH value drops below 7.4, the person is said to have *acidosis;* and if it rises above 7.4, the condition is called *alkalosis.* Usually a person cannot survive if the pH drops to 7.0 or rises to 7.8 for more than a few minutes without medical intervention.

1. What is meant by a formula?
2. Describe three kinds of chemical reactions.
3. Compare the characteristics of an acid with those of a base.
4. What is meant by pH?

Figure 2.14

Approximate pH values of some common substances.

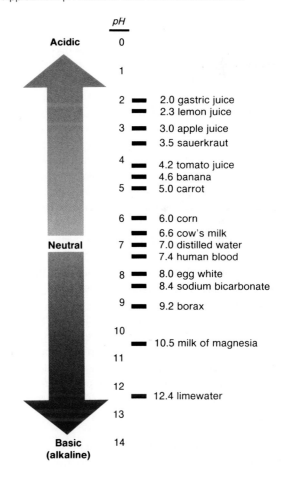

Chemical Constituents of Cells

The chemicals that enter into metabolic reactions or are produced by them can be divided into two large groups. Except for a few simple molecules, those that contain carbon atoms are called **organic,** and those that lack carbon are called **inorganic.**

Generally, inorganic substances will dissolve in water or react with water to produce ions; thus they are electrolytes. Some organic compounds will dissolve in water also, but as a group they are more likely to dissolve in organic liquids like ether or alcohol. Those that will dissolve in water usually do not release ions, and are, therefore, called *nonelectrolytes.*

Inorganic Substances

Among the inorganic substances common in cells are water, oxygen, carbon dioxide, and certain salts.

Water

Water (H_2O) is the most abundant compound in living material and is responsible for about two-thirds of the weight of an adult human. It is the major ingredient of blood and other body fluids, including those within cells.

When substances dissolve in water, relatively large pieces of the material break into smaller ones, and eventually molecular-sized particles or ions result. These tiny particles are much more likely to react with one another than were the original large pieces. Consequently, most metabolic reactions occur in water.

Water also plays an important role in the transportation of chemicals within the body. The aqueous portion of blood, for example, carries many vital substances such as oxygen, sugars, salts, and vitamins, from the organs of digestive and respiratory systems to the body cells. Water also carries waste materials, such as carbon dioxide and urea, from these cells to the kidneys, which remove them from the blood and excrete them to the outside of the body.

In addition, water can absorb and transport heat. For instance, heat released from muscle cells during exercise can be carried by the blood from deeper parts to the surface. At the same time, water released by skin cells in the form of perspiration can carry heat away by evaporation.

Oxygen

Molecules of oxygen gas (O_2) enter the body through the respiratory organs and are transported to cells by the blood. The oxygen is used by cellular organelles in the process of releasing energy from glucose and certain other molecules. The energy is needed to drive the cell's metabolic activities. A continuing supply of oxygen is necessary for cell survival and, ultimately, for the survival of the organism.

Carbon Dioxide

Carbon dioxide (CO_2) is one of the simple, carbon-containing compounds of the inorganic group. It is produced as a waste product when energy is released during respiration. As it moves from the cells into the surrounding body fluids and blood, most of the carbon dioxide reacts with water to form a weak acid (carbonic acid, H_2CO_3). This acid ionizes, releasing hydrogen ions (H^+) and bicarbonate ions (HCO_3^-), which the blood carries to the respiratory organs. Here the chemical reactions reverse, and carbon dioxide gas is given off into the air in the lungs.

Inorganic Salts

Various kinds of inorganic salts are common in body parts and fluids. They are the sources of many necessary ions, including ions of sodium (Na^+), chloride (Cl^-), potassium (K^+), calcium (Ca^{+2}), magnesium (Mg^{+2}), phosphate (PO_4^{-3}), carbonate (CO_3^{-2}), bicarbonate (HCO_3^-), and sulfate (SO_4^{-2}). These ions play important roles in metabolic processes including those involved in maintenance of proper water concentrations in body fluids, blood clotting, bone development, energy transfer within cells, and muscle and nerve functions.

These electrolytes must not only be present, but they must be present in the proper concentrations, both inside and outside cells, if homeostasis is to be maintained. Such a condition is called **electrolyte balance.** In certain diseases, this balance tends to be lost, and modern medical treatment places considerable emphasis on restoring it.

Chart 2.7 summarizes the functions of some of the inorganic substances that commonly occur in cells.

---

1. What is the difference between an organic and an inorganic molecule?
2. What is the difference between an electrolyte and a nonelectrolyte?
3. Define electrolyte balance.

Organic Substances

Important groups of organic substances found in cells include carbohydrates, lipids, proteins, and nucleic acids.

Chart 2.7	Inorganic substances common in cells	
Substance	Symbol or formula	Functions
Water	H_2O	Major ingredient of body fluids; medium in which most biochemical reactions occur; transports various chemical substances; helps regulate body temperature.
Oxygen	O_2	Used in the release of energy from glucose molecules during respiration.
Carbon dioxide	CO_2	Waste product that results from the respiration of glucose molecules; reacts with water to form carbonic acid.
Ions of inorganic salts	Na^+, Cl^-, K^+, Ca^{+2}, Mg^{+2}, PO_4^{-3}, CO_3^{-2}, HCO_3^-, SO_4^{-2}	Play important roles in various metabolic processes including those involved in the maintenance of water balance, blood clotting, bone development, energy transfer within cells, and muscle and nerve actions.

Carbohydrates

Carbohydrates provide much of the energy that cells need. They also supply materials that are needed to build certain cell structures, and they often are stored as reserve energy supplies.

Carbohydrate molecules contain atoms of carbon, hydrogen, and oxygen. These molecules usually have twice as many hydrogen as oxygen atoms, the same ratio of hydrogen to oxygen that occurs in water molecules (H_2O). This ratio is easy to see in the molecular formulas of glucose ($C_6H_{12}O_6$) and sucrose, or table sugar ($C_{12}H_{22}O_{11}$).

The carbon atoms of carbohydrate molecules are joined in chains whose lengths vary with the kind of carbohydrate. For example, those with shorter chains are called **sugars.** Sugars with 6 carbon atoms are known as *simple sugars* or *monosaccharides,* and they are the building blocks of more complex carbohydrate molecules. The simple sugars include glucose, fructose, and galactose. Figures 2.15 and 2.16 illustrate the molecular structure of glucose.

In complex carbohydrates, the simple sugars are bound to form molecules of varying sizes, as shown in figure 2.17. Some, like sucrose, maltose, and lactose, are *double sugars* or *disaccharides*, whose molecules each contain 2 building blocks. Others are made up of many simple sugar units joined together to form *polysaccharides*, such as plant starch. Starch molecules consist of highly branched chains, each containing from 24 to 30 glucose units. Because the number of glucose units within a starch molecule may vary, starch is symbolized by the formula $(C_6H_{10}O_5)_x$, with x equal to the number of glucose units in the molecule.

Animals, including humans, synthesize a polysaccharide similar to starch called *glycogen*. Its molecules also consist of branched chains of sugar units, but each chain contains only about a dozen glucose units.

Figure 2.15
Left, some glucose molecules ($C_6H_{12}O_6$) have a straight chain of carbon atoms; *right*, more commonly, glucose molecules form a ring structure, which can be symbolized with this shape:

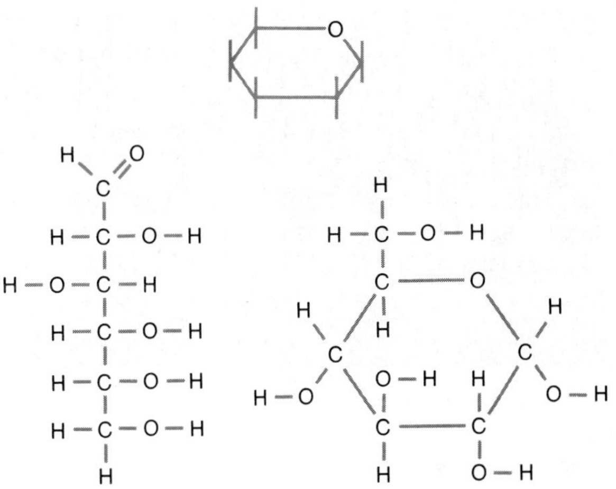

Figure 2.16
Space-filling model of a glucose molecule.

Figure 2.17

(a) A simple sugar molecule consists of one 6-carbon atom building block; (b) a double sugar molecule consists of two building blocks; (c) a complex carbohydrate molecule consists of many building blocks.

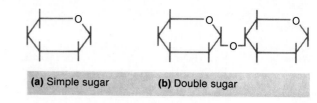

(a) Simple sugar   (b) Double sugar

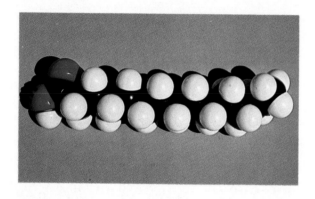

(c) Complex carbohydrate

Figure 2.18

Space-filling model of a fatty acid molecule.

Lipids

Lipids represent a group of organic substances that are insoluble in water, but are soluble in organic solvents such as ether and chloroform. They include a number of compounds, such as phospholipids and cholesterol, that have vital functions in cells and are important constituents of cell membranes (see chapter 3). The most common lipids, however, are *fats.* Fats are used to build cell parts and to supply energy for cellular activities. In fact, fat molecules can supply more energy, gram for gram, than can carbohydrate molecules.

Like carbohydrates, fat molecules are composed of carbon, hydrogen, and oxygen atoms. However, they contain a much smaller proportion of oxygen than do carbohydrates. This fact can be illustrated by the formula for the fat, tristearin, $C_{57}H_{110}O_6$.

The building blocks of fat molecules are **fatty acids** and **glycerol.** These smaller molecules are united so that each glycerol molecule is combined with 3 fatty acid molecules. The result is a single fat molecule or *triglyceride* (figures 2.18 and 2.19).

Although the glycerol portion of every fat molecule is the same, there are many kinds of fatty acids and, therefore, many kinds of fats. Fatty acid molecules differ in the lengths of their carbon atom chains and in the ways the carbon atoms are combined. In some cases, the carbon atoms are all joined by single carbon-carbon bonds. This type of fatty acid is said to be *saturated*; that is, each carbon atom is bound to as many hydrogen atoms as possible and is thus saturated with hydrogen atoms. Other fatty acids have one or more double bonds between carbon atoms. Those with one double bond are called *monounsaturated fatty acids,* and those with two or more double bonds are said to be *polyunsaturated.* Similarly, fat molecules that contain only saturated fatty acids are called **saturated fats,** and those that include unsaturated fatty acids are called **unsaturated fats.** (See figure 2.20.)

Thus each kind of fat molecule has particular kinds of fatty acids joined to its glycerol portion, and each has special properties.

A diet that contains a high proportion of saturated fat seems to increase a person's chance of developing a serious disease of the heart and arteries (atherosclerosis). For this reason, many nutritionists are recommending that polyunsaturated fats be substituted for some dietary saturated fats.

As a rule, saturated fats are more abundant in fatty foods that have higher melting points and thus are solids at room temperature, such as butter, lard, and most other animal fats. Unsaturated fats, on the other hand, are likely to be plentiful in fatty foods that have lower melting points and are, therefore, liquids at room temperature, such as soft margarine and various seed oils, including corn oil and soybean oil. There are exceptions, however, since coconut oil is high in saturated fat.

A *phospholipid* molecule is similar to a fat molecule in that it contains a glycerol portion and fatty acid chains. The phospholipid, however, has only two fatty acid chains, and in place of the third one there is a portion containing a phosphate group. This phosphate portion is soluble in water (hydrophilic) and forms the "head" of the molecule, while the fatty acid portion is insoluble in water (hydrophobic) and forms a "tail." (See figure 2.21.)

Figure 2.19

A fat molecule consists of a glycerol portion and three fatty acid portions.

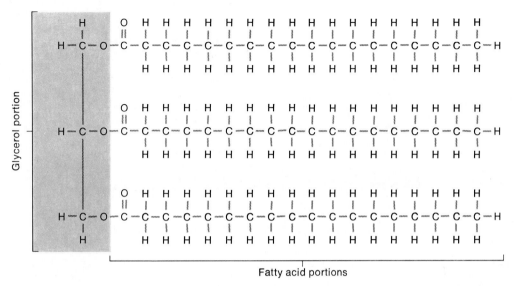

Figure 2.20

What are the major differences between (a) a molecule of a saturated fatty acid and (b) a molecule of unsaturated fatty acid?

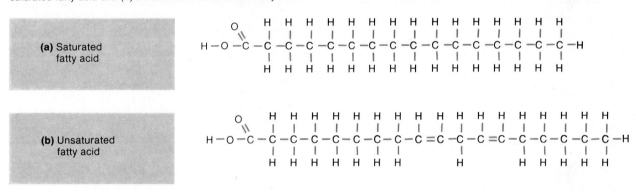

(a) Saturated fatty acid

(b) Unsaturated fatty acid

Figure 2.21

(a) A fat molecule contains a glycerol portion and three fatty acids; (b) in a phospholipid molecule, one fatty acid is replaced by a phosphoric acid-containing group.

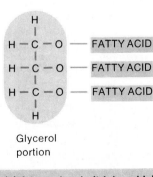

Glycerol portion

(a) A fat molecule (triglyceride)

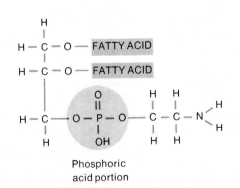

Phosphoric acid portion

(b) A phospholipid molecule (cephalin)

Figure 2.22

Structural formula for cholesterol, a lipid with a relatively complex molecule.

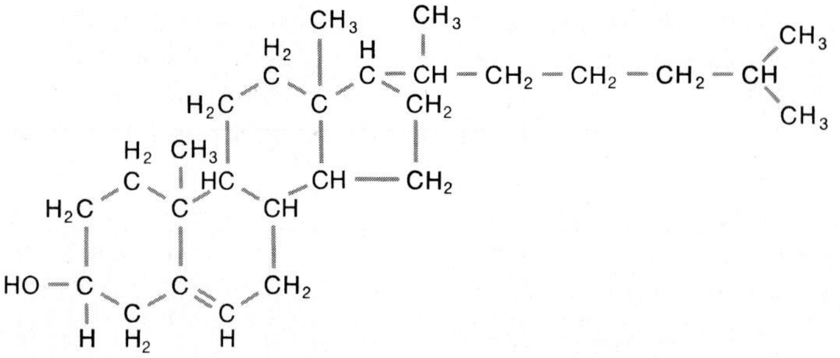

Cholesterol

Figure 2.23

Some representative amino acids and their structural formulas. Each amino acid molecule has a particular shape due to the arrangement of its parts. Note the portion that is common to all amino acid molecules.

Amino acid	Structural formulas	Amino acid	Structural formulas
Alanine		Phenylalanine	
Valine		Tyrosine	
Cysteine		Histidine	

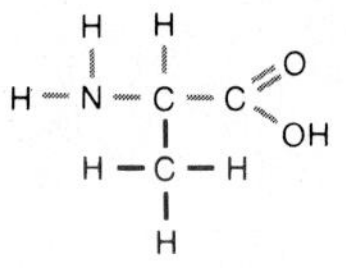

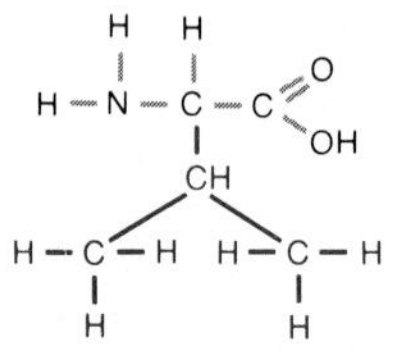

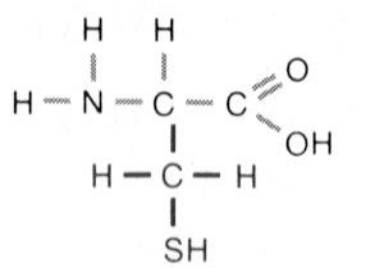

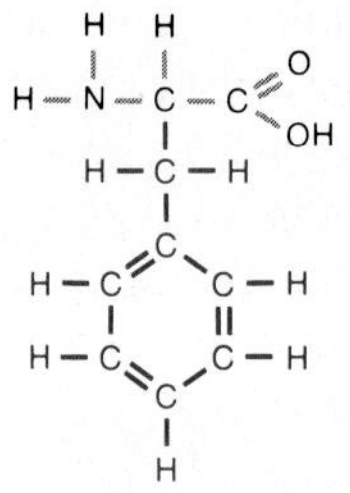

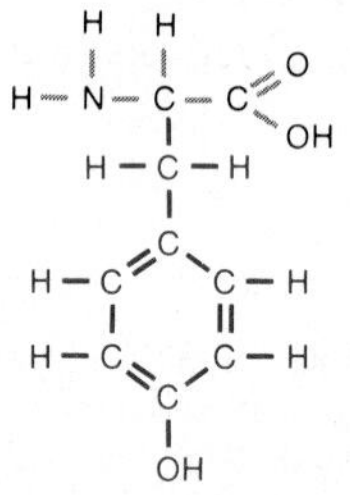

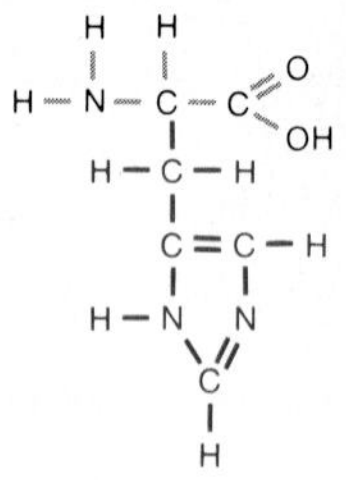

A *cholesterol* molecule is relatively large and complex. Each contains four rings of carbon atoms that are fused together, giving the molecule a somewhat rigid structure. As in the case of a phospholipid, a cholesterol molecule has a water soluble portion and a water insoluble portion. (See figure 2.22.)

Proteins

Proteins serve as structural materials, energy sources, chemical messengers (hormones), and receptors on various surfaces that bond to particular kinds of molecules. Other proteins play vital roles in all metabolic processes by acting as **enzymes.** Enzymes are proteins that act as catalysts. That is, they have special properties and can speed up specific chemical reactions without being changed or used up themselves. (Enzymes are discussed in more detail in chapter 4.)

Like carbohydrates and lipids, proteins contain atoms of carbon, hydrogen, and oxygen. In addition, proteins always contain nitrogen atoms, and sometimes sulfur atoms as well. The building blocks of proteins are smaller molecules called **amino acids** (figures 2.23 and 2.24). About 20 different amino acids occur commonly in proteins. They are joined together in chains, varying in length from less than 100 to more than 50,000 amino acids. The hemoglobin molecules, which carry oxygen and are responsible for the red color of blood, each contain a protein portion that includes 574 amino acid units arranged in four chains. The amino acid sequence of a specific protein (parathyroid hormone) is shown in figure 13.3.

Each type of protein contains specific numbers and kinds of amino acids arranged in particular sequences and folded up into a unique three-dimensional structure. Consequently, different kinds of protein molecules have different shapes that are related to their particular functions (figures 2.25 and 2.26). The molecules of some (fibrous proteins) are long and threadlike. They have special functions, such as binding body parts together or forming the fine threads of a blood clot. Other protein molecules (globular proteins) are coiled or folded into compact masses. Enzyme molecules are usually of this type.

The special shape of a protein molecule is maintained largely by the presence of *hydrogen bonds* by which parts of the molecule are held to other parts of itself. A hydrogen bond is a weak chemical bond that occurs between a hydrogen atom that is covalently bound to a nitrogen or an oxygen atom and at the same time is shared with another nitrogen or oxygen atom nearby. (See figure 2.27.)

Figure 2.24
Space-filling model of the amino acid, glycine. The blue part represents a nitrogen atom.

Figure 2.25
Each different shape in this chain represents a different amino acid molecule. The whole chain represents a portion of a protein molecule.

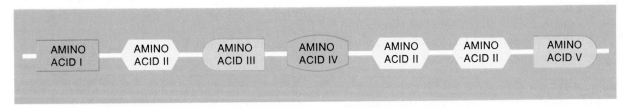

Figure 2.27

Hydrogen bonds, represented by dashed lines in this illustration, help to maintain the special shape of a protein molecule.

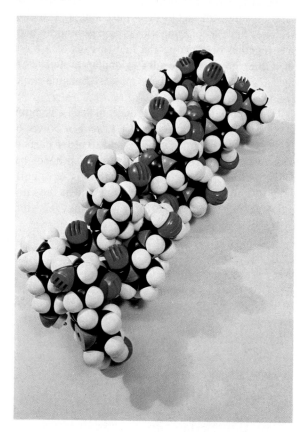

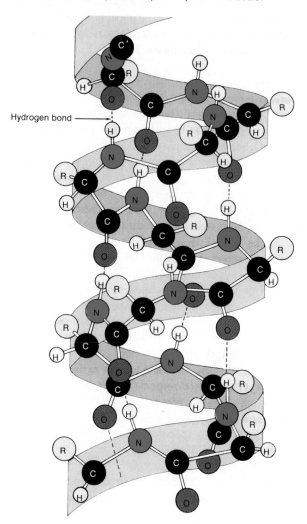

Hydrogen bond →

Portion of a protein molecule

The complex shapes of protein molecules can be altered by exposure to excessive heat, radiation, electricity, or various chemicals. When this occurs, many of the weak hydrogen bonds are broken, the molecules become disorganized, and they are said to be *denatured.* At the same time, they lose their unique properties and behave differently. For example, the protein in egg white is denatured when it is heated. Heating the egg white causes it to change from a liquid to a solid—a change that cannot be reversed. Similarly, if cellular proteins are denatured, they are permanently changed and become nonfunctional. Such changes may threaten the life of the cells.

Nucleic Acids

Of all the compounds present in cells, the most fundamental are the **nucleic acids.** These organic substances control cellular activities in very special ways.

Nucleic acid molecules are generally very large and complex. They contain atoms of carbon, hydrogen, oxygen, nitrogen, and phosphorus, which are bound into building blocks called **nucleotides.** Each nucleotide consists of a 5-carbon *sugar* (ribose or deoxyribose), a *phosphate group*, and one of several *organic bases* (figure 2.28). A nucleic acid molecule consists of many nucleotides united in a chain, as shown in figure 2.29.

There are two major types of nucleic acids. One type, composed of molecules whose nucleotides contain ribose sugar, is called **RNA** (ribonucleic acid). The nucleotides of the second type contain deoxyribose sugar; these are called **DNA** (deoxyribonucleic acid). Figure 2.30 compares the structure of these two 5-carbon sugars.

DNA molecules store information in a kind of molecular code. This information tells cell parts how to construct specific protein molecules, and these proteins, many acting as enzymes, are responsible for all of the metabolic reactions that occur in cells. RNA molecules help to synthesize these proteins.

Chart 2.8 Organic compounds in cells				
Compound	Elements present	Building blocks	Functions	Examples
Carbohydrates	C,H,O	Simple sugar	Energy, cell structure	Glucose, starch
Fats	C,H,O	Glycerol, fatty acids	Energy, cell structure	Tristearin, tripalmitin
Proteins	C,H,O,N (often S)	Amino acids	Cell structure, enzymes, energy	Albumins, hemoglobin
Nucleic acids	C,H,O,N,P	Nucleotides	Store information, synthesize proteins, control cell activities	RNA, DNA

Figure 2.28

A nucleotide consists of a 5-carbon sugar (S), a phosphate group (P), and an organic base (B).

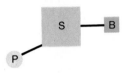

Figure 2.29

A nucleic acid molecule consists of nucleotides joined in a chain.

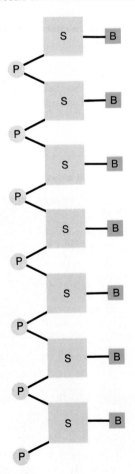

Figure 2.30

How are the molecules of ribose and deoxyribose different? How are they similar to a molecule of glucose?

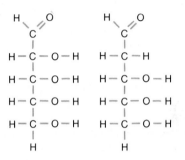

Ribose   **Deoxyribose**

DNA molecules also have a unique ability to duplicate or replicate themselves. They replicate prior to cell reproduction, and each newly formed cell receives an exact copy of its parent's DNA molecules. These molecules contain the instructions for carrying on vital life processes.

Chapter 4 discusses storage of this information in nucleic acid molecules, its use in the manufacture of protein molecules, and the function of these proteins in metabolic reactions.

Chart 2.8 summarizes these four groups of organic compounds.

1. Compare the chemical composition of carbohydrates, lipids, proteins, and nucleic acids.
2. How does an enzyme affect a chemical reaction?
3. What is likely to happen to a protein molecule that is exposed to excessive heat or radiation?
4. What are the functions of DNA and RNA?

PET and NMR Imaging
A Clinical Application

Two of the more recently developed techniques for producing images of internal structures without invading the body are PET and NMR imaging.

PET, or positron emission tomography, makes use of a special group of radioactive isotopes that naturally give rise to *positrons*—positively charged electrons. These elements include carbon-11, nitrogen-13, oxygen-15, and fluorine-18. Whenever a positron is released by one of these substances, it interacts with a nearby negatively charged electron, and the two particles are destroyed (annihilated). At the moment of destruction, two gamma rays appear and travel away from each other in opposite directions. As was mentioned in the previous discussion of radioactive isotopes, such gamma radiation can be detected by using special equipment.

To produce an image of a body part, a person is given a metabolically active compound to which a positron-emitting isotope has been bound chemically. If the brain is being studied, for example, a form of glucose containing fluorine-18 (flurodeoxyglucose) might be used. After the isotope-tagged compound is taken up by the brain tissues, the head is placed within a circular array of radiation detectors.

Whenever the radiation detectors register two gamma rays being emitted simultaneously and traveling in exactly opposite directions (the result of positron-electron annihilation), they record the event. Such data are collected and combined by a computer, which in turn generates a cross-sectional image. This image indicates the location and relative concentration of the radioactive isotope in various regions of the brain, and thus it can be used to study the particular parts metabolizing glucose. Such information, of course, is useful for investigating the physiology of the brain as well as its structure. (See figure 2.31.)

PET imaging also is used to evaluate various brain abnormalities and diseases and to study blood flow in vessels supplying the brain and heart.

In the procedure called **NMR, or nuclear magnetic resonance,** imaging, the body or part being examined is placed in a chamber surrounded by a powerful magnet and a special radio antenna. When the device is operating, the magnetic field created by the magnet affects the alignment and spin of certain types of atoms within the living materials. At the same time, a second rotating magnetic field is adjusted to cause particular kinds of atoms (such as the hydrogen atoms in body fluids and organic compounds) to release weak radio waves with characteristic frequencies. The radio waves are received by the nearby antenna and amplified. The amplified signals are then processed by a computer, and it generates a cross-sectional image that is based upon the locations and concentrations of the particular atoms being investigated. (See figure 2.32.)

Unlike CAT scanning and PET imaging, which depend upon X ray or gamma rays, NMR images are obtained without exposing a patient to ionizing radiation. Also, in addition to visualizing anatomical structures, NMR imaging can provide information concerning the physiology of body parts by revealing something about their chemical composition.

NMR imaging is particularly useful for distinguishing normal and cancerous tissues, depicting blood vessels, and assessing damage sustained by the heart muscle as a result of a heart attack.

Figure 2.31
PET image of a human brain.

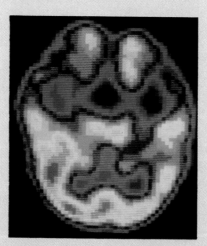

Figure 2.32
NMR image of a human brain (transverse section).

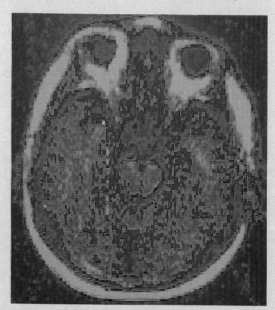

Clinical Terms Related to the Use of Radioactive Isotopes

alpha particle (al'fah par'tĭ-kl) positively charged particle, consisting of two protons and two neutrons, given off by certain radioactive isotopes.

beta particle (ba'tah par'tĭ-kl) negatively charged particle (electron) given off by certain radioactive isotopes; sometimes used to treat diseased tissues on or near the surface of organs.

cobalt 60 (ko'bawlt) radioactive isotope of the element cobalt; used as a source of gamma radiation in the treatment of cancer.

cobalt unit (ko'bawlt u'nit) therapy apparatus that contains cobalt 60.

curie (ku're) a unit of radioactivity used to measure the amount of radiation emitted from a radioactive isotope.

dosimetry (do-sim'ĕ-tre) the process of measuring radiation exposure or dose.

gamma ray (gam'ah ra) a form of radiation, similar to X ray, that is given off by certain radioactive isotopes; the most penetrating form of radiation produced by isotopes.

half-life (haf' lĭf) the time required for a radioactive isotope to lose one-half of its radioactivity.

iodine 131 (i'o-dĭn) a radioactive isotope of the element iodine; sometimes used in testing or treating the thyroid gland.

ionizing radiation (i'on-ī-zing ra-de-a'shun) any form of radiation capable of causing atoms to become electrically charged ions.

iron 59 (i'ern) a radioactive isotope of the element iron; sometimes used to test bone marrow.

irradiation (i-ra''de-a'shun) the process of exposing something to radiation.

nuclear medicine (nu'kle-ar med'ĭ-sin) the branch of medicine concerned with the use of radiation.

potassium 42 (po-tas'e-um) a radioactive isotope of the element potassium; sometimes used to study brain tumors.

rad (rad) a unit used in measuring the amount of radiation exposure; *r*adiation *a*bsorbed *d*ose.

radiation sickness (ra''de-a'shun sik'nes) a disease usually resulting from the effects of radiation on the digestive organs.

rem (rem) a unit used to measure ionizing radiation; quantity that produces the same biological effect as one rad of X ray.

scan (skan) a recording of the radiation emitted by radioactive isotopes within certain tissues.

Chapter Summary

Introduction

Chemistry deals with the composition of substances and changes in their composition.

Biochemistry is the chemistry of living things.

Structure of Matter

Matter is anything that has weight and takes up space.

1. Elements and atoms
   a. Naturally occurring matter on earth is composed of 91 elements.
   b. Elements occur most frequently in mixtures or chemically combined in compounds.
   c. Elements are composed of atoms.
   d. Atoms of different elements vary in size, weight, and ways of interacting.

2. Atomic structure
   a. An atom consists of electrons surrounding an atomic nucleus, which contains protons and neutrons.
   b. Electrons are negatively charged, protons positively charged, and neutrons uncharged.
   c. A complete atom is electrically neutral.
   d. The atomic number of an element is equal to the number of protons in each atom; the atomic weight is equal to the number of protons plus the number of neutrons in each atom.

3. Isotopes
   a. Isotopes are atoms with the same atomic number but different atomic weights (due to differing numbers of neutrons).
   b. All the isotopes of an element react chemically in the same manner.
   c. Some isotopes are radioactive and release radiation.

4. Bonding of atoms
   a. When atoms combine, they gain, lose, or share electrons.
   b. Electrons are arranged in shells around atomic nuclei.
   c. Atoms with filled outer shells are inactive, while atoms with incompletely filled outer shells tend to gain, lose, or share electrons and thus achieve a stable structure.
   d. Atoms that lose electrons become positively charged; atoms that gain electrons become negatively charged.
   e. Ions with opposite charges are attracted to one another and become bound by electrovalent bonds; atoms that share electrons become bound by covalent bonds.

5. Molecules and compounds
   a. When 2 or more atoms of the same element are united, a molecule of that element is formed; when atoms of different kinds are united, a molecule of a compound is formed.
   b. Molecules contain definite kinds and numbers of atoms.
6. Formulas
   a. The numbers and kinds of atoms in a molecule can be represented by a molecular formula.
   b. The arrangement of atoms within a molecule can be represented by a structural formula.
7. Chemical reactions
   a. When a chemical reaction occurs, bonds between atoms are broken or created.
   b. Three kinds of chemical reaction are synthesis, in which larger molecules are formed from smaller particles; decomposition, in which smaller particles are formed from larger molecules; and exchange reactions, in which the parts of 2 different molecules trade positions.
   c. Many reactions are reversible, and the direction depends upon the proportion of reactants and end products, the energy available, and the presence or absence of catalysts.
8. Acids, bases, and salts
   a. Compounds that ionize when they dissolve in water are electrolytes.
   b. Electrolytes that release hydrogen ions are acids, and those that release hydroxyl or other ions that react with hydrogen ions are bases.
   c. Acids and bases that release hydroxyl ions react together to form water and electrolytes called salts.
9. Acid and base concentrations
   a. The concentration of hydrogen (H^+) and hydroxyl (OH^-) ions in a solution can be represented by pH.
   b. A solution with equal numbers of H^+ and OH^- is neutral and has a pH of 7.0; a solution with more H^+ than OH^- is acidic (pH less than 7.0); a solution with fewer H^+ than OH^- is basic (pH more than 7.0).
   c. There is a ten-fold difference in hydrogen ion concentration between each whole number in the pH scale.

Chemical Constituents of Cells

Molecules containing carbon atoms are organic and are usually nonelectrolytes; those lacking carbon atoms are inorganic and are usually electrolytes.

1. Inorganic substances
   a. Water is the most abundant compound in cells and serves as a substance in which chemical reactions occur; it also transports chemicals and heat and helps to release excess body heat.
   b. Oxygen is used in releasing energy from glucose and other molecules.
   c. Carbon dioxide is produced when energy is released during respiration.
   d. Inorganic salts provide ions needed in metabolic processes and must be balanced for effective functioning.
   e. Electrolytes must be present in proper concentrations inside and outside of cells.
2. Organic substances
   a. Carbohydrates provide much of the energy needed by cells; their basic building blocks are simple sugar molecules.
   b. Lipids, such as fats, phospholipids, and cholesterol, supply energy and are used to build cell parts; their basic building blocks are molecules of glycerol and fatty acids.
   c. Proteins serve as structural materials, energy sources, chemical messengers, surface receptors, and enzymes.
      (1) Enzymes initiate or speed chemical reactions without being changed themselves.
      (2) The building blocks of proteins are amino acids.
      (3) Different kinds of proteins vary in the numbers and kinds of amino acids they contain, the sequences in which these amino acids are arranged, and their three-dimensional structure.
      (4) Protein molecules can be denatured by exposure to excessive heat, radiation, electricity, or various chemicals.

d. Nucleic acids are the most fundamental chemicals in cells, because they control cell activities.
   (1) The 2 major kinds are RNA and DNA.
   (2) They are composed of building blocks called nucleotides.
   (3) DNA molecules store information that tells a cell how to construct protein molecules, including enzymes.
   (4) RNA molecules help synthesize proteins.
   (5) DNA molecules are duplicated and passed from parent to offspring cells during cell reproduction.

Application of Knowledge

1. What acidic and alkaline substances do you encounter in your living activities? What foods do you eat regularly that are acidic? What alkaline foods do you eat?
2. Using the information on page 56 to distinguish saturated from unsaturated fats, try to list all of the sources of saturated and unsaturated fats you have eaten during the past 24 hours.
3. How would you reassure a patient who will undergo NMR imaging for evaluation of a tumor, and who fears becoming a radiation hazard to family members?

Review Activities

1. Distinguish between chemistry and biochemistry.
2. Define *matter*.
3. Explain the relationship between elements and atoms.
4. Define *compound*.
5. List the 4 most abundant elements in the human body.
6. Describe the major parts of an atom.
7. Distinguish between protons and neutrons.
8. Explain why a complete atom is electrically neutral.
9. Distinguish between atomic number and atomic weight.
10. Define *isotope*.
11. Explain what is meant by atomic radiation.
12. Describe how electrons are arranged within atoms.
13. Explain why some atoms are chemically inert.
14. Distinguish between an electrovalent bond and a covalent bond.
15. Distinguish between a single covalent bond and a double covalent bond.
16. Explain the relationship between molecules and compounds.
17. Distinguish between a molecular formula and a structural formula.
18. Describe three major types of chemical reactions.
19. Explain what is meant by reversible reaction.
20. Define *catalyst*.
21. Define *acid, base, salt,* and *electrolyte*.
22. Explain what is meant by pH, and describe the pH scale.
23. Distinguish between organic and inorganic substances.
24. Describe the roles played by water and by oxygen in the human body.
25. Explain how carbon dioxide is transported within the body.
26. List several ions needed by cells, and describe their general functions.
27. Define *electrolyte balance*.
28. Describe the general characteristics of carbohydrates.
29. Distinguish between simple and complex carbohydrates.
30. Describe the general characteristics of lipids.
31. Distinguish between saturated and unsaturated fats.
32. Describe the general characteristics of proteins.
33. Define *enzyme*.
34. Explain how protein molecules may become denatured.
35. Describe the general characteristics of nucleic acids.
36. Explain the general functions of nucleic acids.

3

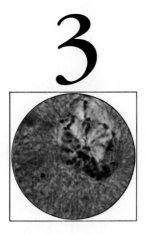

The Cell

The human body is composed entirely of cells, the products of cells, and various fluids. These cells represent the basic structural units of the body in that they are the building blocks from which all larger parts are formed. They are also the functional units, since whatever a body part can do is the result of activities within its cells.

Cells account for the shape, organization, and construction of the body and carry on its life processes. In addition, they can reproduce and thus provide new cells needed for growth and development and for replacement of worn and injured tissues.

Chapter Outline

Chapter Objectives

After you have studied this chapter, you should be able to

1. Explain how cells vary from one another.

2. Describe the general characteristics of a composite cell.

3. Explain how the structure of a cell membrane is related to its function.

4. Describe each kind of cytoplasmic organelle and explain its function.

5. Describe the cell nucleus and its parts.

6. Explain how substances move through cell membranes by physical and physiological processes.

7. Describe the life cycle of a cell.

8. Explain how a cell reproduces.

9. Complete the review activities at the end of this chapter. Note that the items are worded in the form of specific learning objectives. You may want to refer to them before reading the chapter.

Key Terms

active transport (ak′tiv trans′port)

centrosome (sen′tro-sōm)

chromosome (kro′mo-sōm)

cytoplasm (si′to-plazm)

differentiation (dif′′er-en′′she-a′shun)

diffusion (dĭ-fu′zhun)

endocytosis (en′′do-si-to′sis)

endoplasmic reticulum (en′do-plaz′mik rĕ-tik′u-lum)

equilibrium (e′′kwĭ-lib′re-um)

facilitated diffusion (fah-sil′′ĭ-tat′-ed dĭ-fu′zhun)

filtration (fil-tra′shun)

Golgi apparatus (gol′je ap′′ah-ra′tus)

lysosome (li′so-sōm)

micrometer (mi′kro-me′′ter)

mitochondrion (mi′′to-kon′dre-on); plural,
     mitochondria (mi′′to-kon′dre-ah)

mitosis (mi-to′sis)

nucleolus (nu-kle′o-lus)

nucleus (nu′kle-us)

osmosis (oz-mo′sis)

permeable (per′me-ah-bl)

phagocytosis (fag′′o-si-to′sis)

pinocytosis (pi′′no-si-to′sis)

ribosome (ri′bo-sōm)

Aids to Understanding Words

cyt-, cell: *cyt*oplasm—fluid that occupies the space between the cell membrane and the nuclear membrane.

endo-, within: *endo*plasmic reticulum—network of membranes within the cytoplasm.

hyper-, above: *hyper*tonic—a solution that has a greater concentration of dissolved particles than another solution.

hypo-, below: *hypo*tonic—a solution that has a lesser concentration of dissolved particles than another solution.

inter-, between: *inter*phase—stage that occurs between mitotic divisions of a cell.

iso-, equal: *iso*tonic—a solution that has a concentration of dissolved particles equal to that of another solution.

lys-, to break up: *lys*osome—organelle that contains enzymes capable of breaking up molecules of protein and carbohydrate.

mit-, thread: *mit*osis—process during which threadlike chromosomes appear within a cell.

phag-, to eat: *phag*ocytosis—process by which a cell takes in solid particles.

pino-, to drink: *pino*cytosis—process by which a cell takes in tiny droplets of liquid.

-som, body: ribo*som*e—tiny, spherical organelle.

I T IS ESTIMATED that an adult human body consists of about 75 trillion cells. These **cells** have much in common, yet those in different tissues vary in a number of ways. For example, they vary considerably in size.

Cell sizes are measured in units called *micrometers*. A micrometer equals 1/1000th of a millimeter and is symbolized μm. Measured in micrometers, a human egg cell is about 140 μm in diameter and is just barely visible to an unaided eye. This is large when compared to a red blood cell, which is about 7.5 μm in diameter, or the most common white blood cells, which vary from 10–12 μm in diameter. On the other hand, smooth muscle cells can be between 20–500 μm long.

Cells also vary in shape, and typically their shapes are closely related to their functions (figure 3.1). For instance, nerve cells often have long, threadlike extensions that transmit nerve impulses from one part of the body to another. The epithelial cells that line the inside of the mouth serve to shield underlying cells. These protective cells are thin, flattened, and tightly packed, somewhat like the tiles of a floor. Muscle cells, which function to pull parts closer together, are slender and rodlike, with their ends attached to the parts they move.

A Composite Cell

It is not possible to describe a typical cell, because cells vary so greatly in size, shape, and function. For purposes of discussion, however, it is convenient to imagine that one exists. Such a composite cell would contain parts observed in many kinds of cells, even though some of these cells lack parts included in the imagined structure. Thus, the cell shown in figure 3.2 and described in the following sections is not real. Instead, it is a composite cell—one that includes many known cell structures.

Commonly a cell consists of two major parts, one within the other and each surrounded by a thin membrane. The inner portion is called the *cell nucleus,* and it is enclosed by a *nuclear envelope.* A mass of fluid called *cytoplasm* surrounds the nucleus and is encircled by a cytoplasmic or *cell membrane.*

Within the cytoplasm are other membranes that separate it into small subdivisions. These include networks of membranes and membranes that mark off tiny, distinct parts called *cytoplasmic organelles.* These organelles perform specific metabolic functions necessary for cell survival. The nucleus, on the other hand, directs the overall activities of the cell.

Figure 3.1
Cells vary in structure and function. (*a*) A nerve cell transmits impulses from one body part to another, (*b*) epithelial cells protect underlying cells, and (*c*) muscle cells pull parts closer together.

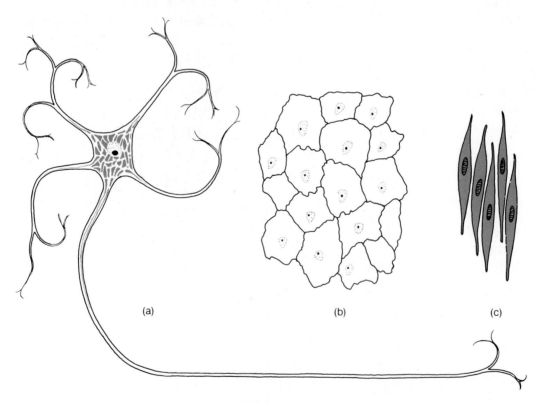

(a) (b) (c)

Figure 3.2
A composite cell and its major organelles.

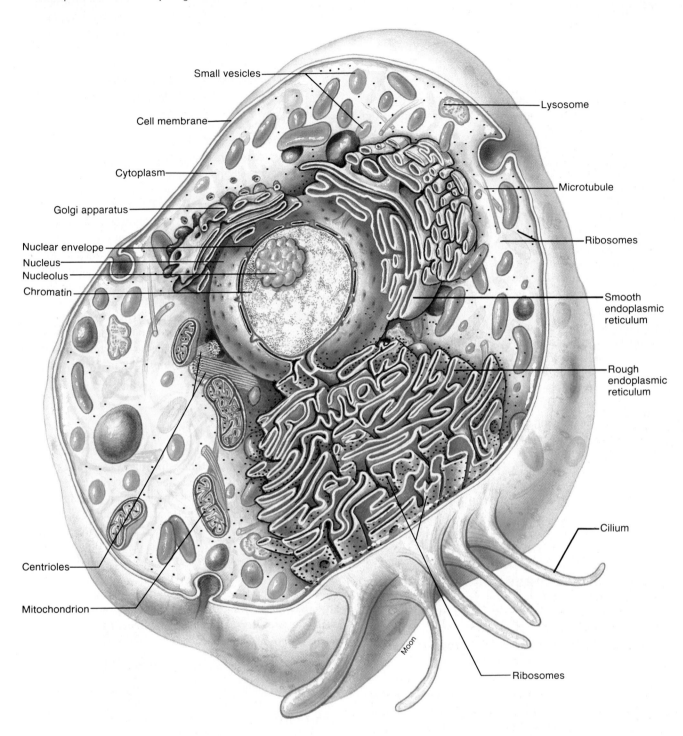

Small vesicles

Cell membrane

Cytoplasm

Golgi apparatus

Nuclear envelope
Nucleus
Nucleolus
Chromatin

Centrioles

Mitochondrion

Lysosome

Microtubule

Ribosomes

Smooth endoplasmic reticulum

Rough endoplasmic reticulum

Cilium

Ribosomes

Moon

1. Give two examples to illustrate the idea that the shape of a cell is related to its function.
2. Name the two major parts of a cell.
3. What are the general functions of these two parts?

Cell Membrane

The **cell membrane** is the outermost limit of the cell, but it is more than a simple envelope surrounding the cellular contents. It is an actively functioning part of the living material and many important metabolic reactions take place on its surface.

General Characteristics

The membrane is extremely thin—visible only with the aid of an electron microscope—but it is flexible and somewhat elastic. It typically has a complex surface with many outpouchings and infoldings that provide extra surface area. The membrane quickly seals minute breaks, but if it is extensively damaged, the cell contents escape, and the cell dies.

In addition to maintaining the wholeness of the cell, the membrane serves to control the entrance and exit of substances. That is, it allows some substances to pass through the membrane and excludes others. A membrane that functions in this manner is called *selectively permeable*. A *permeable membrane,* on the other hand, is a membrane that allows all materials to pass through.

Membrane Structure

Chemically, the cell membrane is composed mainly of lipids and proteins, although it also contains a small quantity of carbohydrate. The basic framework of a membrane consists of a double layer (bilayer) of phospholipid molecules. (See chapter 2.) These molecules are arranged so that their water soluble (hydrophilic) "heads," containing phosphate groups, form the outer surfaces of the membrane, and their water insoluble (hydrophobic) "tails," consisting of fatty acid chains, make up the interior.

The interior of the membrane consists largely of fatty acids and thus has an oily characteristic. Molecules that are soluble in lipids can pass through this layer easily; however, it forms an impermeable barrier to water soluble molecules such as amino acids, sugars, and proteins. Many cholesterol molecules embedded in the interior of the membrane also help to make the membrane less permeable to water soluble substances. In addition, the relatively rigid structure of the cholesterol molecules helps to make the membrane more stable than it would be otherwise.

Although there seems to be only a few types of lipid molecules in the membrane, it contains many kinds of proteins. These proteins are responsible for the special functions of the membrane, and they can be classified according to their shapes. One group of proteins, for example, consists of tightly coiled rodlike molecules embedded in the bilayer of phospholipids. These rodlike proteins may completely span the membrane. That is, they may extend outward from its surface on one side, while the opposite ends of the molecules communicate with the cell's interior. Such proteins often function as receptors that are specialized to combine with molecules such as hormones.

The second group of proteins have molecules with more compact, globular shapes. These proteins also are embedded in the interior of the phospholipid bilayer. Typically, they span the membrane and create narrow passageways or channels through which various ions and molecules can cross the bilayer. For example, some of these proteins form "pores" in the membrane that allow water molecules to pass freely. Others serve as selective channels that allow only particular substances to pass through. In muscle and nerve cells, such selective channels control the movements of sodium and potassium ions which play important roles in the functions of these cells. (See figure 3.3.)

Intercellular Junctions

Many cells, such as blood cells, are not in direct contact with one another because fluid-filled space (intercellular space) separates them. In other cases, however, the cells are tightly packed, and their cell membranes commonly are connected by **intercellular junctions.** In one type, called a *tight junction,* the membranes of adjacent cells converge and become fused together. This fusion surrounds the cell like a belt, and the junction functions to close the intercellular space between the cells.

Another type of intercellular junction called a *desmosome* serves to rivet or "spot weld" adjacent skin cells so they form a reinforced structural unit. The membranes of certain other cells, such as those in heart muscle, are interconnected by *gap junctions* in the form of tubular channels. These channels link the cytoplasm of adjacent cells. They allow ions and nutrients, such as sugars, amino acids, and nucleotides, and certain other relatively small molecules to be exchanged between them (figure 3.4).

Cytoplasm

When viewed through an ordinary light microscope, **cytoplasm** usually appears as a structureless substance with specks scattered throughout. However, a transmission electron microscope (figure 3.5), which produces much greater magnification and resolution,

Figure 3.3

The cell membrane is composed primarily of phospholipids with proteins scattered throughout the lipid layer and associated with its surfaces.

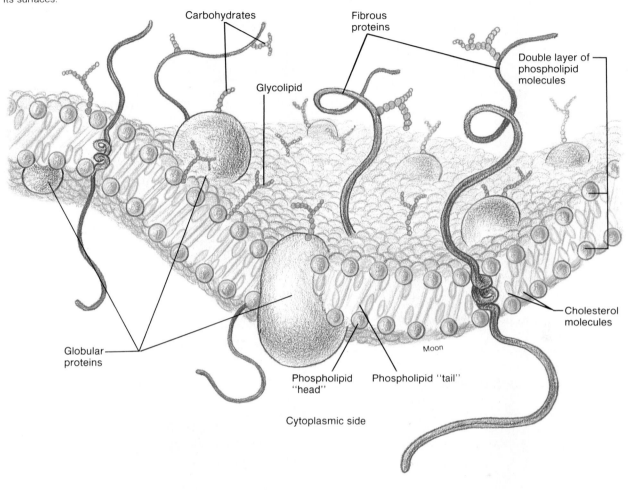

Carbohydrates

Fibrous proteins

Double layer of phospholipid molecules

Glycolipid

Cholesterol molecules

Moon

Globular proteins

Phospholipid "head"

Phospholipid "tail"

Cytoplasmic side

Figure 3.4

Some cells are joined by intercellular junctions, such as desmosomes that serve as "spot welds," or gap junctions that allow movement between the cytoplasm of adjacent cells.

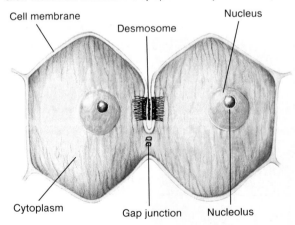

Cell membrane

Desmosome

Nucleus

Cytoplasm

Gap junction

Nucleolus

Figure 3.5

A transmission electron microscope reveals the cytoplasm of cells to be highly structured.

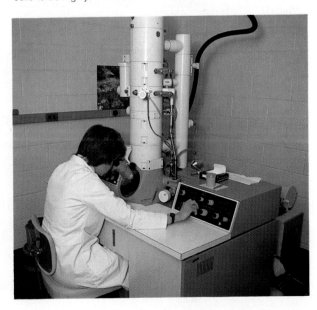

Figure 3.6
A human red blood cell as viewed using (*a*) a light microscope,
(*b*) a transmission electron microscope, and (*c*) a scanning
electron microscope.

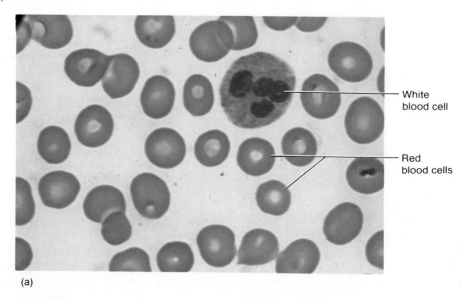

White
blood cell

Red
blood cells

(a)

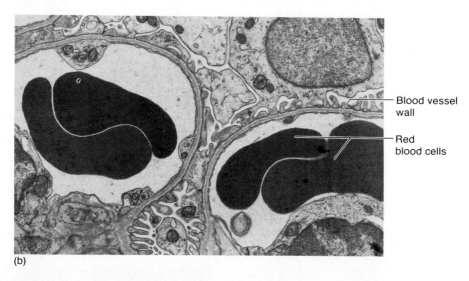

Blood vessel
wall

Red
blood cells

(b)

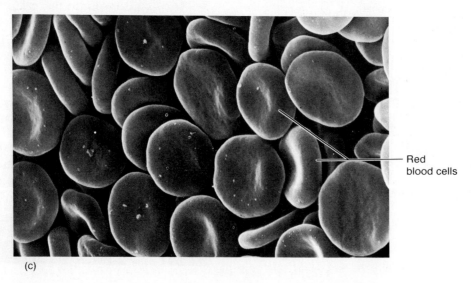

Red
blood cells

(c)

Figure 3.7
(a) A transmission electron micrograph of rough endoplasmic reticulum magnified about 100,000 times. (b) Rough endoplasmic reticulum has ribosomes attached to its surface, while (c) smooth endoplasmic reticulum lacks ribosomes.

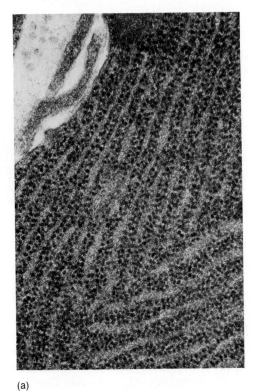

(a)

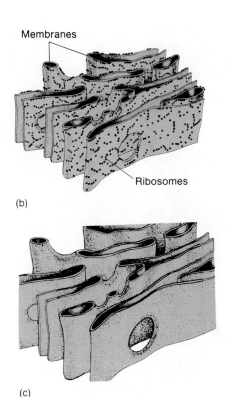

(b)

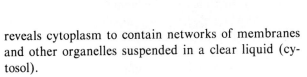

(c)

reveals cytoplasm to contain networks of membranes and other organelles suspended in a clear liquid (cytosol).

The maximum effective magnification that can be achieved using a light microscope is about 1200× (diameters). A transmission electron microscope (TEM) can achieve an effective magnification of nearly 1,000,000×; while another type of electron microscope, the scanning electron microscope (SEM), can provide about 50,000×.

Although the micrographs produced using the light microscope and the transmission electron microscope are typically two-dimensional, those obtained with the scanning electron microscope have a three-dimensional quality (figure 3.6).

The activities of a cell occur largely in its cytoplasm. It is there that food molecules are received, processed, and used. In other words, cytoplasm is a site of metabolic reactions, in which the following **cytoplasmic organelles** play specific roles.

1. **Endoplasmic reticulum.** The endoplasmic reticulum is a complex network of interconnected membranes that form flattened sacs, elongated canals, and fluid-filled vesicles.

This membranous network is connected to the cell membrane, the nuclear envelope, and certain cytoplasmic organelles. Thus, it is widely distributed through the cytoplasm. It seems to function as a tubular communication system through which molecules can be transported from one cell part to another.

The endoplasmic reticulum also functions in the synthesis of protein and lipid molecules. Often a large proportion of its outer membranous surface has numerous tiny, spherical organelles called *ribosomes* attached to it. For this reason it is termed *rough* endoplasmic reticulum. The remaining surface, which lacks ribosomes, is called *smooth* endoplasmic reticulum (figure 3.7).

The ribosomes of rough endoplasmic reticulum synthesize proteins. The resulting molecules may be transported through tubules of the endoplasmic reticulum to the Golgi apparatus and secreted to the outside of the cell. Smooth endoplasmic reticulum contains enzymes involved in the manufacture of various lipid molecules, including certain hormones (steroids).

Figure 3.8

(*a*) A transmission electron micrograph of a Golgi apparatus magnified about 40,000 times; (*b*) the Golgi apparatus consists of membranous sacs that are continuous with the endoplasmic reticulum.

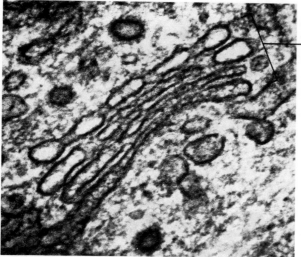

(a)

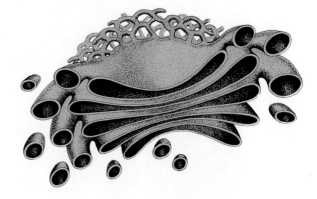

Golgi apparatus

(b)

2. **Ribosomes.** Although many ribosomes are attached to membranes of the endoplasmic reticulum, others occur as free particles scattered throughout the cytoplasm. In both cases, these tiny spherical particles are composed of protein and RNA molecules, and they function in the synthesis of protein molecules. Some of the proteins synthesized by free ribosomes are used for building cell structure, while others act as enzymes. This function is described in more detail in chapter 4.

3. **Golgi apparatus.** The Golgi apparatus is usually located near the nucleus. It is composed of a stack of half a dozen or so flattened, membranous sacs and is involved in the refining and "packaging" of proteins synthesized by the endoplasmic reticulum. (See figure 3.8.)

   These proteins arrive at the Golgi apparatus enclosed in tiny sacs composed of membrane from the endoplasmic reticulum. Also, the protein molecules have been combined with sugar molecules and thus can be called *glycoproteins.* The glycoproteins pass from layer to layer through the Golgi stacks and are modified chemically. They may, for example, have some sugar molecules added or removed. When they reach the outermost layer, the altered glycoproteins are packaged in bits of Golgi apparatus membrane that buds off and forms transport vesicles. Such a vesicle may then move to the cell membrane, where it fuses with the membrane and releases its contents to the outside as a secretion. Other vesicles may transport glycoproteins to various cytoplasmic organelles within the cell. (See figure 3.9.)

   In some cells, including certain liver cells and white blood cells (lymphocytes), glycoprotein molecules are secreted as rapidly as they are synthesized. In other cells, however, such as those that manufacture protein hormones, the vesicles containing proteins are stored in the cytoplasm and are released only when the cells receive particular stimuli from the outside. (Hormone secretion is discussed in chapter 13.)

---

1. What is meant by a selectively permeable membrane?
2. Describe the chemical structure of a cell membrane.
3. What are the functions of the endoplasmic reticulum?
4. Describe the way in which the Golgi apparatus functions.

---

4. **Mitochondria.** Mitochondria are elongated, fluid-filled sacs two to five micrometers long. They often move about slowly in the cytoplasm and can produce others like themselves by dividing. A mitochondrion also contains a supply of DNA that seems to encode information for making a few kinds of RNA and protein molecules. Most of the structural and functional proteins of mitochondria are encoded in the nuclear chromosomes.

Figure 3.9

Figure 3.10

Portions of the Golgi apparatus function in packaging glycoproteins, which then may be moved to the cell membrane and released as a secretion.

(a) A transmission electron micrograph of a mitochondrion magnified about 40,000 times. (b) What is the function of the cristae that form partitions within this saclike organelle?

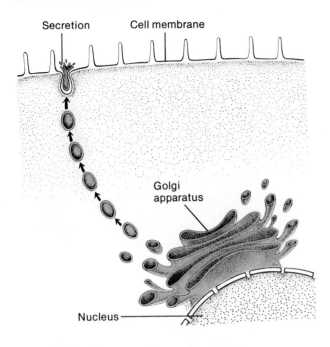

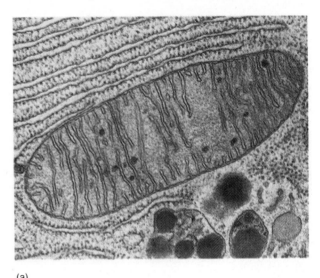

(a)

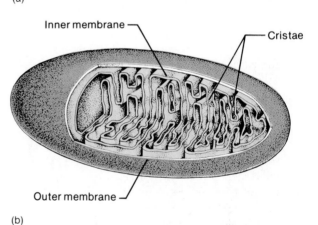

(b)

The membrane of a mitochondrion has two layers—an outer and an inner membrane. The inner membrane is folded extensively to form shelflike partitions called *cristae*. Small, stalked particles that contain enzymes are connected to the cristae. These enzymes and others dissolved in fluid inside the mitochondrion control some of the chemical reactions by which energy is released from glucose and other organic molecules. The mitochondria also function in transforming this energy into a form that is usable by cell parts. For this reason mitochondria are sometimes called the "powerhouses" of cells. This energy-releasing function, involving molecules of the substance called *ATP,* is described in more detail in chapter 4 (figure 3.10).

5. **Lysosomes.** Lysosomes are sometimes difficult to identify because their shapes vary so greatly. However, they commonly appear as tiny, membranous sacs. These sacs contain powerful enzymes that are capable of breaking down molecules of protein, carbohydrates, and nucleic acids. The enzymes function to digest various particles that enter cells. Certain white blood cells, for example, can engulf bacteria that are then digested by the lysosomal enzymes. Consequently, white blood cells aid in preventing bacterial infections.

Lysosomes also function in the destruction of worn cell parts. In fact, sometimes they destroy entire, injured cells that have been engulfed by scavenger cells. How the lysosomal membrane is able to withstand being digested itself is not understood.

Lysosomal digestive activity also seems to be responsible for decreasing the size of body tissues at certain times. Such regression in size occurs in the uterus following birth, in the breasts after the weaning of an infant, and in muscles during periods of prolonged inactivity.

Figure 3.11

(a) A transmission electron micrograph of the two centrioles in a centrosome (142,000×). (b) Note that the centrioles lie at right angles to one another.

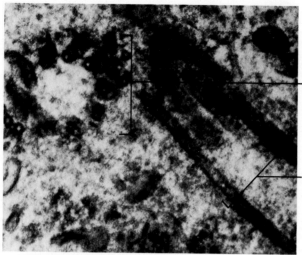

Centriole (cross section)

Centriole (longitudinal section)

(a)

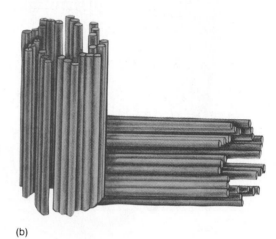

(b)

Figure 3.12

Cilia, such as these (arrow), are common on the surface of certain cells that form the inner lining of respiratory tubes (about 10,000×).

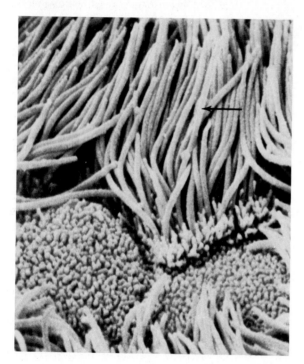

6. **Peroxisomes.** Peroxisomes are membranous sacs that resemble lysosomes in size and shape. They occur most commonly in cells of the liver and kidneys and contain enzymes called *peroxidases*. These enzymes promote certain metabolic reactions that give rise to hydrogen peroxide (H_2O_2) as a by-product. Peroxisomes also contain an enzyme called *catalase* that acts to decompose hydrogen peroxide, which is toxic to cells. Enzymes are discussed in more detail in chapter 4.

   Although it is assumed that peroxisomes have some metabolic function, their specific role is not understood.

7. **Centrosome.** A centrosome (central body) is located in the cytoplasm near the Golgi apparatus and nucleus. It is nonmembranous and consists of two hollow cylinders called *centrioles,* which in turn contain tiny tubelike parts (microtubules). The centrioles usually lie at right angles to each other and function in cell reproduction. During this process, the centrioles move away from one another and take positions on either side of the nucleus. There they aid in the distribution of the chromosomes, which carry genetic information, to the newly forming daughter cells (figure 3.11).

   Centrioles also function in initiating the formation of hairlike cellular projections called *cilia* and *flagella.*

Figure 3.13
(a) Microtubules, such as those shown in this transmission electron micrograph, aid in the distribution of chromosomes during cellular reproduction. (b) Each microtubule is composed of globular proteins.

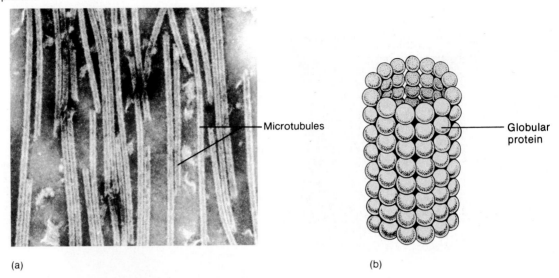

—Microtubules

—Globular protein

(a)

(b)

8. **Cilia** and **flagella.** Cilia and flagella are motile processes that extend outward from the surfaces of certain cells. They are structurally similar and differ mainly in length and number present. Both contain a constant number of microtubules arranged in a distinct pattern.

   Cilia are present in large numbers on the free surfaces of some epithelial cells. Each cilium is a tiny, hairlike structure about 10 μm long and is attached just beneath the cell membrane to a modified centriole called a *basal body.* These projections are arranged in very precise patterns, and they have a "to-and-fro" type of movement. This movement is coordinated so that rows of cilia beat one after the other, creating a wave of motion that sweeps across the ciliated surface. Ciliary action serves to propel fluids like mucus over the surface of certain tissues, such as those lining the tubes of the respiratory tract. (See figure 3.12.)

   Flagella are considerably longer than cilia, and usually there is a single flagellum on a cell. These projections have an undulating, wavelike motion that begins at the base. The flagellum that forms the tail of a sperm cell is responsible for its swimming movements. (See chapter 22.) Flagella also occur in certain cells of the kidneys, testes, and various glands.

9. **Vesicles.** Vesicles (vacuoles) are membranous sacs that vary in size. They may be formed by an action of the cell membrane in which a portion of the membrane folds inward and pinches off. As a result, a small, bubblelike vesicle, containing some liquid or solid material that was outside the cell a moment before, appears in the cytoplasm.

   The process of endocytosis, which involves such vesicle formation, is discussed in a subsequent section of this chapter.

10. **Microfilaments** and **microtubules.** Two types of thin, threadlike structures found in the cytoplasm are microfilaments and microtubules.

    Microfilaments are tiny rods of protein (actin), arranged in meshworks or bundles. They seem to cause various kinds of cellular movements. They are most highly developed in muscle cells as *myofibrils,* which help these cells to shorten or contract. In other cells, microfilaments are usually associated with the inner surface of the cell membrane and seem to aid in cell motility.

    Microtubules are long, slender tubes composed of globular protein (tubulin). They are usually stiff and form an "internal skeleton" within a cell, which helps to maintain the shape of the cell or its parts. For example, they provide structural strength within cilia and flagella.

    Microtubules also aid in moving organelles from one place to another. For instance, they appear in the cytoplasm during cellular reproduction and are involved in the distribution of chromosomes to the newly forming cells. This process is described in more detail in a subsequent section of this chapter (figure 3.13).

Figure 3.14
(*a*) A transmission electron micrograph of a cell nucleus. It contains a nucleolus and masses of chromatin. (*b*) The nuclear envelope is porous and allows substances to pass between the nucleus and the cytoplasm.

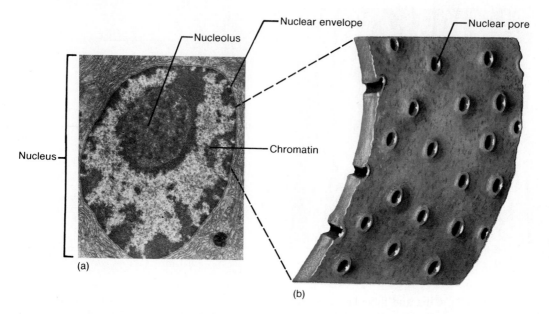

(a)

(b)

In addition to these functional organelles, cytoplasm may also contain masses of lifeless chemical substances called *inclusions*. Most commonly, inclusions consist of cellular products that remain in a cell only temporarily. Inclusions include stored nutrients, such as glycogen and various lipids, and pigments, such as melanin in the skin.

---

1. Why are mitochondria sometimes called "powerhouses" of cells?
2. How do lysosomes within certain white blood cells function?
3. Describe the functions of microfilaments and microtubules.
4. Distinguish between organelles and inclusions.

Cell Nucleus

A **cell nucleus** is usually located near the center of the cytoplasm. It is a relatively large, spherical structure and functions to direct the activities of the cell.

The nucleus is enclosed in a **nuclear envelope** that consists of an inner and an outer membrane. These two membranes have a narrow space between them but are joined at various places that surround relatively large openings or "pores." The pores in the nuclear envelope allow certain dissolved substances to move between the nucleus and the cytoplasm (figure 3.14).

The nucleus contains a fluid (nucleoplasm) in which other structures float. They include the following:

1. **Nucleolus.** A nucleolus ("little nucleus") is a small, dense body composed largely of RNA and protein. It has no surrounding membrane and is formed in specialized regions of certain chromosomes. It functions in the production of ribosomes, and once the ribosomes are formed, they migrate through the pores in the nuclear envelope and enter the cytoplasm. The nuclei of cells, such as those of glands, that synthesize large amounts of protein may contain especially large nucleoli.

2. **Chromatin.** Chromatin consists of loosely coiled fibers present in the nuclear fluid. When the cell begins to undergo the reproductive process, these fibers become more tightly coiled and are transformed into tiny, rodlike *chromosomes*. Chromatin fibers are composed of protein (histones) and DNA molecules, which in turn are organized into tiny, beadlike particles (nucleosomes). The DNA molecules contain the information that directs the cell in carrying out its life processes.

Chart 3.1	Structure and function of cellular organelles	
Organelle	Structure	Function
Cell membrane	Membrane composed mainly of protein and lipid molecules.	Maintains wholeness of cell and controls passage of materials in and out of cell.
Endoplasmic reticulum	Network of interconnected membrane forming sacs and canals.	Transports materials within the cell, provides attachment for ribosomes, and synthesizes lipids.
Ribosomes	Particles composed of protein and RNA molecules.	Synthesize proteins.
Golgi apparatus	Group of flattened, membranous sacs.	Packages protein molecules for secretion.
Mitochondria	Membranous sacs with inner partitions.	Release energy from food molecules and transform energy into usable form.
Lysosomes	Membranous sacs.	Contains enzymes capable of digesting substances that enter cells.
Peroxisomes	Membranous vesicles.	Contain enzymes called peroxidases.
Centrosome	Nonmembranous structure composed of two rodlike centrioles.	Helps distribute chromosomes to daughter cells during cell reproduction and initiates formation of cilia.
Cilia and flagella	Hairlike projections attached to basal bodies beneath cell membrane.	Propels fluids over cellular surface and enables sperm cells to move.
Vesicles	Membranous sacs.	Contain various substances that recently entered the cell.
Microfilaments and microtubules	Thin rods and tubules.	Provide support to cytoplasm and help move objects within the cytoplasm.
Nuclear envelope	Porous double membrane that separates nuclear contents from cytoplasm.	Maintains wholeness of the nucleus and controls passage of materials between nucleus and cytoplasm.
Nucleolus	Dense, nonmembranous body composed of protein and RNA molecules.	Forms ribosomes.
Chromatin	Fibers composed of protein and DNA molecules.	Contains cellular information for carrying on life processes.

Chart 3.1 summarizes the structures and functions of the cellular organelles.

Aging of cells. Although the effects of aging on cells are poorly understood, studies of aged tissues indicate that certain organelles may be altered with age. For example, within the nucleus, the chromatin and chromosomes may show changes such as clumping, shrinking, or fragmenting, and the number and size of the nucleoli may increase. In the cytoplasm, the mitochondria may undergo changes in shape and number, and the Golgi apparatus may become fragmented. Lipid inclusions tend to accumulate in the cytoplasm, while inclusions containing glycogen tend to disappear.

1. How are the nuclear contents separated from the cytoplasm of a cell?
2. What is the function of the nucleolus?
3. What is chromatin?

Movements through Cell Membranes

The cell membrane provides a surface through which various substances enter and leave the cell. More specifically, oxygen and food molecules enter a cell through this membrane, while carbon dioxide and other wastes leave through it. These movements involve physical (or nonliving) processes such as diffusion, facilitated diffusion, osmosis, and filtration; and physiological (or living) mechanisms such as active transport and endocytosis.

Diffusion

Diffusion is the process by which molecules or ions become scattered or are spread from regions where they are in higher concentrations toward regions where they are in lower concentrations.

These molecules and ions are constantly moving at high speeds. Each particle travels in a separate path along a straight line until it collides and bounces off some other particle. Then it moves in another direction, only to collide again and change direction once

Figure 3.15

An example of diffusion. (*a*, *b*, *c*,) If a sugar cube is placed in water, it slowly disappears. As this happens, the sugar molecules diffuse from regions where they are more concentrated toward regions where they are less concentrated. (*d*) Eventually, the sugar molecules are distributed evenly throughout the water.

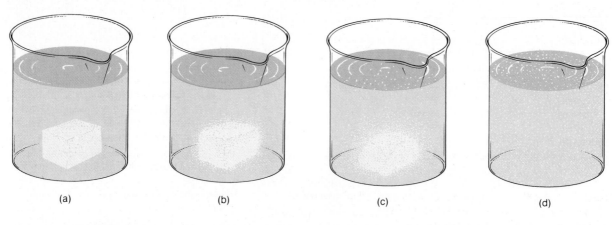

(a) (b) (c) (d)

Figure 3.16

(*1*) The container is separated into two compartments by a permeable membrane. Compartment *A* contains water and sugar molecules, while compartment *B* contains only water molecules. (*2*) As a result of molecular motions, sugar molecules will tend to diffuse from compartment *A* into compartment *B*. Water molecules will tend to diffuse from compartment *B* into compartment *A*. (*3*) Eventually, equilibrium will be achieved.

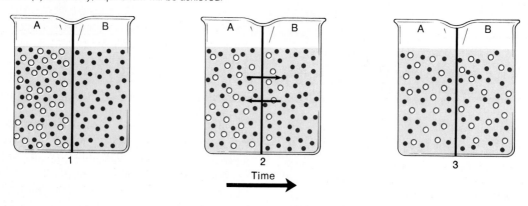

1   2   3

Time

more. Such motion is haphazard, but it accounts for the mixing of molecules that commonly occurs when different kinds of substances are put together.

For example, when some sugar (a solute) is put into a glass of water (a solvent), as illustrated in figure 3.15, the sugar will seem to remain at the bottom for a while. Then it slowly disappears into solution. As this happens, the moving water and sugar molecules are colliding with one another, and in time the sugar and water molecules will be evenly mixed. This mixing occurs by diffusion—the sugar molecules spread from where they are in higher concentration toward the regions where they are less concentrated. Eventually the sugar becomes uniformly distributed in the water. This

condition is called *equilibrium,* and although the molecules continue to move after equilibrium is achieved, their concentrations no longer change.

To better understand how diffusion accounts for the movement of various molecules through a cell membrane, imagine a container of water that is separated into two compartments by a permeable membrane (figure 3.16). This membrane has numerous pores that are large enough for water and sugar molecules to pass through. Sugar molecules are placed in one compartment (*A*) but not in the other (*B*). Although the sugar molecules will move in all directions, more will diffuse from compartment *A* (where they are in greater concentration) through the pores in the membrane and into compartment *B* (where they are in lesser concentration) than will move in the other direction. At the same time, water molecules will tend to diffuse from compartment *B* (where they are in greater concentration) through the pores into compartment *A* (where they are in lesser concentration). Eventually, equilibrium will be achieved when there are equal numbers of water and of sugar molecules in each compartment.

Similarly, oxygen molecules diffuse through cell membranes and enter cells if these molecules are more highly concentrated on the outside than on the inside. Carbon dioxide molecules, too, diffuse through cell membranes and leave cells if they are more concentrated on the inside than on the outside. It is by diffusion that oxygen and carbon dioxide molecules are exchanged between the air and the blood in the lungs and between the blood and the cells of various tissues.

Several factors influence the rate at which diffusion occurs. They include the distance involved, the concentration of the substance, the weight of the molecules, and the temperature. Generally, diffusion occurs more rapidly over a shorter distance, when the concentration of the diffusing substance is greater, when the molecular weight is lower, and when the temperature is higher.

Facilitated Diffusion

Since a cell membrane is composed largely of lipid, molecules that are soluble in lipid, such as oxygen and carbon dioxide, can diffuse readily through it. Other substances that are not soluble in lipid, such as water and certain ions, cannot diffuse easily through the membrane but may pass through membrane pores.

On the other hand, most sugars are insoluble in lipid, and they have molecular sizes that prevent them from passing through membrane pores. Some of these, including glucose, can still enter through the lipid portion of the membrane by a process called *facilitated diffusion*. In this process, the glucose combines with a special carrier molecule at the surface of the membrane. This union of glucose and carrier molecule creates a compound that is soluble in lipid, and it can diffuse to the other side. There, the glucose portion is released and the carrier molecule can return to the opposite side of the membrane and pick up another glucose molecule.

Figure 3.17

Some substances are moved through a cell membrane by facilitated diffusion in which carrier molecules transport the substance from a region of higher concentration to one of lower concentration.

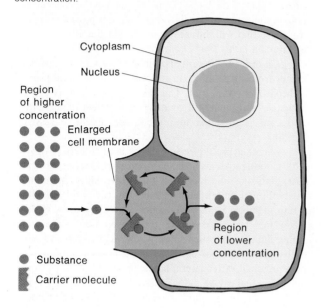

Figure 3.18

An example of osmosis. (*1*) The container is separated into two compartments by a selectively permeable membrane. At first, compartment *A* contains water and sugar molecules, while compartment *B* contains only water molecules. As a result of molecular motions, water molecules will tend to diffuse by osmosis from compartment *B* into compartment *A*. The sugar molecules will remain in compartment *A* because they are too large to pass through the pores of the membrane. (*2*) Since more water is entering compartment *A* than is leaving it, water will accumulate in this compartment, and the level of the liquid will rise on this side.

• Water molecule
○ Sugar molecule

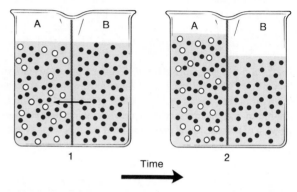

1   Time   2

Osmosis

Osmosis is a special case of diffusion. It occurs whenever *water* molecules diffuse from a region of higher water concentration (where the solute concentration is lower) to a region of lower water concentration (where the solute concentration is higher) through a selectively permeable membrane, such as a cell membrane. In the following examples, it is assumed that the membranes involved are permeable to water molecules but impermeable to glucose molecules.

Ordinarily, the concentrations of water molecules are equal on either side of a cell membrane. Sometimes, however, the water on one side has more solute dissolved in it than the water on the other side. For example, if there is a greater concentration of glucose (solute) in the water outside a cell, there must be a lesser concentration of water there, because the glucose molecules occupy space that would otherwise contain water molecules. Under such conditions, water

Facilitated diffusion is similar to simple diffusion in that it only can cause movement of molecules from regions of higher concentration toward regions of lower concentration. The rate at which facilitated diffusion can occur, however, is limited by the number of carrier molecules in the cell membrane (figure 3.17).

molecules diffuse from inside the cell (where they are in higher concentration) toward the outside (where they are in lesser concentration).

This process, shown in figure 3.18, is similar to the one in figure 3.16, illustrating the diffusion of water and glucose molecules through a permeable membrane. In the case of osmosis, however, the membrane is selectively permeable—water molecules pass through readily, but glucose molecules do not.

Note in figure 3.18 that as osmosis occurs the volume of water on side *A* increases. This increase in volume would be resisted if pressure were applied to the surface of the liquid on side *A*. The amount of pressure needed to stop osmosis in such a case is called *osmotic pressure*. Thus, the osmotic pressure of a solution is a potential pressure and is due to the presence of nondiffusible solute particles in that solution. Furthermore, the greater the number of solute particles in the solution, the greater the osmotic pressure of that solution.

When osmosis occurs, water tends to move toward the region of greater osmotic pressure. Because the amount of osmotic pressure depends upon the difference in concentration of solute particles on opposite sides of the membrane, the greater the difference, the greater the tendency for water to move toward the region of higher solute concentration.

If some cells were put into a water solution that had a greater concentration of solute particles (higher osmotic pressure) than did the cells, there would be a net loss of water from the cells. Consequently, the cells would begin to shrink, and such cells are said to become *crenated*. In time, equilibrium might be achieved, and the shrinking would cease. A solution of this type, in which more water leaves a cell than enters it because the concentration of solute particles is greater outside the cell, is called **hypertonic.**

Conversely, if there is a greater concentration of solute particles inside a cell than in the water around it, water will diffuse into the cell. As this happens, water accumulates within the cell, and it begins to swell. Although cell membranes are somewhat elastic, they may swell so much that they burst. A solution in which more water enters a cell than leaves it because of a lesser concentration of solute particles outside the cell is called **hypotonic.**

A solution that contains the same concentration of solute particles as a cell is said to be **isotonic** to that cell. In such a solution, water enters and leaves a cell in equal amounts, so the cell's size remains unchanged. Figure 3.19 illustrates these three types of solutions.

It is important to control the concentration of solute in solutions that are infused into body tissues or blood. Otherwise, osmosis may cause cells to swell or shrink, and they may be damaged. For instance, if red blood cells are exposed to distilled water (which is hypotonic to them), water will diffuse into the cells and they will burst (hemolyze). On the other hand, if red blood cells are exposed to 0.9% NaCl solution (normal saline), the cells will remain unchanged because this solution is isotonic to human cells. Similarly, a 5% solution of glucose is isotonic to human cells. (The lower percentage is needed with NaCl in part because it ionizes in solution more completely and gives rise to a greater number of solute particles than does glucose.)

Filtration

The passage of substances through membranes by diffusion or osmosis is due to movements of the molecules of those substances. In other instances, molecules are forced through membranes by hydrostatic pressure or blood pressure that is greater on one side of the membrane than on the other. This process is called **filtration**.

Filtration is commonly used to separate solids from a liquid. One method is to pour a mixture onto

Figure 3.19

(*a*) If red blood cells are placed in a hypertonic solution, more water leaves the cells than enters them, and they shrink; (*b*) in a hypotonic solution, more water enters than leaves the cells, and they swell and may burst; (*c*) in an isotonic solution, water enters and leaves the cells in equal amounts, and their size remains unchanged. (Note: the magnification of the cells in (*b*) is lower than in the other two scanning electron micrographs.)

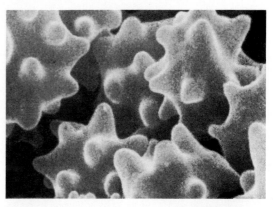

(a)

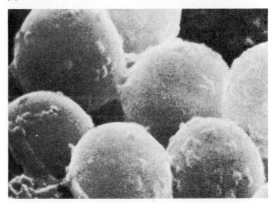

(b)

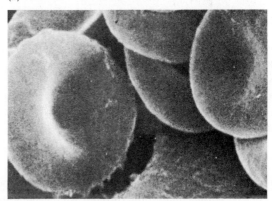

(c)

Figure 3.20

In this example of filtration, gravity provides the force that pulls water through filter paper, while the tiny openings in the paper hold the solids behind.

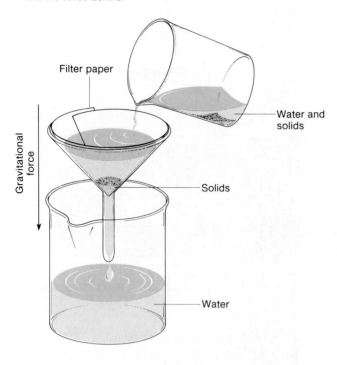

Figure 3.21

In this example of filtration, smaller molecules are forced through the wall of a capillary by blood pressure, while large molecules remain behind.

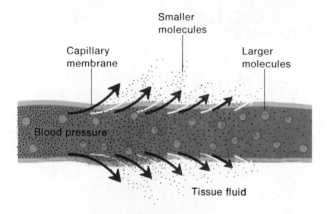

Figure 3.22

During active transport, a molecule or ion combines with a carrier protein, whose shape is altered as a result. This action helps to move the transported particle through the cell membrane.

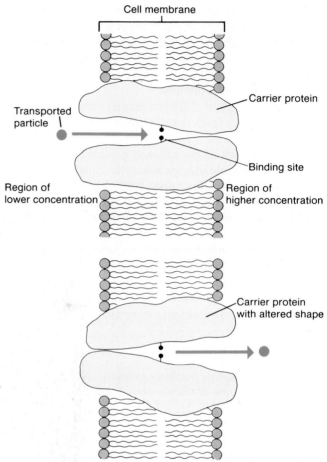

(figure 3.21). The force for this movement comes from blood pressure, created largely by heart action, which is greater within the vessel than outside. Filtration also takes place in the kidneys when water and dissolved wastes are forced out of blood vessels and into kidney tubules by blood pressure. This is the first step in the formation of urine.

1. What kinds of substances move most readily through a cell membrane by diffusion?
2. Explain the differences between diffusion, facilitated diffusion, and osmosis.
3. Distinguish between hypertonic, isotonic, and hypotonic solutions.
4. Explain how filtration occurs within the body.

some filter paper in a funnel (figure 3.20). The paper serves as a porous membrane through which the liquid can pass, leaving the solids behind. Hydrostatic pressure, which is created by the weight of water due to gravity, forces the water through to the other side.

Similarly, in the body, tissue fluid is formed when water and dissolved substances are forced out through the thin walls of blood capillaries, while larger particles such as blood protein molecules are left inside

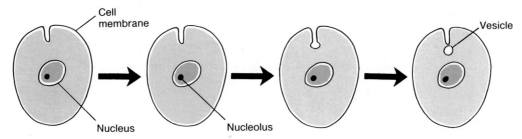

Active Transport

When molecules or ions pass through cell membranes by diffusion, facilitated diffusion, or osmosis, their net movement is from regions of higher concentration to regions of lower concentration. Sometimes, however, the net movement of particles passing through membranes is in the opposite direction; that is, particles move from a region of lower concentration to one of higher concentration.

It is known, for example, that sodium ions can diffuse through cell membranes. Yet the concentration of these ions typically remains many times greater on the outside of cells (in the extracellular fluid) than on the inside (in the intracellular fluid). Furthermore, sodium ions are continually moved from the regions of lower concentration (inside) to regions of higher concentration (outside). Movement of this type is called **active transport.** It depends on life processes within cells and requires energy. In fact, it is estimated that up to 40% of a cell's energy supply may be used for active transport of particles through its membranes.

Although the details are poorly understood, the mechanism of active transport is similar to facilitated diffusion in that it involves specific carrier molecules, probably protein molecules found within cell membranes. As figure 3.22 shows, these carrier molecules have binding sites that combine at the membrane's surface with the particles being transported. Such a union triggers the release of some cellular energy, and it causes the shape of the carrier protein to be altered. As a result, particles are moved through the membrane. Once on the other side, the transported particle is released as a result of an enzyme action, and the carrier molecules can accept another particle at its binding site.

Particles that are carried across cell membranes by active transport include various sugars and amino acids as well as a variety of ions such as those of sodium, potassium, calcium, and hydrogen. Movements of this type are important to cell survival and are involved with the maintenance of homeostasis. Some of these movements are described in detail in subsequent chapters.

Endocytosis

Another physiological process by which substances may move through cell membranes is called **endocytosis.** In this case molecules or other particles that are too large to enter a cell by diffusion or active transport may be conveyed within a tiny vesicle formed by a section of the cell membrane.

There are three forms of endocytosis—pinocytosis, phagocytosis, and receptor-mediated endocytosis.

Pinocytosis

Pinocytosis, which means *cell drinking,* refers to the process by which cells take in tiny droplets of liquid from their surroundings (figure 3.23). When this happens, a small portion of cell membrane becomes indented (invaginated). The open end of the tubelike part thus formed soon seals off, and the result is a small vesicle about 0.1 micrometer in diameter. This tiny sac becomes detached from the surface and moves into the cytoplasm.

For a time, the vesicular membrane, which was part of the cell membrane, separates its contents from the rest of the cell; but eventually the membrane breaks down and the liquid inside becomes part of the cytoplasm. In this way, a cell is able to take in water and dissolved particles, such as proteins, that otherwise might be unable to enter because of large molecular size.

Phagocytosis

Phagocytosis (*cell eating*) is a process that is essentially the same as pinocytosis. In phagocytosis, however, the material taken into the cell is solid rather than liquid.

Certain kinds of cells, including some white blood cells, are called *phagocytes* because they can take in tiny solid particles such as bacterial cells. When a phagocyte first encounters such a particle, the object becomes attached to its cell membrane. This stimulates a portion of the membrane to project outward, surround the object, and slowly draw it inside. The part

Figure 3.24

A cell may take in a solid particle from its surroundings by phagocytosis.

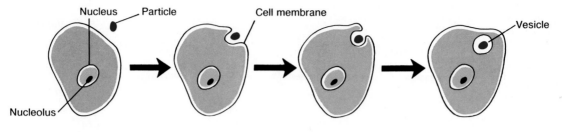

Figure 3.25

When a lysosome combines with a vesicle that contains a phagocytized particle, its digestive enzymes destroy the particle. The products diffuse into the cytoplasm and can be used. Any residue may be expelled.

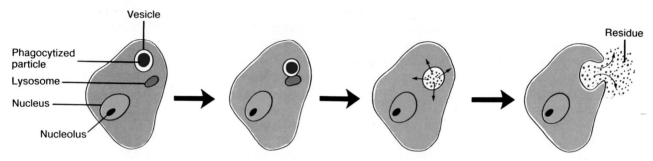

of the cell membrane surrounding the solid detaches from the surface, and a vesicle containing the particle is formed (figure 3.24). Such a vesicle may be several micrometers in diameter.

Commonly, a lysosome soon combines with such a vesicle, and lysosomal digestive enzymes cause the contents to be decomposed (figure 3.25). The products of this decomposition may then diffuse out of the lysosome and into the cell's cytoplasm and may be used as raw materials in metabolic processes. Any remaining residue may be expelled from the cell (exocytosis). In this way phagocytic cells can destroy bacteria that might otherwise cause infections, dispose of foreign objects like dust particles, or remove damaged cells or cell parts that are no longer functional.

Although pinocytosis is thought to provide only a minor route for substances entering cells, phagocytosis is an important line of defense against invasion by disease-causing microorganisms.

As a consequence of an inherited trait, the lysosomes of some individuals fail to produce various digestive enzymes. This leads to one of several "storage diseases," which are characterized by an accumulation of undigested materials in certain types of cells. Tay-Sachs disease is an example of such a condition. In this disease, lipids accumulate abnormally in certain nerve cells.

Receptor-Mediated Endocytosis

Although a variety of substances may enter cells by pinocytosis or phagocytosis, **receptor-mediated endocytosis** moves very specific kinds of particles through membranes. This process involves the presence of particular protein molecules that extend through the cell membrane and are exposed on its outer surface. These proteins serve as *receptor sites* to which specific substances (ligands) from the fluid surroundings of the cell can bind. Thus, when molecules that are capable of binding to the receptor sites are present, only those molecules are selected to enter the cell, leaving other kinds of molecules outside of the cell.

In this process, selected molecules combine with receptor proteins. The formation of such combinations somehow stimulates the membrane to indent and create a tiny vesicle that contains the selected molecules—an action similar to that described previously in the discussion of pinocytosis.

Receptor-mediated endocytosis is of particular importance because it allows a cell to remove specific kinds of substances from its surroundings, even when these substances are present in very low concentrations. (See figure 3.26.)

Chart 3.2 summarizes the types of movement through cell membranes.

Figure 3.26
Receptor-mediated endocytosis. (*a*) A specific substance binds to a receptor site protein; (*b* and *c*) the combination of the substance with the receptor site protein stimulates the cell membrane to indent; (*d*) the resulting vesicle moves the transported substance into the cytoplasm.

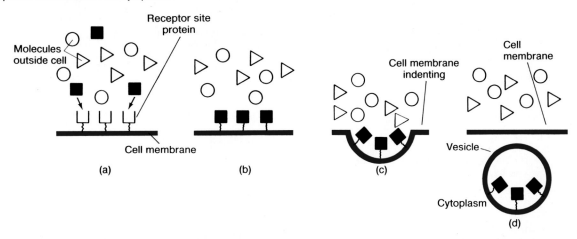

Chart 3.2	Movements through cell membranes		
Process	**Characteristics**	**Source of energy**	**Example**
I. *Physical processes*			
A. Diffusion	Molecules or ions move from regions of higher concentration toward regions of lower concentration.	Molecular motion	Exchange of gases in lungs.
B. Facilitated diffusion	Molecules are moved through a membrane from a region of higher concentration to one of lower concentration by carrier molecules.	Molecular motion	Movement of glucose through a cell membrane.
C. Osmosis	Water molecules move from regions of higher concentration to regions of lower concentration through a selectively permeable membrane.	Molecular motion	Water enters a cell placed in distilled water.
D. Filtration	Molecules are forced from regions of higher pressure to regions of lower pressure.	Blood pressure	Molecules leave blood capillaries.
II. *Physiological processes*			
A. Active transport	Molecules or ions are carried through membranes by other molecules.	Cellular energy	Movement of various ions and amino acids through membranes.
B. Endocytosis			
1. Pinocytosis	Membrane acts to engulf minute droplets of liquid from surroundings.	Cellular energy	Membrane forms vesicle containing liquid and dissolved particles.
2. Phagocytosis	Membrane acts to engulf solid particles from surroundings.	Cellular energy	White blood cell membrane engulfs bacterial cell.
3. Receptor-mediated endocytosis	Membrane acts to engulf selected molecules combined with receptor proteins.	Cellular energy	Cell removes specific kinds of substances from its surroundings.

Figure 3.27
Life cycle of a cell.

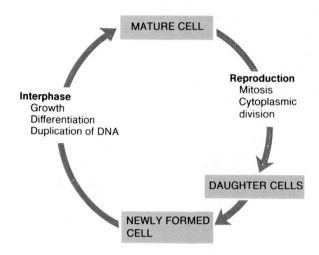

1. What type of mechanism is responsible for maintaining unequal concentrations of ions on opposite sides of a cell membrane?
2. What factors are necessary for the action of this mechanism?
3. What is the difference between pinocytosis and phagocytosis?
4. Describe receptor-mediated endocytosis.

Life Cycle of a Cell

The series of changes that a cell undergoes from the time it is formed until it reproduces is called its *life cycle* (figure 3.27). Superficially, this cycle seems rather simple—a newly formed cell grows for a time and then splits in half to form two daughter cells, which in turn may grow and reproduce. Yet the details of the cycle are quite complex, involving mitosis, cytoplasmic division, interphase, and differentiation.

Mitosis

Cell reproduction involves the dividing of a cell into two portions and includes two separate processes: (1) division of the nuclear parts, which is called **mitosis** (karyokinesis); and (2) division of the cytoplasm (cytokinesis). Another type of cell division called *meiosis* that occurs during the maturation of sex cells is described in chapter 22.

Division of the nuclear parts by mitosis is, of necessity, very precise, because the nucleus contains information, in the form of DNA molecules, that "tells" cell parts how to function. Each daughter cell must have a copy of this information in order to survive. Thus, the DNA molecules of the parent cell must be duplicated, and the duplicate sets must be distributed equally to the daughter cells. Once this has been accomplished, the cytoplasm and its parts can be divided.

Although mitosis is often described in stages, the process is really a continuous one without marked changes between one step and the next (figure 3.28). The idea of stages is useful, however, to indicate the sequence in which major events occur. These stages include the following:

1. **Prophase.** One of the first indications that a cell is going to reproduce is the appearance of *chromosomes*. These structures differentiate from chromatin in the nucleus, as chromatin threads condense into tightly coiled, rodlike parts. The resulting chromosomes contain DNA and protein molecules. Sometime earlier (during interphase) the DNA molecules became replicated, and each chromosome is consequently composed of two identical portions (chromatids). These parts are temporarily fastened together by a region on each called a *centromere* (kinetochore).

Figure 3.28
Mitosis is a continuous process during which the nuclear parts of a cell are divided into two equal portions. After reading about mitosis, identify the phases of the process in this diagram.

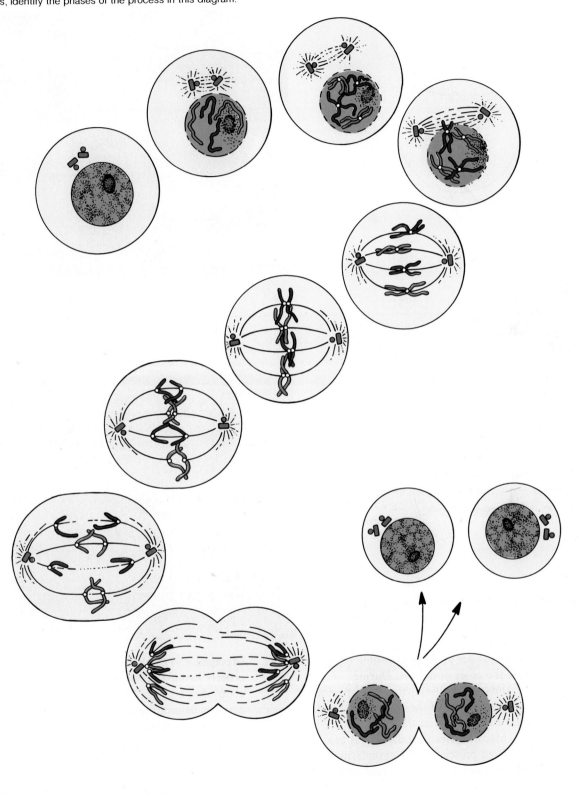

Figure 3.29

In prophase, chromosomes form from chromatin in the nucleus, and the centrioles move to opposite sides of the cell.

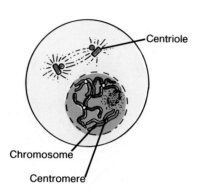

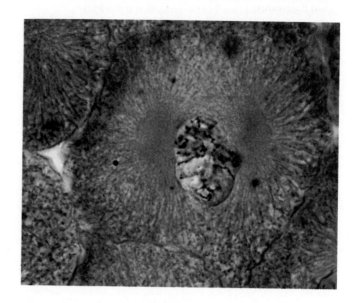

Figure 3.30

In metaphase, the chromosomes become lined up midway between the centrioles.

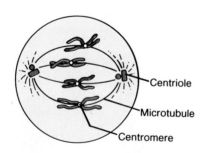

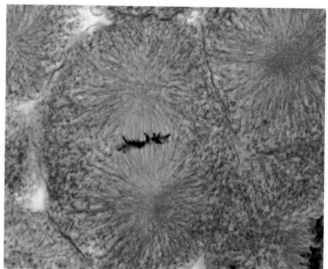

The centrioles of the centrosome replicate just before the onset of mitosis. During prophase, the two resulting pairs of centrioles move to opposite sides of the cytoplasm. Soon the nuclear membrane and the nucleolus disappear. Microtubules (spindle fibers) are assembled from proteins in the cytoplasm, and they become associated with the centrioles and chromosomes (figure 3.29).

2. **Metaphase.** The chromosomes line up in an orderly fashion about midway between the centrioles, apparently as a result of microtubule activity. Microtubules also are attached to the centromeres so that a microtubule accompanying one pair of centrioles is attached to one side of a centromere and a microtubule accompanying the other pair of centrioles is attached to the other side (figure 3.30).

Figure 3.31

In anaphase, the centromeres divide, and the microtubules that have become attached to them pull the chromosome parts toward the centrioles.

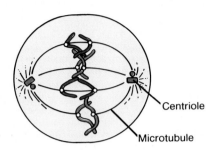

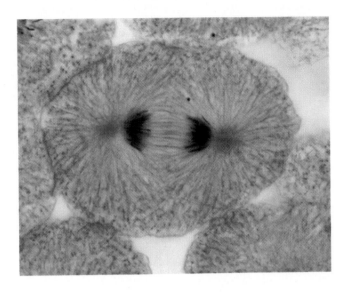

Figure 3.32

In telophase, the chromosomes elongate to become chromatin threads, and the cytoplasm begins to divide.

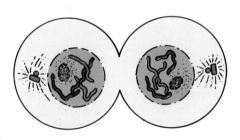

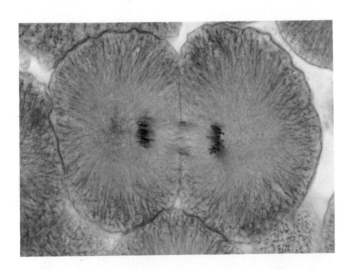

3. **Anaphase.** Soon the centromeres of the chromosome parts separate, and these identical parts become individual chromosomes. The separated chromosomes now move in opposite directions, and once again the movement seems to result from microtubule activity. In this case, the microtubules appear to shorten and pull their attached chromosomes toward the centrioles at opposite sides of the cell (figure 3.31).

4. **Telophase.** The final stage of mitosis is said to begin when the chromosomes complete their migration toward the centrioles. It is much like prophase, but in reverse. As the chromosomes approach the centrioles, they begin to elongate and change from rodlike into threadlike structures. A nuclear membrane forms around each chromosome set, and nucleoli appear within the newly formed nuclei. Finally, the microtubules disappear (figure 3.32). Telophase is accompanied by cytoplasmic division in which the cytoplasm is split into two portions.

Figure 3.33

Following mitosis, the cytoplasm of the parent cell is divided into two portions as seen in these scanning electron micrographs. (From *Scanning Electron Microscopy in Biology*, by R. G. Kessel and C. Y. Shih. © 1976 Springer-Verlag.)

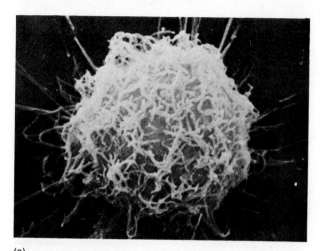

(a)

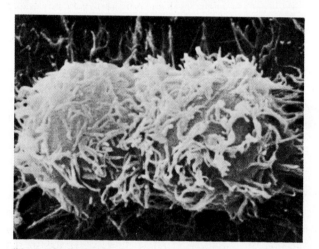

(b)

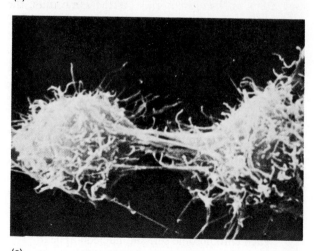

(c)

Chart 3.3	Major events in mitosis
Stage	Major events
Prophase	Chromatin differentiates into chromosomes; centrioles move to opposite sides of cytoplasm; nuclear membrane and nucleolus disappear; microtubules appear and become associated with centrioles and duplicate parts of chromosomes.
Metaphase	Chromosomes become arranged midway between the centrioles; microtubules from the centrioles become attached to the centromeres of each chromosome.
Anaphase	Centromeres separate, and duplicate parts of chromosomes become separated; microtubules shorten and pull individual chromosomes toward centrioles.
Telophase	Chromosomes elongate and form chromatin threads, nuclear membranes appear around each chromosome set, nucleoli appear, microtubules disappear.

Chart 3.3 summarizes the events that take place during these phases in mitosis.

Cytoplasmic Division

The mechanism responsible for cytoplasmic division (cytokinesis) is not well understood, but the cell membrane begins to constrict during anaphase and continues to constrict through telophase. This process involves the musclelike contraction of a ring of microfilaments composed of protein (actin). These filaments are assembled from the cytoplasm and are attached to the inner surface of the cell membrane. The contractile ring is positioned at right angles to the microtubules that pulled the chromosomes to opposite ends of the cell during mitosis. As the ring pinches inward, it separates the two newly formed daughter nuclei and divides about half of the cytoplasmic organelles into each of the new daughter cells.

Although the newly formed cells may differ slightly in size and number of cytoplasmic parts, they have identical chromosomes and thus contain identical DNA information. Except for size, they are copies of the parent cell (figure 3.33).

Interphase

Once formed, the daughter cells usually begin growing. This requires that the young cells obtain nutrients and use them in the manufacture of new living materials and for the synthesis of many vital compounds. At the same time, various cell parts become duplicated. In the nucleus, the chromosomes may be doubling (at least in cells that will soon reproduce); and in the cytoplasm, new ribosomes, lysosomes, mitochondria, and membranes are appearing. This stage in the life cycle is called **interphase,** and it lasts until the cell begins to undergo mitosis.

Many kinds of cells within the body are constantly growing and reproducing and thus increasing the number of cells that are present. Such activity is responsible for the growth and development of an embryo into a child, and a child into an adult, and is necessary to replace cells that have relatively short life spans. It is also responsible for the replacement of cells that are continually worn out or lost due to injury or disease.

Cell Differentiation

Since all body cells are formed by mitosis and contain the same DNA information, it might be expected that they look and act alike; obviously, they do not.

A human begins life as a single cell—a fertilized egg cell. This cell reproduces by mitosis to form 2 daughter cells; they in turn divide into 4 cells, the 4 become 8, and so forth. Then, sometime during development, the cells begin to *specialize.* That is, they develop special structures or begin to function in different ways. Some become skin cells, others become bone cells, and still others become nerve cells (figure 3.34).

The process by which cells develop different characteristics in structure and function is called **differentiation.** The mechanism responsible for this phenomenon is not well understood, but it seems to involve the repression of some of the DNA information. Thus, the DNA information needed for general cell activities may be "switched on" in both nerve and bone cells. The information related to specific bone cell functions, however, may be repressed or "switched off" in the nerve cells. Similarly, the information related to specific nerve cell functions may be repressed in bone cells.

Although the mechanism of differentiation is obscure, the results are obvious. Cells of many kinds are produced, and each kind carries on specialized functions: skin cells protect underlying tissues, red blood cells carry oxygen, and nerve cells transmit impulses. Each type of cell somehow helps the others and aids in the survival of the organism.

1. Why is it important that the division of nuclear materials during mitosis be so precise?
2. Describe the events that occur during mitosis.
3. Name the process by which some cells become muscle cells and others become nerve cells.
4. How does DNA seem to control this process?

Control of Cell Reproduction

The existence of a control mechanism for cell reproduction is evident in the fact that some cells reproduce continually, some occasionally, and some not at all.

Skin cells, blood-forming cells, and cells that line the intestine, for example, reproduce continually throughout life. Those that compose some organs, such as the liver, seem to reproduce until a particular number of cells is present, and then they cease reproducing. Interestingly, if the number of liver cells is reduced by injury or surgery, the remaining cells are somehow stimulated to reproduce again. Still other cells, such as nerve cells, apparently lose their ability to reproduce as they become differentiated; therefore, damage to nerve cells is likely to result in permanent loss of nerve function.

How the reproductive capacities of cells are controlled is not well understood, but the mechanism may involve the release of *growth-inhibiting substances.* Such a substance might slow or stop the growth and reproduction of particular cells when their numbers reach a certain level.

Surface-Volume Relationships

Another factor involved with the control of cell reproduction is the relationship between a cell's membrane surface area and its volume. The quantity of nutrients needed to maintain a cell is directly related to the volume of its living material. At the same time, the quantity of nutrients that can enter the cell is related directly to the surface area of its membrane. As a cell grows, however, its membrane surface area increases proportionately less than its volume. Consequently, the surface area will eventually become inadequate for the needs of the living material inside.

A cell can solve this growth problem by dividing. The resulting daughter cells are smaller than the parent cell and thus have a more favorable surface-volume relationship (figure 3.35).

1. How do cells vary in their rates of reproduction?
2. What factors seem to control the rate at which cells reproduce?

Figure 3.34

During development, the numerous body cells are produced from a single fertilized egg cell by mitosis. As these cells undergo differentiation, they become different kinds of cells that carry on specialized functions.

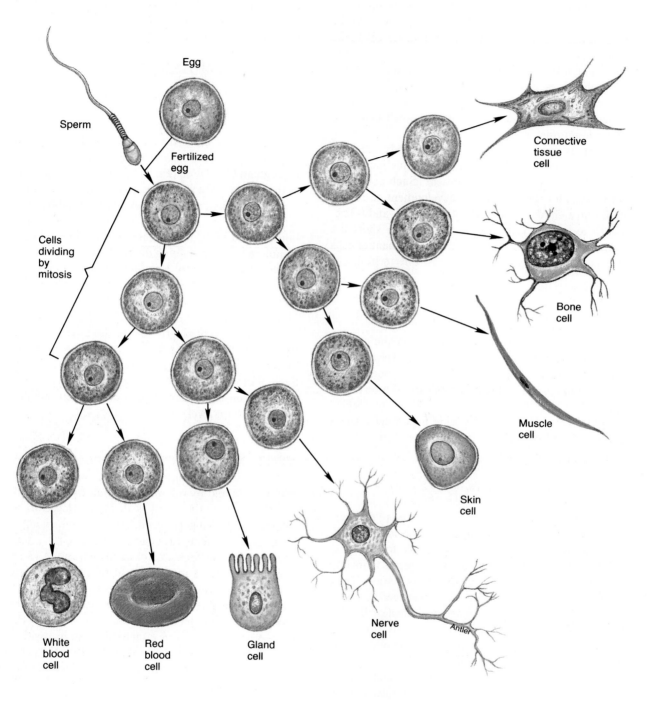

Egg

Sperm

Fertilized egg

Cells dividing by mitosis

Connective tissue cell

Bone cell

Muscle cell

Skin cell

White blood cell

Red blood cell

Gland cell

Nerve cell

Antler

Figure 3.35
When a cell divides, the amount of surface exposed to the surroundings is increased.

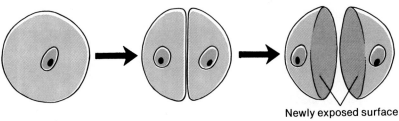

Newly exposed surface

Cancer
A Clinical Application

Cancer is a group of closely related diseases. It can occur in many different tissues and it results from changes in cells that allow them to avoid some of the normal control mechanisms. Cancerous conditions have certain common characteristics, including the following:

1. **Hyperplasia.** Hyperplasia is the uncontrolled reproduction of cells. Although the rate of reproduction among cancer cells is usually unchanged, they are not responsive to normal controls on cell numbers. As a result, cancer cells eventually give rise to large cell populations.

2. **Anaplasia.** The word anaplasia is used to describe the appearance of abnormalities in cellular structure. Typically, cancer cells resemble undifferentiated or primitive cells. That is, they fail to develop the specialized structure of the kind of cell they represent. Also, they fail to function in expected ways. Certain white blood cells, for example, normally function in resisting bacterial infections. When cells of this type become cancerous, they do not function effectively, and the cancer patient becomes more subject to infectious diseases. Cancer cells also are likely to form jumbled masses rather than becoming arranged in orderly groups like normal cells.

3. **Metastasis.** Metastasis is a tendency to spread. Normal cells are usually cohesive; that is, they stick together in groups of similar kinds. Cancer cells often become detached from their cellular mass and may then be carried away from their place of origin and may establish new cancerous growths in other parts of the body. Metastasis is the characteristic most closely associated with the word *malignant*, which suggests a power to threaten life, and this characteristic is often used to distinguish a cancerous growth from a noncancerous (benign) one.

The cause or causes of cancers remain largely unknown. There is, however, increasing evidence that the cause or causes are complex. Among the factors that may be involved are exposure to various chemicals or to harmful radiation, virus infections, changes in DNA structure, the presence of altered genes (oncogenes) that encode abnormal proteins (see chapter 4), or deficiencies in the body's ability to resist disease. Actually, two or more of these factors may be needed to cause a cancer, or perhaps different kinds of cancers are caused by different factors or combinations of factors.

When untreated, cancer cells eventually accumulate in large numbers. They may damage normal cells by effectively competing with them for nutrients. At the same time, cancer cells may invade vital organs and interfere with their functions, obstruct important passageways, or penetrate blood vessels and cause internal bleeding.

If it is detected in its early stages, a cancerous growth may be removed surgically. At other times, cancers are treated with radiation or drugs (chemotherapy), or some combination of these treatments may be used. The drugs most commonly used act on the structure of DNA molecules, affecting the DNA of rapidly reproducing normal cells as well as those of cancer cells. However, the normal cells seem to be able to repair DNA damage more effectively than cancer cells. Thus, the cancer cells are more likely to be destroyed by the drug action.

Chapter Summary

Introduction
Cells vary considerably in size, shape, and function. The shapes of cells are closely related to their functions.

A Composite Cell
1. A cell includes a nucleus, cytoplasm, and a cell membrane.
2. Cytoplasmic organelles perform specific vital functions while the nucleus controls the overall activities of the cell.
3. Cell membrane
   a. The cell membrane forms the outermost limit of the living material.
   b. It acts as a selectively permeable passageway that controls the movements of substances between the cell and its surroundings.
   c. It is composed of protein, lipid, and carbohydrate molecules.
   d. The cell membrane framework consists mainly of a double layer of phospholipid molecules.
   e. Molecules that are soluble in lipids pass through the membrane easily, but the membrane creates a barrier to the passage of water soluble molecules.
   f. Cholesterol molecules help to make the membrane stable.
   g. Proteins are responsible for the special functions of the membrane.
      (1) Rodlike proteins function as receptors on cell surfaces.
      (2) Globular proteins create channels for the passage of various ions and molecules.
   h. Some cells are connected by means of specialized intercellular junctions called tight junctions, desmosomes, and gap junctions.
4. Cytoplasm
   a. Cytoplasm contains networks of membranes and organelles suspended in fluid.
   b. Endoplasmic reticulum is a network of membranes that provides a tubular communication system and an attachment for ribosomes; it also functions in the synthesis of proteins, lipids, and hormones.
   c. Ribosomes are particles of protein and RNA that function in protein synthesis.
   d. Golgi apparatus is composed of a stack of flattened, membranous sacs that package glycoproteins for secretion.
   e. Mitochondria are membranous sacs that contain enzymes involved with releasing energy from food molecules and transforming energy into a usable form.
   f. Lysosomes are membranous sacs containing digestive enzymes that can destroy substances that enter cells.
   g. Peroxisomes are membranous, enzyme-containing vesicles.
   h. Cilia and flagella are motile processes that extend outward from some cell surfaces.
      (1) Cilia are numerous hairlike parts that serve to move fluids across cell surfaces.
      (2) Flagella are longer processes; the tail of a sperm cell is a flagellum that enables the cell to swim.
   i. The centrosome is a nonmembranous structure, consisting of two centrioles, that aids in the distribution of chromosomes during cell reproduction.
   j. Vesicles are membranous sacs containing substances that recently entered the cell.
   k. Microfilaments and microtubules are threadlike processes that aid cellular movements and provide support and stability to cytoplasm.
   l. Cytoplasm may contain nonliving cellular products such as nutrients and pigments called *inclusions*.
5. Cell nucleus
   a. The nucleus is enclosed in a double-layered nuclear envelope that controls the movement of substances between the nucleus and cytoplasm.
   b. A nucleolus is a dense body of protein and RNA that functions in the production of ribosomes.
   c. Chromatin is composed of loosely coiled fibers of protein and DNA that become chromosomes during cell reproduction.

Movements through Cell Membranes
Movement through all membranes may involve physical or physiological processes.
1. Diffusion
   a. Diffusion is a scattering of molecules or ions from regions of higher toward regions of lower concentration.
   b. It is responsible for exchanges of oxygen and carbon dioxide within the body.
   c. Factors that increase the rate of diffusion include short distance, high concentration of diffusing molecules, low molecular weight, and high temperature.
2. Facilitated diffusion
   a. Facilitated diffusion involves the use of carrier molecules in the cell membrane.
   b. This process only moves substances from regions of higher to regions of lower concentration.
3. Osmosis
   a. Osmosis is a special case of diffusion in which water molecules diffuse from regions of higher water concentration to lower water concentration through a selectively permeable membrane.
   b. Osmotic pressure increases as the number of particles dissolved in a solution increases.

c. Cells lose water when placed in hypertonic solutions and gain water when placed in hypotonic solutions.

d. Solutions that contain the same concentration of dissolved particles as a cell are called isotonic.

4. Filtration
   a. Filtration involves movement of molecules from regions of higher toward regions of lower hydrostatic pressure.
   b. Blood pressure causes filtration of water and dissolved substances through capillary walls.

5. Active transport
   a. Active transport is responsible for movement of molecules or ions from regions of lower to regions of higher concentration.
   b. It requires cellular energy and involves action of carrier molecules in the cell membrane.

6. Endocytosis
   a. Pinocytosis is a process by which a cell membrane engulfs tiny droplets of liquid.
   b. Phagocytosis is a process by which a cell membrane engulfs solid particles.
   c. Receptor-mediated endocytosis is a process by which receptor proteins combine with specific kinds of molecules in the cell surroundings and the combinations are engulfed.

Life Cycle of a Cell

1. The life cycle of a cell includes mitosis, cytoplasmic division, interphase, and differentiation.

2. Mitosis
   a. Mitosis is the division and distribution of nuclear parts to daughter cells during cell reproduction.
   b. The stages of mitosis include prophase, metaphase, anaphase, and telophase.

3. Cytoplasmic division is a process by which cytoplasm is divided into two portions following mitosis.

4. Interphase
   a. Interphase is the stage in the life cycle when a cell grows and forms new organelles.
   b. It terminates when the cell begins to undergo mitosis.

5. Cell differentiation involves the development of specialized structures and functions.

Control of Cell Reproduction

1. Cellular reproductive capacities vary greatly.

2. Reproductive capacities of cells may involve the release of growth-inhibiting substances.

3. Surface-volume relationships
   a. As a cell grows, its surface area increases to a lesser degree than its volume, and eventually the area becomes inadequate for the needs of the living material within the cell.
   b. When a cell divides, the daughter cells have more favorable surface-volume relationships.

Application of Knowledge

1. Which process—diffusion, osmosis, or filtration—is most closely related to each of the following situations?
   a. A person is given an injection of isotonic glucose solution.
   b. A person with extremely low blood pressure stops producing urine.
   c. The concentration of urea in the dialyzing fluid of an artificial kidney is kept low.

2. What characteristic of cell membranes may account for the observation that fat-soluble substances like chloroform and ether cause rapid effects upon cells?

3. A person who has been exposed to excessive amounts of X ray may develop a decrease in white blood cell number and an increase in susceptibility to infections. In what way are these effects related?

Review Activities

1. Use specific examples to illustrate how cells vary in size.

2. Describe how the shapes of nerve, epithelial, and muscle cells are related to their functions.

3. Name the major portions of a cell and describe their relationship to one another.

4. Discuss the structure and functions of a cell membrane.

5. Distinguish between organelles and inclusions.

6. Define *selectively permeable*.

7. Describe the chemical structure of a membrane.

8. Explain how the structure of a cell membrane is related to its permeability.

9. Explain the function of membrane proteins.

10. Describe three kinds of intercellular junctions.

11. Describe the structures and functions of each of the following:
    a. endoplasmic reticulum
    b. ribosome
    c. Golgi apparatus
    d. mitochondrion
    e. lysosome
    f. peroxisome
    g. cilium
    h. flagellum
    i. centrosome
    j. vesicle
    k. microfilament
    l. microtubule

12. Define *inclusion*.

13. Describe the structure of the nucleus and the functions of its parts.

14. Distinguish between diffusion and facilitated diffusion.
15. Name four factors that increase the rate of diffusion.
16. Explain how diffusion aids in the exchange of gases within the body.
17. Define *osmosis*.
18. Explain what is meant by *osmotic pressure*.
19. Explain how the number of solute particles in a solution affects its osmotic pressure.
20. Distinguish between solutions that are hypertonic, hypotonic, and isotonic.
21. Define *filtration*.
22. Explain how filtration is involved with the movement of substances through capillary walls.
23. Explain why active transport is called a physiological process while diffusion is a physical process.
24. Explain the function of carrier molecules in active transport.
25. Distinguish between pinocytosis and phagocytosis.
26. Describe the process called *receptor-mediated endocytosis*.
27. List the phases in the life cycle of a cell.
28. Name the two processes included in cell reproduction.
29. Describe the major events of mitosis.
30. Explain what happens during *interphase*.
31. Define *differentiation*.
32. Explain how differentiation may involve the repression of DNA information.
33. Explain how surface-volume relationships may function in the control of cell reproduction.

4

Cellular Metabolism

Cellular metabolism includes all the chemical reactions that occur within cells. These reactions are of two major types and, for the most part, involve the use of food substances. In one type of reaction, molecules of nutrients are converted into simpler forms by changes that are accompanied with the release of energy. In the other type, the nutrient molecules are used in constructive processes that produce the cell's structural and functional materials or molecules that store energy.

In either case, metabolic reactions are controlled by proteins called *enzymes* that are synthesized in cells. The production of enzymes is, in turn, controlled by genetic information held within molecules of DNA, which instructs a cell how to synthesize specific kinds of enzyme molecules.

Chapter Outline

Chapter Objectives

After you have studied this chapter, you should be able to

1. Distinguish between anabolic and catabolic
   metabolism.

2. Explain how enzymes control metabolic processes.

3. Explain how chemical energy is released by cellular
   respiration.

4. Describe how energy is made available
   for cellular activities.

5. Describe the general metabolic pathways of
   carbohydrates, lipids, and proteins.

6. Explain how metabolic pathways are regulated.

7. Describe how genetic information is stored within
   DNA molecules.

8. Explain how this information is used in the synthesis
   of proteins.

9. Describe how DNA molecules are replicated.

10. Explain how genetic information can be altered and
    how such a change may affect an organism.

11. Complete the review activities at the end of this
    chapter. Note that the items are worded in the form of
    specific learning objectives. You may want to refer to
    them before reading the chapter.

Key Terms

aerobic respiration   (a-er-ō'bik res''pĭ-ra'shun)

anabolic metabolism   (an''ah-bol'ik mĕ-tab'o-lizm)

anaerobic respiration   (an''a-er-ō'bik res''pĭ-ra'shun)

catabolic metabolism   (kat''ah-bol'ik mĕ-tab'o-lizm)

coenzyme   (ko-en'zīm)

deamination   (de-am''ĭ-na'shun)

dehydration synthesis   (de''hi-dra'shun sin'thĕ-sis)

DNA

energy   (en'er-je)

enzyme   (en'zīm)

gene   (jēn)

genetic code   (jĕ-net'ik kōd)

glycolysis   (gli-kol'ĭ-sis)

hydrolysis   (hi-drol'ĭ-sis)

mutation   (mu-ta'shun)

oxidation   (ok''sĭ-da'shun)

replication   (re''pli-ka'shun)

RNA

substrate   (sub'strāt)

Aids to Understanding Words

aer-, air: *aer*obic respiration—respiratory process that requires oxygen.

an-, without: *an*aerobic respiration—respiratory process that proceeds without oxygen.

ana-, up: *ana*bolic metabolism—cellular processes in which smaller molecules are used to build larger ones.

cata-, down: *cata*bolic metabolism—cellular processes in which larger molecules are broken down into smaller ones.

co-, with: *co*enzyme—substance that unites with a protein and completes the structure of an enzyme molecule.

de-, undoing: *de*amination—process by which nitrogen-containing portions of amino acid molecules are removed.

mut-, change: *mut*ation—change in the genetic information of a cell.

-strat, spread out: sub*strat*e—substance on which an enzyme acts.

sub-, under: *sub*strate—substance on which an enzyme acts.

zym-, causing to ferment: en*zym*e—protein that initiates or speeds up a chemical reaction without being changed itself.

ALTHOUGH A LIVING cell may appear to be idle, it is actually very active. Numerous metabolic processes needed to maintain its life are occurring all the time.

Metabolic Processes

Metabolic processes, which include all of the chemical reactions that take place in a cell, can be divided into two major types. One type, called *anabolic,* involves the buildup of larger molecules from smaller ones and utilizes energy. The other, called *catabolic,* involves the breakdown of larger molecules into smaller ones and releases energy.

Anabolic Metabolism

Anabolic metabolism includes all the constructive processes used to manufacture substances needed for cellular growth and repair. For example, cells often join simple sugar molecules (monosaccharides) to form larger molecules of glycogen by an anabolic process called *dehydration synthesis.* In this process, the larger carbohydrate molecule is built by bonding monosaccharide molecules together into a complex chain. When adjacent monosaccharide units are joined, an −OH (hydroxyl group) from one monosaccharide molecule and an −H (hydrogen atom) from another are removed. These particles react to produce a water molecule, and the monosaccharides are united by a shared oxygen atom, as shown in figure 4.1. As the process is repeated, the molecular chain becomes longer.

Similarly, glycerol and fatty acid molecules are joined by dehydration synthesis in fat tissue cells to form fat molecules. In this case, 3 hydrogen atoms are removed from a glycerol molecule, and an −OH group is removed from each of 3 fatty acid molecules, as shown in figure 4.2. The result is 3 water molecules and a single fat molecule, whose glycerol and fatty acid portions are bound by shared oxygen atoms.

Figure 4.1
What type of reaction is represented in this diagram?

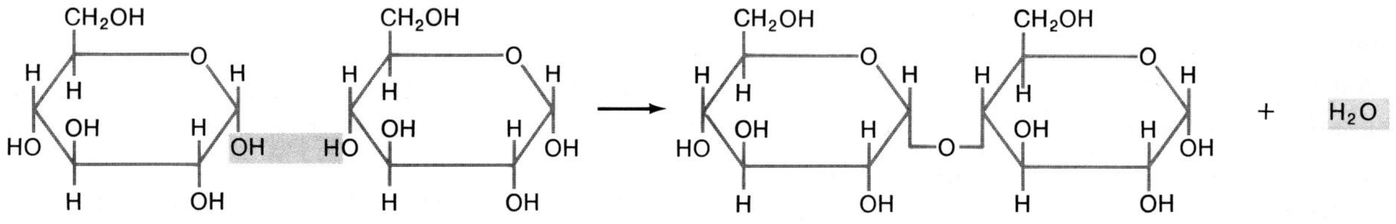

$$C_6H_{12}O_6 \quad + \quad C_6H_{12}O_6 \quad \longrightarrow \quad C_{12}H_{22}O_{11} \quad + \quad H_2O$$

Monosaccharide + Monosaccharide ⟶ Disaccharide + Water

Figure 4.2
A glycerol molecule and three fatty acid molecules may be joined by dehydration synthesis to form a fat molecule.

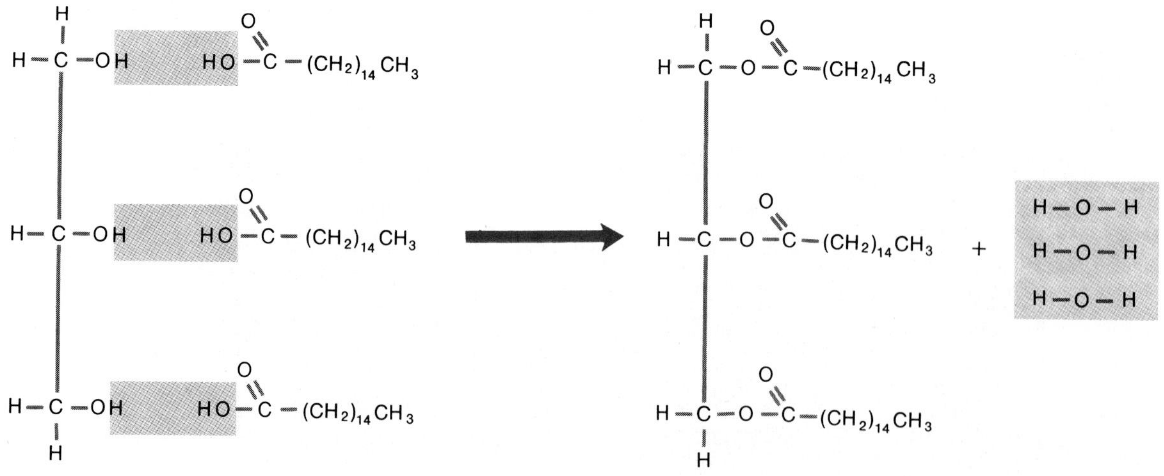

Glycerol + 3 fatty acid molecules ⟶ Fat molecule + 3 water molecules

Cells also build protein molecules by joining amino acids by means of dehydration synthesis. When 2 amino acid molecules are united, an —OH is removed from one and a hydrogen atom from the other. A water molecule is formed, and the amino acid molecules are joined by a bond between a carbon atom and a nitrogen atom (figure 4.3). This type of bond, which is called a *peptide bond,* holds the amino acids together. Two amino acids bound together form a *dipeptide,* while many joined in a chain form a *polypeptide.* Generally, a polypeptide consisting of 100 or more amino acid molecules is called a *protein.*

Catabolic Metabolism

Physiological processes in which larger molecules are broken down into smaller ones constitute **catabolic metabolism.** An example of such a process is *hydrolysis,* which can bring about the decomposition of carbohydrates, lipids, and proteins.

When a molecule of one of these substances is hydrolyzed, a water molecule is used, and the original molecule is split into two simpler parts. The hydrolysis of a disaccharide such as sucrose, for instance, results in molecules of 2 monosaccharides, glucose and fructose.

$$C_{12}H_{22}O_{11} + H_2O \rightarrow C_6H_{12}O_6 + C_6H_{12}O_6$$
(sucrose) (water) (glucose) (fructose)

In this case, the bond between the simple sugars within the sucrose molecule is broken, and the water molecule supplies a hydrogen atom to one sugar molecule and a hydroxyl group to the other. Thus, hydrolysis is the reverse of dehydration synthesis.

Hydrolysis

Disaccharide + Water $\rightleftharpoons$ Monosaccharide + Monosaccharide

Dehydration synthesis

Hydrolysis, more commonly called *digestion,* occurs in various regions of the digestive tract and is responsible for the breakdown of carbohydrates into monosaccharides, fats into glycerol and fatty acids, and proteins into amino acids (figure 4.4). Digestion is discussed in detail in chapter 14.

Both catabolic and anabolic metabolism are carried on continually in cells. However, these activities must be carefully controlled so that the breakdown or energy-releasing reactions occur at rates that are adjusted to the needs of the building-up or energy-utilizing reactions. Any disturbance in this balance is likely to cause abnormalities or death of the cell.

Figure 4.3
When two amino acid molecules are united by dehydration synthesis, a peptide bond is formed between a carbon atom and a nitrogen atom.

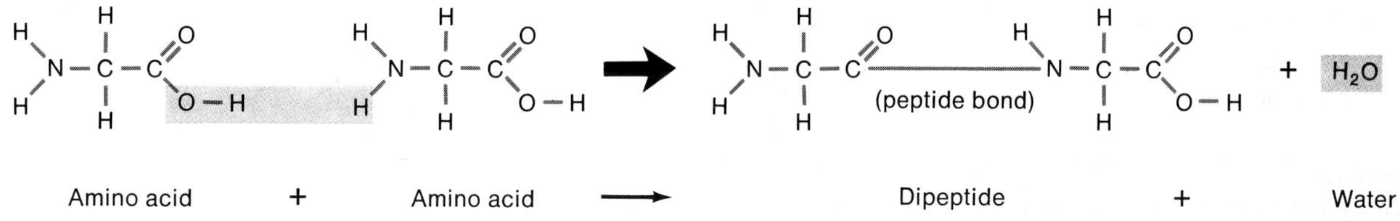

Figure 4.4
Hydrolysis causes the decomposition of (a) carbohydrates into monosaccharides; (b) fats into glycerol and fatty acids; and (c) proteins into amino acids.

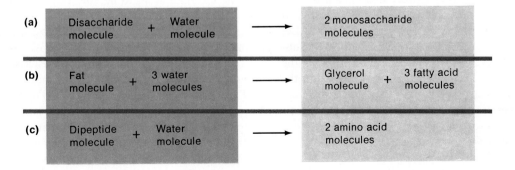

1. What general functions do anabolic metabolism and catabolic metabolism serve?
2. Name a substance formed by the anabolic metabolism of monosaccharides, of amino acids, and of glycerol and fatty acids.
3. Distinguish between dehydration synthesis and hydrolysis.

Control of Metabolic Reactions

Although different kinds of cells may conduct specialized metabolic processes, all cells perform certain basic reactions, such as the buildup and breakdown of carbohydrates, lipids, proteins, and nucleic acids. These reactions actually include hundreds of very specific chemical changes that must occur in an orderly fashion. They are controlled by mechanisms involving the presence of *enzymes*.

Enzymes and Their Actions

Like other chemical reactions, metabolic reactions generally require a certain amount of energy (activation energy) before they will occur. This is why heat is commonly used to increase the rates of chemical reactions in laboratories. The heat energy increases the rate at which molecules move and the frequency of molecular collisions. These collisions increase the likelihood of interactions among the electrons of the molecules that can create new chemical bonds. The temperature conditions that exist naturally in cells are usually too mild to promote the reactions needed to support life, but this is not a problem because cells contain enzymes.

Enzymes are globular proteins (see chapter 2) that promote specific chemical reactions within cells by lowering the amount of energy (activation energy) needed to initiate these reactions. They seem to do this by temporarily binding with the molecules they act upon. This somehow strains or distorts the chemical structure of these molecules and in this way increases the likelihood of a chemical change. Thus, in the presence of enzymes, metabolic reactions are speeded up.

Enzymes are needed in very small quantities because as they function, they are not changed or used up and can, therefore, function again and again. Also, each enzyme acts only on a particular kind of substance, which is called its **substrate.** For example, the substrate of one enzyme called *catalase* (found in the peroxisomes of cells) is hydrogen peroxide, a toxic by-product of certain metabolic reactions. This enzyme's only function is to cause the decomposition of hydrogen peroxide into water and oxygen. In this way it helps to prevent an accumulation of hydrogen peroxide that might damage cells.

You may have noticed the action of the enzyme catalase if you have ever used hydrogen peroxide to cleanse a wound. Injured cells contain catalase, and when hydrogen peroxide contacts them, bubbles of oxygen are released. This is useful since the resulting foam aids in removing debris from inaccessible parts of the wound.

Cellular metabolism includes hundreds of different chemical reactions, and each of these reactions is controlled by a specific kind of enzyme. Thus, there must be hundreds of different kinds of enzymes present in every cell. How each enzyme "recognizes" its specific substrate is not well understood, but this ability seems to involve the shapes of molecules. That is, each enzyme's polypeptide chain is thought to be twisted and coiled into a unique three-dimensional form, that fits the special shape of a substrate molecule. In short, an enzyme molecule is thought to fit a molecule of its substrate much as a key fits a particular lock.

During an enzyme-controlled reaction particular regions of the enzyme molecule called *active sites* temporarily combine with portions of the substrate, creating a substrate-enzyme complex. It is this interaction that seems to strain chemical bonds within the substrate, increasing the likelihood of a change in the substrate molecule. When the substrate is changed, the product of the reaction appears, and the enzyme is released unaltered (figure 4.5).

This activity can be summarized as follows:

$$\text{Substrate molecule} + \text{Enzyme molecule} \rightarrow \text{Substrate-enzyme complex} \rightarrow \text{Product (changed substrate)} + \text{Enzyme molecule}$$

The speed of an enzyme-controlled reaction is related to the number of enzyme and substrate molecules present in the cell. Generally, the reaction occurs more rapidly if the concentration of the enzyme or the concentration of the substrate is increased. Also, the efficiency of different kinds of enzymes varies greatly. Thus some enzymes seem to be able to process only a few substrate molecules per second, while others can process thousands or even millions of substrate molecules in a second.

Commonly, enzymes are named according to the substrates they act upon, with the suffix *-ase* added. Thus, a lipid splitting enzyme is called *lipase,* a protein splitting enzyme is a *protease,* and a starch (amylon) splitting enzyme is an *amylase.* Similarly, *sucrase* is an enzyme that splits the sugar sucrose, *maltase* splits the sugar maltose, and *lactase* splits the sugar lactose.

Figure 4.5
The shape of an enzyme molecule's active site (a) fits the shape of the substrate molecule (b). When the substrate molecule becomes combined temporarily with the enzyme (c) a chemical reaction occurs. The result is product molecules (d) and an unaltered enzyme (e).

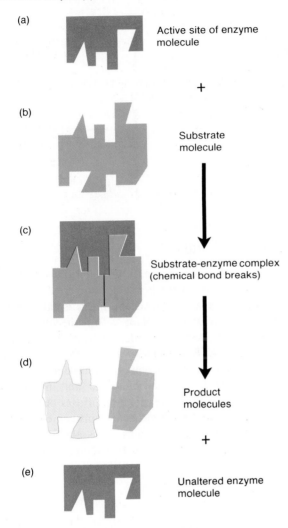

(a)   Active site of enzyme molecule

+

(b)   Substrate molecule

(c)   Substrate-enzyme complex (chemical bond breaks)

(d)   Product molecules

+

(e)   Unaltered enzyme molecule

Cofactors and Coenzymes

Often the protein portion of an enzyme molecule (apoenzyme) is inactive until it is combined with an additional substance. This non-protein part is needed to complete the proper shape of the active site of the enzyme molecule or help bind the enzyme to its substrate. Such a substance is called a **cofactor.** A cofactor may be an ion of an element, such as copper, iron, or zinc, or it may be a relatively small organic molecule, in which case it is called a **coenzyme.** Most coenzymes are obtained from vitamins.

Vitamins are essential substances that cannot be synthesized (or cannot be synthesized in sufficient quantities) by human cells and therefore must be obtained in the diet. Since vitamins provide coenzymes

that can, like enzymes, be used again and again, they are needed by cells in very small quantities. Vitamins are discussed in detail in chapter 15.

Factors That Alter Enzymes

Except for their cofactor portions, enzymes are proteins, and like other proteins, they can be denatured by exposure to excessive heat, radiation, electricity, certain chemicals, or to extreme concentration of acid or base. For example, many enzymes become inactive at 45°C, and nearly all of them are denatured at 55°C. Some poisons are chemicals that cause enzymes to be denatured. Cyanides, for instance, interfere with respiratory enzymes and damage cells by halting their energy-releasing processes.

The antibiotic drug penicillin acts by interfering with enzymes that function in the production of bacterial cell walls. These walls surround and protect normal bacterial cells. In the presence of penicillin, the cell walls are not properly produced, and the affected bacteria are less able to survive. Thus, penicillin may be used to control the growth of bacteria that might otherwise cause an infection.

---

1. What is an enzyme?
2. How can an enzyme control a metabolic reaction?
3. How does an enzyme "recognize" its substrate?
4. What is the role of a cofactor?
5. What factors are likely to denature enzymes?

Energy for Metabolic Reactions

Energy is the capacity to produce changes in matter or the ability to move something, that is, to do work. Therefore, energy is recognized by what it can do. Common forms of energy include heat, light, sound, electrical energy, mechanical energy, and chemical energy.

Although energy cannot be created or destroyed, it can be converted from one form to another. Sunlight is converted to heat when it is absorbed by skin; an ordinary incandescent light bulb changes electrical energy to heat and light; and an automobile engine converts the chemical energy in gasoline to heat and mechanical energy.

Whenever changes take place, energy is being transferred from one part to another. Thus, all metabolic reactions depend on the presence of energy in one form or another.

Release of Chemical Energy

Most metabolic processes use chemical energy. This form of energy is held in the bonds between the atoms of molecules and is released when these bonds are broken. For example, the chemical energy of many substances can be released by burning. Such a reaction must be started, and this is usually accomplished by applying heat to activate the burning process. As the substance burns, molecular bonds are broken, and energy escapes as heat and light.

Similarly, glucose molecules are "burned" in cells, although the process is more correctly called **oxidation.** The energy released by the oxidation of glucose is used to promote cellular metabolism. There are, however, some important differences between the oxidation of substances inside cells and the burning of substances outside them.

Burning usually requires a relatively large amount of energy to activate the process, and most of the energy released escapes as heat or light. In cells, the oxidation process is initiated by enzymes that reduce the amount of energy (activation energy) needed. Also, by transferring energy to special energy-carrying molecules, cells are able to capture about half of the energy released. The rest escapes as heat, which helps maintain body temperature.

The process by which energy is released from molecules such as glucose and is transferred to other molecules is quite complex. The process is called **cellular respiration,** and it involves a number of chemical reactions that must occur in a particular sequence, each one controlled by a separate kind of enzyme. Some of these enzymes are in the cell's cytoplasm, while others are in the mitochondria.

1. What is meant by energy?
2. How does cellular oxidation differ from burning?
3. Define cellular respiration.

Anaerobic Respiration

When a 6-carbon glucose molecule is decomposed in cellular respiration, enzymes control a series of reactions that break it into two 3-carbon pyruvic acid molecules. This phase of respiration (glycolysis) occurs in the cytoplasm, and because it takes place in the absence of oxygen, it is called **anaerobic respiration.**

Although some energy is needed to activate the reactions of anaerobic respiration, more energy is released than is used. A portion of the excess is transferred to molecules of an energy-carrying substance called **ATP** (adenosine triphosphate) (figure 4.6).

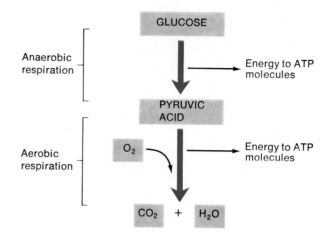

Figure 4.6
Anaerobic respiration occurs in the cytoplasm in the absence of oxygen, while aerobic respiration occurs in the mitochondria in the presence of oxygen.

Aerobic Respiration

Following the anaerobic phase of respiration, oxygen must be available for further molecular breakdown to occur. For this reason, the second phase is called **aerobic respiration.** It takes place within the mitochondria, and as a result, considerably more energy is transferred to ATP molecules.

When the decomposition of a glucose molecule is complete, carbon dioxide molecules and hydrogen atoms remain. The carbon dioxide diffuses out of the cell as a waste, and the hydrogen atoms combine with oxygen to form water molecules. Thus, the final products of glucose oxidation are carbon dioxide, water, and energy.

ATP Molecules

For each glucose molecule that is decomposed, 38 molecules of ATP can be produced. Two of these are the result of anaerobic respiration, while the rest are formed during the aerobic phase.

Each ATP molecule consists of 3 main parts—an adenine portion, a ribose portion, and 3 phosphates in a chain (figure 4.7). (Note the similarity between the structure of an ATP molecule and that of a nucleotide from a nucleic acid molecule described in chapter 2. Actually, ATP represents another type of nucleotide.)

As energy is released during cellular respiration, some of it is captured in the bond of the end phosphate of an ATP molecule. This energy is quickly transferred to another molecule involved in a metabolic process. When this energy transfer occurs, the terminal, high-energy phosphate bond of the ATP is broken, and the energy of this bond is released. Such energy is needed

Figure 4.7
An ATP molecule consists of an adenine portion, a ribose portion, and three phosphates. The wavy lines connecting the last two phosphates represent high energy chemical bonds from which energy can be released quickly.

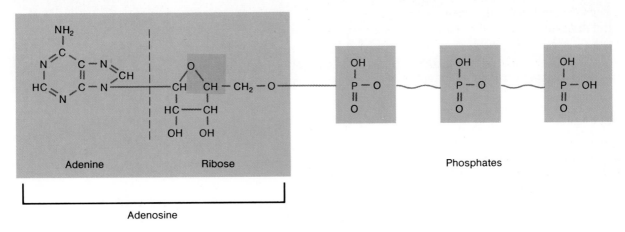

Figure 4.8
What is the significance of this cyclic process?

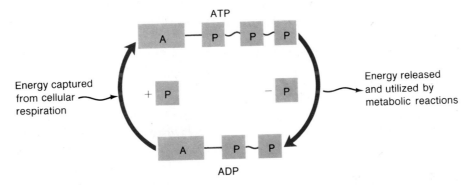

whenever cellular work is performed, as when muscle cells contract, membranes carry on active transport, or cells synthesize substances.

An ATP molecule that has lost its terminal phosphate becomes an **ADP** (adenosine diphosphate) molecule. However, the ADP molecule can be converted back into an ATP by capturing some energy and a phosphate. Thus, as figure 4.8 shows, ATP and ADP molecules shuttle back and forth between the energy-releasing reactions of cellular respiration and the energy-utilizing reactions of the cell.

Although ATP is not the only kind of energy-carrying molecule within a cell, it is the primary one. Without a source of ATP most cells die quickly.

1. What is meant by anaerobic respiration? Aerobic respiration?
2. What happens to the energy released by these processes?
3. What are the final products of these reactions?
4. What is the function of ATP molecules?

Metabolic Pathways

The anabolic and catabolic reactions that occur in cells usually involve a number of different steps that must occur in a particular sequence. For example, the anaerobic phase of cellular respiration, by which glucose is converted to pyruvic acid, involves 10 separate reactions. Since each reaction is controlled by a specific kind of enzyme, these enzymes must act in proper order. Such precision of activity suggests that the enzymes are arranged in precise ways. It is believed that the enzymes responsible for aerobic respiration are located in tiny, stalked particles on the membranes (cristae) within the mitochondria, and that they are positioned in the exact sequence of the reactions they control.

Such a sequence of enzyme-controlled reactions that leads to the production of particular products is called a **metabolic pathway** (figure 4.9). Typically these pathways are interconnected so that molecules of a certain kind of substance may enter more than one pathway as the substance is metabolized. For example,

Figure 4.9
A metabolic pathway consists of a series of enzyme-controlled reactions leading to a product.

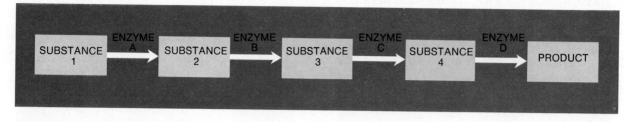

Figure 4.10
Carbohydrates from foods are changed into monosaccharides by hydrolysis. The resulting molecules may enter catabolic pathways and be used as energy sources, or they may enter anabolic pathways and be converted to glycogen or fat.

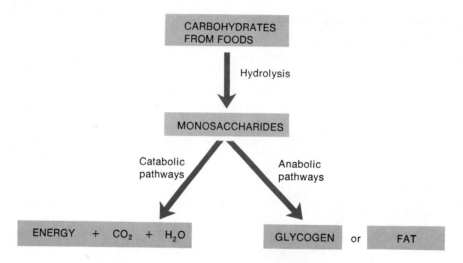

carbohydrate molecules from foods may enter catabolic pathways and be used to supply energy, or they may enter anabolic pathways and be stored as glycogen or fat (figure 4.10).

Carbohydrate Pathways

The average human diet consists largely of carbohydrates, which are changed by digestion to monosaccharides, such as glucose. These substances are used primarily as cellular energy sources, which means they usually enter the catabolic pathways of cellular respiration. As discussed previously, the first phase of this process occurs in the cytoplasm and is anaerobic. This phase is called *glycolysis* and involves the conversion of a 6-carbon glucose into two 3-carbon pyruvic acid molecules. During the second phase of the process the pyruvic acid is transformed into a 2-carbon acetyl group that combines with coenzyme A to form a substance called *acetyl coenzyme A*. It, in turn, is transported into a mitochondrion and is changed into a number of intermediate products by a complex series of chemical

reactions known as the **citric acid cycle** (Kreb's cycle). As these changes occur, energy is released and some of it is transferred to molecules of ATP, while the rest is lost as heat. The end products of the oxidation process are molecules of carbon dioxide and water.

Glycolysis
The chemical reactions of glycolysis are illustrated in figure 4.11. In the early steps of this metabolic pathway, the original glucose molecule is altered by the addition of phosphate groups (a process called *phosphorylation*) and by the rearrangement of its atoms. The phosphate groups and the energy needed for the reactions are supplied by ATP molecules. The result is a molecule of fructose combined with two phosphate groups (fructose-1,6-diphosphate). It is changed into two 3-carbon molecules (glyceraldehyde-3-phosphate), which in turn are converted to pyruvic acid, as follows:

a. An inorganic phosphate group is added to glyceraldehyde-3-phosphate to form 1,3-diphosphoglyceric acid. At the same time, two hydrogen atoms are released.

b. 1,3-diphosphoglyceric acid is changed to 3-phosphoglyceric acid. As this occurs, some energy in the form of a high energy phosphate is transferred from the 1,3-diphosphoglyceric acid to an ADP molecule, converting the ADP to ATP.

c. A slight alteration of 3-phosphoglyceric acid occurs to form 2-phosphoglyceric acid.

d. A change in 2-phosphoglyceric acid converts it into phosphoenolpyruvic acid.

e. Finally, a high energy phosphate is transferred from the phosphoenolpyruvic acid to an ADP molecule, converting it to ATP. A molecule of pyruvic acid remains.

Since two 3-carbon molecules are involved originally, two molecules of pyruvic acid are produced by these reactions. Also a total of four hydrogen atoms are released (step *a*) and four ATP molecules are formed (two in step *b* and two in step *e*). However, because two molecules of ATP are used early in glycolysis, there is a net gain of only two ATP molecules during this phase of cellular respiration.

In the absence of oxygen, the resulting pyruvic acid molecules may be converted into lactic acid. In the presence of oxygen, however, each pyruvic acid molecule is oxidized to an acetyl group which is then combined with a molecule of coenzyme A (obtained from the vitamin, pantothenic acid) to form acetyl coenzyme A. As this occurs, two hydrogen atoms are released for each molecule of acetyl coenzyme A formed. The acetyl coenzyme A is then broken down by means of the citric acid cycle, which is illustrated in figure 4.12.

Citric Acid Cycle

An acetyl coenzyme A molecule enters the citric acid cycle (Kreb's cycle) by combining with a molecule of oxaloacetic acid to form citric acid. The citric acid is then changed by a series of reactions back into oxaloacetic acid, and the cycle is completed.

As citric acid is produced, coenzyme A is released and thus can be used again and again in the formation of acetyl coenzyme A from pyruvic acid molecules.

During various steps in the citric acid cycle, carbon dioxide and hydrogen atoms are released. More specifically, for each glucose molecule metabolized in the presence of oxygen, two molecules of acetyl coenzyme A enter the citric acid cycle. As a result of the cycle, four carbon dioxide molecules and sixteen hydrogen atoms are released. At the same time, two more molecules of ATP are formed.

Figure 4.11
Chemical reactions of glycolysis.

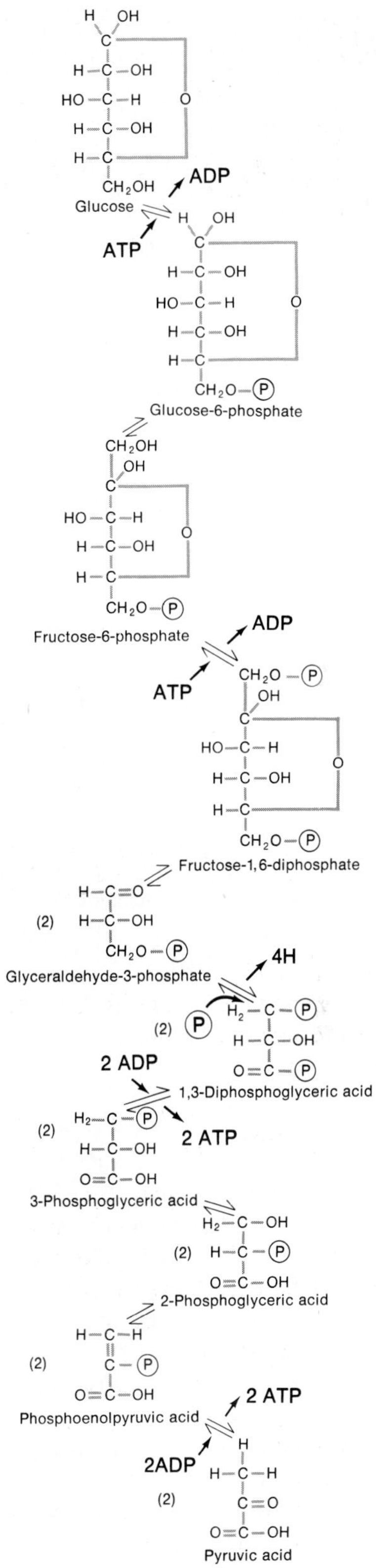

Figure 4.12
Chemical reactions of the citric acid cycle.

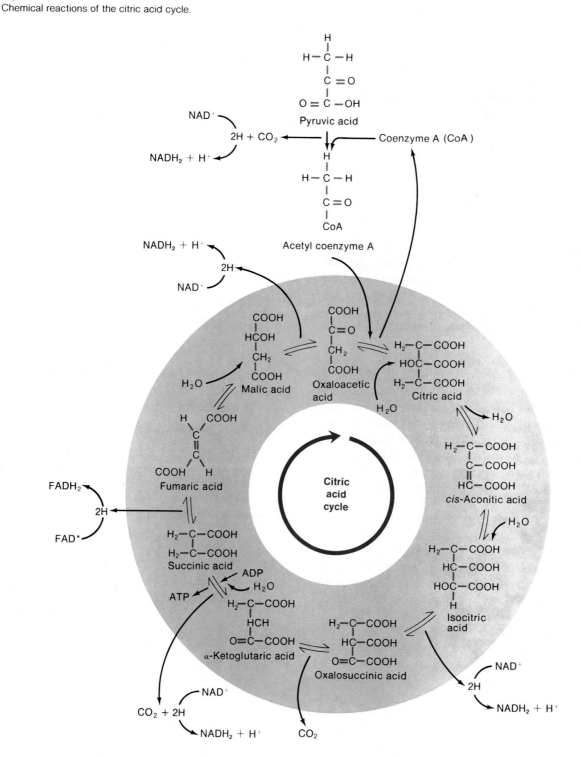

*At this point in the cycle, the hydrogen$_2$ carrier is FAD (flavine adenine dinucleotide).

Figure 4.13
ATP synthesis by oxidative phosphorylation.

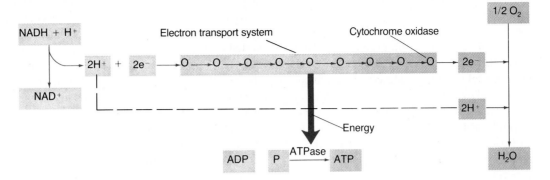

The released carbon dioxide dissolves in the cellular fluid and is transported away by the blood. Most of the hydrogen atoms released from the citric acid cycle and those released during glycolysis and during the formation of acetyl coenzyme A supply electrons that are involved with the production of ATP.

ATP Synthesis

Note in figure 4.12 that the hydrogen atoms released from various metabolic reactions are passed in pairs to hydrogen carriers. One of these carriers is NAD^+ (nicotinamide adenine dinucleotide). When NAD^+ accepts a pair of hydrogen atoms, one of the atoms becomes a hydrogen ion and the other bonds to NAD^+ to form NADH, as follows:

$$NAD^+ + 2H \rightarrow NADH + H^+$$

NAD^+ is a coenzyme obtained from a vitamin (niacin), and when it combines with hydrogen, it is said to be *reduced*. (*Reduction* results from the addition of hydrogen or the gain of electrons; it is the opposite of oxidation.) Another hydrogen acceptor, FAD, acts in a similar manner combining with hydrogen to form $FADH_2$ (figure 4.12).

In their reduced states, the hydrogen carriers NADH and $FADH_2$ hold most of the energy once held by the original glucose molecule.

As shown in figure 4.13, when hydrogen is released from NADH, NAD^+ reappears and can function once again as a hydrogen acceptor. Since this reaction involves the removal of hydrogen, the NAD^+ is said to be oxidized. (*Oxidation* results from the removal of hydrogen or the loss of electrons; it is the opposite of reduction.) At the same time, two electrons from the original pair of hydrogen atoms are passed to a sequence of electron carriers.

The molecules that act as electron carriers comprise an *electron transport system* (cytochrome system). As electrons are passed from one carrier to another, the carriers are alternately reduced and oxidized as they accept or release electrons.

Among the members of the electron transport system are several proteins, including a set of iron-containing molecules called *cytochromes*. The cytochromes are located in the inner membranes of the mitochondria (see chapter 3).

The last cytochrome of the electron transport system (cytochrome oxidase) gives up a pair of electrons and causes two hydrogen ions (formed at the beginning of the sequence) to combine with an atom of oxygen. This process produces a water molecule.

$$2e^- + 2H^+ + \tfrac{1}{2}O_2 \rightarrow H_2O$$

Thus, oxygen is the final acceptor of the electrons.

Note in figure 4.13 that at the same time electrons are being passed through the electron transport system, energy is being released. Some of this energy is used by the enzyme ATPase to combine phosphate and ADP by a high-energy bond (phosphorylation), forming ATP.

Also note in figures 4.11 and 4.12 that twelve pairs of hydrogen atoms are released during the complete breakdown of one glucose molecule—2 pairs from glycolysis, 2 pairs from the conversion of pyruvic acid to acetyl coenzyme A (one pair from each of two pyruvic acid molecules), and 8 pairs from the citric acid cycle (4 pairs for each of 2 acetyl coenzyme A molecules).

As a result of the oxidation (loss of electrons) of 10 pairs of these hydrogen atoms, 30 ATP molecules are produced, while metabolism of the other 2 pairs results in the formation of 4 ATP molecules. Also, there is a net gain of 2 ATP molecules during glycolysis, and 2 ATP molecules are formed as a result of the 2 acetyl coenzyme A molecules entering the citric acid cycle. Thus, a maximum of 38 ATP molecules are formed for each glucose molecule that is metabolized.

Since this process of forming ATP involves both the oxidation of hydrogen atoms and the bonding of phosphate to ATP, it is called *oxidative phosphorylation*.

Storage of Carbohydrates

Whenever excess glucose is present, it may enter anabolic carbohydrate pathways and be converted into storage forms such as glycogen.

Although cells generally can produce glycogen, the liver and muscle cells store the greatest amounts. Following a meal, when the blood glucose level is relatively high, liver cells obtain glucose from the blood and convert it to glycogen. Between meals, when the blood glucose level is lower, the reaction is reversed, and glucose is released into the blood. This mechanism ensures that various tissues will continue to have an adequate supply of glucose molecules to support their respiratory processes.

Glucose can also be converted into fat molecules, which are later deposited in fat tissues. This happens when a person takes in more carbohydrates than can be stored as glycogen or are needed for normal activities. The body has an almost unlimited capacity to perform this type of anabolic metabolism, so an excessive intake of nutrients can result in becoming overweight (obese).

Lipid Pathways

Although foods may contain lipids in the form of phospholipids or cholesterol, the most common dietary lipids are fats. As was explained in chapter 2, a molecule of such a lipid consists of a glycerol portion and 3 fatty acids and is called a *triglyceride.*

The metabolism of lipids is controlled mainly by the liver, which can remove them from the circulating blood and alter their molecular structures. For example, the liver can shorten or lengthen the carbon chains of fatty acid molecules or introduce double bonds into these chains, thus converting fatty acids from one form to another.

Lipids provide for a variety of physiological functions; however, they are used mainly to supply energy. Gram for gram, fats contain more than twice as much chemical energy as carbohydrates or proteins.

Before energy can be released from a triglyceride molecule, it must undergo hydrolysis. As shown in figure 4.14, some of the resulting fatty acid portions can then be converted into molecules of acetyl coenzyme A by a series of reactions called *beta oxidation.*

In the first phase of beta oxidation reactions, the fatty acids are converted into activated forms. This change requires a supply of energy from ATP molecules and the presence of a special group of enzymes (thiokinases). Each of the enzymes in this group can act upon a fatty acid with a particular carbon chain length. Once the fatty acid molecules have been activated, they are broken down by other enzymes called

fatty acid oxidases that are located within mitochondria. In this phase of the reactions, segments of fatty acid chains containing 2 carbon atoms each are removed. Some of these segments are converted into acetyl coenzyme A molecules. Other segments are converted into compounds called *ketone bodies,* which later may be changed to acetyl coenzyme A. In either case, the resulting acetyl coenzyme A can be oxidized by means of the citric acid cycle. The glycerol portions of the triglyceride molecules also can enter metabolic pathways leading to the citric acid cycle, or they can be used to synthesize glucose.

Another possibility for glycerol and fatty acid molecules resulting from the hydrolysis of fats is to be changed back into fat molecules by anabolic processes and stored in fat tissue. Additional fat molecules can be synthesized from excess molecules of glucose or amino acids. Fats cannot be converted back into glucose, however.

When ketone bodies are formed faster than they can be decomposed, some of them are eliminated through the lungs and kidneys. Consequently, the breath and urine may develop a fruity odor due to the presence of a ketone called *acetone.* This sometimes happens when a person fasts or diets, forcing body cells to metabolize fat, in order to lose weight. Persons suffering from *diabetes mellitus* also are likely to metabolize excessive amounts of fats, and they too may have acetone in the breath and urine. At the same time, they may develop a serious imbalance in pH called *acidosis* due to an accumulation of acidic ketone bodies.

---

1. What is meant by a metabolic pathway?
2. How are carbohydrates used within cells?
3. What must happen to fat molecules before they can be used as energy sources?

Protein Pathways

When dietary proteins are digested, the resulting amino acids are absorbed and transported by the blood to various body cells. Many of these amino acids are joined to form protein molecules again, which then may be incorporated into cell parts or serve as enzymes. Still others may be decomposed to supply energy.

When protein molecules are used as energy sources, they must first be broken down into amino acids. The amino acids undergo *deamination,* a process that occurs in the liver and involves removing the nitrogen-containing portions ($-NH_2$ groups) from the amino acids. These $-NH_2$ groups are converted later into a waste substance called *urea.*

Depending upon the amino acids involved, the remaining deaminated portions of the amino acid molecules are decomposed by one of several pathways—

Figure 4.14
Fats from foods are digested into glycerol and fatty acids. These
molecules may enter catabolic pathways and be used as energy
sources.

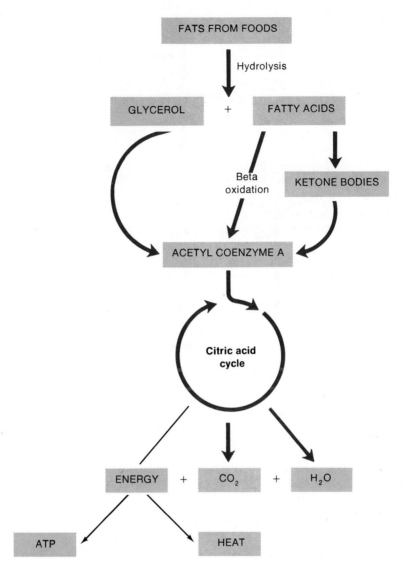

some of which lead to the formation of acetyl coenzyme A, while others lead to various steps of the citric
acid cycle more directly. As energy is released from the
cycle, some of it is captured in molecules of ATP (figure
4.15). If energy is not needed immediately, the deaminated portions of the amino acids may be changed into
glucose or fat molecules by still other metabolic pathways.

Glucose can be changed back into some amino
acids if certain nitrogen-containing molecules are
available. However, about 8 necessary amino acids
cannot be synthesized in human cells and so must
be provided in the diet. For this reason, these are
called *essential amino acids*. They are discussed in
chapter 15.

Urea is created in the liver from amino groups that result
from the deamination of amino acids. It is carried in the
blood to the kidneys, which excrete the urea in urine. As
a result of certain kidney disorders, the ability to remove
urea from the blood is impaired, and the blood urea concentration rises. Thus, a blood test called *blood urea nitrogen* (BUN) that determines the blood urea
concentration is often used to evaluate kidney function.

Regulation of Metabolic Pathways

As mentioned previously, the rate at which an enzyme-
controlled reaction occurs usually increases if either the
number of substrate molecules or enzyme molecules is
increased. However, the rate at which a metabolic
pathway functions is often determined by a regulatory

Figure 4.15

Proteins from foods are digested into amino acids, but before these smaller molecules can be used as energy sources, they must be deaminated.

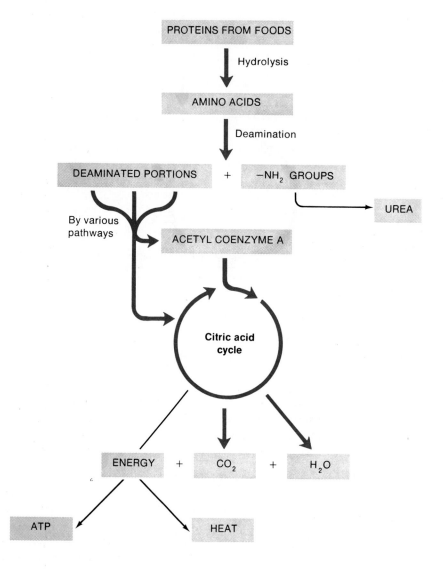

Figure 4.16

A rate-limiting enzyme in a metabolic pathway may be controlled by a negative feedback mechanism in which the product of the pathway inhibits the enzyme.

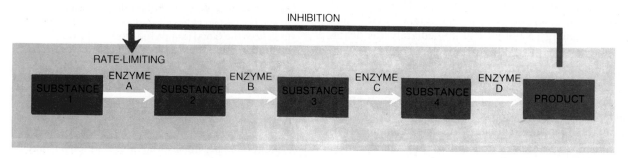

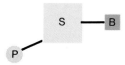

enzyme responsible for one of its steps. This regulatory enzyme is present in limited quantity. Consequently, it can become saturated whenever the substrate concentration increases above a certain amount. Once the enzyme is saturated, increasing the amount of substrate will no longer affect the rate of reaction. In this way, a single enzyme in a pathway can control the whole pathway.

As a rule, such a *rate-limiting enzyme* is the first enzyme in a series. This position is important because some intermediate substance of the pathway might accumulate if an enzyme occupying some other location in the sequence were involved.

Commonly a rate-limiting enzyme is controlled, in turn, by a *negative feedback mechanism*. In this case, the final product of a metabolic pathway acts to inhibit the regulatory enzyme. Thus, as the product increases, the pathway is inhibited and product production decreases; when the amount of product decreases, the pathway is not inhibited and more product is synthesized. Consequently, the rate at which the product is produced remains relatively stable (figure 4.16).

1. How do cells utilize proteins?
2. What is meant by a rate-limiting enzyme?
3. How might a metabolic pathway be controlled by a negative feedback mechanism?

Nucleic Acids and Protein Synthesis

Since enzymes control the metabolic processes that enable cells to survive, the cells must possess information for producing these specialized proteins. Such information is held in **DNA** (deoxyribonucleic acid) molecules in the form of a genetic code. This code "instructs" cells how to synthesize specific protein molecules.

Genetic Information

Children resemble their parents because of inherited traits; but what is actually passed from parents to a child is *genetic information*. This information is received in the form of DNA molecules from the parents' sex cells, and as an offspring develops, the information is passed from cell to cell by mitosis. Genetic information "tells" the cells of the developing body how to

Figure 4.18
The nucleotides of a DNA strand are joined to form a sugar-phosphate backbone (P–S). The organic bases of the nucleotides (B) extend from this backbone and are weakly held to the bases of the second strand by hydrogen bonds (dashed lines).

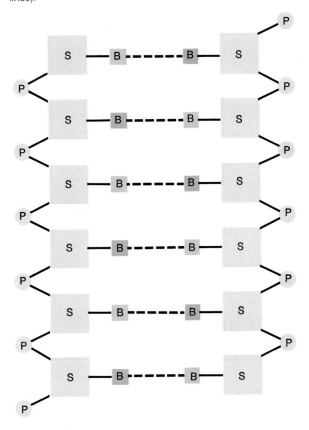

construct specific protein molecules, which in turn function as structural materials, enzymes, or other vital substances.

The portion of a DNA molecule that contains the genetic information for making one kind of protein is called a **gene.** Thus, inherited traits are determined by the genes contained in the parents' sex cells that fused to form the first cell of the offspring's body. These genes instructed the cells to synthesize the particular enzymes needed to control metabolic pathways.

DNA Molecules

As described in chapter 2, the building blocks of nucleic acids (nucleotides) each contain a 5-carbon sugar (ribose or deoxyribose), a phosphate group, and one of several organic bases (figure 4.17).

These nucleotides are joined so that the sugar and phosphate portions alternate and form a long chain, or "backbone." In a DNA molecule, the organic bases project out from this backbone and are bound by weak hydrogen bonds to those of the second strand (figure 4.18). The resulting structure is something like a ladder

Figure 4.19

(a) The molecular ladder of a double-stranded DNA molecule is twisted into the form of a double helix. Each strand contains only four kinds of organic bases—A, T, C, and G; (b) space-filling model of a portion of a DNA molecule.

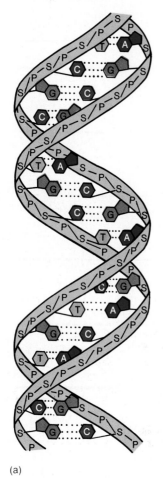

(a)

(b)

in which the uprights represent the sugar and phosphate backbones of the 2 strands, and the crossbars represent the organic bases. In addition, the molecular ladder is twisted to form a *double helix* (figure 4.19).

The organic base of a DNA nucleotide can be one of four kinds: *adenine* (A), *thymine* (T), *cytosine* (C), or *guanine* (G) (figure 4.20). Therefore, there are only four kinds of DNA nucleotides: adenine nucleotide, thymine nucleotide, cytosine nucleotide, and guanine nucleotide.

A strand of DNA consists of nucleotides arranged in a particular sequence (figure 4.21). Moreover, the nucleotides of one strand are paired in a special way with those of the other strand. Only certain organic bases have the molecular shapes needed to fit together so that their nucleotides can bond with one another. Specifically, an adenine will bond only to a thymine, and a cytosine will bond only to a guanine.

As a consequence of such base pairing, a DNA strand possessing the base sequence T, C, A, G would have to be joined to a second strand with the complementary base sequence A, G, T, C (figure 4.22). It is the particular sequence of base pairs that encodes the genetic information held in a DNA molecule.

The Genetic Code

Genetic information contains instructions for synthesizing proteins, and because proteins consist of 20 different amino acids joined in particular sequences, the genetic information must tell how to position the amino acids correctly in a polypeptide chain.

Each of the 20 different amino acids is represented in a DNA molecule by a particular sequence of 3 nucleotides. That is, the sequence G, T, A in a DNA strand represents one kind of amino acid; the sequence G, C, A represents another kind, and T, T, A still another kind (chart 4.1). Other sequences represent instructions for beginning or ending the synthesis of a protein molecule.

Figure 4.20
Each nucleotide of a DNA molecule contains one of four organic bases: adenine, thymine, cytosine, or guanine.

Adenine

Thymine

Cytosine

Guanine

Chart 4.1	Some nucleotide sequences of the genetic code	
Amino acids	DNA Sequence	RNA Sequence
Alanine	GGT	CCA
Arginine	GCA	CGU
Asparagine	TTA	AAU
Aspartic acid	CTA	GAU
Cysteine	ACA	UGU
Glutamic acid	CTT	GAA
Glutamine	GTT	CAA
Glycine	CCG	GGC
Histidine	GTA	CAU
Isoleucine	TAG	AUC
Leucine	GAA	CUU
Lysine	TTT	AAA
Methionine	TAC	AUG
Phenylalanine	AAA	UUU
Proline	GGA	CCU
Serine	AGG	UCC
Threonine	TGC	ACG
Tryptophan	ACC	UGG
Tyrosine	ATA	UAU
Valine	CAA	GUU
Instructions		
Start protein synthesis	TAC	AUG
Stop protein synthesis	ATT	UAA

Figure 4.21
A strand of DNA consists of a chain of nucleotides arranged in a particular sequence. Within the chain there are four kinds of nucleotides: adenine nucleotide (A); thymine nucleotide (T); cytosine nucleotide (C); and guanine nucleotide (G).

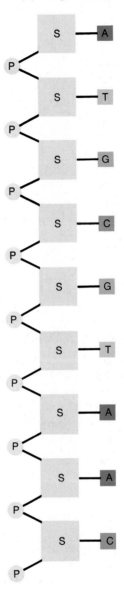

Thus, the sequence in which the nucleotide groups are arranged within a DNA molecule can denote the arrangement of amino acids within a protein molecule, as well as indicating how to start or stop the synthesis of a protein. This method of storing information used for the synthesis of particular protein molecules is termed the **genetic code.**

Although DNA molecules are located in the chromatin within a cell's nucleus, protein synthesis occurs in the cytoplasm. Therefore the genetic information must somehow be transferred from the nucleus into the cytoplasm. This transfer of information is the function of certain RNA molecules.

Figure 4.22

The nucleotides of a double-stranded molecule of DNA are paired so that an adenine nucleotide of one strand is joined to a thymine nucleotide of the other strand, and a guanine nucleotide of one is joined to a cytosine nucleotide of the other. The dotted lines represent weak hydrogen bonds between the paired bases of these nucleotides.

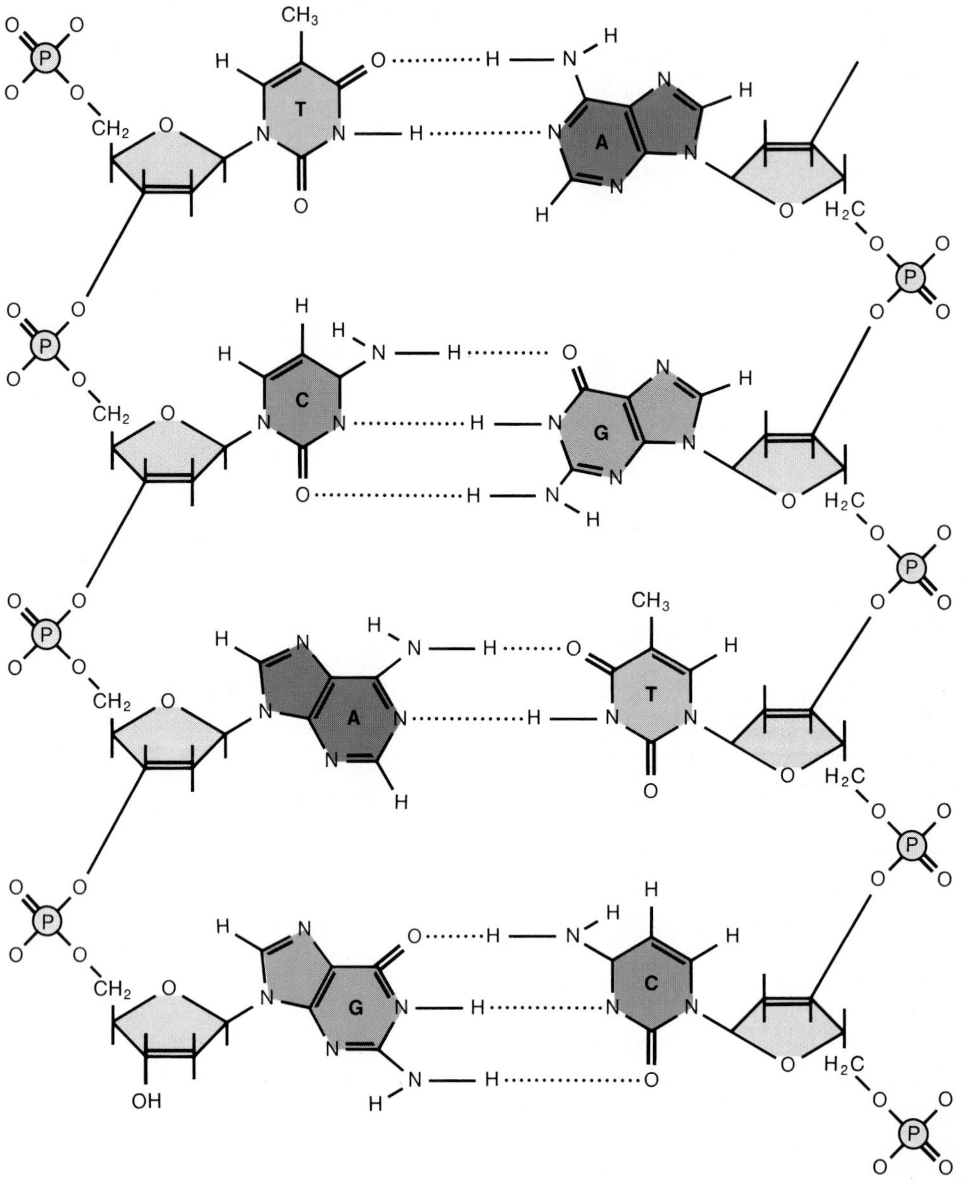

RNA Molecules

RNA (ribonucleic acid) molecules differ from DNA molecules in several ways. For example, RNA molecules are usually single-stranded, and their nucleotides contain ribose rather than deoxyribose sugar. Like DNA, RNA nucleotides each contain one of four organic bases, but while adenine, cytosine, and guanine nucleotides occur in both DNA and RNA, thymine nucleotides are found only in DNA. In its place, RNA molecules contain *uracil* (U) nucleotides (figure 4.23).

One step in the transfer of information from the nucleus to the cytoplasm involves the synthesis of a type of RNA called **messenger RNA** (mRNA).

In the process of producing messenger RNA, an enzyme called *RNA polymerase* becomes associated with the DNA base sequence that represents the beginning of a gene—that is, the instructions for synthesizing a particular protein. Such a DNA base sequence is called a *promoter*. As a result of the enzymatic action, the double-stranded DNA molecule unwinds and pulls apart and exposes the gene. The

Figure 4.23

Each nucleotide of an RNA molecule contains one of four organic bases: adenine, uracil, cytosine, or guanine. How are these nucleotides similar to those found in DNA molecules? How are they different?

Adenine

Uracil

Cytosine

Guanine

RNA polymerase moves along the exposed strand, causing the production of a molecule of messenger RNA that is formed of nucleotides complementary to those arranged along the DNA strand. (The other strand of DNA is not used in this process, but it is important in the replication of the DNA molecule as is explained in a subsequent section of this chapter.) For example, if the sequence of DNA bases is A, T, G, C, G, T, A, A, C, then the complementary bases in the developing messenger RNA molecule would be U, A, C, G, C, A, U, U, G, as shown in figure 4.24.

The RNA polymerase continues to move along the exposed DNA strand until it reaches a special DNA base sequence (termination signal) that represents the end of the gene. At this point, the RNA polymerase releases the newly formed messenger RNA molecule and leaves the DNA. The DNA then rewinds and assumes its former double-helix structure. This process of copying DNA information into the structure of a messenger RNA molecule is called *transcription*.

Since an amino acid is represented by a sequence of 3 nucleotides in a DNA molecule, the same amino acid will be represented in the transcribed messenger RNA by the complementary set of 3 nucleotides. Such a set of nucleotides in a messenger RNA molecule is called a *codon*.

Figure 4.24

When an RNA molecule is synthesized beside a strand of DNA, complementary nucleotides bond as in a double-stranded molecule of DNA with one exception: RNA contains uracil nucleotides (U) in place of thymine nucleotides (T).

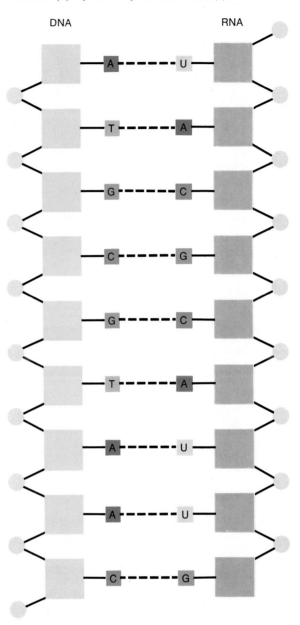

DNA   RNA

Once they are formed, messenger RNA molecules (each of which consists of hundreds or even thousands of nucleotides) can move out of the nucleus through the tiny pores in the nuclear envelope and enter the cytoplasm (figure 4.25). There they become associated with ribosomes and act as patterns or templates for the synthesis of protein molecules—a process called *translation*. (See figure 4.26.)

Figure 4.25

After copying a section of DNA information, a messenger RNA molecule (mRNA) moves out of the nucleus and enters the cytoplasm. There it becomes associated with a ribosome.

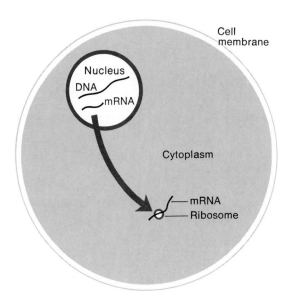

Figure 4.26

Transcription is the process by which DNA information is copied into the structure of messenger RNA. Translation is the process by which messenger RNA information is used in the synthesis of protein.

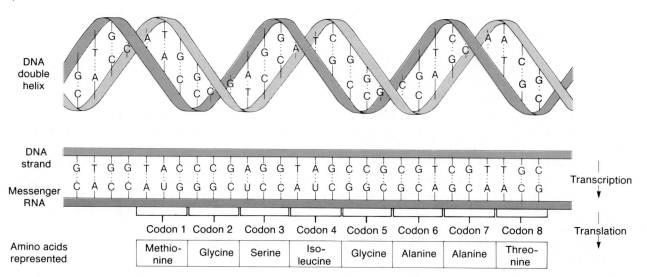

	Codon 1	Codon 2	Codon 3	Codon 4	Codon 5	Codon 6	Codon 7	Codon 8
Amino acids represented	Methio-nine	Glycine	Serine	Iso-leucine	Glycine	Alanine	Alanine	Threo-nine

Protein Synthesis

Before a protein molecule can be synthesized, the correct amino acids must be present in the cytoplasm to serve as building blocks. Furthermore, these amino acids must be positioned in the proper locations along a strand of messenger RNA. The positioning of amino acid molecules is the function of a second kind of RNA molecule that is synthesized in the nucleus and called **transfer RNA** (tRNA). A transfer RNA molecule, however, consists of only 70–80 nucleotides and has a complex three-dimensional shape.

Since 20 different kinds of amino acids are involved in protein synthesis, there must be at least 20 different kinds of transfer RNA molecules to serve as guides. However, before a transfer RNA molecule can pick up its particular kind of amino acid, the amino acid must be activated. This step is influenced by the presence of special enzymes and by ATP, which provides the energy needed to form a bond between the amino acid and its transfer RNA molecule.

Figure 4.27
Molecules of transfer RNA bring specific amino acids and place
them in the sequence determined by the codons of the
messenger RNA molecule. These amino acids bond and form the
polypeptide chain of a protein molecule.

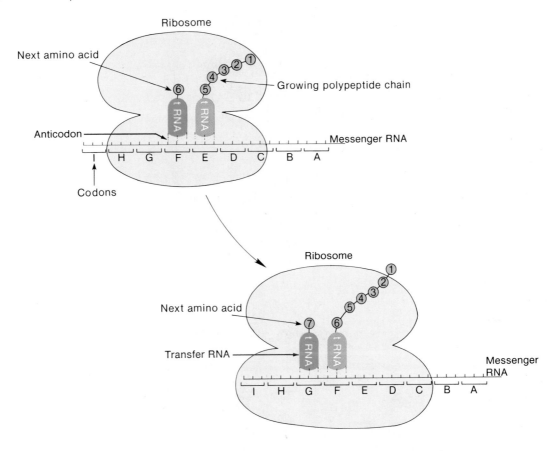

Each type of transfer RNA includes a special region at one end that contains 3 nucleotides in a particular sequence. These nucleotides can only bond to the
complementary set of 3 nucleotides of a messenger
RNA molecule (codon), and thus the set of 3 nucleotides in the transfer RNA molecule is called an *anticodon*. In this way, the transfer RNA carries its amino
acid to a correct position on a messenger RNA strand.
This action occurs on the surface of a ribosome.

A ribosome is a tiny particle of 2 unequal-sized
subunits composed of *ribosomal RNA* (rRNA) and
protein. By its smaller subunit a ribosome binds to a
molecule of messenger RNA near a codon. This action
allows a transfer RNA molecule with the complementary anticodon to bring the amino acid it carries into
position and become temporarily joined to the ribosome. Then the transfer RNA molecule releases its
amino acid and returns to the cytoplasm (figure 4.27).

This process is repeated again and again as the
ribosome moves along the messenger RNA, and the
amino acids, which are released by the transfer RNA
molecules, are added one at a time to a developing protein molecule. The enzymes necessary for the bonding

of the amino acids are found in the larger subunit of
the ribosome. This subunit also seems to hold the
growing chain of amino acids as it develops.

Actually, a molecule of messenger RNA is usually associated with several ribosomes at the same time.
Thus, several protein molecules, each in a different
stage of development, may be present at any given moment.

As the protein molecule develops, it folds into its
unique shape, and when it is completed it is released to
become a separate functional molecule. The transfer
RNA molecules can pick up other amino acids from
the cytoplasm, and the messenger RNA molecules can
function again and again.

ATP molecules provide the energy needed to form
the various chemical bonds in this process. A complete
protein molecule may consist of many hundreds of
amino acids, and the energy from 3 ATP molecules is
required to link each amino acid to the growing chain.
Consequently, a large proportion of a cell's energy
supply is needed to support protein synthesis, especially if the cell is young and growing.

Figure 4.28
When a DNA molecule replicates, its strands pull apart, and a
new strand of complementary nucleotides forms along each of
the old strands.

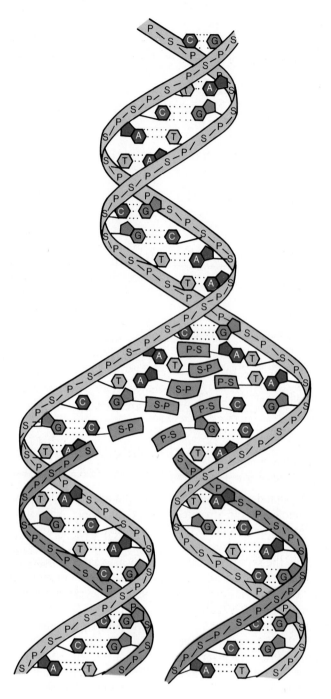

1. What is the function of DNA?
2. How is information carried from the nucleus to the cytoplasm?
3. How are protein molecules synthesized?

Replication of DNA

When a cell reproduces, each daughter cell needs a copy of the parent cell's genetic information so it will be able to synthesize the proteins necessary to build cellular parts and carry on metabolism.

DNA molecules can be replicated and this replication takes place during the interphase of the cell's life cycle.

As the *replication* process begins, bonds are broken between the complementary base pairs of the double strands in each DNA molecule. Then the double-stranded structure pulls apart and unwinds, exposing the organic bases of its nucleotides. New nucleotides of the four types found in DNA pair with the exposed bases, and enzymes cause the complementary bases to join together. In this way a new strand of complementary nucleotides is constructed along each of the old strands. As a result, two complete DNA molecules are produced, each with one old strand of the original molecule and one new strand (figure 4.28).

During mitosis, these two DNA molecules that are held within chromosomes are separated so that one passes to each of the newly forming daughter cells.

Changes in Genetic Information

The amount of genetic information held within a set of human chromosomes is very large. In fact, some investigators estimate that it may equal the amount of the text held within twenty sets of *Encyclopaedia Britannica.*

Since each of the trillions of cells in an adult body resulted from mitosis, this large amount of information had to be replicated many times and with a high degree of accuracy. However, occasionally a mistake occurs or the DNA structure is damaged, and the genetic information is altered. Such a change is called a **mutation.** If the cell in which it occurs survives, the mutation is likely to be passed to future generations of cells.

Nature of Mutations

Mutations can originate in a number of ways. For example, during DNA replication, an organic base may be paired incorrectly within the newly forming strand, or some extra organic bases may be built into its structure. In other instances, sections of DNA strands may be deleted, moved to other regions of the molecule, or even attached to other chromosomes. In any case, the consequences are similar—the genetic information is changed. If a protein is constructed from this information, its molecular structure is likely to be faulty and the protein nonfunctional (figure 4.29).

Fortunately, cells can detect damage in their DNA molecules, and they possess enzymes that often can repair damage occurring in a single strand of DNA. These *repair enzymes* are able to clip out defective nucleotides and cause the resulting gap to be filled with nucleotides complementary to those on the remaining strand of DNA. In this way, the original structure of the double-stranded molecule can be restored.

On the other hand, both strands of the DNA molecule may be damaged in the same region. This type of change is unlikely to be repaired, and if the cell survives and reproduces, its daughter cells are likely to receive copies of the mutation.

Effects of Mutations

The effects of mutations vary greatly. At one extreme there is little or no effect. Perhaps the mutant cell is unable to manufacture an enzyme that is relatively unimportant, and the cell continues to function effectively. At the other extreme, mutations cause cell death. In such a case, the cell may be unable to produce an enzyme needed for energy release, so it cannot survive.

Figure 4.29

(a) The DNA code for the amino acid histidine is GTA. (b) If something happens to change the guanine in this section of the molecule to thymine, the DNA code is changed to TTA, which represents the amino acid, asparagine. Such a change would be a mutation.

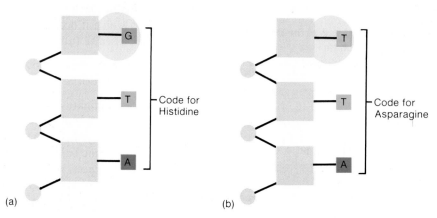

(a) Code for Histidine

(b) Code for Asparagine

Mutagenic Factors
A Clinical Application

Although some mutations arise spontaneously, without known cause, certain factors are known to increase the likelihood of a mutation. These factors are called *mutagens,* and they include various chemicals and forms of ionizing radiation.

Some common forms of ionizing radiation, such as ultraviolet light and X ray, are most likely to damage single strands of DNA by breaking them or by causing adjacent organic bases to link together. Since this type of damage may be repaired by repair enzymes, such radiation is considered to be a weaker mutagen than other more energetic forms. For example, radiation that includes neutrons may damage both strands of a DNA molecule—damage that is unlikely to be repaired.

Ultraviolet light is not very penetrating, so its effect is usually confined to the surfaces of the skin and eyes. X ray on the other hand is very penetrating and can affect cells located in deep tissues and organs.

Chemical mutagens produce a variety of effects on DNA molecules, including removal of bases, production of cross-links between strands, and the linking of themselves to organic bases. Although some of this damage may be repairable, many chemicals are considered to be very powerful mutagens.

It is difficult to avoid contact with mutagenic factors in everyday living. Skin is exposed to ultraviolet light from sunlight; X rays and radioactive isotopes are often used in the diagnosis and treatment of diseases; and mutagenic chemicals occur in a variety of commonly used substances such as tobacco and various petroleum products. Yet with the knowledge that these factors are hazardous to cells of present and future generations, the only wise course is to limit exposure to them whenever possible.

The mutations of most concern are those between these extremes. They result in some decrease in cell efficiency, and although the cell may live and reproduce, it has difficulty functioning normally. In time, the increasing numbers of such mutated cells may produce abnormal effects in the structure and function of the tissues and organs of which these cells are parts.

It also makes a difference whether a mutation occurs in a body cell of an adult or in a cell that is part of a developing embryo. In an adult, the mutant cell might not be noticed because there are many normally functioning cells around it. In the embryo, however, the affected cell might be the ancestor of great numbers of cells that are forming the body of a child. In fact, all the cells of the child's body could be defective if the mutation were present in the fertilized egg.

Well over one hundred human diseases are known to be related to defects in the genetic information of cells caused by mutations. The majority of these involve an inability to produce specific enzymes. As an example, the disease called phenylketonuria (PKU) results from a defective enzyme. The normal enzyme functions in the metabolic breakdown of the amino acid phenylalanine. When the defective enzyme is present, phenylalanine is only partially metabolized, and a toxic substance accumulates. This substance interferes with the normal activity of the child's nervous system. If the condition is untreated, it usually results in severe mental retardation. Treatment for PKU involves restricting the intake of foods containing phenylalanine.

1. How do DNA molecules become replicated?
2. What is a mutation?
3. How do mutations occur?
4. What effects are likely to be produced by mutations?

Chapter Summary

Introduction
A cell continuously carries on metabolic processes.

Metabolic Processes
1. Anabolic metabolism
   a. Anabolic metabolism consists of constructive processes (such as dehydration synthesis) in which smaller molecules are used to build larger ones.
   b. In dehydration synthesis, hydrogen atoms and hydroxyl groups are removed, water is formed, and smaller molecules become bound by shared atoms.
   c. Complex carbohydrates are synthesized from monosaccharides, fats are synthesized from glycerol and fatty acids, and proteins are synthesized from amino acids.
2. Catabolic metabolism
   a. Catabolic metabolism consists of decomposition processes in which larger molecules are broken down into smaller ones.
   b. In hydrolysis, a water molecule is used to supply a hydrogen atom to one portion of a molecule and a hydroxyl group to a second portion; the bond between these two portions is broken.
   c. Complex carbohydrates are decomposed into monosaccharides, fats are decomposed into glycerol and fatty acids, and proteins are decomposed into amino acids.

Control of Metabolic Reactions
Metabolic reactions are controlled by enzymes.

1. Enzymes and their actions
   a. Metabolic reactions require energy to get them started.
   b. Enzymes are proteins that promote metabolic reactions.
   c. An enzyme acts upon a molecule by temporarily combining with it and straining its chemical bonds.
   d. The shape of an enzyme molecule seems to fit the shape of its substrate molecule.
   e. When an enzyme combines with its substrate, the substrate is changed, resulting in a product, while the enzyme is unaltered.
   f. The rate of enzyme-controlled reactions depends upon the number of enzyme and substrate molecules present and the efficiency of the enzyme.
   g. Enzymes may be named according to their substrates with -ase added.

2. Cofactors and coenzymes
   a. Cofactors are necessary parts of some enzyme molecules.
   b. A cofactor may be an ion or a small organic molecule called a coenzyme.
   c. Vitamins, which provide coenzymes, usually cannot be synthesized by human cells in adequate amounts.
3. Factors that alter enzymes
   a. Enzymes are proteins and can be denatured.
   b. Factors that may denature enzymes include excessive heat, radiation, electricity, certain chemicals, and extreme concentration of acid or base.

Energy for Metabolic Reactions
Energy is a capacity to produce change or to do work.

Common forms of energy include heat, light, sound, electrical energy, mechanical energy, and chemical energy.

Whenever changes take place, energy is being transferred from one part to another.

1. Release of chemical energy
   a. Most metabolic processes utilize chemical energy that is released when molecular bonds are broken.
   b. The energy released from glucose during cellular respiration is used to promote metabolism.
   c. Cellular respiration is controlled by enzymes in the cytoplasm and mitochondria.
2. Anaerobic respiration
   a. The first phase of glucose decomposition occurs in the cytoplasm and is anaerobic.
   b. Some of the energy released is transferred to molecules of ATP.
3. Aerobic respiration
   a. The second phase of glucose decomposition occurs within the mitochondria and is aerobic.
   b. Considerably more energy is transferred to ATP molecules during this phase than during the anaerobic phase.
   c. The final products of glucose decomposition are carbon dioxide, water, and energy.
4. ATP molecules
   a. Thirty-eight molecules of ATP can be produced for each glucose molecule that is decomposed.
   b. Energy is captured in the bond of the terminal phosphate of each ATP molecule.
   c. Captured energy is released when the terminal phosphate bond of an ATP molecule is broken.
   d. An ATP molecule that loses its terminal phosphate becomes an ADP molecule.
   e. An ADP can be converted to an ATP by capturing some energy and a phosphate.

Metabolic Pathways

Metabolic processes usually involve a number of steps that must occur in the correct sequence.

A sequence of enzyme-controlled reactions is called a metabolic pathway.

Typically metabolic pathways are interconnected.

1. Carbohydrate pathways
   a. Carbohydrates may enter catabolic pathways and be used as energy sources.
   b. Glycolysis involves the conversion of a 6-carbon glucose molecule into two 3-carbon pyruvic acid molecules and results in a net gain of ATP.
   c. The citric acid cycle is a complex series of reactions in which molecules are decomposed, carbon dioxide and hydrogen atoms are released, and ATP molecules are formed.
   d. Hydrogen atoms from the citric acid cycle are converted to hydrogen ions, which in turn combine with oxygen to form water molecules.
   e. Electrons from hydrogen atoms enter an electron transport system, and as energy is released from the system, some of it is used to form ATP.
   f. For each glucose molecule metabolized, a maximum of 38 ATP molecules are formed.
   g. When present in excess, carbohydrates may enter anabolic pathways and be converted to glycogen or fat.
2. Lipid pathways
   a. Most dietary fats are triglycerides.
   b. Before fats can be used as an energy source they must be converted into glycerol and fatty acids.
   c. Fatty acids are decomposed by beta oxidation.
      (1) Beta oxidation involves activating fatty acids and breaking them down into segments containing two carbon atoms each.
      (2) Fatty acid segments are converted into acetyl coenzyme A, which can be oxidized by means of the citric acid cycle.
   d. Fats can be synthesized from glycerol and fatty acids and from excess glucose or amino acids.
3. Protein pathways
   a. Proteins are used as building materials for cellular parts, as enzymes, and as energy sources.
   b. Before proteins can be used as energy sources, they must be decomposed into amino acids, and the amino acids must be deaminated.
   c. The deaminated portions of amino acids can be broken down into carbon dioxide and water, or can be converted into glucose or fat.
   d. About 8 essential amino acids cannot be synthesized by human cells and must be obtained in foods.

4. Regulation of metabolic pathways
   a. A metabolic pathway may be regulated by a rate-limiting enzyme.
   b. The regulatory enzyme, in turn, may be controlled by a negative feedback mechanism in which the product of the pathway inhibits the enzyme.
   c. The rate of product production usually remains stable.

Nucleic Acids and Protein Synthesis

DNA molecules contain information that instructs a cell how to synthesize the correct enzymes.

1. Genetic information
   a. Inherited traits result from DNA information that is passed from parents to offspring.
   b. A gene is a portion of a DNA molecule that contains the genetic information for making one kind of protein.
2. DNA molecules
   a. The nucleotides of a DNA strand are arranged in a particular sequence.
   b. The nucleotides are paired with those of the second strand in a complementary fashion.
3. The genetic code
   a. The sequence of nucleotides in a DNA molecule represents the sequence of amino acids in a protein molecule.
   b. Genetic information is transferred from the nucleus to the cytoplasm by RNA molecules.
4. RNA molecules
   a. RNA molecules are usually single stranded, contain ribose instead of deoxyribose, and contain uracil nucleotides in place of thymine nucleotides.
   b. Messenger RNA molecules, which are synthesized in the nucleus, contain a nucleotide sequence that is complementary to that of an exposed strand of DNA.
   c. Messenger RNA molecules move into the cytoplasm, become associated with ribosomes, and act as patterns for the synthesis of protein molecules.
5. Protein synthesis
   a. Molecules of transfer RNA serve to position amino acids along a strand of messenger RNA.
   b. A ribosome binds to a messenger RNA molecule and allows a transfer RNA molecule to recognize its correct position on the messenger RNA.

c. The ribosome contains enzymes needed for the synthesis of the developing protein and holds the protein until it is completed.
d. As the protein develops, it folds into a unique shape.
e. ATP provides the energy needed for these events.
6. Replication of DNA
a. Each daughter cell needs a copy of the parent cell's genetic information.
b. DNA molecules are replicated during interphase of the cell's life cycle.
c. Each new DNA molecule contains one old strand and one new strand.

Changes in Genetic Information

The amount of information within a DNA molecule is very large. Occasionally a change or mutation occurs in the genetic information.

1. Nature of mutations
a. Mutations involve a variety of kinds of changes within DNA molecules.
b. A protein synthesized from damaged DNA information is likely to be nonfunctional.
c. Repair enzymes can correct some forms of DNA damage.
2. Effects of mutations
a. The mutations that cause a decrease in cell efficiency are of most concern.
b. A mutation that occurs in a cell of a developing embryo may affect a great number of cells in the developing body.

Application of Knowledge

1. Since enzymes are proteins, they may be denatured. Relate this to the fact that a very high fever accompanying an illness may be life-threatening.
2. Some weight-reducing diets drastically limit the dieter's intake of carbohydrates but allow liberal use of fat and protein foods. What changes would such a diet cause in the cellular metabolism of the dieter? What changes could be noted in the urine of such a person?

Review Activities

1. Define *anabolic* and *catabolic metabolism*.
2. Distinguish between dehydration synthesis and hydrolysis.
3. Explain what is meant by a peptide bond.
4. Define *enzyme*.
5. Describe how an enzyme is thought to interact with its substrate.
6. List three factors that increase the rates of enzyme-controlled reactions.
7. Explain how enzymes are named.
8. Define *cofactor*.
9. Explain why humans need vitamins in their diets.
10. Explain how an enzyme may be denatured.
11. Define *energy*.
12. Explain how oxidation of molecules inside cells differs from burning of substances.
13. Define *cellular respiration*.
14. Distinguish between anaerobic and aerobic respiration.
15. Explain the importance of ATP to cellular processes.
16. Describe the relationship between ATP and ADP.
17. Explain what is meant by a metabolic pathway.
18. Describe the process called *glycolysis*.
19. Review the major events of the citric acid cycle.
20. Explain how carbohydrates are stored.
21. Define *beta oxidation*.
22. Explain how fats may serve as energy sources.
23. Define *deamination* and explain its importance.
24. Explain why some amino acids are essential.
25. Explain how one enzyme can regulate a metabolic pathway.
26. Describe how a negative feedback mechanism can help control a metabolic pathway.
27. Explain what is meant by genetic information.
28. Describe the relationship between a DNA molecule and a gene.
29. Describe the general structure of a DNA molecule.
30. Explain how genetic information is stored in a DNA molecule.
31. Distinguish between messenger RNA and transfer RNA.
32. Distinguish between RNA transcription and translation.
33. Explain the function of a ribosome in protein synthesis.
34. Distinguish between a codon and an anticodon.
35. Explain how a DNA molecule is replicated.
36. Define *mutation* and explain how mutations may originate.
37. Define *repair enzyme*.
38. Explain how a mutation may affect an organism.

Chapter Outline

Chapter Objectives

After you have studied this chapter, you should be able to

1. Describe the general characteristics and functions of epithelial tissue.

2. Name the major types of epithelium and identify an organ in which each is found.

3. Explain how glands are classified.

4. Describe the general characteristics of connective tissue.

5. Describe the major cell types and fibers of connective tissue.

6. List the major types of connective tissues that occur within the body.

7. Describe the major functions of each type of connective tissue.

8. Distinguish between the three types of muscle tissue.

9. Describe the general characteristics and functions of nerve tissue.

10. Complete the review activities at the end of this chapter. Note that the items are worded in the form of specific learning objectives. You may want to refer to them before reading the chapter.

5

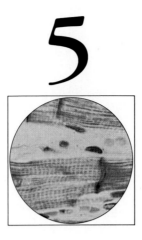

Tissues

Cells, the basic units of structure and function within the human organism, are organized into groups and layers called *tissues*.

Each type of tissue is composed of similar cells that are specialized to carry on particular functions. For example, epithelial tissues form protective coverings and function in secretion and absorption, connective tissues provide support for softer body parts and bind structures together, muscle tissues are responsible for producing body movements, and nerve tissue is specialized to conduct impulses that help to control and coordinate body activities.

Chapter Outline

Chapter Objectives

After you have studied this chapter, you should be able to

1. Describe the general characteristics and functions of epithelial tissue.

2. Name the major types of epithelium and identify an organ in which each is found.

3. Explain how glands are classified.

4. Describe the general characteristics of connective tissue.

5. Describe the major cell types and fibers of connective tissue.

6. List the major types of connective tissues that occur within the body.

7. Describe the major functions of each type of connective tissue.

8. Distinguish between the three types of muscle tissue.

9. Describe the general characteristics and functions of nerve tissue.

10. Complete the review activities at the end of this chapter. Note that the items are worded in the form of specific learning objectives. You may want to refer to them before reading the chapter.

Key Terms

adipose tissue   (ad′ĭ-pōs tish′u)

cartilage   (kar′tĭ-lij)

chondrocyte   (kon′dro-sīt)

connective tissue   (kŏ-nek′tiv tish′u)

epithelial tissue   (ep″ĭ-the′le-al tish′u)

fibroblast   (fi′bro-blast)

fibrous tissue   (fi′brus tish′u)

macrophage   (mak′ro-fāj)

muscle tissue   (mus′el tish′u)

nerve tissue   (nerv tish′u)

neuroglia   (nu-rog′le-ah)

neuron   (nu′ron)

osteocyte   (os″te-o-sīt″)

reticuloendothelial tissue
   (rĕ-tik″u-lo-en″do-the′le-al tish′u)

Aids to Understanding Words

adip-, fat: *adip*ose tissue—tissue that stores fat.

-cyt, cell: osteo*cyt*e—a bone cell.

epi-, upon: *epi*thelial tissue—tissue that covers all free body surfaces.

-glia, glue: neuro*glia*—cells that bind nerve tissue together.

inter-, between: *inter*calated disc—band located between the ends of adjacent cardiac muscle cells.

macro-, large: *macro*phage—a large phagocytic cell.

neuro-, nerve: *neuro*n—a nerve cell.

osseo-, bone: *osseo*us tissue—bone tissue.

phago-, to eat: *phago*cyte—a cell that engulfs and destroys foreign particles.

pseudo-, false: *pseudo*stratified epithelium—tissue whose cells appear to be arranged in layers, but are not.

squam-, scale: *squam*ous epithelium—tissue whose cells appear flattened or scalelike.

stratum, layer: *strat*ified epithelium—tissue whose cells occur in layers.

stria-, groove: *stria*ted muscle—tissue whose cells are characterized by alternating light and dark cross-markings.

IN ALL COMPLEX organisms, cells are organized into layers or groups called **tissues.** Although the cells of different tissues vary in size, shape, arrangement, and function, those within a tissue are quite similar.

Usually tissue cells are separated by nonliving, intercellular materials, which they secrete. These intercellular materials vary in composition from one tissue to another and may take the form of solids, semisolids, or liquids. For example, bone tissue cells are separated by a solid intercellular substance, while blood tissue cells are separated by a liquid.

The tissues of the human body include four major types: *epithelial tissues, connective tissues, muscle tissues,* and *nerve tissue.* These tissues are organized into organs that have specialized functions.

This chapter is concerned primarily with various types of epithelial and connective tissues. Other kinds of tissues will be discussed in subsequent chapters.

1. What is meant by a tissue?
2. List the four major types of tissues.

Epithelial Tissues

Epithelial tissues are widespread throughout the body. They cover all body surfaces—inside and out. They are also the major tissues of glands.

Since epithelium covers organs, forms the inner lining of body cavities, and lines hollow organs, it always has a free surface—one that is exposed to the outside or to an open space internally. The underside of the tissue is anchored to connective tissue by a thin, nonliving layer called the *basement membrane* (basement lamina).

As a rule, epithelial tissues lack blood vessels. However, they are nourished by substances that diffuse from underlying connective tissues, which are well supplied with blood vessels.

Although the cells of some tissues have limited abilities to reproduce, those of epithelium reproduce readily. Injuries to epithelium are likely to heal rapidly as new cells replace lost or damaged ones. For example, skin cells and the cells that line the stomach and intestines are continually being damaged and replaced.

Epithelial cells are tightly packed, and there is little intercellular material between them. Often they are attached to one another by desmosomes (see chapter 3). Consequently, these cells can provide effective protective barriers, such as the outer layer of the skin and the inner lining of the mouth. Other epithelial functions include secretion, absorption, excretion, and sensory reception.

Epithelial tissues are classified according to the specialized shapes, arrangements, and functions of their cells. For example, epithelial tissues that are composed of single layers of cells are called *simple,* those with many layers of cells are said to be *stratified,* those with thin, flattened cells are called *squamous,* those with cubelike cells are called *cuboidal,* and those with elongated cells are called *columnar.* In the following descriptions, note that the free surfaces of various epithelial cells are modified in ways reflecting their specialized functions.

Simple Squamous Epithelium

Simple squamous epithelium consists of a single layer of thin, flattened cells. These cells fit tightly together, somewhat like floor tiles, and their nuclei are usually broad and thin (figure 5.1).

As a rule, substances pass rather easily through this type of tissue, and this tissue often occurs where diffusion and filtration are taking place. For instance, simple squamous epithelium lines the air sacs of the lungs where oxygen and carbon dioxide are exchanged. It also forms the walls of capillaries, lines the insides of blood and lymph vessels, and covers the membranes that line body cavities.

Simple Cuboidal Epithelium

Simple cuboidal epithelium consists of a single layer of cube-shaped cells. These cells usually have centrally located spherical nuclei (figure 5.2).

This tissue covers the ovaries, lines kidney tubules, and lines ducts of various glands, such as salivary glands, the pancreas, and the liver. In the kidneys, it functions in secretion and absorption, while in glands it is concerned with the secretion of glandular products.

Simple Columnar Epithelium

The cells of **simple columnar epithelium** are elongated; that is, they are longer than they are wide. This tissue is composed of a single layer of cells whose nuclei are usually located at about the same level near the basement membrane (figure 5.3).

Simple columnar epithelium occurs in the linings of the uterus and various organs of the digestive tract, including the stomach and intestines. Since its cells are elongated, the tissue is relatively thick, providing protection for underlying tissues. It also functions in the secretion of digestive fluids and in the absorption of nutrient molecules resulting from the digestion of foods.

Columnar cells, whose principal function is absorption, often have numerous tiny, cylindrical processes extending outward from their cell surfaces. These processes, called *microvilli,* are from $0.5-1.0$ μm in

Figure 5.1

(a) Simple squamous epithelium consists of a single layer of tightly packed flattened cells. (b) What features can you identify in this micrograph? (Note: in the micrograph, the cells are viewed on the flattened surface of the tissue.)

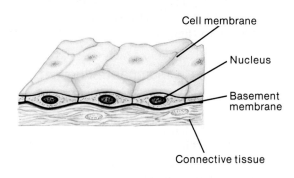

Cell membrane

Nucleus

Basement membrane

Connective tissue

(a)

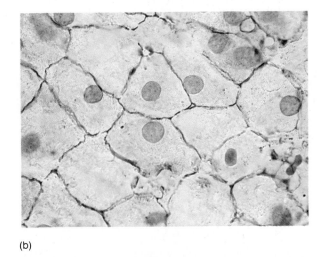

(b)

Figure 5.2

(a) Simple cuboidal epithelium consists of a single layer of tightly packed cube-shaped cells. (b) Note the layer of simple cuboidal cells indicated by the arrow in this micrograph.

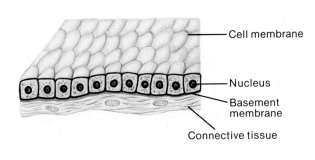

Cell membrane

Nucleus

Basement membrane

Connective tissue

(a)

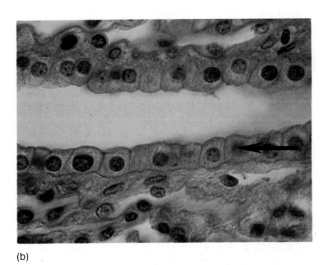

(b)

Figure 5.3

(a) Simple columnar epithelium consists of a single layer of elongated cells. (b) What is the function of the goblet cell indicated by the arrow in this micrograph?

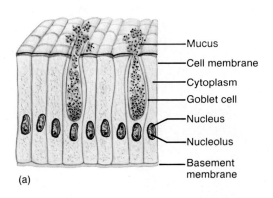

Mucus

Cell membrane

Cytoplasm

Goblet cell

Nucleus

Nucleolus

Basement membrane

(a)

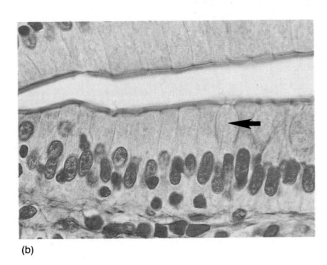

(b)

Figure 5.4

A scanning electron micrograph of microvilli, which occur on the exposed surfaces of some columnar epithelial cells.

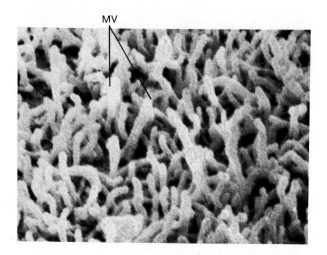

MV

length. They function to increase the surface of the cell membrane where it is exposed to the substances being absorbed (figure 5.4).

Typically, there are specialized, flask-shaped glandular cells scattered along the columnar cells of this tissue. These cells, called *goblet cells,* secrete a thick, protective fluid called *mucus* onto the free surface of the tissue (figure 5.3).

Pseudostratified Columnar Epithelium

The cells of **pseudostratified columnar epithelium** appear to be stratified or layered, but they are not. The layered effect occurs because nuclei are located at two or more levels within the cells of the tissue. However, the cells, which vary in shape, all reach the basement membrane, even though some of them may not contact the free surface.

Figure 5.5

(*a*) Pseudostratified columnar epithelium appears stratified, because the nuclei of various cells are located at different levels. (*b*) What features of the tissue can you identify in this micrograph?

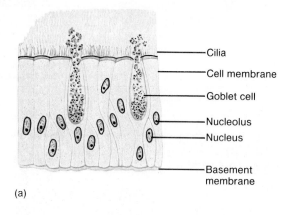

- Cilia
- Cell membrane
- Goblet cell
- Nucleolus
- Nucleus
- Basement membrane

(a)

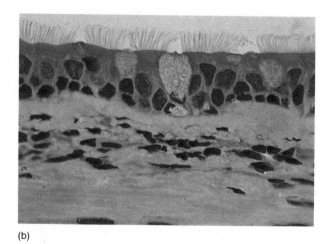

(b)

Figure 5.6

(*a*) Stratified squamous epithelium consists of many layers of cells. (*b*) In this micrograph, note how the cells near the surface of the tissue have become flattened.

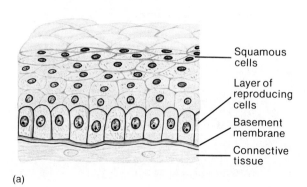

- Squamous cells
- Layer of reproducing cells
- Basement membrane
- Connective tissue

(a)

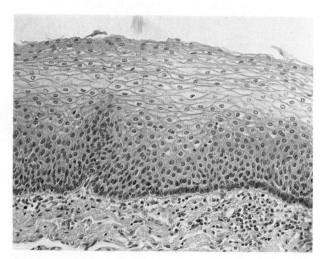

(b)

These cells commonly possess microscopic hair-like projections called *cilia* which measure 7–10 μm in length. They extend from the free surfaces of the cells, and move constantly. Goblet cells also are scattered throughout this tissue, and the mucus they secrete is swept along by the activity of the cilia (figures 5.5 and 3.12).

Pseudostratified columnar epithelium is found lining the passages of the respiratory system and in various tubes of the reproductive systems. In the respiratory passages, the mucus-covered linings are sticky and tend to trap particles of dust and microorganisms that enter with the air. The cilia move the mucus and its captured particles upward and out of the airways. In the reproductive tubes, the cilia aid in moving sex cells from one region to another.

Stratified Squamous Epithelium

Stratified squamous epithelium consists of many layers of cells, making this tissue relatively thick. Only the cells near the free surface, however, are likely to be flattened. Those in the deeper layers, where cellular reproduction occurs, are usually cuboidal or columnar. As the newer cells grow, older ones are pushed further and further outward, and they tend to become flattened (figure 5.6).

This tissue forms the outermost layer of the skin (epidermis). As the older cells are pushed outward, they accumulate a protein called *keratin*. As this happens, the cells become hardened and die. This action produces a covering of dry, tough, protective material that prevents the escape of water from underlying tissues and the entrance of various microorganisms.

Stratified squamous epithelium also lines the mouth cavity, throat, vagina, and anal canal. In these parts, the tissue is not keratinized; it remains moist, and the cells on the free surfaces remain alive.

Transitional Epithelium

Transitional epithelium (uroepithelium) is specialized to undergo changes in tension. It forms the inner lining of the urinary bladder and the passageways of the urinary system. When the wall of one of these organs is contracted, the tissue consists of several layers of cuboidal cells; however, when the organ is distended, the tissue is stretched and may have only a few layers (figure 5.7).

In addition to providing a distensible lining, transitional epithelium is thought to form a barrier that helps to prevent the contents of the urinary tract from diffusing out of its passageways.

Figure 5.7
(*a*) Micrograph of transitional epithelium. (*b*) This tissue consists of many layers when the wall of an organ is contracted; (*c*) the tissue is thin when the wall is stretched.

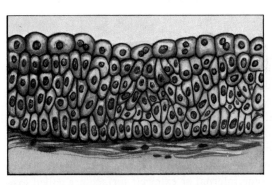

(b)

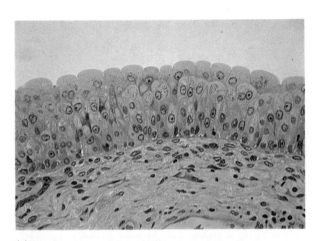

(a)

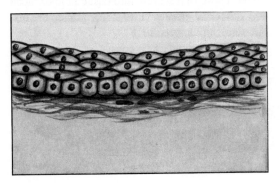

(c)

Chart 5.1 Epithelial tissues		
Type	Function	Location
Simple squamous epithelium	Filtration, diffusion, osmosis	Air sacs of lungs, walls of capillaries, linings of blood and lymph vessels
Simple cuboidal epithelium	Secretion, absorption	Surface of ovaries, linings of kidney tubules, and linings of ducts of various glands
Simple columnar epithelium	Protection, secretion, absorption	Linings of uterus and tubes of the digestive tract
Pseudostratified columnar epithelium	Protection, secretion, movement of mucus and cells	Linings of respiratory passages and various tubes of the reproductive systems
Stratified squamous epithelium	Protection	Outer layer of skin, linings of mouth cavity, throat, vagina, and anal canal
Transitional epithelium	Distensibility, protection	Inner lining of urinary bladder and passageways of urinary tract

Chart 5.1 summarizes the characteristics of the different types of epithelial tissue.

A cancer arising from epithelium is called a *carcinoma,* and it is estimated that up to 90% of all human cancers are of this type. Most of these cancers begin on surfaces that contact the external environment, such as skin, linings of the airways in the respiratory tract, or linings of the stomach or intestines in the digestive tract. This observation suggests that the more common cancer-causing agents may not penetrate tissues very deeply.

1. List the general characteristics of epithelial tissue.
2. Explain how epithelial tissues are classified.
3. Describe the structure of each type of epithelium.
4. Describe the special functions of each type of epithelium.

Glandular Epithelium

Glandular epithelium is composed of cells that are specialized to manufacture and secrete various substances into ducts or into body fluids. Such glandular cells are usually found within columnar or cuboidal epithelium, and one or more of them constitutes a *gland.* Those glands that secrete their products into ducts that open onto some internal or external surface are called **exocrine glands.** Those that secrete into tissue fluid or blood are called **endocrine glands.** (Endocrine glands are discussed in detail in chapter 13.)

An exocrine gland may consist of a single epithelial cell (unicellular gland), such as a mucus-secreting goblet cell, or it may be composed of many cells (multicellular gland). In turn, the multicellular forms can be divided structurally into two groups—simple and compound glands.

A *simple gland* communicates with the surface by means of an unbranched duct, and a *compound gland* has a branched duct. These two types of glands can be further classified according to the shapes of their secretory portions. Thus, those that consist of epithelial-lined tubes are called *tubular glands*; those whose terminal portions form saclike dilations are called *alveolar glands.* Figure 5.8 illustrates several types of exocrine glands.

Chart 5.2 summarizes the types of exocrine glands and their characteristics and provides an example of each type.

Glandular secretions are classified according to whether they consist of cellular products or portions of the glandular cells. Glands that release fluid cellular products through cell membranes without the loss of cytoplasm are called *merocrine* glands. Those of an intermediate type that lose small portions of their glandular cell bodies during secretion are called *apocrine* glands. Glands that release entire cells filled with secretory products are called *holocrine* glands (figure 5.9). Chart 5.3 summarizes these glands and their secretions.

Most secretory cells are merocrine, and they can be further subdivided as either *serous cells* or *mucous cells.* The secretion of serous cells is typically watery and has a high concentration of enzymes. Such cells are common in the glands of the digestive tract. Mucous cells secrete the thicker fluid called *mucus.* This substance is rich in the glycoprotein *mucin* and is secreted abundantly from the inner linings of the digestive and respiratory tubes.

1. Distinguish between exocrine and endocrine glands.
2. Explain how glands are classified according to their structures; according to their secretions.

Figure 5.8
Structural types of exocrine glands.

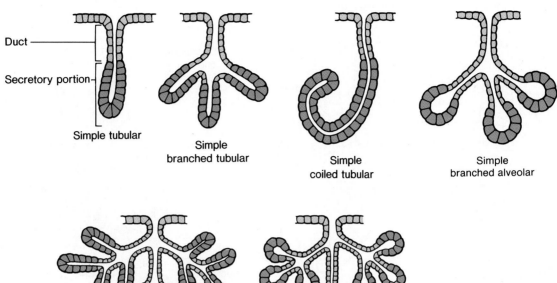

Duct

Secretory portion

Simple tubular

Simple
branched tubular

Simple
coiled tubular

Simple
branched alveolar

Compound tubular

Compound alveolar

Chart 5.2	Types of exocrine glands	
Type	**Characteristics**	**Example**
Unicellular glands	A single secretory cell	Mucus-secreting goblet cell (see figure 5.3)
Multicellular glands	Gland consists of many cells	
Simple glands	Glands communicate with surface by means of unbranched ducts	
1. Simple tubular gland	Straight tubelike gland that opens directly onto surface; no duct present	Intestinal glands of small intestine (see figure 14.2)
2. Simple coiled tubular gland	Long coiled tubelike gland; long duct	Eccrine (sweat) glands of skin (see figure 6.10)
3. Simple branched tubular gland	Branched tubelike gland; duct short or absent	Brunner's glands in small intestine (see figure 14.2)
4. Simple branched alveolar gland	Secretory portions of gland expanded into saclike compartments arranged along duct	Sebaceous gland of skin (see figure 6.7)
Compound glands	Glands communicate with surface by means of branched ducts	
1. Compound tubular gland	Secretory portions coiled tubules, usually branched	Mucous glands of mouth; bulbourethral glands of male (see figure 22.1)
2. Compound alveolar gland	Secretory portions irregularly branched tubules with numerous saclike outgrowths	Salivary glands (see figure 14.10)

Figure 5.9
(a) Merocrine glands release fluid without a loss of cytoplasm;
(b) apocrine glands lose small portions of their cell bodies during
secretion; and (c) holocrine glands release entire cells filled with
secretory products.

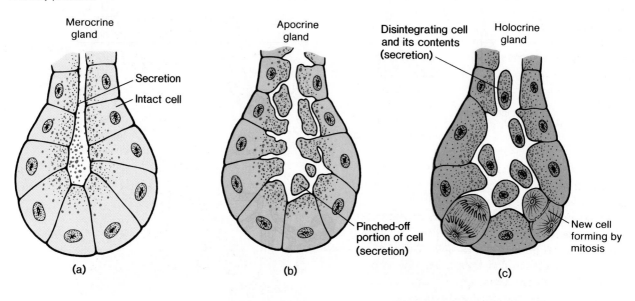

Chart 5.3 Types of glandular secretions		
Type	Description of secretion	Example
Merocrine glands	A fluid cellular product that is released through the cell membrane	Salivary glands, pancreatic glands, certain sweat glands of the skin
Apocrine glands	Cellular product and portions of the free ends of glandular cells that are pinched off during secretion	Mammary glands, certain sweat glands of the skin
Holocrine glands	Entire cells that are laden with secretory products	Sebaceous glands of the skin

Connective Tissues

General Characteristics

Connective tissues occur in all parts of the body. They bind structures together, provide support and protection, serve as frameworks, fill spaces, store fat, produce blood cells, provide protection against infections, and help to repair tissue damage.

Connective tissue cells are usually further apart than epithelial cells, and they have an abundance of intercellular material or *matrix* between them. This matrix consists of *fibers* and a *ground substance* whose consistency varies from fluid to semisolid or solid.

Connective tissue cells are able to reproduce. In most cases, they have good blood supplies and are well nourished. Although some connective tissues, such as bone and cartilage, are quite rigid, loose connective tissue, adipose connective tissue, and fibrous connective tissue are more flexible.

Major Cell Types

Connective tissues contain a variety of cell types. Some of them are called *resident cells* because they are usually present in relatively stable numbers. These include fibroblasts, macrophages, and mast cells. Another group known as *wandering cells* temporarily appears in the tissues, usually in response to an injury or infection. The wandering cells include several types of white blood cells.

The **fibroblast** is the most common kind of cell in connective tissues. It is relatively large and usually star-shaped. Fibroblasts function to produce fibers of protein in the intercellular matrix of connective tissues.

Macrophages (histiocytes) are almost as numerous as fibroblasts in some connective tissues. They are usually attached to fibers but can become detached and actively move about. Macrophages are specialized to carry on phagocytosis. Since they function as scavenger cells that can clear foreign particles from tissues, macrophages represent an important defense against infectious agents. They also play a role in immunity, which is discussed in chapter 19 (figure 5.10).

Mast cells are relatively large cells that are widely distributed in connective tissues and are usually located near blood vessels. Their function is not well understood; however, they are known to contain *heparin,* a compound that prevents blood clotting. They also contain *histamine,* a substance that promotes some of the reactions associated with inflammation and allergies such as asthma and hayfever (see chapter 19 for more information about allergies).

Connective Tissue Fibers

There are three types of connective tissue fibers produced by fibroblasts: *collagenous fibers, elastic fibers,* and *reticular fibers.*

Collagenous fibers are relatively thick threadlike parts composed of the protein *collagen,* which is the major structural protein of the body. (Collagen also is the main ingredient of leather; when this protein is denatured by boiling, it becomes gelatin.) These fibers are grouped in long, parallel bundles, and they are flexible but only slightly elastic. More importantly, they have great tensile strength—that is, they are capable of resisting considerable pulling force. Thus, collagenous fibers are important components of body parts that hold structures together.

When collagenous fibers are present in abundance, the tissue containing them is called *dense connective tissue.* Such a tissue appears white, and for this reason collagenous fibers are sometimes called *white fibers. Loose connective tissue,* on the other hand, is sparsely supplied with collagenous fibers.

Elastic fibers are composed of bundles of microfibrils embedded in a protein called *elastin.* These fibers tend to be branched and form complex networks in various tissues. They have less strength than collagenous fibers, but they are very elastic. That is, they are easily stretched or deformed and will resume their original lengths and shapes when the force acting upon them is removed. They are common in body parts that are normally subjected to stretching, such as the vocal cords and various air passages of the respiratory system. Elastic fibers sometimes are called *yellow fibers,* because tissues well supplied with them appear yellowish.

Reticular fibers are very thin fibers composed of collagen. They are highly branched and form delicate supporting networks in a variety of tissues.

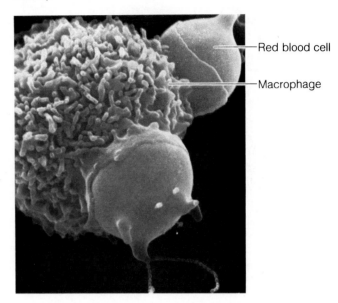

—Red blood cell

—Macrophage

When skin is exposed to sunlight excessively, its connective tissue fibers tend to lose their elasticity, and the skin becomes increasingly stiff and leathery. In time it may sag and wrinkle. On the other hand, skin of a healthy, well-nourished person that remains covered usually shows much less change.

1. What are the general characteristics of connective tissue?
2. What are the major types of resident cells in connective tissue?
3. What is the primary function of fibroblasts?
4. What are the characteristics of collagen and elastin?

Loose Connective Tissue

Loose connective tissue forms delicate, thin membranes throughout the body. The cells of this tissue, which are mainly fibroblasts, are located some distance apart and are separated by a gel-like ground substance that contains many collagenous and elastic fibers (figure 5.11).

This tissue binds the skin to the underlying organs and fills spaces between muscles. It lies beneath most layers of epithelium, where its numerous blood vessels provide nourishment for the epithelial cells.

Figure 5.11

(a) Loose connective tissue contains numerous fibroblasts that produce collagenous and elastic fibers. (b) Micrograph of loose connective tissue.

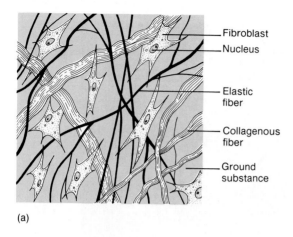

(a)

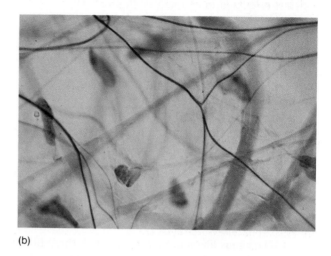

(b)

Figure 5.12

(a) Adipose tissue cells contain large fat droplets that cause the nuclei to be pushed close to the cell membrane. (b) Note the nucleus indicated by the arrow in this micrograph.

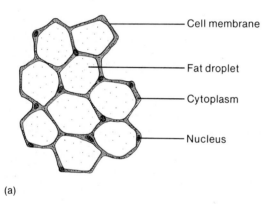

— Cell membrane

— Fat droplet

— Cytoplasm

— Nucleus

(a)

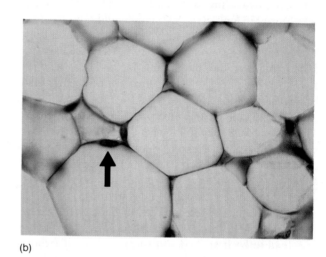

(b)

Adipose Tissue

Adipose tissue, commonly called fat, is really a specialized form of loose connective tissue. Certain cells within loose connective tissue (adipocytes) store fat in droplets within their cytoplasm. At first these cells resemble fibroblasts, but as they accumulate fat, they become swollen, and their nuclei are pushed to one side (figure 5.12). When they occur in such large numbers that other cell types are crowded out, they form adipose tissue.

Adipose tissue is found beneath the skin and in spaces between muscles. It also occurs around the kidneys, behind the eyeballs, in certain abdominal membranes, on the surface of the heart, and around various joints.

Adipose tissue serves as a protective cushion for joints and some organs, such as the kidneys. It also functions as a heat insulator beneath the skin and stores energy in fat molecules.

Because excess food substances are likely to be converted to fat and stored, the amount of adipose tissue present in the body is usually related to a person's diet. During a period of fasting, adipose cells may lose their fat droplets, shrink in size, and become more like fibroblasts again.

In infants and young children, there usually is a continuous layer of adipose tissue just beneath the skin. This layer gives their bodies the rounded figure found in well-nourished youngsters. In adults, this subcutaneous fat tends to become thinner in some regions and thicker in others. For example, in males it usually thickens in the upper back, upper arms, lower back, and buttocks; while in females it is more likely to develop in the breasts, buttocks, and upper legs.

Figure 5.13

(a) Fibrous connective tissue consists largely of tightly packed collagenous fibers. (b) The collagenous fibers are stained red in this micrograph.

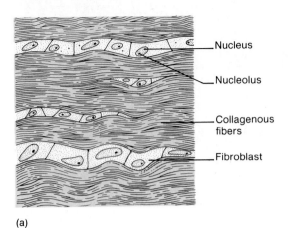

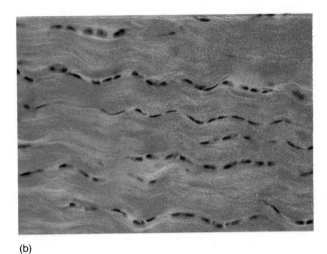

(a)   (b)

Figure 5.14

(a) Elastic connective tissue contains many elastic fibers with collagenous fibers between them. (b) Micrograph of elastic tissue.

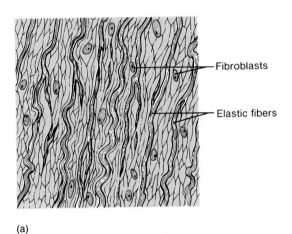

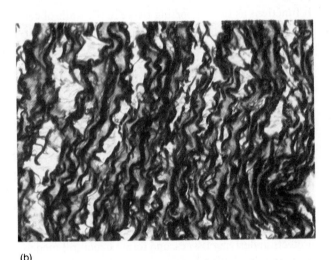

(a)   (b)

Fibrous Connective Tissue

Fibrous connective tissue is a dense tissue that contains many closely packed, thick, collagenous fibers and a fine network of elastic fibers. It has relatively few cells, almost all of which are fibroblasts (figure 5.13).

Since collagenous fibers are very strong, this type of tissue can withstand pulling forces, and it often binds body parts together. For example, **tendons,** which connect muscles to bones, and **ligaments,** which connect bones to bones at joints, are composed of fibrous connective tissue. This type of tissue also occurs in the protective white layer of the eyeball and in the deeper portions of the skin.

The blood supply to fibrous connective tissue is relatively poor, so tissue repair occurs slowly. This is why a sprain, which involves damage to the tissues surrounding a joint, may take some time to heal.

Elastic Connective Tissue

Elastic connective tissue consists mainly of yellow, elastic fibers arranged in parallel strands or in branching networks. Spaces between the fibers contain collagenous fibers and fibroblasts.

This tissue is found in the attachments between adjacent vertebrae of the backbone (ligamenta flava). It also occurs in the layers within the walls of various hollow internal organs, including the larger arteries, some portions of the heart, and the larger airways. It imparts an elastic quality to these structures. (See figure 5.14.)

Figure 5.15

(a) Reticular connective tissue is composed of thin collagenous fibers arranged in a network. (b) Micrograph of reticular tissue.

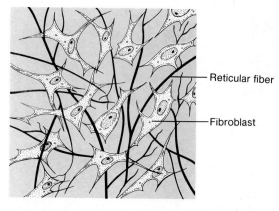

Reticular fiber

Fibroblast

(a)

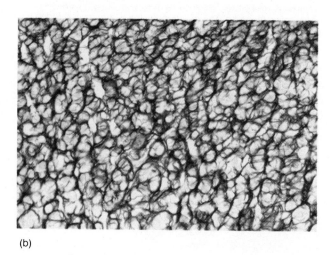

(b)

Reticular Connective Tissue

Reticular connective tissue is composed of thin, collagenous fibers arranged in a three-dimensional network. It functions as the supporting tissue in the walls of certain internal organs, such as the liver, spleen, and various lymphatic organs. (See figure 5.15.)

---

1. How is loose connective tissue related to adipose tissue?
2. What are the functions of adipose tissue?
3. How can fibrous, elastic, and reticular connective tissues be distinguished?

Cartilage

Cartilage is one of the rigid connective tissues. It supports parts, provides frameworks and attachments, protects underlying tissues, and forms structural models for many developing bones.

The matrix of cartilage is abundant and is composed largely of fibers embedded in a gel-like ground substance. This ground substance is rich in a protein-polysaccharide complex (chondromucoprotein) and contains an abundance of water. The cartilage cells or **chondrocytes** occupy small chambers called *lacunae,* which are completely surrounded by matrix.

As a rule, a cartilaginous structure is enclosed in a covering of fibrous connective tissue called the *perichondrium.* Although cartilage tissue lacks a direct blood supply, there are blood vessels in the perichondrium that surround it. The cartilage cells obtain nutrients from these vessels by diffusion that is aided by the water in the matrix. This lack of a direct blood supply is related to the slow rate of cellular reproduction and repair that is characteristic of cartilage.

There are three kinds of cartilage, and each contains a different type of intercellular material. Hyaline cartilage has very fine collagenous fibers in its matrix, elastic cartilage contains a dense network of elastic fibers, and fibrocartilage contains many large collagenous fibers.

Hyaline cartilage (figure 5.16), the most common type of cartilage, looks somewhat like milk glass. It occurs on the ends of bones in many joints, in the soft part of the nose, and in the supporting rings of the respiratory passages. In an embryo, many of the skeletal parts are formed at first of hyaline cartilage, which is later replaced by bone. Hyaline cartilage also plays a role in the growth of most bones and in the repair of bone fractures. (See chapter 7.)

Elastic cartilage (figure 5.17), which is more flexible than hyaline cartilage because it contains many elastic fibers in its matrix, provides the framework for the external ears and parts of the larynx.

Fibrocartilage (figure 5.18), a very tough tissue, contains many collagenous fibers. It often serves as a shock absorber for structures that are subjected to pressure. For example, fibrocartilage forms pads (intervertebral disks) between the individual parts of the backbone. It also forms protective cushions between bones in the knees and the pelvic girdle.

Bone

Bone (osseous tissue) is the most rigid of the connective tissues. Its hardness is due largely to the presence of mineral salts, such as calcium phosphate and calcium carbonate, in the intercellular matrix. This intercellular material also contains a considerable amount of collagen.

Figure 5.16
(a) Hyaline cartilage cells are located in lacunae surrounded by intercellular material containing very fine, collagenous fibers.
(b) Micrograph of hyaline cartilage.

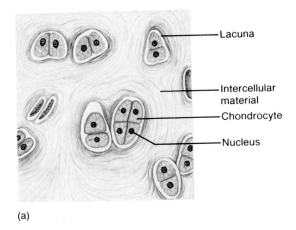

Lacuna

Intercellular material

Chondrocyte

Nucleus

(a)

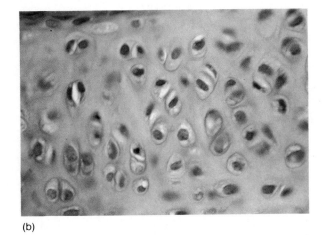

(b)

Figure 5.17
(a) Elastic cartilage contains fine collagenous fibers and many elastic fibers in its intercellular material. (b) Micrograph of elastic cartilage.

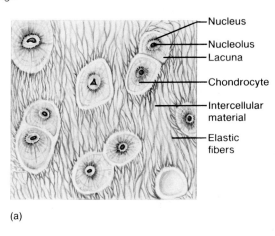

Nucleus

Nucleolus

Lacuna

Chondrocyte

Intercellular material

Elastic fibers

(a)

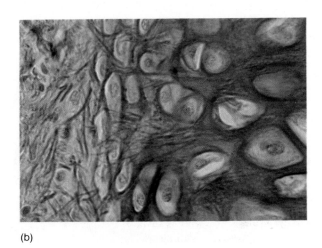

(b)

Figure 5.18
(a) Fibrocartilage contains many large collagenous fibers in its intercellular material. (b) Micrograph of fibrocartilage.

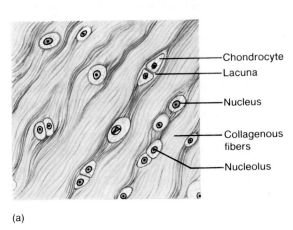

Chondrocyte

Lacuna

Nucleus

Collagenous fibers

Nucleolus

(a)

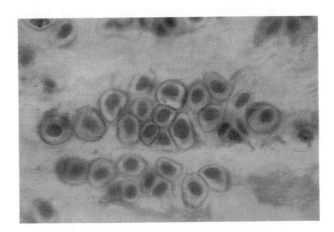

(b)

Figure 5.19

(*a*) Bone matrix is deposited in concentric layers around osteonic canals. (*b*) What features of the tissue can you identify in this micrograph?

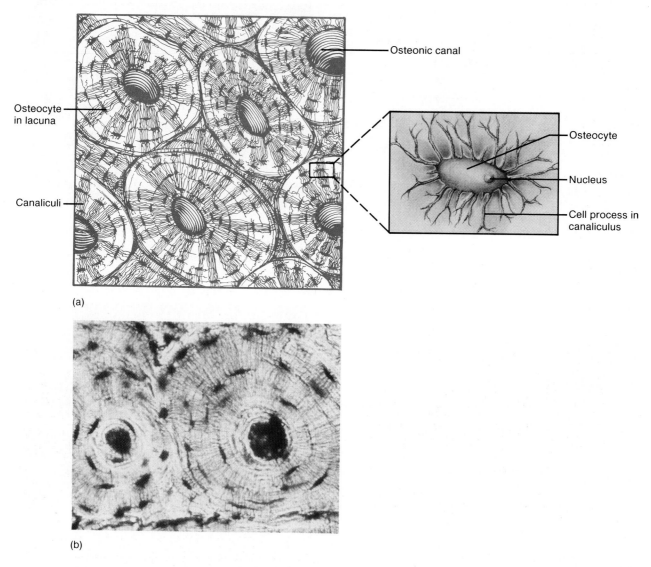

Osteonic canal

Osteocyte in lacuna

Canaliculi

Osteocyte

Nucleus

Cell process in canaliculus

(a)

(b)

Bone provides an internal support for body structures. It protects vital parts in the cranial and thoracic cavities, and serves as an attachment for muscles. It also contains red marrow, which functions in forming blood cells, and it stores various inorganic salts.

Bone matrix is deposited in thin layers called *lamellae,* which most commonly are arranged in concentric patterns around tiny longitudinal tubes called *osteonic* (haversian) *canals.* Bone cells called *osteocytes* are located in lacunae, rather evenly spaced between the lamellae. Consequently, they too are arranged in patterns of concentric circles (see figures 5.19 and 7.4).

Each osteonic canal contains a blood vessel, so that every bone cell is fairly close to a nutrient supply. In addition, the bone cells have numerous cytoplasmic processes that extend outward and pass through minute tubes (*canaliculi*) in the matrix. These cellular processes are attached to the membranes of nearby cells by specialized intercellular junctions (gap junctions). As a result, materials can move rapidly between blood vessels and bone cells. Thus, in spite of its inert appearance, bone is a very active tissue. When it is injured, bone heals much more rapidly than cartilage.

Chart 5.4 lists the characteristics of the major types of connective tissue.

Chart 5.4   Connective tissues

Type	Function	Location
Loose connective tissue	Binds organs together, holds tissue fluids	Beneath the skin, between muscles, beneath most epithelial layers
Adipose tissue	Protection, insulation, and storage of fat	Beneath the skin, around the kidneys, behind the eyeballs, on the surface of the heart
Fibrous connective tissue	Binds organs together	Tendons, ligaments, skin
Elastic connective tissue	Provides elastic quality	Between adjacent vertebrae, in walls of arteries and airways
Reticular connective tissue	Support	Walls of liver, spleen, and lymphatic organs
Hyaline cartilage	Support, protection, provides framework	Ends of bones, nose, and rings in walls of respiratory passages
Elastic cartilage	Support, protection, provides framework	Framework of external ear and part of the larynx
Fibrocartilage	Support, protection	Between bony parts of backbone, pelvic girdle, and knee
Bone	Support, protection, provides framework	Bones of skeleton

Figure 5.20
(a) Blood tissue consists of an intercellular fluid in which red blood cells, white blood cells, and platelets are suspended.
(b) Micrograph of human blood cells.

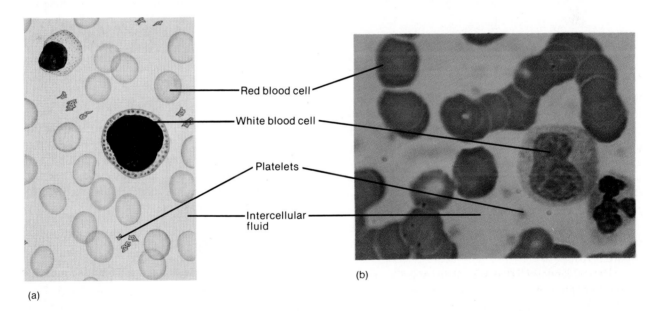

Red blood cell

White blood cell

Platelets

Intercellular fluid

(a)

(b)

Other Connective Tissues

Other important connective tissues include blood (vascular tissue) and reticuloendothelial tissue.

Blood is composed of cells that are suspended in fluid intercellular matrix called *blood plasma*. These cells include red blood cells, white blood cells, and some cellular fragments called *platelets*. Most of the cells are formed by special tissues (hematopoietic tissues) found in the hollow parts of certain bones (figure 5.20). Blood is described in detail in chapter 17.

Of the blood cells, only the red cells function entirely within the blood vessels. The white cells migrate through capillary walls and enter connective tissues where they carry on their major activities. In this way, blood transports cells from the hollow parts of bones to other connective tissues where they usually reside until they die.

Reticuloendothelial tissue is composed of a variety of specialized cells that are widely scattered throughout the body. As a group, these cells are phagocytic; that is, they function to ingest and destroy foreign particles, such as microorganisms, that may invade

Inflammation
A Clinical Application

Inflammation is a localized response to factors that are damaging or potentially damaging to tissues. These factors include traumatic injuries, excessive heat, chemical irritants, bacterial infections, and other stressful conditions.

The inflammation reaction occurs mainly in connective tissues and involves various components of the circulatory system. It helps to prevent the spread of infections, brings about the destruction of foreign substances entering the tissues, and promotes the healing process.

As an example, imagine that some bacteria have entered connective tissue through a break in the skin. In response, chemicals, such as histamine, are released from damaged and stressed cells. These substances, together with others from the invading bacteria, stimulate a variety of changes to take place. Nearby blood vessels become dilated and cause an increase in the blood supply to the affected tissues. This causes the region to appear redder and feel warmer. At the same time, the permeability of the smaller blood vessels increases, and extra fluid filters into the intercellular spaces of the tissue. This causes swelling (a condition called *edema*) and puts pressure on local nerve endings, which causes the region to become painful.

The fluid that enters from the blood soon clots. This seals off the inflamed area by filling tissue spaces and lymphatic vessels, and it helps delay the spread of bacteria into surrounding tissues.

Macrophages residing in the connective tissue become mobilized and begin to phagocytize bacteria almost immediately. The inflamed tissue releases a substance (leukocytosis-inducing factor) that is carried away by the blood and stimulates the release of many white blood cells from bone marrow. Within a few hours, large numbers of these cells (neutrophils) migrate into the inflamed tissues from the blood by passing through capillary walls. They are attracted toward chemicals diffusing out from the affected tissues, and they also act as phagocytes.

Eventually, a fluid-filled cavity filled with a creamy, yellowish substance called *pus* may appear in the inflamed region. The pus represents an accumulation of tissue fluid, bacteria, and dead and dying white blood cells and macrophages. Sometimes such a cavity opens itself to the surface or into a body cavity, and the pus drains out.

As the inflammation subsides, macrophages clean up the cellular debris by phagocytosis and the excess fluid is absorbed by surrounding tissues. The gap is filled with new connective tissue formed by fibroblasts that migrate into the area.

---

the body. Thus, they are particularly important in defending the body against infection.

The reticuloendothelial cells include types found in the blood, lungs, brain, bone marrow, and lymph glands. The most common ones, however, are *macrophages*. Typically, a macrophage remains in a fixed position until it "senses" a foreign particle. Then, the phagocyte becomes motile, moves toward the invader, and may engulf and destroy it. Once the invader has been destroyed, the macrophage becomes fixed again (figure 5.10).

Reticuloendothelial tissue is discussed in more detail in chapter 19.

---

1. Describe the general characteristics of cartilage.
2. Explain why injured bone heals more rapidly than cartilage.
3. Why is blood considered a connective tissue?
4. What do the cells of reticuloendothelial tissue have in common?

Muscle Tissues

Muscle tissues are contractile—their elongated cells or *muscle fibers* can change shape by becoming shorter and thicker. As they contract, the fibers pull at their attached ends and cause body parts to move. The three types of muscle tissue are skeletal muscle, smooth muscle, and cardiac muscle.

Skeletal Muscle Tissue

Skeletal muscle tissue (figure 5.21) is found in muscles that usually are attached to bones and can be controlled by conscious effort. For this reason, it is often called *voluntary* muscle. The cells are long and threadlike, with alternating light and dark cross-markings called *striations*. Each cell, or muscle fiber, has many nuclei, located just beneath its cell membrane. When the cell is stimulated by an action of a nerve fiber, it contracts and then relaxes. If the connecting nerve fiber is damaged, it may not be possible to control the muscle fiber's action, and the muscle fiber is paralyzed.

Skeletal muscles are responsible for moving the head, trunk, and limbs. They also produce the movements involved with facial expressions, writing, talking, and singing, and with such actions as chewing, swallowing, and breathing.

Figure 5.21
(a) Skeletal muscle tissue is composed of striated muscle fibers that contain many nuclei. (b) Micrograph of skeletal muscle tissue.

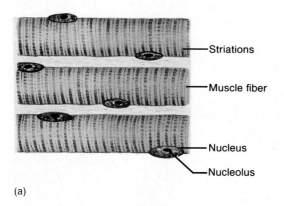

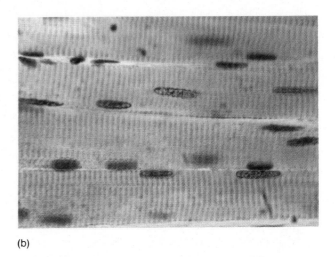

- Striations

- Muscle fiber

- Nucleus

- Nucleolus

(a)

(b)

Figure 5.22
(a) Smooth muscle tissue is formed of spindle-shaped cells, each containing a single nucleus. (b) Note the nucleus indicated by the arrow in this micrograph.

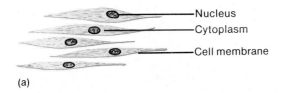

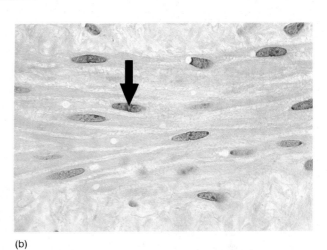

- Nucleus
- Cytoplasm
- Cell membrane

(a)

(b)

Smooth Muscle Tissue

Smooth muscle tissue (figure 5.22) is called smooth because it lacks striations. This tissue is found in the walls of hollow internal organs such as the stomach, intestines, urinary bladder, uterus, and blood vessels. Unlike skeletal muscle, smooth muscle usually cannot be stimulated to contract by conscious efforts. Thus, it is a type of *involuntary* muscle.

Smooth muscle cells are shorter than those of skeletal muscle, and they each have a single, centrally located nucleus. Smooth muscle tissue is responsible for the movements that force food through the digestive tube, constrict blood vessels, and empty the urinary bladder.

Cardiac Muscle Tissue

Cardiac muscle tissue (figure 5.23) occurs only in the heart. Its cells, which are striated, are joined end to end. The resulting fibers are branched and interconnected in complex networks. Each cell has a single nucleus. At its end, where it touches another cell, there is a specialized intercellular junction called an *intercalated disc,* which occurs only in cardiac tissue.

Cardiac muscle, like smooth muscle, is controlled *involuntarily* and, in fact, can continue to function without being stimulated by nerve impulses. This tissue makes up the bulk of the heart and is responsible for pumping blood through the heart chambers and into the blood vessels.

Figure 5.23

(*a*) How is this cardiac muscle tissue similar to skeletal muscle? How is it similar to smooth muscle? (*b*) Note the intercalated disc indicated by the arrow in this micrograph.

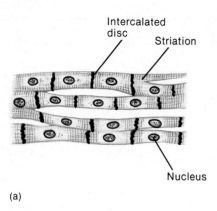

Intercalated disc

Striation

Nucleus

(a)

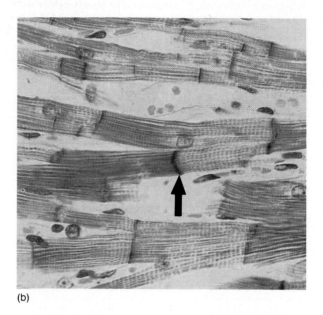

(b)

Figure 5.24

(*a*) Neurons in nerve tissue function to transmit impulses to other neurons or to muscles or glands. (*b*) What features of a neuron can you identify in this micrograph?

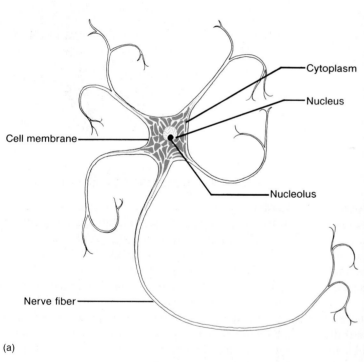

Cytoplasm

Nucleus

Cell membrane

Nucleolus

Nerve fiber

(a)

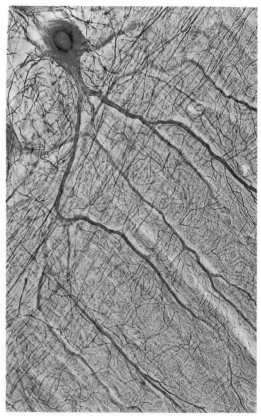

(b)

Chart 5.5 Muscle and nerve tissues					
Type	Function	Location	Type	Function	Location
Skeletal muscle tissue	Voluntary movements of skeletal parts	Muscles attached to bones	Cardiac muscle tissue	Heart movements	Heart muscle
Smooth muscle tissue	Involuntary movements of internal organs	Walls of hollow internal organs	Nerve tissue	Sensitivity and conduction of nerve impulses	Brain, spinal cord, and nerves

The cells of different tissues vary greatly in their abilities to reproduce. For example, the epithelial cells of the skin and the inner lining of the digestive tube, and the connective tissue cells that form blood cells in bones are reproducing continuously. However, striated and cardiac muscle cells and nerve cells do not seem to reproduce at all after becoming differentiated.

Fibroblasts respond rapidly to injuries by increasing in numbers and becoming active in fiber production. Therefore, they often are the principal agents of repair in tissues that have limited abilities to regenerate themselves. For instance, cardiac muscle tissue typically degenerates in the regions damaged by a heart attack. Such tissue may be replaced by connective tissue built by fibroblasts, which later appears as a scar.

1. List the general characteristics of muscle tissue.
2. Distinguish between skeletal, smooth, and cardiac muscle tissue.

Nerve Tissue

Nerve tissue is found in the brain, spinal cord, and associated nerves. The basic cells of this tissue are called *nerve cells,* or *neurons,* and they are among the more highly specialized body cells. Neurons are sensitive to certain types of changes in their surroundings. They respond by transmitting nerve impulses along cytoplasmic extensions to other neurons or to muscles or glands (figure 5.24).

As a result of the extremely complex patterns by which neurons are connected with each other and with various body parts, they are able to coordinate and regulate many body functions.

In addition to neurons, nerve tissue contains **neuroglial cells.** These cells function to support and bind the nerve tissue together, carry on phagocytosis, and aid in supplying nutrients to neurons by connecting them to blood vessels. Nerve tissue is discussed in detail in chapter 10.

Chart 5.5 summarizes the general characteristics of muscle and nerve tissues.

1. Describe the general characteristics of nerve tissue.
2. Distinguish between neurons and neuroglial cells.

Chapter Summary

Introduction
Cells are arranged in layers or groups called tissues. Cells are separated by intercellular materials whose composition varies from solids to liquids.

The four major types of human tissues are epithelial tissue, connective tissue, muscle tissue, and nerve tissue.

Epithelial Tissues
1. General characteristics
   a. Epithelial tissue covers all free body surfaces and is the major tissue of glands.
   b. It is anchored to connective tissue by basement membrane. Epithelial tissue lacks blood vessels, contains little intercellular material, and is replaced continuously.
   c. It functions in protection, secretion, absorption, excretion, and sensory reception.
2. Simple squamous epithelium
   a. This tissue consists of a single layer of thin, flattened cells through which substances pass rather easily.
   b. It functions in the exchange of gases in the lungs and lines blood vessels, lymph vessels, and various membranes within the thorax and abdomen.
3. Simple cuboidal epithelium
   a. This tissue consists of a single layer of cube-shaped cells.
   b. It carries on secretion and absorption in the kidneys and various glands.

4. Simple columnar epithelium
    a. This tissue is composed of elongated cells whose nuclei are located near the basement membrane.
    b. It lines the uterus and digestive tract, where it functions in protection, secretion, and absorption.
    c. Absorbing cells often possess microvilli.
    d. This tissue usually contains goblet cells that secrete mucus.
5. Pseudostratified columnar epithelium
    a. This tissue appears stratified because the nuclei are located at two or more levels within the cells.
    b. Its cells may have cilia that function to move mucus or cells over the surface of the tissue.
    c. It lines various tubes of the respiratory and reproductive systems.
6. Stratified squamous epithelium
    a. This tissue is composed of many cell layers.
    b. It protects underlying cells from harmful environmental effects.
    c. It covers the skin and lines the mouth, throat, vagina, and anal canal.
7. Transitional epithelium
    a. This tissue is specialized to undergo distension.
    b. It occurs in the walls of various organs of the urinary tract.
    c. It helps prevent the contents of the urinary passageways from diffusing outward.
8. Glandular epithelium
    a. Glandular epithelium is composed of cells that are specialized to secrete substances.
    b. One or more cells can constitute a gland.
        (1) Exocrine glands secrete into ducts.
        (2) Endocrine glands secrete into tissue fluid or blood.
    c. Glands are classified according to the arrangement of their cells.
        (1) Simple glands have unbranched ducts.
        (2) Compound glands have branched ducts.
        (3) Tubular glands consist of simple epithelial-lined tubes.
        (4) Alveolar glands consist of saclike dilations connected to the surface by narrowed ducts.

    d. Glandular secretions are classified according to the contents of the secretions.
        (1) Merocrine glands secrete watery fluids without loss of cytoplasm. Most secretory cells are merocrine.
            (a) Serous cells secrete watery fluid with a high enzyme content.
            (b) Mucous cells secrete mucus.
        (2) Apocrine glands lose portions of their cells during secretion.
        (3) Holocrine glands release cells filled with secretory products.

Connective Tissues
1. General characteristics
    a. Connective tissue connects, supports, protects, provides frameworks, fills spaces, stores fat, produces blood cells, provides protection against infection, and helps repair tissues.
    b. It consists of cells, usually some distance apart, and considerable intercellular material.
2. Major cell types
    a. Fibroblasts produce collagenous and elastic fibers.
    b. Macrophages function as phagocytes.
    c. Mast cells contain heparin and histamine.
3. Connective tissue fibers
    a. Collagenous fibers are composed of collagen and have great tensile strength.
    b. Elastic fibers are composed of microfibrils embedded in elastin and are very elastic.
    c. Reticular fibers are very fine collagenous fibers.
4. Loose connective tissue
    a. This tissue forms thin membranes between organs and binds them together.
    b. It is found beneath the skin and between muscles.
    c. It contains tissue fluids in intercellular spaces.
5. Adipose tissue
    a. Adipose tissue is a specialized form of loose connective tissue that stores fat, provides a protective cushion, and functions as a heat insulator.
    b. It is found beneath the skin, in certain abdominal membranes, and around the kidneys, heart, and various joints.
6. Fibrous connective tissue
    a. This tissue is composed largely of strong, collagenous fibers that bind parts together.
    b. It is found in tendons, ligaments, eyes, and skin.
7. Elastic connective tissue
    a. This tissue is composed mainly of elastic fibers.
    b. It imparts an elastic quality to the walls of certain hollow internal organs.

8. Reticular connective tissue
   a. This tissue consists largely of thin, branched collagenous fibers.
   b. It supports the walls of the liver, spleen, and lymphatic organs.
9. Cartilage
   a. Cartilage provides support and framework for various parts.
   b. Its intercellular material is largely composed of fibers and a gel-like ground substance.
   c. It lacks a direct blood supply and is slow to heal following an injury.
   d. Cartilaginous structures are usually enclosed in a perichondrium, which contains blood vessels.
   e. Major types are hyaline cartilage, elastic cartilage, and fibrocartilage.
   f. Cartilage occurs at ends of various bones, in the ear, in the larynx, and in pads between bones of the backbone, pelvic girdle, and knees.
10. Bone
    a. The intercellular matrix of bone contains mineral salts and collagen.
    b. Its cells are arranged in concentric circles around osteonic canals and are interconnected by canaliculi.
    c. It is an active tissue that heals rapidly following an injury.
11. Other connective tissues
    a. Blood
       (1) Blood is composed of cells suspended in fluid.
       (2) The cells are formed by special tissue in the hollow parts of bones.
       (3) It transports white blood cells to connective tissues.
    b. Reticuloendothelial tissue
       (1) This tissue is composed of a variety of phagocytic cells widely distributed in body organs.
       (2) It defends the body against invasion by microorganisms.

Muscle Tissues
1. General characteristics
   a. Muscle tissue is contractile tissue that moves parts attached to it.
   b. Three types are skeletal, smooth, and cardiac muscle tissues.
2. Skeletal muscle tissue
   a. This tissue is attached to bones and is controlled by conscious effort.
   b. Cells or muscle fibers are long and threadlike with alternating light and dark cross-markings.
   c. Muscle fibers contract when stimulated by nerve action and relax immediately.

3. Smooth muscle tissue
   a. This tissue is found in walls of hollow internal organs.
   b. Usually it is controlled by involuntary activity.
4. Cardiac muscle tissue
   a. This tissue is found only in the heart.
   b. Cells are joined by intercalated discs and arranged in branched, interconnecting networks.
   c. It is controlled by involuntary activity.

Nerve Tissue
1. Nerve tissue is found in brain, spinal cord, and nerves.
2. Neurons
   a. These cells are sensitive to changes and respond by transmitting nerve impulses to other neurons or to other body parts.
   b. They function in coordinating and regulating body activities.
3. Neuroglial cells
   a. Some forms of these cells bind and support nerve tissue.
   b. Others carry on phagocytosis.
   c. Still others connect neurons to blood vessels.

Application of Knowledge

1. The sweeping action of the ciliated epithelium that lines the respiratory passages is inhibited by excessive exposure to tobacco smoke or air-borne pollutants. What special problems is this likely to create?
2. Joints, such as the elbow, shoulder, and knee, contain considerable amounts of cartilage and fibrous connective tissue. How is this related to the fact that joint injuries are often very slow to heal?
3. There is a group of disorders called collagenous diseases that are characterized by deterioration of connective tissues. Why would you expect such diseases to produce widely varying symptoms?
4. The cardinal signs of inflammation are swelling, redness, pain, and warmth. How would you explain the reason for each of these signs to a patient?

Review Activities

1. Define *tissue*.
2. Name the four major types of tissue found in the human body.
3. Describe the general characteristics of epithelial tissue.
4. Distinguish between a simple epithelium and a stratified epithelium.
5. Explain how the structure of simple squamous epithelium is related to its function.
6. Name an organ in which each of the following tissues is found, and give the function of the tissue in each case:
   a. Simple squamous epithelium
   b. Simple cuboidal epithelium
   c. Simple columnar epithelium
   d. Pseudostratified columnar epithelium
   e. Stratified squamous epithelium
   f. Transitional epithelium
7. Define *gland*.
8. Distinguish between an exocrine gland and an endocrine gland.
9. Explain how glands are classified according to the structure of their ducts and the arrangement of their cells.
10. Explain how glands are classified according to the nature of their cellular secretions.
11. Distinguish between a serous cell and a mucous cell.
12. Describe the general characteristics of connective tissue.
13. Describe three major types of connective tissue cells.
14. Distinguish between collagen and elastin.
15. Explain the relationship between loose connective tissue and adipose tissue.
16. Explain how the quantity of adipose tissue in the body is related to diet.
17. Distinguish between elastic and reticular connective tissues.
18. Explain why injured fibrous connective tissue and cartilage are usually slow to heal.
19. Name the major types of cartilage, and describe their differences and similarities.
20. Describe how bone cells are arranged in bone tissue.
21. Describe vascular tissue.
22. Describe the solid components of blood.
23. Define *reticuloendothelial tissue*.
24. Describe the general characteristics of muscle tissue.
25. Distinguish between skeletal, smooth, and cardiac muscle tissues.
26. Describe the general characteristics of nerve tissue.
27. Distinguish between neurons and neuroglial cells.

6

The Skin and the Integumentary System

The previous chapters dealt with the lower levels of organization within the human organism—tissues, cells, cellular organelles, and the chemical substances that form these parts.

Chapter 6 explains how tissues are grouped to form organs and how organs comprise organ systems. In this explanation, various membranes, including the skin, are used as examples of organs. Since the skin acts with hair follicles, sebaceous glands, and sweat glands to provide a variety of vital functions, these organs together constitute the integumentary organ system.

Chapter Outline

Chapter Objectives

After you have studied this chapter, you should be able to

1. Describe the four major types of membranes.

2. Describe the structure of the various layers of the skin.

3. List the general functions of each of these layers of the skin.

4. Describe the accessory organs associated with the skin.

5. Explain the functions of each accessory organ.

6. Explain how the skin functions in regulating body temperature.

7. Summarize the factors that determine skin color.

8. Complete the review activities at the end of this chapter. Note that the items are worded in the form of specific learning objectives. You may want to refer to them before reading the chapter.

Key Terms

conduction   (kon-duk'shun)

convection   (kon-vek'shun)

cutaneous membrane   (ku-ta'ne-us mem'brān)

dermis   (der'mis)

epidermis   (ep''ī-der'mis)

evaporation   (e-vap''o-ra'shun)

hair follicle   (hār fol'ĭ-kl)

integumentary   (in-teg-u-men'tar-e)

keratinization   (ker''ah-tin''ī-za'shun)

melanin   (mel'ah-nin)

mucous membrane   (mu'kus mem'brān)

sebaceous gland   (se-ba'shus gland)

serous membrane   (se'rus mem'brān)

subcutaneous   (sub''ku-ta'ne-us)

sweat gland   (swet gland)

Aids to Understanding Words

alb-, white: *alb*ino—condition characterized by a lack of pigment.

cut-, skin: sub*cut*aneous—beneath the skin.

derm-, skin: *derm*is—inner layer of the skin.

epi-, upon: *epi*dermis—outer layer of the skin.

follic-, small bag: hair *follic*le—tubelike depression in which a hair develops.

kerat-, horn: *kerat*in—protein produced as epidermal cells die and harden.

melan-, black: *melan*in—dark pigment produced by certain epidermal cells.

por-, channel: *por*e—opening by which a sweat gland communicates to the surface.

seb-, grease: *seb*aceous gland—gland that secretes an oily substance.

TWO OR MORE kinds of tissues grouped together and performing specialized functions constitute an organ. Thus, thin sheetlike structures called *membranes,* composed of epithelium and connective tissue, that cover body surfaces and line body cavities are organs. One of these membranes, the skin, together with various accessory organs makes up the **integumentary system.**

Types of Membranes

There are four major types of membranes: *serous, mucous, cutaneous,* and *synovial.* Usually these structures are relatively thin. Serous, mucous, and cutaneous membranes are composed of epithelial tissue and some underlying connective tissue; synovial membranes are composed entirely of various connective tissues.

Serous membranes line the body cavities that lack openings to the outside. They form the inner linings of the thorax and abdomen, and they cover the organs within these cavities. Such a membrane consists of a layer of simple squamous epithelium (mesothelium) covering a thin layer of loose connective tissue. It secretes a watery *serous fluid,* which helps to lubricate the surfaces of the membrane. (See figures 1.7 and 1.8.)

Mucous membranes line the cavities and tubes that open to the outside of the body. These include the oral and nasal cavities and the tubes of the digestive, respiratory, urinary, and reproductive systems. Mucous membrane consists of epithelium overlying a layer of loose connective tissue; however, the type of epithelium varies with the location of the membrane. For example, stratified squamous epithelium lines the oral cavity, pseudostratified columnar epithelium lines part of the nasal cavity, and simple columnar epithelium lines the small intestine. Specialized cells within a mucous membrane secrete the thick fluid called *mucus.*

The **cutaneous membrane** is an organ of the integumentary system and is more commonly called *skin.* It is described in detail in the next section of this chapter.

Synovial membranes form the inner linings of joint cavities between the ends of bones at freely movable joints. These membranes usually include fibrous connective tissue overlying loose connective tissue and adipose tissue. They secrete a thick, colorless *synovial fluid* into the joint cavity, which lubricates the ends of the bones within the joint.

The Skin and Its Tissues

The skin is one of the larger and more versatile organs of the body, and it plays vital roles in the maintenance of homeostasis. For example, skin functions as a protective covering which prevents many harmful substances, including microorganisms, from entering the body. At the same time, it retards the loss of water by diffusion from deeper tissues. Skin aids in the regulation of body temperature, houses sensory receptors, synthesizes various chemicals, and excretes small quantities of waste substances. Also, people are recognized and their ages often are judged by their skin characteristics.

The skin includes two distinct layers of tissues. The outer layer, called the **epidermis,** is composed of stratified squamous epithelium. The inner layer, or **dermis,** is thicker than the epidermis, and it includes a variety of tissues, such as fibrous connective tissue, epithelial tissue, smooth muscle tissue, nerve tissue, and blood (figure 6.1).

Beneath the dermis are masses of loose connective and adipose tissues that bind the skin to underlying organs. They form the **subcutaneous layer.**

---

1. Name the four types of membranes and explain how they differ.
2. List the general functions of the skin.
3. Name the tissue (tissues) found in the outer layer of the skin; the inner layer.

The Epidermis

Since the epidermis is composed of epithelium, it lacks blood vessels. However, the deepest cells of the epidermis, which are next to the dermis, are nourished by dermal blood vessels and are capable of reproducing. They form a layer called the *stratum germinativum.* As the cells of this layer divide, older epidermal cells are slowly pushed away from the dermis toward the surface of the skin. The further the cells travel, the poorer their nutrient supply becomes, and in time they die.

Meanwhile, the membranes of the older cells become thickened and develop numerous intercellular junctions (desmosomes) that fasten them to adjacent cells. At the same time the cells begin to undergo a hardening process called **keratinization,** during which strands of tough, fibrous, waterproof protein called *keratin* develop within the cell membranes. As a result, many layers of tough, tightly packed dead cells accumulate in the outer portion of the epidermis. This outermost layer is called the *stratum corneum,* and the dead cells that compose it are easily rubbed away.

Figure 6.1
A section of skin. Why is the skin considered to be an organ?

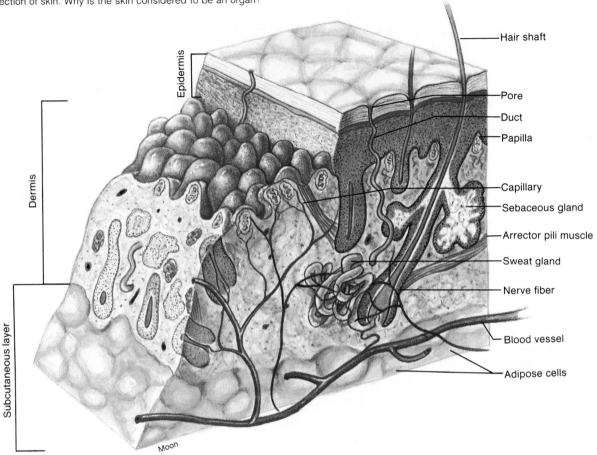

Epidermis

Dermis

Subcutaneous layer

Moon

Hair shaft

Pore

Duct

Papilla

Capillary

Sebaceous gland

Arrector pili muscle

Sweat gland

Nerve fiber

Blood vessel

Adipose cells

The structural organization of the epidermis varies from region to region. For example, it is most highly developed on the palms of the hands and the soles of the feet and may be 0.8 to 1.4 mm thick. In these areas four layers can be distinguished. They include *stratum basale* (stratum germinativum), which is the deepest layer; *stratum spinosum; stratum granulosum,* a granular layer; and *stratum corneum,* a fully keratinized layer. The cells of these layers are characterized by changes they undergo as they are pushed toward the surface (figure 6.2).

In other body regions, the epidermis is usually much thinner, averaging 0.07 to 0.12 mm in most areas. Also, where the epidermis is thin, the stratum granulosum may be missing. Chart 6.1 describes the characteristics of each layer of the epidermis.

In healthy skin, the production of epidermal cells is closely balanced with the loss of stratum corneum, so that skin seldom wears away completely. In fact, the rate of cellular reproduction tends to increase in regions where the skin is being rubbed or pressed regularly. This response causes the growth of calluses on the palms and soles as well as the development of corns on the toes when poorly fitting shoes rub the skin excessively.

Since blood vessels in the dermis supply nutrients to the epidermis, any interference with blood flow is likely to result in the death of epidermal cells. For example, when a person lies in one position for a prolonged period of time, the weight of the body pressing against even a bed interferes with the skin's blood supply. If cells die, the tissues begin to break down (necrosis) and a *pressure ulcer* (also called decubitus ulcer or bedsore) may appear.

Pressure ulcers usually occur in the skin overlying bony projections, such as on the hip, heel, elbow, or shoulder. These ulcers often can be prevented by changing the body position frequently or by massaging the skin to stimulate blood flow in regions associated with bony prominences. Special care must be taken with frequent turning of the body to prevent pressure ulcers in a paralyzed person who lacks sensations of pressure and cannot perform the usual response of frequently shifting body position.

A diet rich in proteins and other necessary nutrients and an adequate intake of fluids to maintain blood volume also helps to prevent this condition.

Figure 6.2
(a) The various layers of the epidermis are characterized by changes that occur in cells as they are pushed toward the surface of the skin. (b) Which layers of the epidermis can you identify in this micrograph from the palm of the hand?

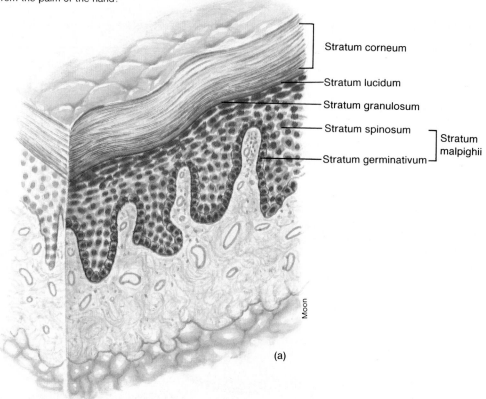

Stratum corneum
Stratum lucidum
Stratum granulosum
Stratum spinosum
Stratum germinativum

Stratum malpighii

Moon

(a)

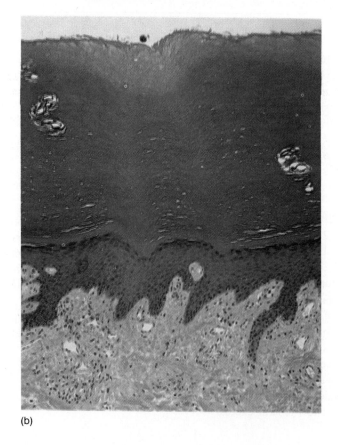

(b)

Chart 6.1	Layers of the epidermis	
Layer	**Location**	**Characteristics**
Stratum corneum (horny layer)	Outermost layer	Many layers of flat keratinized, dead epithelial cells that are flattened and nonnucleated
Stratum granulosum	Beneath the stratum corneum	Three to five layers of flattened granular cells that contain shrunken fibers of keratin and shriveled nuclei
Stratum spinosum	Beneath stratum granulosum	Many layers of cells with centrally located, large oval nuclei and developing fibers of keratin; cells becoming flattened
Stratum basale (basal cell layer)	Deepest layer	A single row of cuboidal or columnar cells that undergo mitosis; this layer also includes pigment-producing melanocytes

Figure 6.3

Melanocytes that occur mainly in the deeper layers of the epidermis produce the pigment called melanin.

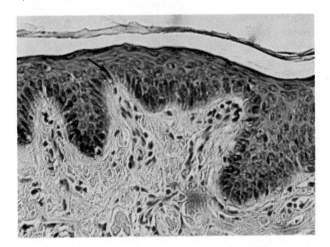

The epidermis has important protective functions. It shields the moist underlying tissues against excess water loss, mechanical injury, and the effects of harmful chemicals. When it is unbroken, the epidermis also prevents the entrance of many disease-causing microorganisms. **Melanin** is a pigment that occurs in the epidermis and is produced by specialized cells known as *melanocytes*. It absorbs light energy (figure 6.3), and in this way, it helps to protect deeper cells from the damaging effects of ultraviolet rays of sunlight.

Melanocytes are found in the deepest portion of the epidermis and in the underlying connective tissue of the dermis. Although they are the only cells that can produce melanin, the pigment also may occur in nearby epidermal cells. This happens because melanocytes have long, pigment-containing cellular extensions that pass upward between neighboring epidermal cells, and they can transfer granules of melanin into these other cells by a process called *cytocrine secretion*. Consequently, nearby epidermal cells may contain more melanin than the melanocytes.

1. Explain how the epidermis is formed.
2. What factors help prevent the loss of body fluids through the skin?
3. What is the function of melanin?

The Dermis

The surface between the epidermis and dermis is usually uneven, because the epidermis has ridges projecting inward and the dermis has finger-like *papillae* passing into the spaces between the ridges (see figure 6.1).

The dermis binds the epidermis to the underlying tissues. It is composed largely of fibrous connective tissue that includes tough collagenous fibers and elastic fibers surrounded by a gel-like substance. Networks of these fibers give the skin the qualities of toughness and elasticity. On the average, the dermis is 1.0 to 2.0 mm thick; however, it may be as thin as 0.5 mm or less on the eyelids or as thick as 3.0 mm on the soles.

As a person ages, the skin is likely to undergo changes, particularly in those regions that are exposed to sunlight. The rate and degree of change will vary from person to person. Typically the skin becomes drier, scaly, and its thickness decreases. Also, there is some degeneration of connective tissue fibers, and this leads to loss of elasticity and tensile strength, followed by wrinkling of the skin.

The dermis also contains muscle fibers. In some regions, such as in the skin that encloses the testes (scrotum), there are numerous smooth muscle cells that can cause the skin to wrinkle when they contract. Other smooth muscles in the dermis are associated with accessory organs such as hair follicles and various glands. Many striated muscle fibers are anchored in this layer in the skin of the face. They help produce the voluntary movements associated with facial expressions.

Skin Cancer

A Practical Application

Cancer of the skin is most likely to arise from nonpigmented epithelial cells within the deep layer of the epidermis or from pigmented melanocytes. Skin cancers originating from epithelial cells are called cutaneous *carcinomas* (basal cell carcinoma or squamous cell carcinoma); those arising from melanocytes are known as cutaneous *melanomas*.

Carcinomas represent the more common type of skin cancer, and they occur most frequently in members of light-skinned populations over forty years of age. These cancers seem to be caused by the effects of the ultraviolet portion of sunlight on the DNA of skin cells, and they usually appear in persons who are exposed to sunlight regularly. Thus, the incidence of cutaneous carcinoma is increased in persons who have spent considerable time outdoors—farmers, sailors, sportspeople, sunbathers, and so forth.

This type of skin cancer often develops from hard, dry, scaly growths (lesions) that have reddish bases. Such a lesion may be flat or raised above the surface and it adheres firmly to the skin. They appear most often in the skin of the neck, face, and scalp. Fortunately, cutaneous carcinomas usually are slow-growing and can be cured completely by being removed surgically or treated with radiation.

Because melanomas develop from melanocytes, they are pigmented with melanin. Such lesions typically have a variety of colored areas—variegated brown, black, gray, or blue—that are arranged haphazardly. They usually also have irregular margins rather than smooth, regular outlines.

This type of skin cancer may appear in young adults as well as older ones and seems to be caused by relatively short, intermittent exposure to high intensity sunlight. Thus, the incidence of melanoma is increased in persons who may work indoors but occasionally experience blistering sunburns during weekend activities or vacations.

Melanomas occur most often in light-skinned persons, particularly those whose skins burn without tanning. They usually appear in the skin of the trunk, especially the back, or in the skin of the limbs. Such a lesion may arise from normal-appearing skin or from a mole (nevus). Typically it enlarges by spreading through the skin horizontally, but eventually it may thicken and grow downward into the skin and invade the deeper tissues. If a melanoma is removed surgically while it is in its horizontal growth phase, it may be cured. Once it has thickened and spread into the deeper tissues, however, it becomes difficult to treat, and the survival rate for persons with such a cancer is very low.

For reasons that are not well understood, the incidence of melanoma within the U.S. population has been increasing rapidly for the past twenty years. To reduce the occurrence, it is advisable to avoid exposing the skin to high intensity sunlight. Whenever people remain outdoors, they should wear protective clothing or make use of sun-screening lotions with high protective factors. In addition, the skin should be self-examined regularly, and any unusual lesions—particularly those undergoing changes in color, shape, or surface texture—should be examined by a physician.

As was discussed, blood vessels in the dermis supply nutrients to the deep living layers of the epidermis, as well as to dermal cells. These vessels also play an important role in the regulation of body temperature—a function that is explained in a subsequent section of this chapter.

There are numerous nerve fibers scattered throughout the dermis. Some of them (motor fibers) carry impulses to dermal muscles and glands, causing these structures to react. Others (sensory fibers) carry impulses away from specialized sensory receptors located within the dermis (figure 6.4).

One set of dermal receptors (pacinian corpuscles) is stimulated by heavy pressure, while another set (Meissner's corpuscles) is sensitive to light touch. Still other receptors are stimulated by temperature changes or by factors that can damage tissues. Sensory receptors are discussed in chapter 12.

Hair follicles, sebaceous glands, and sweat glands also occur at various depths in the dermis. These parts that are composed largely of epithelial tissue are discussed in subsequent sections of this chapter.

Skin cells can play an important role in the production of vitamin D, which is necessary for the normal development of bones and teeth. This vitamin can be formed from a substance (dehydrocholesterol) that is synthesized by cells in the digestive system or obtained in the diet. When it reaches the skin by means of the blood and is exposed to ultraviolet light from the sun, this compound is converted to a substance that becomes vitamin D. (See chapter 13.)

The Subcutaneous Layer

As was mentioned, the subcutaneous layer (hypodermis) lies beneath the dermis and consists largely of loose connective and adipose tissues (see figure 6.1). The collagenous and elastic fibers of this layer are continuous with those of the dermis, and although most run parallel to the surface of the skin, they travel in all directions. As a result there is no sharp boundary between the dermis and the subcutaneous layer.

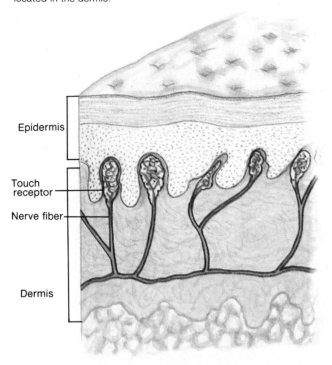

Epidermis

Touch receptor

Nerve fiber

Dermis

Accessory Organs of the Skin

Hair Follicles

Hair is present on all skin surfaces except the palms, soles, lips, nipples, and various parts of the external reproductive organs; however, it is not always well developed. For example, it is very fine on the forehead and the anterior surface of the arm.

Each hair develops from a group of epidermal cells at the base of a tubelike depression called a **hair follicle**. This follicle extends from the surface into the dermis and may pass into the subcutaneous layer. The epidermal cells at its base receive nourishment from dermal blood vessels that occur in a projection of connective tissue (dermal papilla) at the base of the follicle. As these epidermal cells divide and grow, older cells are pushed toward the surface. The cells that move upward and away from the nutrient supply become keratinized and die. Their remains constitute the shaft of a developing hair. In other words, a hair is composed of dead epidermal cells (figures 6.5 and 6.6).

Usually a hair grows for a time and then undergoes a resting period during which it remains anchored in its follicle. Later a new hair begins to grow from the base of the follicle, and the old hair is pushed outward and drops off. Sometimes, however, the hairs are not replaced. In the scalp, the result is baldness (alopecia). This condition is commonly due to heredity and is most likely to occur in males.

Hair color is determined by the type and amount of pigment produced by epidermal melanocytes at the base of the hair follicles. For example, dark hair contains an abundance of melanin, while blond hair contains an intermediate quantity; white hair of an albino person lacks this pigment. Bright red hair contains an iron pigment that does not occur in hair of any other color. A mixture of pigmented and unpigmented hair usually appears gray.

A bundle of smooth muscle cells, forming the *arrector pili muscle* (figure 6.1), is attached to each hair follicle. This muscle is positioned so that the hair within the follicle stands on end when the muscle contracts. If a person is emotionally upset or very cold, nerve impulses may stimulate the arrector pili muscles to contract, causing gooseflesh or goose bumps. Each hair follicle also has one or more sebaceous glands associated with it.

Sebaceous Glands

Sebaceous glands (figure 6.1) are holocrine glands. Their cells produce globules of a fatty material that accumulates, causing the cells to swell and burst. The resulting mixture of fatty material and cellular debris is called *sebum*.

The adipose tissue of the subcutaneous layer functions as a heat insulator—helping to conserve body heat and impeding the entrance of heat from the outside. The amount of adipose tissue varies greatly with each individual's nutritional condition. It also varies in thickness from one region to another. For example, it is usually quite thick over the abdomen and is absent in the eyelids.

The subcutaneous layer also contains the major blood vessels that supply the skin. Branches of these vessels form a network (rete cutaneum) between the dermis and the subcutaneous layer. They, in turn, give off smaller vessels that supply the dermis above and the underlying adipose tissue.

Subcutaneous injections are administered into the layer beneath the skin. *Intradermal injections,* on the other hand, are injected between layers of tissues within the skin. Subcutaneous injections and intramuscular injections, which are administered into muscles, are sometimes called hypodermic injections.

---

1. What kinds of tissues make up the dermis?
2. What are the functions of these tissues?
3. What is the function of the subcutaneous layer?

Figure 6.5
(*a*) A hair grows from the base of a hair follicle when epidermal cells undergo cell division and older cells move outward and become keratinized. (*b*) What features can you identify in this micrograph of a hair follicle?

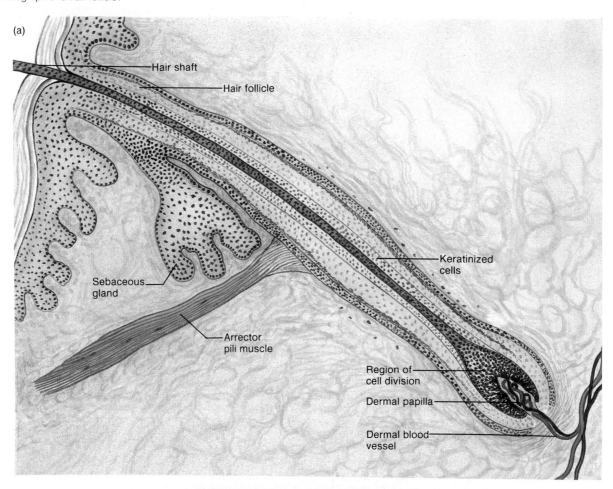

(a)

Hair shaft

Hair follicle

Keratinized cells

Sebaceous gland

Arrector pili muscle

Region of cell division

Dermal papilla

Dermal blood vessel

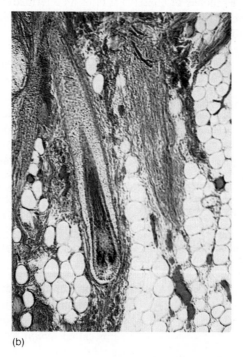

(b)

Figure 6.6
Note how the hairs penetrate the surface layer in this scanning electron micrograph of skin.

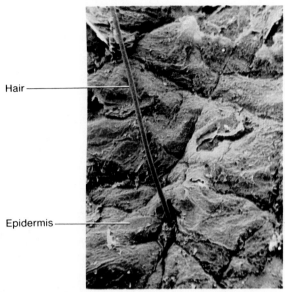

Hair

Epidermis

Figure 6.7
Sebaceous glands secrete sebum through short ducts into a hair follicle (shown here in cross section).

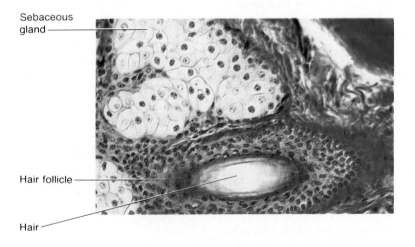

Sebaceous gland

Hair follicle

Hair

Sebum is secreted into hair follicles through short ducts and helps to keep the hairs and the skin soft, pliable, and relatively waterproof (figure 6.7).

Sebaceous glands are found scattered throughout the skin, except on the palms and soles where they are lacking. In some regions, such as on the lips, in the corners of the mouth, and on various parts of the external reproductive organs, they open directly to the surface of the skin rather than being connected to hair follicles.

A disorder of the sebaceous glands is responsible for the condition called *acne* (acne vulgaris), which is common in adolescents. In this condition, the glands become overactive and inflamed in some body regions. At the same time, their ducts may become plugged, and the inflamed glands may be surrounded by small red elevations containing blackheads (comedones) or pimples (pustules).

Figure 6.8

A nail is produced by epithelial cells that reproduce and undergo keratinization in the lunula region of the nail.

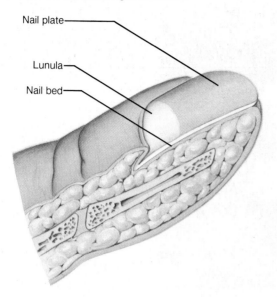

Nails

Nails are protective coverings on the ends of fingers and toes. Each nail consists of a *nail plate* that overlies a surface of skin called the *nail bed*. The nail plate is produced by specialized epithelial cells that are continuous with the epithelium of the skin. The whitish, half-moon-shaped region (lunula) at the base of a nail plate is its most active growing region. The epithelial cells in this thickened region reproduce, and the newly formed cells undergo keratinization. This gives rise to tiny, horny scales that become part of the nail plate, pushing it forward over the nail bed. In time the plate extends beyond the end of the nail bed and is gradually worn away (figure 6.8).

Sweat Glands

Sweat glands (figure 6.1) occur in nearly all regions of the skin, but are most numerous in the palms and soles. Each gland consists of a tiny tube that originates as a ball-shaped coil in the dermis or subcutaneous layer. The coiled portion of the gland is closed at its deep end and is lined with sweat-secreting epithelial cells.

Some sweat glands, the *apocrine glands,* respond to emotional stress. Apocrine secretions typically have odors, and the glands are considered to be scent glands. They begin to function at puberty and are responsible for some skin regions becoming moist when a person is emotionally upset, frightened, or experiencing pain. They also become active when a person is sexually stimulated.

Figure 6.9

How do the functions of eccrine sweat glands and apocrine sweat glands differ?

In adults the apocrine glands are most numerous in the armpits (axillary regions), groin, and in the regions around the nipples. They are usually associated with hair follicles.

Other sweat glands, the *eccrine glands,* are not connected to hair follicles. They function throughout life by responding to elevated body temperature due to environmental heat or physical exercise. These glands are common on the forehead, neck, and back, where they produce profuse sweating on hot days and when a person is physically active. They also are responsible for the moisture that may appear on the palms and soles when a person is emotionally stressed.

Figure 6.10

The tubular portion of the eccrine gland in this micrograph follows a very irregular path.

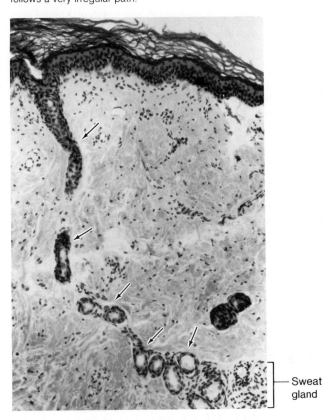

Sweat gland

The fluid secreted by these glands is carried away by a tubular part that opens at the surface as a pore. Although it is mostly water, the fluid contains small quantities of salts and certain wastes, such as urea and uric acid. Thus, the secretion of sweat is, to a limited degree, an excretory function (figures 6.9 and 6.10).

With advancing age, there is a reduction in sweat gland activity, and in the very old, sweat glands may be replaced by fibrous tissues. Similarly, there is a decrease in sebaceous gland activity with age, so that the skin of elderly people tends to be dry and lack oils.

---

1. Explain how a hair forms.
2. What causes gooseflesh?
3. What is the function of the sebaceous glands?
4. How does the composition of a fingernail differ from that of a hair?
5. Describe the functions of the sweat glands.

Regulation of Body Temperature

The regulation of body temperature is vitally important because metabolic processes involve chemical reactions, whose rates are affected by heat conditions. As a result, even slight shifts in body temperature can disrupt the rates of such reactions and produce metabolic disorders.

Normally, the temperature of deeper body parts remains close to 37°C (98.6°F). The maintenance of a stable temperature requires that the amount of heat lost by the body be balanced by the amount it produces. The skin plays a key role in the homeostatic mechanism associated with the regulation of body temperature.

Heat Production and Loss

Heat is a by-product of cellular respiration. Thus, the more active cells are the major heat producers. These include muscle cells and the cells of certain glands such as the liver.

When heat production is excessive, nerve impulses stimulate structures in the skin and other organs to react in ways that promote heat loss. For example, during physical exercise, active muscles release heat, and the blood carries away the excess. When the warmed blood reaches the part of the brain that functions as the body's thermostat (hypothalamus), this part signals muscles in the walls of specialized dermal blood vessels to relax. As the vessels dilate (vasodilation), more blood enters them and some of the heat it carries is likely to escape to the outside. At the same time various deeper blood vessels are caused to contract (vasoconstriction), diverting blood to the surface, and the heart is stimulated to beat faster, causing more blood to move out of the deeper regions.

The primary means of body heat loss is **radiation,** by which heat in the form of *infrared heat rays* escapes from warmer surfaces to cooler surroundings. These rays radiate in all directions, much like those from the bulb of a heat lamp.

Lesser amounts of heat are lost by conduction and convection. In **conduction,** heat moves from the body directly into the molecules of cooler objects in contact with its surface. For example, heat is lost by conduction into the seat of a chair when a person sits down. The heat loss continues as long as the chair is cooler than the body surface in contact with it.

Figure 6.11
Some actions involved in body temperature regulation.

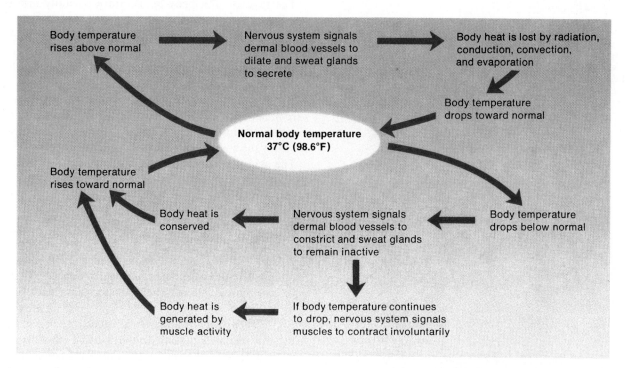

Heat is also lost by conduction to the air molecules that contact the body. As air becomes heated it moves away from the body, carrying heat with it, and is replaced by cooler air moving inward. This type of continuous circulation of air over a warm surface is called **convection.**

Still another means of body heat loss is **evaporation.** When the body temperature is rising above normal, the nervous system stimulates eccrine sweat glands to become active and to release fluid onto the surface of the skin. As this fluid evaporates (changes from a liquid to a gas), it carries heat away from the surface, cooling the skin. The cooling effect of evaporation can be experienced by rubbing some water on the skin and waiting for the water to disappear. Cooling by evaporation is also used when a feverish patient receives a sponge bath to help lower the body temperature.

Whenever sweat is produced profusely, there is danger that an *electrolyte imbalance* may develop. This can occur because sweat contains certain essential electrolytes, and if they are lost in excess, a person may experience an unpleasant condition called heat exhaustion. Its symptoms include dizziness, headache, muscle cramps, and nausea, and they are treated by providing needed salts to restore electrolyte balance.

If heat is lost excessively, as may occur in a very cold environment, the brain triggers different responses in the skin structures. For example, muscles in the walls of dermal blood vessels are stimulated to contract; this decreases the flow of heat-carrying blood through the skin and helps to reduce heat loss by radiation, conduction, and convection. At the same time, the sweat glands remain inactive, decreasing heat loss by evaporation. If the body temperature continues to drop, the nervous system may stimulate muscle fibers in the skeletal muscles throughout the body to contract slightly. This action requires an increase in the rate of cellular respiration and produces heat as a by-product. If this response does not raise the body temperature to normal, small groups of muscles may be stimulated to contract rhythmically with still greater force, and the person begins to shiver, generating more heat.

Figure 6.11 summarizes some of the actions involved in the body temperature-regulating mechanism.

Problems in Temperature Regulation

Unfortunately, the body's temperature-regulating mechanism does not always operate satisfactorily, and the consequences may be unpleasant and dangerous. For example, since air can hold only a limited amount of water vapor, on a hot, humid day the air may become saturated with it. At such times, the sweat glands may be activated, but the sweat will not evaporate. The skin becomes wet, but the person remains hot and uncomfortable. In addition, if the air temperature is high, heat is lost less effectively by radiation. In fact, if the air temperature is higher than the body temperature, the person may gain heat from the surroundings, and the body temperature may rise even higher.

Figure 6.12
An albino lacks the ability to produce the skin pigment melanin
because of the presence of a mutant gene.

1. Why is the regulation of body temperature so
   important?
2. How is body heat produced?
3. By what means is heat lost from the body?
4. How does the skin help to regulate body
   temperature?

Skin Color

Skin color, like hair color, is due largely to the presence
or absence of pigment produced by epidermal melanocytes. The amount of pigment synthesized by these
cells is influenced by genetic, environmental, and physiological factors.

Genetic Factors

Regardless of racial origin, all humans possess about
the same concentrations of melanocytes in their skins.
Differences in skin coloration are due largely to differences in the amount of melanin these cells produce,
which is directed by the genes each person inherits.
Those whose genes cause a relatively large amount of
pigment to be produced have dark skins, while those
whose genes cause less pigment to be formed have
lighter complexions. Still others inherit mutant genes,
and their cells are unable to manufacture melanin. As
a consequence, their skins remain nonpigmented, and
they are termed *albinos* (figure 6.12).

Skin color also is influenced by the distribution and the
size of pigment granules within the melanocytes. The
granules of very dark skin tend to occur singly and are
relatively large; those in lighter skin tend to occur in clusters of 2–4 granules and are somewhat smaller.

Environmental Factors

Environmental factors such as sunlight, ultraviolet light
from sunlamps, or X rays also affect skin color. These
factors cause existing melanin to darken rapidly, and
they stimulate melanocytes to produce more pigment
and transfer it to nearby epidermal cells within a few
days. This is why sunbathing results in skin tanning.
Unless the exposure to sunlight is continued, however,
the tan is eventually lost, as pigmented epidermal cells
become keratinized and are worn away.

Physiological Factors

Blood in the dermal vessels causes additional skin color.
For example, when the blood is well oxygenated, the
blood pigment *hemoglobin* is bright red, making the
skins of light-complexioned persons and albinos appear
pinkish. On the other hand, when the blood oxygen
concentration is low, hemoglobin is dark red and the
skin appears bluish—a condition called *cyanosis*.

The state of the blood vessels also has an effect
on skin color. If the vessels are dilated, more blood enters the dermis, and the skin becomes redder than usual.
This may happen when a person is overheated, embar-
rassed, or under the influence of alcohol. Conversely,
conditions that produce blood vessel constriction cause
blanching of the skin. Thus, if the body temperature is
dropping abnormally or if a person is frightened, the
skin may appear pale.

The presence of a yellow-orange plant pigment
called *carotene*, which is especially common in yellow
vegetables, may cause the skin to have a yellowish color.
The yellowish color results from the accumulation of
carotene in the adipose tissue of the subcutaneous layer.

Various diseases may affect skin color. In certain
forms of liver disease, bile pigments may occur in the
skin and cause it to appear yellowish (jaundiced). A
person with anemia may have pale skin due to a decrease in the concentrations of hemoglobin in blood
cells.

Some middle-aged and older people have a chronic skin
disorder called *acne rosacea*. This condition is characterized by redness of the nose, cheeks, chin, and forehead that results from persistently dilated dermal blood
vessels. It is sometimes accompanied by eruptions similar to those associated with adolescent acne. Although
acne rosacea seems to be inherited, the problem may
be aggravated by certain foods and beverages, including alcohol.

1. What factors influence skin color?
2. Which of these factors are genetic? Which are
   environmental?

Healing of Wounds and Burns
A Clinical Application

As described in chapter 5, **inflammation** is a normal response to injury or stress in which blood vessels in the affected tissues become dilated. At the same time, the permeability of these vessels increases, and there is a tendency for excessive amounts of fluids to leave the blood vessels and enter the damaged tissues. Thus, when skin is injured, it may become reddened, swollen, warm, and painful to touch. On the other hand, the dilated blood vessels provide more nutrients and oxygen to the tissues, which aid the healing process.

The process by which damaged skin heals depends on the extent of the injury. If a break in the skin is shallow, epithelial cells along its margin are stimulated to reproduce more rapidly than usual, and the newly formed cells simply fill the gap.

If the injury extends into the dermis or subcutaneous layer, blood vessels are broken, and the blood that escapes forms a clot in the wound. Tissue fluids seep into the area and become dried. The blood clot and the dried fluids form a scab that covers and protects the damaged tissues. Before long, fibroblasts from connective tissue at the wound margins migrate into the region and begin forming new collagenous fibers. These fibers tend to bind the edges of the wound together. The closer the edges of the wound, the sooner the gap can be filled; this is one reason for suturing or otherwise closing a large break in the skin.

As the healing process continues, blood vessels send out new branches that grow into the area beneath the scab. Phagocytic cells remove dead cells and other debris. Eventually the damaged tissues are replaced, and the scab sloughs off. If the wound was extensive, the newly formed connective tissue may appear on the surface as a scar.

In large, open wounds, the healing process may be accompanied by the formation of *granulations* that develop in the exposed tissues as small, rounded masses. Each of these granulations consists of a new branch of a blood vessel and a cluster of connective tissue cells that are being nourished by the vessel. Figure 6.13 shows the stages in the healing of a wound.

As with other types of injuries, the skin's response to a burn depends upon the amount of damage it sustains. If the skin is only slightly burned, for example, from a minor sunburn, the skin may become warm and reddened (erythema) as a result of dermal blood vessel dilation. This response may be accompanied by mild edema, and in time the surface layer of skin may be shed. Such a burn that involves injury to the epidermis alone is called a *superficial partial-thickness* (first degree) *burn.* Healing of the injury usually occurs within a few days to two weeks, and there is no scarring of the skin.

A burn that involves destruction of some epidermis as well as some underlying dermis is called a *deep partial-thickness* (second degree) *burn.* In this condition, fluid escapes from damaged dermal capillaries, and as the fluid accumulates beneath the outer layer of epidermal cells, blisters appear. The injured region becomes moist and firm, and it may appear dark red to waxy white. Such a burn most commonly occurs as a result of exposure to hot objects, hot liquids, flames, or burning clothing.

The healing of a deep partial-thickness burn involves accessory organs of the skin that survive the injury because they are located in deeper portions of the dermis. These organs, which include hair follicles, sweat glands, and sebaceous glands, contain epithelial cells. During healing these cells may be able to grow out onto the surface of the dermis, spread over it, and form a new epidermis. In time recovery of the skin usually is complete, and scar tissue does not develop unless an infection occurs.

A burn that destroys the epidermis, dermis, and the accessory organs of the skin is called a *full-thickness* (third degree) *burn.* In this instance, the injured skin becomes dry and leathery, and its color may vary from red to black to white.

A full-thickness burn usually occurs as a result of prolonged exposure to hot objects, flames, or corrosive chemicals, or to immersion in hot liquids. Since most of the epithelial cells in the affected region are likely to be destroyed, spontaneous healing can occur only by the growth of epithelial cells inward from its margin. If the injury is large, it may be necessary to remove a thin layer of skin from an unburned region of the body and transplant it in the injured area to speed healing. This procedure is called an *autograft.*

If the burn is too extensive to allow removal of skin from the patient, cadaveric skin from a skin bank or various skin substitutes may be used to cover the injury. In this case the transplant is called a *homograft,* and it serves as a temporary covering that decreases the size of the wound, helps to prevent infection, and aids in preserving deeper tissues. In time, after healing has begun, the temporary covering may be removed and replaced with an autograft, as skin becomes available in areas that have healed. The healing of wounds covered by grafts, however, is likely to be accompanied by extensive scarring.

Figure 6.13

(a) If normal skin is (b) injured deeply, (c) blood escapes from dermal blood vessels and (d) a blood clot soon forms. (e) The blood clot and dried tissue fluid form a scab that protects the damaged region. (f) Later, blood vessels send out branches, and fibroblasts migrate into the area. (g) The fibroblasts produce new connective tissue fibers, and when the skin is largely repaired, the scab sloughs off.

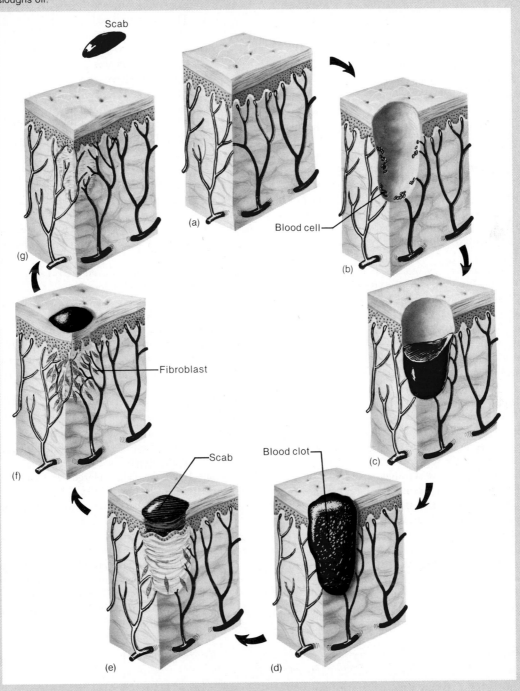

Scab

Blood cell

(a)

(b)

Blood clot

Blood clot

(c)

Fibroblast

Scab

(f)

(g)

(e)

(d)

Common Skin Disorders

acne (ak'ne) a disease of the sebaceous glands accompanied by blackheads and pimples.

alopecia (al″o-pe'she-ah) loss of hair.

athlete's foot (ath'-lētz foot) a fungus infection (tinea pedis) usually involving the skin of the toes and soles.

birthmark (berth' mark) vascular tumor involving the skin and subcutaneous tissues.

boil (boil) a bacterial infection (furuncle) involving a hair follicle and/or sebaceous glands.

carbuncle (kar' bung-kl) a bacterial infection similar to a boil, spreading into the subcutaneous tissues.

cyst (sist) a liquid-filled sac or capsule.

dermatitis (der″mah-ti'tis) an inflammation of the skin.

eczema (ek'zĕ-mah) a noncontagious skin rash often accompanied by itching, blistering, and scaling.

erythema (er″ĭ-the'mah) reddening of the skin, due to dilation of dermal blood vessels, in response to injury or inflammation.

herpes (her'pēz) an infectious disease of the skin usually caused by the virus herpes simplex and characterized by recurring formations of small clusters of vesicles.

impetigo (im″pĕ-ti'go) a contagious disease of bacterial origin characterized by pustules that rupture and become covered with loosely held crusts.

keloids (ke'loid) fibrous tumors usually initiated by an injury.

mole (mol) skin tumor (nevi) that usually is pigmented; colors range from yellow to brown or black.

pediculosis (pĕ-dik″u-lo'sis) a disease produced by infestation of lice.

pruritus (proo-ri'tus) itching of the skin without eruptions.

psoriasis (so-ri'ah-sis) a chronic skin disease characterized by red patches covered with silvery scales.

pustule (pus'tūl) elevated, pus-filled area.

scabies (ska'bēz) a disease resulting from an infestation of mites.

seborrhea (seb″o-re'ah) a disease characterized by hyperactivity of the sebaceous glands and accompanied by greasy skin and dandruff.

shingles (shing'gelz) disease caused by a viral infection (herpes zoster) of certain spinal nerves, characterized by severe localized pain and blisters in the skin areas supplied by the affected nerves.

ulcer (ul'ser) an open sore.

wart (wort) a flesh-colored raised area caused by a virus infection.

Chapter Summary

Introduction

Organs are composed of two or more kinds of tissues. The skin together with its accessory organs constitutes the integumentary system.

Types of Membranes

1. Serous membranes
   a. Serous membranes are organs that line body cavities lacking openings to the outside.
   b. They are composed of epithelium and loose connective tissue.
   c. They secrete watery serous fluid that lubricates membrane surfaces.
2. Mucous membranes
   a. Mucous membranes are organs that line cavities and tubes opening to the outside.
   b. They are composed of various kinds of epithelium and loose connective tissue.
   c. They secrete mucus.
3. The cutaneous membrane is the skin.
4. Synovial membranes
   a. Synovial membranes are organs that line joint cavities.
   b. They are composed of fibrous connective tissue overlying loose connective tissue and adipose tissue.
   c. They secrete synovial fluid that lubricates the ends of bones at joints.

The Skin and Its Tissues

Skin functions as a protective covering, aids in regulating body temperature, houses sensory receptors, synthesizes various chemicals, and excretes wastes. It is composed of an epidermis and a dermis, and has a subcutaneous layer beneath.

1. The epidermis
   a. The epidermis is a layer composed of stratified squamous epithelium, which lacks blood vessels.
   b. The deepest layer, called stratum basale, contains cells undergoing mitosis.
   c. Epidermal cells undergo keratinization as they are pushed toward the surface.
   d. The outermost layer, stratum corneum, is composed of dead epidermal cells.
   e. The production of epidermal cells is balanced with the rate at which they are lost.
   f. Epidermis functions to protect underlying tissues against water loss, mechanical injury, and the effects of harmful chemicals.
   g. Melanin protects underlying cells from the effects of ultraviolet light.
   h. Melanocytes transfer melanin to nearby epidermal cells.

2. The dermis
   a. The dermis is a layer composed largely of fibrous connective tissue that binds the epidermis to underlying tissues.
   b. It also contains muscle fibers, blood vessels, and nerve fibers.
   c. Dermal blood vessels supply nutrients to all skin cells and help regulate body temperature.
   d. Nerve tissue is scattered through the dermis.
      (1) Some dermal nerve fibers carry impulses to muscles and glands of the skin.
      (2) Other dermal nerve fibers are associated with various sensory receptors in the skin.
3. The subcutaneous layer
   a. The subcutaneous layer is composed of loose connective tissue and adipose tissue.
   b. Adipose tissue helps to conserve body heat.
   c. This layer contains blood vessels that supply the skin.

Accessory Organs of the Skin

1. Hair follicles
   a. Hair occurs in nearly all regions of the skin.
   b. Each hair develops from epidermal cells at the base of a tubelike hair follicle.
   c. As newly formed cells develop and grow, older cells are pushed toward the surface and undergo keratinization.
   d. A hair usually grows for a while, undergoes a resting period, and is replaced by a new hair.
   e. Hair color is determined by the type and amount of pigment in its cells.
   f. A bundle of smooth muscle cells and one or more sebaceous glands are attached to each hair follicle.
2. Sebaceous glands
   a. Sebaceous glands secrete sebum, which helps keep skin and hair soft and waterproof.
   b. In some regions they open directly to the skin surface.
3. Nails
   a. Nails are protective covers on the ends of fingers and toes.
   b. They are produced by epidermal cells that undergo keratinization.
4. Sweat glands
   a. Sweat glands are located in nearly all regions of the skin.
   b. Each gland consists of a coiled tube.
   c. Apocrine glands respond to emotional stress, while eccrine glands respond to elevated body temperature.
   d. Sweat is primarily water, but also contains salts and waste products.

Regulation of Body Temperature

Regulation of body temperature is vital because heat affects the rates of metabolic reactions. Normal temperature of deeper body parts is about 37°C (98.6°F).

1. Heat production and loss
   a. Heat is a by-product of cellular respiration.
   b. When the body temperature rises above normal, more blood is caused to enter dermal blood vessels.
   c. Heat is lost to the outside by radiation, conduction, convection, and evaporation.
   d. Sweat gland activity increases heat loss by evaporation.
   e. If the body temperature drops below normal, dermal blood vessels constrict and sweat glands become inactive.
   f. During excessive heat loss, skeletal muscles are stimulated to contract involuntarily; this increases cellular respiration and produces additional heat.
2. Problems in temperature regulation
   a. Air holds a limited amount of water vapor.
   b. When the air is saturated with water, sweat may fail to evaporate and the body temperature may remain elevated.

Skin Color

All humans possess about the same concentration of melanocytes. Skin color is due largely to the amount of melanin in the epidermis.

1. Genetic factors
   a. Each person inherits genes for melanin production.
   b. Dark skin is due to genes that cause large amounts of melanin to be produced; lighter skin is due to genes that cause lesser amounts to form.
   c. Mutant genes may cause a lack of melanin in the skin.
2. Environmental factors
   a. Environmental factors that influence skin color include sunlight, ultraviolet light, and X ray.
   b. These factors cause existing melanin to darken and stimulate melanin production.
3. Physiological factors
   a. Blood in dermal vessels may cause the skin of light-complexioned persons to appear pinkish or bluish.
   b. Carotene in the subcutaneous layer may cause the skin to appear yellowish.
   c. Various diseases may affect skin color.

Application of Knowledge

1. What special problems would result from a loss of 50% of a person's functional skin surface? How might this person's environment be modified to compensate partially for such a loss?

2. A premature infant typically lacks subcutaneous adipose tissue. Also, the surface area of an infant's small body is relatively large compared to its volume. How do you think these factors influence such an infant's ability to regulate its body temperature?

Review Activities

1. Explain why a membrane is an organ.
2. Define *integumentary system*.
3. Distinguish between serous and mucous membranes.
4. Explain the function of synovial membranes.
5. List six functions of skin.
6. Distinguish between the epidermis and the dermis.
7. Describe the subcutaneous layer.
8. Explain what happens to epidermal cells as they undergo keratinization.
9. List the layers of the epidermis.
10. Describe the function of melanocytes.
11. Describe the structure of the dermis.
12. Review the functions of dermal nerve tissue.
13. Explain the function of the subcutaneous layer.
14. Distinguish between a hair and a hair follicle.
15. Review how hair color is determined.
16. Explain the function of sebaceous glands.
17. Describe how nails are formed.
18. Distinguish between apocrine and eccrine glands.
19. Explain the importance of body temperature regulation.
20. Describe the role of the skin in promoting the loss of excess body heat.
21. Explain how body heat is lost by radiation.
22. Distinguish between conduction and convection.
23. Describe the body's responses to decreasing body temperature.
24. Review how the humidity of the air may interfere with body temperature regulation.
25. Explain how environmental factors affect skin coloration.
26. Describe three physiological factors that affect the color of skin.

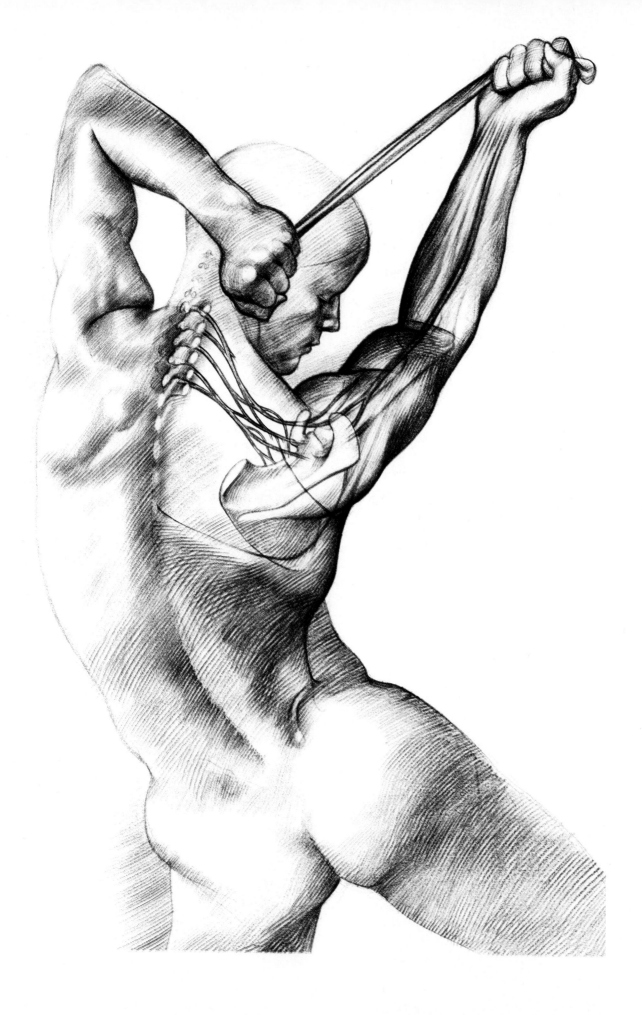

Unit 2

Support and Movement

The chapters of unit 2 deal with structures and functions of the skeletal and muscular systems. They describe how organs of the skeletal system support and protect other body parts and how they function with organs of the muscular system to enable body parts to move. They also describe how skeletal structures participate in the formation of blood and in the storage of inorganic salts, and how muscular tissues act to produce body heat and to move body fluids.

7
The Skeletal System

8
Joints of the Skeletal System

9
The Muscular System

7

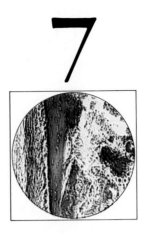

The Skeletal System

The bones of the skeleton are composed of several kinds of tissues and, thus, are the organs of the *skeletal system*.

Since bones are rigid structures, they provide support and protection for softer tissues, and they act together with skeletal muscles to make body movements possible. They also house the tissue that produces blood cells, and they store inorganic salts.

The shapes of individual bones are closely related to their functions. Projections provide places for the attachments of muscles, tendons, and ligaments; openings serve as passageways for blood vessels and nerves, and the ends of bones are modified to form joints with other bones.

Chapter Outline

Chapter Objectives

After you have studied this chapter, you should be able to

1. Classify bones according to their shapes and name
   an example from each group.

2. Describe the general structure of a bone
   and list the functions of its parts.

3. Distinguish between intramembranous and
   endochondral bones and explain how such bones
   grow and develop.

4. Describe the effects of sunlight, nutrition, hormonal
   secretions, and exercise on bone development.

5. Discuss the major functions of bones.

6. Distinguish between the axial and appendicular
   skeletons, and name the major parts of each.

7. Locate and identify the bones and the major features
   of the bones that comprise the skull, vertebral
   column, thoracic cage, pectoral girdle, upper limb,
   pelvic girdle, and lower limb.

8. Complete the review activities at the end of this
   chapter. Note that the items are worded in the form of
   specific learning objectives. You may want to refer to
   them before reading the chapter.

Key Terms

articular cartilage   (ar-tik′u-lar kar′tĭ-lij)

compact bone   (kom′pakt bōn)

diaphysis   (di-af′ĭ-sis)

endochondral bone   (en′do-kon′dral bōn)

epiphyseal disk   (ep′′ĭ-fiz′e-al disk)

epiphysis   (e-pif′ĭ-sis)

hematopoiesis   (hem′′ah-to-poi-e′sis)

intramembranous bone   (in′trah-mem′brah-nus bōn)

lever   (lev′er)

marrow   (mar′o)

medullary cavity   (med′u-lār′′e kav′ĭ-te)

osteoblast   (os′te-o-blast)

osteoclast   (os′te-o-klast)

osteocyte   (os′te-o-sīt)

osteon   (os′te-on)

periosteum   (per′′e-os′te-um)

spongy bone   (spun′je bōn)

Aids to Understanding Words

acetabul-, vinegar cup: *acetabul*um—depression of the coxal bone that articulates with the head of the femur.

ax-, an axis: *ax*ial skeleton—upright portion of the skeleton that supports the head, neck, and trunk.

-blast, budding or developing: osteo*blast*—cell that forms bone tissue.

carp-, wrist: *carp*als—bones of the wrist.

-clast, broken: osteo*clast*—cell that breaks down bone tissue.

condyl-, knob: *condyl*e—a rounded, bony process.

corac-, beaklike: *corac*oid process—beaklike process of the scapula.

cribr-, sievelike: *cribr*iform plate—portion of the ethmoid bone with many small openings.

crist-, ridge: *crist*a galli—a bony ridge that projects upward into the cranial cavity.

fov-, pit: *fov*ea capitis—a pit in the head of a femur.

gladi-, sword: *gladi*olus—middle portion of the bladelike sternum.

glen-, joint socket: *glen*oid cavity—a depression in the scapula that articulates with the head of the humerus.

meat-, passage: auditory *meat*us—canal of the temporal bone that leads inward to parts of the ear.

odont-, tooth: *odont*oid process—a toothlike process of the second cervical vertebra.

poie-, making: hemato*poie*sis—process by which blood cells are formed.

A N INDIVIDUAL BONE is composed of a variety of tissues, including bone tissue, cartilage, fibrous connective tissue, blood, and nerve tissue. Because there is so much nonliving material present in the matrix of bone tissue, the whole organ may appear to be inert. A bone, however, contains very active tissues.

Bone Structure

Although the various bones of the skeletal system vary greatly in size and shape, they are similar in their structure, the way they develop, and the functions they provide.

Classification of Bones

Bones can be classified according to their shapes as long, short, flat, and irregular (figure 7.1).

1. **Long bones** have long longitudinal axes and expanded ends, such as those of arm and leg bones.
2. **Short bones** are somewhat cubelike with their lengths and widths roughly equal. The bones of the wrists and ankles, for example, are of this type.
3. **Flat bones** are platelike structures with broad surfaces, such as the ribs, the scapulae, and the bones of the skull that form a protective wall around the brain.
4. **Irregular bones** have a variety of shapes and are usually connected to several other bones. Irregular bones include the vertebrae that comprise the backbone and many of the facial bones.

In addition to these four groups of bones, some authorities recognize a fifth group called the *round* or *sesamoid* bones. The members of this group are usually small, and they often occur in tendons adjacent to joints, where tendons undergo compression. The kneecap (patella) is an example of a very large sesamoid bone.

Parts of a Long Bone

In describing the structure of bone, a long bone, such as one in an arm or leg, will be used as an example (figure 7.2). At each end of such a bone there is an expanded portion called an **epiphysis,** which articulates

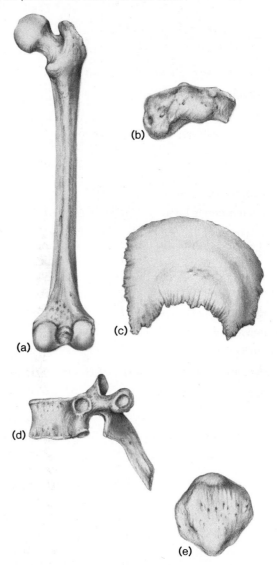

Figure 7.1
(*a*) A femur of the leg is a long bone, (*b*) a tarsal bone of the ankle is a short bone, (*c*) a parietal bone of the skull is a flat bone, (*d*) a vertebra of the backbone is an irregular bone, and (*e*) the patella of the knee is a round bone.

or forms a joint with another bone. On its outer surface, the articulating portion of the epiphysis is coated with a layer of hyaline cartilage called **articular cartilage.** The shaft of the bone, which is located between the epiphyses, is called the **diaphysis.**

Except for the articular cartilage on its ends, the bone is completely enclosed by a tough, vascular covering of fibrous tissue called the **periosteum.** This membrane is firmly attached to the bone, and its fibers are continuous with various ligaments and tendons that are connected to it. The periosteum also functions in the formation and repair of bone tissue.

Figure 7.2

The major parts of a long bone.

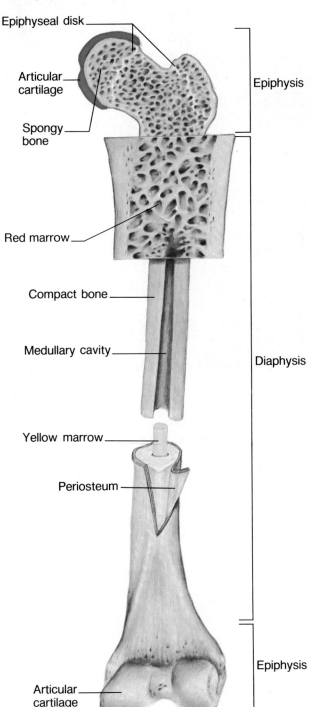

Epiphyseal disk

Articular cartilage

Spongy bone

Red marrow

Compact bone

Medullary cavity

Yellow marrow

Periosteum

Articular cartilage

Epiphysis

Diaphysis

Epiphysis

Each bone has a shape closely related to its functions. Bony projections called *processes*, for example, provide sites for the attachment of ligaments and tendons; grooves and openings serve as passageways for blood vessels and nerves; a depression of one bone might articulate with a process of another.

The wall of the diaphysis is composed of tightly packed tissue called *compact bone*. This type of bone is solid, strong, and resistant to bending.

The epiphyses, on the other hand, are composed largely of *spongy* (cancellous) *bone* with thin layers of compact bone on their surfaces. Spongy bone consists of numerous branching bony plates called *trabeculae*. Irregular interconnecting spaces occur between these plates and help reduce the weight of the bone. Spongy bone provides strength and its bony plates are most highly developed in the regions of the epiphyses that are subjected to forces of compression.

Both compact and spongy tissues usually are present in each bone. Short, flat, and irregular bones typically consist of a mass of spongy bone that is either covered by a layer of compact bone or sandwiched between plates of compact bone.

Compact bone in the diaphysis of a long bone forms a rigid tube with a hollow chamber called the **medullary cavity.** This cavity is continuous with the spaces of the spongy bone. All of these areas are lined with a thin layer of squamous cells called *endosteum* and are filled with a specialized type of soft connective tissue called *marrow*.

Microscopic Structure

As was discussed in chapter 5, bone cells (osteocytes) are located in minute, bony chambers called *lacunae* which are arranged in concentric circles around *osteonic canals* (haversian canals). These cells communicate with nearby cells by means of cellular processes passing through canaliculi. The intercellular material of bone tissue is largely collagen, which gives it strength and resilience, and inorganic salts, which make it hard and resistant to crushing.

In compact bone, the osteocytes and layers of intercellular material clustered concentrically around an osteonic canal form a cylinder-shaped unit called an **osteon** (haversian system). Many of these units cemented together form the substance of compact bone (figure 7.3).

Figure 7.3

What features do you recognize in this scanning electron micrograph of a single osteon in compact bone (about 1300×)? (*Tissues and Organs: A Text-Atlas of Scanning Electron Microscopy,* by R. G. Kessel and R. H. Kardon. © 1979 W. H. Freeman and Company.)

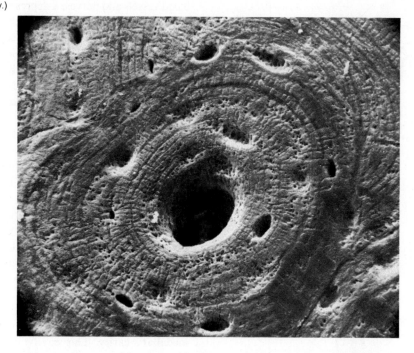

Figure 7.4

Compact bone is composed of osteons cemented together.

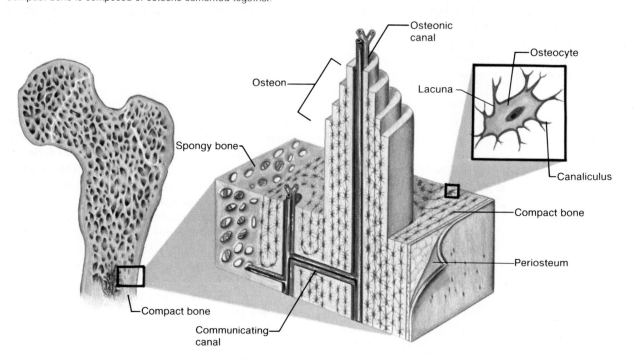

Each osteonic canal contains one or two small blood vessels (usually capillaries) surrounded by some loose connective tissue. Blood in these vessels provides nourishment for the cells associated with the osteonic canal.

Osteonic canals travel longitudinally through bone tissue. They are interconnected by transverse communicating canals (*Volkmann's canals*). These canals contain larger blood vessels by which the vessels in the osteonic canals communicate with the surface of the bone and the medullary cavity (figure 7.4).

Although spongy bone is also composed of osteocytes and intercellular material, the trabeculae are usually very thin, and the cells are not arranged around osteonic canals. They are nourished by diffusion of substances into canaliculi that lead from the bone cells to the surface of the bony plates.

1. Explain how bones are classified.
2. List five major parts of a long bone.
3. How do compact and spongy bone differ in structure?
4. Describe the microscopic structure of a bone.

Bone Development and Growth

Although parts of the skeletal system begin to form during the first few weeks of prenatal development, bony structures continue to grow and develop into adulthood. These bones form by the replacement of existing connective tissue, in one of two ways. Some first appear between sheetlike layers of connective tissues, and are called intramembranous bones. Others begin as masses of cartilage that are later replaced by bone tissue. They are called endochondral bones (figure 7.5).

Intramembranous Bones

The **intramembranous bones** include the broad, flat bones of the skull. During their development (osteogenesis), membranelike layers of primitive connective tissues appear in the sites of the future bones. These layers are supplied with dense networks of blood vessels, and some of the connective tissue cells arrange themselves around these vessels. These primitive cells enlarge and become differentiated into bone-forming cells, called **osteoblasts,** which in turn deposit bony matrix around themselves. As a result, spongy bone is produced in all directions along the blood vessels within the layers of primitive connective tissues. Later, some of the spongy bone may be converted to compact bone as spaces become filled with bone matrix.

Figure 7.5
The tissues of this miscarried fetus have been cleared, and the developing bones have been stained selectively.

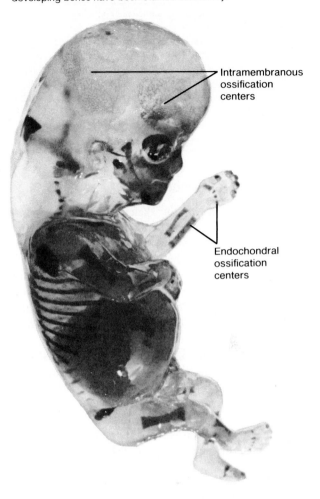

Intramembranous ossification centers

Endochondral ossification centers

As development continues, the osteoblasts may become completely surrounded by matrix, and in this manner they become secluded within lacunae. At the same time, matrix enclosing the cellular processes of the osteoblasts gives rise to canaliculi. Once they are isolated in lacunae, the bone cells are called **osteocytes.**

The cells of the primitive connective tissue that persist outside the developing bone give rise to the periosteum. Osteoblasts on the inside of the periosteum create a layer of compact bone over the surface of the newly formed spongy bone.

This process of forming an intramembranous bone by the replacement of connective tissue is called *intramembranous ossification.* Chart 7.1 lists the major steps of the process.

Chart 7.1	Major steps in bone development	
Intramembranous ossification	**Endochondral ossification**
1. Membranelike layers of primitive connective tissue appear in sites of future bones.	1. Masses of hyaline cartilage form models of future bones.
2. Primitive connective tissue cells become arranged around the blood vessels in these layers.	2. Cartilage tissue breaks down and disappears.
3. Connective tissue cells differentiate into osteoblasts, which form spongy bone.	3. Blood vessels and differentiating osteoblasts from periosteum invade the disintegrating tissue.
4. Osteoblasts become osteocytes when they are completely surrounded by bony matrix.	4. Osteoblasts form spongy bone in space occupied by cartilage.
5. Connective tissue on the surface of each developing structure forms a periosteum.	

Endochondral Bones

Most of the bones of the skeleton are **endochondral bones.** Their development (osteogenesis) proceeds from masses of hyaline cartilage with shapes similar to future bony structures. These cartilaginous models grow rapidly for a time, and then begin to undergo extensive changes. For example, cartilage cells enlarge and increase the sizes of their respective lacunae. This is accompanied by destruction of the surrounding matrix, and soon the cartilage cells die and degenerate.

About the same time, a periosteum forms from connective tissue on the outside surface of the developing structure, and as the cartilage breaks down, blood vessels and undifferentiated connective tissue cells invade the disintegrating tissue. Some of the invading cells differentiate into osteoblasts and begin to form spongy bone in the space previously occupied by the cartilage.

This process of forming an endochondral bone by the replacement of hyaline cartilage is called *endochondral ossification*. Its major steps are listed in chart 7.1.

Growth of an Endochondral Bone

In a long bone, replacement of hyaline cartilage by bony tissue begins in the center of the diaphysis. This region is called the *primary ossification center*, and bone develops from it toward the ends of the cartilaginous structure. Meanwhile, osteoblasts from the periosteum deposit a thin layer of compact bone around the primary ossification center by intramembranous ossification. The epiphyses of the developing bone remain cartilaginous and continue to grow. Later, *secondary ossification centers* appear in the epiphyses, and spongy bone forms in all directions from them. As spongy bone is deposited in the diaphysis and in the epiphysis, a band of cartilage, called the **epiphyseal disk,** is left between the two ossification centers.

The cartilaginous cells of the epiphyseal disk are arranged in four layers, each of which may be several cells thick, as shown in figure 7.6. The first layer, closest to the end of the epiphysis, is composed of resting cells. Although these cells are not actively participating in the growing process, this layer does anchor the epiphyseal disk to the bony tissue of the epiphysis.

The second layer contains rows of numerous young cells that are undergoing mitosis. As new daughter cells are produced, and as intercellular material is formed around them, the cartilaginous disk thickens.

The rows of older cells, which are left behind when new cells appear, form the third layer. These cells enlarge and cause the epiphyseal disk to thicken still more. Consequently, the length of the entire bone increases. At the same time, calcium salts accumulate in the intercellular substance of this layer, and the oldest of the enlarged cells begin to die.

The fourth layer of the epiphyseal disk is quite thin and is composed largely of dead cells and calcified intercellular substance. The intercellular substance nearest the diaphysis disintegrates. The resulting space is invaded by blood vessels and differentiating osteoblasts, which deposit bone tissue in place of the calcified cartilage.

A long bone will grow in length while the cartilaginous cells of the epiphyseal disk are active. However, once ossification centers of the diaphysis and epiphysis come together and the epiphyseal disk becomes ossified, growth in length is no longer possible in that end of the bone.

Growth in thickness, on the other hand, is due to deposition of compact bone by intramembranous ossification occurring on the outside, just beneath the periosteum. As this compact bone is forming on the surface of the structure, other bone tissue is being eroded away on the inside by the action of large, specialized cells called **osteoclasts.** These cells apparently resorb bone by secreting hydrolyzing enzymes (see

Figure 7.6
(a) The cartilaginous cells of an epiphyseal disk are arranged in four layers, each of which may be several cells thick. (b) A micrograph of an epiphyseal disk.

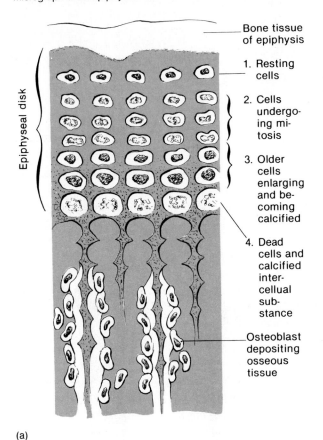

Bone tissue of epiphysis

1. Resting cells

2. Cells undergo-ing mi-tosis

3. Older cells enlarging and be-coming calcified

4. Dead cells and calcified inter-cellular sub-stance

Osteoblast depositing osseous tissue

Epiphyseal disk

(a)

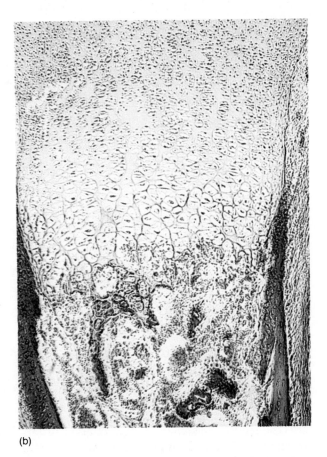

(b)

chapter 4). This action is responsible for creating the space that becomes the medullary cavity in the diaphysis and later fills with marrow.

The bone in the central regions of the epiphyses remains spongy, and hyaline cartilage on the ends of the epiphyses persists throughout life as articular cartilage. Figure 7.7 illustrates the stages in the development and growth of an endochondral bone. Chart 7.2 lists the ages at which various bones become ossified.

It is possible to determine whether a child's long bones are still growing by examining an X-ray photograph to see if the epiphyseal disks are present (figure 7.8). If a disk is damaged as a result of a fracture before it becomes ossified, elongation of the long bone may cease prematurely, or if growth continues, it may be uneven. For this reason, injuries to the epiphyses of a young person's bones are of special concern. On the other hand, an epiphysis is sometimes altered surgically in order to equalize growth of bones that are developing at very different rates.

Factors Affecting Bone Development and Growth

Bone development and growth are influenced by a number of factors, including nutrition, exposure to sunlight, hormonal secretions, and physical exercise. For example, vitamin D is necessary for proper absorption of calcium in the small intestine. In the absence of this vitamin, calcium is poorly absorbed, the inorganic salt portion of bone matrix is deficient in calcium, and the bones are likely to be deformed. In children this condition is called *rickets,* and in adults it is called *osteomalacia.*

Although vitamin D is relatively uncommon in natural foods, it is readily available in eggs and in milk and other dairy products that have been fortified with vitamin D. It also can be formed from a substance (dehydrocholesterol) that is produced by cells in the digestive tract or obtained in the diet. This substance is carried by the blood to the skin, and when it is exposed to ultraviolet light from the sun, it is converted to a compound that becomes vitamin D.

Chart 7.2 Ossification timetable

Age	Occurrence	Age	Occurrence
Third month of prenatal development	Ossification in long bones beginning	17 to 20 years	Bones of the upper limbs and scapulae become completely ossified.
Fourth month	Most primary ossification centers have appeared in the diaphyses of bones.	18 to 23 years	Bones of the lower limbs and coxal bones become completely ossified.
Birth to 5 years	Secondary ossification centers appear in the epiphyses.	23 to 25 years	Bones of the sternum, clavicles, and vertebrae become completely ossified.
5 to 12 years in females, or 5 to 14 years in males	Ossification is spreading rapidly from the ossification centers and various bones are becoming ossified.	By 25 years	Nearly all bones are completely ossified.

Figure 7.7

Major stages in the development of an endochondral bone.

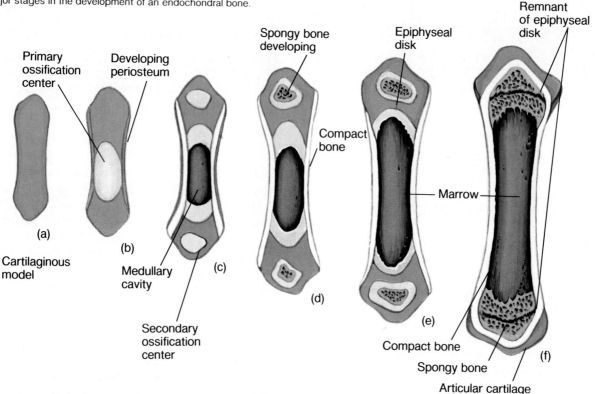

Vitamins A and C also are needed for normal bone development and growth. Vitamin A is necessary for bone resorption that occurs during normal development. Thus, a deficiency of vitamin A may result in retardation of bone development. Vitamin C is needed for the synthesis of collagen, so its lack also may inhibit bone development. In this case, osteoblasts produce less collagen in the intercellular material of the bone tissue, and the resulting bones are abnormally slender and fragile.

Hormones that affect bone development and growth include those secreted by the pituitary gland, thyroid gland, parathyroid glands, and ovaries or testes. The pituitary gland, for instance, secretes a chemical called **growth hormone,** which stimulates reproduction of cartilage cells in the epiphyseal disks. In the absence of this hormone, the long bones of the limbs fail to develop normally, and the child may become a pituitary dwarf. Such a person is very short, but has normal body proportions. If excessive amounts of growth hormone are released before the epiphyseal disks are ossified, the affected person may become a pituitary giant.

Figure 7.8
The presence of an epiphyseal disk (arrow) in a child's femur is
an indication that the bone is still growing in length.

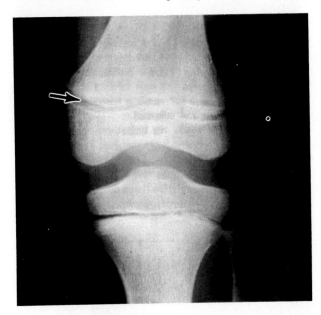

In an adult, excessive secretion of growth hormone causes a condition called *acromegaly*, in which the hands, feet, and jaw enlarge (see chapter 13).

Thyroid hormone stimulates the replacement of cartilage in the epiphyseal disks by bone tissue. Thus, an excessive secretion of this hormone can halt bone growth by causing premature ossification of the disks. A deficiency of thyroid hormone also may produce stunted growth, because it is necessary for the normal function of the pituitary gland. Without the normal effect of the thyroid hormone, the pituitary gland fails to secrete enough growth hormone, and the result is a person of small stature (see chapter 13).

Both male and female sex hormones (called androgens and estrogens, respectively) from the testes, ovaries, and adrenal glands promote the formation of bone tissue. Following puberty, when these hormones are being secreted abundantly, bones tend to grow considerably. However, sex hormones also stimulate ossification of the epiphyseal disks, and consequently they cause bones to stop growing in length at a relatively early age. The effect of estrogens on the disks is somewhat stronger than that of androgens. For this reason, females typically reach their maximum heights earlier than males.

Physical stress also has a stimulating effect on bone growth. For example, when skeletal muscles contract, they pull at their attachments on bones, and the resulting stress stimulates the bone tissue to thicken and strengthen (hypertrophy). Conversely, with lack of exercise, the same bone tissue undergoes a wasting process and tends to become thinner and weaker (atrophy). This is the reason that the bones of athletes usually are stronger and heavier than those of nonathletes and that the bones of casted limbs may decrease in size.

1. Describe the development of an intramembranous bone.
2. Explain how an endochondral bone develops.
3. List the steps in the growth of a long bone.
4. Explain how nutritional factors affect bone development.
5. What effects do hormones have on bone growth?
6. How does physical exercise affect bone structure?

Functions of Bones

Skeletal parts provide shape, support, and protection for body structures. They also act as levers that aid body movements, house tissues that produce blood cells, and store various inorganic salts.

Support and Protection

Bones give shape to structures such as the head, face, thorax, and limbs. They also provide support and protection. For example, the bones of the feet, legs, pelvis, and backbone support the weight of the body. The bones of the skull protect the eyes, ears, and brain. Those of the rib cage and shoulder girdle protect the heart and lungs, while bones of the pelvic girdle protect the lower abdominal and internal reproductive organs.

Lever Actions

Whenever limbs or other body parts are moved, bones and muscles function together as simple mechanical devices called *levers*. Such a lever has four basic components: (a) a rigid bar or rod, (b) a pivot or fulcrum on which the bar turns, (c) an object or weight that is moved, and (d) a force that supplies energy for the movement of the bar.

A playground seesaw is a lever. The board of the seesaw serves as a rigid bar that rocks on a pivot near its center. The person on one end of the board represents the weight that is moved, while the person at the opposite end supplies the force needed for moving the board and its rider.

Fractures
A Clinical Application

Although a **fracture** may involve injury to cartilaginous structures, it is usually defined as a break in a bone. A fracture can be classified according to its cause and the nature of the break sustained. For example, a break due to injury is a *traumatic* fracture, while one resulting from disease is a *spontaneous* or *pathologic* fracture.

If a broken bone is exposed to the outside by an opening in the skin, the injury is termed a *compound fracture*. Such a fracture is accompanied by the added danger of infection, since microorganisms almost surely enter through the broken skin. On the other hand, if the break is protected by uninjured skin, it is called a *simple fracture*. Figure 7.9 shows several types of fractures.

Repair of a Fracture

Whenever a bone is broken, blood vessels within the bone and its periosteum are ruptured, and the periosteum is likely to be torn. Blood escaping from the broken vessels spreads through the damaged area and soon forms a blood clot, or *hematoma*. As vessels in surrounding tissues dilate, these tissues become swollen and inflamed.

Within days or weeks the hematoma is invaded by developing blood vessels and large numbers of osteoblasts, originating from the periosteum. The osteoblasts multiply rapidly in the regions close to the new blood vessels, building spongy bone nearby. Granulation tissue develops, and in regions further from a blood supply, fibroblasts produce masses of fibrocartilage.

Figure 7.9
Various types of traumatic fractures.

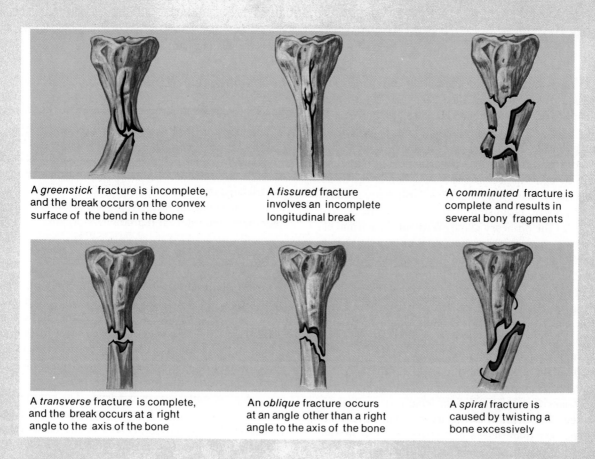

A *greenstick* fracture is incomplete, and the break occurs on the convex surface of the bend in the bone

A *fissured* fracture involves an incomplete longitudinal break

A *comminuted* fracture is complete and results in several bony fragments

A *transverse* fracture is complete, and the break occurs at a right angle to the axis of the bone

An *oblique* fracture occurs at an angle other than a right angle to the axis of the bone

A *spiral* fracture is caused by twisting a bone excessively

Meanwhile, phagocytic cells begin to remove the blood clot as well as any dead or damaged cells in the affected area. Osteoclasts also appear and resorb bone fragments, thus aiding in "cleaning up" debris.

In time, a large amount of fibrocartilage fills the gap between the ends of the broken bone, and this mass is termed a cartilaginous *callus*. The callus is later replaced by bone tissue in much the same way as the hyaline cartilage of a developing endochondral bone is replaced. That is, the cartilaginous callus is broken down, the area is invaded by blood vessels and osteoblasts, and the space is filled with a bony callus.

Usually more bone is produced at the site of a healing fracture than is needed to replace the damaged tissues. However, osteoclasts are able to remove the excess, and the final result of the repair process is a bone shaped very much like the original one. Figure 7.10 shows the steps in the healing of a fracture.

The rate at which a fracture is repaired depends on several factors. For instance, if the ends of the broken bone are close together, healing is more rapid than if they are far apart. This is the reason for setting fractured bones and for using casts or metal pins to keep the broken ends together. Also, some bones naturally heal more rapidly than others. The long bones of the arms, for example, may heal in half the time required by the leg bones. Furthermore, as age increases, so does the time required for healing.

Figure 7.10
Major steps in the repair of a fracture.

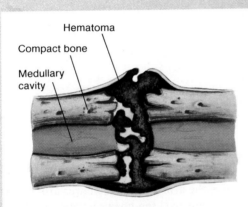

(a) Blood escapes from ruptured blood vessels and forms a hematoma

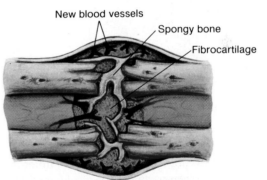

(b) Spongy bone forms in regions close to developing blood vessels, and fibrocartilage forms in more distant regions

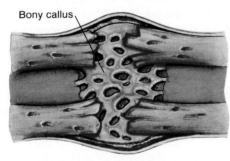

(c) Fibrocartilage is replaced by a bony callus

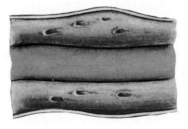

(d) Osteoclasts remove excess bony tissue, making new bone structure much like the original

Figure 7.11

Three types of levers: (a) A first-class lever is used in a pair of scissors, (b) a second-class lever is used in a wheelbarrow, and (c) a third-class lever is used in a pair of forceps.

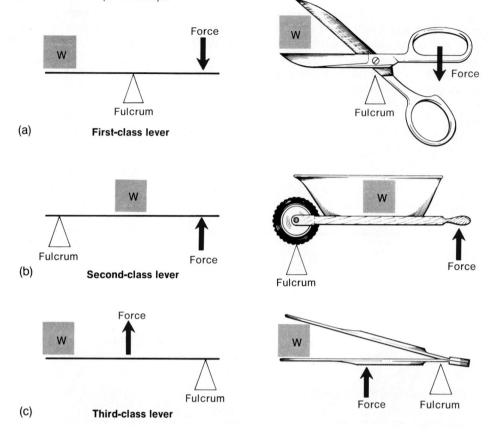

(a) **First-class lever**

(b) **Second-class lever**

(c) **Third-class lever**

There are three kinds of levers, and they differ in the arrangements of their parts, as shown in figure 7.11. A first-class lever is one whose parts are arranged like those of the seesaw. Its pivot is located between the weight and the force, making the sequence of parts weight-pivot-force. Other examples of first-class levers are scissors and hemostats (used to clamp blood vessels closed).

The parts of a second-class lever are arranged in the sequence pivot-weight-force, as in a wheelbarrow.

The parts of a third-class lever are arranged in the sequence pivot-force-weight. This type of lever is illustrated when eyebrow tweezers or forceps are used to grasp an object.

The actions of bending and straightening the arm at the elbow, for example, involve bones and muscles functioning together as levers, as illustrated in figure

7.12. When the arm is bent, the lower arm bones represent the rigid bar, the elbow joint is the pivot, the hand is the weight that is moved, and the force is supplied by muscles on the anterior side of the upper arm. One of these muscles, the *biceps brachii,* is attached by a tendon to a process on the *radius* bone in the lower arm, a short distance below the elbow. Since the parts of this lever are arranged in the sequence pivot-force-weight, it is an example of a third-class lever.

When the arm is straightened at the elbow, the lower arm bones again serve as a rigid bar, the hand as the weight, and the elbow joint as the pivot. However, this time the force is supplied by the *triceps brachii,* a muscle located on the posterior side of the upper arm. A tendon of this muscle is attached to a process of the *ulna* bone at the point of the elbow. Thus, the parts of the lever are arranged weight-pivot-force, and it is an example of a first-class lever.

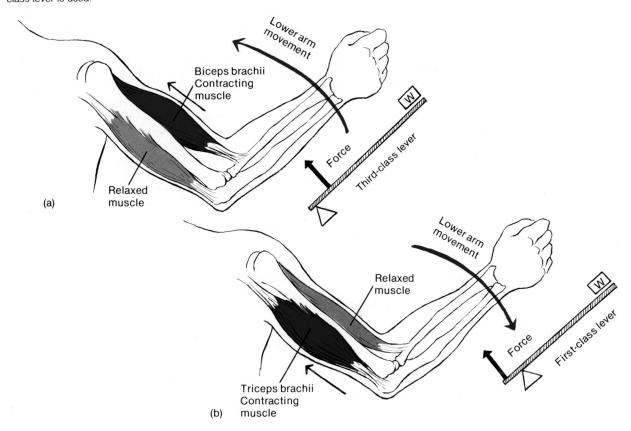

Although many lever arrangements occur throughout the skeletal-muscular systems, they are not always easy to identify. Nevertheless, these levers provide advantages in movements. The parts of some levers, such as those that function in moving the limbs, are arranged in ways that produce rapid motions, while others, such as those that move the head, aid in maintaining posture with minimal effort.

Blood Cell Formation

The process of blood cell formation is called **hematopoiesis.** Very early in life it occurs in a structure called a *yolk sac*, which lies outside the body of a human embryo. (See chapter 23.) Later in development, blood cells are manufactured in the liver and spleen, and still later they are formed in the marrow within bones.

Marrow is a soft, netlike mass of connective tissue found in the medullary cavities of long bones, in the irregular spaces of spongy bone, and in the larger osteonic canals of compact bone tissue.

There are two kinds of marrow—red and yellow. *Red marrow* functions in the formation of red blood cells (erythrocytes), certain white blood cells (leukocytes), and blood platelets (thrombocytes). It is red because of the red, oxygen-carrying pigment called **hemoglobin** contained within the red blood cells.

Red marrow occupies the cavities of most bones in an infant. With age, however, more and more of it is replaced by yellow marrow that functions as fat storage tissue and is inactive in blood cell production.

Osteoporosis
A Practical Application

Osteoporosis is a disorder of the skeletal system in which there is an excessive loss of bone volume and mineral content. This disorder is associated with aging. The affected bones develop spaces and canals that become enlarged and filled with fibrous and fatty tissues. Such bones are easily fractured and may break spontaneously because they are no longer able to support body weight. A person with osteoporosis may, for example, suffer a spontaneous fracture of the upper leg bone (femur) at the hip or the collapse of sections of the backbone (vertebrae). Similarly, the distal portion of a lower arm bone (radius) near the wrist may fracture as a result of a minor stress.

This condition is responsible for a large proportion of fractures occurring in persons over forty-five years of age. Although it may affect persons of either sex, it is most common in light-skinned females after menopause (see chapter 22).

Factors that increase the risk of osteoporosis include a low intake of dietary calcium, a lack of physical exercise, and, in females, a decrease in the blood concentration of the sex hormone called *estrogen*. (This hormone is produced by the ovaries, which become inactive at menopause.) Heavy use of alcohol and cigarette smoking also seem to increase the risk of developing this condition.

Fortunately, osteoporosis may be preventable provided steps are taken early enough. It is known, for example, that bone mass usually reaches a maximum at about age thirty-five. Thereafter bone loss may exceed bone formation in both sexes. To reduce such loss, people in their mid-twenties and older are advised to ensure that their dietary calcium intake reaches at least the U.S. government's recommended daily allowance of 800 milligrams. (Some nutritionists believe that 1000–1500 milligrams of calcium is needed daily to control bone loss.) An 8-ounce glass of nonfat milk, for example, contains about 275 milligrams of calcium. It also is recommended that people engage in regular physical exercises in which their bones support their body weight, such as walking or jogging. Additionally, post-menopausal females may require estrogen replacement therapy, which should be carried out under the supervision of a physician. As a rule, women have about 30% less bone than men, and after menopause, women typically lose bone mass twice as fast as men do.

In an adult, red marrow is found primarily in the spongy bone of the skull, ribs, sternum, clavicles, vertebrae, and pelvis. If the blood cell supply is deficient, some yellow marrow may change back into red marrow and become active in blood cell production. Blood cell formation is described in more detail in chapter 17.

Storage of Inorganic Salts

As mentioned in chapter 5, the intercellular matrix of bone tissue contains both collagen and inorganic mineral salts. Actually, the salts are responsible for about 70% of the matrix by weight, and are mostly in the form of tiny crystals of a type of calcium phosphate called *hydroxyapatite.*

Calcium is needed for a number of vital metabolic processes, including blood clot formation, nerve impulse conduction, and muscle cell contraction. When a low blood calcium ion concentration exists, osteoclasts are stimulated to break down bone tissue and calcium salts are released from the matrix into the blood. On the other hand, if the blood calcium ion concentration is excessively high, osteoclast activity is inhibited and osteoblasts are stimulated to form bone tissue. As a result, the excessive calcium is stored in the matrix. The details of this mechanism are presented in chapter 13.

In addition to storing calcium and phosphorus, bone tissue contains lesser amounts of magnesium, sodium, potassium, and carbonate ions. Bones also tend to accumulate certain metallic elements such as lead, radium, and strontium, which are not normally present in the body but are sometimes ingested accidentally.

A radioactive isotope of strontium, *strontium-90,* is a by-product of nuclear reactions, such as those that occur in atomic explosions and nuclear power plants. If strontium-90 is released into the environment, it may be taken in by plants and animals because it is chemically similar to calcium and can be used metabolically by organisms in the same ways they use calcium. Humans may ingest strontium-90 by drinking milk from cows that fed upon such plants. If this happens, some of the strontium-90 may accumulate in the human bones, and nearby cells would be subjected to its radiation. Because radiation from strontium-90 can cause mutations, such exposure may result in the development of abnormal cells associated with bone cancers or leukemias.

Figure 7.13
Wormian bones are extra bones that sometimes develop in sutures between the flat bones of the skull.

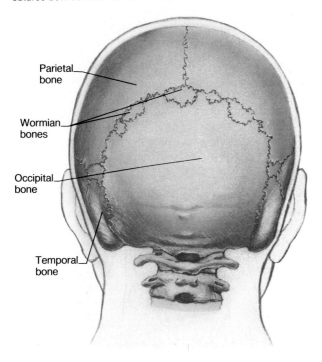

Parietal bone

Wormian bones

Occipital bone

Temporal bone

1. Name three major functions of bones.
2. Explain how body parts form a first-class lever; a third-class lever.
3. Explain how the concentration of blood calcium is controlled.
4. What inorganic elements may be stored in bone tissue?

Organization of the Skeleton

Number of Bones

Although the number of bones in a human skeleton is often reported to be 206, the number actually varies from person to person. Some people may lack certain bones, while others have extra ones. For example, the flat bones of the skull usually grow together and become tightly joined along irregular lines called **sutures.** Occasionally extra bones called *wormian* or *sutural bones* develop in these sutures (figure 7.13). Also extra small, round sesamoid bones may develop in tendons, where they apparently function to reduce friction in places where tendons pass over bony prominences (chart 7.3).

Chart 7.3 Bones of the adult skeleton

1. Axial Skeleton
   a. Skull — 22 bones

      8 cranial bones
      frontal 1
      parietal 2
      occipital 1
      temporal 2
      sphenoid 1
      ethmoid 1

      13 facial bones
      maxilla 2
      palatine 2
      zygomatic 2
      lacrimal 2
      nasal 2
      vomer 1
      inferior nasal concha 2

      1 mandible

   b. Middle ear bones — 6 bones
      malleus 2
      incus 2
      stapes 2

   c. Hyoid — 1 bone
      hyoid bone 1

   d. Vertebral column — 26 bones
      cervical vertebra 7
      thoracic vertebra 12
      lumbar vertebra 5
      sacrum 1
      coccyx 1

   e. Thoracic cage — 25 bones
      rib 24
      sternum 1

2. Appendicular Skeleton
   a. Pectoral girdle — 4 bones
      scapula 2
      clavicle 2

   b. Upper limbs — 60 bones
      humerus 2
      radius 2
      ulna 2
      carpal 16
      metacarpal 10
      phalanx 28

   c. Pelvic girdle — 2 bones
      coxal bones 2

   d. Lower limbs — 60 bones
      femur 2
      tibia 2
      fibula 2
      patella 2
      tarsal 14
      metatarsal 10
      phalanx 28

      Total — 206 bones

Figure 7.14

Major bones of the skeleton. (*a*) Anterior view; (*b*) posterior view. (Note that the axial and appendicular portions are distinguished with color.)

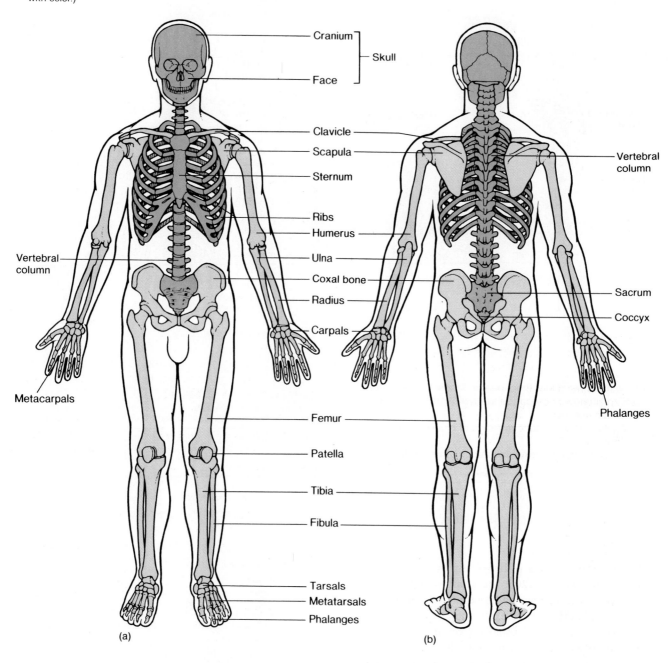

Cranium
Skull
Face

Clavicle
Scapula
Sternum

Ribs
Humerus
Ulna
Coxal bone
Radius

Carpals

Vertebral column

Metacarpals

Femur

Patella

Tibia

Fibula

Tarsals
Metatarsals
Phalanges

(a)

Vertebral column

Sacrum

Coccyx

Phalanges

(b)

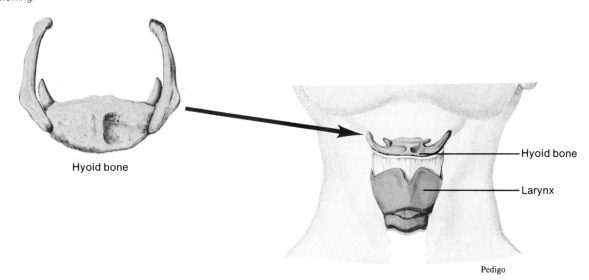

Hyoid bone

Hyoid bone

Larynx

Pedigo

Divisions of the Skeleton

For purposes of study, it is convenient to divide the skeleton into two major portions—an axial skeleton and an appendicular skeleton (figure 7.14).

The **axial skeleton** consists of the bony and cartilaginous parts that support and protect the organs of the head, neck, and trunk. These parts include the following:

1. **Skull.** The skull is composed of the brain case, or *cranium*, and the facial bones.
2. **Hyoid bone.** The hyoid (hi'oid) bone (figure 7.15) is located in the neck between the lower jaw and the larynx. It does not articulate with any other bones, but is fixed in position by muscles and ligaments. The hyoid bone supports the tongue and serves as an attachment for certain muscles that help to move the tongue and function in swallowing. It can be felt approximately a finger's width above the anterior prominence of the larynx.
3. **Vertebral column.** The vertebral column, or backbone, consists of many vertebrae separated by cartilaginous *intervertebral disks*. This column forms the central axis of the skeleton. Near its distal end, several vertebrae are fused to form the **sacrum**, which is part of the pelvis. A small, rudimentary tailbone called the **coccyx** is attached to the end of the sacrum.
4. **Thoracic cage.** The thoracic cage protects the organs of the thorax and the upper abdomen. It is composed of twelve pairs of ribs that articulate posteriorly with thoracic vertebrae. It

also includes the **sternum** (ster'num), or breastbone, to which most of the ribs are attached anteriorly.

The **appendicular skeleton** consists of the bones of the limbs and those that anchor the limbs to the axial skeleton. It includes the following:

1. **Pectoral girdle.** The pectoral girdle is formed by a **scapula** (scap'u-lah) or shoulder blade and a **clavicle** (klav'ĭ-k'l) or collarbone on both sides of the body. The pectoral girdle connects the bones of the arms to the axial skeleton and aids in arm movements.
2. **Upper limbs** (arms). Each upper limb consists of a **humerus** (hu'mer-us) or upper arm bone, and two lower arm bones—a **radius** (ra'de-us) and an **ulna** (ul'nah). These three bones articulate with each other at the elbow joint. At the distal end of the radius and ulna, there are eight **carpals** (kar'pals), or wrist bones. The bones of the palm are called **metacarpals,** and the finger bones are **phalanges** (fah-lan'jēz).
3. **Pelvic girdle.** The pelvic girdle is formed by two **coxal** (kok'sal) or **innominate** (ĭ-nom'ĭ-nāt) bones (hip bones), which are attached to each other anteriorly and to the sacrum posteriorly. They connect the bones of the legs to the axial skeleton and, with the sacrum and coccyx, form the **pelvis,** which protects the lower abdominal and internal reproductive organs.
4. **Lower limbs** (legs). Each lower limb consists of a **femur** (fe'mur), or thigh bone, and two lower leg bones—a large **tibia** (tib'e-ah), or shinbone, and a slender **fibula** (fib'u-lah), or calf bone.

The Skeletal System / 197

Chart 7.4 Terms used to describe skeletal structures

Term	Definition	Example
Condyle (kon'dĭl)	A rounded process that usually articulates with another bone	Occipital condyle of the occipital bone (figure 7.19)
Crest (krest)	A narrow, ridgelike projection	Iliac crest of the ilium (figure 7.51)
Epicondyle (ep''ĭ-kon'dĭl)	A projection situated above a condyle	Medial epicondyle of the humerus (figure 7.46)
Facet (fas'et)	A small, nearly flat surface	Facet of a thoracic vertebra (figure 7.36)
Fissure (fish'-ūr)	A cleft or groove	Inferior orbital fissure in the orbit of the eye (figure 7.17)
Fontanel (fon''tah-nel')	A soft spot in the skull where membranes cover the space between bones	Anterior fontanel between the frontal and parietal bones (figure 7.31)
Foramen (fo-ra'men)	An opening through a bone that usually serves as a passageway for blood vessels, nerves, or ligaments	Foramen magnum of the occipital bone (figure 7.19)
Fossa (fos'ah)	A relatively deep pit or depression	Olecranon fossa of the humerus (figure 7.46)
Fovea (fo've-ah)	A tiny pit or depression	Fovea capitis of the femur (figure 7.56)
Head (hed)	An enlargement on the end of a bone	Head of the humerus (figure 7.46)
Linea (lin'e-ah)	A narrow ridge	Linea aspera of the femur (figure 7.56)
Meatus (me-a'tus)	A tubelike passageway within a bone	Auditory meatus of the ear (figure 7.18)
Process (pros'es)	A prominent projection on a bone	Mastoid process of the temporal bone (figure 7.18)
Ramus (ra'mus)	A structure given off from another larger one	Ramus of the mandible (figure 7.29)
Sinus (si'nus)	A cavity or hollow space within a bone	Frontal sinus of the frontal bone (figure 7.23)
Spine (spīn)	A thornlike projection	Spine of the scapula (figure 7.43)
Suture (soo'cher)	A line of union between bones	Lambdoidal suture between the occipital and parietal bones (figure 7.18)
Trochanter (tro-kan'ter)	A relatively large process	Greater trochanter of the femur (figure 7.56)
Tubercle (tu'ber-kl)	A small, knoblike process	Tubercle of a rib (figure 7.40)
Tuberosity (tu''bē-ros'ĭ-te)	A knoblike process usually larger than a tubercle	Radial tuberosity of the radius (figure 7.47)

These three bones articulate with each other at the knee joint, where the **patella** (pah-tel'ah), or kneecap, covers the anterior surface. At the distal ends of the tibia and fibula, there are seven **tarsals** (tahr'sals) or ankle bones. The bones of the foot are called **metatarsals,** while those of the toes (like the fingers) are **phalanges.** Chart 7.4 defines some terms used to describe skeletal structures.

1. Distinguish between the axial and the appendicular skeletons.
2. List the bones that each includes.

The Skull

A human skull usually consists of twenty-two bones that, except for the lower jaw, are firmly interlocked along sutures. Eight of these immovable bones make up the cranium, and thirteen form the facial skeleton. The **mandible** (man'dĭ-b'l), or lower jawbone, is a movable bone held to the cranium by ligaments (figures 7.16, 7.17, 7.18, and 7.19).

The Cranium

The **cranium** (kra'ne-um) encloses and protects the brain, and its surface provides attachments for various muscles that make chewing and head movements possible. Some of the cranial bones contain air-filled cavities called *sinuses,* which are lined with mucous membranes and are connected by passageways to the nasal cavity. Sinuses reduce the weight of the skull and increase the intensity of the voice by serving as resonant sound chambers.

The eight bones of the cranium (chart 7.5) are as follows:

1. **Frontal bone.** The frontal bone forms the anterior portion of the skull above the eyes and includes the forehead, the roof of the nasal cavity, and the roofs of the orbits (bony sockets) of the eyes. On the upper margin of each orbit, the frontal bone is marked by a *supraorbital foramen* (or *supraorbital notch* in some skulls), through which blood vessels and nerves pass to the tissues of the forehead.

Figure 7.16
Anterior view of the skull.

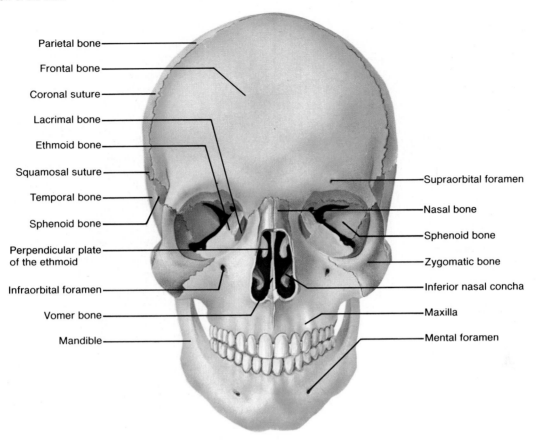

Parietal bone

Frontal bone

Coronal suture

Lacrimal bone

Ethmoid bone

Squamosal suture

Temporal bone

Sphenoid bone

Perpendicular plate
of the ethmoid

Infraorbital foramen

Vomer bone

Mandible

Supraorbital foramen

Nasal bone

Sphenoid bone

Zygomatic bone

Inferior nasal concha

Maxilla

Mental foramen

Figure 7.17
The orbit of the eye includes both cranial and facial bones.

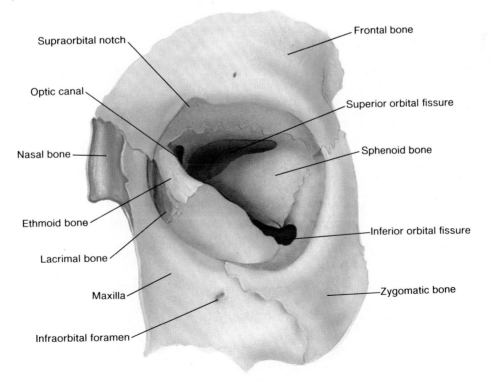

Supraorbital notch

Optic canal

Nasal bone

Ethmoid bone

Lacrimal bone

Maxilla

Infraorbital foramen

Frontal bone

Superior orbital fissure

Sphenoid bone

Inferior orbital fissure

Zygomatic bone

Figure 7.18

Lateral view of the skull.

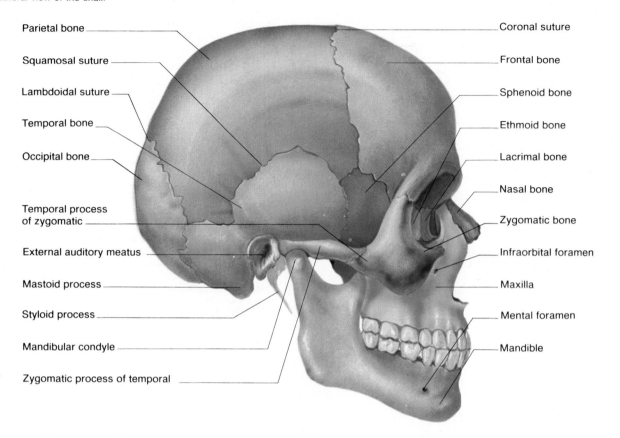

Parietal bone

Squamosal suture

Lambdoidal suture

Temporal bone

Occipital bone

Temporal process
of zygomatic

External auditory meatus

Mastoid process

Styloid process

Mandibular condyle

Zygomatic process of temporal

Coronal suture

Frontal bone

Sphenoid bone

Ethmoid bone

Lacrimal bone

Nasal bone

Zygomatic bone

Infraorbital foramen

Maxilla

Mental foramen

Mandible

Figure 7.19

Inferior view of the skull.

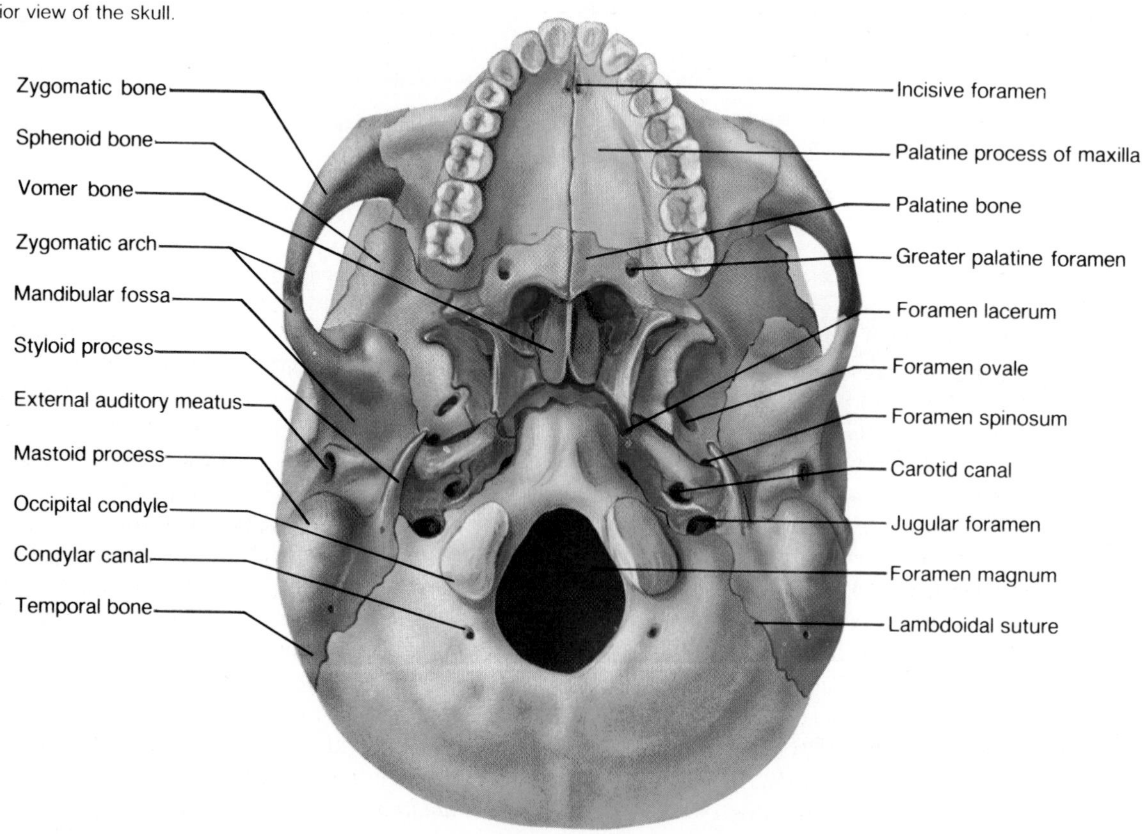

Zygomatic bone

Sphenoid bone

Vomer bone

Zygomatic arch

Mandibular fossa

Styloid process

External auditory meatus

Mastoid process

Occipital condyle

Condylar canal

Temporal bone

Incisive foramen

Palatine process of maxilla

Palatine bone

Greater palatine foramen

Foramen lacerum

Foramen ovale

Foramen spinosum

Carotid canal

Jugular foramen

Foramen magnum

Lambdoidal suture

Chart 7.5 Cranial bones

Name and number	Description	Special features
Frontal 1	Forms forehead, roof of nasal cavity, and roofs of orbits	Supraorbital foramen, frontal sinuses
Parietal 2	Form side walls and roof of cranium	
Occipital 1	Forms back of skull and base of cranium	Foramen magnum, occipital condyles
Temporal 2	Form side walls and floor of cranium	External auditory meatus, mandibular fossa, mastoid process, styloid process, zygomatic process
Sphenoid 1	Forms parts of base of cranium, sides of skull, and floors and sides of orbits	Sella turcica, sphenoidal sinuses
Ethmoid 1	Forms parts of roof and walls of nasal cavity, floor of cranium, and walls of orbits	Cribriform plates, perpendicular plate, superior and middle nasal conchae, ethmoidal sinuses, crista galli

Figure 7.20
Lateral surface of the right temporal bone. What sensory structures are located within this bone?

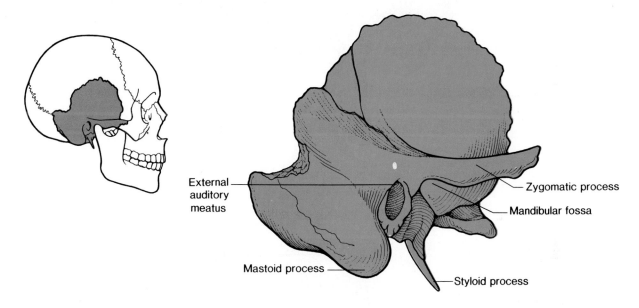

External auditory meatus

Mastoid process

Zygomatic process

Mandibular fossa

Styloid process

Within the frontal bone are two *frontal sinuses*, one above each eye near the midline. Although it is a single bone in adults, the frontal bone develops in two parts. These halves grow together and are usually completely fused by the fifth or sixth year of age.

2. **Parietal bones.** One *parietal bone* is located on each side of the skull just posterior to the frontal bone. It is shaped like a curved plate and has four borders. Together, the parietal bones form the bulging sides and roof of the cranium. They are fused in the midline along the *sagittal suture*, and they meet the frontal bone along the *coronal suture*.

3. **Occipital bone.** The occipital bone joins the parietal bones along the *lambdoidal suture*. It forms the back of the skull and the base of the cranium. There is a large opening on its lower surface called the *foramen magnum*, through which nerve fibers from the brain pass and enter the vertebral canal to become part of the spinal cord. Rounded processes called *occipital condyles,* which are located on each side of the foramen magnum, articulate with the first vertebra of the vertebral column.

4. **Temporal bones.** A temporal (tem′por-al) bone (figure 7.20) on each side of the skull joins the parietal bone along a *squamosal suture*. The temporal bones form parts of the sides and the base of the cranium. Located near the inferior margin is an opening, the *external auditory meatus*, which leads inward to parts of the ear. The temporal bones also house the internal ear structures and have depressions, the *mandibular fossae* (glenoid fossae), that articulate with processes of the mandible.

Figure 7.21

(a) The sphenoid bone as viewed from above; (b) posterior view.

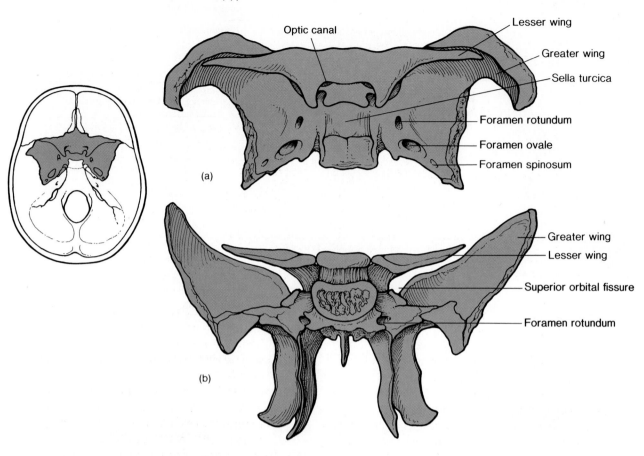

Optic canal

Lesser wing

Greater wing

Sella turcica

Foramen rotundum

Foramen ovale

Foramen spinosum

(a)

Greater wing

Lesser wing

Superior orbital fissure

Foramen rotundum

(b)

Figure 7.22

(a) Ethmoid bone viewed from above and (b) from behind.

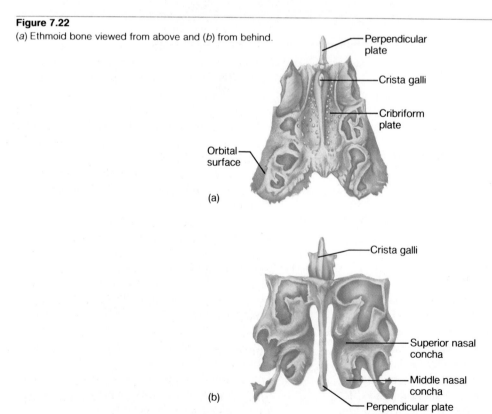

Perpendicular plate

Crista galli

Cribriform plate

Orbital surface

(a)

Crista galli

Superior nasal concha

Middle nasal concha

Perpendicular plate

(b)

Figure 7.23
Lateral wall of the nasal cavity.

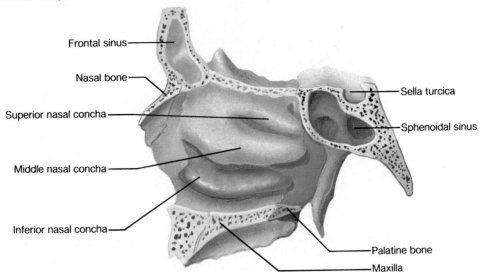

Frontal sinus

Nasal bone

Superior nasal concha

Middle nasal concha

Inferior nasal concha

Sella turcica

Sphenoidal sinus

Palatine bone

Maxilla

Below each external auditory meatus, there are two projections—a rounded *mastoid process* and a long, pointed *styloid process*. The mastoid process provides an attachment for certain muscles of the neck, while the styloid process serves as an anchorage for muscles associated with the tongue and pharynx. An opening near the mastoid process, the *carotid canal*, transmits a branch of the carotid artery, while an opening between the temporal and occipital bones, the *jugular foramen*, accommodates the jugular vein (figure 7.19).

The mastoid process is of clinical interest because it may become infected. The tissues in this region of the temporal bone contain a number of interconnected air cells, lined with mucous membranes, that communicate with the middle ear. These spaces sometimes become inflamed when microorganisms spread from an infected middle ear (*otitis media*). The resulting mastoid infection, called *mastoiditis*, is of particular concern because the membranes that surround the brain are close by and may also become infected.

A *zygomatic process* projects anteriorly from the temporal bone in the region of the external auditory meatus. It joins the *zygomatic bone* and helps form the prominence of the cheek.

5. **Sphenoid bone.** The sphenoid (sfe'noid) bone (figure 7.21) is wedged between several other bones in the anterior portion of the cranium. It consists of a central part and two winglike structures that extend laterally toward each side of the skull. This bone helps form the base of the cranium, the sides of the skull, and the floors and sides of the orbits. Along the midline within the cranial cavity, a portion of the

sphenoid bone rises up and forms a saddle-shaped mass called *sella turcica* (Turk's saddle). The depression of this saddle is occupied by the pituitary gland, which hangs from the base of the brain by a stalk.

The sphenoid bone also contains two *sphenoidal sinuses*, which lie side by side and are separated by a bony septum that projects downward into the nasal cavity.

6. **Ethmoid bone.** The ethmoid (eth'moid) bone (figure 7.22) is located in front of the sphenoid bone. It consists of two masses, one on each side of the nasal cavity, which are joined horizontally by thin *cribriform plates*. These plates form part of the roof of the nasal cavity, and nerves associated with the sense of smell pass through tiny openings (olfactory foramina) in them. Portions of the ethmoid bone also form sections of the cranial floor, orbital walls, and nasal cavity walls. A *perpendicular plate* projects downward in the midline from the cribriform plates to form the bulk of the nasal septum.

Delicate scroll-shaped plates called *superior* and *middle nasal conchae* project inward from the lateral portions of the ethmoid bone toward the perpendicular plate. These bones, which are also called *turbinate bones*, support mucous membranes that line the nasal cavity. The mucous membranes, in turn, begin the processes of moistening, warming, and filtering air as it enters the respiratory tract. The lateral portions of the ethmoid bone contain many small air spaces, the *ethmoidal sinuses*. Various structures in the nasal cavity are shown in figure 7.23.

Figure 7.24

Floor of the cranial cavity viewed from above.

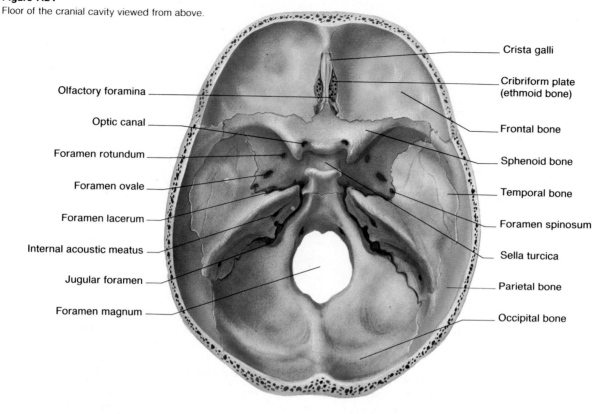

Crista galli

Cribriform plate (ethmoid bone)

Olfactory foramina

Optic canal

Foramen rotundum

Foramen ovale

Foramen lacerum

Internal acoustic meatus

Jugular foramen

Foramen magnum

Frontal bone

Sphenoid bone

Temporal bone

Foramen spinosum

Sella turcica

Parietal bone

Occipital bone

Projecting upward into the cranial cavity between the cribriform plates is a triangular process of the ethmoid bone called the *crista galli* (cock's comb). This process serves as an attachment for membranes that enclose the brain. A view of the cranial cavity is shown in figure 7.24.

The Facial Skeleton

The **facial skeleton** consists of thirteen immovable bones and a movable lower jawbone. In addition to forming the basic shape of the face, these bones provide attachments for various muscles that move the jaw and control facial expressions.

The bones of the facial skeleton are as follows:

1. **Maxillary bones.** The maxillary (mak'si-ler''e) bones (maxillae) form the upper jaw; together, they are the keystone of the face, for all the other immovable facial bones articulate with them.

    Portions of these bones comprise the anterior roof of the mouth (*hard palate*), the floors of the orbits, and the sides and floor of the nasal cavity. They also contain the sockets of the upper teeth. Inside the maxillae, lateral

to the nasal cavity, are *maxillary sinuses* (antrum of Highmore). These spaces are the largest of the sinuses, and they extend from the floor of the orbits to the roots of the upper teeth. Figure 7.25 shows the maxillary and other sinuses. See chart 7.6 for a summary of the sinuses.

During development, portions of the maxillary bones called *palatine processes* grow together and fuse along the midline to form the anterior section of the hard palate.

Sometimes the fusion of the palatine processes of the maxillae is incomplete at the time of birth; the result is called a *cleft palate*. Infants with this deformity may have trouble sucking because of the opening that remains between the oral and nasal cavities.

The inferior border of each maxillary bone projects downward forming an *alveolar process.* Together these processes create a horseshoe-shaped *alveolar arch* (dental arch). Cavities in this arch are occupied by the teeth, which are attached to these bony sockets by connective tissue (see chapter 14).

Figure 7.25
Location of the sinuses.

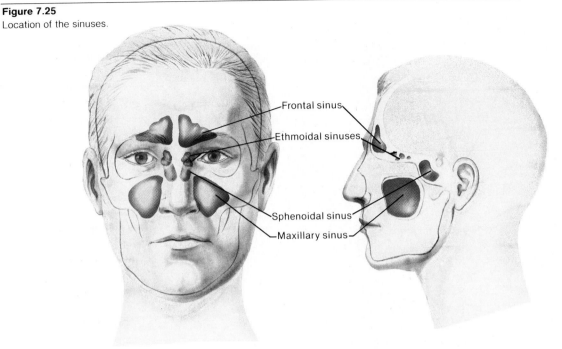

2. **Palatine bones.** The palatine (pal′ah-tīn) bones (figure 7.26) are located behind the maxillae. Each bone is roughly L-shaped. The horizontal portions serve as both the posterior section of the hard palate and the floor of the nasal cavity. The perpendicular portions help form the lateral walls of the nasal cavity.

3. **Zygomatic bones.** The zygomatic (zi″go-mat′ik) bones (malar bones) are responsible for the prominences of the cheeks below and to the sides of the eyes. These bones also help form the lateral walls and the floors of the orbits. Each bone has a *temporal process*, which extends posteriorly to join the zygomatic process of a temporal bone. Together these processes form a *zygomatic arch* (figures 7.18 and 7.19).

4. **Lacrimal bones.** A lacrimal (lak′rĭ-mal) bone is a thin, scalelike structure located in the medial wall of each orbit between the ethmoid bone and the maxilla. A groove in its anterior portion leads from the orbit to the nasal cavity, providing a pathway for a tube that carries tears from the eye to the nasal cavity.

5. **Nasal bones.** The nasal (na′zal) bones are long, thin, and nearly rectangular. They lie side by side and are fused at the midline, where they form the bridge of the nose. These bones serve as attachments for the cartilaginous tissues that are largely responsible for the shape of the nose.

Chart 7.6	Sinuses of the cranial and facial bones	
Sinuses	**Number**	**Location**
Frontal sinuses	2	Frontal bone above each eye and near the midline
Sphenoidal sinuses	2	Sphenoid bone above the posterior portion of the nasal cavity
Ethmoidal sinuses	2 groups of small air cells	Ethmoid bone on either side of the upper portion of the nasal cavity
Maxillary sinuses	2	Maxillary bones lateral to the nasal cavity and extending from the floor of the orbits to the roots of the upper teeth

Figure 7.26
The horizontal portions of the palatine bones form the posterior section of the hard palate, and the perpendicular portions help form the lateral walls of the nasal cavity.

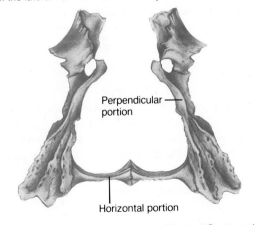

Perpendicular portion

Horizontal portion

The Skeletal System / 205

Figure 7.27

Sagittal section of the skull.

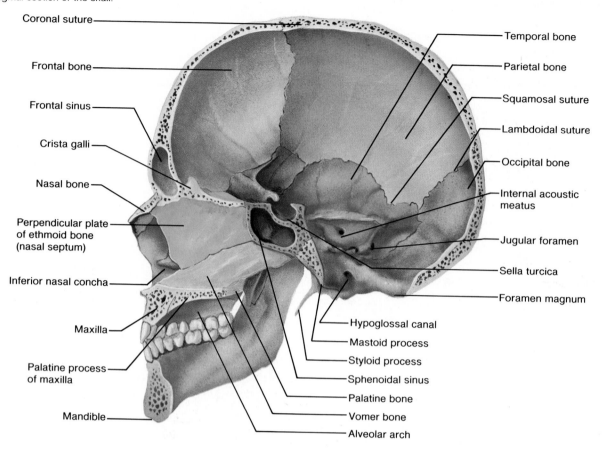

Coronal suture

Frontal bone

Frontal sinus

Crista galli

Nasal bone

Perpendicular plate of ethmoid bone (nasal septum)

Inferior nasal concha

Maxilla

Palatine process of maxilla

Mandible

Temporal bone

Parietal bone

Squamosal suture

Lambdoidal suture

Occipital bone

Internal acoustic meatus

Jugular foramen

Sella turcica

Foramen magnum

Hypoglossal canal

Mastoid process

Styloid process

Sphenoidal sinus

Palatine bone

Vomer bone

Alveolar arch

Figure 7.28

A frontal section of the skull.

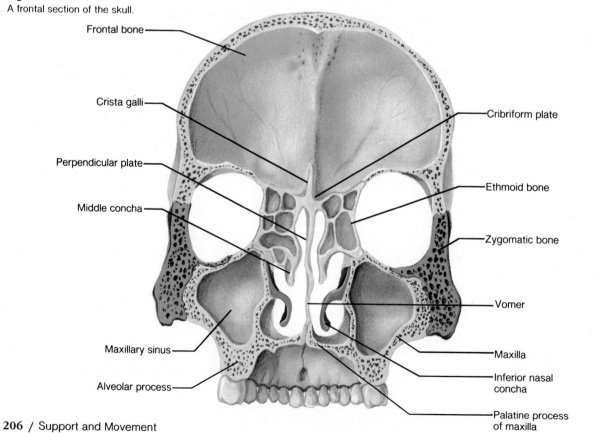

Frontal bone

Crista galli

Perpendicular plate

Middle concha

Maxillary sinus

Alveolar process

Cribriform plate

Ethmoid bone

Zygomatic bone

Vomer

Maxilla

Inferior nasal concha

Palatine process of maxilla

Figure 7.29
(a) Lateral view of the mandible; (b) inferior view.

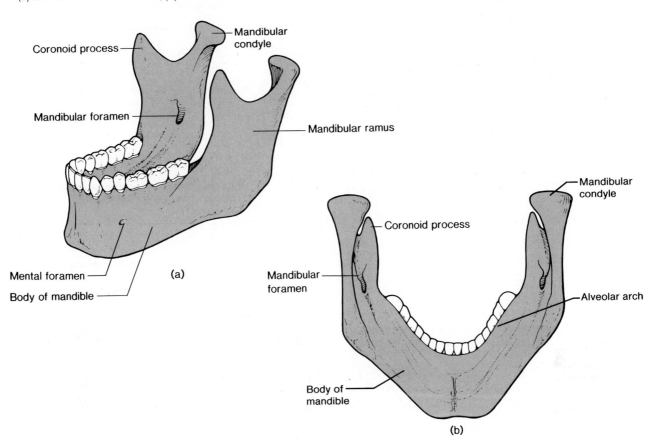

Coronoid process — Mandibular condyle

Mandibular foramen — Mandibular ramus

Mental foramen — (a)

Body of mandible

Coronoid process — Mandibular condyle

Mandibular foramen — Alveolar arch

Body of mandible

(b)

6. **Vomer bone.** The thin, flat vomer (vo'mer) bone is located along the midline within the nasal cavity. Posteriorly it joins the perpendicular plate of the ethmoid bone, and together they form the nasal septum (figures 7.27 and 7.28).

7. **Inferior nasal conchae.** The inferior nasal conchae (kong'ke) are fragile, scroll-shaped bones attached to the lateral walls of the nasal cavity. They are the largest of the conchae and are positioned below the conchae of the ethmoid bone (figures 7.16 and 7.23). Like the superior and middle conchae, the inferior conchae provide support for mucous membranes within the nasal cavity.

8. **Mandible.** The mandible (lower jawbone) consists of a horizontal, horseshoe-shaped body with a flat *ramus* projecting upward at each end. The rami are divided into two processes—a posterior *mandibular condyle* and an anterior *coronoid process* (figure 7.29). The mandibular condyles articulate with the mandibular fossae of the temporal bones, while the coronoid processes serve as attachments for muscles used

in chewing. Other large chewing muscles are inserted on the lateral surfaces of the rami.

A curved bar of bone on the superior border of the mandible, the *alveolar arch,* contains the sockets of the lower teeth.

On the medial side of the mandible, near the center of each ramus, is a *mandibular foramen.* This opening admits blood vessels and a nerve, which supply the roots of the lower teeth. Dentists commonly inject anesthetic into the tissues near this foramen to temporarily block nerve impulse conduction and cause the teeth on that side of the jaw to become insensitive. Branches of these vessels and nerve emerge from the mandible through the *mental foramen,* which opens on the outside near the point of the jaw. They supply the tissues of the chin and lower lip.

Chart 7.7 contains a descriptive summary of the fourteen facial bones. Various features of these bones can be seen in the X-ray films in figure 7.30.

Chart 7.8 lists the major foramina and passageways through bones of the skull, as well as their general locations and the structures they transmit.

Chart 7.7 Facial skeleton

Name and number	Description	Special features
Maxilla 2	Form upper jaw, anterior roof of mouth, floors of orbits, and sides and floor of nasal cavity	Sockets of upper teeth, maxillary sinuses, palatine process
Palatine 2	Form posterior roof of mouth, and floor and lateral walls of nasal cavity	
Zygomatic 2	Form prominences of cheeks, and lateral walls and floors of orbits	Temporal process
Lacrimal 2	Form part of medial walls of orbits	Groove that leads from orbit to nasal cavity
Nasal 2	Form bridge of nose	
Vomer 1	Forms part of nasal septum	
Inferior nasal concha 2	Extend into nasal cavity from its lateral walls	
Mandible 1	Forms lower jaw	Body, ramus, mandibular condyle, coronoid process, mandibular foramen, mental foramen

Figure 7.30

(a) X-ray films of the skull from the front and (b) from the side.
What features of the cranium and facial skeleton do you
recognize?

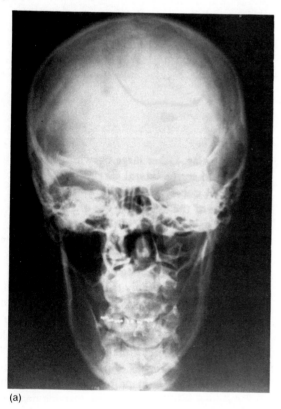

(a)

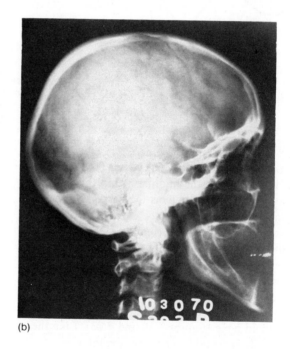

(b)

Chart 7.8   Passageways through bones of the skull

Passageway	Location	Major structures transmitted
Carotid canal (figure 7.19)	Inferior surface of the temporal bone	Internal carotid artery, veins, and nerves
Condylar canal (figure 7.19)	Base of skull in occipital bone	Veins between the skull and the neck
Foramen lacerum (figure 7.19)	Floor of cranial cavity between temporal and sphenoid bones	Branch of pharyngeal artery (in life opening is largely covered by fibrocartilage)
Foramen magnum (figure 7.19)	Base of skull in occipital bone	Nerve fibers between brain and spinal cord
Foramen ovale (figure 7.19)	Floor of cranial cavity in sphenoid bone	Mandibular division of trigeminal nerve and veins
Foramen rotundum (figure 7.24)	Floor of cranial cavity in sphenoid bone	Maxillary division of trigeminal nerve
Foramen spinosum (figure 7.24)	Floor of cranial cavity in sphenoid bone	Middle meningeal blood vessels and branch of mandibular nerve
Greater palatine foramen (figure 7.19)	Posterior portion of hard palate in palatine bone	Palatine blood vessels and nerves
Hypoglossal canal (figure 7.27)	Near margin of foramen magnum in occipital bone	Hypoglossal nerve
Incisive foramen (figure 7.19)	Anterior portion of hard palate	Nasopalatine nerves
Inferior orbital fissure (figure 7.17)	Floor of the orbit	Maxillary nerve and blood vessels
Infraorbital foramen (figure 7.17)	Below the orbit in maxillary bone	Infraorbital blood vessels and nerves
Internal acoustic meatus (figure 7.24)	Floor of cranial cavity in temporal bone	Branches of facial, vestibular, and cochlear nerves and blood vessels
Jugular foramen (figure 7.19)	Base of the skull between temporal and occipital bones	Glossopharyngeal, vagus and accessory nerves, and blood vessels
Mandibular foramen (figure 7.29)	Inner surface of ramus of mandible	Inferior alveolar blood vessels and nerves
Mental foramen (figure 7.29)	Near point of jaw in mandible	Mental nerve and blood vessels
Optic canal (figure 7.17)	Posterior portion of orbit in sphenoid bone	Optic nerve and ophthalmic artery
Supraorbital foramen (figure 7.16)	Upper margin of orbit in frontal bone	Supraorbital blood vessels and nerves
Superior orbital fissure (figure 7.17)	Lateral wall of orbit	Oculomotor, trochlear, abducent, and ophthalmic division of trigeminal nerves

The Infantile Skull

At birth the skull is incompletely developed, and the cranial bones are separated by fibrous membranes. These membranous areas are called **fontanels** or, more commonly, soft spots. They permit some movement between the bones, so that the developing skull is partially compressible and can change shape slightly. This action enables an infant's skull to pass more easily through the birth canal. Eventually the fontanels close as the cranial bones grow together. The posterior fontanel usually closes about two months after birth, the sphenoid fontanel closes at about three months, the mastoid fontanel closes near the end of the first year, but the anterior one may not close until the middle or end of the second year.

Other characteristics of an infantile skull (figure 7.31) include a relatively small face with a prominent forehead and large orbits. The jaw and nasal cavity are small, the sinuses are incompletely formed, and the frontal bone is in two parts. The skull bones are thin, but they are also somewhat flexible and thus are less easily fractured than adult bones.

---

1. Locate and name each of the bones of the cranium.
2. Locate and name each of the bones of the face.
3. Explain how an adult skull differs from that of an infant.

Figure 7.31
(a) Lateral view and (b) superior view of the infantile skull.

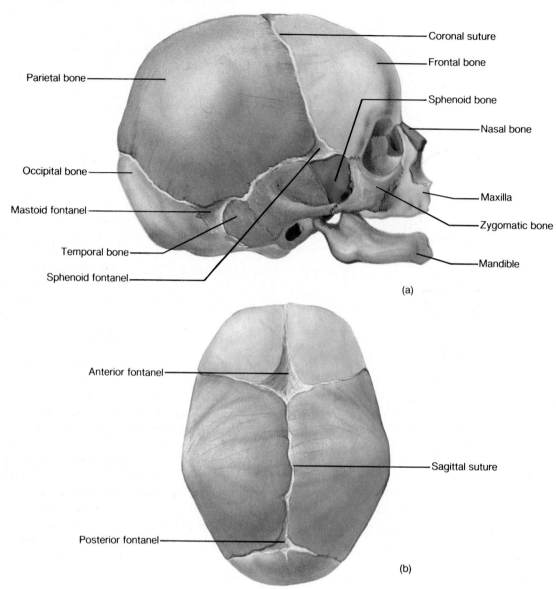

Parietal bone

Occipital bone

Mastoid fontanel

Temporal bone

Sphenoid fontanel

Coronal suture

Frontal bone

Sphenoid bone

Nasal bone

Maxilla

Zygomatic bone

Mandible

(a)

Anterior fontanel

Sagittal suture

Posterior fontanel

(b)

The Vertebral Column

The **vertebral column** extends from the skull to the pelvis and forms the vertical axis of the skeleton. It is composed of many bony parts called **vertebrae.** These are separated by masses of fibrocartilage called *intervertebral discs* and are connected to one another by ligaments. The vertebral column supports the head and the trunk of the body, yet is flexible enough to permit movements, such as bending forward, backward, or to the side, and turning or rotating on the central axis. It also protects the spinal cord, which passes through a *vertebral canal* formed by openings in the vertebrae.

In an infant there are thirty-three separate bones in the vertebral column. Five of these bones eventually fuse to form the sacrum, and four others join to become the coccyx. As a result, an adult vertebral column has twenty-six bones.

Normally the vertebral column has four curvatures that give it a degree of resiliency. The names of the curves correspond to the regions in which they occur, as shown in figure 7.32. The *thoracic* and *pelvic curvatures* are concave anteriorly and are called primary curves. The *cervical curvature* in the neck and the *lumbar curvature* in the lower back are convex anteriorly and are secondary curves.

Figure 7.32
The curved vertebral column consists of many vertebrae separated by intervertebral disks.

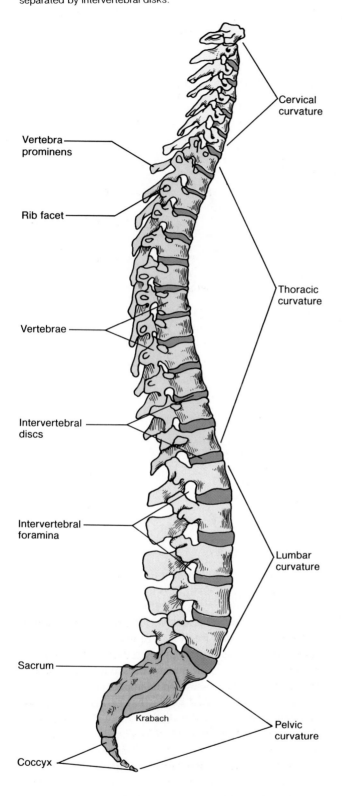

Vertebra prominens

Rib facet

Vertebrae

Intervertebral discs

Intervertebral foramina

Sacrum

Krabach

Coccyx

Cervical curvature

Thoracic curvature

Lumbar curvature

Pelvic curvature

A Typical Vertebra

Although the vertebrae in different regions of the vertebral column have special characteristics, they also have features in common. Thus, a typical vertebra (figure 7.33) has a drum-shaped *body* (centrum) that forms a thick, anterior portion of the bone. A longitudinal row of these vertebral bodies supports the weight of the head and trunk. The intervertebral discs, which separate adjacent vertebrae, are fastened to the roughened upper and lower surfaces of the bodies. These disks cushion and soften the forces created by such movements as walking and jumping, which might otherwise fracture vertebrae or jar the brain.

The bodies of adjacent vertebrae are joined on their anterior surfaces by *anterior ligaments* and on their posterior surfaces by *posterior ligaments*.

Projecting posteriorly from each vertebral body are two short stalks called *pedicles*. They form the sides of the *vertebral foramen*. Two plates called *laminae* arise from the pedicles and fuse in the back to become a *spinous process*. The pedicles, laminae, and spinous process together complete a bony *vertebral arch* around the vertebral foramen, through which the spinal cord passes.

Between the pedicles and laminae of a typical vertebra is a *transverse process* that projects laterally and toward the back. Various ligaments and muscles are attached to the dorsal spinous process and the transverse processes. Projecting upward and downward from each vertebral arch are *superior* and *inferior articulating processes*. These processes bear cartilage-covered facets by which each vertebra is joined to the one above and the one below it.

On the lower surfaces of the vertebral pedicles are notches that align to create openings, called *intervertebral foramina*. These openings provide passageways for spinal nerves that proceed between adjacent vertebrae and connect to the spinal cord.

Athletes, such as gymnasts, jumpers, and pole vaulters, who repeatedly land on hard surfaces, sometimes develop breaks or cracks in their vertebrae. These fractures commonly involve the articulating processes of these bones. Such a fracturing of a vertebra is called *spondylolysis*. It may be prevented by limiting the number of landings experienced during individual practice periods and by padding the landing areas with floor mats.

The Skeletal System / 211

Figure 7.33

(a) Lateral view of a typical vertebra; (b) adjacent vertebrae are joined at their articulating processes; (c) superior view of a typical vertebra.

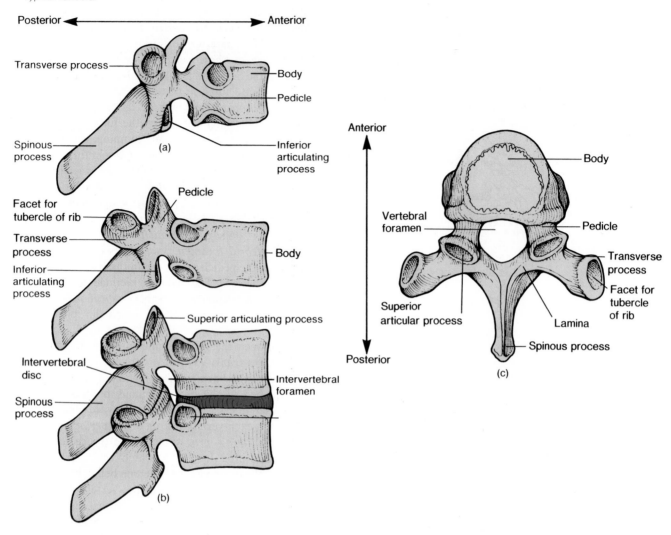

Cervical Vertebrae

Seven **cervical vertebrae** comprise the bony axis of the neck. Although these are the smallest of the vertebrae, their bone tissues are denser than those in any other region of the vertebral column.

The transverse processes of the cervical vertebrae are distinctive because they have *transverse foramina*, which serve as passageways for arteries leading to the brain. Also, the spinous processes of the second through the fifth cervical vertebrae are uniquely forked (bifid). These processes provide attachments for various muscles.

The spinous process of the seventh vertebra is longer and protrudes beyond the other cervical spines.

It is called the *vertebra prominens* and, because it can be felt through the skin, it is a useful landmark for locating other vertebral parts (figure 7.32).

Two of the cervical vertebrae, shown in figure 7.34, are of special interest. The first vertebra, or **atlas** (at′las), supports and balances the head. It has practically no body or spine and appears as a bony ring with two transverse processes. On its upper surface, the atlas has two kidney-shaped facets that articulate with the occipital condyles of the skull.

The second cervical vertebra, or **axis** (ak′sis), bears a toothlike *odontoid process* on its body. This process projects upward and lies in the ring of the atlas. As the head is turned from side to side, the atlas pivots around the odontoid process. (See figure 7.35.)

Figure 7.34
How do the structures of the (a) atlas and (b) axis function together to allow movement of the head? (c) Lateral view of the axis.

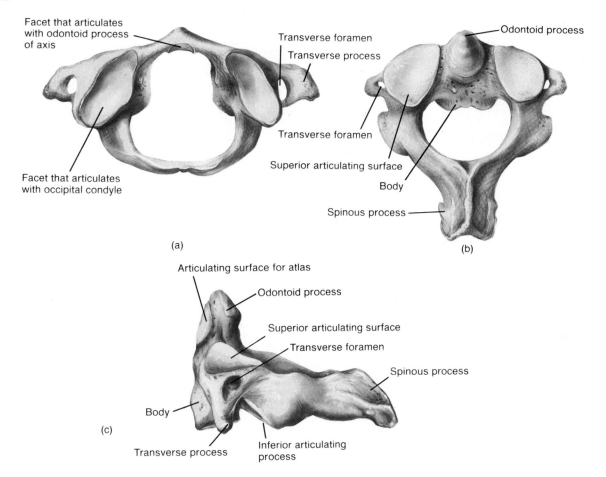

Facet that articulates with odontoid process of axis

Transverse foramen

Transverse process

Odontoid process

Transverse foramen

Superior articulating surface

Body

Spinous process

Facet that articulates with occipital condyle

(a)

(b)

Articulating surface for atlas

Odontoid process

Superior articulating surface

Transverse foramen

Spinous process

Body

(c)

Transverse process

Inferior articulating process

Thoracic Vertebrae

The twelve **thoracic vertebrae** are larger than those in the cervical region. They have long, pointed spinous processes that slope downward, and have facets on the sides of their bodies that articulate with ribs.

Beginning with the third thoracic vertebra and moving downward, the bodies of these bones increase in size. This reflects the stress placed on them by the increasing amounts of body weight they bear.

Lumbar Vertebrae

There are five **lumbar vertebrae** in the small of the back (loins). Since the lumbars must support more weight than the vertebrae above them, they have developed larger and stronger bodies. The transverse processes of these vertebrae project backward at relatively sharp angles, while their short, thick spinous processes are directed nearly horizontally.

Figure 7.35
What features can you identify in this x-ray film of the neck?

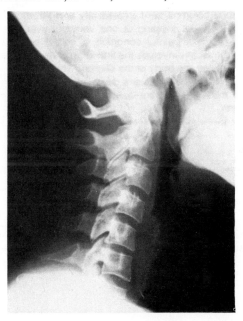

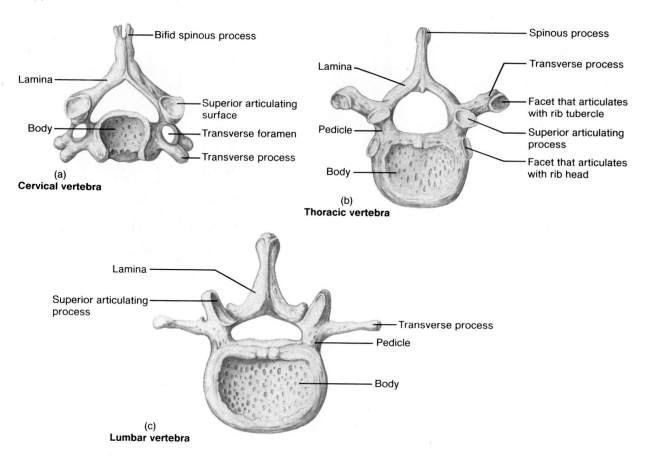

(a)
Cervical vertebra

(b)
Thoracic vertebra

(c)
Lumbar vertebra

Figure 7.36 compares the structures of the lumbar, thoracic, and cervical vertebrae.

Gymnasts, football players, and others who have their vertebral columns bent excessively and forcefully may experience the slipping of one vertebra over the one below it. This painful condition is called *spondylolisthesis*. It usually involves the fifth lumbar vertebra sliding forward over the body of the sacrum. The condition may be prevented by exercises designed to strengthen the back muscles that are associated with the vertebral column.

The Sacrum

The **sacrum** (sa′krum) is a triangular structure at the base of the vertebral column. It is composed of five vertebrae that are separated early in life, but gradually become fused together between the eighteenth and thirtieth years. The spinous processes of these fused bones are represented by a ridge of tubercles. To the sides of the tubercles are rows of openings, the *dorsal sacral foramina*, through which nerves and blood vessels pass (figure 7.37).

The sacrum is wedged between the coxal bones of the pelvis and is united to them by fibrocartilage at the *sacroiliac joints*. The weight of the body is transmitted to the legs through the pelvic girdle at these joints.

Figure 7.37
(a) Anterior view of the sacrum and coccyx; (b) posterior view.

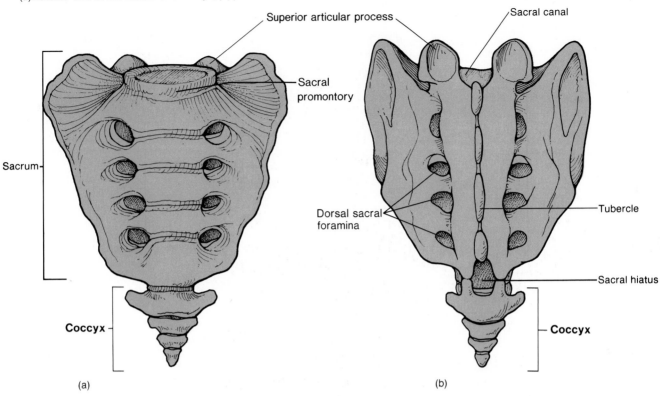

Chart 7.9	Bones of the vertebral column				
Bones	Number	Special features	Bones	Number	Special features
Cervical vertebrae	7	Transverse foramina, facets of atlas articulate with occipital condyles of skull, odontoid process of axis articulates with atlas, spinous processes of second through fifth vertebrae are bifid	Lumbar vertebrae	5	Large bodies; transverse processes that project backward at sharp angles; short, thick spinous processes directed nearly horizontally
			Sacrum	5 vertebrae fused into 1 bone	Dorsal sacral foramina, sacral promontory, sacral canal, sacral hiatus
Thoracic vertebrae	12	Pointed spinous processes that slope downward, facets that articulate with ribs	Coccyx	4 vertebrae fused into 1 bone	

The sacrum forms the posterior wall of the pelvic cavity. The upper anterior margin of the sacrum, which represents the body of the first sacral vertebra, is called the *sacral promontory*. This projection can be felt during a vaginal examination and is used as a guide in determining the size of the pelvis. This measurement is helpful, because an infant must pass through a woman's pelvic cavity during birth.

The vertebral foramina of the sacral vertebrae form the *sacral canal*, which continues through the sacrum to an opening of variable size at the tip called the *sacral hiatus*. This foramen exists because the laminae of the last sacral vertebra are not fused.

The Coccyx

The **coccyx** (kok'siks), or tailbone, is the lowest part of the vertebral column and is composed of four vertebrae that fuse together by the twenty-fifth year. It is attached by ligaments to the margins of the sacral hiatus. When a person is sitting, pressure is exerted upon the coccyx, and it moves forward acting somewhat like a shock absorber. Sitting down with great force sometimes causes the coccyx to be fractured or dislocated.

Chart 7.9 summarizes the bones of the vertebral column.

Figure 7.38

The thoracic cage includes the thoracic vertebrae, the sternum, the ribs, and the costal cartilages that attach the ribs to the sternum.

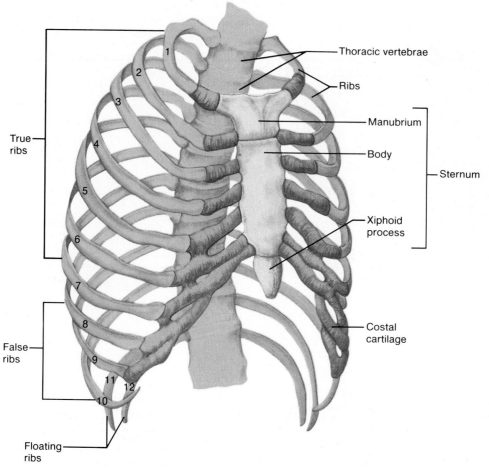

Figure 7.39

X-ray film of the thoracic cage viewed from the front. Note the shadow of the heart behind the sternum and above the diaphragm.

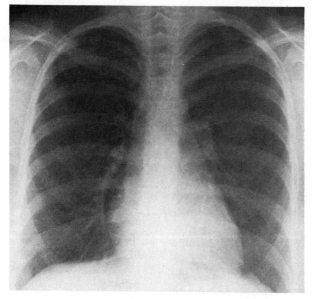

1. Describe the structure of the vertebral column.
2. Explain the difference between the vertebral column of an adult and that of an infant.
3. Describe a typical vertebra.
4. How do the structures of a cervical, a thoracic, and a lumbar vertebra differ?

The Thoracic Cage

The **thoracic cage** includes the ribs, the thoracic vertebrae, the sternum, and the costal cartilages by which the ribs are attached to the sternum. These parts support the shoulder girdle and arms, protect the visceral organs in the thoracic and upper abdominal cavities, and play a role in breathing (figures 7.38 and 7.39).

The Ribs

Regardless of sex, each person usually has twelve pairs of ribs—one pair attached to each of the twelve thoracic vertebrae. Occasionally, however, there may be extra ribs associated with the cervical or lumbar vertebrae.

The first seven rib pairs, the *true ribs* (vertebrosternal ribs), join the sternum directly by their costal cartilages. The remaining five pairs are called *false ribs* because their cartilages do not reach the sternum directly. Instead, the cartilages of the upper three false ribs (vertebrochondral ribs) join the cartilages of the ribs next above, while the last two rib pairs have no cartilaginous attachments to the sternum. These last two pairs are sometimes called *floating ribs* (vertebral ribs).

A typical rib (figure 7.40) has a long, slender shaft that curves around the chest and slopes downward. On the posterior end is an enlarged *head* by which the rib articulates with a facet on the body of its own vertebra and usually with the body of the next higher vertebra. The neck of the rib is a flattened part, lateral to the head, to which various ligaments are attached. Near the neck is a *tubercle* that articulates with the transverse process of the vertebra.

The costal cartilages are composed of hyaline cartilage. They are attached to the anterior ends of the ribs and continue in line with them toward the sternum.

Figure 7.40

A typical rib.

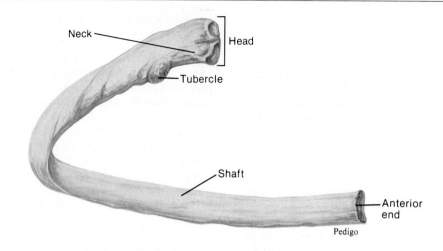

Neck

Head

Tubercle

Shaft

Anterior end

Pedigo

Figure 7.41

The pectoral girdle, to which the arms are attached, consists of a clavicle and a scapula on either side.

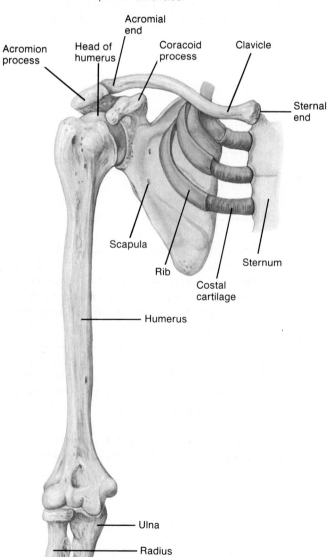

Acromial end

Head of humerus

Coracoid process

Clavicle

Acromion process

Sternal end

Scapula

Rib

Sternum

Costal cartilage

Humerus

Ulna

Radius

The Sternum

The **sternum** (breastbone) is located along the midline in the anterior portion of the thoracic cage. It is a flat, elongated bone that develops in three parts—an upper *manubrium*, a middle *body* (gladiolus), and a lower *xiphoid process* that projects downward (figure 7.38).

The sides of the manubrium and the body are notched where they articulate with costal cartilages. The manubrium also articulates with the clavicles by facets on its superior border. It usually remains as a separate bone until middle age or later, when it fuses to the body of the sternum.

The xiphoid process begins as a piece of cartilage. It slowly ossifies, and by middle life usually it is fused to the body of the sternum also.

The red marrow within the spongy bone of the sternum functions in blood-cell formation into adulthood. Since the sternum has a thin covering of compact bone and is easy to reach, samples of its blood-cell-forming tissue may be removed for use in diagnosing diseases. This procedure, a *sternal puncture*, involves suctioning (aspirating) some marrow through a hollow needle.

---

1. What bones make up the thoracic cage?
2. Describe a typical rib.
3. What are the differences between true, false, and floating ribs?

The Pectoral Girdle

The **pectoral girdle** (shoulder girdle) is composed of four parts—two clavicles (collarbones) and two scapulae (shoulder blades). Although the word *girdle* suggests

Figure 7.42

X-ray film of the right shoulder region viewed from the front. What features can you identify?

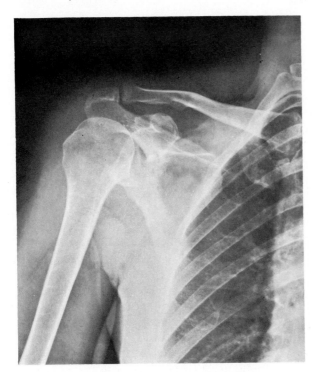

Figure 7.43

(a) Posterior surface of the scapula; (b) lateral view showing the glenoid cavity that articulates with the head of the humerus.

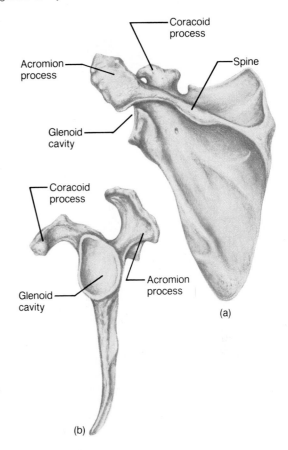

a ring-shaped structure, the pectoral girdle is an incomplete ring. It is open in the back between the scapulae, and its bones are separated in front by the sternum. However, the pectoral girdle supports the arms and serves as an attachment for several muscles that move the arms (figures 7.41 and 7.42).

The Clavicles

The **clavicles** are slender, rodlike bones with elongated S-shapes (figure 7.41). They are located at the base of the neck and run horizontally between the sternum and shoulders. The medial (or sternal) ends of the clavicles articulate with the manubrium, while the lateral (or acromial) ends join processes of the scapulae.

The clavicles act as braces for the freely movable scapulae, and thus help to hold the shoulders in place. They also provide attachments for muscles of the arms, chest, and back. As a result of its elongated double curve, the clavicle is structurally weak. If it is compressed from the end due to abnormal pressure on the shoulder, it is likely to fracture.

The Scapulae

The **scapulae** are broad, somewhat triangular bones located on either side of the upper back. They have flat bodies with concave anterior surfaces. The posterior surface of each scapula is divided into unequal portions by a *spine*. This spine leads to a *head*, which bears two processes—an *acromion process* that forms the tip of the shoulder and a *coracoid process* that curves forward and downward below the clavicle (figure 7.43). The acromion process articulates with a clavicle and provides attachments for muscles of the arm and chest. The coracoid process also provides attachments for arm and chest muscles.

On the head of the scapula, between the processes, is a depression called the *glenoid cavity*. It articulates with the head of the upper arm bone (humerus).

1. What bones form the pectoral girdle?
2. What is the function of the pectoral girdle?

Figure 7.44
(*a*) Frontal view of the left arm in the anatomical position;
(*b*) posterior view of the elbow.

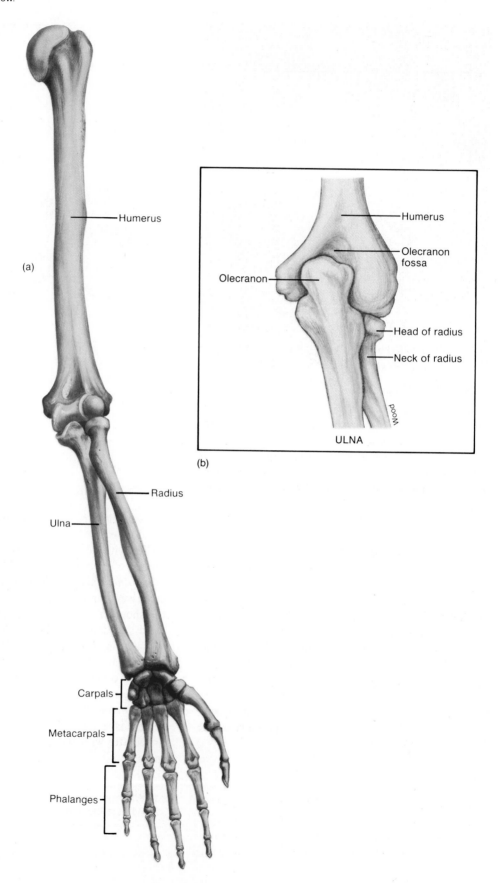

(a)

Humerus

Radius

Ulna

Carpals

Metacarpals

Phalanges

Humerus

Olecranon
fossa

Olecranon

Head of radius

Neck of radius

Wood

ULNA

(b)

Figure 7.45
X-ray film of the left elbow and lower arm viewed from the front.

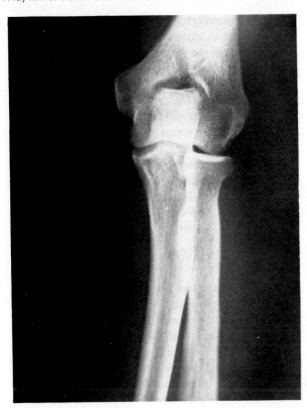

The Upper Limb

The bones of the upper limb form the framework of the arm, wrist, palm, and fingers. They also provide attachments for muscles and function in levers for moving limb parts. These bones include a humerus, a radius, an ulna, and several carpals, metacarpals, and phalanges (figures 7.44 and 7.45).

The Humerus

The **humerus** (figure 7.46) is a heavy bone that extends from the scapula to the elbow. At its upper end, it has a smooth, rounded *head* that fits into the glenoid cavity of the scapula. Just below the head, there are two processes—a *greater tubercle* on the lateral side and a *lesser tubercle* anteriorly. These tubercles provide attachments for muscles that move the arm at the shoulder. Between them is a narrow furrow, the *intertubercular groove*, through which a tendon passes from a muscle in the upper arm (biceps brachii) to the shoulder.

The narrow groove along the lower margin of the head that separates it from the tubercles is called the *anatomical neck*. Just below the head and the tubercles of the humerus is a tapering region called the *surgical neck*, so named because fractures commonly occur

Figure 7.46
(*a*) Posterior surface and (*b*) anterior surface of the left humerus.

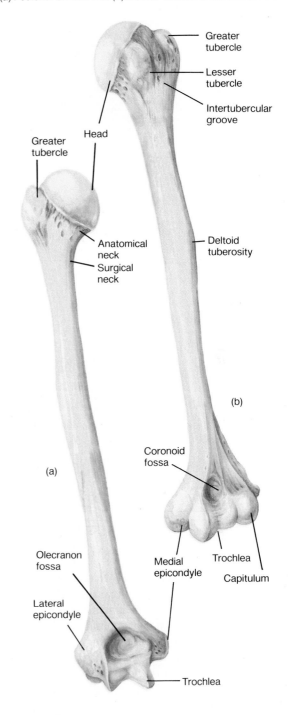

Figure 7.47
The head of the radius articulates with the radial notch of the ulna, and the head of the ulna articulates with the ulnar notch of the radius.

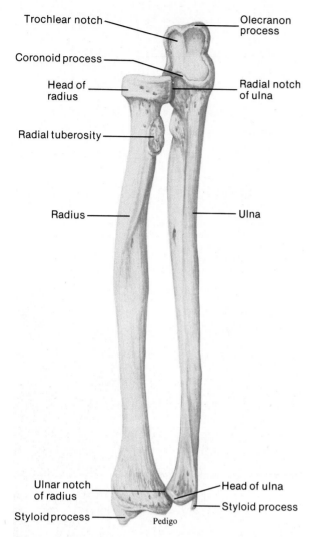

Trochlear notch

Olecranon process

Coronoid process

Head of radius

Radial notch of ulna

Radial tuberosity

Radius

Ulna

Ulnar notch of radius

Head of ulna

Styloid process

Styloid process

Pedigo

there. Near the middle of the bony shaft on the lateral side, there is a rough V-shaped area called the *deltoid tuberosity*. It provides an attachment for the muscle (deltoid) that raises the arm horizontally to the side.

At the lower end of the humerus, there are two smooth *condyles*—a knoblike *capitulum* on the lateral side and a pulley-shaped *trochlea* on the medial side. The capitulum articulates with the radius at the elbow, while the trochlea joins the ulna.

Above the condyles on either side are *epicondyles*, which provide attachments for muscles and ligaments of the elbow. Between the epicondyles anteriorly there is a depression, the *coronoid fossa*, that receives a process of the ulna (coronoid process) when the elbow is bent. Another depression on the posterior surface, the *olecranon fossa*, receives an ulnar process (olecranon process) when the arm is straightened at the elbow.

The Radius

The **radius,** located on the thumb side of the lower arm, is somewhat shorter than its companion, the ulna (figure 7.47). The radius extends from the elbow to the wrist and crosses over the ulna when the hand is turned so that the palm faces backward.

A thick, disklike *head* at the upper end of the radius articulates with the capitulum of the humerus and a notch of the ulna (radial notch). This arrangement allows the radius to rotate freely.

On the radial shaft, just below the head, is a process called the *radial tuberosity*. It serves as an attachment for a muscle (biceps brachii) that functions to bend the arm at the elbow. At the lower end of the radius, a lateral *styloid process* provides attachments for ligaments of the wrist.

The Ulna

The **ulna** is longer than the radius and overlaps the end of the humerus posteriorly. At its upper end, the ulna has a wrenchlike opening, the *trochlear notch* (semilunar notch), that articulates with the trochlea of the humerus. There is a process on either side of this notch. The one above it, the *olecranon process*, provides an attachment for the muscle (triceps brachii) that straightens the arm at the elbow. During this movement, the olecranon process of the ulna fits into the olecranon fossa of the humerus. Similarly, the coronoid process, just below the trochlear notch, fits into the coronoid fossa of the humerus when the elbow is bent.

At the lower end of the ulna, its knoblike *head* articulates with a notch of the radius (ulnar notch) laterally and with a disk of fibrocartilage inferiorly (figure 7.47). This disk, in turn, joins a wrist bone (triquetrum). A medial *styloid process* at the distal end of the ulna provides attachments for ligaments of the wrist.

The Hand

The hand is composed of a wrist, a palm, and five fingers (figures 7.48 and 7.49). The skeleton of the wrist consists of eight small **carpal bones** that are firmly bound in two rows of four bones each. The resulting compact mass is called a *carpus*.

The carpus is rounded on its proximal surface, where it articulates with the radius and with the fibrocartilaginous disk on the ulnar side. The carpus is concave anteriorly, forming a canal through which tendons and nerves extend to the palm. Its distal surface articulates with the metacarpal bones. The individual bones of the carpus are named in figure 7.48.

Five **metacarpal bones,** one in line with each finger, form the framework of the palm. These bones are cylindrical, with rounded distal ends that make the knuckles on a clenched fist. The metacarpals articulate proximally with the carpals and distally with the pha-

Chart 7.10 Bones of the pectoral girdle and upper limbs

Name and number	Location	Special features
Clavicle 2	Base of neck, between sternum and scapula	Sternal end, acromial end
Scapula 2	Upper back, forming part of the shoulder	Body, spine, head, acromion process, coracoid process
Humerus 2	Upper arm, between scapula and elbow	Head, greater tubercle, lesser tubercle, intertubercular groove, surgical neck, deltoid tuberosity, capitulum, trochlea, medial epicondyle, lateral epicondyle, coronoid fossa, olecranon fossa
Radius 2	Lateral side of lower arm, between elbow and wrist	Head, radial tuberosity, styloid process
Ulna 2	Medial side of lower arm, between elbow and wrist	Trochlear notch, olecranon process, head, styloid process
Carpal 16	Wrist	Arranged in two rows of four bones each
Metacarpal 10	Palm	One in line with each finger
Phalanx 28	Finger	Three in each finger, two in each thumb

Figure 7.48
The left hand viewed from the back.

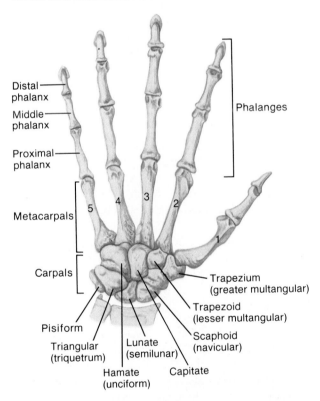

- Distal phalanx
- Middle phalanx
- Proximal phalanx
- Phalanges
- Metacarpals
- Carpals
- Pisiform
- Triangular (triquetrum)
- Hamate (unciform)
- Lunate (semilunar)
- Capitate
- Scaphoid (navicular)
- Trapezoid (lesser multangular)
- Trapezium (greater multangular)

Figure 7.49
X-ray film of the left hand. Note the small sesamoid bone associated with the joint at the base of the thumb (arrow).

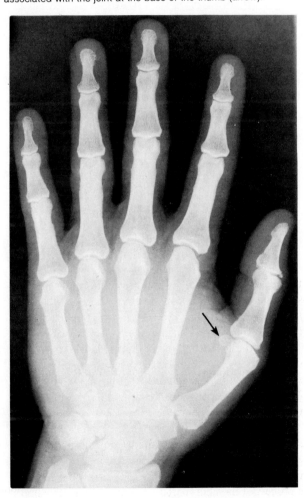

langes. The metacarpal on the lateral side is the most freely movable; it permits the thumb to oppose the fingers when something is grasped in the hand. These bones are numbered 1 to 5, beginning with the metacarpal of the thumb.

The **phalanges** are the bones of the fingers. There are three in each finger—a proximal, a middle, and a distal phalanx—and two in the thumb. (It lacks a middle phalanx.) Thus, there are fourteen finger bones in each hand.

Chart 7.10 summarizes the bones of the pectoral girdle and upper limbs.

Figure 7.50

(*a*) Anterior view and (*b*) posterior view of the pelvic girdle. This girdle provides an attachment for the legs and together with the sacrum and coccyx forms the pelvis.

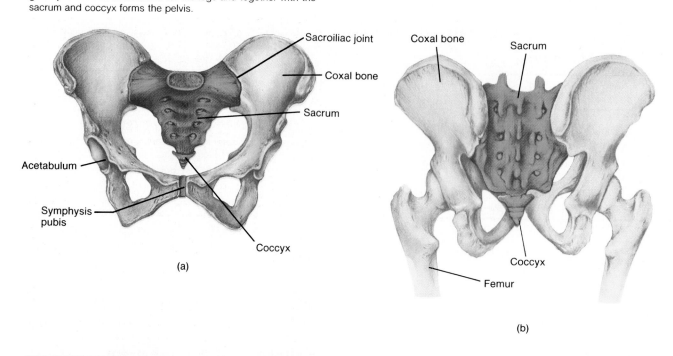

(a)

(b)

Figure 7.51

(*a*) Lateral surface of the right coxal bone; (*b*) medial view.

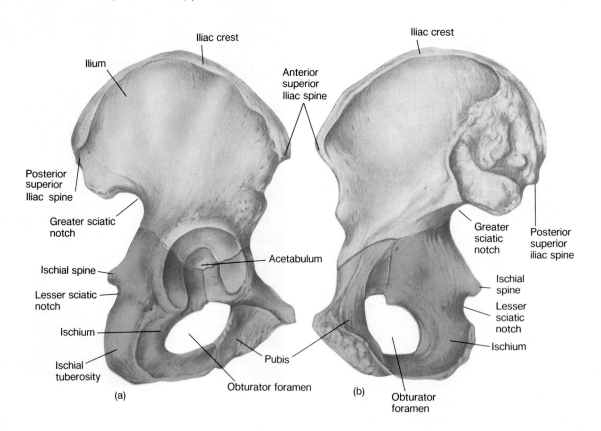

(a)

(b)

1. Locate and name each of the bones of the upper limb.
2. Explain how these bones articulate with one another.

The Pelvic Girdle

The **pelvic girdle** consists of the two coxal bones (hip bones) that articulate with each other anteriorly and with the sacrum posteriorly (figure 7.50). With the sacrum and coccyx, the pelvic girdle forms the ringlike pelvis, which provides stable support for the trunk of the body and attachments for the legs. The weight of the body is transmitted through the pelvis to the legs and then onto the ground. The pelvis also protects the urinary bladder, the distal end of the large intestine, and the internal reproductive organs.

The Coxal Bones

Each **coxal bone** (os coxa) develops from three parts—an ilium, an ischium, and a pubis. These parts fuse in the region of a cup-shaped cavity called the *acetabulum*. This depression is on the lateral surface of the hip bone, and it receives the rounded head of the femur or thigh bone (figure 7.51).

The **ilium** (il'e-um), which is the largest and uppermost portion of the coxal bone, flares outward to form the prominence of the hip. The margin of this prominence is called the *iliac crest*.

Posteriorly, the ilium joins the sacrum at the *sacroiliac joint*. Anteriorly, a projection of the ilium, the *anterior superior iliac spine*, can be felt lateral to the groin. This spine provides attachments for ligaments and muscles and is an important surgical landmark.

A common injury that occurs in contact sports, such as football, involves bruising the soft tissues and bone associated with the anterior superior iliac spine. This painful injury is called a *hip pointer* and may be prevented by wearing protective padding.

On the posterior border of the ilium, there is a *posterior superior iliac spine*. Below this spine there is a deep indentation, the *greater sciatic notch*, through which a number of nerves and blood vessels pass.

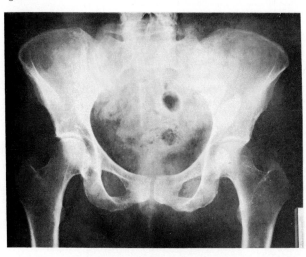

The **ischium** (is'ke-um), which forms the lowest portion of the coxal bone, is L-shaped with its angle, the *ischial tuberosity*, pointing posteriorly and downward. This tuberosity has a rough surface that provides attachments for ligaments and leg muscles. It also supports the weight of the body when a person is sitting. Above the ischial tuberosity, near the junction of the ilium and ischium, is a sharp projection called the *ischial spine*. This spine, which can be felt during a vaginal examination, is used as a guide for determining the size of the pelvis. The distance between the ischial spines represents the shortest diameter of the pelvic outlet.

The **pubis** (pu'bis) constitutes the anterior portion of the coxal bone. The two pubic bones come together in the midline to create a joint called the *symphysis pubis*. The angle formed by these bones below the symphysis is the *pubic arch*.

A portion of each pubis passes posteriorly and downward to join an ischium. Between the bodies of these bones on either side there is a large opening, the *obturator foramen*, which is the largest foramen in the skeleton. In life, this foramen is covered and nearly closed by an obturator membrane. (See figure 7.52.)

Figure 7.53

The female pelvis is usually wider in all diameters and roomier than that of the male. (*a*) Male pelvis; (*b*) female pelvis.

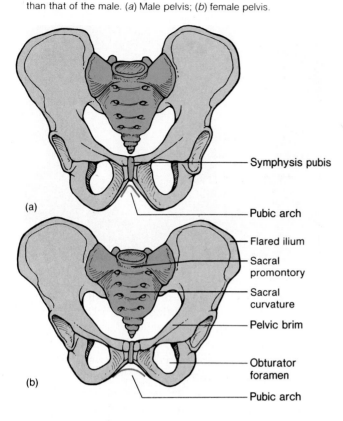

(a)

(b)

- Symphysis pubis
- Pubic arch
- Flared ilium
- Sacral promontory
- Sacral curvature
- Pelvic brim
- Obturator foramen
- Pubic arch

The Greater and Lesser Pelves

If a line is drawn along each side of the pelvis from the sacral promontory downward and anteriorly to the upper margin of the symphysis pubis, it marks the *pelvic brim* (linea terminalis). This margin, which can be traced in figure 7.53, separates the lower, or lesser (true), pelvis from the upper, or greater (false), pelvis.

The *greater pelvis* is bounded posteriorly by the lumbar vertebrae, laterally by the flared parts of the iliac bones, and anteriorly by the abdominal wall. Although the false pelvis helps to support abdominal organs, it has no particular importance in childbirth.

The *lesser pelvis* is bounded posteriorly by the sacrum and coccyx, and laterally and anteriorly by the lower ilium, ischium, and pubis bones. This portion of the pelvis surrounds a short, canal-like cavity that has an upper inlet and a lower outlet. This cavity is of special interest because an infant passes through it during childbirth.

Chart 7.11	Some sexual differences of the skeleton
Part	**Sexual differences**
Skull	Female skull is relatively smaller and lighter, and its muscular attachments are less conspicuous. The female forehead is longer vertically, the facial area is rounder, the jaw is smaller, and the mastoid process is less prominent than that of a male.
Pelvis	Female pelvic bones are lighter, thinner, and have less obvious muscular attachments. The obturator foramina and the acetabula are smaller, and farther apart than those of a male.
Pelvic cavity	Female pelvic cavity is wider in all diameters and is shorter, roomier, and less funnel-shaped. The distances between the ischial spines and between the ischial tuberosities are greater than in the male.
Sacrum	Female sacrum is relatively wider, the sacral promontory projects forward to a lesser degree, and the sacral curvature is bent more sharply posteriorly than in a male.
Coccyx	Female coccyx is more movable than that of a male.

Sexual Differences in Pelves

Although it is difficult to find examples of adult male and female pelves that have all of the male or female characteristics, some differences usually are present. These differences are related to the function of the female pelvis as a birth canal. Usually the female iliac bones are more flared than those of the male, and consequently the female hips usually are broader. The angle of the female pubic arch may be greater; there may be more distance between the ischial spines and the ischial tuberosities; and the sacral curvature may be shorter and flatter. Thus, the female pelvic cavity is usually wider in all diameters and is roomier than that of the male. Also, the bones of the female pelvis are usually lighter, more delicate, and show less evidence of muscle attachments (figure 7.53). Chart 7.11 summarizes some of the sexual differences of the skeleton.

1. Locate and name each of the bones of the pelvis.
2. Explain what is meant by the greater pelvis and the lesser pelvis.
3. How can a male and female pelvis be distinguished?

The Lower Limb

The bones of the lower limb form the frameworks of the leg, ankle, instep, and toes. They include a femur, a tibia, a fibula, and several tarsals, metatarsals, and phalanges (figures 7.54 and 7.55).

Figure 7.54
(*a*) The left lower leg viewed from the front; (*b*) posterior view of the knee.

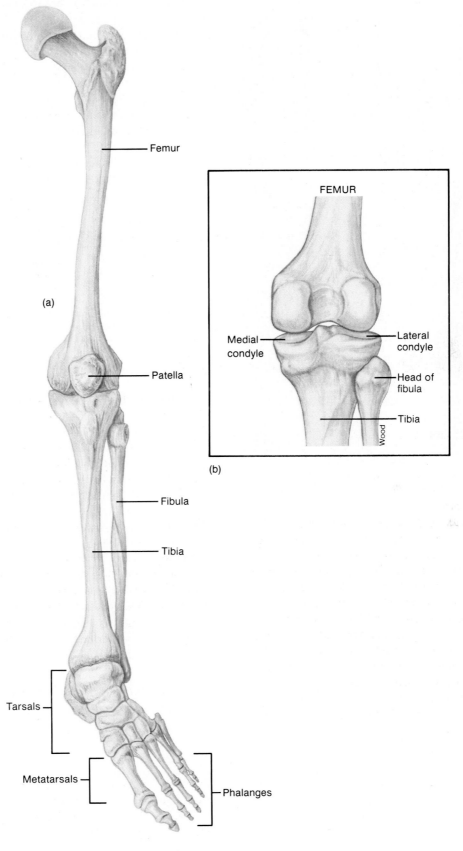

Femur

(a)

Patella

Fibula

Tibia

Tarsals

Metatarsals

Phalanges

FEMUR

Medial condyle

Lateral condyle

Head of fibula

Tibia

Wood

(b)

Figure 7.55

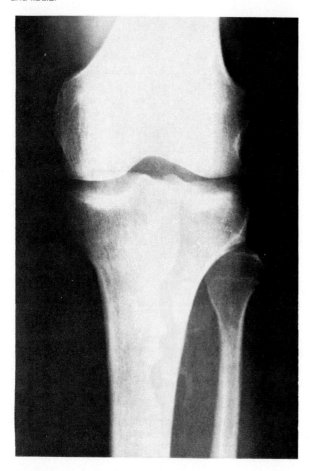

X-ray film of the left knee showing the ends of the femur, tibia, and fibula.

The Femur

The **femur,** or thigh bone, is the longest bone in the body and extends from the hip to the knee. A large, rounded *head* at its upper end projects medially into the acetabulum of the coxal bone. On the head, a pit called the *fovea capitis* marks the attachment of a ligament. Just below the head, there is a constriction, or *neck,* and two large processes—an upper, lateral *greater trochanter* and a lower, medial *lesser trochanter.* These processes provide attachments for muscles of the legs and buttocks. On the posterior surface in the middle third of the shaft, there is a longitudinal crest called the *linea aspera.* This rough strip serves as an attachment for several muscles (figure 7.56).

At the lower end of the femur, two rounded processes, the *lateral* and *medial condyles,* articulate with the tibia of the lower leg. A **patella** (kneecap) also articulates with the femur on its distal anterior surface.

On the medial surface at its distal end, there is a prominent *medial epicondyle,* and on the lateral surface there is a *lateral epicondyle.* These projections provide attachments for various muscles and ligaments.

The patella is a flat sesamoid bone located in a tendon that passes anteriorly over the knee (see figure 7.54). Because of its position, the patella controls the angle at which this tendon continues toward the tibia, and so it functions in lever actions associated with lower leg movements.

As a result of a blow to the knee or a forceful unnatural movement of the leg, the patella sometimes slips out of position to one side. This painful condition is called a *patellar dislocation.* Such a displacement may be prevented by exercises that strengthen muscles associated with the knee and by wearing protective padding. Unfortunately, once the soft tissues that hold the patella in place are stretched, patellar dislocation tends to recur.

The Tibia

The **tibia** (shinbone) is the larger of the two lower leg bones and is located on the medial side. Its upper end is expanded into *medial* and *lateral condyles,* which have concave surfaces and articulate with the condyles of the femur. Below the condyles, on the anterior surface, is a process called the *tibial tuberosity,* which provides an attachment for the *patellar ligament*—a continuation of the patella-bearing tendon. A prominent *anterior crest* extends downward from the tuberosity and serves as an attachment for connective tissues in the lower leg.

At its lower end, the tibia expands to form a prominence on the inner ankle called the *medial malleolus,* which serves as an attachment for ligaments. On its lateral side is a depression that articulates with the fibula. The inferior surface of its distal end articulates with a large bone (the talus) in the foot (figure 7.57).

Figure 7.56
(a) Anterior surface and (b) posterior surface of the left femur.

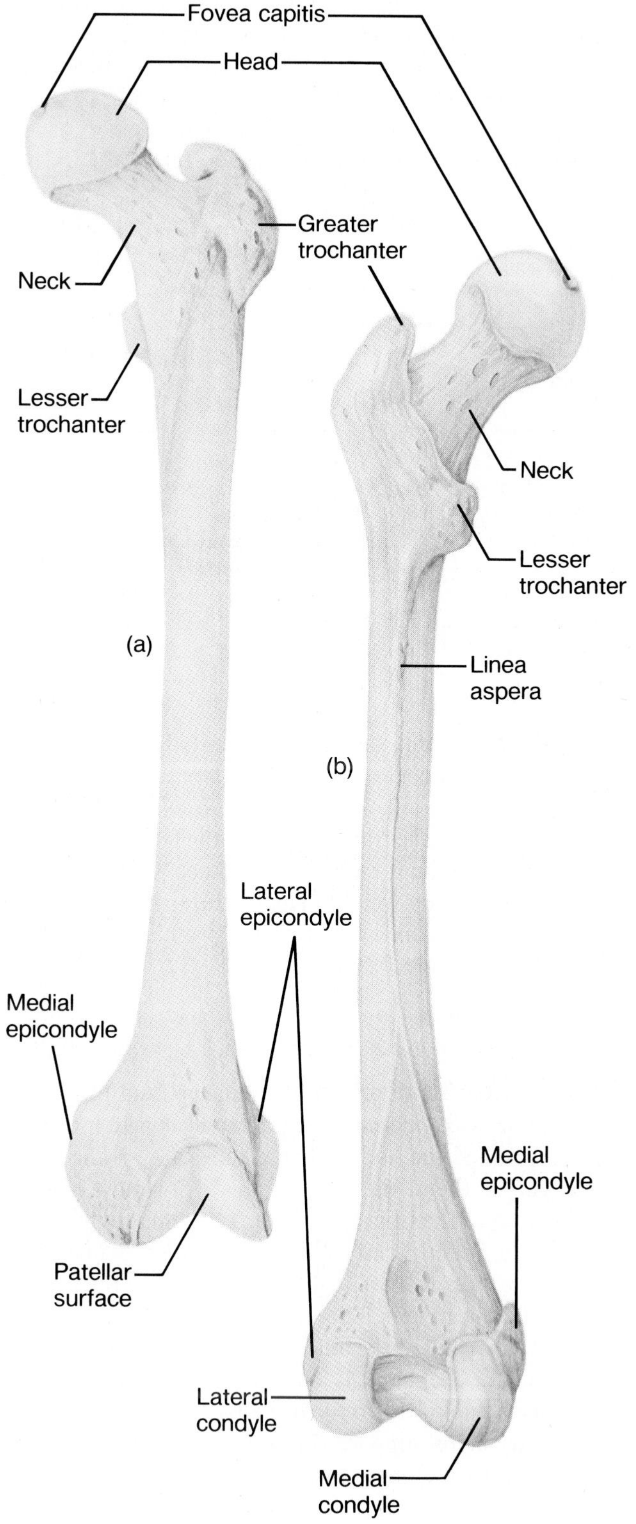

Fovea capitis

Head

Neck

Greater trochanter

Lesser trochanter

Neck

Lesser trochanter

(a)

(b)

Linea aspera

Medial epicondyle

Lateral epicondyle

Patellar surface

Lateral condyle

Medial condyle

Medial epicondyle

Figure 7.57
Bones of the left lower leg viewed from the front.

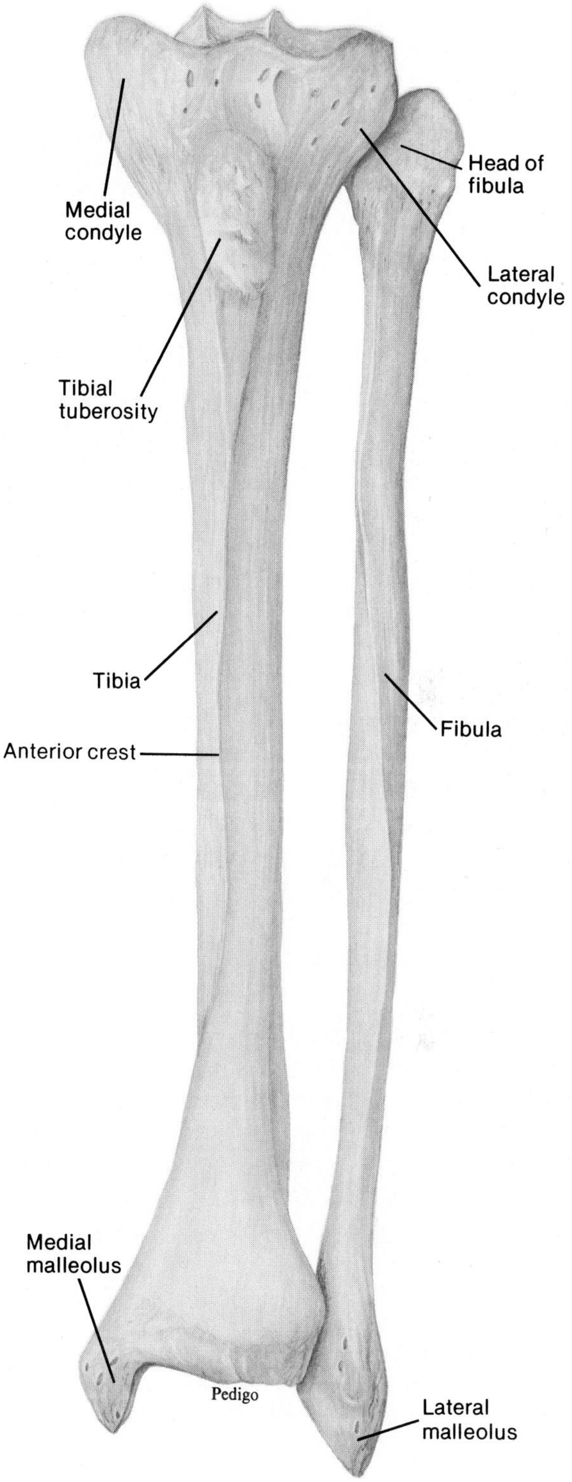

Medial condyle

Head of fibula

Lateral condyle

Tibial tuberosity

Tibia

Anterior crest

Fibula

Medial malleolus

Pedigo

Lateral malleolus

Figure 7.58

The talus moves freely where it articulates with the tibia and fibula.

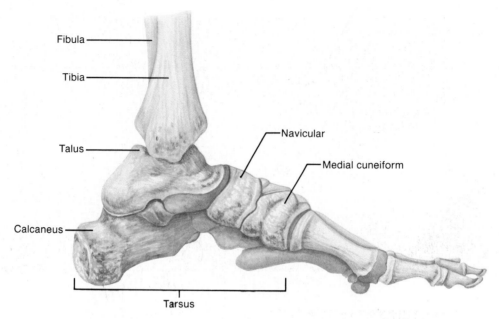

Fibula

Tibia

Talus

Calcaneus

Navicular

Medial cuneiform

Tarsus

Figure 7.59

The left foot viewed from above.

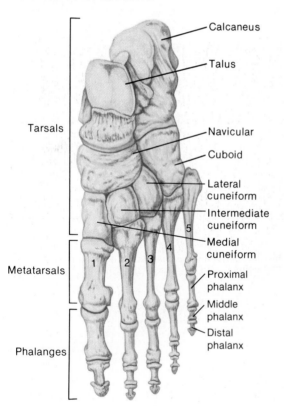

Tarsals

Metatarsals

Phalanges

Calcaneus

Talus

Navicular

Cuboid

Lateral cuneiform

Intermediate cuneiform

Medial cuneiform

Proximal phalanx

Middle phalanx

Distal phalanx

1 2 3 4 5

The Fibula

The **fibula** is a long, slender bone located on the lateral side of the tibia. Its ends are slightly enlarged into an upper *head* and a lower *lateral malleolus*. The head articulates with the tibia just below the lateral condyle; however, it does not enter into the knee joint and does not bear any body weight. The lateral malleolus articulates with the ankle and forms a prominence on the lateral side (figure 7.57).

The Foot

The foot consists of an ankle, an instep, and five toes. The ankle is composed of seven **tarsal bones,** forming a group called the *tarsus.* These bones are arranged so that one of them, the **talus** (ta′lus), can move freely where it joins the tibia and fibula. The remaining tarsal bones are bound firmly together, forming a mass on which the talus rests. The individual bones of the tarsus are named in figures 7.58 and 7.59.

The largest of the ankle bones, the **calcaneus** (kal-ka′ne-us), or heel bone, is located below the talus where it projects backward to form the base of the heel. The calcaneus helps support the weight of the body and provides an attachment for muscles that move the foot.

Figure 7.60

(a) X-ray film of the left foot viewed from the medial side; (b) from above. What features can you identify?

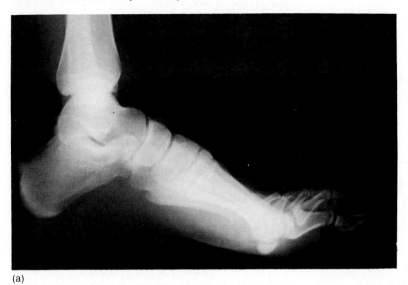

(a)

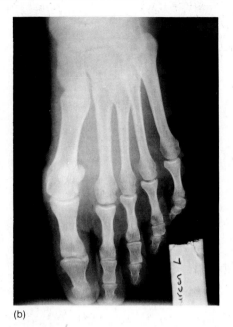

(b)

Chart 7.12 Bones of the pelvic girdle and lower limbs		
Name and number	Location	Special features
Coxal bone 2	Hip, articulating with each other anteriorly and with the sacrum posteriorly	Ilium, iliac crest, anterior superior iliac spine, ischium, ischial tuberosity, ischial spine, obturator foramen, acetabulum, pubis
Femur 2	Upper leg, between the hip and knee	Head, fovea capitis, neck, greater trochanter, lesser trochanter, linea aspera, lateral condyle, medial condyle
Patella 2	Anterior surface of knee	
Tibia 2	Medial side of lower leg, between knee and ankle	Medial condyle, lateral condyle, tibial tuberosity, anterior crest, medial malleolus
Fibula 2	Lateral side of lower leg, between knee and ankle	Head, lateral malleolus
Tarsal 14	Ankle	Freely movable talus that articulates with lower leg bones and six other bones bound firmly together
Metatarsal 10	Instep	One in line with each toe, arranged and bound by ligaments to form arches
Phalanx 28	Toe	Three in each toe, two in the great toe

The instep consists of five, elongated **metatarsal bones** that articulate with the tarsus. They are numbered 1 to 5 beginning on the medial side (figure 7.59). The heads at the distal ends of these bones form the ball of the foot. The tarsals and metatarsals are arranged and bound by ligaments to form the arches of the foot. A longitudinal arch extends from the heel to the toe, and a transverse arch stretches across the foot. These arches provide a stable, springy base for the body. Sometimes, however, the tissues that bind the metatarsals become weakened, producing fallen arches or flat feet.

The **phalanges** of the toes are similar to those of the fingers. They are in line with the metatarsals and articulate with them. Each toe has three phalanges—a proximal, a middle, and a distal phalanx—except the great toe, which has only two and lacks the middle phalanx. Chart 7.12 summarizes the bones of the pelvic girdle and lower limbs. (See figure 7.60.)

1. Locate and name each of the bones of the lower limb.
2. Explain how these bones articulate with one another.
3. Describe how the foot is adapted to support the body.

Clinical Terms Related to the Skeletal System

achondroplasia (a-kon″dro-pla′ze-ah)   an inherited condition in which the formation of cartilaginous bone is retarded. The result is a type of dwarfism.

acromegaly (ak″ro-meg′ah-le)   a condition due to an overproduction of growth hormone in adults and characterized by abnormal enlargement of facial features, hands, and feet.

Colles fracture (kol′ez frak′tūre)   a fracture at the distal end of the radius in which the smaller fragment is displaced posteriorly.

epiphysiolysis (ep″ĭ-fiz″e-ol′ĭ-sis)   a separation or loosening of the epiphysis from the diaphysis of a bone.

laminectomy (lam″ĭ-nek′to-me)   surgical removal of the posterior arch of a vertebra, usually to relieve the symptoms of a ruptured intervertebral disk.

lumbago (lum-ba′go)   a dull ache in the lumbar region of the back.

orthopedics (or″tho-pe′diks)   the science of prevention, diagnosis, and treatment of diseases and abnormalities involving the skeletal and muscular systems.

ostalgia (os-tal′je-ah)   pain in a bone.

ostectomy (os-tek′to-me)   surgical removal of a bone.

osteitis (os″te-i′tis)   inflammation of bone tissue.

osteochondritis (os″te-o-kon-dri′tis)   inflammation of bone and cartilage tissues.

osteogenesis (os″te-o-jen′ĕ-sis)   the development of bone.

osteogenesis imperfecta (os″te-o-jen′ĕ-sis im-per-fek′ta)   a congenital condition characterized by the development of deformed and abnormally brittle bones.

osteoma (os″te-o′mah)   a tumor composed of bone tissue.

osteomalacia (os″te-o-mah-la′she-ah)   a softening of adult bone due to a disorder in calcium and phosphorus metabolism, usually caused by a deficiency of vitamin D.

osteomyelitis (os″te-o-mi″ĕ-li′tis)   inflammation of bone caused by the action of bacteria or fungi.

osteonecrosis (os″te-o-ne-kro′sis)   death of bone tissue. This condition occurs most commonly in the head of the femur in elderly persons and may be due to obstructions in arteries that supply the bone.

osteopathology (os″te-o-pah-thol′o-je)   the study of bone diseases.

osteotomy (os″te-ot′o-me)   the cutting of a bone.

Roentgenogram (rent-gen′o-gram″)   an X-ray film.

Chapter Summary

Introduction

Individual bones are the organs of the skeleton system. Bone contains very active tissues.

Bone Structure

Bone structure reflects its function.

1. Classification of bones
    Bones are grouped according to their shapes—long, short, flat, or irregular.
2. Parts of a long bone
    a. Epiphyses are covered with articular cartilage and articulate with other bones.
    b. The shaft of a bone is called the diaphysis.
    c. Except for the articular cartilage, a bone is covered by a periosteum.
    d. Compact bone provides strength and resistance to bending.
    e. Spongy bone provides strength where needed and reduces the weight of bone.
    f. The diaphysis contains a medullary cavity filled with marrow.
3. Microscopic structure
    a. Compact bone contains osteons cemented together.
    b. Osteonic canals contain blood vessels that nourish the cells of osteons.
    c. Communicating canals connect osteonic canals transversely and communicate with the bone's surface and the medullary cavity.
    d. Cells of spongy bone are nourished by diffusion from the surface of the bony plates.

Bone Development and Growth

1. Intramembranous bones
    a. Intramembranous bones include certain flat bones of the skull.
    b. They develop from layers of connective tissues.
    c. Bone tissue is formed by osteoblasts within the membranous layers.
    d. Mature bone cells are called osteocytes.
    e. Membranous tissues give rise to an inner and outer periosteum.
2. Endochondral bones
    a. Endochondral bones include most of the bones of the skeletal system.
    b. They develop first as hyaline cartilage that is later replaced by bone tissue.
3. Growth of an endochondral bone
    a. Primary ossification center appears in the diaphysis, while secondary ossification centers appear in the epiphyses.
    b. An epiphyseal disk remains between the primary and secondary ossification centers.

c. An epiphyseal disk consists of layers of cells: resting cells, young reproducing cells, older enlarging cells, and dying cells.

d. The epiphyseal disk is responsible for growth in length.

e. Long bones continue to growth in length until the epiphyseal disks are ossified.

f. Growth in thickness is due to intramembranous ossification occurring beneath the periosteum.

g. The medullary cavity is created by the action of osteoclasts.

4. Factors affecting bone development and growth

a. Deficiencies of vitamin A, C, or D result in abnormal development.

b. Insufficient secretion of pituitary growth hormone may result in dwarfism; excessive secretion may result in giantism or acromegaly.

c. Deficiency of thyroid hormone delays bone growth.

d. Male and female sex hormones promote bone formation and stimulate ossification of the epiphyseal disks.

Functions of Bones

1. Support and protection

a. Skeletal parts provide shape and form for body structures.

b. They support and protect softer, underlying tissues.

2. Lever actions

a. Bones and muscles function together as levers.

b. A lever consists of a rod, pivot (fulcrum), weight that is moved, and a force that supplies energy.

c. Parts of a first-class lever are arranged weight-pivot-force; parts of a second-class lever are arranged pivot-weight-force; parts of a third-class lever are arranged pivot-force-weight.

3. Blood cell formation

a. At different ages, hematopoiesis occurs in the yolk sac, liver and spleen, and red bone marrow.

b. Red marrow functions in the production of red blood cells, white blood cells, and blood platelets.

4. Storage of inorganic salts

a. The intercellular material of bone tissue contains large quantities of calcium phosphate in the form of hydroxyapatite.

b. When blood calcium ion concentration is low, osteoclasts resorb bone, thus releasing calcium salts.

c. When blood calcium ion concentration is high, osteoblasts are stimulated to store calcium salts in bones.

d. Bone tissues may accumulate lead, radium, or strontium.

e. Bone stores lesser amounts of sodium, magnesium, potassium, and carbonate ions.

Organization of the Skeleton

1. Number of bones

a. Usually there are 206 bones in the human skeleton, but the number may vary.

b. Extra bones in sutures are called wormian bones.

2. Divisions of the skeleton

a. The skeleton can be divided into axial and appendicular portions.

b. The axial skeleton consists of the skull, hyoid bone, vertebral column, and thoracic cage.

c. The appendicular skeleton consists of the pectoral girdle, upper limbs, pelvic girdle, and lower limbs.

The Skull

The skull consists of 22 bones, which include 8 cranial bones, 13 facial bones, and 1 mandible.

1. The cranium

a. The cranium encloses and protects the brain and provides attachments for muscles.

b. Some cranial bones contain air-filled sinuses that help to reduce the weight of the skull.

c. Cranial bones include frontal bone, parietal bones, occipital bone, temporal bones, sphenoid bone, and ethmoid bone.

2. The facial skeleton

a. Facial bones provide the basic shape of the face and attachments for muscles.

b. Facial bones include maxillary bones, palatine bones, zygomatic bones, lacrimal bones, nasal bones, vomer bone, inferior nasal conchae, and mandible.

3. The infantile skull

a. Incompletely developed bones are separated by fontanels that enable the infantile skull to change shape slightly during birth.

b. Proportions of the infantile skull are different from those of an adult skull, and its bones are less easily fractured.

The Vertebral Column

The vertebral column extends from the skull to the pelvis and protects the spinal cord. It is composed of vertebrae separated by intervertebral disks. It consists of 33 bones in an infant and 26 bones in an adult. It has 4 curvatures—cervical, thoracic, lumbar, and pelvic.

1. A typical vertebra

a. A typical vertebra consists of a body, pedicles, laminae, spinous process, transverse processes, and superior and inferior articulating processes.

b. Notches on the lower surfaces of the pedicles provide intervertebral foramina through which spinal nerves pass.

2. Cervical vertebrae
   a. Cervical vertebrae comprise the bones of the neck.
   b. Transverse processes bear transverse foramina.
   c. Atlas (first vertebra) supports and balances the head.
   d. Odontoid process of the axis (second vertebra) provides a pivot for the atlas when the head is turned from side to side.
3. Thoracic vertebrae
   a. Thoracic vertebrae are larger than cervical vertebrae.
   b. Their long spinous processes slope downward, and facets on the sides of bodies articulate with the ribs.
4. Lumbar vertebrae
   a. Vertebral bodies of lumbar vertebrae are large and strong.
   b. Their transverse processes project back at sharp angles, and spinous processes are directed horizontally.
5. The sacrum
   a. The sacrum is a triangular structure that bears rows of dorsal sacral foramina.
   b. It is united with the coxal bones at the sacroiliac joints.
   c. Sacral promontory provides a guide for determining the size of the pelvis.
6. The coccyx
   a. The coccyx forms the lowest part of the vertebral column.
   b. It acts as a shock absorber when a person sits.

The Thoracic Cage
The thoracic cage includes the ribs, thoracic vertebrae, sternum, and costal cartilages.

It supports the shoulder girdle and arms, protects visceral organs, and functions in breathing.

1. The ribs
   a. Twelve pairs of ribs are attached to the thoracic vertebrae.
   b. Costal cartilages of true ribs join the sternum directly; those of the false ribs join indirectly or lack cartilaginous attachments, as do the floating ribs.
   c. A typical rib bears a shaft, head, and tubercles that articulate with the vertebrae.
2. The sternum
   a. The sternum consists of a manubrium, body, and xiphoid process.
   b. It articulates with costal cartilages and clavicles.

The Pectoral Girdle
The pectoral girdle is composed of two clavicles and two scapulae. It forms an incomplete ring that supports the arms and provides attachments for muscles that move the arms.

1. The clavicles
   a. Clavicles are rodlike bones that run horizontally between the sternum and shoulder.
   b. They function to hold the shoulders in place and provide attachments for muscles.
2. The scapulae
   a. The scapulae are broad, triangular bones with bodies, spines, heads, acromion processes, coracoid processes, and glenoid cavities.
   b. They articulate with the humerus of each arm, and provide attachments for muscles of the arms and chest.

The Upper Limb
Bones of the limb provide frameworks and attachments for muscles and function in levers that move the limb and its parts.

1. The humerus
   a. The humerus extends from the scapula to the elbow.
   b. It bears a head, greater tubercle, lesser tubercle, intertubercular groove, surgical neck, deltoid tuberosity, capitulum, trochlea, epicondyles, coronoid fossa, and olecranon fossa.
2. The radius
   a. The radius is located on the thumb side of the lower arm between the elbow and wrist.
   b. It bears a head, radial tuberosity, and styloid process.
3. The ulna
   a. The ulna is longer than the radius and overlaps the humerus posteriorly.
   b. It bears a trochlear notch, olecranon process, head, and styloid process.
   c. It articulates with the radius laterally and with a disk of fibrocartilage inferiorly.
4. The hand
   a. The hand is composed of a wrist, palm, and 5 fingers.
   b. It includes 8 carpals that form a carpus, 5 metacarpals, and 14 phalanges.

The Pelvic Girdle
The pelvic girdle consists of two coxal bones that articulate with each other anteriorly and with the sacrum posteriorly.

The sacrum, coccyx, and pelvic girdle form the pelvis.

The girdle provides support for body weight and attachments for muscles, and protects visceral organs.

1. The coxal bones
Each coxal bone consists of an ilium, ischium, and pubis, which are fused in the region of the acetabulum.
   a. The ilium
      (1) The ilium, the largest portion of the coxal bone, joins the sacrum at the sacroiliac joint.
      (2) It bears an iliac crest with anterior and posterior superior iliac spines.
   b. The ischium
      (1) The ischium is the lowest portion of the coxal bone.
      (2) It bears an ischial tuberosity and ischial spine.
   c. The pubis
      (1) The pubis is the anterior portion of the coxal bone.
      (2) Pubis bones are fused anteriorly at the symphysis pubis.
2. The greater and lesser pelves
   a. Lesser pelvis is below the pelvic brim; greater pelvis is above it.
   b. Lesser pelvis functions as a birth canal; greater pelvis helps to support abdominal organs.
3. Sexual differences in pelves
   a. Differences between male and female pelves are related to the function of the female pelvis as a birth canal.
   b. Usually the female pelvis is more flared; pubic arch is broader; distance between the ischial spines and the ischial tuberosities is greater; and sacral curvature is shorter.

The Lower Limb
1. The femur
   a. The femur extends from the knee to the hip.
   b. It bears a head, fovea capitis, neck, greater trochanter, lesser trochanter, linea aspera, lateral condyle, and medial condyle.
   c. Patella articulates with its anterior surface.
2. The tibia
   a. The tibia is located on the medial side of the lower leg.
   b. It bears medial and lateral condyles, tibial tuberosity, anterior crest, and medial malleolus.
   c. It articulates with the talus of the ankle.
3. The fibula
   a. The fibula is located on the lateral side of the tibia.
   b. It bears a head and lateral malleolus that articulates with the ankle.
4. The foot
   a. The foot consists of an ankle, instep, and 5 toes.
   b. It includes 7 tarsals that form the tarsus, 5 metatarsals, and 14 phalanges.

Application of Knowledge
1. What steps do you think should be taken to reduce the chances of persons accumulating abnormal metallic elements such as lead, radium, and strontium in their bones?
2. Why do you think incomplete, longitudinal fractures of bone shafts (greenstick fractures) are more common in children than in adults?
3. When a child's bone is fractured, growth may be stimulated at the epiphyseal disk of that bone. What problems might this extra growth create in an arm or leg before the growth of the other limb compensates for the difference in length?
4. How would you explain the observation that elderly persons often develop "bowed backs" and appear shorter than they were in earlier years?

Review Activities

Part A
1. List four groups of bones, based upon their shapes, and name an example from each group.
2. Sketch a typical long bone and label its epiphyses, diaphysis, medullary cavity, periosteum, and articular cartilages.
3. Distinguish between spongy and compact bone.
4. Explain how osteonic canals and communicating canals are related.
5. Explain how the development of intramembranous bone differs from that of endochondral bone.
6. Distinguish between osteoblasts and osteocytes.
7. Explain the function of an epiphyseal disk.
8. Explain how a bone grows in thickness.
9. Define *osteoclast*.
10. Describe the effects of vitamin deficiencies on bone development.
11. Explain the causes of pituitary dwarfism and giantism.
12. Describe the effects of thyroid and sex hormones on bone development.
13. Explain the effects of exercise on bone structure.
14. Provide several examples to illustrate how bones support and protect body parts.
15. Describe a lever, and explain how its parts may be arranged to form first-, second-, and third-class levers.
16. Describe the functions of red and yellow bone marrow.
17. Explain the mechanism that regulates the concentration of blood calcium.

18. List three substances that may be stored in bone abnormally.
19. Distinguish between the axial and appendicular skeletons.
20. List the bones that form the pectoral and the pelvic girdles.
21. Name the bones of the cranium and the facial skeleton.
22. Explain the importance of fontanels.
23. Describe a typical vertebra.
24. Explain the differences between cervical, thoracic, and lumbar vertebrae.
25. Describe the locations of the sacroiliac joint, the sacral promontory, and the sacral hiatus.
26. Name the bones that comprise the thoracic cage.
27. Name the bones of the upper limb.
28. Define *coxal bone.*
29. List the major differences that may occur between the male and female pelves.
30. List the bones of the lower limb.

Part B

Match the parts listed in column I with the bones listed in column II.

I	II
1. Coronoid process	A. Ethmoid bone
2. Cribriform plate	B. Frontal bone
3. Foramen magnum	C. Mandible
4. Mastoid process	D. Maxillary bone
5. Palatine process	E. Occipital bone
6. Sella turcica	F. Temporal bone
7. Supraorbital notch	G. Sphenoid bone
8. Temporal process	H. Zygomatic bone
9. Acromion process	I. Femur
10. Deltoid tuberosity	J. Fibula
11. Greater trochanter	K. Humerus
12. Lateral malleolus	L. Radius
13. Medial malleolus	M. Scapula
14. Olecranon process	N. Sternum
15. Radial tuberosity	O. Tibia
16. Xiphoid process	P. Ulna

8

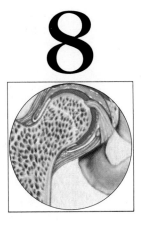

Joints of the Skeletal System

Joints are the junctions between bones of the skeletal system. That is, a joint occurs wherever two or more bones come together. Most joints function in lever systems that make movements possible by bending or straightening. Others, however, are relatively rigid structures that help to hold bones in place or function in the growth process.

This chapter describes how various types of joints are classified, how they differ in structure, and how they permit different movements. It also describes in detail the structure of several large, freely movable joints.

Chapter Outline

Chapter Objectives

After you have studied this chapter, you should be able to

1. Explain how joints can be classified according to the amount of movement they permit.

2. Describe how bones of immovable joints are held together.

3. Describe how bones of slightly movable joints are held together.

4. Describe the general structure of a freely movable joint.

5. List six types of freely movable joints and name an example of each type.

6. Explain how skeletal muscles produce movements at joints and identify several types of joint movements.

7. Describe the shoulder joint and explain how its articulating parts are held together.

8. Describe the elbow joint and explain how its articulating parts are held together.

9. Describe the hip joint and explain how its articulating parts are held together.

10. Describe the knee joint and explain how its articulating parts are held together.

11. Complete the review activities at the end of this chapter. Note that they are worded in the form of specific learning objectives. You may want to refer to them before reading the chapter.

Key Terms

articulation   (ar-tik'u-la''shun)

bursa   (ber'sah)

fontanel   (fon'tah-nel)

gomphosis   (gom-fo'sis)

ligament   (lig'ah-ment)

meniscus   (mĕ-nis'kus)

suture   (su'chur)

symphysis   (sim'fi-sis)

synchondrosis   (sin''kon-dro'sis)

syndesmosis   (sin''des-mo'sis)

synovial   (sĭ-no've-al)

Aids to Understanding Words

acetabul-, vinegar cup: *acetabul*um—depression of the coxal bone that articulates with the head of the femur.

annul-, a ring: *annul*ar ligament—a ring-shaped band of connective tissue within the elbow that encircles the head of the radius.

burs-, a pouch: subcutaneous prepatellar *burs*a—a fluid-filled sac between the skin and the patella.

condyl-, a knob: medial *condyl*e—-a rounded bony process at the distal end of the femur.

fov-, a pit: *fov*ea capitis—a pit in the head of the femur to which a ligament is attached.

glen-, joint socket: *glen*oid cavity—a depression in the scapula that articulates with the head of the humerus.

labr-, a lip: glenoidal *labr*um—a rim of fibrocartilage attached to the margin of the glenoid cavity.

ov-, egg-like: syn*ov*ial fluid—thick fluid within a joint cavity that resembles eggwhite.

sutur-, a sewing: *sutur*e—a type of joint in which flat bones are interlocked by a set of tiny bony processes.

syndesm-, binding together: *syndesm*osis—a type of joint in which the bones are held together by long fibers of connective tissue.

WHEREVER TWO OR more bones meet, a **joint,** or **articulation,** is formed. Such joints represent the functional junctions between bones. They serve to bind various parts of the skeletal system together, allow bone growth to occur, permit certain parts of the skeleton to change shape during childbirth, and enable body parts to move in response to skeletal muscle contractions.

Classification of Joints

Although joints vary considerably in structure and function, they can be classified according to the amount of movement that can occur at each bony junction. On this basis, three general groups can be identified—immovable joints, slightly movable joints, and freely movable joints.

Immovable Joints

Immovable joints (synarthroses) are found between bones that come into close contact with one another. The bones at such joints are fastened tightly together by a thin layer of fibrous connective tissue. As a rule, no appreciable movement occurs between these bones; sometimes, however, a very small degree of motion is possible. There are three types of immovable joints. They include the following:

1. **Syndesmosis.** In this type of joint the bones are bound together by relatively long fibers of connective tissue that form an *interosseous ligament.* Since this ligament is flexible and may be twisted, the joint may permit some slight movement. An example of a syndesmosis is found at the distal ends of the tibia and fibula, where these bones are joined to form the tibiofibular articulation. (See figure 8.1.)

2. **Suture.** Sutures occur only between flat bones of the skull, where the relatively broad margins of adjacent bones grow together and become united by a thin layer of fibrous connective tissue called a *sutural ligament.* In the infantile skull, as is described in chapter 7, the skull is incompletely developed and many of the bones are separated by membranous areas called *fontanels.* (See figure 7.31.) These areas allow the skull to change shape slightly during childbirth, but as the bones continue to grow, the fontanels close and are replaced by sutures. With time, some of the bones at sutures become interlocked by a set of tiny bony processes, and the sutural ligament itself may be changed to bone. An example of such a suture can be found between the parietal and occipital bones of an adult skull, where they meet to form the lambdoidal suture. (See figure 8.2.)

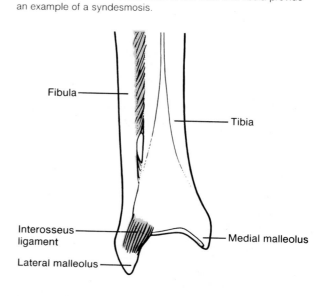

Figure 8.1
The articulation at the distal ends of the tibia and fibula provide an example of a syndesmosis.

Fibula

Tibia

Interosseus ligament

Medial malleolus

Lateral malleolus

The obliteration of the sutures begins in early adulthood, usually before the thirtieth year. The process occurs first in the deeper parts of the sutures and extends slowly to the external surface. Since the obliteration usually is not complete until advanced age, the condition of the sutures may be used to judge the age of a human skull.

3. **Gomphosis.** A gomphosis is a joint created by the union of a cone-shaped bony process in a bony socket. Such an articulation is formed by the peglike root of a tooth fastened to a jawbone by a *periodontal ligament.* This ligament surrounds the root and attaches it firmly to the jaw with bundles of thick collagenous fibers. (See figure 8.3.)

1. What is a joint?
2. Describe three types of immovable joints.
3. What is the function of a fontanel?

Slightly Movable Joints

The bones of **slightly movable joints** (amphiarthroses) are connected by hyaline cartilage or fibrocartilage. There are two types, as follows:

1. **Synchondrosis.** In a synchondrosis the bones are united by bands of hyaline cartilage. Many of these joints are temporary structures which disappear as a result of the growth process. An example is found in an immature long bone where an epiphysis is connected to a diaphysis by a band of hyaline cartilage (the epiphyseal disk). As is described in chapter 7, this band functions in the growth of the bone and in time

Figure 8.2
(a) The joints between the bones of the cranium are immovable and are called sutures; (b) the bones at a suture are separated by a sutural ligament.

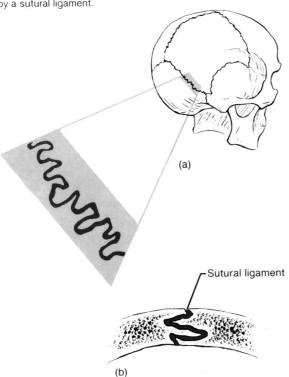

(a)

Sutural ligament

(b)

Figure 8.3
The articulation between the root of a tooth and the jawbone is a gomphosis.

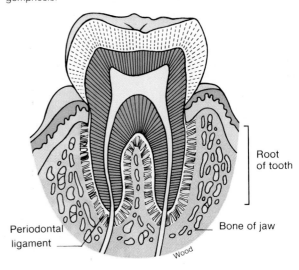

Root of tooth

Bone of jaw

Periodontal ligament

Wood

is converted from cartilage into bone. When this ossification is complete, usually before the age of twenty-five years, movement no longer occurs at the joint. (See figure 7.8.)

Another example of a synchondrosis occurs between the sternum and the first rib, which are united by costal cartilage. In this instance, the joint is permanent. It also allows some movement, since the costal cartilage can bend and twist. (See figure 8.4.)

2. **Symphysis.** The articular surfaces of the bones at a symphysis are covered by a thin layer of hyaline cartilage, and the cartilage, in turn, is attached to a pad of resilient fibrocartilage. A limited amount of movement occurs at such a joint whenever forces cause the cartilaginous pad to become compressed or deformed. The joint formed by the bodies of adjacent vertebrae separated by an *intervertebral disk* is an example of a symphysis. (See figure 8.5.)

Each intervertebral disk is composed of a band of fibrous fibrocartilage (anulus fibrosus) that surrounds a gelatinous core (nucleus pulposus). The disk acts as a shock absorber and helps to equalize pressures between the vertebrae when the body moves. Since each disk is slightly flexible, the combined movement of all of the joints in the vertebral column

allows the limited motion that occurs when the back is bent forward, bent to the side, or is twisted. Another example of this type of joint is the symphysis pubis in the pelvic girdle, which allows some movement as an infant passes through the birth canal during childbirth.

Freely Movable Joints

Most joints of the skeletal system are **freely movable** (diarthroses), and they have much more complex structures than the immovable or slightly movable types. For example, these joints contain articular cartilage, a joint capsule, and a synovial membrane.

General Structure of a Freely Movable Joint

The articular ends of bones in a freely movable joint are covered with a thin layer of hyaline cartilage. (See figure 8.6.) This layer, which is called the **articular cartilage,** is resistant to wear and produces a minimum of friction when it is compressed as the joint is moved.

The bones are held together by a tubular **joint capsule** that has two distinct layers. The outer layer consists largely of dense, white, fibrous connective tissue, whose fibers are attached to the periosteum around the circumference of each bone of the joint near its articular end. Thus, the outer fibrous layer of the capsule completely encloses the other parts of the joint. It is, however, flexible enough to permit movement and strong enough to help prevent the articular surfaces from being pulled apart.

Bundles of strong, tough collagenous fibers called **ligaments** reinforce the joint capsule and help to bind the articular ends of the bones together. Some ligaments appear as thickenings in the fibrous layer of the

Figure 8.4

The articulation between the first rib and the sternum is an example of a synchondrosis.

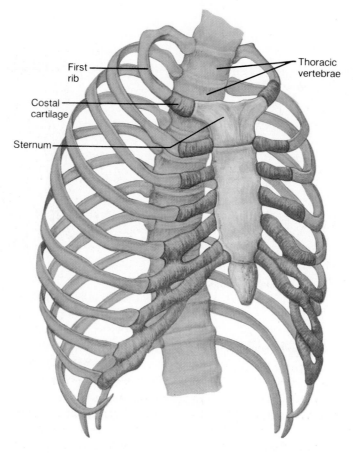

Figure 8.5

(a) The intervertebral disks that separate the bodies of adjacent vertebrae and *(b)* the disk that separates the pubic bones of the pelvic girdle are composed of fibrocartilage.

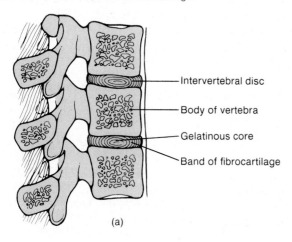

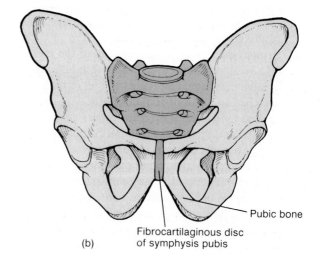

Figure 8.6

The generalized structure of a freely movable joint. What is the function of the synovial fluid within this type of joint?

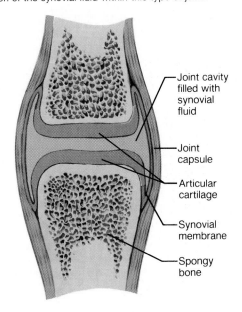

Joint cavity filled with synovial fluid

Joint capsule

Articular cartilage

Synovial membrane

Spongy bone

Figure 8.7

The articulating surfaces of the femur and tibia are separated by menisci; also there are several bursae associated with the knee joint.

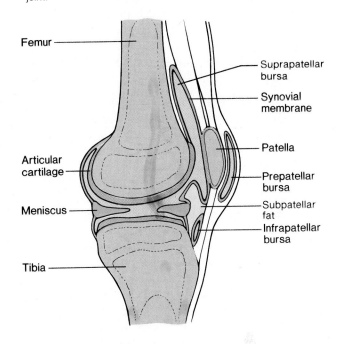

Femur

Suprapatellar bursa

Synovial membrane

Patella

Articular cartilage

Prepatellar bursa

Subpatellar fat

Meniscus

Infrapatellar bursa

Tibia

capsule, while others are *accessory structures* located outside the capsule. In either case these structures also prevent excessive movement at the joint. That is, the ligament is relatively inelastic, and it becomes drawn tightly whenever a normal limit of movement has been achieved in the joint.

The inner layer of the joint capsule consists of a shiny, vascular lining of loose connective tissue called the **synovial membrane.** This membrane covers all of the surfaces within the joint capsule, except the areas covered by cartilage. The synovial membrane surrounds a closed sac called the *synovial cavity,* and into this joint cavity the membrane secretes a clear, viscous fluid called **synovial fluid.** In some regions the surface of the synovial membrane possesses villi as well as larger folds and projections that extend into the cavity. Besides filling spaces and irregularities of the joint cavity, these extensions increase the surface area of the synovial membrane. In addition to secreting fluid, the membrane may store adipose tissue and may form movable fatty pads within the joint. The synovial membrane also functions to reabsorb fluid. Thus it may help to remove substances from a joint cavity that is injured or infected.

Synovial fluid has a consistency similar to egg white, and it serves to moisten and lubricate the smooth cartilaginous surfaces within the joint. It also helps to supply articular cartilage with nutrients that are obtained from blood vessels of the synovial membrane. However, the volume of synovial fluid present in a joint cavity is relatively small. Usually there is just enough

to cover the articulating surfaces with a thin film of fluid. For example, the amount in the cavity of the knee is 0.5 ml. or less.

Some freely movable joints are partially or completely divided into two compartments by disks of fibrocartilage called **menisci** (sing. **meniscus**) located between the articular surfaces. Such a disk is attached to the fibrous layer of the joint capsule peripherally, and its free surface projects into the joint cavity. In the case of the knee joint, crescent-shaped menisci cushion the articulating surfaces and help to distribute the body weight onto these surfaces. (See figure 8.7.)

Certain freely movable joints also have closed, fluid-filled sacs called **bursae** associated with them. Each bursa has an inner lining of synovial membrane, which may be continuous with the synovial membrane of a nearby joint cavity. These sacs contain synovial fluid and are commonly located between the skin and underlying bony prominences, as in the case of the patella of the knee or the olecranon process of the elbow. Bursae act as cushions and aid the movement of tendons that glide over such bony parts or over other tendons. The names of bursae indicate their locations. Thus, there is a *suprapatellar bursa,* a *prepatellar bursa,* and an *infrapatellar bursa,* shown in figure 8.7.

---

1. Describe two types of slightly movable joints.
2. What is the function of an intervertebral disk?
3. Describe the structure of a freely movable joint.

Types of Freely Movable Joints

The articulating bones of freely movable joints have a variety of shapes that allow different kinds of movement. Based upon the shapes of their parts and the movements they permit, these joints can be classified into six major types—ball and socket joints, condyloid joints, gliding joints, hinge joints, pivot joints, and saddle joints.

1. *Ball and socket joints.* A **ball and socket joint** (spheroidal) consists of a bone with a globular or slightly egg-shaped head that articulates with a cup-shaped cavity of another bone. Such a joint allows for a wider range of motion than does any other kind. Movements in all planes, as well as rotational movement around a central axis, are possible. The hip and shoulder contain joints of this type. (See figure 8.8(a).)

2. *Condyloid joints.* In a **condyloid joint** (ellipsoidal), an ovoid condyle of one bone fits into an elliptical cavity of another bone, as in the joints between the metacarpals and phalanges. This type of joint permits a variety of movements in different planes; rotational movement, however, is not possible. (See figure 8.8(b).)

3. *Gliding joints.* The articulating surfaces of **gliding joints** (arthrodial) are nearly flat or only slightly curved. These joints allow only sliding or back-and-forth motion. Most of the joints within the wrist and ankle as well as those between the articular processes of adjacent vertebrae belong to this group. (See figure 8.8(c).)

4. *Hinge joints.* In a **hinge joint** (ginglymoid), the convex surface of one bone fits into the concave surface of another, as in the case of the elbow and the joints of the phalanges. Such a joint resembles the hinge of a door in that it permits movement in one plane only. (See figure 8.8(d).)

5. *Pivot joints.* In a **pivot joint** (trochoid), a cylindrical surface of one bone rotates within a ring formed of bone and fibrous tissue of a ligament. The movement at such a joint is limited to rotation about a central axis. The joint between the proximal ends of the radius and the ulna, where the head of the radius rotates in a ring formed by the radial notch of the ulna and a ligament (annular ligament), is of this type. Similarly, a pivot joint functions as the head is turned from side to side. In this case, the ring formed by a ligament (transverse ligament) and the anterior arch of the atlas rotates around the odontoid process of the axis. (See figure 8.8(e).)

Figure 8.8
Types and examples of freely movable joints.

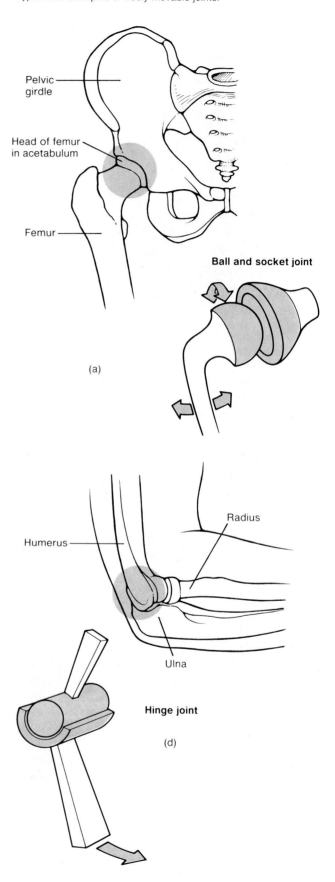

Pelvic girdle

Head of femur in acetabulum

Femur

Ball and socket joint

(a)

Humerus

Radius

Ulna

Hinge joint

(d)

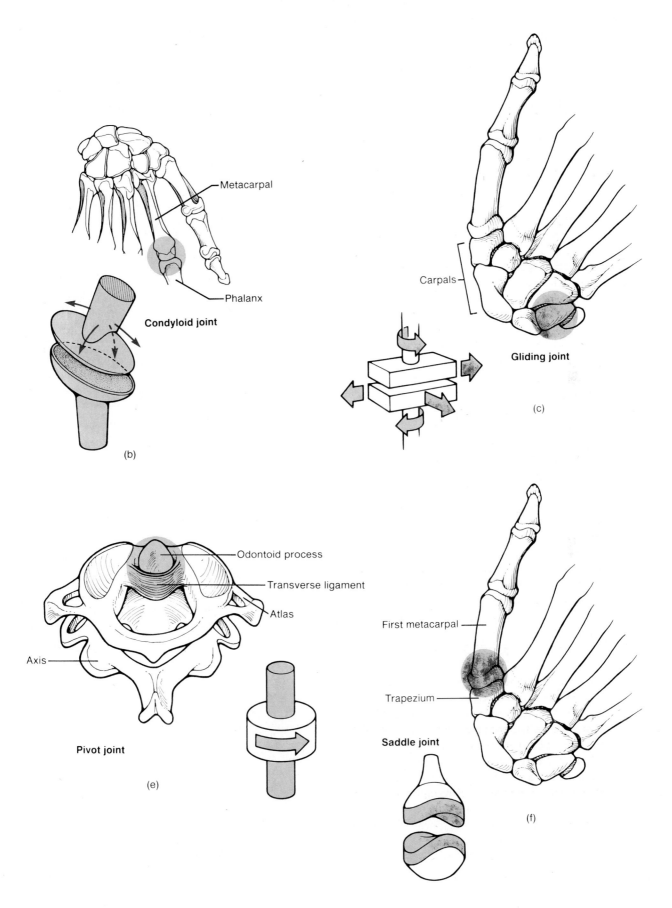

Metacarpal

Phalanx

Condyloid joint

(b)

Carpals

Gliding joint

(c)

Odontoid process

Transverse ligament

Atlas

Axis

Pivot joint

(e)

First metacarpal

Trapezium

Saddle joint

(f)

Chart 8.1 Types of joints

Type of joint	Description	Possible movements	Example
Immovable (synarthrosis)	Articulating bones fastened together by thin layer of fibrous connective tissue		
1. *Syndesmosis*	Bones bound by interosseous ligament	Joint flexible and may be twisted	Tibiofibular articulation
2. *Suture*	Flat bones united by sutural ligament	Bones grow together and become interlocked	Parietal bones articulate at sagittal suture of skull
3. *Gomphosis*	Cone-shaped process fastened in bony socket by periodontal ligament	None	Root of tooth united with jawbone
Slightly movable (amphiarthrosis)	Articulating bones connected by hyaline cartilage or fibrocartilage		
1. *Synchondrosis*	Bones united by bands of hyaline cartilage	Movement occurs during growth process until ossification occurs	Joint between epiphysis and diaphysis of a long bone
2. *Symphysis*	Articular surfaces separated by thin layers of hyaline cartilage attached to band of fibrocartilage	Limited movement as when back is bent or twisted	Joints between bodies of adjacent vertebrae
Freely movable (diarthrosis)	Articulating bones surrounded by joint capsule of ligaments and synovial membranes; ends of articulating bones covered by hyaline cartilage and separated by synovial fluid		
1. *Ball and socket*	Ball-shaped head of one bone articulates with cup-shaped socket of another	Movements in all planes and rotation	Shoulder, hip
2. *Condyloid*	Oval-shaped condyle of one bone articulates with elliptical cavity of another	Variety of movements in different planes, but no rotation	Joints between metacarpals and phalanges
3. *Gliding*	Articulating surfaces are nearly flat or slightly curved	Sliding or twisting	Joints between various bones of wrist and ankle
4. *Hinge*	Convex surface of one bone articulates with concave surface of another	Up and down motion in one plane	Elbow and joints of phalanges
5. *Pivot*	Cylindrical surface of one bone articulates with ring of bone and fibrous tissue	Rotation	Joint between proximal ends of radius and ulna
6. *Saddle*	Articulating surfaces have both concave and convex regions; surface of one bone fits complementary surface of another	Variety of movements	Joint between carpal and metacarpal of thumb

6. *Saddle joints.* A **saddle joint** (sellar) is formed between bones whose articulating surfaces have both concave and convex regions. The surface of one bone fits the complementary surface of the other. This arrangement permits a variety of movements, as in the case of the joint between the carpal (trapezium) and the metacarpal of the thumb. (See figure 8.8(f).)

Chart 8.1 summarizes the types of joints.

Types of Joint Movements

Movements at synovial joints are produced by actions of skeletal muscles. Typically, one end of a muscle is attached to a relatively immovable or fixed part on one side of a joint, and the other end of the muscle is fastened to a movable part on the other side. When the muscle contracts, its fibers pull its movable end (insertion) toward its fixed end (origin), and a movement occurs at the joint.

Figure 8.9
Joint movements illustrating flexion, extension, hyperextension, dorsiflexion, plantar flexion, abduction, and adduction.

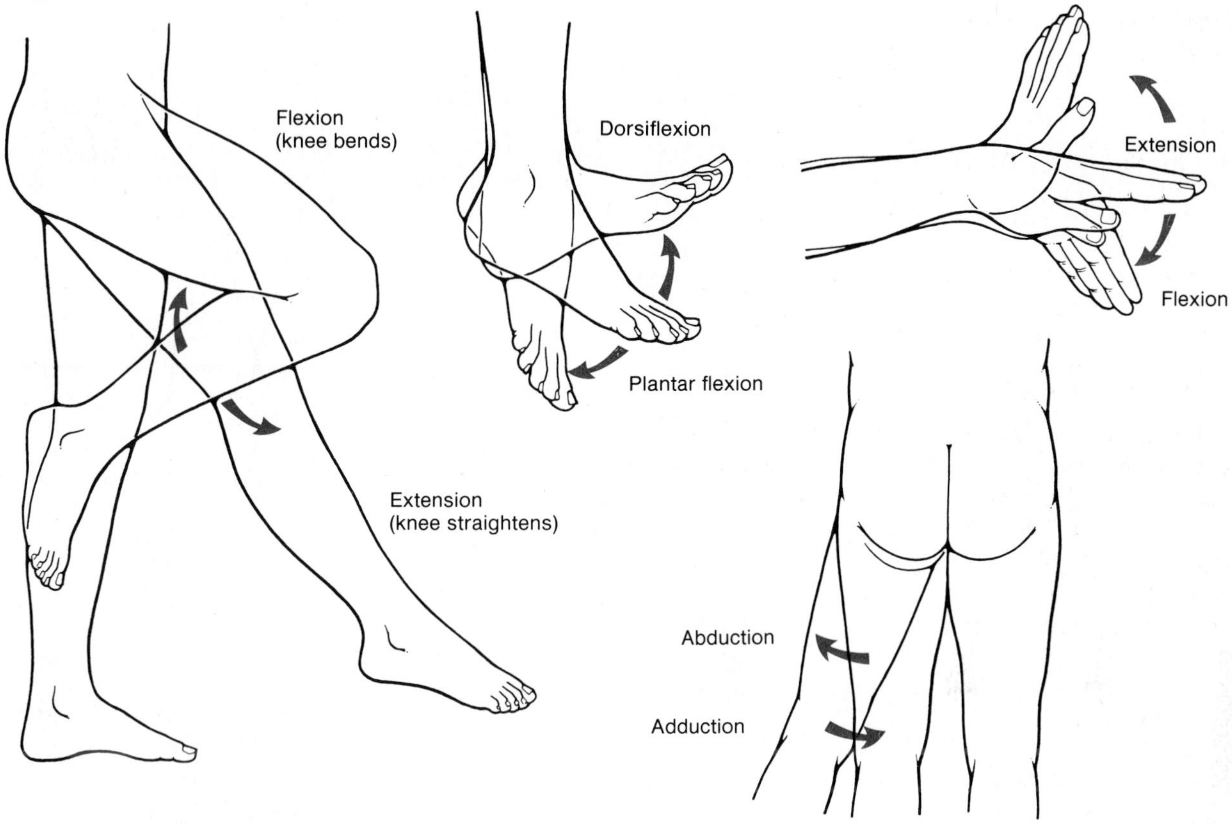

The following terms are used to describe movements of parts at joints that occur in different directions and in different planes (figures 8.9, 8.10, and 8.11):

Flexion (flek'shun)   bending parts at a joint so that the angle between them is decreased and the parts come closer together (bending the leg at the knee).

Extension (ek-sten'shun)   straightening parts at a joint so that the angle between them is increased and the parts move further apart (straightening the leg at the knee).

Hyperextension (hi''per-ek-sten'shun)   excessive extension of the parts at a joint, beyond the anatomical position (bending the head back beyond the upright position).

Dorsiflexion (dor''si-flek'shun)   flexing the foot at the ankle (bending the foot upward).

Plantar flexion (plan'tar flek'shun)   extending the foot at the ankle (bending the foot downward).

Abduction (ab-duk'shun)   moving a part away from the midline (lifting the arm horizontally to form a right angle with the side of the body).

Adduction (ah-duk'shun)   moving a part toward the midline (returning the arm from the horizontal position to the side of the body).

Rotation (ro-ta'shun)   moving a part around an axis (twisting the head from side to side).

Circumduction (ser''kum-duk'shun)   moving a part so that its end follows a circular path (moving the finger in a circular motion without moving the hand).

Supination (soo''pĭ-na'shun)   turning the hand so the palm is upward.

Pronation (pro-na'shun)   turning the hand so the palm is downward.

Eversion (e-ver'zhun)   turning the foot so the sole is outward.

Inversion (in-ver'zhun)   turning the foot so the sole is inward.

Protraction (pro-trak'shun)   moving a part forward (thrusting the chin forward).

Retraction (re-trak'shun)   moving a part backward (pulling the chin backward).

Elevation (el''ĕ-va'shun)   raising a part (shrugging the shoulders).

Depression (de-presh'un)   lowering a part (drooping the shoulders).

---

1. Name six types of freely movable joints.
2. Describe the structure of each of these types of joints.
3. What terms are used to describe various movements that occur at freely movable joints?

Figure 8.10

Joint movements illustrating rotation, circumduction, supination, and pronation.

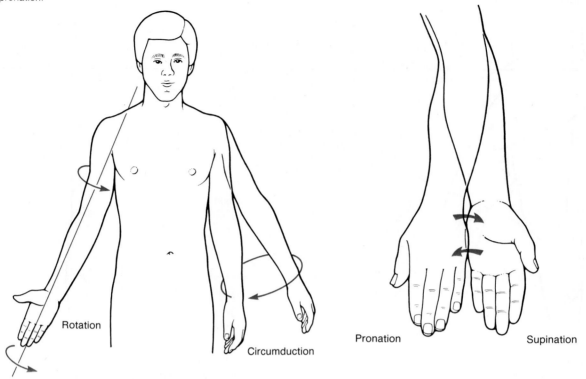

Rotation

Circumduction

Pronation

Supination

Figure 8.11

Joint movements illustrating eversion, inversion, elevation, depression, protraction, and retraction.

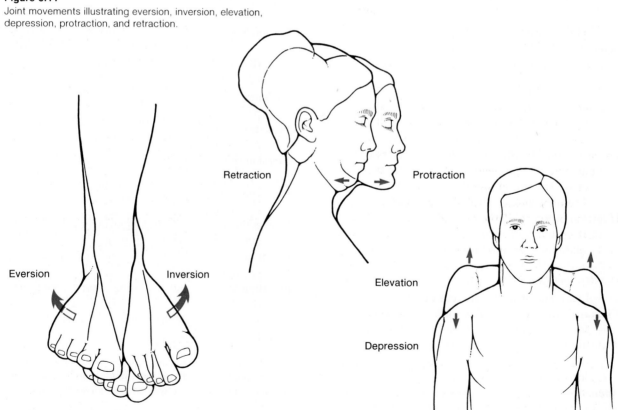

Eversion

Inversion

Retraction

Protraction

Elevation

Depression

Figure 8.12

The shoulder joint allows movements in all directions. Note the bursa associated with this joint.

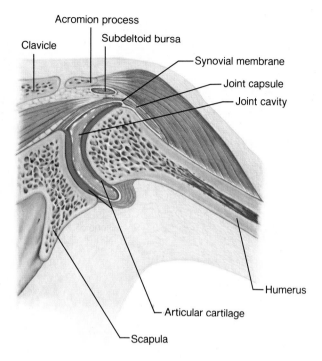

Acromion process

Clavicle

Subdeltoid bursa

Synovial membrane

Joint capsule

Joint cavity

Humerus

Articular cartilage

Scapula

Examples of Freely Movable Joints

The shoulder, elbow, hip, and knee are examples of large freely movable joints. Although these joints have much in common, each has a unique structure that is related to its specific function.

The Shoulder Joint

The **shoulder joint** is a ball and socket joint that consists of the rounded head of the humerus and the shallow glenoid cavity of the scapula. These parts are protected above by the coracoid and acromion processes of the scapula, and they are held together by various fibrous connective tissues and muscles.

The **joint capsule** of the shoulder is attached along the circumference of the glenoid cavity and the anatomical neck of the humerus. Although it completely envelopes the joint, the capsule is very loose and by itself is unable to keep the bones of the joint in close contact. The capsule, however, is surrounded and reinforced by muscles and tendons, and these structures are largely responsible for keeping the articulating parts together. (See figure 8.12.)

The tendons of several muscles are intimately blended with the fibrous layer of the shoulder joint capsule and they reinforce and support the joint. This structure is called the *rotator cuff* of the shoulder. It is sometimes injured as a result of the centrifugal forces created when the shoulder joint is used for throwing objects.

The ligaments that help to prevent displacement of the articulating surfaces of the shoulder joint include the following:

1. **Coracohumeral** (kor″ah-ko-hu′mer-al) **ligament.** This ligament is composed of a broad band of connective tissue that connects the coracoid process of the scapula to the greater tubercle of the humerus. It functions to strengthen the superior portion of the joint capsule.
2. **Glenohumeral** (gle″no-hu′mer-al) **ligaments.** These include three bands of fibers that appear as thickenings in the ventral wall of the joint capsule. They extend from the edge of the glenoid cavity to the lesser tubercle and the anatomical neck of the humerus.
3. **Transverse humeral ligament.** This ligament consists of a narrow sheet of connective tissue fibers that runs between the lesser and the greater tubercles of the humerus. Together with the intertubercular groove of the humerus, the ligament creates a canal (retinaculum) through which the long head of the biceps brachii muscle passes.
4. **Glenoidal labrum** (gle′noid-al la′brum). This ligament is composed of fibrocartilage. It is attached along the margin of the glenoid cavity and forms a rim with a thin, free edge that serves to deepen the cavity. (See figure 8.13.)

Figure 8.13

(a) The articulating surfaces of the shoulder are held together by ligaments; *(b)* the glenoidal labrum is a ligament composed of fibrocartilage.

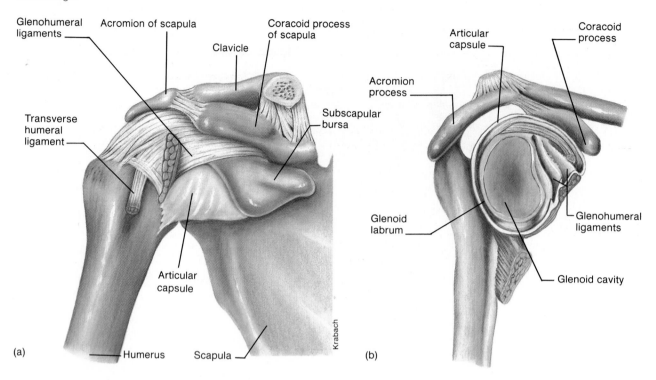

There are several bursae associated with the shoulder joint. The major ones include the *subscapular bursa* located between the joint capsule and the tendon of the subscapularis muscle, the *subdeltoid bursa* between the joint capsule and the deep surface of the deltoid muscle, the *subacromial bursa* between the joint capsule and the under surface of the acromion process of the scapula, and the *subcoracoid bursa* between the joint capsule and the coracoid process of the scapula. Of these, the subscapular bursa usually is continuous with the synovial cavity of the joint cavity, and although the others do not communicate with the joint cavity, they may be connected to each other. (See figures 8.12 and 8.13.)

Due to the looseness of its attachments and to the relatively large articular surface of the humerus compared to the shallow depth of the glenoid cavity, the shoulder joint is capable of a very wide range of movement. These include flexion, extension, abduction, adduction, rotation, and circumduction. Such movements also may be aided by motion occurring simultaneously in the joint formed between the scapula and clavicle.

Since the bones of the shoulder joint are held together mainly by supporting muscles rather than by bony structures and strong ligaments, the joint is somewhat weak. Consequently, the articulating surfaces may become displaced or dislocated rather easily. Such a *dislocation* most commonly occurs during forceful abduction, as may happen when a person falls on an outstretched arm. This movement may cause the head of the humerus to press against the lower part of the joint capsule where its wall is relatively thin and poorly supported by ligaments.

The Elbow Joint

The **elbow joint** is a complex structure that includes two articulations—a hinge joint between the trochlea of the humerus and the trochlear notch of the ulna, and a gliding joint between the capitulum of the humerus and a shallow depression (fovea) on the head of the radius. These unions are completely enclosed and held together by a joint capsule, whose sides are thickened by ulnar and radial collateral ligaments and whose anterior surface is reinforced by fibers from a muscle (brachialis) in the upper arm. (See figure 8.14.)

The **ulnar collateral ligament,** which consists of a thick band of fibrous connective tissue, is located in the medial wall of the capsule. The anterior portion of this

Figure 8.14
The elbow joint allows hinge movements as well as pronation and supination of the hand.

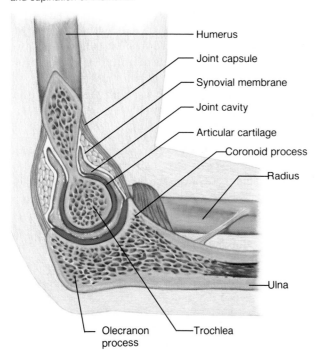

Humerus

Joint capsule

Synovial membrane

Joint cavity

Articular cartilage

Coronoid process

Radius

Ulna

Olecranon process

Trochlea

Figure 8.15
The ulnar collateral ligament (a) and the radial collateral ligament (b) strengthen the capsular wall of the elbow joint.

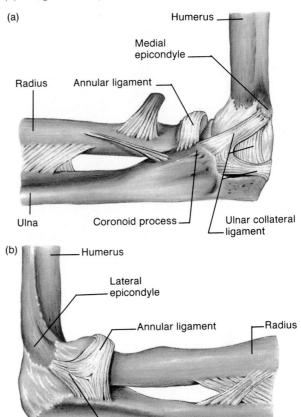

(a)

Humerus

Medial epicondyle

Radius

Annular ligament

Ulna

Coronoid process

Ulnar collateral ligament

(b)

Humerus

Lateral epicondyle

Annular ligament

Radius

Olecranon process

Radial collateral ligament

Ulna

ligament connects the medial epicondyle of the humerus to the medial margin of the coronoid process of the ulna. Its posterior part is attached to the medial epicondyle of the humerus and to the olecranon process of the ulna.

The **radial collateral ligament,** which strengthens the lateral wall of the joint capsule, is composed of a fibrous band extending between the lateral epicondyle of the humerus and the *annular ligament* of the radius. The annular ligament, in turn, is attached to the margin of the trochlear notch of the ulna, and it encircles the head of the radius, functioning to keep the head in contact with the radial notch of the ulna. The resulting radioulnar joint also is enclosed by the elbow joint capsule, so that its function is closely associated with the elbow. (See figure 8.15.)

The *synovial membrane* that forms the inner lining of the elbow capsule projects into the joint cavity between the radius and ulna and partially divides the joint into humerus-ulnar and humerus-radial portions. Also, varying amounts of adipose tissue form fatty pads between the synovial membrane and the fibrous layer of the joint capsule. These pads help to protect nonarticular bony areas during joint movements.

The only movements that can occur at the elbow between the humerus and ulna are hinge-type movements—flexion and extension. The head of the radius, however, is free to rotate in the annular ligament, and this movement is responsible for pronation and supination of the hand.

Injuries to the elbow, shoulder, and knee commonly are diagnosed and treated using a procedure called *arthroscopy*. This procedure makes use of a thin, tubular instrument about 25 centimeters long called an *arthroscope*. The instrument, which contains optical fibers, can be inserted through a small incision in the joint capsule. A surgeon can then view the interior of the joint directly or observe an image of the joint on a television screen. In either case, the surgeon can use the arthroscope to explore the joint cavity and to guide other instruments inserted through the capsule that are needed to repair or remove injured parts.

1. What parts help to keep the articulating surfaces of the shoulder joint together?
2. What factors allow an especially wide range of motion in the shoulder?
3. What parts make up the hinge joint of the elbow?
4. What parts of the elbow permit pronation and supination of the hand?

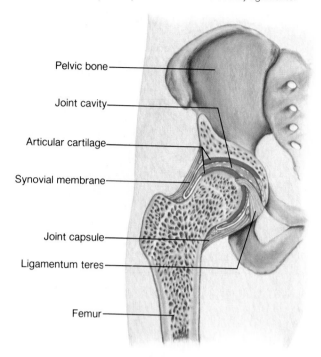

Pelvic bone

Joint cavity

Articular cartilage

Synovial membrane

Joint capsule

Ligamentum teres

Femur

The Hip Joint

The **hip joint** is a ball and socket joint that consists of the head of the femur and the cup-shaped acetabulum of the coxal bone. A ligament (ligamentum capitis) is attached to a pit (fovea capitis) on the head of the femur and to connective tissue within the acetabulum. This attachment, however, seems to have little importance in holding the articulating bones together. Instead, it serves to carry blood vessels to the head of the femur.

A horseshoe-shaped ring of fibrocartilage (acetabular labrum) at the rim of the acetabulum deepens the cavity of the acetabulum. It encloses the head of the femur and helps to hold it securely in place. In addition, a heavy, cylindrical joint capsule that is reinforced with still other ligaments surrounds the articulating structures and connects the neck of the femur to the margin of the acetabulum. (See figure 8.16.)

The major ligaments of the hip joint include the following:

1. **Iliofemoral** (il"e-o-fem'o-ral) **ligament.** This ligament consists of a Y-shaped band of very strong fibers that connects the anterior inferior iliac spine of the coxal bone to a bony line (intertrochanteric line) extending between the greater and lesser trochanters of the femur. The ileofemoral ligament is the strongest ligament in the body, and it helps to prevent extension of the femur when the body is standing erect.

2. **Pubofemoral** (pu"bo-fem'o-ral) **ligament.** The pubofemoral ligament extends between the superior portion of the pubis and the iliofemoral ligament. Its fibers also blend with the fibers of the joint capsule.

3. **Ischiofemoral** (is"ke-o-fem'o-ral) **ligament.** This ligament consists of a band of strong fibers that originate on the ischium just posterior to the acetabulum and blend with the fibers of the joint capsule. (See figure 8.17.)

As in the case of the shoulder, the joint capsule of the hip is surrounded by muscles. On the other hand, the articulating parts of the hip are held more closely together than those of the shoulder. Consequently there is considerably less freedom of movement at the hip joint. The structure of the hip joint, however, permits a wide variety of movements, including extension, flexion, abduction, adduction, rotation, and circumduction.

To help correct problems created by joint injury or disease, it is sometimes desirable to replace the articulating parts with artificial (prosthetic) devices. The hip joint, for example, can be totally replaced. In this procedure, called *total hip arthroplasty,* the acetabulum is replaced by a low-friction polyethylene socket that is fastened into the coxal bone using surgical bone cement. The head of the femur is replaced by a metallic, ball-shaped part.

The Knee Joint

The **knee joint** is the largest and the most complex of the synovial joints. It consists of the medial and lateral condyles at the distal end of the femur and the medial and lateral condyles at the proximal end of the tibia. In addition, the femur articulates anteriorly with the patella. Although the knee is sometimes considered a modified hinge joint, the articulations between femur and tibia are the condyloid type, and the joint between the femur and patella is a gliding joint.

The *joint capsule* of the knee is relatively thin, but it is greatly strengthened by ligaments and the tendons of several muscles. Anteriorly, for example, the capsule is covered by the fused tendons of several muscles in the thigh. Fibers from these tendons descend to the patella, partially enclose it, and continue downward to the tibia. The capsule is attached to the margins of the femoral and tibial condyles as well as to the areas between these condyles. (See figure 8.18.)

The ligaments associated with the joint capsule that help to keep the articulating surfaces of the knee joint in contact include the following:

1. **Patellar** (pah-tel'ar) **ligament.** This ligament represents a continuation of a tendon from a large muscle group in the thigh (quadriceps

Figure 8.17
The major ligaments of the hip joint. *(a)* Anterior view;
(b) posterior view.

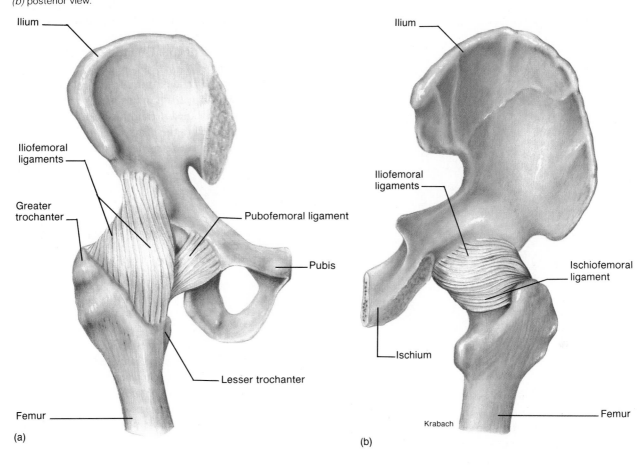

Ilium

Iliofemoral
ligaments

Greater
trochanter

Pubofemoral ligament

Pubis

Lesser trochanter

Femur

(a)

Ilium

Iliofemoral
ligaments

Ischiofemoral
ligament

Ischium

Femur

Krabach

(b)

femoris). It consists of a strong flat band that
extends from the margin of the patella to the
tibial tuberosity.

2. **Oblique popliteal** (ŏ′blēk pop-lit′e-al) **ligament.**
This ligament connects the lateral condyle of the
femur to the margin of the head of the tibia.

3. **Arcuate** (ar′ku-āt) **popliteal ligament.**   This
ligament appears as a Y-shaped system of fibers
that extends from the lateral condyle of the femur
to the head of the fibula.

4. **Tibial collateral** (tib′e-al kŏ-lat′er-al) **ligament.**
This ligament is a broad, flat band of tissue
that connects the medial condyle of the femur
to the medial condyle of the tibia.

5. **Fibular** (fib′u-lar) **collateral ligament.**   This
ligament consists of a strong, round cord
located between the lateral condyle of the
femur and the head of the fibula.

   In addition to the ligaments that strengthen the
joint capsule, there are two called **cruciate** (kroo′she-
āt) **ligaments** within the joint that help to prevent displacement of the articulating surfaces. These strong
bands of fibrous tissue stretch upward between the tibia
and femur, crossing each other on the way. They are

Figure 8.18
The knee joint is the most complex of the synovial joints.

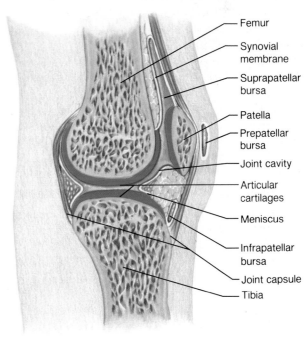

Femur

Synovial
membrane

Suprapatellar
bursa

Patella

Prepatellar
bursa

Joint cavity

Articular
cartilages

Meniscus

Infrapatellar
bursa

Joint capsule

Tibia

Figure 8.19

Ligaments within the knee joint help to strengthen it. *(a)* Anterior view; *(b)* posterior view.

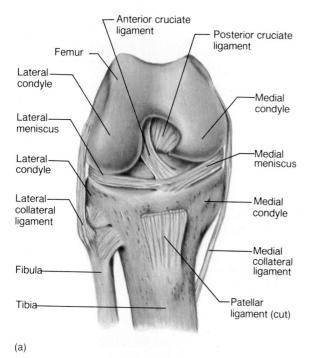

Anterior cruciate ligament

Posterior cruciate ligament

Femur

Lateral condyle

Medial condyle

Lateral meniscus

Lateral condyle

Medial meniscus

Lateral collateral ligament

Medial condyle

Fibula

Medial collateral ligament

Tibia

Patellar ligament (cut)

(a)

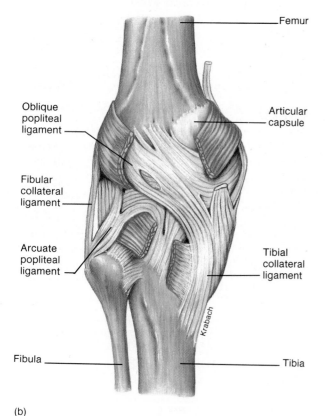

Femur

Oblique popliteal ligament

Articular capsule

Fibular collateral ligament

Arcuate popliteal ligament

Tibial collateral ligament

Fibula

Tibia

(b)

named according to their positions of attachment to the tibia. Thus, the *anterior cruciate ligament* originates from the anterior intercondylar area of the tibia and extends to the lateral condyle of the femur. The *posterior cruciate ligament* connects the posterior intercondylar area of the tibia to the medial condyle of the femur. (See figure 8.19.)

Separating the articulating surfaces of the femur and tibia are two fibrocartilaginous *menisci.* Each meniscus is roughly C-shaped with a thick rim and thinner center and is attached to the head of the tibia. The medial and lateral menisci form depressions that fit the corresponding condyles of the femur, thus compensating for the differences in shapes between the surfaces of the femur and tibia (figure 8.18).

A common type of injury to the knee involves tearing or displacing a meniscus. Usually this occurs as a result of forcefully twisting the knee when the leg is flexed. Since the meniscus is composed of fibrocartilage, such an injury is likely to heal very slowly. Also, if a torn and displaced portion of cartilage becomes jammed between the articulating surfaces, movement of the joint may be prevented (ankylosis).

There are several bursae associated with the knee joint. These include a large extension of the knee joint cavity called the *suprapatellar bursa* located between the anterior surface of the lower part of the femur and the muscle group (quadriceps femoris) above it; a large *prepatellar bursa* between the patella and the skin; and a smaller, *infrapatellar bursa* between the upper part of the tibia and the patellar ligament. (See figure 8.18.)

As with a hinge joint, the basic structure of the knee joint permits flexion and extension. However, when the knee is flexed, rotation also is possible.

Following an injury to the knee, the synovial membrane may become inflamed (acute synovitis) and secrete fluid excessively. As a result, the joint cavity may become distended. In this condition, the knee usually appears enlarged above and on the sides of the patella.

1. What parts help to keep the articulating surfaces of the hip together?
2. What types of movement does the structure of the hip permit?
3. What types of joints are included within the knee?
4. What parts help to hold the articulating surfaces of the knee together?

Disorders of Joints
A Clinical Application

Joints are subjected to considerable stress, because they are used so frequently. Also, they must provide for a great variety of body movements, and some of them must support body weight. Injuries such as dislocations and sprains occur commonly during strenuous physical activity. In addition, joints may be affected by inflammation as well as a number of degenerative diseases such as arthritis.

Dislocation

A *dislocation* involves the displacement of the articulating bones of a joint. This condition usually occurs as a result of a fall or some other unusual body movement. The joints of the shoulders, knees, elbows, fingers, and jaw are common sites for such an injury. A dislocation is characterized by an obvious deformity of the joint, some loss of ability to move the parts involved, localized pain, and swelling. It presents a severe physical problem and requires medical attention.

Sprains

Sprains are the result of overstretching or tearing the connective tissues, ligaments, and tendons associated with a joint, but without dislocating the articular bones. Usually sprains are caused by forceful wrenching or twisting movements involving joints of the wrists or ankles. For example, the ankle may be sprained if it is inverted excessively when running or jumping, causing the ligaments on its lateral surface to be stretched. In severe injuries, these tissues may be pulled loose from their attachments. A sprained joint is likely to be painful, swollen, and movement at the joint may be restricted. The immediate treatment for a sprain is rest; more serious cases require medical attention.

Bursitis

Bursitis is an inflammation of a bursa that may be caused by excessive use of a joint or by stress on a bursa. For example, there is a bursa between the heel bone (calcaneus) and the Achilles tendon, which may become inflamed as a result of a sudden increase in physical activity involving use of the feet. Similarly a form of bursitis called "tennis elbow" involves the bursa between the olecranon process and the

skin, and a condition called "housemaid's knee" is bursitis of the prepatellar bursae. Generally, bursitis is treated with rest, although severe cases may require medication prescribed by a physician.

Arthritis

Arthritis is a disease condition that causes inflamed, swollen, and painful joints. Although there are several types of arthritis, the most prevalent forms are rheumatoid arthritis and osteoarthritis.

Rheumatoid arthritis is the most painful and potentially crippling of the arthritic diseases. In this type of arthritis the synovial membrane of a freely movable joint becomes inflamed and grows thicker, forming a mass called a pannus. This change usually is followed by damage to the articular cartilages of the joint and by an invasion of the joint by fibrous tissues. These tissues interfere increasingly with joint movements, and in time they may become ossified so that the articulating bones become fused together (bony ankylosis).

Rheumatoid arthritis may affect only a few or many joints. It may be accompanied by the development of other disorders, including anemia, osteoporosis, and muscular atrophy as well as abnormal changes in the skin, eyes, lungs, blood vessels, and heart.

Osteoarthritis is the most common type of arthritis. It is a degenerative disease that occurs as a result of aging and affects a large proportion of the population over sixty years of age. In this condition, the articular cartilages soften and disintegrate gradually so that the articular surfaces become roughened. Consequently, the joints involved are painful, and there is some restriction of movement at the joints. Osteoarthritis is most likely to affect joints that have received the greatest use over the years, such as those of the fingers, hips, knees, and the lower regions of the vertebral column. Other factors that seem to increase the chance of osteoarthritis include joint injuries, excess body weight, and certain metabolic disorders.

As a rule, osteoarthritis develops relatively slowly, and its symptoms may be controlled by the use of medications. In more severe cases joint functions are disrupted and parts of joints may need to be replaced surgically.

Clinical Terms Related to Joints

ankylosis (ang''ki-lo'sis)   loss of mobility of a joint.

arthralgia (ar-thral'je-ah)   pain in a joint.

arthrocentesis (ar''thro-sen-te'sis)   puncture and removal of fluid from a joint cavity.

arthrodesis (ar''thro-de'sis)   surgery performed to fuse the bones at a joint.

arthrogram (ar'thro-gram)   x-ray film of a joint after an injection of radio-opaque fluid into the joint cavity.

arthrology (ar-throl'o-je)   study of joints and diseases involving them.

arthropathy (ar-throp'ah-the)   any joint disease.

arthroplasty (ar'thro-plas''te)   surgery performed to make a joint more movable.

arthroscopy (ar-thros'ko-pe)   examination of the interior of a joint using a tubular instrument called an arthroscope.

arthrostomy (ar-thros'to-me)   surgical opening of a joint to allow fluid drainage.

arthrotomy (ar-throt'o-me)   surgical incision of a joint.

gout (gowt)   metabolic disease in which excessive uric acid in the blood is deposited in joints, causing them to become inflamed, swollen, and painful.

hemarthrosis (hem''ar-thros'sis))   blood in a joint cavity.

hydrarthrosis (hi''drar-thro'sis)   accumulation of fluid within a joint cavity.

luxation (luk-sa'shun)   dislocation of a joint.

pyarthrosis (pi''ar-thro'sis)   presence of pus within a joint cavity.

subluxation (sub''luk-sa'shun)   partial dislocation of a joint.

Chapter Summary

Introduction
1. A joint is formed wherever two or more bones meet.
2. Joints are the functional junctions between bones.

Classification of Joints
Joints can be classified according to the amount of movement they permit.

1. Immovable joints
   a. Bones at immovable joints are fastened tightly together by a layer of fibrous connective tissue.
   b. Little or no movement occurs at such a joint.
   c. There are three types of immovable joints.
      (1) A syndesmosis is characterized by bones bound by relatively long fibers of connective tissue.
      (2) A suture occurs where flat bones are united by a thin layer of connective tissue and become interlocked by a set of bony processes.
      (3) A gomphosis is created by the union of a cone-shaped bony process in a bony socket.
2. Slightly movable joints
   a. Bones of slightly movable joints are held together by a layer of cartilage.
   b. There are two types.
      (1) A synchondrosis is characterized by bones united by hyaline cartilage that disappears as a result of growth.
      (2) A symphysis is a joint whose articular surfaces are covered by hyaline cartilage and attached to a pad of fibrocartilage.
3. Freely movable joints
   a. Freely movable joints have a more complex structure than other types of joints.
   b. These joints include articular cartilage, a joint capsule, and a synovial membrane.

General Structure of a Freely Movable Joint
1. The articular ends of bones are covered by a layer of cartilage called *articular cartilage*.
2. The bones are held together by a joint capsule that is strengthened by ligaments.
3. The inner layer of the joint capsule is lined by a synovial membrane that secretes synovial fluid.
4. Synovial fluid moistens and lubricates the articular surfaces.
5. Some freely movable joints are divided into compartments by menisci.
6. Some freely movable joints have fluid-filled bursae associated with them.
   a. Bursae are usually located between the skin and underlying bony prominences.
   b. Bursae act as cushions and aid the movement of tendons over bony parts.
   c. Bursae are named according to their locations.

Types of Freely Movable Joints
1. Ball and socket joints
   a. In this type of joint, a globular head of a bone fits into a cup-shaped cavity of another.
   b. These joints permit a wide variety of movements.
   c. The hip and shoulder are ball and socket joints.
2. Condyloid joints
   a. A condyloid joint consists of an ovoid condyle of one bone that fits into an elliptical cavity of another.

b. This joint permits a variety of movements.

c. The joints between the metacarpals and phalanges are condyloid.

3. Gliding joints
a. Articular surfaces of gliding joints are nearly flat.
b. These joints permit the articular surfaces to slide back and forth.
c. Most of the joints of the wrist and ankle are gliding joints.

4. Hinge joints
a. In a hinge joint a convex surface of one bone fits into a concave surface of another.
b. This joint permits movement in one plane only.
c. The elbow and the joints of the phalanges are hinge type.

5. Pivot joints
a. In a pivot joint a cylindrical surface of one bone rotates within a ring of bone or fibrous tissue.
b. This joint permits rotational movement.
c. The articulation between the proximal ends of the radius and ulna is a pivot joint.

6. Saddle joints
a. A saddle joint is formed between bones that have complementary surfaces with both concave and convex regions.
b. This joint permits a variety of movements.
c. The articulation between the carpal and metacarpal of the thumb is a saddle joint.

Types of Joint Movements

1. Muscles acting at freely movable joints produce movements in different directions and in different planes.

2. Movements include flexion, extension, hyperextension, dorsiflexion, plantar flexion, abduction, adduction, rotation, circumduction, supination, pronation, eversion, inversion, elevation, depression, protraction, and retraction.

Examples of Freely Movable Joints

1. The shoulder joint
a. The shoulder joint is a ball and socket joint that consists of the head of the humerus and the glenoid cavity of the scapula.
b. A cylindrical joint capsule envelopes the joint.
   (1) The capsule is loose and by itself cannot keep the articular surfaces together.
   (2) It is reinforced by surrounding muscles and tendons.
c. Several ligaments help to prevent displacement of bones.
d. There are several bursae associated with the shoulder joint.
e. Since its parts are loosely attached, the shoulder joint permits a wide range of movements.

2. The elbow joint
a. The elbow includes a hinge joint between the humerus and the ulna and a gliding joint between the humerus and radius.
b. The joint capsule is reinforced by collateral ligaments.
c. A synovial membrane partially divides the joint cavity into two portions.
d. The joint between the humerus and ulna permits flexion and extension only.

3. The hip joint
a. The hip joint is a ball and socket joint between the femur and the coxal bone.
b. A ring of fibrocartilage deepens the cavity of the acetabulum.
c. The articular surfaces are held together by a heavy joint capsule that is reinforced by ligaments.
d. The hip joint permits a wide variety of movements.

4. The knee joint
a. The knee joint includes two condyloid joints between the femur and tibia, and a gliding joint between the femur and patella.
b. The joint capsule is relatively thin, but is strengthened by ligaments and tendons.
c. Several ligaments, some of which are within the joint capsule, help to keep the articular surfaces together.
d. Two menisci separate the articulating surfaces of the femur and tibia and help to compensate for differences in the shapes of these surfaces.
e. There are several bursae associated with the knee joint.
f. The knee joint permits flexion and extension; when the leg is flexed at the knee, some rotation is possible.

Application of Knowledge

1. How would you explain to an athlete the reason damaged ligaments and cartilages of joints are so slow to heal following an injury?

2. Compared to the shoulder and hip joints, in what way is the knee joint poorly protected and thus especially vulnerable to injuries?

Review Activities

Part A

1. Define *joint*.
2. Explain how joints are classified.
3. Compare the structure of an immovable joint with that of a slightly movable joint.
4. Distinguish between a syndesmosis and a suture.
5. Describe a gomphosis and name an example of this type of joint.
6. Compare the structures of a synchondrosis and a symphysis.
7. Explain how the joints between adjacent vertebrae permit movement.
8. Describe the general structure of a freely movable joint.
9. Describe how a joint capsule may be reinforced.
10. Explain the function of the synovial membrane.
11. Explain the function of synovial fluid.
12. Define *meniscus*.
13. Define *bursa*.
14. List six types of freely movable joints and name an example of each type.
15. Describe the movements permitted by each type of freely movable joint.
16. Name the parts that comprise the shoulder joint.
17. Name the major ligaments associated with the shoulder joint.
18. Explain why the shoulder joint permits such a wide range of movements.
19. Name the parts that comprise the elbow joint.
20. Describe the major ligaments associated with the elbow joint.
21. Name the movements permitted by the elbow joint.
22. Name the parts that comprise the hip joint.
23. Describe how the articular surfaces of the hip joint are held together.
24. Explain why there is less freedom of movement in the hip joint than in the shoulder joint.
25. Name the parts that comprise the knee joint.
26. Describe the major ligaments associated with the knee joint.
27. Explain the function of the menisci of the knee.
28. Describe the locations of the bursae associated with the knee joint.

Part B

Match the movements in column I with the descriptions in column II.

I	II
1. Rotation	A. Turning palm upward
2. Supination	B. Decreasing angle between parts
3. Extension	C. Moving part forward
4. Eversion	D. Moving part around an axis
5. Protraction	E. Turning sole of foot outward
6. Flexion	F. Increasing angle between parts
7. Pronation	G. Lowering a part
8. Abduction	H. Turning palm downward
9. Depression	I. Moving part away from midline

9

The Muscular System

Muscles, the organs of the *muscular system,* consist largely of muscle cells that are specialized to undergo contractions. During these contractions, chemical energy of some nutrients is converted into mechanical energy or movement.

When muscle cells contract, they pull on the body parts to which they are attached. This action usually causes some movement, as when joints of the legs are flexed and extended during walking. At other times, muscular contractions resist motion, as when they help to hold body parts in postural positions. Muscles are also responsible for the movement of body fluids such as blood and urine. In addition they function in heat production, which helps maintain body temperature.

Chapter Outline

Chapter Objectives

After you have studied this chapter, you should be able to

1. Describe how connective tissue is included in the structure of a skeletal muscle.

2. Name the major parts of a skeletal muscle fiber and describe the function of each part.

3. Explain the major events that occur during muscle fiber contraction.

4. Explain how energy is supplied to the muscle fiber contraction mechanism, how oxygen debt develops, and how a muscle may become fatigued.

5. Distinguish between fast and slow muscles.

6. Distinguish between a twitch and a sustained contraction.

7. Describe how skeletal muscles are affected by exercise.

8. Explain how various types of muscular contractions are used to produce body movements and maintain posture.

9. Distinguish between the structure and function of a multiunit smooth muscle and a visceral smooth muscle.

10. Compare the fiber contraction mechanisms of skeletal, smooth, and cardiac muscles.

11. Explain how the locations of skeletal muscles are related to the movements they produce and how muscles interact in producing such movements.

12. Identify and describe the location of the major skeletal muscles of each body region and describe the action of each muscle.

13. Complete the review activities at the end of this chapter. Note that the items are worded in the form of specific learning objectives. You may want to refer to them before reading the chapter.

Key Terms

actin   (ak'tin)

antagonist   (an-tag'o-nist)

aponeurosis   (ap''o-nu-ro'sēz)

fascia   (fash'e-ah)

insertion   (in-ser'shun)

motor neuron   (mo'tor nu'ron)

motor unit   (mo'tor u'nit)

muscle impulse   (mus'el im'puls)

myofibril   (mi''o-fi'bril)

myogram   (mi'o-gram)

myosin   (mi'o-sin)

neurotransmitter   (nu''ro-trans'mit-er)

origin   (or'ĭ-jin)

oxygen debt   (ok'sĭ-jen det)

prime mover   (prim mōōv'er)

recruitment   (re-krōōt'ment)

sarcomere   (sar'ko-mēr)

synergist   (sin'er-jist)

threshold stimulus   (thresh'old stim'u-lus)

Aids to Understanding Words

calat-, something inserted: inter*calat*ed disk— membranous band that separates adjacent cardiac muscle cells.

erg-, work: syn*erg*ist—muscle that works together with a prime mover to produce a movement.

fasc-, a bundle: *fasc*iculus—a bundle of muscle fibers.

-gram, something written: myo*gram*—recording of a muscular contraction.

hyper-, over, more: muscular *hyper*trophy— enlargement of muscle fibers.

inter-, between: *inter*calated disk—membranous band that separates adjacent cardiac muscle cells.

iso-, equal: *iso*tonic contraction—contraction during which the tension in a muscle remains unchanged.

laten-, hidden: *laten*t period—period between the time a stimulus is applied and the beginning of a muscle contraction.

myo-, muscle: *myo*fibril—contractile fiber of a muscle cell.

reticul-, a net: sarcoplasmic *reticul*um—network of membranous channels within a muscle fiber.

syn-, together: *syn*ergist—muscle that works together with a prime mover to produce a movement.

tetan-, stiff: *tetan*ic contraction—sustained muscular contraction.

-tonic, stretched: iso*tonic* contraction—contraction during which the tension in a muscle remains unchanged.

-troph, well fed: muscular hyper*troph*y—enlargement of muscle fibers.

voluntar-, of one's free will: *voluntar*y muscle—muscle that can be controlled by conscious effort.

THERE ARE THREE types of muscle tissues—
skeletal muscle, smooth muscle, and cardiac
muscle, as described in chapter 5. This chapter, however, is primarily concerned with skeletal muscle, the
type found in muscles that are attached to bones and
are under conscious control.

Structure of a Skeletal Muscle

A skeletal muscle is an organ of the muscular system
and is composed of several kinds of tissue. These include skeletal muscle tissue, nerve tissue, blood, and
various connective tissues.

Connective Tissue Coverings

An individual skeletal muscle is separated from adjacent muscles and held into position by layers of fibrous
connective tissue called **fascia.** This connective tissue
surrounds each muscle and may project beyond the end
of its muscle fibers to form a cordlike **tendon.** Fibers in
a tendon intertwine with those in the periosteum of a
bone, thus attaching the muscle to the bone. In other
cases, the connective tissues associated with a muscle
form broad fibrous sheets called **aponeuroses,** which
may be attached to the coverings of adjacent muscles
(figure 9.1).

The layer of connective tissue that closely surrounds a skeletal muscle is called the *epimysium.* Other
layers of connective tissue called the *perimysium* extend inward from the epimysium and separate the
muscle tissue into small compartments. These compartments contain bundles of skeletal muscle fibers
called *fascicles.* Each muscle fiber within a fascicle is
surrounded by a layer of connective tissue in the form
of a thin, delicate covering called *endomysium* (figure
9.2).

Thus, all parts of a skeletal muscle are enclosed
in layers of connective tissue. This arrangement allows
the parts to have some independent movement. Also,
numerous blood vessels and nerves pass through these
layers (figure 9.3).

The fascia associated with the individual organs
of the muscular system is part of a complex network of
fasciae that extends throughout the body. The portion
of the network that surrounds and penetrates the muscles is called *deep fascia.* It is continuous with the *subcutaneous fascia* that lies beneath the skin, forming the
subcutaneous layer described in chapter 6. The network also is continuous with the *subserous fascia* that
forms the connective tissue layer of the serous membranes covering organs in various body cavities and
lining these cavities (see chapter 6).

Figure 9.1
Various connective tissues attach skeletal muscles to bones or to
other muscles.

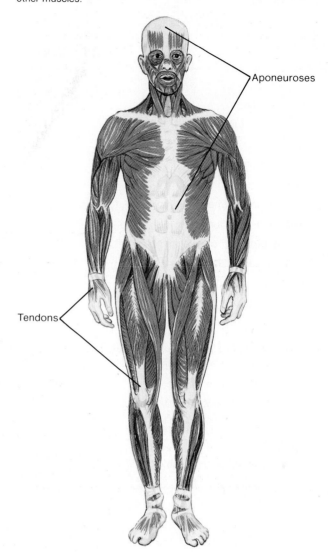

Aponeuroses

Tendons

Skeletal Muscle Fibers

As mentioned in chapter 5, a skeletal muscle fiber represents a single cell of a muscle. This fiber is responsive
to stimulation and when it responds, it contracts and
then relaxes.

Each skeletal muscle fiber is a thin, elongated
cylinder with rounded ends that are attached to connective tissues associated with a muscle. Just beneath
its cell membrane or *sarcolemma,* the cytoplasm or
sarcoplasm of the fiber contains many small, oval nuclei and mitochondria. Also, within the sarcoplasm are
numerous threadlike **myofibrils** that lie parallel to one
another (figure 9.2).

Figure 9.2

(a) A skeletal muscle is composed of a variety of tissues, including layers of connective tissue. *(b)* Fascia covers the surface of the muscle, epimysium lies beneath the fascia, and perimysium extends into the structure of the muscle and separates muscle cells into fasciculi. *(c)* Individual muscle fibers are separated by endomysium. *(d)* A single muscle fiber.

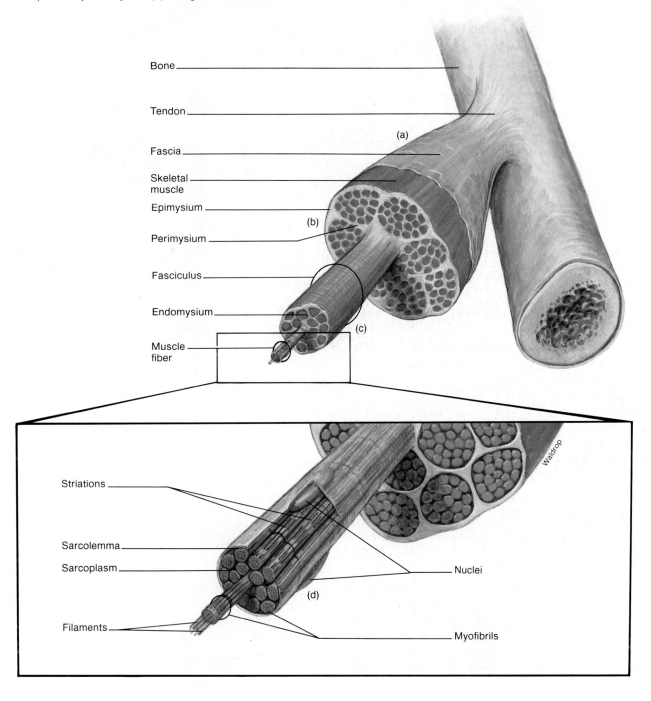

Figure 9.3

Scanning electron micrograph of a muscle surrounded by its connective tissue sheath, the epimysium (215 X). (*Tissues and Organs: A Text-Atlas of Scanning Electron Microscopy,* by R. G. Kessel and R. H. Kardon. © 1979 W. H. Freeman and Company.)

Epimysium

Muscle

Blood vessel

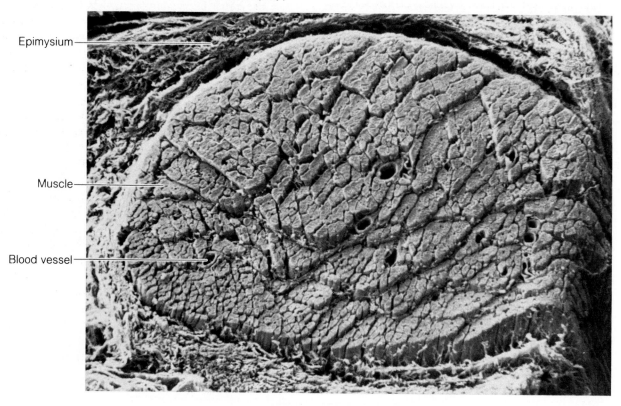

Figure 9.4

(a) A skeletal muscle fiber contains numerous myofibrils, each consisting of *(b)* units called sarcomeres. *(c)* The characteristic striations of a sarcomere are due to the arrangement of actin and myosin filaments.

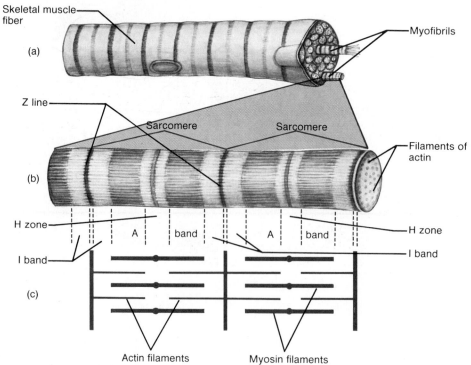

Skeletal muscle fiber

(a)

Myofibrils

Z line

Sarcomere

Sarcomere

Filaments of actin

(b)

H zone

A band

A band

H zone

I band

I band

(c)

Actin filaments

Myosin filaments

The myofibrils play a fundamental role in the muscle contraction mechanism. They contain two kinds of protein filaments—thick ones composed of the protein **myosin** and thin ones composed of the protein **actin.** The arrangement of these filaments produces the characteristic alternating light and dark striations of muscle fiber (figures 9.4 and 9.5).

Myosin filaments are located primarily within the dark portions, or *A bands,* of the striations, and actin filaments occur primarily in the light areas, or *I bands.* The actin filaments, however, also extend into the A bands, and when the muscle fiber contracts, the actin filaments slide further into these bands. Note in figure 9.4 that only myosin filaments occur between the ends of the actin filaments within the A bands. As a result, light H zones appear in the centers of the A bands.

The actin filaments are attached to Z lines at the ends of the I bands. These Z lines extend across the muscle fiber so that those of adjacent myofibrils lie side by side. The segment of the myofibril between two successive Z lines is called a **sarcomere** (figures 9.4 and 9.5). The consequent regular arrangement of the sarcomeres causes the muscle fiber to appear striated.

Within the cytoplasm of a muscle fiber is a network of membranous channels that surrounds each myofibril and runs parallel to it. This is the **sarcoplasmic reticulum,** and it corresponds to the endoplasmic reticulum of other cells. Another set of membranous channels called **transverse tubules** (T-tubules) extends inward, as invaginations from the fiber's membrane, and passes all the way through the fiber. Thus, each of these tubules opens to the outside of the muscle fiber and contains extracellular fluid. Furthermore, each transverse tubule lies between two enlarged portions of the sarcoplasmic reticulum called *cisternae* near the region where the actin and myosin filaments overlap. The sarcoplasmic reticulum and transverse tubules function in activating the muscle contraction mechanism when the fiber is stimulated (figure 9.6).

Figure 9.5
Identify the bands of the striations in this transmission electron micrograph of myofibrils.

Figure 9.6
Within the sarcoplasm of a skeletal muscle fiber are a network of sarcoplasmic reticulum and a system of transverse tubules.

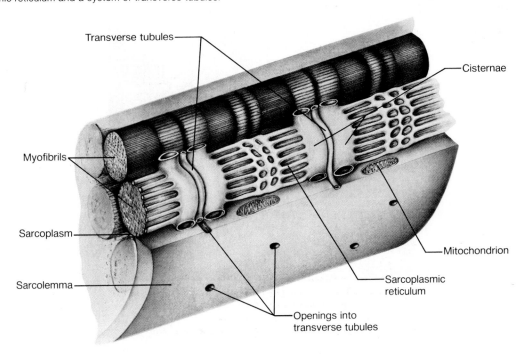

Transverse tubules

Cisternae

Myofibrils

Sarcoplasm

Sarcolemma

Mitochondrion

Sarcoplasmic reticulum

Openings into transverse tubules

Figure 9.7

(a) A neuromuscular junction includes the end of a motor neuron and the motor end plate of a muscle fiber; *(b)* micrograph of a neuromuscular junction.

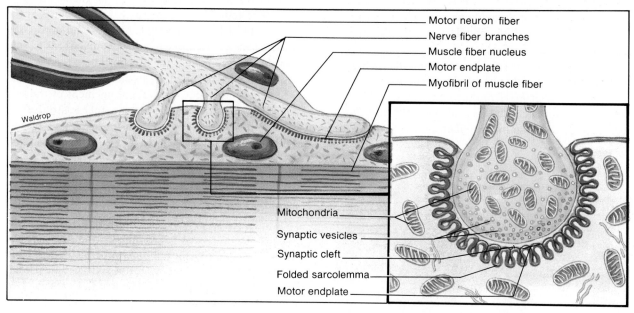

Motor neuron fiber
Nerve fiber branches
Muscle fiber nucleus
Motor endplate
Myofibril of muscle fiber

Waldrop

Mitochondria
Synaptic vesicles
Synaptic cleft
Folded sarcolemma
Motor endplate

(a)

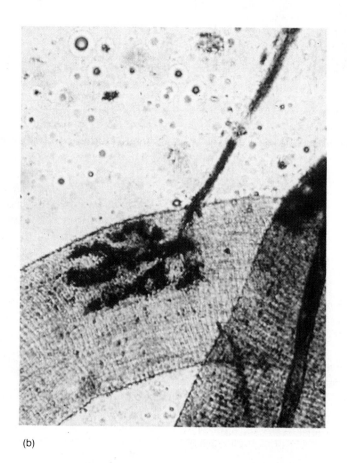

(b)

Although muscle fibers and the connective tissues associated with them are flexible and somewhat elastic, they can be torn if they are overstretched. This type of injury is common in athletes and is called a *muscle strain* or *muscle pull*. The seriousness of the injury depends on the degree of damage sustained by the tissues. In a mild strain, for example, only a few muscle fibers may be injured, the fascia remains intact, and there is little loss of function. In a severe strain, many muscle fibers as well as fascia are torn, and muscle function may be lost completely. A severe strain is very painful and is accompanied by discoloration and swelling of tissues due to the rupture of blood vessels. Such an injury may require surgery to reconnect the separated tissues.

1. Describe how connective tissue is associated with a skeletal muscle.
2. Describe the general structure of a skeletal muscle fiber.
3. Explain why skeletal muscle fibers appear striated.
4. Explain the relationship between the sarcoplasmic reticulum and transverse tubules.

The Neuromuscular Junction

Each skeletal muscle fiber is connected to a fiber of a nerve cell. Such a nerve fiber is an extension of a **motor neuron** that passes outward from the brain or spinal cord. Usually a skeletal muscle fiber will contract only when it is stimulated by the action of a motor neuron.

The site where the nerve fiber and muscle fiber meet is called a **neuromuscular junction** (myoneural junction). At this junction the muscle fiber membrane is specialized to form a **motor end plate**. In this region, muscle fiber nuclei and mitochondria are abundant and the sarcolemma is highly folded (figure 9.7).

The end of the motor nerve fiber is branched, and the ends of these branches project into recesses (synaptic clefts) of the muscle fiber membrane. The cytoplasm at the ends of the nerve fibers is rich in mitochondria and contains many tiny vesicles (synaptic vesicles) that store chemicals called **neurotransmitters.**

When a nerve impulse traveling from the brain or spinal cord reaches the end of a motor nerve fiber, some of the vesicles release a neurotransmitter into the gap between the nerve and the motor end plate. This action stimulates the muscle fiber to contract.

Motor Units

Although a muscle fiber usually has a single motor end plate, the nerve fibers are densely branched, and one motor fiber usually has connections to many muscle fibers. Furthermore, when a motor nerve fiber transmits an impulse, all of the muscle fibers connected to it are

Figure 9.8
A motor unit consists of one motor neuron and all the muscle fibers with which it communicates.

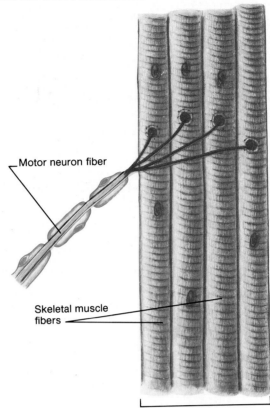

Motor neuron fiber

Skeletal muscle fibers

Motor unit

stimulated to contract simultaneously. Together, a motor neuron and the muscle fibers that it controls constitute a **motor unit** (figure 9.8).

The number of muscle fibers in a motor unit varies considerably. The fewer muscle fibers in the motor units, however, the finer the movements that can be produced in a particular muscle. For example, the motor units of the muscles that move the eyes may contain fewer than ten muscle fibers per unit and can produce very slight movements. Conversely, the motor units of the large muscles in the back may include a hundred or more muscle fibers. When these units are stimulated, the movements produced are coarse in comparison with those of the eye muscles.

If some of the motor nerve fibers innervating a muscle are lost due to injury or disease, their motor units become paralyzed. However, branches from remaining uninjured nerve fibers may grow out to join the paralyzed muscle fibers, creating new motor units with several times the usual number of muscle fibers. As a result, some muscular functions may be restored, although the degree of control over the muscle fibers is decreased.

Figure 9.9
Myosin molecules have projections that extend toward nearby actin filaments.

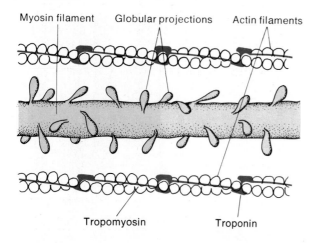

Myosin filament · Globular projections · Actin filaments

Tropomyosin · Troponin

Figure 9.10
According to the ratchet theory, (a) when calcium ions are present, active sites on an actin filament are exposed; (b) globular projections on a myosin filament form cross-bridges at the active sites; (c) a myosin cross-bridge bends slightly, pulling an actin filament; (d) the cross-bridge is broken; and (e) the myosin projection forms a cross-bridge with the next active site.

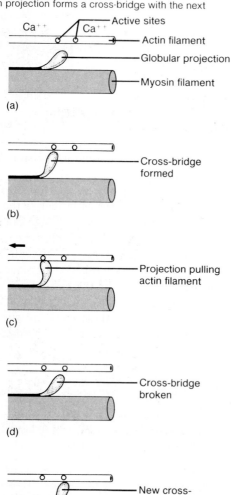

Ca⁺⁺ · Ca⁺⁺ · Active sites · Actin filament · Globular projection · Myosin filament

(a)

Cross-bridge formed

(b)

Projection pulling actin filament

(c)

Cross-bridge broken

(d)

New cross-bridge formed

(e)

Skeletal Muscle Contraction

A muscle fiber contraction is a complex action involving a number of cell parts and chemical substances. The final result is a sliding movement within the myofibrils in which the filaments of actin and myosin merge. When this happens, the muscle fiber is shortened, and it pulls on its attachments.

Role of Actin and Myosin

A myosin molecule is composed of two twisted protein strands with globular protein parts projecting outward along their lengths. In the presence of calcium ions, these globular parts can react with actin filaments and form cross-bridges with them. This reaction between the actin and myosin filaments generates the force involved with shortening the myofibrils during a contraction.

An actin molecule also consists of double twisted strands of protein, and it has ADP molecules attached to its surface. These ADP molecules serve as active sites with which the globular parts of myosin molecules interact to form cross-bridges.

Two other proteins, **tropomyosin** and **troponin,** are also associated with actin. The tropomyosin molecules occupy the longitudinal grooves of the actin helix, and each has a troponin molecule attached to its surface (figure 9.9).

When a muscle fiber is at rest, these tropomyosin-troponin complexes seem to inhibit the active sites on the actin molecules and, thus, prevent the formation of cross-bridges. If a high concentration of *calcium ions* is present, however, the ions bind to the troponin, and this apparently modifies the position of the tropomyosin. As the tropomyosin molecules move, the active sites on the actin filaments are exposed, and cross-bridges can form between the actin and myosin filaments.

It is not known how the formation of cross-bridges results in shortening myofibrils. One theory (ratchet theory) suggests that the head of a myosin cross-bridge can attach to an actin active site and bend slightly, pulling the actin filament with it. Then the head may release, straighten itself, and combine with another active site further down the actin filament. Presumably this cycle can be repeated again and again as the actin filament is moved toward the center of the sarcomere (figures 9.10 and 9.11).

Stimulus for Contraction

A skeletal muscle fiber normally does not contract until it is stimulated by a motor nerve fiber releasing a neurotransmitter at the motor end plate. In skeletal muscle,

Chart 9.1 Major events of muscle contraction and relaxation

Muscle fiber contraction	Muscle fiber relaxation
1. Stimulation occurs when acetylcholine is released from the end of a motor neuron.	1. Cholinesterase causes acetylcholine to decompose, and muscle fiber membrane is no longer stimulated.
2. Acetylcholine diffuses across gap at neuromuscular junction.	2. Calcium ions are actively transported into the sarcoplasmic reticulum.
3. Muscle fiber membrane is stimulated and a muscle impulse travels deep into the fiber through the transverse tubules.	3. Cross-bridges between actin and myosin filaments are broken.
4. Calcium ions diffuse from sarcoplasmic reticulum into the sarcoplasm and bind to troponin molecules.	4. Actin and myosin filaments slide apart.
5. Tropomyosin molecules move and expose specific sites on actin filaments.	5. Muscle fiber lengthens as it relaxes and its resting state is reestablished.
6. Cross-bridges form between actin and myosin filaments.	6. Troponin and tropomyosin molecules inhibit the interaction between actin and myosin filaments.
7. Actin filaments slide inward along myosin filaments.	
8. Muscle fiber shortens as contraction occurs.	

the neurotransmitter is a compound called **acetylcholine.** This substance is synthesized in the cytoplasm at the distal end of the motor neuron and stored in its vesicles. When a nerve impulse reaches the end of the nerve fiber, many of these vesicles discharge their acetylcholine into the gap between the nerve fiber and the motor end plate (figure 9.7).

The acetylcholine diffuses rapidly across the gap, combines with certain protein molecules (receptors) in the muscle fiber membrane, and thus stimulates the membrane. As a result of this stimulus, a **muscle impulse** (action potential), very much like a nerve impulse (described in chapter 10), passes in all directions over the surface of the muscle fiber membrane. It also travels through the transverse tubules, deep into the fiber.

The sarcoplasmic reticulum contains a high concentration of calcium ions. In response to a muscle impulse, the membranes of the cisternae seem to become more permeable to these ions, and the ions diffuse into the sarcoplasm of the muscle fiber.

When calcium ions are present in the sarcoplasm in a relatively high concentration, cross-bridges form between the actin and myosin filaments, and a muscle contraction occurs. The contraction continues while the calcium ions are present, but they are quickly moved back into the sarcoplasmic reticulum by an active transport mechanism (calcium pump). Consequently, the calcium concentration of the sarcoplasm is lowered, the cross-bridges are broken, and within a fraction of a second the muscle fiber relaxes.

Meanwhile, the acetylcholine that stimulated the muscle fiber in the first place is rapidly decomposed by the action of an enzyme called **cholinesterase,** which is present at the neuromuscular junction within the membranes of the motor end plate. This action prevents a single nerve impulse from causing a continued stimulation of the muscle fiber.

Chart 9.1 summarizes the major events leading to muscle contraction and relaxation.

Figure 9.11

(a) As cross-bridges form between filaments of actin and myosin, *(b, c)* the actin filaments are pulled toward the center of the A band, causing the myofibril to contract.

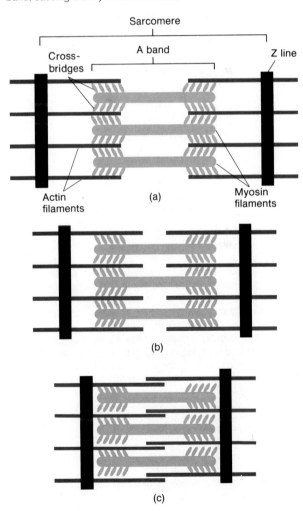

Figure 9.12
Energy released by cellular respiration may be used to promote the synthesis of ATP or to synthesize creatine phosphate. Later, energy from creatine phosphate may be used to promote ATP synthesis.

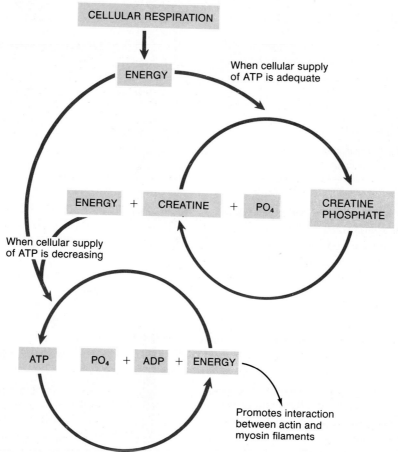

A substance called botulinus toxin, produced by a bacterium (*Clostridium botulinum*), can prevent the release of acetylcholine from motor nerve fibers. It is responsible for a very serious form of food poisoning called *botulism*. This condition is most likely to result from eating home-processed food that has not been heated adequately to kill the bacteria or to inactivate the toxin.

When botulinus toxin is present in the body, muscle fibers fail to be stimulated, and muscles, including those responsible for breathing, may be paralyzed. As a consequence, botulism may cause death.

1. Describe a neuromuscular junction.
2. Define a motor unit.
3. List four proteins associated with myofibrils and explain their relationships.
4. Explain how the filaments of a myofibril interact during muscle contraction.
5. Explain how a motor nerve impulse can trigger a muscle contraction.

Energy Sources for Contraction

The energy used in muscle fiber contraction comes from ATP molecules, which are supplied by numerous mitochondria positioned close to the myofibrils. The globular portions of the myosin filaments contain an enzyme called **ATPase** that causes ATP to decompose into ADP and phosphate and, at the same time, to release some energy (see chapter 4). This energy makes possible the reaction between the actin and myosin filaments. However, there is only enough ATP in a muscle fiber to operate the contraction mechanism for a very short time. Consequently, when a fiber is active ATP must be regenerated.

The primary source of energy available to regenerate ATP from ADP and phosphate is a substance called **creatine phosphate.** Like ATP, creatine phosphate contains high-energy phosphate bonds, and it is actually four to six times more abundant than ATP in muscle fibers. Creatine phosphate, however, cannot directly supply energy for a cell's energy-utilizing reactions. Instead, it acts to store energy released from mitochondria whenever sufficient amounts of ATP are already present (figure 9.12).

Figure 9.13
The oxygen needed to support aerobic respiration is carried in
the blood and stored in myoglobin.

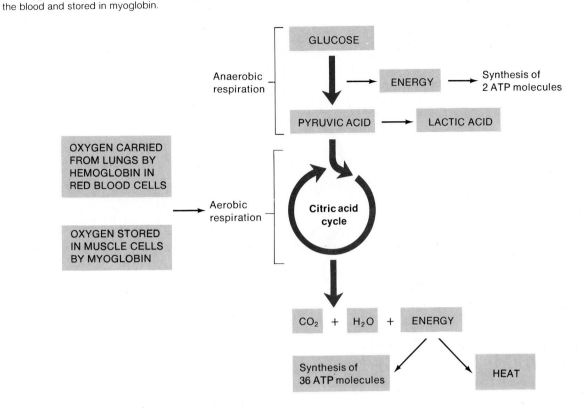

Oxygen Supply and Cellular Respiration

At times when ATP is being decomposed, energy from creatine phosphate can be transferred to ADP molecules, which are quickly converted back into ATP. The amount of ATP and creatine phosphate present in a skeletal muscle, however, is usually not sufficient to support maximal muscle activity for more than a few seconds. Consequently, the muscle fibers in an active muscle soon become dependent upon cellular respiration of glucose as a source of energy for synthesizing ATP. Typically a muscle has a supply of glucose in the form of stored glycogen molecules.

Oxygen Supply and Cellular Respiration

As is described in chapter 4, the early phase of cellular respiration occurs in the cytoplasm and is *anaerobic*, taking place in the absence of oxygen. This phase involves a partial breakdown of energy-supplying molecules such as glucose and results in a gain of only a few ATP molecules. The complete breakdown of glucose occurs in the mitochondria and is *aerobic*. This process, which includes the complex series of reactions of the *citric acid cycle*, produces a relatively large number of ATP molecules.

The oxygen needed to support aerobic respiration is carried from the lungs to body cells by the blood. It is transported within the red blood cells loosely bound to molecules of hemoglobin, the pigment responsible for the red color of blood. In regions of the body where the oxygen concentration is relatively low, oxygen is released from hemoglobin and becomes available for cellular respiration.

Another pigment, **myoglobin,** is synthesized in the muscle cells and is responsible for the reddish brown color of skeletal muscle tissue. It has properties similar to hemoglobin in that it can combine loosely with oxygen. In fact, myoglobin has a greater attraction for oxygen than does hemoglobin, and it functions to store oxygen in muscle tissue, at least temporarily. This ability to store oxygen reduces a muscle's need for a continuous blood supply during muscular contraction, which may be accompanied by a decreased blood flow. Such a decreased blood flow can result from the compression of blood vessels passing through muscle tissue when its fibers contract and thicken (figure 9.13).

Oxygen Debt

When a person is resting or is moderately active, the ability of the respiratory and circulatory systems to supply oxygen to the skeletal muscles is usually adequate to support aerobic respiration. When skeletal muscles are used strenuously for even a minute or two, however, these systems usually cannot supply oxygen

Figure 9.14

The lactic acid that accumulates in muscles as a result of anaerobic respiration can be converted back into glucose by liver cells.

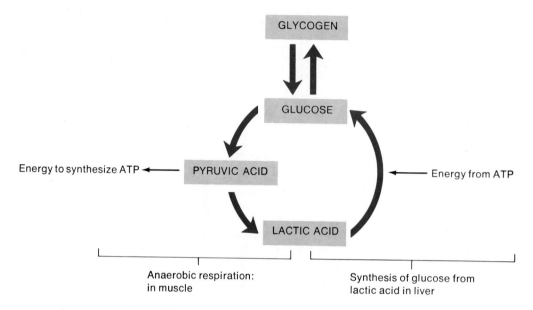

efficiently enough to meet the needs of aerobic respiration. Consequently the muscle fibers must depend more and more on the anaerobic phase of respiration to obtain energy.

In anaerobic respiration, glucose molecules are changed to *pyruvic acid.* If the oxygen supply is low, the pyruvic acid is converted to *lactic acid,* which diffuses out of the muscle fibers and is carried to the liver by the blood.

Liver cells change the lactic acid into *glucose;* however, this conversion also requires energy from ATP. During strenuous exercise, the available oxygen is used primarily to synthesize the ATP needed for the muscle fiber contraction rather than to make ATP for changing lactic acid into glucose. Consequently, as lactic acid accumulates, a person develops an **oxygen debt** that must be paid back at a later time. The amount of this oxygen debt is equal to the amount of oxygen needed by the liver cells to convert the accumulated lactic acid into glucose plus the amount needed by muscle cells to resynthesize ATP and creatine phosphate and return these substances to their original concentrations (figure 9.14).

The conversion of lactic acid back into glucose is a relatively slow process. It may require several hours to repay an oxygen debt following strenuous exercise.

Muscle Fatigue

If a muscle is exercised strenuously for a prolonged period, it may lose its ability to contract and is said to be *fatigued.* This condition may result from an interruption in the muscle's blood supply or rarely from an exhaustion of the supply of acetylcholine in its motor nerve fibers. However, muscle fatigue is more likely to arise from an accumulation of lactic acid in the muscle as a result of anaerobic respiration. The lactic acid causes factors, such as pH, to change so that muscle fibers are no longer responsive to stimulation.

Occasionally a muscle becomes fatigued and cramps at the same time. A cramp is a painful condition in which the muscle contracts spasmodically, but does not relax completely. This condition seems to be due to a lack of ATP, which is needed to move calcium ions back into the sarcoplasmic reticulum and break the connections between actin and myosin filaments before relaxation of muscle fibers can occur.

Tolerance to the effects of lactic acid varies. Therefore, some persons experience muscular fatigue more quickly than others. Resistance to this effect also varies among athletes; as a group, however, they can exercise and produce less lactic acid than nonathletes. In part, this is because the strenuous exercise of physical training stimulates new capillaries to grow within the muscles. Thus, more oxygen and nutrients can be supplied to the muscle fibers of athletes.

A few hours after death, the skeletal muscles undergo a partial contraction that causes the joints to become fixed. This condition, *rigor mortis*, may continue for seventy-two hours or more. It seems to result from a lack of ATP in the muscle fibers, which prevents relaxation. Thus, the actin and myosin filaments of the muscle fibers remain bound together until the muscles begin to decompose.

Fast and Slow Muscles

Muscles vary in the speeds at which they contract. Those that move the eyes contract about ten times faster than those responsible for maintaining posture, and the muscles that move the limbs contract at intermediate rates. Thus, the speed of contraction is related to the special function of a muscle.

The slow contracting postural muscles, such as the long muscles of the back, often are called *red muscles* because most of their fibers contain the red, oxygen-storing pigment, myoglobin. These fibers also are well supplied with blood containing the red, oxygen-carrying pigment, hemoglobin. In addition, red muscle fibers contain many mitochondria so that they are well adapted to carry on aerobic respiration. Consequently, red muscle fibers can generate ATP fast enough to keep up with the ATP breakdown that occurs when they contract. For this reason, they can contract for prolonged periods of time without undergoing fatigue.

Fast contracting muscles are called *white muscles* because they contain less myoglobin and have a poorer blood supply than red muscles. They include certain hand muscles as well as those that move the eyes. Most of the fibers of these muscles have fewer mitochondria and thus have a reduced capacity to carry on aerobic respiration. However, the white fibers have more extensive sacroplasmic reticulum, which stores and reabsorbs calcium ions, and their ATPase has higher activity than that of red fibers. Because of these factors, white muscle fibers can contract rapidly, though they tend to fatigue as ATP and the substances needed to regenerate it are depleted and as lactic acid accumulates.

Although some muscles contain a predominate number of slow contracting red fibers or fast contracting white fibers, most skeletal muscles contain a mixture of the two types.

Heat Production

Heat is produced as a by-product of cellular respiration, so it is generated by all active cells. Since muscle tissue represents such a large proportion of the total body mass, it is a particularly important source of heat.

Only about 25% of the energy released by cellular respiration is available for use in metabolic processes; the rest is lost as heat. Thus, whenever muscles are active, large amounts of heat are released. This heat is transported to other tissues by the blood and helps to maintain body temperature. If heat is present in excess, mechanisms that promote heat loss (described in chapter 6) are activated, and homeostasis is maintained.

1. What substances provide energy used to regenerate ATP?
2. What are the sources of oxygen needed for aerobic respiration?
3. What is the relationship between lactic acid and oxygen debt?
4. Distinguish between fast and slow muscles.
5. What is the relationship between cellular respiration and heat production?

Muscular Responses

One way to observe muscle contraction is to remove a single muscle fiber from a skeletal muscle and connect it to a device that senses and records changes in the fiber's length. In such experiments, the muscle fiber is usually stimulated with an electrical stimulator that is capable of producing stimuli of varying strength and frequency.

Threshold Stimulus

By exposing an isolated muscle fiber to a series of stimuli of increasing strength, it can be shown that the fiber remains unresponsive until a certain strength of stimulation is applied. This minimal strength needed to elicit a contraction is called the **threshold stimulus.**

All-or-None Response

When a muscle fiber is exposed to a stimulus of threshold strength (or above), it responds to its fullest extent. Increasing the strength of the stimulus does not affect the degree to which the fiber contracts. In other words, there are no partial contractions of a muscle fiber—if it contracts at all, it will contract completely even though it may not shorten completely. This phenomenon is called the **all-or-none response.**

To record how a whole muscle responds to stimulation, a skeletal muscle can be removed from a frog or other small animal and mounted in a special laboratory apparatus. The muscle is stimulated electrically, and when it contracts, it pulls on a lever. The lever's movement is recorded, and the resulting pattern is called a **myogram.**

If a muscle is exposed to a single stimulus of sufficient strength to activate some of its motor units, the muscle will contract and then relax. This action—a single contraction that lasts only a fraction of a second—is called a **twitch.** A twitch produces a myogram like that in figure 9.15. It is apparent from this record that the muscle response did not begin immediately following the stimulation. There is a lag between the time that the stimulus was applied and the time that the muscle responded. This time lag is called the **latent period.** In a frog muscle, the latent period lasts for about 0.01 second, and it is even shorter in a human muscle.

The latent period is followed by a *period of contraction* during which the muscle pulls at its attachments, and a *period of relaxation* during which it returns to its former length (figure 9.15).

If a muscle is exposed to two stimuli (of threshold strength or above) in quick succession, it may respond with a twitch to the first stimulus but not to the second. This is

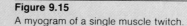
Figure 9.15
A myogram of a single muscle twitch.

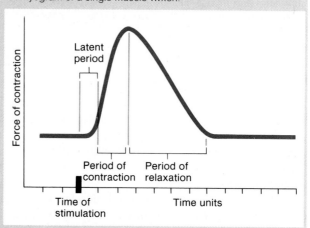

because it takes an instant following a contraction for the muscle fibers to reestablish the electrolyte concentrations necessary to conduct another muscle impulse. Thus, there is a very brief moment following stimulation during which a muscle remains unresponsive. This time is called the **refractory period.**

Recruitment of Motor Units

Since the muscle fibers within a muscle are organized into motor units, and each motor unit is controlled by a single motor neuron, all the muscle fibers in a motor unit are stimulated at the same time. Consequently, a motor unit also responds in an all-or-none manner. A whole muscle, however, does not behave like this, because it is composed of many motor units controlled by different motor neurons responding to different thresholds of stimulation. Thus, if only the motor neurons with low thresholds of stimulation are stimulated, a relatively small number of motor units contract. At higher intensities of stimulation, other motor neurons respond and more motor units are activated. Such an increase in the number of motor units being activated is called **recruitment.** As the intensity of stimulation increases, recruitment of motor units continues until finally all possible motor units are activated and the muscle is contracting with maximal tension.

Staircase Effect

If a muscle that has been inactive is subjected to a series of stimuli, it undergoes a series of contractions. However, the strength of each successive contraction increases until a maximum is reached. This phenomenon is called the **staircase effect** (treppe).

Although the cause of the staircase effect is unknown, it is believed to involve a net increase in the concentration of calcium ions available in the sarcoplasm of the muscle fibers. This might occur if each stimulus in the series causes the release of some calcium ions and if there is a failure of the sarcoplasmic reticulum to recapture the ions immediately (figure 9.16).

Sustained Contractions

If a muscle is exposed to a series of stimuli of increasing frequency, a point is reached when the muscle is unable to complete its relaxation period before the next stimulus in the series arrives. When this happens, the contractions begin to combine and the muscle contraction is sustained. Such a combination of twitches is called *summation of twitches* or *wave summation.*

At the same time that twitches are combining, the strength of the contractions may be increasing. This is due to the recruitment of motor units. The smaller motor units, which have finer fibers, tend to respond earlier in the series of stimuli. The larger motor units, which contain thicker fibers, respond later and produce more forceful contractions. The production of such a sustained contraction of increasing strength is called *multiple motor unit summation.* When the resulting

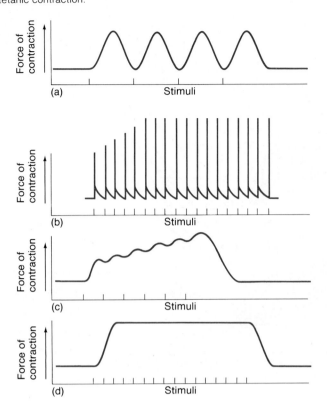

forceful, sustained contraction lacks even partial relaxation, it is termed a **tetanic** (tě-tan'ik) **contraction** (tetanus) (figure 9.16).

Although twitches may occur occasionally in human skeletal muscles, as when an eyelid twitches, such contractions are of limited use. More commonly, muscular contractions are sustained, and they are smooth rather than irregular or jerky because a mechanism operates within the spinal cord to stimulate contractions in different sets of motor units at different moments. Thus, while some motor units are contracting, others are relaxing.

Tetanic contractions occur frequently in skeletal muscles during everyday activities. In many cases the condition occurs only in a portion of a muscle. When a person lifts a weight or walks, for example, sustained contractions are maintained in arm or leg muscles for varying lengths of time. These contractions are responses to a rapid series of stimuli transmitted from the brain and spinal cord on motor neuron fibers.

Actually, even when a muscle appears to be at rest, a certain amount of sustained contraction is occurring in its fibers. This is called **muscle tone** (tonus). It is a response to nerve impulses originating repeatedly in the spinal cord that travel to small numbers of muscle fibers. The result is a continuous state of partial contraction.

Muscle tone is particularly important in maintaining posture. Tautness in the muscles of the neck, trunk, and legs enable a person to stand or sit. If tone is suddenly lost, such as when a person loses consciousness, the body will collapse. Although muscle tone is maintained in health, it is lost if motor nerve fibers are cut or if diseases interfere with the conduction of nerve impulses.

When skeletal muscles are contracted very forcefully they may generate up to 50 pounds of pull for each square inch of muscle cross section. Consequently, large muscles such as those in the thigh can pull with several hundred pounds of force. Occasionally this force is so great that the tendons of muscles are torn away from their attachments on bones.

Isometric and Isotonic Contractions

Sometimes a skeletal muscle contracts but the parts to which it is attached do not move. This happens, for instance, when a person pushes against the wall of a building. Tension within the muscles increases, but the wall does not move, and the muscles remain the same length. Contractions of this type are called **isometric.** Isometric contractions occur continuously in postural muscles that function to stabilize skeletal parts and hold the body upright.

Figure 9.17

(a) Isotonic contractions occur when a muscle contracts and shortens; (b) isometric contractions occur when a muscle contracts, but does not shorten.

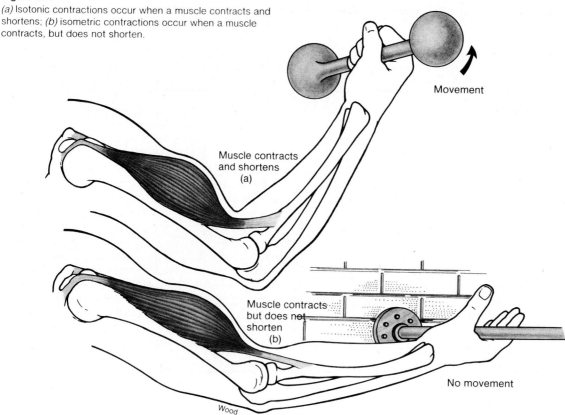

Movement

Muscle contracts
and shortens
(a)

Muscle contracts
but does not
shorten
(b)

No movement

Wood

At other times, muscles shorten when they contract. For example, if a person lifts an object, the tautness in the muscles remains unchanged, their attached ends pull closer together, and the object is moved. This type of contraction is termed **isotonic.** Figure 9.17 illustrates these two types of contraction.

Skeletal muscles can contract either isometrically or isotonically, and most body actions involve both types of contraction. In walking, for instance, certain leg muscles contract isometrically and keep the limb stiff as it touches the ground, while other muscles contract isotonically, causing the leg to bend and lift upward.

1. Define threshold stimulus.
2. What is meant by an all-or-none response?
3. Distinguish between a twitch and a sustained contraction.
4. Define muscle tone.
5. Explain the difference between isometric and isotonic contractions.

Smooth Muscles

The contractile mechanisms of smooth and cardiac muscles are essentially the same as that of skeletal muscles. However, the cells of these tissues have important structural and functional differences.

Smooth Muscle Fibers

As is discussed in chapter 5, smooth muscle cells are shorter than those of skeletal muscle, and they have single, centrally located nuclei. These cells are elongated with tapering ends. Smooth muscle cells contain filaments of *actin* and *myosin* in myofibrils that extend the lengths of the cells. However, these filaments are very thin and more randomly arranged than in skeletal muscle. Consequently, smooth muscle cells lack striations. They also apparently lack transverse tubules, and their sarcoplasmic reticula are not well developed.

There are two major types of smooth muscles. In one type, called **multiunit smooth muscle,** the muscle fibers are less well organized and occur as separate fibers rather than in sheets. Smooth muscle of this type is found in the irises of the eyes and in the walls of blood vessels. Typically it contracts only after stimulation by motor nerve impulses.

The second type of smooth muscle is called **visceral.** It is composed of sheets of spindle-shaped cells that are in close contact and possess gap junctions between one another. These cells are positioned so the thick portion of each cell is next to the thin parts of adjacent cells. Visceral smooth muscle is the more common type and is found in the walls of hollow visceral organs such as the stomach, intestines, urinary

Skeletal muscles are very responsive to use and disuse. For example, those that are forcefully exercised tend to enlarge. This phenomenon is called *muscular hypertrophy*. Conversely, a muscle that is not used undergoes *atrophy*—that is, it decreases in size and strength.

The way a muscle responds to use also depends on the type of exercise involved. For instance, when a muscle contracts relatively weakly as occurs during swimming and running exercises, its slow, fatigue-resistant red fibers are most likely to be activated. As a result, these fibers develop more mitochondria, and more extensive capillary networks develop in their surroundings. Such changes increase the fibers' abilities to resist fatigue during prolonged periods of exercise, although their sizes and strengths may remain unchanged.

Forceful exercise, such as weight lifting, in which a muscle exerts more than 75% of its maximum tension involves the muscle's fast, fatigable white fibers. In response, existing muscle fibers develop new filaments of actin and myosin, and as their diameters increase the whole muscle enlarges. There is, however, no production of new muscle fibers during hypertrophy.

Since the strength of a contraction is directly related to the diameter of the muscle fibers, an enlarged muscle is capable of producing stronger contractions than before. However, such a change does not increase the muscle's ability to resist fatigue during activities such as running or swimming.

If exercise is discontinued, there is a reduction in the capillary networks and in the number of mitochondria within the muscle fibers. Also the size of the actin and myosin filaments decreases and the entire muscle atrophies. Such atrophy commonly occurs when limbs are immobilized by casts or when accidents or diseases interfere with motor nerve impulses. A muscle that cannot be exercised may decrease to less than one-half its usual size within a few months.

The fibers of muscles whose motor neurons are severed not only decrease in size, but also may become fragmented and, in time, be replaced by fat or fibrous connective tissue. However, if such a muscle is reinnervated within the first few months following an injury, its function may be restored. Meanwhile, atrophy may be delayed by treatments in which electrical stimulation is used to cause muscular contractions against loads.

---

bladder, and uterus. Usually there are two thicknesses of smooth muscle in the walls of these organs. The fibers of the outer coats are directed longitudinally, while those of the inner coats are arranged circularly. These muscular layers are responsible for the changes in size and shape that occur in visceral organs as they carry on their special functions.

The fibers of visceral smooth muscles are capable of stimulating each other. Consequently, when one fiber is stimulated, the impulse moving over its surface may excite adjacent fibers, which in turn stimulate still others. Visceral smooth muscles also display *rhythmicity*—a pattern of repeated contractions. This phenomenon is due to the presence of self-exciting fibers from which spontaneous impulses travel periodically into the surrounding muscle tissue.

These two features of visceral smooth muscle—transmission of impulses from cell to cell and rhythmicity—are largely responsible for the wavelike motion called **peristalsis** that occurs in various tubular organs (see chapter 14). Peristalsis involves alternate contractions and relaxations of the longitudinal and circular muscles. These movements help force the contents of a tube along its length. In the intestines, for example, peristaltic waves move masses of food substances through the tube and mix them with digestive fluids. Similar activity in the ureters moves urine from the kidneys to the urinary bladder.

Smooth Muscle Contraction

Smooth muscle contraction resembles skeletal muscle contraction in a number of ways. Both mechanisms involve reactions of actin and myosin, both are triggered by membrane impulses and the release of calcium ions, and both use energy from ATP molecules. There are, however, significant differences. For example, smooth muscle fibers seem to lack troponin, the protein that binds to calcium ions in skeletal muscle. Instead, the smooth muscle seems to make use of a protein called *calmodulin,* which binds to calcium ions released when its fibers are stimulated, thus activating the actin-myosin contraction mechanism.

Although acetylcholine is the neurotransmitter substance in skeletal muscle, two neurotransmitters affect smooth muscle—acetylcholine and norepinephrine. Each of these substances stimulates contractions in some smooth muscles and inhibits it in others. These actions are described in more detail in chapter 11 in the section dealing with the autonomic nervous system.

Smooth muscles are also affected by a number of hormones that stimulate contraction in some cases and alter the degree of response to neurotransmitters in others. For example, during the later stages of the birth process, the smooth muscles in the wall of the uterus are stimulated to contract by the hormone called oxytocin (see chapter 22).

Figure 9.18
The intercalated disks of cardiac muscle, shown in this
transmission electron micrograph, hold adjacent cells together.

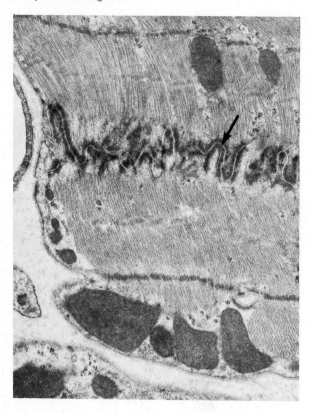

Stretching of smooth muscle fibers can also trigger contractions. This response is particularly important to the function of visceral smooth muscle in the walls of various hollow organs, such as the urinary bladder and intestines. For example, when the wall of the intestine is stretched by its contents, an automatic contraction can move the contents away.

Smooth muscle is slower to contract and slower to relax than skeletal muscle. On the other hand, it can maintain a forceful contraction for a longer time with a given amount of ATP. Unlike skeletal muscle, smooth muscle fibers can change length without changing tautness; because of this, smooth muscles in the stomach and intestinal walls can stretch as these organs become filled, while the pressure inside remains unchanged.

1. Describe the two major types of smooth muscle.
2. What special characteristics of visceral smooth muscle make peristalsis possible?
3. How is smooth muscle contraction similar to that of skeletal muscle?
4. How do their contraction mechanisms differ?

Cardiac Muscle

Cardiac muscle appears only in the heart. It is composed of striated cells that are joined end to end forming fibers. These fibers are interconnected in branching, three-dimensional networks. Each cell contains a single nucleus and numerous filaments of actin and myosin similar to those in skeletal muscle. Cardiac muscle also has a well-developed sarcoplasmic reticulum, a system of transverse tubules, and many mitochondria. However, the cisternae of the sarcoplasmic reticulum of a cardiac muscle fiber are less well developed and store less calcium than those of a skeletal muscle fiber. On the other hand, the transverse tubules of cardiac muscle fibers are larger than those in skeletal muscle, and they release large quantities of calcium ions into the sarcoplasm in response to a muscle impulse. This extra calcium from the transverse tubules makes it possible for cardiac muscle to maintain a contraction for a longer time than skeletal muscle.

The calcium ions provided by transverse tubules are obtained from the fluid outside of the muscle fiber. Consequently, the strength of the cardiac muscle contraction is controlled partially by the extracellular calcium ion concentration.

Chart 9.2	Types of muscle tissue		
	Skeletal	**Smooth**	**Cardiac**
Major location	Skeletal muscles	Walls of hollow visceral organs	Wall of heart
Major function	Movement of bones at joints, maintenance of posture	Movement of visceral organs, peristalsis	Pumping action of heart
Cellular characteristics			
Striations	Present	Absent	Present
Nucleus	Multiple nuclei	Single nucleus	Single nucleus
Special features	Transverse tubule system well developed	Lacks transverse tubules	Transverse tubule system well developed, adjacent cells separated by intercalated disks
Mode of control	Voluntary	Involuntary	Involuntary
Contraction characteristics	Contracts and relaxes relatively rapidly	Contracts and relaxes relatively slowly, self-exciting, rhythmic	Network of fibers contracts as a unit, self-exciting, rhythmic, remains refractory until contraction is completed

The opposing ends of cardiac muscle cells are separated by cross-bands called *intercalated disks*. These bands are the result of elaborate junctions of membranes at the cell's boundary. They help to hold adjacent cells together and transmit the force of contraction from cell to cell. Intercellular junctions between the fused membranes of the intercalated disks allow diffusion of ions between the cells. This makes it possible for muscle impulses to travel rapidly from cell to cell (see figures 5.23 and 9.18).

When one portion of the cardiac muscle network is stimulated, the impulse passes to the other fibers of the network, and the whole structure contracts as a unit; that is, the network responds to stimulation in an all-or-none manner. Cardiac muscle is also self-exciting and rhythmic. Consequently, a pattern of contraction and relaxation is repeated again and again and is responsible for the rhythmic contractions of the heart. Also unlike skeletal muscle, cardiac muscle remains refractory until a contraction is completed, so that sustained or tetanic contractions do not occur in the heart muscle.

Chart 9.2 summarizes the characteristics of the three types of muscles.

1. How is cardiac muscle similar to skeletal muscle?
2. How does it differ from skeletal muscle?
3. What is the function of intercalated disks?

Skeletal Muscle Actions

Skeletal muscles are responsible for a great variety of body movements. The action of each muscle—that is, the movement it causes—depends largely upon the kind of joint it is associated with and the way it is attached on either side of the joint.

Origin and Insertion

As is mentioned in chapter 7, one end of a skeletal muscle is usually fastened to a relatively immovable or fixed part, and the other end is connected to a movable part on the other side of a joint. The immovable end is called the **origin** of the muscle, and the movable one is its **insertion.** When a muscle contracts, its insertion is pulled toward its origin (figure 9.19).

Some muscles have more than one origin or insertion. The *biceps brachii* in the upper arm, for example, has two origins. This is reflected in its name, *biceps,* which means *two heads.* As figure 9.19 shows, one head is attached to the coracoid process of the scapula, and the other one arises from a tubercle above the glenoid cavity of the scapula. The muscle extends along the front surface of the humerus and is inserted by means of a tendon on the radial tuberosity of the radius. When the biceps brachii contracts, its insertion is pulled toward its origin, and the arm bends at the elbow.

Sometimes the end of a muscle changes its function in different body movements; that is, the origin may become the insertion or vice versa. For instance, a large muscle in the chest, *pectoralis major*, connects the humerus to bones of the thorax. If the arm moves across the chest, the end of the muscle attached to the humerus acts as the insertion. When a person hangs from a bar to do chinning exercises, however, the arm becomes fixed and the body is pulled up. In this movement, the end attached to the humerus acts as the origin.

Figure 9.19
The biceps brachii has two heads that originate on the scapula.
This muscle is inserted on the radius by means of a tendon.
What movement results as this muscle contracts?

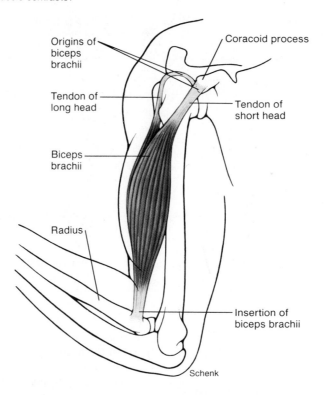

Origins of biceps brachii

Coracoid process

Tendon of long head

Tendon of short head

Biceps brachii

Radius

Insertion of biceps brachii

Schenk

Similarly, the *latissimus dorsi* muscle in the back has its origin on the pelvic girdle and vertebral column and its insertion on the proximal end of the humerus. Usually it functions to pull the arm toward the back as in rowing a boat or swimming. During chinning exercises, however, the end attached to the humerus becomes the origin of the muscle, and the end attached to the pelvis and vertebral column moves.

Interaction of Skeletal Muscles

Skeletal muscles almost always function in groups rather than singly. Consequently, when a particular body movement occurs, a person must do more than command a single muscle to contract; instead that person wills the movement to occur, and the appropriate group of muscles responds to the decision.

By carefully observing body movements, it is possible to determine the particular roles of various muscles. For instance, when the arm is lifted horizontally away from the side, a *deltoid* muscle is responsible for most of the movement, and so it is said to be the **prime mover.** However, while the prime mover is acting, certain nearby muscles also are contracting. They help to hold the shoulder steady and in this way make the action of the prime mover more effective. Muscles that assist the prime mover are called **synergists.**

Still other muscles act as **antagonists** to prime movers. These muscles are capable of resisting a prime mover's action and are responsible for movement in the opposite direction—the antagonist of the prime mover that raises the arm can lower the arm, or the antagonist of the prime mover that bends the arm can straighten it. If both a prime mover and its antagonist contract simultaneously, the part they act upon remains rigid. Consequently, smooth body movements depend upon the antagonists' giving way to the actions of the prime movers whenever the prime movers are contracting. These complex actions are controlled by the nervous system, as described in chapter 11.

1. Distinguish between the origin and insertion of a muscle.
2. Define prime mover.
3. What is the function of a synergist? An antagonist?

Figure 9.20
Anterior view of superficial skeletal muscles.

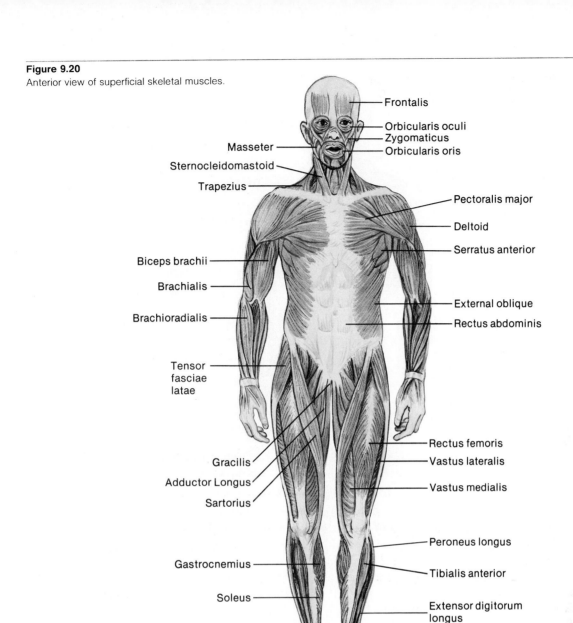

Frontalis
Orbicularis oculi
Zygomaticus
Orbicularis oris
Masseter
Sternocleidomastoid
Trapezius
Pectoralis major
Deltoid
Serratus anterior
Biceps brachii
Brachialis
Brachioradialis
External oblique
Rectus abdominis
Tensor fasciae latae
Gracilis
Adductor Longus
Sartorius
Rectus femoris
Vastus lateralis
Vastus medialis
Peroneus longus
Gastrocnemius
Tibialis anterior
Soleus
Extensor digitorum longus

Nelson

Major Skeletal Muscles

The following section concerns the locations, actions, and attachments of some of the major skeletal muscles of the body. (Figures 9.20 and 9.21 show the locations of superficial skeletal muscles—that is, those near the surface.) Notice that the names of these muscles often describe them in some way. A name may indicate a muscle's size, shape, location, action, number of attachments, or the direction of its fibers, as in the following examples:

pectoralis major   a muscle of large size (*major*) located in the pectoral region or chest.
deltoid   shaped like a delta or triangle.
extensor digitorum   acts to extend the digits (fingers or toes).
biceps brachii   a muscle with two heads (*biceps*), or points of origin, and located in the brachium or arm.
sternocleidomastoid   attached to the sternum, clavicle, and mastoid process.
external oblique   located near the outside, with fibers that run obliquely or in a slanting direction.

The Muscular System / **281**

Figure 9.21

Posterior view of superficial skeletal muscles.

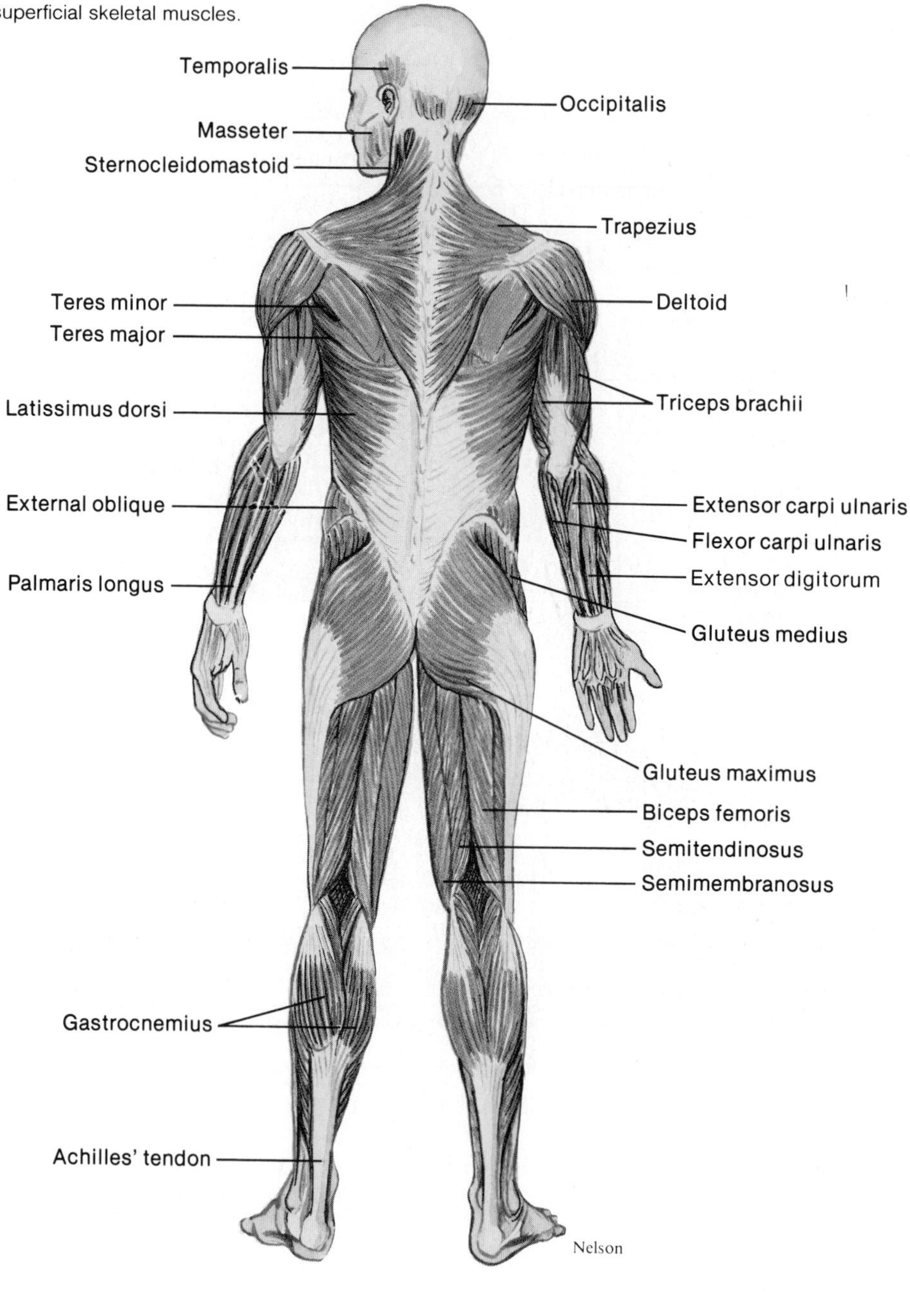

Temporalis

Masseter

Sternocleidomastoid

Occipitalis

Trapezius

Teres minor

Teres major

Deltoid

Latissimus dorsi

Triceps brachii

External oblique

Extensor carpi ulnaris

Flexor carpi ulnaris

Palmaris longus

Extensor digitorum

Gluteus medius

Gluteus maximus

Biceps femoris

Semitendinosus

Semimembranosus

Gastrocnemius

Achilles' tendon

Nelson

Muscles of Facial Expression

A number of small muscles that lie beneath the skin of the face and scalp enable us to communicate feelings through facial expression. Many of these muscles are located around the eyes and mouth, and they are responsible for such expressions as surprise, sadness, anger, fear, disgust, and pain. As a group, the muscles of facial expression connect the bones of the skull to connective tissue in various regions of the overlying skin. They include the following (figure 9.22 and chart 9.3):

Epicranius   *Buccinator*
Orbicularis oculi   *Zygomaticus*
Orbicularis oris   *Platysma*

The **epicranius** (ep″ĭ-kra′ne-us) covers the upper part of the cranium and consists of two muscular parts—the *frontalis,* which lies over the frontal bone, and the *occipitalis,* which lies over the occipital bone. These parts are united by a broad, tendinous membrane called the *epicranial aponeurosis,* which covers the cranium like a cap. Contraction of the epicranius raises the eyebrows and causes the skin of the forehead to wrinkle horizontally, as when a person expresses surprise. Headaches often result from sustained contraction of this muscle.

The **orbicularis oculi** (or-bik′u-la-rus ok′u-li) is a ringlike band of muscle, called a *sphincter muscle,* that surrounds the eye. It lies in the subcutaneous tissue of

Figure 9.22
Muscles of expression and mastication.

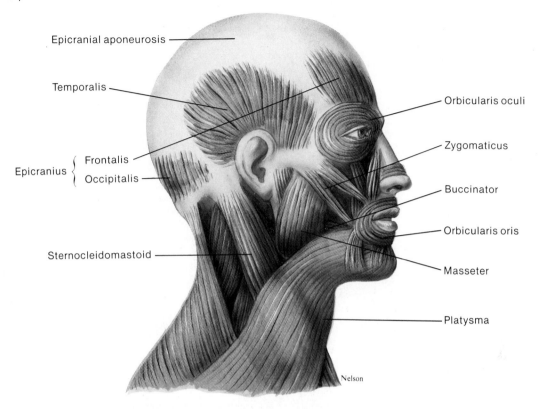

- Epicranial aponeurosis
- Temporalis
- Epicranius { Frontalis / Occipitalis }
- Sternocleidomastoid
- Orbicularis oculi
- Zygomaticus
- Buccinator
- Orbicularis oris
- Masseter
- Platysma

Nelson

Chart 9.3	Muscles of facial expression		
Muscle	Origin	Insertion	Action
Epicranius	Occipital bone	Skin and muscles around eye	Raises eyebrow
Orbicularis oculi	Maxillary and frontal bones	Skin around eye	Closes eye
Orbicularis oris	Muscles near the mouth	Skin of central lip	Closes lips, protrudes lips
Buccinator	Outer surfaces of maxilla and mandible	Orbicularis oris	Compresses cheeks inward
Zygomaticus	Zygomatic bone	Orbicularis oris	Raises corner of mouth
Platysma	Fascia in upper chest	Lower border of mandible	Draws angle of mouth downward

the eyelid and causes the eye to close or blink. At the same time, it compresses the nearby tear gland, or *lacrimal gland,* aiding the flow of tears over the surface of the eye. Contraction of the orbicularis oculi also causes the appearance of folds or crow's feet that radiate laterally from the corner of the eye.

The **orbicularis oris** (or-bik'u-la-rus o'ris) is a sphincter muscle that encircles the mouth. It lies between the skin and the mucous membranes of the lips, extending upward to the nose and downward to the region between the lower lip and chin. The orbicularis oris is sometimes called the kissing muscle because it causes the lips to close and pucker.

The **buccinator** (buk'si-na''tor) is located in the wall of the cheek. Its fibers are directed forward from the bones of the jaws to the angle of the mouth, and

when they contract the cheek is compressed inward. This action helps to hold food in contact with the teeth when a person is chewing. The buccinator also aids in blowing air out of the mouth, and for this reason, it is sometimes called the trumpeter muscle.

The **zygomaticus** (zi''go-mat'ik-us) extends from the zygomatic arch downward to the corner of the mouth. When it contracts, the corner of the mouth is drawn up, as in smiling or laughing.

The **platysma** (plah-tiz'mah) is a thin, sheetlike muscle whose fibers extend from the chest upward over the neck to the face. It functions to pull the angle of the mouth downward, as in pouting. The platysma also helps to lower the mandible.

The muscles that move the eye are described in chapter 12.

Figure 9.23
The medial pterygoid muscles help close the jaw. The lateral pterygoid muscles help open the mouth and can protrude the jaw.

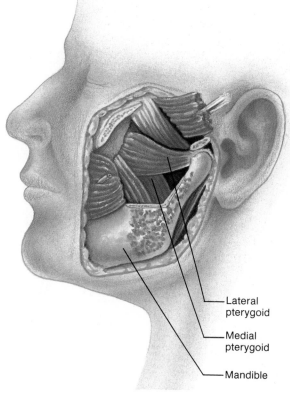

— Lateral pterygoid

— Medial pterygoid

— Mandible

Chart 9.4	Muscles of mastication		
Muscle	Origin	Insertion	Action
Masseter	Lower border of zygomatic arch	Lateral surface of mandible	Closes jaw
Temporalis	Temporal bone	Coronoid process and anterior ramus of mandible	Closes jaw
Medial pterygoid	Sphenoid, palatine, and maxilla bones	Medial surface of mandible	Closes jaw
Lateral pterygoid	Sphenoid bone	Anterior surface of mandibular condyle	Lowers jaw, pulls mandible forward, and moves it from side to side

Muscles of Mastication

Chewing movements are produced by four pairs of muscles that are attached to the mandible. Three pairs of these muscles act to close the lower jaw, as in biting; the fourth pair can lower the jaw, cause side-to-side grinding motions of the mandible, and pull it forward, causing it to protrude. The muscles of mastication include the following (figures 9.22 and 9.23, and chart 9.4):

Masseter   *Medial pterygoid*
Temporalis   *Lateral pterygoid*

The **masseter** (mas-se′ter) is a thick, flattened muscle that can be felt just in front of the ear when the teeth are clenched. Its fibers extend downward from the zygomatic arch to the mandible. The masseter functions primarily to raise the jaw, but also can control the rate at which the jaw falls open in response to gravity.

The **temporalis** (tem-po-ra′lis) is a fan-shaped muscle located on the side of the skull above and in front of the ear. Its fibers, which also raise the jaw, pass downward beneath the zygomatic arch to the mandible.

The **medial pterygoid** (ter′ĭ-goid) extends back and downward from the sphenoid, palatine, and maxillary bones to the ramus of the mandible. It closes the jaw.

Figure 9.24
Deep muscles in the back of the neck help to move the head.
(The splenius capitis is removed on the left to show the
underlying muscles.)

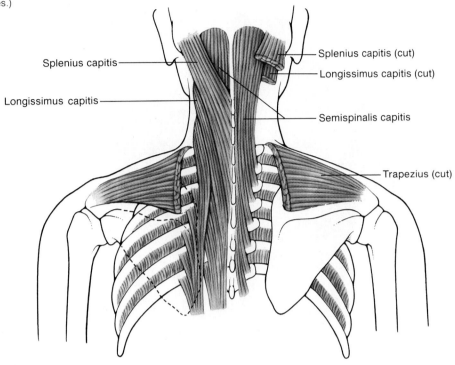

<table>
</table>

Muscle	Origin	Insertion	Action
		Chart 9.5 Muscles that move the head	
Sternocleido-mastoid	Anterior surface of sternum and upper surface of clavicle	Mastoid process of temporal bone	Pulls head to one side, pulls head toward chest, or raises sternum
Splenius capitis	Spinous processes of lower cervical and upper thoracic vertebrae	Mastoid process of temporal bone	Rotates head, bends head to one side, or brings head into upright position
Semispinalis capitis	Processes of lower cervical and upper thoracic vertebrae	Occipital bone	Extends head, bends head to one side, or rotates head
Longissimus capitis	Processes of lower cervical and upper thoracic vertebrae	Mastoid process of temporal bone	Extends head, bends head to one side, or rotates head

The fibers of the **lateral pterygoid** are directed forward from the region just below the mandibular condyle to the sphenoid bone. This muscle can open the mouth, pull the mandible forward making it protrude, and move the mandible from side to side.

Sometimes an individual who is emotionally tense or who has a poor bite (malocclusion) uses his or her jaw muscles to grind or clench the teeth excessively. This action may create stress on the temporomandibular joint—the articulation between the mandibular condyle of the mandible and the mandibular fossa of the temporal bone. As a result the person may experience a variety of unpleasant symptoms, including headache, earache, and pain in the jaw, neck, or shoulder. This condition is called *temporomandibular joint syndrome* (TMJ syndrome).

Muscles That Move the Head

Head movements result from the actions of paired muscles in the neck and upper back. These muscles are responsible for flexing, extending, and rotating the head. They include the following (figures 9.24, 9.26, and chart 9.5):

Sternocleidomastoid   *Semispinalis capitis*
Splenius capitis   *Longissimus capitis*

The **sternocleidomastoid** (ster″no-kli″do-mas′toid) is a long muscle in the side of the neck that extends upward from the thorax to the base of the skull behind the ear. When the sternocleidomastoid on one

Figure 9.25

Muscles of the posterior shoulder. (The trapezius is removed on the right to show underlying muscles.)

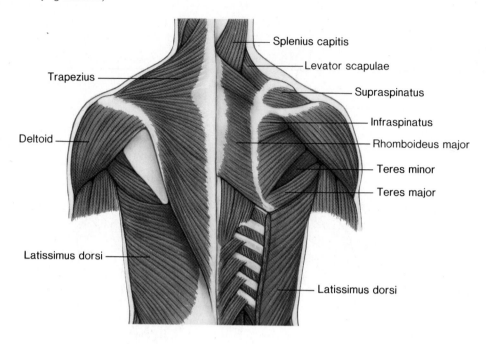

Splenius capitis

Levator scapulae

Trapezius

Supraspinatus

Infraspinatus

Deltoid

Rhomboideus major

Teres minor

Teres major

Latissimus dorsi

Latissimus dorsi

Figure 9.26

Muscles of the anterior chest and abdominal wall. (The left pectoralis major is removed to show the pectoralis minor.)

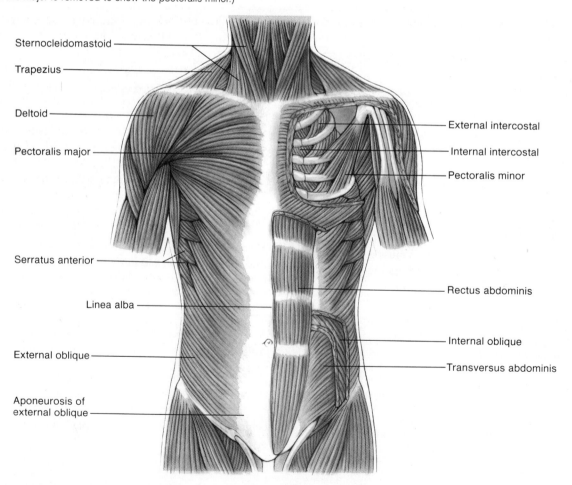

Sternocleidomastoid

Trapezius

Deltoid

External intercostal

Pectoralis major

Internal intercostal

Pectoralis minor

Serratus anterior

Linea alba

Rectus abdominis

External oblique

Internal oblique

Transversus abdominis

Aponeurosis of external oblique

Chart 9.6   Muscles that move the pectoral girdle

Muscle	Origin	Insertion	Action
Trapezius	Occipital bone and spines of cervical and thoracic vertebrae	Clavicle, spine, and acromion process of scapula	Rotates scapula; various fibers raise scapula, pull scapula medially, or pull scapula and shoulder downward
Rhomboideus major	Spines of upper thoracic vertebrae	Medial border of scapula	Raises and adducts scapula
Levator scapulae	Transverse processes of cervical vertebrae	Medial margin of scapula	Elevates scapula
Serratus anterior	Outer surfaces of upper ribs	Ventral surface of scapula	Pulls scapula anteriorly and downward
Pectoralis minor	Sternal ends of upper ribs	Coracoid process of scapula	Pulls scapula forward and downward or raises ribs

side contracts, the face is turned to the opposite side. When both muscles contract, the head is bent toward the chest. If the head is fixed in position by other muscles, the sternocleidomastoids can raise the sternum—an action that aids forceful inhalation.

The **splenius capitis** (sple'ne-us kap'ĭ-tis) is a broad, straplike muscle located in the back of the neck. It connects the base of the skull to the vertebrae in the neck and upper thorax. A splenius capitus acting singly causes the head to rotate and bend toward one side. Acting together, these muscles bring the head into an upright position.

The **semispinalis capitis** (sem''e-spi-na'lis kap'ĭ-tis) is a broad, sheetlike muscle extending upward from the vertebrae in the neck and thorax to the occipital bone. It functions to extend the head, bend it to one side, or rotate it.

The **longissimus capitis** (lon-jis'ĭ-mus kap'ĭ-tis) is a narrow band of muscle that ascends from the vertebrae of the neck and thorax to the temporal bone. It also functions to extend the head, bend it to one side, or rotate it.

Muscles That Move the Pectoral Girdle

The muscles that move the pectoral girdle are closely associated with those that move the upper arm. A number of these chest and shoulder muscles connect the scapula to nearby bones and act to move the scapula upward, downward, forward, and backward. They include the following (figures 9.25, 9.26, and chart 9.6):

Trapezius   *Serratus anterior*
Rhomboideus major   *Pectoralis minor*
Levator scapulae

The **trapezius** (trah-pe'ze-us) is a large, triangular muscle in the upper back that extends horizontally from the base of the skull and vertebral column to the shoulder. Its fibers are arranged in three groups—upper, middle, and lower. Together these fibers act to rotate the scapula. The upper fibers acting alone raise the scapula and shoulder, as when the shoulders are shrugged to express a feeling of indifference. The middle fibers pull the scapula toward the vertebral column, and the lower fibers draw the scapula and shoulder downward. When the shoulder is fixed in position by other muscles, the trapezius can pull the head backward or to one side.

The **rhomboideus** (rom'boid-e-us) **major** connects the upper thoracic vertebrae to the scapula. It acts to raise the scapula and to adduct it.

The **levator scapulae** (le-va'tor scap'u-lē) is a straplike muscle that runs almost vertically through the neck, connecting the cervical vertebrae to the scapula. It functions to elevate the scapula.

The **serratus anterior** (ser-ra'tus an-te're-or) is a broad, curved muscle located on the side of the chest. It arises as fleshy slips on the upper ribs and extends along the medial wall of the axilla to the ventral surface of the scapula. It functions to pull the scapula downward and anteriorly and is used to thrust the shoulder forward, as when pushing something.

The **pectoralis** (pek''to-ra'lis) **minor** is a thin, flat muscle that lies beneath the larger pectoralis major. It extends laterally and upward from the ribs to the scapula and serves to pull the scapula forward and downward. When other muscles fix the scapula in position, the pectoralis minor can raise the ribs and thus aid forceful inhalation.

Figure 9.27

Muscles of the posterior surface of the scapula and the upper arm.

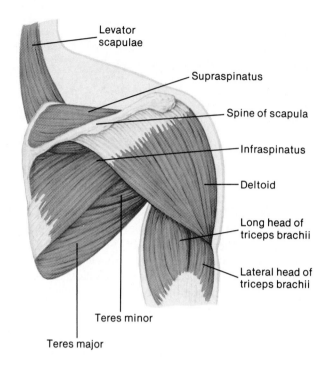

Levator scapulae

Supraspinatus

Spine of scapula

Infraspinatus

Deltoid

Long head of triceps brachii

Lateral head of triceps brachii

Teres minor

Teres major

Muscles That Move the Upper Arm

The upper arm is one of the more freely movable parts of the body. Its many movements are made possible by muscles that connect the humerus to various regions of the pectoral girdle, ribs, and vertebral column (figures 9.25, 9.26, 9.27, 9.28, and 9.29). These muscles can be grouped according to their primary actions—flexion, extension, abduction, and rotation—as follows (chart 9.7):

Flexors	Abductors
Coracobrachialis	*Supraspinatus*
Pectoralis major	*Deltoid*
Extensors	Rotators
Teres major	*Subscapularis*
Latissimus dorsi	*Infraspinatus*
	Teres minor

Flexors

The **coracobrachialis** (kor″ah-ko-bra′ke-al-is) extends from the scapula to the middle of the humerus along its medial surface. It acts to flex and adduct the upper arm.

The **pectoralis major** is a thick, fan-shaped muscle located in the upper chest. Its fibers extend from the center of the thorax through the armpit to the humerus. This muscle functions primarily to pull the upper arm forward and across the chest. It can also rotate the humerus medially and adduct the arm from a raised position.

Extensors

The **teres** (te′rēz) **major** connects the scapula to the humerus. It extends the humerus and can also adduct and rotate the upper arm medially.

The **latissimus dorsi** (lah-tis′ĭ-mus dor′si) is a wide, triangular muscle that curves upward from the lower back, around the side, and to the armpit. It can extend and adduct the arm and rotate the humerus inwardly. It also acts to pull the shoulder downward and back. This muscle is used to pull the arm back in swimming, climbing, and rowing movements.

Abductors

The **supraspinatus** (su″prah-spi′na-tus) is located in the depression above the spine of the scapula on its posterior surface. It connects the scapula to the greater tubercle of the humerus and acts to abduct the upper arm.

The **deltoid** (del′toid) is a thick, triangular muscle that covers the shoulder joint. It connects the clavicle and scapula to the lateral side of the humerus and acts to abduct the upper arm. The deltoid's posterior fibers can extend the humerus, and its anterior fibers can flex the humerus.

Rotators

The **subscapularis** (sub-scap′u-lar-is) is a large, triangular muscle that covers the anterior surface of the scapula. It connects the scapula to the humerus and functions to rotate the upper arm medially.

Figure 9.28
Cross section of the upper arm.

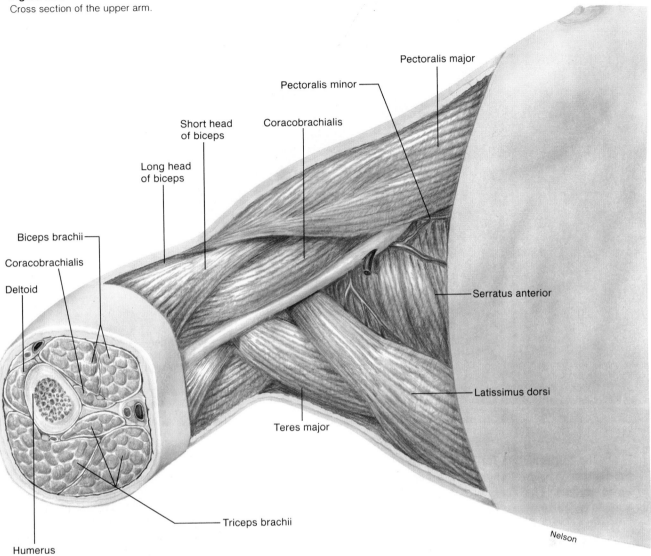

Pectoralis major

Pectoralis minor

Coracobrachialis

Short head
of biceps

Long head
of biceps

Biceps brachii

Coracobrachialis

Deltoid

Serratus anterior

Latissimus dorsi

Teres major

Triceps brachii

Humerus

Nelson

Chart 9.7 Muscles that move the upper arm			
Muscle	**Origin**	**Insertion**	**Action**
Coracobrachialis	Coracoid process of scapula	Shaft of humerus	Flexes and adducts the upper arm
Pectoralis major	Clavicle, sternum, and costal cartilages of upper ribs	Intertubercular groove of humerus	Pulls arm forward and across chest, rotates humerus, or adducts arm
Teres major	Lateral border of scapula	Intertubercular groove of humerus	Extends humerus, or adducts and rotates arm medially
Latissimus dorsi	Spines of sacral, lumbar, and lower thoracic vertebrae, iliac crest, and lower ribs	Intertubercular groove of humerus	Extends and adducts the arm and rotates humerus inwardly, or pulls the shoulder downward and back
Supraspinatus	Posterior surface of scapula	Greater tubercle of humerus	Abducts the upper arm
Deltoid	Acromion process, spine of the scapula, and the clavicle	Deltoid tuberosity of humerus	Abducts upper arm, extends humerus, or flexes humerus
Subscapularis	Anterior surface of scapula	Lesser tubercle of humerus	Rotates arm medially
Infraspinatus	Posterior surface of scapula	Greater tubercle of humerus	Rotates arm laterally
Teres minor	Lateral border of scapula	Greater tubercle of humerus	Rotates arm laterally

Figure 9.29
Muscles of the anterior shoulder and the upper arm.

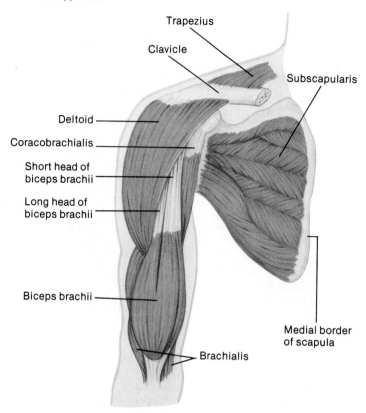

The **infraspinatus** (in″frah-spi′na-tus) occupies the depression below the spine of the scapula on its posterior surface. The fibers of this muscle attach the scapula to the humerus and act to rotate the upper arm laterally.

The **teres minor** is a small muscle connecting the scapula to the humerus. It acts to rotate the upper arm laterally.

Muscles That Move the Forearm

Most forearm movements are produced by muscles that connect the radius or ulna to the humerus or pectoral girdle. A group of muscles located along the anterior surface of the humerus acts to flex the elbow, while a single posterior muscle serves to extend this joint. Other muscles cause movements at the radioulnar joint and are responsible for rotating the forearm.

The muscles that move the forearm include the following (figures 9.27, 9.28, 9.29, 9.30, and chart 9.8).

Flexors
  Biceps brachii
  Brachialis
  Brachioradialis

Extensor
  Triceps brachii
Rotators
  Supinator
  Pronator teres
  Pronator quadratus

Flexors

The **biceps brachii** (bi′seps bra′ke-i) is a fleshy muscle that forms a long, rounded mass on the anterior side of the upper arm. It connects the scapula to the radius and functions to flex the arm at the elbow and to rotate the hand laterally (supination), as when a person turns a doorknob or screwdriver.

The **brachialis** (bra′ke-al-is) is a large muscle beneath the biceps brachii. It connects the shaft of the humerus to the ulna and is the strongest flexor of the elbow.

The **brachioradialis** (bra″ke-o-ra″de-a′lis) connects the humerus to the radius and aids in flexing the elbow.

Extensor

The **triceps brachii** (tri′seps bra′ke-i) has three heads and is the only muscle on the back of the upper arm. It connects the humerus and scapula to the ulna and is the primary extensor of the elbow.

Rotators

The **supinator** (su′pi-na-tor) is a short muscle whose fibers run from the ulna and the lateral end of the humerus to the radius. It assists the biceps brachii in rotating the forearm laterally (supination).

Figure 9.30
Muscles of the anterior forearm.

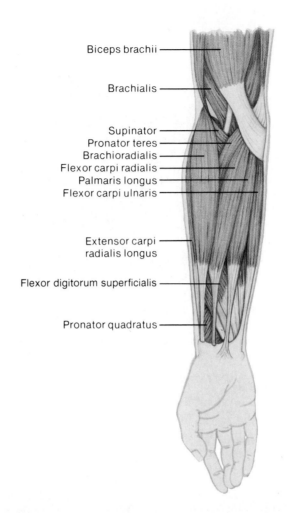

Biceps brachii ——

Brachialis ——

Supinator ——
Pronator teres ——
Brachioradialis ——
Flexor carpi radialis ——
Palmaris longus ——
Flexor carpi ulnaris ——

Extensor carpi ——
radialis longus

Flexor digitorum superficialis ——

Pronator quadratus ——

Chart 9.8	Muscles that move the forearm		
Muscle	Origin	Insertion	Action
Biceps brachii	Coracoid process and tubercle above glenoid cavity of scapula	Radial tuberosity of radius	Flexes arm at elbow and rotates the hand laterally
Brachialis	Anterior shaft of humerus	Coronoid process of ulna	Flexes arm at elbow
Brachioradialis	Distal lateral end of humerus	Lateral surface of radius above styloid process	Flexes arm at elbow
Triceps brachii	Tubercle below glenoid cavity and lateral and medial surfaces of humerus	Olecranon process of ulna	Extends arm at elbow
Supinator	Lateral epicondyle of humerus and crest of ulna	Lateral surface of radius	Rotates forearm laterally
Pronator teres	Medial epicondyle of humerus and coronoid process of ulna	Lateral surface of radius	Rotates arm medially
Pronator quadratus	Anterior distal end of ulna	Anterior distal end of radius	Rotates arm medially

Figure 9.31
Muscles of the posterior forearm.

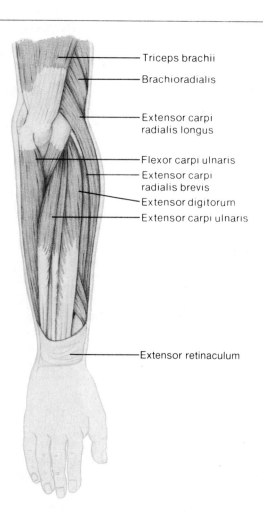

Triceps brachii

Brachioradialis

Extensor carpi radialis longus

Flexor carpi ulnaris

Extensor carpi radialis brevis

Extensor digitorum

Extensor carpi ulnaris

Extensor retinaculum

Chart 9.9 Muscles that move the wrist, hand, and fingers			
Muscle	Origin	Insertion	Action
Flexor carpi radialis	Medial epicondyle of humerus	Base of second and third metacarpals	Flexes and abducts wrist
Flexor carpi ulnaris	Medial epicondyle of humerus and olecranon process	Carpal and metacarpal bones	Flexes and adducts wrist
Palmaris longus	Medial epicondyle of humerus	Fascia of palm	Flexes wrist
Flexor digitorum profundus	Anterior surface of ulna	Bases of distal phalanges in fingers two through five	Flexes distal joints of fingers
Flexor digitorum superficialis	Medial epicondyle of humerus, coronoid process of ulna, and radius	Tendons of fingers	Flexes fingers and hand
Extensor carpi radialis longus	Distal end of humerus	Base of second metacarpal	Extends wrist and abducts hand
Extensor carpi radialis brevis	Lateral epicondyle of humerus	Base of second and third metacarpals	Extends wrist and abducts hand
Extensor carpi ulnaris	Lateral epicondyle of humerus	Base of fifth metacarpal	Extends and adducts wrist
Extensor digitorum	Lateral epicondyle of humerus	Posterior surface of phalanges in fingers two through five	Extends fingers

The **pronator teres** (pro-na′tor te′rēz) is a short muscle connecting the ends of the humerus and ulna to the radius. It functions to rotate the arm medially, as when the hand is turned so the palm is facing downward (pronation).

The **pronator quadratus** (pro-na′tor kwod-ra′tus) runs from the distal end of the ulna to the distal end of the radius. It assists the pronator teres in rotating the arm medially.

Muscles That Move the Wrist, Hand, and Fingers

Many muscles are responsible for wrist, hand, and finger movements. They originate from the distal end of the humerus and from the radius and ulna. The two major groups of these muscles are flexors on the anterior side of the forearm and extensors on the posterior side. These muscles include the following (figures 9.30, 9.31, and chart 9.9).

Flexors
Flexor carpi radialis
Flexor carpi ulnaris
Palmaris longus
Flexor digitorum profundus
Flexor digitorum superficialis
Extensors
Extensor carpi radialis longus
Extensor carpi radialis brevis
Extensor carpi ulnaris
Extensor digitorum

Flexors

The **flexor carpi radialis** (flek′sor kar-pi′ ra″de-a′lis) is a fleshy muscle that runs medially on the anterior side of the forearm. It extends from the distal end of the humerus into the hand, where it is attached to metacarpal bones. The flexor carpi radialis functions to flex and abduct the wrist.

The **flexor carpi ulnaris** (flek′sor kar-pi′ ul-na′ris) is located along the medial border of the forearm. It connects the distal end of the humerus and the proximal end of the ulna to carpal and metacarpal bones. It serves to flex and adduct the wrist.

The **palmaris longus** (pal-ma′ris long′gus) is a slender muscle located on the medial side of the forearm between the flexor carpi radialis and the flexor carpi ulnaris. It connects the distal end of the humerus to fascia of the palm and functions to flex the wrist.

The **flexor digitorum profundus** (flek′sor dij″ĭ-to′rum pro-fun′dus) is a large muscle that connects the ulna to the distal phalanges. It acts to flex the distal joints of the fingers, as when a fist is made.

The **flexor digitorum superficialis** (flek′sor dij″ĭ-to′rum su″per-fish″e-a-lis) is a large muscle located beneath the flexor carpi ulnaris. It arises by three heads—one from the medial epicondyle of the humerus, one from the medial side of the ulna, and one from the radius. It is inserted in the tendons of the fingers and acts to flex the fingers and, by a combined action, to flex the hand.

Extensors

The **extensor carpi radialis longus** (eks-ten′sor kar-pi′ ra″de-a′lis long′gus) runs along the lateral side of the forearm, connecting the humerus to the hand. It functions to extend the wrist and assists in abducting the hand.

The **extensor carpi radialis brevis** (eks-ten′sor kar-pi′ ra″de-a′lis bre′is) is a companion of the extensor carpi radialis longus and is located medially to it. This muscle runs from the humerus to metacarpal bones and functions to extend the wrist. It also assists in abducting the hand.

The **extensor carpi ulnaris** (eks-ten′sor kar-pi′ ul-na′ris) is located along the posterior surface of the ulna and connects the humerus to the hand. It acts to extend the wrist and assists in adducting it.

The **extensor digitorum** (eks-ten′sor dij″ĭ-to′rum) runs medially along the back of the forearm. It connects the humerus to the posterior surface of the phalanges and functions to extend the fingers.

A structure called the *extensor retinaculum* consists of a group of heavy connective tissue fibers in the fascia of the wrist. It connects the lateral margin of the radius with the medial border of the styloid process of the ulna and certain bones of the wrist. The retinaculum gives off branches of connective tissue to the underlying wrist bones, creating a series of sheath-like compartments through which the tendons of the extensor muscles pass to the wrist and fingers (figure 9.31).

Muscles of the Abdominal Wall

Although the walls of the chest and pelvic regions are supported directly by bone, those of the abdomen are not. Instead, the anterior and lateral walls of the abdomen are composed of broad, flattened muscles arranged in layers. These muscles connect the rib cage and vertebral column to the pelvic girdle. A band of tough connective tissue, called the **linea alba,** extends from the xiphoid process of the sternum to the symphysis pubis. It serves as an attachment for some of the abdominal wall muscles.

Chart 9.10 Muscles of the abdominal wall			
Muscle	Origin	Insertion	Action
External oblique	Outer surfaces of lower ribs	Outer lip of iliac crest and linea alba	Tenses abdominal wall and compresses abdominal contents
Internal oblique	Crest of ilium and inguinal ligament	Cartilages of lower ribs, linea alba, and crest of pubis	Same as above
Transversus abdominis	Costal cartilages of lower ribs, processes of lumbar vertebrae, lip of iliac crest, and inguinal ligament	Linea alba and crest of pubis	Same as above
Rectus abdominis	Crest of pubis and symphysis pubis	Xiphoid process of sternum and costal cartilages	Same as above, also flexes vertebral column

Contraction of these muscles decreases the size of the abdominal cavity and increases the pressure inside. This action helps to press air out of the lungs during forceful exhalation, and also aids in defecation, urination, vomiting, and childbirth.

The abdominal wall muscles (figure 9.26 and chart 9.10) include the following:

External oblique
Internal oblique
Transversus abdominis
Rectus abdominis

The **external oblique** (eks-ter′nal ŏ-blēk) is a broad, thin sheet of muscle whose fibers slant downward from the lower ribs to the pelvic girdle and the linea alba. When this muscle contracts, it tenses the abdominal wall and compresses the contents of the abdominal cavity.

Similarly, the **internal oblique** (in-ter′nal ŏ-blēk) is a broad, thin sheet of muscle located beneath the external oblique. Its fibers run up and forward from the pelvic girdle to the lower ribs. Its function is similar to that of the external oblique.

The **transversus abdominis** (trans-ver′sus ab-dom′i-nis) forms a third layer of muscle beneath the external and internal obliques. Its fibers run horizontally from the lower ribs, lumbar vertebrae, and ilium to the linea alba and pubic bones. It functions in the same manner as the external and internal obliques.

The **rectus abdominis** (rek′tus ab-dom′i-nis) is a long, straplike muscle that connects the pubic bones to the ribs and sternum. It is crossed transversely by three or more fibrous bands that give it a segmented appearance. The muscle functions with other abdominal wall muscles to compress the contents of the abdominal cavity, and it also helps to flex the vertebral column.

Muscles of the Pelvic Outlet

The outlet of the pelvis is spanned by two muscular sheets—a deeper **pelvic diaphragm** and a more superficial **urogenital diaphragm.** The pelvic diaphragm forms the floor of the pelvic cavity, and the urogenital diaphragm fills the space within the pubic arch. The muscles of the male and female pelvic outlets include the following (figure 9.32, and chart 9.11).

Pelvic diaphragm
    Levator ani
    Coccygeus
Urogenital diaphragm
    Superficial transversus perinei
    Bulbospongiosus
    Ischiocavernosus
    Sphincter urethrae

Pelvic Diaphragm

The **levator ani** (le-va′tor ah-ni′) muscles form a thin sheet across the pelvic outlet. They are connected at the midline posteriorly by a ligament that extends from the tip of the coccyx to the anal canal. Anteriorly, they are separated in the male by the urethra and the anal canal; and in the female by the urethra, vagina, and anal canal. These muscles help to support the pelvic viscera and provide sphincterlike action in the anal canal and vagina.

An *external anal sphincter* that is under voluntary control and an *internal anal sphincter* that is formed of involuntary muscle fibers of the intestine encircle the anal canal and keep it closed.

The **coccygeus** (kok-sij′e-us) is a fan-shaped muscle that extends from the ischial spine to the coccyx and sacrum. It aids the levator ani in its functions.

Chart 9.11　Muscles of the pelvic outlet

Muscle	Origin	Insertion	Action
Levator ani	Pubic bone and ischial spine	Coccyx	Supports pelvic viscera, and provides sphincterlike action in anal canal and vagina
Coccygeus	Ischial spine	Sacrum and coccyx	Same as above
Superficial transversus perinei	Ischial tuberosity	Central tendon	Supports pelvic viscera
Bulbospongiosus	Central tendon	Males: urogenital diaphragm and fascia of penis Females: pubic arch and root of clitoris	Males: assists emptying urethra Females: constricts vagina
Ischiocavernosus	Ischial tuberosity	Pubic arch	Assists the function of the bulbospongiosus
Sphincter urethrae	Margins of pubis and ischium	Fibers of each unite with those from other side	Opens and closes urethra

Figure 9.32

(a) Muscles of the male pelvic outlet, and (b) female pelvic outlet.

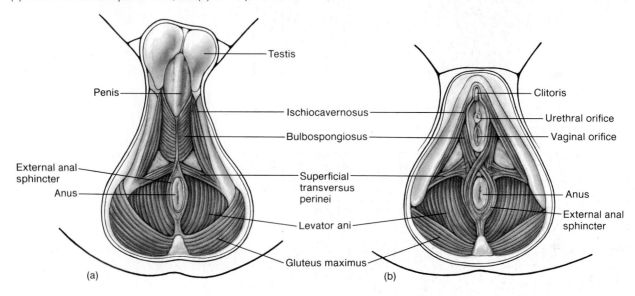

Urogenital Diaphragm

The **superficial transversus perinei** (su''per-fish'al trans-ver'sus per''ĭ-ne'i) consists of a small bundle of muscle fibers that passes medially from the ischial tuberosity along the posterior border of the urogenital diaphragm. It assists other muscles in supporting the pelvic viscera.

In males, the **bulbospongiosus** (bul''bo-spon''je-o'sus) muscles are united and surround the base of the penis. They assist in emptying the urethra. In females, these muscles are separated medially by the vagina and act to constrict the vaginal opening. They also can retard the flow of blood in veins, which helps to maintain an erection in the penis of the male and the clitoris of the female.

The **ischiocavernosus** (is''ke-o-kav''er-no'sus) **muscle** is a tendinous structure that extends from the ischial tuberosity to the margin of the pubic arch. It assists the function of the bulbospongiosus muscle.

The **sphincter urethrae** (sfingk'ter u-re'thrē) are muscles that arise from the margins of the pubic and ischial bones. Each arches around the urethra and unites with the one on the other side. Together they act as a sphincter that closes the urethra by compression and opens it by relaxation, thus helping to control the flow of urine.

Figure 9.33
Muscles of the anterior right thigh.

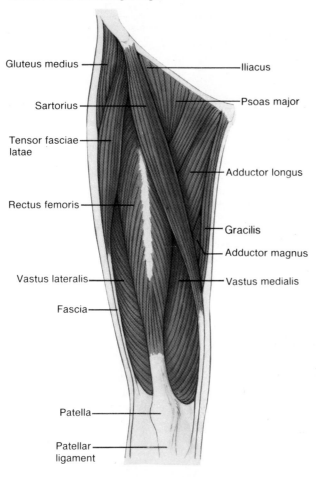

- Gluteus medius
- Iliacus
- Sartorius
- Psoas major
- Tensor fasciae latae
- Adductor longus
- Rectus femoris
- Gracilis
- Adductor magnus
- Vastus lateralis
- Vastus medialis
- Fascia
- Patella
- Patellar ligament

Figure 9.34
Muscles of the lateral right thigh.

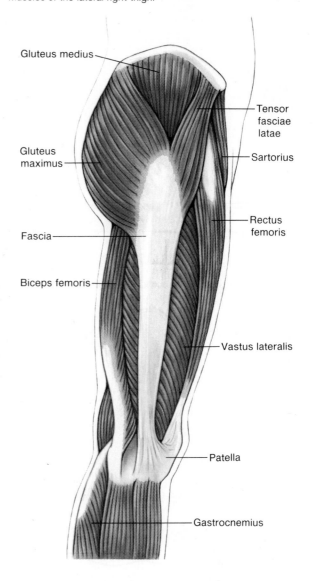

- Gluteus medius
- Tensor fasciae latae
- Gluteus maximus
- Sartorius
- Rectus femoris
- Fascia
- Biceps femoris
- Vastus lateralis
- Patella
- Gastrocnemius

Muscles That Move the Thigh

The muscles that move the thigh are attached to the femur and to some part of the pelvic girdle. They can be separated into anterior and posterior groups. The muscles of the anterior group act primarily to flex the thigh; those of the posterior group extend, abduct, or rotate it. The muscles in these groups (figures 9.33, 9.34, 9.35, 9.36, and chart 9.12) include the following:

Anterior group	Posterior group
Psoas major	*Gluteus maximus*
Iliacus	*Gluteus medius*
	Gluteus minimus
	Tensor fasciae latae

Still another group of muscles attached to the femur and pelvic girdle functions to adduct the thigh. This group includes:

Adductor longus
Adductor magnus
Gracilis

Anterior Group

The **psoas** (so'as) **major** is a long, thick muscle that connects the lumbar vertebrae to the femur. It functions to flex the thigh.

The **iliacus** (il'e-ak-us), a large, fan-shaped muscle, lies along the lateral side of the psoas major. The iliacus and the psoas major are the primary flex of the thigh, and they serve to advance the leg in w movements.

Posterior Group

The **gluteus maximus** (gloo'te-us ̄ largest muscle in the body and ́ the buttock. It connects the il ̄ to the femur by fascia of ̄'

Figure 9.35
Muscles of the posterior right thigh.

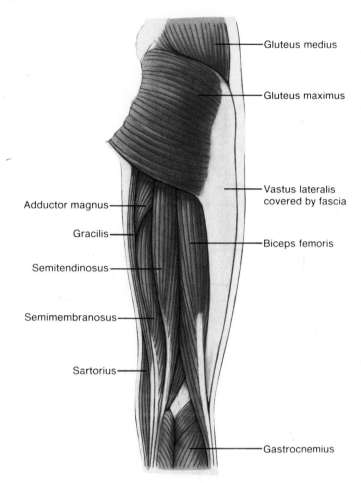

- Gluteus medius
- Gluteus maximus
- Vastus lateralis covered by fascia
- Adductor magnus
- Gracilis
- Biceps femoris
- Semitendinosus
- Semimembranosus
- Sartorius
- Gastrocnemius

Chart 9.12 Muscles that move the thigh			
Muscle	**Origin**	**Insertion**	**Action**
Psoas major	Lumbar intervertebral disks, bodies and transverse processes of lumbar vertebrae	Lesser trochanter of femur	Flexes thigh
Iliacus	Iliac fossa of ilium	Lesser trochanter of femur	Flexes thigh
Gluteus maximus	Sacrum, coccyx, and posterior surface of ilium	Posterior surface of femur and fascia of thigh	Extends leg at hip
Gluteus medius	Lateral surface of ilium	Greater trochanter of femur	Abducts and rotates thigh medially
Gluteus minimus	Lateral surface of ilium	Greater trochanter of femur	Same as gluteus medius
Tensor fasciae latae	Anterior iliac crest	Fascia of thigh	Abducts, flexes, and rotates thigh medially
Adductor longus	Pubic bone near symphysis pubis	Posterior surface of femur	Adducts, flexes, and rotates thigh laterally
Adductor magnus	Ischial tuberosity	Posterior surface of femur	Adducts, extends, and rotates thigh laterally
Gracilis	Lower edge of symphysis pubis	Medial surface of tibia	Adducts thigh, flexes and rotates leg medially at knee

Figure 9.36
A cross section of the thigh.

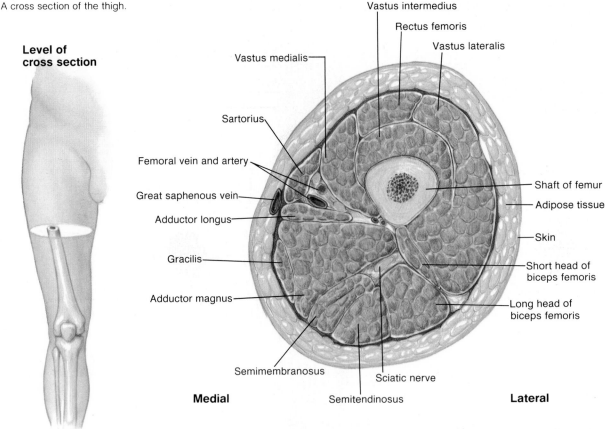

Level of cross section

Vastus intermedius
Rectus femoris
Vastus lateralis
Vastus medialis
Sartorius
Femoral vein and artery
Great saphenous vein
Adductor longus
Gracilis
Adductor magnus
Shaft of femur
Adipose tissue
Skin
Short head of biceps femoris
Long head of biceps femoris
Semimembranosus
Sciatic nerve
Semitendinosus

Medial   **Lateral**

the thigh. The gluteus maximus causes the leg to straighten at the hip when a person walks, runs, or climbs. It is also used to raise the body from a sitting position.

The **gluteus medius** (gloo'te-us me'de-us) is partly covered by the gluteus maximus. Its fibers extend from the ilium to the femur, and they function to abduct the thigh and rotate it medially.

The **gluteus minimus** (gloo'te-us min'ĭ-mus) lies beneath the gluteus medius and is its companion in attachments and functions.

The **tensor fasciae latae** (ten'sor fash'e-e lah-tē) connects the ilium to the fascia of the thigh, which continues downward to the tibia. This muscle functions to abduct and flex the thigh and to rotate it medially.

Thigh Adductors

The **adductor longus** (ah-duk'tor long'gus) is a long, triangular muscle that runs from the pubic bone to the femur. It functions to adduct the thigh and assists in flexing and rotating it laterally.

The **adductor magnus** (ah-duk'tor mag'nus) is the largest adductor of the thigh. It is a triangular muscle that connects the ischium to the femur. It adducts the thigh and assists in extending and rotating it laterally.

The **gracilis** (gras'il-is) is a long, straplike muscle that passes from the pubic bone to the tibia. It functions to adduct the thigh and to flex and rotate the leg medially at the knee.

Muscles That Move the Lower Leg

The muscles that move the lower leg connect the tibia or fibula to the femur or the pelvic girdle (figures 9.33, 9.34, 9.35, 9.36, and chart 9.13). They can be separated into two major groups—those that cause flexion at the knee and those that cause extension at the knee. The muscles of these groups include the following:

Flexors
  Biceps femoris   *Semimembranosus*
  Semitendinosus   *Sartorius*
Extensor
  Quadriceps femoris

Chart 9.13   Muscles that move the lower leg

Muscle	Origin	Insertion	Action
Hamstring group			
Biceps femoris	Ischial tuberosity and linea aspera of femur	Head of fibula and lateral condyle of tibia	Flexes and rotates leg laterally and extends thigh
Semitendinosus	Ischial tuberosity	Medial surface of tibia	Flexes and rotates leg medially and extends thigh
Semimembranosus	Ischial tuberosity	Medial condyle of tibia	Flexes and rotates leg medially and extends thigh
Sartorius	Spine of ilium	Medial surface of tibia	Flexes leg and thigh, abducts thigh, rotates thigh laterally, and rotates leg medially
Quadriceps femoris group		Patella by common tendon, which continues as patellar ligament to tibial tuberosity	Extends leg at knee
Rectus femoris	Spine of ilium and margin of acetabulum		
Vastus lateralis	Greater trochanter and posterior surface of femur		
Vastus medialis	Medial surface of femur		
Vastus intermedius	Anterior and lateral surfaces of femur		

Flexors

As its name implies, the **biceps femoris** (bi'seps fem'or-is) has two heads, one attached to the ischium and the other attached to the femur. This muscle passes along the back of the thigh on the lateral side and connects to the proximal ends of the fibula and tibia. The biceps femoris is one of the hamstring muscles, and its tendon (hamstring) can be felt as a lateral ridge behind the knee. This muscle functions to flex and rotate the leg laterally and to extend the thigh.

The **semitendinosus** (sem''e-ten'di-no-sus) is another of the hamstring muscles. It is a long, bandlike muscle on the back of the thigh toward the medial side, connecting the ischium to the proximal end of the tibia. The semitendinosus is so named because it becomes tendinous in the middle of the thigh, continuing to its insertion as a long, cordlike tendon. It functions to flex and rotate the leg medially and to extend the thigh.

The **semimembranosus** (sem''e-mem'brah-no-sus) is the third hamstring muscle and is the most medially located muscle in the back of the thigh. It connects the ischium to the tibia and functions to flex and rotate the leg medially and to extend the thigh.

The **sartorius** (sar-to're-us) is an elongated straplike muscle that passes obliquely across the front of the thigh and then descends over the medial side of the knee. It connects the ilium to the tibia and functions to flex the leg and the thigh. It can also abduct the thigh, rotate it laterally, and assist in rotating the leg medially.

Extensor

The large, fleshy muscle group called **quadriceps femoris** (kwod'ri-seps fem'or-is) occupies the front and sides of the thigh and is the primary extensor of the knee. It is composed of four parts—*rectus femoris, vastus lateralis, vastus medialis,* and *vastus intermedius.* These parts connect the ilium and femur to a common *patellar tendon,* which passes over the front of the knee and attaches to the patella. This tendon then continues as the *patellar ligament* to the tibia.

Occasionally, as a result of traumatic injury in which a muscle, such as the quadriceps femoris, is compressed against an underlying bone, new bone tissue may begin to develop within the damaged muscle. This condition is called *myositis ossificans*. When the bone tissue has matured (several months following the injury), the newly formed bone can be removed surgically.

Figure 9.37
Muscles of the anterior right lower leg.

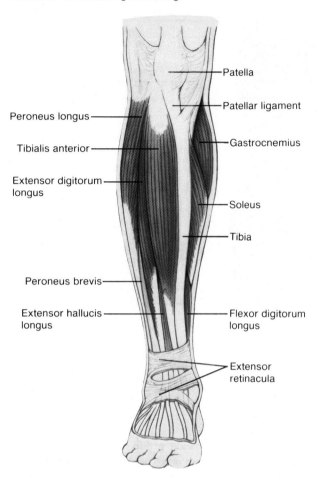

Figure 9.38
Muscles of the lateral right lower leg.

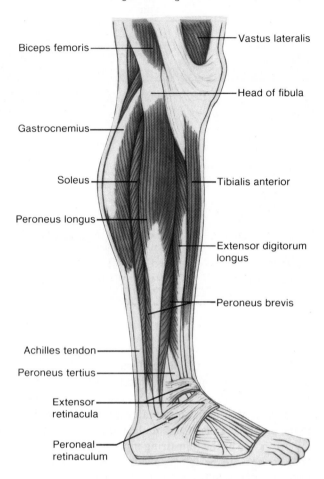

Muscles That Move the Ankle, Foot, and Toes

A number of muscles that function to move the ankle, foot, and toes are located in the lower leg (figures 9.37, 9.38, 9.39, and chart 9.14). They attach the femur, tibia, and fibula to various bones of the foot and are responsible for a variety of movements—moving the foot upward (dorsiflexion) or downward (plantar flexion), and turning the sole of the foot inward (inversion) or outward (eversion). These muscles include the following:

Dorsal flexors
   Tibialis anterior
   Peroneus tertius
   Extensor digitorum longus
Plantar flexors
   Gastrocnemius
   Soleus
   Flexor digitorum longus
Invertor
   Tibialis posterior
Evertor
   Peroneus longus

Dorsal Flexors

The **tibialis anterior** (tib″e-a′lis an-te′re-or) is an elongated, spindle-shaped muscle located on the front of the lower leg. It arises from the surface of the tibia, passes medially over the distal end of the tibia, and attaches to bones of the ankle and foot. Contraction of the tibialis anterior causes dorsiflexion and inversion of the foot.

The **peroneus tertius** (per″o-ne′us ter′shus) is a muscle of variable size that connects the fibula to the lateral side of the foot. It functions in dorsiflexion and eversion of the foot.

The **extensor digitorum longus** (eks-ten′sor dij″i-to′rum long′gus) is situated along the lateral side of the lower leg just behind the tibialis anterior. It arises from the proximal end of the tibia and the shaft of the fibula. Its tendon divides into four parts as it passes over the front of the ankle. These parts continue over the surface of the foot and attach to the four lateral toes. The actions of the extensor digitorum longus include dorsiflexion of the foot, eversion of the foot, and extension of the toes.

Figure 9.39
Muscles of the medial right lower leg.

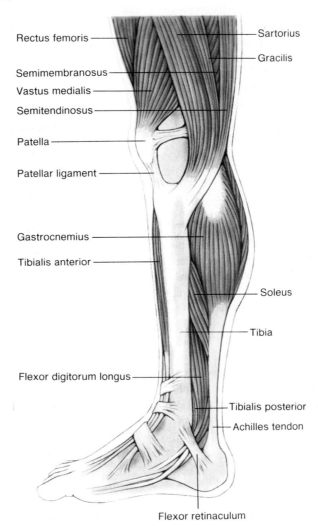

Rectus femoris ————————— Sartorius

————————— Gracilis

Semimembranosus —————

Vastus medialis —————

Semitendinosus —————

Patella ——————

Patellar ligament ——————

Gastrocnemius ——————

Tibialis anterior ——————

————————— Soleus

————————— Tibia

Flexor digitorum longus ——————

————————— Tibialis posterior

————————— Achilles tendon

Flexor retinaculum

Chart 9.14	Muscles that move the ankle, foot, and toes		
Muscle	Origin	Insertion	Action
Tibialis anterior	Lateral condyle and lateral surface of tibia	Tarsal bone (cuneiform) and first metatarsal	Dorsiflexion and inversion of foot
Peroneus tertius	Anterior surface of fibula	Dorsal surface of fifth metatarsal	Dorsiflexion and eversion of foot
Extensor digitorum longus	Lateral condyle of tibia and anterior surface of fibula	Dorsal surfaces of second and third phalanges of four lateral toes	Dorsiflexion and eversion of foot and extension of toes
Gastrocnemius	Lateral and medial condyles of femur	Posterior surface of calcaneus	Plantar flexion of foot and flexion of leg at knee
Soleus	Head and shaft of fibula and posterior surface of tibia	Posterior surface of calcaneus	Plantar flexion of foot
Flexor digitorum longus	Posterior surface of tibia	Distal phalanges of four lateral toes	Plantar flexion and inversion of foot and flexion of four lateral toes
Tibialis posterior	Lateral condyle and posterior surface of tibia and posterior surface of fibula	Tarsal and metatarsal bones	Plantar flexion and inversion of foot
Peroneus longus	Lateral condyle of tibia and head and shaft of fibula	Tarsal and metatarsal bones	Plantar flexion and eversion of foot; also supports arch

Plantar Flexors

The **gastrocnemius** (gas''trok-ne'me-us), on the back of the lower leg, forms part of the calf. It arises by two heads from the femur. The distal end of this muscle joins the strong *Achilles tendon,* which descends to the heel and attaches to the calcaneus. The gastrocnemius is a powerful plantar flexor of the foot, that aids in pushing the body forward when a person walks or runs. It also functions to flex the leg at the knee.

As a result of strenuous athletic activity, the Achilles tendon may be partially or completely torn. This injury occurs most frequently in middle-aged athletes, who run or play games that involve quick movements and directional changes. As a rule, a torn Achilles tendon requires surgical treatment.

The **soleus** (so'le-us) is a thick, flat muscle located beneath the gastrocnemius. These two muscles make up the calf of the leg. The soleus arises from the tibia and fibula, and it extends to the heel by way of the Achilles tendon. It acts with the gastrocnemius to cause plantar flexion of the foot.

The **flexor digitorum longus** (flek'sor dij''i-to'rum long'gus) extends from the posterior surface of the tibia to the foot. Its tendon passes along the plantar surface of the foot. There it divides into four parts that attach to the terminal bones of the four lateral toes. This muscle assists in plantar flexion of the foot, flexion of the four lateral toes, and inversion of the foot.

Invertor

The **tibialis posterior** (tib''e-a'lis pos-tēr'e-or) is the deepest of the muscles on the back of the lower leg. It connects the fibula and tibia to ankle bones by means of a tendon that curves under the medial malleolus. This muscle assists in inversion and plantar flexion of the foot.

Evertor

The **peroneus longus** (per''o-ne'us long'gus) is a long, straplike muscle located on the lateral side of the lower leg. It connects the tibia and the fibula to the foot by means of a stout tendon that passes behind the lateral malleolus. It functions in eversion of the foot, assists in plantar flexion, and helps support the arch of the foot.

As in the case of the wrist, fascia in various regions of the ankle is thickened to form retinacula. Anteriorly, for example, *extensor retinacula* connect the tibia and fibula as well as the calcaneus and fascia of the sole. These retinacula form sheaths for tendons crossing the front of the ankle.

Posteriorly, on the inside, a *flexor retinaculum* runs between the medial malleolus and the calcaneus and forms sheaths for tendons passing beneath the foot. *Peroneal retinacula* connect the lateral malleolus and the calcaneus, providing sheaths for tendons on the lateral side of the ankle (figures 9.38 and 9.39).

Clinical Terms Related to the Muscular System

convulsion (kun-vul'shun)   involuntary contraction of muscles.

electromyography (e-lek''tro-mi-og'rah-fe)   technique for recording the electrical changes that occur in muscle tissues.

fibrillation (fi''bri-la'shun)   spontaneous contractions of individual muscle fibers, producing rapid and uncoordinated activity within a muscle.

fibrosis (fi-bro'sis)   degenerative disease in which skeletal muscle tissue is replaced by fibrous connective tissue.

fibrositis (fi''bro-si'tis)   inflammatory condition of fibrous connective tissues, especially in the muscle fascia. This disease is also called muscular rheumatism.

muscular dystrophy (mus'ku-lar dis'tro-fe)   a progressively crippling disease of unknown cause in which the muscles gradually weaken and atrophy.

myalgia (mi-al'je-ah)   pain resulting from any muscular disease or disorder.

myasthenia gravis (mi''as-the'ne-ah gra'vis)   chronic disease characterized by muscles that are weak and easily fatigued. It results from a disorder at some of the neuromuscular junctions so that a stimulus is not transmitted from the motor neuron to the muscle fiber.

myokymia (mi''o-ki'me-ah)   persistent quivering of a muscle.

myology (mi-ol'o-je)   the study of muscles.

myoma (mi-o'mah)   a tumor composed of muscle tissue.

myopathy (mi-op'ah-the)   any muscular disease.

myositis (mi''o-si'tis)   inflammation of skeletal muscle tissue.

myotomy (mi-ot'o-me)   the cutting of muscle tissue.

myotonia (mi''o-to'ne-ah)   prolonged muscular spasm.

paralysis (pah-ral'i-sis)   loss of ability to move a body part.

shin splints (shin splints)   a soreness on the front of the lower leg due to straining the flexor digitorum longus, often as a result of walking up and down hills.

torticollis (tor''ti-kol'is)   condition in which the neck muscles, such as the sternocleidomastoids, contract involuntarily. It is more commonly called wryneck.

Chapter Summary

Introduction

There are three types of muscle tissue—skeletal, smooth, and cardiac.

Structure of a Skeletal Muscle

Skeletal muscles are composed of nerve, vascular, and various connective tissues as well as skeletal muscle tissue.

1. Connective tissue coverings
   a. Skeletal muscles are covered with fascia.
   b. Other connective tissues invest cells and groups of cells within the muscle's structure.
   c. Fascia is part of a complex network of connective tissue that extends throughout the body.
2. Skeletal muscle fibers
   a. Each fiber represents a single muscle cell, which is the unit of contraction.
   b. Muscle fibers are cylindrical cells with numerous nuclei.
   c. The cytoplasm contains mitochondria, a sarcoplasmic reticulum, and myofibrils composed of actin and myosin.
   d. Striations are produced by the arrangement of the actin and myosin filaments.
   e. Transverse tubules extend from the cell membrane into the cytoplasm and are associated with the cisternae of the sarcoplasmic reticulum.
3. The neuromuscular junction
   a. Muscle fibers are stimulated to contract by motor neurons.
   b. The motor end plate of a muscle fiber lies on one side of a neuromuscular junction.
   c. In response to a nerve impulse, the end of the motor nerve fiber secretes a neurotransmitter, which diffuses across the junction and stimulates the muscle fiber.
4. Motor units
   a. One motor neuron and the muscle fibers associated with it constitute a motor unit.
   b. Finer movements can be produced in muscles whose motor units contain small numbers of muscle fibers.

Skeletal Muscle Contraction

Muscle fiber contraction results from a sliding movement within the myofibrils in which actin and myosin filaments merge.

1. Role of actin and myosin
   a. Globular portions of myosin filaments can form cross-bridges with actin filaments.
   b. The reaction between actin and myosin filaments generates the force of contraction.
   c. When a fiber is at rest, troponin and tropomyosin molecules interfere with cross-bridge formation.
   d. When calcium ions are present, the inhibition is removed.

2. Stimulus for contraction
   a. Muscle fiber is usually stimulated by acetylcholine released from the end of a motor nerve fiber.
   b. Acetylcholine causes muscle fiber to conduct an impulse that reaches the deep parts of the fiber by means of the transverse tubules.
   c. A muscle impulse signals the sarcoplasmic reticulum to release calcium ions.
   d. Cross-bridges form between actin and myosin, and the actin filaments move inward.
   e. Muscle fiber relaxes when calcium ions are transported back into the sarcoplasmic reticulum.
   f. Cholinesterase decomposes acetylcholine.
3. Energy sources for contraction
   a. ATP supplies the energy for muscle fiber contraction.
   b. Creatine phosphate stores energy that can be used to synthesize ATP as it is decomposed.
   c. Active muscles depend upon cellular respiration.
4. Oxygen supply and cellular respiration
   a. Anaerobic respiration results in a gain of only a few ATP molecules, while aerobic respiration results in the gain of many more ATP molecules.
   b. Oxygen is carried from the lungs to body cells by hemoglobin in red blood cells.
   c. Myoglobin in muscle cells stores some oxygen temporarily.
5. Oxygen debt
   a. During rest or moderate exercise, oxygen is supplied to muscles in sufficient concentration to support aerobic respiration.
   b. During strenuous exercise, an oxygen deficiency may develop, and lactic acid may accumulate as a result of anaerobic respiration.
   c. The amount of oxygen needed to convert accumulated lactic acid to glucose and to restore the supplies of ATP and creatine phosphate is called oxygen debt.
6. Muscle fatigue
   a. A fatigued muscle loses its ability to contract.
   b. Muscle fatigue is usually due to the effects of an accumulation of lactic acid.
   c. Athletes usually produce less lactic acid than nonathletes because of their increased ability to supply oxygen and nutrients to muscles.
7. Fast and slow muscles
   a. The speed of contraction is related to a muscle's special function.
   b. Slow or red muscles can generate ATP fast enough to keep up with ATP breakdown and can contract for prolonged periods of time.
   c. Fast or white muscles have reduced ability to carry on aerobic respiration and tend to fatigue relatively rapidly.

8. Heat production
   a. Muscles represent an important source of body heat.
   b. Most of the energy released by cellular respiration is lost as heat.

Muscular Responses

1. Threshold stimulus is the minimal stimulus needed to elicit a muscular contraction.
2. All-or-none response
   a. If a muscle fiber contracts at all, it will contract completely.
   b. Motor units respond in an all-or-none manner.
3. Recruitment of motor units
   a. At low intensity of stimulation, relatively small numbers of motor units contract.
   b. At high intensities of stimulation, other motor units are recruited until the muscle contracts with maximal tension.
4. Staircase effect
   a. An inactive muscle undergoes a series of contractions of increasing strength when subjected to a series of stimuli.
   b. This staircase effect seems to be due to failure to remove calcium ions from the sarcoplasm rapidly enough.
5. Sustained contractions
   a. A rapid series of stimuli may produce a summation of twitches and a sustained contraction.
   b. When contractions fuse, the strength of contraction may increase due to recruitment of fibers.
   c. Even when a muscle is at rest, its fibers usually maintain tone—that is, remain partially contracted.
6. Isometric and isotonic contractions
   a. When a muscle contracts but its attachments do not move, the contraction is called isometric.
   b. When a muscle contracts and its ends are pulled closer together, the contraction is called isotonic.
   c. Most body movements involve both isometric and isotonic contractions.

Smooth Muscles

The contractile mechanisms of smooth and cardiac muscles are similar to that of skeletal muscle.

1. Smooth muscle fibers
   a. Cells contain filaments of actin and myosin.
   b. They lack transverse tubules and the sarcoplasmic reticula are not well developed.
   c. Types include multiunit smooth muscle and visceral smooth muscle.
   d. Visceral smooth muscle displays rhythmicity and is self-exciting.
   e. Movement of material through visceral organs is often aided by peristalsis.
2. Smooth muscle contraction
   a. In smooth muscles, calmodulin binds to calcium ions and activates the contraction mechanism.
   b. Both acetylcholine and norepinephrine serve as neurotransmitter substances for smooth muscles.
   c. Other factors that affect smooth muscle contractions include hormones and stretching.
   d. Smooth muscle can maintain a contraction for a longer time with a given amount of energy than can skeletal muscle.
   e. Smooth muscles can change their lengths without changing their tautness.

Cardiac Muscle

1. Cardiac muscle remains contracted for a longer time than skeletal muscle, because transverse tubules supply extra calcium ions.
2. The ends of adjacent cells are separated by intercalated disks, which hold the cells together.
3. A network of fibers contracts as a unit and responds to stimulation in an all-or-none manner.
4. Cardiac muscle is self-exciting, rhythmic, and remains refractory until a contraction is completed.

Skeletal Muscle Actions

The type of movement produced by a muscle depends on the way it is attached on either side of a joint.

1. Origin and insertion
   a. The movable end of a muscle is its insertion, and the immovable end is its origin.
   b. Some muscles have more than one origin or insertion.
   c. Sometimes the end of a muscle changes its function in different body movements.
2. Interaction of skeletal muscles
   a. Skeletal muscles function in groups.
   b. A prime mover is responsible for most of a movement. Synergists aid prime movers. Antagonists can resist the movement of a prime mover.
   c. Smooth movements depend upon antagonists' giving way to the actions of prime movers.

Major Skeletal Muscles

Muscle names often describe sizes, shapes, locations, actions, number of attachments, or direction of fibers.

1. Muscles of facial expression
   a. These muscles lie beneath the skin of the face and scalp and are used to communicate feelings through facial expression.
   b. They include the epicranius, orbicularis oculi, orbicularis oris, buccinator, zygomaticus, and platysma.

2. Muscles of mastication
   a. These muscles are attached to the mandible and are used in chewing.
   b. They include the masseter, temporalis, medial pterygoid, and lateral pterygoid.
3. Muscles that move the head
   a. Head movements are produced by muscles in the neck and upper back.
   b. They include the sternocleidomastoid, splenius capitis, semispinalis capitis, and longissimus capitis.
4. Muscles that move the pectoral girdle
   a. Most of these muscles connect the scapula to nearby bones and are closely associated with muscles that move the upper arm.
   b. They include the trapezius, rhomboideus major, levator scapulae, serratus anterior, and pectoralis minor.
5. Muscles that move the upper arm
   a. These muscles connect the humerus to various regions of the pectoral girdle, ribs, and vertebral column.
   b. They include the coracobrachialis, pectoralis major, teres major, latissimus dorsi, supraspinatus, deltoid, subscapularis, infraspinatus, and teres minor.
6. Muscles that move the forearm
   a. These muscles connect the radius and ulna to the humerus or pectoral girdle.
   b. They include the biceps brachii, brachialis, brachioradialis, triceps brachii, supinator, pronator teres, and pronator quadratus.
7. Muscles that move the wrist, hand, and fingers
   a. These muscles arise from the distal end of the humerus and from the radius and ulna.
   b. They include the flexor carpi radialis, flexor carpi ulnaris, palmaris longus, flexor digitorum profundus, flexor digitorum superficialis, extensor carpi radialis longus, extensor carpi radialis brevis, extensor carpi ulnaris, and extensor digitorum.
   c. An extensor retinaculum forms sheaths for tendons of the extensor muscles.
8. Muscles of the abdominal wall
   a. These muscles connect the rib cage and vertebral column to the pelvic girdle.
   b. They include the external oblique, internal oblique, transversus abdominus, and rectus abdominis.
9. Muscles of the pelvic outlet
   a. These muscles form the floor of the pelvic cavity and fill the space of the pubic arch.
   b. They include the levator ani, coccygeus, superficial transversus perinei, bulbospongiosus, ischiocavernosus, and sphincter urethrae.
10. Muscles that move the thigh
    a. These muscles are attached to the femur and to some part of the pelvic girdle.
    b. They include the psoas major, iliacus, gluteus maximus, gluteus medius, gluteus minimus, tensor fasciae latae, adductor longus, adductor magnus, and gracilis.
11. Muscles that move the lower leg
    a. These muscles connect the tibia or fibula to the femur or pelvic girdle.
    b. They include the biceps femoris, semitendinosus, semimembranosus, sartorius, and quadriceps femoris.
12. Muscles that move the ankle, foot, and toes
    a. These muscles attach the femur, tibia, and fibula to various bones of the foot.
    b. They include the tibialis anterior, peroneus tertius, extensor digitorum longus, gastrocnemius, soleus, flexor digitorum longus, tibialis posterior, and peroneus longus.
    c. Retinacula form sheaths for tendons passing to the foot.

Application of Knowledge

1. Why do you think athletes generally perform better if they warm up by exercising lightly before a competitive event?
2. Following childbirth, a woman may experience urinary incontinence when sneezing or coughing. What muscles of the pelvic floor should be strengthened by exercise to help control this problem?
3. What steps might be taken to minimize atrophy of skeletal muscles in patients who are confined to bed for prolonged times?
4. As lactic acid and other substances accumulate in an active muscle, they tend to stimulate pain receptors and the muscle may feel sore. How might the application of heat or substances that cause blood vessels to dilate help to relieve such soreness?

Review Activities

Part A

1. List the three types of muscle tissue.
2. Distinguish between tendon and aponeurosis.
3. Describe the connective tissue coverings of a skeletal muscle.
4. Distinguish between deep fascia, subcutaneous fascia, and subserous fascia.
5. List the major parts of a skeletal muscle fiber and describe the function of each part.
6. Describe a neuromuscular junction and a neurotransmitter substance.
7. Define *motor unit* and explain how the number of fibers within a unit affects muscular contractions.
8. Describe the major events that occur when a muscle fiber contracts.
9. Explain how ATP and creatine phosphate are related and how these substances function in the muscle fiber contraction mechanism.
10. Describe how oxygen is supplied to muscles.
11. Describe how oxygen debt may develop.
12. Explain how muscles may become fatigued and how a person's physical condition may affect tolerance to fatigue.
13. Distinguish between fast and slow muscles.
14. Explain how the maintenance of body temperature is related to the action of skeletal muscles.
15. Define *threshold stimulus.*
16. Explain what is meant by an all-or-none response.
17. Explain what is meant by motor unit recruitment.
18. Describe the staircase effect.
19. Explain how a skeletal muscle can be stimulated to produce a sustained contraction.
20. Distinguish between tetanic contraction and muscle tone.
21. Distinguish between isometric and isotonic contractions and explain how each is used in body movements.
22. Compare the structures of a smooth and a skeletal muscle fiber.
23. Distinguish between multiunit and visceral smooth muscles.
24. Define *peristalsis* and explain its function.
25. Compare the characteristics of smooth and skeletal muscle contractions.
26. Compare the structures of cardiac and skeletal muscles.
27. Compare the characteristics of cardiac and skeletal muscle contraction.
28. Distinguish between a muscle's origin and its insertion.
29. Define *prime mover, synergist,* and *antagonist.*

Part B

Match the muscles in column I with the descriptions and functions in column II.

I	II
1. Buccinator	A. Inserted on the coronoid process of the mandible.
2. Epicranius	B. Draws the corner of the mouth upward.
3. Medial pterygoid	C. Can raise and adduct the scapula.
4. Platysma	D. Can pull the head into an upright position.
5. Rhomboideus major	E. Consists of two parts—the frontalis and the occipitalis.
6. Splenius capitus	F. Compresses the cheeks.
7. Temporalis	G. Extends over the neck from the chest to the face.
8. Zygomaticus	H. Pulls the jaw to the side in grinding movements.

I	II
9. Biceps brachii	I. Primary extensor of the elbow.
10. Brachialis	J. Pulls the shoulder back and downward.
11. Deltoid	K. Abducts the arm.
12. Latissimus dorsi	L. Rotates the arm laterally.
13. Pectoralis major	M. Pulls the arm forward and across the chest.
14. Pronator teres	N. Rotates the arm medially.
15. Teres minor	O. Strongest flexor of the elbow.
16. Triceps brachii	P. Strongest supinator of the forearm.

I	II
17. Biceps femoris	Q. Inverts the foot.
18. External oblique	R. A member of the quadriceps femoris group.
19. Gastrocnemius	S. A plantar flexor of the foot.
20. Gluteus maximus	T. Compresses the contents of the abdominal cavity.
21. Gluteus medius	U. Largest muscle in the body.
22. Gracilis	V. A hamstring muscle.
23. Rectus femoris	W. Adducts the thigh.
24. Tibialis anterior	X. Abducts the thigh.

Part C
What muscles can you identify in the bodies of these models whose muscles are enlarged by exercise?

Reference Plates
Surface Anatomy

The following set of reference plates is presented to help you locate some of the more prominent surface features in various regions of the body. For the most part the labeled structures are easily seen or palpated through the skin. Locate as many of these features as possible on your body as a review.

Plate 8
Surface anatomy of head and neck, lateral view.

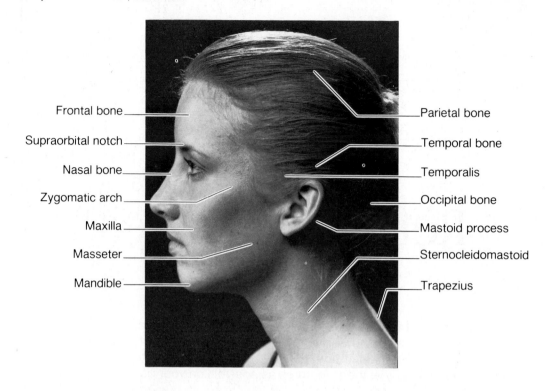

Frontal bone

Supraorbital notch

Nasal bone

Zygomatic arch

Maxilla

Masseter

Mandible

Parietal bone

Temporal bone

Temporalis

Occipital bone

Mastoid process

Sternocleidomastoid

Trapezius

Plate 9
Surface anatomy of arm and thorax, lateral view.

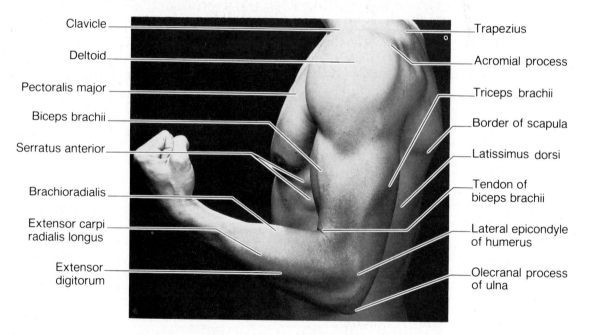

Clavicle

Deltoid

Pectoralis major

Biceps brachii

Serratus anterior

Brachioradialis

Extensor carpi
radialis longus

Extensor
digitorum

Trapezius

Acromial process

Triceps brachii

Border of scapula

Latissimus dorsi

Tendon of
biceps brachii

Lateral epicondyle
of humerus

Olecranal process
of ulna

Plate 10
Surface anatomy of back and arms, posterior view.

Biceps brachii
Triceps brachii
Deltoid

Trapezius

Teres major

Infraspinatus

Border of scapula

Vertebral spine

Latissimus dorsi

Spinous processes
of vertebrae

Plate 11
Surface anatomy of torso and arms, anterior view.

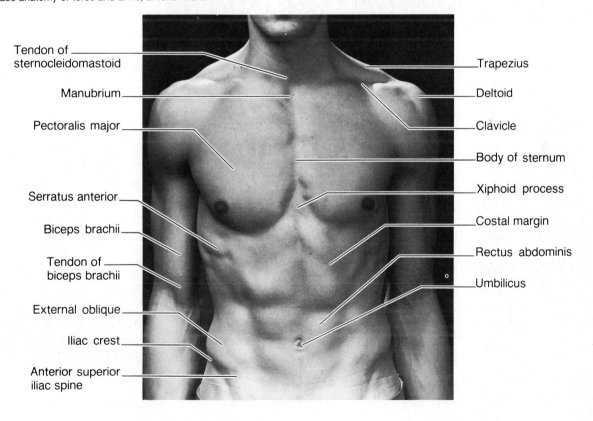

Tendon of
sternocleidomastoid

Manubrium

Pectoralis major

Serratus anterior

Biceps brachii

Tendon of
biceps brachii

External oblique

Iliac crest

Anterior superior
iliac spine

Trapezius

Deltoid

Clavicle

Body of sternum

Xiphoid process

Costal margin

Rectus abdominis

Umbilicus

Plate 12
Surface anatomy of torso and upper legs, posterior view.

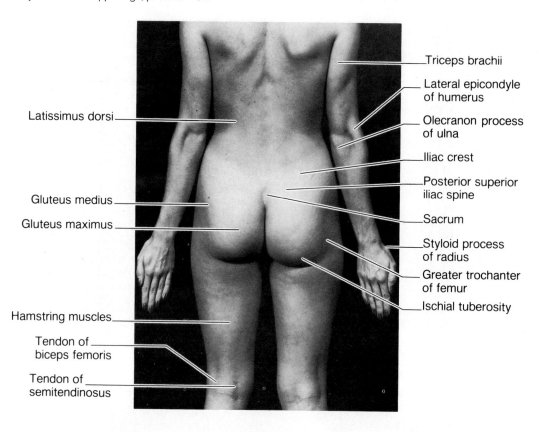

Triceps brachii

Lateral epicondyle of humerus

Olecranon process of ulna

Iliac crest

Posterior superior iliac spine

Sacrum

Styloid process of radius

Greater trochanter of femur

Ischial tuberosity

Latissimus dorsi

Gluteus medius

Gluteus maximus

Hamstring muscles

Tendon of biceps femoris

Tendon of semitendinosus

Plate 13
Surface anatomy of arm, anterior view.

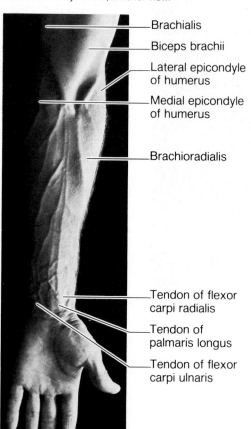

Brachialis

Biceps brachii

Lateral epicondyle of humerus

Medial epicondyle of humerus

Brachioradialis

Tendon of flexor carpi radialis

Tendon of palmaris longus

Tendon of flexor carpi ulnaris

Plate 14
Surface anatomy of the hand.

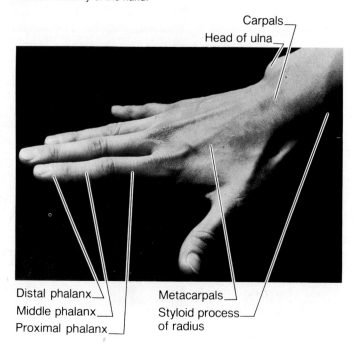

Carpals

Head of ulna

Distal phalanx

Middle phalanx

Proximal phalanx

Metacarpals

Styloid process of radius

Plate 15
Surface anatomy of knee and surrounding area, anterior view.

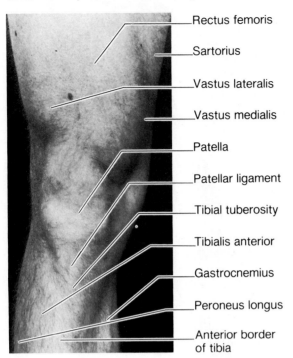

— Rectus femoris

— Sartorius

— Vastus lateralis

— Vastus medialis

— Patella

— Patellar ligament

— Tibial tuberosity

— Tibialis anterior

— Gastrocnemius

— Peroneus longus

— Anterior border of tibia

Plate 16
Surface anatomy of knee and surrounding area, lateral view.

Rectus femoris —

Biceps femoris —

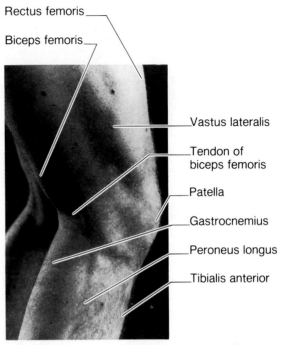

— Vastus lateralis

— Tendon of biceps femoris

— Patella

— Gastrocnemius

— Peroneus longus

— Tibialis anterior

Plate 17
Surface anatomy of ankle and lower leg, medial view.

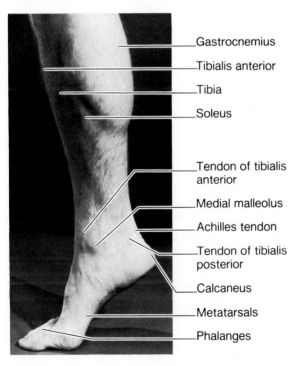

— Gastrocnemius

— Tibialis anterior

— Tibia

— Soleus

— Tendon of tibialis anterior

— Medial malleolus

— Achilles tendon

— Tendon of tibialis posterior

— Calcaneus

— Metatarsals

— Phalanges

Plate 18
Surface anatomy of ankle and foot.

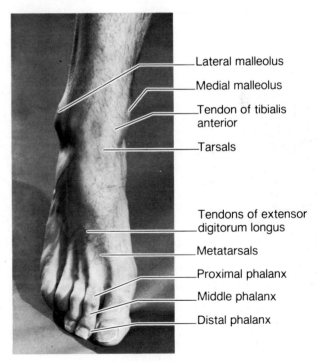

— Lateral malleolus

— Medial malleolus

— Tendon of tibialis anterior

— Tarsals

— Tendons of extensor digitorum longus

— Metatarsals

— Proximal phalanx

— Middle phalanx

— Distal phalanx

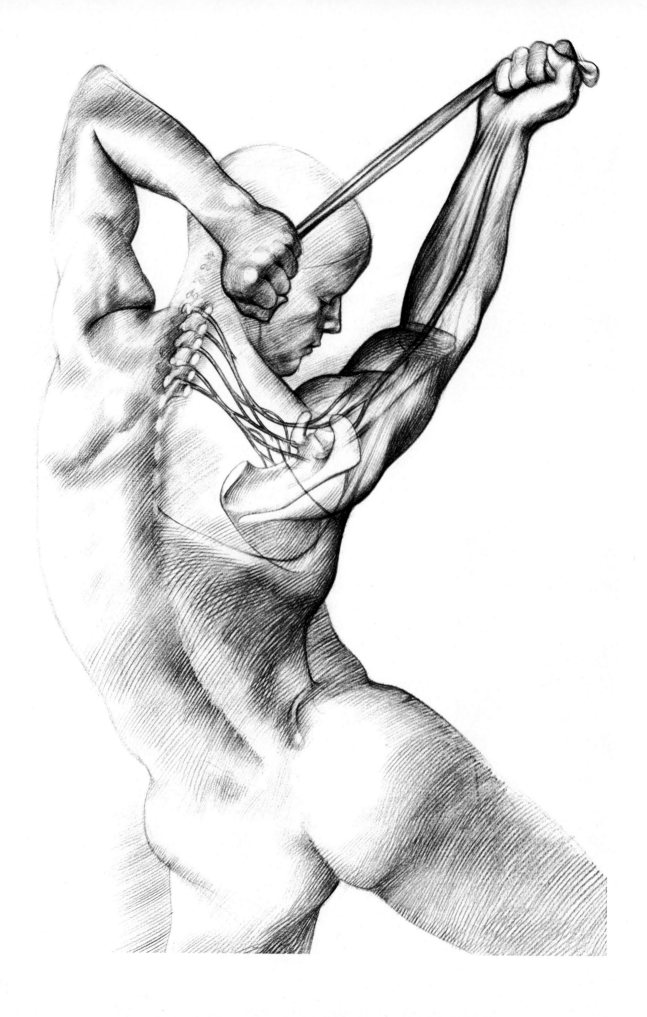

Unit 3

Integration and Coordination

The chapters of unit 3 are concerned with the structures and functions of the nervous and endocrine systems. They describe how the organs of these systems serve to keep the parts of the human body functioning together as a whole and how these organs help to maintain a stable internal environment, which is vital to the survival of the organism.

10

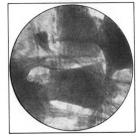

The Nervous System I
Basic Structure and Function

Neurons, the structural and functional units of the nervous system, are able to communicate with one another by means of nerve impulses. Some of these cells are specialized to detect changes that occur inside and outside the body and can transmit impulses to processing centers within the brain or spinal cord. The processing cells gather the incoming information, integrate it, and act on it in various ways. Still other neurons can transmit signals to muscles and glands, causing them to respond.

These functions are essential in the control and coordination of body parts so they are able to work together in ways that help to maintain a stable internal environment.

Chapter Outline

Chapter Objectives

After you have studied this chapter, you should be able to

1. Explain the general functions of the nervous system.

2. Describe the general structure of a neuron.

3. Name four types of neuroglial cells and describe the functions of each.

4. Explain how an injured nerve fiber may regenerate.

5. Explain how a membrane becomes polarized.

6. Describe the events that lead to the conduction of a nerve impulse.

7. Explain how a nerve impulse is transmitted from one neuron to another.

8. Distinguish between excitatory and inhibitory postsynaptic potentials.

9. Explain two ways impulses are processed in neuronal pools.

10. Explain how neurons are classified.

11. Describe how nerve fibers in peripheral nerves are classified.

12. Describe a reflex arc.

13. Explain what is meant by reflex behavior.

14. Complete the review activities at the end of this chapter. Note that the items are worded in the form of specific learning objectives. You may want to refer to them before reading the chapter.

Key Terms

action potential   (ak'shun po-ten'shal)

axon   (ak'son)

central nervous system   (sen'tral ner'vus sis'tem)

convergence   (kon-ver'jens)

dendrite   (den'drit)

divergence   (di-ver'jens)

effector   (ĕ-fek'tor)

facilitation   (fah-sil''ĭ-ta'shun)

myelin   (mi'ĕ-lin)

neuroglia   (nu-rog'le-ah)

neurolemma   (nu''ro-lem'ah)

neuron   (nu'ron)

neurotransmitter   (nu''ro-trans-mit'er)

peripheral nervous system   (pĕ'rif'er-al ner'vus sis'tem)

receptor   (re-sep'tor)

reflex   (re'fleks)

summation   (sum-ma'shun)

synapse   (sin'aps)

threshold   (thresh'old)

Aids to Understanding Words

astr-, starlike: *astr*ocyte—a star-shaped neuroglial cell.

ax-, axle: *ax*on—a cylindrical nerve fiber that carries impulses away from a neuron cell body.

dendr-, tree: *dendr*ite—branched nerve fibers that serve as receptor surfaces of a neuron.

ependym-, tunic: *ependym*a—neuroglial cells that line spaces within the brain and spinal cord.

-lemm, rind or peel: neuro*lemm*a—a sheath that surrounds the myelin of a nerve fiber.

moto-, moving: *moto*r neuron—neuron that stimulates a muscle to contract or a gland to release a secretion.

oligo-, few: *oligo*dendrocyte—small neuroglial cell with few cellular processes.

peri-, all around: *peri*pheral nervous system—portion of the nervous system that consists of the nerves branching from the brain and spinal cord.

saltator-, a dancer: *saltator*y conduction—nerve impulse conduction in which the impulse seems to jump from node to node along the nerve fiber.

sens-, feeling: *sens*ory neuron—neuron that can be stimulated by a sensory receptor and conducts impulses into the brain or spinal cord.

syn-, together: *syn*apse—junction between two neurons.

ORGANS OF THE nervous system, like other organs, are composed of various tissues, including nerve tissue, blood, and connective tissues. These organs can be divided into two groups. One group, consisting of the brain and spinal cord, forms the **central nervous system,** while the other includes the nerves (peripheral nerves) that connect the central nervous system to other body parts and is called the **peripheral nervous system.** Together these parts provide three general functions—a sensory function, an integrative function, and a motor function (figure 10.1).

General Functions of the Nervous System

The sensory function of the nervous system involves sensory *receptors* at the ends of peripheral nerves. These receptors are specialized to gather information by detecting changes that occur within and around the body. They monitor such factors as light and sound intensities outside the body as well as temperature and oxygen concentration of its internal fluids.

The information that is gathered is converted to signals in the form of *nerve impulses,* and they are transmitted over peripheral nerves to the central nervous system. There the signals are integrated—that is, they are brought together creating sensations, adding to memory, or being used to produce thoughts. As a

result of this integrative function, conscious or subconscious decisions are made and then acted upon by using motor functions.

The motor functions of the nervous system employ peripheral nerves that carry impulses from the central nervous system to responsive parts called *effectors.* These effectors are outside the nervous system and include muscles that may contract when they are stimulated by nerve impulses, and glands that may produce a secretion when they are stimulated.

Thus, the nervous system can detect changes occurring in the body, make decisions on the basis of the information received, and cause muscles or glands to respond. Typically these responses are directed toward counteracting the effects of the changes and, in this way, the nervous system helps to maintain stable internal conditions.

Nerve Tissue

The nerve tissue of the brain and spinal cord consists of masses of nerve cells, or **neurons,** that are the structural and functional units of the nervous system. These cells are specialized to react to physical and chemical changes occurring in their surroundings. They also function to conduct nerve impulses to other neurons and to cells outside the nervous system. (See figure 10.2.)

As mentioned in chapter 5, **neuroglial cells** are accessory cells within nerve tissue. They greatly outnumber the neurons and function much like connective tissue cells in other body structures; that is, they fill spaces and surround or support various parts.

Neuron Structure

Although neurons vary considerably in size and shape, they have certain features in common. These include a cell body and tubular processes filled with cytoplasm that conduct nerve impulses to or from the cell body (figure 10.3).

Figure 10.1

The nervous system consists of the brain, spinal cord, and numerous peripheral nerves.

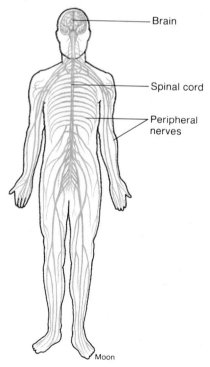

Brain

Spinal cord

Peripheral nerves

Moon

Figure 10.2

Neurons are the structural and functional units of the nervous system.

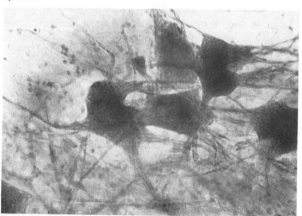

Figure 10.3
Two types of neurons are (a) motor neuron, and (b) sensory
neuron.

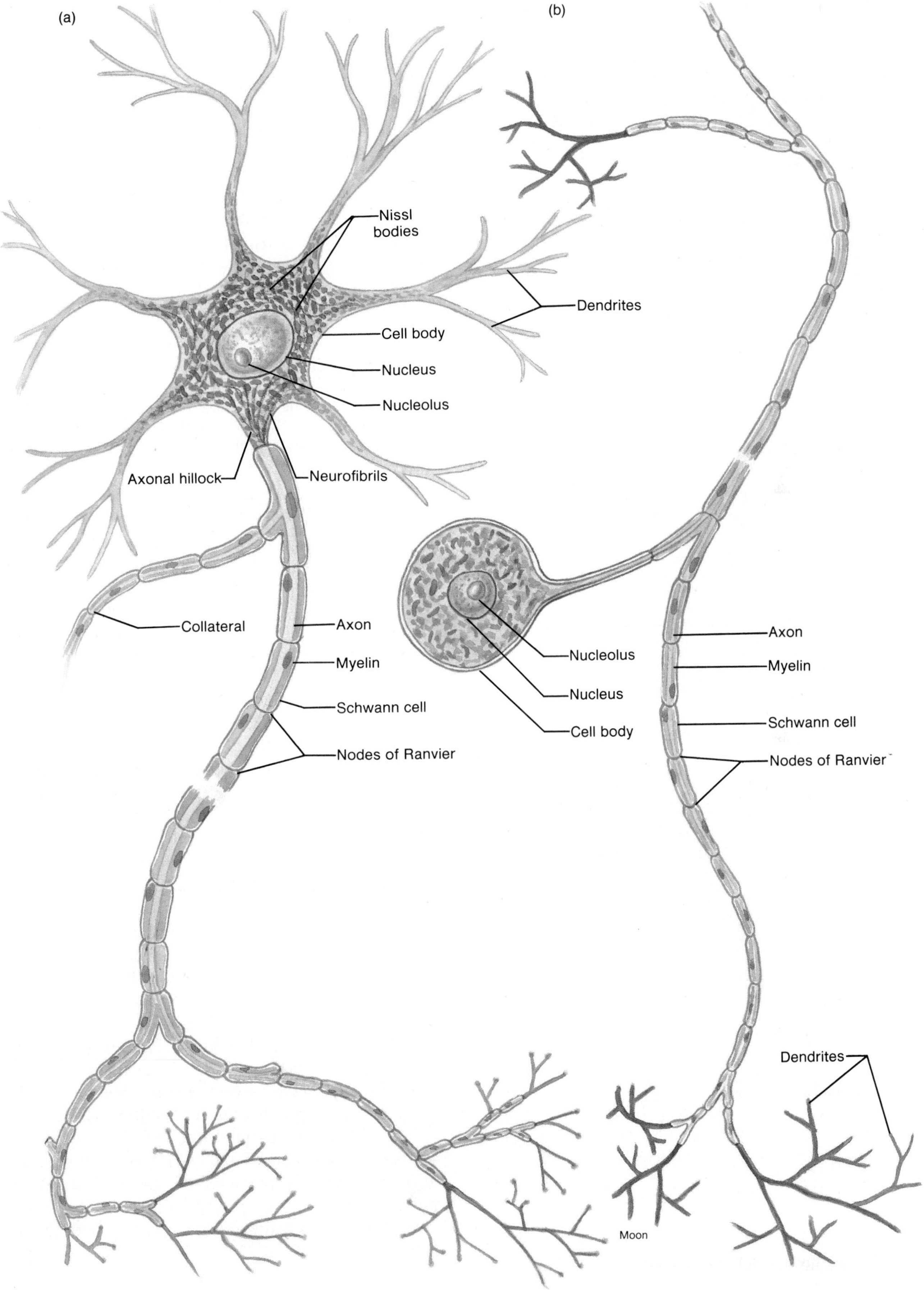

(a)

(b)

Nissl bodies

Dendrites

Cell body

Nucleus

Nucleolus

Axonal hillock

Neurofibrils

Collateral

Axon

Myelin

Schwann cell

Nodes of Ranvier

Nucleolus

Nucleus

Cell body

Axon

Myelin

Schwann cell

Nodes of Ranvier

Dendrites

Moon

Figure 10.4
(a) The portion of a Schwann cell that is tightly wound around an axon forms the myelin sheath; (b) the cytoplasm and nucleus of the Schwann cell, remaining on the outside, form the neurolemmal sheath.

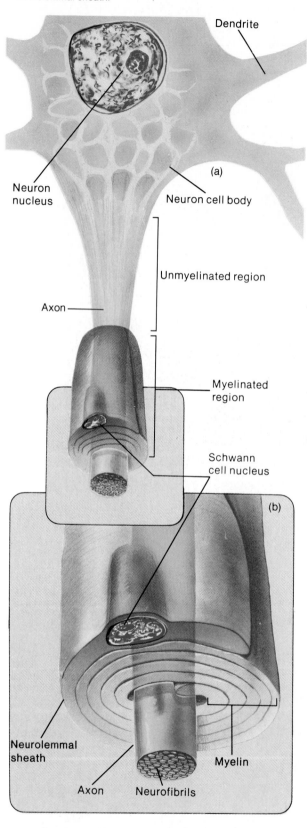

Dendrite

Neuron nucleus

Neuron cell body

(a)

Unmyelinated region

Axon

Myelinated region

Schwann cell nucleus

(b)

Neurolemmal sheath

Myelin

Axon   Neurofibrils

The **cell body** (perikaryon) contains a mass of granular cytoplasm, a cell membrane, and various other organelles usually found in cells, although some of these have special names. Inside the cell body, for example, are mitochondria, lysosomes, a Golgi apparatus, numerous microtubules, and a network of fine threads called **neurofibrils,** which extend into the cell processes. Scattered throughout the cytoplasm are many membranous sacs called **nissl bodies** that are similar to the rough endoplasmic reticulum in other cells. Ribosomes attached to the surfaces of these membranous parts function in the manufacture of vital protein molecules. Centrosomes are usually missing; cytoplasmic inclusions, however, are common. Some of these contain glycogen, lipids, or melanin and other pigments.

Neurons commonly contain inclusions of a golden brown substance called *lipofuscin,* which is thought to be a by-product of lysosomal activity. This pigment tends to accumulate in neurons and may become so abundant that the nuclei and other organelles are pushed to the sides of the cell bodies. Some investigators believe this to be a consequence of the aging process.

Near the center of the cell body there is a large, spherical nucleus with a conspicuous nucleolus. This nucleus, however, apparently does not undergo mitosis after the nervous system is developed and consequently mature neurons seem to be incapable of reproduction.

Two kinds of nerve fibers, called **dendrites** and **axons,** extend from the cell bodies of most neurons. Although a neuron usually has many dendrites, it has a single axon.

In most neurons, the dendrites are relatively short and highly branched. These processes, together with the membrane of the cell body, provide the main receptive surfaces of the neuron to which processes from other neurons communicate. Often the dendrites possess tiny, thornlike spines (dendritic spines) on their surfaces, which serve as contact points for parts of other neurons.

The axon, which usually arises from a slight elevation of the cell body (axon hillock), is a slender, cylindrical process with a nearly smooth surface and uniform diameter. It is specialized to conduct nerve impulses away from the region of the cell body. Many mitochondria, microtubules, and neurofibrils occur within its cytoplasm. Although it begins as a single fiber, the axon may give off branches, called *collaterals.* Near its end, it may have many fine branches, each with a specialized ending called a *presynaptic terminal,* which contacts the receptive surface of another cell.

Larger axons passing through peripheral nerves commonly are enclosed in sheaths composed of neuroglial cells called **Schwann cells.** These cells are tightly wound around the axons, somewhat like insulation on a wire. As a result, the axons are coated with many

Figure 10.5
A transmission electron micrograph of myelinated and
unmyelinated axons in cross section.

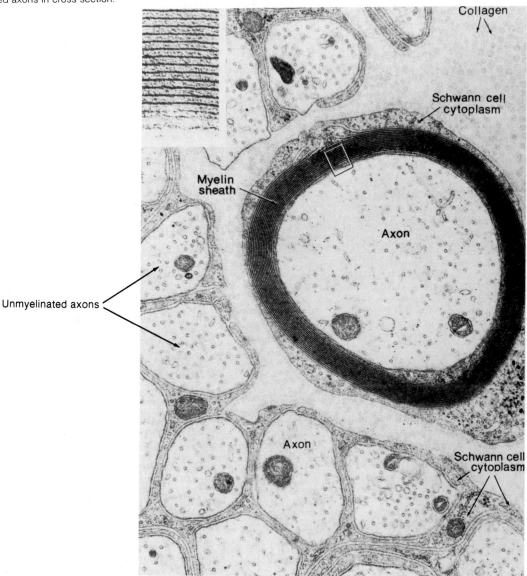

layers of cell membranes that have little or no cytoplasm between them. These membranes are composed
largely of a lipid-protein (lipoprotein) that has a higher
proportion of lipid than other cell surface membranes.
This lipid-protein is called **myelin,** and it forms a *myelin sheath* on the outside of an axon. In addition, the
portions of the Schwann cells that contain cytoplasm
and nuclei remain outside the myelin sheath and comprise a **neurolemma** (neurolemmal sheath), which surrounds the myelin sheath (figure 10.4).

The smallest axons also are enclosed by Schwann
cells, but they are not wound around these axons. Consequently such axons lack myelin sheaths.

Axons that possess myelin sheaths are called *myelinated* (medullated) nerve fibers, while those that lack
these sheaths are *unmyelinated* nerve fibers (figure
10.5). Groups of myelinated fibers appear white, and
masses of such fibers are responsible for the white
matter in the brain and spinal cord. In this instance,
however, the myelin is produced by another kind of
neuroglial cell (oligodendrocyte) rather than by
Schwann cells. Furthermore, nerve fibers in the brain
and spinal cord lack neurolemmal sheaths. The gray
matter of these parts consists of unmyelinated nerve
fibers and neuron cell bodies.

Figure 10.6
Types of neuroglial cells found within the central nervous system
include *(a)* microglial cell, *(b)* oligodendrocyte, *(c)* astrocyte, and
(d) ependymal cell.

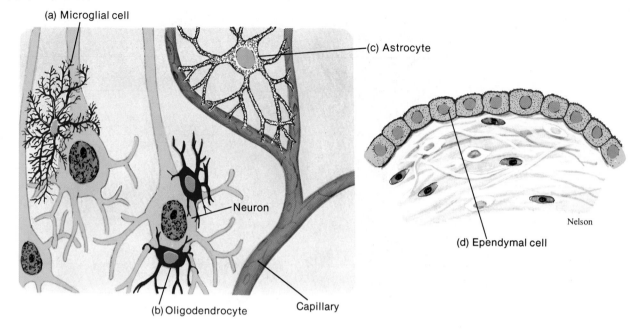

(a) Microglial cell

(c) Astrocyte

Neuron

Nelson

(d) Ependymal cell

(b) Oligodendrocyte   Capillary

When neurons are deprived of oxygen, they undergo a
series of irreversible structural changes in which their
shapes are altered and their nuclei shrink. This phenomenon is called *ischemic cell change,* and in time the affected cells disintegrate. Such an oxygen deficiency can
result from lack of blood flow through nerve tissue (ischemia), an abnormally low blood oxygen concentration, or
the presence of toxins that block aerobic respiration.

1. List the general functions of the nervous system.
2. Describe a neuron.
3. Explain how myelin is formed in the peripheral
   nervous system.

Neuroglial Cells

As mentioned earlier, neuroglial cells fill spaces, support neurons, and provide frameworks within the organs of the nervous system. They also seem to enclose
neurons in ways that prevent contact between nearby
cells except at particular sites (figures 10.6 and 10.7).

Within the peripheral nervous system, the neuroglial cells include the Schwann cells previously described. In the central nervous system, the neuroglial
cells are several times more numerous than neurons.

They include the following types:

1. **Astrocytes.** As their name implies, astrocytes
   are star-shaped cells. They are commonly found
   between nerve tissues and blood vessels where
   they seem to function in providing structural
   support and in holding parts together by means
   of numerous cellular processes. Astrocytes may
   have a nutritive function involving the transport
   of substances from blood vessels to neurons.
   They also respond to injury of brain tissue and
   are responsible for the formation of scar tissue
   that fills spaces and closes gaps following such
   injuries.
2. **Oligodendrocytes.** Oligodendrocytes resemble
   astrocytes, but are smaller and have fewer
   processes. They are commonly arranged in rows
   along nerve fibers, and they function in the
   formation of myelin within the brain and spinal
   cord.

   Unlike the Schwann cells of the
   peripheral nervous system, oligodendrocytes can
   send out a number of cellular processes, each of
   which wraps tightly around a nearby axon. In
   this way a single oligodendrocyte may provide
   some myelin for up to 40 axons. However, these
   cells do not form neurolemmal sheaths.

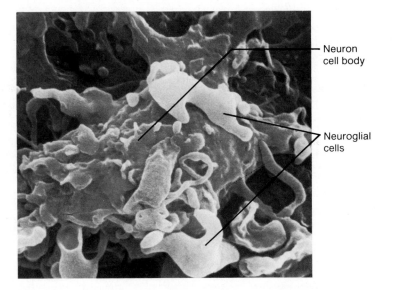

Neuron cell body

Neuroglial cells

3. **Microglia.** Microglial cells are relatively small and have fewer processes than other types of neuroglial cells. These cells are scattered throughout the central nervous system where they help support neurons and are able to phagocytize bacterial cells and cellular debris. They usually increase in number whenever the brain or spinal cord is inflamed because of injury or disease.

4. **Ependyma.** Ependymal cells are cuboidal or columnar in shape and may have cilia. They form an epithelial-like membrane that is one cell thick and covers the inside of spaces within the brain, called *ventricles;* the inside of the *central canal* that extends downward through the spinal cord; and the specialized capillaries called *choroid plexuses,* that are associated with the ventricles of the brain.

Although myelin begins to form on nerve fibers during the fourteenth week of development, many of the nerve fibers in a newborn infant are not completely myelinated. Consequently, the nervous system is unable to function as effectively as that of an older child or adult. An infant's responses to stimuli are therefore coarse and undifferentiated, and may involve its whole body. Essentially all myelinated fibers have begun to develop sheaths by the time a child begins to walk. Myelination continues into adolescence.

Any interference with the supply of essential nutrients during the developmental years may result in an insufficient amount of myelin formation. This, in turn, may be reflected in impaired function of the nervous system.

Regeneration of Nerve Fibers

Injury to a cell body is most likely to cause the death of the neuron; however, a damaged axon may be regenerated. For example, if an axon in a peripheral nerve is separated from its cell body by injury or disease, the distal portion of the axon and its myelin sheath deteriorates within a few weeks. Macrophages remove the fragments of myelin and other cellular debris. Although some Schwann cells may also degenerate, a thin basement membrane and a layer of connective tissue (endoneurium) surrounding the Schwann cells will remain. These parts form a tube that leads back to the original connection of the axon.

The proximal end of the injured axon develops sprouts shortly after the injury, and one of these sprouts or branches may grow into the tube formed by the basement membrane and connective tissue. At the same time, remaining Schwann cells proliferate along the length of the degenerating fiber, and form new myelin around the growing axon.

The growth of such a regenerating fiber is slow (3 to 4 millimeters per day), but eventually the new fiber may reestablish the former connection (figure 10.8).

A similar type of degeneration occurs in the distal portion of injured axons in the central nervous system, although the rate of degeneration is somewhat slower. However, such axons lack connective tissue sheaths, and the myelin-producing oligodendrocytes fail to proliferate. Consequently, if the proximal end of a damaged

Figure 10.8

If a myelinated axon is injured *(a)* the proximal portion of the fiber may survive, but *(b)* the portion distal to the injury degenerates. *(c and d)* In time, the proximal portion may develop extensions that grow into the tube of basement membrane and connective tissue cells previously occupied by the fiber and *(e)* the former connection may be reestablished.

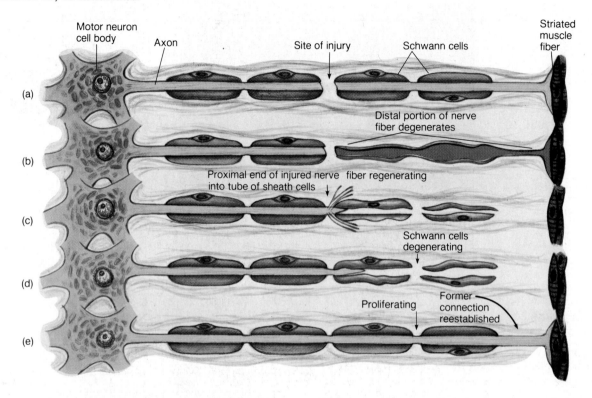

axon begins to grow, there is no tube of sheath cells to guide it, and a functionally significant regeneration is unlikely to take place.

If a peripheral nerve is severed, it is very important that the two cut ends be closely connected as soon as possible so that the regenerating sprouts of the nerve fibers can more easily reach the tubes formed by the basement membranes and connective tissues on the other side of the gap.

When the gap exceeds 3 millimeters, the regenerating fibers tend to form a tangled mass called a *neuroma*. A neuroma composed of sensory nerve fibers is likely to be painfully sensitive to pressure. The development of such growths sometimes complicates a patient's recovery following the amputation of a limb.

1. What is a neuroglial cell?
2. Name and describe four types of neuroglial cells.
3. Explain how an injured peripheral nerve fiber might regenerate.
4. Explain why functionally significant regeneration is unlikely to occur in the central nervous system.

Cell Membrane Potential

A cell membrane is usually electrically charged or *polarized* so that the outside is positively charged with respect to the inside. This polarization is due to an unequal distribution of ions on either side of the membrane, and it is of particular importance to the conduction of muscle and nerve impulses.

Distribution of Ions

The distribution of ions is determined in part by the presence of pores or channels in cell membranes that were discussed in chapter 3. The size and shape of these channels are related to the ways in which proteins and lipids are arranged in the membranes. Some channels are always open, while others can be opened or closed, as if by a gate. Furthermore, they can be selective; that is, a particular kind of channel may allow one kind of ion to pass through and exclude all other kinds. The effective size and electrical charge of an ion also affect the way it passes through a membrane (figure 10.9).

Figure 10.9
Some of the channels in cell membranes through which ions pass can be closed *(a)* or opened *(b)* by a gate-like mechanism.

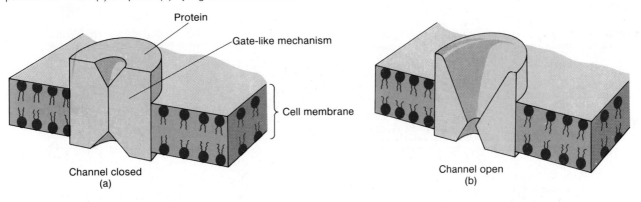

Protein

Gate-like mechanism

Cell membrane

Channel closed
(a)

Channel open
(b)

Figure 10.10
A nerve fiber at rest is polarized as a result of an unequal distribution of ions on either side of its membrane.

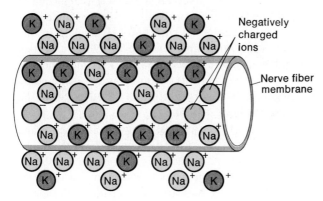

Negatively charged ions

Nerve fiber membrane

As a consequence of such factors, potassium ions tend to pass through cell membranes much more easily than sodium ions, and sodium ions pass through more easily than calcium ions. The relative ease with which potassium ions diffuse through membranes makes them a major contributor to membrane polarization.

Resting Potential

When nerve cells are at rest (that is, not conducting impulses), there is a relatively greater concentration of sodium ions (Na^+) on the outside of their membranes and a relatively greater concentration of potassium ions (K^+) on the inside (figure 10.10). In the cytoplasm of these cells, there are large numbers of negatively charged ions, including those of phosphate, sulfate, and protein, that cannot diffuse through the cell membranes.

However, since the membranes are very permeable to potassium ions and only slightly permeable to sodium ions, potassium ions tend to diffuse freely through the membrane to the outside, and sodium ions diffuse inward more slowly. At the same time, the cell

Figure 10.11
An active transport mechanism in the nerve fiber membrane moves sodium ions outward and potassium ions inward.

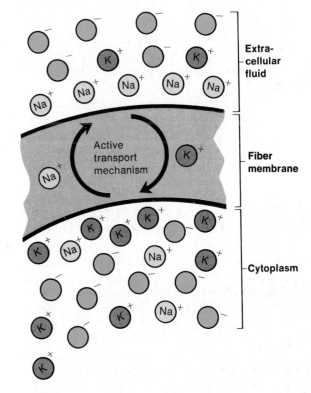

Extra-cellular fluid

Active transport mechanism

Fiber membrane

Cytoplasm

membrane expends energy to actively transport these ions in the opposite directions, preventing them from reaching equilibrium by diffusion. Sodium ions are, therefore, actively transported outward through the membrane, and potassium ions are transported inward. Since potassium ions can readily diffuse out again and sodium can enter only with difficulty, the net effect is for more positively charged ions to leave the cell than to enter it. As a result, the outside of the membrane becomes positively charged with respect to the inside, which is negative (figure 10.11).

Figure 10.12

When a polarized nerve fiber is stimulated, sodium channels open, some of the ions diffuse inward, and the membrane is said to depolarize.

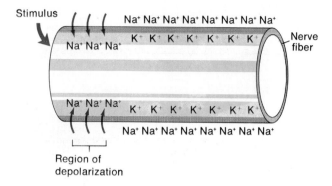

Figure 10.13

When the potassium channels open, potassium ions diffuse outward, and the membrane is repolarized.

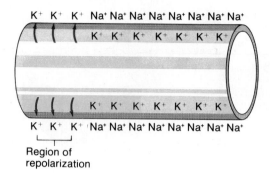

The difference in electrical charge between two points is called the *potential difference* and can be measured in units called *volts*. In the case of a resting nerve cell, the difference between the inside and the outside of the membrane is about −70 millivolts and is called the **resting potential.** As long as the nerve cell membrane is undisturbed it remains in this polarized state.

Local Potential Changes

Nerve cells are excitable; that is, they can respond to changes in their surroundings. Some nerve cells, for example, are specialized to detect changes in temperature, light, or pressure occurring outside the body, while others are responsive to signals coming from nearby neurons. In either case, the stimulation usually affects the resting potential in a particular local region of the cell membrane. If in response to the stimulus, the membrane's resting potential becomes less negative (moving toward zero), the membrane is said to be *depolarizing;* if the resting potential becomes more negative (moving away from zero), the membrane is *hyperpolarizing.*

Such changes in the local potential of a membrane are *graded.* This means the amount of change in potential is directly related to the intensity of the stimulation. Furthermore, if another stimulus of the same type is received before the effect of the first one subsides, the change in local membrane potential is greater. This additive phenomena is called *summation,* and as a result of summated potentials, a level called the *threshold potential* may be reached. Thus, by this means, many subthreshold potential changes may be combined to reach threshold. At threshold, an action potential is produced in a nerve fiber.

Action Potentials

Once the threshold potential is reached, the portion of the cell membrane that is being stimulated undergoes a sudden change in permeability. Channels in the membrane that are highly selective for sodium ions open and allow the ions to pass inward. This movement is aided by the fact that the sodium ions are attracted by the negative electrical condition on the other side of the membrane (figure 10.12).

As the sodium ions rush inward, the membrane loses its electrical charge and becomes *depolarized*. At almost the same time, channels open in the membrane that allow some of the potassium ions to pass through, and as they diffuse outward, the outside of the membrane becomes positively charged once more. The membrane, then, is said to become *repolarized,* and it remains in this state until it is stimulated again (figure 10.13).

This rapid sequence of changes, involving depolarization and repolarization, takes about 1/1000th second or less and is called an **action potential.** Actually, only a small proportion of the sodium and potassium ions present move through the membrane during an action potential, so that many action potentials can occur before the concentrations of sodium ions and potassium ions on either side of the membrane change significantly. Eventually, however, these ion concentrations may change enough that action potentials cease. When this happens, an active transport mechanism in the membrane reestablishes the original concentrations of sodium and potassium, and the resting potential returns.

When an action potential occurs at one point in a nerve cell membrane, it causes an electric current to flow to adjacent portions of the membrane. This local current stimulates the membrane to its threshold level, triggering other action potentials. These in turn stimulate still other areas, and a wave of action potentials

Figure 10.14

(a) An action potential in one region stimulates the adjacent region, and *(b,c)* a wave of action potentials or a nerve impulse moves along the fiber.

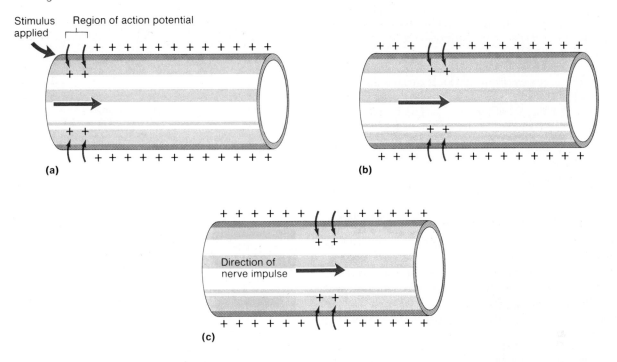

Stimulus applied | Region of action potential

(a)

(b)

Direction of nerve impulse

(c)

Chart 10.1 Events leading to conduction of a nerve impulse	
1. Nerve fiber membrane develops resting potential.	6. Potassium ions diffuse outward.
2. Threshold stimulus is received.	7. Ion channels close and membrane becomes less permeable to sodium and potassium.
3. Sodium channels in membrane open.	8. Wave of action potentials travels the length of the nerve fiber as a nerve impulse.
4. Sodium ions diffuse inward.	
5. Potassium channels in membrane open.	

moves away in all directions from the point of stimulation, traveling to the end of the nerve fiber without decreasing in amplitude. This propagation of action potentials along a fiber constitutes a **nerve impulse** (figure 10.14).

This description of a nerve impulse applies also to the muscle impulse that was mentioned in chapter 9. In the muscle fiber, stimulation that occurs at the motor end plate triggers an impulse to travel over the surface of the fiber and down into its transverse tubules. See chart 10.1 for a summary of the events leading to the conduction of a nerve impulse.

Refractory Period

For a moment following the passage of a nerve impulse, an ordinary stimulus will not be able to trigger another impulse on a nerve fiber. This brief period of time, the **refractory period,** has two parts. During the *absolute refractory period,* which lasts about 1/2500

of a second, the fiber's membrane is changing in sodium permeability and cannot be stimulated. This is followed by a *relative refractory period,* during which the membrane is reestablishing its resting potential. While the membrane is in this relative refractory period, even though polarization is incomplete, an impulse may be triggered by a stimulus of high intensity.

As time passes, however, the fiber gradually becomes more excitable. Simultaneously, the intensity of stimulation needed to trigger an impulse decreases until the fiber's original excitability is restored. This return to the resting state usually takes from 10 to 30 milliseconds.

Because of the refractory period, a nerve fiber cannot be stimulated continuously. Thus, the refractory period acts to limit the rate at which nerve impulses can be conducted. In other words, the time between impulses cannot be less than the absolute refractory period, and the maximum rate of nerve impulses is about 1 impulse per millisecond.

Recording an Action Potential
A Laboratory Application

Since an action potential occurs in about 1/1000 second, an instrument that responds rapidly is needed to record the sequence of changes. The instrument usually used is the **cathode ray oscilloscope.** This instrument includes a *cathode ray tube* that looks and functions somewhat like the picture tube of a television (figure 10.15). It contains an *electron gun,* which fires a concentrated beam of electrons from the neck toward the face of the tube. The face is coated with a thin layer of fluorescent material that produces a spot of light where the electron beam strikes it.

Horizontal deflection plates are located on either side of the electron beam within the cathode ray tube. These plates can be electrically charged by the action of a *sweep circuit* of the oscilloscope. This circuit causes one of the horizontal deflection plates to become positively charged while the other becomes negatively charged. When this happens, the beam of electrons (which are negatively charged) is drawn toward the positive plate and repelled by the negative one. As a result, the electron beam moves across the face of the tube and produces a glowing horizontal line, called a *trace*.

The sweep circuit can be adjusted to move the electron beam across the face of the cathode ray tube from left to right at a known velocity. Each time the beam reaches the right edge of the tube, it almost instantly jumps back to the left side and begins a new horizontal trace.

Other plates called *vertical deflection plates* are located above and below the electron beam within the cathode ray tube. If these plates become electrically charged, the electron beam moves up or down.

When the oscilloscope is being used to record an action potential of a nerve fiber, electrodes from an electronic *amplifier* are placed in contact with the fiber, and the amplifier is connected to the vertical deflection plates. Any change in nerve fiber membrane potential is detected by the electrodes, and a signal is transmitted to the amplifier. The amplifier increases the intensity of the signal in direct proportion to the amount of change occurring in the membrane potential and causes the vertical deflection plates to become charged. If this happens when the electron beam is moving across the face of the cathode ray tube, its horizontal trace will be deflected up or down in direct proportion to the change in membrane potential detected by the electrodes.

Figure 10.16 illustrates an action potential (monophasic) of the type recorded on an oscilloscope when one electrode is placed on the surface of a nerve fiber membrane and the other electrode is inserted into the cytoplasm of the fiber. Note that the fiber's resting potential is about -85 millivolts. As the action potential occurs following stimulation of the fiber, the potential momentarily becomes positive, and then the negative resting potential is reestablished.

Figure 10.15

A cathode ray oscilloscope can be used to record the action potential of a nerve fiber.

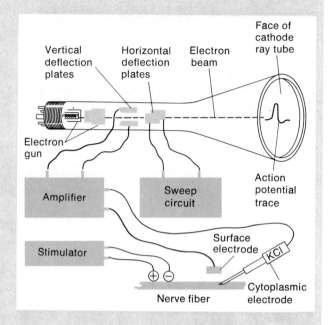

Figure 10.16

An action potential as it might be recorded on the face of a cathode ray tube.

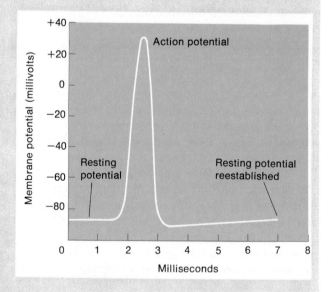

Figure 10.17
On a myelinated fiber, a nerve impulse appears to jump from
node to node.

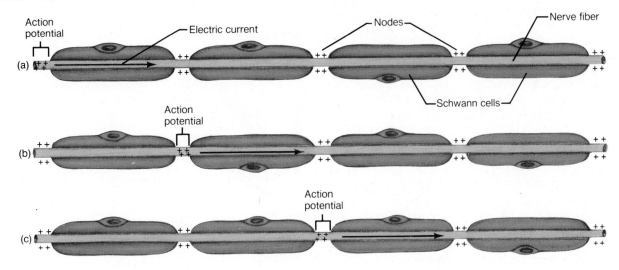

All-or-None Response

Like muscle fiber contraction, nerve impulse conduction is an *all-or-none response*. In other words, if a nerve fiber responds at all, it responds completely. Thus, a nerve impulse is conducted whenever a stimulus of threshold intensity or above is applied to a nerve fiber, and all impulses carried on that fiber will be of the same strength. Greater intensity of stimulation of a nerve fiber does not produce a stronger impulse.

Impulse Conduction

An unmyelinated nerve fiber conducts an impulse over its entire membrane surface. A myelinated fiber functions differently. Myelin contains a high proportion of lipid that excludes water and water soluble substances. Thus, it serves as an insulator by preventing almost all flow of ions through the membrane.

Considering this, it might seem that the myelin sheath would prevent the conduction of a nerve impulse altogether, and this would be true if the sheath were continuous. It is, however, interrupted by constrictions called **nodes of Ranvier,** which occur between adjacent Schwann cells (figure 10.3). At these nodes the fiber membrane is especially permeable to sodium and potassium ions.

When such a fiber is stimulated, an action potential occurs at a node. This causes an electric current to flow away through the cytoplasm of the fiber. As this local current reaches the next node, it stimulates the membrane to its threshold level, and an action potential occurs there. Consequently, a nerve impulse traveling along a myelinated fiber appears to jump from node to node. This type of impulse conduction, called **saltatory conduction,** is many times faster than conduction on an unmyelinated fiber (figure 10.17).

The speed of nerve impulse conduction is also related to the diameter of the fiber—the greater the diameter, the faster the impulse. For example, an impulse on a thick myelinated fiber, such as a motor fiber associated with a skeletal muscle, might travel 120 meters per second, while an impulse on a thin unmyelinated fiber, such as a sensory fiber associated with the skin, might move only 0.5 meter per second.

---

1. How is the resting potential achieved?
2. Explain how a polarized nerve fiber responds to stimulation.
3. List the major events that occur during an action potential.
4. Define refractory period.
5. Explain how impulse conduction differs in myelinated and unmyelinated fibers.

Figure 10.18

For an impulse to continue from one neuron to another, it must cross the synaptic cleft at a synapse. A synapse may occur (a) between an axon and a dendrite, or (b) between an axon and a cell body.

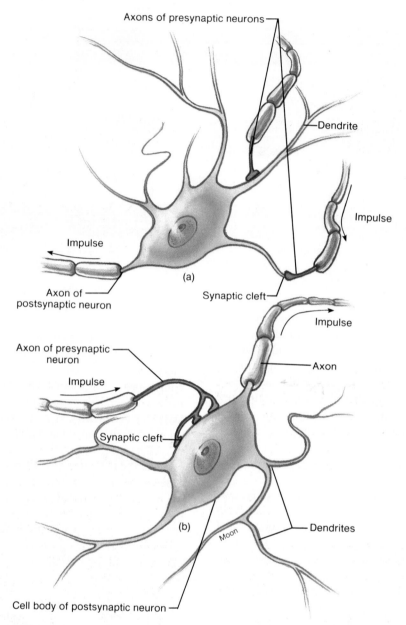

The Synapse

Within the nervous system, nerve impulses travel from neuron to neuron along complex nerve pathways. The junction between the parts of two such neurons is called a **synapse.** Actually, these cells, called *presynaptic* and *postsynaptic neurons,* are not in direct contact at the synapse. There is a gap called a *synaptic cleft* between them, and for an impulse to continue along a nerve pathway it must cross this gap (figure 10.18).

Synaptic Transmission

As was mentioned, a nerve impulse travels in both directions away from the point of stimulation. Within a neuron, however, an impulse will usually travel from a dendrite or cell body and then move along the axon to the presynaptic terminal at its end. There the impulse crosses a synapse and continues to a dendrite or cell body of another neuron. The process of crossing the gap at a synapse is called *synaptic transmission.*

Figure 10.19

(a) When a nerve impulse reaches the synaptic knob at the end of an axon, *(b)* synaptic vesicles release a neurotransmitter substance that diffuses across the synaptic cleft.

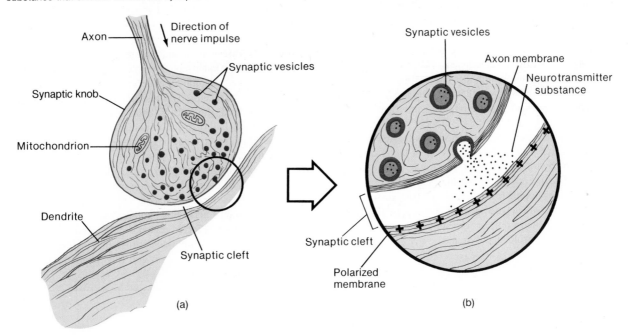

(a)

(b)

The typical one-way transmission from axon to dendrite or cell body is due to the fact that axons usually have rounded *synaptic knobs* at their presynaptic terminals, which dendrites lack. These knobs contain numerous membranous sacs, called *synaptic vesicles*, and when a nerve impulse reaches a knob, some of the vesicles respond by releasing a substance called a **neurotransmitter.**

As figure 10.19 shows, the neurotransmitter diffuses across the synaptic cleft and reacts with specific receptors of the postsynaptic neuron membrane. If a sufficient amount of neurotransmitter is released, the postsynaptic membrane is stimulated, and a nerve impulse may be triggered (figure 10.20).

Neurotransmitter Substances

Many different neurotransmitter substances are produced in the nervous system; however, each neuron seems to release only one or two kinds. These substances include *acetylcholine* that also stimulates skeletal muscle contractions (see chapter 9); a group called *monoamines* (such as epinephrine, norepinephrine, dopamine, and serotonin) that are formed by modifying amino acid molecules; a group of *amino acids* (such as glycine, glutamic acid, aspartic acid, and gamma-aminobutyric acid—GABA); and a large group of *peptides* (such as enkephalins and substance P) that consist of relatively short chains of amino acids.

Figure 10.20

A transmission electron micrograph of a synaptic knob filled with synaptic vesicles.

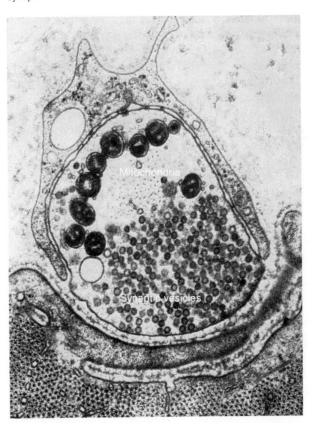

These neurotransmitter substances usually are synthesized in the cytoplasm of the synaptic knobs and generally are stored in the synaptic vesicles.

When an action potential passes over the surface of a synaptic knob, it produces an increase in the membrane's permeability to calcium ions by causing its calcium ion channels to open. Calcium ions diffuse inward, and in response to their presence, some of the synaptic vesicles fuse with the membrane and release their contents into the synaptic cleft. The vesicles may then break away from the membrane and reenter the cytoplasm where they are filled quickly with neurotransmitter substance again. Vesicles that are damaged or lost are replaced by new ones formed in the neuron cell body and transported through the axon to the synaptic knob. (Chart 10.2 summarizes this activity.)

Some neurotransmitters are destroyed through rapid decomposition by enzymes present in the synaptic clefts. Others are removed from the cleft by being transported back into the synaptic knob of the neuron that released them or into nearby neurons or neuroglial cells. Acetylcholine, for example, is decomposed by the enzyme *cholinesterase,* which is present in the membranes at synapses that release this neurotransmitter. Similarly, the monoamine neurotransmitters epinephrine and norepinephrine are inactivated by the enzyme *monoamine oxidase* found in mitochondria. Such destruction or removal of the neurotransmitter is important in preventing a continuous stimulation of the postsynaptic neuron.

Neuropeptides

Neuropeptides are peptides that are synthesized by neurons in the brain or spinal cord. They act as neurotransmitters or as *neuromodulators*—substances that alter a neuron's response to a neurotransmitter or block the release of a neurotransmitter.

Among the neuropeptides are two forms called *enkephalins* that occur throughout the brain and spinal cord. Each enkephalin molecule consists of five amino acids in a chain. The synthesis of these substances seems to increase during periods of painful stress, and they are known to bind to the same receptors in the brain (opiate receptors) as the narcotic morphine. Although they are thought to play a role in relieving pain sensations, enkephalins probably have other functions as well.

Another morphinelike peptide called *beta-endorphin* is found mainly in the pituitary gland. It seems to be decomposed less rapidly and to have a much more potent pain-relieving property than the enkephalins.

Still another neuropeptide, which consists of eleven amino acids and is widely distributed throughout the nervous system, is called **substance P.** It seems to function as a neurotransmitter (or perhaps as a neuromodulator) in the neurons that transmit pain impulses into the spinal cord and on to the brain. Some investigators think the enkephalins and endorphins may relieve pain by inhibiting the release of substance P from pain-transmitting neurons.

As is discussed in chapter 13, some of the neurotransmitter substances also function as hormones.

Chart 10.2   Events leading to the release of neurotransmitter substance

1. Action potential passes along a nerve fiber and over the surface of its synaptic knob.
2. Synaptic knob membrane becomes more permeable to calcium ions, and they diffuse inward.
3. In the presence of calcium ions, synaptic vesicles fuse to synaptic knob membrane.
4. Synaptic vesicles release their neurotransmitter substance contents into the synaptic cleft.
5. Synaptic vesicles reenter the cytoplasm of the nerve fiber and are refilled with neurotransmitter substance.

Synaptic Potentials

The local potentials at a synapse are called *synaptic potentials,* and they provide a means by which one neuron can influence another. As mentioned before, the changes in such potentials are graded and can be depolarizing or hyperpolarizing.

Although the kind of neurotransmitter released by the synaptic knobs of a particular neuron is always the same, other neurons may secrete different neurotransmitters. Furthermore, the effects of neurotransmitters on synaptic potentials vary—some are excitatory and others are inhibitory.

For example, if a neurotransmitter acts on a postsynaptic membrane by combining with specialized receptors and causes its sodium ion channels to open, sodium ions will diffuse inward and cause the membrane to depolarize. As a result, an action potential may be triggered. This kind of membrane change is called an *excitatory postsynaptic potential* (EPSP), and it lasts for about 15 milliseconds.

If, instead, a different neurotransmitter combines with other receptors and causes the permeability of potassium to increase, these ions will diffuse outward, and the membrane becomes hyperpolarized. Since an action potential is now less likely to occur, this change is called an *inhibitory postsynaptic potential* (IPSP).

Within the brain and spinal cord, the synaptic knobs of a thousand or more nerve fibers may communicate with the dendrites and cell body of each neuron. Furthermore, at any moment, some of the postsynaptic potentials may be producing excitatory effects on each neuron, while others are producing inhibitory effects (figure 10.21).

Whether or not an action potential is triggered on a neuron depends upon the integrated sum of the postsynaptic potential effects. That is, if the net effect is more excitatory than inhibitory, threshold may be reached, and an action potential will occur. Conversely, if the net effect is inhibitory, no impulse will be transmitted.

This summation of the excitatory and inhibitory effects of the postsynaptic potentials commonly takes place at a specialized "trigger zone" in the membrane

Figure 10.21
Some of the synaptic knobs that communicate with the dendrites and cell body of a neuron have excitatory functions, while others are inhibitory.

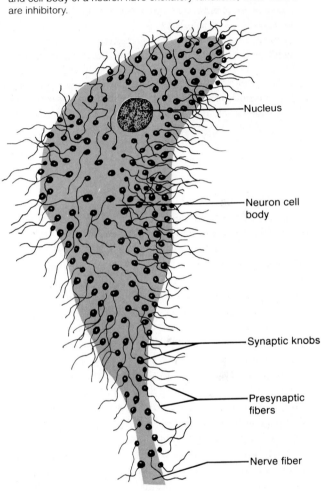

Nucleus

Neuron cell body

Synaptic knobs

Presynaptic fibers

Nerve fiber

of a dendrite or in a proximal region of the axon. This region has an especially low threshold for triggering an action potential; thus, it serves as a decision-making part of the neuron.

1. Describe a synapse.
2. Explain the function of a neurotransmitter.
3. Define neuropeptide.
4. Distinguish between EPSP and IPSP.

The way the nervous system processes nerve impulses and acts upon them reflects, in part, the organization of neurons and their nerve fibers within the brain and spinal cord.

Neuronal Pools

The neurons within the central nervous system are organized into many groups with varying numbers of cells called *neuronal pools*. Each pool receives impulses from input (afferent) nerve fibers. These impulses are processed according to the special characteristics of the pool, and any resulting impulses are conducted away on output (efferent) fibers.

Each input fiber divides many times as it enters, and its branches spread over a certain region of the neuronal pool. The branches give off smaller branches and their terminals form hundreds of synapses with the dendrites and cell bodies of the neurons in the pool.

Facilitation

As a result of incoming impulses, a particular neuron of the pool may receive excitatory and inhibitory stimulation. As before, if the net effect of the stimulation is excitatory, threshold may be reached, and an outgoing impulse will be triggered. If the net effect is excitatory but subthreshold, an impulse will not be triggered; however, in this case the neuron is more excitable to incoming stimulation than before and it is said to be *facilitated*.

Convergence

Any single neuron in a neuronal pool may receive impulses from two or more incoming fibers. Furthermore, these fibers may originate from different parts of the nervous system, and they are said to *converge* when they lead to the same neuron.

Convergence makes it possible for a neuron to summate impulses from different sources. For example, if a neuron is facilitated by receiving subthreshold stimulation from one input fiber, its threshold may be reached if it receives additional stimulation from a second input fiber. Thus, if an output impulse is triggered from this neuron, it reflects a summation of impulses from two different sources. Such an output impulse may travel to a particular effector and cause a special response (figure 10.22).

Incoming impulses often represent information from various sensory receptors that have detected changes taking place. Convergence allows the nervous system to bring together a variety of information, process it, and respond to it in a special way.

Divergence

Impulses leaving a neuron of a neuronal pool often *diverge* by passing into several other output fibers. For example, an impulse from one neuron may stimulate two others; each of these, in turn, may stimulate several others, and so forth. This arrangement of diverging nerve fibers can cause an impulse to be *amplified*—that is, to be spread to increasing numbers of neurons within the pool (figure 10.22).

Figure 10.22

(a) Nerve fibers of neurons 1 and 2 converge to the cell body of neuron 3; *(b)* the nerve fiber of neuron 4 diverges to the cell bodies of neurons 5 and 6.

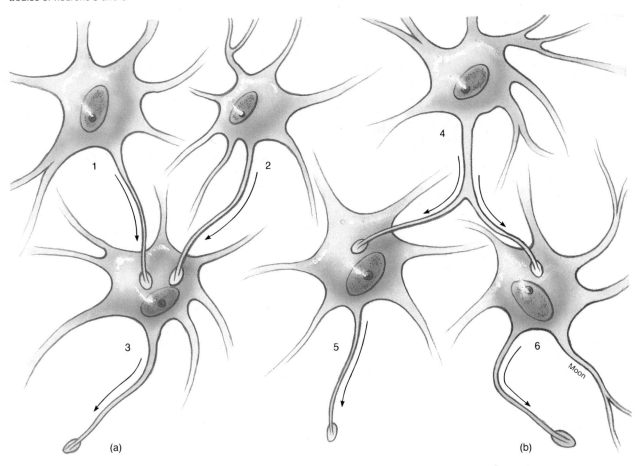

(a)

(b)

As a result of divergence, an impulse originating from a single neuron in the central nervous system may be amplified so that enough impulses reach the motor units within a skeletal muscle to cause a forceful contraction.

Similarly, an impulse originating from a sensory receptor may diverge and reach several different regions of the central nervous system, where the resulting impulses can be processed and acted upon.

---

1. What is a neuronal pool?
2. Define *facilitation*.
3. What is meant by convergence?
4. What is the relationship between divergence and amplification?

Classification of Neurons and Nerve Fibers

Various neurons differ in the structure, size, and shape of their cell bodies. They also vary in the length and size of their axons and dendrites, and in the number of synaptic knobs by which they communicate with other neurons.

Neurons also vary in function. Some carry impulses into the brain or spinal cord, others carry impulses out from the brain or spinal cord, and still others conduct impulses from neuron to neuron within the brain or spinal cord.

Classification of Neurons

On the basis of *structural* differences, neurons can be classified into three major groups, as shown in figure 10.23:

1. **Multipolar neurons.** Multipolar neurons have many processes arising from their cell bodies. Only one process of each neuron is an axon; the rest are dendrites. Most neurons whose cell bodies lie within the brain or spinal cord are of this type.
2. **Bipolar neurons.** The cell body of a bipolar neuron has only two processes, one arising from either end. Although these processes have similar structural characteristics, one serves as an axon and the other as a dendrite. Such neurons are found within specialized parts of the eyes, nose, and ears.

337

Figure 10.23
Structural types of neurons include the *(a)* multipolar neuron,
(b) bipolar neuron, and *(c)* unipolar neuron. Where can examples
of each of these be found in the body?

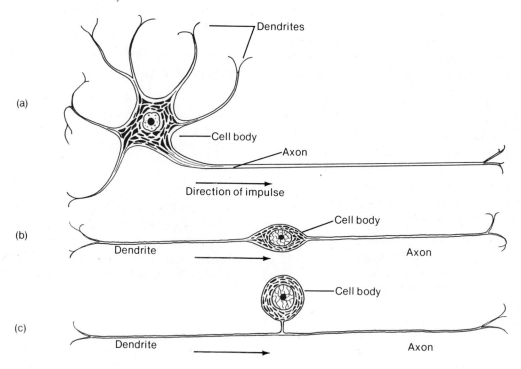

3. **Unipolar neurons.** Each of these neurons has a single process extending from its cell body. A short distance from the cell body, this process divides into two branches: one branch is connected to some peripheral body part and serves as a dendrite, and the other enters the brain or spinal cord and serves as an axon. Unipolar neurons occur in specialized masses of nerve tissue called *ganglia,* which are located outside the brain and spinal cord.

Neurons can also be classified on the basis of their *functional* differences into the following groups:

1. **Sensory neurons.** Sensory neurons (afferent neurons) are those that carry nerve impulses from peripheral body parts into the brain or spinal cord. These neurons have specialized *receptor ends* at the tips of their dendrites, or their dendrites are closely associated with *receptor cells* located in the skin or various sensory organs.

Changes that occur inside or outside the body are likely to stimulate receptor ends or receptor cells, triggering nerve impulses. The impulses travel along the sensory neuron fibers, which lead to the brain or spinal cord, and are processed in these parts by other neurons. Most sensory neurons are unipolar (figure 10.23).

2. **Interneurons.** Interneurons (association or internuncial neurons) lie within the brain or spinal cord. They are multipolar and form links between other neurons. Interneurons function to transmit impulses from one part of the brain or spinal cord to another. That is, they may direct incoming sensory impulses to appropriate parts for processing and interpreting. Other incoming impulses are transferred to motor neurons.

3. **Motor neurons.** Motor neurons (efferent neurons) are multipolar and carry nerve impulses out from the brain or spinal cord to **effectors**—parts of the body capable of responding, such as muscles or glands. When motor impulses reach muscles, for example, these effectors are stimulated to contract; when they reach glands, the glands are stimulated to release secretions.

Two specialized groups of motor neurons supply impulses to smooth and cardiac muscles. One type, the *accelerator neurons,* cause an increase in muscular activities. The second type, *inhibitory neurons,* cause such actions to decrease.

Types of Nerves and Nerve Fibers

While a nerve fiber is an extension of a neuron, a **nerve** is a cordlike bundle (or group of bundles) of nerve fibers held together by layers of connective tissues, as described in chapter 11.

Like nerve fibers, nerves that conduct impulses into the brain or spinal cord are called **sensory nerves,** while those that carry impulses out to muscles or glands are termed **motor nerves.** Most nerves, however, include both sensory and motor fibers, and they are called **mixed nerves.**

Nerves originating from the spinal cord that communicate with other body parts are called *spinal nerves,* while those originating from the brain that communicate with other body parts are *cranial nerves.* The nerve fibers within these structures can be subdivided further into four groups as follows:

1. **General somatic efferent fibers** carry motor impulses outward from the brain or spinal cord to skeletal muscles and cause them to contract.
2. **General visceral efferent fibers** carry motor impulses outward to various smooth muscles and glands associated with internal organs, causing certain muscles to contract or certain glands to release their secretions.
3. **General somatic afferent fibers** carry sensory impulses inward to the brain or spinal cord from receptors in the skin and skeletal muscles.
4. **General visceral afferent fibers** carry sensory impulses to the central nervous system from blood vessels and internal organs.

The term *general* in each of the above categories indicates that the fibers are associated with general structures such as the skin, skeletal muscles, glands, and visceral organs. Three other groups of fibers, found only in cranial nerves, are associated with more specialized structures:

1. **Special visceral efferent fibers** carry motor impulses outward from the brain to the muscles involved with chewing, swallowing, speaking, and facial expressions.
2. **Special visceral afferent fibers** carry sensory impulses inward to the brain from the olfactory and taste receptors.
3. **Special somatic afferent fibers** carry sensory impulses inward from the receptors of sight, hearing, and equilibrium.

1. Explain how neurons are classified according to structure and according to function.
2. How is a neuron related to a nerve?
3. What is a mixed nerve?
4. Explain how cranial and spinal nerve fibers are grouped.

Nerve Pathways

The routes followed by nerve impulses as they travel through the nervous system are called *nerve pathways.* The simplest of these pathways includes only a few neurons and is called a reflex arc.

Reflex Arcs

A **reflex arc** begins with the receptor at the end of a sensory nerve fiber. This fiber usually leads to several interneurons within the central nervous system, which serve as a processing or *reflex center.* Fibers from interneurons may connect with interneurons in other parts of the nervous system. They also communicate with motor neurons, whose fibers pass outward to effectors that respond when they are stimulated. (See figure 10.24.)

Such a reflex arc represents the behavioral unit of the nervous system. That is, it constitutes the structural and functional basis for the simplest acts—the reflexes.

Reflex Behavior

Reflexes are automatic, unconscious responses to changes occurring inside or outside the body. They are mechanisms that help to maintain homeostasis by controlling many of the body's involuntary processes such as heart rate, breathing rate, blood pressure, and digestive activities. Reflexes are also involved in the automatic actions of swallowing, sneezing, coughing, and vomiting.

The *knee-jerk reflex* (patellar reflex) is an example of a simple reflex that may employ only two neurons—a sensory neuron connected directly to a motor neuron. This reflex is initiated by striking the patellar ligament just below the patella. As a result, the quadriceps femoris group of muscles, which is attached to the patella by a tendon, is pulled slightly, and stretch receptors located within the muscles are stimulated. These receptors, in turn, trigger impulses that pass along the fiber of a sensory neuron into the spinal cord. Within the spinal cord, the sensory axon forms a synapse with a dendrite of a motor neuron. The impulse then continues along the axon of the motor neuron and

Figure 10.24

A reflex arc usually includes a receptor, sensory neuron, interneurons, motor neuron, and an effector.

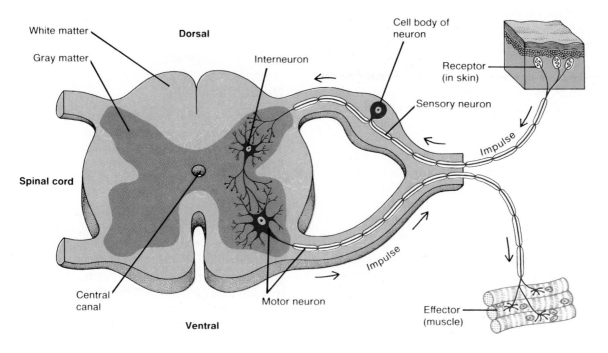

White matter

Gray matter

Dorsal

Interneuron

Cell body of neuron

Receptor (in skin)

Sensory neuron

Impulse

Spinal cord

Central canal

Motor neuron

Ventral

Impulse

Effector (muscle)

Figure 10.25

The knee-jerk reflex involves only two neurons—a sensory neuron and a motor neuron.

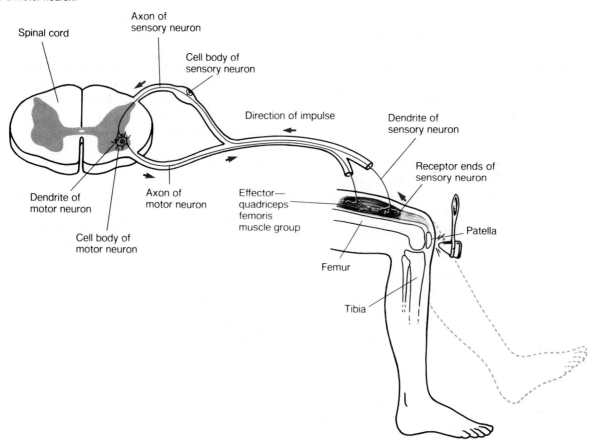

Spinal cord

Axon of sensory neuron

Cell body of sensory neuron

Direction of impulse

Dendrite of sensory neuron

Receptor ends of sensory neuron

Dendrite of motor neuron

Axon of motor neuron

Effector— quadriceps femoris muscle group

Patella

Cell body of motor neuron

Femur

Tibia

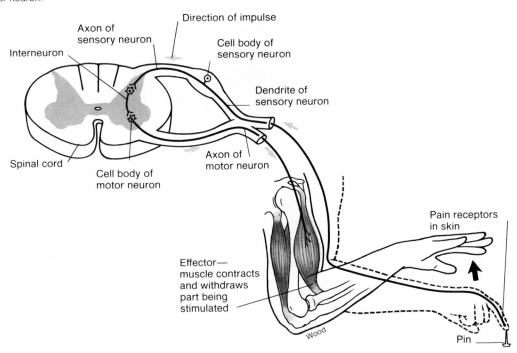

travels back to the quadriceps femoris. The muscles respond by contracting, and the reflex is completed as the lower leg extends (figure 10.25).

This reflex is helpful in maintaining an upright posture. For example, if a person is standing still and the knee begins to bend as a result of gravity pulling downward, the quadriceps femoris is stretched, the reflex is triggered, and the leg straightens again.

Another type of reflex, called a *withdrawal reflex* (figure 10.26), occurs when a person unexpectedly touches a finger to something hot or otherwise painful. As this happens, some of the skin receptors are activated, and sensory impulses travel to the spinal cord. There the impulses pass on to interneurons of the reflex center and are directed to motor neurons. The motor neurons transmit signals to flexor muscles in the arm, and they contract in response, causing the part to be pulled away from the painful stimulus.

At the same time, some of the incoming impulses stimulate interneurons that inhibit the action of the antagonistic extensor muscles. This inhibition (reciprocal innervation) allows flexor muscles to effectively withdraw the hand.

While flexor muscles of the stimulated side (ipsilateral side) are caused to contract, the flexor muscles of the other arm (contralateral side) are being inhibited. Furthermore, the extensor muscles on this other side are caused to contract. This phenomenon, which is called a *crossed extensor reflex,* is due to interneuron pathways within the reflex center of the spinal cord that allow sensory impulses arriving on one side of the cord to pass across to the other side and produce an opposite effect (figure 10.27).

Concurrent with the withdrawal reflex, other interneurons in the spinal cord carry sensory impulses to the brain, and the person becomes aware of the painful experience.

A withdrawal reflex is, of course, protective because it serves to prevent excessive tissue damage when a body part touches something that is potentially harmful. Chart 10.3 summarizes the parts of a reflex arc.

1. What is a nerve pathway?
2. Describe a reflex arc.
3. Define a reflex.
4. Describe the actions that occur during a withdrawal reflex.

Figure 10.27

When the flexor muscle on one side is stimulated to contract in a withdrawal reflex, the extensor on the opposite side also contracts.

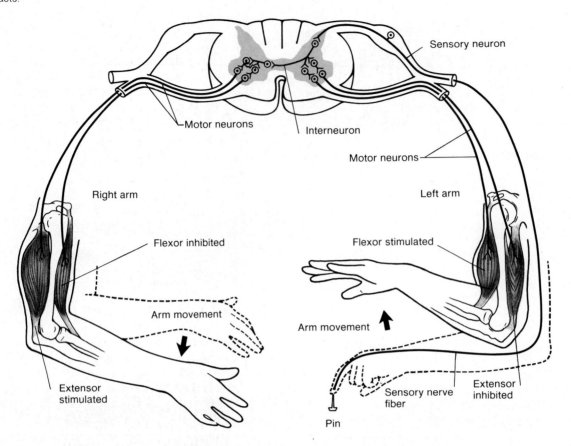

Sensory neuron

Motor neurons

Interneuron

Motor neurons

Right arm

Left arm

Flexor inhibited

Flexor stimulated

Arm movement

Arm movement

Extensor stimulated

Sensory nerve fiber

Extensor inhibited

Pin

Chart 10.3 Parts of a reflex arc		
Part	Description	Function
Receptor	The receptor end of a dendrite or a specialized receptor cell in a sensory organ	Sensitive to a specific type of internal or external change
Sensory neuron	Dendrite, cell body, and axon of a sensory neuron	Transmits nerve impulse from the receptor into the brain or spinal cord
Interneuron	Dendrite, cell body, and axon of a neuron within the brain or spinal cord	Serves as processing center; conducts nerve impulse from the sensory neuron to a motor neuron
Motor neuron	Dendrite, cell body, and axon of a motor neuron	Transmits nerve impulse from the brain or spinal cord out to an effector
Effector	A muscle or gland outside the nervous system	Responds to stimulation by the motor neuron and produces the reflex or behavioral action

Some Uses of Reflexes
A Clinical Application

Since normal reflexes depend on normal neuron functions, reflexes are commonly used to obtain information concerning the condition of the nervous system. An anesthesiologist, for instance, may try to initiate a reflex in a patient who is being anesthetized in order to determine how the anesthetic drug is affecting nerve functions. Also, in the case of injury to some part of the nervous system, various reflexes may be tested to discover the location and extent of the damage.

If any portion of a reflex arc is injured, the normal characteristics of that arc are likely to be altered. For example, a *plantar reflex* is normally initiated by stroking the sole of the foot, and the usual response includes a flexion of the foot and toes. However, in persons who have suffered damage to certain nerve pathways (corticospinal tract) there may be an abnormal response called the *Babinski reflex*. In this case the reflex response is dorsiflexion, in which the great toe extends upward and the smaller toes fan apart. If the injury is minor, the response may consist of plantar flexion with failure of the great toe to flex, or plantar flexion followed by dorsiflexion. The Babinski reflex is, however, present normally in infants up to the age of 12 months and is thought to reflect a degree of immaturity in their corticospinal tracts.

Other reflexes that may be tested during a neurological examination include the following:

1. **Biceps-jerk reflex.** This reflex can be elicited by bending a person's arm at the elbow. The examiner's finger is placed on the inside of the bent elbow over the tendon of the biceps muscle, and the finger is tapped. The biceps contracts in response, and the forearm is flexed at the elbow.

2. **Triceps-jerk reflex.** This reflex can be caused by flexing a person's arm at the elbow and tapping the short tendon of the triceps muscle close to its insertion near the tip of the elbow. The muscle contracts in response, and the forearm is extended slightly.

3. **Abdominal reflexes.** These reflexes occur when the examiner strokes the skin of the abdomen. For example, a dull pin may be drawn from the sides of the abdomen upward toward the midline and above the umbilicus. Normally, the abdominal muscles underlying the skin contract in response, and the umbilicus is moved toward the region that was stimulated.

4. **Ankle-jerk reflex.** This reflex is elicited by tapping the Achilles tendon just above its insertion on the calcaneus. The response is plantar flexion, produced by contraction of the gastrocnemius and soleus muscles.

5. **Cremasteric reflex.** This reflex is obtained in males by stroking the upper inside of the thigh. As a result, the testis on the same side is elevated by contracting muscles.

6. **Anal reflex.** This reflex is elicited by stroking the skin surrounding the anus. The anal sphincter muscles contract in response.

Chapter Summary

Introduction
1. Organs of the nervous system are divided into the central and peripheral nervous systems.
2. These parts provide sensory, integrative, and motor functions.

General Functions of the Nervous System
1. Sensory functions employ receptors that detect changes in internal and external body conditions.
2. Integrative functions bring sensory information together and make decisions that are acted upon by using motor functions.
3. Motor functions make use of effectors that respond when they are stimulated by motor impulses.

Nerve Tissue
1. Neuron structure
   a. A neuron includes a cell body, cell processes, and other organelles usually found in cells.
   b. Dendrites and the cell body provide receptive surfaces.
   c. A single axon arises from the cell body and may be enclosed in a myelin sheath and a neurolemma.
2. Neuroglial cells
   a. Neuroglial cells are accessory cells.
   b. They fill spaces, support neurons, hold nerve tissue together, produce myelin, and carry on phagocytosis.
   c. They include Schwann cells, astrocytes, oligodendrocytes, microglia, and ependyma.

Spinal Cord Injuries
A Clinical Application

Injuries to the spinal cord may be caused indirectly, as by a blow to the head or by a fall, or they may be due to forces applied directly to the cord. The consequences will depend on the amount of damage sustained by the cord.

Less severe injuries to the spinal cord, as may be sustained from a blow to the head, whiplash in an automobile accident, or rupture of an intervertebral disk, causing the cord to be compressed or distorted, are often accompanied by pain, weakness, and muscular atrophy in the regions supplied by the damaged nerve fibers.

Normal spinal reflexes depend on two-way communication between the spinal cord and the brain. When nerve pathways are injured, the cord's reflex activities in sites below the injury are usually depressed. At the same time, there is a lessening of sensations and muscular tone in the parts innervated by the affected fibers. This condition is called *spinal shock*, and it may last for days or weeks; normal reflex activity may return eventually. However, if nerve fibers are severed, some of the cord's functions are likely to be permanently lost.

Among the more common causes of direct injury to the spinal cord are gunshot wounds, stabbings, and fractures or dislocations of vertebrae during a vehicular accident (figure 11.8). Regardless of the cause, if nerve fibers in ascending tracts are cut, sensations arising from receptors below the level of the injury will be lost. If descending tracts are damaged, the result will be a loss of motor functions.

For example, if the right lateral corticospinal tract is severed in the neck near the first cervical vertebra, control of the voluntary muscles in the right arm and leg is lost, and the limbs become paralyzed (hemiplegia). Problems of this type, involving fibers of the descending tracts, are said

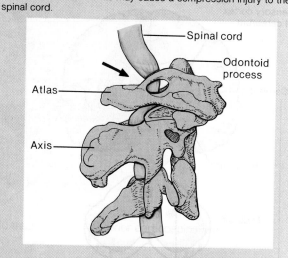

Figure 11.8
A dislocation of the atlas may cause a compression injury to the spinal cord.

Spinal cord

Odontoid process

Atlas

Axis

to produce the *upper motor neuron syndrome*. This condition is characterized by *spastic paralysis* in which muscle tone increases, and there is very little atrophy of the muscles involved. However, there usually is uncoordinated reflex activity (hyperreflexia) during which the flexor and extensor muscles of affected limbs alternately undergo spasms.

If, on the other hand, there is injury to motor neurons or their fibers in the horns of the spinal cord, the resulting condition is called *lower motor neuron syndrome*. It is characterized by *flaccid paralysis*, in which there is total loss of muscle tone, and the muscles undergo extreme atrophy. Reflex activity also is lost.

The corticospinal tracts also are called *pyramidal tracts* after the pyramid-shaped parts in the medulla oblongata through which they pass. Other descending tracts are called *extrapyramidal tracts,* and they include the reticulospinal and rubrospinal tracts.

2. **Reticulospinal** (rĕ-tik″u-lo-spi′nal) **tracts.** The *lateral* reticulospinal tracts are located in the lateral funiculi, while the *anterior* and *medial* reticulospinal tracts are in the anterior funiculi. Some fibers in the lateral tracts cross over, while others remain uncrossed. Those of the anterior and medial tracts remain uncrossed. Motor impulses transmitted on these tracts originate in the brain and function in the control of muscular tone and in the activity of the sweat glands.

3. **Rubrospinal** (roo″bro-spi′nal) **tracts.** The fibers of the rubrospinal tracts cross over in the brain and pass through the lateral funiculi. They carry motor impulses from the brain to skeletal muscles and are involved with muscular coordination and control of posture.

The disease called *poliomyelitis* is caused by viruses that sometimes affect motor neurons, including those in the anterior horns of the spinal cord. When this happens, the person may develop paralysis in the muscles innervated by the affected neurons, but may not suffer any sensory losses in these parts.

In addition to serving as a pathway for various nerve tracts, the spinal cord functions in many reflexes like the knee-jerk and withdrawal reflexes described in chapter 10. Such reflexes are called **spinal reflexes,** because the reflex arcs pass through the cord.

Figure 11.9

(a) The brain develops from a tubular part with three cavities. The cavities persist as the ventricles. (b) The wall of the tube gives rise to various regions of the brain.

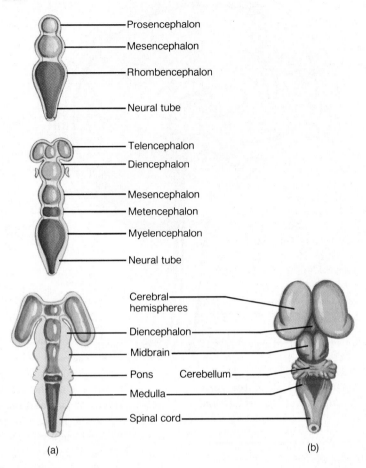

(a)   (b)

---

1. Describe the structure of the spinal cord.
2. What is meant by ascending and descending tracts?
3. What is the consequence of fibers crossing over?
4. Name the major tracts of the spinal cord and list the kinds of impulses each conducts.

The Brain

The **brain** is the largest and most complex part of the nervous system. It occupies the cranial cavity and is composed of about one hundred billion (10^{11}) neurons and innumerable nerve fibers, by which the neurons communicate with one another and with neurons in other parts of the system.

The brain contains nerve centers associated with sensory functions and is responsible for sensations and perceptions. It issues motor commands to skeletal muscles, and carries on higher mental functions, such as memory and reasoning. It also includes centers associated with the coordination of muscular movements, and contains centers and nerve pathways involved in the regulation of visceral activities.

Development of the Brain

The basic structure of the brain reflects the way in which it forms. It develops from a tubelike structure (neural tube) that gives rise to the central nervous system. The portion that becomes the brain has three major cavities or vesicles at one end. They are called the *forebrain, midbrain,* and *hindbrain* (figure 11.9). Later, the forebrain divides into anterior and posterior portions, and the hindbrain partially divides into two parts. The resulting five cavities persist in the mature brain as fluid-filled spaces called *ventricles* and the tubes that connect them. The tissue surrounding the spaces differentiates into various structural and functional regions of the brain.

The wall of the anterior portion of the forebrain gives rise to the *cerebrum* and *basal ganglia,* while the posterior portion forms a section of the brain stem called the *diencephalon.* The region produced by the midbrain continues to be called *midbrain* in the adult

Figure 11.10
The major portions of the brain include the cerebrum,
cerebellum, and brain stem.

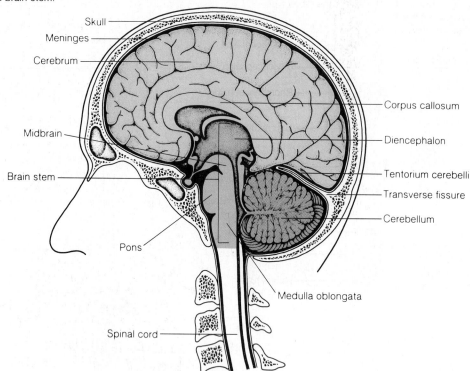

Skull

Meninges

Cerebrum

Midbrain

Brain stem

Pons

Spinal cord

Corpus callosum

Diencephalon

Tentorium cerebelli

Transverse fissure

Cerebellum

Medulla oblongata

structure, and the hindbrain gives rise to the *cerebellum, pons,* and *medulla oblongata* (figure 11.10 and
chart 11.2).

Occasionally during early development, the neural tube
fails to close completely. If this happens in the anterior
end, the brain fails to develop or only a rudimentary one
forms. This condition is called *anencephaly,* and it usually results in the death of the developing offspring.

When the neural tube defect occurs in the posterior portion, the vertebral column remains open and the
spinal cord is exposed to some degree. This condition
is called *spina bifida.* If the spinal cord is covered by the
developing skin, the problem may be corrected surgically, and the child may have only minor problems. However, if at birth the nerve tissue is protruding from the
site of the defect, the child is likely to suffer irreversible
brain damage and some degree of paralysis, and will require extensive medical and surgical treatment.

Structure of the Cerebrum

The cerebrum, which develops from the anterior portion of the forebrain, is the largest part of the mature
brain. It consists of two large masses called **cerebral
hemispheres,** which are essentially mirror images of
each other (figure 11.11). These hemispheres are connected by a deep bridge of nerve fibers called the **corpus
callosum** and are separated by a layer of dura mater
called the *falx cerebri.*

The surface of the cerebrum is marked by numerous ridges or **convolutions** (gyri), which are separated by grooves. A shallow groove is called a **sulcus,**
and a very deep one is a **fissure.** Although the arrangement of these elevations and depressions is complex,
they form fairly distinct patterns in all brains. For example, a *longitudinal fissure* separates the right and
left cerebral hemispheres, a *transverse fissure* separates the cerebrum from the cerebellum, and various
sulci divide each hemisphere into lobes (figures 11.10
and 11.11).

The lobes of the cerebral hemispheres (figure
11.11) are named after the skull bones that they un-
derlie. They include the following:

1. **Frontal lobe.** The frontal lobe forms the
   anterior portion of each cerebral hemisphere. It
   is delimited posteriorly by a *central sulcus*
   (fissure of Rolando) that passes out from the
   longitudinal fissure at a right angle, and
   inferiorly by a *lateral sulcus* (fissure of
   Sylvius) that passes out from the under surface
   of the brain along its sides.
2. **Parietal lobe.** The parietal lobe is posterior to
   the frontal lobe and is separated from it by the
   central sulcus.
3. **Temporal lobe.** The temporal lobe lies below the
   frontal lobe and is separated from it by the
   lateral sulcus.

Chart 11.2 Structural development of the brain

Embryonic vesicle	Ventricle produced	Regions of the brain produced
Forebrain (Prosencephalon)		
Anterior portion (Telencephalon)	Lateral ventricles	Cerebrum Basal ganglia
Posterior portion (Diencephalon)	Third ventricle	Thalamus Hypothalamus Optic nerves Posterior pituitary gland Pineal gland
Midbrain (Mesencephalon)	Cerebral aqueduct (Aqueduct of Sylvius)	Midbrain
Hindbrain (Rhombencephalon)		
Anterior portion (Metencephalon)	Fourth ventricle	Cerebellum Pons
Posterior portion (Myelencephalon)	Fourth ventricle	Medulla oblongata

Figure 11.11
Lobes of the cerebral hemispheres. *(a)* Lateral view of the right hemisphere; *(b)* hemispheres viewed from above.

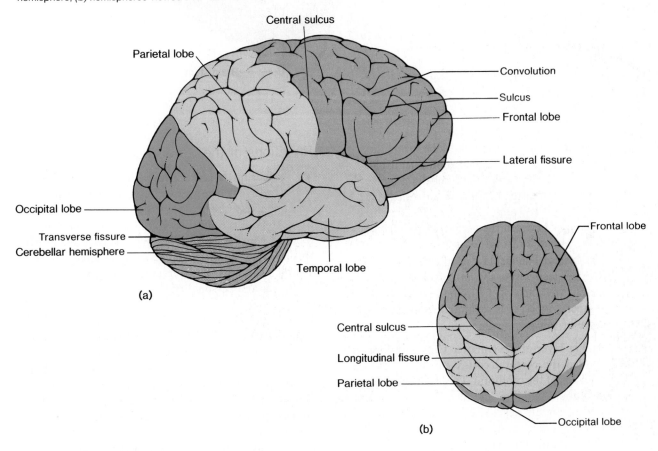

(a)

(b)

Figure 11.12

Some motor, sensory, and association areas of the cerebral cortex.

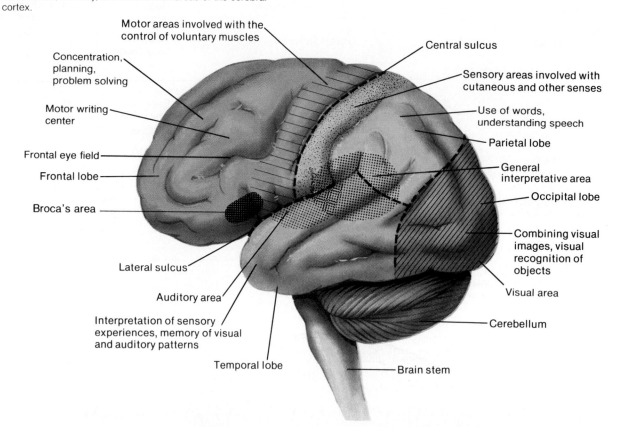

4. **Occipital lobe.** The occipital lobe forms the posterior portion of each cerebral hemisphere and is separated from the cerebellum by a shelflike extension of dura mater called the *tentorium cerebelli.* There is no distinct boundary between the occipital lobe and the parietal and temporal lobes.

5. **The insula.** The insula (island of Reil) is located deep within the lateral sulcus and is covered by parts of the frontal, parietal, and temporal lobes. It is separated from them by a circular sulcus.

A thin layer of gray matter (2 to 5 millimeters thick) called the **cerebral cortex** constitutes the outermost portion of the cerebrum. This layer, which covers the convolutions and dips into the sulci and fissures, is estimated to contain nearly 75% of all the neuron cell bodies in the nervous system.

Just beneath the cerebral cortex are masses of white matter, making up the bulk of the cerebrum. These masses contain bundles of myelinated nerve fibers that connect neuron cell bodies of the cortex with other parts of the nervous system. Some of these fibers pass from one cerebral hemisphere to the other by way of the corpus callosum, while others carry sensory or motor impulses from portions of the cortex to nerve centers in the brain or spinal cord.

Functions of the Cerebrum

The cerebrum is concerned with higher brain functions, in that it contains centers for interpreting sensory impulses arriving from various sense organs as well as centers for initiating voluntary muscular movements. It stores the information of memory and utilizes this information in the processes associated with reasoning. It also functions in determining a person's intelligence and personality.

Functional Regions of the Cortex

The regions of the cerebral cortex that perform specific functions have been located by using a variety of techniques. For example, persons who have suffered brain injuries or who have had portions of their brains removed surgically have been studied. Impaired abilities of these persons provide clues as to the functions of the specific parts.

In other studies, areas of cortices have been exposed surgically and stimulated mechanically or electrically. Stimulation of particular regions of the cortex is followed by responses in certain muscles or by specific sensations.

As a result of such investigations, it is possible to prepare maps that indicate the functional regions of the cerebral cortex. Although there is considerable overlap in these areas, the cortex can be divided into sections known as *motor, sensory,* and *association areas.*

Figure 11.13

(a) Motor areas involved with the control of voluntary muscles; (b) sensory areas involved with cutaneous and certain other senses.

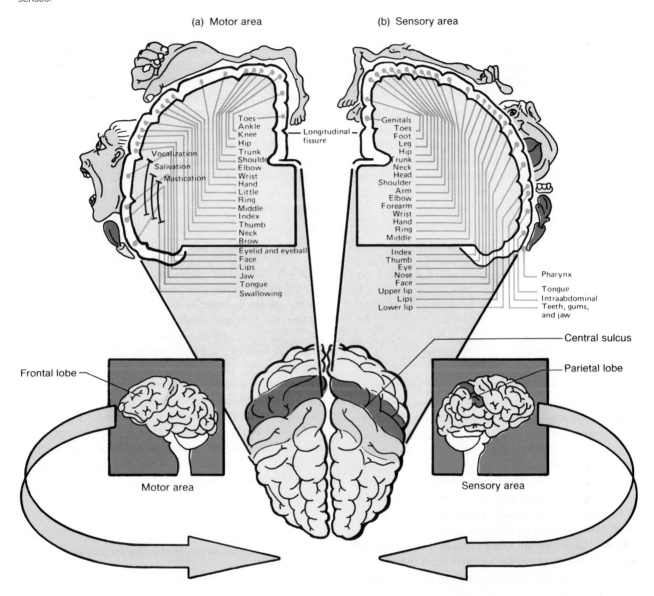

(a) Motor area

(b) Sensory area

Motor area labels (top to bottom): Toes, Ankle, Knee, Hip, Trunk, Shoulder, Elbow, Wrist, Hand, Little, Ring, Middle, Index, Thumb, Neck, Brow, Eyelid and eyeball, Face, Lips, Jaw, Tongue, Swallowing

Vocalization, Salivation, Mastication

Sensory area labels (top to bottom): Genitals, Toes, Foot, Leg, Hip, Trunk, Neck, Head, Shoulder, Arm, Elbow, Forearm, Wrist, Hand, Ring, Middle, Index, Thumb, Eye, Nose, Face, Upper lip, Lips, Lower lip

Pharynx, Tongue, Intraabdominal, Teeth, gums, and jaw

Longitudinal fissure

Central sulcus

Frontal lobe

Parietal lobe

Motor area

Sensory area

Motor Areas

The *primary motor areas* of the cerebral cortex lie in the frontal lobes just anterior to the central sulcus (precentral gyrus), and in the anterior wall of this sulcus. The nerve tissue in these regions contains numerous, large *pyramidal cells* named because of their pyramid-shaped cell bodies (figure 11.12).

As described earlier, impulses from these pyramidal cells travel downward through the brain stem and into the spinal cord on the *corticospinal tracts*. Most of the nerve fibers in these tracts cross over from one side of the brain to the other within the brain stem and descend as lateral corticospinal tracts (figure 11.7).

Within the spinal cord, the corticospinal fibers synapse with motor neurons in the gray matter of the anterior horns. Axons of the motor neurons lead outward through peripheral nerves to various voluntary muscles. Impulses transmitted on these pathways in special patterns and frequencies are responsible for fine movements in skeletal muscles. More specifically, as figure 11.13 shows, cells in the upper portions of the motor areas send impulses to muscles in the legs and thighs; those in the middle portions control muscles in the shoulders and arms; and those in lower portions activate muscles of the head, face, and tongue.

The reticulospinal and rubrospinal tracts function to coordinate and control motor functions involved with the maintenance of balance and posture. Many of these fibers pass into the basal ganglia on the way to the spinal cord. Some of the impulses conducted on these pathways normally inhibit muscular actions.

In addition to the primary motor areas, certain other regions of the frontal lobe are involved with higher motor functions. For example, a region called *Broca's area* is located just anterior to the primary motor cortex and above the lateral sulcus. This area usually occurs in the left cerebral hemisphere. It coordinates the complex muscular actions of the mouth, tongue, and larynx, which make speech possible. A person with an injury to this area may be able to understand spoken words, but cannot speak.

Above Broca's area there is a region called the *frontal eye field*. The motor cortex in this area controls the voluntary movements of the eyes and eyelids. Nearby is the cortex responsible for movements of the head that are made to direct the eyes. Another region, just in front of the primary motor area, controls the muscular movements of the hands and fingers that make skills, such as writing, possible (figure 11.12).

An injury to some part of the motor system may result in a loss of the ability to produce purposeful muscular movements. Such a condition may impair use of the hands, arms, legs, head, or eyes, and is called *apraxia*. Apraxia involving the speech muscles, accompanied by loss of the ability to speak, is called *aphasia*.

Sensory Areas

Sensory areas, which occur in several lobes of the cerebrum, function in interpreting impulses that arrive from various sensory receptors. These interpretations give rise to feelings or sensations. For example, the sensations of temperature, touch, pressure, and pain from all parts of the skin arise in the anterior portions of the parietal lobes along the central sulcus (postcentral gyrus) and in the posterior wall of this sulcus (figure 11.12). The posterior parts of the occipital lobes are concerned with vision, while the posterior, dorsal portions of the temporal lobes contain the centers for hearing. The sensory areas for taste seem to be located near the bases of the central sulci along the lateral sulci, and the sense of smell arises from centers deep within the cerebrum.

Like motor fibers, the sensory fibers cross over so that centers in the right cerebral hemisphere interpret impulses originating from the left side of the body and vice versa (figure 11.6). The sensory areas concerned with vision, however, receive impulses from both eyes and those for hearing receive impulses from both ears.

Association Areas

Association areas occupy the anterior portions of the frontal lobes and are widespread in the lateral portions of the parietal, temporal, and occipital lobes. They function to analyze and interpret sensory experiences and are involved with memory, reasoning, verbalizing, judgment, and emotional feelings (figure 11.12).

The association areas of the frontal lobes are concerned with a number of higher intellectual processes, including those necessary for concentration, planning, and complex problem solving. The anterior and inferior portions of these lobes (prefrontal areas) function in the control of emotional behavior and produce awareness of the possible consequences of behavior.

The association areas of the parietal lobes help make meaning of sensory information and aid in understanding speech and choosing words to express thoughts and feelings. Awareness of body parts stems from the posterior regions of these lobes.

The association areas of the temporal lobes and the regions at the posterior ends of the lateral fissures are concerned with the interpretation of complex sensory experiences, such as those needed to understand speech and read printed words. These regions also are involved with the memory of visual scenes, music, and other complex sensory patterns.

The association areas of the occipital lobes that are adjacent to the visual centers are important in analyzing visual patterns and combining visual images with other sensory experiences—necessary, for instance, when one recognizes another person or an object.

Of particular importance is the region where the parietal, temporal, and occipital association areas come together, near the posterior end of the lateral sulcus. This region is called the *general interpretative area*, and it plays the primary role in complex thought processing. Its function makes it possible for a person to recognize words and arrange them to express a thought, and to read and understand ideas presented in writing. Chart 11.3 summarizes some of the functions of the cerebral lobes.

---

1. How does the brain develop?
2. Describe the cerebrum.
3. List the general functions of the cerebrum.
4. Where are the primary motor and sensory regions located?
5. Explain the functions of association areas.

Hemisphere Dominance

Both cerebral hemispheres participate in basic functions, such as receiving and analyzing sensory impulses, controlling skeletal muscles on opposite sides of the body, and storing memory. However, in most persons, one acts as a **dominant hemisphere** for certain other functions.

In over 90% of the population, the left hemisphere is dominant for the language-related activities of speech, writing, and reading. It is also dominant for complex intellectual functions requiring verbal, analytical, and computational skills. In other persons, the right hemisphere is dominant, and in some, the hemispheres are equally dominant.

Chart 11.3 Functions of the cerebral lobes

Lobe	Functions	Lobe	Functions
Frontal lobes	Motor areas control movements of voluntary skeletal muscles	Temporal lobes	Sensory areas are responsible for hearing
	Association areas carry on higher intellectual processes such as those required for concentration, planning, complex problem-solving, and judging the consequences of behavior		Association areas are used in the interpretation of sensory experiences and in the memory of visual scenes, music, and other complex sensory patterns
Parietal lobes	Sensory areas are responsible for the sensations of temperature, touch, pressure, and pain from the skin	Occipital lobes	Sensory areas are responsible for vision
	Association areas function in the understanding of speech and in using words to express thoughts and feelings		Association areas function in combining visual images with other sensory experiences

Various tests indicate that the left hemisphere is dominant in 90% of the right-handed adults and 64% of the left-handed ones. The right hemisphere is dominant in 10% of the right-handed adults and 20% of the left-handed ones. The hemispheres are equally dominant in the remaining 16% of left-handed persons.

As a consequence of hemisphere dominance, Broca's area on one side almost completely controls the motor activities associated with speech. For this reason, over 90% of the patients with language impairment involving the cerebrum have disorders in the left hemisphere.

In addition to carrying on basic functions, the nondominant hemisphere seems to specialize in nonverbal functions such as those involving motor tasks that require orientation of the body in the surrounding space, understanding and interpreting musical patterns and nonverbal visual experiences. It is also concerned with emotional and intuitive thought processes. For example, although the region in the *nondominant hemisphere* that corresponds to Broca's area is incapable of controlling speech, it seems to influence the emotional aspects of the language expressed in speech.

Nerve fibers of the *corpus callosum,* which connect the cerebral hemispheres, make it possible for the dominant side to control the motor cortex of the nondominant hemisphere. These fibers also allow sensory information reaching the nondominant hemisphere to be transferred to the general interpretative area of the dominant one, where it can be used in decision making.

Storage of Memory

Memory refers to the storage of information and the ability to recall this information when it is needed. Memory is required for all higher brain functions such as learning, reasoning, and adapting behavior to immediate needs.

Although most memory occurs in the cerebral cortex, certain other parts of the nervous system can also store information and are involved with the processes of sorting and retrieving it.

The establishment of memory occurs in stages. In the first stage, sensory information reaches the brain and is held by ongoing neuronal activity for a few seconds or so. This type of temporary storage is called **short-term memory,** and although it may not last for long, it is useful for remembering a few bits of information such as a new telephone number long enough to use it.

Short-term memory seems to be processed mainly in the anterior portion of the frontal lobe and in a part of the brain stem (thalamus). It is easily disrupted and can be displaced and lost as new sensory information arrives. However, it also can be transferred into the next phase of memory storage, which is more permanent.

Information in short-term memory may be recalled quickly, and if it is consciously reviewed, it may be moved into a form called **recent memory.** This phase involves a portion of the temporal lobe (hippocampus) and produces memory that may last for minutes or days. However, recent memory usually is stored relatively weakly and is subject to being lost as a result of disuse.

The most stable form of memory is called **long-term memory.** It results from repeated recall of new information or repeated sensory experiences. Processing this form takes place over large areas of the cerebral cortex, and the information is stored diffusely in both hemispheres. However, bits of information seem to be sorted into groups, so that each new bit is stored in association with other information of the same kind.

The brain has a vast capacity to store long-term memory. The stored information, such as one's name, one's date of birth, letters of the alphabet, and words commonly used in speech, usually can be recalled quickly.

Figure 11.14

A frontal section of the left cerebral hemisphere reveals the basal ganglia.

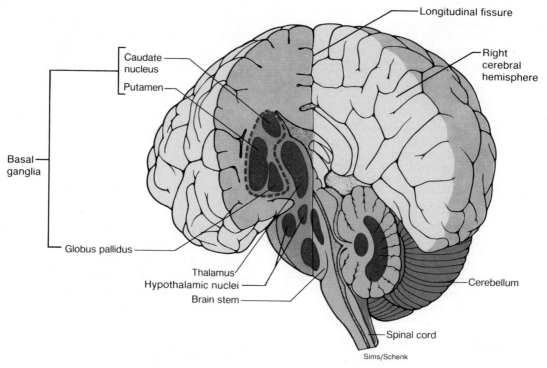

Sims/Schenk

The way in which long-term memory is processed is poorly understood, but some investigators believe it involves neuronal activity that is intense and repetitive enough to cause stable changes in nerve pathways. For example, structural or chemical changes might occur that produce persistent facilitation of certain synapses. If all of the synapses of a particular pathway remain facilitated, an incoming impulse at a later time might excite this pathway, causing impulses to flow through it, thus recalling a bit of memory.

Basal Ganglia

The **basal ganglia** (basal nuclei) are masses of gray matter located deep within the cerebral hemispheres. They include the *caudate nucleus, putamen,* and *globus pallidus,* which develop from the anterior portion of the forebrain. Although the precise function of these parts is poorly understood, the neuron cell bodies they contain are known to serve as relay stations for motor impulses originating in the cerebral cortex and passing into the brain stem and spinal cord. It also is known that most of the inhibitory neurotransmitter dopamine is produced in the basal ganglia. Impulses from these parts normally inhibit motor functions and thus aid in the control of various muscular activities (figure 11.14).

Since the basal ganglia function to inhibit muscular activity, disorders in this portion of the brain may be accompanied by a reduction in mobility (hypokinesia) due to excessive discharge of impulses from the ganglia; or involuntary movements (hyperkinesia) due to lack of inhibiting impulses from them.

Parkinson's disease, for example, is thought to be caused by degeneration of the neurons that synthesize dopamine in the basal ganglia. This condition is characterized by slowness of movements, difficulty in initiating voluntary muscular actions, and by tremors. It often is treated with a drug called L-dopa, which can be converted into dopamine by cellular enzymes. Although this treatment may control some of the symptoms of Parkinson's disease for a time, it usually does not alter the course of the disease.

1. What is meant by hemisphere dominance?
2. What are the functions of the nondominant hemisphere?
3. Distinguish between short-term, recent, and long-term memory.
4. What is the function of the basal ganglia?

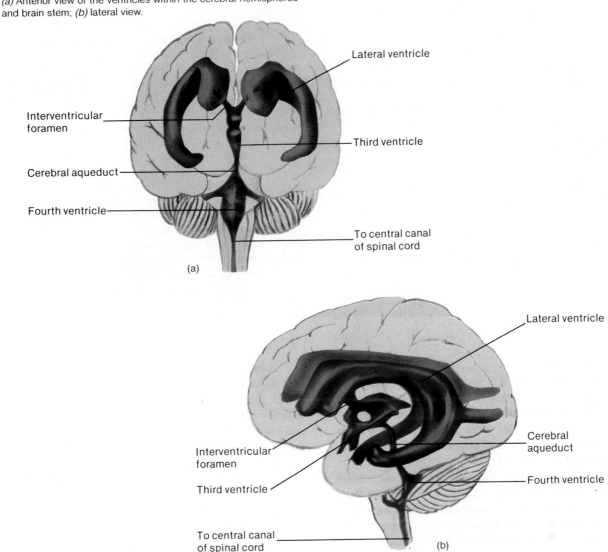

The Ventricles and Cerebrospinal Fluid

Within the cerebral hemispheres and brain stem is a series of interconnected cavities called **ventricles,** shown in figure 11.15. These spaces are continuous with the central canal of the spinal cord and, like it, they are filled with cerebrospinal fluid.

The largest of the ventricles are the *lateral ventricles* (first and second ventricles), which extend into the cerebral hemispheres and occupy portions of the frontal, temporal, and occipital lobes.

A narrow space that constitutes the *third ventricle* is located in the midline of the brain, beneath the corpus callosum. This ventricle communicates with the lateral ventricles through openings *(interventricular foramina)* in its anterior end.

The *fourth ventricle* is located in the brain stem just in front of the cerebellum. It is connected to the third ventricle by a narrow canal, the *cerebral aqueduct* (aqueduct of Sylvius), which passes lengthwise through the brain stem. This ventricle is continuous with the central canal of the spinal cord and has openings in its roof that lead into the subarachnoid space of the meninges.

Cerebral Injuries
A Clinical Application

The effects of injuries to the cerebral cortex depend on which areas are damaged and to what extent. When particular portions of the cortex are injured, the special functions of these portions are likely to be lost or at least depressed.

It is often possible to deduce the location and extent of a brain injury by determining what abilities the patient is missing. For example, if the motor areas of one frontal lobe have been damaged, the person is likely to be partially or completely paralyzed on the opposite side of the body. Similarly, if the visual cortex of one occipital lobe is injured, the person may suffer partial blindness in both eyes; if both visual cortices are damaged, the blindness may be total.

A person with damage to the association areas of the frontal lobes may have difficulty in concentrating on complex mental tasks. Such an individual usually appears disorganized and is easily distracted.

If the general interpretative area of the dominant hemisphere is injured, the person may be unable to interpret sounds as words or understand ideas presented in writing. However, the dominance of one hemisphere usually does not become established until after five or six years of age. Consequently if the general interpretative area is destroyed in a child, the corresponding region on the other side of the brain may be able to take over the functions, and the child's language abilities may develop normally. If such an injury occurs in an adult, the nondominant hemisphere may develop only limited interpretative functions, and the person is likely to have a severe intellectual disability.

Brain damage also may interfere with memory. For example, if the portion of the temporal lobe that functions in the processing of recent memory is lost, the person will be unable to acquire new recent memory or subsequent long term memory. Generally, the amount of stored memory that is lost as a result of injury is roughly proportional to the amount of cerebral cortex damaged.

Cerebrospinal Fluid

The cerebrospinal fluid is secreted by tiny cauliflowerlike masses of specialized capillaries from the pia mater called **choroid plexuses.** These structures project outward from the inner walls of the ventricles. They are covered by epithelial-like *ependymal cells* of the neuroglia (figure 11.16).

Although choroid plexuses occur in the medial walls of the lateral ventricles and the roofs of the third and fourth ventricles, most of the cerebrospinal fluid seems to arise in the lateral ventricles. From there it circulates slowly into the third and fourth ventricles and into the central canal of the spinal cord. It also enters the subarachnoid space of the meninges by passing through the wall of the fourth ventricle near the cerebellum.

Nearly 800 milliliters of cerebrospinal fluid are secreted daily, but only about 140 milliliters are present normally. Thus, almost all the cerebrospinal fluid that is formed completes its circuit by being reabsorbed into the blood. This reabsorption occurs gradually through fingerlike structures called *arachnoid granulations* that project from the subarachnoid space into the blood-filled dural sinuses (figure 11.16).

Cerebrospinal fluid is a clear, slightly viscid liquid that differs slightly in composition from the fluid that leaves the capillaries in other parts of the body. Specifically, it contains a greater concentration of sodium and lesser concentrations of glucose and potassium than do other extracellular fluids. Its primary function, however, seems to be protective.

Since it occupies the subarachnoid space of the meninges, cerebrospinal fluid completely surrounds the brain and spinal cord. In effect, these organs float in the fluid that supports and protects them by absorbing shocks and other forces that might otherwise jar and damage their delicate tissues. Cerebrospinal fluid also aids in maintaining a stable ionic concentration in the central nervous system and provides a pathway to the blood for waste substances.

---

1. Where are the ventricles of the brain located?
2. How is cerebrospinal fluid formed?
3. Describe the pattern of cerebrospinal fluid circulation.

Figure 11.16
Cerebrospinal fluid is secreted by choroid plexuses in the walls of the ventricles. The fluid circulates through the ventricles and central canal, enters the subarachnoid space, and is reabsorbed into the blood of the dural sinuses through arachnoid granulations.

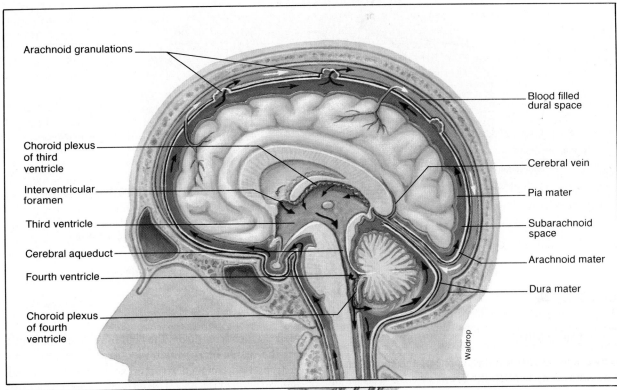

Arachnoid granulations

Choroid plexus of third ventricle

Interventricular foramen

Third ventricle

Cerebral aqueduct

Fourth ventricle

Choroid plexus of fourth ventricle

Blood filled dural space

Cerebral vein

Pia mater

Subarachnoid space

Arachnoid mater

Dura mater

Waldrop

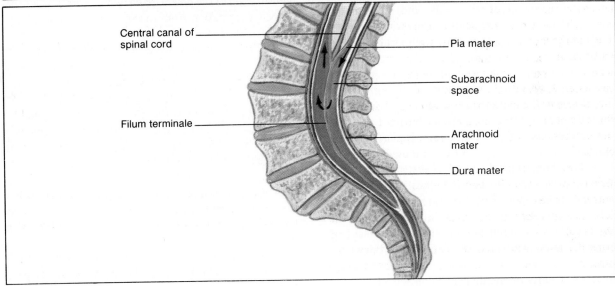

Central canal of spinal cord

Filum terminale

Pia mater

Subarachnoid space

Arachnoid mater

Dura mater

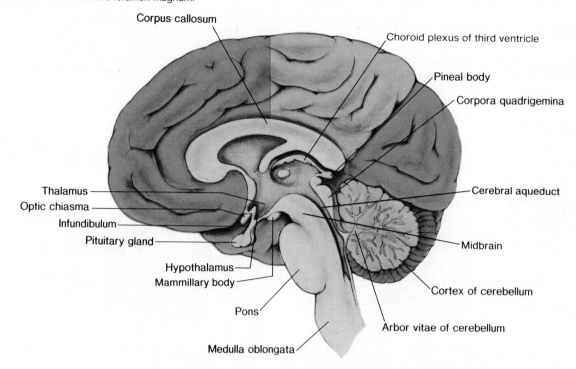

Corpus callosum

Choroid plexus of third ventricle

Pineal body

Corpora quadrigemina

Thalamus

Optic chiasma

Infundibulum

Pituitary gland

Cerebral aqueduct

Midbrain

Hypothalamus

Mammillary body

Cortex of cerebellum

Pons

Arbor vitae of cerebellum

Medulla oblongata

The Brain Stem

The region of the brain that connects the cerebrum to the spinal cord is called the *brain stem.* It consists of the diencephalon, midbrain, pons, and medulla oblongata. These parts include numerous tracts of nerve fibers and several masses of gray matter called *nuclei* (figure 11.17).

The Diencephalon

The *diencephalon* develops from the posterior portion of the forebrain and is located between the cerebral hemispheres and above the midbrain. It generally surrounds the third ventricle and is composed largely of gray matter organized into nuclei. Among these, a dense nucleus, called the *thalamus,* bulges into the third ventricle from each side. Another region of the diencephalon that includes many nuclei is called the *hypothalamus.* It lies below the thalamic nuclei and forms the lower walls and the floor of the third ventricle.

Other parts of the diencephalon include (a) the **optic tracts** and **optic chiasma,** which is formed by the optic nerve fibers crossing over; (b) the **infundibulum,** a conical process behind the optic chiasma to which the pituitary gland is attached; (c) the **posterior pituitary**

gland, which hangs from the floor of the hypothalamus; (d) the **mammillary** (mam'ĭ-ler''e) **bodies,** which appear as two rounded structures behind the infundibulum; and (e) the **pineal gland,** which forms as a cone-shaped evagination from the roof of the diencephalon (see chapter 13).

The **thalamus** serves as a central relay station for sensory impulses ascending from other parts of the nervous system to the cerebral cortex. It receives all sensory impulses (except those associated with the sense of smell) and channels them to appropriate regions of the cortex for interpretation. In addition, all regions of the cerebral cortex can communicate with the thalamus by means of descending fibers, so the two parts function closely together.

Although the cerebral cortex pinpoints the origin of sensory stimulation, the thalamus seems to produce a general awareness of certain sensations such as pain, touch, and temperature.

The **hypothalamus** is interconnected by nerve fibers to the cerebral cortex, thalamus, and other parts of the brain stem so that it can receive impulses from them and send impulses to them. The hypothalamus plays key roles in maintaining homeostasis by regulating a variety of visceral activities and by serving as a link between the nervous and endocrine systems.

Cerebrospinal Fluid Pressure
A Clinical Application

Because cerebrospinal fluid is secreted and reabsorbed continuously, the fluid pressure in the ventricles remains relatively constant. Sometimes, however, an infection, a tumor, or a blood clot interferes with the circulation of this fluid, and the pressure within the ventricles increases. As the intracranial pressure (ICP) increases, there is danger that cerebral blood vessels will be collapsed, retarding blood flow, and that the brain tissues may be injured by being forced against the skull.

The pressure of cerebrospinal fluid can be determined by performing a *lumbar puncture*. This procedure, shown in figure 11.18, involves inserting a fine needle into the subarachnoid space between the third and fourth or the fourth and fifth lumbar vertebrae—a level below the end of the spinal cord. The pressure of the fluid, which is usually about 10 millimeters of mercury, is measured by means of a device called a *manometer*. At the same time, samples of cerebrospinal fluid may be withdrawn and tested for the presence of abnormal constituents. The presence of red blood cells, for example, is abnormal and may indicate the existence of a hemorrhage somewhere in the central nervous system.

If the pressure is found to be excessive, a temporary drain may be employed to decrease the fluid volume and thus lower the pressure. Such a drain may be inserted into the subarachnoid space between the fourth and fifth lumbar vertebrae. In an infant whose cranial sutures have not yet united, increasing intracranial pressure also may cause an enlargement of the cranium called *hydrocephalus* or "water on the brain." Hydrocephalus is often treated by inserting a shunt that drains fluid away from the cranial cavity and into some other body region where it can be reabsorbed into the blood or excreted.

Figure 11.18
(a) A lumbar puncture is performed by inserting a fine needle between the third and fourth lumbar vertebrae and (b) withdrawing a sample of cerebrospinal fluid from the subarachnoid space.

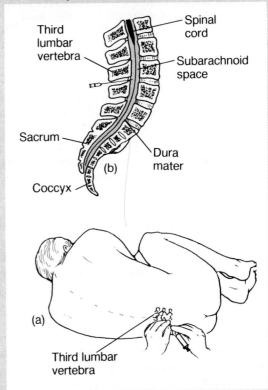

Among the many important functions of the hypothalamus are the following:

1. Regulation of heart rate and arterial blood pressure.
2. Regulation of body temperature.
3. Regulation of water and electrolyte balance.
4. Control of hunger and regulation of body weight.
5. Control of movements and glandular secretions of the stomach and intestines.
6. Production of neurosecretory substances that stimulate the pituitary gland to release various hormones. These hormones help regulate growth, control various glands, and influence reproductive physiology.
7. Regulation of sleep and wakefulness.

In an infant, the hypothalamus is less effective in regulating body temperature than it is in older children and adults. Consequently, changes in environmental temperatures may be reflected in corresponding changes in an infant's body temperature. In a cold environment, for example, an infant's body temperature may drop to a level that threatens its life.

Structures in the general region of the diencephalon also play important roles in the control of emotional responses. For example, portions of the cerebral cortex in the medial parts of the frontal and temporal lobes are interconnected with the hypothalamus, thalamus, basal ganglia, and other deep nuclei. Together these structures comprise a complex called the **limbic system.**

The *limbic system* is involved in emotional experience and expression, and can modify the way a person acts. It functions to produce such feelings as fear, anger, pleasure, and sorrow. More specifically, the limbic system seems to recognize upsets in a person's physical or psychological condition that might threaten survival. By causing pleasant or unpleasant feelings about experiences, the limbic system guides the person into behavior that is likely to increase the chance of survival.

The Midbrain

The **midbrain** (mesencephalon) is a short section of the brain stem located between the diencephalon and the pons. It contains bundles of myelinated nerve fibers that join lower parts of the brain stem and spinal cord with higher parts of the brain. The midbrain includes several masses of gray matter that serve as reflex centers, and the *cerebral aqueduct* that connects the third and fourth ventricles (figure 11.17).

Two prominent bundles of nerve fibers on the underside of the midbrain comprise the *cerebral peduncles*. These fibers include the corticospinal tracts and are the main motor pathways between the cerebrum and lower parts of the nervous system. Beneath the cerebral peduncles are some large bundles of sensory fibers that carry impulses upward to the thalamus.

Two pairs of rounded knobs on the upper surface of the midbrain mark the location of four nuclei known collectively as *corpora quadrigemina*. These masses contain the centers for certain visual reflexes, such as those responsible for moving the eyes to view something as the head is turned. They also contain the auditory reflex centers that operate when it is necessary to move the head so that sounds can be heard more distinctly.

Near the center of the midbrain is a mass of gray matter called the *red nucleus*. This nucleus communicates with the cerebellum and with centers of the spinal cord, and it functions in reflexes concerned with the maintenance of posture.

The Pons

The **pons** appears as a rounded bulge on the underside of the brain stem where it separates the midbrain from the medulla oblongata. The dorsal portion of the pons consists largely of longitudinal nerve fibers, which relay impulses to and from the medulla oblongata and the cerebrum. Its ventral portion contains large bundles of transverse nerve fibers, which transmit impulses from the cerebrum to centers within the cerebellum.

Several nuclei of the pons relay sensory impulses from peripheral nerves to higher brain centers. Other nuclei function with centers of the medulla oblongata in regulating the rate and depth of breathing (figure 11.17).

Medulla Oblongata

The **medulla oblongata** is an enlarged continuation of the spinal cord extending from the level of the foramen magnum to the pons. Its dorsal surface is flattened to form the floor of the fourth ventricle, and its ventral surface is marked by the corticospinal tracts, most of whose fibers cross over at this level. On each side of the medulla oblongata is an oval swelling called the *olive*, from which a large bundle of nerve fibers arises and passes to the cerebellum.

Because of its location, all ascending and descending nerve fibers connecting the brain and spinal cord must pass through the medulla oblongata. As in the spinal cord, the white matter of the medulla surrounds a central mass of gray matter. Here, however, the gray matter is broken up into nuclei that are separated by nerve fibers. Some of these nuclei relay ascending impulses to the other side of the brain stem and then onto higher brain centers. The *nucleus gracilis* and the *nucleus cuneatus,* for example, receive sensory impulses from fibers of the fasciculus gracilis and the fasciculus cuneatus and pass them on to the thalamus or the cerebellum.

Other nuclei within the medulla oblongata function as control centers for vital visceral activities. These centers include the following:

1. **Cardiac center.** Impulses originating in the cardiac center are transmitted to the heart on peripheral nerves. Impulses on these nerves can cause the heart to beat more slowly or more rapidly than it would otherwise.
2. **Vasomotor center.** Certain cells of the vasomotor center initiate impulses that travel to smooth muscles in the walls of blood vessels and stimulate them to contract. This action causes constriction of the vessels (vasoconstriction) and a rise in blood pressure. Other cells of the vasomotor center can produce the opposite effect—dilation of blood vessels and a consequent drop in the blood pressure.
3. **Respiratory center.** The respiratory center functions with centers in the pons to regulate the rate, rhythm, and depth of breathing.

Still other nuclei within the medulla oblongata function as centers for certain nonvital reflexes, including those associated with coughing, sneezing, swallowing, and vomiting. Since the medulla contains vital reflex centers, injuries to this part of the brain stem often are fatal.

Reticular Formation

Scattered throughout the medulla oblongata, pons, and midbrain is a complex network of nerve fibers associated with tiny islands of gray matter. This network, the **reticular formation** (reticular activating system), extends from the upper portion of the spinal cord into the

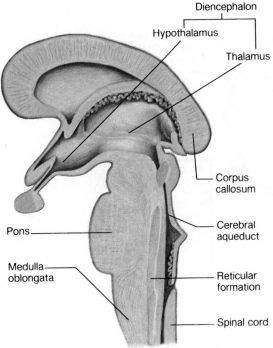

Diencephalon
Hypothalamus
Thalamus
Corpus callosum
Cerebral aqueduct
Reticular formation
Spinal cord
Pons
Medulla oblongata

diencephalon (figure 11.19). Its intricate system of nerve fibers interconnects centers of the hypothalamus, basal ganglia, cerebellum, and cerebrum with fibers in all the major ascending and descending tracts.

When sensory impulses reach the reticular formation, it responds by signaling the cerebral cortex, activating it into a state of wakefulness. Without this arousal, the cortex remains unaware of stimulation and cannot interpret sensory information or carry on thought processes. Thus, sleep results from decreased activity in the reticular formation. If the reticular formation ceases to function, as in certain injuries, the person remains unconscious and cannot be aroused even with strong stimulation (comatose).

The reticular formation also seems to act as a filter for incoming sensory impulses. Those impulses that are judged to be important, such as those originating in pain receptors, are passed onto the cerebral cortex, while others are disregarded. This selective action of the reticular formation frees the cortex from what would otherwise be a continual bombardment of sensory stimulation, and allows the cortex to concentrate upon more significant information. The cerebral cortex also can activate the reticular system, so that intense cerebral activity tends to keep a person awake.

In addition the reticular formation plays a role in regulating various motor activities so that coordinated muscular movements are smooth. It exerts an influence on spinal reflexes so that some are inhibited and others are enhanced.

Types of Sleep

There are two types of sleep—normal and paradoxical. **Normal sleep** (slow wave sleep) occurs when a person is particularly tired. It results from decreasing activity of the reticular formation and is restful, dreamless, and accompanied by a reduced blood pressure and respiratory rate.

Paradoxical sleep is so named because when it occurs, some areas of the brain are active and other areas are asleep. Episodes of this type of sleep usually last from 5 to 20 minutes and occur about every 90 minutes during normal sleep. However, the incidence of these episodes tends to increase as a person becomes rested. This form of sleep seems to result from the channeling of activating impulses to some regions of the brain and not to others. It is accompanied by dreams, irregular heart and respiratory rates, and rapid eye movements (REM).

1. What are the major functions of the thalamus? The hypothalamus?
2. How may the limbic system influence a person's behavior?
3. What vital reflex centers are located in the brain stem?
4. What is the function of the reticular formation?
5. Describe two types of sleep.

Figure 11.20

The cerebellum, which is located below the occipital lobes of the cerebrum, communicates with other parts of the nervous system by means of the cerebellar peduncles.

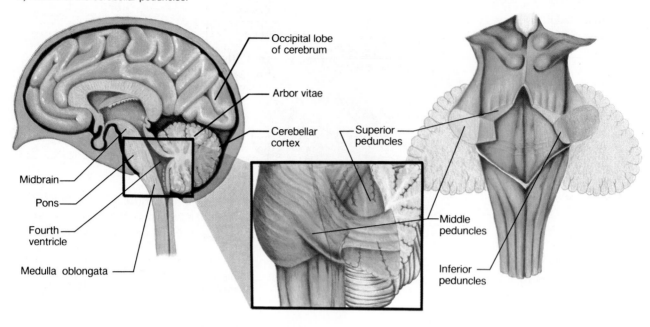

Occipital lobe of cerebrum

Arbor vitae

Cerebellar cortex

Midbrain

Pons

Fourth ventricle

Medulla oblongata

Superior peduncles

Middle peduncles

Inferior peduncles

Figure 11.21

What structures can you identify in this midsagittal section of the brain?

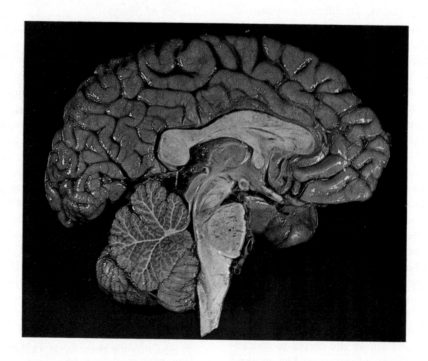

The Cerebellum

The **cerebellum** is a large mass of tissue located below the occipital lobes of the cerebrum and posterior to the pons and medulla oblongata. It consists of two lateral hemispheres partially separated by a layer of dura mater called the *falx cerebelli*. These hemispheres are connected in the midline by a structure called the *vermis* (figures 11.20 and 11.21).

Like the cerebrum, the cerebellum is composed primarily of white matter with a thin layer of gray matter, the **cerebellar cortex,** on its surface. This cortex is doubled over on itself in a series of complex folds with myelinated nerve fibers branching into them. As a result, a cut into the cerebellum reveals a treelike pattern of white matter, called the *arbor vitae,* that is surrounded by gray matter. A number of nuclei lie deep within each cerebellar hemisphere, the largest and most important of which is the *dentate nucleus.*

The cerebellum communicates with other parts of the central nervous system by means of three pairs of nerve tracts called **cerebellar peduncles** (ser''ĕ-bel'ar pe-dung'k'ls). One pair, the *inferior peduncles,* bring sensory information via the spinal cord and medulla oblongata to the cerebellum concerning the actual position of body parts such as limbs and the attitude of joints. The *middle peduncles* transmit information, originating in the cerebral cortex, about the desired position of these parts. After integrating and analyzing the information from these two sources, the cerebellum sends impulses from the dentate nucleus via the *superior peduncles* to the midbrain (figure 11.20). In response, motor impulses are transmitted downward through the pons, medulla oblongata, and spinal cord, and they stimulate or inhibit skeletal muscles at appropriate times to cause movements of body parts into the desired positions. This activity of the cerebellum makes rapid and complex muscular movements possible.

Thus, the cerebellum functions as a center for the control and coordination of skeletal muscles. The sensory impulses it receives come from receptors in muscles, tendons, and joints (proprioceptors) and from special sense organs, such as the eyes and ears (see chapter 12). For example, the cerebellum receives sensory information from the semicircular canals of the inner ears. This information concerns the motion and position of the head, and is used by the cerebellum to help maintain equilibrium.

Chart 11.4 Subdivisions of the nervous system

1. Central nervous system (CNS)
   a. Brain
   b. Spinal cord
2. Peripheral nervous system (PNS)
   a. Cranial nerves arising from the brain
      (1) Somatic fibers connecting to the skin and skeletal muscles
      (2) Autonomic fibers connecting to visceral organs
   b. Spinal nerves arising from the spinal cord
      (1) Somatic fibers connecting to the skin and skeletal muscles
      (2) Autonomic fibers connecting to visceral organs

Damage to the cerebellum is likely to result in tremors, inaccurate movements of voluntary muscles, loss of tone, reeling walk, and loss of equilibrium.

---

1. Where is the cerebellum located?
2. What are the major functions of the cerebellum?
3. What kinds of receptors provide information to the cerebellum?

The Peripheral Nervous System

The **peripheral nervous system** (PNS) consists of the nerves that branch out from the central nervous system (CNS) and connect it to other body parts. It includes the *cranial nerves* that arise from the brain and the *spinal nerves* that arise from the spinal cord.

This portion of the nervous system also can be subdivided into somatic and autonomic nervous systems. Generally, the **somatic system** consists of the cranial and spinal nerve fibers that connect the CNS to the skin and skeletal muscles. The **autonomic nervous system** includes those fibers that connect the CNS to visceral organs such as the heart, stomach, intestines, and various glands. Chart 11.4 outlines the subdivisions of the nervous system.

Structure of Peripheral Nerves

As is described in chapter 10, a peripheral nerve consists of bundles of nerve fibers surrounded by connective tissue. The outermost layer of the connective tissue is called the *epineurium,* which is dense and includes many collagenous fibers. Each bundle of nerve fibers (fascicle) is, in turn, enclosed in a sleeve of less dense

Brain Waves
A Clinical Application

Brain waves are recordings of fluctuating electrical changes that occur in the brain. To obtain such a recording, electrodes are positioned on the surface of a surgically-exposed brain (electrocorticogram, ECoG) or on the outer surface of the head (electroencephalogram, EEG). These electrodes detect electrical changes taking place in the extracellular fluid of the brain in response to changes in potential occurring among large groups of neurons. The resulting signals from the electrodes are amplified and recorded.

The source of the brain waves seems to be the neurons in the cerebral cortex. However, the changes that occur in these waves are also related to activities in other parts of the brain, such as the reticular formation, that influence the cortex.

In general, the intensity of electrical changes is directly related to the degree of neuronal activity. Thus, the brain waves change markedly in amplitude and frequency between sleep and wakefulness.

Brain waves are classified as alpha, beta, theta, and delta waves. *Alpha waves* are recorded most easily from the posterior regions of the head and have a frequency of 8–13 cycles per second. They occur when a person is awake and resting with the eyes closed. These waves disappear during sleep, and if a wakeful person's eyes are opened, the waves are replaced by higher frequency beta waves.

Beta waves have a frequency of more than 13 cycles per second and usually are recorded in the anterior region of the head. They occur when a person is actively engaged in mental activity or is under tension.

Theta waves have a frequency of 4–7 cycles per second and occur mainly in the parietal and temporal regions of the cerebrum. They are produced normally in children and in adults who are in the early stages of sleep, but they are considered to be abnormal in an adult who is awake.

Delta waves have a frequency below 4 cycles per second and occur during sleep. They seem to come from the cerebral cortex when it is not being activated by the reticular formation (figure 11.22).

The patterns of brain waves sometimes are useful for diagnosing disease conditions. For example, they are used to distinguish various types of seizure disorders (epilepsy) and to locate brain tumors. Such abnormal growths as tumors take up space previously occupied by normal nerve tissue, or they compress the normal tissues around them. In either case, abnormal brain waves produced in the affected area may reveal the location of the tumor.

Brain waves also are used to determine when *brain death* has occurred. Such brain death, accompanied by the cessation of neuronal activity, may be verified by the presence of an EEG which lacks waves. However, since certain drugs can greatly depress brain functions, such substances must be excluded as the cause of the flat pattern before brain death can be confirmed.

Figure 11.22
Brain waves are recordings of fluctuating electrical changes that occur in the brain.

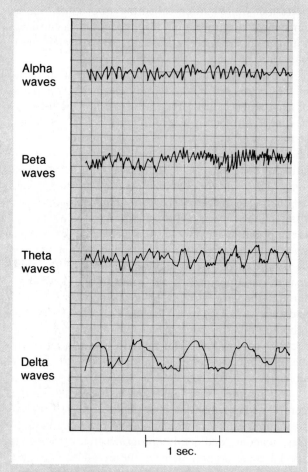

Figure 11.23

A nerve is composed of bundles of nerve fibers held together by connective tissues.

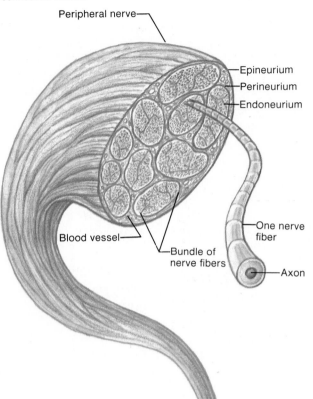

Peripheral nerve

Epineurium

Perineurium

Endoneurium

Blood vessel

Bundle of nerve fibers

One nerve fiber

Axon

connective tissue called the *perineurium*. The individual nerve fibers are surrounded by a small amount of loose connective tissue called *endoneurium* (figures 11.23 and 11.24).

Blood vessels usually are found in the epineurium and perineurium, and they give rise to a network of capillaries in the endoneurium.

The Cranial Nerves

Twelve pairs of **cranial nerves** arise from various locations on the underside of the brain. With the exception of the first pair, which springs from the cerebrum, these nerves originate from the brain stem. They pass from their sites of origin through various foramina of the skull and lead to parts of the head, neck, and trunk.

Although most cranial nerves are mixed, some of those associated with special senses, such as smell and vision, contain only sensory fibers. Others that are closely involved with the activities of muscles and glands are composed primarily of motor fibers and have limited sensory functions.

When sensory fibers are present in cranial nerves, the neuron cell bodies to which the fibers are attached are located outside the brain and are usually in groups called *ganglia* (singular, *ganglion*). On the other hand, motor neuron cell bodies are typically located within the gray matter of the brain.

Figure 11.24

Scanning electron micrograph of the cross section of a nerve (about 1000X). Note the bundles or fascicles of nerve fibers. (*Tissues and Organs: A Text-Atlas of Scanning Electron Microscopy*, by R. G. Kessel and R. H. Kardon. © 1979 W. H. Freeman and Company.)

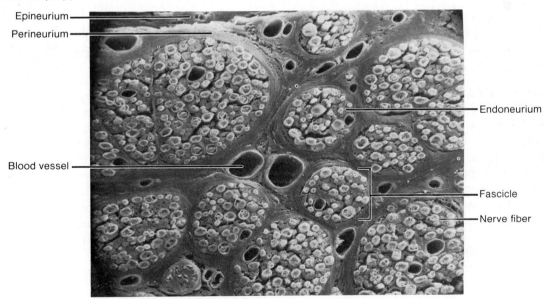

Epineurium

Perineurium

Endoneurium

Blood vessel

Fascicle

Nerve fiber

Figure 11.25
Except for the first pair, the cranial nerves arise from the brain stem. They are identified either by numbers indicating their order, or by names describing their function or the general distribution of their fibers.

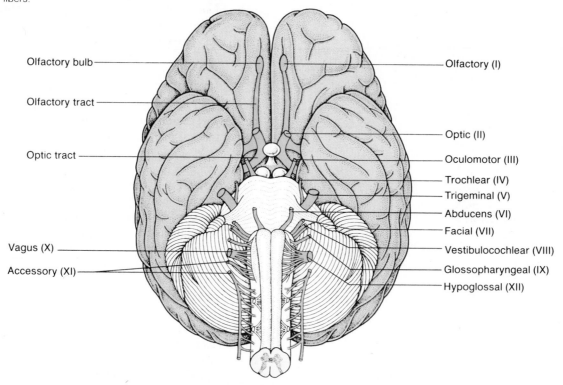

Olfactory bulb — Olfactory (I)

Olfactory tract

Optic tract — Optic (II)

— Oculomotor (III)
— Trochlear (IV)
— Trigeminal (V)
— Abducens (VI)
— Facial (VII)
— Vestibulocochlear (VIII)

Vagus (X) — — Glossopharyngeal (IX)

Accessory (XI) — — Hypoglossal (XII)

Cranial nerves are designated either by a number or a name. The numbers indicate the order in which the nerves arise from the anterior to posterior areas of the brain, and the names describe their primary functions or the general distribution of their fibers (figure 11.25).

The first pair of cranial nerves, the **olfactory nerves** (I), are associated with the sense of smell and contain only sensory neurons. These neurons are located in the lining of the upper nasal cavity where they serve as *olfactory receptor cells*. Axons from these receptors pass upward through the cribriform plates of the ethmoid bone and into *olfactory bulbs* that lie just beneath the frontal lobes of the cerebrum. Sensory impulses travel from these bulbs along *olfactory tracts* to cerebral centers where they are interpreted. The result is the sensation of smell.

The second pair of cranial nerves, the **optic nerves** (II), lead from the eyes to the brain and are associated with the sense of vision. The sensory cell bodies of the nerve fibers occur in ganglia within the eyes, and their axons pass through the *optic foramina* of the orbits and continue into the brain.

Some of the fibers in each optic nerve cross over to the other side of the brain. Specifically, the fibers arising from the medial half of each eye cross over, while those arising from the lateral sides do not. This crossing produces the X-shaped *optic chiasma* in the hypothalamus.

Sensory impulses of the optic nerves continue along *optic tracts* to centers in the thalamus and travel from there to the visual cortices of the occipital lobes.

The third pair of cranial nerves, the **oculomotor** (ok″u-lo-mo′tor) **nerves** (III), arise from the midbrain and pass into the orbits of the eyes. One component of each nerve connects to a number of voluntary muscles. These include the *levator palpebrae superioris* muscles, which function to raise the eyelids, and most of the muscles attached to the eye surfaces that cause them to move—the *superior rectus, medial rectus, inferior rectus,* and *inferior oblique* muscles.

A second portion of each oculomotor nerve is part of the autonomic nervous system, supplying involuntary muscles inside the eyes. These muscles act in adjusting the amount of light that enters the eyes and in focusing the lenses of the eyes.

Figure 11.26
The trigeminal nerve has three large branches that supply
various parts of the head and face.

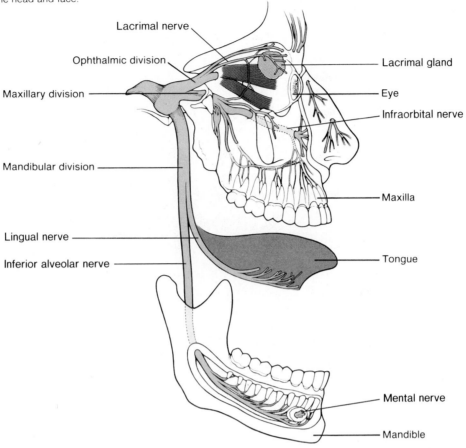

Lacrimal nerve

Ophthalmic division

Maxillary division

Mandibular division

Lingual nerve

Inferior alveolar nerve

Lacrimal gland

Eye

Infraorbital nerve

Maxilla

Tongue

Mental nerve

Mandible

Although the fibers of the oculomotor nerves are primarily motor, some sensory fibers are present. These transmit sensory information to the brain concerning the condition of various muscles.

The fourth pair, the **trochlear** (trok'le-ar) **nerves** (IV), are the smallest cranial nerves. They arise from the midbrain and carry motor impulses to a pair of external eye muscles, the *superior obliques,* which are not supplied by the oculomotor nerves. The trochlear nerves also contain some sensory fibers that transmit information about the condition of certain muscles.

The fifth pair, the **trigeminal** (tri-jem'i-nal) **nerves** (V), are the largest of the cranial nerves and arise from the pons. They are mixed nerves, but their sensory portions are more extensive than their motor portions. Each sensory component includes three large branches, the ophthalmic, maxillary, and mandibular divisions (figure 11.26).

The *ophthalmic division* consists of sensory fibers that bring impulses to the brain from the surface of the eye, the tear gland, and the skin of the anterior scalp, forehead, and upper eyelid. The fibers of the

maxillary division carry sensory impulses from the upper teeth, upper gum, and upper lip, as well as from the mucous lining of the palate and the skin of the face. The *mandibular division* includes both motor and sensory fibers. The sensory branches transmit impulses from the scalp behind the ear, the skin of the jaw, the lower teeth, the lower gum, and the lower lip. The motor branches supply the muscles of mastication and certain muscles in the floor of the mouth.

A disorder of the trigeminal nerve called *tic douloureux* (or trigeminal neuralgia) is characterized by severe recurring pain in the face and forehead on the affected side. If the pain cannot be controlled with drugs, the sensory portion of the nerve is sometimes severed surgically. Although this procedure may relieve the pain, the patient will lose other sensations in the parts supplied by the sensory branch. Consequently, persons who have had such surgery must be cautious when eating or drinking hot foods or liquids, because they may not sense burning. These persons should also inspect their mouths daily for the presence of food particles or damage to their cheeks from biting, which they may not feel.

Figure 11.27
The facial nerve is associated with the taste receptors on the
tongue and with the muscles of facial expression.

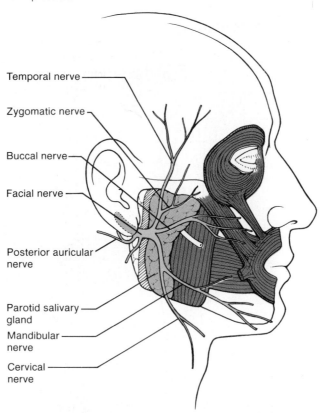

Temporal nerve

Zygomatic nerve

Buccal nerve

Facial nerve

Posterior auricular
nerve

Parotid salivary
gland

Mandibular
nerve

Cervical
nerve

The sixth pair of cranial nerves, the **abducens** (ab-du'senz) **nerves** (VI), are quite small and originate from the pons near the medulla oblongata. They enter the orbits of the eyes and supply motor impulses to a pair of external eye muscles, the *lateral rectus* muscles. The sensory fibers of these nerves provide the brain with information concerning the condition of muscles.

The seventh pair of cranial nerves, the **facial** (fa'shal) **nerves** (VII), arise from the lower part of the pons and emerge on the sides of the face. Their sensory branches are associated with taste receptors on the anterior two-thirds of the tongue, and some of their motor fibers transmit impulses to muscles of facial expression. Still other motor fibers of these nerves function in the autonomic nervous system by stimulating secretions from tear glands and certain salivary glands (submandibular and sublingual glands). (See figure 11.27.)

The eighth pair of cranial nerves, the **vestibulocochlear** (ves-tib"u-lo-kok'le-ar) **nerves** (VIII, acoustic nerves), are sensory nerves and arise from the medulla oblongata. Each of these nerves has two distinct parts—a vestibular branch and a cochlear branch.

The neuron cell bodies of the *vestibular branch* fibers are located in ganglia near the vestibule and semicircular canals of the inner ear. These structures contain receptors that are sensitive to changes in the position of the head. The impulses they initiate pass into the cerebellum, where they are used in reflexes associated with the maintenance of equilibrium.

The neuron cell bodies of the *cochlear branch* fibers are located in a ganglion of the cochlea, a part of the inner ear that houses the hearing receptors. Impulses from this branch pass through the pons and medulla oblongata on their way to the temporal lobe for interpretation.

The ninth pair of cranial nerves, the **glossopharyngeal** (glos"o-fah-rin'je-al) **nerves** (IX), are associated with the tongue and pharynx, as the name implies. These nerves arise from the medulla oblongata, and although they are mixed nerves, their predominant fibers are sensory. These fibers carry impulses from the lining of the pharynx, tonsils, and posterior third of the tongue to the brain. Fibers in the motor component of the glossopharyngeal nerves innervate constrictor muscles in the wall of the pharynx that function in swallowing.

Figure 11.28
The vagus nerve extends from the medulla oblongata downward
into the chest and abdomen to supply many visceral organs.

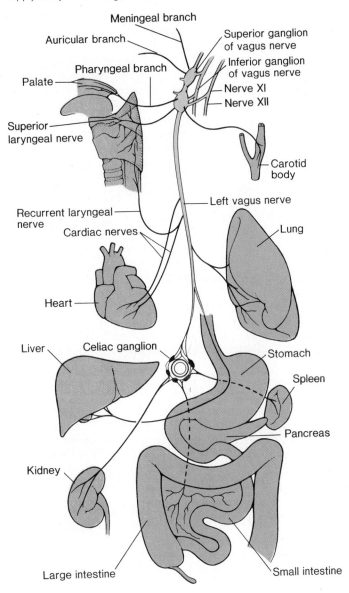

Meningeal branch
Auricular branch
Superior ganglion of vagus nerve
Pharyngeal branch
Inferior ganglion of vagus nerve
Palate
Nerve XI
Nerve XII
Superior laryngeal nerve
Carotid body
Left vagus nerve
Recurrent laryngeal nerve
Lung
Cardiac nerves
Heart
Liver
Celiac ganglion
Stomach
Spleen
Pancreas
Kidney
Large intestine
Small intestine

Other branches of the glossopharyngeal nerves function in the autonomic nervous system. Some sensory fibers, for example, conduct impulses from special receptors in the wall of an artery in the neck (carotid artery) to a center in the medulla oblongata. This information is used in the regulation of blood pressure. Certain autonomic motor fibers within the glossopharyngeal nerves lead to the parotid salivary gland in front of the ear, and can stimulate it to secrete saliva.

The tenth pair of cranial nerves, the **vagus** (va′gus) **nerves** (X), originate in the medulla oblongata and extend downward through the neck into the chest and abdomen. These nerves are mixed, and although they contain both somatic and autonomic branches, the autonomic fibers are the predominant ones.

Among the somatic components are motor fibers that carry impulses to muscles of the larynx. These fibers are associated with speech and with swallowing reflexes that employ muscles in the soft palate and pharynx. Vagal sensory fibers carry impulses from the lining of the pharynx, larynx, esophagus, and the viscera of the thorax and abdomen to the brain.

Autonomic motor fibers of the vagus nerves supply the heart and a variety of smooth muscles and glands in the visceral organs of the thorax and abdomen. (See figure 11.28.)

The eleventh pair of cranial nerves, the **accessory** (ak-ses′o-re) **nerves** (XI, spinal accessory), have origins in the medulla oblongata and the spinal cord; thus, they have both cranial and spinal branches.

Chart 11.5   Functions of cranial nerves

Nerve		Type	Function
I	Olfactory	Sensory	Sensory fibers transmit impulses associated with the sense of smell.
II	Optic	Sensory	Sensory fibers transmit impulses associated with the sense of vision.
III	Oculomotor	Primarily motor	Motor fibers transmit impulses to muscles that raise eyelids, move eyes, adjust amount of light entering eyes, and focus lenses.
			Some sensory fibers transmit impulses associated with the condition of muscles.
IV	Trochlear	Primarily motor	Motor fibers transmit impulses to muscles that move the eyes.
			Some sensory fibers transmit impulses associated with the condition of muscles.
V	Trigeminal	Mixed	
	Ophthalmic division		Sensory fibers transmit impulses from the surface of the eyes, the tear glands, scalp, forehead, and upper eyelids.
	Maxillary division		Sensory fibers transmit impulses from the upper teeth, upper gum, upper lip, lining of the palate, and skin of the face.
	Mandibular division		Sensory fibers transmit impulses from the scalp, skin of the jaw, lower teeth, lower gum, and lower lip.
			Motor fibers transmit impulses to muscles of mastication and muscles in floor of the mouth.
VI	Abducens	Primarily motor	Motor fibers transmit impulses to muscles that move the eyes.
			Some sensory fibers transmit impulses associated with the condition of muscles.
VII	Facial	Mixed	Sensory fibers transmit impulses associated with taste receptors of the anterior tongue.
			Motor fibers transmit impulses to muscles of facial expression, tear glands, and salivary glands.
VIII	Vestibulocochlear	Sensory	
	Vestibular branch		Sensory fibers transmit impulses associated with the sense of equilibrium.
	Cochlear branch		Sensory fibers transmit impulses associated with the sense of hearing.
IX	Glossopharyngeal	Mixed	Sensory fibers transmit impulses from the pharynx, tonsils, posterior tongue, and carotid arteries.
			Motor fibers transmit impulses to muscles of pharynx used in swallowing and to salivary glands.
X	Vagus	Mixed	Motor fibers transmit impulses to muscles associated with speech and swallowing, to the heart, and to smooth muscles of visceral organs in the thorax and abdomen.
			Sensory fibers transmit impulses from the pharynx, larynx, esophagus, and visceral organs of the thorax and abdomen.
XI	Accessory	Motor	
	Cranial branch		Motor fibers transmit impulses to muscles of soft palate, pharynx, and larynx.
	Spinal branch		Motor fibers transmit impulses to muscles of neck and back.
XII	Hypoglossal	Motor	Motor fibers transmit impulses to muscles that move the tongue.

Each *cranial branch* of an accessory nerve joins a vagus nerve and carries impulses to muscles of the soft palate, pharynx, and larynx. The *spinal branch* descends into the neck and supplies motor fibers to the trapezius and sternocleidomastoid muscles.

The twelfth pair of cranial nerves, the **hypoglossal** (hi″po-glos′al) **nerves** (XII), arise from the medulla oblongata and pass into the tongue. They consist primarily of motor fibers that carry impulses to muscles that move the tongue in speaking, chewing, and swallowing.

The functions of the cranial nerves are summarized in chart 11.5.

The consequences of cranial nerve injuries depend on the location and extent of the injuries. For example, if only one member of a nerve pair is damaged, loss of function is limited to the affected side, but if both nerves are injured losses occur on both sides. Also, if a nerve is severed completely, the functional loss is total; if the cut is incomplete, the loss may be partial.

1. Define *peripheral nervous system*.
2. Distinguish between somatic and autonomic nerve fibers.
3. Describe the structure of a peripheral nerve.
4. Name the cranial nerves and list the major functions of each.

Figure 11.29
There are thirty-one pairs of spinal nerves.

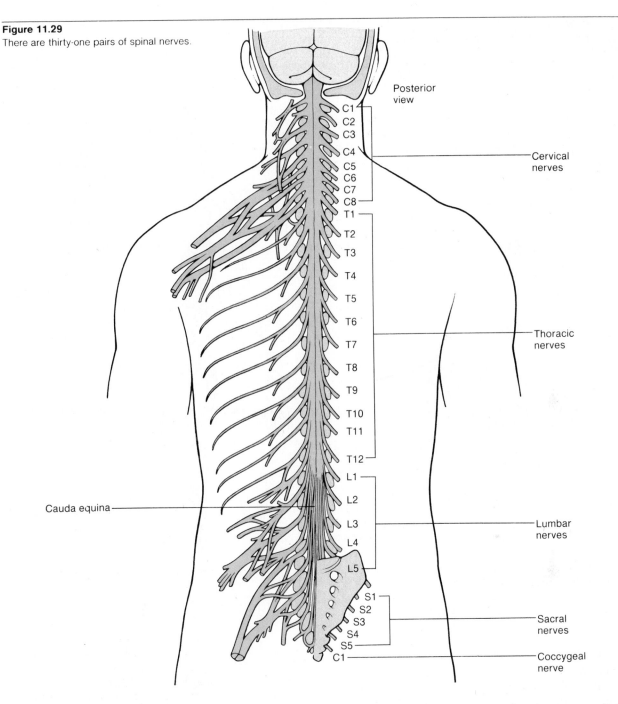

Posterior
view

Cervical
nerves

Thoracic
nerves

Cauda equina

Lumbar
nerves

Sacral
nerves

Coccygeal
nerve

C1
C2
C3
C4
C5
C6
C7
C8
T1
T2
T3
T4
T5
T6
T7
T8
T9
T10
T11
T12
L1
L2
L3
L4
L5
S1
S2
S3
S4
S5
C1

The Spinal Nerves

Thirty-one pairs of **spinal nerves** originate from the spinal cord. They are all mixed nerves, and they provide a two-way communication system between the spinal cord and parts in the arms, legs, neck, and trunk.

Although spinal nerves are not named individually, they are grouped according to the level from which they arise, and each nerve is numbered in sequence (figure 11.29). Thus, there are eight pairs of *cervical nerves* (numbered C1 to C8), twelve pairs of *thoracic nerves* (numbered T1 to T12), five pairs of *lumbar nerves* (numbered L1 to L5), five pairs of *sacral nerves* (numbered S1 to S5), and one pair of *coccygeal nerves.*

The nerves arising from the upper part of the spinal cord pass outward nearly horizontally, while those from the lower portions of the spinal cord descend at sharp angles. This arrangement is a consequence of growth. In early life, the spinal cord extends the entire length of the vertebral column, but with age the column grows more rapidly than the cord. As a result, the adult spinal cord ends at the level between the first and second lumbar vertebrae, so the lumbar, sacral, and coccygeal nerves descend to their exits beyond the end of the cord. These descending nerves form a structure called *cauda equina,* which is shaped somewhat like a horse's tail (figure 11.29).

Figure 11.30
(a) Dermatomes on the anterior body surface and *(b)* on the posterior surface.

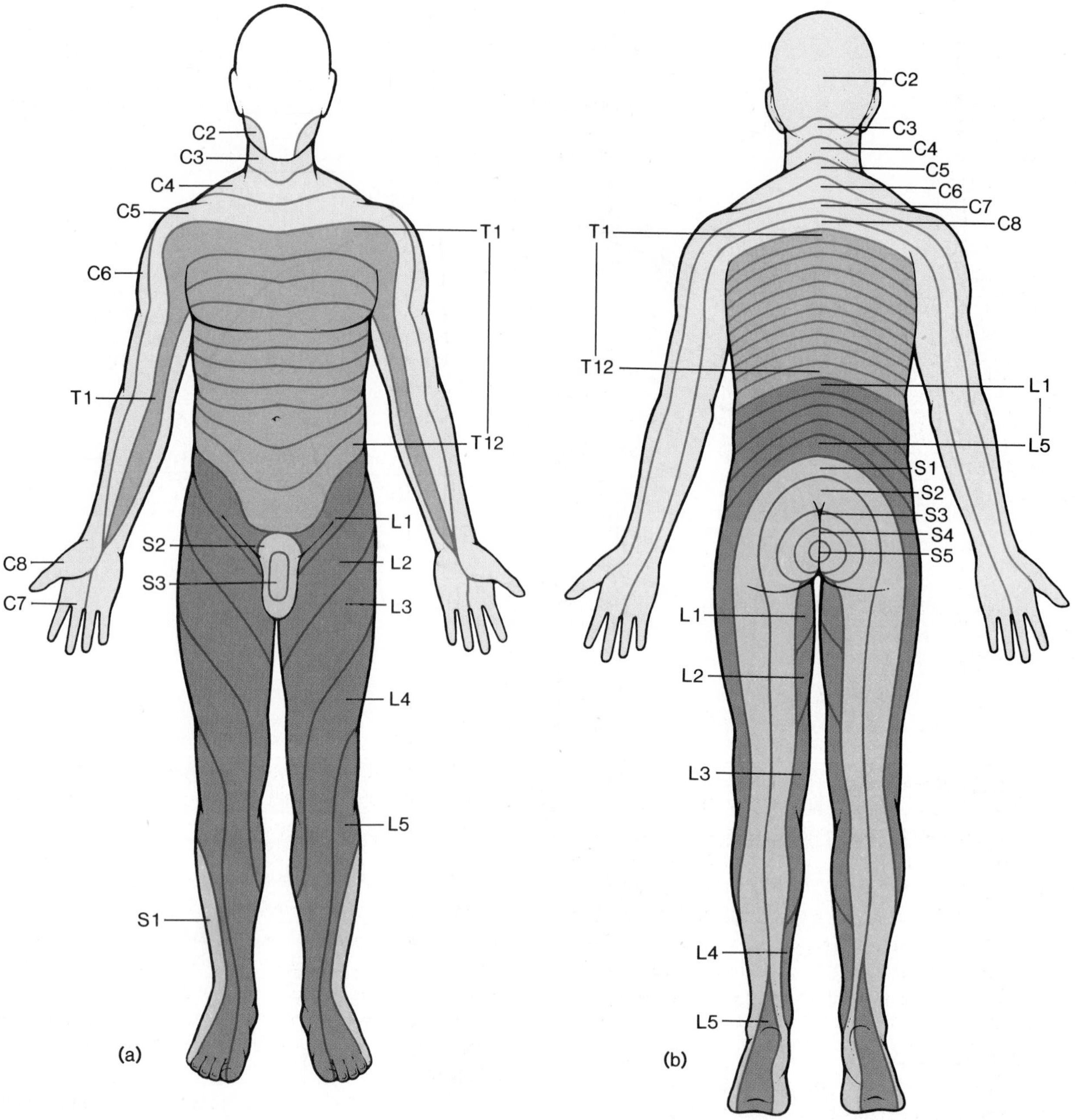

(a)

(b)

Figure 11.31

Each spinal nerve has a posterior and an anterior branch; the thoracic and lumbar spinal nerves also have a visceral branch.

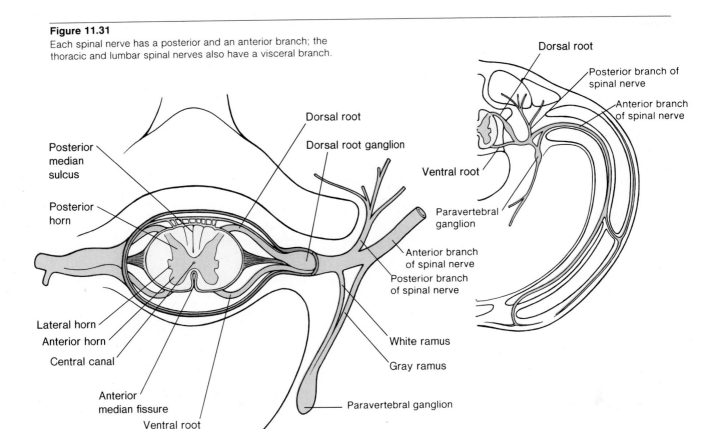

Each spinal nerve emerges from the cord by two short branches, or *roots,* which lie within the vertebral column. The **dorsal root** (sensory root) can be identified by the presence of an enlargement called the *dorsal root ganglion.* This ganglion contains the cell bodies of the sensory neurons whose dendrites conduct impulses inward from peripheral body parts. The axons of these neurons extend through the dorsal root and into the spinal cord, where they form synapses with dendrites of other neurons.

An area of skin where a group of sensory nerve fibers lead to a particular dorsal root is called a *dermatome.* These dermatomes are highly organized, but they vary considerably in size and shape, as indicated in figure 11.30. A map of the dermatomes often is useful in localizing the sites of injuries to dorsal roots or the spinal cord.

The **ventral root** (motor root) of each spinal nerve consists of axons from motor neurons whose cell bodies are located within the gray matter of the cord.

A ventral root and a dorsal root unite to form a spinal nerve, which passes outward from the vertebral canal through an *intervertebral foramen.* Just beyond its foramen, each spinal nerve divides into several parts. One of these parts, the small *meningeal branch,* reenters the vertebral canal through the intervertebral foramen and supplies the meninges and blood vessels of the cord, as well as the intervertebral ligaments and the vertebrae.

As figure 11.31 shows, a *posterior branch* (posterior ramus) of each spinal nerve turns posteriorly and innervates muscles and skin of the back. The main portion of the nerve, the *anterior branch* (anterior ramus), continues forward to supply muscles and skin on the front and sides of the trunk and limbs.

The spinal nerves in the thoracic and lumbar regions have a fourth or *visceral branch,* which is part of the autonomic nervous system.

Figure 11.32
The anterior branches of the spinal nerves in the thoracic region
give rise to intercostal nerves. Those in other regions combine to
form complex networks called plexuses.

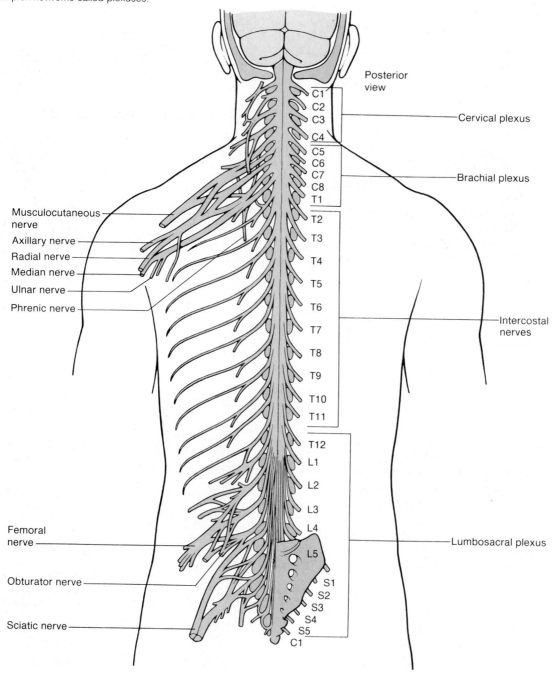

Posterior
view

Cervical plexus

Brachial plexus

C1
C2
C3
C4
C5
C6
C7
C8
T1

Musculocutaneous
nerve

Axillary nerve

Radial nerve

Median nerve

Ulnar nerve

Phrenic nerve

T2
T3
T4
T5
T6
T7
T8
T9
T10
T11

Intercostal
nerves

T12
L1
L2
L3
L4
L5

Femoral
nerve

Lumbosacral plexus

Obturator nerve

S1
S2
S3
S4
S5
C1

Sciatic nerve

Except in the thoracic region, anterior branches
of the spinal nerves combine to form complex networks, called **plexuses,** instead of continuing directly to
the peripheral body parts. In a plexus, the fibers of various spinal nerves are sorted and recombined, so that
fibers associated with a particular peripheral part reach
it in the same nerve, even though the fibers originate
from different spinal nerves (figure 11.32).

Cervical Plexuses

The **cervical plexuses** lie deep in the neck on either side.
They are formed by the anterior branches of the first
four cervical nerves, and fibers from these plexuses
supply the muscles and skin of the neck. In addition,
fibers from the third, fourth, and fifth cervical nerves
pass into the right and left **phrenic** (fren′ik) **nerves,**
which conduct motor impulses to the muscle fibers of
the diaphragm.

Brachial Plexuses

The anterior branches of the lower four cervical nerves and the first thoracic nerve give rise to **brachial plexuses.** These nets of nerve fibers are located deep within the shoulders between the neck and the axillae (armpits). The major branches emerging from the brachial plexuses include the following:

1. *Musculocutaneous nerves,* which supply muscles of the arms on the anterior sides and the skin of the forearms.
2. *Ulnar nerves,* which supply muscles of the forearms and hands, and skin of the hands.
3. *Median nerves,* which supply muscles of the forearms and muscles and skin of the hands.
4. *Radial nerves,* which supply muscles of the arms on the posterior sides, and skin of the forearms and hands.
5. *Axillary nerves,* which supply muscles and skin of the upper, lateral, and posterior regions of the arm.

In contact sports, such as football, a player's shoulder girdle is sometimes depressed at the same time that the head and neck are pushed away from the depressed side. This may result in a painful stretching of the brachial plexus that athletes call a "stinger."

Lumbosacral Plexuses

The **lumbosacral** (lum″bo-sa′kral) **plexuses** are formed on either side by the last thoracic and the lumbar, sacral, and coccygeal nerves. These networks of nerve fibers extend from the lumbar region of the back into the pelvic cavity, giving rise to a number of motor and sensory fibers associated with the lower abdominal wall, external genitalia, buttock, thighs, legs, and feet. The major branches of these plexuses include the following:

1. *Obturator nerves,* which supply the adductor muscles of the thighs.
2. *Femoral nerves,* which divide into many branches, supplying motor impulses to muscles of the thighs and legs, and receiving sensory impulses from the skin of the thighs and lower legs.
3. *Sciatic nerves,* which are the largest and longest nerves in the body. They pass downward into the buttock and descend into the thighs, where they divide into *tibial* and *common peroneal nerves.* The many branches of these nerves supply muscles and skin in the thighs, legs, and feet.

The anterior branches of the thoracic spinal nerves do not enter a plexus. Instead, they travel into spaces between the ribs and become **intercostal** (in″ter-kos′tal) **nerves.** These nerves supply motor impulses to the intercostal muscles and the upper abdominal wall muscles. They also receive sensory impulses from skin of the thorax and abdomen.

---

1. How are spinal nerves grouped?
2. Describe the way a spinal nerve joins the spinal cord.
3. Name and locate the major nerve plexuses.

The Autonomic Nervous System

The *autonomic nervous system* is the portion of the peripheral nervous system that functions independently (autonomously) and continuously without conscious effort. This system controls visceral functions by regulating the actions of smooth muscles, cardiac muscle, and various glands. It is concerned with regulating heart rate, blood pressure, breathing rate, body temperature, and other visceral activities that aid in the maintenance of homeostasis. Portions of the autonomic system also are responsive during times of emotional stress, and they serve to prepare the body to meet the demands of strenuous physical activity.

General Characteristics

Autonomic activities are regulated largely by reflexes in which sensory signals originate from receptors within visceral organs and skin. These signals are transmitted on afferent nerve fibers to nerve centers within the brain or spinal cord. Motor impulses then travel out from these centers on efferent nerve fibers within cranial and spinal nerves.

Typically these efferent fibers lead to ganglia outside the central nervous system. The impulses they carry are integrated within the ganglia and are relayed to various visceral organs—muscles or glands—which respond by contracting, secreting, or being inhibited. The integrative function of the ganglia provides a degree of independence from the brain and spinal cord, and the visceral efferent nerve fibers associated with these ganglia comprise the autonomic nervous system.

The autonomic nervous system is subdivided into two sections called the **sympathetic** and **parasympathetic divisions,** which act together. For example, some visceral organs are supplied with nerve fibers from each of the divisions. In such cases, impulses on one set of fibers tend to activate an organ, while impulses on the other set inhibit it. Thus, the divisions may act antagonistically, so that the actions of some visceral organs are regulated by alternately being activated or inhibited.

Although the functions of the autonomic divisions are mixed—each activates some organs and inhibits others—they have general functional differences. The sympathetic division is concerned primarily with preparing the body for energy-expending, stressful, or emergency situations. The parasympathetic division, however, is most active under ordinary, restful conditions. It also counterbalances the effects of the sympathetic division, and restores the body to a resting state following a stressful experience. For example, during an emergency the sympathetic division will cause the heart and breathing rates to increase, and following the emergency, the parasympathetic division will slow these activities.

Autonomic Nerve Fibers

All of the nerve fibers of the autonomic nervous system are efferent or motor fibers. However, unlike the motor pathways of the somatic nervous system, which usually include a single neuron between the brain or spinal cord and a skeletal muscle, those of the autonomic system involve two neurons, as shown in figure 11.33. The cell body of one neuron is located in the brain or spinal cord. Its axon, the **preganglionic fiber,** leaves the CNS and forms a synapse with one or more nerve fibers whose cell bodies are housed within an autonomic ganglion. The axon of such a second neuron is called a **postganglionic fiber,** and it extends to a visceral effector.

Sympathetic Division

In the sympathetic division (thoracolumbar division) the preganglionic fibers originate from neurons within the lateral horn of the spinal cord. These neurons are found in all of the thoracic and the upper two or three lumbar segments of the cord. Their axons exit in the ventral roots of spinal nerves along with various somatic motor fibers.

After traveling a short distance, preganglionic fibers leave the spinal nerves through branches called *white rami* (sing. ramus) and enter sympathetic ganglia. Two groups of such ganglia, called **paravertebral ganglia,** are located in chains along the sides of the vertebral column. These ganglia, together with the fibers that interconnect them, comprise the **sympathetic trunks.** (See figure 11.34.)

The paravertebral ganglia are found just beneath the parietal pleura in the thorax and the parietal peritoneum in the abdomen. (See chapter 1.) Although these ganglia are located some distance from the visceral organs they help to control, other sympathetic ganglia are positioned nearer to the viscera. The *collateral sympathetic ganglia,* for example, are found within the thorax closely associated with certain large blood vessels. (See figure 11.36.)

Some of the preganglionic fibers that enter paravertebral ganglia synapse with neurons within these ganglia. Other fibers travel through the ganglia and pass

Figure 11.33

(a) Somatic neuron pathways usually have a single neuron between the central nervous system and an effector.
(b) Autonomic neuron pathways involve two neurons between the central nervous system and an effector.

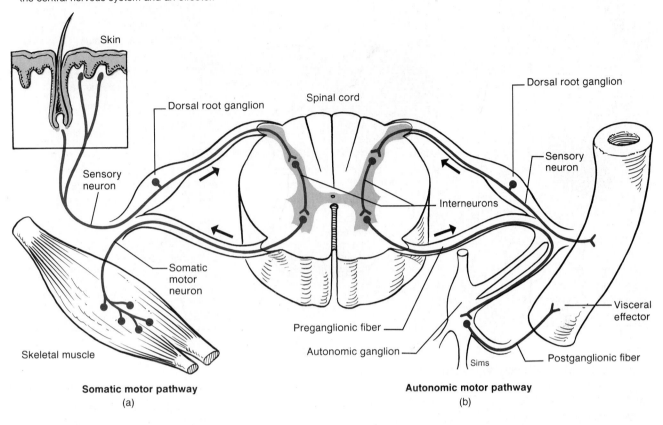

Skin

Dorsal root ganglion

Spinal cord

Dorsal root ganglion

Sensory neuron

Sensory neuron

Interneurons

Somatic motor neuron

Preganglionic fiber

Visceral effector

Skeletal muscle

Autonomic ganglion

Sims

Postganglionic fiber

Somatic motor pathway
(a)

Autonomic motor pathway
(b)

Figure 11.34

A chain of paravertebral ganglia extends along each side of the vertebral column.

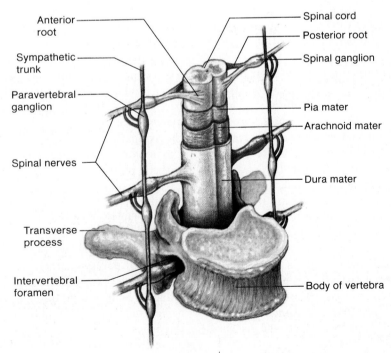

Anterior root

Spinal cord

Posterior root

Sympathetic trunk

Spinal ganglion

Paravertebral ganglion

Pia mater

Arachnoid mater

Spinal nerves

Dura mater

Transverse process

Intervertebral foramen

Body of vertebra

Figure 11.35

Sympathetic fibers leave the spinal cord in the ventral roots of spinal nerves, enter sympathetic ganglia, and synapse with other neurons that extend to visceral effectors.

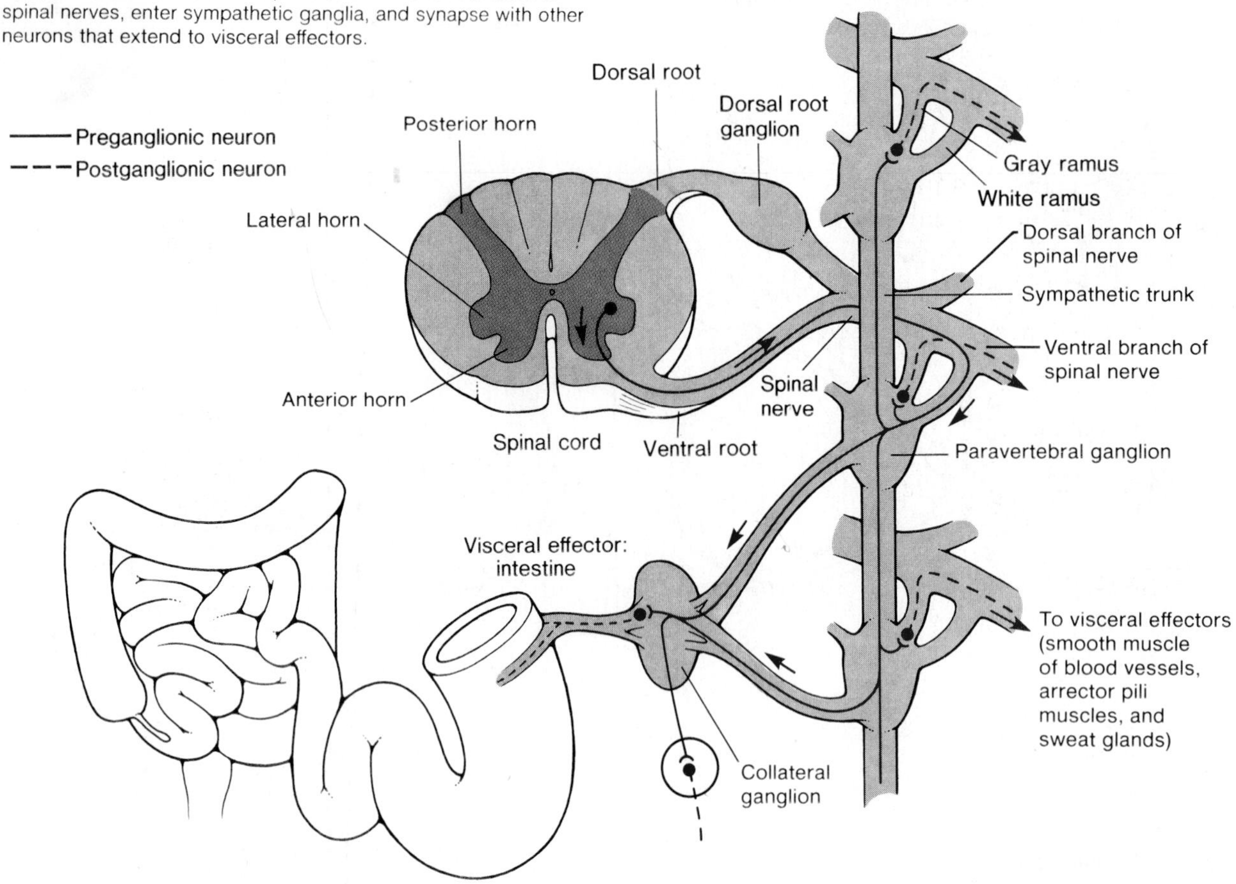

—— Preganglionic neuron
- - - Postganglionic neuron

Posterior horn
Dorsal root
Dorsal root ganglion
Gray ramus
White ramus
Dorsal branch of spinal nerve
Sympathetic trunk
Ventral branch of spinal nerve
Paravertebral ganglion
Lateral horn
Anterior horn
Spinal cord
Ventral root
Spinal nerve
Visceral effector: intestine
To visceral effectors (smooth muscle of blood vessels, arrector pili muscles, and sweat glands)
Collateral ganglion

up or down the sympathetic trunk and synapse with neurons in ganglia at higher or lower levels within the chain. Still other fibers pass through to collateral ganglia before they synapse. Typically, a preganglionic axon will synapse with several other neurons within a sympathetic ganglion.

The axons of the second neurons in sympathetic pathways, the postganglionic fibers, extend out from the sympathetic ganglia to visceral effectors. Those leaving paravertebral ganglia usually pass through branches called **gray rami** and return to a spinal nerve before proceeding to an effector. (See figure 11.35.) These branches appear gray because the postganglionic axons generally are unmyelinated, whereas the preganglionic axons in the white rami are nearly all myelinated. An important exception to the usual arrangement of sympathetic fibers occurs in a set of preganglionic fibers that pass through the sympathetic ganglia and extend out to the medulla of each adrenal gland. These fibers terminate within the glands on special hormone-secreting cells that release norepinephrine or epinephrine when they are stimulated. The functions of the adrenal medulla gland and its hormones are discussed in chapter 13.

Parasympathetic Division

The preganglionic fibers of the parasympathetic division (craniosacral division) arise from neurons in the midbrain and medulla oblongata of the *brain stem* and from the *sacral region* of the spinal cord. From there, they lead outward on cranial or sacral nerves to ganglia located near or within various visceral organs (terminal ganglia). The relatively short postganglionic fibers continue from the ganglia to specific muscles or glands within these visceral organs. As in the case of the sympathetic fibers, the parasympathetic preganglionic axons usually are myelinated, and the parasympathetic postganglionic fibers are unmyelinated.

The parasympathetic preganglionic fibers associated with parts of the head are included in the oculomotor, facial, and glossopharyngeal nerves. Those that innervate organs of the thorax and upper abdomen are parts of the vagus nerves. (The vagus nerves carry about 75% of all parasympathetic fibers.) Preganglionic fibers arising from the sacrum are found within the branches of the second through the fourth sacral spinal nerves, and they carry impulses to visceral organs within the pelvis. Figure 11.36 compares the distribution of nerve fibers in the sympathetic division with that in the parasympathetic division.

Figure 11.36

The preganglionic fibers of the sympathetic division arise from the thoracic and lumbar regions of the spinal cord, while those of the parasympathetic division arise from the brain and sacral region of the cord.

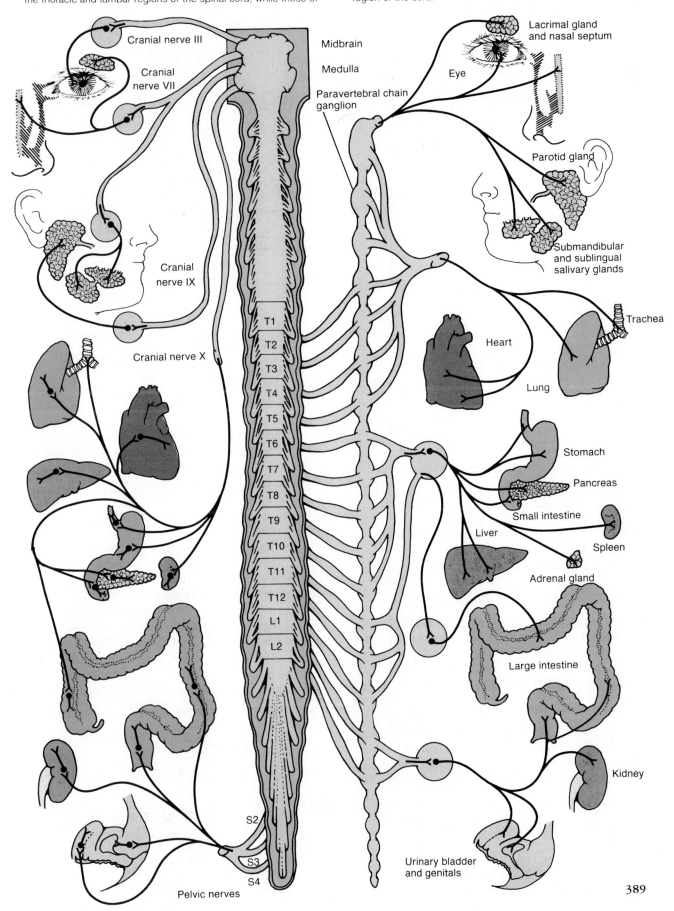

Figure 11.37

Most sympathetic fibers are adrenergic and secrete norepinephrine at the ends of the postganglionic fibers; parasympathetic fibers are cholinergic and secrete acetylcholine at the ends of the postganglionic fibers.

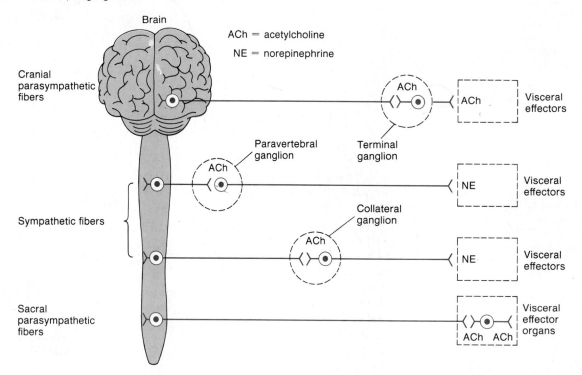

1. What parts of the nervous system are included in the autonomic nervous system?
2. How are the subdivisions of the autonomic system distinguished?
3. What functions are generally controlled by the sympathetic division? By the parasympathetic division?
4. Describe a sympathetic nerve pathway and a parasympathetic nerve pathway.

Autonomic Transmitter Substances

The preganglionic fibers of the sympathetic and parasympathetic divisions all secrete *acetylcholine;* their postganglionic fibers, however, use different transmitter substances. The parasympathetic postganglionic fibers secrete acetylcholine and are called **cholinergic fibers;** however, most sympathetic postganglionic fibers secrete norepinephrine (noradrenalin) and are called **adrenergic fibers.** Exceptions to this include the sympathetic postganglionic fibers leading to sweat glands and to blood vessels in muscles that secrete acetylcholine and, therefore, are cholinergic (figure 11.37).

These different postganglionic transmitter substances (mediators) are responsible for the different effects that the sympathetic and parasympathetic divisions have on visceral organs.

Although each division can activate some effectors and inhibit others, most visceral organs are controlled primarily by one division. In other words, the divisions usually are not actively antagonistic. For example, the diameter of most blood vessels, which lack parasympathetic innervation, is regulated by the sympathetic division. Smooth muscles in the walls of these vessels are continuously stimulated by sympathetic impulses; thus, they are maintained in a state of partial contraction called *sympathetic tone.* The diameter of such a blood vessel can be increased (dilated) by decreasing the amount of sympathetic stimulation, which allows the muscular wall to relax. Conversely, the vessel can be constricted by increasing the sympathetic stimulation.

Similarly, the parasympathetic division is dominant in controlling movement in the digestive system. Parasympathetic impulses stimulate stomach and intestinal motility, and when these impulses decrease, movement is reduced.

The effects of adrenergic and cholinergic fibers on various visceral effectors are summarized in chart 11.6.

Chart 11.6 Effects of autonomic stimulation on various effectors

Effector location	Response to sympathetic stimulation	Response to parasympathetic stimulation
Integumentary system		
Apocrine glands	Increased secretion	No action
Eccrine glands	Increased secretion (cholinergic effect)	No action
Special senses		
Iris of eye	Dilation	Constriction
Tear gland	Slight increase in secretion	Great increase in secretion
Endocrine system		
Adrenal cortex	Increased secretion	No action
Adrenal medulla	Increased secretion	No action
Digestive system		
Muscle of gallbladder wall	Relaxation	Contraction
Muscle of intestinal wall	Decreases peristaltic action	Increases peristaltic action
Muscle of internal anal sphincter	Contraction	Relaxation
Pancreatic glands	Reduced secretion	Great increase in secretion
Salivary glands	Reduced secretion	Great increase in secretion
Respiratory system		
Muscles in walls of bronchioles	Dilation	Constriction
Cardiovascular system		
Blood vessels supplying muscles	Constriction (alpha adrenergic) Dilation (beta adrenergic) Dilation (cholinergic)	No action
Blood vessels supplying skin	Constricted	No action
Blood vessels supplying heart (coronary arteries)	Dilation (beta adrenergic) Constriction (alpha adrenergic)	Dilation
Muscles in wall of heart	Increased contraction rate	Decreased contraction rate
Urinary system		
Muscle of bladder wall	Relaxation	Contraction
Muscle of internal urethral sphincter	Contraction	Relaxation
Reproductive systems		
Blood vessels to clitoris and penis		Dilation leading to erection
Muscles associated with male internal reproductive organs	Ejaculation	

Actions of Autonomic Transmitters

As in the case of stimulation occurring at neuromuscular junctions (chapter 9) and at synapses (chapter 10), the actions of autonomic transmitters result from combining of the transmitters with receptors. These receptors are proteins located in the membranes of effector cells, and when a transmitter combines with a receptor, there is a change in the membrane. For example, the membrane's permeability to certain ions may increase. In smooth muscle cells, an action potential followed by muscular contraction results from such a change. Similarly, a gland cell may respond to a change in its membrane by secreting a special product.

Acetylcholine is known to combine with two types of cholinergic receptors called muscarinic and nicotinic. The *muscarinic receptors* are located in the membranes of effector cells at the ends of all postganglionic parasympathetic nerve fibers and at the ends of the cholinergic sympathetic fibers. Responses from these receptors are excitatory and occur relatively slowly. The *nicotinic receptors* occur in the synapses between the preganglionic and postganglionic neurons of the parasympathetic and sympathetic pathways. They produce rapid excitatory responses (figure 11.38). (Receptors at neuromuscular junctions of skeletal muscles are also nicotinic.)

Both norepinephrine and epinephrine from the adrenal glands combine with adrenergic receptors of effector cells. There are two major types of these receptors called alpha-adrenergic and beta-adrenergic. When norepinephrine is present, it stimulates the effector cells by combining mainly with the *alpha receptors,* and has only a slight effect on the *beta receptors*. However, epinephrine can combine with either type of receptor. Consequently, the way each of these adrenergic substances influences effector cells depends upon the relative numbers of alpha and beta receptors present in the cell membranes. For example, an effector cell

Figure 11.38

(a) and (b) Muscarinic receptors occur in the membranes of effector cells at the ends of cholinergic fibers; (c) nicotinic receptors are found in the membranes of skeletal muscle fibers.

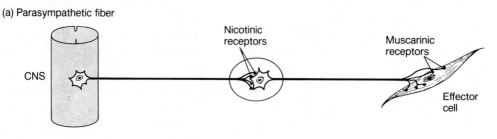

(a) Parasympathetic fiber

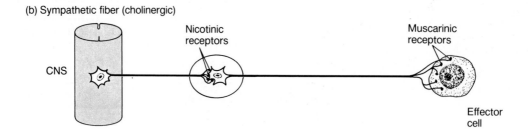

(b) Sympathetic fiber (cholinergic)

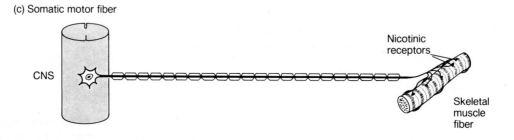

(c) Somatic motor fiber

which contains both types of receptors will be affected by both substances, but one with only beta receptors will be influenced mainly by epinephrine (figure 11.39).

The actions resulting from adrenergic receptors are sometimes excitatory and sometimes inhibitory. Therefore, when norepinephrine combines with the alpha receptors of smooth muscles, the action usually is excitatory and the cells contract, but when epinephrine combines with the beta receptors of these cells, the action is inhibitory and the cells relax.

The acetylcholine released by cholinergic fibers is rapidly decomposed by the action of the enzyme *cholinesterase*. (This decomposition also occurs in the neuromuscular junctions of skeletal muscle.) Thus, acetylcholine usually produces an effect for only a fraction of a second. Much of the norepinephrine released from adrenergic fibers, however, is taken back into the nerve endings by an active transport mechanism. It can be inactivated by the enzyme, *monoamine oxidase*, found in mitochondria. This may take a few seconds, and during that time some molecules may diffuse into nearby tissues and be decomposed by other enzymes. On the other hand, some norepinephrine molecules may enter the blood and remain active until they diffuse into tissues containing the inactivating enzymes. For these reasons, norepinephrine is likely to produce a more prolonged effect than acetylcholine. In fact, when the adrenal medulla releases norepinephrine and epinephrine into the blood in response to sympathetic stimulation, these substances may trigger sympathetic responses in organs throughout the body that last up to thirty seconds.

Until recently it was believed that activities of the autonomic nervous system were beyond conscious control. It is now known that some people can learn to alter their visceral responses consciously. To accomplish this, a person needs to receive information concerning the state of the visceral action to be controlled. Normally a person is unaware of such activities, but electronic equipment can measure blood pressure, for example, and continuously feed back this information to the person by means of a signal. Then the person can consciously try to produce a visceral response that will alter the blood pressure in a desired way.

This procedure, called **biofeedback,** has been used successfully to control heart rates and blood pressures, and to alter the diameters of blood vessels and the amount of blood flowing into certain regions of the body. It also has been used to control migraine headaches, which are associated with changes in the muscular tone of certain blood vessels, and to control epileptic seizures.

Figure 11.39
Norepinephrine combines mainly with alpha-adrenergic receptors, while epinephrine combines with both alpha- and beta-adrenergic receptors.

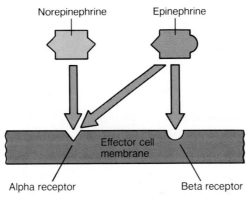

Many drugs influence autonomic functions. Some, like ephedrine, enhance sympathetic effects by stimulating the release of norepinephrine from sympathetic nerve endings. Others, like reserpine, inhibit sympathetic activity by preventing the synthesis of norepinephrine. Another group of drugs, which includes pilocarpine, produce parasympathetic effects, and some, like atropine, block the action of acetylcholine in visceral effectors.

Still other drugs, such as the tricyclic antidepressants used to treat depression, act by blocking the mechanism responsible for the uptake of norepinephrine. This allows norepinephrine to accumulate and to continue its effect by combining with receptor sites. Another group of antidepressant drugs produces a similar effect by inhibiting the action of monoamine oxidase, the enzyme that normally inactivates norepinephrine.

Control of Autonomic Activity

Although the autonomic nervous system has some degree of autonomy because of the integration of impulses that occurs within its ganglia, it is controlled largely by the brain and spinal cord. For example, as discussed previously, there are control centers in the medulla oblongata for cardiac, vasomotor, and respiratory activities. These reflex centers receive sensory impulses from visceral organs by means of vagus nerve fibers, and they make use of autonomic nerve pathways to stimulate motor responses in muscles and glands. Thus, they exert control over the autonomic nervous system. Similarly, the hypothalamus helps regulate body temperature, water and electrolyte balance, hunger, and thirst by employing and controlling autonomic pathways.

Still higher levels within the brain, including the limbic system and cerebral cortex control the autonomic nervous system when a person is stressed emotionally. These parts of the central nervous system make use of autonomic pathways to regulate a person's emotional expression and behavior.

The regulation of particular visceral organs is discussed in subsequent chapters that deal with individual organs and organ systems.

1. Distinguish between cholinergic and adrenergic fibers.
2. Explain how the fibers of one division can control the actions of an organ.
3. What transmitter substances are used in the autonomic nervous system?
4. Describe two types of cholinergic receptors and two types of adrenergic receptors.

Clinical Terms Related to the Nervous System

analgesia (an″al-je′ze-ah)   loss or reduction in the ability to sense pain, but without loss of consciousness.

analgesic (an″al-je′sik)   a pain-relieving drug.

anesthesia (an″es-the′ze-ah)   a loss of feeling.

aphasia (ah-fa′ze-ah)   a disturbance or loss in the ability to use words or to understand them, usually due to damage to cerebral association areas.

apraxia (ah-prak′se-ah)   an impairment in a person's ability to make correct use of objects.

ataxia (ah-tak′se-ah)   a partial or complete inability to coordinate voluntary movements.

cerebral palsy (ser′ĕ-bral pawl′ze)   a condition characterized by partial paralysis and lack of muscular coordination.

coma (ko′mah)   an unconscious condition in which there is an absence of responses to stimulation.

cordotomy (kor-dot′o-me)   a surgical procedure in which a nerve tract within the spinal cord is severed, usually to relieve intractable pain.

craniotomy (kra″ne-ot′o-me)   a surgical procedure in which part of the skull is opened.

electroencephalogram (EEG) (e-lek″tro-en-sef′ah-lo-gram″)   a recording of the electrical activity of the brain.

encephalitis (en″sef-ah-li′tis)   an inflammation of the brain and meninges characterized by drowsiness and apathy.

epilepsy (ep′ĭ-lep″se)   a disorder of the central nervous system that is characterized by temporary disturbances in normal brain impulses; it may be accompanied by convulsive seizures and loss of consciousness.

hemiplegia (hem″ĭ-ple′je-ah)   paralysis on one side of the body and the limbs on that side.

Huntington's chorea (hunt′ing-tunz ko-re′ah)   a rare hereditary disorder of the brain characterized by involuntary convulsive movements and mental deterioration.

laminectomy (lam″ĭ-nek′to-me)   surgical removal of the posterior arch of a vertebra, usually to relieve the symptoms of a ruptured intervertebral disk.

monoplegia (mon″o-ple′je-ah)   paralysis of a single limb.

multiple sclerosis (mul′tĭ-pl skle-ro′sis)   a disease of the central nervous system characterized by loss of myelin and the appearance of scarlike patches throughout the brain and spinal cord or both.

neuralgia (nu-ral′je-ah)   a sharp, recurring pain associated with a nerve, usually caused by inflammation or injury.

neuritis (nu-ri′tis)   an inflammation of a nerve.

paraplegia (par″ah-ple′je-ah)   paralysis of both legs.

quadriplegia (kwod″rĭ-ple′je-ah)   paralysis of all four limbs.

vagotomy (va-got′o-me)   severing of a vagus nerve.

Chapter Summary

Introduction

The brain and spinal cord are surrounded by bone and protective membranes called meninges.

The Meninges

1. The meninges consist of a dura mater, arachnoid mater, and pia mater.
2. Cerebrospinal fluid occupies the space between the arachnoid and pia maters.

The Spinal Cord

The spinal cord is a nerve column that extends from the brain into the vertebral canal.

It terminates at the level between the first and second lumbar vertebrae.

1. Structure of the spinal cord
   a. The spinal cord is composed of thirty-one segments, each of which gives rise to a pair of spinal nerves.
   b. It is characterized by a cervical enlargement, a lumbar enlargement, and two deep longitudinal grooves that divide it into right and left halves.
   c. It has a central core of gray matter that is surrounded by white matter.
   d. The white matter is composed of bundles of myelinated nerve fibers.
2. Functions of the spinal cord
   a. The cord provides a two-way communication system between the brain and body parts outside the nervous system.
   b. Ascending tracts carry sensory impulses to the brain; descending tracts carry motor impulses to muscles and glands.
   c. Many of the fibers in the ascending and descending tracts cross over in the spinal cord or brain.

The Brain

The brain is the largest and most complex part of the nervous system.

It contains nerve centers that are associated with sensations, issue motor commands, and carry on higher mental functions.

1. Development of the brain
   a. Brain structure reflects the way it forms.
   b. It develops from a tubular part with three cavities—the forebrain, midbrain, and hindbrain.
   c. The cavities persist as ventricles, and the walls give rise to structural and functional regions.
2. Structure of the cerebrum
   a. The cerebrum consists of two cerebral hemispheres connected by the corpus callosum.
   b. Its surface is marked by ridges and grooves; sulci divide each hemisphere into lobes.
   c. The cerebral cortex is a thin layer of gray matter near the surface.
   d. White matter consists of myelinated nerve fibers that interconnect neurons within the nervous system and with other body parts.

3. Functions of the cerebrum
   a. The cerebrum is concerned with higher brain functions, such as interpretation of sensory impulses, control of voluntary muscles, storage of memory, thought, and reasoning.
   b. The cerebral cortex can be divided into sensory, motor, and association areas.
   c. The primary motor regions lie in the frontal lobes near the central sulcus and are aided by other areas of the frontal lobes that control special motor functions.
   d. Areas responsible for interpreting sensory impulses from the skin are located in the parietal lobes near the central sulcus; other specialized sensory areas occur in the temporal and occipital lobes.
   e. Association areas function to analyze and interpret sensory impulses and are involved with memory, reasoning, verbalizing, judgment, and emotional feelings.
   f. In most persons, one cerebral hemisphere is dominant for certain intellectual functions.
   g. Memory is established in phases and is stored diffusely in both hemispheres.
4. Basal ganglia
   a. Basal ganglia are masses of gray matter located deep within the cerebral hemispheres.
   b. They function as relay stations for motor impulses originating in the cerebral cortex, and they aid in the control of motor activities.
5. The ventricles and cerebrospinal fluid
   a. Ventricles are interconnected cavities within the cerebral hemispheres and brain stem.
   b. These spaces are filled with cerebrospinal fluid.
   c. Cerebrospinal fluid is secreted by choroid plexuses in the walls of the ventricles; it circulates through the ventricles and is reabsorbed into the blood of the dural sinuses.
6. The brain stem
   a. The brain stem extends from the base of the cerebrum to the spinal cord.
   b. It consists of the diencephalon, midbrain, pons, and medulla oblongata.
   c. The diencephalon contains the thalamus, which serves as a central relay station for incoming sensory impulses, and the hypothalamus, which plays important roles in maintaining homeostasis.
   d. The limbic system functions to produce emotional feelings and to modify behavior.
   e. The midbrain contains reflex centers associated with eye and head movements.
   f. The pons transmits impulses between the cerebrum and other parts of the nervous system and contains centers that help to regulate the rate and depth of breathing.
   g. The medulla oblongata transmits all ascending and descending impulses and contains several vital and nonvital reflex centers.
   h. The reticular formation acts to filter incoming sensory impulses, arousing the cerebral cortex into wakefulness whenever significant impulses are received.
   i. Normal sleep results from decreasing activity of the reticular formation, and paradoxical sleep occurs when activating impulses are received by some parts of the brain, but not by others.
7. The cerebellum
   a. The cerebellum consists of two hemispheres connected by the vermis.
   b. It is composed of white matter surrounded by a thin cortex of gray matter.
   c. The cerebellum functions primarily as a reflex center in the coordination of skeletal muscle movements and the maintenance of equilibrium.

The Peripheral Nervous System

The peripheral nervous system consists of cranial and spinal nerves that branch out from the brain and spinal cord to all body parts.

It can be divided into somatic and autonomic portions.

1. Structure of peripheral nerves
   a. A nerve consists of a bundle of nerve fibers surrounded by connective tissues.
   b. The connective tissues form an outer epineurium, a perineurium enclosing bundles of nerve fibers, and an endoneurium surrounding each fiber.
2. The cranial nerves
   a. Twelve pairs of cranial nerves connect the brain to parts in the head, neck, and trunk.
   b. Although most are mixed, some are pure sensory, and others are primarily motor.
   c. The names of cranial nerves indicate their primary functions or the general distributions of their fibers.
   d. Some cranial nerve fibers are somatic and others are autonomic.
3. The spinal nerves
   a. Thirty-one pairs of spinal nerves originate from the spinal cord.
   b. These mixed nerves provide a two-way communication system between the spinal cord and parts in the arms, legs, neck, and trunk.
   c. Spinal nerves are grouped according to the levels from which they arise, and they are numbered in sequence.
   d. Each nerve emerges by a dorsal and a ventral root.
      (1) A dorsal root contains sensory fibers and is characterized by the presence of a dorsal root ganglion.
      (2) A ventral root contains motor fibers.
   e. Just beyond its foramen, each spinal nerve divides into several branches.
   f. Most spinal nerves combine to form plexuses in which nerve fibers are sorted and recombined so that those fibers associated with a particular part reach it together.

The Autonomic Nervous System
The autonomic nervous system consists of the portions of the nervous system that function without conscious effort.

It is concerned primarily with the regulation of visceral activities that aid in maintaining homeostasis.

1. General characteristics
    a. Autonomic functions operate as reflex actions controlled from centers in the hypothalamus, brain stem, and spinal cord.
    b. Autonomic nerve fibers are associated with ganglia in which impulses are integrated before passing out to effectors.
    c. The integrative function of the ganglia provides a degree of independence from the central nervous system.
    d. The autonomic nervous system consists of the visceral efferent fibers associated with these ganglia.
    e. The autonomic nervous system is subdivided into two divisions—sympathetic and parasympathetic.
    f. The sympathetic division functions in meeting stressful and emergency conditions.
    g. The parasympathetic division is most active under ordinary conditions.
2. Autonomic nerve fibers
    a. The autonomic fibers are efferent or motor.
    b. Sympathetic fibers leave the spinal cord and synapse in ganglia.
        (1) Preganglionic fibers pass through white rami to reach paravertebral ganglia.
        (2) Paravertebral ganglia and interconnecting fibers comprise the sympathetic trunks.
        (3) Preganglionic fibers synapse within paravertebral or collateral ganglia.
        (4) Postganglionic fibers usually pass through gray rami to reach spinal nerves before passing to effectors.
        (5) A special set of sympathetic preganglionic fibers passes through ganglia and extends to the adrenal medulla.
    c. Parasympathetic fibers begin in the brain stem and sacral region of the spinal cord and synapse in ganglia near visceral organs.

3. Autonomic transmitter substances
    a. Preganglionic sympathetic and parasympathetic fibers secrete acetylcholine.
    b. Most postganglionic sympathetic fibers secrete norepinephrine and are adrenergic; postganglionic parasympathetic fibers secrete acetylcholine and are cholinergic.
    c. The different effects of the autonomic divisions are due to different transmitter substances released by the postganglionic fibers.
    d. Most visceral organs are controlled mainly by one division.
4. Actions of autonomic transmitters
    a. Transmitters combine with receptors and cause changes in cell membranes.
    b. There are two types of cholinergic and two types of adrenergic receptors.
    c. The way cells respond to transmitters depends upon the number and type of receptors present in their membranes.
    d. Acetylcholine acts very briefly; norepinephrine and epinephrine may have more prolonged effects.
5. Control of autonomic activity
    The autonomic nervous system is controlled largely by the central nervous system.
    a. The medulla oblongata employs autonomic fibers to regulate cardiac, vasomotor, and respiratory activities.
    b. The hypothalamus employs autonomic fibers in the regulation of various visceral functions.
    c. The limbic system and cerebral cortex control emotional responses through the autonomic nervous system.

Application of Knowledge

1. If a physician plans to obtain a sample of spinal fluid from a patient, what anatomical site can be safely used, and how should the patient be positioned to facilitate this procedure?
2. What functional losses would you expect to observe in a patient who has suffered injury to the right occipital lobe of the cerebral cortex? the right temporal lobe?

3. Based on your knowledge of the cranial nerves, devise a set of tests to assess the normal functions of each of these nerves.
4. The Brown-Sequard syndrome is due to an injury on one side of the spinal cord. It is characterized by paralysis below the injury and on the same side as the injury and by loss of sensations of temperature and pain from the opposite side. How would you explain these symptoms?

Review Activities

1. Name the layers of the meninges and explain their functions.
2. Describe the location of cerebrospinal fluid within the meninges.
3. Describe the structure of the spinal cord.
4. Name the major ascending and descending tracts of the spinal cord and list the functions of each.
5. Explain the consequences of nerve fibers crossing over.
6. Describe how the brain develops.
7. Describe the structure of the cerebrum.
8. Define *cerebral cortex.*
9. Describe the location and function of the primary motor areas of the cortex.
10. Describe the location and function of Broca's area.
11. Describe the location and function of the sensory areas of the cortex.
12. Explain the function of the associations areas of the lobes of the cerebrum.
13. Define *hemisphere dominance.*
14. Explain the function of the corpus callosum.
15. Distinguish between short-term, recent, and long-term memory.
16. Describe the location and function of the basal ganglia.
17. Describe the location of the ventricles of the brain.
18. Explain how cerebrospinal fluid is produced and how it functions.
19. Name the parts of the brain stem and describe the general functions of each.
20. Define the limbic system and explain its functions.
21. Name the parts of the midbrain and describe the general functions of each.
22. Describe the pons and its functions.
23. Describe the medulla oblongata and its functions.
24. Describe the location and function of the reticular formation.
25. Distinguish between normal and paradoxical sleep.
26. Describe the functions of the cerebellum.
27. Distinguish between somatic and autonomic nervous systems.
28. Describe the structure of a peripheral nerve.
29. Name, locate, and describe the major functions of each pair of cranial nerves.
30. Explain how the spinal nerves are grouped and numbered.
31. Define *cauda equina.*
32. Describe the structure of a spinal nerve.
33. Define *plexus* and locate the major plexuses of the spinal nerves.
34. Distinguish between the sympathetic and parasympathetic divisions of the autonomic nervous system.
35. Explain how autonomic ganglia provide a degree of independence from the central nervous system.
36. Distinguish between a preganglionic fiber and a postganglionic fiber.
37. Define *paravertebral ganglion.*
38. Trace a sympathetic nerve pathway through a ganglion to an effector.
39. Explain why the effects of the sympathetic and parasympathetic divisions differ.
40. Distinguish between cholinergic and adrenergic nerve fibers.
41. Define *sympathetic tone.*
42. Explain how autonomic transmitters influence the actions of effector cells.
43. Distinguish between alpha and beta adrenergic receptors.
44. Describe three examples in which the central nervous system employs autonomic nerve pathways.

12

Somatic and Special Senses

Before parts of the nervous system can act to control body functions, they must detect what is happening inside and outside the body. This information is gathered by sensory receptors that are sensitive to changes occurring in their surroundings.

Although receptors vary greatly in their individual characteristics, they can be grouped into two major categories. The members of one group are widely distributed throughout the skin and deeper tissues and generally have simple forms. These receptors are associated with the *somatic senses* of touch, pressure, temperature, and pain. Members of the second group function as parts of complex, specialized sensory organs that are responsible for the *special senses* of smell, taste, hearing, equilibrium, and vision.

Chapter Outline

Chapter Objectives

After you have studied this chapter, you should be able to:

1. Name five kinds of receptors and explain the function of each kind.

2. Explain how receptors stimulate sensory impulses.

3. Explain how a sensation is produced.

4. Distinguish between somatic and special senses.

5. Describe the receptors associated with the senses of touch and pressure, temperature, and pain.

6. Describe how the sense of pain is produced.

7. Explain the importance of stretch receptors in muscles and tendons.

8. Explain the relationship between the senses of smell and taste.

9. Name the parts of the ear and explain the function of each part.

10. Distinguish between static and dynamic equilibrium.

11. Name the parts of the eye and explain the function of each part.

12. Explain how light is refracted by the eye.

13. Explain how depth and distance are perceived.

14. Describe the visual nerve pathway.

15. Complete the review activities at the end of this chapter. Note that the items are worded in the form of specific learning objectives. You may want to refer to them before reading the chapter.

Key Terms

accommodation   (ah-kom''o-da'shun)

auditory   (aw'di-to''re)

chemoreceptor   (ke''mo-re-sep'tor)

cochlea   (kok'le-ah)

dynamic equilibrium   (di-nam'ik e''kwĭ-lib're-um)

labyrinth   (lab'ĭ-rinth)

mechanoreceptor   (mek''ah-no-re-sep'tor)

olfactory   (ol-fak'to-re)

optic   (op'tik)

photoreceptor   (fo''to-re-sep'tor)

projection   (pro-jek'shun)

proprioceptor   (pro''pre-o-sep'tor)

referred pain   (re-furd' pān)

refraction   (re-frak'shun)

sensory adaptation   (sen'so-re ad''ap-ta'shun)

static equilibrium   (stat'ik e''kwĭ-lib're-um)

thermoreceptor   (ther''mo-re-sep'tor)

Aids to Understanding Words

aud-, to hear: *aud*itory—pertaining to hearing.

choroid, skinlike: *choroid* coat—middle, vascular layer of the eye.

cochlea, snail: *cochlea*—coiled tube within the inner ear.

corn-, horn: *corn*ea—transparent outer layer in the anterior portion of the eye.

iris, rainbow: *iris*—colored, muscular part of the eye.

labyrinth, maze: *labyrinth*—complex system of interconnecting chambers and tubes of the inner ear.

lacri-, tears: *lacri*mal gland—tear gland.

lut-, yellow: macula *lut*ea—yellowish spot on the retina.

macula, spot: *macula* lutea—yellowish spot on the retina.

oculi-, eye: orbicularis *oculi*—muscle associated with the eyelid.

olfact-, to smell: *olfact*ory—pertaining to the sense of smell.

palpebra, eyelid: levator *palpebra*e superioris—muscle associated with the eyelid.

scler-, hard: *scler*a—tough, outer protective layer of the eye.

therm-, heat: *therm*oreceptor—receptor sensitive to changes in temperature.

tympano-, drum: *tympan*ic membrane—the eardrum.

vitre-, glass: *vitre*ous humor—clear, jellylike substance within the eye.

As CHANGES OCCUR within the body and in its surroundings, *sensory receptors* are stimulated, and they trigger nerve impulses. These impulses travel on sensory pathways into the central nervous system to be processed and interpreted. As a result the person often experiences a particular type of feeling or sensation.

Receptors and Sensations

Although there are many kinds of sensory receptors, they have features in common. For example, each type of receptor is particularly sensitive—that is, it has a low threshold—to a distinct kind of environmental change and is much less sensitive to other forms of stimulation.

Types of Receptors

On the basis of their sensitivities, it is possible to identify five general groups of sensory receptors. The members of each group are sensitive to changes in one of the following factors:

1. *Chemical concentration.* Receptors that are stimulated by changes in concentration of chemical substances are called **chemoreceptors.** Those associated with the senses of smell and taste are of this type. Chemoreceptors in various internal organs can detect changes in the blood concentrations of such substances as oxygen, hydrogen ions, and glucose.
2. *Tissue damage.* Whenever tissues are damaged in any way, **pain receptors** (nociceptors) are likely to be stimulated.
3. *Temperature change.* Receptors sensitive to temperature change are called **thermoreceptors,** and there are two different sets. One set, the *heat receptors,* responds to temperatures above body temperature; the other set, the *cold receptors,* is sensitive to temperatures below that of the body.
4. *Mechanical forces.* A number of sensory receptors are sensitive to mechanical forces, such as changes in pressure or movement of fluids. As a rule, these **mechanoreceptors** detect changes that cause them to become deformed. For example, those called *proprioceptors* are sensitive to changes in the tensions of muscles and tendons.
5. *Light intensity.* Light or **photoreceptors** occur only in the eyes, and they respond whenever they are exposed to sufficient intensity of light energy.

Sensory Impulses

Although the ends of nerve fibers often serve as sensory receptors, other kinds of cells located close to nerve fiber endings may also serve as receptors. In either case, when receptors are stimulated, local changes occur in their membrane potentials (receptor potential). As a result, an electric current is generated that is graded and, thus, reflects the intensity of stimulation (see chapter 10).

If a receptor is a nerve fiber and the change in its membrane potential is sufficient to reach threshold, an action potential is generated, and a sensory impulse travels away on the fiber. However, if the receptor is another type of cell, its receptor potential must be transferred to a nerve fiber before an action potential can be triggered.

Such sensory impulses enter the central nervous system by means of peripheral nerves and are analyzed and interpreted by the brain (see chapter 11).

Sensations

A **sensation** is a feeling that occurs when sensory impulses are interpreted by the brain. Because all the nerve impulses that travel from sensory receptors to the central nervous system are alike, different kinds of sensations must be due to the way the brain interprets the impulses rather than to differences in the receptors. In other words, when a receptor is stimulated, the resulting sensation depends on what region of the cerebral cortex receives the impulse. For example, impulses reaching one region always are interpreted as sounds, and those reaching another portion always are sensed as touch. Impulses reaching still other regions of the brain always are interpreted as pain.

It makes little difference how the impulses are stimulated. Pain receptors, for example, can be stimulated by excessive heat, cold, or pressure, but the sensation is always the same since the impulses are interpreted by the same part of the brain. Similarly, nerve impulses are sometimes triggered in visual receptors by factors other than light. When this happens, the person may "see" lights, even though no light is entering the eye, since any impulses reaching the visual cortex are interpreted as light.

At the same time a sensation is created, the brain causes it to seem to come from the receptors being stimulated. This process is called **projection,** because the brain projects the sensation back to its apparent source. Projection allows the person to pinpoint the region of stimulation. Thus, the eyes appear to see, the ears appear to hear, and so forth.

Figure 12.1
Touch and pressure receptors include *(a)* free ends of sensory nerve fibers, *(b)* Meissner's corpuscle, and *(c)* Pacinian corpuscle.

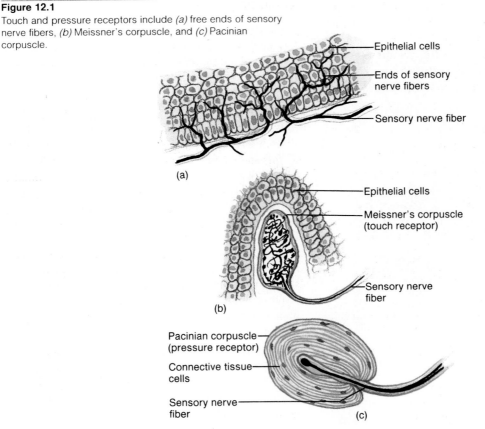

Epithelial cells

Ends of sensory nerve fibers

Sensory nerve fiber

(a)

Epithelial cells

Meissner's corpuscle (touch receptor)

Sensory nerve fiber

(b)

Pacinian corpuscle (pressure receptor)

Connective tissue cells

Sensory nerve fiber

(c)

Sensory Adaptation

When they are subjected to continuous stimulation, many receptors undergo an adjustment called **sensory adaptation.** As the receptors adapt, impulses leave them at lower and lower rates, until finally they may completely fail to send signals. Once receptors have adapted, impulses can be triggered only if the strength of the stimulus is changed.

Sensory adaptation is experienced when a person enters a room where there is a strong odor. At first the scent seems intense, but it becomes less and less noticeable as the smell receptors adapt. If the person remains in the room for a minute or more, he or she may become totally adapted to the odor and quite unaware of its presence. If the person leaves the room, however, and then reenters it, the receptors probably will be stimulated once again.

1. List five general types of sensory receptors.
2. What do these receptors have in common?
3. Explain how a sensation occurs.
4. What is meant by sensory adaptation?

Somatic Senses

Somatic senses are those that involve receptors associated with the skin, muscles, joints, and visceral organs. These senses can be divided into three groups, as follows:

1. Those associated with changes occurring at the body surface (exteroceptive senses), which include the senses of touch, pressure, and temperature.
2. Those associated with changes occurring in muscles and tendons, and in body position (proprioceptive senses).
3. Those associated with changes occurring in visceral organs (visceroceptive senses). Except for visceral pain, these visceral senses will be discussed in later chapters.

Touch and Pressure Senses

The senses of touch and pressure employ several kinds of receptors. As a group, these receptors are sensitive to mechanical forces that cause tissues to be deformed or displaced (figure 12.1). They include the following:

1. **Free ends of sensory nerve fibers.** These receptors occur commonly in epithelial tissues

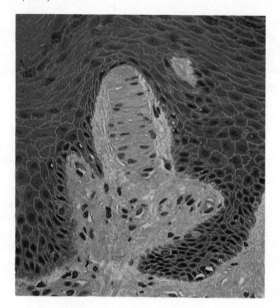

where they end between epithelial cells. They are associated with sensations of touch and pressure.

2. **Meissner's corpuscles.** These structures consist of small, oval masses of flattened connective tissue cells surrounded by connective tissue sheaths. Two or more nerve fibers branch into each corpuscle and end within it as tiny knobs.

   Meissner's corpuscles are especially numerous in the hairless portions of the skin, such as the lips, fingertips, palms, soles, nipples, and external genital organs. They are sensitive to the motion of objects that barely contact the skin, and impulses from them are interpreted as the sensation of light touch. They are also used when a person touches something to judge its texture. (See figure 12.2.)

3. **Pacinian corpuscles.** These sensory bodies are relatively large, ellipsoidal structures composed of connective tissue fibers and cells. They are commonly found in the deeper subcutaneous tissues of such organs as the hands, feet, penis, clitoris, urethra, and breasts, and also occur in tendons and the ligaments of joints.

   Pacinian corpuscles are stimulated by heavy pressure and are thought to be associated with the sensation of deep pressure. They may also serve to detect vibrations in tissues.

Temperature Senses

As mentioned earlier, the temperature senses employ two kinds of thermoreceptors. Although there is some question about the identity of these receptors, they seem to include two types of *free nerve endings* located in the skin called *heat receptors* and *cold receptors*. The heat receptors are most sensitive to temperatures above 25°C (77°F) and become unresponsive at temperatures above 45°C (113°F). As 45°C is approached, pain receptors also are triggered, producing a burning sensation.

Cold receptors are most sensitive to temperatures between 10°C (50°F) and 20°C (68°F). If the temperature drops below 10°C, pain receptors are stimulated, and the person feels a freezing sensation.

Both heat and cold receptors demonstrate rapid adaptation, so that within about a minute following stimulation, the sensation of hot or cold begins to fade. A person experiences this form of sensory adaptation upon entering a tub of hot water or wading into a cold lake or stream. At first the temperature sensation may be unpleasant, but within a few moments the temperature receptors adapt, at least partially, and the person feels more comfortable.

Sense of Pain

The sense of pain also involves receptors that consist of free nerve fiber endings. These receptors are widely distributed throughout the skin and internal tissues, except for the nerve tissue in the brain, which lacks pain receptors.

Pain receptors have a protective function in that they are stimulated whenever tissues are being damaged. The resulting sensory impulses are likely to stimulate the reticular formation and, thus, activate the brain. The pain sensation usually is perceived as unpleasant, and it serves as a signal that something should be done to remove the source of the stimulation.

Although most pain receptors can be stimulated by more than one type of change, some are most sensitive to mechanical damage while others are particularly sensitive to extremes in temperature. Still other pain receptors are most responsive to the presence of various chemicals such as hydrogen ions, potassium ions, or polypeptides (kinins) from the breakdown of proteins, histamine, and acetylcholine. A deficiency of blood (ischemia) and, thus, a deficiency of oxygen (hypoxia) in a tissue or the stimulation of mechanical-sensitive receptors also triggers pain sensation. Pain elicited during a muscle cramp, for example, seems to be related to an interruption of blood flow that occurs

Headache
A Practical Application

One of the more commonly experienced forms of pain is headache. Although the brain tissue itself lacks pain receptors, nearly all other tissues of the head, including meninges and blood vessels associated with the brain, are well supplied with them.

Most headaches seem to be related to stressful life situations that result in fatigue, emotional tension, anxiety, or frustration. These conditions are reflected in various physiological changes. For example, they may cause prolonged contraction of skeletal muscles in the forehead, sides of the head, or back of the neck. Such contractions may stimulate pain receptors and produce what is often called a *tension headache*. In other cases, people suffer more severe *vascular headaches* that accompany constriction or dilation of cranial blood vessels. The throbbing headache of a "hangover" following excessive consumption of alcohol may be due to blood pulsating through cranial vessels. Similar headaches may be experienced by persons who are sensitive to certain food additives, such as monosodium glutamate (MSG) that sometimes is used to enhance the flavor of foods.

Still another form of vascular headache is called *migraine*. In this disorder, certain cranial blood vessels seem to constrict, producing a localized cerebral blood deficiency. As a result, the person may experience a variety of symptoms, such as seeing patterns of bright light that may interfere with vision or feeling numbness in the limbs or face. Typically, the vasoconstriction is followed by vasodilation of the cranial vessels and a severe headache, which is usually limited to one side of the head and lasts for several hours. Migraine headaches tend to run in families and generally begin in adolescence. They also are more common in women than in men.

Even though the pain of some headaches originates inside the cranium, the sensation often is referred to the surface. Other causes of headaches include high blood pressure, increased intracranial pressure due to tumor or blood escaping from a ruptured vessel, temporomandibular joint syndrome (see chapter 9), decreased cerebrospinal fluid pressure following a lumbar puncture (see chapter 11), or sensitivity to or withdrawal from various drugs.

---

as the sustained contraction squeezes capillaries, as well as to the stimulation of mechanical-sensitive pain receptors. Also, when blood flow is interrupted, pain-stimulating chemicals accumulate excessively. Increasing the blood flow through the sore tissue may relieve the resulting pain, and this is why heat is sometimes applied to reduce muscle soreness. Heat causes blood vessels to dilate and, thus, promotes blood flow, which helps to reduce the concentration of the pain-stimulating substances. In some conditions, accumulating chemicals cause the thresholds of pain receptors to be lowered. Consequently, inflamed tissues may become more sensitive to heat or pressure than before.

Pain receptors adapt very little, if at all, and once such a receptor has been activated, even by a single stimulus, it may continue to send impulses into the central nervous system for some time.

A painful condition may occur in the heart of a person suffering from the artery disease called *atherosclerosis*. In this disease, solid materials accumulate abnormally inside arteries, interfering with the flow of blood. If this happens in the arteries that supply blood to the heart muscle, the victim may feel chest pain, called *angina pectoris*, during times of physical exertion. This pain seems to be related to a decreased oxygen supply in the heart muscle and a buildup of pain-stimulating chemicals. The pain is relieved by rest, which provides time for blood to supply oxygen and to remove the offending substances.

Visceral Pain

As a rule, pain receptors are the only receptors in visceral organs whose stimulation produces sensations. Pain receptors in these organs seem to respond differently to stimulation than those associated with surface tissues. For example, localized damage to intestinal tissue, as may occur during surgical procedures, may not elicit any pain sensations even in a conscious person. However, when visceral tissues are subjected to more widespread stimulation, as when intestinal tissues are stretched or when the smooth muscles in the intestinal walls undergo spasms, a strong pain sensation may follow. Once again, the resulting pain seems to be related to the stimulation of mechanical-sensitive receptors and to a decreased blood flow accompanied by a lower tissue oxygen concentration and an accumulation of pain-stimulating chemicals.

Another characteristic of visceral pain is that it may feel as if it is coming from some part of the body other than the part being stimulated—a phenomenon called **referred pain.** Pain originating from the heart, for example, may be referred to the left shoulder or the inside of the left arm. Pain from the lower esophagus, stomach, or small intestine may seem to be coming from the upper central (epigastric) region of the abdomen; pain from the urogenital tract may be referred to the lower central (hypogastric) region of the abdomen or to the sides between the ribs and the hip (figure 12.3).

Figure 12.3

Surface regions to which visceral pain may be referred.

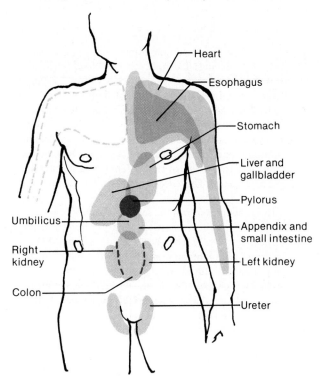

Heart

Esophagus

Stomach

Liver and gallbladder

Pylorus

Umbilicus

Appendix and small intestine

Right kidney

Left kidney

Colon

Ureter

The occurrence of referred pain seems to be related to *common nerve pathways* that are used by sensory impulses coming both from skin areas and from visceral organs (see chapter 11). In other words, pain impulses from the heart seem to be conducted over the same nerve paths as those coming from the skin of the shoulder and the inside of the left arm, as shown in figure 12.4. Consequently, the cerebral cortex may incorrectly interpret the source of the impulses as the shoulder or arm rather than the heart.

Pain originating from the parietal layers of thoracic and abdominal membranes—parietal pleura, pericardium, or peritoneum—usually is not referred; instead such pain is likely to be felt over the area being stimulated.

Role of Neuropeptides

Neuropeptides are substances synthesized in the brain and spinal cord that serve as neurotransmitters and neuromodulators (see chapter 10). These substances include a group, called **enkephalins,** that are released by axons in the parts of the spinal cord that carry pain impulses and in parts of the brain within the limbic system. The limbic system, which includes the thalamus, also serves as the pathway for pain impulses. Enkephalins act to inhibit such impulses; thus, they relieve pain sensations much as morphine and certain other

Figure 12.4

Pain originating in the heart may feel as if it is coming from the skin because sensory impulses from these two regions follow common nerve pathways to the brain.

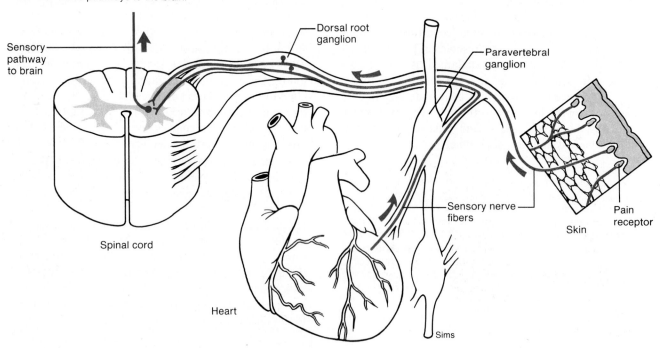

Sensory pathway to brain

Dorsal root ganglion

Paravertebral ganglion

Sensory nerve fibers

Pain receptor

Skin

Spinal cord

Heart

Sims

narcotic drugs do. In fact, enkephalins seem to function by binding to the same receptor sites on neuron membranes as does morphine.

Another group of neuropeptides, called **endorphins,** have pain-suppressing, morphinelike actions. They occur in the pituitary gland and in various regions of the nervous system involved with the transmission of pain impulses.

It is believed that enkephalins and endorphins may be released at times when pain impulses are triggered excessively, and that in this way they provide natural pain control.

---

1. Describe three types of touch and pressure receptors.
2. Describe the thermoreceptors.
3. What types of stimuli excite pain receptors?
4. What is referred pain?
5. Explain how neuropeptides may help to control pain.

Stretch Receptors

Stretch receptors are proprioceptors that provide information to the spinal cord and brain concerning the lengths and tensions of muscles. There are two main kinds of these receptors: muscle spindles and Golgi tendon organs; however, no sensation results when they are stimulated.

Figure 12.5
Muscle spindles, which are modified muscle fibers, are stimulated by changes in muscle length.

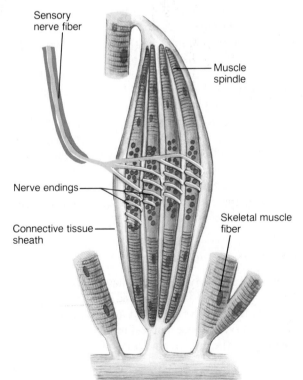

Sensory nerve fiber

Muscle spindle

Nerve endings

Connective tissue sheath

Skeletal muscle fiber

Muscle spindles are located in skeletal muscles near the junctions of these muscles and their tendons. Each spindle consists of one or more small, modified skeletal muscle fibers (intrafusal fibers) enclosed in a connective tissue sheath. Near its center, each fiber has a specialized nonstriated region with the end of a sensory nerve fiber wrapped around it (figure 12.5).

The striated portions of the muscle fiber can contract, and they keep the spindle taut. However, if the whole muscle becomes relaxed and is stretched longer than usual, the muscle spindle detects this change and sensory nerve impulses may be triggered on its nerve fiber. Such sensory impulses travel into the spinal cord and onto motor fibers leading back to the same muscle, causing it to contract. This action, which is called a **stretch reflex,** opposes the lengthening of the muscle and helps a person to maintain the desired position of a limb in spite of gravitational or other forces tending to move it (see chapter 10).

Golgi tendon organs are found in tendons close to their attachments to muscles. Each is connected to a set of skeletal muscle fibers and is innervated by a sensory neuron.

These receptors have relatively high thresholds and are stimulated by increased tension. Sensory impulses from these receptors produce a reflex that inhibits contraction of the muscle whose tendon they occupy. Thus, the Golgi tendon organs stimulate a reflex with an effect opposite that of a stretch reflex. This reflex also helps to maintain posture, and it protects muscle attachments from being pulled away from their insertions by excessive tension. Chart 12.1 summarizes the somatic receptors and their functions.

Chart 12.1	Somatic receptors	
Type	Function	Sensation
Free nerve endings (mechanoreceptors)	Detect changes in pressure	Touch, pressure
Meissner's corpuscles (mechanoreceptors)	Detect objects moving over the skin	Touch, texture
Pacinian corpuscles (mechanoreceptors)	Detect changes in pressure	Deep pressure, vibrations
Free nerve endings (thermoreceptors for heat)	Detect changes in temperature	Heat
Free nerve endings (thermoreceptors for cold)	Detect changes in temperature	Cold
Free nerve endings (pain receptors)	Detect tissue damage	Pain
Muscle spindle (mechanoreceptor)	Detect changes in muscle length	None
Golgi tendon organ (mechanoreceptor)	Detect changes in muscle tension	None

Control of Pain
A Practical Application

Most people suffer from pains that arise from injuries and diseases, and few are able to remain indifferent to these sensations.

Interestingly, the intensity of stimulation needed to trigger pain receptors seems to be about the same for all individuals, yet people vary greatly in the ways they perceive pain and respond to it. These differences seem to be related to past experiences, cultural conditioning, and the circumstances under which the painful stimuli are encountered.

Acute pain, which is short-term and often due to a known cause, usually can be endured. However, *chronic pain,* which persists for a long period of time, tends to weaken its victim and become the dominant feature of that person's life.

When pain is persistent and interferes with a person's ability to function effectively, a way to control it is usually sought. For example, the person may take a pain-reducing drug (analgesic). Other methods of relieving pain include mechanically rubbing the affected part or a nearby region of the skin, applying heat or a liniment that produces a local irritation of the skin, or treatment with acupuncture—the insertion of needles into various tissues. Although acupuncture is not well understood, such procedures are known to trigger impulses on large sensory peripheral nerve fibers and they, in turn, seem to inhibit the transmission of pain impulses originating in the same areas. Some investigators believe that the sensory impulses cause the release of enkephalins and endorphins within the spinal cord or brain, and that these neuropeptides produce an analgesic effect.

Treatment of chronic pain often requires the use of narcotic drugs. In some difficult cases, electrodes are applied to the skin or implanted in appropriate regions of the spinal cord and are used to stimulate certain nerve fibers using a battery-powered device. Such electrical stimulation may produce an analgesic effect over relatively large areas of the body. In other cases, various sensory nerve pathways may be sectioned surgically. This procedure will prevent pain impulses from reaching the pain-interpreting centers of the brain, but it will also result in a permanent loss of sensation from body parts innervated by the sectioned nerve fibers.

---

1. Describe a muscle spindle.
2. Explain how this receptor helps to maintain posture.
3. Where are Golgi tendon organs located?
4. What is the function of this receptor?

Sense of Smell

The sense of smell, like the other special senses, is associated with complex sensory structures in the head.

Olfactory Receptors

The smell or olfactory receptors are similar to those for taste (discussed in a subsequent section of this chapter) in that they are chemoreceptors stimulated by chemicals dissolved in liquids. These two senses function closely together and aid in food selection, since food is often smelled at the same time that it is tasted. It is often difficult to tell what part of a food sensation is due to smell and what part is really taste. For this reason, an onion tastes quite different when sampled with the nostrils closed, because much of the usual onion sensation is due to odor. Similarly, during a head cold, excessive mucus secretions may cover the olfactory receptors, and food may seem to lose its taste.

Olfactory Organs

The **olfactory organs,** which contain the olfactory receptors, also include various epithelial supporting cells. These organs appear as yellowish-brown masses surrounded by pinkish mucous membrane. They cover the upper parts of the nasal cavity, the superior nasal conchae, and a portion of the nasal septum (figure 12.6).

The **olfactory receptor cells** are bipolar neurons surrounded by columnar epithelial cells. The neurons have knobs at the distal ends of their dendrites that are covered by hairlike cilia. The cilia project into the nasal cavity and are thought to be the sensitive portions of the receptors.

Chemicals that stimulate olfactory receptors enter the nasal cavity as gases, but they must dissolve at least partially in the watery fluids that surround the cilia before they can be detected. It is believed that such chemicals must also be soluble in lipids before they can stimulate the receptors, because cilia are composed primarily of lipid materials.

Olfactory Nerve Pathways

Once the olfactory receptors have been stimulated, nerve impulses are triggered, and they travel along the axons of the receptor cells that are the fibers of the olfactory nerves. These nerve fibers pass through tiny openings in the cribriform plates of the ethmoid bone

Figure 12.6
The olfactory receptor cells, which have cilia at their distal ends, are supported by columnar epithelial cells.

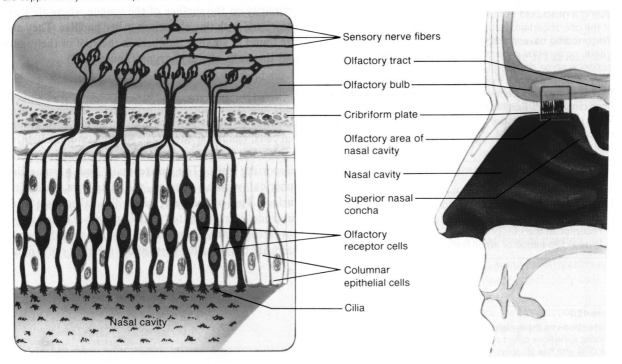

Sensory nerve fibers
Olfactory tract
Olfactory bulb
Cribriform plate
Olfactory area of nasal cavity
Nasal cavity
Superior nasal concha
Olfactory receptor cells
Columnar epithelial cells
Cilia
Nasal cavity

and lead to neurons located in enlargements called **olfactory bulbs.** These structures lie on either side of the crista galli of the ethmoid bone. (See figure 7.24.) From the olfactory bulbs, the impulses travel along the **olfactory tracts** to interpreting centers located in the base of the frontal lobes (olfactory cortex).

Olfactory Stimulation

The way in which various substances stimulate the olfactory receptors is poorly understood. One hypothesis suggests that the shapes of gaseous molecules may fit receptor sites on the cilia that have complementary shapes. A nerve impulse, according to this idea, is triggered when a molecule binds to its particular receptor site.

Attempts to classify the sensations associated with odors have met with only limited success. Many researchers, however, agree that there are at least seven groups of primary odors:

1. *Camphoraceous,* like the scent of camphor.
2. *Musky,* like the scent of musk.
3. *Floral,* like the scent of flowers.
4. *Pepperminty,* like the scent of oil of peppermint.
5. *Ethereal,* like the scent of ether.
6. *Pungent,* like the scent of spices.
7. *Putrid,* like the scent of decaying meat.

Any particular odor might be described as one of the primary odors or some combination of two or more of them.

Since the olfactory organs are located high in the nasal cavity above the usual pathway of inhaled air, a person may have to sniff and force air over the receptor areas to smell something that has a faint odor. Also, olfactory receptors undergo sensory adaptation rather rapidly, so the intensity of an olfactory sensation drops about 50% within a second following stimulation. Within a minute, the receptors may become almost insensitive to a given odor, but even though they have adapted to one scent, their sensitivity to other odors remains unchanged.

The neurons that function as olfactory receptors are the only parts of the nervous system that are in direct contact with the outside environment. Because of their exposed positions, these neurons are subject to damage; and because damaged neurons usually are not replaced, persons are likely to experience a progressive diminishing of olfactory sense with age. In fact, some investigators have estimated that a person loses about 1% of the olfactory receptors every year.

stimulate these receptors are usually *organic compounds,* such as sugars and polysaccharides. A few inorganic substances, including some salts of lead and beryllium, also elicit sweet sensations.

Sour receptors occur primarily along the margins of the tongue. They are stimulated mainly by acids. The intensity of a sour sensation is roughly proportional to the concentration of the *hydrogen ions* in the substance being tasted.

Salt receptors are most common in the tip and the upper front portion of the tongue. They are stimulated mainly by ionized *inorganic salts.* The quality of the sensation each produces is related to the kind of positively charged ion, such as Na^+ from table salt, that it releases into solution.

Bitter receptors are located toward the back of the tongue. They are stimulated by a variety of chemical substances, most of which are organic compounds, although some inorganic salts of magnesium and calcium produce bitter sensations, too.

One group of bitter compounds of particular interest are the *alkaloids,* which include a number of poisons such as strychnine, nicotine, and morphine. The fact that persons commonly reject bitter substances may be related to a kind of protective mechanism that causes them to avoid alkaloids in foods.

Taste receptors, like olfactory receptors, undergo sensory adaptation relatively rapidly. The resulting loss of taste can be circumvented by moving bits of food over the surface of the tongue to stimulate different receptors at different times.

Although the taste cells are located very close to the surface of the tongue and are somewhat exposed to damage by environmental factors, the sense of taste is not as likely to diminish with age as the sense of smell. This is because taste cells are reproduced continually, so that any one of these cells functions for only about a week before it is replaced.

Taste Nerve Pathways

Sensory impulses from taste receptors located in the anterior two-thirds of the tongue travel on fibers of the facial nerve (VII); those from receptors in the posterior one-third of the tongue and the back of the mouth pass along the glossopharyngeal nerve (IX); and those from receptors at the base of the tongue and the pharynx travel on the vagus nerve (X) (see chapter 11).

These cranial nerves conduct the impulses into the medulla oblongata. From there the impulses ascend to the thalamus and are directed to the gustatory cortex, located in the parietal lobe along a deep portion of the lateral sulcus.

---

1. Why is saliva necessary for a sense of taste?
2. Name the four primary taste sensations.
3. What characteristic of taste receptors helps to maintain a sense of taste with age?
4. Trace a sensory impulse from a taste receptor to the cerebral cortex.

Sense of Hearing

The organ of hearing, the **ear,** has external, middle, and inner parts. In addition to making hearing possible, the ear functions in the sense of equilibrium, which will be discussed in a subsequent section of this chapter.

External Ear

The external ear consists of an outer, funnel-like structure called the **auricle** and an S-shaped tube, the **external auditory meatus,** that leads inward for about 2.5 cm (figure 12.10).

The external auditory meatus passes into the temporal bone. Near its opening the tube is guarded by hairs. It is lined with skin that contains numerous modified sweat glands called *ceruminous glands,* which secrete wax (cerumen). The hairs and wax help to keep relatively large foreign objects, such as insects, from entering the ear.

Sounds generally are created by vibrations of objects that are transmitted through matter in the form of sound waves. For example, the sounds of some musical instruments are produced by vibrating strings or reeds, and the sounds of the voice are created by vibrating vocal folds in the larynx. The auricle of the ear helps to collect sound waves traveling through air and directs them into the auditory meatus.

After entering the meatus, the sound waves pass to the end of the tube and cause pressure changes on the eardrum. The eardrum moves back and forth in response and thus reproduces the vibrations of the sound wave source.

Middle Ear

The middle ear consists of an air-filled space in the temporal bone called the **tympanic cavity** that separates the external and internal ears, an eardrum or tympanic membrane, and three small bones called auditory ossicles.

The **tympanic membrane** is a semitransparent membrane covered by a thin layer of skin on its outer surface and by mucous membrane on the inside. It has an oval margin and is cone-shaped, with the apex of the cone directed inward. Its cone shape is maintained by the attachment of one of the auditory ossicles (malleus).

Figure 12.10
Major parts of the ear.

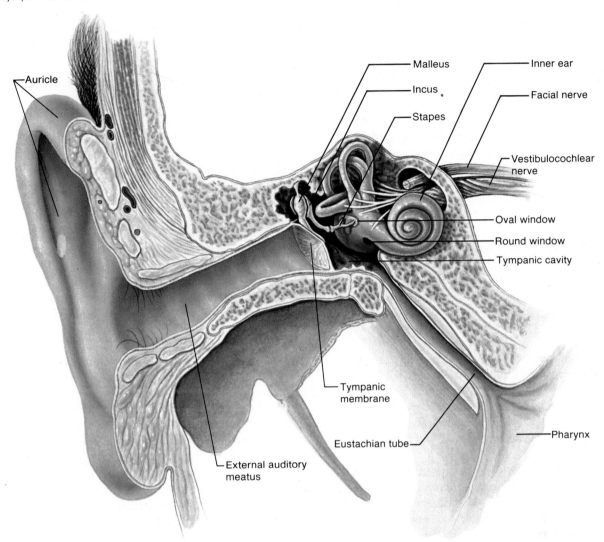

The **auditory ossicles**—the *malleus* (hammer), *incus* (anvil), and *stapes* (stirrup)—are attached to the wall of the tympanic cavity by tiny ligaments and are covered by mucous membrane. These bones form a bridge connecting the eardrum to the inner ear and function to transmit vibrations between these parts. Specifically, the malleus is attached to the eardrum, and when the eardrum vibrates, the malleus vibrates in unison with it. The malleus causes the incus to vibrate, and it passes the movement onto the stapes. The stapes is held by ligaments to an opening in the wall of the tympanic cavity called the **oval window.** Vibration of the stapes at the oval window causes motion in a fluid within the inner ear. These vibrations are responsible for stimulating the receptors of hearing (figure 12.10).

In addition to transmitting vibrations, the auditory ossicles form a lever system that helps increase (amplify) the force of the vibrations as they are passed from the eardrum to the oval window. Also, because the ossicles transmit vibrations from the relatively large surface of the eardrum to a much smaller area at the oval window, the vibration force becomes concentrated as it travels from the external to the inner ear. As a result of these two factors, the pressure (per mm^2) applied by the stapes at the oval window is about twenty-two times greater than that exerted on the eardrum by sound waves.

The middle ear also contains two small skeletal muscles that are attached to the auditory ossicles. One of them, the *tensor tympani,* is inserted on the medial surface of the malleus, and when it contracts it pulls the bone inward. The other muscle, the *stapedius,* is attached to the posterior side of the stapes and serves to pull it outward (figure 12.11). These muscles are the effectors in the **tympanic reflex,** which is elicited by loud sounds. When the reflex occurs, the muscles contract and the malleus and stapes are moved. As a result, the bridge of ossicles in the middle ear becomes more rigid, and its effectiveness in transmitting vibrations to the inner ear is reduced.

The tympanic reflex is a protective mechanism that reduces pressure from loud sounds that might otherwise damage the hearing receptors. The tensor tympani muscle also functions to maintain a steady pull on the eardrum. This is important because a loose tympanic membrane would not be able to transmit vibrations effectively to the auditory ossicles.

Eustachian Tube

An **eustachian tube** (auditory tube) connects each middle ear to the throat. This tube allows air to pass between the tympanic cavity and the outside of the body by way of the throat and mouth. It is important in maintaining equal air pressure on both sides of the eardrum, which is necessary for normal hearing (figure 12.10).

The function of the eustachian tube can be experienced during rapid altitude change. For example, as a person moves from a high altitude to a lower one, the air pressure on the outside of the membrane becomes greater and greater. The eardrum may be pushed inward, out of its normal position, and hearing may be impaired.

When the air pressure difference is great enough, some air may force its way up through the eustachian tube into the middle ear. At the same time, the pressure on both sides of the eardrum is equalized, and the drum moves back into its regular position. The person usually hears a popping sound at this moment, and normal hearing is restored. A reverse movement of air ordinarily occurs when a person moves from low altitude into a higher one.

The eustachian tube is usually closed by valvelike flaps in the throat, which may inhibit air movements into the middle ear. Swallowing, yawning, or chewing may aid in opening the valves, and these actions can hasten the equalization of air pressure if discomfort is experienced during altitude changes.

The mucous membranes that line the eustachian tubes are continuous with the linings of the middle ears. Consequently, the tubes provide a route by which mucous membrane infections of the throat (pharyngitis) may pass to the ear and cause an infection of the middle ear (otitis media). For this reason, it is poor practice to pinch one nostril when blowing the nose, because pressure in the nasal cavity may then force material from the throat up the eustachian tube and into the middle ear.

Inner Ear

The inner ear consists of a complex system of intercommunicating chambers and tubes called a **labyrinth.** In fact, there are two such structures in each ear—the osseous and membranous labyrinths.

The *osseous labyrinth* is a bony canal in the temporal bone; the *membranous labyrinth* lies within the osseous one and has a similar shape (figure 12.12). Between the osseous and membranous labyrinths is a fluid, called *perilymph,* that is secreted by cells in the wall of the bony canal. The membranous labyrinth contains another fluid, called *endolymph,* whose composition is slightly different.

The parts of the labyrinths include a **cochlea** that functions in hearing, and three **semicircular canals** that function in providing a sense of equilibrium. A bony chamber, called the **vestibule,** which is located between the cochlea and the semicircular canals, contains membranous structures that serve both hearing and equilibrium.

The **cochlea,** as its name suggests, is shaped like the coiled shell of a snail. Inside, it contains a bony core (modiolus) and a thin bony shelf (spiral lamina) that winds around the core like the threads of a screw. The shelf divides the bony labyrinth of the cochlea into upper and lower compartments. The upper compartment, called the *scala vestibuli,* leads from the oval window to the apex of the spiral. The lower compartment, the *scala tympani,* extends from the apex of the cochlea to a membrane-covered opening in the wall of the inner ear called the **round window.** These compartments constitute the bony labyrinth of the cochlea, and they are filled with perilymph. At the apex of the cochlea, the fluids in the chambers can flow together through a small opening (helicotrema).

The membranous labyrinth of the cochlea is represented by the *cochlear duct* (scala media), which is filled with endolymph. It lies between the two bony

Figure 12.11

Medial view of the middle ear. Two small muscles that are attached to the malleus and stapes serve as effectors in the tympanic reflex.

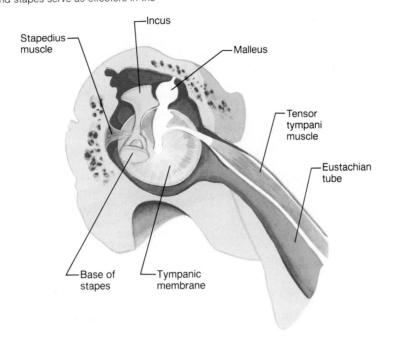

Incus

Stapedius muscle

Malleus

Tensor tympani muscle

Eustachian tube

Base of stapes

Tympanic membrane

Figure 12.12

The osseous labyrinth of the inner ear is separated from the membranous labyrinth by perilymph.

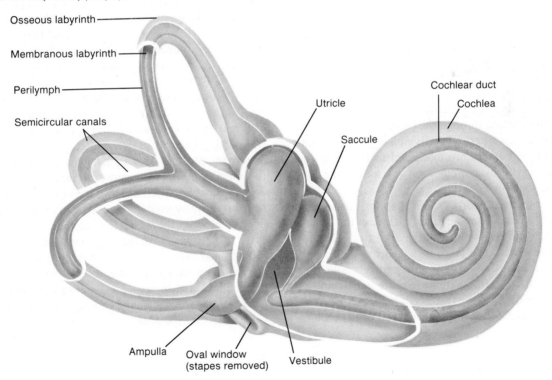

Osseous labyrinth

Membranous labyrinth

Perilymph

Semicircular canals

Utricle

Saccule

Cochlear duct

Cochlea

Ampulla

Oval window (stapes removed)

Vestibule

Figure 12.13

(a) The cochlea consists of a coiled, bony canal with a membranous tube inside. (b) If the cochlea is unwound, the membranous tube is seen ending as a closed sac at the apex of the bony canal.

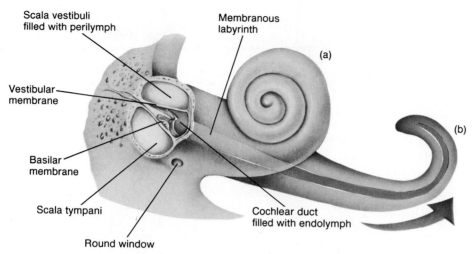

Scala vestibuli filled with perilymph

Membranous labyrinth

Vestibular membrane

Basilar membrane

Scala tympani

Round window

Cochlear duct filled with endolymph

(a)

(b)

compartments and ends as a closed sac at the apex of the cochlea. The cochlear duct is separated from the scala vestibuli by a *vestibular membrane* (Reissner's membrane) and from the scala tympani by a *basilar membrane* (figure 12.13).

The basilar membrane extends from the bony shelf of the cochlea and forms the floor of the cochlear duct. It contains many thousands of stiff, elastic fibers, whose lengths vary becoming progressively longer from the base of the cochlea to its apex. Vibrations entering the perilymph at the oval window travel along the scala vestibuli and pass through the vestibular membrane to enter the endolymph of the cochlear duct, where they cause movements in the basilar membrane.

After passing through the basilar membrane, the sound vibrations enter the perilymph of the scala tympani, and their forces are dissipated to the air in the tympanic cavity by movement of the membrane covering the round window.

The **organ of Corti,** which contains the hearing receptors, is located on the upper surface of the basilar membrane and stretches from the apex to the base of the cochlea. Its receptor cells, which are called **hair cells,** are arranged in rows, and they possess numerous hairlike processes that extend into the endolymph of the cochlear duct. Above these cells is a *tectorial membrane* that is attached to the bony shelf of the cochlea

and passes like a roof over the receptor cells, making contact with the tips of their hairs. (See figures 12.14 and 12.15.)

As sound vibrations pass through the inner ear, the hairs shear back and forth against the tectorial membrane, and the mechanical deformation of the hairs stimulates the receptor cells. Various receptor cells, however, have slightly different sensitivities to such deformation of their hairs. Thus, a sound that produces a particular frequency of vibration will excite certain receptor cells, while a sound involving another frequency will stimulate a different set of cells.

Although the receptor cells are epithelial cells, they act somewhat like neurons (see chapter 10). For example, when such a cell is at rest, its membrane is polarized. When it is stimulated appropriately its membrane becomes depolarized, ion channels open, and the membrane becomes more permeable to calcium ions. Although the cell has no axon or dendrites, it has neurotransmitter-containing vesicles in the cytoplasm near its base. In the presence of calcium ions, some of these vesicles fuse with the cell membrane and release neurotransmitter substance to the outside. The neurotransmitter stimulates the ends of nearby sensory nerve fibers, and in response they transmit nerve impulses along the cochlear branch of the vestibulocochlear nerve to the brain.

Figure 12.14
(a) Cross section of the cochlea; *(b)* organ of Corti.
(*a* from McClintock, J. R., *Physiology of the Human Body,* 2d ed. Copyright
© 1978 John Wiley & Sons, Inc., New York. Reprinted by permission.)

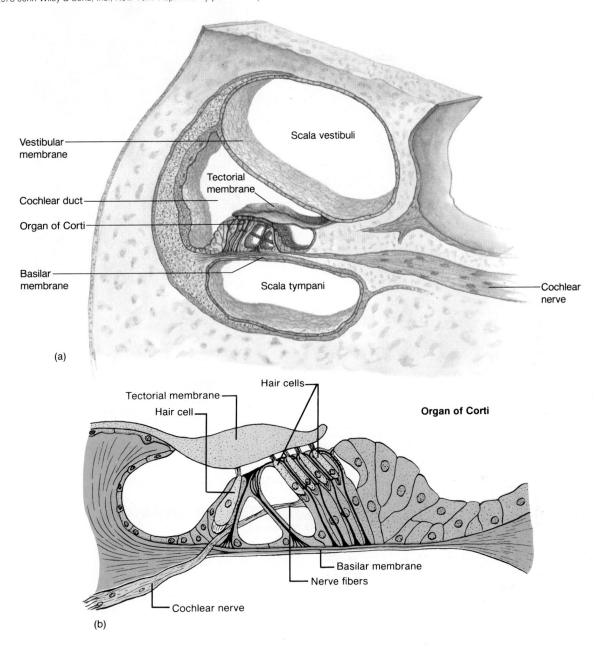

Figure 12.15

(a) A micrograph of the organ of Corti; (b) a scanning electron micrograph of hair cells in the organ of Corti.

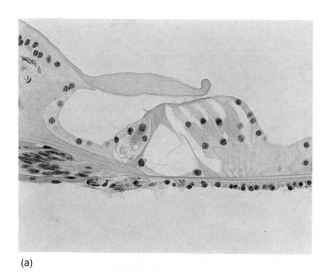

(a)

(b)

Figure 12.16

Receptors in various regions of the cochlear duct are sensitive to different frequencies of vibration (cps).

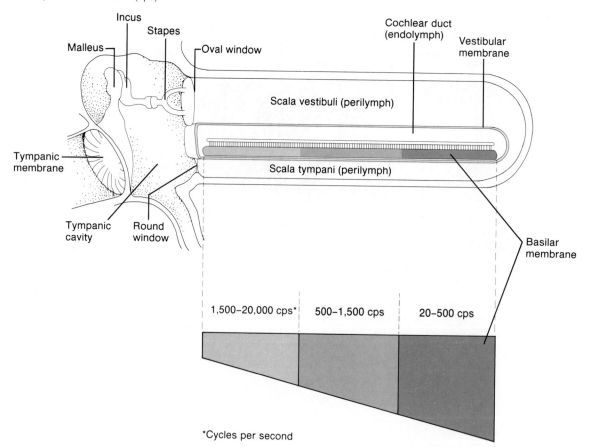

*Cycles per second

Figure 12.17
The auditory nerve pathway extends into the medulla oblongata,
proceeds through the midbrain to the thalamus, and passes into
the auditory cortex of the cerebrum.
(*At right,* from McClintock, J. R., *Physiology of the Human Body*, 2d ed.
Copyright © 1978 John Wiley & Sons, Inc., New York. Reprinted by
permission.)

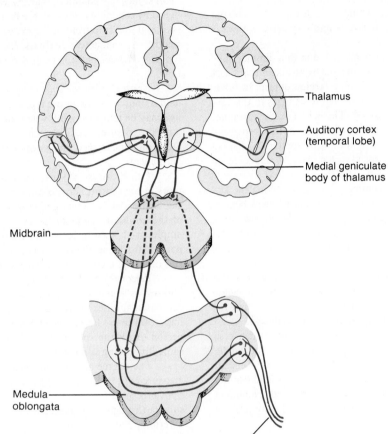

Although the human ear is able to detect sound waves with frequencies varying from about 20 to 20,000 vibrations per second, the range of greatest sensitivity is between 2,000 and 3,000 vibrations per second. (See figure 12.16.)

Auditory Nerve Pathways

The cochlear branches of the vestibulocochlear nerves enter the auditory nerve pathways that extend into the medulla oblongata and proceed through the midbrain to the thalamic regions. From there they pass into the auditory cortices of the temporal lobes of the cerebrum where they are interpreted. On the way, some of these fibers cross over, so that impulses arising from each ear

are interpreted on both sides of the brain. Consequently, damage to a temporal lobe on one side of the brain is not necessarily accompanied by complete hearing loss in the ear on that side. (See figure 12.17.)

Chart 12.2 summarizes the pathway of vibrations through the parts of the middle and inner ears.

1. Describe the external, middle, and inner ears.
2. Explain how sound waves are transmitted through the parts of the ear.
3. Describe the tympanic reflex.
4. Distinguish between the osseous and membranous labyrinths.
5. Explain the function of the organ of Corti.

Deafness
A Clinical Application

Partial or complete hearing loss can be caused by a variety of factors, including interference with the transmission of vibrations to the inner ear (conductive deafness), or damage to the cochlea, auditory nerve, or auditory nerve pathways (sensorineural deafness).

Conductive deafness may be due to a simple plugging of the external auditory meatus by an accumulation of dry wax or the presence of some foreign object. In other instances, the impairment is related to changes in the eardrum or the auditory ossicles. The eardrum, for example, may harden as a result of disease and thus be less responsive to sound waves, or it may be torn or perforated by disease or injury. The auditory ossicles, too, may be damaged or destroyed by disease or may lose their mobility.

One of the more common diseases involving the auditory ossicles is *otosclerosis*. In this condition, new bone is deposited abnormally around the base of the stapes, interfering with the motion of the ossicle needed to transmit vibrations effectively to the inner ear. Although the cause of otosclerosis is unknown, a victim's hearing often can be restored by surgical procedures. For instance, the base of the stapes may be freed by chipping away the bone that holds it fixed in position, or the stapes may be removed and replaced with a wire or plastic substitute (prosthesis).

Two tests used to help diagnose conductive deafness are the Weber and Rinne tests. In the Weber test, the handle of a vibrating tuning fork is pressed against the forehead. A person with normal hearing perceives the sound coming from directly in front, while a person with sound conduction blockage in one middle ear hears the sound coming from the impaired side.

In the Rinne test, a vibrating tuning fork is held against the bone behind the ear. After the sound is no longer heard by conduction through the bones of the skull, the fork is moved to just in front of the external auditory meatus. In middle ear conductive deafness, the vibrating fork can no longer be heard, but a normal ear will continue to hear its tone.

Sensorineural deafness can be caused by excessively loud sounds. If the exposure is of short duration, the hearing loss may be temporary, but when there is repeated and prolonged exposure, such as occurs in some foundries, near jackhammers, or on a firing range, the impairment may be permanent.

Other causes of sensorineural deafness include tumors in the central nervous system, brain damage as a result of vascular accidents, and the use of certain drugs. Some of the antibiotic drugs in the streptomycin group, for example, are known to be able to cause damage to inner ear parts. Also, some people experience hearing losses as a result of aging, a condition called *presbycusis*.

Persons over 65 years of age commonly have lost their abilities to hear high frequency sounds and they also have decreased abilities to discriminate speech sounds. Thus it is important to use lower voice tones in speaking to an elderly person. Also, the speaker should face the listener so that some lip reading is possible.

The degree of a person's hearing impairment can be measured with an *audiometer*. This device produces vibrations of known frequencies that are transmitted to the test subject through earphones. During the test, it is possible to determine the percentage of hearing loss a person has suffered at each vibration frequency.

Chart 12.2   Steps in the generation of sensory impulses from the ear

1. Sound waves enter external auditory meatus.
2. Waves of changing pressures cause the eardrum to reproduce the vibrations coming from the sound wave source.
3. Auditory ossicles amplify and transmit vibrations to the end of the stapes.
4. Movement of the stapes at the oval window transmits vibrations to the perilymph in the scala vestibuli.
5. Vibrations pass through the vestibular membrane and enter the endolymph of the cochlear duct.
6. Different frequencies of vibration in endolymph stimulate different sets of receptor cells.
7. Receptor cell becomes depolarized; its membrane becomes more permeable to calcium ions.
8. In the presence of calcium ions, vesicles at the base of the receptor cell release neurotransmitter.
9. Neurotransmitter stimulates the ends of nearby sensory neurons.
10. Sensory impulses are triggered on fibers of the cochlear branch of the vestibulocochlear nerve.
11. Auditory cortex of the temporal lobe interprets the sensory impulses.

Figure 12.18

The saccule and utricle, which are expanded portions of the membranous labyrinth, are located within the bony chamber of the vestibule.

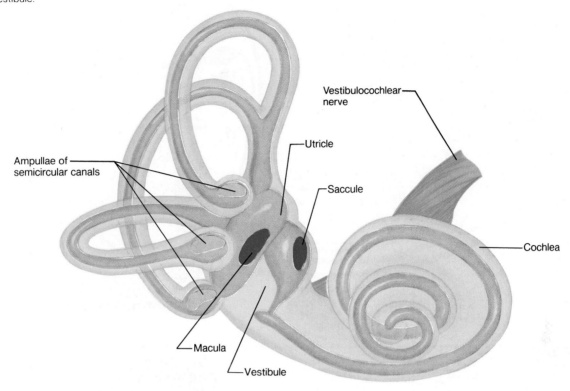

Vestibulocochlear nerve

Utricle

Ampullae of semicircular canals

Saccule

Cochlea

Macula

Vestibule

Sense of Equilibrium

The sense of equilibrium actually involves two senses—a sense of *static equilibrium* and a sense of *dynamic equilibrium*—that result from the actions of different sensory organs. The organs associated with static equilibrium function in maintaining the stability of the head and body when these parts are motionless. When the head and body are suddenly moved or rotated, the organs concerned with dynamic equilibrium aid in maintaining balance.

Static Equilibrium

The organs of **static equilibrium** are located within the vestibule, the bony chamber between the semicircular canals and the cochlea. More specifically, the membranous labyrinth inside the vestibule consists of two expanded chambers—an **utricle** and a **saccule.** The larger utricle communicates with the saccule and the membranous portions of the semicircular canals; the saccule in turn communicates with the cochlear duct (figure 12.18).

Each of these chambers has a small patch of hair cells and supporting cells called a **macula** on its wall. When the head is upright, the hairs of the macula in the utricle project vertically, those in the saccule project horizontally. In each case, the hairs are in contact with a sheet of gelatinous material (otolithic membrane) that has crystals of calcium carbonate (otoconia) embedded on its surface. These particles increase the weight of the gelatinous sheet, making it more responsive to changes in position. The hair cells, which serve as sensory receptors, have nerve fibers wrapped around their bases. These fibers are associated with the vestibular portion of the vestibulocochlear nerve.

The usual stimulus to the hair cells occurs when the head is bent forward, backward, or to one side. Such movements may cause the gelatinous mass of one or more maculae to be tilted, and as it sags in response to gravity, the hairs projecting into it are bent. This action stimulates the hair cells, and they signal the nerve fibers associated with them in a manner similar to that of the hearing receptors. The resulting nerve impulses

Figure 12.19

The macula is responsive to changes in the position of the head. *(a)* Macula with the head in an upright position; *(b)* with the head bent forward.

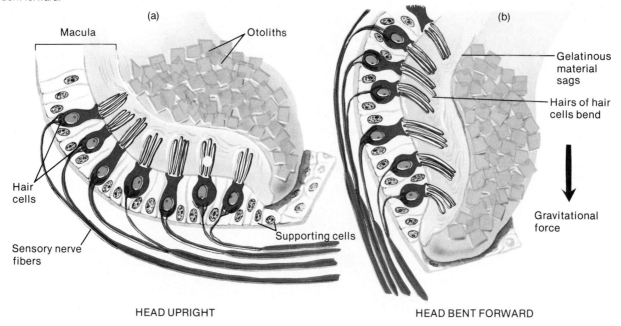

HEAD UPRIGHT   HEAD BENT FORWARD

Figure 12.20

Scanning electron micrograph of "hairs" of hair cells, such as those found in the utricle and saccule.

Figure 12.21

A crista ampullaris is located within the ampulla of each semicircular canal.

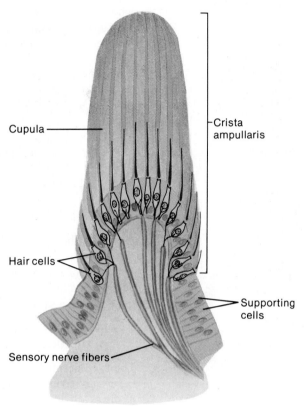

travel into the central nervous system, informing the brain as to the position of the head. The brain may act on this information by sending motor impulses to skeletal muscles, and they may contract or relax appropriately so that balance is maintained (figure 12.19).

The maculae also function in the sense of dynamic equilibrium. For example, if the head or body is thrust forward or backward abruptly, the gelatinous mass of the maculae lags slightly behind, and the hair cells are stimulated. In this way, the maculae aid the brain in detecting movements such as falling and in maintaining posture while walking (figure 12.20).

Dynamic Equilibrium

Each semicircular canal follows a circular path about six millimeters in diameter. The three bony **semicircular canals** lie at right angles to each other, and each occupies a different plane in space. Two of them, the *superior* and the *posterior canals,* stand vertically, while the third or *lateral canal* is horizontal.

Suspended in the perilymph of each bony canal is a membranous semicircular canal that ends in a swelling, called an **ampulla.** The ampullae communicate with the utricle of the vestibule.

An ampulla contains a septum that crosses the tube and houses a sensory organ. Each of these organs, called a **crista ampullaris,** contains a number of sensory hair cells and supporting cells. As in the maculae, the hair cells have hairs that extend upward into a dome-shaped gelatinous mass, called the *cupula.* Also, the hair cells are connected at their bases to nerve fibers that make up part of the vestibular branch of the vestibulocochlear nerve (figure 12.21).

The hair cells of the crista are ordinarily stimulated by rapid turns of the head or body. At such times, the semicircular canals move with the head or body, but the fluid inside the membranous canals remains stationary because of inertia. This causes the cupula in one or more of the canals to be bent in a direction opposite to that of the head or body movement, and the hairs embedded in it are also bent. This bending of the sensory hairs stimulates the hair cells to signal their

Figure 12.22
(a) When the head is stationary, the cupula of the crista ampullaris remains upright. (b) When the head is moving rapidly, the cupula is bent and sensory receptors are stimulated.

(a)

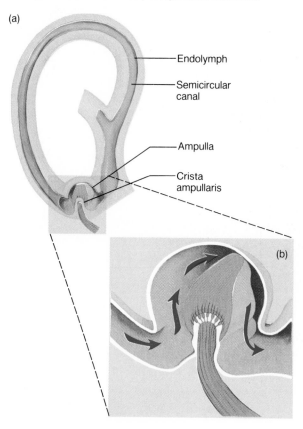

- Endolymph
- Semicircular canal
- Ampulla
- Crista ampullaris

(b)

associated nerve fibers and, as a result, impulses travel to the brain (figure 12.22).

Parts of the cerebellum are particularly important in interpreting impulses from the semicircular canals. Analysis of such information allows the nervous system to predict the consequences of rapid body movements. The brain can then send motor impulses to stimulate appropriate skeletal muscles so that loss of balance can be prevented.

Some people experience a discomfort called *motion sickness* when they are moving in a boat, airplane, or automobile. This discomfort seems to be caused by abnormal and irregular body motions that disturb the organs of equilibrium. Symptoms of motion sickness include nausea, vomiting, dizziness, headache, and prostration.

Other sensory structures also aid in the maintenance of equilibrium. For example, various proprioceptors, particularly those associated with the joints of the neck, supply the brain with information concerning the position of body parts. In addition, the eyes can detect changes in posture that result from body movements. Visual information is so important that a person who has suffered damage to the organs of equilibrium may be able to maintain normal balance as long as the eyes remain open and body movements are performed slowly.

---

1. Distinguish between the senses of static and dynamic equilibrium.
2. What structures function to provide the sense of static equilibrium? Dynamic equilibrium?
3. How does sensory information from other receptors help maintain equilibrium?

Sense of Sight

Although the eye contains the visual receptors and is the primary organ of sight, its functions are assisted by a number of *accessory organs*. These include the eyelids and the lacrimal apparatus that aid in protecting the eye, and a set of extrinsic muscles that move it.

Visual Accessory Organs

The eye, lacrimal gland, and the extrinsic muscles of the eye are housed within the pear-shaped orbital cavity of the skull. The orbit, which is lined with the periosteums of various bones, also contains fat, blood vessels, nerves, and a variety of connective tissues.

Each **eyelid** (palpebra) is composed of four layers—skin, muscle, connective tissue, and conjunctiva. The skin of the eyelid, which is the thinnest skin of the body, covers the lid's outer surface and fuses with its inner lining, the *conjunctiva,* near the margin of the lid (figure 12.23).

The muscles of the eyelids include the *orbicularis oculi,* whose fibers encircle the opening between the lids and spread out onto the cheek and forehead. This muscle acts as a sphincter and closes the lids when it contracts.

Fibers of the *levator palpebrae superioris* muscle arise from the roof of the orbit and are inserted in the connective tissue of the upper lid. When these fibers contract, the upper lid is raised and the eye opens.

The connective tissue layer of the eyelid, which helps give it form, contains numerous modified sebaceous glands (tarsal glands). The oily secretions of these glands are carried by ducts to openings along the borders of the lids. This secretion aids in preventing the lids from sticking together.

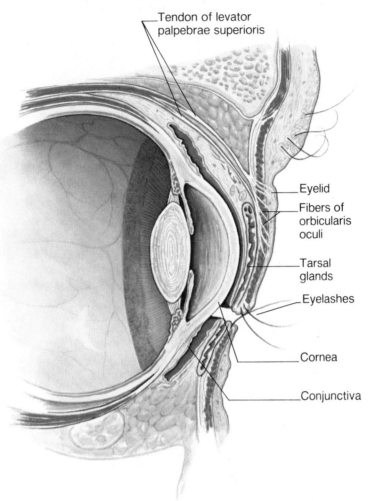

Tendon of levator palpebrae superioris

Eyelid

Fibers of orbicularis oculi

Tarsal glands

Eyelashes

Cornea

Conjunctiva

The **conjunctiva** is a mucous membrane that lines the inner surfaces of the eyelids and folds back to cover the anterior surface of the eyeball, except for its central portion (cornea). Although the conjunctiva that lines the eyelids is relatively thick, the part on the eyeball is very thin. It is also freely movable and quite transparent, so that blood vessels are clearly visible beneath it.

The *lacrimal apparatus* (figure 12.24) consists of the **lacrimal gland,** which secretes tears, and a series of *ducts,* which carry the tears into the nasal cavity. The gland is located in the orbit, above and to the lateral side of the eye. It secretes tears continuously, and they pass out through tiny tubules, and flow downward and medially across the eye.

Tears are collected by two small ducts (superior and inferior canaliculi) whose openings (puncta) can be seen on the medial borders of the eyelids. From these ducts, the fluid moves into the *lacrimal sac,* which lies in a deep groove of the lacrimal bone, and then into the *nasolacrimal duct,* which empties into the nasal cavity.

Glandular cells of the conjunctiva also secrete a tearlike liquid that, together with the secretion of the lacrimal gland, keeps the surface of the eye and the lining of the lids moist and lubricated. Tears contain an enzyme, called *lysozyme,* that functions as an antibacterial agent, reducing the chance of eye infections.

When a person is emotionally upset by grief or disappointment, or when the conjunctiva is irritated, the tear glands are likely to secrete excessive fluids. Tears may spill over the edges of the eyelids, and the nose may fill with fluid. This response involves motor impulses carried to the lacrimal glands on parasympathetic nerve fibers.

The **extrinsic muscles** of the eye arise from the bones of the orbit and are inserted by broad tendons on the eye's tough outer surface. There are six such muscles that function to move the eye in various directions

Figure 12.24
The lacrimal apparatus consists of a tear-secreting
gland and a series of ducts.

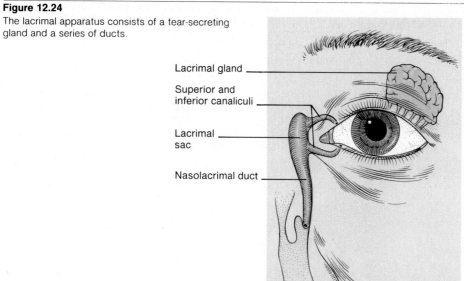

Lacrimal gland _____

Superior and
inferior canaliculi _____

Lacrimal _____
sac

Nasolacrimal duct _____

Figure 12.25
The extrinsic muscles of the eye.

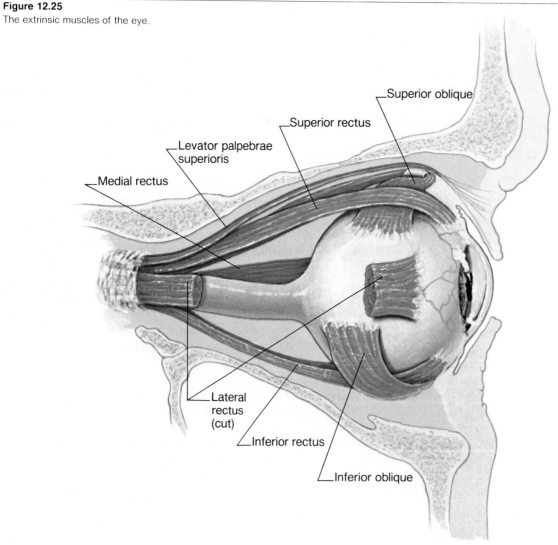

Superior oblique

Superior rectus

Levator palpebrae
superioris

Medial rectus

Lateral
rectus
(cut)

Inferior rectus

Inferior oblique

Chart 12.3 Muscles associated with the eye

Skeletal muscles			Smooth muscles		
Name	*Innervation*	*Function*	*Name*	*Innervation*	*Function*
Orbicularis oculi	Facial nerve (VII)	Closes eye	Ciliary muscles	Oculomotor nerve (III) parasympathetic fibers	Causes suspensory ligaments to relax
Levator palpebrae superioris	Oculomotor nerve (III)	Opens eye			
Superior rectus	Oculomotor nerve (III)	Rotates eye upward and toward midline	Iris, circular muscles	Oculomotor nerve (III) parasympathetic fibers	Causes size of pupil to decrease
Inferior rectus	Oculomotor nerve (III)	Rotates eye downward and toward midline	Iris, radial muscles	Sympathetic fibers	Causes size of pupil to increase
Medial rectus	Oculomotor nerve (III)	Rotates eye toward midline			
Lateral rectus	Abducens nerve (VI)	Rotates eye away from midline			
Superior oblique	Trochlear nerve (IV)	Rotates eye downward and away from midline			
Inferior oblique	Oculomotor nerve (III)	Rotates eye upward and away from midline			

(figure 12.25). Although any given eye movement may involve more than one of them, each muscle is associated with one primary action, as follows:

1. **Superior rectus**—rotates the eye upward and toward the midline.
2. **Inferior rectus**—rotates the eye downward and toward the midline.
3. **Medial rectus**—rotates the eye toward the midline.
4. **Lateral rectus**—rotates the eye away from the midline.
5. **Superior oblique**—rotates the eye downward and away from the midline.
6. **Inferior oblique**—rotates the eye upward and away from the midline.

The motor units of the extrinsic eye muscles contain the smallest number of muscle fibers (five to ten) of any muscles in the body. Because of this, the eyes can be moved with great precision. Also, the eyes move together so that they are aligned when looking at something. Such alignment involves complex motor adjustments that result in the contraction of certain eye muscles while their antagonists are relaxed. For example, when the eyes move to the right, the lateral rectus of the right eye and the medial rectus of the left eye must contract. At the same time, the medial rectus of the right eye and the lateral rectus of the left eye must relax. A person whose eyes are not coordinated well enough to produce alignment is said to have *strabismus,* or squint.

Chart 12.3 summarizes the muscles associated with the eye.

When one eye deviates from the line of vision, the person has double vision (diplopia). If this condition persists, there is danger that changes will occur in the brain to suppress the image from the deviated eye. As a result, the turning eye may become blind (suppression amblyopia). Such monocular blindness can often be prevented if the eye deviation is treated early in life with exercises, glasses, and surgery. For this reason, vision screening programs for preschool children are very important.

---

1. Explain how the eyelid is moved.
2. Describe the conjunctiva.
3. What is the function of the lacrimal apparatus?
4. Describe the function of each extrinsic eye muscle.

Figure 12.26
Sagittal section of the eye.

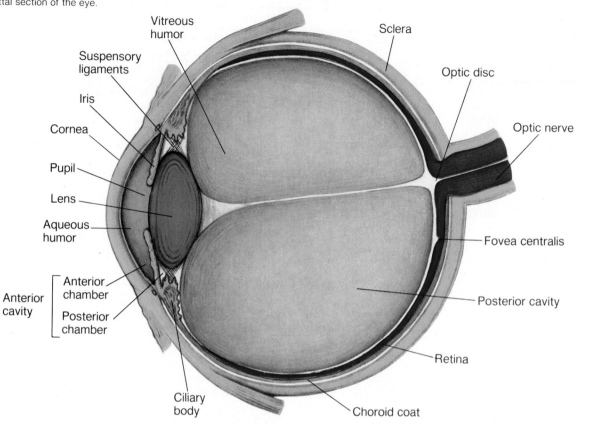

Vitreous humor

Suspensory ligaments

Iris

Cornea

Pupil

Lens

Aqueous humor

Anterior cavity

Anterior chamber

Posterior chamber

Ciliary body

Sclera

Optic disc

Optic nerve

Fovea centralis

Posterior cavity

Retina

Choroid coat

Figure 12.27
Anterior portion of the eye.

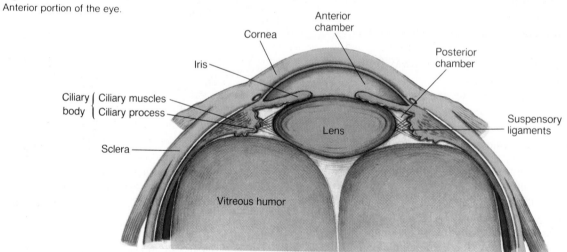

Cornea

Anterior chamber

Iris

Posterior chamber

Ciliary body

Ciliary muscles

Ciliary process

Lens

Sclera

Vitreous humor

Suspensory ligaments

Structure of the Eye

The eye is a hollow, spherical structure about 2.5 cm in diameter. Its wall has three distinct layers—an outer *fibrous tunic,* a middle *vascular tunic,* and an inner *nervous tunic.* The spaces within the eye are filled with fluids that provide support for its wall and internal parts, and help maintain its shape. Figure 12.26 shows the major parts of the eye.

The Outer Tunic

The anterior 1/6 of the outer tunic bulges forward as the transparent **cornea,** which serves as the window of the eye and helps focus entering light rays. It is composed largely of connective tissue with a thin layer of epithelium on the surface. The transparency of the cornea is due to the fact that it contains relatively few cells and no blood vessels. Also, the cells and collagenous fibers are arranged in unusually regular patterns.

On the other hand, the cornea is well supplied with nerve fibers that enter its margin and radiate toward its center. These fibers are associated with numerous pain receptors that have very low thresholds. Cold receptors are also abundant in the cornea, but heat and touch receptors seem to be lacking.

Along its circumference, the cornea is continuous with the **sclera,** the white portion of the eye. This part makes up the posterior 5/6 of the outer tunic and is opaque due to the presence of many large, haphazardly arranged, collagenous and elastic fibers. The sclera provides protection and serves as an attachment for the extrinsic muscles of the eye.

In the back of the eye, the sclera is pierced by the **optic nerve** and certain blood vessels and is attached to the dura mater that encloses these structures and other parts of the central nervous system.

The Middle Tunic

The middle or vascular tunic of the eyeball (uveal layer) includes the choroid coat, the ciliary body, and the iris.

The **choroid coat,** in the posterior 5/6 of the globe, is loosely joined to the sclera and is honeycombed with blood vessels that provide nourishment to surrounding tissues. The choroid also contains numerous pigment-producing melanocytes that give it a brownish black appearance. The melanin of these cells absorbs excess light and helps keep the inside of the eye dark.

The **ciliary body,** which is the thickest part of the middle tunic, extends forward from the choroid and forms an internal ring around the front of the eye. Within the ciliary body there are many radiating folds called *ciliary processes* and two distinct groups of muscle fibers that constitute the *ciliary muscles.* These structures are shown in figure 12.27.

Figure 12.28
The lens and ciliary body viewed from behind.

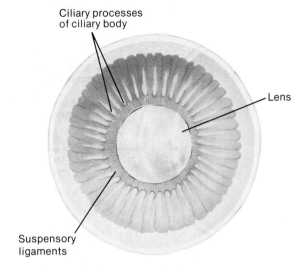

Ciliary processes of ciliary body

Lens

Suspensory ligaments

The transparent **lens** is held in position by a large number of strong but delicate fibers, called *suspensory ligaments,* that extend inward from the ciliary processes. The distal ends of these fibers are attached along the margin of a thin capsule that surrounds the lens. The body of the lens lies directly behind the iris and is composed of "fibers" that arise from epithelial cells. In fact, the cytoplasm of these cells makes up the transparent substance of the lens.

The lens capsule is a clear, membranelike structure composed largely of intercellular material. It is quite elastic, and this quality keeps it under constant tension. As a result, the lens can assume a globular shape. However, the suspensory ligaments attached to the margin of the capsule are also under tension, and as they pull outward, the capsule and the lens inside are kept somewhat flattened (figure 12.28).

If the tension on the suspensory ligaments is relaxed, the elastic capsule rebounds, and the lens surface becomes more convex. Such a change occurs in the lens when the eye is focused to view a close object. This adjustment is called **accommodation.**

The relaxation of the suspensory ligaments during accommodation is a function of the ciliary muscles. One set of these muscle fibers is arranged in a circular pattern forming a sphincterlike structure around the ciliary processes. The fibers of the other set extend back from fixed points in the sclera to the choroid. When the circular muscle fibers contract, the diameter of the ring formed by the ciliary processes is lessened; when the other fibers contract, the choroid is pulled forward and the ciliary body is shortened. Each of these actions causes the suspensory ligaments to become relaxed, and

Figure 12.29

(a) How is the focus of the eye affected when the lens becomes thin as the ciliary muscle fibers relax? (b) How is it affected when the lens thickens as the ciliary muscle fibers contract?

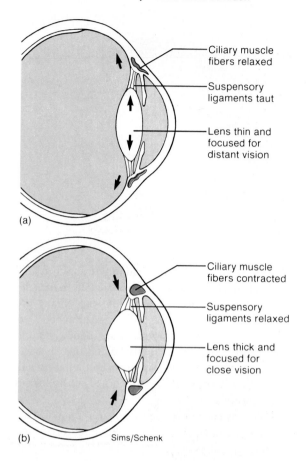

(a)

(b)   Sims/Schenk

1. Describe the three layers of the eye.
2. What factors contribute to the transparency of the cornea?
3. How is the shape of the lens changed during accommodation?

The **iris** is a thin diaphragm composed largely of connective tissue and smooth muscle fibers that is seen from the outside as the colored portion of the eye. It extends forward from the periphery of the ciliary body and lies between the cornea and the lens. The iris divides the space separating these parts, which is called the *anterior cavity,* into an *anterior chamber* (between the cornea and the iris) and a *posterior chamber* (between the iris and the lens).

The epithelium on the inner surface of the ciliary body continuously secretes a watery fluid called **aqueous humor** into the posterior chamber. The fluid circulates from this chamber through the **pupil,** a circular opening in the center of the iris, and into the anterior chamber (figure 12.30). It fills the space between the cornea and the lens, helps nourish these parts, and aids in maintaining the shape of the front of the eye. The aqueous humor subsequently leaves the anterior chamber through veins and a special drainage canal (canal of Schlemm) located in its wall.

A disorder called *glaucoma* sometimes develops in the eyes as a person ages. This condition occurs when the rate of aqueous humor formation exceeds the rate of its removal. As a result, fluid accumulates in the anterior chamber of the eye, and the fluid pressure rises.

Since liquids cannot be compressed, the increasing pressure from the anterior chamber is transmitted to all parts of the eye, and in time the blood vessels that supply the receptor cells of the retina may be squeezed closed. If this happens, cells that fail to receive needed nutrients and oxygen may die, and permanent blindness may result.

Glaucoma often can be controlled. It is relatively easy to measure the pressure of the fluids within the eyes, and if the pressure is found to be reaching a dangerous level, steps can be taken to reduce it. For example, drugs that reduce the formation of aqueous humor by the ciliary body may be used, or an opening may be created surgically to allow excess aqueous humor to drain into the tissues beneath the conjunctiva.

The smooth muscle fibers of the iris are arranged into two groups, a circular set and a radial set. These muscles function to control the size of the pupil, which

the lens thickens in response. In this thickened state, the lens is focused for viewing closer objects than before (figure 12.29).

To focus on a more distant object, the ciliary muscles are relaxed, tension on the suspensory ligaments increases, and the lens becomes thinner again.

A relatively common eye disorder, particularly in older people, is called *cataract.* In this condition, the lens or its capsule slowly loses its transparency and becomes cloudy and opaque. As a result, clear images cannot be focused on the retina, and in time the person may become blind.

Cataract is usually treated by surgically removing the lens. The lens may then be replaced by an artificial one, or the loss of refractive power of the eye may be corrected with eyeglasses or contact lenses.

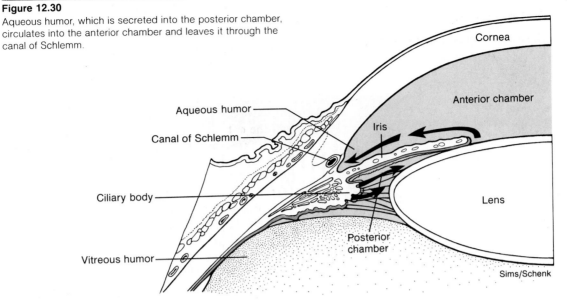

is the opening that light passes through as it enters the eye. The circular set acts as a sphincter, and when it contracts, the pupil gets smaller and the intensity of the light entering decreases. When the radial muscle fibers contract, the diameter of the pupil increases and the intensity of the light entering increases.

The sizes of the pupils change constantly in response to pupillary reflexes that are triggered by such factors as light intensity, gaze, and variations in emotional state. For example, bright light elicits a reflex, and impulses travel along parasympathetic nerve fibers to the *circular muscles* of the irises. The pupils become constricted in response. Conversely, in dim light, impulses travel on sympathetic nerve fibers to the *radial muscles* of the irises, and the pupils become dilated (figure 12.31).

The color of the eyes is determined largely by the amount and distribution of melanin in the irises. If melanin is present only in the epithelial cells that cover an iris's posterior surface, the iris appears blue. When this condition exists together with denser than usual tissue within the iris, it looks gray, and when melanin is present within the body of the iris as well as in the epithelial covering, it is brown.

The Inner Tunic

The inner tunic of the eye consists of the **retina,** which contains the visual receptor cells (photoreceptors).This nearly transparent sheet of tissue is continuous with the optic nerve in the back of the eye and extends forward as the inner lining of the eyeball. It ends just behind the margin of the ciliary body.

Figure 12.31
In dim light, the radial muscles of the iris are stimulated to contract, and the pupil becomes dilated. In bright light, the circular muscles of the iris are stimulated to contract, and the pupil becomes smaller.

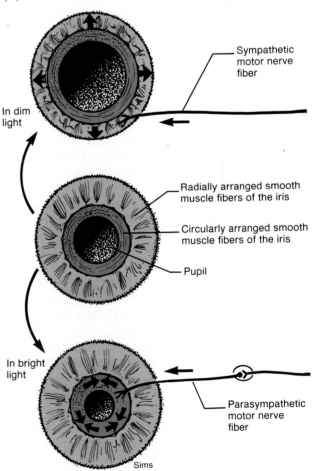

Figure 12.32
The retina consists of several cell layers.

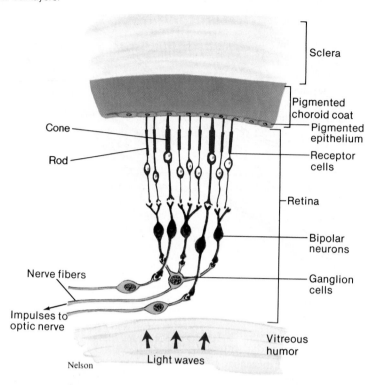

Sclera
Pigmented choroid coat
Pigmented epithelium
Cone
Rod
Receptor cells
Retina
Bipolar neurons
Nerve fibers
Ganglion cells
Impulses to optic nerve
Vitreous humor
Nelson
Light waves

Although the retina is thin and delicate, its structure is quite complex. It has a number of distinct layers, including pigmented epithelium, receptor cells, nerve fibers, ganglion cells, and limiting membranes (figures 12.32 and 12.33).

In the central region of the retina there is a yellowish spot called the **macula lutea** that occupies about one square millimeter. It has a depression in its center called the **fovea centralis.** This depression is in the region of the retina that produces the sharpest vision.

Just medial to the fovea centralis is an area called the **optic disk** (figure 12.34). Here the nerve fibers from the retina leave the eye and become parts of the optic nerve. A central artery and vein also pass through at the optic disk. These vessels are continuous with capillary networks of the retina and, together with vessels in the underlying choroid coat, they supply blood to the cells of the inner tunic.

Because there are no receptor cells in the region of the optic disk, it is commonly referred to as the blind spot of the eye.

The space bounded by the lens, ciliary body, and retina is the largest compartment of the eye and is called the *posterior cavity.* It is filled with a clear, jellylike fluid called **vitreous humor.** This fluid functions to support the internal parts of the eye and helps maintain its shape.

In summary, light waves entering the eye must pass through the cornea, aqueous humor, lens, vitreous humor, and several layers of the retina before they reach the photoreceptors. The layers of the eye are summarized in chart 12.4.

Although the eye is well protected by the bony orbit, sometimes it is injured in sports by a ball or a forceful blow. If the eye is jarred sufficiently, some of its contents may be displaced. For example, the suspensory ligaments may be torn and the lens dislocated into the posterior cavity. Similarly, a blow to the eye (or to the head) may cause the retina to pull away from the underlying vascular choroid coat. Once the retina is detached, there is danger that photoreceptor cells will die because of lack of oxygen and nutrients. Unless such a *detached retina* can be repaired surgically, this injury may result in varying degrees of visual loss or blindness. Athletic injuries to the eye usually can be prevented by wearing protective devices such as face masks or goggles.

1. Explain the origin of aqueous humor and trace its path through the eye.
2. How is the size of the pupil regulated?
3. Describe the structure of the retina.

Figure 12.33

Note the layers of cells and nerve fibers in this light micrograph of the retina.

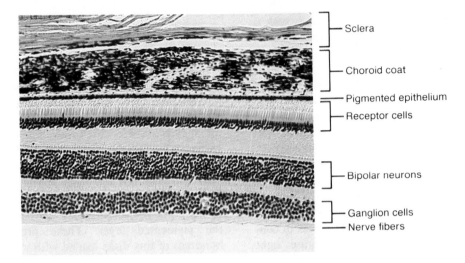

- Sclera
- Choroid coat
- Pigmented epithelium
- Receptor cells
- Bipolar neurons
- Ganglion cells
- Nerve fibers

Figure 12.34

Nerve fibers leave the eye in the area of the optic disk (arrow) to form the optic nerve.

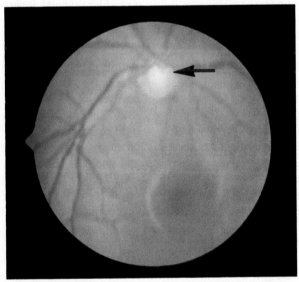

Chart 12.4 Layers of the eye				
Tunic	Posterior portion	Function	Anterior portion	Function
Outer layer	Sclera	Protection	Cornea	Light transmission
Middle layer	Choroid coat	Blood supply, pigment prevents reflection	Ciliary body, iris	Accommodation, controls light intensity
Inner layer	Retina	Photoreception, impulse transmission	None	

Refraction of Light

When a person sees something, it is either giving off light or light waves are being reflected from it. These waves enter the eye, and an image of what is seen is focused upon the retina. This focusing process involves bending of light waves—a phenomenon called **refraction.**

Refraction occurs when light waves pass at an oblique angle from a medium of one optical density into a medium of a different optical density. For example, as figure 12.35 shows, when light passes obliquely from a less dense medium such as air into a denser medium such as glass, or from air into the cornea of the eye, the light is bent toward a line perpendicular to the surface between these substances. When the surface between such refracting media is curved, a lens is formed. A lens with a *convex* surface causes light waves to converge, and a lens with a *concave* surface causes light waves to diverge (figure 12.36).

When light arrives from objects outside the eye, the light waves are refracted primarily by the convex surface of the cornea. Then the light is refracted again by the convex surface of the lens and to a lesser extent by the surfaces of the fluids within the chambers of the eye.

If the shape of the eye is normal, light waves are focused sharply upon the retina, much as a motion picture image is focused on a screen for viewing. Unlike the motion picture image, however, the one formed on the retina is upside down and reversed from left to right (figure 12.37). When the visual cortex interprets such an image, it somehow corrects this, and things are seen in their proper positions.

Light waves coming from objects more than 20 feet away are traveling in nearly parallel lines, and they are focused on the retina by the cornea and by the lens in its more flattened or "at rest" condition. Light waves arriving from objects less than 20 feet away, however, reach the eye along more divergent lines—in fact, the closer the object, the more divergent the lines.

Divergent light waves tend to come into focus behind the retina unless something is done to increase the refracting power of the eye. This increase is accomplished by accommodation, which results in a thickening of the lens. As the lens thickens, light waves are converged more strongly, so that diverging light waves coming from close objects are focused on the retina.

1. What is meant by refraction?
2. What parts of the eye provide refracting surfaces?
3. Why is it necessary to accommodate for viewing close objects?

Visual Receptors

The photoreceptors of the eye are actually modified neurons, and there are two distinct kinds. One group of receptors have long, thin projections at their terminal ends and are called **rods.** The cells of the other group have short, blunt projections and are called **cones.**

Instead of being located in the surface layer of the retina, the rods and cones are found in a deep portion, closely associated with a layer of pigmented epithelium. The projections from the receptors extend into the pigmented layer. These projections contain hundreds of tiny disks loaded with visual pigments.

The epithelial pigment of the retina functions to absorb light waves that are not absorbed by the receptor cells, and with the pigment of the choroid coat, it serves to keep light from reflecting about inside the eye. The pigment layer also stores vitamin A, which can be used to synthesize visual pigments. Albinos, who lack melanin in all parts of their bodies, including their eyes, suffer a considerable loss in visual sharpness (acuity). This is because light reflects from inside their eyes and tends to stimulate their visual receptors excessively.

The visual receptors are only stimulated when light reaches them. Thus, when a light image is focused on an area of the retina, some receptors are stimulated, and impulses travel away from them to the brain.

However, the impulse leaving each activated receptor provides only a fragment of the information needed for the brain to interpret a total scene.

Rods and cones function differently. For example, rods are hundreds of times more sensitive to light than cones, and, as a result, they enable persons to see in relatively dim light. In addition, the rods produce colorless vision, while cones can detect colors.

Still another difference involves visual acuity—the sharpness of the images perceived. Cones allow a person to see sharp images, while rods enable one to see more general outlines of objects. This characteristic is related to the fact that nerve fibers from many rods may converge, and their impulses may be transmitted to the brain on the same nerve fiber (see chapter

Figure 12.35
When light passes at an oblique angle from air into glass, the light waves are bent toward a line perpendicular to the surface of the glass.

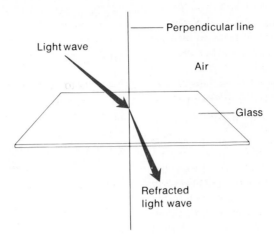

Figure 12.36
(a) A lens with a convex surface causes light waves to converge; *(b)* a lens with a concave surface causes them to diverge.

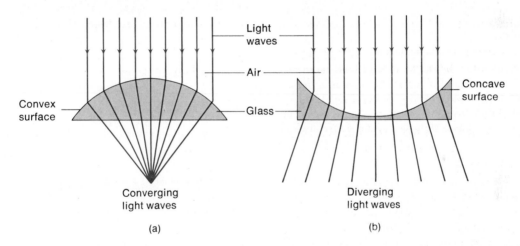

Figure 12.37
The image of an object formed on the retina is upside-down.

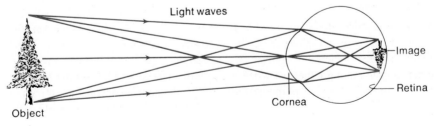

Refraction Disorders
A Practical Application

Unfortunately, the elastic quality of the lens capsule tends to lessen with age, and persons over forty-five years of age often are unable to accommodate sufficiently to read the fine print in books and newspapers. Their eyes remain focused for distant vision. This condition is termed *presbyopia*, or farsightedness of age, and it can usually be corrected by using eyeglasses or contact lenses that make up for the eye's loss in refracting power.

Other persons suffer from problems that result when their eyeballs are too short or too long for sharp focusing. For example, if an eye is too short, light waves are not focused sharply on the retina since their point of focus lies some distance behind it. A person with this condition may be able to bring the image of distant objects into focus by accommodation, but this requires contraction of the ciliary muscles at times when these muscles are at rest in a normal eye. Still more accommodation is needed to view closer objects, and the victim may suffer from ciliary muscle fatigue, pain, and headache whenever it is necessary to do close work.

Since people with short eyeballs commonly are unable to accommodate enough to focus on very close objects, they are said to be *farsighted*. Treatment for this condition (hyperopia) involves the use of glasses or contact lenses with *convex* surfaces that cause images to be focused closer to the front of the eye.

If an eyeball is too long, light waves tend to be focused in front of the retina, and the image produced on it is blurred. In other words, the refracting power of the eye, even when the lens is flattened, is too great. Although a person with this problem may be able to focus on close objects by accommodation, distance vision is invariably poor. For this reason, the person is said to be *nearsighted*. Treatment for nearsightedness (myopia) makes use of glasses or contact lenses with *concave* surfaces that cause images to be focused further from the front of the eye (figures 12.38 and 12.39).

Still another refraction problem is termed *astigmatism*. In this condition, there is a defect in the curvature of the cornea or, sometimes, in the curvature of the lens. The normal cornea has a spherical curvature, like the inside of

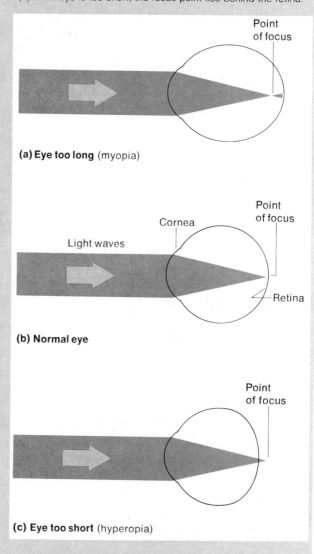

Figure 12.38

(a) If an eye is too long, the focus point of images lies in front of the retina; *(b)* in a normal eye, the focus point is on the retina; *(c)* if the eye is too short, the focus point lies behind the retina.

(a) Eye too long (myopia)

(b) Normal eye

Point of focus

Cornea

Light waves

Retina

(c) Eye too short (hyperopia)

Point of focus

---

10). Thus, if a point of light stimulates a rod, the brain cannot tell which one of many receptors has actually been stimulated. Such a convergence of impulses occurs to a much lesser degree among cones, so when a cone is stimulated, the brain is able to pinpoint the stimulation more accurately (figure 12.40).

As was mentioned, the area of sharpest vision is the fovea centralis in the macula lutea. This area lacks rods, but contains densely packed cones with few or no converging fibers. Also, the overlying layers of the retina, as well as the retinal blood vessels, are displaced to the sides. This displacement more fully exposes the receptors to incoming light. Consequently, to view something in detail, one moves the eyes so the important part of an image falls upon the fovea.

The concentration of cones decreases in areas further away from the macula, while the concentration of rods increases in these areas. Also, the amount of convergence among the rods and cones increases toward the periphery of the retina. As a result of these factors, the visual sensations from images focused on the sides of the retina tend to be blurred compared with those focused on the central portion of the retina.

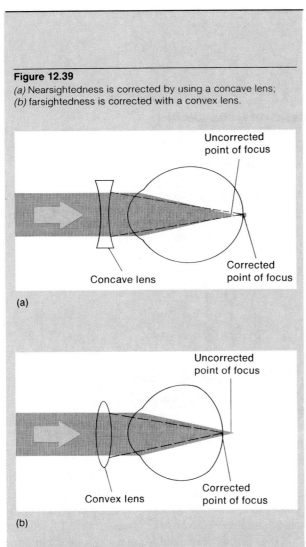

Figure 12.39
(a) Nearsightedness is corrected by using a concave lens;
(b) farsightedness is corrected with a convex lens.

At the same time that the rhodopsin molecules are decomposing, changes occur in the ionic permeability of the rod cell membrane, causing receptor potentials to be generated. These receptor potentials are proportional to the intensity of the light stimulating the photoreceptors, and they are transmitted through the receptor cells to adjacent layers of neurons in the retina. Action potentials are triggered in these neurons, and, as a result, nerve impulses travel away from the retina, through the optic nerve, and into the brain.

In bright light, nearly all of the rhodopsin in the rods of the retina is decomposed, and the sensitivity of these receptors is greatly reduced. In dim light, however, rhodopsin can be regenerated from scotopsin and retinene faster than it is broken down. This regeneration process requires cellular energy, which is provided by energy-carrying molecules of ATP.

The light sensitivity of an eye whose rods have converted the available scotopsin and retinene to rhodopsin increases about 100,000 times, and the eye is said to be *dark adapted*.

A person needs a dark-adapted eye to see in dim light. For example, when a person moves from daylight into a darkened theater it may be difficult to see well enough to locate a seat, but soon the eyes become adapted to the dim light, and the vision improves.

Later, when leaving the theater and entering the sunlight, the person may feel some discomfort or even pain. This occurs at the moment that most of the rhodopsin decomposes in response to the bright light. At the same time, the light sensitivity of the eyes decreases greatly, and they become *light adapted*.

Many people, particularly children, suffer from vitamin A deficiency due to improper diet. In such cases, the quantity of retinene available for the manufacture of rhodopsin may be reduced and consequently the sensitivity of their rods may be low. This condition, called *night blindness,* is characterized by poor vision in dim light. Fortunately, the problem is usually easy to correct by adding vitamin A to the diet or by providing the victim with injections of vitamin A. Vitamin A deficiency remains one of the principal nutritional problems throughout the world, even though most countries have ample supplies of the plant foods that provide this needed substance.

a ball; an astigmatic cornea usually has an elliptical curvature, like the bowl of a spoon. As a result, some portions of an image are in focus on the retina, but other portions are blurred, and vision is distorted.

Without corrective lenses, astigmatic eyes tend to accommodate back and forth reflexly in an attempt to sharpen focus. The consequence of this continual action is likely to be ciliary muscle fatigue and headache.

Visual Pigments

Both rods and cones contain light-sensitive pigments that decompose when they absorb light energy. The light-sensitive substance in rods is called **rhodopsin,** or visual purple, and it is embedded in discs of membrane that are stacked within these receptor cells. (See figure 12.41.) In the presence of light, rhodopsin molecules break down into molecules of a colorless protein called *scotopsin* and a yellowish substance called *retinene* (retinaldehyde), which is synthesized from vitamin A.

The light-sensitive pigments of cones are similar to rhodopsin in that they are composed of retinene combined with a protein; the protein, however, differs from that in the rods. In fact, there are three sets of cones, each containing an abundance of one of the three different visual pigments.

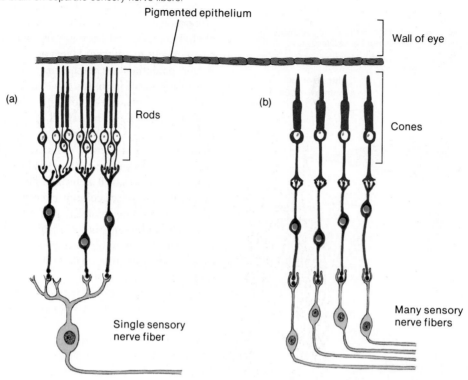

The wavelength of a particular kind of light determines the color perceived from it. For example, the shortest wavelengths of visible light are perceived as violet, while the longest visible wavelengths are sensed as red. As far as cone pigments are concerned, one type (erythrolabe) is most sensitive to red light waves, another (chlorolabe) to green light waves, and a third (cyanolabe) to blue light waves. The color a person perceives depends upon which set of cones or combination of sets of cones is stimulated by the light in a given image. If all three sets of cones are stimulated, the person senses the light as white, and if none are stimulated, the person senses black.

Some persons have defective color vision, due to a decreased sensitivity in one or more cone sets. Such people perceive colors differently from those with normal vision and are said to be *color blind.*

Stereoscopic Vision

Stereoscopic vision (stereopsis) is vision that involves the perception of distance and depth as well as the height and width of objects. Such vision is due largely to the fact that the pupils of the eyes are 6–7 cm apart. Consequently, objects that are relatively close (less than 20 feet away) produce slightly different retinal images.

That is, the right eye sees a little more of one side of an object, while the left eye sees a little more of the other side. These two images are somehow superimposed and interpreted by the visual cortex of the brain, and the result is the perception of a single object in three dimensions (figure 12.42).

Since this type of depth perception depends on vision with two eyes (binocular vision), it follows that a one-eyed person has a lessened ability to judge distance and depth accurately. To compensate when making such judgments, a person with one eye can use clues provided by the relative sizes and relative positions of familiar objects.

Visual Nerve Pathways

As is mentioned in chapter 10, the axons of the ganglion cells in the retina leave the eyes to form the *optic nerves.* Just anterior to the pituitary gland, these nerves give rise to the X-shaped *optic chiasma,* and within the chiasma some of the fibers cross over. More specifically, the fibers from the nasal half of each retina cross over, while those from the temporal sides do not. Thus, fibers from the nasal half of the left eye and the temporal half of the right eye form the right *optic tract;* and fibers from the nasal half of the right eye and the temporal half of the left form the left optic tract.

Figure 12.41
Rhodopsin is embedded in discs of membrane that are stacked within the rod cells.

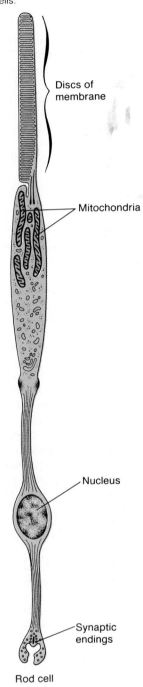

Rod cell

Figure 12.42
Stereoscopic vision results from the formation of two slightly different retinal images.

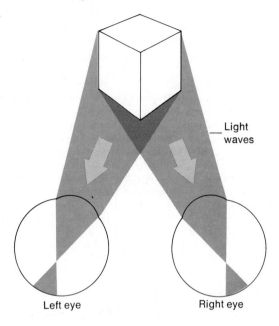

Figure 12.43
The visual pathway includes the optic nerve, optic chiasma, optic tract, and optic radiations.

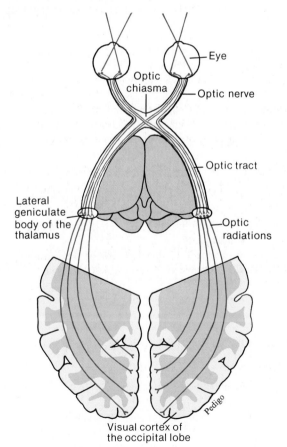

The nerve fibers continue in the optic tracts, and just before they reach the thalamus, a few of them leave to enter nuclei that function in various visual reflexes. Most of the fibers, however, enter the thalamus and synapse in its posterior portion (lateral geniculate body). From this region the visual impulses enter nerve pathways called *optic radiations,* and they lead to the visual cortex of the occipital lobes (figure 12.43).

Other fibers conducting visual impulses pass downward into various regions of the brainstem. These impulses are important for controlling head and eye movements associated with following an object visually, for controlling the simultaneous movements of both eyes, and for controlling certain visual reflexes, such as those involved with movements of the iris muscles.

Since each visual cortex receives impulses from each eye, a person may develop partial blindness in both eyes if either visual cortex is injured. For example, if the right visual cortex (or the right optic tract) is injured, sight may be lost in the temporal side of the right eye and the nasal side of the left eye. Similarly, if the central portion of the optic chiasma, where fibers from the nasal sides of the eyes cross over, is damaged, the nasal sides of both eyes are blinded.

1. Distinguish between the rods and the cones of the retina.
2. Explain the roles of visual pigments.
3. What factors make stereoscopic vision possible?
4. Trace the pathway of visual impulses from the retina to the occipital cortex.

Clinical Terms Related to the Senses

amblyopia (am″ble-o′pe-ah) dimness of vision due to a cause other than a refractive disorder or lesion.

ametropia (am″ĕ-tro′pe-ah) an eye condition characterized by inability to focus images sharply on the retina.

anopia (an-o′pe-ah) absence of the eye.

audiometry (aw″de-om′ĕ-tre) the measurement of auditory acuity for various frequencies of sound waves.

blepharitis (blef″ah-ri′tis) an inflammation of the margins of the eyelids.

causalgia (kaw-zal′je-ah) a persistent, burning pain usually associated with injury to a limb.

conjunctivitis (kon-junk″ti-vi′tis) an inflammation of the conjunctiva.

diplopia (di-plo′pe-ah) double vision, or the sensation of seeing two objects when only one is viewed.

emmetropia (em″ĕ-tro′pe-ah) normal condition of the eyes; eyes with no refractive defects.

enucleation (e-nu″kle-a′shun) removal of the eyeball.

exophthalmos (ek″sof-thal′mos) condition in which the eyes protrude abnormally.

hemianopia (hem″e-ah-no′pe-ah) defective vision affecting half of the visual field.

hyperalgesia (hi″per-al-je′ze-ah) an abnormally increased sensitivity to pain.

iridectomy (ir″i-dek′to-me) the surgical removal of part of the iris.

iritis (i-ri′tis) an inflammation of the iris.

keratitis (ker″ah-ti′tis) an inflammation of the cornea.

labyrinthectomy (lab″i-rin-thek′to-me) the surgical removal of the labyrinth.

labyrinthitis (lab″i-rin-thi′tis) an inflammation of the labyrinth.

Ménière's disease (men″e-ārz′ di-zez) an inner ear disorder characterized by ringing in the ears, increased sensitivity to sounds, dizziness, and loss of hearing.

neuralgia (nu-ral′je-ah) pain resulting from inflammation of a nerve or a group of nerves.

neuritis (nu-ri′tis) an inflammation of a nerve.

nystagmus (nis-tag′mus) an involuntary oscillation of the eyes.

otitis media (o-ti′tis me′de-ah) an inflammation of the middle ear.

otosclerosis (o″to-skle-ro′sis) a formation of spongy bone in the inner ear, which often causes deafness by fixing the stapes to the oval window.

pterygium (tĕ-rij′e-um) an abnormally thickened patch of conjunctiva that extends over part of the cornea.

retinitis pigmentosa (ret″i-ni′tis pig″men-to′sa) a progressive retinal sclerosis characterized by deposits of pigment in the retina and atrophy of the retina.

retinoblastoma (ret″i-no-blas-to′mah) an inherited, highly malignant tumor arising from immature retinal cells.

tinnitus (ti-ni′tus) a ringing or buzzing noise in the ears.

tonometry (to-nom′ĕ-tre) the measurement of fluid pressure within the eyeball.

trachoma (trah-ko′mah) a virus-caused disease of the eye, characterized by conjunctivitis, that may lead to blindness.

tympanoplasty (tim″pah-no-plas′te) the surgical reconstruction of the middle ear bones and establishment of continuity from the tympanic membrane to the oval window.

uveitis (u″ve-i′tis) an inflammation of the uvea, which includes the iris, ciliary body, and the choroid coat.

vertigo (ver′ti-go) a sensation of dizziness.

Chapter Summary

Introduction
Sensory receptors are sensitive to environmental changes and initiate impulses to the brain and spinal cord.

Receptors and Sensations
1. Types of receptors
   a. Each type of receptor is sensitive to a distinct type of stimulus.
   b. The major types of receptors include the following:
      (1) Chemoreceptors, sensitive to changes in chemical concentration.
      (2) Pain receptors, sensitive to tissue damage.
      (3) Thermoreceptors, sensitive to temperature changes.
      (4) Mechanoreceptors, sensitive to mechanical forces.
      (5) Photoreceptors, sensitive to light.
2. Sensory impulses
   a. When receptors are stimulated, changes occur in their membrane potentials.
   b. Receptor potentials are transferred to nerve fibers and action potentials are triggered.
3. Sensations
   a. Sensations are feelings resulting from sensory stimulation.
   b. A particular part of the sensory cortex always interprets impulses reaching it in the same way.
   c. The brain projects a sensation back to the region of stimulation.
4. Sensory adaptations are adjustments made by sensory receptors to continuous stimulation in which impulses are triggered at slower and slower rates.

Somatic Senses
Somatic senses are those involved with receptors in skin, muscles, joints, and visceral organs.

They can be grouped as exteroceptive, proprioceptive, and visceroceptive senses.

1. Touch and pressure senses
   a. Free ends of sensory nerve fibers are responsible for the sensations of touch and pressure.
   b. Meissner's corpuscles are responsible for the sensations of light touch.
   c. Pacinian corpuscles are responsible for the sensations of heavy pressure and vibrations.
2. Thermoreceptors include two sets of free nerve endings that serve as heat and cold receptors.
3. Sense of pain
   a. Pain receptors are free nerve endings stimulated by tissue damage. They provide a protective function, do not adapt rapidly, and can be stimulated by changes in temperature, mechanical force, and chemical concentration.
   b. The only receptors in visceral organs that provide sensations are pain receptors. These receptors are most sensitive to lack of blood flow and the presence of certain chemicals. The sensations they produce are likely to feel as if they were coming from some other part (referred pain).
   c. Certain neuropeptides, synthesized in the brain and spinal cord, seem to relieve pain by inhibiting pain impulses.
4. Stretch receptors
   a. Stretch receptors provide information about the condition of muscles and tendons.
   b. Muscle spindles are stimulated when a muscle is relaxed, and they initiate a reflex that causes the muscle to contract.
   c. Golgi tendon organs are stimulated when muscle tension increases, and they initiate a reflex that causes muscle relaxation.

Sense of Smell
1. Olfactory receptors
   a. Olfactory receptors are chemoreceptors that are stimulated by chemicals dissolved in liquid.
   b. They function together with taste receptors and aid in food selection.
2. Olfactory organs
   a. The olfactory organs consist of receptors and supporting cells in the nasal cavity.
   b. Olfactory receptors are neurons with cilia that are sensitive to lipid-soluble chemicals.
3. Olfactory nerve pathways. Nerve impulses travel from the olfactory receptors through the olfactory nerves, olfactory bulbs, and olfactory tracts to interpreting centers in the frontal lobes of the cerebrum.
4. Olfactory stimulation
   a. Olfactory impulses may result when various gaseous molecules combine with specific sites on the cilia of the receptor cells.
   b. Primary odors include camphoraceous, musky, floral, pepperminty, ethereal, pungent, and putrid.
   c. Olfactory receptors adapt rapidly.
   d. Olfactory receptors are often damaged by environmental factors, but are not replaced.

Sense of Taste

1. Taste receptors
   a. Taste buds consist of receptor cells and supporting cells.
   b. Taste cells have taste hairs that are sensitive to particular chemicals dissolved in water.
   c. Taste hair surfaces seem to have receptor sites to which chemicals combine and trigger impulses to the brain.
   d. There are four primary kinds of taste cells, each particularly sensitive to a certain group of chemicals.
2. Taste sensations
   a. The four primary taste sensations are sweet, sour, salty, and bitter.
   b. Various taste sensations result from the stimulation of one or more sets of taste receptors.
   c. Sweet receptors are most plentiful near the tip of the tongue, sour receptors along the margins, salt receptors in the tip and upper front, and bitter receptors toward the back.
3. Taste nerve pathways
   a. Sensory impulses from taste receptors travel on fibers of the facial, glossopharyngeal, and vagus nerves.
   b. These impulses are carried to the medulla and ascend to the thalamus, from which they are directed to the gustatory cortex in the parietal lobes.

Sense of Hearing

1. The external ear collects sound waves created by vibrating objects.
2. Middle ear
   a. Auditory ossicles of the middle ear conduct sound waves from the tympanic membrane to the oval window of the inner ear. They also increase the force of these waves.
   b. Skeletal muscles attached to the auditory ossicles act in the tympanic reflex to protect the inner ear from the effects of loud sounds.
3. Eustachian tubes connect the middle ears to the throat and function to help maintain equal air pressure on both sides of the eardrums.
4. Inner ear
   a. The inner ear consists of a complex system of interconnected tubes and chambers—the osseous and membranous labyrinths. It includes the cochlea, which in turn houses the organ of Corti.
   b. The organ of Corti contains the hearing receptors that are stimulated by vibrations in the fluids of the inner ear.
   c. Different frequencies of vibrations stimulate different sets of receptor cells; the human ear can detect sound frequencies from about 20 to 20,000 vibrations per second.

5. Auditory nerve pathways
   a. The nerve fibers from hearing receptors travel in the cochlear branch of the vestibulocochlear nerves.
   b. Auditory impulses travel into the medulla oblongata, midbrain, and thalamus, and are interpreted in the temporal lobes of the cerebrum.

Sense of Equilibrium

1. Static equilibrium is concerned with maintaining the stability of the head and body when these parts are motionless. The organs of static equilibrium are located in the vestibule.
2. Dynamic equilibrium is concerned with balancing the head and body when they are moved or rotated suddenly. The organs of this sense are located in the ampullae of the semicircular canals.
3. Other parts that help with the maintenance of equilibrium include the eyes and the proprioceptors associated with certain joints.

Sense of Sight

1. Visual accessory organs include the eyelids and lacrimal apparatus that function to protect the eye, and the extrinsic muscles that move the eye.
2. Structure of the eye
   a. The wall of the eye has an outer, a middle, and an inner layer that function as follows:
      (1) The outer layer (sclera) is protective, and its transparent anterior portion (cornea) refracts light entering the eye.
      (2) The middle layer (choroid) is vascular and contains pigments that help to keep the inside of the eye dark.
      (3) The inner layer (retina) contains the visual receptor cells.
   b. The lens is a transparent, elastic structure whose shape is controlled by the action of ciliary muscles.
   c. The iris is a muscular diaphragm that controls the amount of light entering the eye; the pupil is an opening in the iris.
   d. Spaces within the eye are filled with fluids (aqueous and vitreous humors) that help to maintain its shape.
3. Refraction of light
   a. Light waves are refracted primarily by the cornea and lens to focus an image on the retina.
   b. The lens must be thickened to focus on close objects.
4. Visual receptors
   a. The visual receptors are called rods and cones.
   b. Rods are responsible for colorless vision in relatively dim light, and cones are responsible for color vision.

5. Visual pigments
   a. A light-sensitive pigment in rods (rhodopsin) decomposes in the presence of light and triggers nerve impulses.
   b. Color vision seems to be related to the presence of three sets of cones containing different light-sensitive pigments, and each is sensitive to a different wavelength of light; the color perceived depends on which set or sets of cones are stimulated.
6. Stereoscopic vision
   a. Stereoscopic vision involves the perception of distance and depth.
   b. Stereoscopic vision occurs because of the formation of two slightly different retinal images that the brain superimposes and interprets as one image in three dimensions.
   c. A one-eyed person uses relative sizes and positions of familiar objects to judge distance and depth.
7. Visual nerve pathways
   a. Nerve fibers from the retina form the optic nerves.
   b. Some fibers cross over in the optic chiasma.
   c. Most of the fibers enter the thalamus and synapse with others that continue to the visual cortex.
   d. Other impulses pass into the brain stem and function in various visual reflexes.

Application of Knowledge

1. How would you explain the following observation? A person enters a tub of water and reports that it is uncomfortably warm, yet a few moments later says the water feels comfortable, even though the water temperature remains unchanged.
2. How would you explain the fact that some serious injuries, such as those produced by a bullet entering the abdomen, may be relatively painless, while others such as those involving a crushing of the skin may produce considerable discomfort?
3. Labyrinthitis is a condition in which the tissues of the inner ear are inflamed. What symptoms would you expect to observe in a patient with this disorder?
4. Sometimes, as a result of an injury to the eye, the retina becomes detached from its pigmented epithelium. Assuming that the retinal tissues remain functional, what is likely to happen to the person's vision if the retina moves unevenly toward the interior of the eye?

Review Activities

1. List five groups of sensory receptors and name the kind of change to which each is sensitive.
2. Explain how sensory impulses are stimulated by receptors.
3. Define *sensation*.
4. Explain what is meant by the projection of a sensation.
5. Define *sensory adaptation*.
6. Explain how somatic senses can be grouped.
7. Describe the functions of free nerve endings, Meissner's corpuscles, and Pacinian corpuscles.
8. Explain how thermoreceptors function.
9. Compare pain receptors with other types of somatic receptors.
10. List the factors that are likely to stimulate visceral pain receptors.
11. Define *referred pain*.
12. Explain how neuropeptides may function to relieve pain.
13. Distinguish between muscle spindles and Golgi tendon organs.
14. Explain how the senses of smell and taste function together to create the flavors of foods.
15. Describe the olfactory organ and its function.
16. Trace a nerve impulse from the olfactory receptor to the interpreting center of the cerebrum.
17. List the seven primary olfactory sensations.
18. Explain how the salivary glands aid the function of the taste receptors.
19. Name the four primary taste sensations and describe the patterns in which the taste receptors are distributed on the tongue.
20. Explain why taste sensation is less likely to diminish with age than olfactory sensation.
21. Trace the pathway of a taste impulse from the receptor to the cerebral cortex.
22. Distinguish between the external, middle, and inner ears.
23. Trace the path of a sound vibration from the tympanic membrane to the hearing receptors.
24. Describe the functions of the auditory ossicles.
25. Describe the tympanic reflex, and explain its importance.
26. Explain the function of the eustachian tube.
27. Distinguish between the osseous and the membranous labyrinths.
28. Describe the cochlea and its function.
29. Describe a hearing receptor.
30. Explain how a hearing receptor stimulates a sensory neuron.

31. Trace a nerve impulse from the organ of Corti to the interpreting centers of the cerebrum.
32. Describe the organs of static and dynamic equilibrium and their functions.
33. Explain how the sense of vision helps to maintain equilibrium.
34. List the visual accessory organs and describe the functions of each.
35. Name the three layers of the eye wall and describe the functions of each.
36. Describe how accommodation is accomplished.
37. Explain how the iris functions.
38. Distinguish between aqueous and vitreous humor.
39. Distinguish between the macula lutea and the optic disk.
40. Explain how light waves are focused on the retina.
41. Distinguish between rods and cones.
42. Explain why cone vision is generally more acute than rod vision.
43. Describe the function of rhodopsin.
44. Explain how the eye becomes adapted to light and dark.
45. Describe the relationship between light wavelengths and color vision.
46. Define *stereoscopic vision.*
47. Explain why a person with binocular vision is able to judge distance and depth of close objects more accurately than a one-eyed person.
48. Trace a nerve impulse from the retina to the visual cortex.

13

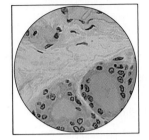

The Endocrine System

The endocrine system consists of a variety of loosely related cells, tissues, and organs that act together with parts of the nervous system to control body activities and maintain homeostasis. Each of these systems provides a means by which body parts can communicate with one another and adjust to changing needs. Whereas the parts of the nervous system communicate by means of nerve impulses carried on nerve fibers, the parts of the endocrine system make use of hormones that act as chemical messengers.

Chapter Outline

Chapter Objectives

After you have studied this chapter, you should be able to:

1. Distinguish between endocrine and exocrine glands.

2. Describe how hormones can be classified according to their chemical composition.

3. Explain how steroid and nonsteroid hormones produce effects on target cells.

4. Discuss how hormone secretions are regulated by feedback mechanisms.

5. Explain how hormone secretions may be controlled by the nervous system.

6. Name and describe the location of the major endocrine glands of the body, and list the hormones they secrete.

7. Describe the general functions of these hormones.

8. Explain how the secretion of each of these hormones is regulated.

9. Distinguish between physical and psychological stress.

10. Describe the general stress response.

11. Complete the review activities at the end of this chapter. Note that the items are worded in the form of specific learning objectives. You may want to refer to them before reading the chapter.

Key Terms

adenylate cyclase   (ah-den′ĭ-lat sĭ′klās)

adrenal cortex   (ah-dre′nal kor′teks)

adrenal medulla   (ah-dre′nal me-dul′ah)

anterior pituitary   (an-ter′e-or pĭ-tu′ĭ-tar″e)

cyclic AMP   (sik′lik)

kinase   (ki′nās)

metabolic rate   (met″ah-bol′ik rāt)

negative feedback   (neg′ah-tiv fēd′bak)

pancreas   (pan′kre-as)

parathyroid gland   (par″ah-thi′roid gland)

pineal gland   (pin′e-al gland)

posterior pituitary   (pos-ter′e-or pĭ-tu′ĭ-tar″e)

prostaglandin   (pros″tah-glan′din)

steroid   (ste′roid)

target cell   (tar′get sel)

thymus gland   (thi′mus gland)

thyroid gland   (thi′roid gland)

Aids to Understanding Words

-crin, to secrete: endo*crin*e—pertaining to internal secretions.

diuret-, to pass urine: *diuret*ic—a substance that promotes the production of urine.

endo-, within: *endo*crine gland—a gland that releases its secretion internally into a body fluid.

exo-, outside: *exo*crine gland—a gland that releases its secretion to the outside through a duct.

hyper-, above: *hyper*thyroidism—condition resulting from above normal secretion of thyroid hormone.

hypo-, below: *hypo*thyroidism—condition resulting from below normal secretion of thyroid hormone.

lact-, milk: pro*lact*in—a hormone that promotes milk production.

para-, beside: *para*thyroid glands—a set of glands located on the surface of the thyroid gland.

toc-, birth: oxy*toc*in—a hormone that stimulates uterine muscles to contract during childbirth.

tropic-, influencing: adrenocortico*tropic* hormone—a hormone secreted by the anterior pituitary gland that stimulates the adrenal cortex.

vas-, vessel: *vas*opressin—a substance that causes blood vessel walls to contract.

THE TERM *ENDOCRINE* is used to describe cells, tissues, and organs that secrete hormones into body fluids. In contrast, the term *exocrine* refers to parts whose secretory products are carried by tubes or ducts to some internal or external body surface. Thus, the thyroid and parathyroid glands that secrete hormones into the blood are endocrine glands (ductless glands), while sweat glands and salivary glands are exocrine (figure 13.1).

General Characteristics of the Endocrine System

As a group, endocrine structures function to help regulate metabolic processes. They control the rates of certain chemical reactions, aid in the transport of substances through membranes, and help regulate water and electrolyte balances. They also play vital roles in reproductive processes and in development and growth.

Although many hormones have localized effects, those described in this chapter have more widespread actions and are produced by larger endocrine glands. These glands include the pituitary gland, thyroid gland, parathyroid glands, adrenal glands, and pancreas. Several other hormone-secreting glands and tissues, including those involved in the processes of digestion and reproduction, are discussed in subsequent chapters (figure 13.2).

Hormones and Their Actions

A **hormone** is a substance secreted by a cell that has an effect on the metabolic activity of another cell or tissue. Such substances are released into the extracellular spaces surrounding the endocrine cells. Some of them travel only short distances and produce their effects locally. Other hormones are transported in the blood to all parts of the body and may produce rather general effects. In either case, the physiological action of a particular hormone is restricted to its *target cells*—those that possess specific receptors for the hormone molecules.

Chemistry of Hormones

Hormones are organic substances with special molecular structures. They are very potent and thus can stimulate cellular changes when they are present in extremely low concentrations. Chemically, most of them are steroids or steroidlike substances that are synthesized from cholesterol (see chapter 2), or they are amines, peptides, proteins, or glycoproteins that are synthesized from amino acids.

Steroids are compounds whose molecules contain complex rings of carbon and hydrogen atoms. The difference between one steroid and another is due to the types and numbers of atoms attached to these rings and

Figure 13.1

Endocrine glands, such as *(a)* the thyroid gland, release hormones into body fluids; while exocrine glands, such as *(b)* sweat glands, release their secretions into ducts that lead to body surfaces.

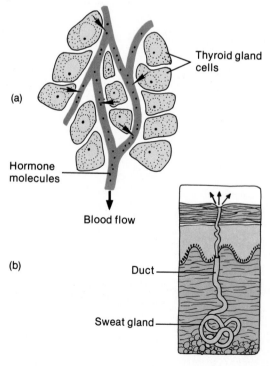

Figure 13.2

Locations of major endocrine glands.

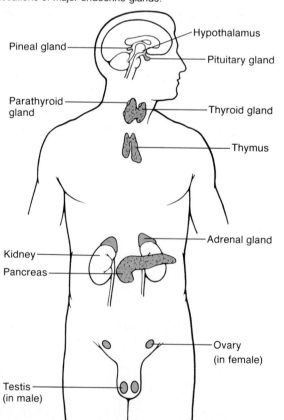

the ways they are joined. The steroid hormones include sex hormones (such as estrogens and androgens) and those secreted by the adrenal cortex (including aldosterone and cortisol). The substance commonly called vitamin D is a modified steroid and can be converted into a hormone, as is discussed later in this chapter (see chapter 15).

Hormones that are *amines* include norepinephrine and epinephrine. As discussed in chapter 11, these substances are produced by certain neurons. They are also synthesized in the adrenal medulla (the inner portion of the adrenal gland) from a single amino acid (tyrosine).

The *peptide* hormones are composed of relatively short chains of amino acids. This group includes those associated with the posterior pituitary gland (ADH and oxytocin) and some produced in the hypothalamus (including TRH, GnRH, and GIH).

Protein hormones are composed of many amino acids united to form intricate molecular structures (see chapter 2). They include the hormone secreted by the parathyroid gland (PTH) and some of those from the anterior pituitary gland (such as GH and PRL). Certain other hormones from the anterior pituitary (FSH, LH, and TSH) are *glycoproteins*. Their molecules include a protein joined to a carbohydrate portion (figure 13.3).

Figure 13.3
(a) Amino acid sequence of a protein hormone; structural formulas of *(b)* a steroid hormone and *(c)* an amine hormone.

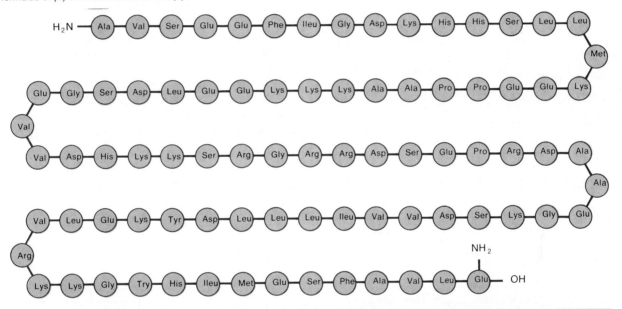

(a) Parathyroid hormone (PTH)

(b) Cortisol

(c) Norepinephrine

Chart 13.1 Hormone names and abbreviations

Source	Name	Abbreviation	Synonym
Hypothalamus	Corticotropin-releasing hormone	CRH	
	Gonadotropin-releasing hormone	GnRH	Luteinizing hormone-releasing hormone (LHRH)
	Growth hormone release-inhibiting hormone	GIH	Somatostatin (SS)
	Growth hormone-releasing hormone	GRH	
	Prolactin release-inhibiting factor	PRIF	
	Prolactin-releasing factor	PRF	
	Thyrotropin-releasing hormone	TRH	
Anterior pituitary gland	Adrenocorticotropic hormone	ACTH	Corticotropin
	Follicle stimulating hormone	FSH	Gonadotropin
	Growth hormone	GH	Somatotropin (STH)
	Luteinizing hormone	LH	Gonadotropin, interstitial cell stimulating hormone (ICSH)
	Prolactin	PRL	
	Thyroid stimulating hormone	TSH	Thyrotropin
Posterior pituitary gland	Antidiuretic hormone	ADH	Vasopressin
	Oxytocin	OT	
Thyroid gland	Calcitonin		
	Thyroxine	T_4	
	Triiodothyronine	T_3	
Parathyroid gland	Parathyroid hormone	PTH	Parathormone
Adrenal medulla	Epinephrine		Adrenalin
	Norepinephrine		Noradrenalin
Adrenal cortex	Aldosterone		
	Cortisol		Hydrocortisone
Pancreas	Glucagon		Hyperglycemic factor
	Insulin		

Chart 13.2 Types of hormones

Type of compound	Precursor molecule	Examples
Amines	Amino acids	Norepinephrine, epinephrine
Peptides	Amino acids	ADH, oxytocin, TRH, GIH, GnRH
Protein	Amino acids	PTH, GH, PRL
Glycoprotein	Protein and carbohydrate	FSH, LH, TSH
Steroid	Cholesterol	Estrogen, androgen, aldosterone, cortisol

1. What is the difference between an endocrine gland and an exocrine gland?
2. What is a hormone?
3. How are hormones chemically classified?

Actions of Hormones

Different kinds of hormones act in different ways. Some alter the activities of enzymes that, in turn, cause substances to be synthesized; others alter the rates at which chemicals are transported through cell membranes. The ability of a hormone to influence a particular kind of cell depends on the presence of receptor molecules in the cell or on its membranes. In other words, a hormone's target cells have certain receptors that other cells do not possess.

Steroid Hormones

Steroid hormones, unlike proteins, peptides, and amines, are soluble in lipids such as those that make up the bulk of cell membranes. For this reason, steroid

Chart 13.1 lists the names and abbreviations of some of the hormones discussed in this chapter. Chart 13.2 summarizes the chemical composition of various hormones.

Figure 13.4

(a) A steroid hormone passes through a cell membrane and *(b)* combines with a receptor protein in the cytoplasm. *(c)* The steroid-protein complex enters the nucleus and *(d)* activates the synthesis of messenger RNA. *(e)* The messenger RNA leaves the nucleus and *(f)* functions in the manufacture of protein molecules.

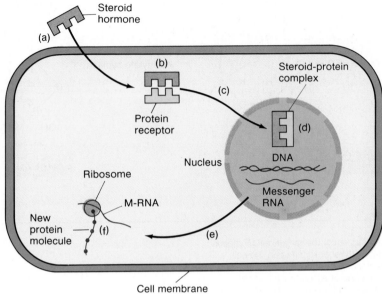

molecules can enter a target cell relatively easily by diffusion. Once they are inside they deliver their messages by combining with specific cytoplasmic protein molecules—the receptors. The steroid-protein complex may then enter the nucleus, bind to particular regions of DNA molecules, and thus activate certain genes, bringing about the synthesis of particular kinds of messenger RNA molecules.

As is discussed in chapter 4, messenger RNA can leave the nucleus and enter the cytoplasm where it functions in manufacturing specific proteins. Thus, steroid hormones influence cells by causing special proteins to be synthesized—proteins that may act as enzymes and alter the rates of cellular processes, or act as parts of membrane transport systems, or cause enzymes to be activated or inhibited (figure 13.4 and chart 13.3).

The degree of cellular response induced by a steroid hormone is proportional to the number of hormone-receptor complexes formed.

Nonsteroid Hormones

The nonsteroid hormones, such as amines, peptides, and proteins, act by combining with specific receptors located in the target-cell membranes. Each receptor is a protein molecule that has a *binding site* and an *activity site*. A hormone delivers its message to the cell by uniting with the binding-site of its receptor. This combination stimulates the receptor's activity site to interact with other membrane proteins. As a result, the

---

Chart 13.3 Sequence of steroid hormone action

1. Endocrine gland secretes steroid hormone.
2. Hormone enters target cell by diffusing through the cell membrane.
3. Hormone combines with a receptor protein molecule.
4. Hormone-protein complex enters the nucleus and promotes the synthesis of messenger RNA.
5. Messenger RNA enters the cytoplasm and functions in manufacturing protein molecules.
6. Proteins may function as enzymes and cause reactions that are recognized as the hormone's action.

---

actions of membrane-bound enzymes or membrane transport mechanisms are altered, causing changes in the concentrations of still other cellular substances. These substances serve as "second messengers" in the hormonal stimulation mechanism in that they, in turn, induce cellular changes that appear as responses to the original hormone.

The second messenger employed by one group of hormones is **cyclic adenosine monophosphate** (cAMP). In this mechanism a hormone binds to its receptor, and the resulting hormone-receptor complex activates an enzyme called **adenylate cyclase** that is bound to the inside of the cell membrane. The activated enzyme causes ATP molecules within the cytoplasm to become

Figure 13.5

ATP molecules are converted into cyclic AMP by the action of the enzyme, adenylate cyclase. What cellular changes may take place as a result of this reaction?

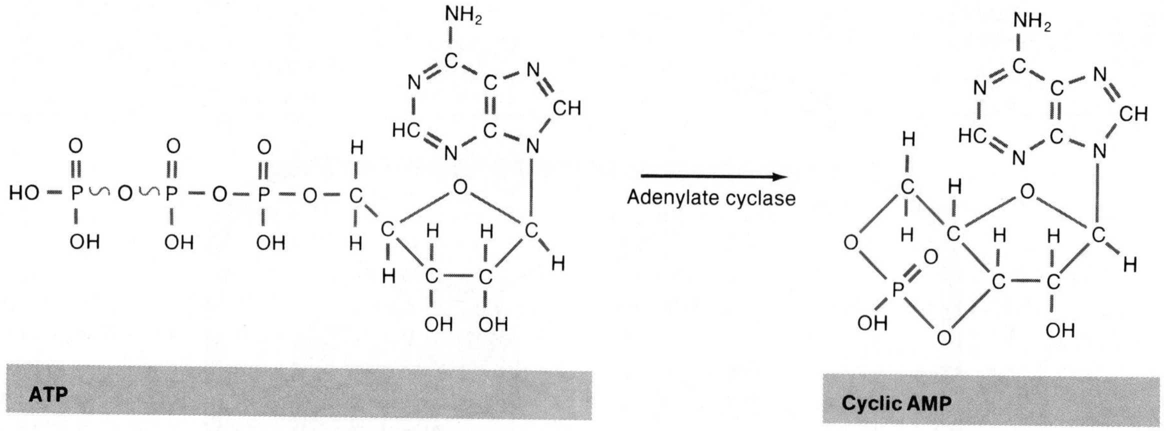

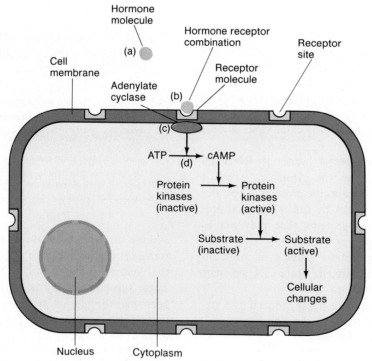

ATP

Adenylate cyclase

Cyclic AMP

Figure 13.6

(a) Nonsteroid hormone molecules reach the target cell by means of body fluids and *(b)* combine with receptor sites on the cell membrane. *(c)* As a result, molecules of adenylate cyclase are activated, and *(d)* cause the change of ATP into cyclic AMP. *(e)* Cyclic AMP promotes a series of reactions leading to various cellular changes.

Hormone molecule

Hormone receptor combination

Receptor site

Cell membrane

Receptor molecule

Adenylate cyclase

(a)

(b)

(c)

ATP → cAMP
(d)

Protein kinases (inactive) → Protein kinases (active)

Substrate (inactive) → Substrate (active)

Cellular changes

Nucleus   Cytoplasm

Chart 13.4   Sequence of some nonsteroid hormone actions

1. Endocrine gland secretes hormone.	5. Acting as an enzyme, adenylate cyclase causes ATP molecules to be converted into cyclic AMP molecules.
2. Hormone is carried to its target cell by body fluid.	6. Cyclic AMP activates various protein kinases.
3. Hormone combines with receptor site on membrane of its target cell.	7. These enzymes activate protein substrates in cell to induce changes in metabolic processes.
4. Adenylate cyclase molecules are activated within target cell's membrane.	8. Cellular changes are recognized as the hormone's action.

molecules of cAMP (figure 13.5). The cAMP, in turn, activates another set of enzymes called **protein kinases.** Protein kinases bring about the transfer of phosphate groups from ATP molecules to various protein substrate molecules. This phosphorylation alters the shapes of the substrate molecules and converts them from inactive forms into active ones. The activated protein molecules then induce changes in various cellular processes. (See figure 13.6.) Thus, the response of any particular cell to such a hormone is determined not only by the type of membrane receptors present, but also by the kinds of protein substrate molecules it contains. Chart 13.4 summarizes these actions.

Cellular responses to this second messenger mechanism include altering membrane permeabilities, activating various enzymes, promoting the synthesis of certain proteins, stimulating or inhibiting specific metabolic pathways, promoting cellular movements, and initiating the secretion of hormones and other substances.

The cyclic AMP produced by this mechanism is quickly inactivated by still another enzyme (phosphodiesterase) so that its action is short-lived. For this reason, a continuing response in a cell will depend upon a continuing signal from hormone molecules combining with receptors in the cell membrane.

Hormones whose actions involve the cyclic AMP mechanism include releasing hormones from the hypothalamus; TSH, ACTH, FSH, and LH from the anterior pituitary gland; ADH from the posterior pituitary gland; PTH from the parathyroid glands; norepinephrine and epinephrine from the adrenal glands; calcitonin from the thyroid gland; and glucagon from the pancreas.

Another group of nonsteroid hormones employs second messengers other than cAMP. These hormones include GnRH and GIH from the hypothalamus, GH and PRL from the anterior pituitary gland, oxytocin from the posterior pituitary gland, and insulin from the pancreas. The mechanisms that produce responses to these hormones are less well understood than the cAMP mechanism.

In one such mechanism, the binding of a hormone to its receptor causes an increase in calcium ion concentration within the target cell by stimulating the transport of calcium ions inward through the cell membrane or by inducing a release of calcium ions from cellular storage sites. The calcium combines with the protein *calmodulin* (see chapter 9), causing its molecular structure to be changed and activated. The activated calmodulin can interact with various cellular enzymes, altering their activities and thus producing a variety of responses.

Still another mechanism involves a substance called *cyclic guanosine monophosphate* (cGMP). This substance is similar to cAMP, and it is thought to function as a second messenger in much the same manner as cAMP. Thus a hormone-receptor combination activates the enzyme guanylate cyclase, which is responsible for the formation of cGMP, and the cGMP, in turn, induces changes in certain protein kinases. The precise role of cGMP, however, is not well understood.

As was mentioned, the cellular response to a steroid hormone is proportional to the number of hormone-receptor complexes formed; the response to a hormone operating through a second messenger, however, is greatly amplified. This amplification occurs when even a small number of hormone-receptor combinations are produced, because each combination can activate a membrane-bound enzyme. This enzyme, in turn, can cause a large number of second messenger molecules to appear, thus inducing a relatively strong cellular response. As a consequence of such amplification, cells are highly sensitive to changes in the concentrations of nonsteroid hormones.

Although it is not known how they accomplish this, some protein hormones enter cells. The significance of this movement is not clear; investigators believe that the act of binding to receptor sites produces an immediate, short-term effect on a cell. Long-lasting effects may be produced by actions taking place after the hormone, often with its receptor, has entered the cell.

Figure 13.7

A prostaglandin molecule contains a 20-carbon fatty acid with a 5-carbon ring.

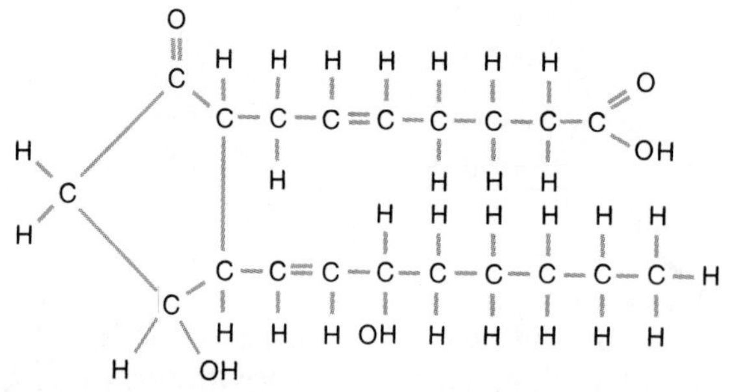

Prostaglandin E₂ (PGE₂)

Among the protein hormones that enter cells are growth hormone, prolactin, and gonadotropins from the anterior pituitary gland, PTH from the parathyroid glands, and insulin from the pancreas.

---

1. How do steroid hormones seem to act on their target cells?
2. How do nonsteroid hormones seem to act on their target cells?
3. What is meant by a "second messenger"?

Prostaglandins

Another group of compounds called **prostaglandins** have regulating effects on cells. These substances are lipids (20-carbon fatty acids that include 5-carbon rings) and are synthesized from fatty acids (arachidonic acid) found in cell membranes. They occur in a wide variety of cells including those of the liver, kidneys, heart, lungs, thymus gland, pancreas, brain, and various reproductive organs (figure 13.7).

Prostaglandins are very potent substances and are present in very small quantities. They are not stored in cells. Instead, they are synthesized just before they are released and are rapidly inactivated.

Although their modes of action are not well understood, some prostaglandins seem to function as modulators of hormones. For example, they seem to influence the activation of adenylate cyclase in cell membranes, and thus are able to regulate the formation of cyclic AMP. In some cases, this action increases the formation of cyclic AMP and in others it decreases its formation. In either case, by controlling the amount of cyclic AMP produced, the prostaglandins are able to alter the cell's response to a hormone.

Prostaglandins produce a variety of effects. For example, they can cause the smooth muscles in the airways of the lungs and in blood vessels to relax; however, they also can cause smooth muscle in the walls of the uterus and intestine to contract. They stimulate the secretions of hormones from the adrenal cortex and inhibit the secretion of hydrochloric acid from the wall of the stomach. They also influence the movements of sodium ions and water in the kidneys, help regulate blood pressure, and have powerful effects on both male and female reproductive physiology. When tissues are injured or stressed, prostaglandins promote the inflammation response (see chapter 5).

Prostaglandins are usually present in relatively high concentration in the synovial fluid of joints affected by rheumatoid arthritis. Certain anti-inflammatory drugs, including aspirin, seem to inhibit the production of these substances, and these are useful for treating the symptoms of arthritis.

---

1. What are prostaglandins?
2. Describe one possible function of these substances.
3. What kinds of effects do they produce?

Control of Hormonal Secretions

Since hormones are very potent substances that function to regulate metabolic processes, the amounts of hormones released by endocrine cells also must be regulated. One mechanism commonly employed for this purpose is a *feedback system*.

Figure 13.8

An example of a negative feedback system: *(1)* gland *A* secretes a hormone that stimulates gland *B* to release another hormone; *(2)* the hormone from gland *B* causes changes in its target cells and inhibits the activity of gland *A*.

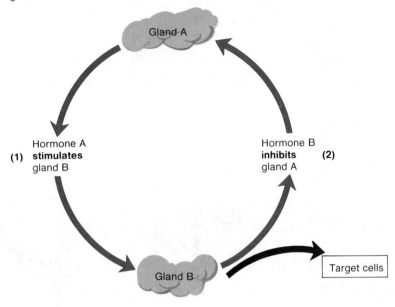

Figure 13.9

As a result of negative feedback systems, some hormone concentrations remain relatively stable, although they may fluctuate slightly above and below the average concentrations.

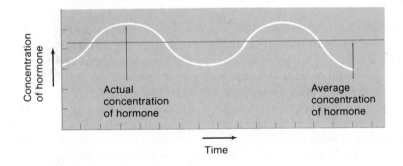

Negative Feedback Systems

In a feedback system, the hormone-secreting tissue continuously receives information (feedback) in the form of chemical signals concerning the cellular process it controls. If conditions change, the tissue can act upon the information and adjust its rate of secretion.

Commonly the control of hormonal secretions involves a **negative feedback system,** as described in chapter 1. In such a system, an endocrine gland is sensitive to the concentration of a substance it regulates or to the concentration of a product of a process it regulates. Whenever this concentration reaches a certain level, the endocrine gland is inhibited (a negative effect), and its secretory activity decreases. Then, as the concentration of its hormone decreases, the concentration of the regulated substance decreases, too, and the inhibition on the gland is released. When the gland is no longer inhibited, it begins to secrete its hormone again (figure 13.8).

As a result of such negative feedback systems, the concentrations of some hormones remain relatively stable, although they may slightly fluctuate within normal ranges (figure 13.9).

Figure 13.10
This diagram represents a positive feedback system. How does
it differ from a negative feedback system?

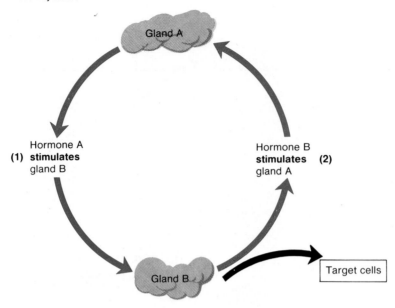

Gland A

Hormone A
(1) stimulates
gland B

Hormone B
stimulates (2)
gland A

Gland B

Target cells

Positive Feedback Systems

A **positive feedback system** is operating when an endocrine gland is stimulated to increase its rate of hormonal secretion (a positive effect) by a substance it
causes to be produced. Consequently, still more product
forms, and then more hormone is secreted (figure
13.10).

A positive feedback system is unstable in that it
tends to produce extreme changes in conditions, and
such positive feedback systems occur rarely in organisms. However, one example of this type of system involves the sex hormone estrogen. This steroid is secreted
by cells in the ovary in response to an anterior pituitary
hormone (FSH). When estrogen is present in a certain
blood concentration, it causes an increase in the secretion of some anterior pituitary hormones (including
FSH), so that positive feedback exists for a short time
(see chapter 22).

Excess hormones that are not used in the interactions
with target cells usually are inactivated by the liver and
kidneys. For this reason, a person with a liver or kidney
disease may experience effects resulting from increasing concentrations of certain hormones.

Nerve Control

Still other types of hormone control mechanisms involve the nervous system. For example, some endocrine
glands, such as the adrenal medulla, secrete their hormones in response to nerve impulses, and no other stimulus seems able to cause a secretion. In the case of the
adrenal medulla, hormones (norepinephrine and epinephrine) are secreted in response to stimulation from
preganglionic sympathetic nerve fibers originating in
the hypothalamus of the brain (see chapter 11).

Still another kind of control system involves interaction between an endocrine gland and the hypothalamus. In this system, neurosecretory cells in the
hypothalamus secrete substances, called **releasing** (or
inhibiting) **hormones** (or factors), whose target cells are
in the anterior pituitary gland. The gland responds to
the substance by secreting its own hormone. Then, as
the gland's hormone reaches a certain concentration in
the body fluids, a negative feedback system operates to
inhibit the hypothalamus, and its secretion of the releasing hormone decreases.

---

1. What is a feedback system?
2. How does a negative feedback system operate to
   control hormonal secretions?
3. Describe a positive feedback mechanism.
4. How does the nervous system help to regulate
   hormonal secretions?

Figure 13.11
The pituitary gland is attached to the hypothalamus and lies in
the sella turcica of the sphenoid bone.

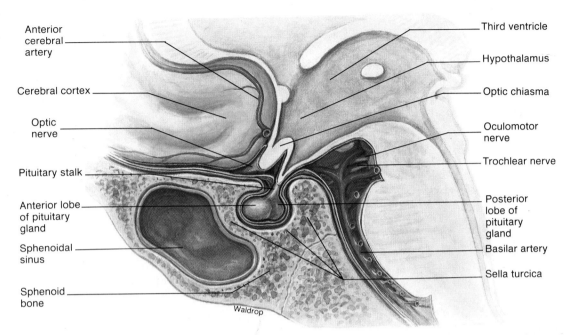

Anterior cerebral artery — Cerebral cortex — Optic nerve — Pituitary stalk — Anterior lobe of pituitary gland — Sphenoidal sinus — Sphenoid bone — Waldrop — Third ventricle — Hypothalamus — Optic chiasma — Oculomotor nerve — Trochlear nerve — Posterior lobe of pituitary gland — Basilar artery — Sella turcica

The Pituitary Gland

The **pituitary gland** (hypophysis) is about 1 cm in diameter and is located at the base of the brain. It is attached to the hypothalamus by the pituitary stalk, or *infundibulum,* and lies in the sella turcica of the sphenoid bone, as shown in figure 13.11.

This gland consists of two distinct portions: an anterior lobe (adenohypophysis) and a posterior lobe (neurohypophysis). The *anterior lobe* secretes a number of hormones, including growth hormone (GH), thyroid-stimulating hormone (TSH), adrenocorticotropic hormone (ACTH), follicle-stimulating hormone (FSH), luteinizing hormone (LH), and prolactin (PRL). Although the *posterior lobe* does not synthesize any hormones, two important ones, antidiuretic hormone (ADH) and oxytocin, are secreted by nerve fibers within its tissues.

During fetal development, a narrow region appears between the anterior and posterior lobes of the pituitary gland. This part is called the *intermediate lobe* (pars intermedia). It produces melanocyte-stimulating hormone (MSH), which regulates the formation of melanin—the pigment found in the skin—and portions of the eyes and brain. This region atrophies during prenatal development, and appears only as a vestige in adults.

Most of the pituitary activities are controlled by the brain. The release of hormones from the posterior lobe, for example, occurs when nerve impulses from the *hypothalamus* signal the axon ends of neurosecretory cells in this lobe. Secretions from the anterior lobe typically are controlled by releasing hormones produced by the hypothalamus (figure 13.12). These releasing hormones are transmitted by blood, in the vessels of a capillary net in the region of the hypothalamus. These vessels merge to form the **hypophyseal portal veins** that pass downward along the pituitary stalk and give rise to a capillary net in the anterior lobe. Thus, substances released into the blood from the hypothalamus are carried directly to the anterior lobe.

The hypothalamus (described in chapter 11) receives information from nearly all parts of the nervous system. This information includes data concerning a person's emotional state, body temperature, blood nutrient concentrations, and so forth. The hypothalamus sometimes acts on such information by signaling the pituitary gland to release hormones.

1. Where is the pituitary gland located?
2. List the hormones secreted by the anterior lobe and the posterior lobe of the pituitary gland.
3. Explain how the hypothalamus is related to the actions of the anterior and posterior lobes of the pituitary gland.

Figure 13.12

Axons in the posterior lobe of the pituitary gland are stimulated to release hormones by nerve impulses originating in the hypothalamus; cells of the anterior lobe are stimulated by releasing hormones secreted by hypothalamic neurons.

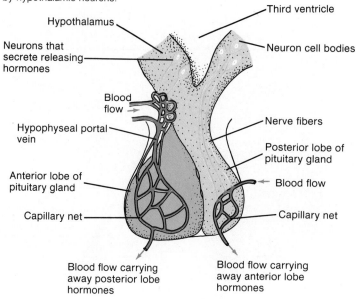

Anterior Lobe Hormones

The anterior lobe of the pituitary gland is enclosed in a dense capsule of collagenous connective tissue and consists largely of epithelial tissue arranged in blocks around many thin-walled blood vessels. Within the epithelial tissue five types of secretory cells have been identified. They include *somatotropes* that secrete GH, *mammatropes* that secrete PRL, *thyrotropes* that secrete TSH, *corticotropes* that secrete ACTH, and *gonadotropes* that secrete FSH and LH. (See figure 13.13.)

Growth hormone (GH), which also is called *somatotropin* (STH), is a protein that generally stimulates body cells to increase in size and undergo more rapid cell division than usual. It enhances the movement of amino acids through cell membranes and causes an increase in the rate at which cells convert these molecules into proteins. GH also causes cells to decrease the rate at which they utilize carbohydrates and to increase the rate at which they use fats. The hormone's effect on amino acids, however, seems to be the more important one.

Although the exact mechanism for controlling growth hormone secretion is unknown, it appears to involve two substances from the hypothalamus called *growth hormone-releasing hormone* (GRH) and *growth hormone release-inhibiting hormone* (GIH), which is also called *somatostatin*. A person's nutritional state also seems to play a role in the control of GH, for more of it is released during periods of protein

deficiency and of abnormally low blood glucose concentration. Conversely, when blood protein and glucose levels are increased, there is a resulting decrease in growth hormone secretion. Apparently the hypothalamus is able to sense changes in the concentrations of certain blood nutrients, and it releases GRH in response to some of them.

One of the more obvious effects of growth hormone results from its ability to stimulate the growth of cartilage in the epiphyseal discs of long bones and thus promote their elongation. As is mentioned in chapter 8, if growth hormone is not secreted in sufficient amounts during childhood, body growth is limited, and a type of *dwarfism* (hypopituitary dwarfism) results. In this condition, body parts usually are correctly proportioned and mental development is normal. However, an abnormally low secretion of growth hormone is usually accompanied by lessened secretions from other anterior lobe hormones, leading to additional hormone deficiency symptoms. For example, a hypopituitary dwarf often fails to develop adult sexual features unless hormone therapy is provided.

Hypopituitary dwarfism is sometimes treated by administering growth hormone, and this treatment may stimulate a rapid increase in height. The procedure, however, must be started before the epiphyseal disks of the person's long bones have become ossified. Otherwise growth in height is not possible.

An oversecretion of growth hormone during childhood may result in *gigantism*—a condition in which the person's height may exceed 8 feet. Gigantism, which is relatively rare, is usually accompanied

Figure 13.13
Hormones released from the anterior lobe of the pituitary gland and their target organs.

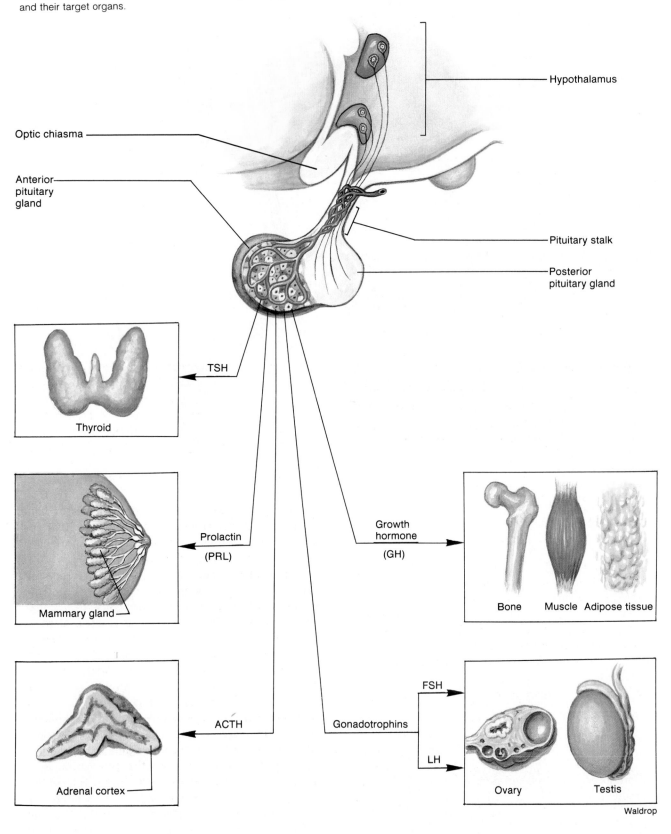

Hypothalamus

Optic chiasma

Anterior pituitary gland

Pituitary stalk

Posterior pituitary gland

TSH

Thyroid

Prolactin (PRL)

Mammary gland

Growth hormone (GH)

Bone   Muscle   Adipose tissue

ACTH

Adrenal cortex

Gonadotrophins

FSH

LH

Ovary   Testis

Waldrop

Figure 13.14

Acromegaly is caused by an oversecretion of growth hormone in adulthood. Note the changes in facial features of this individual from ages nine (a), sixteen (b), thirty-three (c), and fifty-two (d).

(a)   (b)

(c)   (d)

Figure 13.15

TRH from the hypothalamus stimulates the anterior pituitary gland to release TSH. TSH stimulates the thyroid gland to release hormones, which in turn cause the hypothalamus to reduce its secretion of TRH.

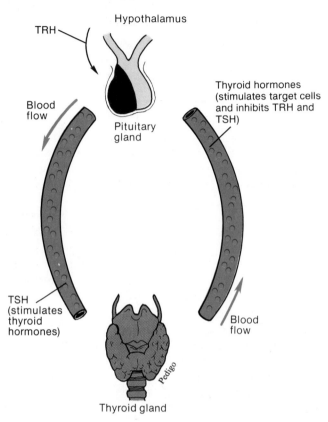

Hypothalamus

TRH

Blood flow

Pituitary gland

Thyroid hormones (stimulates target cells and inhibits TRH and TSH)

TSH (stimulates thyroid hormones)

Blood flow

Thyroid gland

by a tumor of the pituitary gland. In such cases, various pituitary hormones in addition to GH are likely to be secreted excessively, so that a giant often suffers from a variety of metabolic disturbances and has a shortened life expectancy.

If growth hormone is secreted excessively in an adult, after the epiphyses of the long bones have ossified, the person does not grow taller. The soft tissues, however, may continue to enlarge and the bones may become thicker. As a consequence, an affected individual may develop greatly enlarged hands and feet, a protruding jaw, and a large tongue and nose. This condition is called *acromegaly*, and like gigantism, it is often associated with a pituitary tumor (figure 13.14).

Prolactin (PRL) is a protein, and as its name suggests, it functions in milk production in females. More specifically, prolactin stimulates and sustains milk production following the birth of an infant.

Although this hormone seems to have little effect in males by itself, it amplifies the effect of the hormone LH. LH stimulates the production of male sex hormones (androgens); thus, when prolactin is present, increased amounts of androgens are secreted in response to LH (see chapter 22).

Although the regulation of prolactin secretion is poorly understood, two substances from the hypothalamus seem to be involved. One substance, called *prolactin release-inhibiting factor* (PRIF), acts to restrain the secretion of prolactin, while the other, *prolactin-releasing factor* (PRF), stimulates its secretion.

Thyroid-stimulating hormone (TSH), which is also called *thyrotropin,* is a glycoprotein—a protein bound to a carbohydrate. Its major function is to control the secretion of hormones from the thyroid gland, which are described in a later section of this chapter.

TSH secretion is partially regulated by the hypothalamus, which secretes *thyrotropin-releasing hormone* (TRH). TSH secretion is also regulated by circulating thyroid hormones that exert an inhibiting effect on the release of TRH and TSH; therefore, as the blood concentration of thyroid hormones increases, the secretions of TRH and TSH are reduced (figure 13.15).

Certain external factors influence the release of these hormones. These factors include exposure to extreme cold, which is accompanied by increased hormonal secretions, and emotional stress, which sometimes triggers increased hormonal secretions and other times causes decreased secretions.

---

1. How does growth hormone affect the utilization of carbohydrates, fats, and proteins?
2. What are the functions of prolactin?
3. How is TSH secretion regulated?

Adrenocorticotropic hormone (ACTH) is a protein that controls the manufacture and secretion of certain hormones from the outer layer (cortex) of an adrenal gland. These adrenal hormones are discussed in a later section of this chapter.

The secretion of ACTH may be regulated in part by *corticotropin-releasing hormone* (CRH), which is released from the hypothalamus in response to decreased concentrations of adrenal cortical hormones. Also, as in the case of TSH, various forms of stress serve to stimulate the secretion of ACTH.

Both **follicle-stimulating hormone** (FSH) and **luteinizing hormone** (LH) are glycoproteins and are called *gonadotropins,* which means they exert their actions on the gonads or reproductive organs. FSH, for example, is responsible for growth and development of egg-containing follicles in the female ovaries. It also stimulates the follicle cells to secrete a group of female sex hormones, collectively called *estrogen.*

In males, FSH stimulates the initial production of sperm cells in the testes at puberty. LH, which in males is also called *interstitial cell-stimulating hormone* (ICSH) since it acts upon the interstitial cells of the testes, promotes the secretion of sex hormones in both males and females, and is essential for the release of egg cells from the female ovaries. Other functions of these gonadotropins and the ways they interact are discussed in chapter 22.

The mechanism that regulates the secretion of gonadotropins is not well understood. It is known, however, that the hypothalamus secretes a *gonadotropin-releasing hormone* (GnRH). The hypothalamus apparently fails to secrete this hormone until the age of puberty because gonadotropins are virtually absent in the body fluids of infants and children.

---

1. What is the function of ACTH?
2. Describe the functions of FSH and LH in a male and in a female.
3. What is a gonadotropin?

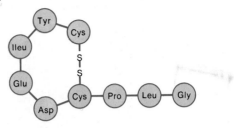

Oxytocin

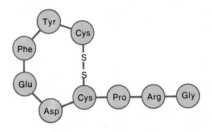

Antidiuretic hormone

Posterior Lobe Hormones

Unlike the anterior lobe of the pituitary gland, which is composed primarily of glandular epithelial cells, the posterior lobe consists largely of nerve fibers and neuroglial cells (*pituicytes*). The neuroglial cells function to support the nerve fibers that originate in the hypothalamus.

As mentioned earlier, the two hormones associated with the posterior lobe—antidiuretic hormone (ADH) and oxytocin (OT)—are actually produced by specialized neurons in the hypothalamus (figure 13.12). These substances travel down axons through the pituitary stalk to the posterior lobe and are stored in vesicles (secretory granules) near the ends of the axons. The hormones are released into the blood in response to nerve impulses coming from the hypothalamus.

Antidiuretic hormone (ADH) is a polypeptide consisting of a relatively short chain of amino acids. This is also true of oxytocin, and as figure 13.16 shows, the molecules of these substances are similar except for amino acids in two locations.

A *diuretic* is a substance that acts to increase urine production. An *antidiuretic,* then, is a chemical that inhibits urine formation. ADH produces its antidiuretic effect by acting on the kidneys and causing them to reduce the amount of water they excrete. In this way, ADH is important in regulating the water concentration of body fluids (see chapter 20).

Chart 13.5 Hormones of the pituitary gland

Anterior lobe

Hormone	Action	Source of control
Growth hormone (GH)	Stimulates increase in size and rate of reproduction of body cells; enhances movement of amino acids through membranes; promotes growth of long bones	Growth hormone-releasing hormone (GRH) and growth hormone release-inhibiting hormone (GIH) from the hypothalamus
Prolactin (PRL)	Sustains milk production after birth; amplifies effect of LH in males.	Secretion restrained by prolactin release-inhibiting factor (PRIF) and stimulated by prolactin-releasing factor (PRF) from the hypothalamus
Thyroid-stimulating hormone (TSH)	Controls secretion of hormones from the thyroid gland	Thyrotropin-releasing hormone (TRH) from the hypothalamus
Adrenocorticotropic hormone (ACTH)	Controls secretion of certain hormones from the adrenal cortex	Corticotropin-releasing hormone (CRH) from the hypothalamus
Follicle-stimulating hormone (FSH)	Responsible for the development of egg-containing follicles in ovaries; stimulates follicle cells to secrete estrogen; in male stimulates the production of sperm cells	Gonadotropin-releasing hormone (GnRH) from the hypothalamus
Luteinizing hormone (LH or ICSH in males)	Promotes secretion of sex hormones; plays role in release of egg cell in females	Gonadotropin-releasing hormone (GnRH) from the hypothalamus

Posterior Lobe

Hormone	Action	Source of control
Antidiuretic hormone (ADH)	Causes kidneys to reduce water excretion; in high concentration, causes blood pressure to rise	Hypothalamus in response to changes in blood water concentration and blood volume
Oxytocin (OT)	Causes contractions of muscles in uterine wall; causes muscles associated with milk-secreting glands to contract	Hypothalamus in response to stretch in uterine and vaginal walls and stimulation of breasts

---

Beverage alcohol (ethyl alcohol) is thought to inhibit the normal secretion of ADH. Consequently, an abnormally large volume of urine may be produced after a person drinks alcoholic beverages. The lost body fluid must later be replaced if water balance is to be maintained.

When it is present in high concentrations, ADH causes contractions of certain smooth muscles, including those in the walls of some blood vessels. As a result, the blood pressure in these vessels increases. For this reason, ADH is also called *vasopressin*. Although ADH is seldom present in sufficient quantities to elevate blood pressure, it may be released following a severe loss of blood. Blood pressure may drop as a consequence of excessive bleeding, and in this situation ADH's vasopressor effect may help to restore normal pressure.

The secretion of ADH is regulated by the hypothalamus. Certain neurons in this part of the brain called *osmoreceptors* are sensitive to changes in the water concentration of body fluids. If, for example, a person is dehydrating due to lack of water intake, the solutes in blood become more and more concentrated. The osmoreceptors can sense this change and signal the posterior lobe to release ADH. The ADH is transported in the blood to the kidneys, and as a result of its effects, less urine is produced. This action conserves water.

If, on the other hand, a person drinks an excess amount of water, the body fluids become more dilute and the release of ADH is inhibited. Consequently, the kidneys excrete more dilute urine until the water concentration of the body fluids returns to normal.

Another factor that affects ADH secretion is blood volume. The normal volume causes pressure that stretches the walls of blood vessels. Pressure receptors in some of the walls are stimulated by such stretching, and they signal the hypothalamus with impulses that inhibit the release of ADH. However, if the blood volume is decreased due to a hemorrhage, the pressure drops, and the receptors send fewer inhibiting impulses. As a result, the secretion of ADH increases and, as before, this hormone causes the kidneys to conserve water. This action helps to compensate for some of the decreased blood volume.

If any parts involved in the ADH regulating mechanism are damaged due to an injury or a tumor, the hormone may not be synthesized or released normally. The resulting ADH deficiency produces a condition called *diabetes insipidus*. This disease is characterized by an output of as much as 25 to 30 liters of very dilute urine per day (polyuria) and a rise in the concentration of dissolved substances in the body fluids. It also is accompanied by a sensation of great thirst (polydipsia).

Figure 13.17
(a) The thyroid gland consists of two lobes connected anteriorly
by an isthmus; (b) thyroid hormones are secreted by follicle cells.

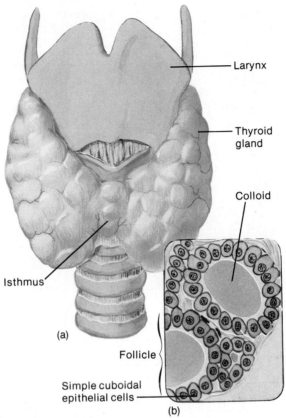

Larynx

Thyroid gland

Colloid

Isthmus

(a)

Follicle

Simple cuboidal epithelial cells

(b)

Oxytocin also has an antidiuretic action, but it is weaker in this respect than ADH. In addition, it can cause contractions of the smooth muscles in the uterine wall and may play a role in the later stages of childbirth by stimulating uterine contractions. The mechanism that triggers the release of oxytocin is not clearly understood. It is known, however, that the uterus becomes more and more sensitive to oxytocin's effects during pregnancy. Also, it is believed that stretching of uterine and vaginal tissues late in pregnancy, caused by the growing fetus, may initiate nerve impulses to the hypothalamus. The hypothalamus may then signal the posterior lobe to release oxytocin, which, in turn, may enhance uterine wall contractions during labor.

Oxytocin also has an effect upon the breasts, causing contractions in certain cells associated with milk-producing glands and their ducts. In lactating breasts, this action forces liquid from the milk glands into the milk ducts and causes the milk to be ejected from the breasts—an effect that is necessary before milk can be removed.

The mechanical stimulation provided by sucking the nipple of a breast initiates nerve impulses that travel to the mother's hypothalamus. The hypothalamus responds by signaling the posterior lobe to release oxytocin, which in turn stimulates the release of milk. Thus,

milk is normally not ejected from the milk glands and ducts until it is needed. Oxytocin has no established function in males, although it is present in the male posterior pituitary.

Chart 13.5 reviews the hormones of the pituitary gland.

If the uterus is not contracting sufficiently to expel a fully developed fetus, commercial preparations of oxytocin are sometimes used to stimulate uterine contractions, thus inducing labor. Also, such preparations are often administered to the mother following childbirth to ensure that the uterine muscles contract enough to squeeze broken blood vessels closed, minimizing the danger of hemorrhage.

---

1. What is the function of ADH?
2. What effects does oxytocin produce?

The Thyroid Gland

The **thyroid gland,** shown in figure 13.17, is a very vascular structure that consists of two large lobes connected by a broad isthmus. It is located just below the larynx on either side and in front of the trachea. It has a special ability to remove iodine from the blood.

Figure 13.18

A micrograph of thyroid gland tissue. The open spaces
surrounded by follicle cells are filled with colloid.

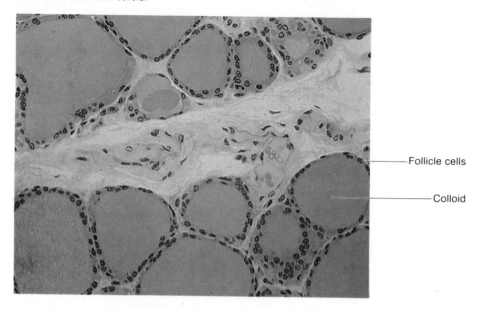

Follicle cells

Colloid

Structure of the Gland

The thyroid gland is covered by a capsule of connective
tissue and is made up of many secretory parts called
follicles. The cavities of the follicles are lined with a
single layer of cuboidal epithelial cells and are filled
with a clear, viscous glycoprotein called *colloid*. The
follicle cells produce and secrete hormones that may
be stored in the colloid or released into the blood of nearby capillaries (figure 13.18). Some other hormone-
secreting cells called extrafollicular cells (C cells) occur
outside the follicles.

Thyroid Hormones

The thyroid gland produces several hormones. Some
that are synthesized in the follicles have marked effects
on the metabolic rates of body cells. Another hormone
that is produced by the extrafollicular cells influences
the levels of blood calcium and phosphate.

Of the thyroid hormones that affect metabolic
rates, the most important are thyroxine (T_4) and triiodothyronine (T_3). They help regulate the metabolism
of carbohydrates, lipids, and proteins. For example, they
increase the rate at which cells release energy from
carbohydrates; enhance the rate of protein synthesis;
and stimulate the breakdown and mobilization of lipids.
As a result of their actions, these hormones are needed
for normal growth and development of children, and
they are essential for the maturation of the nervous
system (figure 13.19).

Before follicle cells can produce thyroxine and
triiodothyronine, they must be supplied with iodine salts
(iodides). Such salts are normally obtained from foods,
and after they have been absorbed from the intestine,
they are carried by the blood to the thyroid gland. An
efficient active transport mechanism called the *iodine
pump* moves the iodides into the follicle cells, where
they are concentrated. The iodides, together with an
amino acid (tyrosine), are used to synthesize hormones.

Follicle cells also secrete a protein called *thyroglobulin,* which is the main ingredient of thyroid colloid. Thyroglobulin is used to store thyroid hormones
whenever they are produced in excess. The stored hormones are bound to the thyroglobulin until the hormone concentration of the body fluids drops below a
certain level; then enzymes cause the hormones to be
released from the colloid, and they diffuse into the
blood. Once they are in the blood, thyroid hormones

Figure 13.19
The hormones thyroxine and triiodothyronine have very similar molecular structures.

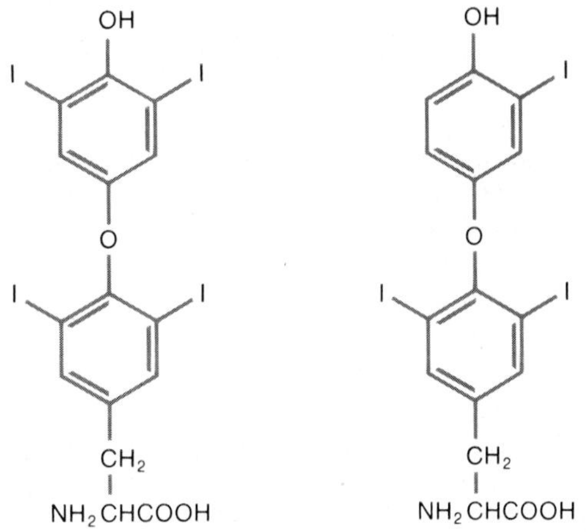

Thyroxine (T$_4$)   Triiodothyronine (T$_3$)

Chart 13.6 Hormones of the thyroid gland

Hormone	Action	Source of control
Thyroxine (T$_4$)	Increases rate of energy release from carbohydrates; increases rate of protein synthesis; accelerates growth; stimulates activity in the nervous system	TSH from the anterior pituitary gland
Triiodothyronine (T$_3$)	Same as above	Same as above
Calcitonin	Lowers blood calcium level by inhibiting the release of calcium from bones	Elevated blood calcium, digestive hormones

combine with blood proteins (alpha globulins) and are transported to body cells. Although triiodothyronine is nearly five times more potent, thyroxine accounts for at least 95% of the circulating thyroid hormones.

The thyroid hormone that influences blood calcium and phosphate concentrations is a polypeptide called **calcitonin**. This substance helps regulate the concentration of these ions by inhibiting the rate at which they leave the bones and enter the extracellular fluids. This is accomplished by inhibiting the bone resorbing activity of osteoclasts (see chapter 8). At the same time, calcitonin causes an increase in the rate at which calcium and phosphate are deposited in bone matrix by stimulating the activity of osteoblasts. It also increases the excretion of calcium ions and phosphate ions by the kidneys. Thus, calcitonin acts to lower the blood calcium and phosphate concentrations.

The secretion of calcitonin is stimulated by a high blood calcium ion concentration, as may occur following the absorption of calcium ions from a recent meal. Its secretion also is stimulated by certain hormones, such as gastrin, that are released from digestive organs when they are active. For these reasons, it is thought that calcitonin may function to prevent a prolonged elevation in blood calcium after eating.

Chart 13.6 summarizes the thyroid gland's hormones and the hormones' actions and sources of control.

1. Where is the thyroid gland located?
2. What hormones of the thyroid gland affect carbohydrate metabolism and protein synthesis?
3. What substance is essential for the production of these hormones?
4. How does calcitonin influence the concentration of blood calcium and phosphate?

Disorders of the Thyroid Gland
A Clinical Application

Most functional disorders of the thyroid gland are characterized by *overactivity* (hyperthyroidism) or *underactivity* (hypothyroidism) of the gland cells.

A variety of laboratory tests are available to help a physician judge the functional condition of a patient's thyroid gland. These include tests that measure the blood concentration of triiodothyronine (T_3 test), the blood concentration of thyroxine (T_4 test), and the rate at which iodine is taken up by the gland.

In the iodine uptake test, the patient drinks some distilled water containing a small amount of *radioactive iodine* (I-131). Some hours later, the radioactivity of the thyroid gland is measured using a scintillation counter (see figure 2.2). The results indicate how effectively the gland was able to remove iodine from the blood and concentrate it, thus giving an indication of the gland's condition (figure 2.3).

A thyroid disorder may develop at any time during a person's life as a result of a developmental problem, an injury, a disease, or a dietary deficiency. One form of **hypothyroidism** appears in infants when their thyroid glands fail to function normally. An affected child may appear normal at birth because it has received an adequate supply of thyroid hormones from its mother during pregnancy. When its own thyroid gland fails to produce sufficient quantities of hormones, the child soon develops a condition called *cretinism*. Cretinism is characterized by severe symptoms including stunted growth, abnormal bone formation, retarded mental development, low body temperature, and sluggishness. Without treatment with thyroid hormones within a month or so following birth, the child is likely to suffer from permanent mental retardation (figure 13.20).

If a person develops hypothyroidism later in life, the symptoms include an abnormally low metabolic rate, abnormal sensitivity to cold, physical sluggishness, and poor appetite. The person also may appear mentally dull and may develop swollen tissues due to an accumulation of body fluid in the subcutaneous tissue—a condition called *myxedema*.

On the other hand, a person with **hyperthyroidism** has an abnormally high metabolic rate, is sensitive to heat, is restless or overactive, eats excessively, and appears mentally alert. Also, the person's eyes are likely to protrude

Figure 13.20
Cretinism is due to an underactive thyroid gland during infancy and childhood. How can this condition be prevented?

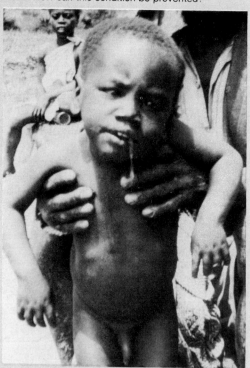

Figure 13.21
Hyperthyroidism may produce protrusion of the eyes.

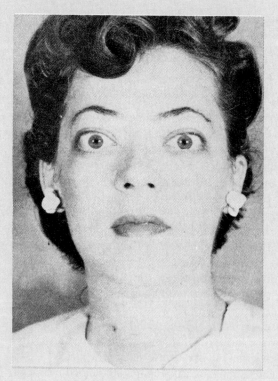

(exophthalmos) because of edematous swelling in the tissues behind them (figure 13.21). At the same time, the thyroid gland is likely to enlarge, producing a bulge in the neck called a *goiter*. Such a goiter that is associated with hyperthyroidism is said to be a *toxic goiter*.

Another type of goiter, called *simple* or *endemic goiter*, sometimes affects persons who live in regions where iodine is lacking in the soil and drinking water. Such a person is likely to develop an iodine deficiency, which is reflected in an inability to produce thyroid hormones. Since these hormones normally exert an inhibiting effect on the secretion of TSH, without such inhibition, the anterior lobe of the pituitary gland releases TSH excessively. The resulting overstimulation of the thyroid gland causes it to enlarge, but since the gland is unable to manufacture hormones, the condition is accompanied by the symptoms of hypothyroidism (figure 13.22).

Simple goiter usually can be prevented in regions where iodine deficiencies occur if people use table salt containing iodides—*iodized salt*.

Figure 13.22
Endemic goiter is caused by an iodine deficiency.

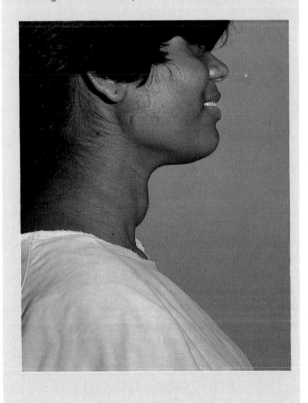

The Parathyroid Glands

The **parathyroid glands** are located on the posterior surface of the thyroid gland, as shown in figure 13.23. Usually there are four of them—two associated with each of the thyroid's lateral lobes. These glands secrete a hormone that functions in the regulation of blood calcium and phosphate levels.

Structure of the Glands

Each parathyroid gland is a small, yellowish brown structure covered by a thin capsule of connective tissue. The body of the gland consists of numerous tightly packed secretory cells that are closely associated with capillary networks.

Parathyroid Hormone

The only hormone known to be secreted by the parathyroid glands is a protein called **parathyroid hormone** (PTH) or *parathormone* (figure 13.3). This substance causes an increase in the blood calcium concentration

Figure 13.23
The parathyroid glands are embedded in the posterior surface of the thyroid gland.

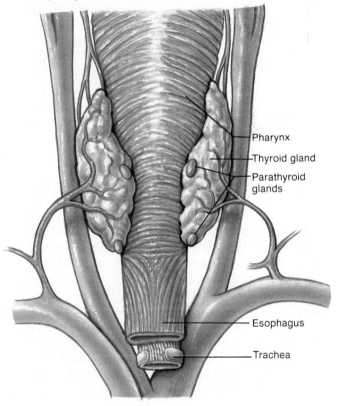

Pharynx

Thyroid gland

Parathyroid glands

Esophagus

Trachea

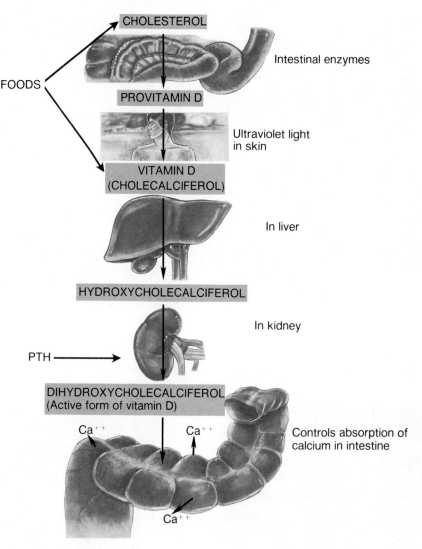

CHOLESTEROL

Intestinal enzymes

FOODS

PROVITAMIN D

Ultraviolet light
in skin

VITAMIN D
(CHOLECALCIFEROL)

In liver

HYDROXYCHOLECALCIFEROL

In kidney

PTH →

DIHYDROXYCHOLECALCIFEROL
(Active form of vitamin D)

Ca^{++} Ca^{++} Controls absorption of
calcium in intestine

Ca^{++}

and a decrease in the blood phosphate level. It does this
by influencing actions in bones, kidneys, and the intestines.

As is described in chapter 8, the intercellular matrix of bone tissue contains a considerable amount of
calcium phosphate and calcium carbonate. PTH seems
to stimulate bone resorption by osteocytes and osteoclasts and inhibit the activity of osteoblasts (see chapter
8). As a result of increased resorption, calcium and
phosphate ions are released from the bones, and the
blood concentrations of these substances increase. At
the same time, PTH causes the kidneys to conserve
blood calcium and to increase the excretion of phosphate in the urine. It also stimulates the absorption of

calcium from food in the intestine, but it does this indirectly by influencing the metabolism of vitamin D.

Vitamin D (cholecalciferol) can be obtained in
some foods. It also can be synthesized in the body from
dietary cholesterol, which is converted into provitamin
D (7-dehydrocholesterol) by intestinal enzymes. This
provitamin is stored largely in the skin, and it can be
changed to vitamin D by exposure to ultraviolet light,
as occurs in sunlight. The resulting vitamin D molecules then can be stored in various tissues or converted
by the liver into another form of vitamin D (hydroxycholecalciferol), which can be transported by the blood.

When PTH is present, hydroxycholecalciferol can
be changed by the kidneys into an active form of vitamin D (dihydroxycholecalciferol). This active form

controls the transport mechanism by which calcium is absorbed from the contents of the intestine. Thus, PTH indirectly regulates calcium absorption in the intestine by causing the kidneys to activate vitamin D molecules (figure 13.24).

The parathyroid glands seem to be free of control by the hypothalamus. Instead, the secretion of PTH is regulated by a *negative feedback mechanism* operating between the glands and the blood calcium concentration (figure 13.25). As the concentration of blood calcium rises, less PTH is secreted; as the concentration of blood calcium drops, more PTH is released.

The homeostasis of calcium is important in a number of physiological processes. For example, as the blood calcium concentration drops (hypocalcemia), the nervous system becomes abnormally excitable, and impulses may be triggered spontaneously. As a result, muscles may undergo tetanic contractions, and the person may die due to a failure of respiratory movements. With an abnormally high concentration of blood calcium (hypercalcemia), the nervous system becomes depressed. Consequently, muscle contractions are weak and reflexes are sluggish.

1. Where are the parathyroid glands located?
2. How does parathyroid hormone help to regulate the concentrations of blood calcium and phosphate?
3. How does the negative feedback system of the parathyroid glands differ from that of the thyroid gland?

Figure 13.25

Parathyroid hormone stimulates the release of calcium from bone and the conservation of calcium by the kidneys. It indirectly stimulates the absorption of calcium by the intestine. The resulting increase in blood calcium concentration inhibits the secretion of this hormone.

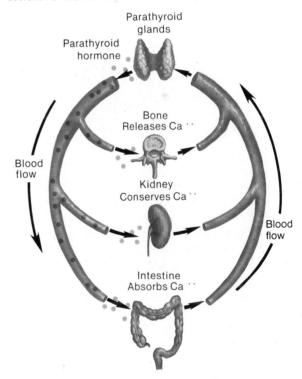

Figure 13.26
(a) An adrenal gland consists of an outer cortex and an inner
medulla; (b) the cortex consists of three layers or zones of cells.

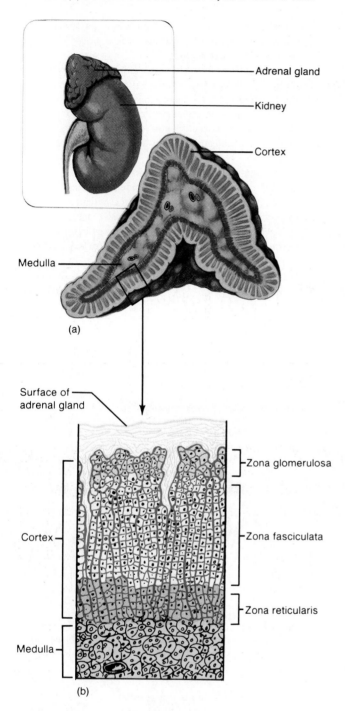

Adrenal gland

Kidney

Cortex

Medulla

(a)

Surface of
adrenal gland

Cortex

Zona glomerulosa

Zona fasciculata

Zona reticularis

Medulla

(b)

The Adrenal Glands

The **adrenal glands** (suprarenal glands) are located in close association with the kidneys. A gland sits atop each kidney like a cap and is embedded in the mass of fat that encloses the kidney.

Structure of the Glands

Although the adrenal glands usually differ somewhat in size and shape, they are generally pyramidal. Each gland is very vascular and consists of two parts, as shown in figure 13.26. The central portion is termed the adrenal medulla, and the outer part is the adrenal cortex. Although these regions are not sharply divided, they represent distinct glands that secrete different hormones.

The **adrenal medulla** consists of irregularly shaped cells that are arranged in groups around blood vessels. These cells are intimately connected with the sympathetic division of the autonomic nervous system. They are modified postganglionic cells, and preganglionic autonomic nerve fibers lead to them from the central nervous system without synapsing (see chapter 11).

The **adrenal cortex,** which makes up the bulk of the gland, is composed of closely packed masses of epithelial cells that are arranged in layers. These layers form an outer, a middle, and an inner zone of the cortex. As in the case of the medulla, the cells of the adrenal cortex are well supplied with blood vessels.

Hormones of the Adrenal Medulla

Cells of the adrenal medulla (chromaffin cells) secrete two closely related hormones, **epinephrine** (adrenalin) and **norepinephrine** (noradrenalin). Both of these substances are amines (catecholamines), and they have similar molecular structures and physiological functions (figure 13.27). In fact, epinephrine is produced from norepinephrine.

The synthesis of these substances begins with the amino acid *tyrosine*. In the first step of the process, tyrosine is converted into a substance called *dopa* by an enzyme in the secretory cells. Dopa is changed to *dopamine* by a second enzyme, and the dopamine is changed to norepinephrine by a third enzyme. In about 10% of the medullary cells, the process ends with norepinephrine; in the rest, still another enzyme converts the norepinephrine to epinephrine.

The effects of the medullary hormones generally resemble those that result when sympathetic nerve fibers stimulate their effectors. The hormonal effects, however, last up to ten times longer because the hormones are removed from the tissues relatively slowly.

Figure 13.27
Epinephrine and norepinephrine have similar molecular structures
and similar functions.

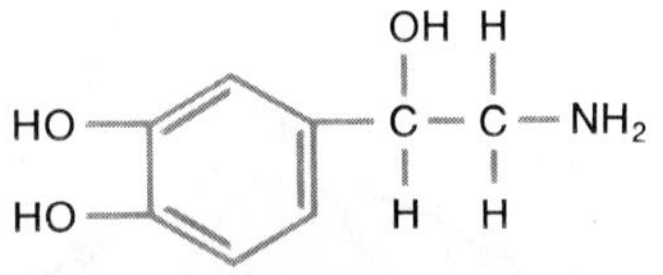

Norepinephrine

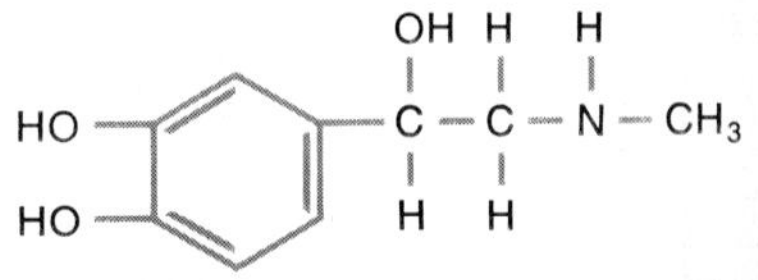

Epinephrine

Chart 13.7 Comparative effects of epinephrine and norepinephrine		
Part or function affected	Epinephrine	Norepinephrine
Heart	Rate increases	Rate increases
	Force of contraction increases	Little or no effect on force of contraction
Blood vessels	Vessels in skeletal muscle vasodilate, decreasing resistance to blood flow	Vessels in skeletal muscles vasoconstrict, increasing resistance to blood flow
Systemic blood pressure	Some increase due to increased cardiac output	Great increase due to vasoconstriction
Airways	Dilation	Less effect
Reticular formation of brain	Activated	Little effect
Liver	Promotes change of glycogen to glucose increasing blood sugar	Little effect on blood sugar
Metabolic rate	Increases	Little or no effect

These effects include increased heart rate and increased force of cardiac muscle contraction, elevated blood pressure, increased breathing rate, and decreased activity in the digestive system (see chart 11.6).

The ratio of the two hormones in the medullary secretion varies with different physiological conditions, but usually is about 80% epinephrine and 20% norepinephrine. Although their effects generally are similar, certain effectors respond differently to them. These differences are due to the relative numbers of alpha and beta receptors in the membranes of the effector cells. As is described in chapter 11, epinephrine combines with either alpha or beta receptors, while norepinephrine combines mainly with alpha receptors. Chart 13.7 compares some of the differences in the effects of these hormones.

Impulses arriving by way of sympathetic nerve fibers stimulate the adrenal medulla to release its hormones at the same time other effectors are being stimulated by sympathetic impulses. As a rule, these impulses originate in the hypothalamus in response to various types of stress. Thus the medullary secretions function together with the sympathetic division of the autonomic nervous system in preparing the body for energy-expending action—fight or flight.

Although a lack of medullary hormones produces no significant effects, tumors in the adrenal medulla (pheochromocytoma) are sometimes accompanied by excessive hormone secretions. Usually the secretion of norepinephrine predominates, and the affected persons show signs of prolonged sympathetic responses—high blood pressure, increased heart rate, elevated blood sugar, and so forth. Treatment for this condition generally involves surgical removal of the tumorous growth causing the problem.

The Pancreas

The pancreas secretes digestive juices as well as hormones.

1. Structure of the gland
   a. The pancreas is located in back of the stomach and is attached to the small intestine.
   b. The endocrine portion, which is called the islets of Langerhans, secretes glucagon, insulin, and somatostatin.
2. Hormones of the islets of Langerhans
   a. Glucagon stimulates the liver to produce glucose, causing an increase in the concentration of blood glucose, and promotes the breakdown of fats.
   b. Insulin promotes the movement of glucose through cell membranes, stimulates the storage of glucose, promotes the synthesis of proteins, and stimulates the storage of fats.

Other Endocrine Glands

1. The pineal gland
   a. The pineal gland is attached to the thalamus near the roof of the third ventricle.
   b. It is innervated by postganglionic sympathetic nerve fibers.
   c. It secretes melatonin, which seems to inhibit the secretion of gonadotropins.
   d. It may help to regulate the female reproductive cycle.
2. The thymus gland
   a. The thymus lies behind the sternum and between the lungs.
   b. Its size diminishes with age.
   c. It secretes thymosin that affects the production of lymphocytes.
3. The reproductive glands
   a. The ovaries secrete estrogens and progesterone.
   b. The placenta secretes estrogens, progesterone, and a gonadotropin.
   c. The testes secrete testosterone.
4. The digestive glands include certain glands of the stomach and small intestine that secrete hormones.

Stress and Its Effects

Stress is the condition produced when the body responds to stressors that threaten the maintenance of homeostasis.

Stress responses involve increased activity of the sympathetic nervous system and increased secretion of adrenal hormones.

1. Types of stress
   a. Physical stress results from environmental factors that are harmful or potentially harmful to tissues.
   b. Psychological stress results from thoughts about real or imagined dangers.
   c. Factors that produce psychological stress vary from person to person and situation to situation.
2. Responses to stress
   a. Responses to stress are directed toward maintaining homeostasis.
   b. Usually these responses involve a general stress syndrome that is controlled by the hypothalamus.

Application of Knowledge

1. Based on your understanding of the actions of glucagon and insulin, would a person with diabetes mellitus be likely to require more insulin or more sugar following strenuous exercise? Why?
2. What problems might result from the prolonged administration of cortisol to a person with a severe inflammatory reaction?
3. How might the environment of a patient with hyperthyroidism be modified to minimize the drain on body energy resources?
4. What hormones would need to be administered to an adult whose anterior pituitary gland had been removed? Why?

Review Activities

1. Explain what is meant by an endocrine gland.
2. Define *hormone* and *target cell*.
3. Explain how hormones can be grouped on the basis of their chemical composition.
4. Explain how steroid hormones exert their influence.
5. Distinguish between a binding site and an activity site of a receptor molecule.
6. Explain how nonsteroid hormones may function through the formation of cAMP.
7. Explain how nonsteroid hormones may function through the increase in intracellular calcium concentration.
8. Explain how the cellular response to a hormone operating through a second messenger is amplified.
9. Define *prostaglandins* and explain their general function.
10. Distinguish between a negative and a positive feedback system.
11. Define *releasing hormone* and provide an example of such a substance.
12. Describe the location and structure of the pituitary gland.
13. List the hormones secreted by the anterior lobe of the pituitary gland.
14. Explain how pituitary gland activity is controlled by the brain.
15. Explain how growth hormone produces its effects.
16. List the major factors that affect the secretion of growth hormone.
17. Explain the causes of pituitary dwarfism and gigantism.
18. Summarize the functions of prolactin.
19. Describe the mechanism that regulates the concentrations of circulating thyroid hormones.
20. Explain how the secretion of ACTH is controlled.
21. List the major gonadotropins and explain the general functions of each.
22. Compare the cellular structures of the anterior and posterior lobes of the pituitary gland.
23. Name the hormones associated with the posterior pituitary and explain their functions.
24. Explain how the release of ADH is regulated.
25. Describe the location and structure of the thyroid gland.
26. Name the hormones secreted by the thyroid gland and list the general functions of each.
27. Define *iodine pump*.
28. Describe the location and structure of the parathyroid glands.
29. Explain the general functions of parathyroid hormone.
30. Describe the mechanism that regulates the secretion of parathyroid hormone.
31. Distinguish between the adrenal medulla and the adrenal cortex.
32. List the hormones produced by the adrenal medulla and describe their general functions.
33. List the steps in the synthesis of medullary hormones.
34. Name the most important hormones of the adrenal cortex and describe the general functions of each.
35. Describe how the secretion of aldosterone is regulated.
36. Describe how the secretion of cortisol is regulated.
37. Describe the location and structure of the pancreas.
38. List the hormones secreted by the pancreas and describe the general functions of each.
39. Summarize how the secretion of hormones from the pancreas is regulated.
40. Describe the location and general function of the pineal gland.
41. Describe the location and general function of the thymus gland.
42. Distinguish between a stressor and stress.
43. List several factors that cause physical and psychological stress.
44. Describe the general stress syndrome.

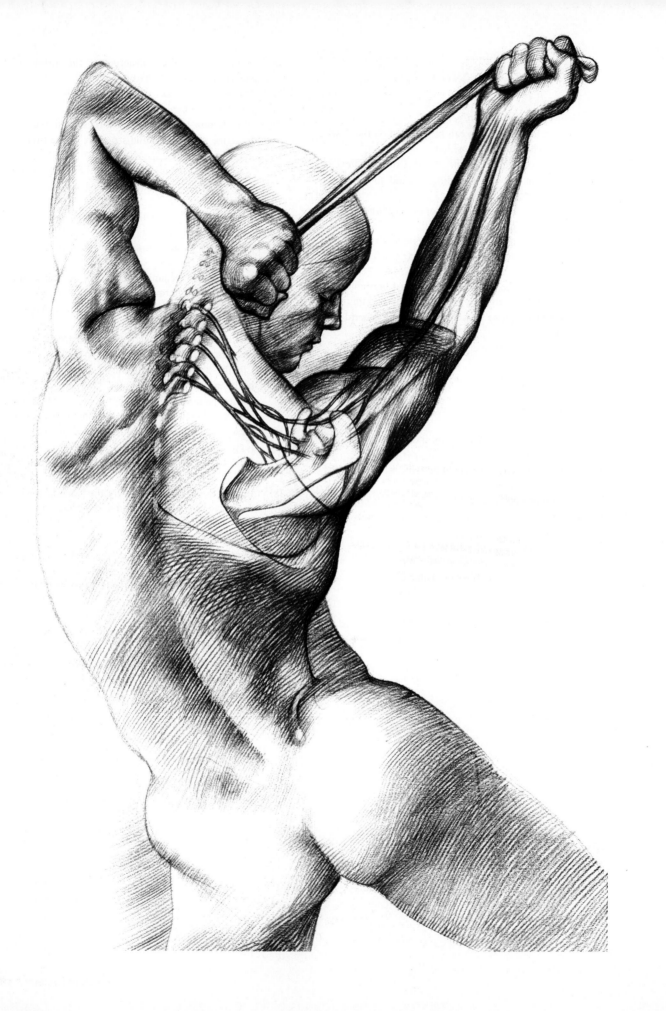

Unit 4

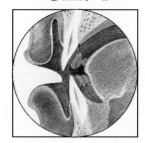

Processing and Transporting

The chapters of unit 4 are concerned with the digestive, respiratory, circulatory, lymphatic, and urinary systems. They describe how organs of these systems obtain nutrients and oxygen from outside the body, how the nutrients are altered chemically and absorbed into body fluids, and how the nutrients and oxygen are transported to body cells. They also explain processes by which these substances are utilized by cells, how resulting wastes are transported and excreted, and how stable concentrations of various substances in body fluids are maintained.

14

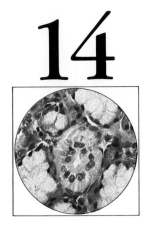

The Digestive System

Most food substances are composed of chemicals whose molecules cannot pass easily through cell membranes and so cannot be absorbed effectively by cells. The *digestive system* functions to solve this problem. Its parts are adapted to ingest foods, to break large particles into smaller ones, to secrete enzymes that decompose food molecules, to absorb the products of this digestive action, and to eliminate the unused residues.

Foods are moved through the digestive tract by muscular contractions in the wall of the tubular alimentary canal, and digestive juices, which contain digestive enzymes, are secreted into the canal by various glands and accessory organs. These functions are controlled largely by interactions between the digestive, nervous, and endocrine systems.

Chapter Outline

Chapter Objectives

After you have studied this chapter, you should be able to

1. Name and describe the location of the organs of the digestive system and their major parts.

2. Describe the general functions of each digestive organ and the liver.

3. Describe the structure of the wall of the alimentary canal.

4. Explain how the contents of the alimentary canal are mixed and moved.

5. List the enzymes secreted by the various digestive organs and describe the function of each enzyme.

6. Describe how digestive secretions are regulated.

7. Explain how digestive reflexes function to control the movement of material through the alimentary canal.

8. Describe the mechanisms of swallowing, vomiting, and defecating.

9. Explain how the products of digestion are absorbed.

10. Complete the review activities at the end of this chapter. Note that the items are worded in the form of specific learning objectives . You may want to refer to them before reading the chapter.

Key Terms

absorption (ab-sorp'shun)

accessory organ (ak-ses'o-re or'gan)

alimentary canal (al''ĭ-men'tar-e kah-nal')

bile (bīl)

chyme (kīm)

circular muscle (ser'ku-lar mus'el)

deciduous (de-sid'u-us)

gastric juice (gas'trik jōos)

intestinal juice (in-tes'tĭ-nal jōos)

intrinsic (in-trin'sik)

longitudinal muscle (lon''jĭ-tu'dĭ-nal mus'el)

mucous membrane (mu'kus mem'brān)

pancreatic juice (pan''kre-at'ik jōos)

peristalsis (per''ĭ-stal'sis)

serous layer (se'rus la'er)

sphincter muscle (sfingk'ter mus'el)

villi; singular, villus (vil'ĭ, vil'us)

Aids to Understanding Words

aliment-, food: *aliment*ary canal—the tubelike portion of the digestive system.

cari-, decay: dental *caries*—tooth decay.

cec-, blindness: *cec*um—blind-ended sac at the beginning of the large intestine.

chym-, juice: *chym*e—semifluid paste of food particles and gastric juice formed in the stomach.

decidu-, falling off: *decidu*ous teeth—those that are shed during childhood.

entero-, intestine: *entero*gastrone—a hormone secreted by cells in the wall of the small intestine.

frenul-, a restraint: *frenul*um—membranous fold that anchors the tongue to the floor of the mouth.

gastr-, stomach: *gastr*ic gland—portion of the stomach that secretes gastric juice.

hepat-, liver: *hepat*ic duct—duct that carries bile from the liver to the common bile duct.

hiat-, an opening: esophageal *hiat*us—opening through which the esophagus penetrates the diaphragm.

lingu-, the tongue: *lingu*al tonsil—mass of lymphatic tissue at the root of the tongue.

peri-, around: *peri*stalsis—wavelike ring of contraction that moves material along the alimentary canal.

pylor-, gatekeeper: *pylor*ic sphincter—muscle that serves as a valve between the stomach and small intestine.

vill-, hairy: *vill*i—tiny projections of mucous membrane in the small intestine.

DIGESTION IS THE process by which food substances are changed into forms that can be absorbed through cell membranes. The **digestive system** includes the organs that promote this process. It consists of an *alimentary canal* that extends from the mouth to the anus, and several *accessory organs* that release secretions into the canal. The alimentary canal includes the mouth, pharynx, esophagus, stomach, small intestine, and large intestine; while the accessory organs include the salivary glands, liver, gallbladder, and pancreas. Major organs of this system are shown in figure 14.1.

General Characteristics of the Alimentary Canal

The **alimentary canal** is a muscular tube about 9 meters long that passes through the body's ventral cavity. Although it is specialized in various regions to carry on particular functions, the structure of its wall, the method by which it moves food, and its type of innervation are similar throughout its length.

Structure of the Wall

The wall of the alimentary canal consists of four distinct layers, although the degree to which they are developed varies from region to region. Beginning with the innermost tissues, these layers, shown in figure 14.2, include the following:

1. **Mucous membrane** or **mucosa.** This layer is formed of surface epithelium, underlying connective tissue (lamina propria), and a small amount of smooth muscle. In some regions, it develops folds and tiny projections that extend into the lumen of the digestive tube and increase its absorptive surface area. It may also contain glands that are tubular invaginations into which the lining cells secrete mucus and digestive enzymes. Therefore, the mucosa functions to protect the tissues beneath it and carry on absorption and secretion.
2. **Submucosa.** The submucosa contains considerable loose connective tissue as well as blood vessels, lymphatic vessels, and nerves. Its vessels serve to nourish surrounding tissues and to carry away absorbed materials.
3. **Muscular layer.** This layer, which is responsible for the movements of the tube, consists of two coats of smooth muscle tissue. The fibers of the inner coat are arranged so that they encircle the tube, and when these *circular fibers* contract, the diameter of the tube is decreased. The fibers of the outer muscular coat run lengthwise, and when these *longitudinal fibers* contract, the tube is shortened.

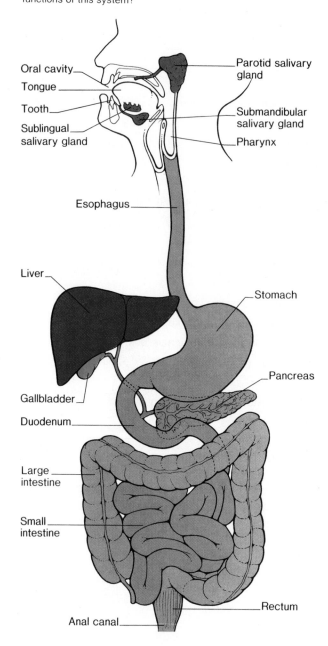

Figure 14.1

Major organs of the digestive system. What are the general functions of this system?

Oral cavity

Tongue

Tooth

Sublingual salivary gland

Parotid salivary gland

Submandibular salivary gland

Pharynx

Esophagus

Liver

Stomach

Gallbladder

Pancreas

Duodenum

Large intestine

Small intestine

Rectum

Anal canal

4. **Serous layer** or **serosa.** The serous or outer covering of the tube is composed of the *visceral peritoneum*, which is formed of epithelium on the outside and connective tissue beneath. The cells of the serosa secrete serous fluid that keeps the tube's outer surface moist. This lubricates the surface so that the organs within the abdominal cavity slide freely against one another.

The characteristics of these layers are summarized in chart 14.1.

Figure 14.2
The wall of the small intestine, as in other portions of the
alimentary canal, includes four layers: mucous membrane,
submucosa, muscular layer, and serous layer.

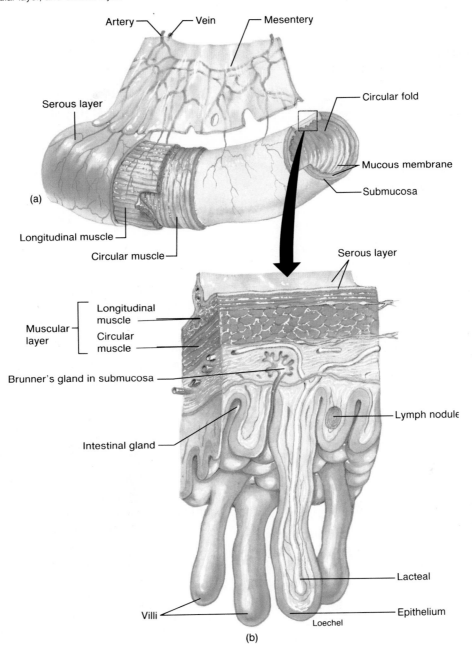

Chart 14.1	Layers in the wall of the alimentary canal	
Layer	**Composition**	**Function**
Mucous membrane	Epithelium, connective tissue, smooth muscle	Protection, absorption, secretion
Submucosa	Loose connective tissue, blood vessels, lymphatic vessels, nerves	Nourishes surrounding tissues, transports absorbed materials
Muscular layer	Smooth muscle fibers arranged in circular and longitudinal groups	Movements of the tube and its contents
Serous layer	Epithelium, connective tissue	Protection

Figure 14.3
(a) Mixing movements occur when small segments of the muscular wall of the alimentary canal undergo rhythmic contractions. *(b)* Peristaltic waves cause the contents to be moved along the canal.

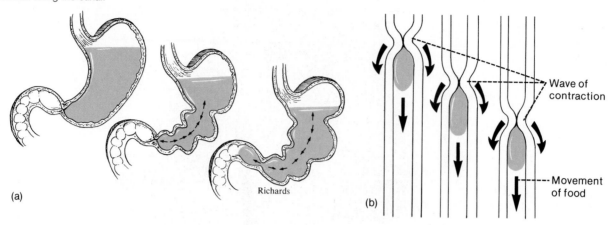

(a)   Richards

(b)   — Wave of contraction

— Movement of food

Movements of the Tube

The motor functions of the alimentary canal are of two basic types—mixing movements and propelling movements (figure 14.3). Mixing occurs when smooth muscles in relatively small segments of the tube undergo rhythmic contractions. When the stomach is full, for example, waves of muscular contractions move along its wall from one end to the other. These waves occur every 20 seconds or so, and their action tends to mix food substances with digestive juices secreted by the mucosa.

Propelling movements include a wavelike motion called **peristalsis.** When peristalsis occurs, a ring of contraction appears in the wall of the tube. At the same time, the muscular wall just ahead of the ring relaxes—a phenomenon called *receptive relaxation.* As the wave moves along, it pushes the tubular contents ahead of it. The usual stimulus for peristalsis is an expansion of the tube due to accumulation of food inside. Such movements create sounds that can be heard through a stethoscope applied to the abdominal wall.

Innervation of the Tube

The alimentary canal is innervated extensively by branches of the sympathetic and parasympathetic divisions of the autonomic nervous system. These nerve fibers are associated mainly with the tube's muscular layer, and they are responsible for maintaining muscle tone and for regulating the strength, rate, and velocity of muscular contractions.

Parasympathetic impulses generally cause an increase in the activities of the digestive system. Some of these impulses originate in the brain and are conducted on branches of the vagus nerves to the esophagus, stomach, pancreas, gallbladder, small intestine, and proximal half of the large intestine. Other parasympathetic impulses arise in the sacral region of the spinal cord and supply the distal half of the large intestine.

The effects produced by sympathetic nerve impulses usually are opposite those of the parasympathetic division. In other words, they inhibit various digestive actions. Sympathetic impulses are responsible for the contraction of certain sphincter muscles in the wall of the alimentary canal, and when they are contracted, these muscles effectively block the movement of materials through the tube.

1. What organs constitute the digestive system?
2. Describe the wall of the alimentary canal.
3. Name the two types of movements that occur in the alimentary canal.
4. What effect do parasympathetic nerve impulses have on digestive actions? What effect do sympathetic nerve impulses have?

The Mouth

The mouth, which is the first portion of the alimentary canal, is adapted to receive food and prepare it for digestion by mechanically reducing the size of solid particles and mixing them with saliva (mastication). It also functions as an organ of speech and an organ of pleasure. The mouth is surrounded by the lips, cheeks,

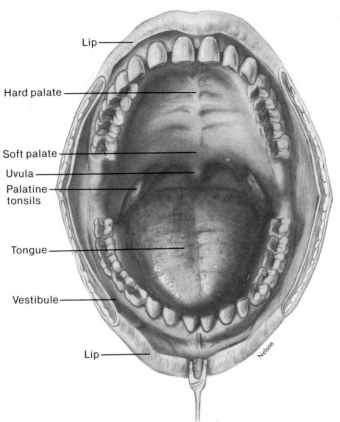

Lip

Hard palate

Soft palate

Uvula

Palatine tonsils

Tongue

Vestibule

Lip

tongue, and palate, and includes a chamber between the palate and tongue called the *oral cavity,* as well as a narrow space between the teeth, cheeks, and lips called the *vestibule* (figure 14.4).

The Cheeks and Lips

The cheeks form the lateral walls of the mouth and consist of outer layers of skin, pads of subcutaneous fat, certain muscles associated with expression and chewing, and inner linings of stratified squamous epithelium.

The lips are highly mobile structures that surround the mouth opening. They contain skeletal muscles and a variety of sensory receptors that are useful in judging the temperature and texture of foods. Their normal reddish color is due to an abundance of blood vessels near their surfaces. The external borders of the lips mark the boundaries between the skin of the face and the mucous membrane that lines the alimentary canal.

The Tongue

The **tongue** is a thick, muscular organ that occupies the floor of the mouth and nearly fills the oral cavity when the mouth is closed. It is covered by mucous membrane and is anchored in the midline to the floor of the mouth by a membranous fold called the **frenulum**.

A person whose frenulum is too short is said to be tongue-tied. Infants with this condition may have difficulty sucking, and older children may be unable to make the tongue movements needed for normal speech. A short frenulum is sometimes corrected in early infancy by cutting it.

The body of the tongue is composed largely of skeletal muscle whose fibers run in several directions. These muscles aid in mixing food particles with saliva during chewing and in moving food toward the pharynx during swallowing. Rough projections, called **papillae,**

Figure 14.5

The surface of the tongue as viewed from above.

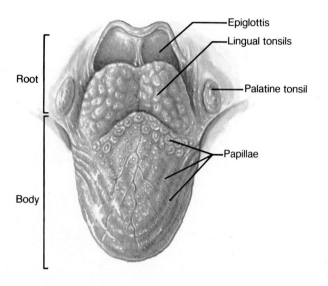

Figure 14.6

A sagittal section of the mouth, nasal cavity, and pharynx.

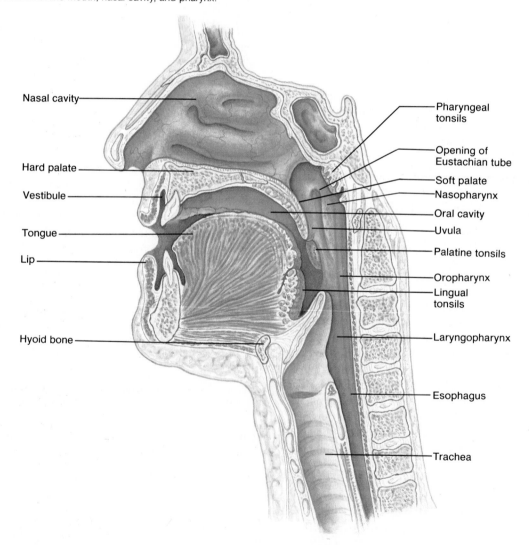

on the surface of the tongue provide friction that is useful in handling food. These papillae also contain taste buds (figure 14.5).

The posterior region, or *root*, of the tongue is anchored to the hyoid bone and is covered with rounded masses of lymphatic tissue called **lingual tonsils.**

The Palate

The **palate** forms the roof of the oral cavity and consists of a hard anterior part and a soft posterior part. The *hard palate* is formed by the palatine processes of the maxillary bones in front and the horizontal portions of the palatine bones in back. The *soft palate* forms a muscular arch that extends posteriorly and downward as a cone-shaped projection called the **uvula.**

During swallowing, muscles draw the soft palate and the uvula upward. This action closes the opening between the nasal cavity and the pharynx, preventing food from entering the nasal cavity.

In the back of the mouth, on either side of the tongue and closely associated with the palate, are masses of lymphatic tissue called **palatine tonsils.** These structures lie beneath the epithelial lining of the mouth and, like other lymphatic tissues, they help protect the body against infections. (See chapter 19 for more detail concerning lymphatic functions.)

The palatine tonsils themselves are common sites of infections, and if they become inflamed, the condition is termed *tonsillitis.* Infected tonsils may become so swollen that they block the passageways of the pharynx and interfere with breathing and swallowing. Since the mucous membranes of the pharynx, eustachian tubes, and middle ears are continuous, there is always danger that such an infection may travel from the throat into the middle ears (otitis media).

When tonsillitis occurs repeatedly and the condition remains unresponsive to antibiotic treatment, the tonsils are often removed. This surgical procedure is called *tonsillectomy.*

Still other masses of lymphatic tissue, called **pharyngeal tonsils** or *adenoids,* occur on the posterior wall of the pharynx, above the border of the soft palate. If these parts become enlarged and block the passage between the pharynx and the nasal cavity, they also may be removed surgically (figure 14.6).

---

1. What is the function of the mouth?
2. How does the tongue contribute to the function of the digestive system?
3. What is the role of the soft palate in swallowing?
4. Where are the tonsils located?

Figure 14.7
(a) The permanent teeth of the upper jaw and (b) the lower jaw.

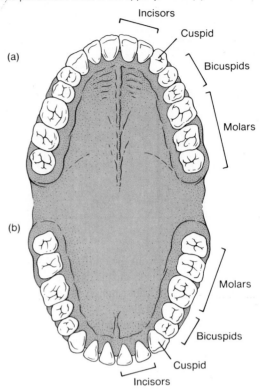

The Teeth

The teeth develop in sockets within the alveolar processes of the mandibular and maxillary bones. Teeth are unique structures in that two sets form during development. The members of the first set, the *primary* or *deciduous teeth,* usually erupt through the gums (gingiva) at regular intervals between the ages of 6 months and 2½ years. There are twenty deciduous teeth—ten in each jaw—and they occur from the midline toward the sides in the following sequence: central incisor, lateral incisor, cuspid (canine), first molar, and second molar.

The deciduous teeth usually are shed in the same order they appeared. Before this happens, though, their roots are resorbed. Then the teeth are pushed out of their sockets by pressure from the developing *secondary* or *permanent* teeth (figure 14.7). This second set consists of thirty-two teeth—sixteen in each jaw—

Chart 14.2 Primary and secondary teeth			
Primary teeth (deciduous)		**Secondary teeth (permanent)**	
Type	*Number*	*Type*	*Number*
Incisor		Incisor	
central	4	central	4
lateral	4	lateral	4
Cuspid	4	Cuspid	4
		Bicuspid	
		first	4
		second	4
Molar		Molar	
first	4	first	4
second	4	second	4
		third	4
Total	20	Total	32

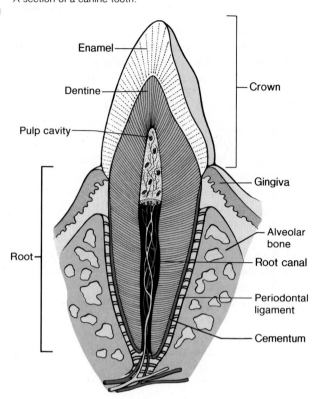

Figure 14.8
A section of a canine tooth.

Enamel

Dentine

Pulp cavity

Root

Crown

Gingiva

Alveolar bone

Root canal

Periodontal ligament

Cementum

and they are arranged from the midline as follows: central incisor, lateral incisor, cuspid, first bicuspid (premolar), second bicuspid, first molar, second molar, and third molar (chart 14.2).

The permanent teeth usually begin to appear at age 6 years, but the set may not be completed until the third molars appear between 17 and 25 years of age. Sometimes these molars, which are also called wisdom teeth, become wedged in abnormal positions within the jaws and fail to erupt. Such teeth are said to be *impacted.*

Teeth function mechanically to break pieces of food into smaller pieces. This action increases the surface area of the food particles and thus makes it possible for digestive enzymes to react more effectively with food molecules.

Different teeth are adapted to handle food in different ways. *Incisors* (front teeth) are chisel-shaped, and their sharp edges bite off relatively large pieces of food. The *cuspids* (canine teeth) are cone-shaped, and they are useful in grasping or tearing food. The *bicuspids* and *molars* have somewhat flattened surfaces and are specialized for grinding particles.

Each tooth consists of two main portions—the *crown,* which projects beyond the gum, and the *root,* which is anchored to the alveolar bone of the jaw. The region where these portions meet is called the *neck* of the tooth. The crown is covered by glossy, white *enamel.* Enamel consists mainly of calcium salts and is the hardest substance in the body. Unfortunately, if enamel is damaged by abrasive action or injury, it is not replaced. It also tends to wear away with age.

The bulk of a tooth beneath the enamel is composed of *dentine,* a substance much like bone, but somewhat harder. The dentine, in turn, surrounds the

tooth's central cavity (pulp cavity), which contains blood vessels, nerves, and connective tissue (pulp). The blood vessels and nerves reach this cavity through tubular *root canals* that extend upward into the root.

The root is enclosed by a thin layer of bone-like material called *cementum,* which is surrounded by a *periodontal ligament* (membrane). This ligament contains bundles of thick collagenous fibers that pass between the cementum and the alveolar bone, firmly attaching the tooth to the jaw. It also contains blood vessels and nerves near the surface of the cementum-covered root (figure 14.8).

The mouth parts and their functions are summarized in chart 14.3.

---

1. How do deciduous teeth differ from permanent teeth?
2. How are various types of teeth adapted to provide specialized functions?
3. Describe the structure of a tooth.
4. Explain how a tooth is attached to the bone of the jaw.

Dental Caries
A Practical Application

Dental caries (decay) involves decalcification of tooth enamel and is usually followed by destruction of the enamel and its underlying dentine. The result is a cavity, which must be cleaned and filled to prevent further erosion of the tooth.

Although the cause or causes of dental caries are not well understood, lack of dental cleanliness and a diet high in sugar and starch seem to promote the problem. Accumulations of food particles on the surfaces and between the teeth are thought to aid the growth of certain kinds of bacteria. These microorganisms utilize carbohydrates in food particles and produce acid by-products. The acids then begin the process of destroying tooth enamel.

Preventing dental caries requires brushing the teeth at least once a day, using dental floss or tape regularly to remove debris from between the teeth, and limiting the intake of sugar and starch, especially between meals. The use of fluoridated drinking water or the application of fluoride solution to children's teeth also helps prevent dental decay.

Loss of teeth is most commonly associated with diseases of the gums and the dental pulp (endodontitis). Such diseases can usually be avoided by practicing good oral hygiene and obtaining regular dental treatment.

Part	Location	Function	Part	Location	Function
Cheeks	Form lateral walls of mouth	Hold food in mouth, muscles function in chewing	Tongue	Occupies floor of mouth	Aids in mixing food with saliva, moves food toward pharynx, contains taste receptors
Lips	Surround mouth opening	Contain sensory receptors used to judge characteristics of foods	Palate	Forms roof of mouth	Holds food in mouth, directs food to pharynx
			Teeth	In sockets of mandibular and maxillary bones	Break food particles into smaller pieces, help mix food with saliva during chewing

Chart 14.3 Mouth parts and their functions

The Salivary Glands

The **salivary glands** function to secrete saliva. This fluid moistens food particles, helps bind them together, and begins the digestion of carbohydrates. Saliva also acts as a solvent, dissolving various food chemicals so they can be tasted, and it helps to cleanse the mouth and teeth. Bicarbonate ions (HCO_3^-) in saliva aid in regulating (buffering) its acid concentration so that its pH usually remains near neutral, between 6.5 and 7.5. This is a favorable range for the action of the salivary enzyme and is important in protecting the teeth against dissolving in an excessively acid environment.

Many small salivary glands are scattered throughout the mucosa of the tongue, palate, and cheeks. They secrete fluid continuously so that the lining of the mouth remains moist. In addition, there are three pairs of major salivary glands: the parotid, submandibular, and sublingual glands.

Salivary Secretions

Within a salivary gland there are two types of secretory cells, called *serous* and *mucous* cells. They occur in varying proportions within different glands. The serous cells produce a watery fluid that contains a digestive enzyme, called **amylase.** This enzyme functions to split starch and glycogen molecules into disaccharides—the first step in the digestion of carbohydrates. Mucous cells secrete the thick, stringy liquid called **mucus** that binds food particles together and acts as a lubricant during swallowing.

Like other digestive structures, the salivary glands are innervated by branches of both sympathetic and parasympathetic nerves. Impulses arriving on sympathetic fibers stimulate the gland cells to secrete a small quantity of viscous saliva. Parasympathetic impulses, on the other hand, elicit the secretion of a large volume of watery saliva. Such parasympathetic impulses are activated reflexly when a person sees, smells, tastes, or

Figure 14.9
Locations of the major salivary glands.

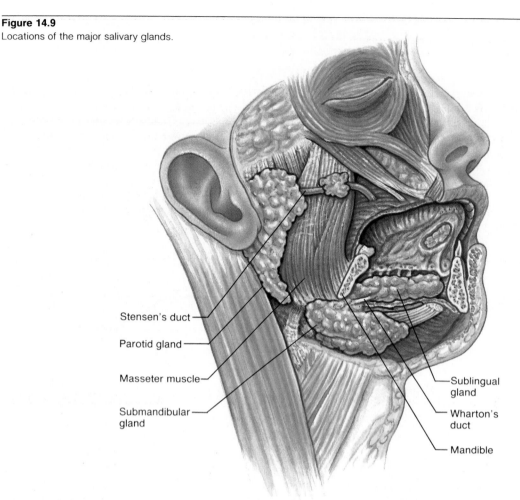

Stensen's duct

Parotid gland

Masseter muscle

Submandibular gland

Sublingual gland

Wharton's duct

Mandible

even thinks about pleasant foods. Conversely, if food looks, smells, or tastes unpleasant, parasympathetic activity is inhibited so that less saliva is produced, and swallowing may become difficult.

Major Salivary Glands

The **parotid glands** are the largest of the major salivary glands. One lies in front of and somewhat below each ear, between the skin of the cheek and the masseter muscle. A *parotid duct* (Stensen's duct) passes from the gland inward through the buccinator muscle, entering the mouth just opposite the upper second molar on either side of the jaw. These glands secrete a clear, watery fluid that is rich in amylase (figure 14.9).

A person who has *mumps,* which is caused by a virus infection, typically develops swellings in one or both of the parotid glands. Occasionally this disease also affects the submandibular glands as well as the pancreas and the gonads.

The **submandibular** (submaxillary) **glands** are located in the floor of the mouth on the inside surface of the jaw (mandible). The secretory cells of these glands are predominantly serous, although some mucous cells are present. Consequently, the submandibular glands secrete a more viscous fluid than the parotid glands. The ducts of the submandibular glands (Wharton's ducts) open under the tongue, near the frenulum (figure 14.10).

The **sublingual glands** are the smallest of the major salivary glands. They are found on the floor of the mouth under the tongue. Their cells are primarily the mucous type, and as a result their secretions, which enter the mouth through many separate ducts, tend to be thick and stringy. (See figure 14.10.)

Chart 14.4 summarizes the characteristics of these glands.

1. What stimulates the salivary glands to secrete saliva?
2. What is the function of saliva?
3. Where are the major salivary glands located?

Figure 14.10
Light micrographs of (a) the parotid salivary gland, (b) the submandibular salivary gland, and (c) the sublingual salivary gland.

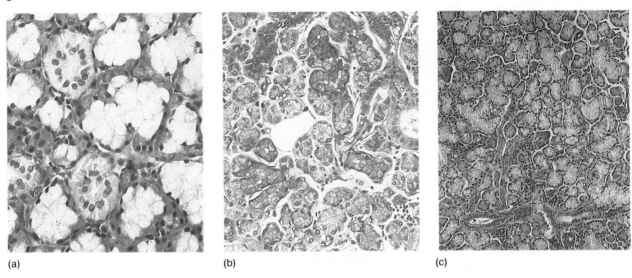

(a)   (b)   (c)

Chart 14.4 The major salivary glands			
Gland	Location	Duct	Type of secretion
Parotid glands	In front and somewhat below the ears, between the skin of the cheeks and the masseter muscles	Parotid ducts pass through the buccinator muscles and enter the mouth opposite the upper second molars	Clear, watery serous fluid rich in amylase
Submandibular glands	In the floor of the mouth on the inside surface of the mandible	Ducts open beneath the tongue near the frenulum	Primarily serous fluid, but with some mucus; more viscous than parotid secretion
Sublingual glands	In the floor of the mouth beneath the tongue	Many separate ducts	Primarily thick, stringy mucus

The Pharynx and Esophagus

The pharynx is a cavity behind the mouth from which the tubular esophagus leads to the stomach. Although neither the pharynx nor the esophagus contributes to the digestive process, they are important passageways, and their muscular walls function in swallowing.

Structure of the Pharynx

The **pharynx** connects the nasal and oral cavities with the larynx and esophagus. (See figure 14.6.) It can be divided into the following parts:

1. The **nasopharynx** is located above the soft palate. It communicates with the nasal cavity and provides a passageway for air during breathing. The eustachian tubes, which connect the pharynx with the middle ears, open through the walls of the nasopharynx.

2. The **oropharynx** is behind the mouth. It opens behind the soft palate into the nasopharynx and projects downward to the upper border of the epiglottis. This portion functions as a passageway for food moving downward from the mouth, and for air moving to and from the nasal cavity.

3. The **laryngopharynx** is located just below the oropharynx. It extends from the upper border of the epiglottis downward to the lower border of the cricoid cartilage of the larynx, where it is continuous with the esophagus.

Figure 14.11
Muscles of the pharyngeal wall as viewed from behind.

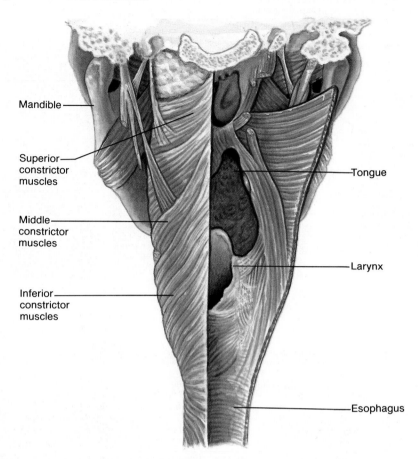

Mandible

Superior constrictor muscles

Middle constrictor muscles

Inferior constrictor muscles

Tongue

Larynx

Esophagus

The muscles in the walls of the pharynx are arranged in inner circular and outer longitudinal groups (figure 14.11). The circular muscles, called *constrictors,* serve to pull the walls inward during swallowing. The *superior* constrictor muscles, which are attached to bony processes of the skull and mandible, curve around the upper part of the pharynx. The *middle* constrictor muscles arise from projections on the hyoid bone and fan around the middle of the pharynx. The *inferior* constrictor muscles originate from cartilage of the larynx and pass around the lower portion of the cavity. Some of the fibers in the lower part of the inferior constrictor muscles remain contracted most of the time, and this action prevents air from entering the esophagus during breathing.

Although the pharyngeal muscles are skeletal muscles, they generally are not under voluntary control. Instead, they function involuntarily in the swallowing reflex.

The Swallowing Mechanism

The act of swallowing (deglutition), which involves a set of complex reflexes, can be divided into three stages. In the first, which is initiated voluntarily, food is chewed and mixed with saliva. Then, it is rolled into a mass (bolus) and forced into the pharynx by the tongue. The second stage begins as the food reaches the pharynx and stimulates sensory receptors located around the pharyngeal opening. This triggers the swallowing reflex, illustrated in figure 14.12, which includes the following actions:

1. The soft palate is raised, preventing food from entering the nasal cavity.
2. The hyoid bone and the larynx are elevated, so that food is less likely to enter the trachea.
3. The tongue is pressed against the soft palate, sealing off the oral cavity from the pharynx.
4. The longitudinal muscles in the pharyngeal wall contract, pulling the pharynx upward toward the food.
5. The lower portion of the inferior constrictor muscles relaxes, opening the esophagus.

Figure 14.12

Steps in the swallowing reflex: *(a)* the tongue forces food into the pharynx; *(b)* the soft palate, hyoid bone, and larynx are raised; the tongue is pressed against the palate; and inferior constrictor muscles relax so that the esophagus opens; *(c)* superior constrictor muscles contract and force food into the esophagus; *(d)* peristaltic waves move food through the esophagus to the stomach.

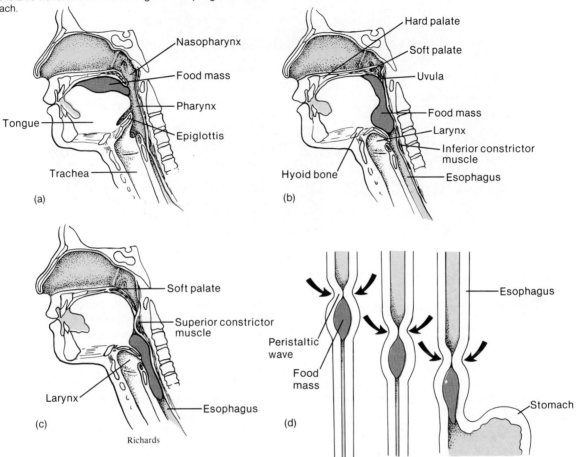

6. The superior constrictor muscles contract, stimulating a peristaltic wave to begin in other pharyngeal muscles, and this wave forces the food into the esophagus.

During the third stage of swallowing, the food enters the esophagus and is transported to the stomach by peristalsis.

The Esophagus

The **esophagus** is a straight, collapsible tube about 25 cm long. It provides a passageway for substances from the pharynx to the stomach. It descends through the thorax behind the trachea, passing through the mediastinum. The esophagus penetrates the diaphragm through an opening, the *esophageal hiatus,* and is continuous with the stomach on the abdominal side of the diaphragm (figure 14.13).

Figure 14.13

The esophagus functions as a passageway between the pharynx and stomach.

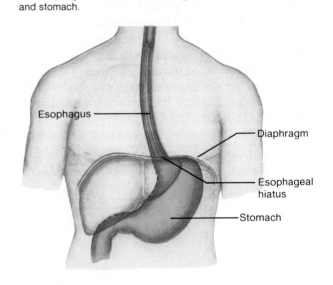

The Digestive System / 503

Figure 14.14
This cross section of the esophagus shows its muscular wall.

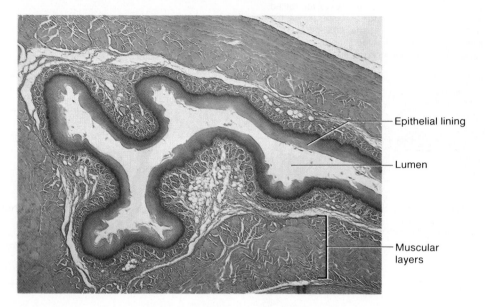

Epithelial lining

Lumen

Muscular layers

There are mucous glands scattered throughout the mucosa of the esophagus, and their secretions keep the inner lining of the tube moist and lubricated.

Occasionally there is a weak place in the diaphragm due to a congenital defect or an injury. As a result of such weakness, a portion of the stomach, large intestine, or some other abdominal organ may protrude upward through the esophageal hiatus and into the thorax. This condition is called *hiatal hernia.*

If gastric juice from the stomach enters the esophagus as a result of such a hernia, the esophageal mucosa may become inflamed. This may lead to the discomfort commonly called "heartburn," difficulty in swallowing, and ulceration accompanied by loss of blood.

Just above the point where the esophagus joins the stomach, some of the circular muscle fibers in its wall are thickened. These fibers are usually contracted and function to close the entrance to the stomach. In this way, they help prevent regurgitation of the stomach contents into the esophagus.

When peristaltic waves reach the stomach, the muscle fibers that guard its entrance relax and allow the food to enter (figure 14.14).

---

1. Describe the regions of the pharynx.
2. List the major events that occur during swallowing.
3. What is the function of the esophagus?

The Stomach

The stomach is a J-shaped, pouchlike organ, about 25–30 cm long, that hangs under the diaphragm in the upper left portion of the abdominal cavity. It has a capacity of about one liter or more, and its inner lining is marked by thick folds (rugae) that tend to disappear when its wall is distended. The stomach functions to receive food from the esophagus, mix it with gastric juice, initiate the digestion of proteins, carry on a limited amount of absorption, and move food into the small intestine.

In addition to the two layers of smooth muscle—an inner circular and an outer longitudinal layer—found in other regions of the alimentary canal, some parts of the stomach have another inner layer of oblique fibers. This third muscular layer is most highly developed near the opening of the esophagus and in the body of the stomach.

Parts of the Stomach

The stomach, shown in figures 14.15 and 14.16, can be divided into cardiac, fundic, body, and pyloric regions. The *cardiac region* is a small area near the esophageal opening (cardia). The *fundic region,* which balloons above the cardiac portion, acts as a temporary storage area and sometimes becomes filled with swallowed air. This produces a gastric air bubble, which may be used as a landmark on an x-ray film of the abdomen. The dilated *body region* is the main part of the stomach and

Figure 14.15
Major regions of the stomach.

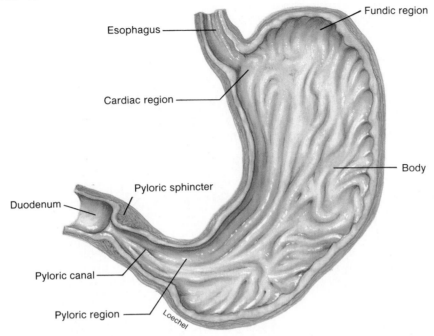

Figure 14.16
X-ray film of a stomach.

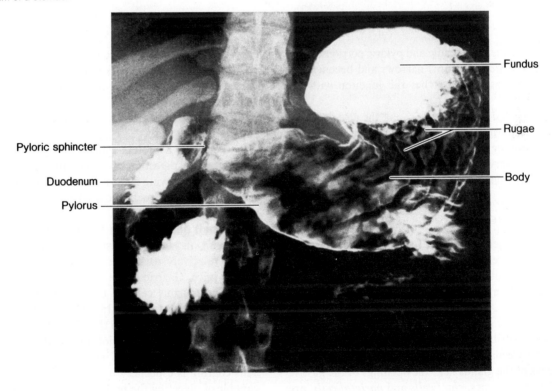

Figure 14.17
(a) The mucosa of the stomach is studded with gastric pits that
are the openings of the gastric glands; *(b)* gastric glands include
mucous cells, parietal cells, and chief cells, each type producing
a different secretion.

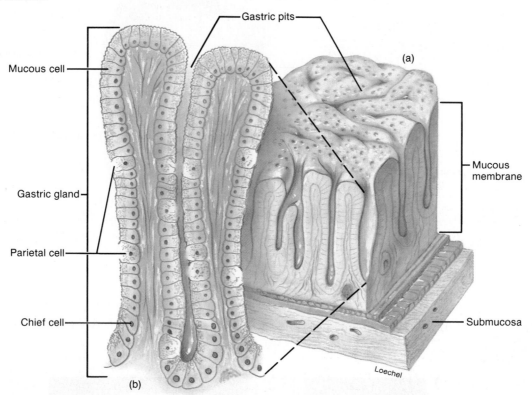

Gastric pits

(a)

Mucous cell

Mucous membrane

Gastric gland

Parietal cell

Chief cell

Submucosa

Loechel

(b)

is located between the fundic and pyloric portions, while the *pyloric region* (antrum) narrows and becomes the *pyloric canal* as it approaches the junction with the small intestine.

At the end of the pyloric canal, the circular layer of fibers in its muscular wall is thickened, forming a powerful muscle called the **pyloric sphincter** (pylorus). This muscle serves as a valve that prevents regurgitation of food from the intestine back into the stomach.

Gastric Secretions

The mucous membrane that forms the inner lining of the stomach is relatively thick, and its surface is studded with many small openings. These openings, called *gastric pits,* are located at the ends of tubular **gastric glands** (figure 14.17). Although their structure and the composition of their secretions vary in different parts of the stomach, gastric glands generally contain three types of secretory cells. One type, the *mucous cell,* occurs in the necks of the glands near the openings of the gastric pits. The other types, *chief cells* and *parietal cells,* are found in the deeper parts of the glands (figures 14.17 and 14.18). The chief cells secrete *digestive enzymes,* and the parietal cells release *hydrochloric acid.* The products of the mucous cells, chief cells, and parietal cells together form **gastric juice.**

Although gastric juice contains several digestive enzymes, **pepsin** is by far the most important. It is secreted by the chief cells as an inactive, nonerosive substance called **pepsinogen.** When pepsinogen contacts the hydrochloric acid of gastric juice, however, it is changed rapidly into pepsin.

Pepsin is a protein-splitting enzyme capable of beginning the digestion of nearly all types of dietary protein. This enzyme is most active in an acid environment, and the hydrochloric acid in gastric juice provides such an environment.

Small quantities of a fat-splitting enzyme, *gastric lipase,* also occur in gastric juice. However, its action is relatively weak due in part to the low pH of gastric juice. Gastric lipase acts mainly on butterfat.

The mucous cells of the gastric glands secrete large quantities of mucus. In addition, the cells of the mucous membrane between these glands release a more viscous and alkaline secretion that is thought to form a protective coating on the inside of the stomach wall. This coating is especially important because pepsin is capable of digesting the proteins of the stomach tissues as well as those in foods. Thus, the coating normally prevents the stomach from digesting itself.

Still another component of gastric juice is **intrinsic factor.** This substance, which is secreted by the parietal cells of the gastric glands, aids in absorption

Figure 14.18
A light micrograph of the gastric mucosa.

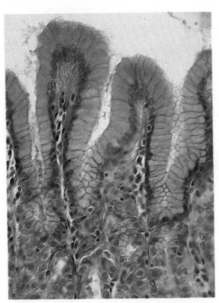

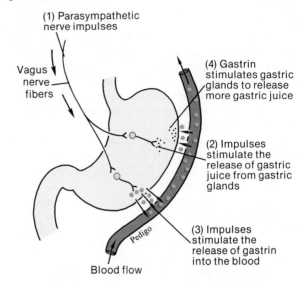

(1) Parasympathetic nerve impulses

Vagus nerve fibers

(4) Gastrin stimulates gastric glands to release more gastric juice

(2) Impulses stimulate the release of gastric juice from gastric glands

(3) Impulses stimulate the release of gastrin into the blood

Pedigo

Blood flow

Chart 14.5	Major components of gastric juice	
Component	Source	Function
Pepsinogen	Chief cells of the gastric glands	An inactive form of pepsin
Pepsin	Formed from pepsinogen in the presence of hydrochloric acid	A protein-splitting enzyme capable of digesting nearly all types of protein
Hydrochloric acid	Parietal cells of the gastric glands	Provides acid environment, needed for the conversion of pepsinogen into pepsin
Mucus	Goblet cells and mucous glands	Provides viscous, alkaline protective layer on the stomach wall
Intrinsic factor	Parietal cells of the gastric glands	Aids the absorption of vitamin B_{12}

of vitamin B_{12} from the small intestine, as is explained in chapter 15.

The substances in gastric juice are summarized in chart 14.5.

1. Where is the stomach located?
2. What is secreted by the chief cells of the gastric glands? By the parietal cells?
3. What is the most important digestive enzyme of gastric juice?
4. How is the stomach prevented from digesting itself?

Regulation of Gastric Secretions

Although gastric juice is produced continuously, the rate of production varies considerably from time to time and is under the control of neural and hormonal mechanisms. More specifically, parasympathetic impulses arriving on vagus nerve fibers stimulate gastric glands to secrete large amounts of gastric juice that is rich in hydrochloric acid and pepsin. These impulses also stimulate certain stomach cells to release a hormone,

called **gastrin,** that causes the gastric glands to increase their secretory activity (figure 14.19). On the other hand, sympathetic impulses bring about a decrease in gastric gland activity.

As a result of the ways these factors interact, three stages of gastric secretion can be recognized. They are called the cephalic, gastric, and intestinal phases.

The *cephalic phase* of gastric secretion begins before any food reaches the stomach and may begin before any food is eaten. In this stage, gastric secretions are stimulated by parasympathetic reflexes operating through the vagus nerves, whenever a person tastes, smells, sees, or even thinks about food. Furthermore, the hungrier the person is, the greater the amount of gastric secretion.

The *gastric phase* starts when food enters the stomach. The presence of food chemicals and distension of the wall triggers the release of gastrin from the stomach, and the gastrin in turn stimulates the production of still more gastric juice. Actually, the release of gastrin is promoted by several factors, including

Peptic Ulcers
A Clinical Application

An *ulcer* is an open sore in the mucous membrane resulting from a localized breakdown of the tissues. Although ulcers may occur in various parts of the alimentary canal, they often occur in the stomach. Such *gastric ulcers* are most likely to develop in the wall of the lesser curvature, which is the concave margin of the stomach near the liver.

Gastric ulcers commonly develop in people who have *gastritis*—an inflammation of the gastric mucosa. Such an inflammation may be caused by irritating foods, bacterial infections, or excessive exposure to alcohol, aspirin, or other drugs. As a result of gastritis, the mucosa is weakened and more likely to be damaged by the digestive action of pepsin in gastric juice (figure 14.20).

Ulcers are also common in the first portion of the small intestine, the duodenum. Such *duodenal ulcers* occur in regions that are exposed to the action of pepsin as the contents of the stomach enter the intestine. They often develop in people who are emotionally stressed and whose stomachs secrete increased amounts of acidic gastric juice between meals, when the stomach is essentially empty. There also is an increased frequency of gastric and duodenal ulcers among smokers.

Fortunately for ulcer victims, the cells of the mucous membrane are able to reproduce rapidly. In fact the entire lining of the stomach is thought to be replaced every few days. Peptic ulcer treatment attempts to eliminate the factors that aggravate the condition and gives the damaged

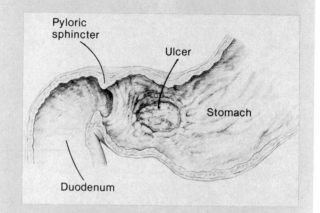

Figure 14.20
Gastric ulcers are usually created by the digestive action of pepsin.

Pyloric sphincter

Ulcer

Stomach

Duodenum

tissue a chance to heal. The treatment commonly includes drugs that reduce the acidity of gastric secretions.

If such measures are unsuccessful, an ulcer patient may be treated surgically. For example, a portion of the stomach may be removed (gastrectomy) so that gastric secretions are greatly reduced, or branches of the vagus nerves may be cut (vagotomy), preventing parasympathetic impulses from reaching the gastric glands.

---

parasympathetic impulses, distension of the stomach wall, and the presence of certain food substances (secretagogues) in the stomach. Such chemicals as meat extracts, partially digested proteins, various spices, caffeine, and alcohol, for example, are especially stimulating to gastrin release. Food in the stomach also elicits additional parasympathetic reflexes that promote the secretion of still more gastric juice.

The *intestinal phase* begins when food leaves the stomach and enters the small intestine. When the food first contacts the intestinal wall, it stimulates intestinal cells to release a hormone that again enhances gastric gland secretions. Although the actual nature of this hormone is unknown, some investigators believe it is identical to gastrin, and they call it *intestinal gastrin*.

As more food moves into the small intestine, however, the secretion of gastric juice from the stomach wall is inhibited. Apparently this inhibition involves a sympathetic reflex triggered by the presence of acids in the upper part of the small intestine. Also, the presence of fats in this region causes the release of the hormone *cholecystokinin* from the intestinal wall, which causes a decrease in gastric motility. Thus, these actions bring about a decrease in gastric secretion and motility as the small intestine fills with food.

Chart 14.6 summarizes the phases of gastric secretion.

---

1. How is the secretion of gastric juice controlled?
2. Distinguish between the cephalic, gastric, and intestinal phases of gastric secretion.
3. What is the function of cholecystokinin?

Chart 14.6	Phases of gastric secretion
Phase	**Action**
Cephalic phase	Parasympathetic reflexes are triggered by sight, taste, smell, or thought of food; gastric juice is secreted in response
Gastric phase	Food in stomach chemically and mechanically stimulates the release of gastrin, which in turn stimulates the secretion of gastric juice; reflex responses also stimulate gastric juice secretion
Intestinal phase	As food enters the small intestine, it stimulates intestinal cells to release intestinal gastrin, which in turn promotes the secretion of gastric juice from the stomach wall

Figure 14.21

(a) As the stomach fills, its muscular wall becomes stretched, but the pyloric sphincter remains closed; (b) mixing movements mix food and gastric juice, creating chyme; (c) peristaltic waves move the chyme toward the pyloric sphincter, which relaxes and allows some chyme to enter the duodenum.

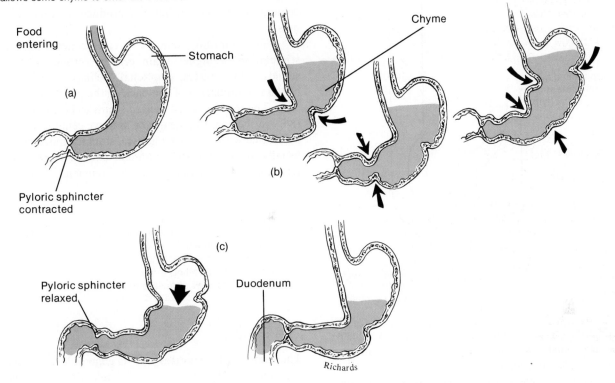

Gastric Absorption

Although gastric enzymes begin the breakdown of proteins, the stomach wall is not well adapted to carry on the absorption of digestive products. Small quantities of water, glucose, certain salts, alcohol, and various lipid-soluble drugs, however, may be absorbed by the stomach.

Mixing and Emptying Actions

As more and more food enters the stomach, the smooth muscles in its wall become stretched. Although the stomach may enlarge, its muscles maintain their tone, and internal pressure normally remains unchanged. A person can eat more than the stomach can comfortably hold, and when this happens the internal pressure may rise enough so that pain receptors are stimulated. The result is a stomachache.

Following a meal, the mixing movements of the stomach wall aid in producing a semifluid paste of food particles and gastric juice called **chyme.** Peristaltic waves push the chyme toward the pyloric region of the stomach, and as it accumulates near the pyloric sphincter, this muscle begins to relax. The muscular pyloric region then pumps the chyme a little at a time (5–15 ml) into the small intestine. This process is illustrated in figure 14.21.

The rate at which the stomach empties depends on several factors, including the fluidity of the chyme and the type of food present. For example, liquids usually pass through the stomach quite rapidly, while solids remain until they are well mixed with gastric juice. Fatty foods may remain in the stomach 3 to 6 hours; foods high in proteins tend to be moved through more quickly; and carbohydrates usually pass through more rapidly than either fats or proteins.

The rate at which the stomach empties is also related to age. An infant's stomach, which has a small capacity, empties very rapidly. As the person grows, the stomach enlarges, and the emptying time increases.

When the duodenum becomes filled with chyme, its internal pressure increases, and the intestinal wall is stretched. This action stimulates sensory receptors in the wall, and an **enterogastric reflex** is triggered. The name of this reflex, like those of other digestive reflexes, describes the origin and termination of reflex impulses. Thus, the enterogastric reflex begins in the small intestine (*entero*) and ends in the stomach (*gastro*).

As a result of the enterogastric reflex, fewer parasympathetic impulses travel to the stomach, and

Figure 14.22
The rate at which chyme leaves the stomach is regulated in part by the enterogastric reflex.

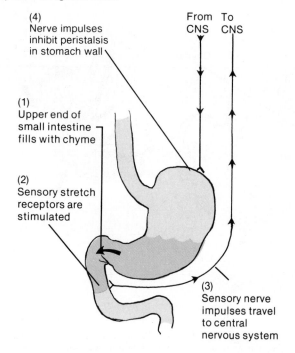

(4)
Nerve impulses inhibit peristalsis in stomach wall

From To
CNS CNS

(1)
Upper end of small intestine fills with chyme

(2)
Sensory stretch receptors are stimulated

(3)
Sensory nerve impulses travel to central nervous system

peristaltic waves are inhibited. Consequently, the intestine is filled less rapidly. Also, if the chyme entering the intestine has a high fat content, the hormone, **cholecystokinin,** is released from the intestinal wall, and causes peristalsis to be inhibited even more (figure 14.22).

As the stomach contents enter the duodenum, accessory organs add their secretions to the chyme. These organs include the pancreas, liver, and gallbladder.

Vomiting results from a complex reflex that empties the stomach another way. This action is usually triggered by irritation or distension in some part of the alimentary canal, such as the stomach or intestines. Sensory impulses travel from the site of stimulation to the *vomiting center* in the medulla oblongata, and a number of motor responses follow. These include taking a deep breath, raising the soft palate and thus closing the nasal cavity, closing the opening to the trachea (glottis), relaxing the sphincterlike muscles at the base of the esophagus, contracting the diaphragm so it moves downward over the stomach, and contracting the abdominal wall muscles so the pressure inside the abdominal cavity is increased. As a result, the stomach is squeezed from all sides, and its contents are forced upward and out through the esophagus, pharynx, and mouth.

Activity in the vomiting center can be stimulated by various drugs (emetics), by toxins in contaminated foods, and sometimes by rapid changes in body motion. In this last situation, sensory impulses from the labyrinths of the inner ears apparently reach the vomiting center, and in some people this produces motion sickness.

The vomiting center also can be activated by stimulation of higher brain centers through sights, sounds, odors, tastes, emotional feelings, or by mechanical stimulation of the back of the pharynx.

The feeling called *nausea* seems to be due to activity in the vomiting center or in nerve centers closely associated with it. During nausea, stomach movements usually are diminished or absent, and duodenal contents may be moved back into the stomach.

1. How is chyme produced?
2. What factors influence the speed with which chyme leaves the stomach?
3. Describe the enterogastric reflex.
4. Describe the vomiting reflex.
5. What factors may stimulate the vomiting reflex?

The Pancreas

The shape of the **pancreas** and its general location are described in chapter 13, as are its endocrine functions. The pancreas also has an exocrine function—the secretion of digestive juice.

Structure of the Pancreas

The pancreas is closely associated with the small intestine and is located behind the parietal peritoneum. It extends horizontally across the posterior abdominal wall with its head in the C-shaped curve of the duodenum and its tail against the spleen (figure 14.23).

The cells that produce pancreatic juice are called *pancreatic acinar cells,* and they make up the bulk of the pancreas. These cells are clustered around tiny tubes, into which they release their secretions. The smaller tubes unite to form larger ones, which in turn give rise to a *pancreatic duct* extending the length of the pancreas. This duct usually connects with the duodenum at the same place where the bile duct from the liver and gallbladder joins the duodenum (figure 13.31 and 14.23).

Figure 14.23
The pancreas is closely associated with the duodenum.

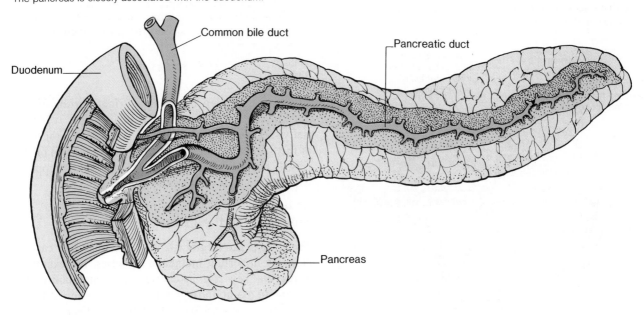

Pancreatic Juice

Pancreatic juice contains enzymes capable of digesting carbohydrates, fats, proteins, and nucleic acids.

The carbohydrate-digesting enzyme is called **pancreatic amylase.** It splits molecules of starch or glycogen into double sugars (disaccharides); the fat-digesting enzyme, **pancreatic lipase,** breaks triglyceride molecules into fatty acids and monoglycerides. (A monoglyceride molecule consists of one fatty acid bound to glycerol.)

The protein-splitting enzymes (proteinases) are **trypsin, chymotrypsin,** and **carboxypeptidase.** Each of these acts to split the bonds between particular combinations of amino acids in proteins. Since no single enzyme can split all possible combinations, the presence of several enzymes is necessary for the complete digestion of protein molecules.

These protein-splitting enzymes are stored in inactive forms within tiny cellular structures called *zymogen granules*. They, as gastric pepsin, are secreted in inactive forms and must be activated by other enzymes after they reach the small intestine. For example, the pancreatic cells release inactive **trypsinogen.** This substance becomes active trypsin when it contacts an enzyme called *enterokinase,* which is secreted by the mucosa of the small intestine. Chymotrypsin and carboxypeptidase are activated, in turn, by the presence of trypsin. This mechanism prevents the enzymatic digestion of proteins within the secreting cells and their ducts.

If something happens to block the release of pancreatic juice, it may accumulate in the duct system of the pancreas, and the trypsinogen may become activated. As a result, portions of the pancreas may become digested, causing a painful condition called *acute pancreatitis.*

The symptoms of acute pancreatitis commonly include severe and steady pain centered in the epigastric region of the abdomen with nausea, vomiting, and abdominal distension. Most commonly, this condition occurs in persons who use alcohol excessively or who have gallstones.

In addition to the above enzymes, pancreatic juice contains two **nucleases** that break down nucleic acid molecules into nucleotides. Pancreatic juice also contains a high concentration of bicarbonate ions that makes it alkaline. This provides a favorable environment for the actions of the digestive enzymes and helps to neutralize the acidic chyme as it arrives from the stomach. At the same time, the alkaline condition in the small intestine blocks the action of pepsin, which might otherwise damage the duodenal wall.

Regulation of Pancreatic Secretion

As in the case of gastric and small intestinal secretions, the release of pancreatic juice is regulated by nerve actions as well as hormones. For example, during the cephalic and gastric phases of gastric secretion, parasympathetic impulses reach the pancreas and

Figure 14.24
Acidic chyme entering the duodenum from the stomach stimulates the release of secretin, which in turn stimulates the release of pancreatic juice.

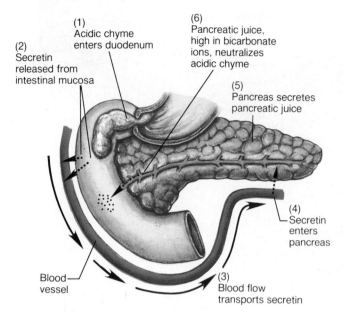

(1)
Acidic chyme enters duodenum

(2)
Secretin released from intestinal mucosa

(6)
Pancreatic juice, high in bicarbonate ions, neutralizes acidic chyme

(5)
Pancreas secretes pancreatic juice

(4)
Secretin enters pancreas

Blood vessel

(3)
Blood flow transports secretin

1. Where is the pancreas located?
2. List the enzymes found in pancreatic juice.
3. What are the functions of these enzymes?
4. How is the secretion of pancreatic juice regulated?

The Liver

The liver is located in the upper right portion of the abdominal cavity, just below the diaphragm. It is partially surrounded by the ribs and extends from the level of the fifth intercostal space to the lower margin of the ribs. It has a reddish brown color and is well supplied with blood vessels (figures 14.25 and 14.26).

Functions of the Liver

The liver is the largest gland in the body and carries on many important metabolic activities. For example, it plays a key role in carbohydrate metabolism by helping to maintain the normal concentration of blood glucose. As is described in chapter 13, liver cells responding to various hormones can decrease blood glucose by converting glucose to glycogen; they also can increase blood glucose by changing glycogen to glucose or by converting noncarbohydrates into glucose.

The liver's effects on lipid metabolism include the oxidation of fatty acids at an especially high rate (see chapter 4); the synthesis of lipoproteins, phospholipids, and cholesterol; and the conversion of carbohydrates into proteins and fats. Fats synthesized in the liver are transported by the blood to adipose tissue for storage.

The most vital liver functions probably are those related to protein metabolism. They include the deamination of amino acids; the formation of urea (see chapter 4); the synthesis of various blood proteins, including several that are necessary for blood clotting (see chapter 17); and the conversion of various amino acids to other amino acids.

Ammonia formed by bacteria in the intestine is normally removed from the blood by liver cells and changed into urea. In the absence of this liver function, the concentration of blood ammonia rises excessively, causing hepatic coma, a condition that can lead to death.

stimulate it to begin releasing digestive enzymes. Also, when acidic chyme enters the duodenum, the pancreas is stimulated to secrete a large quantity of fluid. This time, however, the stimulus comes from a hormone called **secretin,** which is released into the blood from the duodenal mucous membrane in response to the acid in chyme. The pancreatic juice secreted at this time contains few, if any, digestive enzymes. Instead, it has a high concentration of bicarbonate ions that function to neutralize the acid (figure 14.24).

The presence of chyme in the duodenum also stimulates the release of cholecystokinin from the intestinal wall. As before, this hormone reaches the pancreas by way of the blood. However, it causes the secretion of pancreatic juice with a high concentration of digestive enzymes.

Cystic fibrosis is an inherited condition characterized by the production of very thick, sticky mucus that adversely affects various exocrine glands. For example, the viscid mucus tends to clog the ducts of the pancreas. This interferes with the secretion of pancreatic juice, prevents the pancreatic digestive enzymes from reaching the duodenum, and leaves the person vulnerable to malnutrition.

The liver also stores a variety of substances, including glycogen; vitamins A, D, and B_{12} (see chapter 15); and iron. Iron storage occurs when the concentration of blood iron is excessive. The extra iron is combined with a protein (apoferritin) in liver cells, and as a result a substance called *ferritin* is formed. The iron remains stored in this form until the blood iron concentration reaches a certain low level. Then, some of the iron is released. Thus, the liver plays an important role in the homeostasis of iron.

Figure 14.25
The liver is located in the upper right portion of the abdominal cavity and is partially surrounded by ribs.

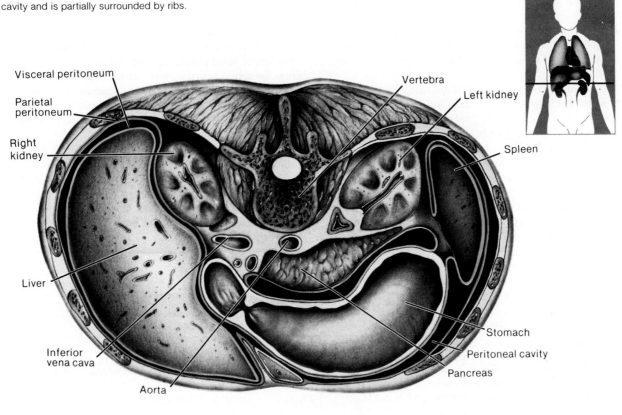

Figure 14.26
(a) Lobes of the liver as viewed from the front and *(b)* from below.

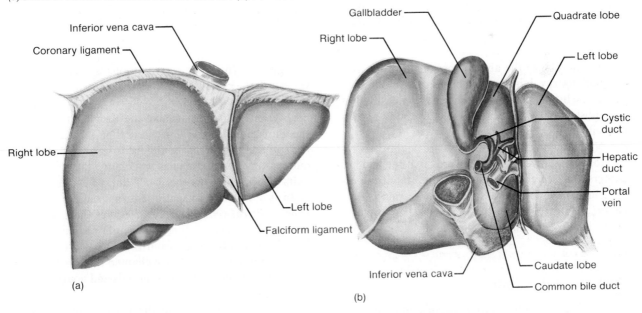

Chart 14.7	Major functions of the liver			
General function	**Specific function**	**General function**	**Specific function**	
Carbohydrate metabolism	Conversion of glucose to glycogen, conversion of glycogen to glucose, conversion of noncarbohydrates to glucose	Protein metabolism	Deamination of amino acids, synthesis of urea, synthesis of blood proteins, interconversion of amino acids	
Lipid metabolism	Oxidation of fatty acids; synthesis of lipoproteins, phospholipids, and cholesterol; conversion of carbohydrates and proteins into fats	Storage	Stores glycogen; vitamins A, D, and B_{12}; and iron	
		Blood filtering	Removes damaged red blood cells and foreign substances by phagocytosis	
		Detoxification	Alters composition of toxic substances	
		Secretion	Secretes bile	

In addition to the above functions, various liver cells help destroy damaged red blood cells and foreign substances by phagocytosis, alter the composition of toxic substances (detoxification) in body fluids, and secrete bile.

Since many of these functions are not directly related to the digestive system, they are discussed in other chapters. Bile secretion, however, is important to digestion and is explained in a subsequent section of this chapter.

Chart 14.7 summarizes the major functions of the liver.

Structure of the Liver

The liver is enclosed in a fibrous capsule and is divided by connective tissue into *lobes*—a large right lobe and a smaller left lobe. These lobes are separated by the *falciform ligament,* a fold of visceral peritoneum that also fastens the liver to the abdominal wall, anteriorly.

As figure 14.26 shows, the right lobe is further subdivided into the *right lobe proper,* the *quadrate lobe,* and the *caudate lobe.*

On its superior surface, the liver is attached to the diaphragm by a fold of visceral peritoneum called the *coronary ligament.* Each lobe is separated into numerous tiny **hepatic lobules,** which are the functional units of the gland (figures 14.27 and 14.28). A lobule consists of numerous hepatic cells that radiate outward from a *central vein.* Platelike groups of these cells are separated from each other by vascular channels called **hepatic sinusoids.** Blood from the digestive tract that is carried in portal veins (see chapter 18) brings newly absorbed nutrients into the sinusoids and nourishes the hepatic cells.

Usually the blood in the portal veins contains many bacterial cells that have entered through the intestinal wall. However, large **Kupffer cells** that are fixed to the inner lining (endothelium) of the hepatic sinusoids remove most of the bacteria from the blood by phagocytosis. Then the blood passes into the central veins of the hepatic lobules and moves out of the liver.

Figure 14.27

A cross section of a hepatic lobule, which is the functional unit of the liver.

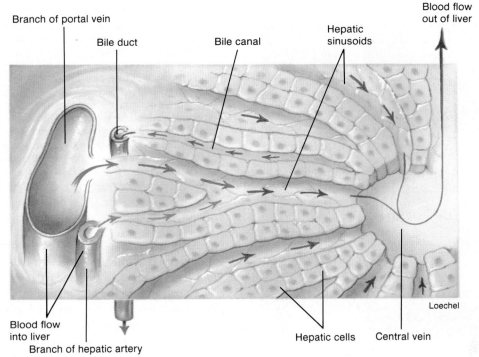

Within the liver lobules, there are many fine *bile canals* that receive secretions from the hepatic cells. The canals of neighboring lobules unite to form larger ducts, and these converge to become the **hepatic ducts.** They merge, in turn, to form the *common hepatic duct.*

1. Describe the location of the liver.
2. Review the functions of the liver.
3. What liver function is directly related to digestion?
4. Describe an hepatic lobule.

Composition of Bile

Bile is a yellowish green liquid that is secreted continuously by hepatic cells. In addition to water, it contains *bile salts, bile pigments* (bilirubin and biliverdin), cholesterol, and various electrolytes. Of these, the bile salts are the most abundant, and they are the only substances in bile that have a digestive function.

Hepatic cells use cholesterol in the production of bile salts and release some cholesterol into the bile in the process of secreting these salts. The cholesterol apparently has no special function in bile or in the alimentary canal.

The bile pigments are products of red blood cell breakdown and are normally excreted in the bile (see chapter 17).

Figure 14.28

A light micrograph of hepatic lobules. What features can you identify?

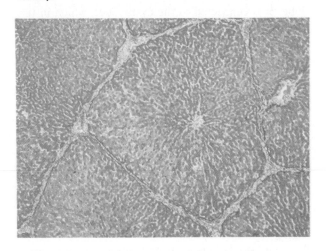

If the excretion of bile pigments is prevented due to obstruction of ducts, they tend to accumulate in the blood and tissues, causing a yellowish tinge in the sclera of the eye and other light-colored tissues. This condition is called *obstructive jaundice.* Jaundice also may occur with liver diseases in which the hepatic cells are unable to secrete ordinary amounts of bile pigments. This form is called *hepatocellular jaundice.* If there is excessive destruction of red blood cells, accompanied by rapid release of pigments, *hemolytic jaundice* results.

Figure 14.29

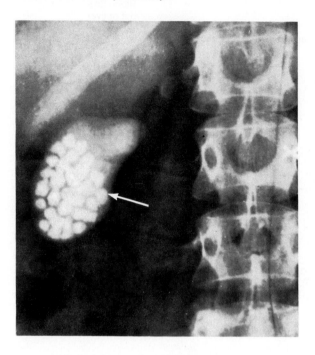

The Gallbladder and Its Functions

The **gallbladder** is a pear-shaped sac located in a depression on the inferior surface of the right hepatic lobe. It is connected to the **cystic duct,** which in turn joins the hepatic duct. The gallbladder has a capacity of 30–50 ml, is lined with columnar epithelial cells, and has a strong muscular layer in its wall (figure 14.29). It stores bile between meals, concentrates bile by reabsorbing water, and releases bile into the duodenum when stimulated by *cholecystokinin* from the small intestine.

The **common bile duct** is formed by the union of the common hepatic and cystic ducts. It leads to the duodenum, where its exit is guarded by a sphincter muscle (sphincter of Oddi). This sphincter normally remains contracted, so that bile collects in the common bile duct and backs up to the cystic duct. When this happens, the bile flows into the gallbladder and is stored there.

While the bile is in the gallbladder, its composition is altered because the lining reabsorbs some of the water and electrolytes. As these substances are removed, the bile salts, bile pigments, and cholesterol become increasingly concentrated. Although the cholesterol normally remains in solution, under certain conditions it may precipitate and form solid crystals.

If cholesterol continues to come out of solution, these crystals become larger and larger, forming *gallstones.*

Gallstones may form if the bile is concentrated excessively, if the hepatic cells secrete too much cholesterol, or if there is an inflammation in the gallbladder (cholecystitis). If such stones get into the bile duct, they may block the flow of bile, causing obstructive jaundice and considerable pain (figure 14.29). Generally gallstones that cause obstructions are surgically removed. At the same time, the gallbladder is removed by a surgical procedure called *cholecystectomy.* Consequently, the person is unable to produce gallstones or store bile. Following surgery, bile continues to reach the intestine by means of the hepatic and common bile ducts.

Regulation of Bile Release

Normally bile does not enter the duodenum until the gallbladder is stimulated to contract by *cholecystokinin.* This hormone is released from the intestinal mucosa in response to the presence of fats in the contents of the small intestine. The sphincter at the base of the common bile duct usually remains contracted until a peristaltic wave in the duodenal wall approaches it. Then just before the wave reaches it, the sphincter undergoes receptive relaxation and a squirt of bile enters the duodenum (figure 14.30).

The hormones that help control digestive functions are summarized in chart 14.8.

Functions of Bile Salts

The bile salts are the only components of bile with a digestive function. Although they do not act as digestive enzymes, these salts aid the actions of enzymes and enhance the absorption of fatty acids and certain fat-soluble vitamins.

Molecules of fats tend to clump together, forming masses called *fat globules.* Bile salts affect fat globules much like a soap or detergent would affect them. That is, they cause fat globules to break up into smaller droplets, an action called **emulsification.** Emulsification also is aided by the presence of monoglycerides, resulting from the action of pancreatic lipase on triglyceride molecules. As a result of emulsification, the total surface area of the fatty substance is greatly increased, and the tiny droplets mix with water. The fat-splitting enzymes (lipases) can then act on the fat molecules more effectively.

Bile salts aid in the absorption of fatty acids and cholesterol by forming complexes (micelles) that are very soluble in chyme and more easily absorbed by ep-

Figure 14.30
The gallbladder is stimulated to release bile when fat-containing chyme enters the duodenum.

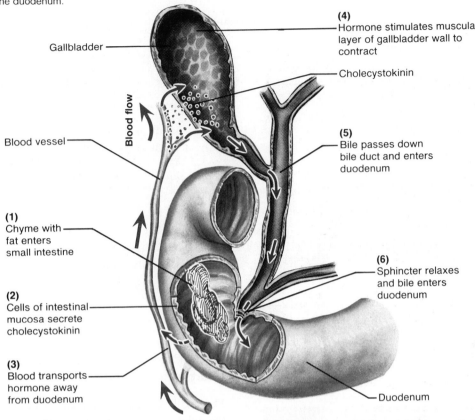

Gallbladder

(4) Hormone stimulates muscular layer of gallbladder wall to contract

Cholecystokinin

Blood flow

Blood vessel

(5) Bile passes down bile duct and enters duodenum

(1) Chyme with fat enters small intestine

(2) Cells of intestinal mucosa secrete cholecystokinin

(6) Sphincter relaxes and bile enters duodenum

(3) Blood transports hormone away from duodenum

Duodenum

Chart 14.8 Hormones of the digestive tract		
Hormone	Source	Function
Gastrin	Gastric cells, in response to the presence of food	Causes gastric glands to increase their secretory activity
Intestinal gastrin	Cells of small intestine, in response to the presence of chyme	Causes gastric glands to increase their secretory activity
Cholecystokinin	Intestinal wall cells, in response to the presence of fats in the small intestine	Causes gastric glands to decrease their secretory activity and inhibits gastric motility; stimulates pancreas to secrete fluid with a high digestive enzyme concentration; stimulates gallbladder to contract and release bile
Secretin	Cells in the duodenal wall, in response to chyme entering the small intestine	Stimulates pancreas to secrete fluid with a high bicarbonate ion concentration

ithelial cells. Along with these lipids, various fat-soluble vitamins, such as vitamins A, D, E, and K, are also absorbed. Thus, if bile salts are lacking, lipids may be poorly absorbed and the person is likely to develop vitamin deficiencies.

Nearly all the bile salts are reabsorbed by the mucous membrane of the small intestine along with the fatty acids. They enter the blood and are carried to the liver where they are resecreted into the bile ducts by

the hepatic cells. The small quantities that are lost in the feces are replaced by synthesis of bile salts in liver cells.

1. Explain how bile originates.
2. Describe the function of the gallbladder.
3. How is the secretion of bile regulated?
4. How does bile function in digestion?

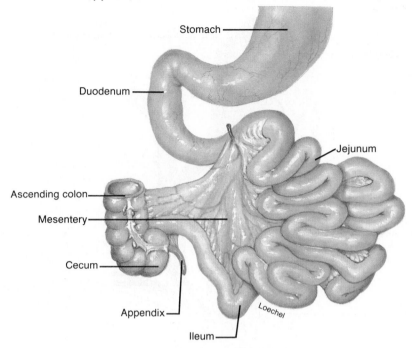

Figure 14.32
X-ray film of a normal small intestine.

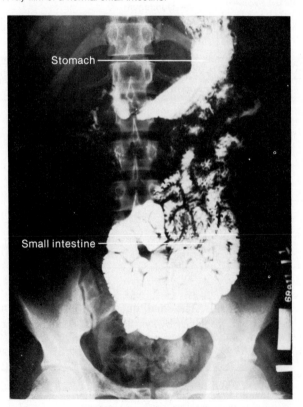

The Small Intestine

The small intestine is a tubular organ that extends from the pyloric sphincter to the beginning of the large intestine. With its many loops and coils, it fills much of the abdominal cavity. Although it is 5.5–6.0 meters (18–20 feet) long in a cadaver when the muscular wall lacks tone, the small intestine may be only half this long in a living person.

As mentioned, this portion of the alimentary canal receives secretions from the pancreas and liver. It also completes the digestion of the nutrients in chyme, absorbs the various products of digestion, and transports the remaining residues to the large intestine.

Parts of the Small Intestine

The small intestine, shown in figures 14.31 and 14.32, consists of three portions: the duodenum, jejunum, and ileum.

The **duodenum,** which is about 25 centimeters long and 5 centimeters in diameter, lies behind the parietal peritoneum (retroperitoneal), and is the most fixed portion of the small intestine. It follows a C-shaped path as it passes in front of the right kidney and the upper three lumbar vertebrae.

The rest of the small intestine lies free in the peritoneal cavity and is mobile. The proximal 2/5 of this portion is called the **jejunum,** and the remainder is the **ileum.** These portions are suspended from the posterior abdominal wall by a double-layered fold of peritoneum

Figure 14.33
Portions of the small intestine are suspended from the posterior
abdominal wall by mesentery formed by folds of the peritoneal
membrane.

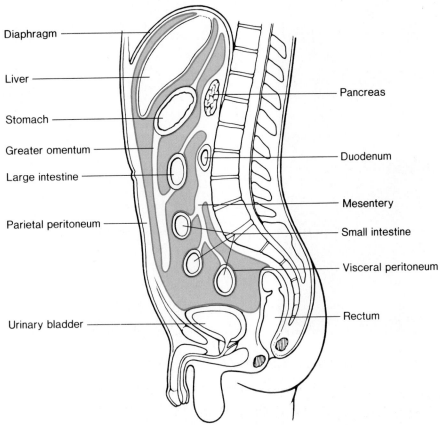

Diaphragm

Liver

Stomach

Greater omentum

Large intestine

Parietal peritoneum

Urinary bladder

Pancreas

Duodenum

Mesentery

Small intestine

Visceral peritoneum

Rectum

called **mesentery** (figure 14.33). This supporting tissue
contains the blood vessels, nerves, and lymphatic vessels that supply the intestinal wall.

A filmy fold of peritoneal membrane called the *greater
omentum* drapes like an apron from the stomach over
the transverse colon and the folds of the small intestine.
If infections occur in the wall of the alimentary canal, cells
from the omentum may adhere to the inflamed region
and help to wall it off so that the infection is less likely
to enter the peritoneal cavity.

Although there is no distinct separation between
the jejunum and ileum, the diameter of the jejunum
tends to be greater, and its wall is thicker, more vascular, and more active than that of the ileum.

Structure of the Small Intestinal Wall

Throughout its length, the inner wall of the small intestine has a velvety appearance. This is due to the
presence of innumerable tiny projections of mucous
membrane called **intestinal villi** (figures 14.2 and 14.34).
These structures are most numerous in the duodenum
and the proximal portion of the jejunum. They project
into the passageway or **lumen** of the alimentary canal

Figure 14.34
Structure of a single intestinal villus. What is the function of such
projections?

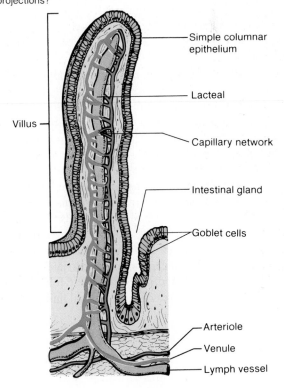

Villus

Simple columnar
epithelium

Lacteal

Capillary network

Intestinal gland

Goblet cells

Arteriole

Venule

Lymph vessel

Figure 14.35

Light micrograph of intestinal villi from the wall of the duodenum.

Figure 14.36

Microvilli at the free surface of the columnar epithelial cells that line the small intestine and greatly increase the surface area of these cells.

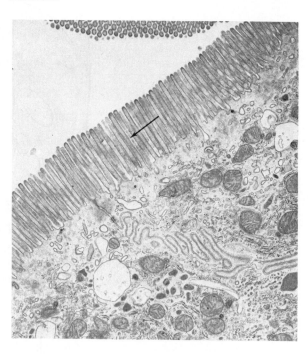

contacting the intestinal contents. They increase the surface area of the intestinal lining and play an important role in the absorption of digestive products.

Each villus consists of a layer of simple columnar epithelium and a core of connective tissue containing blood capillaries, a lymphatic capillary called a **lacteal,** and nerve fibers. At their free surfaces, the epithelial cells possess many fine extensions, called *microvilli,* that create a brushlike border and greatly increase the surface area of the intestinal cells (figures 14.35 and 14.36). The presence of microvilli enhances the process of absorption.

The blood and lymph capillaries function to carry away substances absorbed by a villus, while the nerve fibers act to stimulate or inhibit its activities.

Between the bases of the villi are tubular **intestinal glands** (crypts of Lieberkühn) that extend downward into the mucous membrane. The deeper layers of the small intestinal wall are much like those of other parts of the alimentary canal in that they include a submucosa, a muscular layer, and a serous layer.

The lining of the small intestine also is characterized by the presence of numerous circular folds, called *plicae circulares,* that are especially well developed in the lower duodenum and upper jejunum. Together with the villi and microvilli, these folds help to increase the surface area of the intestinal lining (figure 14.37).

Secretions of the Small Intestine

In addition to mucous-secreting goblet cells that occur extensively throughout the mucosa of the small intestine, there are many specialized *mucous-secreting glands* (Brunner's glands) that occur in the submucosa within the proximal portion of the duodenum. These glands secrete large quantities of viscid, alkaline mucus in response to various stimuli.

The intestinal glands at the bases of the villi secrete great amounts of a watery fluid. The villi rapidly reabsorb this fluid, and it provides a vehicle for moving digestive products into the villi. The fluid secreted by the intestinal glands has a pH that is nearly neutral (6.5–7.5), and seems to lack digestive enzymes. However, the epithelial cells of the intestinal mucosa have digestive enzymes embedded in the surfaces of their microvilli. These enzymes can break down food molecules just before absorption takes place. They include **peptidases,** which split peptides into amino acids; **sucrase, maltase,** and **lactase,** which split the double sugars (disaccharides) sucrose, maltose, and lactose respectively into the simple sugars (monosaccharides) glucose, fructose, and galactose; and **intestinal lipase,** which splits fats into fatty acids and glycerol.

Chart 14.9 summarizes the sources and actions of the major digestive enzymes.

Figure 14.37
(a) The inner lining of the small intestine contains many circular
folds, the plicae circulares; (b) a longitudinal section through
some of these folds.

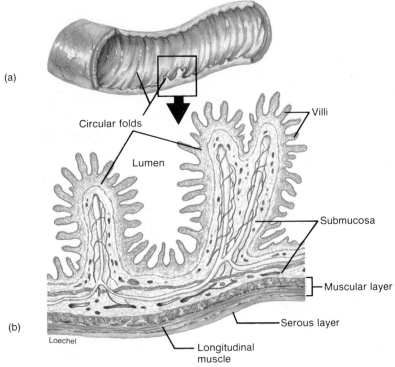

(a)

Circular folds

Lumen

Villi

Submucosa

Muscular layer

Serous layer

(b)

Loechel

Longitudinal
muscle

Chart 14.9	A summary of the digestive enzymes	
Enzyme	Source	Digestive action
Salivary enzyme		
Amylase	Salivary glands	Begins carbohydrate digestion by converting starch and glycogen to disaccharides
Gastric enzymes		
Pepsin	Gastric glands	Begins the digestion of proteins
Lipase	Gastric glands	Begins the digestion of butterfat
Intestinal enzymes		
Peptidase	Mucosal cells	Converts peptides into amino acids
Sucrase, maltase, lactase	Mucosal cells	Converts disaccharides into monosaccharides
Lipase	Mucosal cells	Converts fats into fatty acids and glycerol
Enterokinase	Mucosal cells	Activates trypsinogen
Pancreatic enzymes		
Amylase	Pancreas	Converts starch and glycogen into disaccharides
Lipase	Pancreas	Converts fats into fatty acids and glycerol
Proteinases a. Trypsin b. Chymotrypsin c. Carboxypeptidase	Pancreas	Converts proteins or partially digested proteins into peptides
Nuclease	Pancreas	Converts nucleic acids into nucleotides

Figure 14.38
Digestion changes complex carbohydrates into disaccharides.
The disaccharides are then converted into monosaccharides that
are absorbed by intestinal villi and enter the blood.

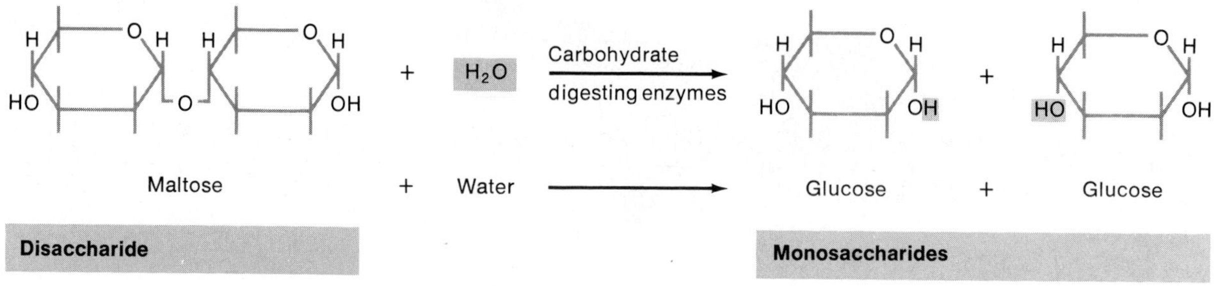

Maltose + Water ⟶ Glucose + Glucose

Disaccharide   **Monosaccharides**

The amount of lactase produced in the small intestine usually reaches a maximum shortly after birth and thereafter tends to decrease. As a result, some adults produce insufficient quantities of lactase to break down the lactose or milk sugar in their diets. When this happens, the lactose from milk or certain milk products remains undigested and causes an increase in the osmotic pressure of the intestinal contents. Consequently, water is drawn from the tissues into the intestine. At the same time, intestinal bacteria may act upon the undigested sugar and produce organic acids and gases. As a result, the person may feel bloated and suffer from intestinal cramps and diarrhea.

Regulation of Small Intestinal Secretions

Since mucus functions to protect the intestinal wall in the same way it protects the stomach lining, it is not surprising that mucus secretions are enhanced by mechanical stimulation and the presence of irritants, such as gastric juice. Consequently, as the stomach contents enter the small intestine, the duodenal mucous glands are stimulated to release large quantities of mucus.

Secretions from goblet cells and intestinal glands are similarly stimulated by direct contact with chyme, which provides both chemical and mechanical stimuli, and by reflexes triggered by distension of the intestinal wall. These reflex actions involve parasympathetic motor impulses that cause secretory cells to increase their activities. Some investigators believe that a hormone (enterocrinin) may also aid in the regulation of intestinal secretions. This hormone apparently is released from cells in the intestinal wall and stimulates the intestinal glands to increase their secretions.

1. Describe the parts of the small intestine.
2. Distinguish between intestinal villi and microvilli.
3. What is the function of the intestinal glands?
4. List the digestive enzymes formed by intestinal cells.

Absorption in the Small Intestine

Because villi greatly increase the surface area of the intestinal mucosa, the small intestine is the most important absorbing organ of the alimentary canal. In fact, the small intestine is so effective at absorbing digestive products, water, and electrolytes, that very little absorbable material reaches its distal end.

Carbohydrate digestion begins in the mouth as a result of the activity of salivary amylase, and is completed in the small intestine by enzymes from the intestinal mucosa and pancreas (figure 14.38). The resulting monosaccharides are absorbed by the villi and enter blood capillaries. Although small quantities may pass into the villi by diffusion, most simple sugars are absorbed by active transport or facilitated diffusion (see chapter 3). The exact mechanisms, however, are poorly understood.

Protein digestion begins in the stomach as a result of pepsin activity and is completed in the small intestine by enzymes from the intestinal mucosa and the pancreas. During this process, large protein molecules are converted into amino acids, as shown in figure 14.39. These smaller particles are then absorbed into the villi by active transport and are carried away by the blood.

Fat molecules are digested almost entirely by enzymes from the intestinal mucosa and pancreas (figure 14.40). The absorption mechanism for the resulting fatty acid molecules involves several steps, including the following: (1) the fatty acid molecules are dissolved in the epithelial cell membranes of the villi and diffuse into these cells, (2) the endoplasmic reticula of the cells use the fatty acids to resynthesize fat molecules similar to those previously digested, and (3) these fats collect in clusters that become encased in protein. The resulting large molecules of lipoprotein are called *chylomicrons,* and they make their ways to the lacteal of the villus. Periodic contractions of smooth muscles in the villus help to empty the lacteal and the lymph carries the chylomicrons to the blood. (See figure 14.41.)

Figure 14.39

The amino acids that result from protein digestion are absorbed by intestinal villi and enter the blood.

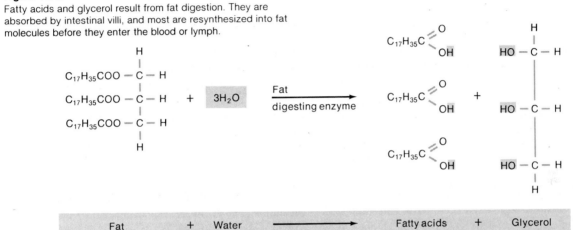

| Peptide (portion of protein molecule) | + | Water | ⟶ | Amino acid | + | Amino acid |

Figure 14.40

Fatty acids and glycerol result from fat digestion. They are absorbed by intestinal villi, and most are resynthesized into fat molecules before they enter the blood or lymph.

| Fat | + | Water | ⟶ | Fatty acids | + | Glycerol |

Figure 14.41

Fatty acid absorption involves several steps.

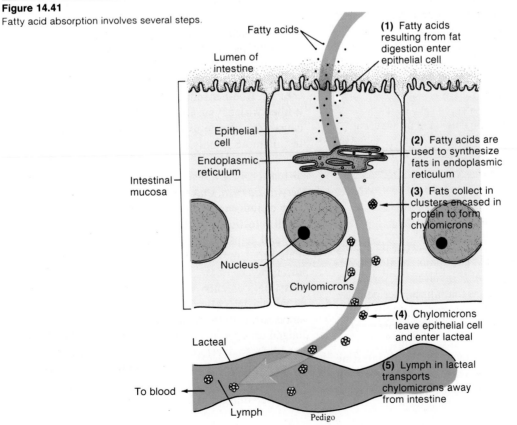

Fatty acids

Lumen of intestine

(1) Fatty acids resulting from fat digestion enter epithelial cell

Epithelial cell

Endoplasmic reticulum

Intestinal mucosa

(2) Fatty acids are used to synthesize fats in endoplasmic reticulum

(3) Fats collect in clusters encased in protein to form chylomicrons

Nucleus

Chylomicrons

(4) Chylomicrons leave epithelial cell and enter lacteal

Lacteal

To blood ←

Lymph

Pedigo

(5) Lymph in lacteal transports chylomicrons away from intestine

The Digestive System / 523

Chylomicrons are transported by the blood to capillaries of muscle and adipose tissues. A specific type of protein (apoprotein) associated with the surface of chylomicrons activates an enzyme (lipoprotein lipase) that is attached to the inner lining of such capillaries. As a result of the enzyme's action, fatty acids and monoglycerides are released from the chylomicrons, and they enter muscle or adipose cells to be utilized as energy sources or to be stored. The remnants of the chylomicrons travel in the blood to the liver, where they bind to receptors on the surface of liver cells. The remnants quickly enter liver cells by receptor-mediated endocytosis (see chapter 3) and are destroyed by lysosomal activity.

Some fatty acids with relatively short carbon chains may be absorbed directly into the blood capillary of the villus without being converted back into fat.

In addition to absorbing the products of carbohydrate, protein, and fat digestion, the intestinal villi function in the absorption of various electrolytes and water. Certain ions, such as those of sodium, potassium, chloride, nitrate, and bicarbonate, are readily absorbed; but others, including calcium, magnesium, and sulfate, are poorly absorbed.

The absorption of electrolytes usually involves active transport, and water is absorbed by osmosis. Thus, even though the intestinal contents may be hypertonic to the epithelial cells at first, as nutrients and electrolytes are absorbed, the intestinal contents tend to become hypotonic to the cells. Then, water follows the nutrients and electrolytes into the villi by osmosis.

The absorption process is summarized in chart 14.10.

---

1. What substances resulting from the digestion of carbohydrate, protein, and fat molecules are absorbed by the small intestine?
2. What ions are absorbed by the small intestine?
3. What transport mechanisms are used by intestinal villi?
4. Describe how fatty acids are absorbed and transported.

Movements of the Small Intestine

Like the stomach, the small intestine carries on *mixing movements* and *peristalsis*. The major mixing movement is called *segmentation*. It involves the formation of small, ringlike contractions that appear periodically, cutting the chyme into segments and moving it to and fro. Segmentation also serves to slow the movement of chyme through the intestine.

Chyme is propelled through the small intestine by peristaltic waves. These waves are usually weak, and they stop after pushing the chyme a short distance.

Chart 14.10	Intestinal absorption of nutrients	
Nutrient	Absorption mechanism	Means of transport
Monosaccharides	Diffusion and active transport	Blood in capillaries
Amino acids	Active transport	Blood in capillaries
Fatty acids and glycerol	Diffusion into cells	
	(a) Most fatty acids are converted back into fats and incorporated in chylomicrons for transport	Lymph in lacteals
	(b) Some fatty acids with relatively short carbon chains are transported without being converted back into fats	Blood in capillaries
Electrolytes	Diffusion and active transport	Blood in capillaries
Water	Osmosis	Blood in capillaries

Consequently, food materials move relatively slowly through the small intestine, taking from 3 to 10 hours to travel its length.

As might be expected, parasympathetic impulses enhance both mixing and peristaltic movements, and sympathetic impulses inhibit them. Reflexes involving parasympathetic impulses to the small intestine sometimes originate in the stomach. For example, as the stomach fills with food and its wall becomes distended, a reflex (gastroenteric reflex) is triggered, and peristaltic activity in the small intestine is greatly increased. Another reflex is initiated when the duodenum is filled with chyme and its wall is stretched. This reflex causes the chyme to be moved through the small intestine more rapidly.

Stimulation of the small intestinal wall by overdistension or by severe irritation may elicit a strong *peristaltic rush,* that passes along its entire length. This type of movement serves to sweep the contents of the small intestine into the large intestine relatively rapidly and helps relieve the small intestine of its problem. This rapid movement of chyme prevents the normal absorption of water, nutrients, and electrolytes from the intestinal contents. The result is *diarrhea,* a condition in which defecation becomes more frequent and the stools are watery. If the diarrhea continues for a prolonged time, problems in water and electrolyte balance are likely to develop.

At the distal end of the small intestine, where the ileum joins the cecum of the large intestine, there is a sphincter muscle called the **ileocecal valve.** Normally

Figure 14.42
Parts of the large intestine.

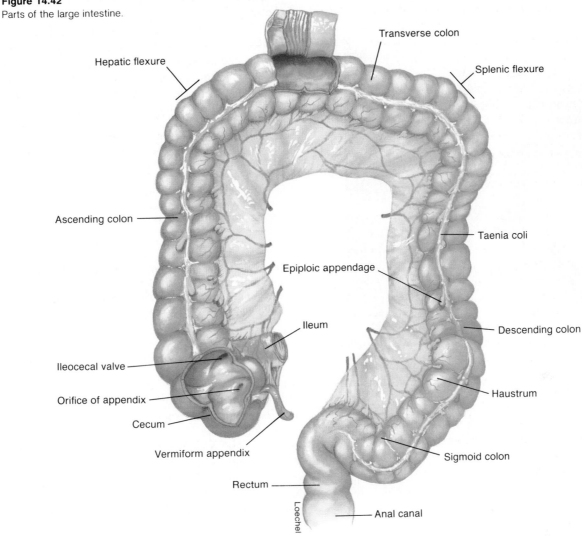

this sphincter remains constricted, preventing the contents of the small intestine from entering. At the same time, it prevents the contents of the large intestine from backing up into the ileum. However, after a meal a reflex (gastroileal reflex) is elicited, and peristalsis in the ileum is increased. This action forces some of the contents of the small intestine into the cecum.

1. Describe the movements of the small intestine.
2. How are these movements initiated?
3. What is meant by a peristaltic rush?
4. What stimulus causes the ileocecal valve to relax?

The Large Intestine

The large intestine is so named because its diameter is greater than that of the small intestine. This portion of the alimentary canal is about 1.5 meters long and begins in the lower right side of the abdominal cavity, where the ileum joins the cecum. From there the large intestine (colon) travels upward on the right side,

crosses obliquely to the left, and descends into the pelvis. At its distal end it opens to the outside of the body as the anus.

The large intestine reabsorbs water and electrolytes from the chyme remaining in the alimentary canal. It also forms and stores the feces until defecation occurs.

Parts of the Large Intestine

The large intestine, shown in figures 14.42 and 14.43, consists of the cecum, colon, rectum, and anal canal.

The **cecum,** which represents the beginning of the large intestine, is a dilated, pouchlike structure that hangs slightly below the ileocecal opening. Projecting downward from it is a narrow tube with a closed end called the **vermiform appendix.** Although the human appendix has no known digestive function, it does contain lymphatic tissue that may serve to resist infections.

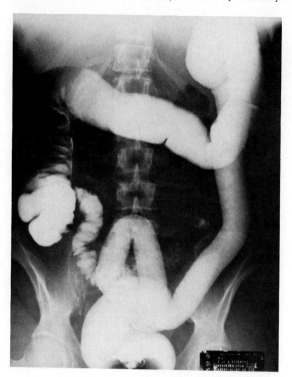

Occasionally the appendix itself may become infected and inflamed, causing the condition called *appendicitis*. When this happens, the appendix is often removed surgically to prevent its rupture. If it does break open, the contents of the large intestine may enter the abdominal cavity and cause a serious infection of the peritoneum called *peritonitis*.

The **colon** can be divided into four portions—ascending, transverse, descending, and sigmoid colons. The **ascending colon** begins at the cecum and travels upward against the posterior abdominal wall to a point just below the liver. There it turns sharply to the left (as the right colic or hepatic flexure) and becomes the **transverse colon.** The transverse colon is the longest and the most mobile part of the large intestine. It is suspended by a fold of peritoneum and tends to sag in the middle below the stomach. As the transverse colon approaches the spleen, it turns abruptly downward (as the left colic or splenic flexure) and becomes the **descending colon.** At the brim of the pelvis, the descending colon makes an S-shaped curve, called the **sigmoid colon,** and then becomes the rectum.

The **rectum** lies next to the sacrum and generally follows its curvature. It is firmly attached to the sacrum by the peritoneum, and it ends about 5 cm below the tip of the coccyx, where it becomes the anal canal (figure 14.44).

The **anal canal** is formed by the last 2.5 or 4.0 cm of the large intestine. The mucous membrane in the canal is folded into a series of six to eight longitudinal *anal columns.* At its distal end, the canal opens to the outside as the **anus.** This opening is guarded by two sphincter muscles, an *internal anal sphincter* composed of smooth muscle, and an *external anal sphincter* of skeletal muscle.

Each anal column contains a branch of the rectal vein, and if something interferes with the blood flow in these vessels, the anal columns may become enlarged and inflamed. This condition, called *hemorrhoids,* may be aggravated by bowel movements and may be accompanied by discomfort and bleeding.

1. What is the general function of the large intestine?
2. Describe the parts of the large intestine.
3. Distinguish between the internal sphincter and the external sphincter of the anus.

Structure of the Large Intestinal Wall

Although the wall of the large intestine includes the same types of tissues found in other parts of the alimentary canal, it has some unique features. For example, it lacks the villi that are characteristic of the small intestine. Also, the layer of longitudinal muscle fibers does not cover its wall uniformly. Instead, the fibers are arranged in three distinct bands (teniae coli) that extend the entire length of the colon. These bands exert tension on the wall, creating a series of pouches (haustra). The large intestinal wall also is characterized by small collections of fat (epiploic appendages) in the serous layer on its outer surface (figure 14.42).

Functions of the Large Intestine

Unlike the small intestine, which functions to secrete digestive enzymes and to absorb the products of digestion, the large intestine has little or no digestive function. The mucous membrane that forms the inner lining of the large intestine, however, contains many tubular glands. Structurally these glands are similar to those of the small intestine, but they are composed almost entirely of goblet cells. Consequently, mucus is the only significant secretion of this portion of the alimentary canal (figures 14.45 and 14.46).

The rate of mucus secretion is controlled largely by mechanical stimulation from chyme and by parasympathetic impulses. In both cases, the goblet cells respond by increasing their production of mucus, which in turn protects the intestinal wall against the abrasive action of materials passing through it. Mucus also aids

Figure 14.44
The rectum and anal canal are located at the distal end of the alimentary canal.

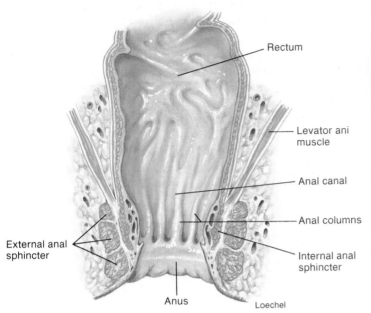

Rectum

Levator ani muscle

Anal canal

Anal columns

External anal sphincter

Internal anal sphincter

Anus

Loechel

Figure 14.45
A scanning electron micrograph of the large intestine mucosa. Notice the openings of the goblet cells (arrow). (*Tissues and Organs: A Text-Atlas of Scanning Electron Microscopy,* by R. G. Kessel and R. H. Kardon. © 1979 W. H. Freeman and Company.)

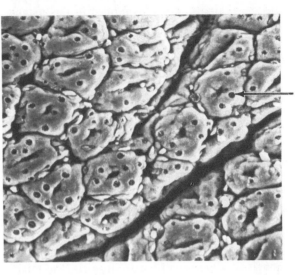

Figure 14.46
Light micrograph of the mucosa of the human large intestine.

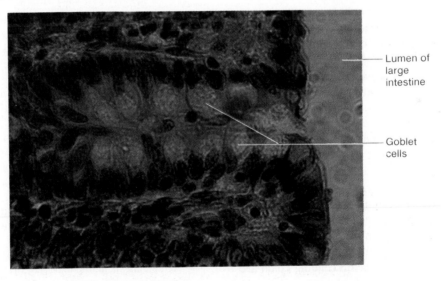

Lumen of large intestine

Goblet cells

in holding particles of fecal matter together, and because it is alkaline, mucus helps to control the pH of the large intestinal contents. This is important because acids are sometimes released from the feces as a result of bacterial activity.

The chyme entering the large intestine usually has few nutrients remaining in it. It contains materials that could not be digested or absorbed by the small intestine, in addition to water, various electrolytes, mucus, and bacteria.

Absorption in the large intestine is normally limited to water and electrolytes, which usually are absorbed in the proximal half of the tube. Electrolytes, such as sodium ions, can be absorbed by active transport, while water enters the mucosa by osmosis. As a result of these processes, little sodium or water is lost in the feces.

Many bacteria (colon bacilli) normally inhabit the large intestine, and these organisms can break down some of the substances that escape the actions of the human enzymes. For instance, cellulose passes through the alimentary canal almost unchanged, but colon bacteria can break down this carbohydrate and use it as an energy source. At the same time, these bacteria synthesize certain vitamins that can be absorbed by the intestinal mucosa and used to supplement the supply of dietary vitamins. Vitamins formed in this manner include vitamins K, B_{12}, thiamine, and riboflavin. Bacterial actions in the large intestine may give rise to gases (flatus), which sometimes cause discomfort and pain.

---

1. How does the structure of the large intestine differ from that of the small intestine?
2. What substances can be absorbed by the large intestine?
3. What useful substances are produced as a result of bacteria inhabiting the large intestine?

Movements of the Large Intestine

The movements of the large intestine—mixing and peristalsis—are similar to those of the small intestine, although they are usually more sluggish. The mixing movements break the fecal matter into segments and turn it so that all portions are exposed to the mucosa. This helps in water and electrolyte absorption.

The peristaltic waves of the large intestine are different from those in the small intestine. Instead of occurring frequently, they come only two or three times each day. These waves produce *mass movements* in which a relatively large section of the colon constricts vigorously, forcing its contents to move toward the rectum. Typically, mass movements occur following a meal, as a result of a reflex that is initiated in the small intestine (duodenocolic reflex). Abnormal irritations of the mucosa can also trigger such movements. For instance, a person suffering from an inflamed colon (colitis) may experience frequent mass movements.

When it is appropriate to defecate, a person usually can initiate a *defecation reflex* by holding a deep breath and contracting abdominal wall muscles. This action increases the internal abdominal pressure and forces feces into the rectum. As the rectum fills, its wall is distended and the defecation reflex is triggered. As a result, peristaltic waves in the descending colon are stimulated, and the internal anal sphincter relaxes. At the same time, other reflexes involving the sacral region of the spinal cord, cause the peristaltic waves to strengthen, the diaphragm to lower, the glottis to close, and the abdominal wall muscles to contract. These actions cause an additional increase in the internal abdominal pressure and assist in squeezing the rectum.

The external anal sphincter is signaled to relax, and the feces are forced to the outside. A person can inhibit defecation voluntarily by keeping the external anal sphincter contracted.

The Feces

As was mentioned, **feces** are composed largely of materials that could not be digested, together with water, electrolytes, mucus, and bacteria. Usually they are about 75% water, and the color is normally due to the presence of bile pigments that have been altered somewhat by bacterial action.

The pungent odor of feces results from a variety of compounds produced by bacteria. These include phenol, hydrogen sulfide, indole, skatole, and ammonia.

---

1. How does peristalsis in the large intestine differ from peristalsis in the small intestine?
2. List the major events that occur during defecation.
3. Describe the composition of feces.

Clinical Terms Related to the Digestive System

achalasia (ak″ah-la′ze-ah)   failure of the smooth muscle to relax at some junction in the digestive tube, such as that between the esophagus and stomach.

achlorhydria (ah″klōr-hi′dre-ah)   a lack of hydrochloric acid in gastric secretions.

aphagia (ah-fa′je-ah)   an inability to swallow.

cholecystitis (ko″le-sis-ti′tis)   inflammation of the gallbladder.

cholecystolithiasis (ko″le-sis″to-li-thi′ah-sis)   the presence of stones in the gallbladder.

cirrhosis (si-ro′sis)   a liver condition in which the hepatic cells degenerate and the surrounding connective tissues thicken.

diverticulitis (di″ver-tik″u-li′tis)   inflammation of small pouches (diverticula) that sometimes form in the lining and wall of the colon.

dumping syndrome (dum′ping sin′drōm)   a set of symptoms, including diarrhea, that often occur following a gastrectomy.

dysentery (dis′en-ter″e)   an intestinal infection, caused by viruses, bacteria, or protozoans, that is accompanied by diarrhea and cramps.

dyspepsia (dis-pep′se-ah)   indigestion; difficulty in digesting a meal.

dysphagia (dis-fa′ze-ah)   difficulty in swallowing.

enteritis (en″tĕ-ri′tis)   inflammation of the intestine.

esophagitis (e-sof″ah-ji′tis)   inflammation of the esophagus.

gastrectomy (gas-trek′to-me)   partial or complete removal of the stomach.

Disorders of the Large Intestine
A Clinical Application

If feces are allowed to remain in the colon too long, the material tends to become hard and dry as more and more water is removed from it. This action may produce a condition called *constipation*, in which the feces become so dry that defecation is difficult.

The causes of constipation include prolonged inhibition of the defecation reflex, lack of tone in the muscular layer of the large intestine, and muscular spasms in the intestinal wall that block the movement of feces. Constipation is sometimes treated with laxatives (cathartics), which are substances that can stimulate fecal movements.

Some laxatives act by causing an irritating effect on the intestinal mucosa, which stimulates peristaltic waves. Others prevent reabsorption of water by coating the feces with oil, or keeping the feces moist by increasing the osmotic pressure within the contents of the large intestine.

Constipation may also be treated by enemas, by increasing the fluid intake, by adding bulk foods to the diet, or by resensitizing the person to the defecation reflex.

Diarrhea creates an effect opposite to that of constipation. It occurs when the feces move more rapidly than usual through the colon and consequently remain very liquid. Among the more common causes of diarrhea are hypersecretion of fluids and abnormally frequent peristaltic waves.

Such actions often accompany emotional stress or intestinal irritations produced by bacterial infections, viral infections, or various chemicals.

Sometimes the mucosa of the large intestine becomes extensively ulcerated, and the person is said to have *ulcerative colitis*. The cause of this condition is unknown, but it is often associated with prolonged periods of emotional stress. In any case, the colon becomes very active, mass movements occur frequently, and the person has diarrhea most of the time.

Treatment for ulcerative colitis involves reducing nervous tension, changing the diet, and providing medication. If such measures are unsuccessful, a surgeon may perform an *ileostomy*, in which the colon is removed and a portion of the ileum is connected to an opening in the abdominal wall. As a result, the small intestinal contents can move to the outside of the body without passing through the colon.

A related procedure, called *colostomy*, is sometimes performed in cases of abscesses of the colon or intestinal obstructions, or following removal of the rectum. In this procedure, the colon is connected to an opening on the surface of the abdomen so that the rectum and anus are bypassed.

---

gastritis (gas-tri′tis)   inflammation of the stomach lining.

gastrostomy (gas-tros′to-me)   the creation of an opening in the stomach wall to allow the administration of food and liquids when swallowing is not possible.

gingivitis (jin″ji-vi′tis)   inflammation of the gums.

glossitis (glŏ-si′tis)   inflammation of the tongue.

hemorrhoidectomy (hem″ŏ-roi-dek′to-me)   removal of hemorrhoids.

hemorrhoids (hem′ŏ-roids)   a condition in which veins associated with the lining of the anal canal become enlarged.

hepatitis (hep″ah-ti′tis)   inflammation of the liver.

ileitis (il″e-i′tis)   inflammation of the ileum.

pharyngitis (far″in-ji′tis)   inflammation of the pharynx.

proctitis (prok-ti′tis)   inflammation of the rectum.

pyloric stenosis (pi-lor′ik stě-no′sis)   a congenital obstruction at the pyloric sphincter due to an enlargement of the pyloric muscle.

pylorospasm (pi-lor′o-spazm)   a spasm of the pyloric portion of the stomach or of the pyloric sphincter.

pyorrhea (pi″o-re′ah)   an inflammation of the dental periosteum accompanied by the formation of pus.

stomatitis (sto″mah-ti′tis)   inflammation of the lining of the mouth.

vagotomy (va-got′o-me)   sectioning of the vagus nerve fibers.

Chapter Summary

Introduction

Digestion is the process of changing food substances into forms that can be absorbed.

The digestive system consists of an alimentary canal and several accessory organs.

General Characteristics of the Alimentary Canal

Various regions of the canal are specialized to perform specific functions.

1. Structure of the wall
   a. The wall consists of four layers.
   b. These layers include mucous membrane, submucosa, muscular layer, and serous layer.
2. Movements of the tube
   a. Motor functions include mixing and propelling movements.
   b. Peristalsis is responsible for propelling movements.
   c. The wall of the tube undergoes receptive relaxation just ahead of a peristaltic wave.

3. Innervation of the tube
   a. The tube is innervated by branches of the sympathetic and parasympathetic divisions of the autonomic nervous system.
   b. Parasympathetic impulses generally cause an increase in digestive activities; sympathetic impulses generally inhibit digestive activities.
   c. Sympathetic impulses are responsible for the contraction of certain sphincter muscles that control movement through the alimentary canal.

The Mouth

The mouth is adapted to receive food and begin preparing it for digestion. It also serves as an organ of speech and pleasure.

1. The cheeks and lips
   a. Cheeks form the lateral walls of the mouth.
   b. Lips are highly mobile and possess a variety of sensory receptors useful in judging the characteristics of food.
2. The tongue
   a. The tongue is a thick, muscular organ that aids in mixing food with saliva and moving it toward the pharynx.
   b. Its rough surface aids in handling food and contains taste buds.
   c. Lingual tonsils are located on the root of the tongue.
3. The palate
   a. The palate comprises the roof of the mouth and includes hard and soft portions.
   b. The soft palate closes the opening to the nasal cavity during swallowing.
   c. Palatine tonsils are located on either side of the tongue in the back of the mouth.
   d. Tonsils consist of lymphatic tissues, but are common sites of infections and may become enlarged so that they interfere with swallowing and breathing.
4. The teeth
   a. Two sets develop in sockets of the mandibular and maxillary bones.
   b. There are twenty deciduous and thirty-two permanent teeth.
   c. They break food into smaller pieces, increasing the surface area of food that is exposed to digestive actions.
   d. Different kinds are adapted to handle foods in different ways, such as biting, grasping, or grinding.
   e. Each tooth consists of a crown and root and is composed of enamel, dentine, pulp, nerves, and blood vessels.
   f. A tooth is attached to the alveolar bone by collagenous fibers of the periodontal ligament.

The Salivary Glands

Salivary glands secrete saliva, which moistens food, helps bind food particles together, begins digestion of carbohydrates, makes taste possible, helps cleanse the mouth, and regulates pH in the mouth.

1. Salivary secretions
   a. Salivary glands include serous cells that secrete digestive enzymes and mucous cells that secrete mucus.
   b. Parasympathetic impulses stimulate the secretion of serous fluid.
2. Major salivary glands
   a. The parotid glands are the largest, and they secrete saliva that is rich in amylase.
   b. The submandibular glands in the floor of the mouth produce viscid saliva.
   c. The sublingual glands in the floor of the mouth primarily secrete mucus.

The Pharynx and Esophagus

The pharynx and esophagus serve only as passageways.

1. Structure of the pharynx
   a. The pharynx is divided into a nasopharynx, oropharynx, and laryngopharynx.
   b. Its muscular walls contain fibers arranged in circular and longitudinal groups.
2. The swallowing mechanism
   a. The act of swallowing occurs in three stages.
      (1) Food is mixed with saliva and forced into the pharynx.
      (2) Involuntary reflexes move the food into the esophagus.
      (3) Food is transported to the stomach.
3. The esophagus
   a. The esophagus passes through the mediastinum and penetrates the diaphragm.
   b. Some circular muscle fibers at the end of the esophagus help to prevent the regurgitation of food from the stomach.

The Stomach

The stomach receives food, mixes it with gastric juice, carries on a limited amount of absorption, and moves food into the small intestine.

1. Parts of the stomach
   a. The stomach is divided into cardiac, fundic, body, and pyloric regions.
   b. The pyloric sphincter serves as a valve between the stomach and the small intestine.
2. Gastric secretions
   a. Gastric glands secrete gastric juice.
   b. Gastric juice contains pepsin, hydrochloric acid, lipase, and intrinsic factor.
3. Regulation of gastric secretions
   a. Gastric secretions are enhanced by parasympathetic impulses and the hormone, gastrin.
   b. Three stages of gastric secretion include cephalic, gastric, and intestinal phases.
   c. Presence of food in the small intestine reflexly inhibits gastric secretions.

4. Gastric absorption
   a. The stomach is not well adapted for absorption.
   b. A few substances such as water and other small molecules may be absorbed through its wall.
5. Mixing and emptying actions
   a. As the stomach fills, its wall stretches, but its internal pressure remains unchanged.
   b. Mixing movements aid in producing chyme; peristaltic waves move the chyme into the pyloric region.
   c. Muscular wall of pyloric region pumps chyme into small intestine.
   d. The rate of emptying depends on the fluidity of the chyme and the type of food present.
   e. The upper part of the small intestine fills, and an enterogastric reflex causes the peristaltic waves in the stomach to be inhibited.
   f. Vomiting results from a complex reflex that can be stimulated by a variety of factors.

The Pancreas

1. Structure of the pancreas
   a. The pancreas is closely associated with the duodenum.
   b. It produces pancreatic juice that is secreted into a pancreatic duct.
   c. The pancreatic duct leads to the duodenum.
2. Pancreatic juice
   a. Pancreatic juice contains enzymes that can split carbohydrates, proteins, fats, and nucleic acids.
   b. It has a high bicarbonate ion concentration that helps to neutralize chyme and causes the intestinal contents to be alkaline.
3. Regulation of pancreatic secretion
   a. Secretin from the duodenum stimulates the release of pancreatic juice that contains few digestive enzymes, but has a high bicarbonate ion concentration.
   b. Cholecystokinin from the intestinal wall stimulates the release of pancreatic juice that has a high concentration of digestive enzymes.

The Liver

1. Functions of the liver
   a. The liver is the largest gland in the body.
   b. It carries on a variety of important functions involving the metabolism of carbohydrates, lipids, and proteins; the storage of substances, the filtering of blood, the destruction of toxic chemicals, and the secretion of bile.
   c. Bile is the only liver secretion that directly affects digestion.
2. Structure of the liver
   a. The liver is a highly vascular organ, enclosed in a fibrous capsule, and divided into lobes.
   b. Each lobe contains hepatic lobules, the functional units of the liver.
   c. Bile from the lobules is carried by bile canals to hepatic ducts that unite to form the common bile duct.

3. Composition of bile
   a. Bile contains bile salts, bile pigments, cholesterol, and various electrolytes.
   b. Only the bile salts have digestive functions.
   c. Bile pigments are products of red blood cell breakdown.
4. The gallbladder and its functions
   a. The gallbladder stores bile between meals.
   b. Release of bile from the common bile duct is controlled by a sphincter muscle.
   c. Gallstones may sometimes form within the gallbladder.
5. Regulation of bile release
   a. Release of bile is stimulated by cholecystokinin from the small intestine.
   b. Sphincter muscle at the base of the common bile duct relaxes as a peristaltic wave in the duodenal wall approaches.
6. Digestive functions of bile salts
   a. Bile salts emulsify fats and aid in the absorption of fatty acids, cholesterol, and certain vitamins.
   b. Bile salts are reabsorbed in the small intestine.

The Small Intestine

The small intestine extends from the pyloric sphincter to the large intestine.

It receives secretions from the pancreas and liver, completes the digestion of nutrients, absorbs the products of digestion, and transports the residues to the large intestine.

1. Parts of the small intestine
   a. The small intestine consists of the duodenum, jejunum, and ileum.
   b. It is suspended from the posterior abdominal wall by mesentery.
2. Structure of the small intestinal wall
   a. The wall is lined with villi that increase surface area and aid in mixing and absorption.
   b. Microvilli on the free ends of epithelial cells greatly increase the surface area.
   c. Intestinal glands are located between the villi.
   d. Circular folds in the lining of the intestinal wall also increase its surface area.
3. Secretions of the small intestine
   a. Intestinal glands primarily secrete a watery fluid that lacks digestive enzymes, but provides a vehicle for moving food substances to the villi.
   b. Digestive enzymes embedded in the surfaces of microvilli can split molecules of sugars, proteins, and fats.
4. Regulation of small intestinal secretions
   a. Secretions are enhanced by the presence of gastric juice and chyme and by the mechanical stimulation of distension.
   b. A hormone from the intestinal wall may also stimulate secretions.

5. Absorption in the small intestine
   a. Monosaccharides, amino acids, fatty acids, and glycerol are absorbed by the villi.
   b. Villi also absorb water and electrolytes.
   c. Fat molecules with longer chains of carbon atoms enter the lacteals of the villi; other products of digestion enter the blood capillaries of the villi.
6. Movements of the small intestine
   a. Movements include mixing by segmentation and peristalsis.
   b. Overdistension or irritation may stimulate a peristaltic rush and result in diarrhea.
   c. The ileocecal valve controls movement from the small intestine into the large intestine.

The Large Intestine
The large intestine functions to reabsorb water and electrolytes and to form and store feces.

1. Parts of the large intestine
   a. The large intestine consists of the cecum, colon, rectum, and anal canal.
   b. The colon is divided into ascending, transverse, descending, and sigmoid portions.
2. Structure of the large intestinal wall
   a. Basically the large intestine wall is like the wall in other parts of the alimentary canal.
   b. Unique features include a layer of longitudinal muscle fibers that are arranged in distinct bands.
3. Functions of the large intestine
   a. The large intestine has little or no digestive function, although it secretes mucus.
   b. The rate of mucus secretion is controlled by mechanical stimulation and parasympathetic impulses.

   c. Absorption in the large intestine is generally limited to water and electrolytes.
   d. Many bacteria inhabit the large intestine and may aid the body by synthesizing certain vitamins.
4. Movements of the large intestine
   a. Movements are similar to those in the small intestine.
   b. Mass movements occur two to three times each day.
   c. Defecation is stimulated by a defecation reflex.
5. The feces
   a. Feces is formed and stored in the large intestine.
   b. It consists largely of water, undigested material, mucus, and bacteria.
   c. The color of feces is due to bile salts that have been altered by bacterial action.

Application of Knowledge

1. If a patient has 95% of the stomach removed (subtotal gastrectomy) as treatment for severe ulcers or cancer, how would the digestion and absorption of foods be affected? How would the patient's eating habits have to be altered? Why?
2. Why is it that a person with inflammation of the gallbladder (cholecystitis) may also develop an inflammation of the pancreas (pancreatitis)?
3. What effect is a before-dinner cocktail likely to have on digestion? Why are such beverages inadvisable for persons with ulcers?
4. What type of acid-alkaline disorder is likely to develop if the stomach contents are lost repeatedly by vomiting over a prolonged period of time? What acid-alkaline disorder is likely to develop as a result of prolonged diarrhea?

Review Activities

1. List and describe the location of the major parts of the alimentary canal.
2. List and describe the location of the accessory organs of the digestive system.
3. Name the four layers of the wall of the alimentary canal.
4. Distinguish between mixing movements and propelling movements.
5. Define *peristalsis*.
6. Explain the relationship between peristalsis and receptive relaxation.
7. Describe the general effects of parasympathetic and sympathetic impulses on the alimentary canal.
8. Discuss the functions of the mouth and its parts.
9. Distinguish between lingual, palatine, and pharyngeal tonsils.
10. Compare the deciduous and permanent teeth.
11. Explain how the various types of teeth are adapted to perform specialized functions.
12. Describe the structure of a tooth.
13. Explain how a tooth is anchored in its socket.
14. List and describe the location of the major salivary glands.
15. Explain how the secretions of the salivary glands differ.
16. Discuss the digestive functions of saliva.
17. Name and locate the three major regions of the pharynx.
18. Describe the mechanism of swallowing.
19. Explain the function of the esophagus.
20. Describe the structure of the stomach.
21. List the enzymes in gastric juice and explain the function of each enzyme.
22. Explain how gastric secretions are regulated.
23. Describe the mechanism that controls the emptying of the stomach.
24. Describe the enterogastric reflex.
25. Explain the mechanism of vomiting.
26. Describe the location of the pancreas and the pancreatic duct.
27. List the enzymes found in pancreatic juice and explain the function of each enzyme.
28. Explain how pancreatic secretions are regulated.
29. List the major functions of the liver.
30. Describe the structure of the liver.
31. Describe the composition of bile.
32. Explain how gallstones may form.
33. Define *cholecystokinin*.
34. Explain the functions of bile salts.
35. List and describe the location of the parts of the small intestine.
36. Name the enzymes of the intestinal mucosa and explain the function of each enzyme.
37. Explain how the secretions of the small intestine are regulated.
38. Describe the functions of the intestinal villi.
39. Summarize how each major type of digestive product is absorbed.
40. Explain how the movement of the intestinal contents is controlled.
41. List and describe the location of the parts of the large intestine.
42. Explain the general functions of the large intestine.
43. Describe the defecation reflex.

15

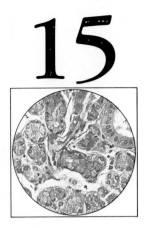

Nutrition and Metabolism

To maintain health, the human body must obtain a variety of chemicals in adequate amounts from its environment. These substances, which include carbohydrates, lipids, proteins, vitamins, and minerals, are called *nutrients.* They are needed to supply energy for cellular processes, to play vital roles in metabolic reactions, and to provide building materials for growth, maintenance, and repair of tissues. The process by which nutrients are taken in and utilized by the body is called *nutrition.*

Chapter Outline

Chapter Objectives

After you have studied this chapter, you should be able to

1. Define *nutrition, nutrients,* and *essential nutrients.*

2. List the major sources of carbohydrates, lipids, and proteins.

3. Describe how carbohydrates are utilized by cells.

4. Describe how lipids are utilized by cells.

5. Describe how amino acids are utilized by cells.

6. Define *nitrogen balance.*

7. Explain how the energy values of foods are determined.

8. Explain how various factors affect an individual's energy needs.

9. Define *energy balance.*

10. Explain what is meant by desirable weight.

11. List the fat-soluble and water-soluble vitamins.

12. Describe the general functions of each vitamin.

13. Distinguish between a vitamin and a mineral.

14. List the major minerals and trace elements.

15. Describe the general functions of each mineral and trace element.

16. Describe an adequate diet.

17. Distinguish between primary and secondary malnutrition.

18. Complete the review activities at the end of this chapter. Note that the items are worded in the form of specific learning objectives. You may want to refer to them before reading the chapter.

Key Terms

acetyl coenzyme A   (as'ĕ-til ko-en'zīm)

antioxidant   (an''te-ok'si-dant)

basal metabolic rate   (ba'sal met''ah-bol'ik rāt)

calorie   (kal'o-re)

calorimeter   (kal''o-rim'ĕ-ter)

dynamic equilibrium   (di-nam'ik e''kwĭ-lib're-um)

energy balance   (en'er-je bal'ans)

malnutrition   (mal''nu-trish'un)

mineral   (min'er-al)

nitrogen balance   (ni'tro-jen bal'ans)

nutrient   (nu'tre-ent)

triglyceride   (tri-glis'er-īd)

vitamin   (vi'tah-min)

Aids to Understanding Words

bas-, base: *bas*al metabolic rate—metabolic rate of body under resting (basal) conditions.

calor-, heat: *calor*ie—unit used in measurement of heat or energy content of foods.

carot-, carrot: *carot*ene—yellowish plant pigment responsible for the color of carrots and other yellowish plant parts.

mal-, bad: *mal*nutrition—poor nutrition resulting from lack of food or failure to use available foods to best advantage.

meter-, a measure: calori*meter*—instrument used to measure the caloric contents of foods.

nutri-, nourishing: *nutri*ent—chemical substance needed to nourish body cells.

obes-, fat: *obes*ity—condition in which body contains excessive fat.

pell-, skin: *pell*agra—vitamin deficiency condition that is characterized by dermatitis and other symptoms.

THE NUTRIENTS NEEDED by the body include the carbohydrates, lipids, proteins, vitamins, minerals, and water that are introduced in chapters 2 and 4.

As a group, these substances are obtained from foods, and some of them are altered by digestive processes after being ingested. Once they are processed by the digestive system, they are absorbed and transported to body cells by the blood. The ways in which the nutrients are changed chemically and used anabolically and catabolically to support life processes constitute **metabolism.** (The metabolic pathways of carbohydrates, lipids, and proteins are discussed in chapter 4.)

Some nutrients, such as certain amino acids and fatty acids, are particularly important because they are necessary for health and cannot be synthesized in adequate amounts by human cells. These are called **essential nutrients** since it is essential that they be provided in the diet.

Carbohydrates

Carbohydrates are organic compounds, such as sugars and starch, that are used primarily to supply energy for cellular processes.

Sources of Carbohydrates

Carbohydrates are ingested in a variety of forms, including starch from grains and certain vegetables; glycogen from meats and seafoods; disaccharides from cane sugar, beet sugar, and molasses; and simple sugars from honey and various fruits. During digestion, the more complex carbohydrates, such as starch and glycogen, are changed to monosaccharides—a form that can be absorbed and transported to body cells.

An exception to this is provided by the polysaccharide called *cellulose,* which is abundant in plant foods. Although it is composed of glucose units arranged in chains, somewhat like plant starch, cellulose cannot be broken down by human digestive enzymes; consequently, it passes through the alimentary canal largely unchanged. The presence of cellulose in foods, however, is important to the function of the digestive system. It provides bulk (sometimes called fiber or roughage) in the digestive tube against which the muscular wall can push. Thus, cellulose facilitates the movement of food substances through the digestive tract. (Other carbohydrates that provide fiber include hemicellulose, pectin, and lignin.)

Figure 15.1

The monosaccharides, fructose and galactose, are converted into glucose by the liver.

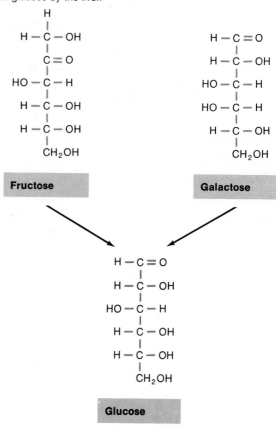

Utilization of Carbohydrates

The monosaccharides that are absorbed from the digestive tract include *fructose, galactose,* and *glucose.* Fructose and galactose are normally converted into glucose by the action of the liver (figure 15.1), and glucose is the form of carbohydrate that is most commonly oxidized by cells as fuel.

If glucose is present in excessive amounts, some of it is converted to *glycogen* (glycogenesis) and stored as a glucose reserve in the liver and muscles. Glucose can be mobilized rapidly from glycogen (glycogenolysis) when it is needed to supply energy. Only a certain amount of glycogen can be stored, however, and excess glucose is usually converted into fat and stored in adipose tissue (figure 15.2).

It is estimated that about 100 grams of glycogen can be stored in an adult liver and that another 200 grams can be stored in various muscular tissues. This amount of glycogen is sufficient to meet the energy demands of body cells for about 12 hours when the body is at rest.

Figure 15.2
Monosaccharides from foods are used to supply energy, are stored as glycogen, or are converted to fat.

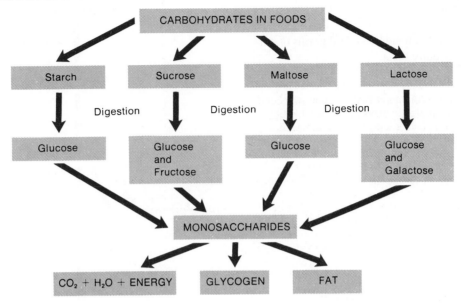

1. List several common sources of carbohydrates.
2. In what form are carbohydrates utilized as a cellular fuel?
3. Explain what happens to excessive glucose in the body.

Figure 15.3
Whenever the supply of glucose is inadequate, liver cells may convert noncarbohydrates, such as amino acids, into glucose.

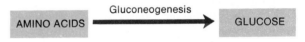

Carbohydrate Requirements

Although most carbohydrates are used to supply energy, some are used to produce vital cellular substances. These include the 5-carbon sugars *ribose* and *deoxyribose* that are needed for the synthesis of the nucleic acids RNA and DNA, as well as the disaccharide *lactose* (milk sugar) that is produced when the breasts are actively secreting milk.

Many cells also can obtain energy by oxidizing fatty acids. Some cells, however, such as the neurons of the central nervous system, are dependent on a continuous supply of glucose for survival. Even a temporary drop in the glucose supply may produce a serious functional disorder of the nervous system or the death of nerve cells. Consequently, the presence of some carbohydrates in the body is essential; if an adequate supply is not received from foods, the liver may convert noncarbohydrates, such as amino acids from proteins, into glucose (gluconeogenesis). Thus, the need for glucose has priority over the need to manufacture proteins from available amino acids (figure 15.3).

Humans survive on widely varying amounts of carbohydrates, as a result of differences in the carbohydrate content of available or preferred foods. In the United States, for example, it is estimated that a typical adult diet supplies about 50% of total body energy in the form of carbohydrates. Since carbohydrate foods are generally inexpensive, this percentage is likely to be higher among members of lower economic groups. It is possible to maintain adequate nutrition, however, with relatively high or relatively low carbohydrate intake provided the needs for all *essential nutrients* are met. Furthermore, poor nutrition among people with limited economic resources is more likely to be related to low intake of essential amino acids, vitamins, and minerals than to an excessive use of carbohydrate foods.

Since carbohydrates provide the primary source of fuel for cellular processes, the need for carbohydrates varies with individual energy requirements. Persons who are physically active require more fuel than those who are sedentary. The minimal requirement for carbohydrates in the human diet is unknown. It is estimated, however, that an intake of at least 100 grams daily is necessary to avoid excessive breakdown of protein and to avoid metabolic disorders that sometimes accompany excessive utilization of fats. Persons in the United States typically include 200–300 grams of carbohydrates in their daily diets.

A nutritionist might recommend that an adult diet contain enough carbohydrates to maintain a desirable body weight, enough proteins to supply the cellular needs for essential amino acids, enough fats to supply essential fatty acids and fat-soluble vitamins, and sufficient amounts of other essential vitamins and minerals.

1. Name two uses of carbohydrates other than supplying energy.
2. What cells are particularly dependent on a continuous supply of glucose?
3. How does the body obtain glucose when insufficient amounts of carbohydrates are ingested?
4. What is the daily minimum requirement of carbohydrates?

Lipids

As is described in chapter 2, **lipids** comprise a group of organic compounds. This group includes fats, oils, and fatlike substances that are used to supply energy for cellular processes and to build structures such as cell membranes. Although lipids include fats, phospholipids, and cholesterol, the most common dietary lipids are fats called *triglycerides,* such as those described in chapter 4 (see figure 2.19).

Sources of Lipids

Triglycerides are contained in foods of both plant and animal origin. They occur, for example, in meats, eggs, milk, and lard, as well as in various nuts and plant oils such as corn oil, peanut oil, and olive oil.

Cholesterol is obtained in relatively high concentrations from foods such as liver, egg yolk, and brain, and is present in lesser amounts in whole milk, butter, cheese, and meats. It does not occur in foods of plant origin.

Utilization of Lipids

During digestion, triglycerides are broken down into fatty acids and glycerol, and after being absorbed, these products are transported by the lymph and blood to various tissues. The metabolism of these substances is controlled mainly by the liver and the adipose tissues.

Figure 15.4
Fatty acids are used by the liver to synthesize a variety of lipids.

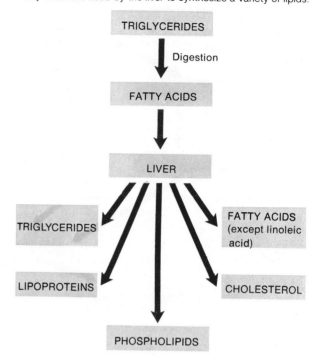

As explained in chapter 4, the liver can convert fatty acids from one form to another, but it cannot synthesize one type of fatty acid called *linoleic acid.* This substance is an **essential fatty acid** needed for the production of certain phospholipids, which in turn are necessary for the formation of cell membranes and the transport of circulating lipids. Good sources of linoleic acid include corn oil, cottonseed oil, and soy oil.

The liver uses free fatty acids in the synthesis of triglycerides, phospholipids, and lipoproteins that may then be released into the blood. Thus, the liver is largely responsible for the control of circulating lipids (figure 15.4). In addition, it is thought to regulate the total amount of cholesterol in the body by synthesizing cholesterol and releasing it into the blood, or by removing cholesterol from the blood and excreting it into the bile. The liver also uses cholesterol in the production of bile salts.

Cholesterol is not used as an energy source, but it does provide structural material for a variety of cell parts and furnishes molecular components for the synthesis of various sex hormones and hormones produced by the adrenal cortex.

Excessive triglycerides are stored in adipose tissue, and if the blood lipid concentration drops (in response to fasting, for example), some of these triglycerides are hydrolyzed into free fatty acids and glycerol, and released into the blood.

In addition to storing energy materials, adipose tissues beneath the skin function as an insulating layer. Such insulation helps reduce the rate of body heat loss during cold weather. On the other hand, it may also interfere with heat loss during warm weather and cause considerable discomfort.

Lipid Requirements

As in the case of carbohydrates, the lipid content of human diets varies widely. Thus, one person's diet may provide 10% of the body's energy in the form of fats, while another's provides 40%.

The amounts and types of fats needed for health are unknown. Linoleic acid, however, is an essential fatty acid, and to prevent deficiency conditions from developing, nutritionists recommend that infants receive formulas in which 3% of the energy intake is in the form of linoleic acid. Fatty acid deficiencies have not been observed in adults, so it has been concluded that a typical adult diet consisting of a variety of foods provides an adequate supply of this essential nutrient.

Since fats contain fat-soluble vitamins, the intake of fats also must be sufficient to supply needed amounts of these nutrients.

1. What foods commonly supply lipids?
2. Which fatty acid is an essential nutrient?
3. What is the role of the liver in the utilization of lipids?
4. What is the function of cholesterol?
5. How much lipid should be included in the diet?

Proteins

As is described in chapter 2, **proteins** are organic compounds that serve as structural materials in cells, act as enzymes that regulate metabolic reactions, and are often used to supply energy.

Sources of Proteins

Foods rich in proteins include meats, fish, poultry, cheese, nuts, milk, eggs, cereals, and various legumes such as beans and peas that contain lesser amounts of protein.

During digestion, proteins are broken down into their component amino acids, and these smaller molecules are absorbed and transported to the body cells in the blood.

As is mentioned in chapter 4, the human body can synthesize many amino acids (nonessential amino acids). However, eight amino acids needed by the adult body (ten needed by growing children) cannot be synthesized, or are not synthesized in adequate amounts. They are called **essential amino acids** because it is essential that they are present in the diet.

Chart 15.1	Amino acids found in proteins
Alanine	Leucine (e)
Arginine (ch)	Lysine (e)
Aspartic acid	Methionine (e)
Asparagine	Phenylalanine (e)
Cysteine	Proline
Glutamic acid	Serine
Glutamine	Threonine (e)
Glycine	Tryptophan (e)
Histidine (ch)	Tyrosine
Isoleucine (e)	Valine (e)
Hydroxyproline	

Eight essential amino acids (e) cannot be synthesized by human cells and must be provided in the diet. Two additional amino acids (ch) are essential in growing children.

All the essential amino acids must be present in the body at the same time for growth and tissue repair to occur. If one essential molecule is missing, the process of protein synthesis cannot take place. Since essential amino acids are not stored, those that are present and are not used in protein synthesis are oxidized as energy sources or are converted into carbohydrates or fats.

Plant proteins typically contain less than adequate amounts of one or more of the essential amino acids. However, protein-containing plant foods can be combined in a meal so that one supplies the amino acids that another lacks. For example, beans are deficient in the essential amino acid *methionine,* but they contain adequate amounts of *lysine.* Rice is deficient in lysine but contains adequate amounts of *methionine.* Thus, if beans and rice are eaten together, they complement one another, and the meal provides a suitable combination of essential amino acids.

Chart 15.1 lists the amino acids found in foods and indicates those that are essential.

On the basis of the kinds of amino acids that they provide, proteins can be classified as complete or incomplete. The **complete proteins,** which include those available in milk, meats, fish, poultry, and eggs, contain adequate amounts of the essential amino acids to maintain body tissues and promote normal growth and development. **Incomplete proteins,** such as *zein* in corn, which lacks the essential amino acids tryptophan and lysine, and *gelatin,* which lacks tryptophan, are unable by themselves to support tissue maintenance or normal growth and development.

A protein called *gliadin,* which occurs in wheat, is an example of a *partially complete protein.* It is deficient in the essential amino acid lysine, and although it does not contain enough lysine to promote growth, it does contain enough to maintain life.

Figure 15.5

Amino acids resulting from the digestion of proteins are used to synthesize nonessential amino acids and proteins, and to supply energy.

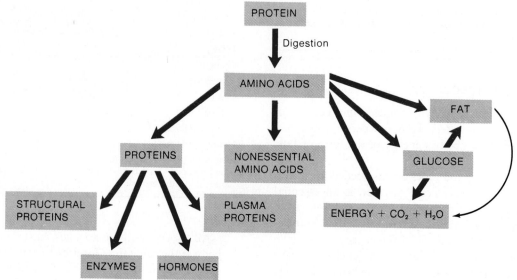

1. What foods provide rich sources of proteins?
2. Why are some amino acids called essential?
3. Distinguish between a complete protein and an incomplete protein.

Utilization of Amino Acids

As is discussed in chapter 4, amino acids are used by body cells in a variety of ways. For example, some amino acids are incorporated into protein molecules that provide cellular structure, as in the *actin* and *myosin* of muscle fibers, the *collagen* of connective tissue fibers, and the *keratin* of skin. Other amino acids are used to synthesize proteins that function in various body processes: hemoglobin, which transports oxygen; plasma proteins, which help to regulate water balance and control pH; enzymes, which catalyze metabolic reactions; and hormones, which help regulate metabolic activities.

Amino acids also represent potential sources of energy. When they are present in excess or when supplies of carbohydrates and fats are insufficient to provide needed energy, amino acids are oxidized by various metabolic pathways (figure 15.5).

Body cells use carbohydrates preferentially as an energy source, and so a diet that supplies an adequate amount of this nutrient is said to have a *protein-sparing effect*. In other words, when sufficient carbohydrates are present, proteins remain available for tissue building and repair rather than being converted into carbohydrates for use as an energy source.

A relatively high proportion of tissue proteins may be used as energy sources during times of starvation. These molecules are drawn from structural parts of cells and, consequently, such utilization of structural proteins is accompanied by wasting tissues.

Nitrogen Balance

In a healthy adult, proteins are continuously being built up and broken down. Although these processes occur at different rates in different tissues, the overall gain of body proteins equals the loss, so that a state of *dynamic equilibrium* exists. Since proteins contain a relatively high percentage of nitrogen, dynamic equilibrium is also characterized by a **nitrogen balance**—a condition in which the amount of nitrogen taken in is equal to the amount excreted.

A person who is starving, however, will have a *negative nitrogen balance* because the amount of nitrogen excreted as a result of amino acid oxidation will exceed the amount replaced by the diet. A growing child, a pregnant woman, or an athlete in training is likely to have a *positive nitrogen balance,* since the amount of protein being built into new tissue probably exceeds the amount being used for energy.

Protein Requirements

In addition to supplying essential amino acids, proteins are needed to provide molecular parts and nitrogen for the synthesis of nonessential amino acids and various nonprotein nitrogenous substances. Consequently, the amount of protein required by individuals varies according to body size, metabolic rate, and nitrogen balance condition.

Chart 15.2 Protein, lipid, and carbohydrate nutrients

Nutrient	Sources	Calories per gram	Utilization	Adult daily requirement
Protein	Meats, fish, poultry, cheese, nuts, milk, eggs, cereals, legumes	4.1	Production of protein molecules used to build cell structure and to function as enzymes or hormones; used in the transport of oxygen, regulation of water balance, control of pH; amino acids may be broken down and oxidized for energy or converted to carbohydrates or fats for storage	0.8 grams per kilogram of body weight; diet must include the essential amino acids
Lipid	Meats, eggs, milk, lard, plant oils	9.5	Oxidized for energy; production of triglycerides, phospholipids, lipoproteins, and cholesterol; stored in adipose tissue; glycerol portions of fat molecules may be used to synthesize glucose	Amount and type needed for health unknown; diet must supply an adequate amount of essential fatty acids and fat-soluble vitamins
Carbohydrates	Primarily from starch and sugars in foods of plant origin, and from glycogen in meats and seafoods	4.1	Oxidized for energy; used in production of ribose, deoxyribose, and lactose; stored in liver and muscles as glycogen; converted to fats and stored in adipose tissue	Varies with individual energy needs; at least 100 grams per day to prevent excessive breakdown of proteins

For an average adult, nutritionists recommend a daily protein intake of about 0.8 grams per kilogram (0.4 grams per pound) of body weight. For a pregnant woman, who needs to maintain a positive nitrogen balance, the recommendation is increased by an additional 30 grams of protein per day. Similarly, a nursing mother requires an additional 20 grams of protein per day to maintain a high level of milk production.

The consequences of protein deficiencies are severe, particularly among growing children. As mentioned previously, a decreased protein intake is likely to result in a negative nitrogen balance and tissue wasting. It may also be accompanied by a decrease in the level of plasma proteins, and if this occurs, the osmotic pressure of the blood decreases. As a result of this osmotic change, fluids collect excessively in the tissues, producing a condition called *nutritional edema*.

Chart 15.2 summarizes the sources, uses, and requirements of protein, lipid, and carbohydrate nutrients.

Protein deficiencies in pregnant women are likely to result in anemia, miscarriages, or premature births. Children whose diets are inadequate in proteins tend to develop a condition called *kwashiorkor*, which is characterized by nutritional edema and failure to grow (figure 15.6). The nervous systems of such children may also fail to develop, so they may be mentally retarded.

Figure 15.6
How can the condition known as kwashiorkor be prevented?

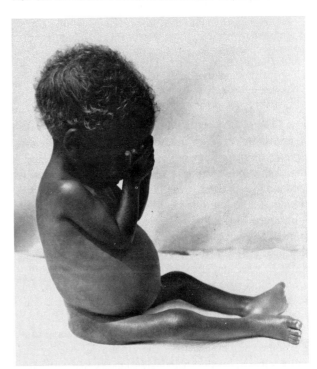

1. What are the physiological functions of proteins?
2. What is meant by a negative nitrogen balance? A positive nitrogen balance?
3. How can edema result from inadequate nutrition?

Energy Expenditures

Energy can be obtained from carbohydrates, fats, and proteins. Since it is required for all metabolic processes, the supply of energy is of prime importance to cell survival. If the diet is deficient in energy-supplying materials, structural molecules may gradually be consumed; and if this continues, death may result.

Energy Values of Foods

The amount of potential energy contained in a food can be expressed as **calories,** which are units of heat.

Although a calorie is defined as the amount of heat needed to raise the temperature of a gram of water by 1 degree Celsius (°C), the calorie used in the measurements of food energy is 1000 times greater. This *large calorie* (Cal.) is equal to the amount of heat needed to raise the temperature of a kilogram (1000 gm) of water by 1°C (actually from 15°C to 16°C). This unit is also called a *kilocalorie,* but it is customary in nutritional studies to refer to it simply as a calorie.

The American Institute of Nutrition and the International Congress of Nutrition have recommended that a unit of energy called the *joule* (J.) be used in place of the calorie. Therefore, an increasing use of this unit to express the energy values of foods is expected to occur. One kilocalorie is equal to 4.184 kilojoules (kJ.).

The caloric contents of various foods can be determined by using an instrument called a *bomb calorimeter,* shown in figure 15.7, that consists of a metal chamber submerged in a known volume of water. A food sample is dried, weighed, and placed inside the chamber. The chamber is filled with oxygen gas and submerged in the water. Then, the food is ignited and allowed to oxidize completely. As heat is released from the food, it causes the temperature of the surrounding water to rise, and the change in temperature is measured. Since the volume of the water is known, the amount of heat released from the food can be calculated in calories.

The caloric values determined this way are usually somewhat higher than the amount of energy actually released by metabolic oxidation because nutrients

Figure 15.7

A bomb calorimeter can be used to measure the caloric content of a food sample.

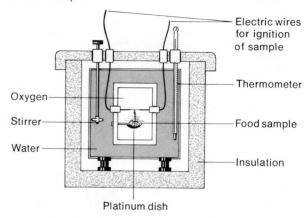

generally are not completely absorbed from the digestive tract. Also, molecules of amino acids in proteins are not completely oxidized in the body. Portions of these molecules, which still contain some energy, are excreted in urea or are transformed into other nitrogenous substances. When such losses are taken into account, cellular oxidation yields on the average about 4.1 calories from 1 gram of carbohydrates, 4.1 calories from 1 gram of protein, and 9.5 calories from 1 gram of fat.

---

1. What term is used to designate the potential energy in a food substance?
2. How can the energy value of a food be determined?
3. What is the energy value of a gram of carbohydrate? A gram of protein? A gram of fat?

Energy Requirements

The amount of energy required to support metabolic activities for 24 hours varies from person to person. The factors that influence energy needs include the individual's basal metabolic rate, degree of muscular activity, body temperature, and rate of growth.

The **basal metabolic rate** (BMR) is a measurement of the rate at which the body expends energy under *basal conditions*—the conditions that exist when the body is at rest, after an overnight fast, and in a comfortable environment. The apparatus used to determine this rate measures the amount of oxygen taken in by a resting person as well as the amount of carbon dioxide given off during a certain period of time (figure 15.8).

The amount of oxygen consumed by the body is directly proportional to the amount of energy released by cellular respiration. The BMR, therefore, reveals the

Figure 15.8
The basal metabolic rate is determined by measuring the rate at which a person uses oxygen.

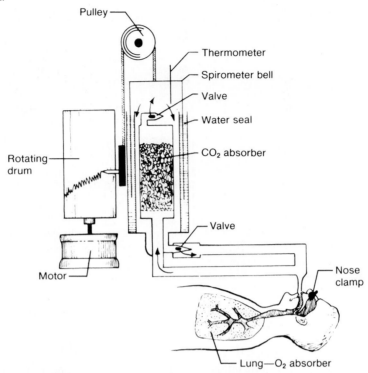

total amount of energy expended in a given time period to support the maintenance activities of such organs as the brain, heart, lungs, liver, and kidneys.

Although the average adult basal metabolic rate indicates a need for approximately one calorie of energy per hour for each kilogram of body weight, this need varies with such factors as sex, body size, body temperature, and level of endocrine gland activity. For example, since heat loss is directly proportional to the body surface area, and a smaller person has a relatively larger surface area, such a person will have a higher BMR. Males tend to have higher rates than females; as body temperature increases, the BMR increases; and as the blood level of thyroxine or epinephrine increases, so does the BMR.

Maintaining the basal metabolic rate usually requires the body's greatest expenditure of energy. The energy required to support voluntary muscular activity comes next, though this amount varies greatly with the type of activity (chart 15.3). For example, energy needed to maintain posture while sitting at a desk might amount to 100 Cal. per hour above the basal need, while running or swimming might require 500–600 Cal. per hour.

Chart 15.3 Calories used during various activities	
Kinds of activity	Calories (per hour)
Walking up stairs	1100
Running (jogging)	570
Swimming	500
Vigorous exercise	450
Slow walking	200
Dressing and undressing	118
Sitting at rest	100

The maintenance of body temperature may require additional energy expenditure, particularly in cold weather. In this case, extra energy may be released by involuntary muscular contractions, such as shivering, or through voluntary muscular actions.

As mentioned in the discussion of nitrogen balance, growing children and pregnant women have special needs for nutrients because their bodies are actively engaged in the production of new tissues. They also require increased caloric intake, since tissue building uses energy.

Chart 15.4 Recommended daily dietary allowances of calories for maintenance of good nutrition

	Years (from–to)	Weight (kg)	(lb)	Height (cm)	(in)	Energy (kcal)
Infants	0–6 mos	6	13	60	24	kg × 115
	6 mos–1	9	20	71	28	kg × 105
Children	1–3	13	29	90	35	1300
	4–6	20	44	112	44	1700
	7–10	28	62	132	52	2400
Males	11–14	45	99	157	62	2700
	15–18	66	145	176	69	2800
	19–22	70	154	177	70	2900
	23–50	70	154	178	70	2700
	51–75	70	154	178	70	2400
	76+	70	154	178	70	2050
Females	11–14	46	101	157	62	2200
	15–18	55	120	163	64	2100
	19–22	55	120	163	64	2100
	23–50	55	120	163	64	2000
	51–75	55	120	163	64	1800
	76+	55	120	163	64	1600
Pregnant						+300
Lactating						+500

Source: Food and Nutrition Board, *Recommended Dietary Allowances,* 9th ed., National Academy of Sciences—National Research Council, Washington, D.C., 1980, p. 23.

Energy Balance

A state of **energy balance** exists when the caloric intake in the form of foods is equal to the caloric output resulting from the basal metabolic rate, muscular activities, and so forth. Under these conditions, the body weight remains constant, except perhaps for slight variations due to changes in water content.

If, on the other hand, the caloric intake exceeds the output, a *positive* energy balance occurs and the excessive nutrients are likely to be stored in the tissues. At the same time, the body weight increases, since an excess of 3500 Cal. can be stored as a pound of fat. Conversely, if the caloric output exceeds the input, the energy balance is *negative,* and stored materials are mobilized from the tissues for oxidation, causing body weight loss. Chart 15.4 gives the number of Cal. recommended (by height, weight, and age) for maintaining energy balance.

---

1. What is meant by basal metabolic rate?
2. What factors influence the BMR?
3. What is meant by *energy balance?*

Desirable Weight

The most obvious and common nutritional disorders involve caloric imbalances. Persons in underdeveloped countries, for example, are often undernourished, and obesity is rare. Conversely, in highly developed countries, such as the United States, people are likely to be overnourished and overweight.

It is difficult to determine what is a desirable body weight. In the past, weight standards have been based on *average* weights and heights within a certain population, and the degrees of underweight and overweight have been expressed as percentage deviations from these averages. These standards reflected the gradual gain in weight that usually occurs in members of the United States population as they grow older. Later, it was recognized that such an increase in weight after the ages of 25 to 30 years is not necessary and may not be conducive to health. Consequently, standards of *desirable weights* were prepared, as shown in charts 15.5 and 15.6.

These figures were based on the assumption that the average body weights for persons 25 to 30 years of age are desirable weights for all older persons, and it was recommended that each individual maintain this desirable weight throughout life.

Chart 15.5 Desirable weights for women 25 years of age and over*†
Weight in pounds according to frame (in indoor clothing and two-inch heels)

Height		Small frame	Medium frame	Large frame
Feet	*Inches*			
4	10	92–98	96–107	104–119
4	11	94–101	98–110	106–122
5	0	96–104	101–113	109–125
5	1	99–107	104–116	112–128
5	2	102–110	107–119	115–131
5	3	105–113	110–122	118–134
5	4	108–116	113–126	121–138
5	5	111–119	116–130	125–142
5	6	114–123	120–135	129–146
5	7	118–127	124–139	133–150
5	8	122–131	128–143	137–154
5	9	126–135	132–147	141–158
5	10	130–140	136–151	145–163
5	11	134–144	140–155	149–168
6	0	138–148	144–159	153–173

*Metropolitan Life Insurance Company, New York, 1959.
†For women between 18 and 25, subtract 1 pound for each year under 25.

Chart 15.6 Desirable weights for men 25 years of age and over*
Weight in pounds according to frame (in indoor clothing and one-inch heels)

Height		Small frame	Medium frame	Large frame
Feet	*Inches*			
5	2	112–120	118–129	126–141
5	3	115–123	121–133	129–144
5	4	118–126	124–136	132–148
5	5	121–129	127–139	135–152
5	6	124–133	130–143	138–156
5	7	128–137	134–147	142–161
5	8	132–141	138–152	147–166
5	9	136–145	142–156	151–170
5	10	140–150	146–160	155–174
5	11	144–154	150–165	159–179
6	0	148–158	154–170	164–184
6	1	152–162	158–175	168–189
6	2	156–167	162–180	173–194
6	3	160–171	167–185	178–199
6	4	164–175	172–190	182–204

*Metropolitan Life Insurance Company, New York, 1959.

Chart 15.7 Height-weight guidelines for adult women based on longevity*
Weight in pounds according to frame (in 3 pounds of clothes and shoes with one-inch heels)

Height		Small frame	Medium frame	Large frame
Feet	Inches			
4	10	102–111	109–121	118–131
4	11	103–113	111–123	120–134
5	0	104–115	113–126	122–137
5	1	106–118	115–129	125–140
5	2	108–121	118–132	128–143
5	3	111–124	121–135	131–147
5	4	114–127	124–138	134–151
5	5	117–130	127–141	137–155
5	6	120–133	130–144	140–159
5	7	123–136	133–147	143–163
5	8	126–139	136–150	146–167
5	9	129–142	139–153	149–170
5	10	132–145	142–156	152–173
5	11	135–148	145–159	155–176
6	0	138–151	148–162	158–179

*Courtesy of the Metropolitan Life Insurance Company, New York, 1983. Source of basic data: 1979 Build Study, Society of Actuaries and Association of Life Insurance Medical Directors of America, 1980.

Chart 15.8 Height-weight guidelines for adult men based on longevity*
Weight in pounds according to frame (in 5 pounds of clothes and shoes with one-inch heels)

Height		Small frame	Medium frame	Large frame
Feet	Inches			
5	2	128–134	131–141	138–150
5	3	130–136	133–143	140–153
5	4	132–138	135–145	142–156
5	5	134–140	137–148	144–160
5	6	136–142	139–151	146–164
5	7	138–145	142–154	149–168
5	8	140–148	145–157	152–172
5	9	142–151	148–160	155–176
5	10	144–154	151–163	158–180
5	11	146–157	154–166	161–184
6	0	149–160	157–170	164–188
6	1	152–164	160–174	168–192
6	2	155–168	164–178	172–197
6	3	158–172	167–182	176–202
6	4	162–176	171–187	181–207

*Courtesy of the Metropolitan Life Insurance Company, New York, 1983. Source of basic data: 1979 Build Study, Society of Actuaries and Association of Life Insurance Medical Directors of America, 1980.

Obesity
A Practical Application

One of the more serious and growing nutritional problems in the United States is obesity—the accumulation of excessive body fat. Although the indirect causes of this condition are complex and may involve physiological as well as psychological and genetic factors, the direct cause usually is overeating. That is, the person has maintained a positive energy balance by taking in more nutrients than were needed to supply the body's energy requirements. The excessive nutrients were stored as fat, and over a period of time the person became obese.

Successful treatment of obesity involves maintaining a negative energy balance—taking in less food than is needed to supply the body's energy requirements. Since a pound of fat represents the storage of about 3500 Cal., the person's diet must have a deficit of 3500 Cal. in order for a pound of fat to be lost. Thus, if the daily intake provides a deficit of 500 Cal., one pound of fat should be lost in a week.

Some fad diets cause more rapid reduction of weight by promoting the loss of body water. Unfortunately for the obese person, such a weight loss has little or no effect on stored fat and usually lasts for only a short time.

Obesity is associated with a number of adverse health conditions. These include heart disease, high blood pressure, strokes, noninsulin-dependent diabetes mellitus, damage to weight-bearing bones and associated joints, arthritis, gallstones, and certain types of cancer. Obese persons also tend to have shortened life spans.

Figure 15.9

(a) An obese person is overweight and has excess adipose tissue. *(b)* An athlete may be overweight due to muscular hypertrophy, but would not be considered obese.

(a)

(b)

More recently, height-weight guidelines have been prepared that are based upon the characteristics of people who live the longest. The weights in charts 15.7 and 15.8 tend to be somewhat more lenient than those in the desirable weight charts.

Using desirable weight as the standard, *overweight* can be defined as the condition in which a person exceeds the desirable weight by 10 to 20%. A person who exceeds this standard by more than 20% is usually *obese,* though **obesity** is more correctly defined as the presence of excessive adipose tissue. In other words, being overweight and being obese are not the same. For example, as figure 15.9 shows, an athlete or a person whose work involves heavy muscular activity may be overweight, but not obese. Conversely, a sedentary person may be excessively fat, and therefore obese, but not overweight.

Thus, determining what body weight a particular person should maintain for optimal health is a complex problem and, in fact, the average weights and heights for clinically healthy persons in the United States are not known.

1. What is meant by *desirable weight?*
2. Distinguish between overweight and obesity.

Figure 15.10
A molecule of beta-carotene can be converted into two
molecules of retinal, which in turn can be changed into retinol.

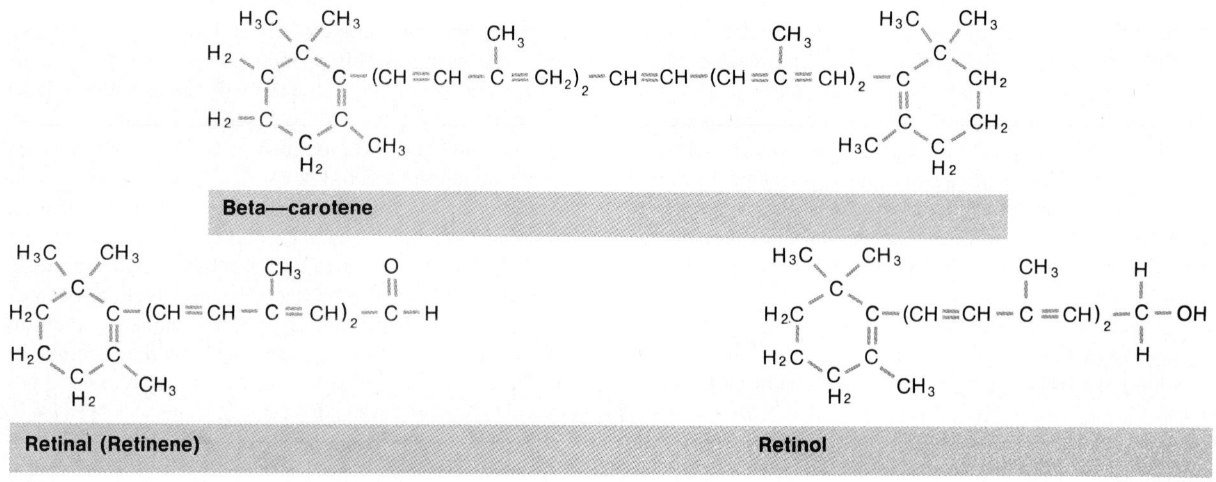

Beta—carotene

Retinal (Retinene)

Retinol

Vitamins

Vitamins are organic compounds (other than carbohydrates, lipids, and proteins) that must be present in
small amounts for normal metabolic processes, but
cannot be synthesized in adequate amounts by body
cells. They are, then, essential nutrients that must be
supplied in foods.

For convenience, vitamins are often grouped on
the basis of their solubilities, since some are soluble in
fats (or fat solvents) and others are soluble in water.
Those that are *fat-soluble* include vitamins A, D, E,
and K; the *water-soluble* group includes the B vitamins
and vitamin C.

Fat-Soluble Vitamins

Since the fat-soluble vitamins dissolve in fats, they
occur in association with lipids and are influenced by
the same factors that affect lipid absorption. For example, the presence of bile salts in the intestine promotes the absorption of these compounds. As a group,
the fat-soluble vitamins are stored in moderate quantities within various tissues, and because they are fairly
resistant to the effects of heat, they usually are not destroyed by cooking or food processing.

1. What are vitamins?
2. Distinguish between fat-soluble vitamins and water-
   soluble vitamins.
3. How do bile salts affect the absorption of vitamins?

Vitamin A occurs in several forms, including retinol and retinal (retinene). This vitamin can be synthesized by body cells from a group of yellowish plant
pigments called *carotenes*. Whenever vitamin A or its
precursors are present in excess, they are stored mainly
in the liver, which functions to regulate the concentrations of these substances within the body (figure 15.10).

It is estimated that the liver of a healthy adult stores
enough vitamin A to supply the body needs for a year.
Infants and children, on the other hand, usually lack such
reserves, and consequently they are more likely to develop vitamin A deficiencies if their diets are inadequate.

Vitamin A is relatively stable to the effects of
heat, acids, and alkalis, but it is readily destroyed by
oxidation and is unstable in the presence of light.

Although functions of vitamin A are not fully
understood, it is known that retinal is used in the synthesis of *rhodopsin* (visual purple) in the rods of the
retina. Vitamin A is thought to be required for the production of light-sensitive pigments in the cones as well.
Thus, this vitamin is needed to maintain normal vision.
It is also believed to function in the synthesis of mu-
coproteins and mucopolysaccharides of mucus, in the
development of normal bones and teeth, and in the
maintenance of epithelial cells in skin and mucous
membranes.

Vitamin A as such is obtained only from foods of
animal origin such as liver, fish, whole milk, butter, and
eggs. However, its precursor, carotene, is widespread in
leafy green vegetables and in yellow or orange vegetables and fruits.

A deficiency of vitamin A is characterized by a condition called *night blindness,* in which a person is unable to see normally in dim light. Also, various epithelial tissues tend to undergo degenerative changes, and as a result, the body becomes more susceptible to infections by microorganisms.

An overdose of vitamin A may produce a serious condition, called *hypervitaminosis A,* that is characterized by peeling of the skin, loss of hair, nausea, headache, and dizziness. In chronic cases, growth may be inhibited, and the bones and joints may undergo degenerative changes.

1. What substance is used by body cells to synthesize vitamin A?
2. What conditions will destroy vitamin A?
3. What foods are good sources of vitamin A?

Vitamin D is a group of sterol compounds that have similar properties. One of these substances, vitamin D_3 (cholecalciferol), is found in foods such as milk, egg yolk, and fish liver oils. Vitamin D_2 (ergocalciferol) is produced commercially by exposing a sterol obtained from yeasts (ergosterol) to ultraviolet light.

As explained in chapter 13, vitamin D also can be synthesized from dietary cholesterol that has been converted to provitamin D by intestinal enzymes, stored in the skin, and exposed to ultraviolet light.

Like other fat-soluble vitamins, vitamin D is resistant to the effects of heat and is also resistant to oxidation, acids, and alkalis. It is stored primarily in the liver and occurs to a lesser extent in the skin, brain, spleen, and bones.

When it is needed, stored vitamin D can be changed by the liver into a form (hydroxycholecalciferol) that is transported in the blood. When parathyroid hormone is present, this form of vitamin D is converted in the kidneys into an active form of the vitamin (dihydroxycholecalciferol). It, in turn, is carried in the blood as a hormone to the intestine where it promotes the absorption of calcium and phosphorus and thus, helps to ensure that adequate amounts of these minerals are available in the blood for tooth and bone formation and various metabolic processes.

Although natural foods are generally poor sources of vitamin D, it is often added to food during processing. Homogenized milk, nonfat milk, and evaporated milk, for example, are typically fortified with vitamin D. (Note: *fortified* means essential nutrients have been added to a food in which they originally were absent or present in lesser amounts; *enriched* means essential nutrients have been at least partially replaced in a food that has lost nutrients as a result of processing.)

Figure 15.11
Rickets, which is characterized by failure of the bones and teeth to develop normally, is caused by a deficiency of vitamin D.

The presence of excessive amounts of vitamin D produces a toxic condition, called *hypervitaminosis D,* that is characterized by diarrhea, nausea, and weight loss. Over prolonged periods, overdoses of vitamin D can cause calcification of various soft tissues and irreversible damage to the kidneys.

In children, a deficiency of vitamin D results in the condition called *rickets,* in which the bones and teeth fail to develop normally (figure 15.11). In adults or in elderly persons who have a minimal exposure to sunlight, such a deficiency may be accompanied by *osteomalacia,* a condition in which the bones undergo decalcification and become weakened due to disturbances in calcium and phosphorus metabolism.

1. Where is vitamin D stored?
2. What are the functions of vitamin D?
3. What foods are good sources of vitamin D?

Vitamin E includes a group of compounds. The most active of these, metabolically, is *alphatocopherol.* This substance is resistant to the effects of heat, acids, and visible light, but is unstable in alkalis and in the presence of ultraviolet light or oxygen. In

Chart 15.9 Fat-soluble vitamins

Vitamin	Characteristics	Functions	Sources
Vitamin A	Occurs in several forms; synthesized from carotenes; stored in liver; stable in heat, acids, and alkalis; unstable in light	Necessary for synthesis of visual pigments, mucoproteins, and mucopolysaccharides; for normal development of bones and teeth; and for maintenance of epithelial cells	Liver, fish, whole milk, butter, eggs, leafy green vegetables, and yellow and orange vegetables and fruits
Vitamin D	A group of sterols; resistant to heat, oxidation, acids, and alkalis; stored in liver, skin, brain, spleen, and bones	Promotes absorption of calcium and phosphorus; promotes development of teeth and bones	Produced in skin exposed to ultraviolet light; in milk, egg yolk, fish liver oils, fortified foods
Vitamin E	A group of compounds; resistant to heat and visible light; unstable in presence of oxygen and ultraviolet light; stored in muscles and adipose tissue	An antioxidant; prevents oxidation of vitamin A and polyunsaturated fatty acids; may help maintain stability of cell membranes	Oils from cereal seeds, salad oils, margarine, shortenings, fruits, nuts, and vegetables
Vitamin K	Occurs in several forms; resistant to heat, but destroyed by acids, alkalis, and light; stored in liver	Needed for synthesis of prothrombin; needed for blood clotting	Leafy green vegetables, egg yolk, pork liver, soy oil, tomatoes, cauliflower

fact, it combines so readily with oxygen that it may prevent the oxidation of other compounds, and for this reason it is said to be an **antioxidant.**

Although vitamin E is present in all tissues, it is stored primarily in the muscles and adipose tissues. It also occurs in relatively high concentrations in the pituitary and adrenal glands.

The precise functions of vitamin E are unknown, but it apparently plays a role as an antioxidant by inhibiting the oxidation of vitamin A in the digestive tract and of polyunsaturated fatty acids in the tissues. It is also thought to help maintain the stability of cell membranes.

While vitamin E is widely distributed among foods, its richest sources are oils from cereal seeds such as wheat germ oil. Other good sources are salad oils, margarine, shortenings, fruits, nuts, and vegetables. Since this vitamin is so easily obtained, deficiency conditions are rare.

Vitamin E does not pass readily through the placental membranes that separate the blood of a fetus from the blood of its mother. Consequently, a newborn child usually has a relatively low reserve of this nutrient.

---

1. Where is vitamin E stored?
2. What are the functions of vitamin E?
3. What foods are good sources of vitamin E?

Vitamin K, like the other fat-soluble vitamins, occurs in several chemical forms. One of these, called vitamin K_1 (phylloquinone), is found in foods, while another, called vitamin K_2, is produced by bacteria (*Escherichia coli*) that normally inhabit the human intestinal tract.

These vitamins are resistant to the effects of heat, but are destroyed by oxidation or by exposure to acids, alkalis, or light. They are stored to a limited degree in the liver.

Vitamin K functions primarily in the liver, where it is necessary for the formation of *prothrombin,* a protein required for blood clotting. Consequently, a deficiency of vitamin K is characterized by prolonged clotting time and a tendency to hemorrhage.

The richest sources of vitamin K are leafy green vegetables. Other good sources are egg yolk, pork liver, soy oil, tomatoes, and cauliflower.

The fat-soluble vitamins and their properties are summarized in chart 15.9.

Newborns sometimes develop vitamin K deficiencies and experience abnormal bleeding because their intestines lack bacteria that can synthesize this vitamin. Similarly, adults who have been treated with antibiotic drugs may develop a vitamin K deficiency since such drugs often interfere with the activities of intestinal bacteria.

---

1. Where in the body is vitamin K synthesized?
2. What is the function of vitamin K?
3. What foods are good sources of vitamin K?

and various amino acids, the conversion of tryptophan to niacin, the production of antibodies, and the synthesis of nucleic acids.

Since the functions of vitamin B_6 are generally related to the metabolism of nitrogen-containing substances, the requirement for this vitamin varies with the protein content of the diet rather than with the caloric intake.

The recommended daily allowance of vitamin B_6 is 2.0 mg, but because this substance is so widespread in foods, deficiency conditions are quite rare.

Good sources of vitamin B_6 include liver, meats, fish, poultry, bananas, avocados, beans, peanuts, whole grain cereals, and egg yolk.

5. **Pantothenic acid.** Pantothenic acid is a yellowish oil that is destroyed by heat, acids, and alkalis. It functions as part of a complex molecule called *coenzyme A,* which, in turn, reacts with intermediate products of carbohydrate and fat metabolism to become *acetyl coenzyme A.* Pantothenic acid is therefore essential to the energy-releasing mechanisms of cells.

A daily adult intake of 4–7 mg seems to be adequate. Most diets apparently provide sufficient amounts, since deficiencies are rare, and no clearly defined set of deficiency symptoms is known.

Good sources of pantothenic acid include meats, fish, whole grain cereals, legumes, milk, fruits, and vegetables.

6. **Cyanocobalamin** or **vitamin B_{12}.** Cyanocobalamin has a complex molecular structure that contains a single atom of the element *cobalt* (figure 15.15). In its pure form, this vitamin is red. It is stable to the effects of heat, but is inactivated by exposure to light, or by the presence of strong acids or alkalis.

The absorption of cyanocobalamin is regulated by the secretion of *intrinsic factor* from the parietal cells of the gastric glands, as explained in chapter 14. Intrinsic factor is thought to combine with cyanocobalamin and to facilitate its transfer through the epithelial lining of the small intestine and into the blood. Although the details of this transport mechanism are not well understood, it is known that calcium ions must be present for the process to take place.

Cyanocobalamin is stored in various tissues, particularly those of the liver, and an average adult body is thought to contain a reserve sufficient to supply cellular needs for 3 to 5 years.

This vitamin is essential for the functions of all cells. It serves as a part of coenzymes needed for the synthesis of nucleic acids and the metabolism of carbohydrates and fats. It also seems to play a role in the formation of myelin within the central nervous system.

Figure 15.15
Vitamin B_{12}, which has the most complex molecular structure of the vitamins, contains cobalt (Co). What is the function of this vitamin?

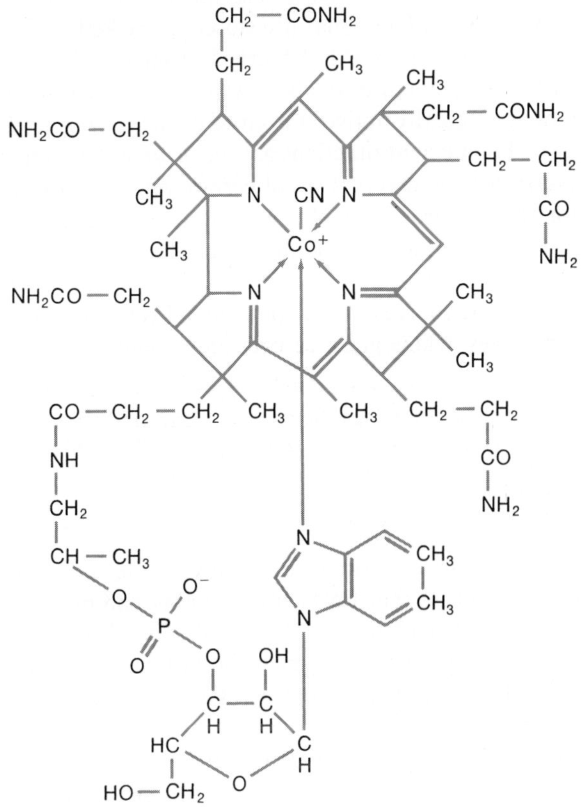

Vitamin B_{12} (cyanocobalamin)

Cyanocobalamin is found only in foods of animal origin, and good sources include liver, meats, poultry, fish, milk, cheese, and eggs. Since persons in the United States consume relatively large amounts of such foods, a dietary lack of this vitamin seldom occurs, although strict vegetarians may develop a deficiency.

A cyanocobalamin deficiency may develop in association with a disorder of the absorption mechanism. Specifically, the gastric glands of some individuals fail to secrete adequate amounts of intrinsic factor. As a result, the vitamin is poorly absorbed. This leads to a condition called *pernicious anemia* that is characterized by the presence of abnormally large red blood cells called *macrocytes.* These cells are produced when bone marrow cells fail to divide properly because of defective synthesis of DNA.

7. **Folacin** or **folic acid.** Folacin is a yellow crystalline compound that exists in several forms. It is easily oxidized in an acid environment and is destroyed by heat in alkaline solutions; consequently, this vitamin may be lost in foods that are stored or cooked.

Folacin is readily absorbed from the digestive tract and is stored in the liver, where it is converted to a physiologically active substance called *folinic acid.*

Folinic acid functions as a coenzyme that is necessary for the metabolism of certain amino acids and for the synthesis of DNA. It also acts together with cyanocobalamin in promoting the production of normal red blood cells.

Good sources of folacin include liver, leafy green vegetables, whole grain cereals, and legumes.

A deficiency of folacin leads to *megaloblastic anemia,* which is characterized by a reduction in the number of normal red blood cells and the presence of large, nucleated red cells.

---

8. **Biotin.** Biotin is a relatively simple compound that is stable to the effects of heat, acids, and light, but may be destroyed by oxidation or alkalis. (The molecular structure of biotin is included in figure 15.18.)

Biotin functions as a necessary coenzyme in a number of metabolic pathways, including those involved with the metabolism of amino acids and fatty acids. It also plays a role in the synthesis of *purines,* which are essential in the synthesis of nucleic acids.

Small quantities of biotin are stored in the tissues of metabolically active organs such as the brain, liver, and kidneys. It is believed to be manufactured by bacteria that inhabit the intestinal tract.

Biotin is widely distributed in foods, and dietary deficiencies are relatively rare. Good sources include liver, egg yolk, nuts, legumes, and mushrooms.

---

1. What substances make up the vitamin B complex?
2. What foods are good sources of vitamin B complex?
3. Which of the B-complex vitamins can be synthesized from tryptophan?
4. What is the general function of each member of the B complex?

Vitamin C or **ascorbic acid.** Ascorbic acid is a crystalline compound that contains six carbon atoms. Chemically, it is closely related to the monosaccharides (figure 15.16). It is one of the least stable of the vitamins, in that it can be destroyed by oxidation, heat, light, or alkalis. It is fairly stable in acids.

The functions of ascorbic acid are poorly understood. It is known, however, to be necessary for the production of the protein, *collagen,* which is a vital part of various connective tissues such as bone, cartilage, and the fibrous connective tissue of skin. It is needed

Figure 15.16
Vitamin C is closely related chemically to the monosaccharides.

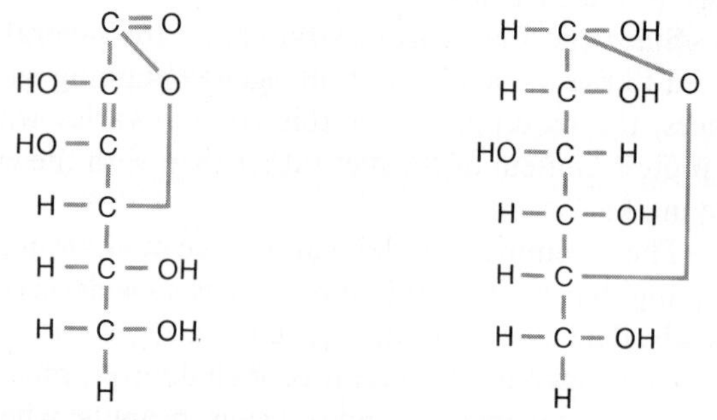

Vitamin C (ascorbic acid)

Glucose (a monosaccharide)

in the conversion of folacin to folinic acid and in the metabolism of certain amino acids. It also promotes the absorption of iron and the synthesis of various hormones from cholesterol.

Although this vitamin is not stored in any great amount, tissues such as the adrenal cortex, pituitary gland, and intestinal glands contain relatively high concentrations. Excessive quantities are either excreted in the urine or oxidized.

There is considerable controversy as to how much ascorbic acid is required to maintain optimal health, and individual needs may vary. It is believed that 10 mg per day is sufficient to prevent deficiency symptoms and that 80 mg per day will saturate the tissues within a few weeks. Consequently, many nutritionists generally recommend a daily adult intake of 60 mg, which is thought to be enough to replenish normal losses and to provide a satisfactory level for cellular needs.

Ascorbic acid is fairly widespread in plant foods. Those that provide particularly high concentrations include citrus fruits, citrus juices, tomatoes, and cabbage. Potatoes, leafy green vegetables, and fresh fruits are also good sources.

Chart 15.10 summarizes the water-soluble vitamins and their characteristics.

A prolonged deficiency of ascorbic acid leads to a disease, called *scurvy,* that occurs more frequently in infants and children than in adults. This condition is characterized by abnormal bone development and swollen, painful joints. There is also a tendency for the cells to pull apart and, consequently, the gums may swell and bleed easily, resistance to infection is lowered, and wounds heal slowly.

---

1. What factors cause the destruction of vitamin C?
2. What is the function of vitamin C?
3. What foods are good sources of vitamin C?

Chart 15.10	Water-soluble vitamins		
Vitamin	Characteristics	Functions	Sources
Thiamine (Vitamin B₁)	Destroyed by heat and oxygen, especially in alkaline environment	Part of coenzyme needed for oxidation of carbohydrates, and coenzyme needed in synthesis of ribose	Lean meats, liver, eggs, whole grain cereals, leafy green vegetables, legumes
Riboflavin (Vitamin B₂)	Stable to heat, acids, and oxidation; destroyed by alkalis and light	Part of enzymes and coenzymes needed for oxidation of glucose and fatty acids and for cellular growth	Meats, dairy products, leafy green vegetables, whole grain cereals
Niacin (Nicotinic Acid)	Stable to heat, acids, and alkalis; converted to niacinamide by cells; synthesized from tryptophan	Part of coenzymes needed for oxidation of glucose and synthesis of proteins, fats, and nucleic acids	Liver, lean meats, poultry, peanuts, legumes
Vitamin B₆	Group of three compounds; stable to heat and acids; destroyed by oxidation, alkalis, and ultraviolet light	Coenzyme needed for synthesis of proteins and various amino acids, for conversion of tryptophan to niacin, for production of antibodies, and for synthesis of nucleic acids	Liver, meats, fish, poultry, bananas, avocados, beans, peanuts, whole grain cereals, egg yolk
Pantothenic Acid	Destroyed by heat, acids, and alkalis	Part of coenzyme needed for oxidation of carbohydrates and fats	Meats, fish, whole grain cereals, legumes, milk, fruits, vegetables
Cyanocobalamin (Vitamin B₁₂)	Complex, cobalt-containing compound; stable to heat; inactivated by light, strong acids, and strong alkalis; absorption regulated by intrinsic factor from gastric glands; stored in liver	Part of coenzyme needed for synthesis of nucleic acids and for metabolism of carbohydrates; plays role in synthesis of myelin	Liver, meats, poultry, fish, milk, cheese, eggs
Folacin (Folic Acid)	Occurs in several forms; destroyed by oxidation in acid environment or by heat in alkaline environment; stored in liver where it is converted into folinic acid	Coenzyme needed for metabolism of certain amino acids and for synthesis of DNA; promotes production of normal red blood cells	Liver, leafy green vegetables, whole grain cereals, legumes
Biotin	Stable to heat, acids, and light; destroyed by oxidation and alkalis	Coenzyme needed for metabolism of amino acids and fatty acids and for synthesis of nucleic acids	Liver, egg yolk, nuts, legumes, mushrooms
Ascorbic Acid (Vitamin C)	Closely related to monosaccharides; stable in acids, but destroyed by oxidation, heat, light, and alkalis	Needed for production of collagen, conversion of folacin to folinic acid, and metabolism of certain amino acids; promotes absorption of iron and synthesis of hormones from cholesterol	Citrus fruits, citrus juices, tomatoes, cabbage, potatoes, leafy green vegetables, fresh fruits

Minerals

The nutrients that have been discussed—carbohydrates, lipids, proteins, and vitamins—are all organic substances. Dietary **minerals,** however, are inorganic elements that play essential roles in human metabolism. These elements usually are extracted from the soil by plants. Humans, in turn, obtain them from plant foods or from animals that have eaten plants.

Characteristics of Minerals

Minerals are responsible for about 4% of the body weight and are most concentrated in the bones and teeth. In fact, the minerals *calcium* and *phosphorus,* which are very abundant in these tissues, account for nearly 75% of the body's minerals.

Minerals usually are incorporated into organic molecules. For example, phosphorus occurs in phospholipids, iron in hemoglobin, and iodine in thyroxine.

However, some occur in inorganic compounds, such as the calcium phosphate of bone; others occur as free ions, such as the sodium, chloride, and calcium ions in blood plasma.

Minerals are present in all body cells, where they comprise parts of structural materials. They function as portions of enzyme molecules, help create the osmotic pressure of body fluids, and play vital roles in the conduction of nerve impulses, the contraction of muscle fibers, the coagulation of blood, and the maintenance of pH.

The concentrations of various minerals in body fluids are regulated by homeostatic mechanisms, which ensure that the excretion of these substances will be balanced with the dietary intake. Thus, toxic excesses are avoided, while minerals that are present in limited amounts are conserved.

---

1. How do minerals differ from other nutrients?
2. What are the major functions of minerals?
3. What are the most abundant minerals in the body?

Major Minerals

Calcium and phosphorus account for nearly 75% of the mineral elements in the body; thus, they are **major minerals.** Other major minerals, each of which accounts for 0.05% or more of the body weight, include potassium, sulfur, sodium, chlorine, and magnesium. Descriptions of the major minerals follow:

1. **Calcium.** Calcium (Ca) is widely distributed in cells and body fluids, even though 99% of the body's supply occurs in the inorganic salts of the bones and teeth. It is essential for nerve impulse conduction, muscle fiber contraction, and blood coagulation. It also serves to decrease the permeability of cell membranes and to activate certain enzymes.

The amount of calcium that is absorbed varies with a number of factors. For example, the proportion of calcium absorbed increases as the body's need for calcium increases. Vitamin D and high protein intake promote calcium absorption; increased motility of the digestive tract or an excessive intake of fats causes a decreased absorption. Consequently, the amount of calcium needed to provide an adequate cellular supply may vary from time to time. However, nutritionists believe a daily intake of 800 mg is sufficient to cover adult needs in spite of variations in absorption.

Unfortunately, only a few foods contain significant amounts of calcium. Milk and milk products and fish with bones, such as salmon or sardines, are the richest sources. Leafy green vegetables such as mustard greens, turnip greens, and kale are good sources.

Since these vegetables are not very popular foods in the United States, it is difficult for most people to maintain an adequate intake of calcium unless milk or milk products such as cheese, cottage cheese, and ice cream are regularly incorporated into the diet.

A deficiency of calcium in children may be accompanied by stunted growth, misshapen bones, and enlarged wrists and ankles. In adults such a deficiency may result in removal of calcium from the bones. As this occurs, the bones may become thinner and more fragile and may fracture easily.

2. **Phosphorus.** Phosphorus (P) is responsible for about 1% of the total body weight, and most of it is incorporated in the calcium phosphate of the bones and teeth. The remainder is distributed throughout the body cells, where it serves as a structural component and plays important roles in nearly all metabolic reactions. More specifically, phosphorus is a constituent of nucleic acids, many proteins, some enzymes, and some vitamins. It also occurs in the phospholipids of cell membranes, the energy-carrying molecules of ATP, and the phosphates of body fluids that function in the regulation of pH. (The molecular structure of ATP is shown in figure 4.7.)

The recommended daily adult intake of phosphorus is 800 mg, and since this mineral is abundant in protein foods, diets adequate in proteins are also adequate in phosphorus. In addition to being found in protein foods such as meats, poultry, fish, cheese, and nuts, phosphorus occurs in whole grain cereals, milk, and legumes.

---

1. What are the functions of calcium?
2. What are the functions of phosphorus?
3. What foods are good sources of these minerals?

---

3. **Potassium.** Potassium (K) is widely distributed throughout the body and tends to be concentrated inside cells rather than in extracellular fluids. On the other hand, *sodium,* which has similar chemical properties, tends to be concentrated outside the cells. More specifically, the ratio of potassium to sodium within a cell is 10:1, while the ratio outside the cell is 1:28 (figure 15.17).

Within cells, potassium functions to help maintain the intercellular osmotic pressure and to regulate pH. It promotes reactions involved in carbohydrate and protein metabolism and plays a vital role in the membrane polarization that occurs in nerve impulse conduction and muscle fiber contraction.

Nutritionists recommend a daily adult intake of 2.5 grams (2500 mg) of potassium, and since this mineral is widely distributed in foods, it is estimated that a typical adult diet provides between 2 and 6 grams each day. Although deficiencies of potassium due to an

Figure 15.17

Potassium ions tend to be more concentrated inside cells, and sodium ions tend to be more concentrated in the body fluids outside cells.

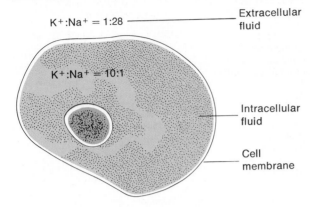

inadequate diet are relatively rare, they may occur for other reasons. For example, when a person has diarrhea, the intestinal contents may pass through the digestive tract so rapidly that potassium absorption is greatly reduced. Vomiting or using diuretic drugs also may lead to potassium depletion. The consequences of such losses may include muscular weakness, cardiac abnormalities, and edema.

Foods rich in potassium are avocados, dried apricots, meats, milk, peanut butter, potatoes, and bananas. Citrus fruits, citrus juices, apples, carrots, and tomatoes provide lesser amounts.

---

1. How is potassium distributed in the body?
2. What is the function of potassium?
3. What foods are good sources of this mineral?

---

4. Sulfur. Sulfur (S) is responsible for about 0.25% of the body weight and is widely distributed through the tissues. It is particularly abundant in the skin, hair, and nails. Most of it is incorporated in molecules of the amino acids *methionine* and *cysteine*. Other sulfur-containing compounds include thiamine, insulin, and biotin (figure 15.18). In addition, sulfur is a constituent of mucopolysaccharides that are found in cartilage, tendons, and bones, and of sulfolipids that occur in the liver, kidneys, salivary glands, and brain.

No daily requirement for sulfur has been established. It is thought, however, that a diet providing adequate amounts of protein will also meet the body's need for sulfur. Good food sources of this mineral include meats, milk, eggs, and legumes.

5. Sodium. It is estimated that about 0.15% of the adult weight is due to sodium (Na), which is widely distributed throughout the body. Only about 10% of this mineral occurs inside the cells, and about 40% is found within the extracellular fluids. The remainder is bound to the inorganic salts of bones.

Figure 15.18

Three examples of essential sulfur-containing nutrients.

Methionine

Thiamine hydrochloride (vitamin B₁)

Biotin

Sodium is readily absorbed from foods by active transport, and the blood concentration of this element is regulated by the kidneys under the influence of the adrenal hormone *aldosterone*.

Sodium helps to maintain the osmotic concentrations of extracellular fluids and thus serves to regulate the water balance between cells and their surroundings. Its presence is necessary for nerve impulse conduction and for muscle fiber contraction, and it aids in regulating pH and transporting various substances across cell membranes.

Although a daily requirement for sodium has not been established, it is believed that the usual human diet provides more than enough to meet the body's needs. On the other hand, sodium may be lost as a result of diarrhea, vomiting, kidney disorders, sweating, or using diuretics. Such losses may cause a variety of symptoms including edema, muscular cramps, and convulsions.

The amount of sodium naturally present in foods varies greatly, and it is commonly added to foods in the form of table salt (sodium chloride). In some geographic regions, there are significant concentrations of sodium in the drinking water.

Chart 15.11 Major minerals

Mineral	Distribution	Functions	Sources
Calcium (Ca)	Mostly in the inorganic salts of bones and teeth	Structure of bones and teeth; essential for nerve impulse conduction, muscle fiber contraction, and blood coagulation; increases permeability of cell membranes; activates certain enzymes	Milk, milk products, leafy green vegetables
Phosphorus (P)	Mostly in the inorganic salts of bones and teeth	Structure of bones and teeth; component in nearly all metabolic reactions; constituent of nucleic acids, many proteins, some enzymes, and some vitamins; occurs in cell membrane, ATP, and phosphates of body fluids	Meats, poultry, fish, cheese, nuts, whole grain cereals, milk, legumes
Potassium (K)	Widely distributed; tends to be concentrated inside cells	Helps maintain intracellular osmotic pressure and regulate pH; promotes metabolism; needed for nerve impulse conduction and muscle fiber contraction	Avocados, dried apricots, meats, nuts, potatoes, bananas
Sulfur (S)	Widely distributed	Essential part of various amino acids, thiamine, insulin, biotin, and mucopolysaccharides	Meats, milk, eggs, legumes
Sodium (Na)	Widely distributed; large proportion occurs in extracellular fluids and bonded to inorganic salts of bone	Helps maintain osmotic pressure of extracellular fluids and regulate water balance; needed for conduction of nerve impulses and contraction of muscle fibers; aids in regulation of pH and in transport of substances across cell membranes	Table salt, cured ham, sauerkraut, cheese, graham crackers
Chlorine (Cl)	Closely associated with sodium; most highly concentrated in cerebrospinal fluid and gastric juice	Helps maintain osmotic pressure of extracellular fluids, regulate pH, and maintain electrolyte balance; essential in formation of hydrochloric acid; aids transport of carbon dioxide by red blood cells	Same as for sodium
Magnesium (Mg)	Abundant in bones	Needed in metabolic reactions that occur in mitrochondria and are associated with the production of ATP; plays role in conversion of ATP to ADP	Milk, dairy products, legumes, nuts, leafy green vegetables

Foods that have relatively high sodium contents include cured ham, sauerkraut, cheese, and graham crackers.

---

1. In what compounds and tissues of the body is sulfur found?
2. What hormone regulates the blood concentration of sodium?
3. What are the functions of sodium?

---

6. **Chlorine.** Chlorine (Cl) in the form of chloride ions is closely associated with sodium. Like sodium, it is widely distributed throughout the body, although it is most highly concentrated in cerebrospinal fluid and in gastric juice.

Together with sodium, chlorine helps to maintain the osmotic concentrations of extracellular fluids, regulate pH, and maintain electrolyte balance. It is also essential for the formation of hydrochloric acid in gastric juice, and it functions in the transport of carbon dioxide by red blood cells.

Chlorine and sodium are usually obtained together in the form of table salt (sodium chloride), and as in the case of sodium, an ordinary diet usually provides considerably more chlorine than the body needs. It too can be lost excessively as a result of vomiting, diarrhea, kidney disorders, sweating, or using diuretics.

7. **Magnesium.** Magnesium (Mg) is responsible for about 0.05% of the body weight and is found in all cells. It is particularly abundant in bones in the form of phosphates and carbonates.

This mineral functions in a number of metabolic reactions including many that occur within mitochondria and are associated with the production of ATP. It also plays a role in the conversion of ATP to ADP; therefore, it is important in providing energy for cellular processes.

When the intake of magnesium is high, a smaller percentage is absorbed from the intestinal tract, and when the intake is low, a larger percentage is absorbed. Absorption increases as the protein intake increases, and it decreases as the calcium and vitamin D intake increases. A reserve supply of magnesium is stored in bone tissue, and excess amounts are excreted in the urine under the influence of *aldosterone*.

The recommended daily allowance of magnesium is 300 mg for females and 350 mg for males. Because a typical diet usually provides only about 120 mg of magnesium for every 1000 calories, it may barely meet the body's needs.

Good sources of magnesium include milk and dairy products (except butter), legumes, nuts, and leafy green vegetables.

The major minerals are summarized in chart 15.11.

---

1. Where are chloride ions most highly concentrated in the body?
2. Where is magnesium stored?
3. What factors influence the absorption of magnesium from the intestinal tract?

Trace Elements

Trace elements are essential minerals that occur in minute amounts, each one making up less than 0.005% of the adult body weight. They include iron, manganese, copper, iodine, cobalt, and zinc.

Iron (Fe) is found primarily in the blood, although some is present in all body cells, and a reserve supply is stored in the liver, spleen, and bone marrow. It functions as part of the *hemoglobin* molecules in red blood cells and is responsible for the ability of these molecules to carry oxygen (figure 15.19). Iron also occurs in *myoglobin* molecules, which are synthesized in muscle cells and act to store oxygen temporarily. In addition, iron functions to catalyze the formation of vitamin A and is incorporated into a number of enzymes.

It has been estimated that an adult male requires from 0.7 to 1 mg of iron daily, while females need 1.2

Figure 15.19
A hemoglobin molecule contains four heme portions, each of which contains a single iron atom (Fe) that can combine with oxygen.

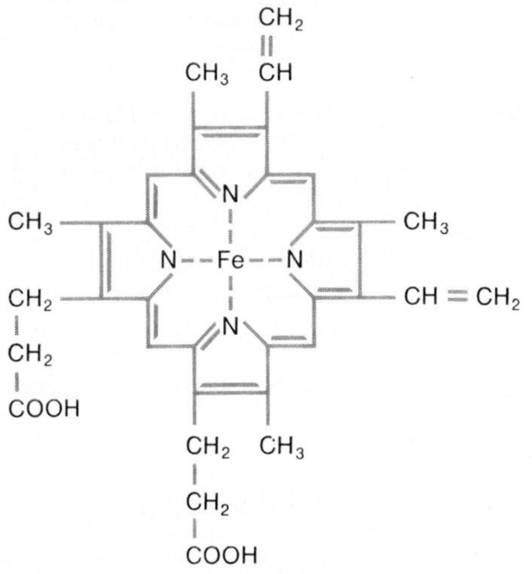

Heme

to 2 mg. It is also believed that although a typical diet supplies about 10–18 mg of iron each day, only 2%–10% of this iron is absorbed. (See chapter 17.) Such a diet provides a narrow margin of safety as far as iron intake is concerned, and it may fail to provide the amount needed by some individuals.

Pregnant women require additional quantities of iron in order to support the formation of a placenta, and the growth and development of a fetus. Iron also is needed for the synthesis of hemoglobin in the fetus as well as in the mother, whose blood volume increases during pregnancy.

Liver is the only really rich source of dietary iron, and since liver is not a very popular food, iron is one of the more difficult nutrients to obtain from natural sources in adequate amounts. Foods that contain some iron include lean meats; dried apricots, raisins, and prunes; enriched whole grain cereals; legumes; and molasses.

Manganese (Mn) is most concentrated in the liver, kidneys, and pancreas. It is necessary for normal growth and development of skeletal structures and other connective tissues, and occurs in enzymes that are essential for the synthesis of fatty acids and cholesterol, for the formation of urea, and for the normal functions of the nervous system.

The daily requirement for manganese is unknown. The richest sources include nuts, legumes, and whole grain cereals; leafy green vegetables and fruits are good sources.

Chart 15.12 Trace elements

Trace element	Distribution	Functions	Sources
Iron (Fe)	Primarily in blood; stored in liver, spleen, and bone marrow	Part of hemoglobin molecule; catalyzes formation of vitamin A; incorporated into a number of enzymes	Liver, lean meats, dried apricots, raisins, enriched whole grain cereals, legumes, molasses
Manganese (Mn)	Most concentrated in liver, kidneys, and pancreas	Occurs in enzymes needed for synthesis of fatty acids and cholesterol, formation of urea, and normal functioning of the nervous system	Nuts, legumes, whole grain cereals, leafy green vegetables, fruits
Copper (Cu)	Most highly concentrated in liver, heart, and brain	Essential for synthesis of hemoglobin, development of bone, production of melanin, and formation of myelin	Liver, oysters, crabmeat, nuts, whole grain cereals, legumes
Iodine (I)	Concentrated in thyroid gland	Essential component for synthesis of thyroid hormones	Food content varies with soil content in different geographic regions; iodized table salt
Cobalt (Co)	Widely distributed	Component of cyanocobalamin; needed for synthesis of several enzymes	Liver, lean meats, poultry, fish, milk
Zinc (Zn)	Most concentrated in liver, kidneys, and brain	Constituent of several enzymes involved in digestion, respiration, bone metabolism, liver metabolism; necessary for normal wound healing and maintaining integrity of the skin	Seafoods, meats, cereals, legumes, nuts, vegetables

1. What is the primary function of iron?
2. Why does the usual diet provide only a narrow margin of safety in supplying iron?
3. How is manganese utilized?
4. What foods are good sources of manganese?

Copper (Cu) is found in all body tissues, but is most highly concentrated in the liver, heart, and brain. It is essential for the synthesis of hemoglobin, the normal development of bone, the production of melanin, and the formation of myelin within the nervous system.

It is believed that a daily intake of 2 mg of copper is sufficient to provide the needs of body cells and that a typical adult diet supplies about 2–5 mg of this mineral. Adults seldom develop copper deficiencies.

Foods rich in copper include liver, oysters, crabmeat, nuts, whole grain cereals, and legumes.

Iodine (I) occurs in minute quantities in all tissues, but is highly concentrated within the thyroid gland. Its only known function is to provide an essential component for the synthesis of *thyroid hormones*. (The molecular structures of two of these hormones are shown in figure 13.19.)

An intake of 1 microgram (.001 mg) of iodine per kilogram of body weight daily is thought to provide an adequate amount for most adults. Since the iodine content of foods varies with the iodine content of soils in different geographic regions, it is sometimes necessary to artificially increase the iodine in foods to prevent deficiencies from developing. The use of iodized table salt, which contains added *potassium iodide* (KI), has proved to be an effective means of accomplishing this.

Cobalt (Co) is widely distributed throughout the body, since it is an essential part of *cyanocobalamin* (vitamin B_{12}) molecules. It is also thought to be necessary for the synthesis of several important enzymes.

The amount of cobalt required in the daily diet is unknown. This mineral is found in a great variety of foods, and the quantity present in the average diet is apparently sufficient to meet the body's need. Good sources of cobalt include liver, lean meats, poultry, fish, and milk.

Zinc (Zn) generally occurs in body tissues, but is most concentrated in the liver, kidneys, and brain. It is a constituent of a large number of enzymes involved in digestion, respiration, bone metabolism, and liver metabolism. It is also necessary for normal wound healing and for maintaining the integrity of the skin.

The daily requirement for zinc is believed to be about 15 mg, and most diets are thought to provide 10–15 mg. Since only a portion of this amount may be absorbed, zinc deficiencies may be relatively common.

The richest sources of zinc are seafoods and meats; cereals, legumes, nuts, and vegetables provide lesser amounts.

Characteristics of these trace elements are summarized in chart 15.12.

Chart 15.13 Recommended daily dietary allowances for various age groups in the United States

	Infants *0–6 mos*	Children *4–6 yrs*	Adolescents *15–18 yrs*		Adults *23–50 yrs*		Adults *51 yrs and over*	
			Males	*Females*	*Males*	*Females*	*Males*	*Females*
Weight, kg (lb)	6 (13)	20 (44)	66 (145)	55 (120)	70 (154)	55 (120)	70 (154)	55 (120)
Height, cm (in)	60 (24)	112 (44)	176 (69)	163 (64)	178 (70)	163 (64)	178 (70)	163 (64)
Protein, g	kg × 2.2	30	56	46	56	44	56	44
Fat-soluble vitamins								
Vitamin A, $\mu g^\dagger$	420	500	1000	800	1000	800	1000	800
Vitamin D, μg	10	10	10	10	5	5	5	5
Vitamin E activity, mg	3	6	10	8	10	8	10	8
Water-soluble vitamins								
Ascorbic acid, mg	35	45	60	60	60	60	60	60
Folacin, μg	30	200	400	400	400	400	400	400
Niacin, mg	6	11	18	14	18	13	16	13
Riboflavin, mg	0.4	1.0	1.7	1.3	1.6	1.2	1.4	1.2
Thiamine, mg	0.3	0.9	1.4	1.1	1.4	1.0	1.2	1.0
Vitamin B_6, mg	0.3	1.3	2.0	2.0	2.2	2.0	2.2	2.0
Vitamin B_{12}, μg	0.5	2.5	3.0	3.0	3.0	3.0	3.0	3.0
Minerals								
Calcium, mg	360	800	1200	1200	800	800	800	800
Phosphorus, mg	240	800	1200	1200	800	800	800	800
Iodine, μg	40	90	150	150	150	150	150	150
Iron, mg	10	10	18	18	10	18	10	10
Magnesium, mg	50	200	400	300	350	300	350	300
Zinc, mg	3	10	15	15	15	15	15	15

Source: Food and Nutrition Board, *Recommended Dietary Allowances,* 9th ed., National Academy of Sciences—National Research Council, Washington, D.C., 1980.

†Microgram

1. How is copper used?
2. What is the function of iodine?
3. Why are zinc deficiencies thought to be relatively common?

Adequate Diets

An adequate diet is one that provides sufficient *energy* (calories), *essential fatty acids, essential amino acids, vitamins,* and *minerals* to support optimal growth and to maintain and repair body tissues. However, since individual needs for nutrients vary greatly with age, sex, growth rate, amount of physical activity, and level of stress, as well as with genetic and environmental factors, it is not possible to design a diet that will be adequate for everyone.

On the other hand, nutrients are so widely distributed in foods that satisfactory amounts and combinations of essential substances usually can be obtained in spite of individual food preferences that are related to cultural backgrounds, life-styles, and emotional attitudes. Chart 15.13 lists the recommended daily allowance for each of the major nutrients.

Malnutrition

Malnutrition is poor nutrition that results from a lack of essential nutrients or a failure to use available foods to best advantage. It may involve *undernutrition* and include the symptoms of deficiency diseases, or it may be due to *overnutrition* arising from an excessive intake of nutrients.

The factors leading to malnutrition are varied. A deficiency condition may, for example, stem from lack of availability or poor quality of food. On the other hand, malnutrition may result from excessive intake of vitamin supplements, which causes hypervitaminosis, or from excessive caloric intake. Malnutrition from causes that involve the diet alone is called *primary malnutrition.*

Secondary malnutrition occurs when a person's individual characteristics make a normally adequate diet unsuitable. A person who secretes low quantities of bile salts, for example, may develop a deficiency of fat-soluble vitamins because the presence of bile salts promotes the absorption of these substances. Likewise, severe and prolonged emotional stress may lead to secondary malnutrition because stress can cause changes in hormonal concentrations, and such changes may result in abnormal breakdown of amino acids or in excessive excretion of various nutrients.

Food Selection
A Practical Application

Although recommended daily allowances of essential nutrients are often used by nutritionists and dietitians as guides in planning adequate diets, these values are usually of limited use to the average person. A more useful aid for most persons is the idea of *basic food groups*.

A basic food group is a class of foods that will supply sufficient amounts of certain essential nutrients when a given quantity of food from the group is included in the diet. For example, one basic food plan includes four food groups, each of which provides a unique contribution toward achieving an adequate diet. In this plan, the food groups and the quantities of food required from each are as follows:

Group 1: Milk and Dairy Products. This group includes milk, cottage cheese, cream cheese, natural and processed cheese, and ice cream. It provides calcium, phosphorus, magnesium, protein, vitamin B_6, vitamin B_{12}, and riboflavin. It is recommended that an adult have two 8-ounce glasses of milk or the equivalent in other dairy foods each day.

Group 2: Fruits and Vegetables. This group includes all fruits and vegetables, and it provides vitamin C, vitamin A, iron, magnesium, and vitamin B_6. It is recommended that an adult diet contain four servings from this group each day, and that one of these be citrus fruit or some other fruit or vegetable that is particularly high in vitamin C content. At least every other day, one of the servings should be a dark green, yellow, or orange vegetable.

Group 3: Meat, Poultry, and Fish. This group includes all meats, poultry, and fish as well as meat substitutes such as eggs, beans, peas, and nuts. It provides protein, vitamin A, thiamine, niacin, vitamin B_6, vitamin B_{12}, riboflavin, phosphorus, magnesium, and iron. For an adult, two or more servings from this group are recommended each day.

Group 4: Breads and Cereals. This group includes all whole grain or enriched breads and cereals. It provides iron, thiamine, niacin, riboflavin, protein, phosphorus, and magnesium. Four or more servings per day from this group are recommended for adults.

Chart 15.14 provides a food guide based on these basic food groups.

In response to emotional problems and an intense fear of becoming overweight, teenagers (particularly girls) sometimes develop a condition called *anorexia nervosa*. In this condition, the person severely restricts food intake and may exercise excessively. As a result, 25% or more of the body weight is lost. The person also may engage in periods of secretive, unrestrained eating (binge-eating) followed by self-induced vomiting and the use of laxatives and diuretics (bulimia). Other symptoms of anorexia nervosa that accompany the malnutrition associated with the chronic starvation include cessation of menstruation, decreased heart rate, inability to maintain normal body temperature, impaired judgement, and hallucinations. In perhaps 3% or more of the cases, anorexics die suddenly, usually as a result of electrolyte imbalances and cardiac disorders.

---

1. What is meant by an adequate diet?
2. What factors influence individual needs for nutrients?
3. What is meant by *primary malnutrition*? By *secondary malnutrition*?

Clinical Terms Related to Nutrients and Nutrition

anorexia (an″o-rek′se-ah)   a loss of appetite.

casein (ka′se-in)   the primary protein found in milk.

celiac disease (se′le-ak di-zēz′)   a digestive disorder characterized by the inability to digest or use fats and carbohydrates.

emaciation (e-ma″se-a′shun)   excessive leanness of the body due to tissue wasting.

extrinsic factor (ek-strin′sik fak′tor)   vitamin B_{12}.

hyperalimentation (hi″per-al″ĭ men-ta″shun)   prolonged intravenous nutrition provided for patients with severe digestive disorders.

hypercalcemia (hi″per-kal-se′me-ah)   an excessive level of calcium in the blood.

hypercalciuria (hi″per-kal″se-u′re-ah)   an excessive excretion of calcium in the urine.

hyperglycemia (hi″per-gli-se′me-ah)   an excessive level of glucose in the blood.

hyperkalemia (hi″per-kah-le′me-ah)   an excessive level of potassium in the blood.

hypernatremia (hi″per-nah-tre′me-ah)   an excessive level of blood sodium.

hypoalbuminemia (hi″po-al-bu″mĭ-ne′me-ah)   a low level of albumin in the blood.

hypoglycemia (hi″po-gli-se′me-ah)   a low level of glucose in the blood.

hypokalemia (hi″po-kah-le′me-ah)   a low level of potassium in the blood.

hyponatremia (hi″po-nah-tre′me-ah)   a low level of blood sodium.

isocaloric (i″so-kah-lo′rik)   containing equal amounts of heat energy.

lipogenesis (lip″o-jen′ĕ-sis)   the formation of fat.

marasmus (mah-raz′mus)   an extreme form of protein and calorie malnutrition.

nyctalopia (nik″tah-lo′pe-ah)   night blindness.

pica (pi′kah)   a hunger for substances that are not suitable foods.

polyphagia (pol″e-fa′je-ah)   excessive intake of food.

proteinuria (pro″te-i-nu′re-ah)   presence of protein in the urine.

provitamin (pro-vi′tah-min)   a substance used as a precursor in vitamin synthesis.

Chart 15.14 A daily food guide

Milk group (8-ounce cups)

2 to 3 cups for children under 9 years

3 or more cups for children 9 to 12 years

4 cups or more for teenagers

2 cups or more for adults

3 cups or more for pregnant women

4 cups or more for nursing mothers

Meat group

2 or more servings. Count as one serving:

   2 to 3 ounces lean, cooked beef, veal, pork, lamb, poultry, fish—without bone

   2 eggs

   1 cup cooked dry beans, dry peas, lentils

   4 tablespoons peanut butter

Vegetable and fruit group (½ cup serving, or 1 piece fruit, etc.)

4 or more servings per day, including:

   1 serving of citrus fruit, or other fruit or vegetable as a good source of vitamin C, or 2 servings of a fair source

   1 serving, at least every other day, of a dark green or deep yellow vegetable for vitamin A

   2 or more servings of other vegetables and fruits, including potatoes

Bread and cereals group

4 or more servings daily (whole grain, enriched, or restored). Count as one serving:

   1 slice bread

   1 ounce ready-to-eat cereal

   ½ to ¾ cup cooked cereal, corn meal, grits, macaroni noodles, rice, or spaghetti

Source: "A Daily Food Guide" in *Consumers All.* Yearbook of Agriculture, 1965, U.S. Department of Agriculture, Washington, D.C., 1965, p. 394.

Chapter Summary

Introduction

Nutrients include carbohydrates, lipids, proteins, vitamins, and minerals.

The ways nutrients are used to support life processes constitute metabolism.

Essential nutrients are needed for health and cannot be synthesized by body cells.

Carbohydrates

Carbohydrates are organic compounds that are used primarily to supply cellular energy.

1. Sources of carbohydrates
   a. Carbohydrates are ingested in a variety of forms.
   b. Starch, glycogen, disaccharides, and monosaccharides are carbohydrates.
   c. Cellulose is a polysaccharide that cannot be digested by human enzymes, but is important in providing bulk that facilitates the movement of intestinal contents.
2. Utilization of carbohydrates
   a. Carbohydrates are absorbed as monosaccharides.
   b. Fructose and galactose are converted to glucose by the liver.
   c. Energy is released from glucose by oxidation.
   d. Excessive glucose is stored as glycogen or converted to fat.
3. Carbohydrate requirements
   a. Most carbohydrates are used to supply energy, although some are used to produce important sugars.
   b. Some cells depend on a continuous supply of glucose to survive.
   c. If inadequate amounts of glucose are available, amino acids may be converted to glucose.

   d. Humans survive with a wide range of carbohydrate intakes.
   e. Poor nutritional status is usually related to low intake of nutrients other than carbohydrates.

Lipids

Lipids are organic compounds that supply energy and are used to build cell structures. They include fats, phospholipids, and cholesterol.

1. Sources of lipids
   a. Triglycerides are obtained from foods of plant and animal origins.
   b. Cholesterol is obtained in foods of animal origin only.
2. Utilization of lipids
   a. Metabolism of triglycerides is controlled mainly by the liver and adipose tissues.
   b. The liver can alter the molecular structures of fatty acids.
   c. Linoleic acid is an essential fatty acid.
   d. The liver also regulates the amount of cholesterol by synthesizing or excreting it.
3. Lipid requirements
   a. Humans survive with a wide range of lipid intakes.
   b. The amounts and types of lipids needed for health are unknown.
   c. Some fats contain fat-soluble vitamins, and the intake of fats must be sufficient to supply these essential nutrients.

Proteins

Proteins are organic compounds that serve as structural materials, act as enzymes, and provide energy.

1. Sources of proteins
   a. Proteins are obtained mainly from meats, dairy products, cereals, and legumes.
   b. During digestion they are broken down into amino acids.

c. Eight amino acids are essential for adults, while ten are essential in growing children.

d. All essential amino acids must be present at the same time for growth and repair of tissues.

e. Complete proteins contain adequate amounts of all the essential amino acids to maintain the tissues and to promote growth.

f. Incomplete proteins lack one or more essential amino acids.

2. Utilization of amino acids

a. Amino acids are incorporated into various structural and functional proteins, including enzymes.

b. During starvation, tissue proteins may be used as energy sources; thus, the tissues waste away.

3. Nitrogen balance

a. In healthy adults, the gain of protein equals the loss of protein, and a nitrogen balance exists.

b. A starving person has a negative nitrogen balance; while a growing child, a pregnant woman, or an athlete in training usually has a positive nitrogen balance.

4. Protein requirements

a. Proteins and amino acids are needed to supply essential amino acids and nitrogen for the synthesis of various nitrogen-containing molecules.

b. The consequences of protein deficiencies are particularly severe among growing children.

Energy Expenditures

Energy is of prime importance to survival and may be obtained from carbohydrates, fats, or proteins.

1. Energy values of foods

a. The potential energy values of foods are expressed in calories.

b. When energy losses due to incomplete absorption and incomplete oxidation are taken into account, 1 gram of carbohydrate or 1 gram of protein yields about 4 calories, while 1 gram of fat yields about 9 calories.

2. Energy requirements

a. The amount of energy required varies from person to person.

b. Factors that influence energy requirements include basal metabolic rate, muscular activity, body temperature, and nitrogen balance.

3. Energy balance

a. Energy balance exists when caloric intake equals caloric output.

b. If balance is positive, body weight increases; if balance is negative, body weight decreases.

4. Desirable weight

a. The most common nutritional disorders involve caloric imbalances.

b. Average weights of persons 25–30 years of age are thought to be desirable for all older persons.

c. Recently, height-weight guidelines have been prepared that are based on longevity.

d. A person who exceeds the desirable weight by 10–20% is called overweight.

e. A person whose body contains an excess of fatty tissue is said to be obese.

Vitamins

Vitamins are organic compounds (other than carbohydrates, lipids, and proteins) that are essential for normal metabolic processes and cannot be synthesized by body cells in adequate amounts.

1. Fat-soluble vitamins

a. General characteristics

(1) Fat-soluble vitamins occur in association with lipids and are influenced by the same factors that affect lipid absorption.

(2) They are fairly resistant to the effects of heat; thus, are not destroyed by cooking or food processing.

b. Vitamin A

(1) Vitamin A occurs in several forms, is synthesized from carotenes, and is stored in the liver.

(2) It functions in the production of pigments necessary for vision.

c. Vitamin D

(1) Vitamin D is represented by a group of related sterols.

(2) It is found in certain foods and is produced commercially; it also can be synthesized in the skin.

(3) When needed, vitamin D is converted by the kidneys to an active form that functions as a hormone and promotes the absorption of calcium and phosphorus by the intestine.

d. Vitamin E

(1) Vitamin E is represented by a group of compounds that are antioxidants.

(2) It is stored in muscles and adipose tissue.

(3) Its precise functions are unknown, but it seems to prevent oxidation of vitamin A and polyunsaturated fatty acids, and to stabilize cell membranes.

e. Vitamin K

(1) Vitamin K_1 occurs in foods; vitamin K_2 is produced by certain bacteria that normally inhabit the intestinal tract.

(2) It is stored to a limited degree in the liver.

(3) It is used in the production of prothrombin that is needed for normal blood clotting.

2. Water-soluble vitamins
    a. General characteristics
        (1) Water-soluble vitamins include the B vitamins and vitamin C.
        (2) B vitamins make up a group called the B complex and are generally involved with the oxidation of carbohydrates, lipids, and proteins.
    b. Vitamin B complex
        (1) Thiamine
            (a) Thiamine functions as parts of coenzymes that act in the oxidation of carbohydrates and in the synthesis of essential sugars.
            (b) Small amounts are stored in the tissues; excess is excreted in the urine.
            (c) Quantities needed vary with caloric intake.
        (2) Riboflavin
            (a) Riboflavin functions as parts of several enzymes and coenzymes that are essential to the oxidation of glucose and fatty acids.
            (b) Its absorption is regulated by an active transport system; excess is excreted in the urine.
            (c) Quantities needed vary with caloric intake.
        (3) Niacin
            (a) Niacin functions as parts of coenzymes needed for the oxidation of glucose and for the synthesis of proteins and fats.
            (b) It can be synthesized from tryptophan; daily requirement varies with the tryptophan intake.
        (4) Vitamin B_6
            (a) Vitamin B_6 is a group of compounds that function as coenzymes needed by a variety of metabolic pathways involved in the synthesis of proteins, various amino acids, antibodies, and nucleic acids.
            (b) Its requirement varies with protein intake.
        (5) Pantothenic acid
            (a) Pantothenic acid functions as part of coenzyme A; thus, it is essential for energy-releasing mechanisms.
            (b) Its daily requirement is not known.
        (6) Cyanocobalamin
            (a) The cyanocobalamin molecule contains cobalt.
            (b) Its absorption is regulated by the secretion of intrinsic factor from the gastric glands.
            (c) It functions as part of coenzymes needed for the synthesis of nucleic acids and for the metabolism of carbohydrates and fats.
        (7) Folacin
            (a) Folacin is converted by the liver to physiologically active folinic acid.
            (b) It functions as a coenzyme needed for the metabolism of certain amino acids, the synthesis of DNA, and the normal production of red blood cells.
        (8) Biotin
            (a) Biotin functions as a coenzyme needed for the metabolism of amino acids and fatty acids, and for the synthesis of purines.
            (b) It is stored in metabolically active organs.
    c. Ascorbic acid (vitamin C)
        (1) Vitamin C is closely related chemically to monosaccharides.
        (2) Its functions are poorly understood, but is thought to be needed for the production of collagen, the metabolism of certain amino acids, and the absorption of iron.
        (3) It is not stored in any great amounts; excess is excreted in the urine.

Minerals
1. Characteristics of minerals
    a. Minerals are responsible for about 4% of body weight.
    b. About 75% of the minerals are found in bones and teeth as calcium and phosphorus.
    c. Minerals are usually incorporated into organic molecules, although some occur in inorganic compounds or as free ions.
    d. They comprise structural materials, function in enzymes, and play vital roles in various metabolic processes.
    e. Mineral concentrations are generally regulated by homeostatic mechanisms.
2. Major minerals
    a. Calcium
        (1) Calcium is essential for the formation of bones and teeth, the conduction of nerve impulses, the contraction of muscle fibers, the coagulation of blood, and the activation of various enzymes.
        (2) Its absorption is affected by existing calcium concentration, vitamin D, protein intake, and motility of the digestive tract.
    b. Phosphorus
        (1) Phosphorus is incorporated for the most part into the salts of bones and teeth.
        (2) It plays roles in nearly all metabolic reactions as a constituent of nucleic acids, proteins, enzymes, and some vitamins.
        (3) It also occurs in the phospholipids of cell membranes, in ATP, and in phosphates of body fluids.

c. Potassium
    (1) Potassium tends to be concentrated inside cells.
    (2) It functions in maintenance of osmotic pressure, regulation of pH, metabolism of carbohydrates and proteins, the conduction of nerve impulses, and the contraction of muscle fibers.
d. Sulfur
    (1) Sulfur is incorporated for the most part into the molecular structures of certain amino acids.
    (2) It is also included in thiamine, insulin, biotin, and mucopolysaccharides.
e. Sodium
    (1) Most sodium occurs in extracellular fluids or is bonded to the inorganic salts of bone.
    (2) Blood concentration of sodium is regulated by the kidneys under the influence of aldosterone.
    (3) It helps maintain osmotic concentrations and regulate water balance and pH.
    (4) It is essential for the conduction of nerve impulses, the contraction of muscle fibers, and the movement of substances across cell membranes.
f. Chlorine
    (1) Chlorine is closely associated with sodium in the form of chloride ions.
    (2) It acts with sodium to help maintain osmotic pressure, regulate pH, and maintain electrolyte balance.
    (3) It is essential for the formation of hydrochloric acid and for the transport of carbon dioxide by red blood cells.
g. Magnesium
    (1) Magnesium is particularly abundant in the bones as phosphates and carbonates.
    (2) It functions in the production of ATP and in the conversion of ATP to ADP.
    (3) A reserve supply of magnesium is stored in the bones; excesses are excreted in the urine.

3. Trace elements
a. Iron
    (1) Iron occurs primarily in hemoglobin of red blood cells and in myoglobin of muscles.
    (2) A reserve supply of iron is stored in the liver, spleen, and bone marrow.
    (3) It is needed to catalyze the formation of vitamin A; incorporated into various enzymes.
b. Manganese
    (1) Most manganese is concentrated in the liver, kidneys, and pancreas.
    (2) It is necessary for normal growth and development of skeletal structures and other connective tissues; it is essential for the synthesis of fatty acids, cholesterol, and urea.
c. Copper
    (1) Most copper is concentrated in the liver, heart, and brain.
    (2) It is needed for synthesis of hemoglobin, development of bones, production of melanin, and formation of myelin.
d. Iodine
    (1) Iodine is most highly concentrated in the thyroid gland.
    (2) It provides an essential component for the synthesis of thyroid hormones.
    (3) It is often added to foods in the form of iodized table salt.
e. Cobalt
    (1) Cobalt is widely distributed throughout the body.
    (2) It is an essential part of cyanocobalamin and probably is needed for the synthesis of several important enzymes.
f. Zinc
    (1) Zinc is most concentrated in the liver, kidneys, and brain.
    (2) It is a constituent of several enzymes involved with digestion, respiration, bone metabolism, and liver metabolism.

Adequate Diets

1. An adequate diet provides sufficient energy and essential nutrients to support optimal growth, as well as maintenance and repair of tissues.
2. Individual needs vary so greatly that it is not possible to design a diet that is adequate for everyone.
3. Malnutrition
   a. Poor nutrition is due to lack of foods or failure to make best use of available foods.
   b. Primary malnutrition is due to poor diet.
   c. Secondary malnutrition is due to an individual characteristic that makes a normal diet unsuitable.

Application of Knowledge

1. How would you explain the fact that the blood sugar level of a person whose diet is relatively low in carbohydrates remains stable?
2. Calculate the carbohydrate, lipid, protein, and caloric content of your diet for a 24-hour period. Assuming that this 24-hour sample is representative of your normal eating habits, what improvements could be made in its composition?
3. Examine the label information on the packages of a variety of dry breakfast cereals. Which types of cereals provide the best sources of vitamins and minerals? Which major nutrients are lacking?

Review Activities

1. Define *essential nutrient*.
2. List some common sources of carbohydrates.
3. Explain the importance of cellulose in the diet.
4. Explain what happens to excessive amounts of glucose in the body.
5. Explain why a temporary drop in the glucose level may produce functional disorders of the nervous system.
6. List some of the factors that affect an individual's need for carbohydrates.
7. Define *triglyceride*.
8. List some common sources of lipids.
9. Describe the role of the liver in fat metabolism.
10. Discuss the functions of cholesterol.
11. List some common sources of proteins.
12. Distinguish between essential and nonessential amino acids.
13. Explain why all of the essential amino acids must be present at the same time.
14. Distinguish between complete and incomplete proteins.
15. Review the major functions of amino acids.
16. Define *nitrogen balance*.
17. Explain why a protein deficiency may be accompanied by edema.
18. Define *calorie*.
19. Explain how the caloric values of foods are determined.
20. Define *basal metabolic rate*.
21. List some of the factors that affect the BMR.
22. Define *energy balance*.
23. Explain what is meant by desirable weight.
24. Distinguish between overweight and obesity.
25. Discuss the general characteristics of fat-soluble vitamins.
26. List the fat-soluble vitamins and describe the major functions of each vitamin.
27. List some good sources for each of the fat-soluble vitamins.
28. Explain what is meant by the vitamin B complex.
29. List the water-soluble vitamins and describe the major functions of each vitamin.
30. List some good sources for each of the water-soluble vitamins.
31. Discuss the general characteristics of the mineral nutrients.
32. List the major minerals and describe the major functions of each mineral.
33. List some good sources for each of the major minerals.
34. Distinguish between a major mineral and a trace element.
35. List the trace elements and describe the major functions of each element.
36. List some good sources for each of the trace elements.
37. Define *adequate diet*.
38. Define *malnutrition*.
39. Distinguish between primary and secondary malnutrition.

16

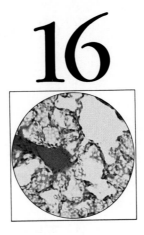

The Respiratory System

Before body cells can oxidize nutrients and release energy, they must be supplied with oxygen, and the carbon dioxide that results from oxidation must be excreted. These two general processes—obtaining oxygen and removing carbon dioxide—are the primary functions of the *respiratory system.*

In addition, the respiratory organs filter particles from incoming air, help to control the temperature and water content of the air, aid in producing sounds used in speech, and play important roles in the sense of smell and the regulation of pH.

Chapter Outline

Chapter Objectives

After you have studied this chapter, you should be able to

1. List the general functions of the respiratory system.

2. Name and describe the location of the organs of the respiratory system.

3. Describe the functions of each organ of the respiratory system.

4. Explain how inspiration and expiration are accomplished.

5. Name and define each of the respiratory air volumes.

6. Explain how the alveolar ventilation rate is calculated.

7. List several nonrespiratory air movements and explain how each occurs.

8. Locate the respiratory center and explain how it controls normal breathing.

9. Discuss how various factors affect the respiratory center.

10. Describe the structure and function of the respiratory membrane.

11. Explain how oxygen and carbon dioxide are transported in the blood.

12. Review the major events that occur during cellular respiration.

13. Explain how oxygen is used by cells.

14. Complete the review activities at the end of this chapter. Note that the items are worded in the form of specific learning objectives. You may want to refer to them before reading the chapter.

Key Terms

alveolus   (al-ve′o-lus)

bronchial tree   (brong′ke-al tre)

carbonic anhydrase   (kar-bon′ik an-hi′drās)

cellular respiration
   (sel′u-lar res′′pi-ra′ shun)

citric acid cycle   (sit′rik as′id si′kl)

expiration   (ek′′spi-ra′shun)

glottis   (glot′is)

hemoglobin   (he′′mo-glo′bin)

hyperventilation   (hi′′per-ven′′ti-la′shun)

inspiration   (in′′spi-ra′shun)

partial pressure   (par′shil presh′ur)

pleural cavity   (ploo′ral kav′i-te)

respiratory center   (re-spi′rah-to′′re sen′ter)

respiratory membrane
   (re-spi′rah-to′′re mem′brān)

respiratory volume   (re-spi′rah-to′′re vol′ūm)

surface tension   (ser′fas ten′shun)

surfactant   (ser-fak′tant)

Aids to Understanding Words

alveol-, a small cavity: *alveol*us—a microscopic air sac within a lung.

bronch-, the windpipe: *bronch*us—a primary branch of the trachea.

carcin-, a spreading sore: *carcin*oma—a type of cancer.

carin-, keellike: *carin*a—a ridge of cartilage between the right and left bronchi.

cric-, a ring: *cric*oid cartilage—a ring-shaped mass of cartilage at the base of the larynx.

epi-, upon: *epi*glottis—a flaplike structure that partially covers the opening into the larynx during swallowing.

hem-, blood: *hem*oglobin—pigment in red blood cells that serves to transport oxygen and carbon dioxide.

tuber-, a swelling: *tuber*culosis—a disease characterized by the formation of fibrous masses within the lungs.

THE **RESPIRATORY SYSTEM** consists of a group of passages that filter incoming air and transport it from the outside of the body into the lungs, and numerous microscopic air sacs in which gas exchanges take place. Although the entire process of exchanging gases between the atmosphere and the body cells is called **respiration,** the process actually involves several events. These include the movement of air in and out of the lungs—commonly called breathing or pulmonary ventilation; the exchange of gases between the air in the lungs and the blood; the transport of gases by the blood between the lungs and body cells; and the exchange of gases between the blood and the body cells. The utilization of oxygen and the production of carbon dioxide by the body cells is called **cellular respiration.**

Organs of the Respiratory System

The organs of the respiratory system include the nose, nasal cavity, sinuses, pharynx, larynx, trachea, bronchial tree, and lungs.

The parts of the respiratory system, shown in figure 16.1, can be divided into two sets or tracts. Those organs outside the thorax constitute the *upper respiratory tract,* and those within the thorax comprise the *lower respiratory tract.*

The Nose

The nose is covered with skin and is supported internally by bone and cartilage. Its two *nostrils* (external nares) provide openings through which air can enter and leave the nasal cavity. These openings are guarded by numerous internal hairs that help prevent the entrance of relatively coarse particles sometimes carried by the air.

The Nasal Cavity

The **nasal cavity,** which is a hollow space behind the nose, is divided medially into right and left portions by the **nasal septum.** This cavity is separated from the cranial cavity by the cribriform plate of the ethmoid bone and from the mouth by the hard palate.

The nasal septum is usually straight at birth, although it is sometimes bent as a result of a birth injury. It remains straight throughout early childhood, but with age it tends to bend toward one side or the other. Such a *deviated septum* may create an obstruction in the nasal cavity that makes breathing difficult.

Figure 16.1
Organs of the respiratory system.

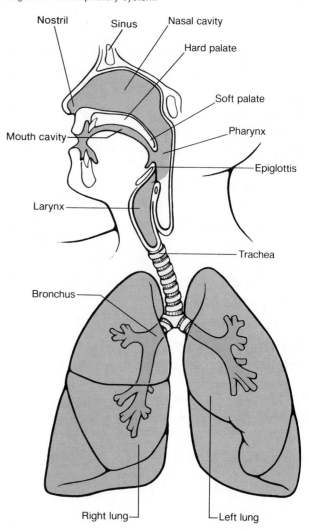

As figure 16.2 shows, **nasal conchae** (turbinate bones) curl out from the lateral walls of the nasal cavity on each side, dividing the cavity into passageways called the *superior, middle,* and *inferior meatuses* (see chapter 7). They also support the mucous membrane that lines the nasal cavity and help to increase its surface area.

The upper posterior portion of the nasal cavity, below the cribriform plate, is slitlike, and its lining contains the olfactory receptors that function in the sense of smell. The remainder of the cavity serves to conduct air to and from the nasopharynx.

The mucous membrane lining the nasal cavity contains pseudostratified ciliated epithelium that is rich in mucous-secreting goblet cells (see chapter 5). It also includes an extensive network of blood vessels and normally appears pinkish. As air passes over the membrane, heat radiates from the blood and warms the air.

Figure 16.2
Major features of the upper respiratory tract.

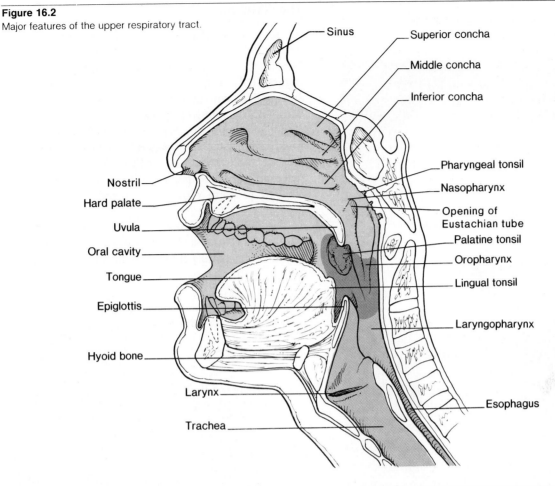

Figure 16.3
Cilia move mucus and trapped particles from the nasal cavity to
the pharynx, where they are swallowed.

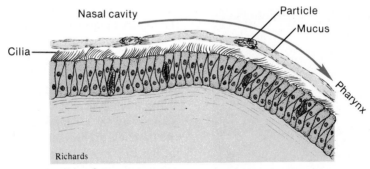

In this way, the temperature of the incoming air quickly adjusts to that of the body. In addition, the incoming air tends to become moistened by evaporation of water from the mucous lining. The sticky mucus secreted by the mucous membrane entraps dust and other small particles entering with the air.

As the cilia of the epithelial cells move, a thin layer of mucus and any entrapped particles are pushed toward the pharynx (figure 16.3). When the mucus reaches the pharynx, it is swallowed. In the stomach, any microorganisms in the mucus, including disease-causing forms, are likely to be destroyed by the action of gastric juice. Thus, the filtering mechanism provided by the mucous membrane not only prevents particles from reaching the lower air passages, but also helps to prevent respiratory infections.

1. What is meant by respiration?
2. What organs constitute the respiratory system?
3. What is the function of the mucous membrane that lines the nasal cavity?
4. What is the function of the cilia in the cells of this lining?

Figure 16.4

(a) X-ray film of a skull from the front and (b) from the side, showing air-filled sinuses within the bones. Which of these sinuses can you identify?

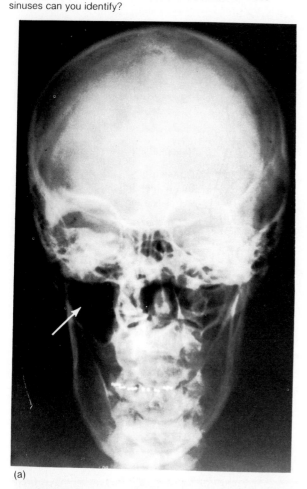

(a)

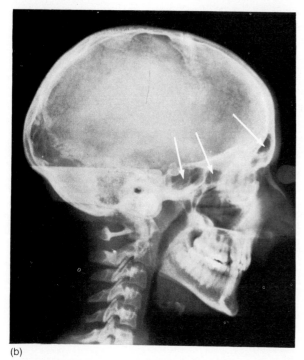

(b)

The Sinuses

As discussed in chapter 7, the **sinuses** (paranasal sinuses) are air-filled spaces located within the *maxillary, frontal, ethmoid,* and *sphenoid bones* of the skull (figure 16.4). These spaces open into the nasal cavity and are lined with mucous membranes that are continuous with the lining of the nasal cavity. Consequently, mucus secretions can drain from the sinuses into the nasal cavity. If this drainage is blocked by membranes that are swollen because of nasal infections or allergic reactions, the accumulating fluids may cause increasing pressure within a sinus and a painful sinus headache.

Although the sinuses function mainly to reduce the weight of the skull, they also serve as resonant chambers that affect the quality of the voice.

It is possible to illuminate a person's frontal sinus in a darkened room by holding a small flashlight just beneath the eyebrow. Similarly, the maxillary sinuses can be illuminated by holding the flashlight in the mouth.

---

1. Where are the sinuses located?
2. What are the functions of the sinuses?

The Pharynx

The **pharynx** (throat) is located behind the mouth cavity and between the nasal cavity and larynx. It functions as a passageway for food traveling from the oral cavity to the esophagus and for air passing between the nasal cavity and larynx (figure 16.2). It also aids in creating the sounds of speech (see chapter 14).

The Larynx

The **larynx** is an enlargement in the airway at the top of the trachea and below the pharynx. It serves as a passageway for air moving in and out of the trachea and functions to prevent foreign objects from entering the trachea. In addition, it houses the *vocal cords* and, because of this, is commonly called the voice box.

The larynx is composed primarily of muscles and cartilages that are bound together by elastic tissue. The largest of the cartilages (shown in figure 16.5) are the thyroid, cricoid, and epiglottic cartilages. These structures occur singly, and the other laryngeal cartilages—the arytenoid, corniculate, and cuneiform cartilages—are paired.

The **thyroid cartilage** was named for the thyroid gland that covers its lower part. This cartilage is the shieldlike structure that protrudes in the front of the neck and is sometimes called the Adam's apple. The

Figure 16.5

(a) Anterior view and *(b)* posterior view of the larynx.

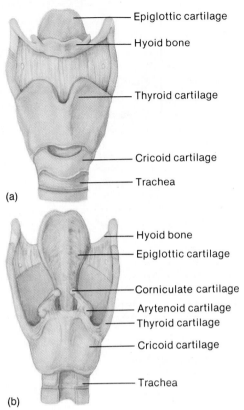

Figure 16.6

(a) Frontal section and *(b)* sagittal section of the larynx.

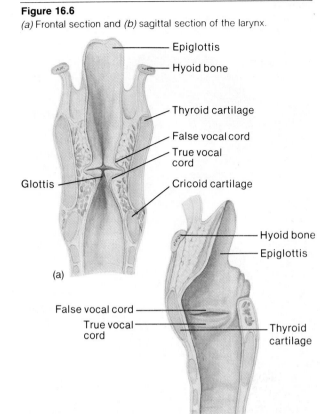

protrusion typically is more prominent in males than in females because of an effect of male sex hormones on the development of the larynx.

The **cricoid cartilage** lies below the thyroid cartilage and marks the lowermost portion of the larynx.

The **epiglottic cartilage** is attached to the upper border of the thyroid cartilage and supports a flaplike structure called the **epiglottis.** The epiglottis usually stands upright and allows air to enter the larynx. During swallowing, however, the larynx is raised by muscular contractions, and the epiglottis is pressed downward by the base of the tongue. As a result, the epiglottis partially covers the opening into the larynx, helping to prevent foods and liquids from entering the air passages.

The pyramid-shaped **arytenoid cartilages** are located above and on either side of the cricoid cartilage. Attached to the tips of the arytenoid cartilages are the tiny, conelike **corniculate cartilages.** These cartilages serve as attachments for muscles that help to regulate tension on the vocal cords during speech and aid in closing the larynx during swallowing.

The **cuneiform cartilages** are small cylindrical parts found in the mucous membrane between the epiglottic and the arytenoid cartilages. They function to stiffen the soft tissues in this region.

Inside the larynx, two pairs of horizontal folds in the mucous membrane extend inward from the lateral walls. The upper folds (vestibular folds) are called *false vocal cords* because they do not function in the production of sounds. Muscle fibers within these folds help to close the larynx during swallowing.

The lower folds are the *true vocal cords.* (Both pairs of cords are shown in figure 16.6.) They contain elastic fibers and are responsible for vocal sounds, which are created by forcing air between the vocal cords, causing them to vibrate from side to side. This action generates sound waves that can be formed into words by changing the shapes of the pharynx and oral cavity and by using the tongue and lips.

The *pitch* of the vocal sound is controlled by changing the tension on the cords. This is accomplished by contracting or relaxing various laryngeal muscles. Increasing the tension produces a higher pitched sound, and decreasing the tension creates a lower pitch.

The *intensity* (loudness) of a vocal sound is related to the force of the air passing over the vocal cords. Louder sounds are produced by using stronger blasts of air to cause the vocal cords to vibrate.

Figure 16.7
Vocal cords as viewed from above (a) with the glottis closed and (b) with the glottis open.

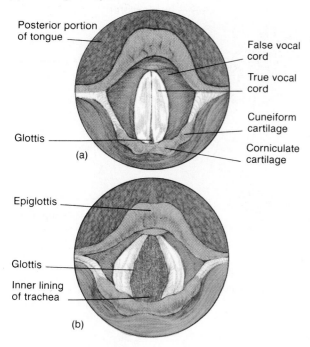

(a)

(b)

Figure 16.8
The trachea transports air between the larynx and the bronchi.

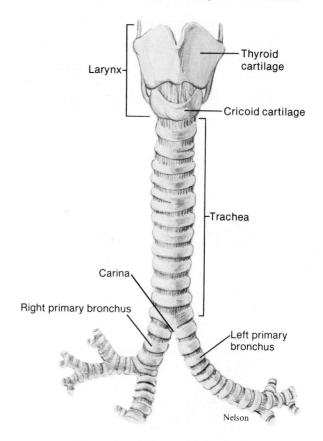

Nelson

During normal breathing, the vocal cords remain relaxed, and the opening between them, called the **glottis,** appears as a triangular slit. When food or liquid is swallowed, however, the glottis is closed by muscles within the larynx, and this prevents foreign substances from entering the trachea (figure 16.7).

The mucous membrane that lines the larynx continues to filter the incoming air by entrapping particles and moving them toward the pharynx by ciliary action.

Occasionally the mucous membrane of the larynx becomes inflamed and swollen as a result of an infection or an irritation from inhaled vapors. When this happens, the vocal cords may not vibrate as freely as before, and the voice may become hoarse. This condition is called *laryngitis,* and although it is usually mild, laryngitis is potentially dangerous because the swollen tissues may obstruct the airway and interfere with breathing. In such cases it may be necessary to provide a passageway by inserting a tube into the trachea through the nose, mouth, or surgical opening in the trachea.

---

1. What part of the respiratory tract is shared with the alimentary canal?
2. Describe the structure of the larynx.
3. How do the vocal cords function to produce sounds?
4. What is the function of the glottis? the epiglottis?

The Trachea

The **trachea** (windpipe) is a flexible cylindrical tube about 2.5 cm in diameter and 12.5 cm in length. It extends downward in front of the esophagus and into the thoracic cavity, where it splits into right and left bronchi, as shown in figure 16.8.

The inner wall of the trachea is lined with a ciliated mucous membrane that contains many goblet cells. This membrane continues to filter the incoming air and to move entrapped particles upward to the pharynx.

Within the tracheal wall there are about twenty C-shaped pieces of hyaline cartilage arranged one above the other. The open ends of these incomplete rings are directed posteriorly, and the gaps between their ends are filled with smooth muscle and connective tissues (figures 16.9 and 16.10). These cartilaginous rings prevent the trachea from collapsing and blocking the airway. At the same time, the soft tissues that complete the rings in the back allow the nearby esophagus to expand as it carries food to the stomach.

Figure 16.9
Cross section of the trachea. What is the significance of the
C-shaped rings of hyaline cartilage in its wall?

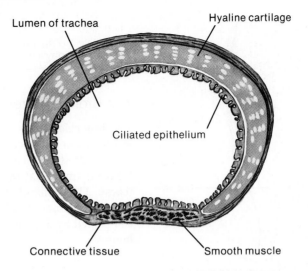

Lumen of trachea

Hyaline cartilage

Ciliated epithelium

Connective tissue

Smooth muscle

Figure 16.10
Light micrograph of a section of the tracheal wall.

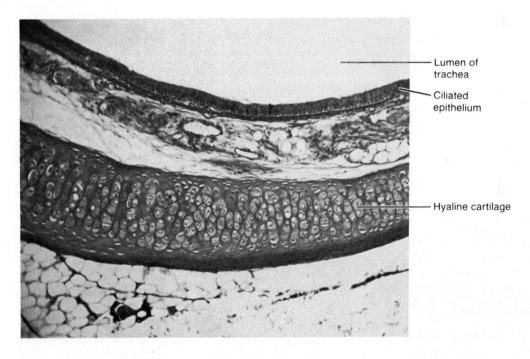

Lumen of trachea

Ciliated epithelium

Hyaline cartilage

Figure 16.11

A tracheostomy may be performed to allow air to bypass an obstruction within the larynx.

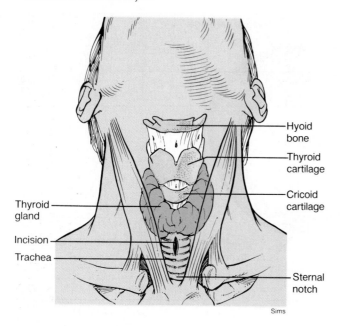

Thyroid gland

Incision

Trachea

Hyoid bone

Thyroid cartilage

Cricoid cartilage

Sternal notch

Sims

If the trachea becomes obstructed as a result of swollen tissues, excessive secretions, or the presence of foreign objects, it may be necessary to create an external opening in this tube so that air can bypass the obstruction (figure 16.11). This is called a *tracheostomy*. It may be a life-saving procedure because if the trachea is blocked, asphyxiation can occur within a few minutes.

The Bronchial Tree

The **bronchial tree** consists of the branched airways leading from the trachea to the microscopic air sacs. It begins with the right and left **primary bronchi** that arise from the trachea at the level of the fifth thoracic vertebrae. The openings of these tubes are separated by a ridge of cartilage called the *carina*. Each bronchus, accompanied by large blood vessels, enters its respective lung.

Figure 16.12

The bronchial tree consists of the passageways that connect the trachea and the alveoli.

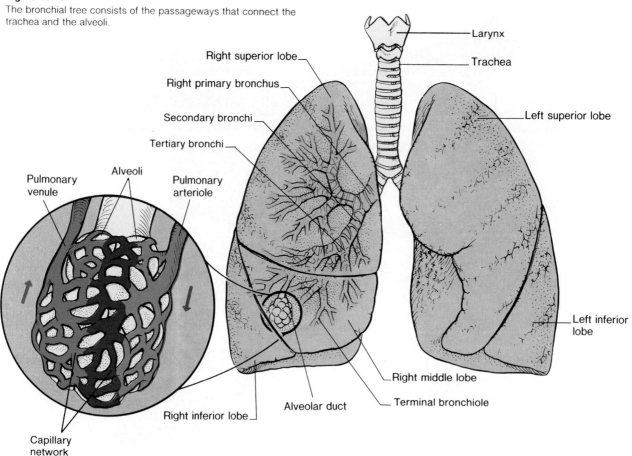

Right superior lobe

Right primary bronchus

Secondary bronchi

Tertiary bronchi

Pulmonary venule

Alveoli

Pulmonary arteriole

Larynx

Trachea

Left superior lobe

Left inferior lobe

Right middle lobe

Alveolar duct

Terminal bronchiole

Right inferior lobe

Capillary network

Branches of the Bronchial Tree

A short distance from its origin, each primary bronchus divides into **secondary** or **lobar bronchi** (two on the left and three on the right), which, in turn, branch into finer and finer tubes (figures 16.12, 16.13, and 16.14). The successive divisions of these branches from the lobar bronchus to the microscopic air sacs follow:

1. *Tertiary* or *segmental bronchi.* Each of these branches supplies a portion of the lung called a *bronchopulmonary segment.* Usually there are ten such segments in the right lung and eight in the left.

2. *Bronchioles.* These small branches of the segmental bronchi enter the basic units of the lung—the *lobules.*

3. *Terminal bronchioles.* These tubes branch from a bronchiole. There are 50 to 80 terminal bronchioles within a lobule of the lung.

4. *Respiratory bronchioles.* Two or more respiratory bronchioles branch from each terminal bronchiole. They are relatively short and have diameters of about 0.5 mm. They are called "respiratory" because a few air sacs bud from their sides; thus, they are the first

Figure 16.13
A plastic cast of the bronchial tree.

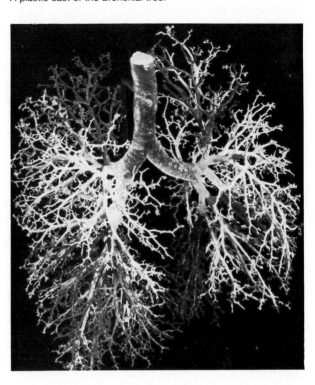

Figure 16.14
This bronchogram X-ray film reveals many branches of the bronchial tree. What other anatomic features can you identify?

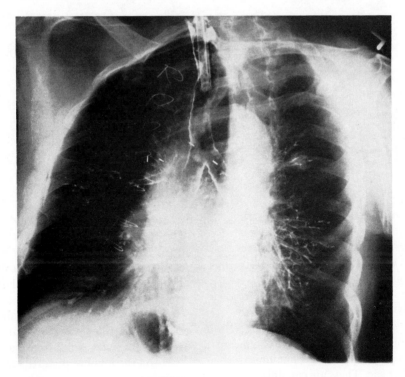

Figure 16.15

The respiratory tubes end in tiny alveoli, each of which is surrounded by a capillary network.

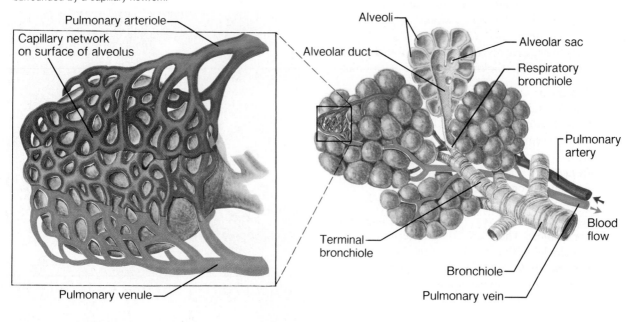

Figure 16.16

Alveoli appear as open spaces in this light micrograph of human lung tissue.

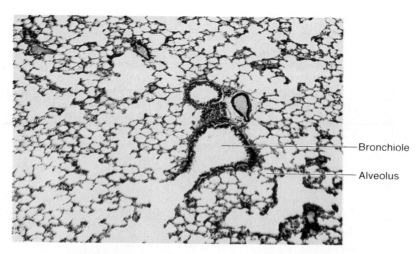

structures in the sequence that can engage in gas exchange.

5. *Alveolar ducts.* Two to ten long, branching alveolar ducts extend from each respiratory bronchiole (figure 16.15).

6. *Alveolar sacs.* Alveolar sacs are thin-walled, closely packed outpouchings of the alveolar ducts.

7. *Alveoli.* Alveoli are thin-walled, microscopic air sacs that open only on the side communicating with an alveolar sac. Thus, air can diffuse freely from the alveolar ducts, through the alveolar sacs, and into the alveoli (figures 16.16 and 16.17).

Structure of the Respiratory Tubes

The structure of a bronchus is similar to that of the trachea, but the cartilaginous rings are replaced with cartilaginous plates where the bronchus enters the lung. These plates are irregularly shaped, and they completely surround the tube giving it a cylindrical form. However, as finer and finer branch tubes appear, the amount of cartilage decreases, and it finally disappears in the bronchioles, which have diameters of about 1 mm.

As the amount of cartilage decreases, a layer of smooth muscle that surrounds the tube just beneath the mucosa becomes more prominent. This muscular layer

Figure 16.17
Scanning electron micrograph of lung tissue.

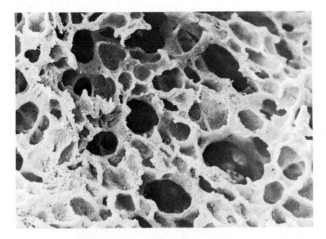

Figure 16.18
Oxygen diffuses from the air within the alveolus into the capillary, while carbon dioxide diffuses from the blood in the capillary into the alveolus.

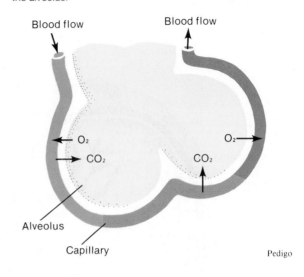

Pedigo

remains in the wall to the ends of the respiratory bronchioles, and only a few muscle fibers occur in the walls of the alveolar ducts.

Elastic fibers are scattered among the smooth muscle cells and are abundant in the connective tissue that surrounds the respiratory tubes. These fibers play an important role in the breathing mechanism, as explained in a subsequent section of this chapter.

As the tubes become finer, there is also a change in the type of cells that line them. For example, the lining of the larger tubes consists of pseudostratified, ciliated columnar epithelium; in the finer tubes, beginning with the respiratory bronchioles, the lining is cuboidal epithelium; and in the alveoli, it is simple squamous epithelium closely associated with a dense network of capillaries.

The trachea and bronchial tree can be examined directly using a flexible, optical instrument called a *fiberoptic bronchoscope.* This procedure (bronchoscopy) sometimes is useful in diagnosing tumors or other pulmonary diseases. It also may be employed to locate and remove aspirated foreign bodies in the air passages.

Functions of the Respiratory Tubes
The branches of the bronchial tree serve as air passages that continue to filter incoming air and distribute it to the alveoli in all parts of the lungs. The alveoli, in turn, provide a large surface area of thin epithelial cells through which gas exchanges can occur. During these exchanges, oxygen diffuses through the alveolar walls and enters the blood in nearby capillaries, while carbon dioxide diffuses from the blood through these walls and enters the alveoli (figures 16.18 and 16.19).

Figure 16.19
Scanning electron micrograph of casts of alveoli and associated capillary networks. These casts were prepared by filling the alveoli and blood vessels with resin and later removing the soft tissues by digestion, leaving only the resin casts (415X). (*Tissues and Organs: A Text-Atlas of Scanning Electron Microscopy,* by Richard D. Kessel and Randy Kardon. © 1979 W. H. Freeman and Company.)

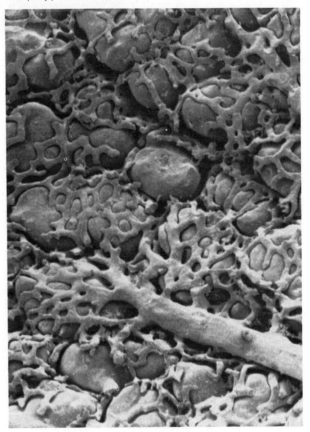

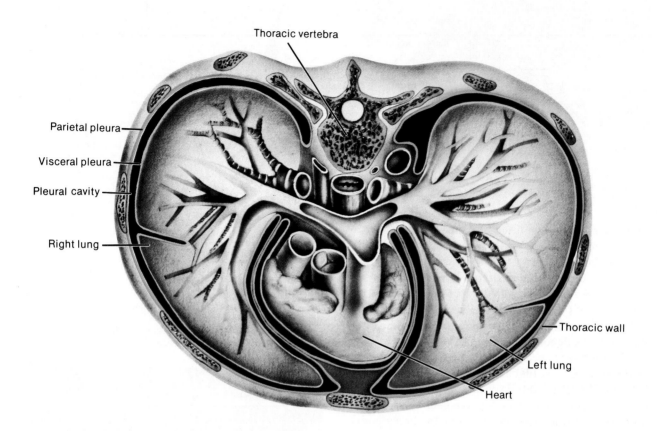

Thoracic vertebra

Parietal pleura

Visceral pleura

Pleural cavity

Right lung

Thoracic wall

Left lung

Heart

It is estimated that there are about 300 million alveoli in an adult lung and that these spaces have a total surface area of between 70 and 80 square meters. (This is roughly equivalent to the area of the floor in a room measuring 25 feet by 30 feet.)

1. What is the function of the cartilaginous rings in the tracheal wall?
2. How do the right and left bronchi differ in structure?
3. List the branches of the bronchial tree.
4. Describe the changes in structure that occur in the respiratory tubes as they become smaller and smaller.
5. How are gases exchanged in the alveoli?

The Lungs

The lungs are soft, spongy, cone-shaped organs located in the thoracic cavity. The right and left lungs are separated medially by the heart and the mediastinum, and they are enclosed by the diaphragm and the thoracic cage.

Each lung occupies most of the thoracic space on its side and is suspended in this cavity by its attachments, which include a bronchus and some large blood vessels. These tubular parts enter the lung on its medial surface through a region called the **hilus.** A layer of serous membrane, the **visceral pleura,** is firmly attached to the surface of each lung, and this membrane folds back at the hilus to become the **parietal pleura.** The parietal pleura, in turn, forms part of the mediastinum and lines the inner wall of the thoracic cavity.

The potential space between the visceral and parietal pleurae is called the **pleural cavity,** and it contains a thin film of serous fluid. This fluid lubricates the adjacent pleural surfaces, reducing friction as they move against one another during breathing. It also helps to hold the pleural membranes together, as explained in the next section of this chapter (figure 16.20).

As figure 16.12 shows, the right lung is larger than the left one, and it is divided into three parts called the superior, middle, and inferior *lobes.* The left lung consists of two parts, a superior and an inferior lobe.

Each lobe is supplied by a lobar bronchus of the bronchial tree. It also has connections to blood and lymphatic vessels and is enclosed by connective tissues.

Chart 16.1   Parts of the respiratory system

Part	Description	Function
Nose	Part of face centered above the mouth and below the space between the eyes	Nostrils provide entrance to nasal cavity; internal hairs begin to filter incoming air
Nasal cavity	Hollow space behind nose	Conducts air to pharynx; mucous lining filters, warms, and moistens air
Sinuses	Hollow spaces in various bones of the skull	Reduce weight of the skull; serve as resonant chambers
Pharynx	Chamber behind mouth cavity and between nasal cavity and larynx	Passageway for air moving from nasal cavity to larynx and for food moving from mouth cavity to esophagus
Larynx	Enlargement at the top of the trachea	Passageway for air; prevents foreign objects from entering trachea; houses vocal cords
Trachea	Flexible tube that connects larynx with bronchial tree	Passageway for air; mucous lining continues to filter air
Bronchial tree	Branched tubes that lead from the trachea to the alveoli	Conducts air from the trachea to the alveoli; mucous lining continues to filter air
Lungs	Soft, cone-shaped organs that occupy a large portion of the thoracic cavity	Contain the air passages, alveoli, blood vessels, connective tissues, lymphatic vessels, and nerves of the lower respiratory tract

A lobe is further subdivided by connective tissue into **lobules,** and each of these units contains terminal bronchioles together with their alveolar ducts, alveolar sacs, alveoli, nerves, and associated blood and lymphatic vessels. Thus, the substance of a lung includes air passages, alveoli, blood vessels, connective tissues, lymphatic vessels, and nerves.

Chart 16.1 summarizes the characteristics of the major parts of the respiratory system.

1. Where are the lungs located?
2. What is the function of the serous fluid within the pleural cavity?
3. How does the structure of the right lung differ from that of the left lung?
4. What kinds of structures make up a lung?

Mechanism for Breathing

Breathing, which is also called *pulmonary ventilation,* entails the movement of air from outside the body into the bronchial tree and alveoli, followed by a reversal of this air movement. These actions are termed **inspiration** (inhalation) and **expiration** (exhalation).

Inspiration

Atmospheric pressure, due to the weight of air, is the force that causes air to move into the lungs. At sea level, this pressure is sufficient to support a column of mercury about 760 mm high in a tube (figure 16.21). Thus, normal air pressure is said to equal 760 mm of mercury (Hg).

Figure 16.21

Atmospheric pressure is sufficient to support a column of mercury 760 mm high.

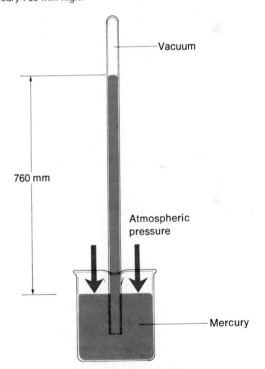

Vacuum

760 mm

Atmospheric pressure

Mercury

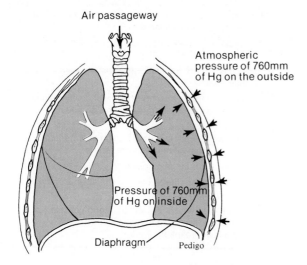

Air passageway

Atmospheric pressure of 760mm of Hg on the outside

Pressure of 760mm of Hg on inside

Diaphragm   Pedigo

Air pressure is exerted on all surfaces in contact with air, and since people breathe air, the inside surfaces of their lungs also are subjected to pressure. In other words, the pressures on the inside of the lungs and alveoli and on the outside of the thoracic wall are about the same (figure 16.22).

If the pressure on the inside of the lungs and alveoli (intra-alveolar pressure) decreases, air from the outside will then be pushed into the airways by atmospheric pressure. This is what happens during normal inspiration. Muscle fibers in the dome-shaped *diaphragm* below the lungs are stimulated to contract by impulses carried on the phrenic nerves. As this happens, the diaphragm moves downward, the size of the thoracic cavity is enlarged, and the intra-alveolar pressure is reduced about 1mm Hg below that of atmospheric pressure. In response to this lower pressure, air is forced into the airways by atmospheric pressure, and the lungs expand.

While the diaphragm is contracting and moving downward, the *external intercostal muscles* between the ribs and certain thoracic muscles may be stimulated to contract. This action raises the ribs and elevates the sternum, so that the size of the thoracic cavity increases even more. As a result, the pressure inside is further reduced and more air is forced into the airways by atmospheric pressure (figure 16.23).

The expansion of the lungs is aided by the fact that the parietal pleura, on the inner wall of the thoracic cavity, and the visceral pleura, attached to the surface of the lungs, are separated only by a thin film of serous fluid. The *water molecules* in this fluid have

a great attraction for one another, and the resulting force, called **surface tension,** holds the moist surfaces of the pleural membranes tightly together. Consequently, when the thoracic wall is moved upward and outward by the action of the intercostal muscles, the parietal pleura is moved too, and the visceral pleura follows it. This helps to expand the lungs in all directions. The steps in inspiration are summarized in chart 16.2.

The attraction between adjacent moist membranes is sufficient to cause the collapse of the tiny air sacs, which also have moist surfaces. The alveolar cells, however, secrete a mixture of lipoproteins, called **surfactant,** that acts to reduce surface tension and decrease this tendency to collapse.

Sometimes the lungs of a newborn fail to produce enough surfactant, and the newborn's breathing mechanism may be unable to overcome the force of surface tension. Consequently, the lungs cannot be ventilated and the newborn is likely to die of suffocation. This condition is called *respiratory distress syndrome* and it is the primary cause of respiratory difficulty in immature newborns and those born to mothers with certain forms of diabetes mellitus.

If a person needs to take a deeper than normal breath, the diaphragm and external intercostal muscles usually can be contracted to a greater degree. Additional muscles, such as the pectoralis minors and sternocleidomastoids, can be used to pull the thoracic cage further upward and outward (figure 16.24).

The ease with which the lungs can be expanded as a result of pressure changes occurring during breathing is called *compliance* (distensibility). In a normal lung, compliance decreases as the lung volume increases, because an inflated lung is more difficult to expand than a deflated one. Conditions that tend to obstruct air passages, destroy lung tissue, or impede lung expansion in other ways also decrease compliance.

Expiration

The forces responsible for normal expiration come from *elastic recoil* of tissues and surface tension. The lungs and thoracic wall, for example, contain a considerable amount of elastic tissue, and as the lungs expand during inspiration, these tissues are stretched. When the diaphragm lowers, the abdominal organs beneath it are compressed. As the diaphragm and the external intercostal muscles relax following inspiration, the elastic tissues cause the lungs and thoracic cage to recoil, and they return to their original positions. Similarly, elastic tissues within the abdominal organs cause them to

Chart 16.2 Major events in inspiration	
1. Nerve impulses are carried on phrenic nerves to muscle fibers in the diaphragm causing them to contract. 2. As the dome-shaped diaphragm moves downward, the size of the thoracic cavity is increased. 3. At the same time, the external intercostal muscles may contract, raising the ribs and causing the size of the thoracic cavity to increase still more.	4. As the size of the thoracic cavity is increased, the intra-alveolar pressure is decreased. 5. Atmospheric pressure, which is relatively greater on the outside, forces air into the respiratory tract through the air passages. 6. The lungs become inflated.

Figure 16.23

The size of the thoracic cavity enlarges as the diaphragm and external intercostal muscles contract.

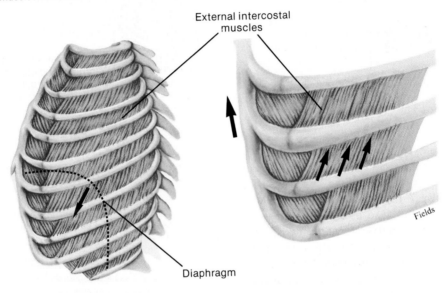

Figure 16.24

(a) Shape of the thorax at the end of normal inspiration.
(b) Shape of the thorax at the end of maximal inspiration, aided by contraction of the sternocleidomastoid and pectoralis minor muscles.

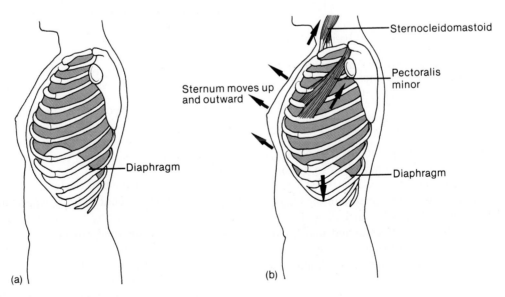

Figure 16.25

(a) Normal expiration is due to elastic recoil of the thoracic wall and abdominal organs; (b) maximal expiration is aided by contraction of the abdominal wall and posterior internal intercostal muscles.

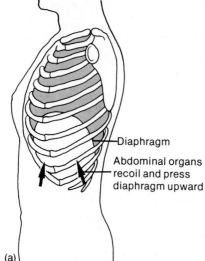

Diaphragm

Abdominal organs recoil and press diaphragm upward

(a)

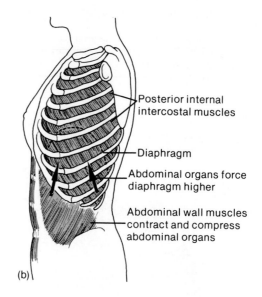

Posterior internal intercostal muscles

Diaphragm

Abdominal organs force diaphragm higher

Abdominal wall muscles contract and compress abdominal organs

(b)

spring back into their previous locations, pushing the diaphragm upward. At the same time, surface tension that develops between the moist surfaces of the alveolar linings tends to cause alveoli to decrease in diameter. Each of these factors tends to increase the intra-alveolar pressure about 1mm Hg above atmospheric pressure, so that the air inside the lungs is forced out through the respiratory passages. Thus, normal expiration is a passive process.

The recoil of the elastic fibers within the lung tissues tends to reduce the pressure in the pleural cavity. Consequently the pressure between the pleural membranes (intrapleural pressure) is usually about 4mm Hg less than atmospheric pressure.

Since the visceral and parietal pleural membranes are held together by surface tension, there is normally no actual space in the pleural cavity between them. However, if the thoracic wall is punctured, atmospheric air may enter the pleural cavity and create a real space between the membranes. This condition is called *pneumothorax,* and when it occurs, the lung on the affected side may collapse because of its elasticity.

If a person needs to exhale more air than normal, the posterior *internal intercostal muscles* can be contracted. These muscles pull the ribs and the sternum down and inward increasing the pressure in the lungs. Also, the *abdominal wall muscles,* including the external and internal obliques, transversus abdominis, and rectus abdominis, can be used to squeeze the abdominal organs inward. Thus, the abdominal wall muscles

Chart 16.3   Major events in expiration

1. The diaphragm and external respiratory muscles relax.
2. Elastic tissues of the lungs, thoracic cage, and abdominal organs, which were stretched during inspiration, suddenly recoil and surface tension causes alveolar walls to collapse.
3. Tissues recoiling around the lungs cause the intra-alveolar pressure to increase.
4. Air is squeezed out of the lungs and into the air passages.

can cause the pressure in the abdominal cavity to increase and force the diaphragm still higher against the lungs (figure 16.25). As a result of these actions, additional air may be squeezed out of the lungs.

Chart 16.3 summarizes the steps in expiration.

1. Describe the events in inspiration.
2. How does surface tension aid expansion of the lungs during inspiration?
3. What forces are responsible for normal expiration?

Respiratory Air Volumes

The amount of air that enters the lungs during a normal, quiet inspiration is about 500 cc. Approximately the same volume leaves during a normal expiration. This volume is termed the **tidal volume.**

During forced inspiration, a quantity of air in addition to the tidal volume enters the lungs. This additional volume is called the *inspiratory reserve volume* (complemental air), and it equals about 3000 cc.

Figure 16.26
Respiratory air volumes.

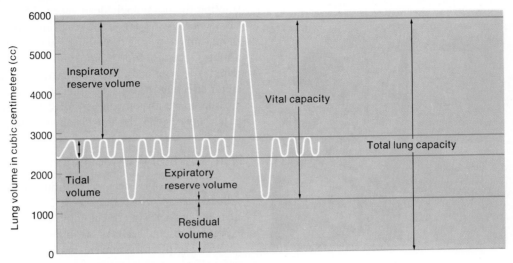

Volume	Quantity of air	Description	Volume	Quantity of air	Description
Tidal volume (TV)	500 cc	Volume moved in or out of the lungs during quiet breathing	Vital capacity (VC)	4600 cc	Maximum volume of air that can be exhaled after taking the deepest breath possible: VC=TV+IRV+ERV
Inspiratory reserve volume (IRV)	3000 cc	Volume that can be inhaled during forced breathing in addition to tidal volume	Residual volume (RV)	1200 cc	Volume that remains in the lungs at all times
Expiratory reserve volume (ERV)	1100 cc	Volume that can be exhaled during forced breathing in addition to tidal volume	Total lung capacity (TLC)	5800 cc	Total volume of air that the lungs can hold: TLC=VC+RV

Chart 16.4 Respiratory air volumes

During forced expiration, about 1100 cc of air in addition to the tidal volume can be expelled from the lungs. This quantity is called the *expiratory reserve volume* (supplemental air). However, even after the most forceful expiration, about 1200 cc of air remains in the lungs. This volume is called *residual volume.*

Residual air remains in the lungs at all times and, consequently, newly inhaled air is always mixed with air that is already in the lungs. This prevents the oxygen and carbon dioxide concentrations in the lungs from fluctuating excessively with each breath.

If the *inspiratory reserve volume* (3000 cc) is combined with the *tidal volume* (500 cc) and the *expiratory reserve volume* (1100 cc), the total is termed the **vital capacity** (4600 cc). This volume is the maximum amount of air a person can exhale after taking the deepest breath possible.

The *vital capacity* plus the *residual volume* equals the *total lung capacity,* which is about 5800 cc.

This total varies with age, sex, and body size (figure 16.26).

Some of the air that enters the respiratory tract during breathing fails to reach the alveoli. This volume (about 150 cc) remains in the passageways of the trachea, bronchi, and bronchioles. Since gas exchanges do not occur through the walls of these passages, this air is said to occupy *anatomic dead space.*

Occasionally, air sacs in some regions of the lungs are nonfunctional due to poor blood flow in the adjacent capillaries. This creates *alveolar dead space.* If the anatomic and alveolar dead space volumes are combined, the total is called *physiologic dead space.* In a normal lung, however, the anatomic and physiologic dead spaces are essentially the same (150 cc).

The respiratory air volumes are summarized in chart 16.4.

Figure 16.27

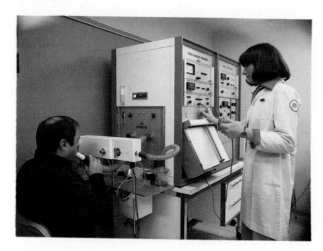

A spirometer, such as this, can be used to measure respiratory air volumes.

With the exception of the residual volume, which is measured using special techniques, respiratory air volumes can be determined by using an instrument, called a *spirometer,* shown in figure 16.27. The results of such measurements may be useful in evaluating the courses of various diseases like emphysema, pneumonia, or lung cancer in which there are losses in functional lung tissue. The results also may be valuable in studying the progress of bronchial asthma and other diseases accompanied by obstructions of the air passages.

1. What is meant by tidal volume?
2. Distinguish between inspiratory and expiratory reserve volumes.
3. How is vital capacity measured?
4. How is the total lung capacity calculated?

Alveolar Ventilation

The amount of new atmospheric air that is moved into the respiratory passages each minute is called the *minute respiratory volume.* This volume can be calculated by multiplying the tidal volume by the breathing rate. Thus, if the tidal volume is 500 cc and the breathing rate is 12 breaths per minute, the minute respiratory volume is 500 cc × 12 or 6000 cc per minute. However, not all of this new air reaches the alveoli. Instead, much of it remains in the air passages, occupying the physiologic dead space.

The amount of new air that does reach the alveoli and is available for gas exchange is calculated by subtracting the physiologic dead space (150 cc) from the tidal volume (500 cc). If the resulting volume (350 cc) is multiplied by the breathing rate (12 breaths per minute), the *alveolar ventilation rate* (4200 cc per minute) is obtained. This ventilation rate is a major factor affecting the concentration of oxygen and carbon dioxide in the alveoli; thus, it affects the exchange of gases between the alveolar air and the blood.

Nonrespiratory Air Movements

Air movements that occur in addition to breathing are called *nonrespiratory movements.* They are used to clear air passages, as in coughing and sneezing, or to express emotional feelings, as in laughing and crying.

Nonrespiratory movements usually result from *reflexes,* although sometimes they are initiated voluntarily. A cough, for example, can be produced through conscious effort or may be triggered by the presence of a foreign object in an air passage.

The act of *coughing* involves taking a deep breath, closing the glottis, and forcing air upward from the lungs against the closure. Then the glottis is suddenly opened, and a blast of air is forced upward from the lower respiratory tract. Usually this rapid rush of air will remove the substance that triggered the reflex.

The most sensitive areas of the air passages are in the larynx and regions near the branches of the major bronchi. The distal portions of the bronchioles (respiratory bronchioles), alveolar ducts, and alveoli lack a nerve supply. Consequently, before any material in these parts can trigger a cough reflex, it must be moved into the larger passages of the respiratory tract.

A *sneeze* is much like a cough, but it clears the upper respiratory passages rather than lower ones. This reflex is usually initiated by a mild irritation in the lining of the nasal cavity and, in response, a blast of air is forced up through the glottis. This time, the air is directed into the nasal passages by depressing the uvula, thus closing the opening between the pharynx and the oral cavity.

Laughing involves taking a breath and releasing it in a series of short expirations. *Crying* consists of very similar movements, and sometimes it is necessary to note a person's facial expressions in order to distinguish laughing from crying.

A *hiccup* is caused by sudden inspiration due to a spasmodic contraction of the diaphragm while the glottis is closed. The sound of the hiccup is created by

Chart 16.5	Nonrespiratory air movements	
Air movement	Mechanism	Function
Coughing	Deep breath is taken, glottis is closed, and air is forced against the closure; suddenly the glottis is opened and a blast of air passes upward	Clears lower respiratory passages
Sneezing	Same as coughing, except air moving upward is directed into the nasal cavity by depressing the uvula	Clears upper respiratory passages
Laughing	Deep breath is released in a series of short expirations	Expresses emotional happiness
Crying	Same as laughing	Expresses emotional sadness
Hiccuping	Diaphragm contracts spasmodically while the glottis is closed	No useful function
Yawning	Deep breath taken	Ventilates a large proportion of the alveoli and aids oxygenation of the blood
Speech	Air is forced through the larynx, causing vocal cords to vibrate; words are formed by lips, tongue, and soft palate	Communication

air striking the vocal folds. The reason for hiccups is not well understood, and this movement apparently serves no useful function.

Yawning is thought to aid respiration by providing an occasional deep breath. During normal, quiet breathing, not all of the alveoli are ventilated and some blood may pass through the lungs without becoming well oxygenated. It is believed that this low blood oxygen concentration somehow triggers the yawn reflex, and in response, a very deep breath is taken. The deep breath ventilates a large proportion of the alveoli and the blood oxygen level rises.

As described earlier, *speech* results from sounds that are produced when air is forced through the larynx causing the vocal cords to vibrate. Words are formed from such sounds by actions of the lips, tongue, and soft palate.

Chart 16.5 summarizes nonrespiratory air movements.

---

1. How is the minute respiratory volume calculated? the alveolar ventilation rate?
2. What nonrespiratory air movements help to clear the air passages?
3. What nonrespiratory air movements are used to express emotions?
4. What seems to be the function of a yawn?

Control of Breathing

Although respiratory muscles can be controlled voluntarily, normal breathing is a rhythmic, involuntary act that continues when a person is unconscious.

The Respiratory Center

Breathing is controlled by a poorly defined collection of neurons in the brain stem called the **respiratory center.** This center initiates impulses periodically that travel on cranial and spinal nerves to various breathing muscles causing inspiration and expiration. The center also is able to adjust the rate and depth of breathing. As a result, cellular needs for a supply of oxygen and the removal of carbon dioxide are met, even during periods of strenuous physical exercise.

Components of the respiratory center are widely scattered throughout the pons and medulla oblongata. However, two areas of the center are of special interest. They are the rhythmicity area of the medulla and the pneumotaxic area of the pons, shown in figure 16.30.

The **medullary rhythmicity area** includes two groups of neurons that extend throughout the length of the medulla oblongata. They are called the dorsal respiratory group and the ventral respiratory group.

The *dorsal respiratory group* is responsible for the basic rhythm of breathing. Neurons of this group emit bursts of impulses that signal the diaphragm and other inspiratory muscles to contract. The impulses of each burst begin weakly, then increase in strength for about two seconds, and cease abruptly. The breathing muscles that contract in response to the impulses cause the volume of air entering the lungs to increase steadily. The neurons remain inactive while expiration occurs passively, and then they emit another burst of inspiratory impulses so that the inspiration-expiration cycle is repeated.

Some Respiratory Disorders
A Clinical Application

Although respiratory disorders are caused by a variety of factors, some are characterized by inadequate ventilation. This group includes paralysis of various breathing muscles, bronchial asthma, emphysema, and lung cancer.

Paralysis of breathing muscles is sometimes caused by injuries to the respiratory center or to spinal nerve tracts that transmit motor impulses. In other instances, paralysis may be due to a disease, such as *poliomyelitis,* that affects parts of the central nervous system and injures motor neurons. The consequences of such paralysis depend on which muscles are affected. Sometimes, by increasing their responses, other muscles are able to compensate for functional losses of a paralyzed muscle. If unaffected muscles are unable to ventilate the lungs adequately, a person must be provided with some type of mechanical breathing device in order to survive.

Bronchial asthma is a condition commonly caused by an *allergic reaction* to foreign substances in the respiratory tract. Typically the foreign substance is a plant pollen that enters with inhaled air. As a result of this reaction, the walls of the small bronchioles become edematous, the cells lining the respiratory tubes secrete abnormally large amounts of thick mucus, and the smooth muscles in these tubes contract. These muscles cause the bronchioles to constrict, reducing the diameters of the air passages. As these changes occur, the person finds it increasingly difficult to breathe and produces a characteristic wheezing sound, as air moves through narrowed passages.

An asthmatic person usually finds it harder to force air out of the lungs than to bring it in. This is because inspiration involves powerful breathing muscles, and as they contract, the lungs expand. This expansion of the tissues helps to open the air passages. Expiration, on the other hand, is usually a passive process due to elastic recoil of stretched tissues. Also, it causes compression of the tissues and decreases the diameters of the bronchioles, and this adds to the problem of moving air through the narrowed air passages.

Emphysema is a progressive, degenerative disease characterized by the destruction of many alveolar walls. As a result, clusters of small air sacs merge to form larger chambers, so that the total surface area of the respiratory membrane decreases. At the same time, the alveolar walls tend to lose their elasticity and the capillary networks associated with the alveoli become less abundant (figure 16.28).

Because of the loss of tissue elasticity, the person with emphysema finds it increasingly difficult to force air out of the lungs. As was mentioned, normal expiration involves an elastic recoil of inflated tissues; in emphysema, abnormal muscular efforts are required to produce this movement.

The cause of emphysema is not clear, but some investigators believe it develops in response to prolonged exposure to respiratory irritants, such as those in tobacco smoke and polluted air. Emphysema is becoming one of the more common respiratory disorders among older persons, although it is not limited to this group.

Lung cancer, like other cancers, involves an uncontrolled growth of abnormal cells. These cells develop in and around the normal tissues and deprive them of nutrients. In effect, the cancer cells cause the death of normal cells by crowding them out.

Some cancerous growths in the lungs result from cancer cells that spread (metastasize) from other parts of the body, such as the breasts, alimentary tract, liver, or kidneys.

Primary pulmonary cancer, which begins in the lungs, is the most common form of cancer in males today. It also is becoming increasingly common among females and is rapidly replacing breast cancer as the leading cause of death from cancer among women. Primary pulmonary cancer may arise from epithelial cells, connective tissue cells, or various blood cells. The most common form arises from epithelium and is called *bronchogenic carcinoma,* which means it originates from the cells that line the tubes of the bronchial tree. This type of cancer seems to occur in response to excessive irritation such as that produced by prolonged exposure to tobacco smoke (figure 16.29).

Once lung cancer cells have appeared, they are likely to produce masses that obstruct air passages and reduce the amount of alveolar surface available for gas exchanges. Furthermore, bronchogenic carcinoma is likely to spread to other tissues relatively quickly and establish secondary cancers. The sites of such secondary cancer commonly include lymph nodes, liver, bones, brain, and kidneys.

Lung cancer is often difficult to control. Usually it is treated by surgical removal of the diseased portions of the lungs, by exposure to ionizing radiation, and by the use of drugs (chemotherapy). Despite these treatments, the survival rate among lung cancer victims remains extremely low.

Figure 16.28

(a) As emphysema develops, the alveoli tend to merge, (b) forming larger chambers.

Figure 16.29

Lung cancer usually starts in the lining (epithelium) of the bronchus. (a) The normal lining shows (4) columnar cells with (2) hairlike cilia, (3) goblet cells that secrete (1) mucus, and (5) basal cells from which new columnar cells arise. (6) A basement membrane separates the epithelial cells from (7) the underlying connective tissue. (b) In the first stage of lung cancer, the basal cells divide repeatedly. The goblet cells secrete excessive mucus, and the cilia function less efficiently in moving the heavy mucus secretion. (c) With the continued multiplication of basal cells, the columnar and goblet cells are displaced. The basal cells penetrate the basement membrane and invade the deeper connective tissue.

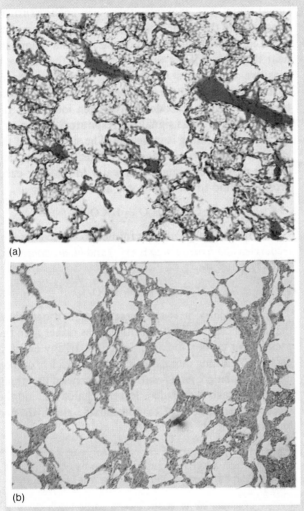

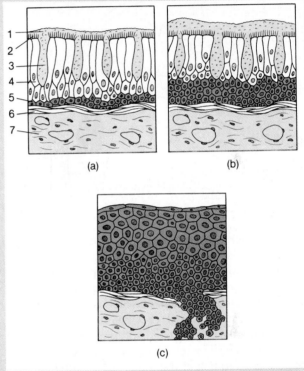

Figure 16.30

The respiratory center is located in the pons and the medulla oblongata.

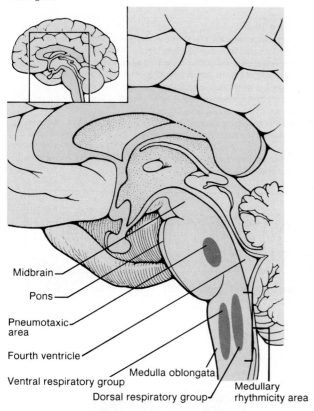

Midbrain

Pons

Pneumotaxic area

Fourth ventricle

Ventral respiratory group

Dorsal respiratory group

Medulla oblongata

Medullary rhythmicity area

The *ventral respiratory group* is quiescent during normal breathing. When there is need for more forceful breathing, however, neurons in this group generate impulses that increase inspiratory movement. Other neurons in the group activate muscles associated with forceful expiration, as well. (See figure 16.30.)

Neurons in the *pneumotaxic area* transmit impulses to the dorsal respiratory group continuously and regulate the duration of inspiratory bursts originating from the dorsal group. In this way the pneumotaxic neurons control the rate of breathing. More specifically, when pneumotaxic signals are strong, the inspiratory bursts have shorter durations, and the rate of breathing is increased; when the pneumotaxic signals are weak, the inspiratory bursts have longer durations, and the rate of breathing is decreased.

---

1. Where is the respiratory center located?
2. Describe how the respiratory center functions to maintain a normal breathing pattern.
3. Explain how the breathing pattern may be changed.

Factors Affecting Breathing

In addition to the controls exerted by the respiratory center, breathing rate and depth are influenced by a variety of other factors. These include the presence of certain chemicals in body fluids, the degree to which the lung tissues are stretched, and the person's emotional state. For example, there are *chemosensitive areas* within the respiratory center. These areas are located in the ventral portion of the medulla oblongata near the origins of the vagus nerves, and they are very sensitive to changes in the blood concentrations of carbon dioxide and hydrogen ions. Thus, if the concentration of carbon dioxide or hydrogen ions rises, the chemosensitive areas signal the respiratory center, and the rate of breathing is increased. The similarity of the effects of these two substances is related to the fact that carbon dioxide combines with water in blood or cerebrospinal fluid to form carbonic acid (H_2CO_3):

$$CO_2 + H_2O \rightleftarrows H_2CO_3$$

The carbonic acid thus formed soon becomes ionized, releasing *hydrogen ions* (H^+) and *bicarbonate ions* (HCO_3^-).

$$H_2CO_3 \rightleftarrows H^+ + HCO_3^-$$

Apparently, it is the presence of hydrogen ions that influences the chemosensitive areas rather than the presence of carbon dioxide molecules. In any event, a person's breathing rate increases when air rich in carbon dioxide is inhaled. As a result of the increased breathing rate, more carbon dioxide is lost in exhaled air, and the blood concentrations of carbon dioxide and hydrogen ions are reduced.

Air to which additional carbon dioxide has been added is sometimes used to stimulate the rate and depth of breathing.

Ordinary air is about 0.04% carbon dioxide. If a patient inhales air containing 4% carbon dioxide, the breathing rate usually doubles.

Low blood oxygen seems to have little direct effect upon the chemosensitive areas associated with the respiratory center. Instead, changes in blood oxygen concentration is sensed by *chemoreceptors* in specialized structures called the *carotid* and *aortic bodies,* which are located in the walls of certain large arteries (carotid arteries and aorta) in the neck and thorax (see chapter 18). When these receptors are stimulated by decreased oxygen concentration, impulses are transmitted to the respiratory center, and the breathing rate is increased. This mechanism usually is not triggered

Figure 16.31

In the process of inspiration, motor impulses travel from the respiratory center to the diaphragm and external intercostal muscles, which contract and cause the lungs to expand. This expansion stimulates stretch receptors in the lungs to send inhibiting impulses to the respiratory center, thus preventing overinflation.

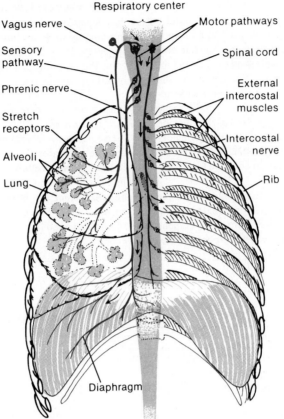

until the blood oxygen concentration reaches a very low level; thus, oxygen seems to play only a minor role in the control of normal respiration.

Although the chemoreceptors of the carotid and aortic bodies also are stimulated by changes in the blood concentrations of carbon dioxide and hydrogen ions, these substances have a much more powerful effect when they act on the chemosensitive areas of the respiratory center. Thus, the effects of carbon dioxide and hydrogen ions on the carotid and aortic bodies are relatively unimportant.

An exception to the normal pattern of chemical control may occur in patients who have chronic obstructive pulmonary diseases (COPD), such as asthma and bronchitis. Over a period of time, these patients seem to adapt to the presence of high levels of carbon dioxide, and in such persons low oxygen levels may serve as effective respiratory stimuli.

An *inflation reflex* (Hering-Breuer reflex) helps to regulate the depth of breathing. This reflex occurs when stretch receptors in the visceral pleura, bronchioles, and alveoli are stimulated as a result of lung tissues being overstretched. The sensory impulses of the reflex travel via the vagus nerves to the pneumotaxic area, and the duration of inspiratory movements is reduced. This action prevents overinflation of the lungs during forceful breathing. (See figure 16.31.)

The normal breathing pattern also may be altered if a person is emotionally upset. Fear, for example, typically causes an increased breathing rate, as does pain; thus, a person may gasp in response to a sudden fright or the chill of a cold shower.

Then, too, because the respiratory muscles are voluntary, the breathing pattern can be altered consciously. In fact, breathing can be stopped altogether for a time.

Exercise and Breathing
A Practical Application

When a person engages in moderate to heavy physical exercise, the amount of oxygen used by the skeletal muscles increases greatly. For example, a young man at rest will utilize about 250 ml of oxygen per minute, but may require 3600 ml per minute during maximal exercise. While oxygen utilization is increasing, the volume of carbon dioxide produced increases also. Since decreased blood oxygen and increased blood carbon dioxide are stimulating to the respiratory center, it is not surprising that exercise is accompanied by an increase in breathing rate. Studies have revealed, however, that the blood oxygen and carbon dioxide concentrations usually remain nearly unchanged during exercise—a reflection of the respiratory system's effectiveness in obtaining oxygen and releasing carbon dioxide to the outside.

This observation has led investigators to search for other factors that might act to increase the breathing rate during exercise. The mechanism that seems to be responsible for most of the increase involves the *cerebral cortex* and *proprioceptors* associated with muscles and joints. (See chapter 12.) Specifically, the cortex seems to transmit stimulating impulses to the respiratory center whenever it signals skeletal muscles to contract. At the same time, muscular movements stimulate proprioceptors, and a *joint reflex* is triggered. In this reflex, sensory impulses are transmitted from the proprioceptors to the respiratory center, and the breathing rate increases.

Whenever an increase in the breathing rate occurs during exercise, there must also be an increase in the blood flow if the needs of body cells are to be met. Physical exercise places an increased demand on both the respiratory and the circulatory systems. If either of these systems fails to keep up with cellular demands, the person will begin to feel out of breath. This feeling, however, is usually due to an inability of the heart and circulatory system to move enough blood between the lungs and the body cells, rather than to an inability of the respiratory system to provide enough air.

Chart 16.6   Factors affecting breathing

Factor	Receptors stimulated	Response	Effect
Stretch of tissues	Stretch receptors in visceral pleura, bronchioles, and alveoli	Inhibits inspiration	Prevents overinflation of lungs during forceful breathing
Low blood oxygen	Chemoreceptors in carotid and aortic bodies	Increases breathing rate	Increases blood oxygen concentration
High blood carbon dioxide	Chemosensitive areas of the respiratory center	Increases breathing rate	Decreases blood carbon dioxide concentration
High blood hydrogen ion	Same as above	Increases breathing rate	Decreases blood hydrogen ion concentration

If a person decides to stop breathing, the blood concentrations of carbon dioxide and hydrogen ions begin to rise, and the concentration of oxygen falls. These changes stimulate the respiratory center, and soon the need to inhale overpowers the desire to hold the breath. On the other hand, a person can increase the breath-holding time by breathing rapidly and deeply in advance. This action, termed **hyperventilation,** causes a lowering of the blood carbon dioxide concentration. Following hyperventilation, it takes longer than usual for the carbon dioxide concentration to reach the level needed to produce an overwhelming effect upon the respiratory center.

Factors affecting breathing are summarized in chart 16.6.

Sometimes a person who is emotionally upset may hyperventilate, become dizzy, and lose consciousness. This effect seems to be due to a lowered carbon dioxide concentration accompanied by a pH change (respiratory alkalosis). Since a normal concentration of carbon dioxide is needed to maintain the tone of smooth muscles in blood vessels, a drop in the carbon dioxide concentration may be followed by dilation of blood vessels and a decrease in blood pressure. As a result, the oxygen supply to the brain cells may be inadequate and the person may faint.

Hyperventilation should never be used to help in holding the breath while swimming, because the person who has hyperventilated may lose consciousness under water and drown.

Figure 16.32
Alveolar pores (arrow) allow air to pass from one alveolus to another.

Figure 16.32
Alveolar pores (arrow) allow air to pass from one alveolus to another.

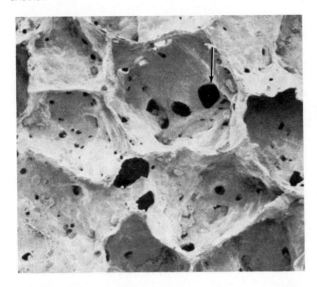

1. Describe the inflation and deflation reflexes.
2. What chemical factors affect breathing?
3. How does hyperventilation result in a decreased respiratory rate?

Alveolar Gas Exchanges

While other parts of the respiratory system move air in and out of the air passages, the alveoli carry on the vital process of exchanging gases between the air and the blood.

The Alveoli

The **alveoli** are microscopic air sacs clustered at the distal ends of the finest respiratory tubes—the alveolar ducts. Each alveolus consists of a tiny space surrounded by a thin wall that separates it from adjacent alveoli. There are minute openings, called **alveolar pores,** in the walls of some alveoli, and although the function of these pores is not clear, they may permit air to pass from one alveolus to another (figure 16.32). This arrangement may provide alternate air pathways if the passages in some portions of the lung become obstructed.

The Respiratory Membrane

The wall of an alveolus consists of an inner lining of simple squamous epithelium and a dense network of capillaries that also are lined with simple squamous cells. Thin basement membranes separate the layers of these flattened cells, and in the spaces between them

Figure 16.33
The respiratory membrane consists of the walls of the alveolus and the capillary.

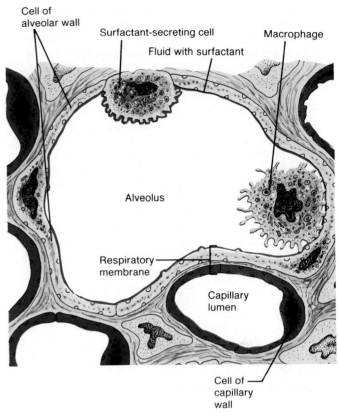

there are elastic and collagenous fibers that help to support the wall. As figure 16.33 shows, there are at least two thicknesses of epithelial cells and basement membranes between the air in an alveolus and the blood in a capillary. These layers make up the **respiratory membrane,** which is of vital importance because it is through this membrane that gas exchange occurs between alveolar air and blood.

Diffusion through the Respiratory Membrane

As described in chapter 3, gas molecules diffuse from regions where they are in higher concentration toward regions where they are in lower concentration. Similarly, gases move from regions of higher pressure toward regions of lower pressure, and it is the *pressure* of a gas that determines the rate at which it will diffuse from one region to another.

Measured by volume, ordinary air is about 78% nitrogen, 21% oxygen, and 0.04% carbon dioxide. Air also contains small amounts of certain other gases that have little or no physiological importance.

In a mixture of gases, such as air, each gas is responsible for a portion of the total weight or pressure produced by the mixture. The amount of pressure each gas creates is called **partial pressure,** and this pressure

As mentioned previously, the disorder called *emphysema* is accompanied by a progressive decrease in the surface area available for gas exchanges because the alveoli tend to merge and form larger chambers. The functional surface of an emphysematous lung may be reduced to ¼ that of a normal lung. This greatly decreases the lung's ability to carry on gas exchanges by diffusion, and its ability to oxygenate blood is further decreased by a loss of capillaries. Similar problems develop in pneumonia, tuberculosis, and a condition called atelectasis.

When a person has *pneumonia,* the alveoli become filled with fluids and blood cells. Pneumonia is usually due to an acute infection by bacterial organisms called *pneumococci* or to the presence of certain viruses.

In response to such infections, the linings of the alveoli become edematous and abnormally permeable. As a result, fluids and white blood cells enter the air sacs in abnormal quantities. As the alveoli fill up, the surface area available for gas exchanges is greatly reduced.

This type of infection is likely to spread to other regions of the respiratory tract and may involve lobes (lobar pneumonia) or an entire lung. The death rate in untreated cases of pneumonia is relatively high.

Tuberculosis is a bacterial disease caused by the *tubercle bacillus.* In this condition, fibrous connective tissue develops around the sites of infection, creating structures called *tubercles.* By walling off the bacteria, the tubercles help to inhibit their spread. Sometimes the protective mechanism fails, and the bacteria become widely distributed throughout the lungs.

In the later stages of tuberculosis, other kinds of bacteria are likely to appear and cause secondary infections in the lungs. There may be extensive destruction of lung tissues that greatly reduces the surface area of the respiratory membrane. In addition, the widespread development of fibrous tissue within a tubercular lung causes an increase in the thickness of the respiratory membrane, which further restricts gas exchange.

Atelectasis refers to the collapse of a lung or some part of it, together with the collapse of the blood vessels that supply the affected region.

Although atelectasis can be caused by a variety of factors, it is commonly due to an obstruction of a respiratory tube. Such a blockage may result from an abnormal secretion of mucus or the presence of some foreign object. In either case, the air in the alveoli beyond the obstruction tends to be absorbed by the lung tissues. As the air pressure in the alveoli decreases, their elastic walls cause them to collapse, and they become nonfunctional.

Fortunately, after a portion of a lung collapses, the functional regions that remain are often able to carry on enough gas exchange to sustain the body cells.

Respiratory distress syndrome, which was discussed previously, is a special form of atelectasis that is characterized by the collapse of alveoli in an infant's lungs due to a lack of surfactant.

---

is directly related to the concentration of the gas in the mixture. For example, because air is 21% oxygen, this gas is responsible for 21% of the atmospheric pressure. Since 21% of 760 mm of Hg is equal to 160 mm of Hg, it is said that the partial pressure of oxygen, symbolized PO_2, in atmospheric air is 160 mm of Hg. Similarly, it can be calculated that the partial pressure of carbon dioxide (PCO_2) in air is 0.3 mm of Hg.

When a mixture of gases dissolves in blood, each gas exerts its own partial pressure in proportion to its dissolved concentration. Furthermore, each gas will diffuse between the liquid and its surroundings and, as figure 16.34 shows, this movement will tend to equalize its partial pressures in the two regions.

For example, the PCO_2 in capillary blood is 46 mm of Hg, while the PCO_2 in alveolar air is 39 mm. As a consequence of the difference between these partial pressures, carbon dioxide diffuses from the blood, where its pressure is higher, through the respiratory membrane and into the alveolar air. When blood leaves the lungs, its PCO_2 is 40 mm of Hg, which is about the same as the PCO_2 of the alveolar air.

Similarly, the PO_2 of capillary blood is 40 mm of Hg, while that of alveolar air is 101 mm. Thus, oxygen diffuses from the alveolar air into the blood, and when the blood leaves the lungs it has a PO_2 of 100 mm of Hg.

---

1. Describe the structure of the respiratory membrane.
2. What is meant by partial pressure?
3. What causes oxygen and carbon dioxide to move across the respiratory membrane?

Transport of Gases

The transport of oxygen and carbon dioxide between the lungs and the body cells is a function of the blood. As these gases enter the blood, they dissolve in the liquid portion (plasma). They soon combine chemically with various blood components, and most are carried in combination with other atoms or molecules.

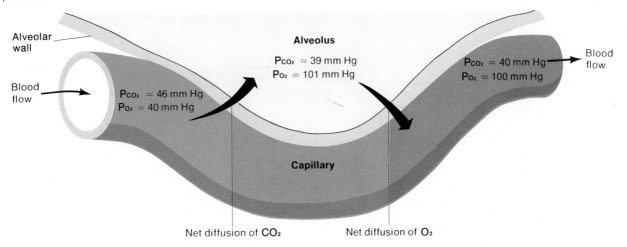

Alveolar wall

Blood flow

Alveolus
P_{CO_2} = 39 mm Hg
P_{O_2} = 101 mm Hg

P_{CO_2} = 46 mm Hg
P_{O_2} = 40 mm Hg

P_{CO_2} = 40 mm Hg
P_{O_2} = 100 mm Hg

Blood flow

Capillary

Net diffusion of CO_2

Net diffusion of O_2

Oxygen Transport

Almost all the oxygen carried in the blood is combined with the compound, **hemoglobin,** that occurs within the red blood cells. It is responsible for the color of these cells.

Hemoglobin has a complex molecular structure that consists of two subunits, called *heme* and *globin.* (See chapter 17.) Globin is a protein that contains 574 amino acids arranged in four polypeptide chains. Each chain is associated with a heme group, and each heme group contains an atom of iron. Each iron atom can combine loosely with an oxygen molecule. Thus, as oxygen dissolves in blood, it combines rapidly with hemoglobin, and the result of this chemical reaction is a new substance called **oxyhemoglobin.**

The amount of oxygen that combines with hemoglobin is determined by the P_{O_2}. The greater the P_{O_2}, the more oxygen will combine with hemoglobin, until the hemoglobin molecules are saturated with oxygen.

Although the proportion of oxygen in air remains the same (about 21%) at high altitudes, the P_{O_2} decreases. Consequently, if a person breathes high altitude air, oxygen diffuses less rapidly from the alveoli into the blood, and the hemoglobin is less likely to become saturated with oxygen. Such a person may develop the symptoms of an oxygen deficiency (hypoxia). For example, at 8000 feet a person may experience anxiety, restlessness, an increased breathing rate, and a rapid pulse. At 12,500 feet, where the P_{O_2} is only 100 mm of Hg rather than 160 mm of Hg at sea level, the symptoms may include drowsiness, mental fatigue, headache, and nausea. As altitude increases, these conditions are likely to increase in intensity until the person loses consciousness and may die at about 23,000 feet.

Figure 16.35
The oxyhemoglobin dissociation curve indicates that increased quantities of oxygen are released as the P_{O_2} decreases.

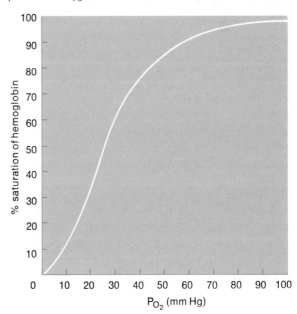

Oxyhemoglobin dissociation at 38°C

The chemical bonds that form between oxygen and hemoglobin molecules are relatively unstable, and as the P_{O_2} decreases, oxygen is released from oxyhemoglobin molecules (figure 16.35). This happens in tissues where cells have used oxygen in their respiratory

Figure 16.36

(a) Oxygen molecules, entering the blood from the alveolus, bond to hemoglobin, forming oxyhemoglobin. (b) In the regions of the body cells, oxyhemoglobin releases oxygen.

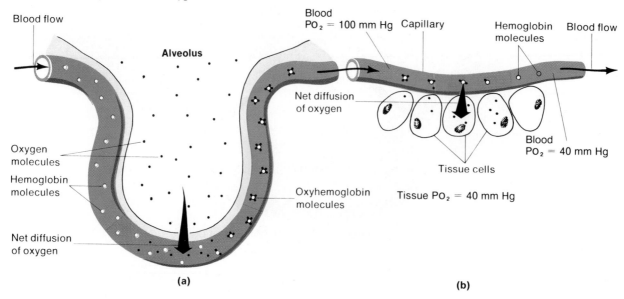

(a)

(b)

processes, and the free oxygen diffuses from the blood into nearby cells, as shown in figure 16.36.

The amount of oxygen that is released from oxyhemoglobin is affected by several other factors, including the blood concentration of carbon dioxide, the pH, and the temperature. Thus, as the concentration of carbon dioxide (PCO_2) increases, oxyhemoglobin tends to release more oxygen (figure 16.37). Also, as the blood becomes more acidic or as the temperature increases, more oxygen is released (figures 16.38 and 16.39).

Because of these factors, more oxygen is released from the blood to skeletal muscle during periods of exercise, since increased muscular activity causes an increase in the PCO_2, a decrease in the pH value, and a rise in the local temperature. At the same time, less active cells receive relatively smaller amounts of oxygen.

Carbon Monoxide

Carbon monoxide (CO) is a toxic gas produced in gasoline engines and some stoves as a result of incomplete combustion of fuel. It is also a component of tobacco smoke. Its toxic effect occurs because it combines with

hemoglobin more effectively than oxygen. Furthermore, carbon monoxide does not dissociate readily from the hemoglobin. Thus, when a person breathes carbon monoxide, increasing quantities of hemoglobin become unavailable for oxygen transport, and the body cells soon begin to suffer.

A victim of carbon monoxide poisoning may be treated by administering oxygen in high concentration, to replace some of the carbon monoxide bound to hemoglobin molecules. Carbon dioxide is usually administered simultaneously to stimulate the respiratory center, which in turn causes an increase in the breathing rate. Rapid breathing is desirable because it helps to reduce the concentration of carbon monoxide in the alveoli.

1. How is oxygen transported from the lungs to the body cells?
2. What factors affect the release of oxygen from oxyhemoglobin?
3. Why is carbon monoxide toxic?

Figure 16.37
The amount of oxygen released from oxyhemoglobin increases as the P_{CO_2} increases.

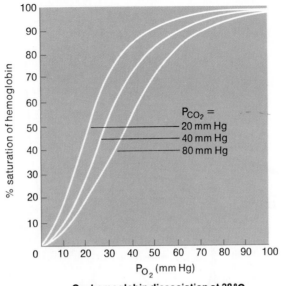

Oxyhemoglobin dissociation at 38 °C

Figure 16.38
The amount of oxygen released from oxyhemoglobin increases as the pH decreases.

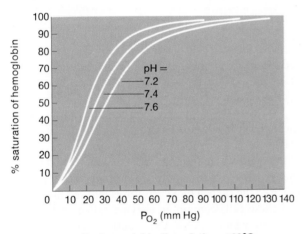

Oxyhemoglobin dissociation at 38°C

Figure 16.39
The amount of oxygen released from oxyhemoglobin increases as the temperature increases.

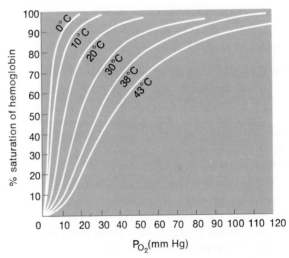

Oxyhemoglobin dissociation at various temperatures

Figure 16.40

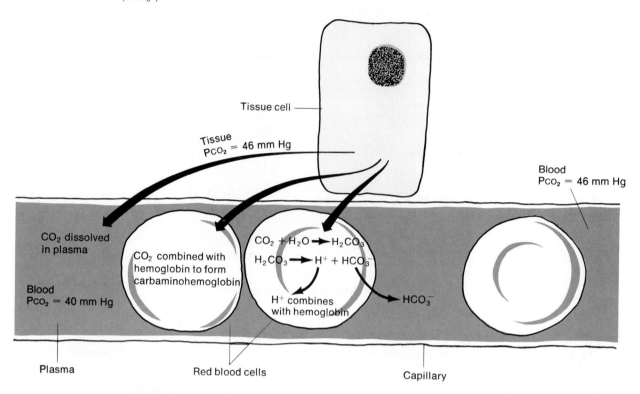

Carbon Dioxide Transport

Blood flowing through the capillaries of body tissues gains carbon dioxide because the tissues have a relatively high PCO_2. This carbon dioxide is transported to the lungs in one of three forms: it may be carried as carbon dioxide dissolved in blood, as part of a compound formed by bonding to hemoglobin, or as part of a bicarbonate ion (figure 16.40).

The amount of carbon dioxide that dissolves in blood is determined by its partial pressure. The higher the PCO_2 of the tissues, the more carbon dioxide will go into solution. However, only about 7% of the carbon dioxide is transported in this form.

Unlike oxygen, which combines with the iron atoms of hemoglobin molecules, carbon dioxide bonds with the amino groups ($-NH_2$) of these molecules. Consequently, oxygen and carbon dioxide do not compete for bonding sites, and both gases can be transported by a hemoglobin molecule at the same time.

When carbon dioxide combines with hemoglobin, a loosely bound compound called **carbamino-hemoglobin** is formed. This substance decomposes readily in regions where the PCO_2 is low, and thus its carbon dioxide is released. Although this method of transporting carbon dioxide is theoretically quite effective, carbaminohemoglobin forms relatively slowly. It is believed that about 23% of the total carbon dioxide is carried this way.

The most important carbon dioxide transport mechanism involves the formation of **bicarbonate ions** (HCO_3^-). As mentioned previously, carbon dioxide reacts with water to form carbonic acid (H_2CO_3). Although this reaction occurs slowly in blood plasma, much of the carbon dioxide diffuses into red blood cells, and these cells contain an enzyme called **carbonic anhydrase,** which speeds the reaction between carbon dioxide and water.

The resulting carbonic acid dissociates almost immediately, releasing hydrogen ions (H^+) and bicarbonate ions (HCO_3^-).

$$H_2CO_3 \rightarrow H^+ + HCO_3^-$$

Most of the hydrogen ions combine quickly with hemoglobin molecules; thus, they are prevented from accumulating and causing a change in the blood pH. The bicarbonate ions tend to diffuse out of the red blood cells and enter the blood plasma. It is estimated that nearly 70% of the carbon dioxide transported in blood is carried in this form.

Figure 16.41

As bicarbonate ions (HCO_3^-) diffuse out of the red blood cell, chloride ions (Cl^-) from the plasma diffuse into the cell, thus maintaining the electrical balance between ions. This exchange of ions is called the chloride shift.

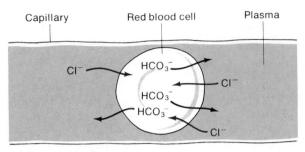

Figure 16.42

In the lungs, carbon dioxide diffuses from the blood into the alveoli.

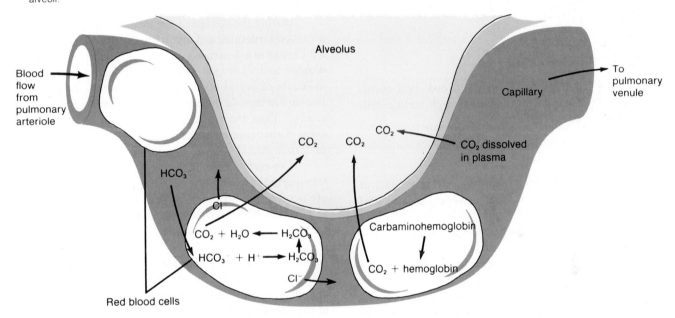

As the bicarbonate ions (HCO_3^-) leave the red blood cells and enter the plasma, *chloride ions* (Cl^-) are repelled electrically, and they move from the plasma into the red cells. This exchange in position of the two negatively charged ions, shown in figure 16.41, maintains the ionic balance between the red cells and the plasma. It is termed the **chloride shift.**

When blood passes through the capillaries of the lungs, it loses its dissolved carbon dioxide by diffusion into the alveoli. This occurs in response to the relatively low P_{CO_2} of the alveolar air. At the same time, hydrogen ions and bicarbonate ions in the red blood cells recombine to form carbonic acid molecules, and under the influence of carbonic anhydrase, the carbonic acid gives rise to carbon dioxide and water.

$$H^+ + HCO_3^- \rightarrow H_2CO_3 \rightarrow CO_2 + H_2O$$

Carbaminohemoglobin also releases its carbon dioxide, and as carbon dioxide continues to diffuse out of the blood, an equilibrium is established between the P_{CO_2} of the blood and the P_{CO_2} of the alveolar air. This process is summarized in figure 16.42.

Chart 16.7 summarizes the transport of blood gases.

1. Describe three ways carbon dioxide can be carried from body cells to the lungs.
2. How is it possible for hemoglobin to carry oxygen and carbon dioxide at the same time?
3. What is meant by the chloride shift?
4. How is carbon dioxide released from the blood into the lungs?

Chart 16.7	Gases transported in blood	
Gas	Reaction involved	Substance transported
Oxygen	Combines with iron atoms of hemoglobin molecules	Oxyhemoglobin
Carbon dioxide	About 7% dissolves in plasma	Carbon dioxide
	About 23% combines with the amino groups of hemoglobin molecules	Carbamino-hemoglobin
	About 70% reacts with water to form carbonic acid; the carbonic acid then dissociates to release hydrogen ions and bicarbonate ions	Bicarbonate ions

Utilization of Oxygen

The most important use of oxygen by body cells occurs during cellular respiration, even though certain cells may use oxygen for other purposes.

Cellular Respiration

As discussed in chapter 4, **cellular respiration** involves the molecular breakdown of substances, such as glucose, and the release of energy from them.

To review, this process involves two phases. The first phase, which takes place in the cytoplasm of cells, occurs in the absence of oxygen and is therefore *anaerobic*. For glucose, this anaerobic phase includes the conversion of glucose to pyruvic acid and is accompanied by the transfer of a relatively small amount of energy to ATP molecules.

The second phase takes place within the mitochondria and is *aerobic*, since it requires oxygen. During this phase, the pyruvic acid is converted into acetyl coenzyme A, and these molecules enter a complex series of chemical reactions that constitute the *citric acid cycle*.

Citric Acid Cycle

A molecule of acetyl coenzyme A contains two carbon atoms, and as it enters the citric acid cycle, it combines with a 4-carbon atom molecule (oxaloacetic acid) to form a 6-carbon molecule of *citric acid*.

The citric acid in turn is oxidized, first to a 5-carbon molecule (α ketoglutaric acid) and then to a 4-carbon molecule (oxaloacetic acid). This 4-carbon molecule can then combine with another molecule of acetyl coenzyme A, and the cycle is completed, as shown in figure 16.43.

During the change of the 6-carbon molecule to a 5-carbon structure, and again during the change from a 5-carbon to a 4-carbon structure, molecules of carbon dioxide and hydrogen atoms are released. For each molecule of acetyl coenzyme A that enters the cycle, two carbon dioxide molecules and eight hydrogen atoms emerge. Thus, the two carbon atoms of the acetyl coenzyme A that entered the citric acid cycle are converted to carbon dioxide.

The carbon dioxide diffuses out of the cell and is transported away by the blood. The hydrogen atoms are accepted by available oxygen under the influence of enzymes called *oxidases,* and water molecules are formed.

While molecules of acetyl coenzyme A are broken down, energy is being released. Some of the energy is transferred to molecules of ATP that can be used to provide energy for various metabolic processes. The rest of the energy is lost as heat. (See chapter 4 for a more detailed discussion of cellular respiration.)

1. What is meant by cellular respiration?
2. Distinguish between the anaerobic and the aerobic phases of cellular respiration.
3. Describe the major events that occur during the citric acid cycle.

Figure 16.43
Oxygen functions to accept hydrogen atoms that are released
from the citric acid cycle, and this action results in the formation
of water molecules.

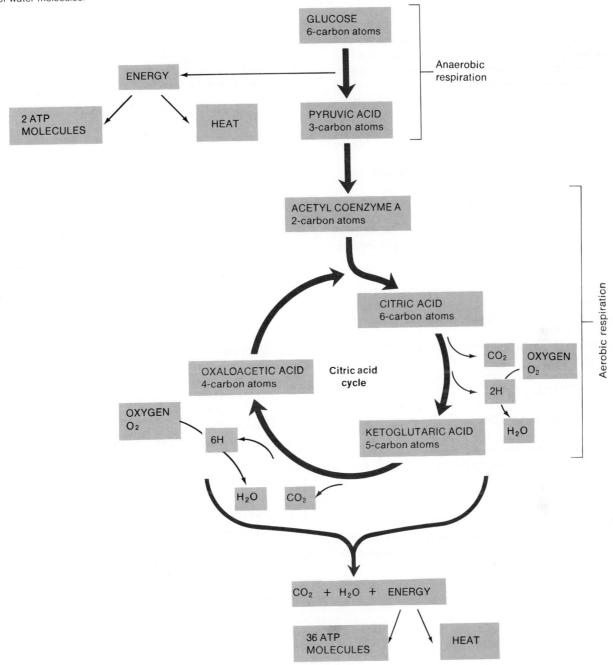

Clinical Terms Related to the Respiratory System

anoxia (ah-nok′se-ah)   an absence or deficiency of oxygen within tissues.

apnea (ap-ne′ah)   temporary absence of breathing.

asphyxia (as-fik′se-ah)   condition in which there is a deficiency of oxygen and an excess of carbon dioxide in the blood and tissues.

atelectasis (at″e-lek′tah-sis)   the collapse of a lung or some portion of it.

bradypnea (brad″e-ne′ah)   abnormally slow breathing.

bronchiolectasis (brong″ke-o-lek′tah-sis)   chronic dilation of the bronchioles.

bronchitis (brong-ki′tis)   inflammation of the bronchial lining.

Cheyne-Stokes respiration (chān stōks res″pi-ra′shun)   irregular breathing characterized by a series of shallow breaths that increase in depth and rate, followed by breaths that decrease in depth and rate.

dyspnea (disp′ne-ah)   difficulty in breathing.

eupnea (up-ne′ah)   normal breathing.

hemothorax (he″mo-tho′raks)   the presence of blood in the pleural cavity.

hypercapnia (hi″per-kap′ne-ah)   excessive carbon dioxide in the blood.

hyperpnea (hi″perp-ne′ah)   increased depth and rate of breathing.

hyperventilation (hi″per-ven″ti-la′-shun)   prolonged, rapid, and deep breathing.

hypoxemia (hi″pok-se′me-ah)   a deficiency in the oxygenation of the blood.

hypoxia (hi-pok′se-ah)   a diminished availability of oxygen in the tissues.

lobar pneumonia (lo′ber nu-mo′ne-ah)   pneumonia that affects an entire lobe of a lung.

pleurisy (ploo′ri-se)   inflammation of the pleural membranes.

pneumoconiosis (nu″mo-ko″ne-o′sis)   a condition characterized by the accumulation of particles from the environment in the lungs and the reaction of the tissues to their presence.

pneumothorax (nu″mo-tho′raks)   entrance of air into the space between the pleural membranes, followed by collapse of the lung.

rhinitis (ri-ni′tis)   inflammation of the nasal cavity lining.

sinusitis (si″nu-si′tis)   inflammation of the sinus cavity lining.

tachypnea (tak″ip-ne′ah)   rapid, shallow breathing.

tracheotomy (tra″ke-ot′o-me)   an incision in the trachea for exploration or the removal of a foreign object.

Chapter Summary

Introduction

The respiratory system includes the passages that transport air to and from the lungs, and the air sacs of the lungs in which gas exchanges occur.

Respiration is the entire process by which gases are exchanged between the atmosphere and the body cells, and it involves several major events.

Organs of the Respiratory System

The respiratory system includes the nose, nasal cavity, sinuses, pharynx, larynx, trachea, bronchial tree, and lungs.

1. The nose
    a. The nose is supported by bone and cartilage.
    b. Nostrils provide entrances for air.
2. The nasal cavity
    a. The nasal cavity is a space behind the nose.
    b. It is divided medially by the nasal septum.
    c. Nasal conchae divide the cavity into passageways and help increase surface area of mucous membranes.
    d. Mucous membrane filters, warms, and moistens incoming air.
    e. Particles trapped in the mucus are carried to the pharynx by ciliary action and swallowed.
3. The sinuses
    a. Sinuses are spaces in the bones of the skull that open into the nasal cavity.
    b. They are lined with mucous membrane that is continuous with the lining of the nasal cavity.
4. The pharynx
    a. The pharynx is located behind the mouth and between the nasal cavity and the larynx.
    b. It functions as a common passage for air and food.
    c. It aids in creating vocal sounds.
5. The larynx
    a. The larynx is an enlargement at the top of the trachea.
    b. It serves as a passageway for air and helps prevent foreign objects from entering the trachea.
    c. It is composed of muscles and cartilages, some of which are single while others are paired.
    d. It contains the vocal cords, which produce sounds by vibrating as air passes over them.
        (1) The pitch of a sound is related to the tension on the cords.
        (2) The intensity of a sound is related to the force of the air passing over the cords.
    e. The glottis and epiglottis help prevent food and liquid from entering the trachea.

6. The trachea
    a. The trachea extends into the thoracic cavity in front of the esophagus.
    b. It divides into right and left bronchi.
    c. The mucous lining continues to filter incoming air.
    d. The wall is supported by cartilaginous rings.
7. The bronchial tree
    a. The bronchial tree consists of branched air passages that lead from the trachea to the air sacs.
    b. The branches of the bronchial tree include primary bronchi, lobar bronchi, segmental bronchi, bronchioles, terminal bronchioles, respiratory bronchioles, alveolar ducts, alveolar sacs, and alveoli.
    c. Structure of the respiratory tubes
        (1) As tubes branch, the amount of cartilage in the walls decreases and the muscular layer becomes more prominent.
        (2) Elastic fibers in the walls aid the breathing mechanism.
        (3) The epithelial lining changes from pseudostratified and ciliated to cuboidal and simple squamous.
    d. Functions of the respiratory tubes include distribution of air and exchange of gases between the alveolar air and the blood.
8. The lungs
    a. The left and right lungs are separated by the mediastinum and enclosed by the diaphragm and the thoracic cage.
    b. The visceral pleura is attached to the surface of the lungs; parietal pleura lines the thoracic cavity.
    c. The right lung has three lobes, and the left lung has two.
    d. Each lobe is composed of lobules that contain alveoli, blood vessels, and supporting tissues.

Mechanism for Breathing

Inspiration and expiration movements are accompanied by changes in the size of the thoracic cavity.

1. Inspiration
    a. Air is forced into the lungs by atmospheric pressure.
    b. Inspiration occurs when the intra-alveolar pressure is reduced.
    c. The intra-alveolar pressure is reduced when the diaphragm moves downward and the thoracic cage moves upward and outward.
    d. Expansion of the lungs is aided by surface tension that holds the pleural membranes together.
    e. Surfactant reduces surface tension within the alveoli.

2. Expiration
    a. The forces of expiration come from the elastic recoil of tissues and surface tension within the alveoli.
    b. Expiration can be aided by thoracic and abdominal wall muscles that pull the thoracic cage downward and inward and compress the abdominal organs.
3. Respiratory air volumes
    a. The amount of air that normally moves in and out during quiet breathing is the tidal volume.
    b. Additional air that can be inhaled is the inspiratory reserve volume; additional air that can be exhaled is the expiratory reserve volume.
    c. Residual air remains in the lungs and is mixed with newly inhaled air.
    d. The vital capacity is the maximum amount of air a person can exhale after taking the deepest breath possible.
    e. The total lung capacity is equal to the vital capacity plus the residual air volume.
    f. Air in the anatomic and alveolar dead spaces is not available for gas exchange.
4. Alveolar ventilation
    a. Minute respiratory volume is calculated by multiplying the tidal volume by the breathing rate.
    b. Alveolar ventilation rate is calculated by subtracting the physiologic dead space from the tidal volume and multiplying the result by the breathing rate.
    c. The alveolar ventilation rate is a major factor affecting the exchange of gases between the alveolar air and the blood.
5. Nonrespiratory air movements
    a. Nonrespiratory air movements are movements that occur in addition to breathing.
    b. They include coughing, sneezing, laughing, crying, hiccuping, and yawning.

Control of Breathing

Normal breathing is rhythmic and involuntary, although the respiratory muscles can be controlled voluntarily.

1. The respiratory center
    a. The respiratory center is located in the brain stem, and includes parts of the medulla oblongata and pons.
    b. The medullary rhythmicity area includes two groups of neurons.
        (1) The dorsal respiratory group is responsible for the basic rhythm of breathing.
        (2) The ventral respiratory group increases inspiratory and expiratory movements during forceful breathing.
    c. The pneumotaxic area regulates the rate of breathing.

2. Factors affecting breathing
   a. Breathing is affected by certain chemicals, stretching of lung tissues, and emotional state.
   b. There are chemosensitive areas associated with the respiratory center.
      (1) Carbon dioxide combines with water to form carbonic acid which, in turn, releases hydrogen ions.
      (2) Chemosensitive areas are influenced mainly by hydrogen ions.
      (3) Stimulation of these areas causes the breathing rate to increase.
   c. There are chemoreceptors in the carotid and aortic bodies of certain arteries.
      (1) These chemoreceptors are sensitive to low oxygen concentration.
      (2) When oxygen concentration is low, breathing rate is increased.
   d. An inflation reflex is triggered by overstretching of lung tissues.
      (1) This reflex causes the duration of inspiratory movements to be reduced.
      (2) This prevents overinflation of the lungs during forceful breathing.
   e. Hyperventilation causes the carbon dioxide concentration to decrease.

Alveolar Gas Exchanges
Alveoli carry on gas exchanges between the air and the blood.

1. The alveoli
   a. The alveoli are tiny air sacs clustered at the distal ends of the alveolar ducts.
   b. Some alveoli have openings into adjacent air sacs that provide alternate pathways for air when passages are obstructed.
2. The respiratory membrane
   a. The respiratory membrane consists of the alveolar and capillary walls.
   b. Gas exchanges take place through these walls.
3. Diffusion through the respiratory membrane
   a. The partial pressure of a gas is determined by the concentration of that gas in a mixture of gases or the concentration dissolved in a liquid.
   b. Gases diffuse from regions of higher partial pressure toward regions of lower partial pressure.
   c. Oxygen diffuses from the alveolar air into the blood; carbon dioxide diffuses from the blood into the alveolar air.

Transport of Gases
Blood transports gases between the lungs and the body cells.

1. Oxygen transport
   a. Oxygen is transported in combination with hemoglobin molecules.
   b. The resulting oxyhemoglobin is relatively unstable and releases its oxygen in regions where the P_{O_2} is low.
2. Carbon monoxide
   a. Carbon monoxide forms as a result of incomplete combustion of fuels.
   b. It combines with hemoglobin more readily than oxygen and forms a stable compound.
   c. Carbon monoxide is toxic because the hemoglobin with which it combines is no longer available for oxygen transport.
3. Carbon dioxide transport
   a. Carbon dioxide may be carried in solution, either bound to hemoglobin or as a bicarbonate ion.
   b. Most carbon dioxide is transported in the form of bicarbonate ions.
   c. The enzyme, carbonic anhydrase, speeds the reaction between carbon dioxide and water to form carbonic acid.
   d. Carbonic acid dissociates to release hydrogen ions and bicarbonate ions.

Utilization of Oxygen
1. Cellular respiration
   a. Cellular respiration occurs in two phases.
      (1) The anaerobic phase takes place in the cytoplasm.
      (2) The aerobic phase takes place in the mitochondria.
   b. During aerobic respiration, a complex series of chemical reactions results in the breakdown of acetyl coenzyme A and the formation of carbon dioxide and water.
2. Citric acid cycle
   a. Carbon atoms of acetyl coenzyme A are converted into carbon dioxide.
   b. Hydrogen atoms that are released are combined with oxygen to form water molecules.
   c. Some of the resulting energy is transferred to molecules of ATP, and the rest escapes as heat.
   d. Oxaloacetic acid combines with another molecule of acetyl coenzyme A to complete the cycle.

Application of Knowledge

1. If the upper respiratory passages are bypassed with a tracheostomy, how might the air entering the trachea be different from air normally passing through this canal? What problems may this create for the patient?

2. In certain respiratory disorders, such as emphysema, the capacity of the lungs to recoil elastically is reduced. Which respiratory air volumes will be affected by a condition of this type? Why?

3. What changes would you expect to occur in the relative concentrations of blood oxygen and carbon dioxide in a patient who breathes rapidly and deeply for a prolonged time? Why?

4. If a person has stopped breathing and is receiving pulmonary resuscitation, would it be better to administer pure oxygen or a mixture of oxygen and carbon dioxide? Why?

Review Activities

1. Describe the general functions of the respiratory system.

2. Explain how the nose and nasal cavity function in filtering incoming air.

3. Name and describe the locations of the major sinuses, and explain how a sinus headache may occur.

4. Distinguish between the pharynx and the larynx.

5. Name and describe the locations and functions of the cartilages of the larynx.

6. Distinguish between the false vocal cords and the true vocal cords.

7. Compare the structure of the trachea and the branches of the bronchial tree.

8. List the successive branches of the bronchial tree from the primary bronchi to the alveoli.

9. Describe how the structure of the respiratory tube changes as the branches become finer.

10. Explain the functions of the respiratory tubes.

11. Distinguish between visceral pleura and parietal pleura.

12. Name and describe the locations of the lobes of the lungs.

13. Explain how normal inspiration and forced inspiration are accomplished.

14. Define *surface tension* and explain how it aids the breathing mechanism.

15. Define *surfactant* and explain its function.

16. Define *compliance*.

17. Explain how normal expiration and forced expiration are accomplished.

18. Distinguish between vital capacity and total lung capacity.

19. Distinguish between anatomic, alveolar, and physiologic dead spaces.

20. Distinguish between minute respiratory volume and alveolar ventilation rate.

21. Compare the mechanisms of coughing and sneezing and explain the function of each.

22. Explain the function of yawning.

23. Describe the location of the respiratory center and name its major components.

24. Describe how the basic rhythm of breathing is controlled.

25. Explain the function of the pneumotaxic area of the respiratory center.

26. Explain why increasing blood concentrations of carbon dioxide and hydrogen ions have similar effects upon the respiratory center.

27. Describe the function of the chemoreceptors in the carotid and aortic bodies.

28. Describe the inflation reflex.

29. Discuss the effects of emotions on breathing.

30. Define *hyperventilation* and explain how it affects the respiratory center.

31. Define *respiratory membrane* and explain its function.

32. Explain the relationship between the partial pressure of a gas and its rate of diffusion.

33. Summarize the gas exchanges that occur through the respiratory membrane.

34. Describe how oxygen is transported in blood.

35. Explain why carbon monoxide is toxic.

36. List three ways that carbon dioxide can be transported in blood.

37. Explain the function of carbonic anhydrase.

38. Define *chloride shift*.

39. Distinguish between anaerobic and aerobic respiration.

40. Explain the major steps in the citric acid cycle.

41. Explain how oxygen is utilized during cellular respiration.

17

The Blood

The blood, heart, and blood vessels constitute the *circulatory system,* and provide a link between the body's internal parts and its external environment. More specifically, the blood transports nutrients from the digestive tract, oxygen from respiratory organs to the body cells, and wastes from these cells to the respiratory and excretory organs. It transports hormones from the endocrine glands to target tissues and bathes the body cells in a liquid of relatively stable composition. It also aids in temperature control by distributing heat from skeletal muscles and other active organs to all body parts. Thus, the blood provides vital support for cellular activities and aids in maintaining a favorable cellular environment.

The heart and the closed system of blood vessels comprise the apparatus that moves blood throughout the body. These organs constitute the *cardiovascular system,* which will be discussed in chapter 18.

Chapter Outline

Chapter Objectives

After you have studied this chapter, you should be able to

1. Describe the general characteristics of blood, and discuss its major functions.

2. Distinguish between the various types of blood cells.

3. Explain how blood cell counts are made and how they are used.

4. Discuss the life cycle of a red blood cell.

5. Explain how red cell production is controlled.

6. List the major components of blood plasma, and describe the functions of each.

7. Define *hemostasis*, and explain the mechanisms that help to achieve it.

8. Review the major steps in blood coagulation.

9. Explain how coagulation can be prevented.

10. Explain the basis for blood typing.

11. Describe how blood reactions may occur between fetal and maternal tissues.

12. Complete the review activities at the end of this chapter. Note that the items are worded in the form of specific learning objectives. You may want to refer to them before reading the chapter.

Key Terms

agglutinin   (ah-gloo'tĭ-nin)

agglutinogen   (ag''loo-tin'o-jen)

albumin   (al-bu'min)

antibody   (an'tĭ-bod''e)

coagulation   (ko-ag''u-la'shun)

embolus   (em'bo-lus)

erythrocyte   (ĕ-rith'ro-sīt)

erythropoietin   (ĕ-rith''ro-poi'ĕ-tin)

fibrinogen   (fi-brin'o-jen)

globulin   (glob'u-lin)

hemostasis   (he''mo-sta'sis)

leukocyte   (lu'ko-sīt)

macrophage   (mak'ro-fāj)

plasma   (plaz'mah)

platelet   (plāt'let)

thrombus   (throm'bus)

Aids to Understanding Words

agglutin-, to glue together: *agglutin*ation—the clumping together of red blood cells.

bil-, bile: *bil*irubin—pigment excreted in bile.

-crit, to separate: hemato*crit*—percentage of cells in a blood sample, determined by separating the cells from the plasma.

embol-, a stopper: *embol*us—an obstruction in a blood vessel.

erythr-, red: *erythr*ocyte—a red blood cell.

hemo-, blood: *hemo*globin—red pigment responsible for the color of blood.

hepar-, liver: *hepar*in—anticoagulant secreted by liver cells.

leuko-, white: *leuko*cyte—a white blood cell.

-lys, to break up: fibrino*lys*in—protein-splitting enzyme that can digest fibrin.

macro-, large: *macro*phage—a large phagocytic cell.

-osis, increased production: leukocyt*osis*—condition in which white blood cells are produced in abnormally large numbers.

-poiet, making: erythro*poiet*in—a hormone that stimulates the production of red blood cells.

poly-, many: *poly*cythemia—condition in which red blood cells appear in abnormally large numbers.

-stas, a halting: hemo*stas*is—arrest of bleeding from damaged blood vessels.

thromb-, a clot: *thromb*ocyte—blood platelet that plays a role in the formation of a blood clot.

A S IS MENTIONED in chapter 5, blood is a type of connective tissue whose cells are suspended in a liquid intercellular material.

Blood and Blood Cells

Whole blood is slightly heavier and three to four times more viscous than water. Its cells, which are formed mostly in red bone marrow, include *red blood cells* and *white blood cells*. Blood also contains some cellular fragments called *platelets* (figure 17.1).

Volume and Composition of Blood

The volume of blood varies with body size. It also varies with changes in fluid and electrolyte concentrations and the amount of fat tissue present. An average-size male (70 kg or 154 lbs), however, will have a blood volume of about 5 liters (5.2 quarts).

If a blood sample is allowed to stand in a tube for a while, the cells, which are over 99% red cells, will become separated from the liquid portion of the blood and settle to the bottom. This separation can be speeded by centrifuging the sample so that the cells quickly become packed into the lower part of the centrifuge tube, as shown in figure 17.2. When the amount of cells and liquid is measured, the percentage of each in the blood sample can be calculated.

A blood sample is usually about 45% red cells. This percentage is called **hematocrit** (HCT). The remaining 55% of a blood sample consists of clear, straw-colored **plasma**. The normal values for this and other common blood tests are provided in Appendix C, page A–6.

In addition to red cells, the solids of the blood include *white blood cells* and *blood platelets*. The plasma is composed of a complex mixture that includes water, amino acids, proteins, carbohydrates, lipids, vitamins, hormones, electrolytes, and cellular wastes. The normal concentrations of plasma components are listed in the appendix.

---

1. What are the major components of blood?
2. What factors affect blood volume?
3. How is hematocrit determined?

Characteristics of Red Blood Cells

Red blood cells or **erythrocytes** are tiny, biconcave disks that are thin near their centers and thicker around their rims. This special shape is related to the red cell's function of transporting gases, in that it provides an increased surface area through which gases can diffuse (figure 17.3). The shape also places the cell membrane closer to various interior parts, where oxygen-carrying *hemoglobin* is found.

Each red blood cell is about ⅓ hemoglobin by volume, and this substance is responsible for the color of the blood. The rest of the cell consists mainly of membrane, water, electrolytes, enzymes, and components of cellular respiratory mechanisms. When the hemoglobin is combined with oxygen, the resulting *oxyhemoglobin* is bright red, and when the oxygen is released, the remaining *deoxyhemoglobin* is darker. In fact, blood rich in deoxyhemoglobin may appear bluish when it is viewed through blood vessel walls.

A person experiencing a prolonged oxygen deficiency (hypoxia) may develop a symptom called *cyanosis*. In this condition, the skin and mucous membranes appear bluish due to an abnormally high concentration of deoxyhemoglobin in the blood. Cyanosis may also occur as a result of exposure to low temperature. In this case, superficial blood vessels become constricted, and more oxygen than usual is removed from the blood flowing through them.

Although red blood cells have nuclei during their early stages of development in the red bone marrow, these nuclei are lost as the cells mature. This characteristic, like the shape of a red blood cell, seems to be related to the function of transporting oxygen, since the space previously occupied by the nucleus is available to hold hemoglobin.

Since their nuclei are missing, red blood cells cannot carry on DNA-directed protein synthesis (see chapter 4) and are unable to reproduce. On the other hand, their cytoplasmic enzymes remain functional, and the cells are able to carry on vital energy-releasing processes. With age, however, the red blood cells become less and less active.

Some individuals produce an abnormal type of hemoglobin molecule that tends to form long chains when exposed to low oxygen concentrations. This causes the red blood cells containing the abnormal molecules to become distorted or sickle-shaped, and the person is said to have *sickle-cell anemia*.

In this condition, the distorted cells may block capillaries and thus interrupt the flow of blood to tissues. The resulting symptoms may include severe joint and abdominal pain, skin ulceration, and chronic kidney disease.

---

1. Describe a red blood cell.
2. What is the function of hemoglobin?
3. What changes occur in a red blood cell as it matures?

Figure 17.1

Blood consists of a liquid portion called plasma and a solid portion that includes red blood cells, white blood cells, and platelets.

Blood sample

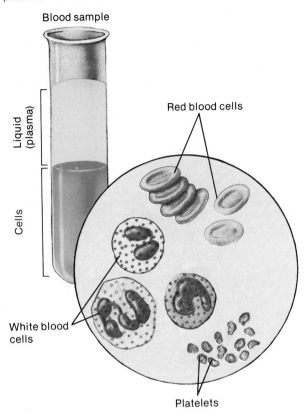

Liquid (plasma)

Cells

Red blood cells

White blood cells

Platelets

Figure 17.2

If a blood-filled capillary tube is centrifuged, the red cells become packed in the lower portion, and the percentage of red cells (hematocrit) can be determined.

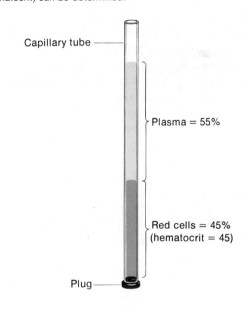

Capillary tube

Plasma = 55%

Red cells = 45%
(hematocrit = 45)

Plug

Figure 17.3

(a) How is the biconcave shape of a red blood cell related to its function? (b) Scanning electron micrograph of human red blood cells.

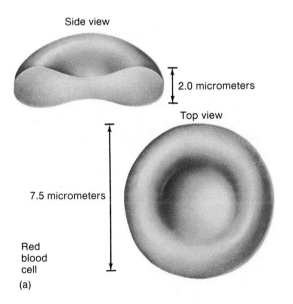

Side view

2.0 micrometers

Top view

7.5 micrometers

Red blood cell

(a)

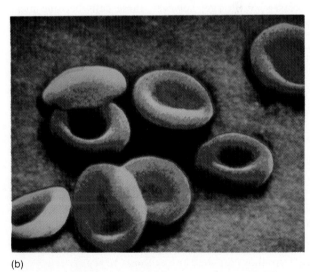

(b)

Red Blood Cell Counts

The number of red blood cells in a cubic millimeter (mm^3) of blood is called the *red cell count*. Although this number varies from time to time even in healthy individuals, the normal range for adult males is 4,600,000–6,200,000 cells per mm^3, while that for adult females is 4,200,000–5,400,000 cells per mm^3. For children, the normal range is 4,500,000–5,100,000 cells per mm^3. (Note: These values may vary slightly with the hospital, physician, and type of equipment used to make counts.) The number of red cells generally increases following exercise, a large meal, a rise in temperature, or an increase in altitude.

Since the number of circulating red blood cells is closely related to the blood's *oxygen-carrying capacity,* any changes in this number may be significant. For this reason, red cell counts are routinely made to help in diagnosing and evaluating the courses of various diseases.

Blood cell counts sometimes are made using a special glass slide called a *hemocytometer.* This device is designed so the cells in a small, but known, volume of blood can be observed and counted with the aid of a microscope. Clinical laboratories often are equipped with electronic counters that detect and record the number of blood cells forced through a tiny opening in the instrument.

Destruction of Red Blood Cells

Red blood cells are quite elastic and flexible, and they readily change shape as they pass through small blood vessels. With age, however, these cells become more fragile, and they are often damaged or ruptured when they squeeze through capillaries, particularly those in active muscles.

Damaged red cells are phagocytized and destroyed, primarily in the liver and spleen, by large reticuloendothelial cells called **macrophages,** which are described in chapter 5. The fragments and contents of ruptured cells are also phagocytized and digested by macrophages.

The hemoglobin molecules from red cells undergoing destruction are broken down into molecular subunits of *heme,* an iron-containing portion, and

Figure 17.4
When a hemoglobin molecule is decomposed, the heme portions are broken down into iron (Fe) and biliverdin. Most of the biliverdin is then converted to bilirubin.

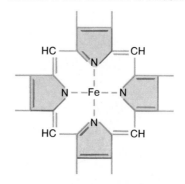

Heme

Biliverdin ($C_{33}H_{34}O_6N_4$)

Bilirubin ($C_{33}H_{36}O_6N_4$)

globin, a protein. The heme is further decomposed into iron and a greenish pigment called **biliverdin.** The iron, combined with a protein called *transferrin,* may be carried by the blood to the blood-cell-forming tissue in the red bone marrow and reused in the synthesis of a new hemoglobin. Otherwise it may be stored in liver cells in the form of an iron-protein complex called *ferritin.* In time, the biliverdin is converted to an orange pigment called **bilirubin.** Biliverdin and bilirubin are excreted in bile as bile pigments (figure 17.4). Chart 17.1 summarizes the process of red cell destruction.

Figure 17.5
Hemocytoblasts in red bone marrow.

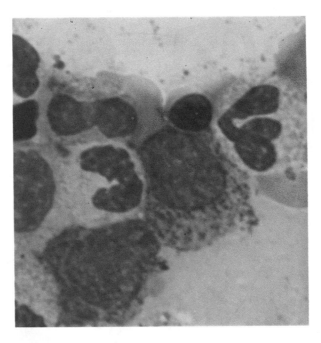

Figure 17.5
Hemocytoblasts in red bone marrow.

Chart 17.1	**Major events in red blood cell destruction**

1. Cells are damaged while squeezing through the capillaries of active tissues.
2. Damaged red cells are phagocytized by macrophages in the liver and spleen.
3. Hemoglobin from the red cells is decomposed into heme and globin.
4. Heme is decomposed into iron and biliverdin.
5. Iron is made available for reuse in the synthesis of new hemoglobin or is stored in the liver as ferritin.
6. Some biliverdin is converted into bilirubin.
7. Biliverdin and bilirubin are excreted in bile as bile pigments.

About one-third of all newborns develop a mild disorder called *physiologic jaundice* within a few days following birth. In this condition, as in other forms of jaundice, the skin and eyes become yellowish due to an accumulation of bilirubin in the tissues.

Physiologic jaundice is thought to be the result of immature liver cells that are somewhat ineffective in excreting bilirubin into the bile. The condition is treated by feedings that promote bowel movements and by exposure to fluorescent light that reduces the concentration of bilirubin in the tissues.

1. What is the normal red blood cell count for a male? For a female?
2. What happens to damaged red blood cells?
3. What are the end products of hemoglobin breakdown?

Red Blood Cell Production and Its Control

As is mentioned in chapter 8, red blood cell formation (hematopoiesis) occurs in the yolk sac, liver, and spleen of an embryo; but after an infant is born, these cells are produced almost exclusively by tissue that lines the vascular sinuses of the red bone marrow.

Within the red marrow, cells that are called **hemocytoblasts** give rise to **erythroblasts** that can synthesize hemoglobin (figure 17.5). The erythroblasts also reproduce and give rise to many daughter cells. The nuclei of these newly formed cells soon shrink and are extruded by being pinched off in thin coverings of cytoplasm and cell membrane. The resulting cells are **erythrocytes.** Some of these young red cells may contain a netlike structure (reticulum) for a day or two.

Figure 17.8

Life cycle of a red blood cell. *(1)* Essential nutrients are absorbed from the intestine; *(2)* nutrients are transported by blood to red bone marrow; *(3)* red blood cells are produced in red bone marrow by mitosis; *(4)* mature red blood cells are released into blood and circulate for about 120 days; *(5)* damaged red blood cells are destroyed in liver by macrophages; *(6)* hemoglobin from red blood cells is decomposed into heme and globin; *(7)* iron from heme is returned to red bone marrow to be reused, and biliverdin is excreted in the bile.

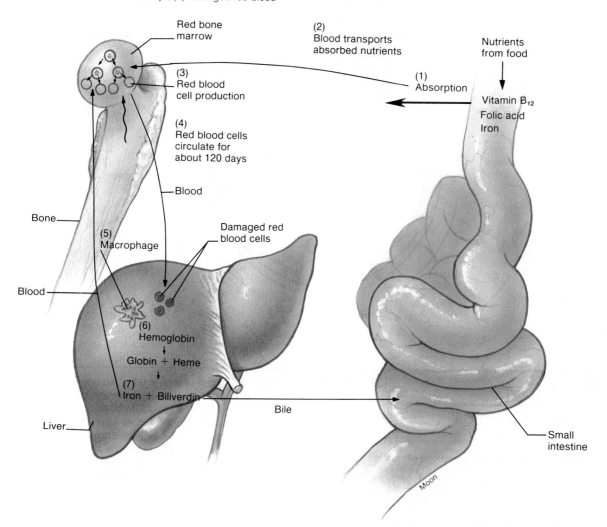

rate of absorption is increased; and when the tissues are becoming saturated with iron, the rate is greatly decreased. Figure 17.8 summarizes the life cycle of a red blood cell.

The dietary factors that affect red blood cell production are summarized in chart 17.3.

During pregnancy, the enlarging uterus and its developing vascular system create a demand for more blood. Consequently, the maternal blood volume usually increases greatly. Since this increase requires accelerated production of red blood cells, a pregnant woman must take additional iron to meet the needs of her own blood-cell-forming tissues as well as those of the developing fetus.

Chart 17.3	Dietary factors affecting red blood cell production	
Substance	Source	Function
Vitamin B$_{12}$	Absorbed from small intestine in the presence of intrinsic factor	Necessary for the synthesis of DNA
Iron	Absorbed from small intestine; conserved during red cell destruction and made available for reuse	Necessary for the synthesis of hemoglobin
Folic Acid	Absorbed from small intestine	Necessary for the synthesis of DNA

Some Red Blood Cell Disorders
A Clinical Application

Whenever a rapid loss or an inadequate production of red cells causes a deficiency of these cells, the resulting condition is termed **anemia**. In disorders of this type, the oxygen-carrying capacity of the blood is reduced, and the person may appear pale and lack energy. Among the more common types of anemia are hemorrhagic anemia, hypochromic anemia, aplastic anemia, and hemolytic anemia.

Hemorrhagic anemia is caused by an abnormal loss of blood and may be mild or severe depending on the amount of bleeding that occurs. Increased production of red cells usually makes up for a blood loss within a few weeks. If the blood loss is prolonged, however, the person may not be able to obtain enough iron from food to replace the lost hemoglobin. As a result, even though the normal number of red cells is reestablished, they may lack a normal concentration of hemoglobin. This condition is called *hypochromic anemia*. It is characterized by the presence of small, pale red blood cells (figure 17.9).

Aplastic anemia is due to a malfunction in the red bone marrow and is characterized by decreased production of red blood cells. This condition is sometimes caused by an overexposure to X rays or some other form of *ionizing radiation*. It can also be caused by the toxic effects of various chemicals or adverse drug reactions.

Hemolytic anemia is characterized by an abnormally high rate of red blood cell rupture (hemolysis). Such cellular breakdown may be caused by a variety of factors including hereditary defects, parasitic infections, and adverse drug reactions.

Another type of red blood cell disorder is called **polycythemia.** In this condition, there is an abnormal increase in the number of circulating red cells. It may result from stimulation of the erythrocyte-forming tissues, as occurs when a person remains at a high altitude or engages in prolonged physical activity. It may occur when the blood becomes concentrated from excessive loss of body fluids that accompanies heavy sweating or shock. Polycythemia also may be caused by a tumorous condition in the red-cell-forming tissues (polycythemia vera), causing the production of abnormally large numbers of cells.

A general effect of polycythemia is to increase the viscosity of the blood, which depends largely on the concentration of red blood cells. (Viscosity is a property related to the ease at which a fluid flows—syrup has a high viscosity and flows sluggishly, while water has a low viscosity and flows easily.) As a result of increased viscosity, blood flow through the blood vessels is decreased, an abnormal amount of hemoglobin may be deoxygenated, and the skin may appear bluish (cyanotic).

Figure 17.9

(a) Normal human erythrocytes; *(b)* erythrocytes from a person with hypochromic anemia.

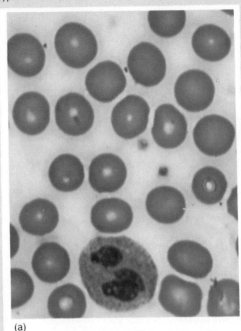

(a)

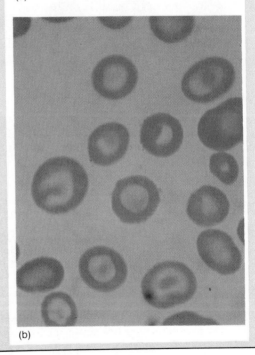

(b)

Figure 17.10
Origin and development of blood cells.

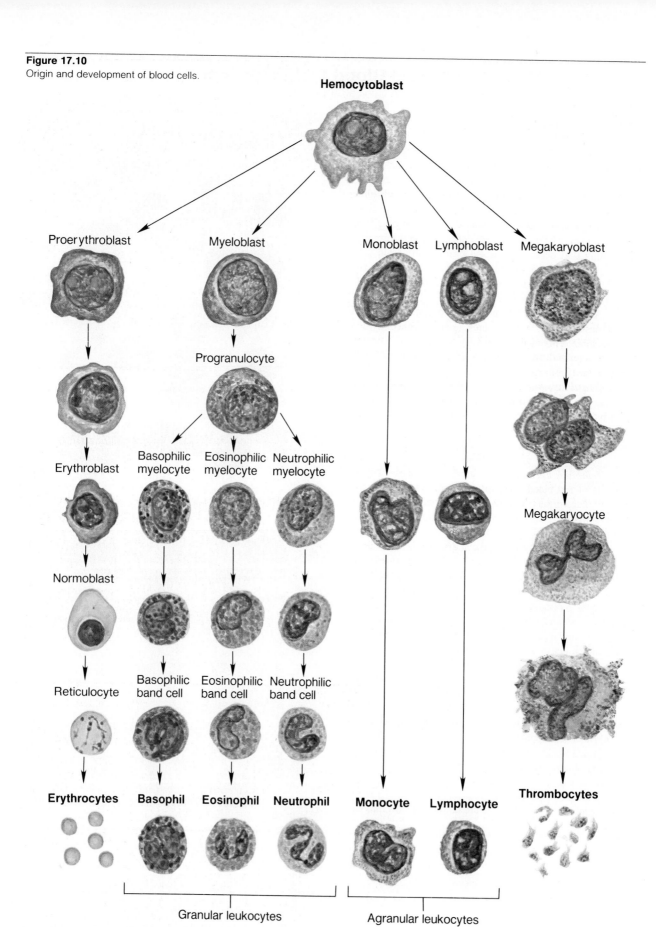

Hemocytoblast

Proerythroblast

Myeloblast

Monoblast

Lymphoblast

Megakaryoblast

Progranulocyte

Erythroblast

Basophilic
myelocyte

Eosinophilic
myelocyte

Neutrophilic
myelocyte

Normoblast

Megakaryocyte

Reticulocyte

Basophilic
band cell

Eosinophilic
band cell

Neutrophilic
band cell

Erythrocytes

Basophil

Eosinophil

Neutrophil

Monocyte

Lymphocyte

Thrombocytes

Granular leukocytes

Agranular leukocytes

1. What vitamins are necessary for red blood cell production?
2. Why is iron needed for the normal development of red blood cells?

Types of White Blood Cells

White blood cells or **leukocytes** function primarily to defend the body against microorganisms. Although these cells do most of their work outside the circulatory system, they use the blood for transportation to sites of infection.

Normally, five types of white cells can be found in circulating blood. They are distinguished by size, the nature of the cytoplasm, the shape of the nucleus, and the staining characteristics. For example, some types have granular cytoplasm and make up a group called *granulocytes,* while others lack cytoplasmic granules and are called *agranulocytes.*

Each **granulocyte** is typically about twice the size of a red cell. Members of this group include three types of white cells—neutrophils, eosinophils, and basophils. These cells develop in the red bone marrow and are produced from *hemocytoblasts* in much the same manner as red cells. They have a relatively short life span, however, averaging about 12 hours. Figure 17.10 illustrates stages in the development of these and other blood cells.

Neutrophils are characterized by the presence of fine cytoplasmic granules that stain pinkish in neutral stain. The nucleus of a neutrophil is lobed and consists of two to five parts connected by thin strands of chromatin. For this reason, neutrophils are also called *polymorphonuclear leukocytes,* which means their nuclei take many forms. Neutrophils account for 54%–62% of the white cells in a normal blood sample (figure 17.11).

Eosinophils contain coarse, uniformly sized cytoplasmic granules that stain deep red in acid stain. The nucleus usually has only two lobes (bilobed). These cells make up 1%–3% of the total number of circulating leukocytes (figure 17.12).

Basophils are similar to eosinophils in size and in the shape of their nuclei. However, they have fewer, more irregularly shaped cytoplasmic granules that stain deep blue in basic stain. This type of leukocyte usually accounts for less than 1% of the white cells (figure 17.13).

The leukocytes of the **agranulocyte** group include monocytes and lymphocytes. While monocytes generally arise from red bone marrow, lymphocytes are formed in organs of the lymphatic system as well as in the red bone marrow.

Figure 17.11
A neutrophil has a lobed nucleus with two to five parts.

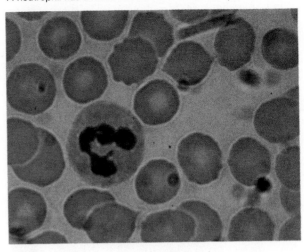

Figure 17.12
An eosinophil is characterized by the presence of red-staining cytoplasmic granules.

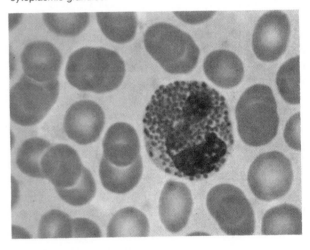

Figure 17.13
A basophil has cytoplasmic granules that stain deep blue.

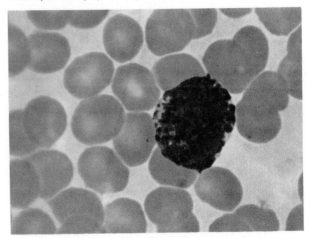

Figure 17.14
A monocyte is the largest type of blood cell.

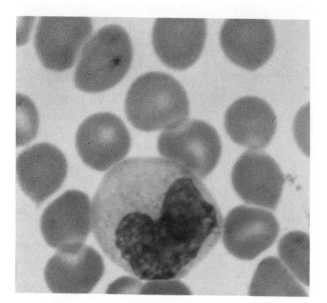

Figure 17.15
The lymphocyte (left) contains a large, round nucleus; neutrophil (right).

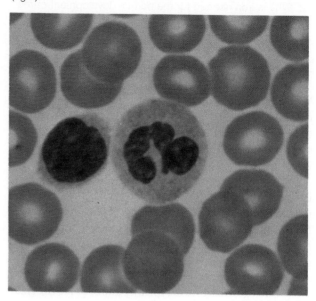

Monocytes are the largest cells found in the blood, having diameters two to three times greater than red cells. Their nuclei vary in shape and are described as round, kidney-shaped, oval, or lobed. They usually make up 3%–9% of the leukocytes in a blood sample and live for several weeks or even months (figure 17.14).

Although large **lymphocytes** are sometimes found in the blood, usually they are only slightly larger than the red cells. Typically, a lymphocyte contains a relatively large, round nucleus with a thin rim of cytoplasm surrounding it. These cells account for 25%–33% of the circulating white blood cells. They have relatively long life spans that may extend for years (figure 17.15).

---

1. Distinguish between granulocytes and agranulocytes.
2. List five types of white blood cells, and explain how they differ from one another.

White Blood Cell Counts

The procedure used to count white blood cells is similar to that used for counting red cells. Before a *white cell count* is made, however, the red cells in the blood sample are destroyed so they will not be mistaken for white cells. Normally there are from 5000 to 10,000 white cells per mm³ of blood.

Since the total number of white blood cells may change in response to abnormal conditions, white blood cell counts are of clinical interest. For example, some infectious diseases are accompanied by a rise in the number of circulating white cells, and if the total

number of white cells exceeds 10,000 per mm³ of blood, the person is said to have **leukocytosis.** This condition occurs during certain acute infections, such as appendicitis. It may also appear following vigorous exercise, emotional disturbances, or excessive loss of body fluids.

If the total white cell count drops below 5000 per mm³ of blood, the condition is called **leukopenia.** Such a deficiency may accompany typhoid fever, influenza, measles, mumps, chicken pox, and poliomyelitis. It may also occur as a result of various anemias or poisoning from lead, arsenic, or mercury.

A *differential white blood cell count* is one in which the percentages of the various types of leukocytes in a blood sample are determined. This test is useful because the relative proportions of white cells may change in particular diseases. Neutrophils, for instance, usually increase during bacterial infections, while eosinophils may increase during certain parasitic infections and allergic reactions.

---

1. What is the normal white blood cell count?
2. Distinguish between leukocytosis and leukopenia.
3. What is a differential white blood cell count?

Functions of White Blood Cells

As was mentioned, white blood cells generally function to protect the body against microorganisms. For example, some leukocytes phagocytize microorganisms that get into the body, and others produce antibodies that react with such foreign particles to destroy or disable them.

Figure 17.16

Leukocytes can squeeze between the cells of a capillary wall and enter the tissue space outside the blood system.

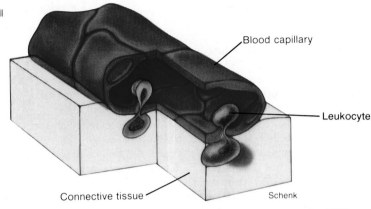

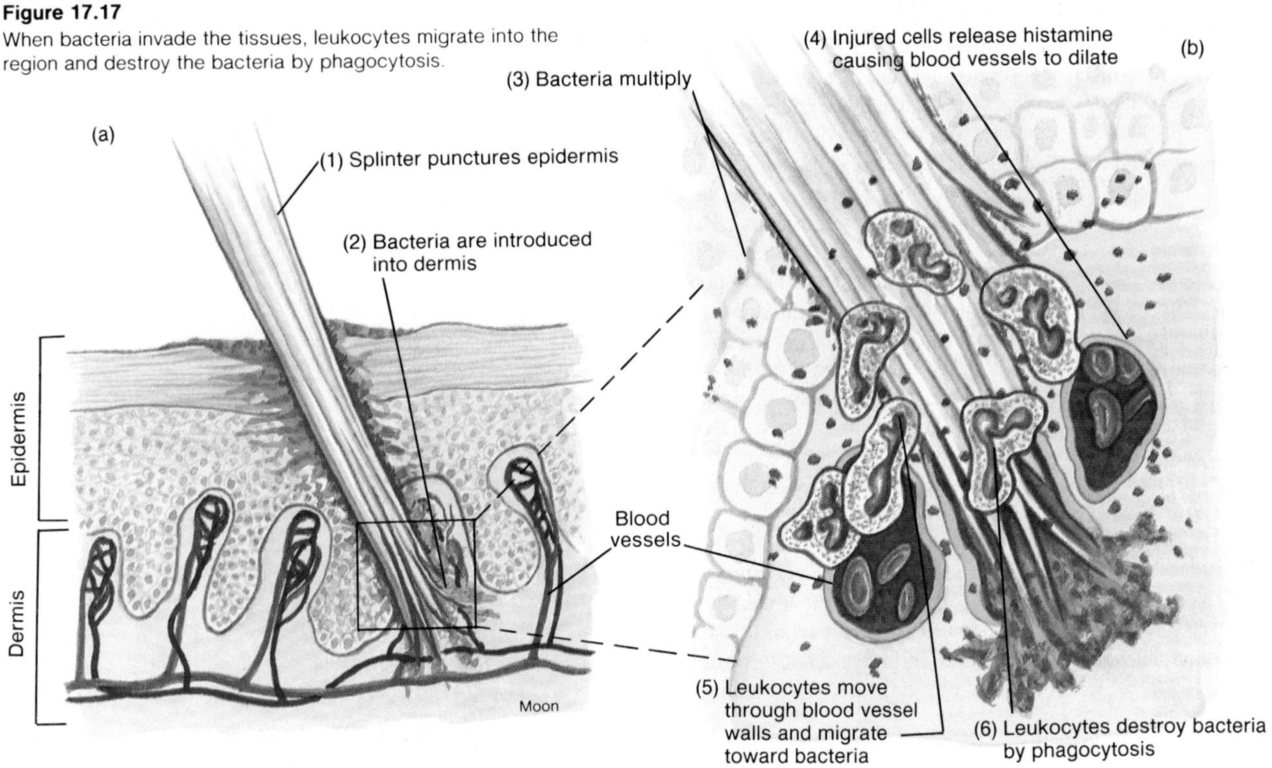

Blood capillary

Leukocyte

Connective tissue

Schenk

Figure 17.17

When bacteria invade the tissues, leukocytes migrate into the region and destroy the bacteria by phagocytosis.

(a)

(1) Splinter punctures epidermis

(2) Bacteria are introduced into dermis

Epidermis

Dermis

Blood vessels

Moon

(3) Bacteria multiply

(4) Injured cells release histamine causing blood vessels to dilate

(b)

(5) Leukocytes move through blood vessel walls and migrate toward bacteria

(6) Leukocytes destroy bacteria by phagocytosis

Leukocytes can squeeze between the cells that make up blood vessel walls. This movement, called **diapedesis,** allows the white cells to leave the circulation. (See figure 17.16.) Once outside the blood, they move through interstitial spaces using a form of self-propulsion, called *ameboid motion.*

The most mobile and active phagocytic leukocytes are the *neutrophils* and the *monocytes.* Although the neutrophils are unable to ingest particles much larger than bacterial cells, monocytes can engulf relatively large objects. Both types of phagocytes contain numerous *lysosomes* that are filled with digestive enzymes capable of breaking down various organic molecules like those in bacterial cell membranes. As a result of their activities, neutrophils and monocytes often become so engorged with digestive products and bacterial toxins that they also die.

When tissues are invaded by microorganisms, the body cells often respond by releasing substances that cause local blood vessels to dilate. One such substance, **histamine,** also causes an increase in the permeability of nearby capillaries and the walls of small veins (venules). As these changes occur, the tissues become reddened and large quantities of fluids leak into the interstitial spaces. The swelling produced by this *inflammatory reaction* tends to delay the spread of the invading microorganisms into other regions (see chapter 5). At the same time, leukocytes are somehow attracted by chemicals released from damaged cells, and they move toward these substances. This phenomenon is called **positive chemotaxis,** and it results in large numbers of white blood cells moving quickly into inflamed areas (figure 17.17).

The Blood / 625

Leukemia
A Clinical Application

Leukemia is a form of cancer characterized by an uncontrolled production of specific types of leukocytes (figure 17.18). There are two major types of leukemia. *Myeloid leukemia* results from an abnormal production of granulocytes by the red bone marrow, while *lymphoid leukemia* is accompanied by increased formation of lymphocytes from lymph nodes. In both types, the cells produced usually fail to mature into functional cells. Thus, even though large numbers of neutrophils may be formed in myeloid leukemia, these cells have little ability to phagocytize bacteria, and the patient has a lowered resistance to infections.

Eventually, the cells responsible for the overproduction of leukocytes tend to spread from the bone marrow or lymph nodes to other parts, and as a result white blood cells are produced abnormally in tissues throughout the body. As with other forms of cancer, the leukemic cells finally appear in such great numbers that they crowd out the normal, functioning cells. For example, leukemic cells originating in red bone marrow may invade other regions of the bone, weakening its structure and stimulating pain receptors. Also, as the normal red marrow is crowded out, the patient is likely to become anemic and develop a deficiency of blood platelets (thrombocytopenia). The lack of platelets usually is reflected in an increasing tendency to bleed.

Leukemias also are classified as *acute* or *chronic*. An acute condition appears suddenly, the symptoms progress rapidly, and death occurs in a few months when the condition is untreated. Chronic forms begin more slowly and may remain undetected for many months. Without treatment, life expectancy is about 3 years.

The greatest success in treatment has been achieved with acute lymphoid leukemia, which is the most common cancerous condition in children. This treatment usually involves counteracting the side effects of the condition, such as anemia, hemorrhaging, and an increased susceptibility to infections, as well as administering chemotherapeutic drugs.

Although acute lymphoid leukemia may occur at any age, the chronic form usually occurs after 50 years of age. Acute myeloid leukemia also may occur at any age, but it is more frequent in adults; chronic myeloid leukemia is primarily a disease of adults between 20 and 50 years of age.

Figure 17.18

(a) Normal blood cells; *(b)* blood cells from a person with leukemia. (Note the increased number of leukocytes.)

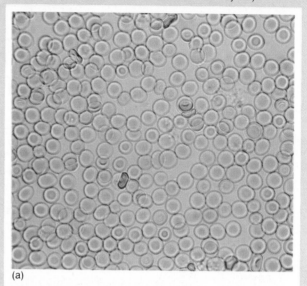

(a)

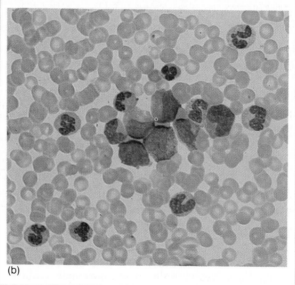

(b)

As bacteria, leukocytes, and damaged cells accumulate in an inflamed area, a thick fluid, called *pus*, often forms, and continues to do so while the invading microorganisms are active. If the pus cannot escape to the outside of the body or into a body cavity, it may remain walled in the tissues for some time. Eventually it may be absorbed by the surrounding cells.

The function of *eosinophils* is uncertain, although there is evidence that they act to detoxify foreign proteins that enter the body through the lungs or intestinal tract. They usually increase in number in parasitic diseases and allergic reactions. They also may release enzymes that break down blood clots and help remove certain products of immune responses (antigen-antibody complexes) described in chapter 19.

Chart 17.4 Cellular components of blood

Component	Description	Number present	Function
Red blood cell (erythrocyte)	Biconcave disk without nucleus, about ⅓ hemoglobin	4,000,000 to 6,000,000 per mm³	Transports oxygen and carbon dioxide
White blood cells (leukocytes)		5000 to 10,000 per mm³	Aids in defense against infections by microorganisms
Granulocytes	About twice the size of red cells, cytoplasmic granules present		
1. Neutrophil	Nucleus with two to five lobes, cytoplasmic granules stain pink in neutral stain	54%–62% of white cells present	Destroys relatively small particles by phagocytosis
2. Eosinophil	Nucleus bilobed, cytoplasmic granules stain red in acid stain	1%–3% of white cells present	Helps to detoxify foreign substances, secretes enzymes that break down clots and removes products of immune reactions
3. Basophil	Nucleus lobed, cytoplasmic granules stain blue in basic stain	Less than 1% of white cells present	Releases anticoagulant, heparin, and histamine
Agranulocytes	Cytoplasmic granules absent		
1. Monocyte	Two to three times larger than red cell, nuclear shape varies from round to lobed	3%–9% of white cells present	Destroys relatively large particles by phagocytosis
2. Lymphocyte	Only slightly larger than red cell, nucleus nearly fills cell	25%–33% of white cells present	Mechanism of immunity
Platelet (thrombocyte)	Cytoplasmic fragment	130,000 to 360,000 per mm³	Helps control blood loss from broken vessels

Some of the cytoplasmic granules of *basophils* contain a blood-clot-inhibiting substance, called *heparin,* while other granules contain *histamine.* Basophils may aid in preventing intravascular blood clot formation by releasing heparin, and may cause an increase in blood flow to injured tissues by releasing histamine. They also may play major roles in certain allergic reactions (see chapter 19).

Lymphocytes play an important role in the mechanism of *immunity.* Some, for example, can form **antibodies** that act against various foreign substances entering the body. This function is discussed in chapter 19.

1. What are the primary functions of white blood cells?
2. How do white blood cells reach microorganisms that are outside blood vessels?
3. Which white blood cells are the most active phagocytes?
4. What are the functions of eosinophils and basophils?

Blood Platelets

Platelets or **thrombocytes** are not complete cells. They arise from very large cells in the red bone marrow, called **megakaryocytes.** These cells give off tiny cytoplasmic fragments, and as they detach and enter the circulation, they become platelets (figure 17.10).

Each platelet is a round disk that lacks a nucleus and is less than half the size of a red blood cell. It is capable of ameboid movement and may live for about 10 days. In normal blood, the platelet count will vary from 130,000 to 360,000 platelets per mm³.

Platelets help close breaks in damaged blood vessels by sticking to the broken surfaces. In the presence of such breaks, platelets release a substance called **serotonin,** which causes smooth muscles in the vessel walls to contract—an action that reduces blood flow. In addition, platelets function to initiate the formation of blood clots, as is explained in a subsequent section of this chapter.

Chart 17.4 summarizes the characteristics of blood cells and platelets.

Figure 17.19
Plasma, which is primarily water, contains about 7% protein and
1.5% of other substances.

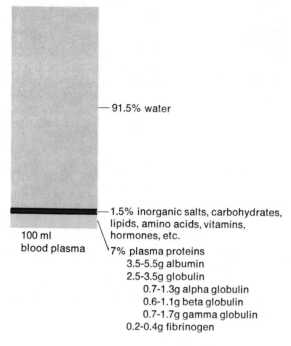

—91.5% water

—1.5% inorganic salts, carbohydrates,
lipids, amino acids, vitamins,
hormones, etc.

100 ml
blood plasma

7% plasma proteins
   3.5-5.5g albumin
   2.5-3.5g globulin
      0.7-1.3g alpha globulin
      0.6-1.1g beta globulin
      0.7-1.7g gamma globulin
   0.2-0.4g fibrinogen

1. What is leukemia?
2. Distinguish between myeloid and lymphoid leukemia.
3. What is the function of blood platelets?
4. What is the normal blood platelet count?

Blood Plasma

Plasma is the clear, straw-colored, liquid portion of the
blood in which the cells and platelets are suspended. It
is approximately 92% water and contains a complex
mixture of organic and inorganic substances that function in a variety of ways. These functions include transporting nutrients, gases, and vitamins; regulating fluid
and electrolyte balances; and maintaining a favorable
pH. Figure 17.19 shows the chemical makeup of
plasma.

Plasma Proteins

The most abundant dissolved substances (solutes) in
plasma are **plasma proteins.** These proteins remain in
the blood and interstitial fluids, and ordinarily are not
used as energy sources. There are three main groups—
albumins, globulins, and fibrinogen. The members of
each group differ in their chemical compositions and in
their physiological functions.

Albumins account for about 60% of the plasma
proteins and have the smallest molecules of these proteins. They are synthesized in the liver, and because
they are so plentiful, they are of particular significance
in maintaining the *osmotic pressure* of the blood.

As is explained in chapter 3, whenever the concentration of dissolved substances changes on either side
of a cell membrane, water is likely to move through the
membrane toward the region where the dissolved molecules are in higher concentration. For this reason, it
is important that the concentration of dissolved substances in plasma remains relatively stable. Otherwise,
water will tend to leave the blood and enter the tissues,
or leave the tissues and enter the blood by *osmosis.* The
presence of albumins (and other plasma proteins) adds
to the osmotic pressure of the plasma and aids in regulating the water balance between the blood and the
tissues. It also helps to control blood volume, which in
turn is directly related to blood pressure.

The concentration of plasma proteins may decrease
significantly if a person is starving and is forced to use
body protein as an energy source, or has a protein-de-
ficient diet. Similarly, the plasma protein level may drop
if liver disease interferes with the synthesis of these proteins. In either case, as the blood protein concentration
decreases the osmotic pressure of the blood decreases,
and water tends to accumulate in the intercellular spaces,
causing tissues to swell (edema).

The **globulins,** which make up about 36% of the
plasma proteins, can be further separated into frac-
tions called *alpha globulins, beta globulins,* and *gamma
globulins.* The alpha and beta globulins are synthesized in the liver, and they have a variety of functions
including the transport of lipids and fat-soluble vitamins. The gamma globulins are produced in lymphatic
tissues, and they include the proteins that function as
antibodies of immunity (see chapter 19).

Fibrinogen plays a primary role in the blood clotting mechanism. It is synthesized in the liver and has
the largest molecules of the plasma proteins. The function of fibrinogen is discussed in a subsequent section
of this chapter.

Chart 17.5 summarizes the characteristics of the
plasma proteins.

1. List three types of plasma proteins.
2. How does albumin help to maintain a water balance
   between the blood and the tissues?
3. Which of the globulins functions in immunity?
4. What is the role of fibrinogen?

Chart 17.5	Plasma proteins						
Protein	Percentage of total	Origin	Function	Protein	Percentage of total	Origin	Function
Albumin	60%	Liver	Helps in the maintenance of blood osmotic pressure	Beta globulins		Liver	Transport of lipids and fat-soluble vitamins
				Gamma globulins		Lymphatic tissues	Constitute antibodies of immunity
Globulin Alpha globulins	36%	Liver	Transport of lipids and fat-soluble vitamins	Fibrinogen	4%	Liver	Plays key role in blood clot formation

Nutrients and Gases

The *plasma nutrients* include amino acids, simple sugars, and various lipids that have been absorbed from the digestive tract. Glucose, for example, is transported by the plasma from the small intestine to the liver, where it may be stored as glycogen or changed into fat. If the blood glucose level drops below the normal range, glycogen may be converted back into glucose, as described in chapter 13.

Recently absorbed amino acids are also carried to the liver, where they may be used in the manufacture of proteins, such as those found in the plasma, or deaminated and used as an energy source. (See chapter 4.)

The lipids of plasma include fats (triglycerides), phospholipids, and cholesterol. Generally these lipids are combined with proteins in complexes called **lipoproteins.** Lipoprotein molecules are relatively large and consist of a core of triglyceride surrounded by a surface layer composed of phospholipid, cholesterol, and protein. Within the protein portion of this layer there are specialized molecules called *apoproteins* that can combine with receptors on the membranes of specific target cells. Lipoprotein molecules also vary in the proportion of the various lipids they contain.

Since fats are less dense than proteins, as the proportion of triglycerides in a lipoprotein increases, the density of the particle decreases. Conversely, as the proportion of triglycerides decreases, the density increases.

On the basis of their densities, which reflect their composition, lipoproteins can be classified as *very low-density lipoproteins* (VLDL), which have a relatively high concentration of triglycerides; *low-density lipoproteins* (LDL), which have a relatively high concentration of cholesterol; and *high-density lipoproteins* (HDL), which have a relatively high concentration of protein and a lower concentration of lipids.

As is discussed in chapter 14, chylomicrons are lipoproteins that function to transport dietary fats to muscle and adipose cells. Similarly, very low-density lipoproteins, which are produced by liver cells, are used to transport triglycerides that the liver has synthesized from excess dietary carbohydrates. After VLDL molecules have delivered their loads of triglycerides to adipose cells, the remnants of the VLDL molecules are converted to low-density lipoproteins. Since most of the triglycerides have been removed, LDL molecules have a relatively higher cholesterol content than do the original VLDL molecules. Various cells, including liver cells, have surface receptors that combine with apoproteins associated with LDL molecules, and these cells slowly remove LDL from plasma by receptor-mediated endocytosis (see chapter 3). This process provides cells with a supply of cholesterol.

Some individuals have an inherited condition in which their cells lack LDL receptors. As a consequence, LDL molecules accumulate in the plasma, and their blood cholesterol concentration becomes abnormally high. This condition, which is called *hypercholestremia,* is chacterized by increased deposition of cholesterol in the walls of arteries and accelerated development of the disease *atherosclerosis.*

After chylomicrons have delivered their triglycerides to cells, the remnants of these molecules are transferred to high-density lipoproteins. These HDL molecules, which are formed in the liver and small intestine, function to transport chylomicron remnants to the liver, where they enter cells rapidly by receptor-mediated endocytosis. The liver disposes of the cholesterol it obtains in this manner by secreting it into bile or by using it to synthesize bile salts.

Much of the cholesterol and bile salts in bile is later reabsorbed by the small intestine and transported back to the liver, and the secretion-reabsorption cycle is repeated. During each cycle, some of the cholesterol and bile salts escapes reabsorption, reaches the large intestine, and is eliminated with feces.

The most important blood gases are oxygen and carbon dioxide. While plasma also contains a considerable amount of dissolved nitrogen, this gas ordinarily has no physiological function.

---

1. What nutrients are found in blood plasma?
2. How are triglycerides transported in plasma?
3. How is cholesterol eliminated from the liver?
4. What gases occur in plasma?

Nonprotein Nitrogenous Substances

Molecules that contain nitrogen atoms, but are not proteins, comprise a group called **nonprotein nitrogenous substances.** Within the plasma, this group includes amino acids, urea, uric acid, creatine, and creatinine. The *amino acids* are present as a result of protein digestion and amino acid absorption. *Urea* and *uric acid* are the products of protein and nucleic acid catabolism, respectively, and *creatinine* results from the metabolism of *creatine.*

As is discussed in chapter 9, creatine occurs as **creatine phosphate** in muscle and brain tissues, and in the blood. This substance functions to store high energy phosphate bonds, much like those of ATP molecules.

Normally, the concentration of nonprotein nitrogenous (NPN) substances remains relatively stable because protein intake and utilization are balanced with the excretion of nitrogenous wastes. Because about half of the NPN is urea, which is ordinarily excreted by the kidneys, a rise in the plasma NPN level may suggest a kidney disorder. Such an increase, however, may also occur as a result of excessive protein catabolism or the presence of an infection.

Plasma Electrolytes

Plasma contains a variety of *electrolytes* that have been absorbed from the intestine or have been released as by-products of cellular metabolism. They include sodium, potassium, calcium, magnesium, chloride, bicarbonate, phosphate, and sulfate ions. Of these, sodium and chloride ions are the most abundant.

Such ions are important in maintaining the osmotic pressure and the pH of the plasma, and like other plasma constituents, they are regulated so that their blood concentrations remain relatively stable. These electrolytes are discussed in chapter 21 in connection with water and electrolyte balance.

---

1. What is meant by a nonprotein nitrogenous substance?
2. Why does kidney disease cause an increase in the blood concentration of these substances?
3. What are the sources of plasma electrolytes?

Hemostasis

The term **hemostasis** refers to the stoppage of bleeding, which is vitally important when blood vessels are damaged. Following such an injury, several things may occur that help to prevent excessive blood loss. These include blood vessel spasm, platelet plug formation, and blood coagulation.

These mechanisms are most effective in minimizing blood losses from relatively small vessels, such as arterioles, capillaries, and venules. Injuries to larger vessels, particularly arteries, may result in a severe hemorrhage that requires special treatment for its control.

Blood Vessel Spasm

When an arteriole or a venule is cut or broken, the smooth muscles in its wall are stimulated to contract, and blood loss is decreased almost immediately. In fact, the ends of a severed vessel may be closed completely by such a *spasm.* This effect seems to result from direct stimulation of the vessel wall as well as from reflexes elicited by pain receptors in the injured tissues.

Although the reflex response may last only a few minutes, the effect of the direct stimulation usually continues for about 30 minutes; and by then the platelet plug and blood coagulation mechanisms normally are operating. Also, as the platelet plug forms, the platelets release a substance called *serotonin* which causes smooth muscles in the blood vessel walls to contract. This vasoconstricting action helps to maintain a prolonged vascular spasm.

Platelet Plug Formation

Platelets tend to stick to the exposed ends of injured blood vessels. Actually they adhere to any rough surfaces, particularly to the *collagen* in connective tissue that underlies the endothelial lining of blood vessels.

Figure 17.20
Steps in platelet plug formation.

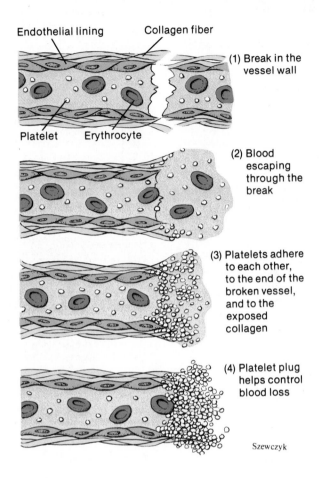

Szewczyk

When platelets contact collagen, their shapes change drastically, and numerous processes begin to protrude from their membranes. At the same time, they tend to stick to each other creating a *platelet plug* in the vascular break. Such a plug may be able to control blood loss if the break is relatively small, but a larger one may require the aid of a blood clot to halt bleeding.

The steps in platelet plug formation are shown in figure 17.20.

1. What is meant by hemostasis?
2. How does blood vessel spasm help to control bleeding?
3. Describe the formation of a platelet plug.

Blood Coagulation

Coagulation, which is the most effective of the hemostatic mechanisms, results in the formation of a *blood clot*. This process may be initiated by a variety of factors and may involve an extrinsic or an intrinsic clotting mechanism.

The *extrinsic clotting mechanism* is triggered by the release of chemical substances from broken blood vessels or damaged tissues. The *intrinsic mechanism* is stimulated by blood contact with foreign surfaces in the absence of tissue damage.

Although the mechanism by which blood coagulates is poorly understood, it is known that many substances called *clotting factors* are involved in the process. Some of these factors promote coagulation (procoagulants) and others inhibit it (anticoagulants). Whether or not the blood coagulates depends on the balance that exists between these two groups of factors. Normally the anticoagulants prevail, and the blood does not clot. As a result of injury (trauma), however, substances that favor coagulation may increase in concentration, and the blood may coagulate.

The basic event in blood clot formation is the conversion of the soluble plasma protein *fibrinogen* into relatively insoluble threads of the protein **fibrin.** This change is triggered by the activation of certain plasma

Chart 17.6 Hemostatic mechanisms

Mechanism	Stimulus	Effect
Blood vessel spasm	Direct stimulus to vessel wall or to pain receptors; serotonin released from platelets	Smooth muscles in vessel wall contract reflexly; vasoconstriction helps to maintain prolonged vessel spasm
Platelet plug formation	Exposure of platelets to rough surfaces or to collagen of connective tissue	Platelets adhere to rough surfaces and to each other, forming a plug
Blood coagulation	Cellular damage and blood contact with foreign surfaces results in the production of substances that favor coagulation	Blood clot forms as a result of a series of reactions terminating in the conversion of fibrinogen into fibrin

Figure 17.21

Hemostasis following tissue damage is likely to involve blood vessel spasm, platelet plug formation, and the extrinsic blood-clotting mechanism.

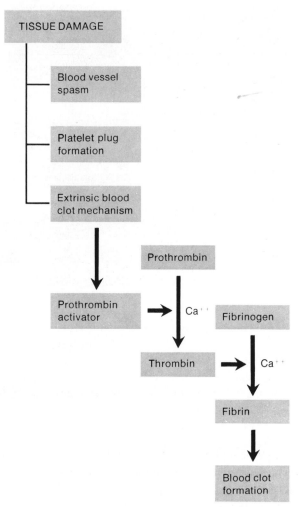

proteins that remain inactive until they are affected by the presence of still other protein factors.

Chart 17.6 and figure 17.21 summarize the three primary hemostatic mechanisms.

Extrinsic Clotting Mechanism

The extrinsic clotting mechanism is triggered when blood contacts damaged blood vessel walls or tissues outside blood vessels. Such damaged tissues release a complex of substances called *tissue thromboplastin,* and its presence initiates a series of reactions involving several clotting factors. These reactions, which also depend on the presence of calcium ions for their completion, lead to the production of *prothrombin activator.*

Prothrombin is an alpha globulin that is continually produced by the liver and thus is normally present in plasma. In the presence of calcium ions, prothrombin is converted into **thrombin** by the action of prothrombin activator. Thrombin, in turn, acts as an enzyme and causes a reaction in molecules of fibrinogen. In this reaction, small portions of the fibrinogen molecules are split off, and the remaining pieces develop attractions for other similar molecules. These activated fibrinogen molecules join, end to end, forming long threads of *fibrin.* The production of fibrin threads also is enhanced by the presence of calcium ions and other protein factors. (See figure 17.21.)

Once threads of fibrin have formed, they tend to stick to the exposed surfaces of damaged blood vessels and create a meshwork in which various blood cells and platelets become entangled (figure 17.22). The resulting mass is the *blood clot,* which may effectively block a vascular break and prevent further loss of blood.

The amount of prothrombin activator that appears in the blood is directly proportional to the degree of tissue damage. Once a blood clot begins to form, it promotes still more clotting. This happens because thrombin also acts directly on blood clotting factors other than fibrinogen, and it can cause prothrombin to form still more thrombin. This type of self-initiating action is an example of a **positive feedback system** (see chapter 13).

Figure 17.22
A scanning electron micrograph of fibrin threads. What factors serve to initiate the formation of fibrin?

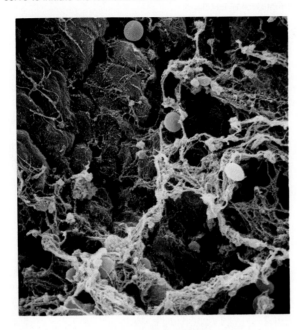

Figure 17.23
The intrinsic blood clotting mechanism may be initiated when blood contacts a foreign surface.

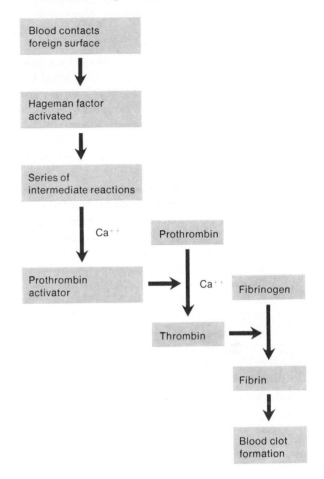

Normally, the formation of a massive clot throughout the blood system is prevented by blood movement, which rapidly carries excessive thrombin away and thus keeps its concentration too low to enhance further clotting. As a result, blood clot formation is usually limited to blood that is standing still, and clotting stops where a clot comes in contact with circulating blood.

Sometimes the clotting mechanism is activated in widespread regions of the circulatory system. This condition, called *disseminated intravascular clotting* (DIC), is usually associated with the presence of bacteria or bacterial toxins in the blood or some disorder causing widespread tissue damage. As a result, many small clots may appear and obstruct blood flow into various tissues and organs, particularly the kidneys. As the plasma clotting factors and platelets become depleted, the patient may develop a serious tendency to bleed.

Intrinsic Clotting Mechanism
A secondary clotting mechanism is initiated by the activation of a substance called the *Hageman factor.* This occurs when blood is exposed to a foreign surface such as collagen or when blood is stored in a glass container. In the presence of *calcium ions,* the activated factor triggers a complex series of changes that involve several clotting factors and lead to the formation of *prothrombin activator.* The subsequent steps of blood clot formation are the same as those described for the extrinsic mechanism (figure 17.23).

Fate of Blood Clots
After a blood clot has formed, it soon begins to retract, apparently due to platelet activity. The tiny processes extending from the platelet membranes adhere to strands of fibrin within the clot, and these processes seem to contract. The blood clot becomes smaller, and the edges of the broken vessel are pulled closer together. At the same time, a fluid, called **serum,** is squeezed from the clot. Serum is essentially plasma minus all of its fibrinogen and most of the other factors involved in the clotting mechanism.

Blood clots that form in ruptured vessels are soon invaded by *fibroblasts*. These cells produce fibrous connective tissue throughout the clots, which helps to strengthen and seal vascular breaks. Many clots, including those that form in tissues as a result of blood leakage (hematomas), disappear in time. This dissolution involves the activation of a plasma protein, called *profibrinolysin,* that is incorporated into blood clots along with other proteins. Profibrinolysin apparently is activated by substances released from the tissues surrounding the clot. When it is activated, profibrinolysin becomes *fibrinolysin* (plasmin), a protein-splitting enzyme that can digest fibrin threads and other proteins associated with blood clots. This action may cause a whole clot to dissolve; however, clots that fill large blood vessels are seldom removed by natural processes.

If a blood clot forms in a vessel abnormally, it is termed a **thrombus.** If the clot becomes dislodged or if a fragment of it breaks loose and is carried away by the blood flow, it is called an **embolus.** Generally, emboli continue to move until they reach narrow places in vessels where they become lodged and interfere with the blood flow.

Such abnormal clot formations are often associated with conditions that cause changes in the endothelial linings of vessels. In the disease *atherosclerosis,* for example, arterial linings are changed by accumulations of fatty deposits. These changes may initiate the clotting mechanism (figure 17.24).

Coagulation also may occur in blood that is flowing too slowly. In this instance, the concentration of clot-promoting substances may increase to a critical level instead of being carried away by more rapidly moving blood, and a clot may form.

Sometimes a blood clot forms in a vessel supplying a vital organ, such as the heart (coronary thrombosis) or brain (cerebral thrombosis). As a consequence of the obstruction, tissues may die (infarction), and this may lead to death of the individual. The blood vessels of the lungs are most likely to be obstructed by emboli arriving through the pulmonary arteries (pulmonary embolism). If a major branch of a pulmonary artery is obstructed, a portion of the lung will become nonfunctional and the life of the patient may be threatened.

---

1. Distinguish between extrinsic and intrinsic clotting mechanisms.
2. What is the basic event in blood clot formation?
3. What prevents the formation of massive clots throughout the blood system?
4. Distinguish between a thrombus and an embolus.

Figure 17.24

(a) A normal artery; *(b)* the inner wall of an artery changed as a result of atherosclerosis. How might this condition promote the formation of blood clots?

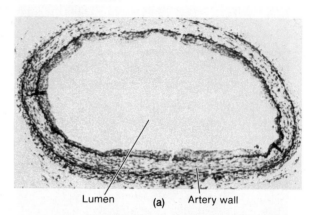

Lumen (a) Artery wall

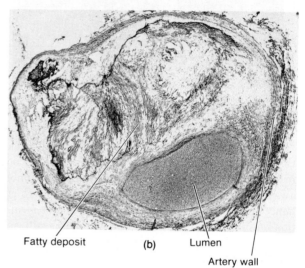

Fatty deposit (b) Lumen

Artery wall

Prevention of Coagulation

In a normal vascular system, spontaneous formation of blood clots is prevented in part by the endothelium of the blood vessels. This smooth lining carries a negative electrical charge that tends to repel platelets and various clotting factors. As long as these vascular characteristics are maintained, the coagulation mechanism is not initiated.

Also, when a clot is forming, fibrin threads adsorb thrombin, thus helping to prevent the spread of the clotting reaction. Any additional thrombin is likely to be inactivated by an alpha globulin, called *antithrombin,* normally present in the plasma. This substance binds to thrombin and blocks its action on fibrinogen.

Coagulation Disorders
A Clinical Application

Coagulation disorders fall into two main groups—those that result in excessive bleeding and those that produce abnormal blood clotting.

Because the liver plays an important role in the synthesis of various plasma proteins, such as prothrombin, it is not surprising that liver diseases are often accompanied by a tendency to bleed. Also, bile salts from the liver are necessary for the efficient absorption of *vitamin K* from the intestine, and this vitamin is essential for the synthesis of prothrombin. If the liver fails to produce enough bile, or if the bile ducts become obstructed, a vitamin K deficiency is likely to develop, and the ability to form blood clots may be diminished. For this reason, vitamin K is often administered to patients with liver diseases or bile duct obstructions before they are treated surgically.

Newborns sometimes have a tendency to hemorrhage because they have relatively low blood levels of certain coagulation factors. These factors usually rise in concentration during the first few weeks of life. Meanwhile, such infants are usually given preventive treatment with vitamin K, which serves to reduce the chance of bleeding.

In other cases, a tendency to bleed is related to abnormally low platelet counts. This condition, called **thrombocytopenia,** is said to occur whenever the platelet count drops below 100,000 platelets per mm³ of blood.

The cause of thrombocytopenia is usually unknown, but various factors that damage red bone marrow, such as excessive exposure to ionizing radiation or adverse drug reactions, may produce this condition. Affected persons bleed easily, numerous capillary hemorrhages usually occur throughout their body tissues, and their skin typically exhibits many small, bruiselike spots.

Hemophilia is a hereditary disease that appears almost exclusively in males. Hemophiliacs are deficient in one of several blood factors necessary for coagulation, and consequently, they usually experience repeated episodes of serious bleeding.

There are several types of hemophilia, and each is caused by the lack of a different blood factor. However, the symptoms of these conditions are so similar that they are difficult to distinguish. In addition to the tendency to hemorrhage severely following minor injuries, hemophilia is characterized by frequent nosebleeds, large intramuscular hematomas, blood in the urine (hematuria), and severe pain and disability from bleeding into joints and body cavities.

Treatment for hemophilia may involve pressure or packing of accessible bleeding sites in an effort to control blood loss. Transfusions are used to replace missing blood factors. For example, a common type of hemophilia caused by a deficiency of a substance, called blood factor VIII, may be treated with fresh plasma, fresh-frozen plasma, or plasma concentrates (cryoprecipitates).

Chart 17.7	Factors that inhibit blood clot formation		
Factor	Action	Factor	Action
Smooth lining of blood vessel	Prevents activation of intrinsic blood clotting mechanism	Fibrin threads	Adsorb thrombin
		Antithrombin in plasma	Interferes with the action of thrombin
Negative charge of blood vessel lining	Repels platelets	Heparin from mast cells and basophils	Interferes with the formation of prothrombin activator

In addition, certain cells including basophils and mast cells, which are common in the connective tissue surrounding capillaries, secrete the anticoagulant *heparin.* This substance interferes with the formation of prothrombin activator, prevents the action of thrombin on fibrinogen, and promotes the removal of thrombin by antithrombin and fibrin adsorption.

Heparin-secreting cells are particularly abundant in the liver and lungs, where capillaries are likely to trap small blood clots that commonly occur in the slow-moving blood of veins. These cells are thought to secrete heparin continually and thus help prevent additional clotting in the circulatory system. Chart 17.7 summarizes the clot-inhibiting factors.

Because of its strong anticoagulant action, heparin is widely used in the treatment of conditions in which blood clots are likely to form abnormally. Administration of heparin by injection almost immediately increases the time required for blood coagulation.

1. How does the lining of a blood vessel help to prevent blood clot formation?
2. What is the function of antithrombin?
3. How does heparin help to prevent blood clot formation?

Blood Groups and Transfusions

Early attempts to transfer blood from one person to another produced varied results. Sometimes the person receiving the transfusion was aided by the procedure. Other times the recipient suffered a blood reaction in which the red blood cells clumped together, obstructing vessels and producing other serious consequences.

Eventually, it was discovered that each individual has a particular combination of substances in his or her blood. Some of these substances reacted with those in another person's blood. These discoveries led to the development of procedures for typing blood. It is now known that safe transfusions of whole blood depend upon properly matching the blood types of the donors and recipients.

Agglutinogens and Agglutinins

The clumping of red cells following a transfusion reaction is called **agglutination.** This phenomenon is due to the presence of substances called **agglutinogens** (antigens) in the red cell membranes and substances called **agglutinins** (antibodies) dissolved in the plasma.

Blood typing involves identifying the agglutinogens that are present in a person's red cells. Although there are many different agglutinogens associated with

human erythrocytes, only a few of them are likely to produce serious transfusion reactions. These include the agglutinogens of the ABO group and those of the Rh group.

Avoiding the mixture of certain kinds of agglutinogens and agglutinins prevents adverse transfusion reactions.

The ABO Blood Group

The *ABO blood group* is based on the presence (or absence) of two major agglutinogens in red cell membranes—*agglutinogen A* and *agglutinogen B*. The erythrocytes of each person contain one of the four following combinations of agglutinogens: only A, only B, both A and B, or neither A nor B.

A person with only agglutinogen A is said to have *type A blood;* a person with only agglutinogen B has *type B blood;* one with both agglutinogen A and B has *type AB blood;* and one with neither agglutinogen A nor B has *type O blood.* Thus, all humans have one of four possible blood types—A, B, AB, or O.

Certain agglutinins in the plasma accompany the agglutinogens in the red cell membranes of each person's blood. Specifically, whenever agglutinogen A is absent, an agglutinin called *anti-A* is present; and whenever agglutinogen B is absent, an agglutinin called

Figure 17.25
Each blood type is characterized by a different combination of agglutinogens and agglutinins.

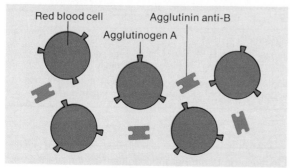

Type A blood

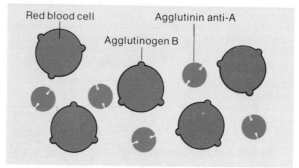

Type B blood

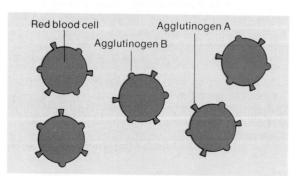

Type AB blood

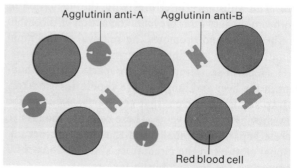

Type O blood

anti-B is present. Therefore, persons with type A blood also have agglutinin anti-B in their plasma; those with B blood have agglutinin anti-A; those with type AB blood have neither agglutinin; and those with type O blood have both agglutinin anti-A and anti-B (figure 17.25).

Chart 17.8 summarizes the agglutinogens and agglutinins of the ABO blood group.

Since an agglutinin of one kind will react with an agglutinogen of the same kind and cause red blood cells to clump together, such combinations are avoided whenever possible. Actually, the major concern in blood transfusion procedures is that the cells in the *transfused blood* not be agglutinated by the agglutinins in the recipient's plasma. For this reason, a person with type A (anti-B) blood should never be given blood of type B or AB, because the red cells of both types would be agglutinated by the anti-B in the recipient's type A blood. Likewise, a person with type B (anti-A) blood should never be given type A or AB blood and a person with type O (anti-A and anti-B) blood should never be given type A, B, or AB blood (figure 17.26).

Since type AB blood lacks both anti-A and anti-B agglutinins, it would appear that an AB person could receive a transfusion of blood of any other type. For this reason, type AB persons are sometimes called *universal recipients*. It should be noted, however, that type A (anti-B) blood, type B (anti-A) blood, and type O (anti-A and anti-B) blood still contain agglutinins (either anti-A or anti-B) that could cause agglutination of type AB cells. Consequently, it is always best to use donor blood (AB) of the same type as the recipient blood (AB). If the matching type is not available and type A, B, or O is used, it should be transfused slowly so that the donor blood is well diluted by the recipient's larger blood volume. This precaution usually avoids serious reactions between the donor's agglutinins and the recipient's agglutinogens.

Similarly, because type O blood lacks agglutinogens A and B, it would seem that this type could be transfused into persons with blood of any other type. Persons with type O blood, therefore, are sometimes called *universal donors*. Type O blood, however, does contain both anti-A and anti-B agglutinins, and if it is given to a person with blood type A, B, or AB, it too should be transfused slowly to minimize the chance of an adverse reaction.

Chart 17.9 summarizes preferred blood type for normal transfusions and permissible blood type for emergency transfusions.

Chart 17.8	Agglutinogens and agglutinins of the ABO blood group	
Blood type	Agglutinogen	Agglutinin
A	A	anti-B
B	B	anti-A
AB	A and B	Neither anti-A nor anti-B
O	Neither A nor B	Both anti-A and anti-B

Chart 17.9	Preferred and permissible blood types for transfusions	
Blood type	Preferred transfusion	Permissible transfusion
A	A	O
B	B	O
AB	AB	A, B, O
O	O	none

Figure 17.26
(a) If red blood cells with agglutinogen A are added to blood containing agglutinin anti-A, *(b)* the agglutinins will react with the agglutinogens of the red blood cells and cause them to clump together.

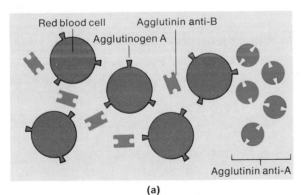

(a)

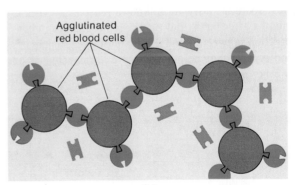

(b)

Agglutination Reactions
A Clinical Application

In any blood reaction involving agglutinogens and agglutinins, the agglutinated red cells usually degenerate or are phagocytized by reticuloendothelial cells. At the same time, hemoglobin and other red cell contents are released, and the blood concentration of *free hemoglobin* increases greatly. Some of this hemoglobin may diffuse out of the vascular system and enter the body tissues, where it is gradually converted into bilirubin. As a result, the tissues may develop a yellowish stain, which is a condition called *jaundice*.

Free hemoglobin also may pass into the kidneys and interfere with the vital functions of these organs, so that a person with a blood transfusion reaction may experience kidney (renal) failure later.

The severity of a blood transfusion reaction depends on a number of factors, including the degree of incompatibility between the blood of the donor and that of the recipient, the quantity of blood that is transfused, and the rate at which the transfusion is administered.

If a reaction occurs, the patient is likely to experience anxiety, breathing difficulty, facial flushing, headache, and severe pain in the neck, chest, and lumbar area.

Infants with erythroblastosis fetalis usually are jaundiced and severely anemic. As their blood-cell-forming tissues respond to the need for more red cells, various immature erythrocytes including *erythroblasts* are released into the blood. (The presence of these immature cells is related to the name of the disease.)

An affected infant may suffer permanent brain damage as a result of bilirubin precipitating in the brain tissues and injuring neurons. This condition is called *kernicterus*, and if the infant survives, it may have motor or sensory losses and exhibit mental deficiencies.

Treatment for erythroblastosis fetalis usually involves exposing the affected infant to bright blue or white *fluorescent light*. Bilirubin is a light-sensitive substance, and this exposure (phototherapy) causes a decrease in the blood bilirubin concentration.

In more severe cases, the infant's Rh-positive blood may be replaced with Rh-negative blood. This procedure is called an *exchange transfusion,* and it reduces the concentration of bilirubin in the infant's tissues in addition to removing the agglutinating red cells, the anti-Rh agglutinins, the free hemoglobin, and the other products of erythrocyte destruction. It also provides a temporary supply of red cells that will not be agglutinated by any remaining anti-Rh agglutinins. In time, the infant's blood-cell-forming tissues will replace the donor's blood cells with Rh-positive cells, but by then the maternal agglutinins will have disappeared.

Erythroblastosis fetalis can be prevented in future offspring by treating Rh-negative mothers with a special blood serum within 72 hours following the birth of each Rh-positive child. This serum is obtained from the blood of another Rh-negative person who has formed anti-Rh agglutinin. The serum antibodies are able to inactivate any Rh-positive cells that may have entered the maternal blood at the time of the infant's birth. The maternal tissues, consequently, are not stimulated to manufacture anti-Rh agglutinins, and the mother's blood should not be a hazard to her next Rh-positive child.

Every person's blood contains a great variety of substances, and although the agglutinogens of the ABO and Rh groups are the most likely to produce serious blood reactions, other factors sometimes cause problems. For this reason, it is a good practice to determine whether samples of recipient and donor blood will produce agglutination of red cells before administering a transfusion to a patient. This procedure, called *crossmatching,* involves mixing a suspension of donor cells in some recipient serum, and then mixing a suspension of recipient cells in some donor serum. If the red cells do not agglutinate in either case, it is probably safe to give the transfusion.

The agglutinogens of the ABO group are *inherited factors* that are present in the red cell membranes at the time of birth. The plasma agglutinins begin to appear spontaneously, for unknown reasons, about 2 to 8 months after birth, and they reach a maximum concentration between 8 and 10 years of age. A person's blood type cannot be changed by a transfusion or by any other procedure.

Blood substitutes, such as isotonic saline solution or isotonic glucose solution, are sometimes administered to persons during emergencies. These substitutes will increase the blood volume only temporarily. If a patient has suffered a severe hemorrhage, a transfusion of whole blood may be necessary for survival.

Figure 17.27
(a) If an Rh-negative woman is pregnant with an Rh-positive fetus, (b) some of the fetal red blood cells with Rh agglutinogens may enter the maternal blood at the time of birth. (c) As a result, the woman's cells may produce anti-Rh agglutinins.

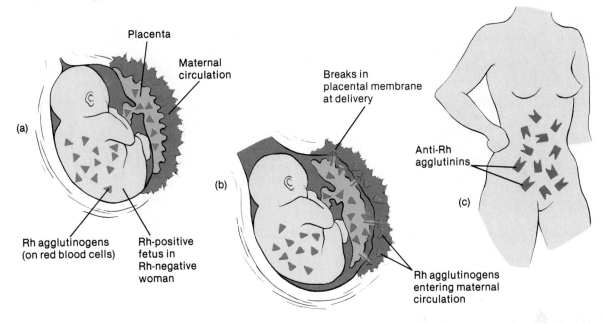

1. Distinguish between agglutinogens and agglutinins.
2. What is meant by blood type?
3. What is the main concern when blood is transfused from one individual to another?
4. Why is a type AB person called a universal recipient?

The Rh Blood Group

The *Rh blood group* was named after the *rhesus monkey,* in which it was first observed. In humans this group is based upon several different Rh agglutinogens (factors). The most important of these is *agglutinogen D;* however, if any of the Rh factors are present in the red cell membranes, the blood is said to be *Rh positive.* Conversely, if the red cells lack Rh agglutinogens, the blood is called *Rh negative.*

As in the case of agglutinogens A and B, the presence (or absence) of an Rh agglutinogen is an inherited trait. Unlike anti-A and anti-B, agglutinins for Rh (*anti-Rh*) do not appear spontaneously. Instead, they form only in Rh-negative persons in response to special stimulation.

If an Rh-negative person receives a transfusion of Rh-positive blood, the recipient's antibody-producing cells will be stimulated by the presence of the Rh agglutinogen and will begin producing *anti-Rh agglutinin.* Generally there are no serious consequences from this initial transfusion, but if the Rh-negative person—who is now sensitized to Rh-positive blood—receives a subsequent transfusion of Rh-positive blood, some months later, the donor's red cells are likely to agglutinate.

A related condition may occur when an Rh-negative woman is pregnant with an Rh-positive fetus for the first time. Such a pregnancy may be uneventful; however, at the time of this infant's birth, (or if a miscarriage occurs) the placental membranes, which separated the maternal blood from the fetal blood during the pregnancy, may be broken, and some of the infant's Rh-positive blood cells may get into the maternal circulation. These Rh-positive cells may then stimulate the maternal tissues to begin producing anti-Rh agglutinins (figure 17.27).

Figure 17.28

(a) If a woman who has developed anti-Rh agglutinins is pregnant with an Rh-positive fetus, (b) agglutinins may pass through the placental membrane and cause the fetal red blood cells to agglutinate.

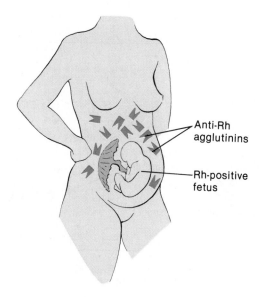

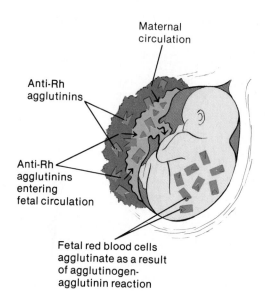

Chart 17.10	Possible events leading to erythroblastosis fetalis

1. Rh-negative woman becomes pregnant with her first Rh-positive child.
2. Pregnancy is uneventful, but at the time of birth some Rh-positive red cells enter the maternal circulation through damaged placental tissues.
3. Maternal tissues produce anti-Rh agglutinins.
4. A second Rh-positive child is conceived.
5. Anti-Rh agglutinins from the maternal circulation pass through the placental membranes and enter the fetal blood.
6. The fetus develops erythroblastosis fetalis as the maternal anti-Rh agglutinins react with the Rh agglutinogens of the fetal red blood cells and cause them to agglutinate.

If the mother, who has already developed anti-Rh agglutinin becomes pregnant with a second Rh-positive fetus, these anti-Rh agglutinins can pass through the placental membrane and react with the fetal red cells, causing them to agglutinate. The fetus then develops a disease called **erythroblastosis fetalis** (hemolytic disease of the newborn) (figure 17.28).

Chart 17.10 reviews the events that can lead to this condition.

Clinical Terms Related to the Blood

anisocytosis (an-i''so-si-to'sis)   condition in which there is an abnormal variation in the size of erythrocytes.

antihemophilic plasma (an''ti-he''mo-fil'ik plaz'mah)   normal blood plasma that has been processed to preserve an antihemophilic factor.

Christmas disease (kris'mas di-zēz')   a hereditary bleeding disease that is due to a deficiency in a clotting factor; also called *hemophilia B*.

citrated whole blood (sit'rāt-ed hōl blud)   normal blood to which a solution of acid citrate has been added to prevent coagulation.

dried plasma (drīd plaz'mah)   normal blood plasma that has been vacuum dried to prevent the growth of microorganisms.

hemochromatosis (he''mo-kro''mah-to'sis)   a disorder in iron metabolism in which excessive iron is deposited in the tissues.

hemorrhagic telangiectasia (hem''o-raj'ik tel-an''je-ek-ta'ze-ah)   a hereditary disorder in which there is a tendency to bleed from localized lesions of capillaries.

heparinized whole blood (hep'er-i-nizd'' hōl blud)   normal blood to which a solution of heparin has been added to prevent coagulation.

macrocytosis (mak''ro-si-to'sis)   condition characterized by the presence of abnormally large erythrocytes.

microcytosis (mi″kro-si-to′sis)   condition characterized by the presence of abnormally small erythrocytes.

neutrophilia (nu″tro-fil′e-ah)   condition in which there is an increase in the number of circulating neutrophils.

normal plasma (nor′mal plaz′mah)   plasma from which the blood cells have been removed by centrifugation or sedimentation.

packed red cells (pakd red selz)   a concentrated suspension of red blood cells from which the plasma has been removed.

pancytopenia (pan″si-to-pe′ne-ah)   condition characterized by an abnormal depression of all the cellular components of blood.

poikilocytosis (poi″ki-lo-si-to′sis)   condition in which the erythrocytes are irregularly shaped.

pseudoagglutination (su″do-ah-gloo″ti-na′shun)   clumping of erythrocytes due to some factor other than an agglutinogen-agglutinin reaction.

purpura (per′pu-rah)   a disease characterized by spontaneous bleeding into the tissues and through the mucous membrane.

septicemia (sep″ti-se′me-ah)   a condition in which disease-causing microorganisms or their toxins are present in the blood.

spherocytosis (sfer″o-si-to′sis)   a hereditary form of hemolytic anemia characterized by the presence of spherical erythrocytes (spherocytes).

thalassemia (thal″ah-se′me-ah)   a group of hereditary hemolytic anemias characterized by the presence of very thin, fragile erythrocytes.

von Willebrand's disease (fon vil′e-brandz di-zez′)   a hereditary condition due to a deficiency of an antihemophilic blood factor and capillary defects, which is characterized by bleeding from the nose, gums, and genitalia.

Chapter Summary

Introduction
Blood is a type of connective tissue with cells that are suspended in a liquid intercellular material.

Blood and Blood Cells
Blood contains red blood cells, white blood cells, and platelets.

1. Volume and composition of blood
   a. Blood volume varies with body size, fluid and electrolyte balance, and fat content.
   b. Blood can be separated into solid and liquid portions.
      (1) The cellular portion is mostly red blood cells.
      (2) Plasma includes water, nutrients, hormones, electrolytes, and cellular wastes.

2. Characteristics of red blood cells
   a. Red blood cells are biconcave disks with shapes that provide increased surface area and place their cell membranes close to internal parts.
   b. They contain hemoglobin that combines loosely with oxygen.
   c. The mature forms lack nuclei, but contain enzymes needed for energy-releasing processes.

3. Red blood cell counts
   a. The red blood cell count equals the number of cells per mm^3 of blood.
   b. The average count is from 4,000,000 to 6,000,000 cells per mm^3.
   c. Red cell count is related to the oxygen-carrying capacity of the blood and is used in diagnosing and evaluating the courses of diseases.

4. Destruction of red blood cells
   a. Red cells are fragile and are damaged while moving through capillaries.
   b. Damaged red cells are phagocytized by macrophages in the liver and spleen.
   c. Hemoglobin molecules are decomposed and the iron they contain is conserved.
   d. The number of red blood cells remains relatively stable.

5. Red blood cell production and its control
   a. During fetal development, red cells are formed in the yolk sac, liver, and spleen; later, red cells are produced by the red bone marrow.
   b. The rate of red cell production is controlled by a negative feedback mechanism that involves a hormone from the kidneys and liver.
      (1) Erythropoietin is released in response to low oxygen levels.
      (2) Low oxygen levels may be caused by high altitude, loss of blood, or chronic lung disease.

6. Dietary factors affecting red blood cell production
   a. Production of red blood cells is affected by the availability of vitamin B_{12} and folic acid that are needed for DNA synthesis.
   b. Iron also is needed for hemoglobin synthesis.
   c. The rate of iron absorption varies with the amount of iron in the body.

7. Types of white blood cells
   a. White blood cells function to defend the body against infections by microorganisms.
   b. Granulocytes include neutrophils, eosinophils, and basophils.
   c. Agranulocytes include monocytes and lymphocytes.

8. White blood cell counts
   a. Normal total white blood cell counts vary from 5000 to 10,000 cells per mm³ of blood.
   b. Number of white cells may change in abnormal conditions such as infections, emotional disturbances, or excessive loss of body fluids.
   c. A differential white cell count indicates the percentages of various types of leukocytes present.
9. Functions of white blood cells
   a. Some white blood cells phagocytize foreign particles; others produce antibodies of immunity.
   b. Leukocytes may be stimulated by the presence of chemicals released by damaged cells and move toward these chemicals.
   c. Eosinophils increase in parasitic diseases and allergies.
   d. Basophils release heparin, which inhibits blood clotting.
10. Blood platelets
   a. Blood platelets are fragments of giant cells that become detached and enter the circulation.
   b. The normal count varies from 130,000 to 360,000 platelets per mm³.
   c. They function to help close breaks in blood vessels.

Blood Plasma

Plasma is the liquid part of the blood and is composed of water and a mixture of organic and inorganic substances.

It functions to transport nutrients and gases, regulate fluid and electrolyte balance, and maintain pH.

1. Plasma proteins
   a. Plasma proteins remain in blood and interstitial fluids, and are not normally used as energy sources.
   b. Three major groups exist.
      (1) Albumins help maintain the osmotic pressure of blood.
      (2) Globulins function to transport lipids and fat-soluble vitamins, and they include the antibodies of immunity.
      (3) Fibrinogen functions in blood clotting.

2. Nutrients and gases
   a. Nutrients include amino acids, simple sugars, and lipids.
      (1) Glucose is stored in the liver as glycogen and is released whenever the blood glucose level falls.
      (2) Amino acids are used to synthesize proteins and are deaminated for use as energy sources.
      (3) Lipids are present as lipoproteins that function in the transport of these lipids.
   b. Gases in plasma include oxygen, carbon dioxide, and nitrogen.
3. Nonprotein nitrogenous substances
   a. Nonprotein nitrogenous substances are composed of molecules that contain nitrogen atoms but are not proteins.
   b. They include amino acids, urea, uric acid, creatine, and creatinine.
      (1) Urea and uric acid are products of catabolic metabolism.
      (2) Creatinine results from the metabolism of creatine.
   c. These substances usually remain stable; an increase may indicate a kidney disorder.
4. Plasma electrolytes
   a. Plasma electrolytes are obtained by absorption from the intestines and are released as by-products of cellular metabolism.
   b. They include ions of sodium, potassium, calcium, magnesium, chlorine, bicarbonate, phosphate, and sulfate.
   c. They are important in the maintenance of osmotic pressure and pH.

Hemostasis

Hemostasis refers to the stoppage of bleeding.

Hemostatic mechanisms are most effective in controlling blood loss from small vessels.

1. Blood vessel spasm
   a. Smooth muscles in walls of arterioles and arteries contract reflexly following injury.
   b. Platelets release serotonin that stimulates vasoconstriction and helps to maintain vessel spasm.
2. Platelet plug formation
   a. Platelets adhere to rough surfaces and exposed collagen.
   b. Platelets stick together at the sites of injuries and form platelet plugs in broken vessels.

3. Blood coagulation
   a. Blood clotting is the most effective means of hemostasis and may be initiated by extrinsic or intrinsic mechanisms.
   b. Clot formation depends on the balance between clotting factors that promote clotting and those that inhibit clotting.
   c. The basic event of coagulation is the conversion of soluble fibrinogen into insoluble fibrin.
   d. Factors that promote clotting include the presence of prothrombin activator, prothrombin, and calcium ions.
   e. After forming, the clot retracts and pulls the edges of a broken vessel closer together.
   f. A thrombus is a blood clot in a vessel; an embolus is a clot or fragment of a clot that has moved in a vessel.
   g. A clot is invaded by fibroblasts that form connective tissue throughout the clot.
   h. The clot eventually may be destroyed by the action of protein-splitting enzymes.
4. Prevention of coagulation
   a. The smooth lining of blood vessels repels platelets.
   b. As a clot forms, fibrin adsorbs thrombin and prevents the reaction from spreading.
   c. Antithrombin interferes with the action of excessive thrombin.
   d. Some cells secrete heparin, an anticoagulant.

Blood Groups and Transfusions

Blood can be typed on the basis of the substances it contains.

Blood substances of certain types will react adversely with other types.

1. Agglutinogens and agglutinins
   a. Red blood cell membranes may contain agglutinogens and blood plasma may contain agglutinins.
   b. Blood typing involves identifying the agglutinogens present in the red cell membranes.
2. The ABO blood group
   a. Blood can be grouped according to the presence or absence of agglutinogens A and B.
   b. Whenever agglutinogen A is absent, agglutinin anti-A is present; whenever agglutinogen B is absent, agglutinin anti-B is present.

c. Adverse transfusion reactions are avoided by preventing the mixing of red cells that contain an agglutinogen with plasma that contains the corresponding agglutinin.
   d. Adverse reactions involve agglutination (clumping) of the red blood cells.
3. The Rh blood group
   a. Rh agglutinogens are present in the red cell membranes of Rh-positive blood; they are absent in Rh-negative blood.
   b. If an Rh-negative person is exposed to Rh-positive blood, anti-Rh agglutinins are produced in response.
   c. Mixing Rh-positive red cells with plasma that contains anti-Rh agglutinins results in agglutination of the positive cells.
   d. If an Rh-negative female is pregnant with an Rh-positive fetus, some of the positive cells may enter the maternal blood at the time of birth and stimulate the maternal tissues to produce anti-Rh agglutinins.
   e. Anti-Rh agglutinins in maternal blood may pass through the placental tissues and react with the red cells of an Rh-positive fetus, causing it to develop erythroblastosis fetalis.

Application of Knowledge

1. What changes would you expect to occur in the hematocrit of a person who is dehydrating? Why?
2. If a patient with an inoperable cancer is treated by using a drug that reduces the rate of cell division, what changes might occur in the patient's white blood cell count? How might the patient's environment be modified to compensate for the effects of these changes?
3. Hypochromic (iron-deficiency) anemia is relatively common among aging persons who are admitted to hospitals for other conditions. What environmental and sociological factors might promote this?

Review Activities

1. List the major components of blood.
2. Define *hematocrit* and explain how it is determined.
3. Describe a red blood cell.
4. Distinguish between oxyhemoglobin and deoxyhemoglobin.
5. Explain what is meant by a red blood cell count.
6. Describe the life cycle of a red blood cell.
7. Distinguish between biliverdin and bilirubin.
8. Define *erythropoietin* and explain its function.
9. Explain how vitamin B_{12} and folic acid deficiencies affect red blood cell production.
10. List two sources of iron that can be used for the synthesis of hemoglobin.
11. Distinguish between granulocytes and agranulocytes.
12. Name five types of leukocytes and list the major functions of each type.
13. Explain the significance of white blood cell counts as aids to diagnosing diseases.
14. Describe a blood platelet and explain its functions.
15. Name three types of plasma proteins and list the major functions of each type.
16. Define *lipoprotein*.
17. Define *apoprotein*.
18. Distinguish between low-density lipoprotein and high-density lipoprotein.
19. Name the sources of VLDL, LDL, HDL, and chylomicrons.
20. Describe how lipoproteins are removed from plasma.
21. Explain how cholesterol is eliminated from plasma and the body.
22. Define *nonprotein nitrogenous substances* and name those commonly present in plasma.
23. Name several plasma electrolytes.
24. Define *hemostasis*.
25. Explain how blood vessel spasms are stimulated following an injury.
26. Explain how a platelet plug forms.
27. List the major steps leading to the formation of a blood clot.
28. Distinguish between fibrinogen and fibrin.
29. Provide an example of a positive feedback system.
30. Define *serum*.
31. Distinguish between a thrombus and an embolus.
32. Explain how a blood clot may be removed naturally from a blood vessel.
33. Describe how blood coagulation may be prevented.
34. Review the function of vitamin K.
35. Distinguish between agglutinogen and agglutinin.
36. Explain the basis of ABO blood types.
37. Explain why a person with blood type AB is sometimes called a universal recipient.
38. Explain why a person with blood type O is sometimes called a universal donor.
39. Distinguish between Rh-positive and Rh-negative blood.
40. Describe how a person may become sensitized to Rh-positive blood.
41. Define *erythroblastosis fetalis* and explain how this condition may develop.

18

The Cardiovascular System

The *cardiovascular system* is the portion of the circulatory system that includes the heart and blood vessels. It functions to move blood between the body cells and organs of the integumentary, digestive, respiratory, and urinary systems, which communicate with the external environment.

In performing this function, the heart acts as a pump that forces blood through the blood vessels. The blood vessels, in turn, form a closed system of ducts that transports blood and allows exchanges of gases, nutrients, and wastes between the blood and the body cells.

Chapter Outline

Chapter Objectives

After you have studied this chapter, you should be able to

1. Name the organs of the cardiovascular system and discuss their functions.

2. Name and describe the location of the major parts of the heart and discuss the function of each part.

3. Trace the pathway of blood through the heart and the vessels of the coronary circulation.

4. Discuss the cardiac cycle and explain how it is controlled.

5. Compare the structures and functions of the major types of blood vessels.

6. Describe the mechanism that aids in the return of venous blood to the heart.

7. Explain how blood pressure is created and controlled.

8. Compare the pulmonary and systemic circuits of the cardiovascular system.

9. Identify and locate the major arteries and veins of the pulmonary and systemic circuits.

10. Complete the review activities at the end of this chapter. Note that the items are worded in the form of specific learning objectives. You may want to refer to them before reading the chapter.

Key Terms

atrium   (a′tre-um)

cardiac conduction system
   (kar′de-ak kon-duk′shun sis′tem)

cardiac cycle   (kar′de-ak si′kl)

cardiac output   (kar′de-ak owt′poot)

diastolic pressure   (di″ah-stol′ik presh′ur)

electrocardiogram   (e-lek″tro-kar′de-o-gram″)

functional syncytium   (funk′shun-al sin-sish′e-um)

myocardium   (mi″o-kar′de-um)

pacemaker   (pās′māk-er)

pericardium   (per″ĭ-kar′de-um)

peripheral resistance   (pe-rif′er-al re-zis′tans)

pulmonary circuit   (pul′mo-ner″e sur′kit)

sphygmomanometer   (sfig″mo-mah-nom′ĕ-ter)

systemic circuit   (sis-tem′ik sur′kit)

systolic pressure   (sis-tol′ik presh′ur)

vasoconstriction   (vas″o-kon-strik′shun)

vasodilation   (vas″o-di-la′shun)

ventricle   (ven′trĭ-kl)

viscosity   (vis-kos′ĭ-te)

Aids to Understanding Words

angio-, a vessel: *angio*tensin—a substance that causes blood vessels to constrict.

brady-, slow: *brady*cardia—an abnormally slow heartbeat.

diastol-, dilation: *diastol*ic pressure—the blood pressure that occurs when the ventricle is relaxed (thus dilated).

ectop-, out of place: *ectop*ic beat—a heartbeat that occurs before it is expected in a normal series of cardiac cycles.

edem-, a swelling: *edem*a—condition in which fluids accumulate in the tissues and cause them to swell.

-gram, something written: electrocardio*gram*—a recording of the electrical changes that occur in the heart muscle during a cardiac cycle.

myo-, muscle: *myo*cardium—the muscle tissue within the wall of the heart.

papill-, nipple: *papill*ary muscle—a small mound of muscle within a chamber of the heart.

phleb-, vein: *phleb*itis—an inflammation of a vein.

scler-, hard: *scler*osis—condition in which a blood vessel wall loses its elasticity and becomes hard.

syn-, together: *syn*cytium—a mass of merging cells that act together.

systol-, contraction: *systol*ic pressure—the blood pressure that occurs during a ventricular contraction.

tachy-, rapid: *tachy*cardia—an abnormally fast heartbeat.

A FUNCTIONAL CARDIOVASCULAR system is vital for survival because without circulation, tissues lack a supply of oxygen and nutrients, and waste substances accumulate. Under such conditions cells soon begin to undergo irreversible changes that quickly lead to death of the organism. The general pattern of the cardiovascular system is shown in figure 18.1.

Structure of the Heart

The heart is a hollow, cone-shaped, muscular pump located within the mediastinum of the thorax and resting upon the diaphragm.

Size and Location of the Heart

Although the size of the heart varies with body size, it is generally about 14 cm long and 9 cm wide in an average adult.

The heart is enclosed laterally by the lungs, posteriorly by the backbone, and anteriorly by the sternum (figure 18.2). Its *base,* which is attached to several large blood vessels, lies beneath the second rib. Its distal end extends downward and to the left, terminating as a bluntly pointed *apex* at the level of the fifth intercostal space. For this reason, it is possible to sense the *apical heartbeat* by feeling or listening to the chest wall between the fifth and sixth ribs, about 7.5 cm to the left of the midline.

Coverings of the Heart

The heart and the proximal ends of the large blood vessels to which it is attached are enclosed by a double-layered **pericardium.** The inner layer of this membrane is called the *visceral pericardium* (epicardium), and it consists of a thin, serous covering closely applied to the

Figure 18.1
The cardiovascular system functions to transport blood between the body cells and various organs that communicate with the external environment.

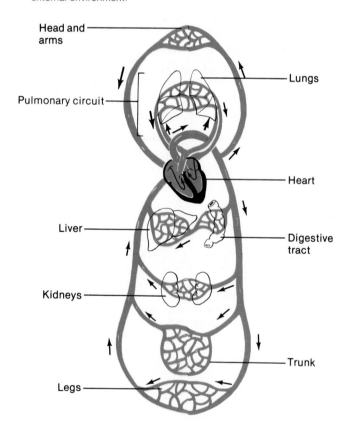

Figure 18.2
The heart is located behind the sternum, where it rests upon the diaphragm.

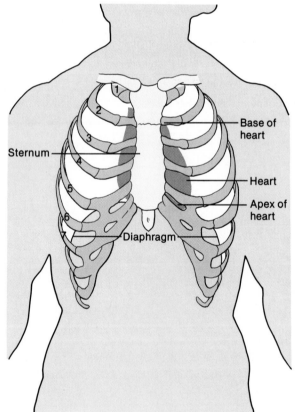

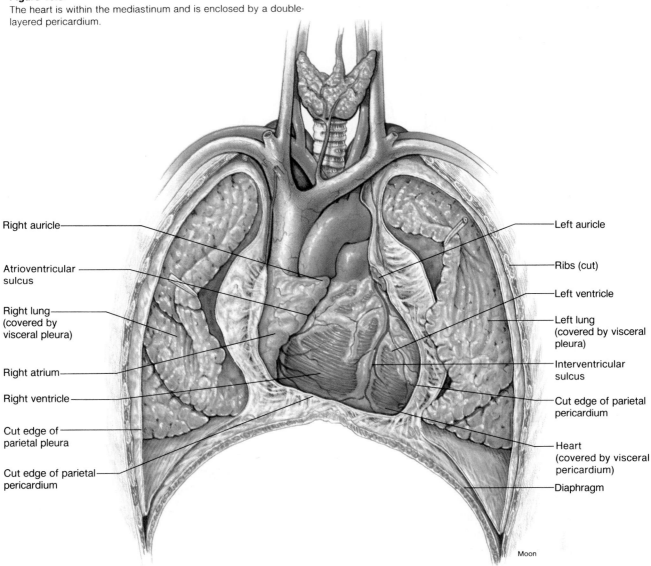

Right auricle

Atrioventricular sulcus

Right lung (covered by visceral pleura)

Right atrium

Right ventricle

Cut edge of parietal pleura

Cut edge of parietal pericardium

Left auricle

Ribs (cut)

Left ventricle

Left lung (covered by visceral pleura)

Interventricular sulcus

Cut edge of parietal pericardium

Heart (covered by visceral pericardium)

Diaphragm

Moon

surface of the heart. At the base of the heart, the visceral pericardium turns back upon itself and becomes the serous part of a loose-fitting *parietal pericardium.*

The parietal pericardium is a tough, protective sac composed largely of white fibrous connective tissue. It is attached to the central portion of the diaphragm, the back of the sternum, the vertebral column, and the large blood vessels emerging from the heart. (See figure 18.3.) Between the parietal and visceral membranes is a potential space, the *pericardial cavity,* that contains a small amount of serous fluid. This fluid serves to reduce friction between the pericardial membranes as the heart moves within them (figure 18.4).

If the pericardium becomes inflamed due to a bacterial or viral infection, the condition is called *pericarditis.* As a result of this inflammation, the layers of the pericardium sometimes become stuck together by adhesions, and this may interfere with heart movements. If this happens, surgery may be required to separate the surfaces and make unrestricted heart actions possible again.

1. Where is the heart located?
2. Where would you listen to hear the apical heartbeat?
3. Distinguish between the visceral pericardium and the parietal pericardium.
4. What is the function of the fluid in the pericardial cavity?

Figure 18.4

The wall of the heart consists of three layers: endocardium, myocardium, and epicardium.

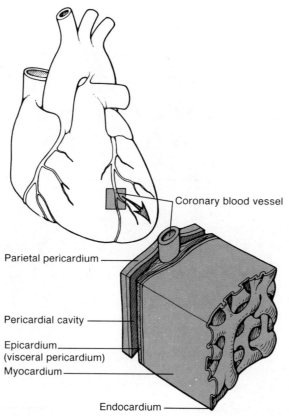

Coronary blood vessel

Parietal pericardium

Pericardial cavity

Epicardium (visceral pericardium)

Myocardium

Endocardium

Wall of the Heart

The wall of the heart is composed of three distinct layers: an outer epicardium, a middle myocardium, and an inner endocardium (figure 18.4).

The outer **epicardium** provides a protective layer and is composed of the visceral pericardium. This serous membrane consists of connective tissue covered by epithelium and includes blood capillaries, lymph capillaries, and nerve fibers. The deeper portion often contains fat, particularly along the paths of larger blood vessels.

The middle layer, or **myocardium,** is relatively thick and consists largely of the cardiac muscle tissue responsible for forcing blood out of the heart chambers. The muscle fibers are arranged in planes separated by connective tissues that are richly supplied with blood capillaries, lymph capillaries, and nerve fibers.

The inner layer, or **endocardium,** consists of epithelium and connective tissue that contains many elastic and collagenous fibers. The connective tissue also contains blood vessels and some specialized cardiac muscle fibers called *Purkinje fibers.* The function of these fibers is described in a subsequent section of this chapter.

The endocardium lines all of the heart chambers and covers structures, such as the heart valves, that project into them. This inner lining is also continuous with the inner linings of the blood vessels (endothelium) attached to the heart.

Chart 18.1 summarizes the characteristics of these three layers of the heart.

An inflammation of the endocardium is termed *endocarditis.* This condition sometimes accompanies bacterial diseases, such as scarlet fever or syphilis, and may produce lasting effects by damaging the valves of the heart.

Heart Chambers and Valves

Internally, the heart is divided into four hollow chambers, two on the left and two on the right. The upper chambers, called **atria** (singular, *atrium*), have relatively thin walls and receive blood from veins. The lower chambers, the **ventricles,** force blood out of the heart into arteries. (Note: veins are blood vessels that carry blood toward the heart; arteries carry blood away from the heart.)

Chart 18.1 Wall of the heart

Layer	Composition	Function
Epicardium (visceral pericardium)	Serous membrane of connective tissue covered with epithelium and including blood capillaries, lymph capillaries, and nerve fibers	Protective outer covering
Myocardium	Cardiac muscle tissue separated by connective tissues and including blood capillaries, lymph capillaries, and nerve fibers	Muscular contractions that force blood from the heart chambers
Endocardium	Membrane of epithelium and connective tissues, including elastic and collagenous fibers, blood vessels, and specialized muscle fibers	Protective inner lining of the chambers and valves

Figure 18.5
What features can you identify in this anterior view of the human heart?

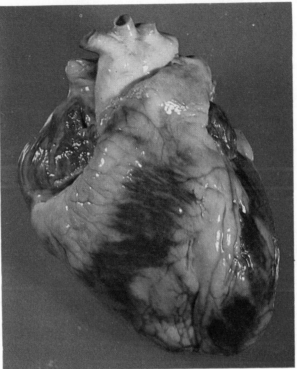

The atrium and ventricle on the right side are separated from those on the left by a *septum.* The atrium on each side communicates with its corresponding ventricle through an opening called the **atrioventricular orifice,** which is guarded by an *atrioventricular valve (A-V valve).*

Grooves on the surface of the heart mark the divisions between its chambers, and they also contain major blood vessels that supply the heart tissues. The deepest of these grooves is the **atrioventricular (coronary) sulcus,** which encircles the heart between the atrial and ventricular portions. Two **interventricular** (anterior and posterior) **sulci** indicate the location of the septum that separates the right and left ventricles.

Small, earlike projections, called **auricles,** extend outward from the atria (figures 18.3 and 18.5).

1. Describe the layers of the heart wall.
2. Name and locate the four chambers of the heart.
3. Name the orifices that occur between the upper and the lower chambers of the heart.
4. Name the structure that separates the right and left sides of the heart.

The right atrium receives blood from two large veins: the *superior vena cava* and *inferior vena cava.* These return blood that is low in oxygen from various body parts. A smaller vein, the *coronary sinus,* also drains blood into the right atrium from the wall of the heart.

Figure 18.6
A frontal section of the heart.

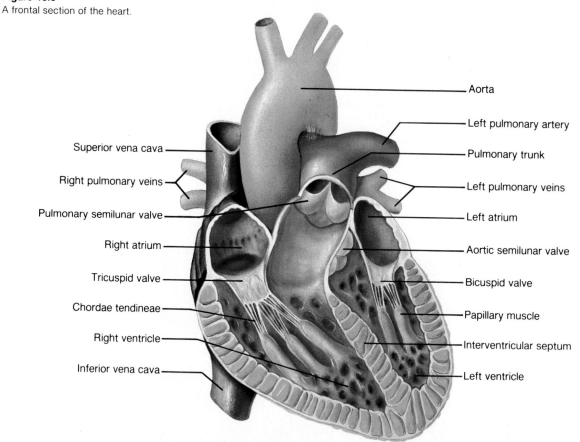

Aorta

Left pulmonary artery

Pulmonary trunk

Left pulmonary veins

Left atrium

Aortic semilunar valve

Bicuspid valve

Papillary muscle

Interventricular septum

Left ventricle

Superior vena cava

Right pulmonary veins

Pulmonary semilunar valve

Right atrium

Tricuspid valve

Chordae tendineae

Right ventricle

Inferior vena cava

The atrioventricular orifice between the right atrium and the right ventricle is guarded by a large **tricuspid valve,** which is composed of three leaflets or cusps as its name implies. This valve permits blood to move from the right atrium into the right ventricle and prevents passage in the opposite direction. The cusps fold passively out of the way when the blood pressure is greater on the atrial side, and they close passively when the pressure is greater on the ventricular side (figure 18.6).

Strong, fibrous strings, called *chordae tendineae,* are attached to the cusps on the ventricular side. These strings originate from small mounds of muscle tissue, the **papillary muscles,** that project inward from the walls of the ventricle. When the cusps close, the chordae tendineae prevent them from swinging back into the atrium.

The right ventricle has a much thinner muscular wall than the left ventricle. This chamber pumps blood a fairly short distance to the lungs against a relatively low resistance to blood flow. The left ventricle, on the other hand, must force blood to all other parts of the body against a much greater resistance to flow.

When the muscular wall of the right ventricle contracts, the blood inside its chamber is put under increasing pressure, and the tricuspid valve closes passively. As a result, the only exit is through the *pulmonary trunk,* which divides to form the left and right *pulmonary arteries.* At the base of this trunk is a **pulmonary semilunar valve** that consists of three cusps. This valve opens as the right ventricle contracts. When the ventricular muscles relax, however, blood begins to back up in the pulmonary trunk. This causes the semilunar valve to close, thus preventing a return flow into the ventricular chamber.

The left atrium receives blood from the lungs through four *pulmonary veins*—two from the right lung and two from the left lung. Blood passes from the left atrium into the left ventricle through the atrioventricular orifice, which is guarded by a valve. This valve consists of two leaflets, and it is appropriately named the **bicuspid** (mitral) **valve.** It prevents blood from flowing back into the left atrium from the ventricle. Like the tricuspid valve, the bicuspid valve is attached to papillary muscles by chordae tendineae.

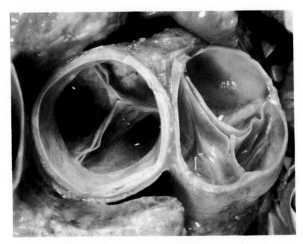

Chart 18.2 Valves of the heart					
Valve	Location	Function	Valve	Location	Function
Tricuspid valve	Right atrioventricular orifice	Prevents blood from moving from right ventricle into right atrium during ventricular contraction	Bicuspid (mitral) valve	Left atrioventricular orifice	Prevents blood from moving from left ventricle into left atrium during ventricular contraction
Pulmonary semilunar valve	Entrance to pulmonary trunk	Prevents blood from moving from pulmonary trunk into right ventricle during ventricular relaxation	Aortic semilunar valve	Entrance to aorta	Prevents blood from moving from aorta into left ventricle during ventricular relaxation

When the left ventricle contracts, the bicuspid valve closes passively, and the only exit is through a large artery called the *aorta*. Its branches distribute blood to all parts of the body.

At the base of the aorta, there is an **aortic semilunar valve** that consists of three cusps. (See figure 18.7.) It opens and allows blood to leave the left ventricle as it contracts. When the ventricular muscles relax, this valve closes and prevents blood from backing up into the ventricle. Chart 18.2 summarizes the heart valves.

A heart disorder that affects up to 6% of the U.S. population is *mitral valve prolapse*. In this condition, one (or both) of the cusps of the bicuspid (mitral) valve is stretched and bulges into the left atrium during ventricular contraction. Although in most cases the valve continues to function adequately, sometimes blood regurgitates into the left atrium causing some degree of disability. The cause of mitral valve prolapse is unknown.

1. Which blood vessels carry blood into the right atrium?
2. Where does the blood go after it leaves the right ventricle?
3. Which blood vessels carry blood into the left atrium?
4. What prevents blood from flowing back into the ventricles when they are relaxed?

Skeleton of the Heart

At their proximal ends, the pulmonary trunk and aorta are surrounded by rings of dense fibrous connective tissue. These rings are continuous with others that encircle the atrioventricular orifices. They provide firm attachments for the heart valves and for various muscle fibers. In addition, they prevent the outlets of the atria and ventricles from dilating during myocardial contraction. The fibrous rings together with other masses

Figure 18.8

The skeleton of the heart consists of fibrous rings to which the heart valves are attached.

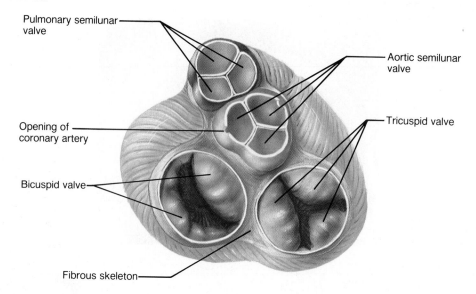

Pulmonary semilunar valve

Aortic semilunar valve

Opening of coronary artery

Tricuspid valve

Bicuspid valve

Fibrous skeleton

of dense fibrous tissue in the upper portion of the interventricular septum constitute the *skeleton of the heart* (figure 18.8).

Path of Blood through the Heart

Blood that is relatively low in oxygen concentration and relatively high in carbon dioxide concentration enters the right atrium through the venae cavae and the coronary sinus. As the right atrial wall contracts, blood passes through the right atrioventricular orifice and enters the chamber of the right ventricle (figure 18.9).

When the right ventricular wall contracts, the tricuspid valve closes the right atrioventricular orifice, and the blood moves into the pulmonary trunk and its branches (pulmonary circuit). From these vessels, blood enters the capillaries associated with the alveoli of the lungs. Gas exchanges occur between the blood in the capillaries and the air in the alveoli. The freshly oxygenated blood, which is now relatively low in carbon dioxide, returns to the heart through the pulmonary veins that lead to the left atrium.

The left atrial wall contracts, and the blood moves through the left atrioventricular orifice and into the chamber of the left ventricle. When the left ventricular wall contracts, the bicuspid valve closes the left atrioventricular orifice, and the blood passes into the aorta and its branches (systemic circuit).

Blood Supply to the Heart

Blood is supplied to the tissues of the heart by the first two branches of the aorta, called the right and left **coronary arteries.** Their openings lie just beyond the aortic semilunar valve (figures 18.8 and 18.10).

One branch of the left coronary artery, the *circumflex artery,* follows the atrioventricular sulcus between the left atrium and the left ventricle. Its branches supply blood to the walls of the left atrium and the left ventricle. Another branch of the left coronary artery, the *anterior interventricular artery,* travels in the anterior interventricular sulcus, and its branches supply the walls of both ventricles.

The right coronary artery passes along the atrioventricular sulcus between the right atrium and the right ventricle. It gives off two major branches—a *posterior interventricular artery,* which travels along the posterior interventricular sulcus and supplies the walls of both ventricles, and a *marginal artery,* which passes along the lower border of the heart. Branches of the marginal artery supply the walls of the right atrium and right ventricle (figures 18.11 and 18.12).

If a branch of a coronary artery becomes abnormally constricted or obstructed by a thrombus or embolus, the myocardial cells it supplies may experience a blood deficiency, called *ischemia.* As a result of ischemia, the person may experience a painful condition called *angina pectoris.* The discomfort of angina pectoris usually occurs during physical activity or an emotional disturbance, and is relieved by rest. It may take the form of

Figure 18.9
The right ventricle forces blood to the lungs, while the left
ventricle forces blood to all other body parts. How does the
composition of the blood in these two chambers differ?

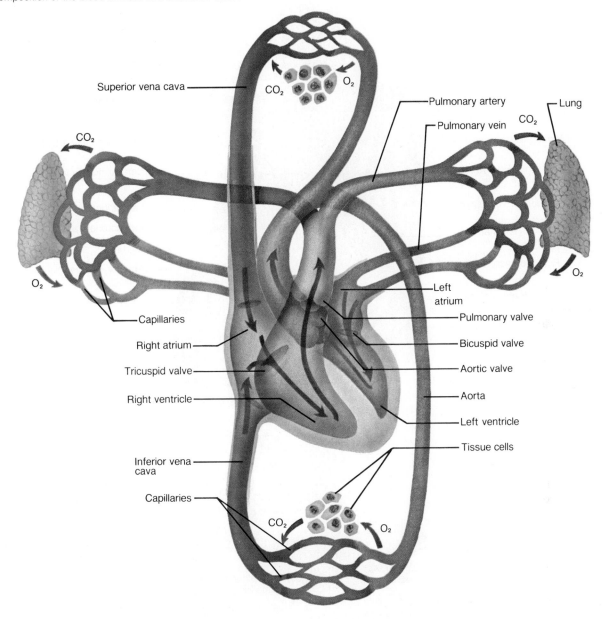

Superior vena cava

CO_2 O_2

Pulmonary artery
Pulmonary vein
Lung
CO_2
CO_2

O_2

O_2

Capillaries
Right atrium
Tricuspid valve
Right ventricle

Left
atrium
Pulmonary valve
Bicuspid valve
Aortic valve
Aorta
Left ventricle
Tissue cells

Inferior vena
cava

Capillaries

CO_2 O_2

pain or a sensation of heavy pressure, tightening, or
squeezing in the chest. Although it is usually felt in the
region behind the sternum or in the anterior portion of
the upper thorax, the pain may radiate to other parts,
including the neck, jaw, throat, arm, shoulder, elbow,
back, or upper abdomen.

Sometimes a portion of the heart dies because of
ischemia, and this condition, a *myocardial infarction,* or
more commonly called a heart attack, is one of the leading
causes of death.

Since the heart must beat continually to supply
blood to the body tissues, the myocardial cells require
a constant supply of freshly oxygenated blood. The
myocardium contains many capillaries fed by branches
of the coronary arteries. The smaller branches of these
arteries have some interconnections (anastomoses) between vessels that provide alternate pathways for blood.

Figure 18.10
Blood vessels associated with the surface of the heart.
(a) Anterior view; *(b)* posterior view.

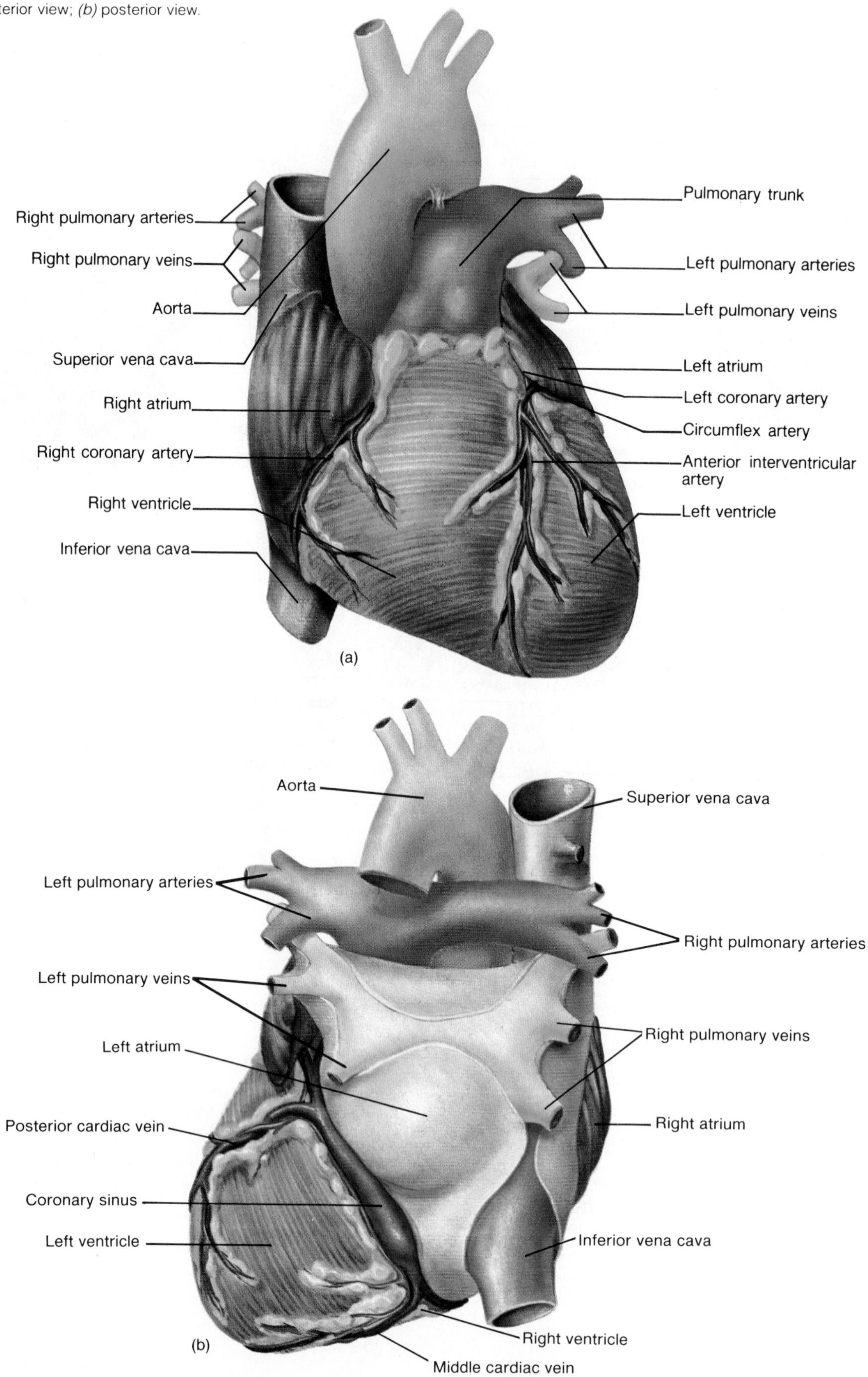

(a)

Right pulmonary arteries
Right pulmonary veins
Aorta
Superior vena cava
Right atrium
Right coronary artery
Right ventricle
Inferior vena cava

Pulmonary trunk
Left pulmonary arteries
Left pulmonary veins
Left atrium
Left coronary artery
Circumflex artery
Anterior interventricular artery
Left ventricle

(b)

Aorta
Left pulmonary arteries
Left pulmonary veins
Left atrium
Posterior cardiac vein
Coronary sinus
Left ventricle

Superior vena cava
Right pulmonary arteries
Right pulmonary veins
Right atrium
Inferior vena cava
Right ventricle
Middle cardiac vein

Figure 18.11

(a) Blood is supplied to heart tissues by branches of the coronary arteries; (b) blood is drained by branches of the cardiac veins.

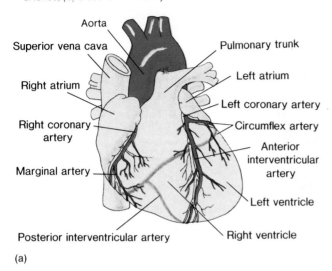

Aorta
Superior vena cava
Right atrium
Right coronary artery
Marginal artery
Posterior interventricular artery
Pulmonary trunk
Left atrium
Left coronary artery
Circumflex artery
Anterior interventricular artery
Left ventricle
Right ventricle

(a)

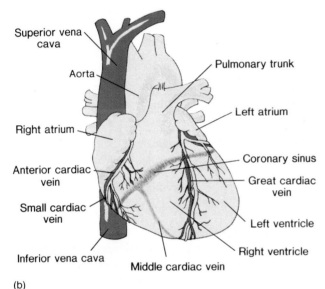

Superior vena cava
Aorta
Right atrium
Anterior cardiac vein
Small cardiac vein
Inferior vena cava
Middle cardiac vein
Pulmonary trunk
Left atrium
Coronary sinus
Great cardiac vein
Left ventricle
Right ventricle

(b)

In most body parts, blood flow in arteries reaches a peak during ventricular contraction. Blood flow in the vessels of the myocardium, however, is poorest during ventricular contraction. This is because the muscle fibers of the myocardium compress nearby vessels as they contract, and this action interferes with blood flow. Also, the openings into the coronary arteries are closed during ventricular contraction. Conversely, during ventricular relaxation, the myocardial vessels are no longer compressed and the orifices of the coronary arteries are open. Consequently, blood flow into the myocardium increases.

Blood that has passed through the capillaries of the myocardium is drained by branches of **cardiac veins,**

Figure 18.12

An angiogram (X-ray film) of the coronary arteries.

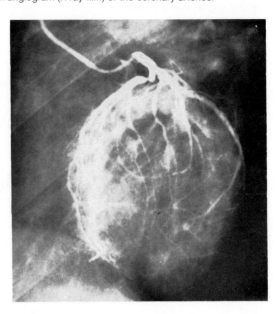

whose paths roughly parallel those of the coronary arteries. As figure 18.11 shows, these veins join an enlarged vessel, the **coronary sinus,** which is on the posterior surface of the heart in the atrioventricular sulcus and empties into the right atrium.

In the *heart transplantation* procedure, the recipient's failing heart is removed, except for the posterior walls of the right and left atria and their connections to the venae cavae and pulmonary veins. The donor heart is prepared similarly and is attached to the atrial cuffs remaining in the recipient's thorax. Finally, the recipient's aorta and pulmonary arteries are connected to those of the donor heart.

1. What structures make up the skeleton of the heart?
2. Review the path of blood through the heart.
3. What vessels supply blood to the heart?
4. How does blood return from the cardiac tissues to the right atrium?

Actions of the Heart

Although the previous discussion described the actions of the heart chambers separately, they do not function independently. Instead, their actions are regulated so that the atrial walls contract while the ventricular walls are relaxed, and ventricular walls contract while the atrial walls are relaxed. Such a series of events, shown

Figure 18.13
(a) The ventricles fill with blood during ventricular diastole and
(b) empty during ventricular systole.

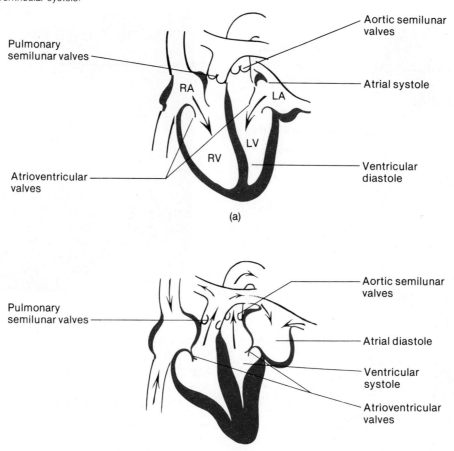

in figure 18.13, constitutes a complete heartbeat or **cardiac cycle.** At the end of each cycle, the atria and the ventricles remain relaxed for a moment and then a new cycle begins.

The Cardiac Cycle

During a cardiac cycle, the pressure within the chambers rises and falls. For example, when the atria are relaxed, blood flows into them from the large, attached veins. As these chambers fill, the pressure inside gradually increases. About 70% of the entering blood flows directly into the ventricles through the atrioventricular orifices before the atrial walls contract. Then, during atrial contraction (atrial systole), the atrial pressure rises suddenly, forcing the remaining 30% of the atrial contents into the ventricles. This is followed by atrial relaxation (atrial diastole) (figure 18.14).

As the ventricles contract (ventricular systole), the A-V valves guarding the atrioventricular orifices close passively and begin to bulge back into the atria, causing the atrial pressure to rise sharply. At the same time, the papillary muscles contract, and by pulling on the chordae tendineae, they prevent the cusps of the A-V valves from bulging too far into the atria. The atrial pressure soon falls, however, as blood flows out of the ventricles into the arteries. During the ventricular contraction, the A-V valves remain closed and the atrial pressure gradually increases as the atria fill with blood. When the ventricles relax (ventricular diastole), the A-V valves open passively, blood flows through them into the ventricles, and the atrial pressure drops to a low point.

Pressure in the ventricles is low while they are filling, but when the atria contract the ventricular pressure increases slightly. Then, as the ventricles contract, the ventricular pressure rises sharply, and as soon as the pressure exceeds that in the atria, the A-V valves

Figure 18.14

Some changes that occur during a cardiac cycle.

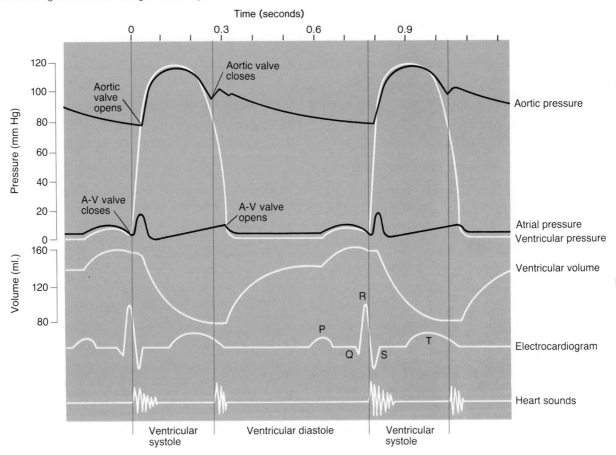

close. The ventricular pressure continues to increase until it exceeds the pressure in the pulmonary trunk and aorta. Then, the semilunar valves open, and blood is ejected from the ventricles into these arteries. When the ventricles are nearly empty, the ventricular pressure begins to drop, and it continues to drop as the ventricles relax. When the ventricular pressure is less than that in the arteries, the semilunar valves are closed by arterial blood flowing back toward the ventricles. As soon as the ventricular pressure falls below that of the atria, the A-V valves open, and the ventricles begin to fill once more.

Heart Sounds

The sounds associated with a heartbeat can be heard with a stethoscope and are described as *lub*-dup sounds. These sounds are due to vibrations in heart tissues that are created as blood flow is suddenly speeded or slowed with the contraction and relaxation of heart chambers, and the opening and closing of valves.

The first part of a heart sound occurs during the ventricular contraction, when the A-V valves are closing. The second part occurs during ventricular relaxation, when the semilunar valves are closing.

Sometimes, during inspiration, the interval between the closure of the pulmonary and the aortic semilunar valves is long enough so that a sound related to each of these events can be heard. In this case, the second heart sound is said to be *split*.

Heart sounds are of particular interest because they provide information concerning the condition of the heart valves. For example, inflammation of the endocardium (endocarditis) may cause changes in the shapes of the valvular cusps (valvular stenosis). Then, when the cusps close, the closure may be incomplete, and some blood may leak back through the valve. If this happens, an abnormal sound called a *murmur* may be heard. The seriousness of a heart murmur depends on the amount of valvular damage. Fortunately for those who have serious problems, it may be possible to repair the damaged valves or to replace them by open heart surgery.

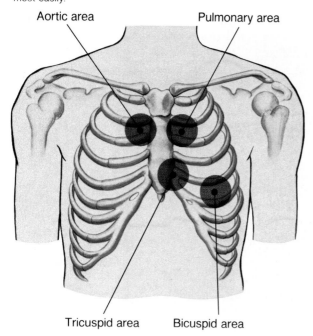

Aortic area   Pulmonary area

Tricuspid area   Bicuspid area

1. Describe the pressure changes that occur in the atria and ventricles during a cardiac cycle.
2. What causes heart sounds?
3. What is meant by a functional syncytium?
4. Where are the functional syncytia of the heart located?

With the aid of a stethoscope, it is possible to hear sounds associated with the aortic and pulmonary semi-lunar valves by listening from the second intercostal space on either side of the sternum. The *aortic sound* is heard on the right, and the *pulmonic sound* is heard on the left.

The sound associated with the bicuspid (mitral) valve can be heard from the fifth intercostal space at the nipple line on the left. The sound of the tricuspid valve can be heard at the tip of the sternum (figure 18.15).

Cardiac Muscle Fibers

As is mentioned in chapter 9, cardiac muscle fibers function much like those of skeletal muscles. In cardiac muscle, however, the fibers are interconnected in branching networks that spread in all directions through the heart. When any portion of this net is stimulated, an impulse travels to all of its parts, and the whole structure contracts as a unit.

A mass of merging cells that act together in this way is called a **functional syncytium.** There are two such structures in the heart—one in the atrial walls and another in the ventricular walls. These masses of muscle fibers are separated from each other by portions of the heart's fibrous skeleton, except for a small area in the right atrial floor. In this region, the *atrial syncytium* and the *ventricular syncytium* are connected by fibers of the cardiac conduction system.

Cardiac Conduction System

Throughout the heart there are clumps and strands of specialized cardiac muscle tissue, whose fibers contain only a few myofibrils. Instead of contracting, these parts function to initiate and distribute impulses (cardiac impulses) throughout the myocardium. They comprise the **cardiac conduction system,** which functions to co-ordinate the events of the cardiac cycle.

A key portion of this conduction system is called the **sinoatrial (S-A) node.** It consists of a small mass of specialized muscle tissue just beneath the epicar-dium. It is located in the posterior wall of the right atrium, below the opening of the superior vena cava, and its fibers are continuous with those of the *atrial syncytium.*

The cells of the S-A node have an ability to excite themselves. That is, without being stimulated by nerve fibers or any other outside agents, these cells initiate impulses that spread into the myocardium and stimu-late cardiac muscle fibers to contract. Furthermore, this activity is rhythmic. The S-A node initiates one im-pulse after another, seventy to eighty times a minute. Thus, it is responsible for the rhythmic contractions of the heart and is often called the **pacemaker.**

As a cardiac impulse travels from the S-A node into the atrial syncytium, the right and left atria con-tract almost simultaneously. Instead of passing directly into the ventricular syncytium, which is separated from the atrial syncytium by the fibrous skeleton of the heart, the cardiac impulse passes along fibers of the conduc-tion system that are continuous with atrial muscle fi-bers. These conducting fibers lead to a mass of specialized muscle tissue called the **atrioventricular node** (A-V node). This node, located in the floor of the right atrium near the interatrial septum and just beneath the endocardium, provides the only normal conduction pathway between the atrial and ventricular syncytia.

The fibers that conduct the cardiac impulse into the A-V node (junctional fibers) have very small di-ameters, and because small fibers conduct impulses slowly, they cause the impulse to be delayed. The im-pulse is delayed still more as it travels through the A-V node, and this delay allows time for the atria to empty and the ventricles to fill with blood.

Once the cardiac impulse reaches the other side of the A-V node, it passes into a group of large fibers that make up the **A-V bundle** (bundle of His), and the

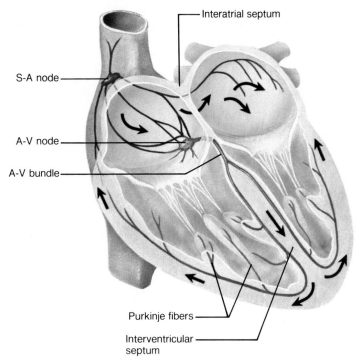

- Interatrial septum

S-A node

A-V node

A-V bundle

Purkinje fibers

Interventricular septum

impulse moves rapidly through them. The A-V bundle enters the upper part of the interventricular septum, and divides into right and left branches that lie just beneath the endocardium. About halfway down the septum, the branches give rise to enlarged **Purkinje fibers.**

The base of the aorta, which contains the aortic valves, is enlarged and protrudes somewhat into the interatrial septum close to the A-V bundle. Consequently, inflammatory conditions, such as bacterial endocarditis, involving the aortic valves (aortic valvulitis) may also affect the A-V bundle.

If a portion of the bundle is damaged, it may no longer conduct impulses normally. As a result, cardiac impulses may reach the two ventricles at different times so that they fail to contract together. This condition is called a *bundle branch block*.

The Purkinje fibers spread from the interventricular septum, into the papillary muscles that project inward from the ventricular walls, and continue downward to the apex of the heart. There they curve around the tips of the ventricles and pass upward over the lateral walls of these chambers. Along the way, the Purkinje fibers give off many small branches that become continuous with cardiac muscle fibers. These parts of the conduction system are shown in figure 18.16.

Figure 18.17
The muscle fibers within the ventricular walls are arranged in patterns of whorls. The fibers of groups *(a)* and *(b)* surround both ventricles.

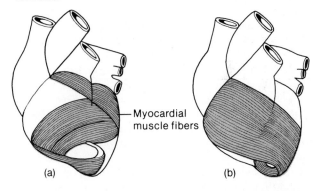

- Myocardial muscle fibers

(a)   (b)

The muscle fibers in the ventricular walls are arranged in irregular whorls, so that when they are stimulated by impulses on Purkinje fibers, the ventricular walls contract with a twisting motion (figure 18.17). This action squeezes or wrings the blood out of the ventricular chambers and forces it into the arteries.

1. What kinds of tissues make up the cardiac conduction system?
2. How is a cardiac impulse initiated?
3. How is this impulse transmitted from the atrium to the ventricles?

The Electrocardiogram
A Laboratory Application

A recording of the electrical changes that occur in the *myocardium* during a cardiac cycle is called an **electrocardiogram** (ECG). These changes result from the depolarization and repolarization associated with the contraction of muscle fibers. Because body fluids can conduct electrical currents, such changes can be detected on the surface of the body.

To record an *ECG*, metal electrodes are placed in certain locations on the skin. These electrodes are connected by wires to an instrument (shown in figure 18.18) that responds to very weak electrical changes by causing a pen or stylus to mark on a moving strip of paper. When the instrument is operating, up-and-down movements of the pen correspond to electrical changes occurring within the body as a result of myocardial activity.

Since the paper moves past the pen at a known rate, the distance between pen deflections can be used to measure the time elapsing between various phases of the cardiac cycle.

As figure 18.19 illustrates, the ECG pattern includes several deflections, or *waves,* during each cardiac cycle. Between cycles, the muscle fibers remain polarized, and no detectable electrical changes occur. Consequently, the pen simply marks along the baseline as the paper moves through the instrument. When the S-A node triggers a cardiac impulse, however, the atrial fibers are stimulated to depolarize, and an electrical change occurs. As a result, the pen is deflected, and when this electrical change is completed, the pen returns to the base position. This pen movement produces a *P wave* that is caused by the depolarization of the atrial fibers just before they contract.

When the cardiac impulse reaches the ventricular fibers, they are stimulated to depolarize rapidly. Because the ventricular walls are much more extensive than those of the atria, the amount of electrical change is greater, and the pen is deflected to a greater degree than before. Once again, when the electrical change is completed, the pen returns to the baseline, leaving a mark called the *QRS complex*. This wave appears just prior to the contraction of the ventricular walls.

Near the end of the ECG pattern, the pen is deflected for a third time, producing a *T wave*. This wave is caused by electrical changes occurring as the ventricular muscle fibers become repolarized relatively slowly. The record of the atrial repolarization is missing from the pattern because the atrial fibers repolarize at the same time that the ventricular fibers depolarize. Thus, the recording of the atrial repolarization is obscured by the QRS complex (figures 18.20 and 18.21).

Figure 18.18
This instrument is used to record an ECG.

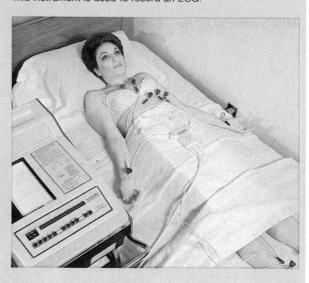

Figure 18.19
A normal ECG.

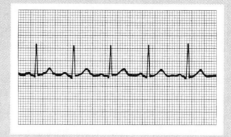

Figure 18.20
In an ECG pattern, the P wave results from depolarization of the atria, the QRS complex results from depolarization of the ventricles, and the T wave results from repolarization of the ventricles.

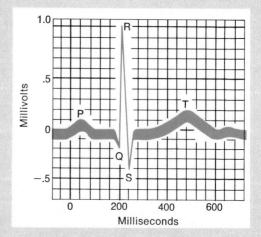

Figure 18.21
In this set of drawings (a–g), the green areas of the hearts
indicate where depolarization is occurring, and the red areas
indicate where tissues are repolarizing; the portion of the ECG
pattern produced at each step is shown by the colored line on
the graph paper.

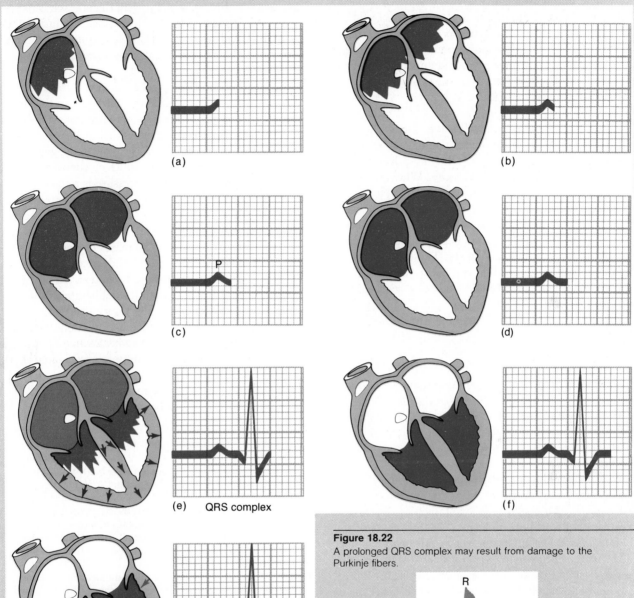

(a)

(b)

(c)   P

(d)

(e)   QRS complex

(f)

(g)

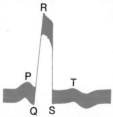

Figure 18.22
A prolonged QRS complex may result from damage to the
Purkinje fibers.

ECG patterns are especially important because they allow a physician to assess the heart's ability to conduct impulses and, therefore, to judge its condition. For example, the time period between the beginning of a P wave and the beginning of a QRS complex (*P-Q interval,* or *P-R interval* if the initial wave is upright) indicates how long it takes for the cardiac impulse to travel from the S-A node through the A-V node and into the ventricular walls. If ischemia or other problems involving the fibers of the A-V conduction pathways are present, this P-Q interval sometimes increases. Similarly, if the Purkinje fibers are injured, the duration of the QRS complex may increase, because it may take longer for an impulse to spread throughout the ventricular walls (figure 18.22).

Regulation of the Cardiac Cycle

The primary function of the heart is to pump blood to the body cells, and when the needs of these cells change, the quantity of blood pumped must change also. For example, during strenuous exercise, the amount of blood required by skeletal muscles increases greatly, and the rate of the heartbeat increases in response to this need. Since the S-A node normally controls the heart rate, changes in this rate often involve factors that affect the pacemaker. These include motor impulses on parasympathetic and sympathetic nerve fibers. (See figure 11.36.)

The parasympathetic fibers that supply the heart arise from neurons in the medulla oblongata and make up parts of the *vagus nerves*. Most of these fibers branch to the S-A node and the A-V node. When nerve impulses reach their endings, these fibers secrete acetylcholine that causes a decrease in S-A and A-V nodal activity. As a result, the rate of heartbeat decreases.

The vagus nerves seem to carry impulses continually to the S-A and A-V nodes. These impulses impose a braking action on the heart. Consequently, parasympathetic activity can cause the heart rate to change in either direction. An increase in impulses causes a slowing of the heart, and a decrease in impulses releases the parasympathetic brake and allows the heartbeat to increase.

Sympathetic fibers reach the heart by means of the *accelerator nerves*, whose branches join the S-A and A-V nodes as well as other areas of the atrial and ventricular myocardium. The endings of these fibers secrete norepinephrine in response to nerve impulses, and this substance causes an increase in the rate and the force of myocardial contractions.

A normal balance between the inhibitory effects of the parasympathetic fibers and the excitatory effects of the sympathetic fibers is maintained by the cardiac center of the medulla oblongata. In this region of the brain, masses of neurons function as *cardioinhibitor* and *cardioaccelerator reflex centers*. These centers receive sensory impulses from various parts of the circulatory system and relay motor impulses to the heart in response.

For example, there are *pressoreceptors* in certain regions of the aorta (aortic sinus and aortic arch) and the carotid arteries (carotid sinuses). These receptors are sensitive to changes in blood pressure, and if the pressure rises, they signal the cardioinhibitor center in the medulla. In response, this reflex center sends *parasympathetic* motor impulses to the heart, causing the heart rate and force of contraction to decrease. This also causes blood pressure to drop toward the normal level (figure 18.23).

Another regulatory reflex involves pressoreceptors in the venae cavae near the entrances to the right atrium. If the blood pressure increases abnormally in these vessels, the receptors signal the cardioaccelerator center, and *sympathetic* impulses flow to the heart. As a result, the heart rate and force of contraction increase, and the venous blood pressure is reduced.

Two other factors that influence heart rate are temperature and various ions. Heart action is increased by a rising body temperature, which accounts for the fact that heart rate usually increases during fever. On the other hand, cardiac activity is decreased by abnormally low body temperature. Consequently, a patient's body temperature is sometimes deliberately lowered (hypothermia) to slow the heart during surgery.

Of the ions that influence heart action, the most important are potassium (K^+) and calcium (Ca^{++}). Although homeostatic mechanisms normally maintain the concentrations of these substances within narrow ranges, these mechanisms sometimes fail, and the consequences can be serious or even fatal.

An excess of *potassium ions* (hyperkalemia) seems to alter the usual polarized state of the cardiac muscle fibers, and the result is a decrease in the rate and force of contractions. In fact, if the potassium ion concentration is very high, the conduction of cardiac impulses may be blocked, and heart action may suddenly stop (cardiac arrest). Conversely, if the potassium concentration falls below normal (hypokalemia), the heart may develop a serious abnormal rhythm (arrhythmia). This condition can also be life threatening.

Excessive *calcium ions* (hypercalcemia) cause increased heart actions, and there is a danger that the heart will undergo a prolonged contraction. Conversely, low calcium (hypocalcemia) depresses heart action. This effect occurs because calcium ions help initiate the muscle contraction mechanism, as described in chapter 9.

Figure 18.23
The activities of the S-A and A-V nodes can be altered by autonomic nerve impulses.

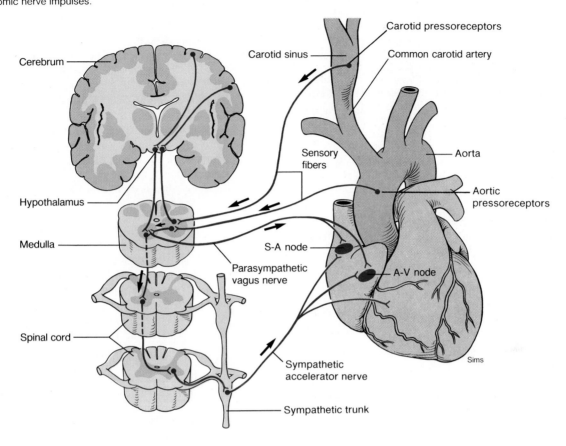

Cerebrum

Carotid pressoreceptors

Carotid sinus

Common carotid artery

Sensory fibers

Aorta

Hypothalamus

Aortic pressoreceptors

Medulla

S-A node

Parasympathetic vagus nerve

A-V node

Spinal cord

Sympathetic accelerator nerve

Sympathetic trunk

Sims

1. What nerves supply parasympathetic fibers to the heart? Sympathetic fibers?
2. How do parasympathetic and sympathetic impulses help control heart rate?
3. How do changes in body temperature affect heart rate?

Blood Vessels

The vessels of the cardiovascular system form a closed circuit of tubes that carry blood from the heart to the body cells and back again. These tubes include arteries, arterioles, capillaries, venules, and veins. The arteries and arterioles conduct blood away from the ventricles of the heart and lead to the capillaries. The capillaries function to exchange substances between the blood and the body cells, and the venules and veins return blood from the capillaries to the atria.

Arteries and Arterioles

Arteries are strong, elastic vessels that are adapted for carrying blood away from the heart under relatively high pressure. These vessels subdivide into progressively thinner tubes and eventually give rise to fine branches called **arterioles.**

Arrhythmias
A Clinical Application

Although slight variations in heart actions sometimes occur normally, marked changes in the usual rate or rhythm may suggest cardiovascular disease. Such abnormal actions, termed cardiac **arrhythmias**, include the following:

1. **Tachycardia.** Tachycardia is characterized by an abnormally fast heartbeat, usually over one hundred beats per minute. This condition may be caused by such factors as an increase in body temperature, nodal stimulation by sympathetic fibers, the presence of various drugs or hormones, heart disease, excitement, exercise, anemia, or shock. Figure 18.24 shows the ECG of a tachycardic heart.

2. **Bradycardia.** Bradycardia means a slow heart rate, usually one that is less than sixty beats per minute. It may be caused by decreased body temperature, nodal stimulation by parasympathetic impulses, or the presence of certain drugs. It also may occur during sleep. Athletes sometimes have unusually slow heartbeats because their hearts have developed the ability to pump a greater than normal volume of blood with each beat (figure 18.25).

3. **Premature heartbeats.** A premature beat is one that occurs before it is expected in a normal series of cardiac cycles (figure 18.26). Such an occurrence is probably caused by cardiac impulses originating from unusual (ectopic) regions of the heart. That is, the impulse originates from a site other than the S-A node. Cardiac impulses may arise from ischemic tissues or from muscle fibers that are irritated by the effects of disease or drugs.

4. **Flutter.** A heart chamber is said to flutter when it is contracting regularly, but at a very rapid rate, such as 250–350 contractions per minute. Although normal hearts may flutter occasionally, this condition is more likely to be due to damage to the myocardium (figure 18.27).

5. **Fibrillation.** Fibrillation is also characterized by rapid heart actions, but unlike flutter, the accompanying contractions are *uncoordinated* because small regions of the myocardium contract and relax independently of all other areas. As a result, the myocardium fails to contract as a whole and the walls of the fibrillating chambers are completely ineffective in pumping blood.

Although a person may survive atrial fibrillation because blood may continue to move through the heart if the ventricles are functional, ventricular fibrillation is very likely to cause death (figure 18.28).

The ventricles can be stimulated to fibrillate by a variety of factors, including ischemia associated with obstruction of a coronary artery (coronary occlusion), electric shock, traumatic injury to the heart or chest wall, or toxic effects of various drugs. Once the ventricles are fibrillating, normal cardiac rhythm is not likely to be restored unless the heart can be *defibrillated*. Under such conditions, if circulation is not reestablished within a few minutes, the person may suffer irreparable damage to the cerebral cortex and may die.

Defibrillation is accomplished by exposing a fibrillating myocardium to a strong electric current for a short time. This action causes all of the muscle fibers within the myocardium to become depolarized simultaneously, so that all contractile activity ceases momentarily. It is hoped that, within the next instant, the S-A node will begin to function, and a normal rhythm will be reestablished.

6. **Arrhythmias due to conduction disorders.** Any interference or block in cardiac impulse conduction may cause arrhythmia. The type of arrhythmia produced, however, varies with the location and extent of the block. Related to this is the fact that certain cardiac tissues other than the S-A node can function as pacemakers.

The S-A node usually initiates seventy to eighty heartbeats per minute. If the S-A node is damaged, impulses originating in the A-V node may travel upward into the atrial myocardium and downward into the ventricular walls, stimulating them to contract. Under the influence of the A-V node acting as a *secondary pacemaker,* the heart may continue to pump blood, but at a rate of forty to sixty beats per minute (figure 18.29). Similarly, the Purkinje fibers can initiate cardiac impulses, causing the heart to contract fifteen to forty times per minute.

Figure 18.24
Tachycardia is characterized by a rapid heartbeat.

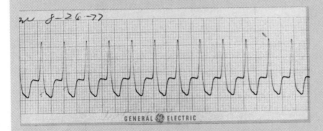

Figure 18.25
Bradycardia is characterized by a slow heartbeat.

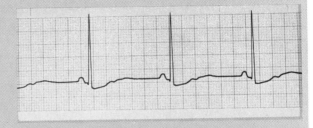

Figure 18.26
Ectopic beats usually are caused by impulses originating from sites other than the S-A node.

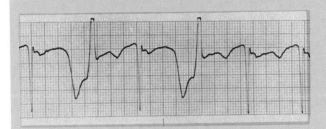

Figure 18.27
Atrial flutter is characterized by an abnormally rapid rate of atrial contraction.

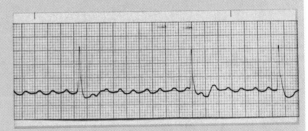

Figure 18.28
Atrial fibrillation is characterized by contractions that are rapid and uncoordinated. Why is this condition less serious than ventricular fibrillation?

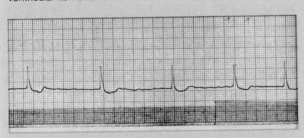

Figure 18.29
When the A-V node is the functional pacemaker, the P wave may be inverted (or absent) and the PR interval may be reduced.

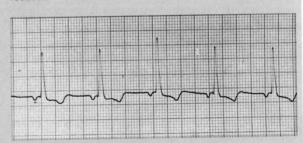

Figure 18.30
(a) The wall of an artery and (b) the wall of a vein.

Artery
Vein

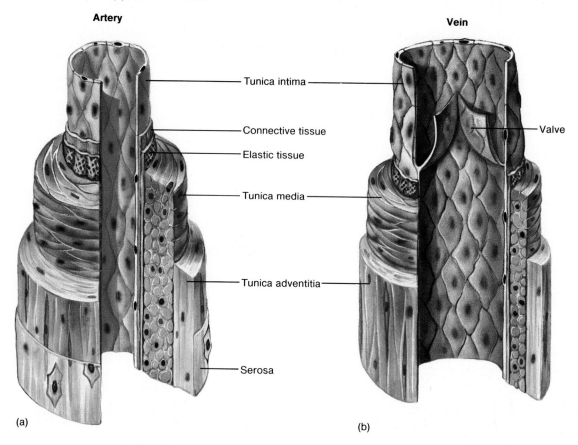

- Tunica intima
- Connective tissue
- Elastic tissue
- Tunica media
- Tunica adventitia
- Serosa
- Valve

(a)
(b)

The wall of an artery consists of three distinct layers, or *tunics,* shown in figure 18.30. The innermost layer (tunica intima) is composed of a layer of epithelium, called *endothelium,* resting on a connective tissue membrane that is rich in elastic and collagenous fibers.

The middle tunic (tunica media) makes up the bulk of the arterial wall. It includes smooth muscle fibers that encircle the tube and a thick layer of elastic connective tissue.

The outer layer (tunica adventitia) is relatively thin and consists chiefly of connective tissue with irregularly arranged elastic and collagenous fibers. This layer attaches the artery to the surrounding tissues. It also contains minute vessels (vasa vasorum) that give rise to capillaries and provide blood to the more external cells of the artery wall.

The endothelial lining provides a smooth surface so that blood cells and platelets can flow without being damaged. The connective tissues give the vessel a tough elasticity that allows it to withstand the force of high blood pressure and, at the same time, to stretch and accommodate the sudden increase in blood volume that accompanies each ventricular contraction.

The smooth muscles in the walls of arteries and arterioles are innervated by sympathetic branches of the autonomic nervous system. Impulses on these *vasomotor* fibers cause the smooth muscles to contract, thus reducing the diameter of the vessel. This action is called **vasoconstriction.** If such vasomotor impulses are inhibited, the muscle fibers relax and the diameter of the vessel increases. In this case, the artery is said to undergo **vasodilation.** Changes in the diameters of arteries greatly influence the flow and pressure of the blood.

Arterioles, which are microscopic continuations of arteries, give off branches called *metarterioles* that, in turn, join capillaries. Although the walls of the larger arterioles have three layers similar to those of arteries, these walls become thinner and thinner as the arterioles approach the capillaries. The wall of a very small arteriole consists only of an endothelial lining and some smooth muscle, surrounded by a small amount of connective tissue (figures 18.31 and 18.32).

Figure 18.31
Small arterioles have some smooth muscle fibers in their walls;
capillaries lack these fibers.

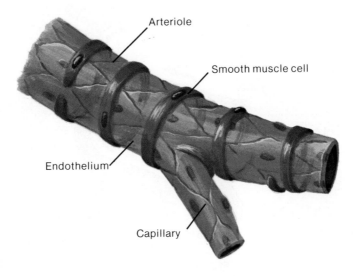

Arteriole

Smooth muscle cell

Endothelium

Capillary

Figure 18.32
Scanning electron micrograph of an arteriole cross section
(3900X). (*Tissues and Organs: A Text-Atlas of Scanning Electron
Microscopy*, by R. G. Kessel and R. H. Kardon. © 1979 W. H.
Freeman and Company.)

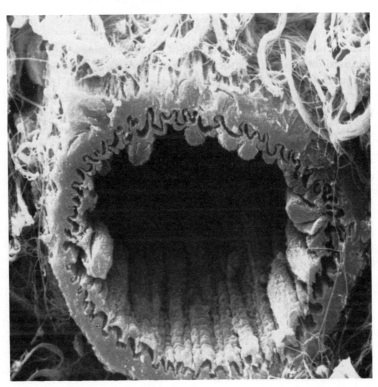

Figure 18.33
Some metarterioles provide arteriovenous shunts by connecting
arterioles directly to venules.

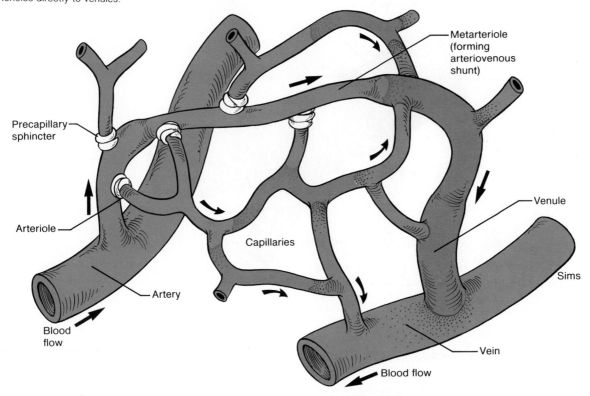

The arteriole and metarteriole walls are adapted
for vasoconstriction and vasodilation in that their
muscle fibers respond to impulses from the autonomic
nervous system by contracting or relaxing. Thus, these
vessels function in helping to control the flow of blood
into the capillaries.

Sometimes metarterioles are connected directly
to venules, and the blood entering them can bypass the
capillaries. These connections between arteriole and
venous pathways, shown in figure 18.33, are called *arteriovenous shunts.*

1. Describe the wall of an artery.
2. What is the function of the smooth muscle in the
   arterial wall?
3. How is the structure of an arteriole different from that
   of an artery?

Capillaries

Capillaries are the smallest blood vessels. They form
the connections between the smallest arterioles and the
smallest venules. Capillaries are essentially extensions
of the inner linings of these larger vessels, in that their
walls consist of endothelium—a single layer of squamous epithelial cells (figure 18.31). These thin walls
form the semipermeable membranes through which
substances in the blood are exchanged for substances
in the tissue fluid surrounding body cells.

Capillary Permeability
The openings or pores in the capillary walls are thin
slits occurring where two adjacent endothelial cells
overlap. The sizes of these pores, and consequently the
permeability of the capillary wall, vary from tissue to
tissue. For example, the pores are relatively small in
the capillaries of smooth, skeletal, and cardiac muscle,
while those in capillaries associated with endocrine
glands and the lining of the small intestine are larger.
Among the capillaries with the largest openings are
those of the liver, spleen, and red bone marrow. The
pores in the walls of these vessels commonly allow large
protein molecules and even intact cells to pass through,
as they enter or leave the blood (figures 18.34 and
18.35).

Figure 18.34

Some substances are exchanged between the tissue fluid and the blood by passing through tiny pores in capillary walls.

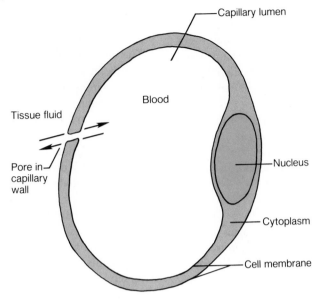

Figure 18.35

Transmission electron micrograph of a capillary cross section. (Note the narrow opening at the cell junction.) The lumen of the capillary is occupied by a red blood cell.

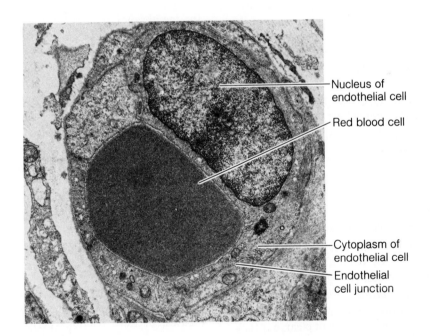

In the brain, the endothelial cells of the capillary walls are more tightly fused (tight junctions) than those in other body regions. Consequently, some substances that readily leave capillaries in other tissues enter brain tissues only slightly or not at all. This resistance to movement is called the *blood-brain barrier,* and it is of particular interest because it prevents certain drugs from entering the brain tissues or cerebrospinal fluid in sufficient concentrations to effectively treat certain diseases.

The capillaries of the hypothalamus in the brain are an exception, since they are more permeable. This allows the hypothalamus to respond to changes in the chemical composition of plasma, as described in chapter 11.

Arrangement of Capillaries

The density of capillaries within tissues varies directly with the tissues' rates of metabolism. Thus, muscle and nerve tissues, which use relatively large quantities of oxygen and nutrients, are richly supplied with capillaries, while hyaline cartilage, the epidermis, and the cornea, in which metabolic rates are very slow, lack capillaries.

The patterns of capillary arrangement also differ in various body parts. For example, some capillaries pass directly from arterioles to venules, while others lead to highly branched networks. Such arrangements make it possible for blood to follow different pathways through a tissue and meet the varying demands of its cells. During periods of exercise, for example, blood can be directed into the capillary networks of the skeletal muscles, where cells are experiencing an increasing need for oxygen and nutrients. At the same time, blood can bypass some of the capillary nets in the tissues of the digestive tract, where the demand for blood is less critical. Conversely, when a person is relaxing after a meal, blood can be shunted from the inactive skeletal muscles into the capillary networks of the digestive organs, where it is needed to support the processes of digestion and absorption.

Regulation of Capillary Blood Flow

The distribution of blood in the various capillary pathways is regulated mainly by smooth muscles that encircle the capillary entrances. As figure 18.33 shows, these muscles form *precapillary sphincters* that may close a capillary by contracting or open it by relaxing. How the sphincters are controlled is not clear, but they seem to respond to the demands of the cells supplied by their individual capillaries. When the cells are low in concentration of oxygen and nutrients, the sphincter relaxes; and when the cellular needs are met, the sphincter may contract again.

1. Describe the wall of a capillary.
2. What is the function of a capillary?
3. How is blood flow into capillaries controlled?

Exchanges in the Capillaries

The vital function of exchanging gases, nutrients, and metabolic by-products between the blood and the tissue fluid surrounding body cells occurs in the capillaries. The substances exchanged move through the capillary walls primarily by the processes of diffusion, filtration, and osmosis described in chapter 3. Of these processes, diffusion provides the most important means of transfer.

It is by diffusion that molecules and ions move from regions where they are more highly concentrated toward regions where they are in lower concentration. Since blood entering capillaries of the systemic circuit generally carries relatively high concentrations of oxygen and nutrients, these substances diffuse through the capillary walls and enter the tissue fluid. Conversely, the concentrations of carbon dioxide and other wastes are generally more highly concentrated in these tissues, and the wastes tend to diffuse into the capillary blood.

The paths followed by these substances depend primarily on their solubilities in lipids. Those that are soluble in lipid, such as oxygen, carbon dioxide, and fatty acids, can diffuse through most areas of the cell membranes that make up the capillary wall because the membranes are largely lipid. Lipid insoluble substances, such as water, sodium ions, and chloride ions, diffuse through pores in the cell membranes and the slitlike openings between the endothelial cells that form the capillary wall (figure 18.34).

Plasma proteins generally remain in the blood because they are not soluble in the lipid portions of the capillary membranes, and their molecular size is too great to permit diffusion through the membrane pores or slitlike openings between the endothelial cells of most capillaries.

As described in chapter 3, *filtration* involves forcing molecules through a membrane by *hydrostatic pressure.* In capillaries, the force is provided by blood pressure generated by contractions of the ventricular walls.

Blood pressure is also responsible for moving blood through the arteries and arterioles. Pressure tends to decrease, however, as the distance from the heart increases, because friction (peripheral resistance) between the blood and the vessel walls slows the flow. For this reason, blood pressure is greater in arteries than in arterioles, and greater in arterioles than in capillaries. It is similarly greater at the arteriole end of a capillary than at the venule end.

Figure 18.36
Substances leave the capillaries because of a net outward filtration pressure; substances enter the capillaries because of a net inward force of osmotic pressure.

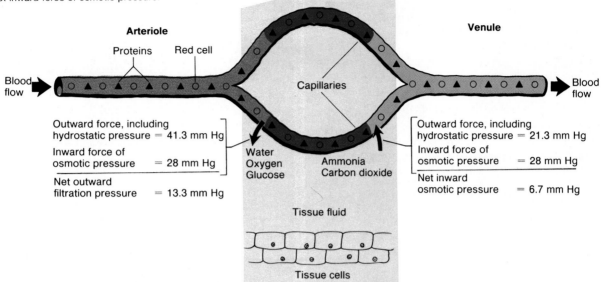

Arteriole   Proteins   Red cell   **Venule**

Blood flow   Capillaries   Blood flow

Outward force, including hydrostatic pressure = 41.3 mm Hg
Inward force of osmotic pressure = 28 mm Hg
Net outward filtration pressure = 13.3 mm Hg

Water Oxygen Glucose   Ammonia Carbon dioxide

Outward force, including hydrostatic pressure = 21.3 mm Hg
Inward force of osmotic pressure = 28 mm Hg
Net inward osmotic pressure = 6.7 mm Hg

Tissue fluid

Tissue cells

Blood components generally fail to pass through the walls of arteries and arterioles because these structures are too thick. When blood reaches the thin-walled capillaries, however, the hydrostatic pressure of the blood serves to increase the rate by which substances diffuse through the capillary wall. This filtration effect occurs primarily at the arteriole ends of capillaries, while diffusion takes place along their entire lengths.

The plasma proteins, which remain in the capillaries, help to make the *osmotic pressure* of the blood greater (hypertonic) than that of the tissue fluid. Although the capillary blood has a greater osmotic attraction for water than does the tissue fluid, this attraction is overcome by the greater force of the blood pressure. As a result, the net movement of water and dissolved substances is outward at the arteriole end of the capillary by filtration.

More specifically the force, including *hydrostatic pressure,* that tends to move fluid outward at the arteriole end of a capillary is 41.3 millimeters of mercury (mm Hg), while the *osmotic pressure* of the blood, which tends to move fluid inward, is only 28 mm Hg. Consequently, there is a net movement of water and dissolved substances outward. However, the blood pressure decreases as the blood moves through the capillary, and at the venule end, the outward force equals 21.3 mm Hg, while the osmotic pressure of the blood remains unchanged at 28 mm Hg. Thus there is a net movement of water and dissolved materials into the venule end of the capillary by osmosis. This process is shown in figure 18.36.

Normally, more fluid leaves the capillaries than returns to them, and the excess is collected and returned to the venous circulation by *lymphatic vessels.* This mechanism is discussed in chapter 19.

Sometimes unusual events cause an increase in the permeability of the capillaries, and an excessive amount of fluid is likely to enter the interstitial spaces. This may occur, for instance, following a traumatic injury to the tissues or in response to certain chemicals such as *histamine,* which increase membrane permeability. In any case, so much fluid may leak out of the capillaries that the lymphatic drainage is overwhelmed, and the affected tissues become swollen (edematous) and painful.

A failing heart—one that is unable to pump blood out of the ventricles as rapidly as it enters—is often accompanied by edema in other parts of the body. This occurs because the blood backs up into the veins, venules, and capillaries, and the blood pressure in these vessels increases. As a result of this increased *back pressure,* the osmotic pressure of the blood in the venule ends of the capillaries is less effective in attracting water from tissue fluid, and the tissues tend to become edematous. This is true particularly of the tissues in the lower extremities if the person is upright, or of the back if the person is supine. In the terminal stages of heart failure, edema may become widespread, and fluid may begin to accumulate in the peritoneal cavity of the abdomen. This condition is called *ascites.*

Figure 18.37
Note the structural differences in the cross sections of *(a)* this
artery and *(b)* this vein.

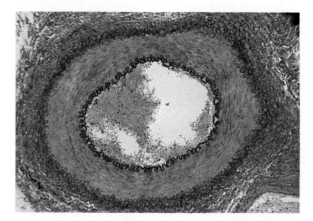

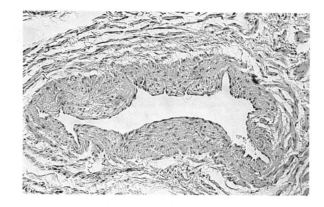

Figure 18.38
(a) Venous valves allow blood to move toward the heart, but
(b) prevent blood from moving away from the heart.

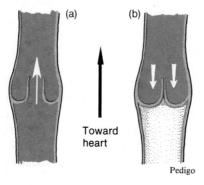

(a)　(b)

Toward
heart

Pedigo

---

1. What forces are responsible for the exchange of
   substances between the blood and tissue fluid?
2. Why is the fluid movement out of a capillary greater at
   its arteriole end than at its venule end?
3. Since more fluid leaves the capillary than returns to it,
   how is the remainder returned to the vascular
   system?

Venules and Veins

Venules are the microscopic vessels that continue from
the capillaries and merge to form **veins.** The veins, which
carry blood back to the atria, follow pathways that
roughly parallel those of the arteries.

The walls of veins are similar to those of arteries
in that they are composed of three distinct layers.
However, the middle layer of the venous wall is poorly
developed. Consequently, veins have thinner walls and
contain less smooth muscle and less elastic tissue than
arteries (figures 18.30 and 18.37).

Many veins, particularly those in the arms and
legs, contain flaplike *valves* that project inward from
their linings. These valves, shown in figure 18.38, are
usually composed of two leaflets that close if the blood
begins to back up in a vein. In other words, the valves
aid in returning blood to the heart, since the valves open
as long as the flow is toward the heart, but close if it is
in the opposite direction.

In addition to providing pathways for blood returning to the heart, the veins function as blood reservoirs that can be drawn on in times of need. For
example, if a hemorrhage accompanied by a drop in
arterial blood pressure occurs, the muscular walls of
veins are stimulated reflexly by sympathetic nerve impulses. The resulting venous constrictions help to raise
the blood pressure. This mechanism ensures a nearly
normal blood flow even when as much as 25% of the
blood volume has been lost. Figure 18.39 illustrates the

Figure 18.39

Most of the blood volume is contained within the veins and venules. (From *Circulation,* by Bjorn Folkow and Eric Neil. Copyright © 1971 by Oxford University Press, Inc. Reprinted by permission.)

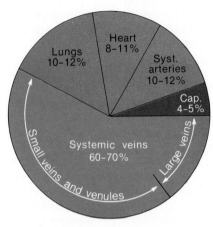

Chart 18.3	Characteristics of blood vessels	
Vessel	Type of wall	Function
Artery	Thick, strong wall with three layers—endothelial lining, middle layer of smooth muscle and elastic tissue, and outer layer of connective tissue	Carries high pressure blood from heart to arterioles
Arteriole	Thinner wall than artery, but with three layers; smaller arterioles have endothelial lining, some smooth muscle tissue, and small amount of connective tissue	Connects artery to capillary; helps to control blood flow into capillary by undergoing vasoconstriction or vasodilation
Capillary	Single layer of squamous epithelium	Provides semipermeable membrane through which nutrients, gases, and wastes are exchanged between blood and tissue cells; connects arteriole to venule
Venule	Thinner wall, less smooth muscle and elastic tissue than arteriole	Connects capillary to vein
Vein	Thinner wall than artery, but with similar layers; middle layer more poorly developed; some with flaplike valves	Carries low pressure blood from venule to heart; valves prevent back flow of blood; serves as blood reservoir

relative volumes of blood in the veins and other blood vessels.

The characteristics of the blood vessels are summarized in chart 18.3.

1. How does the structure of a vein differ from that of an artery?
2. What are the functions of veins and venules?

Blood Pressure

Blood pressure is the force exerted by the blood against the inner walls of blood vessels. Although such a force occurs throughout the vascular system, the term *blood pressure* is most commonly used to refer to systemic arterial pressure.

Arterial Blood Pressure

Arterial blood pressure rises and falls in a pattern corresponding to the phases of the cardiac cycle. That is, when the ventricles contract (ventricular systole), their walls squeeze the blood inside their chambers and force it into the pulmonary trunk and aorta. As a result, the pressures in these arteries rise sharply. The maximum pressure achieved during ventricular contraction is called the **systolic pressure.** When the ventricles relax (ventricular diastole), the arterial pressure drops, and the lowest pressure that remains in the arteries before the next ventricular contraction is termed the **diastolic pressure.**

The surge of blood entering the arterial system during a ventricular contraction causes the elastic walls of the arteries to swell, but the pressure drops almost immediately as the contraction is completed, and the

Some Blood Vessel Disorders
A Clinical Application

It is estimated that nearly half of all deaths in the United States are due to the arterial disease called *atherosclerosis*. This condition is characterized by an accumulation of soft masses of fatty materials, particularly cholesterol, on the inside of the arterial wall. Such deposits are called *plaque*, and as they develop they tend to protrude into the lumens of the vessels and interfere with the blood flow (figure 18.40). Furthermore, plaque often creates a surface that can initiate the formation of a blood clot. As a result, persons with atherosclerosis may develop thrombi or emboli that cause blood deficiency (*ischemia*) or tissue death (*necrosis*) downstream from the obstruction.

The walls of affected arteries also tend to undergo degenerative changes during which they lose their elasticity and become hardened or *sclerotic*. This stage of the disease is called *arteriosclerosis*. When this occurs, there is danger that a sclerotic vessel will rupture under the force of blood pressure.

Although the cause of atherosclerosis is not well understood, the disease seems to be associated with such factors as excessive use of saturated fats and refined carbohydrates in the diet, elevated blood pressure, cigarette smoking, obesity, and lack of physical exercise. Emotional and genetic factors also may increase the susceptibility to atherosclerosis.

Sometimes the wall of an artery is so weakened by the effects of disease that blood pressure causes a region of the artery to become dilated, forming a pulsating sac. This condition is called an *aneurysm* and when it begins to form, the enlargement tends to continue increasing in size. If the resulting sac develops by a longitudinal splitting of the middle layer of the arterial wall, it is called a *dissecting aneurysm*. An aneurysm may cause symptoms by pressing on nearby organs, or it may rupture and produce a great loss of blood.

Although most aneurysms seem to be caused by arteriosclerosis, they may also occur as a consequence of trauma, high blood pressure, infections, or congenital defects in blood vessels. Common sites of aneurysms include the thoracic and abdominal aorta, and the circle of Willis.

Phlebitis, or inflammation of a vein, is a relatively common disorder, and although it may occur in association with an injury or infection or as an aftermath of surgery, it sometimes develops for no apparent reason.

If such an inflammation is restricted to a superficial vein, the blood flow may be rechanneled through other vessels. If it occurs in a deep vein, however, the consequences can be quite serious, particularly if the blood within the affected vessel clots and blocks the normal circulation. This condition is called *thrombophlebitis*.

Varicose veins are distinguished by the presence of abnormal and irregular dilations in superficial veins, particularly those in the lower legs. This condition is usually associated with prolonged, increased back pressure within the affected vessels due to the force of gravity, as occurs when a person stands. The problem can also be aggravated when venous blood flow is obstructed by crossing the legs or by sitting in a chair so that its edge presses against the popliteal area behind the knee.

Excessive back pressure causes the veins to stretch and their diameters to increase. Since the valves within these vessels do not change size, they soon lose their abilities to block the backward flow of blood, and blood tends to accumulate in the enlarged regions.

Increased venous pressure is also accompanied by rising pressure within the venules and capillaries that supply the veins. Consequently, tissues in affected regions typically become edematous and painful.

Although some people seem to inherit a weakness in their venous valves, *varicose veins* are most common in persons who stand or sit for prolonged periods. Pregnancy and obesity also seem to favor the development of this condition. The discomfort associated with varicose veins can sometimes be relieved by elevating the legs or by wearing support hosiery that are put on before a person arises in the morning thus preventing the vessel dilation that occurs upon standing. In other cases, surgical removal of the affected veins may be necessary to reduce the problem.

Figure 18.41
Sites where arterial pulse is most easily detected. *(a. stands for artery.)*

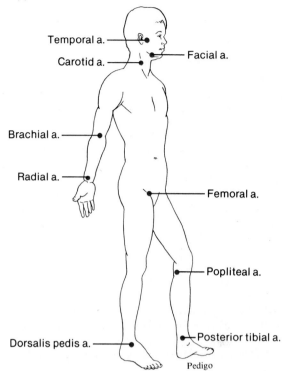

Figure 18.40

As atherosclerosis develops, masses of fatty materials accumulate beneath the inner linings of certain arteries and arterioles, and protrude into their lumens. *(a)* Normal arteriole; *(b), (c),* and *(d)* accumulation of plaque on the inner wall of the arteriole.

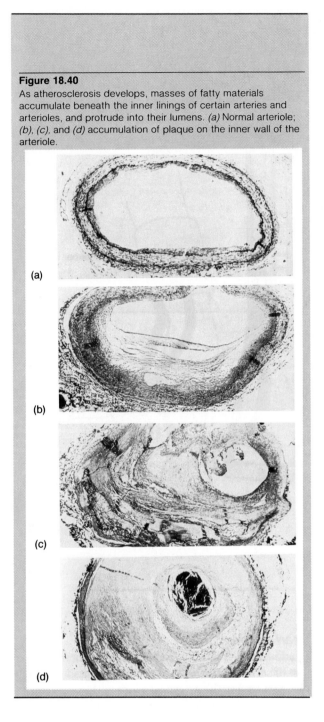

(a)

(b)

(c)

(d)

arterial walls recoil. This alternate expanding and recoiling of an arterial wall can be felt as a *pulse* in an artery that runs close to the surface. Figure 18.41 shows several sites where a pulse can be detected. The radial artery, for example, courses near the surface at the wrist and is commonly used to sense a person's radial pulse.

The radial pulse rate is equal to the rate at which the left ventricle is contracting, and for this reason it can be used to determine the heart rate. A pulse also can reveal something about blood pressure, because an elevated pressure produces a pulse that feels full, while a low pressure is accompanied by a pulse that is easily compressed.

Measurement of Arterial Blood Pressure
A Laboratory Application

Systemic arterial blood pressure can be measured by using an instrument called a *sphygmomanometer* (figure 18.42). This device consists of an inflatable rubber cuff that is connected by tubing to a compressible bulb and a glass tube containing a column of mercury. The bulb is used to pump air into the cuff, and the pressure produced is indicated by a rise in the mercury column. Thus, the pressure in the cuff can be expressed in millimeters of mercury (mm Hg). A pressure of 100 mm Hg, for example, would be enough to force the mercury column upward for a distance of 100 mm.

To measure blood pressure, the cuff of the sphygmomanometer is usually wrapped around the upper arm so it surrounds the brachial artery. Air is pumped into the cuff using the bulb until the cuff pressure exceeds the pressure in the brachial artery. As a result, the vessel is squeezed closed, and its blood flow is stopped. At this moment, if the bell of a stethoscope is placed over the brachial artery at the distal border of the cuff, no sounds can be heard from the vessel because the blood flow has been interrupted. As air is slowly released from the cuff, the air pressure inside it decreases. When the cuff pressure is approximately equal to the systolic blood pressure within the brachial artery, the artery opens enough for a small amount of blood to spurt through. This movement produces a sharp sound that can be heard through the stethoscope, and the height of the mercury column when this first tapping sound is heard represents the *arterial systolic pressure* (SP).

As the cuff pressure continues to drop, a series of increasingly louder sounds can be heard. Then the sounds become abruptly muffled and finally disappear. The height of the mercury column when the sounds become abruptly muffled represents the *arterial diastolic pressure* (DP).

The results of a blood pressure measurement are reported as a fraction, such as 120/80. In this notation the upper number indicates the systolic pressure in mm Hg, and

Figure 18.42
A sphygmomanometer is used to measure arterial blood pressure.

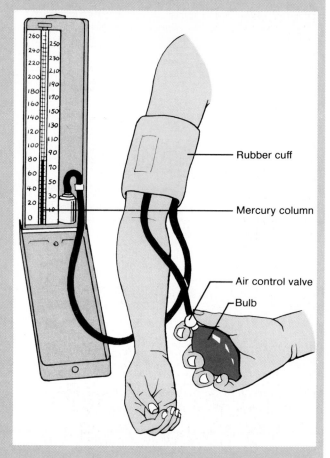

- Rubber cuff
- Mercury column
- Air control valve
- Bulb

---

1. Distinguish between systolic and diastolic blood pressure.
2. What cardiac event is responsible for the systolic pressure? The diastolic pressure?
3. What causes a pulse in an artery?

Factors That Influence Arterial Blood Pressure

The arterial pressure depends on a variety of factors, including heart action, blood volume, resistance to flow, and the viscosity of the blood (figure 18.44).

Heart Action

In addition to creating blood pressure by forcing blood into the arteries, the heart action determines how much blood will enter the arterial system with each ventricular contraction, as well as the rate of this fluid output.

The volume of blood discharged from the ventricle with each contraction is called the **stroke volume** and equals about 70 ml. The volume discharged from the ventricle per minute is called the **cardiac output.** It is calculated by multiplying the stroke volume by the heart rate in beats per minute. (Cardiac output=stroke volume × heart rate.) Thus, if the stroke volume is 70 ml, and the heart rate is 72 beats per minute, the cardiac output is 5040 ml per minute.

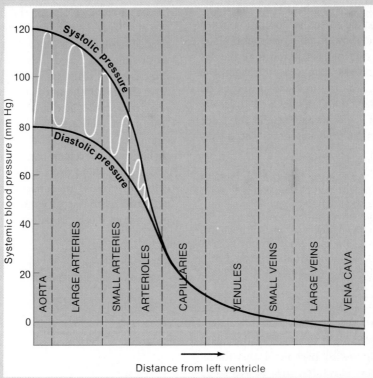

the lower number indicates the diastolic pressure in mm Hg (SP/DP). Figure 18.43 shows how these pressures vary throughout the systemic circuit.

The difference between the systolic and diastolic pressures (SP−DP), which is called the *pulse pressure* (PP), is generally about 40 mm Hg.

The average pressure in the arterial system is also of interest since it represents the force that is effective throughout the cardiac cycle for driving blood to the tissues. This force, called the *mean arterial pressure,* is approximated by adding the diastolic pressure and one-third of the pulse pressure (DP+⅓PP).

Blood pressure varies with the cardiac output. If either the stroke volume or the heart rate increases, so does the cardiac output and, as a result, the blood pressure rises. Conversely, if the stroke volume or the heart rate decreases, so do the cardiac output and the blood pressure.

Blood Volume

The **blood volume** is equal to the sum of the blood cell and plasma volumes in the vascular system. Although the blood volume varies somewhat with age, body size, and sex, for adults it usually remains about 5 liters.

Blood pressure is directly proportional to the volume of blood within the cardiovascular system. Thus, any changes in blood volume are accompanied by

Figure 18.44
Some factors that influence arterial blood pressure.

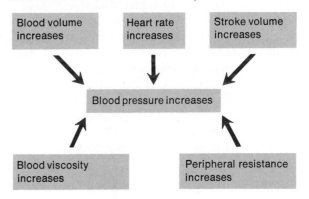

changes in blood pressure. For example, if blood volume is reduced by a hemorrhage, the blood pressure drops. If the normal blood volume is restored by a blood transfusion, normal pressure may be reestablished.

Arterial blood volume, and consequently arterial blood pressure, also vary directly with cardiac output. Thus, if the stroke volume or the heart rate increases, the arterial system may be forced to accept a greater volume of blood than can flow into the peripheral blood vessels. Consequently, the blood volume that the arteries must accommodate is increased even when the ventricles are relaxed. For this reason, an increased cardiac output is reflected in an elevated *diastolic pressure.* On the other hand, an increase in the force of ventricular contraction produces an elevated *systolic pressure.*

Blood volume can be determined by injecting a known volume of an indicator, such as radioactive iodine, into the blood. After a time which allows for thorough mixing, a blood sample is withdrawn, and the concentration of the indicator is measured. The total blood volume is calculated by using the formula: Blood Volume = Amount of Indicator Injected/Concentration of Indicator in Blood Sample.

Peripheral Resistance

Friction between the blood and the walls of the blood vessels creates a force called **peripheral resistance,** which hinders blood flow. This force must be overcome by blood pressure if the blood is to continue flowing. Consequently, factors that alter peripheral resistance cause changes in blood pressure.

For example, if the smooth muscles in the walls of arterioles contract, the peripheral resistance of these constricted vessels increases. Blood tends to back up into the arteries supplying the arterioles, and the arterial pressure rises. Dilation of the arterioles has the opposite effect—peripheral resistance lessens, and the arterial blood pressure drops in response.

As mentioned earlier, the arterial walls are quite elastic, and when the ventricles discharge a surge of blood, the arteries swell. Almost immediately, the elastic tissues recoil, and the vessel walls press against the blood inside. This action helps to force the blood onward against the peripheral resistance created by the arterioles and capillaries. It also tends to convert the intermittent flow of blood, which is characteristic of the arterial system, into a more continuous movement through the capillaries.

Viscosity

The viscosity of a fluid is a physical property that is related to the ease with which its molecules flow past one another. The greater the viscosity, the greater the resistance to flowing.

The presence of blood cells and plasma proteins increases the viscosity of the blood. Since the greater the blood's resistance to flowing, the greater the force needed to move it through the vascular system, it is not surprising that blood pressure rises as blood viscosity increases and drops as viscosity decreases.

Although the viscosity of blood normally remains relatively stable, any condition that alters the concentrations of blood cells or plasma proteins may cause changes in viscosity. For example, anemia and hemorrhage may be accompanied by a decreasing viscosity and a consequent drop in blood pressure. If red blood cells are present in abnormally high numbers (polycythemia), the viscosity increases and there is likely to be a rise in blood pressure.

If blood is flowing very slowly, the red blood cells tend to stick together in groups like stacks of tiny coins called *rouleaux.* The presence of such groups of cells increases the viscosity of blood. Although rouleaux commonly form in veins, blood flow in arteries and arterioles usually is rapid enough to prevent their formation.

---

1. How is cardiac output calculated?
2. What is the relationship between cardiac output and blood pressure?
3. How does blood volume affect blood pressure?
4. What is the relationship between peripheral resistance and blood pressure? Between viscosity and blood pressure?

Control of Blood Pressure

Two important mechanisms for maintaining normal arterial pressure involve the regulation of cardiac output and peripheral resistance.

Cardiac output depends on the volume of blood discharged from the left ventricle with each contraction (stroke volume) and the rate of heartbeat. These actions are affected by mechanical, neural, and chemical factors.

For example, the volume of blood entering the ventricle affects the stroke volume. As the blood enters, myocardial fibers in the ventricular wall are mechanically stretched. Within limits, the greater the length of these fibers, the greater the force with which they

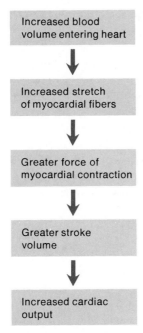

Increased blood volume entering heart

↓

Increased stretch of myocardial fibers

↓

Greater force of myocardial contraction

↓

Greater stroke volume

↓

Increased cardiac output

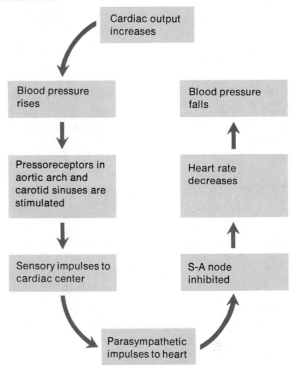

Cardiac output increases

Blood pressure rises

Pressoreceptors in aortic arch and carotid sinuses are stimulated

Sensory impulses to cardiac center

Parasympathetic impulses to heart

S-A node inhibited

Heart rate decreases

Blood pressure falls

contract. This relationship between fiber length and force of contraction is called *Starling's law of the heart.* Because of it, the heart can respond to the demands placed on it by the varying quantities of blood that return from the venous system.

In other words, the more blood that enters the heart from the veins, the stronger the ventricular contraction, the greater the stroke volume, and the greater the cardiac output (figure 18.45). The less blood that returns from the veins, the weaker the ventricular contraction, the lesser the stroke volume, and the lesser the cardiac output.

This mechanism ensures that the volume of blood discharged from the heart is equal to the volume entering its chambers. Consequently, the volume of blood that enters the right atrium from the venae cavae is normally equal to the volume that leaves the left ventricle and enters the aorta.

The neural regulation of heart rate was described previously in this chapter. To review, pressoreceptors (baroreceptors) located in the walls of the aortic arch and carotid sinuses are sensitive to changes in blood pressure. If the arterial pressure rises, nerve impulses travel from the receptors to the *cardiac center* of the medulla oblongata. This center relays parasympathetic impulses to the S-A node in the heart, and its rate decreases in response. As a result of this *cardioinhibitor reflex,* the cardiac output is reduced, and blood pressure falls toward the normal level. Figure 18.46 summarizes this mechanism.

Conversely, if the arterial blood pressure drops, the *cardioaccelerator reflex,* involving sympathetic impulses to the S-A node, is initiated, and the heart beats faster. This response increases the cardiac output, and the arterial pressure rises.

As mentioned in chapter 13, *epinephrine* causes the heart rate to increase and consequently alters cardiac output and blood pressure. Other factors that cause an increase in heart rate and a rise in blood pressure include emotional responses, such as fear and anger; physical exercise; and an increase in body temperature.

Peripheral resistance is regulated primarily by changes in the diameters of arterioles. Since blood vessels with smaller diameters offer greater resistance to blood flow, factors that cause arteriole vasoconstriction bring about increasing peripheral resistance, while those causing vasodilation produce a decrease in resistance.

The *vasomotor center* of the medulla oblongata continually sends sympathetic impulses to the smooth muscles in the arteriole walls. As a result, these muscles are kept in a state of tonic contraction, which helps to maintain the peripheral resistance associated with normal blood pressure. Since the vasomotor center is responsive to changes in blood pressure, it can cause an increase in peripheral resistance by increasing its outflow of sympathetic impulses, or can cause a decrease

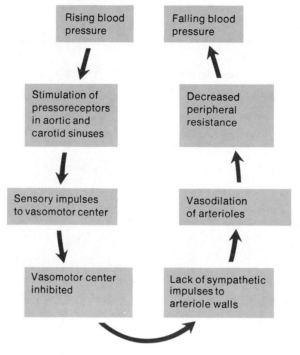

in such resistance by decreasing its sympathetic outflow. In the latter case, the vessels undergo vasodilation
as the sympathetic stimulation is decreased.

For instance, as figure 18.47 illustrates, whenever the arterial blood pressure suddenly rises, the
pressoreceptors in the aortic arch and carotid sinuses
signal the vasomotor center, and the sympathetic outflow to the arteriole walls is inhibited. The resulting
vasodilation causes a decrease in peripheral resistance,
and the blood pressure falls toward the normal level.

Control of vasoconstriction and vasodilation by
the vasomotor center is especially important in the arterioles of the *abdominal viscera* (splanchnic region).
These vessels, if fully dilated, could accept nearly all
the blood of the body and cause the arterial pressure
to approach zero. Thus, control of their diameters is
essential in the regulation of normal peripheral resistance.

Chemical substances, including carbon dioxide,
oxygen, and hydrogen ions, also influence peripheral
resistance by affecting the smooth muscles in the walls
of arterioles and metarterioles and the actions of precapillary sphincters. For example, an increasing Pco_2,

a decreasing Po_2, and a decreasing pH cause relaxation of these muscles and a consequent drop in blood
pressure. In addition, epinephrine and norepinephrine
cause vasoconstriction of many vessels, followed by a
rise in blood pressure; even though epinephrine causes
vasodilation of the vessels within the skeletal muscles.

1. What factors affect cardiac output?
2. Define Starling's law of the heart.
3. What is the function of the pressoreceptors in the
   walls of the aortic arch and carotid sinuses?
4. How does the vasomotor center control the diameter
   of arterioles?

Venous Blood Flow

Blood pressure decreases as the blood moves through
the arterial system and into the capillary networks. In
fact, little pressure remains at the venule ends of capillaries (figure 18.43). Therefore, blood flow through
the venous system is not the direct result of heart action, but depends on other factors, such as skeletal
muscle contraction, breathing movements, and vasoconstriction of veins.

Hypertension
A Clinical Application

Hypertension, or high blood pressure, is characterized by a persistently elevated arterial pressure, and is one of the more common diseases of the cardiovascular system.

Since the cause of high blood pressure is often unknown, it may be called *essential* (or idiopathic) *hypertension.* Sometimes the elevated pressure is related to another problem, such as arteriosclerosis or kidney disease.

Arteriosclerosis, for example, is accompanied by decreasing elasticity of the arterial walls and narrowing of the lumens of these vessels. Both of these effects promote an increase in blood pressure.

Kidney diseases often produce changes that interfere with blood flow to kidney cells. In response, the affected tissues may release an enzyme, called **renin,** that acts on certain plasma proteins to form a substance called **angiotensin.**

Angiotensin is a powerful vasoconstrictor whose action increases the peripheral resistance in the arterial system, and this causes the arterial pressure to rise. Angiotensin also causes *aldosterone* to be released from the adrenal cortex. This hormone promotes the retention of sodium ions and water by the kidneys, and the resulting increase in blood volume causes an additional increase in blood pressure (figure 18.48).

Normally this mechanism ensures that a decrease in blood flow to the kidneys is followed by an increase in arterial pressure, and that, in turn, results in an increase in blood flow to the kidneys. If the decreased blood flow is the result of a disease, such as atherosclerosis, the mechanism may cause high blood pressure and promote further deterioration of the arterial system.

Although the mechanism is not well understood, high sodium intake seems to cause vasoconstriction in some individuals, which leads to increasing blood pressure. Other factors that seem to promote hypertension include obesity, which is accompanied by an increased volume of body fluids and increased heart action needed to provide additional blood flow, and psychological stress, which activates sympathetic nerve impulses that cause generalized vasoconstriction.

The consequences of prolonged, uncontrolled hypertension can be very serious. For example, the left ventricle must contract with greater force than usual in order to discharge a normal volume of blood against increased arterial pressure. As a result of the increased work load, the myocardium tends to thicken, and the heart enlarges. Such hypertrophy may benefit cardiac function at first. However,

Figure 18.48

A mechanism that acts to elevate blood pressure.

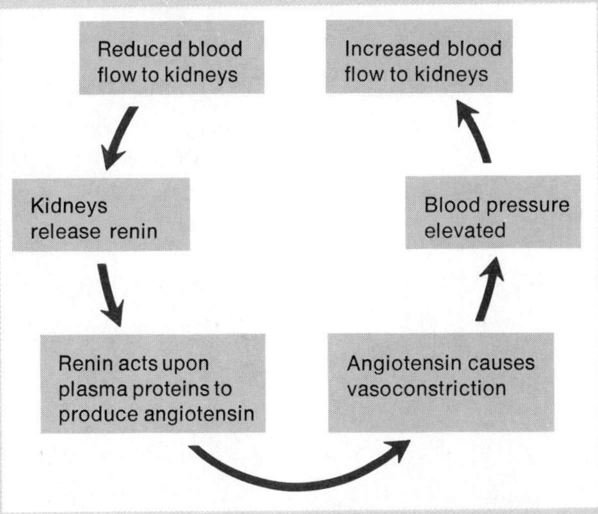

as the muscle tissue increases, the coronary blood vessels may not increase to the same extent. Consequently, the vessels may become less and less able to supply the myocardium with sufficient blood. This change may be accompanied by degeneration of muscle fibers and a replacement of muscle with fibrous tissue. In time, the enlarged heart is weakened and may fail.

Hypertension also enhances the development of atherosclerosis. As a result, various arteries, including those of the coronary circuit, tend to accumulate plaque that may cause arterial occlusions. The patient then may suffer from a *coronary thrombosis* or a *coronary embolism.* Similar changes in the arteries of the brain increase the chances of a *cerebral vascular accident* (CVA)—a cerebral thrombosis, embolism, or hemorrhage—more commonly called a stroke.

The treatment of hypertension may include exercise, control of body weight, reduction of stress, and use of low sodium diets. It also may involve the use of drugs, such as diuretics and inhibitors of sympathetic nerve activity. Diuretics typically increase the urinary excretion of sodium and water, reducing the volume of body fluids. The sympathetic inhibitors may act by blocking the synthesis of neurotransmitters, such as norepinephrine, or by blocking various receptor sites of effector cells.

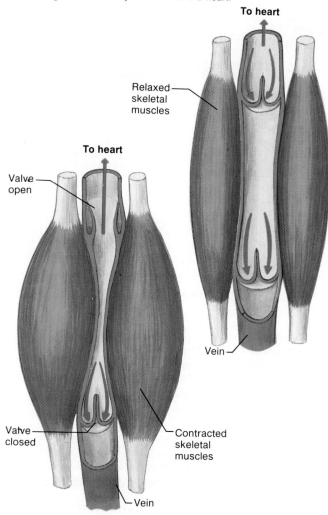

For example, when skeletal muscles contract, they thicken and press on nearby vessels, squeezing the blood inside. As mentioned previously, many veins, particularly those in the arms and legs, contain flaplike *semilunar valves* that project inward from their linings. These valves offer little resistance to blood flowing toward the heart, but they close if blood moves in the opposite direction. Consequently, as *skeletal muscles* exert pressure on veins with valves, some blood is moved from one valve section to another. This massaging action of contracting skeletal muscles helps to push blood through the venous system toward the heart. (See figure 18.49.)

Respiratory movements provide another means of moving venous blood. During inspiration, the pressure within the thoracic cavity is reduced as the diaphragm contracts and the rib cage moves upward and outward. At the same time, the pressure within the abdominal cavity is increased as the diaphragm presses

downward on the abdominal viscera. Consequently, blood tends to be squeezed out of the abdominal veins and into thoracic veins. Back flow into the legs is prevented by valves in the veins of the legs.

During exercise, these respiratory movements act together with skeletal muscle contractions to increase the return of venous blood.

Another mechanism that promotes return of venous blood involves constriction of veins. When venous pressure is low, sympathetic reflexes stimulate smooth muscles in the walls of veins to contract. Such *venoconstriction* serves to increase venous pressure and forces more blood toward the heart.

Also, as was mentioned, the veins provide a blood reservoir that can adapt its capacity to changes in blood volume (figure 18.39). If blood is lost and the blood pressure decreases, venoconstriction can help to return the blood pressure to normal by forcing blood out of this reservoir.

If a person strains with the glottis closed, the intrathoracic pressure is increased, and the return of venous blood to the heart is reduced. This action, which is called the *Valsalva maneuver,* occurs when a person lifts a heavy object, coughs, or strains to defecate or give birth.

Central Venous Pressure

Since all the systemic veins drain into the right atrium, the pressure within this heart chamber is called *central venous pressure.*

This pressure is of special interest because it has an effect on the pressure within the peripheral veins. For example, if the heart is beating weakly, the central venous pressure increases, and blood tends to back up in the venous network, causing its pressure to increase also. If the heart is beating forcefully, however, the central venous pressure and pressure within the venous network decrease.

Other factors that increase the flow of blood into the right atrium, and thus cause the central venous pressure to become elevated, include an increase in blood volume and widespread venoconstriction.

Sometimes, as a result of disease or injury, fluid accumulates rapidly in the pericardial cavity. This condition, called *acute cardiac tamponade,* can be life-threatening because increasing pressure within the pericardial cavity may compress the heart and interfere with the flow of blood into its chambers. An early symptom of acute cardiac tamponade may be an increase in central venous pressure, accompanied by visible engorgement of veins in the neck.

Figure 18.50

The pulmonary circuit consists of the vessels that carry blood between the heart and the lungs; all other vessels are included in the systemic circuit.

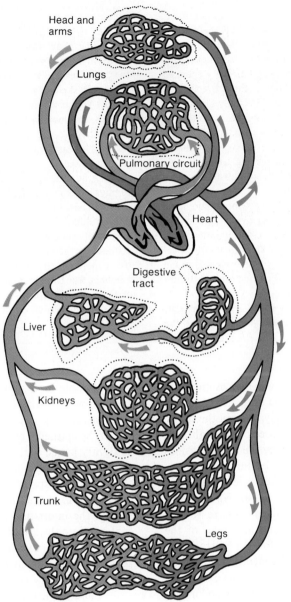

Head and arms

Lungs

Pulmonary circuit

Heart

Digestive tract

Liver

Kidneys

Trunk

Legs

1. What is the function of the venous valves?
2. How do skeletal muscles affect venous blood flow?
3. How do respiratory movements affect venous blood flow?
4. What factors stimulate venoconstriction?

Paths of Circulation

The blood vessels of the cardiovascular system can be divided into two major pathways—a pulmonary circuit and a systemic circuit. The **pulmonary circuit** consists of those vessels that carry blood from the heart to the lungs and back to the heart, while the **systemic circuit** is responsible for carrying blood from the heart to all other parts of the body and back again (figure 18.50).

The circulatory pathways described in the following sections are those of an adult. The fetal pathways, which are somewhat different, are described in chapter 23.

The Pulmonary Circuit

Blood enters the pulmonary circuit as it leaves the right ventricle through the pulmonary trunk. The pulmonary trunk extends upward and posteriorly from the

Cardiovascular Adjustment to Exercise

A Practical Application

When a person engages in strenuous exercise, a number of changes occur in the circulatory system that help to increase blood flow, and therefore oxygen availability, to skeletal muscles. For example, when the skeletal muscles are resting, only about one-fifth of their capillaries are open. As exercise begins, oxygen concentration in the active skeletal muscles decreases and, in response, essentially all of their capillaries open. At the same time, blood flow into the active muscles greatly increases due to vasodilation stimulated by such factors as sympathetic nerve impulses, decreasing oxygen concentration, increasing carbon dioxide concentration, and decreasing pH. Blood flow to the myocardium and skin also increases.

In certain other parts of the body, sympathetic nerve impulses cause vasoconstriction, and as a result, blood flow into the abdominal viscera and inactive skeletal muscles is reduced. Blood flow to the brain and kidneys, however, remains nearly unchanged, so that these vital organs continue to receive oxygen and nutrients at the rate required to support their needs.

When the walls of veins in visceral organs constrict, blood is moved out of these blood reservoirs, adding to the venous return to the heart. This increased venous return also is aided by an increased respiratory rate during exercise.

The heart rate increases in response to decreased parasympathetic stimulation of the S-A and A-V nodes and to sympathetic reflexes triggered by stimulation of proprioceptors in skeletal muscles and of stretch receptors in the lungs. As the volume of blood returning to the heart increases, the ventricular walls are stretched, and this stimulates them to contract with greater force. Consequently, with each heartbeat the stroke volume increases and blood pressure rises.

A number of physiological changes occur in response to physical conditioning that may improve athletic performance. These include increases in heart pumping efficiency, blood volume, blood hemoglobin concentration, and mitochondrial content of muscle fibers. As a result of such changes, oxygen can be delivered to and utilized by muscle tissue more effectively. In the case of athletes, such as marathoners, who train for endurance-type performance, the size of the heart and its chambers may enlarge 40% or more, providing for greater cardiac output.

heart and, about 5 cm above its origin, divides into right and left pulmonary arteries. These branches penetrate the right and left lungs respectively. Within the lungs, they divide into *lobar branches* (three on the right side and two on the left) that accompany the main divisions of the bronchi into the lobes of the lungs. After repeated divisions, the lobar branches give rise to arterioles that continue into the capillary networks associated with the walls of the alveoli (figure 18.51).

The blood in the arteries and arterioles of the pulmonary circuit has a relatively low concentration of oxygen and a relatively high concentration of carbon dioxide. As is explained in chapter 16, gas exchanges occur between the blood and the air as the blood moves through the *pulmonary capillaries.*

Since the right ventricle contracts with less force than the left ventricle, the arterial pressure in the pulmonary circuit is less than that in the systemic circuit. Consequently, the pulmonary capillary pressure is relatively low.

The force tending to move fluid out of a pulmonary capillary is 23 mm Hg, while the force tending to pull fluid into it is 22 mm Hg; thus, such a capillary has a net filtration pressure of 1 mm Hg. This pressure is responsible for a slight, continuous flow of fluid into the narrow interstitial space between the capillary and the alveolus.

The epithelial cells of the alveolar membranes are so tightly joined that ions of sodium, chlorine, and potassium, and molecules of glucose and urea that enter the interstitial space usually fail to enter the alveoli. This helps to maintain a relatively high osmotic pressure in the interstitial fluid. Consequently, any water that gets into the alveoli is rapidly moved back into the interstitial space by osmosis. This mechanism helps to keep the alveoli from filling with fluid, and they are said to remain dry (figure 18.52).

Fluid in the interstitial space may be drawn back into the pulmonary capillaries by the osmotic pressure of the blood, or it may be returned to the circulation by means of lymphatic vessels described in chapter 19.

As a result of the gas exchanges occurring between the blood and the alveolar air, blood entering the venules of the pulmonary circuit is rich in oxygen and low in carbon dioxide.

These venules merge to form small veins, and they in turn converge to form still larger ones. Four *pulmonary veins,* two from each lung, return blood to the left atrium, and this completes the vascular loop of the pulmonary circuit.

The external carotid artery terminates by dividing into *maxillary* and *superficial temporal arteries*. The maxillary artery supplies blood to the teeth, gums, jaws, cheek, nasal cavity, eyelids, and meninges. The temporal artery extends to the parotid salivary gland and to various surface regions of the face and scalp.

The **internal carotid artery** follows a deep course upward along the pharynx to the base of the skull. Entering the cranial cavity, it provides the major blood supply to the brain. Its major branches include the following:

1. *Ophthalmic artery* to the eyeball and various muscles and accessory organs within the orbit.
2. *Posterior communicating artery* that forms part of the circle of Willis.
3. *Anterior choroid artery* to the choroid plexus within the lateral ventricle of the brain and to a variety of nerve structures within the brain.

The internal carotid artery terminates by dividing into *anterior* and *middle cerebral arteries*. The middle cerebral artery passes through the lateral sulcus and supplies the lateral surface of the cerebrum, including the primary motor and sensory areas of the face and arms, the optic radiations, and the speech area (see chapter 11). The anterior cerebral artery extends anteriorly between the cerebral hemispheres and supplies the medial surface of the brain.

Near the base of each internal carotid artery is an enlargement called a **carotid sinus.** Like the aortic sinuses, these structures contain pressoreceptors that function in the reflex control of blood pressure. A number of small epithelial masses, called **carotid bodies,** also occur in the wall of the carotid sinus. These bodies are very vascular and contain chemoreceptors that act with the chemoreceptors of the aortic bodies in regulating various circulatory and respiratory actions.

If a cerebral artery becomes occluded, the person may have a stroke as brain tissue beyond the obstruction becomes ischemic. For example, if a branch of the middle cerebral artery is occluded, the symptoms may include varying degrees of paralysis or sensory loss in the face or arm, or various visual or speech disorders.

Figure 18.56
The main arteries of the shoulder and arm.

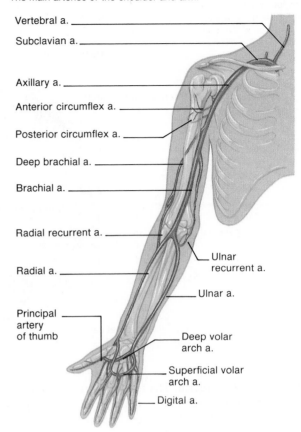

Vertebral a.
Subclavian a.
Axillary a.
Anterior circumflex a.
Posterior circumflex a.
Deep brachial a.
Brachial a.
Radial recurrent a.
Radial a.
Principal artery of thumb
Ulnar recurrent a.
Ulnar a.
Deep volar arch a.
Superficial volar arch a.
Digital a.

Arteries to the Shoulder and Arm

The subclavian artery, after giving off branches to the neck, continues into the upper arm (figure 18.56). It passes between the clavicle and the first rib and becomes the axillary artery.

The **axillary artery** supplies branches to structures in the axilla and the chest wall, including the skin of the shoulder; part of the mammary gland; the upper end of the humerus; the shoulder joint; and various muscles in the back, shoulder, and chest. As this vessel leaves the axilla, it becomes the brachial artery.

The **brachial artery** courses along the humerus to the elbow. It gives rise to a *deep brachial artery* that curves posteriorly around the humerus and supplies the triceps muscle. Shorter branches pass into the muscles on the anterior side of the upper arm, while others descend on each side to the elbow and interconnect with

Figure 18.57

Arteries that supply the thoracic wall. *(a.* stands for *artery;*
m. for *muscle.)*

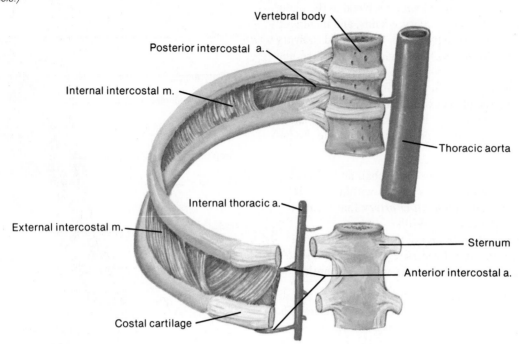

Vertebral body

Posterior intercostal a.

Internal intercostal m.

Thoracic aorta

External intercostal m.

Internal thoracic a.

Sternum

Anterior intercostal a.

Costal cartilage

arteries in the forearm. The resulting arterial network allows blood to reach the lower arm even if some portion of the distal brachial artery becomes obstructed.

Within the elbow, the brachial artery divides into an ulnar and a radial artery. The **ulnar** (ul'nar) **artery** leads downward on the ulnar side of the forearm to the wrist. Some of its branches join the anastomosis around the elbow joint, while others supply blood to flexor and extensor muscles in the lower arm.

The **radial** (ra'de-al) **artery,** a continuation of the brachial artery, travels along the radial side of the forearm to the wrist. As it nears the wrist, it comes close to the surface and provides a convenient vessel for taking the pulse (radial pulse).

Various branches of the radial artery join the anastomosis of the elbow and supply lateral muscles of the forearm.

At the wrist, branches of the ulnar and radial arteries join to form an interconnecting network of vessels. Arteries arising from this network supply blood to structures in the wrist, hand, and fingers.

Arteries to the Thoracic and Abdominal Walls

Blood reaches the thoracic wall through several vessels, including branches from the subclavian artery and the thoracic aorta (figure 18.57).

The subclavian artery contributes to this supply through a branch called the **internal thoracic artery.** This vessel originates in the base of the neck and passes downward on the pleura and behind the cartilages of the upper six ribs. It gives off two *anterior intercostal arteries* to each of the upper six intercostal spaces, which supply the intercostal muscles, other intercostal tissues, and the mammary glands.

Posterior intercostal arteries arise from the thoracic aorta and enter the intercostal spaces between the third through the eleventh ribs. These arteries give off branches that supply the intercostal muscles, the vertebrae, the spinal cord, and various deep muscles of the back.

The blood supply to the anterior abdominal wall is provided primarily by branches of the *internal thoracic* and *external iliac arteries.* Structures in the posterior and lateral abdominal wall are supplied by paired

Figure 18.58
Arteries that supply the pelvic region.

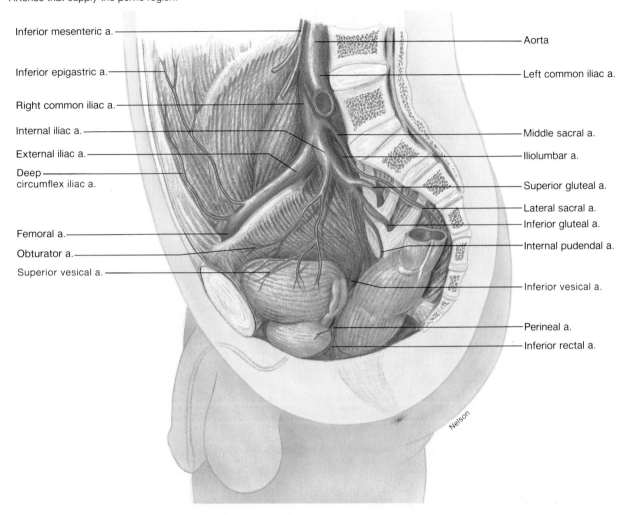

Inferior mesenteric a.
Inferior epigastric a.
Right common iliac a.
Internal iliac a.
External iliac a.
Deep circumflex iliac a.
Femoral a.
Obturator a.
Superior vesical a.

Aorta
Left common iliac a.
Middle sacral a.
Iliolumbar a.
Superior gluteal a.
Lateral sacral a.
Inferior gluteal a.
Internal pudendal a.
Inferior vesical a.
Perineal a.
Inferior rectal a.

Nelson

vessels originating from the abdominal aorta, including the *phrenic* and *lumbar arteries* mentioned previously.

Arteries to the Pelvis and Leg

The abdominal aorta divides to form the **common iliac arteries** at the level of the pelvic brim, and these vessels provide blood to the pelvic organs, gluteal region, and legs.

Each common iliac artery descends a short distance and divides into an internal (hypogastric) and an external branch. The **internal iliac artery** gives off numerous branches to various pelvic muscles and visceral structures, as well as to the gluteal muscles and the external genitalia. Important branches of this vessel include the following (figure 18.58):

1. *Iliolumbar artery* to the ilium and muscles of the back.
2. *Superior* and *inferior gluteal arteries* to the gluteal muscles, pelvic muscles, and skin of the buttocks.
3. *Internal pudendal artery* to muscles in the distal portion of the alimentary canal, the external genitalia, and the hip joint.
4. *Superior* and *inferior vesical arteries* to the urinary bladder. In males these vessels also supply the seminal vesicles and the prostate gland.
5. *Middle rectal artery* to the rectum.
6. *Uterine artery* to the uterus and vagina in females.

Figure 18.59
Main branches of the external iliac artery.

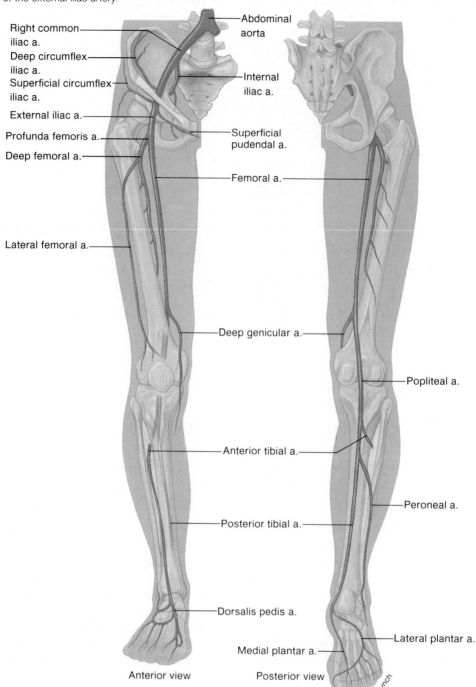

Right common iliac a.

Deep circumflex iliac a.

Superficial circumflex iliac a.

External iliac a.

Profunda femoris a.

Deep femoral a.

Lateral femoral a.

Abdominal aorta

Internal iliac a.

Superficial pudendal a.

Femoral a.

Deep genicular a.

Popliteal a.

Anterior tibial a.

Peroneal a.

Posterior tibial a.

Dorsalis pedis a.

Lateral plantar a.

Medial plantar a.

Anterior view

Posterior view

Lynch

The **external iliac artery** provides the main blood supply to the legs (figure 18.59). It passes downward along the brim of the pelvis and gives off two large branches—an *inferior epigastric artery* and a *deep circumflex iliac artery*. These vessels supply muscles and skin in the lower abdominal wall.

Midway between the pubic symphysis and the anterior superior iliac spine of the ilium, the external iliac artery passes beneath the inguinal ligament and becomes the femoral artery.

The **femoral** (fem′or-al) **artery,** which passes fairly close to the anterior surface of the upper thigh, gives off many branches to muscles and superficial tissues of the thigh. These branches also supply the skin of the groin and lower abdominal wall. Important subdivisions of the femoral artery include the following:

1. *Superficial circumflex iliac artery* to lymph nodes and skin of the groin.
2. *Superficial epigastric artery* to the skin of the lower abdominal wall.
3. *Superficial* and *deep external pudendal arteries* to the skin of the lower abdomen and external genitalia.
4. *Profunda femoris artery,* which is the largest branch of the femoral artery, to the hip joint and various muscles of the thigh.
5. *Deep genicular artery* to distal ends of thigh muscles and to an anastomosis around the knee joint.

As the femoral artery reaches the proximal border of the space behind the knee (popliteal fossa), it becomes the **popliteal** (pop-lit′e-al) **artery.** Branches of this artery supply blood to the knee joint and to certain muscles in the thigh and calf. Also, many of its branches join the anastomosis of the knee and help to provide alternate pathways for blood in the case of arterial obstructions. At the lower border of the popliteal fossa, the popliteal artery divides into the anterior and posterior tibial arteries.

The **anterior tibial** (tib′e-al) **artery** passes downward between the tibia and the fibula, giving off branches to the skin and muscles in anterior and lateral regions of the lower leg. It also communicates with the anastomosis of the knee and a network of arteries around the ankle. This vessel continues into the foot as the *dorsalis pedis artery,* which supplies blood to the foot and toes.

The **posterior tibial artery,** the larger of the two popliteal branches, descends beneath the calf muscles giving off branches to skin, muscles, and other tissues of the lower leg along the way. Some of these vessels join the anastomoses of the knee and ankle.

The largest branch of the posterior tibial artery is the *peroneal artery,* which travels downward along the fibula and contributes to the anastomosis of the ankle. As it passes between the medial malleolus and the heel, this vessel divides into the *medial* and *lateral plantar arteries.* Branches from these arteries supply blood to tissues of the heel, foot, and toes.

The major vessels of the arterial system are shown in figure 18.60.

---

1. Describe the structure of the aorta.
2. Name the vessels that arise from the aortic arch.
3. Name the branches of the thoracic and abdominal aorta.
4. What vessels supply blood to the head? The arm? The abdominal wall? The leg?

The Venous System

Venous circulation is responsible for returning blood to the heart after exchanges of gases, nutrients, and wastes have been made between the blood and the body cells.

Characteristics of Venous Pathways

The vessels of the venous system begin with the merging of capillaries into venules, venules into small veins, and small veins into larger ones. Unlike arterial pathways, however, those of the venous system are difficult to follow. This is because the vessels are commonly interconnected in irregular networks, so that many unnamed tributaries may join to form a relatively large vein.

On the other hand, larger veins typically parallel the courses taken by named arteries, and these veins often have the same names as their companions in the arterial system. Thus, with some exceptions, the name of a major artery also provides the name of the vein next to it. For example, the renal vein parallels the renal artery, the common iliac vein accompanies the common iliac artery, and so forth.

The veins that carry blood from the lungs and myocardium back to the heart have already been described. The veins from all other parts of the body converge into two major pathways that lead to the right atrium. They are the *superior* and *inferior venae cavae.*

Figure 18.60
Major vessels of the arterial system. *(a. stands for artery.)*

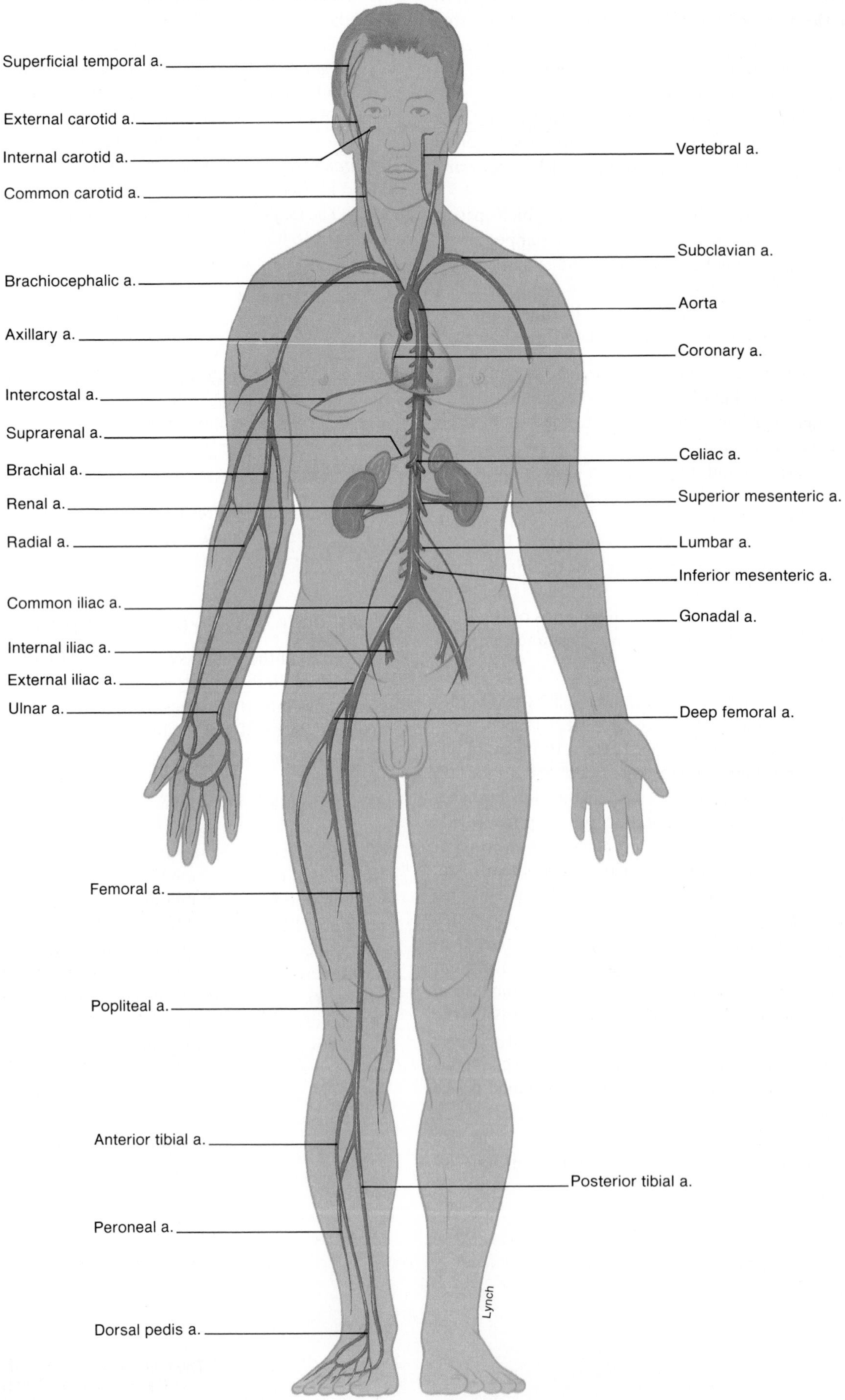

Superficial temporal a.

External carotid a.

Internal carotid a.

Vertebral a.

Common carotid a.

Subclavian a.

Brachiocephalic a.

Aorta

Axillary a.

Coronary a.

Intercostal a.

Suprarenal a.

Celiac a.

Brachial a.

Renal a.

Superior mesenteric a.

Radial a.

Lumbar a.

Inferior mesenteric a.

Common iliac a.

Gonadal a.

Internal iliac a.

External iliac a.

Ulnar a.

Deep femoral a.

Femoral a.

Popliteal a.

Anterior tibial a.

Posterior tibial a.

Peroneal a.

Lynch

Dorsal pedis a.

Figure 18.61

The main veins of the head and neck. (Note that the clavicle has been removed.) *(v. stands for vein.)*

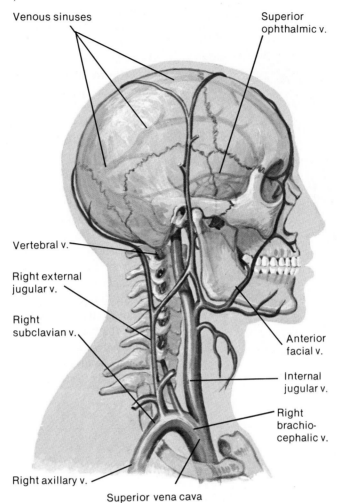

Venous sinuses

Superior ophthalmic v.

Vertebral v.

Right external jugular v.

Right subclavian v.

Anterior facial v.

Internal jugular v.

Right brachio-cephalic v.

Right axillary v.

Superior vena cava

Veins from the Head, Neck, and Brain

Blood from the face, scalp, and superficial regions of the neck is drained by the **external jugular** (jug'u-lar) **veins.** These vessels descend on either side of the neck, passing over the sternocleidomastoid muscles and beneath the platysma. They empty into the *right* and *left subclavian veins* in the base of the neck (figure 18.61).

The **internal jugular veins,** which are somewhat larger than the externals, arise from numerous veins and venous sinuses of the brain and from deep veins in various parts of the face and neck. They pass downward through the neck beside the common carotid arteries and also join the subclavian veins. These unions of the internal jugular and subclavian veins form large **brachiocephalic** (innominate) **veins** on each side. These vessels then merge in the mediastinum and give rise to the **superior vena cava,** which enters the right atrium.

The superior vena cava sometimes becomes compressed by the growth of a lung cancer, the enlargement of a lymph node, or an aortic aneurysm. This condition may interfere with return of blood from the upper body to the heart. It is characterized by pain, shortness of breath, distension of veins draining into the superior vena cava, and swelling of tissues in the face, head, and arms. In severe cases, compression of the superior vena cava may result in such restricted blood flow to the brain that the person's life is threatened.

Veins from the Arm and Shoulder

The arm is drained by a set of deep veins and a set of superficial ones. The deep veins generally parallel the arteries in each region and are given similar names, such as *radial vein, ulnar vein, brachial vein,* and *axillary*

Figure 18.62
The main veins of the arm and shoulder.

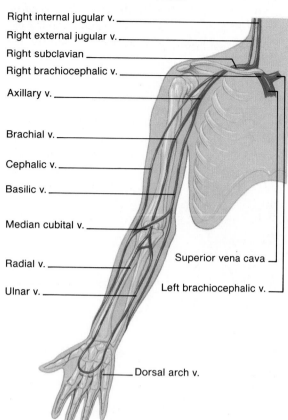

Right internal jugular v.
Right external jugular v.
Right subclavian
Right brachiocephalic v.
Axillary v.
Brachial v.
Cephalic v.
Basilic v.
Median cubital v.
Superior vena cava
Radial v.
Left brachiocephalic v.
Ulnar v.
Dorsal arch v.

vein. The superficial veins are interconnected in complex networks just beneath the skin. They also communicate with the deep vessels of the arm, providing many alternate pathways through which blood can leave the tissues (figure 18.62).

The main vessels of the superficial network are the basilic and cephalic veins. They arise from anastomoses in the hand and wrist on the radial and ulnar sides, respectively.

The **basilic vein** passes along the back of the forearm on the ulnar side for a distance and then curves forward to the anterior surface below the elbow. It continues ascending on the medial side until it reaches the middle of the upper arm. There it penetrates the tissues deeply and joins the *brachial vein.* As the basilic and brachial veins merge, they form the *axillary vein.*

The **cephalic vein** courses upward on the lateral side of the arm from the hand to the shoulder. In the shoulder it pierces the tissues and empties into the axillary vein. Beyond the axilla, the axillary vein becomes the *subclavian vein.*

In the bend of the elbow, a *median cubital vein* ascends from the cephalic vein on the lateral side of the arm to the basilic vein on the medial side. This vein is often used as a site for *venipuncture,* when it is necessary to remove a sample of blood for examination or to add fluids to the blood.

Figure 18.63
Veins that drain the thoracic wall.

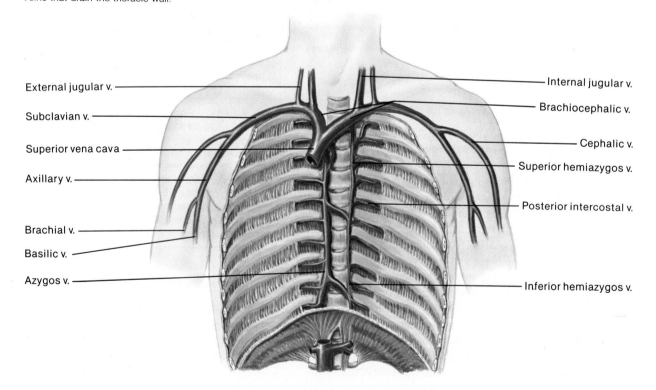

External jugular v.
Subclavian v.
Superior vena cava
Axillary v.
Brachial v.
Basilic v.
Azygos v.

Internal jugular v.
Brachiocephalic v.
Cephalic v.
Superior hemiazygos v.
Posterior intercostal v.
Inferior hemiazygos v.

Veins from the Abdominal and Thoracic Walls

The abdominal and thoracic walls are drained mainly by tributaries of the brachiocephalic and azygos veins. For example, the *brachiocephalic vein* receives blood from the *internal thoracic vein*, which generally drains the tissues supplied by the internal thoracic artery. Some *intercostal veins* also empty into the brachiocephalic vein (figure 18.63).

The **azygos** (az′ĭ-gos) **vein** originates in the dorsal abdominal wall and ascends through the mediastinum on the right side of the vertebral column to join the superior vena cava. It drains most of the muscular tissue in the abdominal and thoracic walls.

Tributaries of the azygos vein include the *posterior intercostal veins,* which drain the intercostal spaces on the right side, and the *superior* and *inferior hemiazygos veins,* which connect to the posterior intercostal veins on the left. Right and left *ascending lumbar veins,* with tributaries that include vessels from the lumbar and sacral regions, also connect to the azygos system.

Veins from the Abdominal Viscera

Although veins usually carry blood directly to the atria of the heart, those that drain the abdominal viscera are exceptions (figure 18.64). They originate in the capillary networks of the stomach, intestines, pancreas, and spleen, and carry blood from these organs through a **portal** (por′tal) **vein** to the liver. There the blood enters capillarylike **hepatic sinusoids** (hĕ-pat′ik si′nŭ-soids).

Figure 18.64
Veins that drain the abdominal viscera.

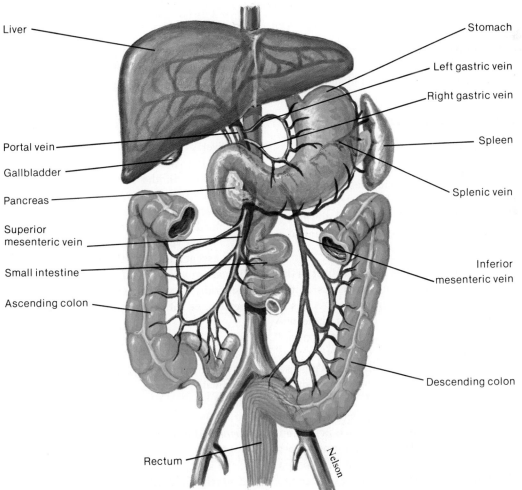

Figure 18.65

In this schematic drawing of the circulatory system, note how the hepatic portal vein drains one set of capillaries and leads to another set.

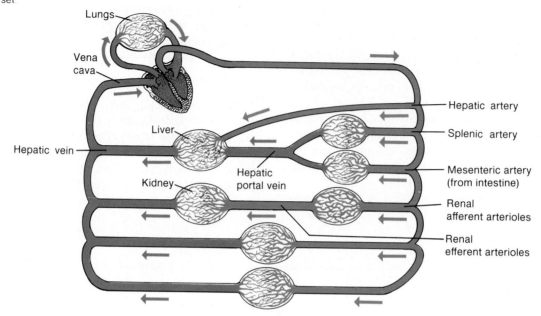

This unique venous pathway is called the **hepatic portal system.** (See figure 18.65.)

The tributaries of the portal vein include the following vessels:

1. Right and left *gastric veins* from the stomach.
2. *Superior mesenteric vein* from the small intestine, ascending colon, and transverse colon.
3. *Splenic vein* from a convergence of several veins draining the spleen, pancreas, and a portion of the stomach. Its largest tributary, the *inferior mesenteric vein,* brings blood upward from the descending colon, sigmoid colon, and rectum.

About 80% of the blood flowing to the liver in the hepatic portal system comes from capillaries in the stomach and intestines and is rich in nutrients. As is discussed in chapter 14, the liver acts on these nutrients in a variety of ways. For example, it monitors the blood glucose concentration and converts excesses into glycogen and fat for storage, or it releases glucose from glycogen if the blood glucose concentration drops below normal.

Similarly, the liver helps to regulate the blood concentrations of recently absorbed amino acids and lipids by modifying their molecules into forms usable by cells, oxidizing them, or changing them into storage forms. The liver also functions to store certain vitamins and to detoxify harmful substances.

The blood in the portal vein nearly always contains bacteria that have entered through intestinal capillaries. These microorganisms are removed by the phagocytic action of large *Kupffer cells* lining the hepatic sinusoids. Consequently, nearly all of the bacteria have been removed from the portal blood before it leaves the liver.

After passing through the hepatic sinusoids of the liver, blood in the hepatic portal system is carried through a series of merging vessels into **hepatic veins.** These veins empty into the *inferior vena cava,* and thus the blood is returned to the general circulation.

Other veins also empty into the inferior vena cava as it ascends through the abdomen. They include the *lumbar, gonadal, renal, suprarenal,* and *phrenic veins.* Generally these vessels drain regions that are supplied by arteries with corresponding names.

Veins from the Leg and Pelvis

As in the arm, veins that drain blood from the leg can be divided into deep and superficial groups (figure 18.66).

The deep veins of the lower leg, such as the *anterior* and *posterior tibial veins,* have names that correspond with the arteries they accompany. At the level of the knee, these vessels form a single trunk, the **popliteal vein.** This vein continues upward through the thigh as the **femoral vein,** which in turn becomes the **external iliac vein** just behind the inguinal ligament.

Figure 18.66
The main veins of the leg and pelvis.

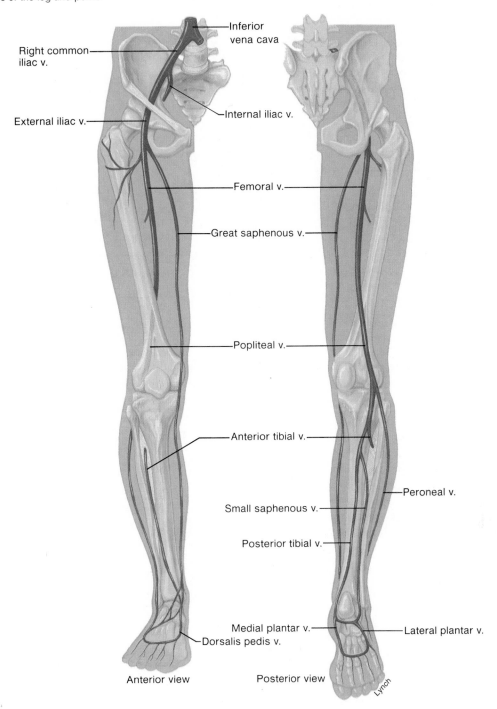

Inferior vena cava

Right common iliac v.

External iliac v.

Internal iliac v.

Femoral v.

Great saphenous v.

Popliteal v.

Anterior tibial v.

Peroneal v.

Small saphenous v.

Posterior tibial v.

Medial plantar v.

Dorsalis pedis v.

Lateral plantar v.

Anterior view

Posterior view

Lynch

Figure 18.67
Major vessels of the venous system.

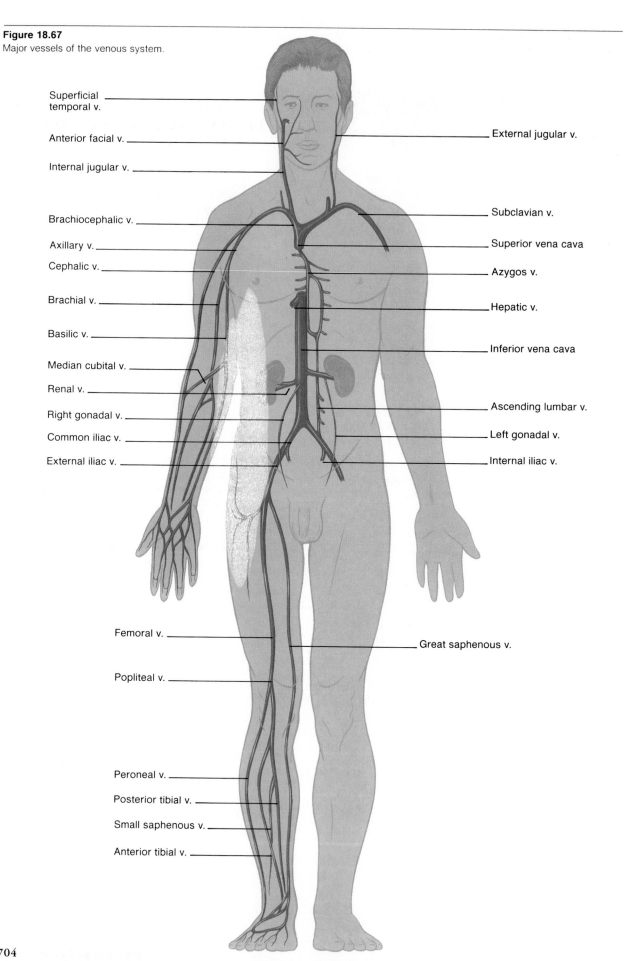

Superficial temporal v.

Anterior facial v.

Internal jugular v.

Brachiocephalic v.

Axillary v.

Cephalic v.

Brachial v.

Basilic v.

Median cubital v.

Renal v.

Right gonadal v.

Common iliac v.

External iliac v.

External jugular v.

Subclavian v.

Superior vena cava

Azygos v.

Hepatic v.

Inferior vena cava

Ascending lumbar v.

Left gonadal v.

Internal iliac v.

Femoral v.

Popliteal v.

Great saphenous v.

Peroneal v.

Posterior tibial v.

Small saphenous v.

Anterior tibial v.

The superficial veins of the foot and leg intercon-nect to form a complex network beneath the skin. These vessels drain into two major trunks: the small and great saphenous veins.

The **small saphenous** (sah-fe′nus) **vein** begins in the lateral portion of the foot and passes upward be-hind the lateral malleolus. It ascends along the back of the calf, enters the popliteal fossa, and joins the *pop-liteal vein.*

The **great saphenous vein**, which is the longest vein in the body, originates on the medial side of the foot. It ascends in front of the medial malleolus and extends upward along the medial side of the leg. In the thigh just below the inguinal ligament, it penetrates deeply and joins the femoral vein. Near its termination, the great saphenous vein receives tributaries from a number of vessels that drain the upper thigh, groin, and lower abdominal wall.

In addition to communicating freely with each other, the saphenous veins communicate extensively with the deep veins of the leg. As a result, there are many pathways by which blood can be returned to the heart from the lower extremities.

Whenever a person stands, the *saphenous veins* are subjected to increased blood pressure because of grav-itational forces. When such pressure is prolonged, these veins are especially prone to developing abnormal dila-tions characteristic of varicose veins. The saphenous veins commonly serve as a source of blood vessel ma-terial for vascular bypass surgery.

In the pelvic region, blood is carried away from organs of the reproductive, urinary, and digestive sys-tems by vessels leading to the **internal iliac vein.** This vein is formed by tributaries corresponding to the branches of the internal iliac artery, such as the *glu-teal, pudendal, vesical, rectal, uterine,* and *vaginal veins.* Typically, these veins have many interconnec-tions and form complex networks (plexuses) in the re-gions of the rectum, urinary bladder, and prostate gland (in the male) or uterus and vagina (in the female).

The internal iliac veins originate deep within the pelvis and ascend to the pelvis brim. There they unite with the right and left *external iliac veins* to form the **common iliac veins.** These vessels in turn merge to pro-duce the *inferior vena cava* at the level of the fifth lumbar vertebra.

Figure 18.67 shows the major vessels of the ve-nous system.

---

1. Name the veins that return blood to the right atrium.
2. What major veins drain blood from the head? The arm? The abdominal viscera? The leg?

Clinical Terms Related to the Cardiovascular System

aneurysm (an′u-rizm)   a saclike swelling in the wall of a blood vessel, usually an artery.

angiocardiography (an″je-o-kar″de-og′rah-fe)   injection of radiopaque solution into the vascular system for X-ray examination of the heart and pulmonary circuit.

angiospasm (an′je-o-spazm″)   a muscular spasm in the wall of a blood vessel.

arteriography (ar″te-re-og′rah-fe)   injection of radiopaque solution into the vascular system for X-ray examination of arteries.

asystole (a-sis′to-le)   condition in which the myocardium fails to contract.

cardiac tamponade (kar′de-ak tam″po-nād′)   compression of the heart by an accumulation of fluid within the pericardial cavity.

congestive heart failure (kon-jes′tiv hart fāl′yer)   condition in which the heart is unable to pump an adequate amount of blood to the body cells.

cor pulmonale (kor pul-mo-na′le)   a heart-lung disorder characterized by pulmonary hypertension and hypertrophy of the right ventricle.

embolectomy (em″bo-lek′to-me)   removal of an embolus through an incision in a blood vessel.

endarterectomy (en″dar-ter-ek′to-me)   removal of the inner wall of an artery to reduce an arterial occlusion.

palpitation (pal″pĭ-ta′shun)   an awareness of a heartbeat that is unusually rapid, strong, or irregular.

pericardiectomy (per″ĭ-kar″de-ek′to-me)   an excision of the pericardium.

phlebitis (flĕ-bi′tis)   inflammation of a vein, usually in the legs.

phlebosclerosis (fleb″o-sklĕ-ro′sis)   abnormal thickening or hardening of the walls of veins.

phlebotomy (flĕ-bot′o-me)   incision of a vein for the purpose of withdrawing blood.

sinus rhythm (si′nus rithm)   normal cardiac rhythm regulated by the S-A node.

thrombophlebitis (throm″bo-flĕ-bi′tis)   formation of a blood clot in a vein in response to inflammation of the venous wall.

valvotomy (val-vot′o-me)   an incision of a valve.

venography (ve-nog′rah-fe)   injection of radiopaque solution into the vascular system for X-ray examination of veins.

Chapter Summary

Introduction

The cardiovascular system is vital for providing oxygen and nutrients to tissues and for removing wastes.

Structure of the Heart

1. Size and location of the heart
   a. The heart is about 14 cm long and 9 cm wide.
   b. It is located within the mediastinum and rests on the diaphragm.
2. Coverings of the heart
   a. The heart is enclosed in a pericardium.
   b. The pericardial cavity is a potential space between the visceral and parietal membranes.
3. Wall of the heart
   a. The wall of the heart is composed of three layers.
   b. These layers include epicardium, myocardium, and endocardium.
4. Heart chambers and valves
   a. The heart is divided into four chambers—two atria and two ventricles—that communicate through atrioventricular orifices on each side.
   b. Right chambers and valves
      (1) The right atrium receives blood from the venae cavae and coronary sinus.
      (2) The right atrioventricular orifice is guarded by the tricuspid valve.
      (3) The right ventricle pumps blood into the pulmonary trunk.
      (4) The base of the pulmonary trunk is guarded by a pulmonary semilunar valve.
   c. Left chambers and valves
      (1) The left atrium receives blood from the pulmonary veins.
      (2) The left atrioventricular orifice is guarded by the bicuspid valve.
      (3) The left ventricle pumps blood into the aorta.
      (4) The base of the aorta is guarded by an aortic semilunar valve.
5. Skeleton of the heart
   a. The skeleton of the heart consists of fibrous rings that enclose bases of the pulmonary artery, aorta, and atrioventricular orifices.
   b. The fibrous rings provide attachments for valves and muscle fibers and prevent the orifices from dilating excessively during ventricular contractions.
6. Path of blood through the heart
   a. Blood that is relatively low in oxygen concentration and high in carbon dioxide concentration enters the right side of the heart from the venae cavae and is pumped into the pulmonary circulation.
   b. After blood is oxygenated in the lungs and some of its carbon dioxide is removed, it returns to the left side of the heart through the pulmonary veins.
   c. From the left ventricle, it moves into the aorta.
7. Blood supply to the heart
   a. Blood is supplied through the coronary arteries.
   b. It is returned to the right atrium through the cardiac veins and coronary sinus.

Actions of the Heart

1. The cardiac cycle
   a. Atria contract while ventricles relax; ventricles contract while atria relax.
   b. Pressure within the chambers rises and falls in repeated cycles.
2. Heart sounds
   a. Heart sounds can be described as *lub*-dup.
   b. Heart sounds are due to vibrations produced as a result of blood and valve movements.
   c. The first part of the sound occurs as A-V valves are closing and the second part is associated with the closing of semilunar valves.
3. Cardiac muscle fibers
   a. Fibers are interconnected to form a functional syncytium.
   b. If any part of the syncytium is stimulated, the whole structure contracts as a unit.
   c. Except for a small region in the floor of the right atrium, the atrial syncytium is separated from the ventricular syncytium by the fibrous skeleton.
4. Cardiac conduction system
   a. This system is composed of specialized muscle tissue and functions to initiate and conduct depolarization waves through the myocardium.
   b. Impulses from the S-A node pass slowly to the A-V node; impulses travel rapidly along the A-V bundle and Purkinje fibers.
   c. Muscle fibers in the ventricular walls are arranged in whorls that squeeze blood out of the ventricles when they contract.
5. Regulation of the cardiac cycle
   a. Heartbeat is affected by physical exercise, body temperature, and concentration of various ions.
   b. S-A and A-V nodes are innervated by branches of sympathetic and parasympathetic nerve fibers.
      (1) Parasympathetic impulses cause heart action to decrease; sympathetic impulses cause heart action to increase.
      (2) Autonomic impulses are regulated by the cardiac center in the medulla oblongata.

Blood Vessels

The blood vessels form a closed circuit of tubes that transport blood between the heart and body cells.

Tubes include arteries, arterioles, capillaries, venules, and veins.

1. Arteries and arterioles
    a. Arteries are adapted to carry high pressure blood away from the heart.
    b. Arterioles are branches of arteries.
    c. Walls of these vessels consist of layers of endothelium, smooth muscle, and connective tissue.
    d. The smooth muscles are innervated by autonomic fibers that can stimulate vasoconstriction or vasodilation.
2. Capillaries
    Capillaries form connections between arterioles and venules.
    The capillary wall consists of a single layer of cells that forms a semipermeable membrane.
    a. Capillary permeability
        (1) The openings in the capillary walls are thin slits between adjacent endothelial cells.
        (2) The sizes of the openings vary from tissue to tissue.
        (3) Endothelial cells of brain capillaries are tightly fused, forming a blood-brain barrier.
    b. Arrangement of capillaries
        Capillary density varies directly with tissue metabolic rates.
    c. Regulation of capillary blood flow
        (1) Capillary blood flow is regulated by precapillary sphincters.
        (2) Precapillary sphincters open when cells are low in oxygen and nutrients, and close when cellular needs are met.
3. Exchanges in capillaries
    a. Gases, nutrients, and metabolic by-products are exchanged between the capillary blood and tissue fluid.
    b. Diffusion provides the most important means of transport.
    c. Diffusion pathways depend on lipid solubilities.
    d. Plasma proteins generally remain in the blood.
    e. Filtration, which is due to the hydrostatic pressure of blood, causes a net outward movement of fluid at the arterial end of a capillary.
    f. Osmosis causes a net inward movement of fluid at the venule end of the capillary.
    g. Some factors cause fluids to accumulate excessively in the tissues.
4. Venules and veins
    a. Venules continue from capillaries and merge to form veins.
    b. Veins carry blood to the heart.
    c. Venous walls are similar to arterial walls, but are thinner and contain less muscle and elastic tissue.

Blood Pressure

Blood pressure is the force exerted by blood against the inside of the blood vessels.

1. Arterial blood pressure
    a. Arterial blood pressure is created primarily by heart action; it rises and falls with phases of the cardiac cycle.
    b. Systolic pressure occurs when the ventricle contracts; diastolic pressure occurs when the ventricle relaxes.
2. Factors that influence arterial blood pressure
    a. Heart action, blood volume, resistance to flow, and blood viscosity influence arterial blood pressure.
    b. Arterial pressure increases as cardiac output, blood volume, peripheral resistance, or blood viscosity increases.
3. Control of blood pressure
    a. Blood pressure is controlled in part by the mechanisms that regulate cardiac output and peripheral resistance.
    b. Cardiac output depends on the volume of blood discharged from the ventricle with each beat and the rate of heartbeat.
        (1) The more blood that enters the heart, the stronger the ventricular contraction, the greater the stroke volume, and the greater the cardiac output.
        (2) Heart rate is regulated by the cardiac center of the medulla oblongata.
    c. Regulation of peripheral resistance involves changes in the diameter of arterioles, which is controlled by the vasomotor center of the medulla oblongata.
4. Venous blood flow
    a. Venous blood flow is not a direct result of heart action; it depends on skeletal muscle contraction, breathing movements, and venoconstriction.
    b. Many veins contain flaplike valves that prevent blood from backing up.
    c. Venous constriction can increase venous pressure and blood flow.
5. Central venous pressure
    a. Central venous pressure is the pressure in the right atrium.
    b. It is influenced by factors that alter inflow of blood to the right atrium.
    c. It affects pressure within the peripheral veins.

Paths of Circulation

1. The pulmonary circuit
    a. The pulmonary circuit is composed of vessels that carry blood from the right ventricle to the lungs, pulmonary capillaries, and vessels that lead back to the left atrium.
    b. Pulmonary capillaries contain lower pressure than those of the systemic circuit.

c. Tightly joined epithelial cells of alveoli walls prevent most substances from entering the alveoli.

d. Water is rapidly drawn out of alveoli into the interstitial fluid by osmotic pressure, so alveoli remain dry.

2. The systemic circuit
   a. The systemic circuit is composed of vessels that lead from the heart to the body cells and back to the heart.
   b. It includes the aorta and its branches.

The Arterial System

1. Principal branches of the aorta
   a. Branches of the ascending aorta include the right and left coronary arteries.
   b. Branches of the aortic arch include the brachiocephalic, left common carotid, and left subclavian arteries.
   c. Branches of the descending aorta include thoracic and abdominal groups.
   d. The descending aorta terminates by dividing into right and left common iliac arteries.

2. Arteries to the neck, head, and brain
   These include branches of the subclavian and common carotid arteries.

3. Arteries to the shoulder and arm
   a. The subclavian artery passes into the upper arm and in various regions is called the axillary and brachial artery.
   b. Branches of the brachial artery include the ulnar and radial arteries.

4. Arteries of the thoracic and abdominal walls
   a. The thoracic wall is supplied by branches of the subclavian artery and thoracic aorta.
   b. The abdominal wall is supplied by branches of the abdominal aorta and other arteries.

5. Arteries of the pelvis and leg
   The common iliac artery supplies the pelvic organs, gluteal region, and leg.

The Venous System

1. Characteristics of venous pathways
   a. Veins are responsible for returning blood to the heart.
   b. Larger veins usually parallel the courses of major arteries.

2. Veins from the head, neck, and brain
   a. These regions are drained by the jugular veins.
   b. Jugular veins unite with subclavian veins to form the brachiocephalic veins.

3. Veins from the arm and shoulder
   a. The arm is drained by sets of superficial and deep veins.
   b. Major superficial veins are the basilic and cephalic veins.
   c. The median cubital vein in the bend of the elbow is often used as a site for venipuncture.

4. Veins from the abdominal and thoracic walls
   These are drained by tributaries of the brachiocephalic and azygos veins.

5. Veins from the abdominal viscera
   a. Blood from the abdominal viscera generally enters the hepatic portal system and is carried to the liver.
   b. Blood in the portal system is rich in nutrients.
   c. The liver functions to regulate the blood concentrations of glucose, amino acids, and lipids.
   d. Phagocytic cells in the liver remove bacteria from the portal blood.
   e. From the liver, the blood is carried by hepatic veins to the inferior vena cava.

6. Veins from the leg and pelvis
   a. These are drained by sets of deep and superficial veins.
   b. Deep veins include the tibial veins, and superficial veins include the saphenous veins.

Application of Knowledge

1. Based on your understanding of the way capillary blood flow is regulated, do you think it is wiser to rest or to exercise following a heavy meal? Give a reason for your answer.

2. If a patient develops a blood clot in the femoral vein of the left leg and a portion of the clot breaks loose, where is the blood flow likely to carry the embolus? What symptoms is this condition likely to produce?

3. When a person strains to lift a heavy object, the intrathoracic pressure is increased. What do you think will happen to the rate of venous blood returning to the heart during such lifting? Why?

Review Activities

1. Describe the general structure, function, and location of the heart.
2. Distinguish between the visceral pericardium and the parietal pericardium.
3. Compare the layers of the cardiac wall.
4. Identify and describe the location of the chambers and the valves of the heart.
5. Describe the skeleton of the heart and explain its function.
6. Trace the path of blood through the heart.
7. Trace the path of blood through the coronary circulation.
8. Describe a cardiac cycle.
9. Describe the pressure changes that occur in the atria and ventricles during a cardiac cycle.
10. Explain the origin of heart sounds.
11. Describe the arrangement of the cardiac muscle fibers.
12. Distinguish between the S-A node and A-V node.
13. Explain how the cardiac conduction system functions in the control of the cardiac cycle.
14. Discuss how the nervous system functions in the regulation of the cardiac cycle.
15. Describe two factors other than the nervous system that affect the cardiac cycle.
16. Distinguish between an artery and an arteriole.
17. Explain how vasoconstriction and vasodilation are controlled.
18. Describe the structure and function of a capillary.
19. Describe the function of the blood-brain barrier.
20. Explain how the blood flow through a capillary is controlled.
21. Explain how diffusion functions in the exchange of substances between plasma and tissue fluid.
22. Explain why water and dissolved substances leave the arteriole end of a capillary and enter the venule end.
23. Describe the effect of histamine on a capillary.
24. Distinguish between a venule and a vein.
25. Explain how veins function as blood reservoirs.
26. Distinguish between systolic and diastolic blood pressures.
27. Name several factors that influence blood pressure and explain how each produces its effect.
28. Describe how blood pressure is controlled.
29. List the major factors that promote the flow of venous blood.
30. Define *central venous pressure*.
31. Distinguish between the pulmonary and systemic circuits of the cardiovascular system.
32. Trace the path of blood through the pulmonary circuit.
33. Explain why the alveoli normally remain dry.
34. Describe the aorta and name its principal branches.
35. On a diagram, locate and identify the major arteries that supply the abdominal visceral organs.
36. On a diagram, locate and identify the major arteries that supply parts in the head, neck, and brain.
37. On a diagram, locate and identify the major arteries that supply parts in the shoulder and arm.
38. On a diagram, locate and identify the major arteries that supply parts in the thoracic and abdominal walls.
39. On a diagram, locate and identify the major arteries that supply parts in the pelvis and leg.
40. Describe the relationship between the major venous pathways and the major arterial pathways.
41. On a diagram, locate and identify the major veins that drain parts in the head, neck, and brain.
42. On a diagram, locate and identify the major veins that drain parts in the arm and shoulder.
43. On a diagram, locate and identify the major veins that drain parts in the abdominal and thoracic walls.
44. On a diagram, locate and identify the major veins that drain parts of the abdominal viscera.
45. Review the actions of the liver on nutrients carried in the portal veins.
46. On a diagram, locate and identify the major veins that drain parts of the leg and pelvis.

19

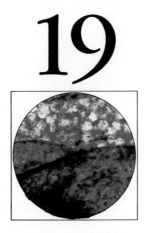

The Lymphatic System

When substances are exchanged between the blood and tissue fluid, more fluid leaves the blood capillaries than returns to them. If the fluid remaining in the interstitial spaces were allowed to accumulate, the hydrostatic pressure in tissues would increase. The *lymphatic system* helps to prevent such an imbalance by providing pathways through which tissue fluid can be transported as lymph from the interstitial spaces to veins, where it becomes part of the blood.

The lymphatic system also helps to defend the tissues against infections by filtering particles from the lymph and by supporting the activities of lymphocytes that furnish immunity against specific disease-causing agents.

Chapter Outline

Chapter Objectives

After you have studied this chapter, you should be able to

1. Describe the general functions of the lymphatic system.

2. Describe the location of the major lymphatic pathways.

3. Describe how tissue fluid and lymph are formed and explain the function of lymph.

4. Explain how lymphatic circulation is maintained and the consequence of lymphatic obstruction.

5. Describe a lymph node and its major functions.

6. Describe the location of the major chains of lymph nodes.

7. Discuss the functions of the thymus and spleen.

8. Distinguish between specific and nonspecific body defenses and provide examples of each defense.

9. Explain how two major types of lymphocytes are formed and how they function in immune mechanisms.

10. Name the major types of immunoglobulins and discuss their origins and actions.

11. Distinguish between primary and secondary immune responses.

12. Distinguish between active and passive immunity.

13. Explain how allergic reactions and tissue rejection reactions are related to immune mechanisms.

14. Complete the review activities at the end of this chapter. Note that these items are worded in the form of specific learning objectives. You may want to refer to them before reading the chapter.

Key Terms

allergen (al′er-jen)

antibody (an′ti-bod′′e)

antigen (an′ti-jen)

clone (klōn)

complement (kom′plĕ-ment)

hapten (hap-ten)

immunity (ĭ-mu′nĭ-te)

immunoglobulin (im′′u-no-glob′u-lin)

interferon (in′′ter-fēr′on)

lymph (limf)

lymphatic pathway (lim-fat′ik path′wa)

lymph node (limf nōd)

lymphocyte (lim′fo-sīt)

macrophage (mak′ro-fāj)

pathogen (path′o-jen)

reticuloendothelial tissue
(rĕ-tik′′u-lo-en′′do-the′le-al tish′u)

spleen (splēn)

thymus (thi′mus)

vaccine (vak′sēn)

Aids to Understanding Words

auto-, self: *auto*immune disease—condition in which the immune system attacks the body's own tissues.

gen-, to be produced: aller*gen*—substance that stimulates an allergic response.

humor-, fluid: *humor*al immunity—immunity resulting from soluble antibodies in body fluids.

immun-, free: *immun*ity—resistance to (freedom from) a specific disease.

inflamm-, setting on fire: *inflamm*ation—a condition characterized by localized redness, heat, swelling, and pain in the tissues.

nod-, knot: *nod*ule—a small mass of lymphocytes surrounded by connective tissue.

patho-, disease: *patho*gen—a disease-causing agent.

THE LYMPHATIC SYSTEM is closely associated with the *cardiovascular system,* since it includes a network of vessels that assist in circulating body fluids. These vessels transport excess fluid away from the interstitial spaces and return it to the bloodstream. The organs of the lymphatic system also help to defend the body against invasion by disease-causing agents (figure 19.1).

Lymphatic Pathways

The lymphatic pathways begin as lymphatic capillaries. These tiny tubes merge to form larger lymphatic vessels, and they, in turn, lead to collecting ducts that unite with veins in the thorax.

Lymphatic Capillaries

Lymphatic capillaries are microscopic, closed-ended tubes. They extend into the interstitial spaces of most tissues, forming complex networks that parallel the networks of blood capillaries (figure 19.2). The walls of the lymphatic capillaries, like those of blood capillaries, consist of a single layer of squamous epithelial cells. This thin wall makes it possible for tissue fluid from the interstitial space to enter the lymphatic capillary. Once the fluid is inside the capillary, it is called **lymph.**

The villi of the small intestine contain specialized lymphatic capillaries called *lacteals,* described in chapter 14. These vessels are responsible for transporting recently absorbed fats away from the digestive tract.

Figure 19.1

Lymphatic vessels transport fluid from interstitial spaces to the bloodstream.

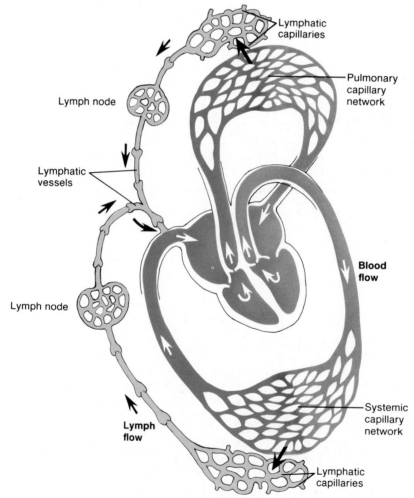

Lymphatic Vessels

Lymphatic vessels, which are formed by the merging of lymphatic capillaries, have walls similar to those of veins. That is, their walls are composed of three layers: an endothelial lining, a middle layer of smooth muscle and elastic fibers, and an outer layer of connective tissue. Also, like veins, the lymphatic vessels have flaplike *valves* that help to prevent the backflow of lymph. Figure 19.3 shows one of these valves.

Typically, lymphatic vessels lead to specialized organs called **lymph nodes,** and after leaving these structures, the vessels merge to form still larger lymphatic trunks.

Lymphatic Trunks and Collecting Ducts

The **lymphatic trunks** drain lymph from relatively large regions of the body, and they are named for the regions they serve. For example, the *lumbar trunk* drains lymph

Figure 19.2
Lymph capillaries are microscopic closed-ended tubes that begin in the interstitial spaces of most tissues.

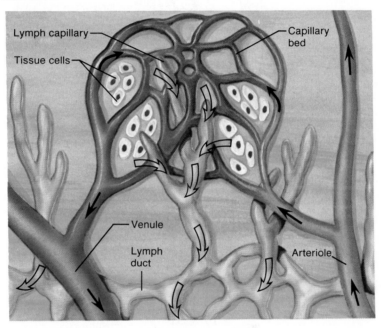

Figure 19.3
A light micrograph of the flaplike valve (arrow) within a lymphatic vessel. What is the function of this valve?

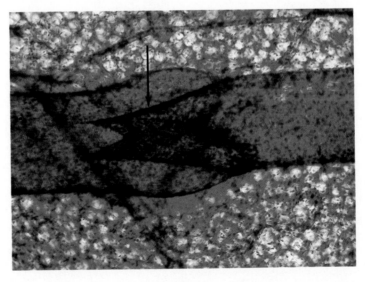

Figure 19.4
Lymphatic vessels merge into larger lymphatic trunks, which in turn drain into collecting ducts.

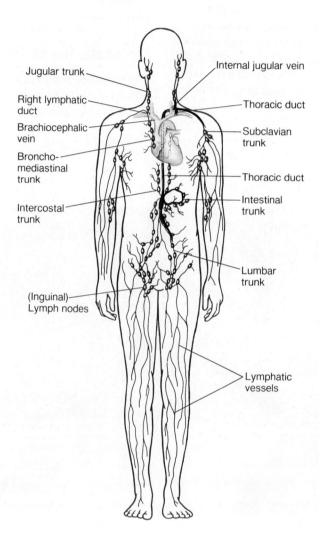

Jugular trunk

Right lymphatic duct

Brachiocephalic vein

Broncho-mediastinal trunk

Intercostal trunk

(Inguinal) Lymph nodes

Internal jugular vein

Thoracic duct

Subclavian trunk

Thoracic duct

Intestinal trunk

Lumbar trunk

Lymphatic vessels

Figure 19.5
A lymphangiogram (X-ray film) of the lymphatic vessels and
lymph nodes of the pelvic region.

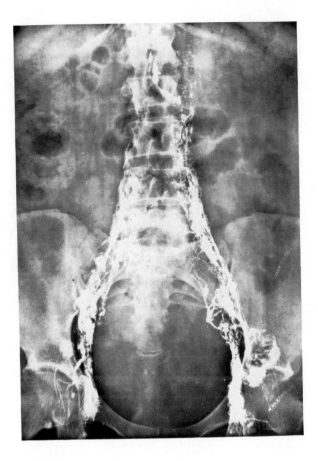

from the legs, lower abdominal wall, and the pelvic organs; the *intestinal trunk* drains organs of the abdominal viscera; the *intercostal* and *bronchomediastinal trunks* receive lymph from portions of the thorax; the *subclavian trunk* drains the arm; and the *jugular trunk* drains portions of the neck and head. These lymphatic trunks then join one of two **collecting ducts**—the thoracic duct or the right lymphatic duct. Figure 19.4 shows the location of the major lymphatic trunks and collecting ducts, and figure 19.5 shows a lymphangiogram or X-ray film of some lymphatic vessels and lymph nodes.

The **thoracic duct** is the larger and longer of the collecting ducts. It begins in the abdomen, passes upward through the diaphragm beside the aorta, ascends in front of the vertebral column through the mediastinum, and empties into the left subclavian vein near the junction of the left jugular vein. This duct drains lymph from the intestinal, lumbar, and intercostal trunks, as well as from the left subclavian, left jugular, and left bronchomediastinal trunks.

The **right lymphatic duct** originates in the right thorax by the union of the right jugular, right subclavian, and right bronchomediastinal trunks. It empties into the right subclavian vein near the junction of the right jugular vein.

After leaving the two collecting ducts, lymph enters the venous system and becomes part of the plasma just before the blood returns to the right atrium.

To summarize, lymph from the lower body regions, left arm, and left side of the head and neck enters the thoracic duct; lymph from the right side of the

Figure 19.6

(a) The right lymphatic duct drains lymph from the upper right side of the body, while the thoracic duct drains lymph from the rest of the body; *(b)* lymph drainage of the right breast.

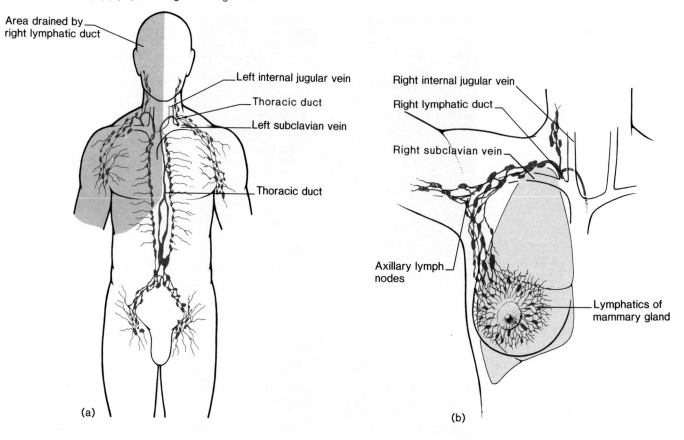

(a)

(b)

head and neck, right arm, and right thorax enters the right lymphatic duct (figure 19.6). Chart 19.1 traces a typical lymphatic pathway.

The skin is richly supplied with lymphatic capillaries. Consequently, if the skin is broken or something is injected into the skin, such as venom from a stinging insect, foreign particles are likely to enter the lymphatic system relatively rapidly.

1. What is the general function of the lymphatic system?
2. Through what lymphatic vessels would lymph pass in traveling from a leg to the bloodstream?

Chart 19.1 Typical lymphatic pathway

Tissue fluid
  leaves the interstitial space and becomes

Lymph
  as it enters the

Lymphatic capillary
  that merges with other capillaries to form the

Afferent lymphatic vessel
  that enters the

Lymph node
  where lymph is filtered and leaves via the

Efferent lymphatic vessel
  that merges with other vessels to form the

Lymphatic trunk
  that merges with other trunks and joins the

Collecting duct
  that empties into the

Subclavian vein
  where lymph is added to the blood.

Tissue Fluid and Lymph

Lymph is essentially tissue fluid that has entered a lymphatic capillary. The formation of lymph is, then, closely associated with the formation of tissue fluid.

Tissue Fluid Formation

As explained in chapter 17, *tissue fluid* originates from blood plasma. This fluid is composed of water and dissolved substances that leave the blood capillaries as a result of diffusion and filtration.

Although tissue fluid contains various nutrients and gases found in plasma, it generally lacks proteins of large molecular size. Some proteins with smaller molecules do leak out of the blood capillaries and enter the interstitial space. Usually, these proteins are not reabsorbed when water and other dissolved substances move back into the venule ends of these capillaries by diffusion and osmosis. As a result, the protein concentration of the tissue fluid tends to rise, causing the *osmotic pressure* of the fluid to rise also.

Lymph Formation

As the osmotic pressure of the tissue fluid rises, it interferes with the osmotic reabsorption of water by the blood capillaries. The volume of fluid in the interstitial spaces then tends to increase, as does the pressure within the spaces.

This increasing interstitial pressure is responsible for forcing some of the tissue fluid into the lymphatic capillaries, where it becomes lymph (figure 19.2).

Function of Lymph

Most of the protein molecules that leak out of the blood capillaries are carried away by lymph and are returned to the bloodstream. At the same time, lymph transports various foreign particles, such as bacterial cells or viruses that may have entered the tissue fluids, to lymph nodes.

Although these proteins and foreign particles cannot easily enter blood capillaries, the lymphatic capillaries are especially adapted to receive them. Specifically, the endothelial cells that form the walls of these vessels are arranged so that the edge of one cell overlaps the edge of an adjacent cell, but is not attached to it. This arrangement, shown in figure 19.7, creates flaplike valves in the lymphatic capillary wall, which are pushed inward when the pressure is greater on the outside of the capillary, but close when the pressure is greater on the inside.

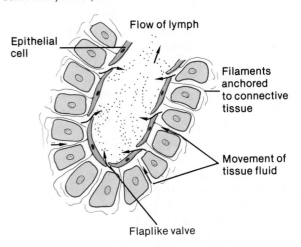

Figure 19.7
Tissue fluid enters lymphatic capillaries through flaplike valves between adjacent epithelial cells.

The epithelial cells of the lymphatic capillary wall are also attached to surrounding connective tissue cells by thin filaments, so that the lumen of a lymphatic capillary remains open even when the outside pressure is increased.

---

1. How would you explain the relationship between tissue fluid and lymph?
2. How does the presence of protein in tissue fluid affect the formation of lymph?
3. What are the major functions of lymph?

Movement of Lymph

Although the entrance of lymph into the lymphatic capillaries is influenced by the *osmotic pressure* of tissue fluid, the movement of lymph through the lymphatic vessels is controlled largely by *muscular activity*.

Flow of Lymph

Lymph, like venous blood, is under relatively low pressure and may not flow readily through the lymphatic vessels without the aid of outside forces. These forces include contraction of skeletal muscles, pressure changes due to the action of breathing muscles, and contraction of smooth muscles in the walls of larger lymphatic vessels.

As *skeletal muscles* contract, they compress lymphatic vessels. This squeezing action causes the lymph inside a vessel to move, but since lymphatic vessels contain valves that prevent backflow, the lymph can only move toward a collecting duct. Similarly, the smooth muscles in the walls of larger lymphatic vessels may contract and compress the lymph inside. This action also helps to force the fluid onward.

Breathing muscles aid the circulation of lymph (as they do that of venous blood) by creating relatively low pressure in the thorax during inhalation. At the same time, the pressure in the abdominal cavity is increased by the contracting diaphragm. Consequently, lymph (and venous blood as well) is squeezed out of the abdominal vessels and into the thoracic vessels. Once again, backflow of lymph (and blood) is prevented by valves within the lymphatic (and blood) vessels.

Since the actions of skeletal and breathing muscles promote the circulation of lymph, it is not surprising that the flow of lymph is greatest during periods of physical exercise. There also is an increase in lymph formation during exercise because of elevated perfusion of fluid from the blood into the interstitial spaces at such times.

Obstruction of Lymph Movement

Because of the continuous movement of fluid from interstitial spaces into blood capillaries and lymphatic capillaries, the volume of fluid in these spaces remains stable. Conditions sometimes occur, however, that interfere with lymph movement, and tissue fluids accumulate in the spaces, causing *edema*.

For example, lymphatic vessels may be obstructed as a result of surgical procedures in which portions of the lymphatic system are removed. Since the affected pathways no longer can drain lymph from the tissues, proteins tend to accumulate in the interstitial spaces. This causes an increase in the osmotic pressure of the tissue fluid, which in turn promotes the accumulation of water within the tissues.

Since lymphatic vessels tend to carry particles away from tissues, these vessels may also transport cancer cells and promote their spread to other sites (metastasis). For this reason, the lymphatic tissues (lymph nodes) in the axillary regions are commonly biopsied during the surgical removal of cancerous breast tissue (mastectomy). If the lymphatic tissue is removed during the procedure, lymphatic drainage from the arm and other nearby tissues may be obstructed, and these parts may become edematous following surgery.

1. What factors promote the flow of lymph?
2. What is the consequence of lymphatic obstruction?

Lymph Nodes

Lymph nodes (lymph glands) are structures located along the lymphatic pathways. They contain large numbers of *lymphocytes* and *macrophages* that are vital in the defense against invasion by microorganisms.

Structure of a Lymph Node

Lymph nodes vary in size and shape; however, they are usually less than 2.5 cm in length and are somewhat bean shaped. A section of a typical lymph node is illustrated in figure 19.8.

The indented region of a bean-shaped node is called the **hilum,** and it is the portion through which blood vessels and nerves enter the structure. The lymphatic vessels leading to a node (afferent vessels) enter separately at various points on its convex surface, while the lymphatic vessels leaving the node (efferent vessels) exit from the hilum.

Each lymph node is enclosed by a capsule of white fibrous connective tissue. This tissue also extends into the node and partially divides it into compartments that contain dense masses of lymphocytes and macrophages. These masses, called **nodules,** represent the structural units of the node.

Spaces within the node, called **lymph sinuses,** provide a complex network of chambers and channels through which lymph circulates as it passes through the node. Lymph enters a lymph node through an *afferent* lymphatic vessel, moves slowly through the lymph sinuses, and leaves through an *efferent* lymphatic vessel (figure 19.9).

Sometimes lymphatic vessels become inflamed due to a bacterial infection. When this happens in superficial vessels, painful reddish streaks may appear beneath the skin—for example, in an arm or leg. This condition is called *lymphangitis,* and it is usually followed by *lymphadenitis,* an inflammation of the lymph nodes. Affected nodes may become greatly enlarged and quite painful.

Nodules also occur singly or in groups associated with the mucous membranes of the respiratory and digestive tracts. The *tonsils,* described in chapter 14, are composed of partially encapsulated lymph nodules. Also, aggregations of nodules, *Peyer's patches,* are scattered throughout the mucosal lining of the ileum of the small intestine.

Figure 19.8

A section of a lymph node. What factors promote the flow of lymph through a node?

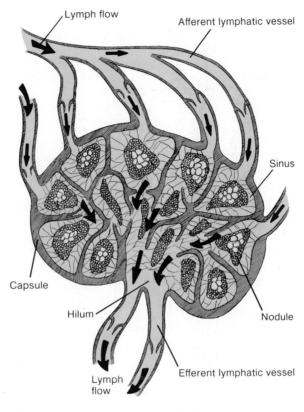

Figure 19.9

Lymph enters and leaves a lymph node through lymphatic vessels.

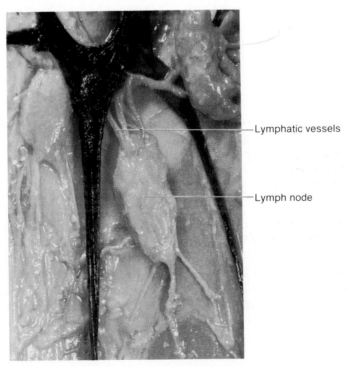

Figure 19.10
Major locations of lymph nodes.

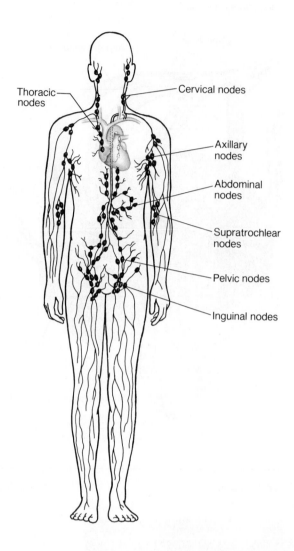

Thoracic nodes

Cervical nodes

Axillary nodes

Abdominal nodes

Supratrochlear nodes

Pelvic nodes

Inguinal nodes

Locations of Lymph Nodes

Lymph nodes generally occur in groups or chains along the larger lymphatic vessels. Although they are widely distributed throughout the body, they are lacking in the tissues of the central nervous system.

The major locations of lymph nodes, shown in figure 19.10, are as follows:

1. **Cervical region.** Nodes in the cervical region occur along the lower border of the mandible, in front of and behind the ears, and deep within the neck along the paths of the larger blood vessels. These nodes are associated with the lymphatic vessels that drain the skin of the scalp and face, as well as the tissues of the nasal cavity and pharynx.

2. **Axillary region.** In the underarm region, nodes receive lymph from vessels that drain the arm, the wall of the thorax, the mammary gland (breast), and the upper wall of the abdomen.

3. **Inguinal region.** Nodes in the inguinal region receive lymph from the legs, the external genitalia, and the lower abdominal wall.

4. **Pelvic cavity.** Within the pelvic cavity, nodes occur primarily along the paths of the iliac blood vessels. They receive lymph from the lymphatic vessels of the pelvic viscera.

5. **Abdominal cavity.** Within this cavity, nodes occur in chains along the main branches of the mesenteric arteries and the abdominal aorta. These nodes receive lymph from the abdominal viscera.

Figure 19.11
The thymus gland is a bilobed organ located between the lungs
and above the heart.

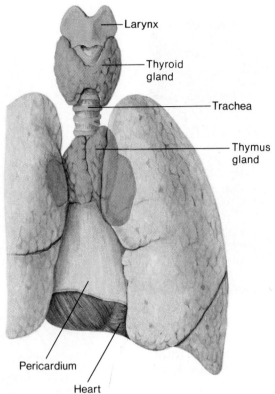

Larynx

Thyroid
gland

Trachea

Thymus
gland

Pericardium

Heart

6. **Thoracic cavity.** Nodes of the thoracic cavity occur within the mediastinum and along the trachea and bronchi. They receive lymph from the thoracic viscera and from the internal wall of the thorax.

The supratrochlear lymph nodes (cubital lymph nodes), which are located superficially on the medial side of the elbow, often become enlarged in children as a result of infections associated with many cuts and scrapes on the hands.

Functions of Lymph Nodes

As mentioned, lymph nodes contain large numbers of *lymphocytes.* Nodes, in fact, are centers for lymphocyte production, although such cells are produced in other tissues as well. Lymphocytes act against foreign particles, such as bacterial cells and viruses, that are carried to the lymph nodes by the lymphatic vessels.

The nodes also contain *macrophages* that engulf and destroy foreign substances, damaged cells, and cellular debris.

Thus, the primary functions of lymph nodes are the production of lymphocytes, which help to defend the body against microorganisms, and the filtration of

potentially harmful foreign particles and cellular debris from lymph before it is returned to the bloodstream.

---

1. How would you distinguish between a lymph node and a lymph nodule?
2. In what body regions are lymph nodes most abundant?
3. What are the major functions of lymph nodes?

Thymus and Spleen

Two other lymphatic organs whose functions are closely related to those of the lymph nodes are the thymus and the spleen.

The Thymus

The **thymus,** shown in figure 19.11, is a soft, bilobed structure whose lobes are surrounded by connective tissue. It is located within the mediastinum, in front of the aortic arch and behind the upper part of the sternum, extending from the root of the neck to the pericardium. Although the thymus varies in size from person to person, it is usually relatively large during

Figure 19.12
A cross section of the thymus gland. Note how the organ is subdivided into lobules.

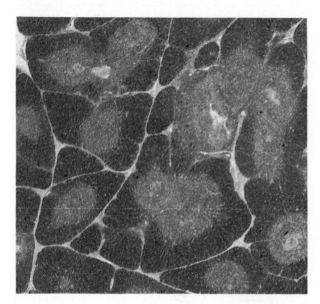

Figure 19.13
The spleen resembles a large lymph node.

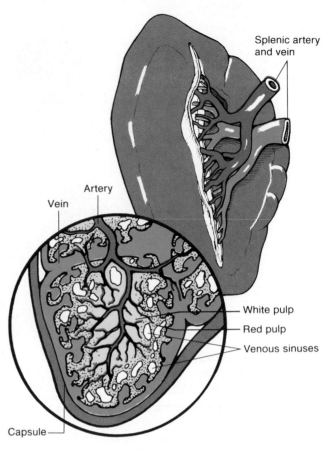

infancy and early childhood. After puberty it tends to decrease in size, and in an adult it may be quite small. In an elderly person, the thymus is often largely replaced by fat and connective tissue.

This organ is composed of lymphatic tissue that is subdivided into *lobules* by connective tissues extending inward from its surface (figure 19.12). The lobules contain large numbers of lymphocytes that developed from precursor cells originating in the bone marrow. The majority of these cells (thymocytes) remain inactive; however, some of them develop into a group (T-lymphocytes) that leaves the thymus and functions in immunity.

Epithelial cells within the thymus may secrete a protein hormone called *thymosin,* which is thought to stimulate the maturation of the T-lymphocytes after they leave the thymus and migrate to other lymphatic tissues.

The Spleen

The *spleen* is the largest of the lymphatic organs. It is located in the upper left portion of the abdominal cavity, just beneath the diaphragm and behind the stomach.

The spleen resembles a large lymph node in some respects. It is, for example, enclosed in connective tissue that extends inward from the surface and partially divides the organ into chambers or *lobules*. It also has a *hilum* on one surface through which blood vessels enter. However, unlike a lymph node, the spaces (venous sinuses) within the chambers of the spleen are filled with *blood* instead of lymph.

In addition to being very vascular, the spleen is soft and elastic, and can be distended by blood filling its venous sinuses. These characteristics are related to the spleen's function as a *blood reservoir.* During times of rest when circulation of blood is decreased, some blood can be stored within the spleen by vasodilation of its blood vessels. Then, during periods of exercise or in response to low oxygen concentration (anoxia) or hemorrhage, these blood vessels are stimulated to undergo vasoconstriction, and some of the stored blood is expelled into the circulation.

Within the lobules of the spleen, the tissues are called *pulp* and are of two types: white pulp and red pulp. The *white pulp* is distributed throughout the spleen in tiny islands. This tissue is composed of nodules (splenic nodules), which are similar to those found in lymph nodes and contain large numbers of lymphocytes. The *red pulp,* which fills the remaining spaces of the lobules, surrounds the venous sinuses. This pulp contains relatively large numbers of red blood cells, which are responsible for its color, along with many lymphocytes and macrophages (figure 19.13).

Chart 19.2 Major organs of the lymphatic system

Organ	Location	Function	Organ	Location	Function
Lymph nodes	In groups or chains along the paths of larger lymphatic vessels	Center for lymphocyte production; house T-lymphocytes and B-lymphocytes that are responsible for immunity; phagocytes filter foreign particles and cellular debris from lymph	Thymus	Within the mediastinum behind the upper portion of the sternum	Houses lymphocytes, changes undifferentiated lymphocytes into T-lymphocytes
			Spleen	In upper left portion of abdominal cavity beneath the diaphragm and behind the stomach	Serves as blood reservoir; phagocytes filter foreign particles, damaged red blood cells, and cellular debris from the blood; houses lymphocytes

The blood capillaries within the red pulp are quite permeable. Red blood cells can squeeze through the pores in these capillary walls and enter the venous sinuses. The older, more fragile red blood cells may rupture as they make this passage and the resulting cellular debris is removed by phagocytic macrophages located within the splenic sinuses.

The macrophages also engulf and destroy foreign particles, such as bacteria, that may be carried in the blood as it flows through the sinuses. Thus, the spleen filters blood much as the lymph nodes filter lymph.

The lymphocytes of the spleen, like those of the thymus, lymph nodes, and nodules, help to defend the body against infections.

Chart 19.2 summarizes the characteristics of the major organs of the lymphatic system.

During fetal development, pulp cells of the spleen function to produce blood cells, much as red bone marrow cells do in later life. As the time of birth approaches, this splenic function ceases. However, in certain diseases, such as *erythroblastosis fetalis*, in which large numbers of red blood cells are destroyed, the splenic pulp cells may resume their blood-cell-forming activity.

---

1. Why are the thymus and spleen considered to be organs of the lymphatic system?
2. What are the major functions of the thymus and the spleen?

Body Defenses against Infection

An infection is a condition caused by the presence of some kind of disease-causing agent. Such agents are termed **pathogens,** and they include viruses and microorganisms such as bacteria, fungi, and protozoans, as well as other parasitic forms of life.

The human body is equipped with a variety of *defense mechanisms* that help to prevent the entrance of pathogens or act to destroy them if they do enter the tissues. Some of these mechanisms are quite general, or *nonspecific,* in that they protect against many types of pathogens. These mechanisms include species resistance, mechanical barriers, actions of enzymes, interferon, inflammation, and phagocytosis.

Other defense mechanisms are very *specific* in their actions, providing protection against particular disease-causing agents. They are responsible for the type of resistance called *immunity.*

Nonspecific Resistance

Species Resistance

Species resistance refers to the fact that a given kind of organism or *species* (such as humans, *Homo sapiens*) develops diseases that are unique to it. At the same time, a species may be resistant to diseases that affect other species. For example, humans are subject to infections by the microorganisms that cause gonorrhea and syphilis, but other animal species are generally resistant to these diseases. Similarly, humans are resistant to certain forms of malaria and tuberculosis that affect various birds.

Mechanical Barriers

The *skin* and the *mucous membranes* lining the tubes of the respiratory, digestive, urinary, and reproductive systems create **mechanical barriers** against the entrance of infectious agents. As long as these barriers remain unbroken, many pathogens are unable to penetrate them. In addition, the mucous-coated ciliated epithelium, described in chapter 16, that lines the respiratory passages acts to entrap and sweep particles out of the airways and into the pharynx, where they are swallowed.

Enzymatic Actions

Enzymatic actions against disease-causing agents are due to the presence of enzymes in various body fluids. Gastric juice, for example, contains the protein-splitting enzyme *pepsin.* It also has a low pH due to the presence of hydrochloric acid, and the combined effect of these substances is lethal to many pathogens that reach the stomach. Similarly, tears contain an enzyme called *lysozyme* that has an antibacterial action against certain pathogens that may get onto the surfaces of the eyes.

Interferon

Interferon is the name given to a group of proteins produced by certain cells, including lymphocytes, in response to the presence of *viruses.* Although the effect of interferon is nonspecific, it interferes with the reproduction (replication) of viruses and thus helps to control the development of the diseases they cause. Furthermore, interferon released from infected cells can be taken in by other, noninfected cells. Consequently, the noninfected cells become protected against the viral infection, and the spread of the pathogenic viruses is inhibited.

Inflammation

Inflammation is a tissue response to injury from mechanical forces, chemical irritants, exposure to extreme temperatures, or exposure to excessive radiation. It may also accompany infections.

The major symptoms of inflammation include localized redness, swelling, heat, and pain. The *redness* is a result of blood vessel dilation and the consequent increase of blood volume within the affected tissues (hyperemia). This effect, coupled with an increase in the permeability of the capillaries, is responsible for tissue *swelling* (edema). The *heat* is due to the presence of blood from deeper body parts, which is generally warmer than that near the surface, while the *pain* results from the stimulation of pain receptors in the affected tissues.

White blood cells tend to accumulate at the sites of inflammation. Some of these cells help to control pathogens by *phagocytosis.* In the case of bacterial infections, the resulting mass of white blood cells, bacterial cells, and damaged tissue cells may create a thick fluid called *pus.*

Chart 19.3   Major actions that may occur during an inflammation response

Blood vessels dilate.

Capillary permeability increases.
   Tissues become red, swollen, warm, and painful.

White blood cells invade the region.
   Pus may form as white blood cells, bacterial cells, and cellular debris accumulate.

Body fluids seep into the area.
   A clot containing threads of fibrin may form.

Fibroblasts appear.
   A connective tissue sac may be formed around the injured tissues.

Phagocytes are active.
   Dead cells and other debris are removed.

Cells reproduce.
   Newly formed cells replace injured ones.

Body fluids (exudate) also tend to collect in inflamed tissues. These fluids contain *fibrinogen* and other clotting elements. As a result of clotting, a network of fibrin threads may develop within a region of damaged tissue. Later, *fibroblasts* may appear and form fibers around the affected area until it is enclosed in a sac of fibrous connective tissue. This action serves to confine an infection and to prevent the spread of pathogens to adjacent tissues.

A *pimple* or *boil* is an example of such a confined infection. An abscess of this type should never be squeezed, because the pressure acting upon the inflamed tissues may force pathogenic microorganisms into adjacent tissues and cause the infection to spread.

Once an infection has been controlled, phagocytic cells tend to remove dead cells and other debris from the site of inflammation. At the same time the dead cells are replaced by cellular reproduction.

The process of inflammation is summarized in chart 19.3.

Phagocytosis

As is mentioned in chapter 17, the most active phagocytic cells in the blood are *neutrophils* and *monocytes.* These wandering cells can leave the bloodstream by squeezing between the cells of blood vessel walls (diapedesis). They are attracted toward sites of inflammation by chemicals released from injured tissues.

Chart 19.4	Types of nonspecific resistance
Type	Description
Species resistance	A species of organism is resistant to certain diseases to which other species are susceptible
Mechanical barriers	Unbroken skin and mucous membranes prevent the entrance of some infectious agents
Enzymatic actions	Enzymes present in various body fluids act against pathogens
Interferon	A group of proteins (produced by cells in response to the presence of viruses) that interferes with the reproduction of the viruses in other cells
Inflammation	A tissue response to injury that helps to prevent the spread of infectious agents into nearby tissues
Phagocytosis	Neutrophils and macrophages engulf and destroy foreign particles and cells

Neutrophils are able to engulf and digest smaller particles, while monocytes can phagocytize somewhat larger objects.

Monocytes give rise to *macrophages* (histiocytes), which become fixed in various tissues or to the inner walls of blood vessels and lymphatic vessels. These relatively nonmotile, phagocytic cells that can divide and produce new macrophages are found in such organs as lymph nodes, spleen, liver, and lungs. This diffuse group of phagocytic cells constitutes the **reticuloendothelial tissue** or **system** (tissue macrophage system).

The name *reticuloendothelium* describes two characteristics of cells that make up this tissue. They are capable of forming fibrous networks or *reticula,* and they are often associated with the linings of blood vessels, the *endothelium.*

The cells that line the vascular sinuses of the red bone marrow are also reticuloendothelial cells. Although these cells are especially concerned with the formation of red and white blood cells, they are also capable of phagocytosis. In fact, they are adapted to remove very small particles, such as foreign protein molecules, from the blood.

As a result of reticuloendothelial activities, foreign particles are removed from lymph as it moves from the interstitial spaces to the bloodstream. Any such particles that reach the blood are likely to be removed by phagocytes located in the vessels and tissues of the spleen, liver, or bone marrow.

Chart 19.4 summarizes the types of nonspecific resistance.

---

1. What is meant by an infection?
2. Explain six nonspecific defense mechanisms.
3. Define reticuloendothelial tissue.

Immunity

Immunity is resistance to specific foreign agents, such as pathogens, or to the toxins they release. It involves a number of *immune mechanisms,* in which certain cells recognize the presence of particular foreign substances and act to eliminate them. The cells that function in immune mechanisms include lymphocytes and macrophages.

Origin of Lymphocytes

Lymphocytes originate from bone marrow cells (stem cells), although later they proliferate in various lymphatic tissues.

Before they become specialized or *differentiated,* developing lymphocytes are released from the marrow and are carried away by the blood. About half of them reach the thymus gland, where they remain for a time. Within the thymus, these undifferentiated cells (thymocytes) undergo special processing, and thereafter they are called *T-lymphocytes* or T-cells (thymus-derived lymphocytes). Later, the T-cells are transported away from the thymus by the blood, where they comprise 70%–80% of the circulating lymphocytes. They tend to reside in various organs of the lymphatic system and are particularly abundant in lymph nodes, the thoracic duct, and the white pulp of the spleen.

Lymphocytes that have been released from the bone marrow and do not reach the thymus gland, differentiate into *B-lymphocytes* (bone marrow-derived lymphocytes). Before they can mature, however, it may be necessary for these B-cells to pass through lymphoid tissue in some presently unknown region of the body, where they are specially processed.

Figure 19.14

Bone marrow releases undifferentiated lymphocytes which, after processing, become T-lymphocytes and B-lymphocytes.

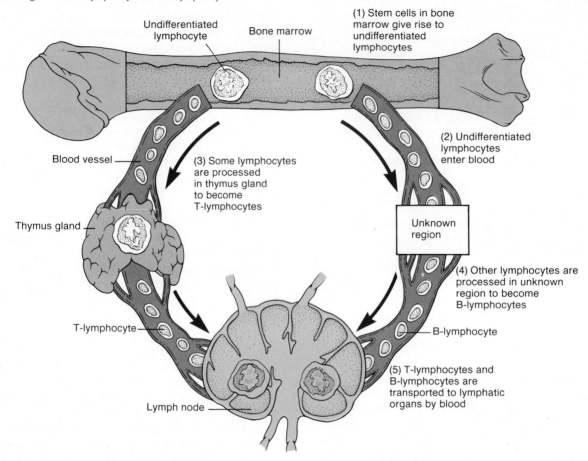

Figure 19.15

Scanning electron micrograph of a human circulating lymphocyte.

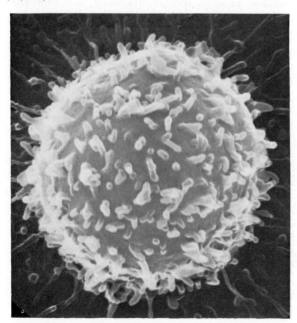

B-lymphocytes are distributed by the blood and constitute 20%–30% of the circulating lymphocytes. They settle in various lymphatic organs with the T-lymphocytes and are abundant in lymph nodes, spleen, bone marrow, secretory glands, intestinal lining, and reticuloendothelial tissue (figures 19.14 and 19.15).

Antigens

Prior to birth, body cells somehow make an inventory of the proteins and various other large molecules that are present in the body. Subsequently, the substances that are present (self-substances) can be distinguished from foreign (nonself) substances. More specifically, T-cells and B-cells have receptors on their surfaces that can recognize foreign substances. When such substances are recognized, lymphocytes can produce immune reactions against them.

Foreign substances, such as proteins, polysaccharides, and some glycolipids to which lymphocytes respond, are called **antigens.** As a group, antigens are large, complex molecules; however, sometimes smaller molecules that cannot stimulate immune responses by themselves combine with larger ones and create a combination that is antigenic. Such a small molecule is

Chart 19.5	A comparison of T-lymphocytes and B-lymphocytes	
Characteristic	T-lymphocytes	B-lymphocytes
Origin of undifferentiated cell	Bone marrow	Bone marrow
Site of differentiation	Thymus gland	Region outside thymus gland
Primary locations	Lymphatic tissues, 70%–80% of circulating lymphocytes	Lymphatic tissues, 20%–30% of circulating lymphocytes
Primary functions	Responsible for cell-mediated immunity, secrete lymphokines, trigger and regulate actions of B-lymphocytes	Responsible for antibody-mediated immunity, secrete antibodies into body fluids

called a **hapten.** When lymphocytes are stimulated by the combined form, they can react either to the hapten or to the larger molecule of the combination. Substances that serve as haptens occur in various drugs such as penicillin, household and industrial chemicals, particles in dust, and products of animal skins (dander).

Functions of Lymphocytes

T-lymphocytes and B-lymphocytes respond to the antigens they recognize in different ways. For example, some T-lymphocytes (cytotoxic T-cells) attach themselves to antigen-bearing cells, such as certain types of bacterial cells, and interact with these cells directly—that is, with cell-to-cell contact. This type of response is called **cell-mediated immunity** (CMI).

T-cells also provide an important defense against many viral infections. Such infections are caused by viruses that grow inside body cells where they are somewhat protected. Most viruses, however, cause antigens to be produced on the membranes of infected cells. T-lymphocytes can detect the antigen-bearing cells, disrupt their membranes and thus destroy them along with the viruses they contain.

T-lymphocytes also can release a group of proteins called *lymphokines.* These proteins, in turn, can cause a variety of effects, such as attracting macrophages and leukocytes into inflamed tissues and retaining them there, or stimulating the production of antibodies by B-cells, as is described in the next section. In addition, T-cells may secrete toxic substances (lymphotoxins) that are lethal to their target cells, growth-inhibiting factors that prevent target cell growth, or interferon that prevents proliferation of viruses.

T-lymphocytes are able to recognize and destroy certain kinds of cancer cells. This has led some investigators to believe that interference with normal T-lymphocyte function may be a factor in the development of some cancers. Furthermore, since T-lymphocyte function seems to decline with age, this may help to explain why some types of cancer appear more commonly in elderly persons.

B-lymphocytes act indirectly against the antigens by producing and secreting globular proteins (immunoglobulins) called **antibodies.** Antibodies are carried by body fluids and react in various ways with specific antigens or antigen-bearing particles to eliminate them. This type of response is called **antibody-mediated immunity** (AMI) or humoral immunity.

T-lymphocytes and B-lymphocytes also interact with each other in complex ways. For example, one group of T-lymphocytes (T-helper cells) must interact with B-cells, either directly or indirectly by means of lymphokines, before the B-cells can differentiate into antibody-secreting cells (plasma cells). Another group of T-lymphocytes (T-suppressor cells) function to regulate the synthesis of antibodies by producing a substance (suppressor factor) that inhibits B-cell activity after the response to an antigen has begun. Thus, both types of lymphocytes are required for normal immune responses. Chart 19.5 compares T- and B-lymphocytes.

1. Define immunity.
2. Explain the difference in origin between a T-lymphocyte and a B-lymphocyte.
3. Explain the difference between an antigen and a hapten.
4. What are the functions of a T-lymphocyte? A B-lymphocyte?

Immunoglobulin Molecules

The antibodies produced and secreted by B-lymphocytes are all soluble globular proteins. These proteins are called **immunoglobulins,** and they constitute the *gamma globulin* fraction of plasma proteins.

Some persons partially or completely lack the ability to form immunoglobulins. As a consequence, they cannot produce antibodies and are very susceptible to developing infectious diseases. This condition is termed *agammaglobulinemia.*

Figure 19.16
An immunoglobulin molecule consists of two identical light chains of amino acids and two identical heavy chains of amino acids.

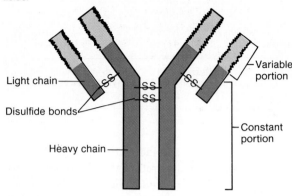

Light chain

Disulfide bonds

Heavy chain

Variable portion

Constant portion

Each immunoglobulin molecule is composed of four chains of amino acids that are linked together by pairs of sulfur atoms (disulfide bonds). Two of these chains are identical light chains (L-chains), and two are identical heavy chains (H-chains). The heavy chains contain about twice as many amino acids as the light ones. Each of the five major types of immunoglobulin molecules contains a particular kind of heavy chain.

As with other proteins, the sequence of amino acids within the peptide chains is responsible for the unique, three-dimensional structure of each immunoglobulin molecule. This special structure, in turn, is related to the physiological properties of the molecule. For example, the terminal portions at one end of each of the heavy and light chains contain a variable sequence of amino acids (variable regions), and it is these parts that are specialized to react with the shape of a specific antigen molecule. The remaining portions of the chains (constant regions) are responsible for other properties of the immunoglobulin, such as its ability to bond to cellular structures or combine with chemical substances (figure 19.16).

Types of Immunoglobulins

Of the five major types of immunoglobulins, three constitute the bulk of the circulating antibodies. They are called immunoglobulin G (IgG), which accounts for about 80% of the antibodies; immunoglobulin A (IgA), which makes up about 13%; and immunoglobulin M (IgM), which is responsible for about 6%. Immunoglobulin D (IgD) and immunoglobulin E (IgE) account for the remainder.

Immunoglobulin G occurs in plasma and tissue fluids and is particularly effective against bacterial cells, viruses, and various toxins.

Immunoglobulin A is found commonly in the secretions of various exocrine glands, and occurs in milk, tears, nasal fluid, gastric juice, intestinal juice, bile, and urine. It also helps to control bacterial and viral infections.

Immunoglobulin A does not pass through the placenta of a pregnant woman to her fetus; however, it can pass from a nursing mother to her baby by means of milk and other secretions of the breasts. These antibodies provide the infant with some protection against various digestive and respiratory disturbances that otherwise might cause serious problems in early infancy.

Immunoglobulin M is the type of antibody that occurs naturally in blood plasma (agglutinins anti-A and/or anti-B), described in chapter 17. Serious reactions sometimes occur following blood transfusions because these substances can cause red blood cells to agglutinate by reacting with the antigens (agglutinogens A and/or B) on their cell membranes.

Immunoglobulin D is found on the surfaces of most B-lymphocytes, especially those of infants; and although little is known about it, the significance of this antibody is thought to be quite limited.

Immunoglobulin E occurs in various exocrine secretions along with IgA. Its best understood action is associated with allergic reactions that are described in a subsequent section of this chapter.

Chart 19.6 summarizes the major immunoglobulins and their functions.

1. What is an immunoglobulin?
2. Describe the structure of an immunoglobulin molecule.
3. Name the five major types of immunoglobulins.

Actions of Antibodies

In general, antibodies react to antigens by attacking them directly, by activating a set of enzymes (complement) that attack the antigens, or by stimulating changes in local areas that help to prevent the spread of the antigens.

In the case of direct attacks, antibodies may combine with antigens and cause them to clump together (agglutination) or to form insoluble substances (precipitation). Such actions make it easier for phagocytic

Chart 19.6	Characteristics of major immunoglobulins	
Type	Occurrence	Major functions
IgG	Tissue fluid and plasma	Acts against bacterial cells, viruses, and toxins
IgA	Secretions of exocrine glands	Acts against bacterial cells and viruses
IgM	Plasma	Reacts with antigens occurring naturally on some red blood cell membranes following certain blood transfusions
IgD	Surface of most B-lymphocytes	Poorly understood, but apparently of limited significance
IgE	Secretions of exocrine glands	Promotes allergic reactions

cells to engulf the antigen-bearing agents and eliminate them. In other instances, antibodies cover the toxic portions of antigen molecules and neutralize their effects (neutralization), or they cause cell membranes to rupture (lysis), thereby destroying invading cells. However, under normal conditions, direct attack by antibodies apparently is of limited value in protecting the body against infections. Of greater importance is the antibody action of activating complement.

Complement is a group of inactive enzymes (complement system) that occurs in plasma and other body fluids. When antibodies combine with antigens, reactive sites on the constant regions of antibody molecules become exposed, and this triggers a series of reactions leading to the activation of the complement enzymes. The activated enzymes produce a variety of effects which include attracting macrophages and neutrophils into the region (chemotaxis), causing antigen-bearing agents to clump together, altering cell membranes in ways that make the cells more susceptible to phagocytosis (opsonization), digesting the membranes of foreign cells so that they rupture (lysis), and altering the molecular structure of viruses thereby making them harmless. Other enzymes promote the inflammation reaction, which helps to prevent the spread of antigens to nearby tissues (see chapter 5).

Immunoglobulin E also promotes inflammation reactions, but in this case the reactions tend to be excessive and harmful to tissues in some individuals. This antibody tends to be attached to the membranes of widely distributed mast cells (see chapter 5). When antigens are present, they combine with the antibodies, and the resulting antigen-antibody complexes stimulate the mast cells to release various substances. These substances, which include histamine, cause the changes associated with inflammation, such as vasodilation and edema.

Chart 19.7 summarizes the actions of antibodies.

Chart 19.7	Actions of antibodies	
General action	Type of effect	Description
Direct attack		
	Agglutination	Causes antigens to clump together
	Precipitation	Causes antigens to form insoluble substances
	Neutralization	Causes antigens to lose toxic properties
	Lysis	Causes cell membranes to rupture
Activation of complement		
	Chemotaxis	Attracts macrophages and neutrophils into region
	Opsonization	Alters cell membranes so cells are more susceptible to being phagocytized
	Inflammation	Promotes local tissue changes that help to prevent the spread of antigens

Immune Responses

Within the populations of T-cells and B-cells of each person, there are millions of varieties. The members of each variety originated from a single early cell, so that the members of a variety are all alike. Also, the members of each variety have a particular type of antigen receptor on their surfaces, and thus are capable of responding only to a specific antigen. Such a group is called a **clone.**

Figure 19.17
When a T-cell encounters an antigen that fits its antigen receptor, it becomes activated and proliferates, thus enlarging its clone.

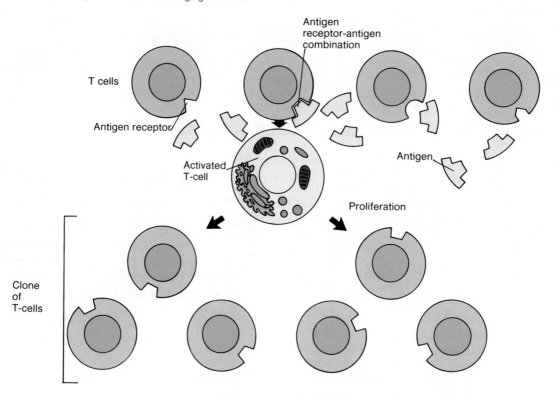

Before a lymphocyte can respond to its antigen, it must be activated. A T-cell, for example, becomes activated when it encounters an antigen that fits its antigen receptors. In response to the antigen receptor-antigen combination, the T-cell proliferates, and its clone is enlarged. (See figure 19.17.) The activation of a B-cell, however, usually requires the presence of a macrophage and a T-helper cell.

A B-cell response usually begins when some antigen-bearing agents that have entered the tissues are phagocytized by a macrophage, and others combine with B-cell antigen receptors. The antigen-bearing agents are digested by macrophage lysosomal activity, and the antigens are displayed on the surface of the macrophage (presenter cell). T-helper cells become activated after they contact displayed antigens that fit their antigen receptors. When such an activated T-cell encounters a B-cell that has combined with the same antigen, the T-helper cell binds to the B-cell and releases lymphokines. The hormonelike action of the lymphokines stimulates the B-cell to proliferate rapidly, thus enlarging its clone. (See figure 19.18.)

Chart 19.8   Steps in the formation of plasma cells

1. Antigen-bearing agents enter tissues; some are phagocytized by macrophages and some combine with surface receptors of B-cells.
2. Lysosomes of macrophages digest agents, and antigens are displayed on surfaces of macrophages.
3. T-helper cells contact displayed antigens and become activated.
4. Activated T-helper cells bind to B-cells that have previously combined with antigen displayed by macrophages.
5. T-helper cells secrete lymphokines that stimulate B-cells to proliferate and enlarge clone.
6. Some of newly formed B-cells of clone become antibody-secreting plasma cells.

Some of the newly formed members of the B-cell clone become **plasma cells,** and they make use of their DNA and protein-synthesizing mechanism (see chapter 4) to produce antibody molecules. These antibodies are similar in structure to the antigen-receptor molecules present on the original B-cell surface, and thus are able to combine with the antigen-bearing agent that has entered the body. Chart 19.8 lists the steps leading to the formation of plasma cells.

Figure 19.18
(a) After digesting antigen-bearing agents, a macrophage displays antigens on its surface; *(b)* T-helper cells become activated when they contact displayed antigens that fit their antigen receptors; *(c)* activated T-helper cell interacts with B-cell and stimulates it to proliferate.

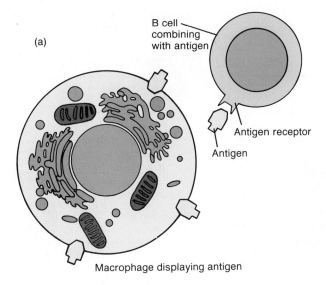

(a)

B cell combining with antigen

Antigen receptor

Antigen

Macrophage displaying antigen

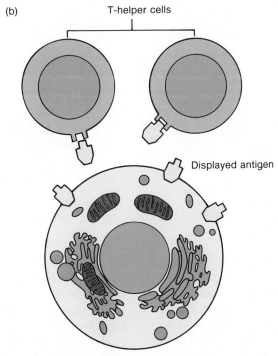

(b)

T-helper cells

Displayed antigen

Macrophage (presenter) cell

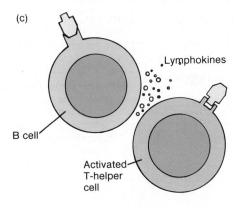

(c)

Lymphokines

B cell

Activated T-helper cell

Figure 19.19
Activated B-cells proliferate, giving rise to antibody-secreting
plasma cells and dormant memory cells.

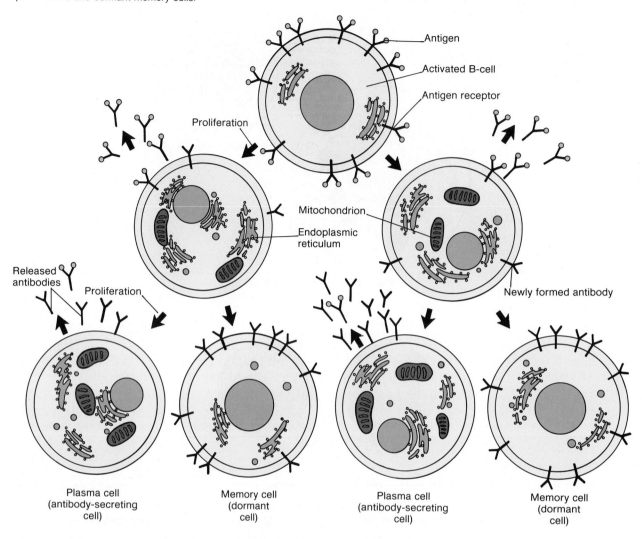

Antigen

Activated B-cell

Antigen receptor

Proliferation

Mitochondrion

Endoplasmic
reticulum

Released
antibodies

Proliferation

Newly formed antibody

Plasma cell
(antibody-secreting
cell)

Memory cell
(dormant
cell)

Plasma cell
(antibody-secreting
cell)

Memory cell
(dormant
cell)

Antibodies are released into lymph and are transported by the blood to all body parts. Antibody production and release may continue for several weeks. Such a response, occurring when an antigen is first encountered, is called a **primary immune response.** Also, following a primary response some of the B-cells and T-cells that are produced during proliferation remain dormant, serving as *memory cells.* In this way, if the same antigens are encountered in the future, the original clones of lymphocytes are increased in numbers and can respond to the antigens to which they were sensitized more rapidly and effectively. A subsequent reaction of this type is called a **secondary immune response** (figure 19.19).

As a result of a primary immune response, detectable concentrations of antibodies usually appear in the body fluids within five to ten days following the exposure to antigens. If the same antigen is encountered some time later, a secondary immune response may produce additional antibodies within a day or two. (See figure 19.20.) Although such newly formed antibodies may persist in the body for only a few months, or perhaps a few years, the memory cells live much longer. Consequently, the ability to produce a secondary immune response may be long-lasting.

1. In what general ways do antibodies act?
2. What is the function of complement?
3. What is meant by a clone of lymphocytes?
4. Distinguish between a primary and a secondary immune response.

Figure 19.20
Antibody concentration increases more rapidly during a
secondary immune response than during a primary response.

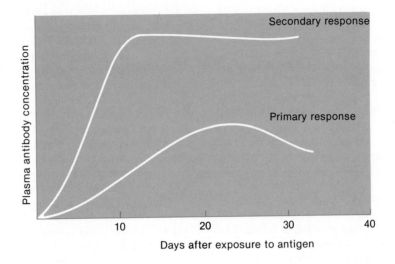

Types of Immunity

One type of immunity is called *naturally acquired active immunity.* It occurs when a person who has been exposed to a live pathogen develops a disease and becomes resistant to that pathogen as a result of a primary immune response.

Another type of active immunity can be produced in response to a **vaccine.** Such a substance contains an antigen that can stimulate a primary immune response against a particular disease-causing agent, but is unable to produce the symptoms of that disease.

A vaccine, for example, might contain bacteria or viruses that have been killed or weakened sufficiently so that they cannot cause a serious infection or produce the severe symptoms of an infection, or it may contain a toxin of an infectious organism that has been chemically altered to destroy its toxic effects. In any case, the antigens present still retain the characteristics needed to stimulate a primary immune response. Thus, a person who has been vaccinated is said to develop *artificially acquired active immunity.*

Vaccines are available to stimulate the development of active immunity against a great variety of diseases. These include such diseases as typhoid fever, cholera, whooping cough, diphtheria, tetanus, polio, measles, and mumps.

Sometimes a person needs protection against a disease-causing microorganism, but lacks the time needed to develop active immunity. In such a case, it may be possible to provide the person with an injection of ready-made antibodies. These antibodies may be obtained from *gamma globulin* separated from the blood of persons who have already developed immunity against the disease in question.

A person who receives an injection of gamma globulin is said to have *artificially acquired passive immunity.* This type of immunity is called passive because the antibodies involved are not produced by the recipient's cells. Such immunity is relatively short-term, seldom lasting more than a few weeks. Furthermore, since the recipient's lymphocytes have not been activated against the pathogens for which the protection was needed, the person continues to be susceptible to those pathogens in the future.

During pregnancy, certain antibodies (immunoglobulin G) are able to pass from the maternal blood into the fetal bloodstream. This transfer is accomplished by receptor-mediated endocytosis (see chapter 3) and involves the presence of receptor sites on cells of the fetal yolk sac (see chapter 23). These receptor sites bind to a region that is common to the molecular structure of IgG molecules. Then, after entering the fetal cells, the antibodies are secreted into the fetal blood. As a result, the fetus acquires some immunity against the pathogens for which the mother has developed active immunities. In this case, the fetus is said to have *naturally acquired passive immunity,* which may remain effective for six months to a year after birth.

Acquired Immune Deficiency Syndrome
A Clinical Application

Acquired immune deficiency syndrome (AIDS) is a disease that was first recognized in 1981 and whose incidence is roughly doubling each year. It is caused by a type of human T-lymphocyte virus (HTLV) that specifically infects T-helper cells—the cells required to activate B-lymphocytes and induce the production of antibodies.

When the virus infects T-helper cells, its genes are incorporated into the cellular DNA. Then, the virus may become dormant, failing to produce any noticeable effects for some time. During this latency period, which may last for several years, the person may show no symptoms of AIDS, but probably can transmit the virus to uninfected persons.

If the viral genes in the T-helper cells are activated by some unknown factor, they induce the production of new viral particles that can infect other T-helper cells and kill them. As a consequence, the person's immune system is greatly impaired, and the group of symptoms that characterize AIDS develops. These symptoms include enlargement of lymph nodes, weight loss, and fever, as well as a variety of severe infections (following depression of the immune system) and the appearance of certain unusual forms of cancer. The AIDS virus also may appear in the nervous system; some patients develop serious neurological problems without impairment of their immune responses.

Transmission of this disease seems to require direct insertion of the AIDS virus into the blood, as may result from certain sexual activities, from use of contaminated hypodermic needles, or from the transfusion of virus-containing blood or blood products. AIDS apparently is not spread by casual contact with AIDS patients.

The persons who are at greatest risk for contracting AIDS are homosexual and bisexual males (who made up 73% of the AIDS patients in recent years), heterosexual drug abusers (17%), and female sexual partners of infected males (1%). The remaining patients (6%) do not belong to one of these high-risk groups and acquired the disease from some other source. The unborn fetuses of infected mothers, for example, acquire the disease in about 50% of the cases.

Chart 19.9 Types of immunity		
Type	**Stimulus**	**Result**
Naturally acquired active immunity	Exposure to live pathogens	Symptoms of a disease and stimulation of an immune response
Artificially acquired active immunity	Exposure to a vaccine containing weakened or dead pathogens	Stimulation of an immune response without the severe symptoms of a disease
Artificially acquired passive immunity	Injection of gamma globulin containing antibodies	Immunity for a short time without stimulating an immune response
Naturally acquired passive immunity	Antibodies passed to fetus from mother with active immunity	Short-term immunity for infant, without stimulating an immune response

The types of immunity are summarized in chart 19.9.

Injections of gamma globulin sometimes are used to provide passive immunity for pregnant women who have been exposed to rubella (German or three-day measles) viruses during the early stages of their pregnancies. Injections of antibodies also are used to provide temporary protection against the effects of poisonous snakebites and certain other toxins.

Allergic Reactions

Allergic reactions are closely related to immune responses in that both may involve the sensitizing of lymphocytes or the combining of antigens with antibodies. Allergic reactions, however, are likely to be excessive and cause damage to tissues.

Some forms of allergic reactions can occur in almost anyone, while another form affects only those people who have inherited from their parents an ability to produce exaggerated immune responses.

Delayed-reaction allergy is an example of a type of allergic response that may occur in anyone. It results from repeated exposure of the skin to certain chemical substances—commonly, household or industrial chemicals or some cosmetics. As a consequence of such repeated contacts, T-lymphocytes eventually become activated by the nonself substance, and a large number of these T-cells enter the skin. As a result of their actions and the actions of macrophages they attract, various chemical factors are released which, in turn, cause eruptions and inflammation of the skin (dermatitis). This reaction is called delayed because it usually takes about 48 hours to occur.

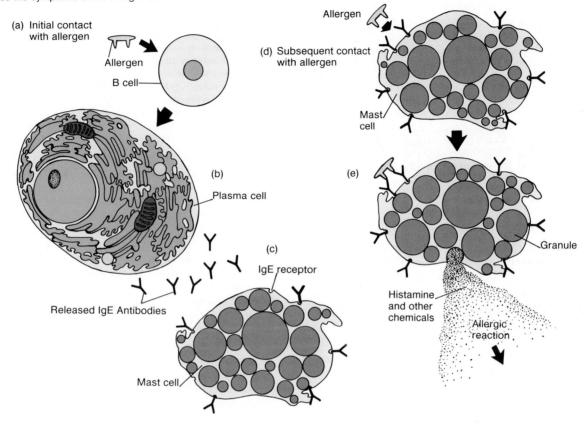

Figure 19.21
In immediate-reaction allergy *(a)* B-cells are activated when they contact an allergen; *(b)* an activated B-cell becomes an antibody-secreting plasma cell; *(c)* the antibodies become attached to the membranes of mast cells; *(d)* subsequently, when the allergen is encountered, it combines with the antibodies of the mast-cell membranes; and *(e)* the mast cell releases substances that cause the symptoms of the allergic reaction.

(a) Initial contact with allergen

Allergen

B cell

(b)

Plasma cell

(c)

IgE receptor

Released IgE Antibodies

Mast cell

(d) Subsequent contact with allergen

Allergen

Mast cell

(e)

Granule

Histamine and other chemicals

Allergic reaction

In other cases, allergic reactions may occur within minutes after contact with a nonself substance. Persons with such *immediate-reaction allergy* have an inherited ability to synthesize abnormally large amounts of immunoglobulin E in response to certain antigens. In this instance, the allergic reaction involves the activation of B-lymphocytes, and the antigen that triggers the reaction is called an **allergen.**

Typically, the B-cells become activated when the allergen is first encountered, and subsequent exposures trigger allergen-antibody reactions. As mentioned previously, immunoglobulin E is attached to the membranes of widely distributed *mast cells;* and when an allergen-antibody reaction occurs, the mast cells release a variety of substances such as *histamine* and *serotonin.* These substances, in turn, cause a variety of physiological effects, including the dilation of blood vessels, swelling of tissues, and contraction of smooth muscles. The result is a severe inflammation reaction that is responsible for the symptoms of the allergy—such as hives, hay fever, asthma, eczema, or gastric disturbances (figure 19.21).

In most individuals, such severe responses to nonself substances are inhibited by a special type of T-lymphocyte called *suppressor cells.* In a person with immediate-reaction allergy, however, this control function of T-cells seems to be less effective. On the other hand, the suppressor cells usually interfere with the production of IgE eventually, and the allergic reaction is terminated.

Transplantation and Tissue Rejection

It is occasionally desirable to transplant some tissue or an organ, such as skin, kidney, heart, or liver, from one person to another to replace a nonfunctional, damaged, or lost body part. In these cases, there is a danger that the recipient's cells may recognize the donor's tissues as being foreign. This triggers the recipient's immune mechanisms, which may act to destroy the donor tissue. Such a response is called a **tissue rejection reaction.**

Tissue rejection involves the activities of lymphocytes and both cell- and antibody-mediated responses—responses similar to those that occur when any

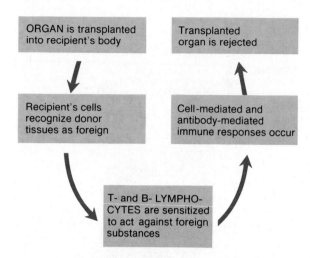

foreign substances are present (figure 19.22). The greater the antigenic difference between the proteins of the recipient and the donor tissues, the more rapid and severe the rejection reaction is likely to be. Thus, the reaction can sometimes be minimized by matching recipient and donor tissues. This means locating a donor whose tissues are antigenically similar to those of the person needing a transplant—a procedure much like matching the blood of a donor with that of a recipient before giving a blood transfusion.

One matching procedure makes use of lymphocytes from circulating blood. These cells contain both self-substances and receptor sites for recognizing foreign antigens. In the test, lymphocytes from a potential donor and a potential recipient are mixed and cultured. The antigenic difference between these cells can then be judged by observing certain cellular responses (mixed lymphocyte reaction), including enlargement, amount of DNA synthesized, and the rate of cell division. Minimal antigenic difference results in the best match.

Another approach to reducing the rejection of transplanted tissue involves the use of *immunosuppressive drugs.* These substances interfere with the recipient's immune mechanisms in various ways. A drug, for example, may suppress the formation of antibodies or the production of T-lymphocytes, thereby reducing the humoral and cellular responses.

Unfortunately, the use of immunosuppressive drugs leaves the recipient relatively unprotected against infections. Although the drug may prevent a tissue rejection reaction, the recipient may develop a serious infectious disease, such as pneumonia, that is difficult to control and may cause death.

1. Explain the difference between active and passive immunities.
2. In what ways is an allergic reaction related to an immune reaction?
3. In what ways is a tissue rejection reaction related to an immune response?

Clinical Terms Related to the Lymphatic System and Immunity

anaphylaxis (an″ah-fi-lak′sis)   hypersensitivity to the presence of a foreign substance.

asplenia (ah-sple′ne-ah)   the absence of a spleen.

autograft (aw′to-graft)   transplantation of tissue from one part of a body to another part of the same body.

histocompatibility (his″to-kom-pat″ĭ-bil′ĭ-te)   compatibility between the tissues of a donor and a recipient based on antigenic similarities.

homograft (ho′mo-graft)   transplantation of tissue from one person to another person.

immunocompetence (im″u-no-kom′pe-tens)   the ability to produce an immune response to the presence of antigens.

immunodeficiency (im″u-no-de-fish′en-se)   a lack of ability to produce an immune response.

lymphadenectomy (lim-fad″ē-nek′to-me)   surgical removal of lymph nodes.

lymphadenopathy (lim-fad″ē-nop′ah-the)   enlargement of the lymph nodes.

lymphadenotomy (lim-fad″ē-not′o-me)   an incision of a lymph node.

lymphocytopenia (lim″fo-si″to-pe′ne-ah)   an abnormally low concentration of lymphocytes in the blood.

lymphocytosis (lim″fo-si″to′sis)   an abnormally high concentration of lymphocytes in the blood.

lymphoma (lim-fo′mah)   a tumor composed of lymphatic tissue.

lymphosarcoma (lim″fo-sar-ko′mah)   a cancer within the lymphatic tissue.

splenectomy (sple-nek′to-me)   surgical removal of the spleen.

splenitis (sple-ni′tis)   an inflammation of the spleen.

splenomegaly (sple″no-meg′ah-le)   an abnormal enlargement of the spleen.

splenotomy (sple-not′o-me)   incision of the spleen.

thymectomy (thi-mek′to-me)   surgical removal of the thymus gland.

thymitis (thi-mi′tis)   an inflammation of the thymus gland.

As mentioned previously, immune responses are usually directed toward nonself or foreign molecules, while self-substances are tolerated by the immune mechanism. Occasionally, something happens to change this, and the tolerance to self-substances is lost. As a result, cell- and antibody-mediated immune responses may be directed toward a person's own tissues. Such a condition is called an **autoimmune disease** (autoallergy).

Although their causal mechanism is not well understood, these diseases are more common in older people, and they may involve viral or bacterial infections. Some investigators believe that the release of abnormally large quantities of antigens may occur when infectious agents cause tissue damage. These agents may also cause body proteins to change into forms that stimulate antibody production. At the same time, the activity of suppressor cells, which normally limits this type of reaction, seems to be repressed.

After they appear, the concentration of the antibodies sometimes fluctuates greatly. Also, they may persist indefinitely, or they may disappear.

Some autoimmune diseases affect specific organs, while others produce more generalized effects. For example, in the disease *autoimmune thyroiditis*, the effects are directed toward the thyroid gland. In this condition, an-tibodies are produced by B-lymphocytes that respond abnormally to self-proteins associated with thyroid hormones (thyroglobulin) and to substances in thyroid epithelial cell membranes. As a result of the antibody actions, the gland may become inflamed and much of its tissue may be destroyed.

Systemic lupus erythematosus (SLE), which occurs most commonly in young women, is an example of an autoimmune disease that affects many organs and tissues. In this case, B-cells produce antibodies that react with cell nuclei and various cytoplasmic substances. The resulting antigen-antibody combinations tend to accumulate in membranes of the kidneys, heart, lungs, blood vessels, and joints, creating functional problems in these organs. Also, antibodies that react with red blood cells, platelets, and lymphocytes may be produced. One of the early symptoms of this disease is an inflammation of the skin over the nose and cheeks.

Other autoimmune diseases include myasthenia gravis, thyrotoxicosis (Grave's disease), chronic atrophic gastritis, primary adrenal atrophy, rheumatoid arthritis, insulin-dependent diabetes mellitus, multiple sclerosis, and systemic sclerosis (scleroderma).

Chapter Summary

Introduction
The lymphatic system is closely associated with the cardiovascular system. It functions to transport excess tissue fluid to the bloodstream and to help defend the body against invasion by disease-causing microorganisms.

Lymphatic Pathways
1. Lymphatic capillaries
   a. Lymphatic capillaries are microscopic closed-ended tubes that extend into interstitial spaces.
   b. They receive lymph through their thin walls.
   c. Lacteals are lymphatic capillaries in the villi of the small intestine.
2. Lymphatic vessels
   a. Lymphatic vessels are formed by the merging of lymphatic capillaries.
   b. They have walls similar to veins and possess valves that prevent backflow of lymph.
   c. They lead to lymph nodes and then merge into lymphatic trunks.
3. Lymphatic trunks and collecting ducts
   a. Lymphatic trunks drain lymph from relatively large body regions.
   b. Trunks lead to two collecting ducts within the thorax.
   c. Collecting ducts join the subclavian veins.

Tissue Fluid and Lymph
1. Tissue fluid formation
   a. Tissue fluid originates from blood plasma and includes water and dissolved substances that passed through the capillary wall.
   b. It generally lacks proteins, but some smaller protein molecules leak into interstitial spaces.
   c. As protein concentration of tissue fluid increases, osmotic pressure increases also.
2. Lymph formation
   a. Rising osmotic pressure in tissue fluid interferes with the return of water to the blood capillaries.
   b. Increasing pressure within interstitial spaces forces some tissue fluid into lymphatic capillaries, and this fluid becomes lymph.
3. Function of lymph
   a. Lymph returns protein molecules to the bloodstream.
   b. It transports foreign particles to the lymph nodes.

Movement of Lymph
1. Flow of lymph
   a. Lymph is under low pressure and may not flow readily without aid from external forces.
   b. Forces that aid movement of lymph include squeezing action of skeletal muscles and low pressure in the thorax created by breathing movements.

2. Obstruction of lymph movement
   a. Any condition that interferes with the flow of lymph results in edema.
   b. Obstruction of lymphatic vessels due to surgery also results in edema.

Lymph Nodes
1. Structure of a lymph node
   a. Lymph nodes are usually bean shaped with blood vessels, nerves, and efferent lymphatic vessels attached to the indented region; afferent lymphatic vessels enter at points on the convex surface.
   b. Lymph nodes are enclosed in connective tissue that extends into the nodes and divides them into nodules.
   c. Nodules contain masses of lymphocytes and macrophages, and spaces through which lymph flows.
2. Locations of lymph nodes
   a. Lymph nodes generally occur in groups or chains along the paths of larger lymphatic vessels.
   b. They occur primarily in cervical, axillary, and inguinal regions and within the pelvic, abdominal, and thoracic cavities.
3. Functions of lymph nodes
   a. Lymph nodes are centers for production of lymphocytes that act against foreign particles.
   b. They contain macrophages that remove foreign particles from lymph.

Thymus and Spleen
1. The thymus
   a. The thymus is a soft, bilobed organ located within the mediastinum.
   b. It tends to decrease in size after puberty.
   c. It is composed of lymphatic tissue that is subdivided into lobules.
   d. Lobules contain lymphocytes, most of which are inactive, that develop from precursor cells in bone marrow.
   e. Some lymphocytes leave the thymus and become T-lymphocytes.
   f. The thymus may secrete a hormone called thymosin, which stimulates lymphocytes that have migrated to other lymphatic tissues.
2. The spleen
   a. The spleen is located in the upper left portion of the abdominal cavity.
   b. It resembles a large lymph node that is encapsulated and subdivided into lobules by connective tissue.
   c. Spaces within lobules are filled with blood.
   d. It acts as a blood reservoir.
   e. It contains numerous macrophages and lymphocytes that filter foreign particles and damaged red blood cells from the blood.

Body Defenses against Infection
Infection is caused by the presence of pathogens.

The body is equipped with specific and nonspecific defenses against infection.

Nonspecific Resistance
1. Species resistance
   Each species of organism is resistant to certain diseases that may affect other species, but is susceptible to diseases that other species may be able to resist.
2. Mechanical barriers
   a. Mechanical barriers include skin and mucous membranes.
   b. They prevent entrance of many pathogens as long as they remain unbroken.
3. Enzymatic actions
   a. Enzymes of gastric juice are lethal to many pathogens.
   b. Enzymes in tears have antibacterial actions.
4. Interferon
   a. Interferon is a group of proteins produced by certain cells in response to the presence of viruses.
   b. It can interfere with the reproduction and spread of viruses.
5. Inflammation
   a. Inflammation is a tissue response to injury or infection.
   b. The response includes localized redness, swelling, heat, and pain.
   c. Chemicals released by damaged tissues attract various white blood cells to the site of inflammation.
   d. Clotting may occur in body fluids that accumulate in affected tissues.
   e. Fibrous connective tissue may form a sac around the injured tissue and thus prevent the spread of infection.
6. Phagocytosis
   a. The most active phagocytes in blood are neutrophils and monocytes; monocytes give rise to macrophages that remain fixed in tissues.
   b. Phagocytic cells associated with the linings of blood vessels in bone marrow, liver, spleen, and lymph nodes constitute the reticuloendothelial tissue.
   c. Phagocytes function to remove foreign particles from tissues and body fluids.

Immunity
1. Origin of lymphocytes
   a. Lymphocytes originate in bone marrow and are released into the blood before they become differentiated.
   b. Some reach the thymus where they become T-lymphocytes.
   c. Others become B-lymphocytes after being processed in some unknown region of the body.
   d. Both T- and B-lymphocytes tend to reside in organs of the lymphatic system.

2. Antigens
   a. Before birth, cells make an inventory of the proteins and other large molecules present in the body.
   b. After the inventory, lymphocytes can differentiate between foreign substances and self-substances.
   c. Antigens are foreign substances that combine with T-cell and B-cell surface receptors and stimulate these cells to respond.
   d. Haptens are small molecules that can combine with larger ones, forming combinations that are antigenic.
3. Functions of lymphocytes
   a. T-lymphocytes attack antigens or antigen-bearing agents directly, providing cell-mediated immunity.
   b. They also release lymphokines that attract phagocytes, stimulate antibody production, and secrete interferon as well as toxins and growth-inhibiting factors that affect target cells.
   c. B-lymphocytes produce antibodies that act against specific antigens, providing antibody-mediated immunity.
   d. Normal immune responses require the interaction of T- and B-lymphocytes.
4. Immunoglobulin molecules
   a. Immunoglobulins are proteins that function as antibodies.
   b. They constitute the gamma globulin fraction of plasma.
   c. Each immunoglobulin molecule consists of four chains of amino acids linked together.
   d. Variable regions at the ends of these chains are specialized to react with antigens.
5. Types of immunoglobulins
   a. There are five major types of immunoglobulins—IgG, IgA, IgM, IgD, and IgE.
   b. IgG, IgA, and IgM make up most of the circulating antibodies.
6. Actions of antibodies
   a. Antibodies attack antigens directly, activate complement, or stimulate local tissue changes that are unfavorable to antigen-bearing agents.
   b. Direct attacks occur by means of agglutination, precipitation, neutralization, or lysis.
   c. Activated enzymes of complement attract phagocytes, alter cells so they become more susceptible to phagocytosis, and promote inflammation.
7. Immune responses
   a. Lymphocyte populations include thousands of clones, each of whose members is capable of responding to a specific antigen.
   b. Before lymphocytes can respond to antigens, they must be activated.
      (1) A T-cell becomes activated and proliferates when an antigen combines with its surface receptors.
      (2) Activation and proliferation of a B-cell usually requires antigen-receptor combination, display of antigens by a macrophage, and release of lymphokines by a T-helper cell.
      (3) Some newly formed B-cells become plasma cells that secrete antibodies into body fluids.
   c. The first response to an antigen is called the primary immune response.
      (1) During this response, antibodies are produced for several weeks.
      (2) Some B-cells and T-cells remain dormant as memory cells.
   d. A secondary immune response occurs rapidly if the same antigen is encountered later.
8. Types of immunity
   a. A person who encounters a pathogen and has a primary immune response develops naturally acquired active immunity.
   b. A person who receives vaccine containing a dead or weakened pathogen develops artificially acquired active immunity.
   c. A person who receives an injection of gamma globulin that contains ready-made antibodies has artificially acquired passive immunity.
   d. When antibodies pass through a placental membrane from a pregnant woman to her fetus, the fetus develops naturally acquired passive immunity.
   e. Active immunity lasts much longer than passive immunity.
9. Allergic reactions
   a. Allergic reactions involve combining antigens with antibodies; reactions are likely to be excessive or violent, and may cause tissue damage.
   b. Delayed-reaction allergy, which can occur in anyone and cause inflammation of the skin, results from repeated exposure to antigenic substances.
   c. A person with immediate-reaction allergy has an inherited ability to produce an abnormally large amount of IgE.
   d. Allergic reactions may damage mast cells, which in turn release histamine and serotonin.
   e. Released chemicals are responsible for the symptoms of the allergic reaction: hives, hay-fever, asthma, eczema, or gastric disturbances.
   f. An allergic reaction usually is terminated by suppressor cells that inhibit the production of IgE.
10. Transplantation and tissue rejection
   a. If tissue is transplanted from one person to another, the recipient's cells may recognize the donor's tissue as foreign and act against it.
   b. Tissue rejection reaction may be reduced by matching the donor and recipient tissues or by using immunosuppressive drugs.
   c. Immunosuppressive drugs interfere with the recipient's immune mechanisms, but cause the recipient to be very susceptible to infection.

Application of Knowledge

1. Based on your understanding of the functions of lymph nodes, how would you explain the fact that enlarged nodes are often removed for microscopic examination as an aid to diagnosing certain disease conditions?
2. Why is it true that an injection into the skin is, to a large extent, an injection into the lymphatic system?
3. Explain why vaccination provides long-lasting protection against a disease, while gamma globulin provides only short-term protection.

Review Activities

1. Explain how the lymphatic system is related to the cardiovascular system.
2. Trace the general pathway of lymph from the interstitial spaces to the bloodstream.
3. Identify and describe the location of the major lymphatic trunks and collecting ducts.
4. Distinguish between tissue fluid and lymph.
5. Describe the primary functions of lymph.
6. Explain how physical exercise promotes lymphatic circulation.
7. Explain how a lymphatic obstruction leads to edema.
8. Describe the structure of a lymph node and list its major functions.
9. Locate the major body regions occupied by lymph nodes.
10. Describe the structure and functions of the thymus gland.
11. Describe the structure and functions of the spleen.
12. Distinguish between specific and nonspecific body defenses against infection.
13. Explain what is meant by species resistance.
14. Name two mechanical barriers to infection.
15. Describe how enzymatic actions function as defense mechanisms.
16. Define *interferon* and explain its action.
17. List the major symptoms of inflammation and explain why each occurs.
18. Identify the major phagocytic cells in the blood and other tissues.
19. Define the *reticuloendothelial tissue* and explain its importance.
20. Review the origin of T-lymphocytes and B-lymphocytes.
21. Distinguish between an antigen and an antibody.
22. Define *hapten*.
23. Explain what is meant by cell-mediated immunity.
24. Explain what is meant by antibody-mediated immunity.
25. Explain two ways in which T-cells and B-cells interact.
26. Describe an immunoglobulin molecule.
27. Distinguish between the variable region and the constant region of an immunoglobulin molecule.
28. List the major types of immunoglobulins and describe their main functions.
29. Explain four mechanisms by which antibodies may attack antigens directly.
30. Explain the function of complement.
31. Define *clone of lymphocytes*.
32. Explain how T-cells become activated.
33. Describe the roles of macrophages and T-helper cells in the activation of B-cells.
34. Explain the function of plasma cells.
35. Distinguish between a primary and a secondary immune response.
36. Distinguish between active and passive immunity.
37. Define *vaccine*.
38. Explain how a vaccine produces its effect.
39. Describe how a fetus may obtain antibodies from the maternal blood.
40. Explain the relationship between an allergic reaction and an immune response.
41. Distinguish between an antigen and an allergen.
42. List the major events leading to a delayed-reaction allergic response.
43. Describe how an immediate-reaction allergic response may occur.
44. Explain the relationship between a tissue rejection and an immune response.
45. Describe two methods used to reduce the severity of a tissue rejection reaction.

20

The Urinary System

Cells form a variety of wastes as by-products of metabolic processes, and if these substances accumulate, their effects are likely to be toxic.

Body fluids, such as blood and lymph, serve to carry wastes away from tissues that produce them. Other parts remove these wastes from the blood and transport them to the outside. The respiratory system, for example, removes carbon dioxide from the blood, and the *urinary system* removes various salts and nitrogenous wastes. In both systems, the wastes are carried to the outside through tubular organs.

The urinary system also helps to maintain the normal concentrations of water and electrolytes within body fluids. It helps to regulate pH, volume of body fluids, red blood cell production, and blood pressure.

Chapter Outline

Chapter Objectives

After you have studied this chapter, you should be able to

1. Name the organs of the urinary system and list their general functions.

2. Describe the locations of the kidneys and the structure of a kidney.

3. List the functions of the kidneys.

4. Trace the pathway of blood through the major vessels within a kidney.

5. Describe a nephron and explain the functions of its major parts.

6. Explain how glomerular filtrate is produced and describe its composition.

7. Explain how various factors affect the rate of glomerular filtration and how this rate is regulated.

8. Discuss the role of tubular reabsorption in the formation of urine.

9. Explain why the osmotic concentration of the glomerular filtrate changes as it passes through a renal tubule.

10. Describe a countercurrent mechanism and explain how it helps concentrate urine.

11. Define *tubular secretion* and explain its role in urine formation.

12. Describe the structure of the ureters, urinary bladder, and urethra.

13. Discuss the process of micturition and explain how it is controlled.

14. Complete the review activities at the end of this chapter. Note that the items are worded in the form of specific learning objectives. You may want to refer to them before reading the chapter.

Key Terms

afferent arteriole   (af′er-ent ar-te′re-ōl)

autoregulation   (aw′′to-reg′′u-la′shun)

Bowman's capsule   (bo′manz kap′sūl)

countercurrent mechanism
   (kown′ter kur′ent mek′ah-nizm)

detrusor muscle   (de-truz′or mus′l)

efferent arteriole   (ef′er-ent ar-te′re-ōl)

glomerulus   (glo-mer′u-lus)

juxtaglomerular apparatus   (juks′′tah-glo-mer′u-lar
   ap′′ah-ra′tus)

micturition   (mik′′tu-rish′un)

nephron   (nef′ron)

peritubular capillary   (per′′i-tū′bu-lar kap′ĭ-ler′′e)

renal corpuscle   (re′nal kor′pusl)

renal cortex   (re′nal kor′teks)

renal medulla   (re′nal mĕ-dul′ah)

renal plasma threshold
   (re′nal plaz′mah thresh′old)

renal tubule   (re′nal tu′būl)

Aids to Understanding Words

calyc-, a small cup: major *calyc*es—cuplike divisions of the renal pelvis.

detrus-, to force away: *detrus*or muscle—muscle within the bladder wall that causes urine to be expelled.

glom-, little ball: *glom*erulus—cluster of capillaries within a renal corpuscle.

juxta-, near to: *juxta*medullary nephron—a nephron located near the renal medulla.

mict-, to pass urine: *mict*urition—process of expelling urine from the bladder.

nephr-, pertaining to the kidney: *nephr*on—functional unit of a kidney.

papill-, nipple: renal *papill*ae—small elevations that project into a renal calyx.

ren-, kidney: *ren*al cortex—outer region of the kidney.

trigon-, a triangular shape: *trigon*e—triangular area on the internal floor of the bladder.

THE URINARY SYSTEM consists of a pair of glandular *kidneys,* which remove substances from the blood, form urine, and help to regulate various metabolic processes; a pair of tubular *ureters,* which transport urine away from the kidneys; a saclike *urinary bladder,* which serves as a urine reservoir; and a tubular *urethra,* which conveys urine to the outside of the body. These organs are shown in figures 20.1 and 20.2.

The Kidney

A **kidney** is a reddish brown, bean-shaped organ with a smooth surface. It is about 12 cm long, 6 cm wide, and 3 cm thick in an adult, and is enclosed in a tough, fibrous capsule (tunic fibrosa).

Location of the Kidneys

The kidneys lie on either side of the vertebral column in a depression high on the posterior wall of the abdominal cavity.

Although the positions of the kidneys may vary slightly with changes in posture and with breathing movements, their upper and lower borders are generally at the levels of the twelfth thoracic and third lumbar vertebrae, respectively. The left kidney usually is about 1.5 to 2 cm higher than the right one.

The kidneys are positioned *retroperitoneally,* which means they are behind the parietal peritoneum and against the deep muscles of the back. As figure 20.3 shows, they are held in position by connective tissue (renal fascia) and masses of adipose tissue (renal fat) that surround them.

Structure of a Kidney

The lateral surface of each kidney is convex, while its medial side is deeply concave. The resulting medial depression leads into a hollow chamber called the **renal sinus.** The entrance to this sinus is termed the *hilum,* and through it pass various blood vessels, nerves, lymphatic vessels, and the ureter (figure 20.4).

The superior end of the ureter is expanded to form a funnel-shaped sac called the **renal pelvis,** which is located inside the renal sinus. The pelvis is divided into two or three tubes called *major calyces* (singular, *calyx*), and they in turn are divided into several (eight to fourteen) *minor calyces.*

A series of small elevations project into the renal sinus from its wall. These projections are called *renal papillae,* and each of them is pierced by tiny openings that lead into a minor calyx.

Figure 20.1

The urinary system includes the kidneys, ureters, urinary bladder, and urethra. What are the general functions of this system?

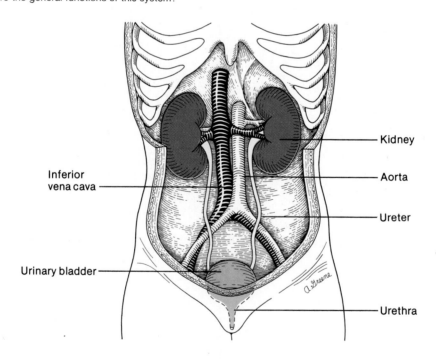

Figure 20.2

What features of the urinary system do you recognize in this
X-ray film?

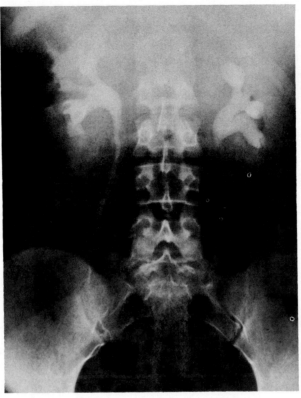

Figure 20.3

The kidneys are behind the parietal peritoneum and are
surrounded and supported by adipose tissue.

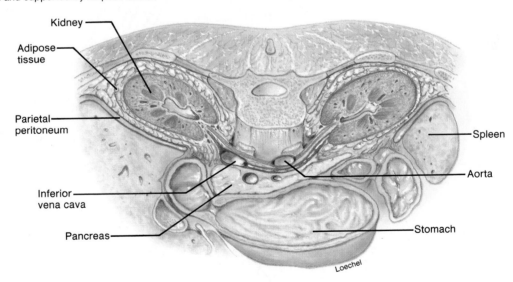

Figure 20.4

(a) Longitudinal section of a kidney; *(b)* a renal pyramid
containing nephrons; *(c)* a single nephron.

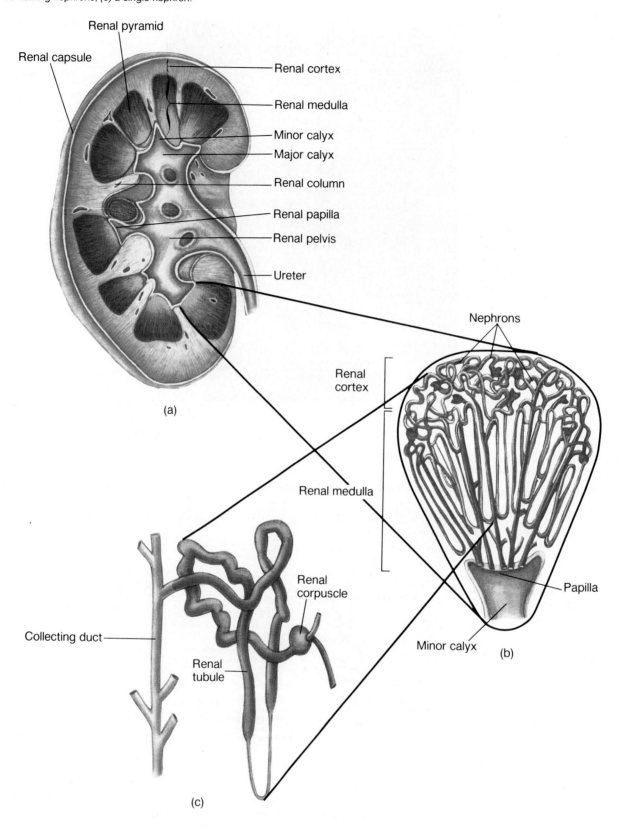

Renal pyramid

Renal capsule

Renal cortex

Renal medulla

Minor calyx

Major calyx

Renal column

Renal papilla

Renal pelvis

Ureter

(a)

Nephrons

Renal cortex

Renal medulla

Papilla

Minor calyx

(b)

Renal corpuscle

Collecting duct

Renal tubule

(c)

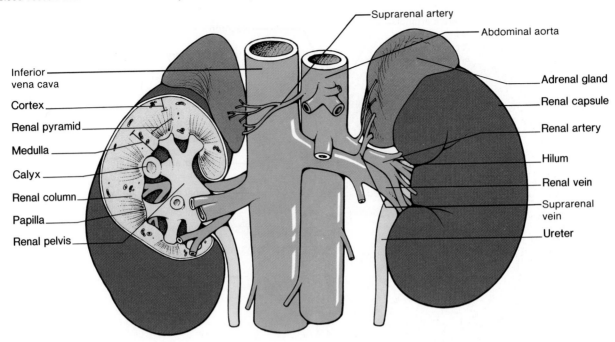

The substance of the kidney is divided into two distinct regions: an inner medulla and an outer cortex. The **renal medulla** is composed of conical masses of tissue called *renal pyramids,* whose bases are directed toward the convex surface of the kidney, and whose apexes form the renal papillae. The tissue of the medulla appears striated due to the presence of microscopic tubules leading from the cortex to the renal papillae.

The **renal cortex,** which appears somewhat granular, forms a shell around the medulla. Its tissue dips into the medulla between adjacent renal pyramids to form *renal columns.* The granular appearance of the cortex is due to the random arrangement of tiny tubules associated with **nephrons,** the functional units of the kidney.

> During development of a human kidney, several pyramidal masses or lobes fuse to form the final structure. Consequently, lobulation is evident on the surface of a fetal kidney, and this condition sometimes persists into adulthood.

Functions of the Kidneys

The kidneys function to remove metabolic wastes from the blood and excrete them to the outside. They also carry on a variety of equally important regulatory activities including helping to control the rate of red blood cell formation by secreting the hormone *erythropoietin* (see chapter 17), helping to regulate blood pressure by secreting the enzyme *renin* (see chapter 18), and helping to regulate the absorption of calcium by activating *vitamin D* (see chapter 13).

The kidneys also help to regulate the volume, composition, and pH of body fluids. These functions involve complex mechanisms that lead to the formation of urine. They are discussed in a subsequent section of this chapter and are explored in still more detail in chapter 21.

1. Where are the kidneys located?
2. Describe the structure of a kidney.
3. Name the functional unit of the kidney.
4. What are the general functions of the kidneys?

Renal Blood Vessels

Blood is supplied to the kidneys by means of **renal arteries** that arise from the abdominal aorta (figure 20.5). These arteries transport a relatively large volume of blood; in fact, when a person is at rest they usually carry from 15%–30% of the total cardiac output into the kidneys.

A renal artery enters a kidney through the hilum and gives off several branches, the *interlobar arteries,* that pass between the renal pyramids. At the junction between the medulla and cortex, the interlobar arteries

Figure 20.6
Main branches of the renal artery and vein.

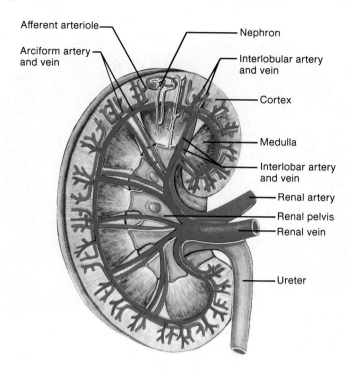

Afferent arteriole

Arciform artery and vein

Nephron

Interlobular artery and vein

Cortex

Medulla

Interlobar artery and vein

Renal artery

Renal pelvis

Renal vein

Ureter

Figure 20.7
A scanning electron micrograph of a cast of the renal blood vessels associated with the glomeruli (260 X). (*Tissues and Organs: A Text-Atlas of Scanning Electron Microscopy,* by R. G. Kessel and R. H. Kardon. © 1979 W. H. Freeman and Company.)

Efferent arteriole

Afferent arteriole

Glomerulus

Peritubular capillary

branch to form a series of incomplete arches, the *arciform arteries* (arcuate arteries), which in turn give rise to *interlobular arteries*. Lateral branches of the interlobular arteries, called **afferent arterioles**, lead to the nephrons.

Venous blood is returned through a series of vessels that correspond generally to the arterial pathways. For example, the venous blood passes through interlobular, arciform, interlobar, and renal veins. The **renal vein** then joins the inferior vena cava as it courses through the abdominal cavity. (Branches of the renal arteries and veins are shown in figures 20.6 and 20.7.)

Patients with end-stage renal disease, who are facing death, sometimes are treated with a *kidney transplant.* In this procedure, a kidney from a living donor or a cadaver, whose tissues are antigenically similar to those of the recipient, is placed surgically in the depression on the medial surface of the right or left ilium (iliac fossa). The renal artery and vein of the donor kidney are connected to the recipient's iliac artery and vein respectively, and its ureter is attached to the dome of the recipient's urinary bladder.

The Nephrons

Structure of a Nephron
A kidney contains about one million nephrons, each consisting of a **renal corpuscle** and a **renal tubule** (figure 20.4).

Figure 20.8
Structure of a nephron and the blood vessels associated with it.

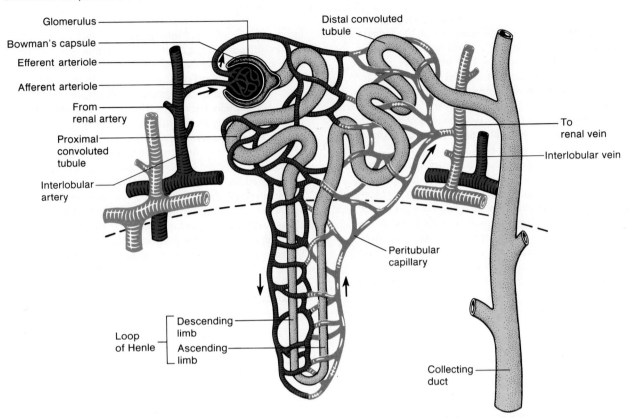

A renal corpuscle (malpighian corpuscle) is composed of a tangled cluster of blood capillaries called a **glomerulus,** and a thin-walled, saclike structure called **Bowman's capsule,** that surrounds the glomerulus.

The Bowman's capsule is an expansion at the closed end of a renal tubule. It is composed of two layers of squamous epithelial cells: a visceral layer that closely covers the glomerulus, and an outer parietal layer that is continuous with the visceral layer and with the wall of the renal tubule.

The cells of the parietal layer are typical squamous epithelial cells; however, those of the visceral layer are highly modified epithelial cells called *podocytes*. Each podocyte has several primary processes extending from its cell body, and these processes, in turn, bear numerous secondary processes or *pedicels*. The pedicels of each cell interdigitate with those of adjacent podocytes, and the clefts between them form a complicated system of *slit pores*. (See figure 20.13.)

The renal tubule leads away from the Bowman's capsule and becomes highly coiled. This coiled portion of the tubule is appropriately named the *proximal convoluted tubule*.

The proximal tubule dips toward the renal pelvis following a straight path into the deeper layers of the cortex to become the *descending limb of the loop of*

Henle. The tubule then curves back toward its renal corpuscle and forms the *ascending limb of the loop of Henle*.

The ascending limb returns in a straight path to the region of the renal corpuscle, where it becomes highly coiled again, and is called the *distal convoluted tubule*. This distal portion is shorter than the proximal tubule, and its convolutions are less complex.

Several distal convoluted tubules merge in the renal cortex to form a *collecting duct*, which in turn passes into the renal medulla, becoming larger and larger as it is joined by other collecting ducts. The resulting tube (papillary duct) empties into a minor calyx through an opening in a renal papilla. The parts of a nephron are shown in figures 20.8 and 20.9.

Juxtaglomerular Apparatus
Near its beginning, the distal convoluted tubule passes between the afferent and efferent arterioles and contacts them. At the point of contact, the epithelial cells of the distal tubule are quite narrow and densely packed. These cells comprise a structure called the *macula densa*.

Glomerulonephritis
A Clinical Application

Nephritis refers to an inflammation of the kidney; *glomerulonephritis* refers to an inflammation affecting the glomeruli. This latter condition may occur in acute or chronic forms and can lead to renal failure.

Acute glomerulonephritis usually results from an abnormal immune reaction that develops one to three weeks following a certain type of bacterial infection (beta-hemolytic streptococci). As a rule, the infectious condition occurs in some other part of the body and does not affect the kidneys directly. Instead, the presence of bacterial antigens triggers an immune reaction, and antibodies are produced against these antigens. As a consequence of antigen-antibody reactions, insoluble immune complexes are formed (see chapter 19), and they are carried by the blood to the kidneys. The antigen-antibody complexes tend to be deposited in the glomerular capillaries, causing them to be blocked.

At the same time, an inflammation reaction is triggered (see chapters 5 and 19), and large numbers of white blood cells are attracted into the region, blocking the capillaries still more. Those capillaries remaining open may become excessively permeable, and if this happens, plasma proteins and red blood cells may be lost into the urine.

In severe cases of glomerulonephritis, the functions of the kidneys may fail completely, and without treatment, the person is likely to die within a week or so.

Chronic glomerulonephritis is a progressive disease in which increasing numbers of nephrons are slowly damaged until finally the kidneys are unable to perform their usual functions. This condition usually is associated with certain diseases other than streptococcal infections and once again, it involves the formation of antigen-antibody complexes that precipitate and accumulate in the glomeruli. The resulting inflammation is prolonged, and it is accompanied by tissue changes in which the glomerular membranes are slowly replaced by fibrous tissue. As this happens, the functions of the nephrons are permanently lost, and eventually the kidneys may fail.

Figure 20.9
Light micrograph of the human renal cortex (about 150 X). What features can you identify?

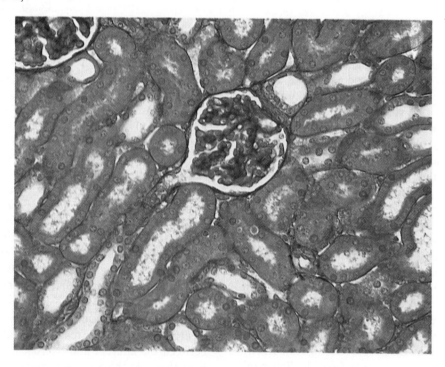

Figure 20.10
The juxtaglomerular apparatus consists of the macula densa and
the juxtaglomerular cells.

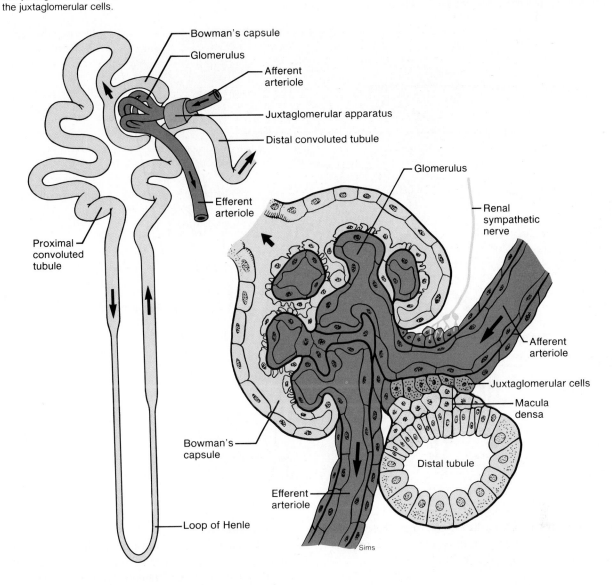

- Bowman's capsule
- Glomerulus
- Afferent arteriole
- Juxtaglomerular apparatus
- Distal convoluted tubule
- Efferent arteriole
- Proximal convoluted tubule
- Glomerulus
- Renal sympathetic nerve
- Afferent arteriole
- Juxtaglomerular cells
- Macula densa
- Bowman's capsule
- Distal tubule
- Efferent arteriole
- Loop of Henle

Sims

Close by, in the walls of the arterioles near their attachments to the glomerulus, some of the smooth muscle cells are enlarged. They are called *juxtaglomerular cells,* and together with the cells of the macula densa, they constitute the **juxtaglomerular apparatus (complex)**. This structure plays an important role in regulating the flow of blood through various renal vessels (figure 20.10).

Cortical and Juxtamedullary Nephrons

Most nephrons have corpuscles located in the renal cortex near the surface of the kidney. They are called *cortical nephrons,* and they have relatively short loops of Henle that usually do not reach the renal medulla.

Another group, called *juxtamedullary nephrons,* have corpuscles close to the renal medulla, and their loops of Henle extend deep into the medulla. Although

Figure 20.11

Cortical nephrons are close to the surface of a kidney; juxtamedullary nephrons are near the renal medulla.

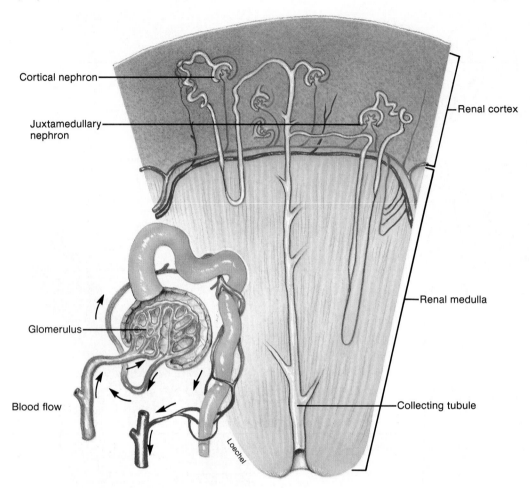

Cortical nephron

Juxtamedullary nephron

Renal cortex

Glomerulus

Renal medulla

Blood flow

Collecting tubule

Loechel

they represent only about 20% of the total, these nephrons play an important role in regulating the process of concentrating urine (figure 20.11).

Blood Supply of a Nephron

The capillary cluster that forms a glomerulus arises from an **afferent arteriole.** After passing through the capillary of the glomerulus, blood enters an **efferent arteriole** (rather than a venule) whose diameter is somewhat less than that of the afferent vessel.

Because of its small diameter, the efferent arteriole creates some resistance to blood flow. This causes blood to back up into the glomerulus, producing a relatively high pressure in the glomerular capillary.

The efferent arteriole branches into a complex, freely interconnecting network of capillaries that surrounds the various portions of the renal tubule. This network is called the **peritubular capillary system** and the blood it contains is under relatively low pressure (figures 20.8 and 20.11).

Special branches of this system, which receive blood primarily from the efferent arterioles of the juxtamedullary nephrons, form capillary loops called *vasa recta.* These loops dip into the renal medulla and are closely associated with the loops of the juxtamedullary nephrons.

After flowing through the vasa recta, blood is returned to the renal cortex, where it joins blood from other branches of the peritubular capillary system and enters the venous system of the kidney (figure 20.12).

Figure 20.12
The capillary loop of the vasa recta is closely associated with the loop of Henle of a juxtamedullary nephron.

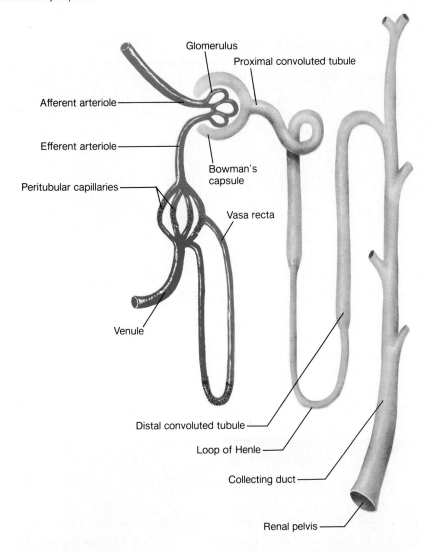

- Glomerulus
- Proximal convoluted tubule
- Afferent arteriole
- Efferent arteriole
- Peritubular capillaries
- Bowman's capsule
- Vasa recta
- Venule
- Distal convoluted tubule
- Loop of Henle
- Collecting duct
- Renal pelvis

1. Describe the characteristics of the blood supply to the kidney.
2. Name the parts of a nephron.
3. What parts comprise the juxtaglomerular apparatus?
4. Distinguish between a cortical and a juxtamedullary nephron.
5. Describe the blood supply of a nephron.

Urine Formation

The functions of the nephrons include the removal of waste substances from the blood and the regulation of water and electrolyte concentrations within the body fluids. The end product of these functions is **urine,** which is excreted to the outside of the body carrying with it wastes, excess water, and excess electrolytes.

Urine formation involves these processes: *filtration,* into renal tubules, of various substances from the plasma within glomerular capillaries; *reabsorption,* into the plasma, of some of these substances; *secretion,* into the renal tubules, of other substances from the plasma within the peritubular capillaries.

Glomerular Filtration

Urine formation begins when water and various dissolved substances are filtered out of the glomerular capillaries and into the Bowman's capsules. The filtration of these materials through the capillary walls is much like the filtration that occurs at the arteriole ends

Figure 20.13

The first step in urine formation is the filtration of substances through the glomerular membrane into Bowman's capsule.

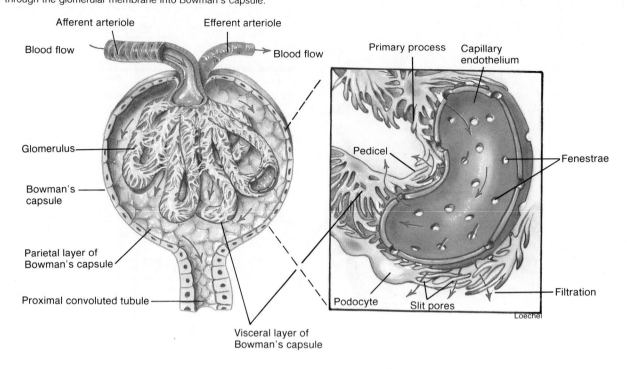

Figure 20.14

Scanning electron micrograph of a glomerulus. Note the slit pores between the pedicels.

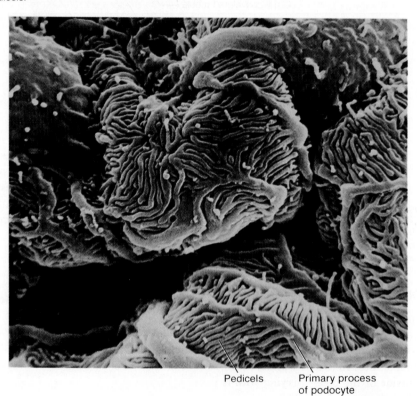

of other capillaries throughout the body. The glomerular capillaries, however, are many times more permeable than the capillaries in other tissues due to the presence of numerous tiny openings (fenestrae) in their walls (figures 20.13 and 20.14).

Filtration Pressure

As in the case of other capillaries, the force mainly responsible for the movement of substances through the glomerular capillary wall is the pressure of the blood inside (glomerular hydrostatic pressure). This movement is also influenced by the osmotic pressure of the plasma in the glomerulus and by the hydrostatic pressure inside the Bowman's capsule. An increase in either of these pressures will oppose movement out of the capillary and thus reduce filtration. The net pressure acting to force substances out of the glomerulus is called **filtration pressure,** and it can be calculated by subtracting the sum of the opposing forces from the hydrostatic pressure of the glomerulus.

$$\begin{matrix}\text{Filtration} \\ \text{pressure}\end{matrix} = \begin{matrix}\text{Glomerular} \\ \text{hydrostatic} \\ \text{pressure}\end{matrix} - \left(\begin{matrix}\text{Glomerular} \\ \text{plasma} \\ \text{osmotic} \\ \text{pressure}\end{matrix} + \begin{matrix}\text{Capsular} \\ \text{hydrostatic} \\ \text{pressure}\end{matrix}\right)$$

The resulting **glomerular filtrate,** which is received by the Bowman's capsule, has about the same composition as the filtrate that becomes tissue fluid elsewhere. That is, glomerular filtrate is largely water and contains essentially the same substances as blood plasma, except for the larger protein molecules, which the filtrate lacks. More specifically, glomerular filtrate contains water; glucose; amino acids; urea; uric acid; creatine; creatinine; and ions of sodium, chlorine, potassium, calcium, bicarbonate, phosphate, and sulfate. The relative concentrations of some substances in plasma, glomerular filtrate, and urine are shown in chart 20.1.

Filtration Rate

The rate of glomerular filtration is directly proportional to the filtration pressure. Consequently, factors that affect the glomerular hydrostatic pressure, the glomerular plasma osmotic pressure, or the hydrostatic pressure in Bowman's capsule will affect the rate of filtration also (figure 20.15).

For example, since the glomerular capillary is located between two arterioles—the *afferent* and *efferent arterioles*—any change in the diameters of these vessels is likely to cause a change in the glomerular hydrostatic pressure, accompanied by a change in the

Chart 20.1 Relative concentrations of certain substances in plasma, glomerular filtrate, and urine

Substance	Concentrations (mEq/l)		
	Plasma	Glomerular filtrate	Urine
Sodium (Na^+)	142	142	128
Potassium (K^+)	5	5	60
Calcium (Ca^{+2})	4	4	5
Magnesium (Mg^{+2})	3	3	15
Chlorine (Cl^-)	103	103	134
Bicarbonate (HCO_3^-)	27	27	14
Sulfate (SO_4^{-2})	1	1	33
Phosphate (PO_4^{-3})	2	2	40

Substance	Concentrations (mg/100 ml)		
	Plasma	Glomerular filtrate	Urine
Glucose	100	100	0
Urea	26	26	1820
Uric acid	4	4	53
Creatinine	1	1	196

Figure 20.15
The rate of glomerular filtration is affected by the hydrostatic pressure of the plasma, the osmotic pressure of the plasma, and the hydrostatic pressure of the fluid in Bowman's capsule.

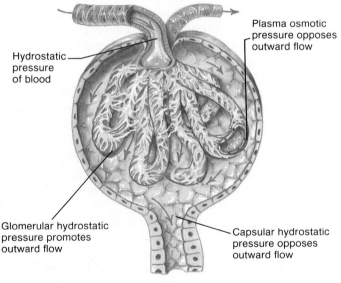

Hydrostatic pressure of blood

Plasma osmotic pressure opposes outward flow

Glomerular hydrostatic pressure promotes outward flow

Capsular hydrostatic pressure opposes outward flow

Loechel

glomerular filtration rate. The afferent arteriole, through which blood enters the glomerulus, may become constricted as a result of mild stimulation by sympathetic nerve impulses. If this occurs, blood flow is diminished, the glomerular hydrostatic pressure is decreased, and the filtration rate drops. If, on the other hand, the efferent arteriole (through which blood leaves the glomerulus) becomes constricted, blood backs up into the glomerulus, the glomerular hydrostatic pressure is increased, and the filtration rate rises. Converse effects are produced by vasodilation of these vessels.

If the arterial blood pressure drops excessively, as may occur in *shock,* the glomerular hydrostatic pressure may decrease below the level required for filtration. At the same time, the epithelial cells of the renal tubules may fail to receive sufficient nutrients to maintain their high rates of metabolism. As a result, cells may die (tubular necrosis) and renal functions may be lost. Changes such as these sometimes cause the effects of shock to be irreversible.

As mentioned earlier, the osmotic pressure of the glomerular plasma is another factor that influences filtration pressure and affects the rate of filtration. In capillaries, the blood hydrostatic pressure, acting to force water and dissolved substances outward, is opposed by the effect of the plasma osmotic pressure that attracts water inward. Actually, as filtration occurs through the capillary wall, proteins remaining in the plasma cause the osmotic pressure within the glomerular capillary to rise. When this pressure reaches a certain high level, filtration ceases. Conversely, conditions that tend to decrease plasma osmotic pressure, such as a decrease in plasma protein concentration, cause an increase in the filtration rate.

For these reasons, the rate of blood flow through the glomerulus also affects the filtration rate. More specifically, if the blood is flowing slowly, a larger proportion of the plasma filters out of the glomerulus, and as the plasma osmotic pressure rises, the filtration rate decreases. If, on the other hand, the flow is rapid, there is less change in the plasma osmotic pressure, and the filtration rate remains higher.

The hydrostatic pressure in Bowman's capsule is still another factor that may affect the filtration pressure and the rate of filtration. This capsular pressure sometimes changes as a result of an obstruction, such as may be caused by a stone in a ureter or an enlarged prostate gland pressing on the urethra. If this occurs, fluids tend to back up into the renal tubules and cause the hydrostatic pressure in Bowman's capsules to rise. Since any increase in capsular pressure opposes glomerular filtration, the rate of filtration may decrease significantly.

Figure 20.16
(a) Relative amounts of glomerular filtrate and (b) urine formed in 24 hours.

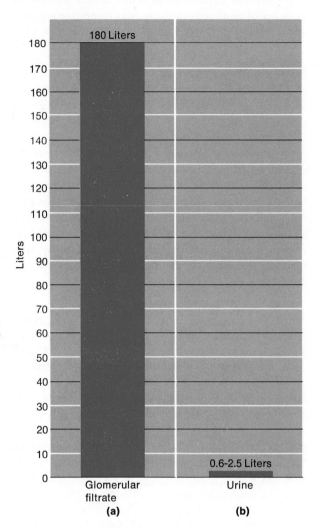

In an average adult, the glomerular filtration rate for the nephrons of both kidneys is about 125 ml per minute, or 180,000 ml (180 liters) in 24 hours. Assuming that the plasma volume is about 3 liters, the production of 180 liters of filtrate in 24 hours means that all of the plasma must be filtered through the glomeruli about sixty times each day (figure 20.16).

Since this 24-hour volume is nearly 45 gallons, it is obvious that not all of it is excreted as urine. Instead, most of the fluid that passes through the renal tubules is reabsorbed and reenters the plasma.

The volume of plasma filtered by the kidneys is also related to the amount of *surface area* within the glomerular capillaries. This surface area is estimated to be about 2 square meters—approximately equal to the surface area of an adult's skin.

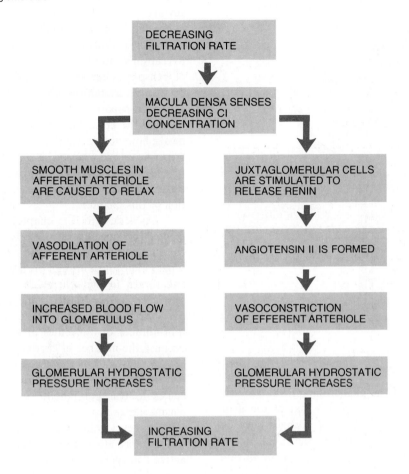

DECREASING FILTRATION RATE

↓

MACULA DENSA SENSES DECREASING Cl CONCENTRATION

SMOOTH MUSCLES IN AFFERENT ARTERIOLE ARE CAUSED TO RELAX

↓

VASODILATION OF AFFERENT ARTERIOLE

↓

INCREASED BLOOD FLOW INTO GLOMERULUS

↓

GLOMERULAR HYDROSTATIC PRESSURE INCREASES

JUXTAGLOMERULAR CELLS ARE STIMULATED TO RELEASE RENIN

↓

ANGIOTENSIN II IS FORMED

↓

VASOCONSTRICTION OF EFFERENT ARTERIOLE

↓

GLOMERULAR HYDROSTATIC PRESSURE INCREASES

INCREASING FILTRATION RATE

1. What general processes are involved with urine formation?
2. How is filtration pressure calculated?
3. What factors influence the rate of glomerular filtration?

Regulation of Filtration Rate

The regulation of the filtration rate involves the *juxtaglomerular apparatus,* which was described previously (figure 20.10), and two negative feedback mechanisms. These mechanisms are triggered whenever the filtration rate is decreasing. For example, as the rate decreases, the concentration of chloride ions reaching the *macula densa* in the distal convoluted tubule also decreases. In response, the macula densa signals the smooth muscles in the wall of the afferent arteriole to relax, and the vessel becomes dilated. This action allows more blood to flow into the glomerulus, increasing the glomerular pressure, and as a consequence, the filtration rate rises toward its previous level.

At the same time that the macula densa signals the afferent arteriole to dilate, it stimulates the *juxtaglomerular cells* to release *renin* (see chapter 13). This enzyme causes a plasma globulin (angiotensinogen) to form a substance called *angiotensin I*. Angiotensin I is converted quickly to *angiotensin II* by an enzyme called *converting enzyme* that is present in the plasma and lungs.

Angiotensin II is a vasoconstrictor, and it stimulates the smooth muscle cells in the wall of the efferent arteriole to contract, constricting the vessel. As a result of this action, blood tends to back up into the glomerulus, and as the glomerular hydrostatic pressure increases, the filtration rate increases also (figure 20.17).

These two mechanisms operate together to ensure a constant blood flow through the glomerulus and a relatively stable glomerular filtration rate, in spite of marked changes occurring in the arterial blood pressure. This phenomenon, by which a mechanism within an organ or tissue maintains a constant blood flow through that part even though the arterial blood pressure is changing, is called **autoregulation.**

Figure 20.18

Reabsorption is the process by which substances are transported from the glomerular filtrate into the blood of the peritubular capillary. What substances are reabsorbed in this manner?

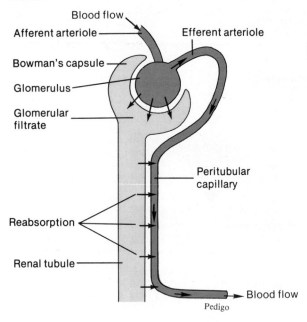

Tubular Reabsorption

If the composition of the glomerular filtrate entering the renal tubule is compared with that of the urine leaving the tubule, it is clear that changes have occurred as the fluid passed through the tubule (chart 20.1). For example, glucose is present in the filtrate, but is absent in the urine. Also, urea and uric acid are considerably more concentrated in urine than they are in the glomerular filtrate. Such changes in fluid composition are largely the result of **tubular reabsorption,** a process by which substances are transported out of the glomerular filtrate, through the epithelium of the renal tubule, and into the blood of the peritubular capillary (figure 20.18).

Since the efferent arteriole is narrower than the peritubular capillary, blood flowing from the former into the latter is under relatively low pressure. Also, the wall of the peritubular capillary is more permeable than that of other capillaries. Both of these factors enhance the rate of fluid reabsorption from the renal tubule.

Although tubular reabsorption occurs throughout the renal tubule, most of it occurs in the proximal convoluted portion. The epithelial cells in this portion have numerous microscopic projections called *microvilli* that form a "brush border" on their free surfaces (see chapter 14). These tiny extensions greatly increase the surface area exposed to glomerular filtrate and enhance the reabsorption process (see figure 14.36).

Various segments of the renal tubule are adapted to reabsorb specific substances, using particular modes of transport. Glucose reabsorption, for example, occurs primarily through the walls of the proximal tubule by active transport. Water also is reabsorbed rapidly through the epithelium of the proximal tubule by osmosis; however, portions of the distal tubule are almost impermeable to water. This characteristic of the distal tubule is important in the regulation of urine concentration and volume, as is described in a subsequent section of this chapter.

As is described in chapter 3, an active transport mechanism depends on the presence of carrier molecules in a cell membrane. These carriers transport passenger molecules through the membrane, release them, and return to the other side to transport more passenger molecules. Such a mechanism has a *limited transport capacity;* that is, it can only transport a certain number of molecules in a given amount of time because the number of carriers is limited.

Usually all of the glucose in the glomerular filtrate is reabsorbed, since there are enough carrier molecules to transport them. Sometimes, however, the plasma glucose concentration increases, and if it reaches a critical level, called the *renal plasma threshold,* there will be more glucose molecules in the filtrate than the active transport mechanism can handle. As a result, some glucose will remain in the filtrate and be excreted in the urine.

The appearance of glucose in the urine is called *glucosuria* (or *glycosuria*). This condition may occur following the administration of glucose intravenously or in a patient with diabetes mellitus. If the cause is diabetes mellitus, the blood glucose concentration rises because of insufficient insulin from the pancreas.

Amino acids also enter the glomerular filtrate and are reabsorbed in the proximal convoluted tubule, apparently by three different active transport mechanisms. Each mechanism is thought to reabsorb a different group of amino acids, whose members have molecular similarities. As a result of their actions, only a trace of amino acids usually remains in urine.

Although the glomerular filtrate is nearly free of protein, some *albumin* may be present. These proteins have relatively small molecules, and they are reabsorbed by *pinocytosis* through the brush border of epithelial cells lining the proximal convoluted tubule. Once they are inside an epithelial cell, the proteins probably are converted to amino acids and moved into the blood of the peritubular capillary.

Other substances reabsorbed by the epithelium of the proximal convoluted tubule include creatine, lactic acid, citric acid, uric acid, ascorbic acid (vitamin C), phosphate ions, sulfate ions, calcium ions, potassium ions, and sodium ions. As a group, these substances are reabsorbed by active transport mechanisms with limited transport capacities (like that of glucose). Such a substance usually does not appear in the urine until its concentration in the glomerular filtrate exceeds its particular threshold.

Sodium and Water Reabsorption

Substances that remain in the renal tubule tend to become more and more concentrated as water is reabsorbed from the filtrate. Most of this water reabsorption occurs *passively* by osmosis in the proximal convoluted tubule and is closely associated with the active reabsorption of sodium ions. In fact, if sodium reabsorption increases, water reabsorption increases; if sodium reabsorption decreases, water reabsorption decreases also.

About 70% of the *sodium ion reabsorption* occurs in the proximal segment of the renal tubule by active transport (sodium pump mechanism). As these positively charged ions (Na^+) are moved through the tubular wall, negatively charged ions including chloride ions (Cl^-), phosphate ions (PO_4^{-3}), and bicarbonate ions (HCO_3^-) accompany them. This movement of negatively charged ions is due to the electrochemical attraction between particles of opposite charge. It is termed **passive transport** because it does not require a direct expenditure of cellular energy.

As more and more sodium ions are actively transported into the peritubular capillary, along with various negatively charged ions, the concentration of solutes within the peritubular blood is increased. Furthermore, since water moves through cell membranes from regions of lesser solute concentration (hypotonic) toward regions of greater solute concentration (hypertonic), water is transported by osmosis from the renal tubule into the peritubular capillary. Because of the movement of solutes and water into the peritubular capillary, the volume of fluid within the renal tubule is greatly reduced (figure 20.19).

Figure 20.19
Water reabsorption by osmosis occurs in response to the reabsorption of sodium by active transport.

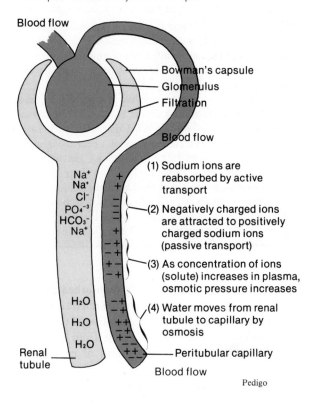

Blood flow

Bowman's capsule
Glomerulus
Filtration

Blood flow

Na+
Na+
Cl-
PO₄⁻³
HCO₃⁻
Na+

(1) Sodium ions are reabsorbed by active transport

(2) Negatively charged ions are attracted to positively charged sodium ions (passive transport)

(3) As concentration of ions (solute) increases in plasma, osmotic pressure increases

H₂O
H₂O
H₂O

(4) Water moves from renal tubule to capillary by osmosis

Renal tubule

Peritubular capillary

Blood flow

Pedigo

1. How is the peritubular capillary adapted for reabsorption?
2. What substances present in glomerular filtrate are not normally present in urine?
3. What mechanisms are responsible for reabsorption of solutes from the glomerular filtrate?
4. Define renal plasma threshold.
5. Describe the role of passive transport in urine formation.

Regulation of Urine Concentration and Volume

Sodium ions continue to be reabsorbed by active transport as the tubular fluid moves through the loop of Henle, the distal convoluted tubule, and the collecting duct. Consequently, almost all the sodium that enters the renal tubule as glomerular filtrate may be reabsorbed before the urine is excreted.

The nephrotic syndrome is a set of symptoms that often appears in patients with renal diseases. This syndrome is characterized by considerable loss of plasma proteins into the urine (proteinuria), widespread edema, and increased susceptibility to infections.

The plasma proteins are lost into the urine because of increased permeability of the glomerular membranes, which accompanies various renal disorders, such as glomerulonephritis. As a consequence of a decreasing plasma protein concentration (hypoproteinemia), the plasma osmotic pressure falls. This may lead to widespread, severe edema as a large volume of fluid accumulates in the interstitial spaces within the tissues and in various body spaces, such as the abdominal cavity, pleural cavity, pericardial cavity, and joint cavities.

Also, as edema develops, the blood volume decreases and the blood pressure drops. These changes may trigger the release of aldosterone from the adrenal cortex (see chapter 13), which in turn stimulates the kidneys to conserve sodium ions and water. This action reduces the urine output and may aggravate the edema.

The nephrotic syndrome sometimes appears in young children who have *lipoid nephrosis*. The cause of this condition is unknown, but it is characterized by changes occurring in the epithelial cells of the glomeruli. As a result of these changes, the cells of the glomerular membranes become enlarged and distorted, allowing proteins to leak through.

Water also continues to be reabsorbed passively by osmosis in various segments of the renal tubule, and as this occurs, the osmotic concentration of the tubular fluid changes. These changes result from a *countercurrent mechanism* and the actions of certain *hormones*.

The **countercurrent mechanism** involves the loops of Henle of the juxtamedullary nephrons. The descending and ascending limbs of these U-shaped structures lie parallel and very close to one another. The mechanism is named for the fact that fluid moving down the descending limb creates a current that is counter to that of the fluid moving up in the ascending limb.

The tubular fluid in the proximal segment is *isotonic* to the plasma of the peritubular capillary blood; however, it becomes *hypertonic* when it moves through the descending limb. Then, as it moves through the ascending limb, the tubular fluid becomes *hypotonic*.

These changes in the osmotic concentration are due primarily to activity in the epithelium of the ascending limb. Specifically, the epithelial lining in the upper portion of this limb is thickened (thick segment) and is impermeable to water. However, the epithelium does carry on the active reabsorption of chloride ions, and sodium ions follow them passively. As NaCl accumulates, the interstitial fluid outside the ascending limb becomes *hypertonic,* while the tubular fluid inside becomes *hypotonic* since it is losing its solute.

The epithelium of the descending limb is thin (thin segment), lacks microvilli, and is quite permeable. Because this segment is surrounded by the hypertonic fluid created by the ascending limb, water tends to leave the descending limb by osmosis. Thus, the contents of the descending limb become more and more concentrated, or *hypertonic* to the plasma of the peritubular capillary.

Also, since the concentration of NaCl in the medullary interstitial fluid is high, NaCl reenters the descending limb by diffusion. This makes the fluid in the descending limb still more concentrated (hypertonic) (figure 20.20).

Thus, NaCl moves from the descending limb into the ascending limb, is reabsorbed into the medullary interstitial fluid, and diffuses back into the descending limb again. Each time this circuit is completed, the concentration of NaCl increases or multiplies. For this reason, the mechanism is called a *countercurrent multiplier.*

As a result of this mechanism, the NaCl concentration of the medullary interstitial fluid is greatest near the tip of the loops of Henle and decreases progressively toward the renal cortex. As is explained later, this concentration gradient is important to the process of concentrating urine (figure 20.21).

The maintenance of the NaCl concentration gradient is aided by another countercurrent mechanism operating in the *vasa recta,* which is described in a previous section of this chapter. In this case, blood flows relatively slowly down the descending portion of the vasa recta, and NaCl enters it by diffusion. Then, as the blood moves back up toward the renal cortex, most

Figure 20.20

(a) Fluid in the ascending limb becomes hypotonic as solute is reabsorbed; (b) fluid in the descending limb becomes hypertonic as it loses water by osmosis and gains solute by diffusion.

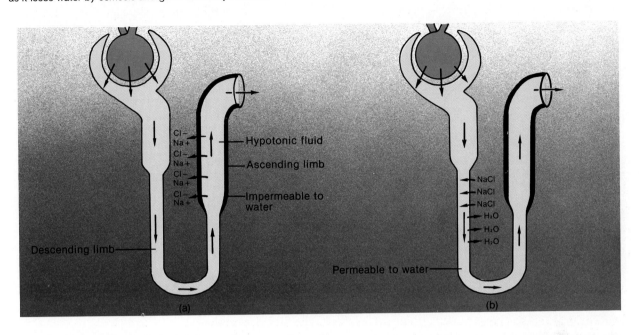

(a)

(b)

Figure 20.21

(a) As NaCl completes the countercurrent circuit again and again, it becomes more concentrated; (b) as a result, an NaCl concentration gradient is established in the medullary interstitial fluid.

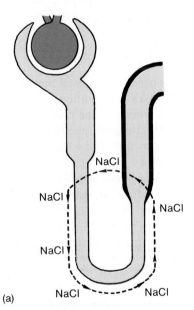

(a)

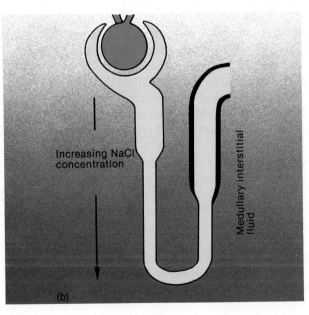

(b)

Figure 20.22

A countercurrent mechanism in the vasa recta helps to maintain the NaCl concentration gradient in the medullary interstitial fluid.

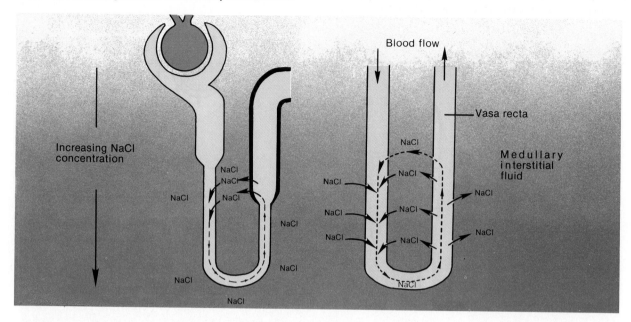

of the NaCl diffuses from the blood and reenters the medullary interstitial fluid. Consequently, little NaCl is carried away from the renal medulla (figure 20.22).

The tubular fluid reaching the distal convoluted tubule is hypotonic to its surroundings. The cells lining this segment and the collecting duct that follows, continue to reabsorb sodium and chloride ions, but these cells are quite impermeable to water. Thus, water tends to accumulate inside the tubule and may be excreted as dilute urine.

As is discussed in chapter 13, antidiuretic hormone (ADH) is produced by specialized neurons in the *hypothalamus.* This hormone is released from the posterior lobe of the pituitary gland in response to a decreasing concentration of water in the blood or to decreasing blood volume and pressure. When it reaches the kidney, ADH causes an increase in the permeability of the epithelial cell linings of the distal convoluted tubule and the collecting duct, and consequently water moves rapidly out of these segments by osmosis. Thus, the urine volume is reduced, and it becomes more concentrated (hypertonic).

This concentrating of the urine also continues as it passes down the *collecting duct,* because the hypertonic interstitial fluid of the renal medulla that surrounds the collecting duct draws water from it by osmosis (figure 20.23).

To summarize, ADH stimulates the production of concentrated urine, which contains soluble wastes and other substances in a minimum of water; it also

inhibits the loss of body fluids whenever there is a danger of dehydration. If the water concentration of the body fluids is excessive, ADH secretion is decreased. In the absence of ADH, the epithelial linings of the distal segment and the collecting duct become less permeable to water, less water is reabsorbed, and the urine tends to be more dilute.

Urea and Uric Acid Excretion

Urea is a by-product of amino acid metabolism. Consequently, its plasma concentration is directly related to the amount of protein in the diet. Urea enters the renal tubule by filtration, and about 50% of it is reabsorbed (passively) by diffusion while the remainder is excreted in the urine.

Urea also is involved in more complex movements within the kidney, including a countercurrent multiplier mechanism that causes it to be concentrated in the medullary interstitial fluid. As a result of this, urea becomes more highly concentrated in the urine than would otherwise be possible.

Uric acid, which results from the metabolism of certain organic bases (purines) in nucleic acids, is reabsorbed by active transport. Although this mechanism seems able to reabsorb all the uric acid normally present in glomerular filtrate, about 10% of the amount filtered is excreted in the urine. This amount is apparently *secreted* into the renal tubule.

Figure 20.23
(a) The distal convoluted tubule and collecting duct are impermeable to water, so water may be excreted as dilute urine; *(b)* if ADH is present, however, these segments become permeable, and water is reabsorbed by osmosis into the hypertonic medullary interstitial fluid.

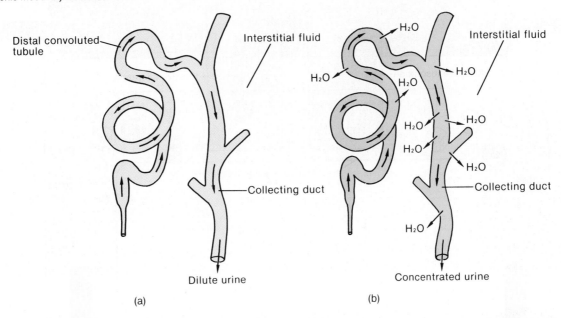

Distal convoluted tubule

Interstitial fluid

Collecting duct

Dilute urine

(a)

H_2O

Interstitial fluid

H_2O

H_2O

H_2O

H_2O

H_2O

H_2O

Collecting duct

H_2O

Concentrated urine

(b)

Gout is a disorder in which the plasma concentration of uric acid becomes abnormally high. Since uric acid is a relatively insoluble substance, it tends to precipitate when it is present in excess. As a result, crystals of uric acid may be deposited in joints and other tissues, where they produce inflammation and extreme pain. The joints of the great toes are affected most commonly, but joints in the hands and feet are often involved.

This condition, which often seems to be inherited, is sometimes treated by administering drugs that inhibit the reabsorption of uric acid and, thus, increase its excretion.

---

1. Describe a countercurrent mechanism.
2. What role does the hypothalamus play in regulating urine concentration and volume?
3. Explain how urea and uric acid are excreted.

Tubular Secretion

Tubular secretion (tubular excretion) is the process by which certain substances are transported from the plasma of the peritubular capillary into the fluid of the renal tubule. As a result, the amount of a particular substance excreted in the urine may be greater than the amount filtered from the plasma in the glomerulus (figure 20.24).

Figure 20.24
Secretory mechanisms act to move substances from the plasma of the peritubular capillary into the fluid of the renal tubule.

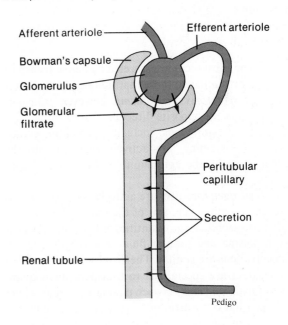

Afferent arteriole

Efferent arteriole

Bowman's capsule

Glomerulus

Glomerular filtrate

Peritubular capillary

Secretion

Renal tubule

Pedigo

Figure 20.25
Passive secretion of potassium ions (or hydrogen ions) may
occur in response to the active reabsorption of sodium ions.

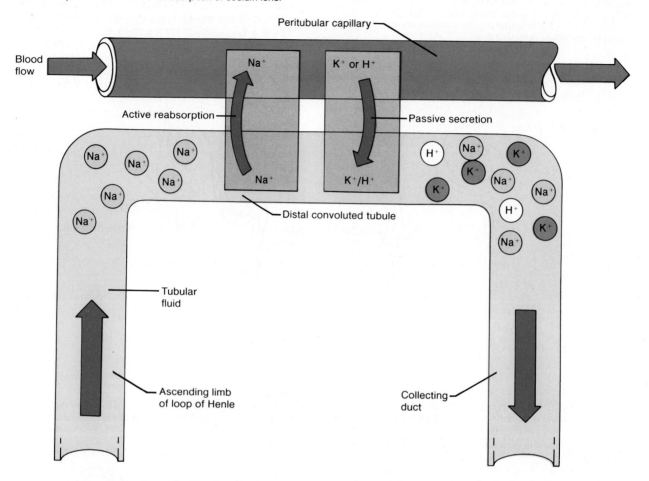

Some substances are secreted by active transport mechanisms similar to those that function in reabsorption. The *secretory mechanisms,* however, transport substances in the opposite direction. For example, certain organic compounds, including penicillin, creatinine, and histamine, are actively secreted into the tubular fluid by the epithelium of the proximal convoluted segment.

Hydrogen ions also are actively secreted. In this case, the proximal segment of the renal tubule is specialized to secrete large quantities of hydrogen ions between plasma and tubular fluid, where hydrogen ion concentrations are similar. The distal segment and collecting duct are specialized to transport small quantities of hydrogen ions between plasma and urine, where hydrogen ion concentrations may vary. The secretion of hydrogen ions plays an important role in the regulation of the pH of body fluids, as is explained in chapter 21.

Although most of the *potassium ions* in the glomerular filtrate are actively reabsorbed in the proximal convoluted tubule, some may be secreted passively in the distal segment and collecting duct. During this process, the active reabsorption of sodium ions out of the tubular fluid creates a negative electrical charge within the tube. Since positively charged potassium ions (K^+) are attracted to regions that are negatively charged, these ions move through the tubular epithelium and enter the tubular fluid. (See figure 20.25.)

To summarize, urine is formed as a result of the following: glomerular filtration of materials from blood plasma; reabsorption of substances, such as glucose, amino acids, proteins, creatine, lactic acid, citric acid, uric acid, ascorbic acid, phosphate ions, sulfate ions, calcium ions, potassium ions, sodium ions, water, and urea; and secretion of substances, such as penicillin, creatinine, histamine, phenobarbital, hydrogen ions, ammonia, and potassium ions (chart 20.2).

Chart 20.2 Functions of nephron parts

Part	Function
Renal Corpuscle	
Glomerulus	Filtration of water and dissolved substances from plasma
Bowman's capsule	Receives glomerular filtrate
Renal Tubule	
Proximal convoluted tubule	Reabsorption of glucose, amino acids, creatine, lactic acid, citric acid, uric acid, ascorbic acid, phosphate ions, sulfate ions, calcium ions, potassium ions, and sodium ions by active transport
	Reabsorption of proteins by pinocytosis
	Reabsorption of water by osmosis
	Reabsorption of chloride ions, and other negatively charged ions by electrochemical attraction
	Active secretion of substances such as penicillin, histamine, and hydrogen ions
Descending limb of loop of Henle	Reabsorption of water by osmosis
Ascending limb of loop of Henle	Reabsorption of chloride ions by active transport and passive reabsorption of sodium ions
Distal convoluted tubule	Reabsorption of sodium ions by active transport
	Reabsorption of water by osmosis
	Active secretion of hydrogen ions
	Passive secretion of potassium ions by electrochemical attraction

1. Define tubular secretion.
2. What substances are actively secreted? Passively secreted?
3. How does the reabsorption of sodium affect the secretion of potassium?

Composition of Urine

The composition of urine varies considerably from time to time because of differences in dietary intake and physical activity. In addition to containing about 95% water, it usually contains *urea* from the catabolic metabolism of amino acids, *uric acid* from the metabolism of nucleic acids, and *creatinine* from the metabolism of creatine. It also may contain a trace of *amino acids,* as well as a variety of *electrolytes* whose concentrations tend to vary directly with the amounts included in the diet (chart 20.1). The normal concentrations of various components of urine are listed in Appendix C on page A–9.

Abnormal constituents of urine include *glucose, proteins, hemoglobin, ketones,* and various *blood cells.* The significance of such substances in urine, however, may depend on the amounts present and on other factors. For example, glucose may appear in urine following a large intake of carbohydrates, proteins may appear following vigorous physical exercise, and ketones may appear following a prolonged fast. Also, some pregnant women have glucose in their urine toward the end of pregnancy.

The volume of urine produced by the kidneys usually varies between 0.6 and 2.5 liters per day. The exact volume is influenced by such factors as the fluid intake, the environmental temperature, the relative humidity of the surrounding air, and a person's emotional condition, respiratory rate, and body temperature. An output of 50–60 cc of urine per hour is considered normal, and an output of less than 30 cc per hour may be an indication of kidney failure.

The kidneys of infants and young children are unable to concentrate urine and conserve water as effectively as those of adults. Consequently, such young persons produce relatively large volumes of urine and tend to lose water rapidly.

1. List the normal constituents of urine.
2. What is the normal hourly output of urine? The minimal hourly output?

Elimination of Urine

After being formed by nephrons, urine passes from the collecting ducts through openings in the renal papillae and enters the major and minor calyces of the kidney. From there it passes through the renal pelvis and is conveyed by a ureter to the urinary bladder. It is excreted to the outside of the body by means of the urethra.

The Ureter

The **ureter** is a tubular organ about 25 cm long, which begins as the funnel-shaped renal pelvis. It extends downward behind the parietal peritoneum and parallel to the vertebral column. Within the pelvic cavity, it courses forward and medially to join the urinary bladder from underneath.

The wall of the ureter is composed of three layers. The inner layer, or *mucous coat,* includes several thicknesses of epithelial cells and is continuous with the linings of the renal tubules and the urinary bladder. The middle layer, or *muscular coat,* consists largely of smooth muscle fibers arranged in circular and longitudinal bundles. The outer layer, or *fibrous coat,* is composed of connective tissue (figure 20.26).

Since the linings of the ureters and the urinary bladder are continuous, infectious agents such as bacteria may ascend from the bladder into the ureters. An inflammation of the bladder, which is called *cystitis,* occurs more commonly in women than men because the female urethral pathway is shorter. An inflammation of the ureter is called *ureteritis.*

Although the ureter is simply a tube leading from the kidney to the urinary bladder, its muscular wall helps to move urine. Muscular peristaltic waves that originate in the renal pelvis force urine along the length of the ureter. These waves are initiated by the presence of urine in the renal pelvis, and their frequency is related to the rate of urine formation. If the rate of urine formation is high, a peristaltic wave may occur every few seconds; if the rate is low, a wave may occur every few minutes.

When such a peristaltic wave reaches the urinary bladder, it causes a jet of urine to spurt into the bladder. The opening through which the urine enters is covered by a flaplike fold of mucous membrane. This fold acts as a valve, allowing urine to move inward from the ureter but preventing it from backing up.

If a ureter becomes obstructed, as when a small kidney stone (renal calculus) is present in its lumen, strong peristaltic waves are initiated in the proximal portion of the tube. Such waves may help to move the stone into the bladder. At the same time, the presence of a stone usually stimulates a sympathetic reflex (ureterorenal reflex) that results in constriction of the renal arterioles and reduces the production of urine in the kidney on the affected side.

Kidney stones, which usually are composed of uric acid, calcium oxalate, calcium phosphate, or magnesium phosphate, sometimes form in the renal pelvis. If such a stone passes into a ureter it may produce severe pain. This pain commonly begins in the region of the kidney and tends to radiate into the abdomen, pelvis, and legs. The pain also may be accompanied by nausea and vomiting.

1. Describe the structure of a ureter.
2. How is urine moved from the renal pelvis?
3. What prevents urine from backing up from the urinary bladder into the ureters?
4. How does an obstruction in a ureter affect urine production?

The Urinary Bladder

The **urinary bladder** is a hollow, distensible, muscular organ. It is located within the pelvic cavity, behind the symphysis pubis, and below the parietal peritoneum (figure 20.27). In a male, it lies against the rectum posteriorly, and in a female it contacts the anterior walls of the uterus and vagina.

Although the bladder is somewhat spherical, its shape is altered by the pressures of surrounding organs. When it is empty, the inner wall of the bladder is thrown into many folds, but as it fills with urine, the wall becomes smoother. At the same time, the superior surface of the bladder expands upward into a dome.

Figure 20.26
Cross section of a ureter.

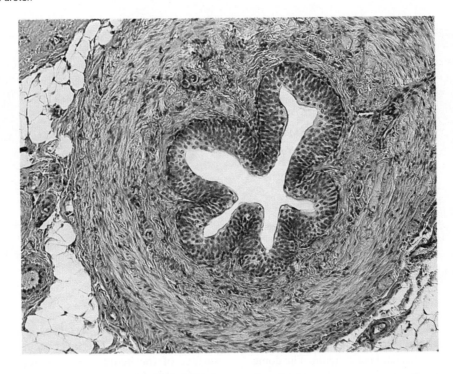

Figure 20.27
The urinary bladder is located within the pelvic cavity and behind the symphysis pubis. In a male, it lies against the rectum.

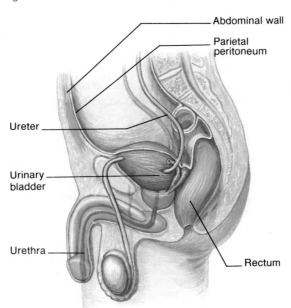

Abdominal wall

Parietal peritoneum

Ureter

Urinary bladder

Urethra

Rectum

Figure 20.28

(a) Frontal section of the male urinary bladder; *(b)* posterior view of the urinary bladder.

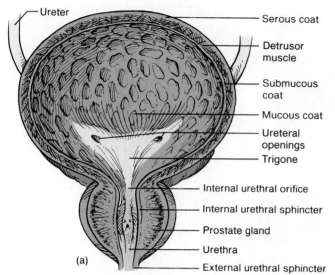

- Ureter
- Serous coat
- Detrusor muscle
- Submucous coat
- Mucous coat
- Ureteral openings
- Trigone
- Internal urethral orifice
- Internal urethral sphincter
- Prostate gland
- Urethra
- External urethral sphincter

(a)

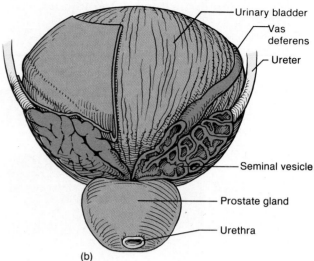

- Urinary bladder
- Vas deferens
- Ureter
- Seminal vesicle
- Prostate gland
- Urethra

(b)

When it is greatly distended, the bladder pushes above the pubic crest and into the region between the abdominal wall and the parietal peritoneum. The dome can reach the level of the umbilicus and press against the coils of the small intestine.

The internal floor of the bladder consists of a triangular area called the *trigone*, which has an opening at each of its three angles (figure 20.28). Posteriorly, at the base of the trigone, the openings are those of the ureters. The *internal urethral orifice*, which opens into the urethra, is located anteriorly at the apex of the trigone. The trigone generally remains in a fixed position even though the rest of the bladder changes shape during distension and contraction, thus preventing a reflux of urine into the ureters.

The wall of the urinary bladder consists of four layers. The inner layer, or *mucous coat*, includes several thicknesses of transitional epithelial cells, similar to those lining the ureters and the upper portion of the urethra. (See chapter 5.)The thickness of this tissue changes as the bladder expands and contracts, so that during distension it appears to be only two or three cells thick, while during contraction it appears to be five or six cells thick (see figure 5.7).

The second layer of the wall is the *submucous coat*. It consists of connective tissue and contains many elastic fibers.

The third layer, or *muscular coat,* is composed primarily of coarse bundles of smooth muscle fibers. These bundles are interlaced in all directions and depths, and together they comprise the **detrusor muscle.** This muscle is supplied with parasympathetic nerve fibers that function in the micturition reflex.

The outer layer, or *serous coat,* consists of the parietal peritoneum. This layer occurs only on the upper surface of the bladder. Elsewhere, the outer coat is composed of fibrous connective tissue.

---

1. Describe the trigone of the urinary bladder.
2. Describe the structure of the bladder wall.
3. What kind of nerve fibers supply the detrusor muscle?

Micturition

Micturition (urination) is the process by which urine is expelled from the urinary bladder. It involves the contraction of the detrusor muscle and may be aided by contractions of muscles in the abdominal wall and pelvic floor, and fixation of the thoracic wall and diaphragm. Micturition also involves the relaxation of the *external urethral sphincter.* This muscle surrounds the urethra about 3 cm from the bladder and is composed of voluntary muscle tissue.

The need to urinate usually is stimulated by distension of the bladder wall as it fills with urine. As the wall expands, stretch receptors are stimulated, and the micturition reflex is triggered.

Although an infant is unable to control the voluntary muscles associated with micturition, voluntary control of micturition becomes possible as various portions of the brain and spinal cord mature.

The *micturition reflex center* is located in the sacral segments of the spinal cord. When it is signalled by sensory impulses from the stretch receptors, *parasympathetic* motor impulses travel out to the detrusor

Figure 20.29
Longitudinal section of the male urinary bladder and urethra.

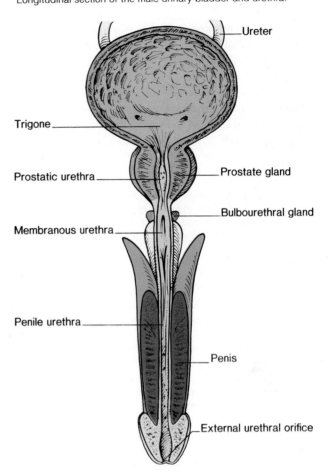

muscle, and it undergoes rhythmic contractions in response. This action is accompanied by a sensation of urgency.

Although the urinary bladder may hold as much as 600 ml of urine, the desire to urinate usually is experienced when it contains about 150 ml. Then, as the volume of urine increases to 300 ml or more, the sensation of fullness becomes increasingly uncomfortable.

Since the external urethral sphincter can be consciously controlled, it ordinarily remains contracted until a decision is made to urinate. This control is aided by nerve centers in the *midbrain* and *cerebral cortex* that are able to inhibit the micturition reflex. When a person decides to urinate, the external urethral sphincter is allowed to relax, and the micturition reflex is no longer inhibited. Nerve centers within the *pons* and *hypothalamus* may function to make the micturition reflex more effective. Consequently, the detrusor muscle contracts, and urine is excreted to the outside through the urethra. Within a few moments, the neurons of the micturition reflex seem to fatigue, the detrusor muscle relaxes, and the bladder begins to fill with urine again.

This process is outlined in chart 20.3.

Damage to the spinal cord above the sacral region may result in loss of voluntary control of urination. If the micturition reflex center and its sensory and motor fibers are uninjured, however, micturition may continue to occur reflexly. In this case, the bladder collects urine until its walls are stretched enough to trigger a micturition reflex, and the detrusor muscle contracts in response. This condition is called an *automatic bladder*.

The Urethra

The **urethra** is a tube that conveys urine from the urinary bladder to the outside of the body. Its wall is lined with mucous membrane and contains a relatively thick layer of smooth muscle tissue, whose fibers are generally directed longitudinally. It also contains numerous mucous glands, called *urethral glands,* that secrete mucus into the urethral canal.

In a female, the urethra is about 4 cm long. It passes forward from the bladder, courses below the symphysis pubis, and empties between the labia minor. Its opening, the *external urethral orifice* (urinary meatus), is located anterior to the vaginal opening and about 2.5 cm posterior to the clitoris.

In a male, the urethra, which functions both as a urinary canal and a passageway for cells and secretions from various reproductive organs, can be divided into three sections: the prostatic urethra, the membranous urethra, and the penile urethra (figure 20.29).

The **prostatic urethra** is about 2.5 cm long and passes from the urinary bladder through the *prostate* *gland,* which is located just below the bladder. Ducts from various reproductive structures join the urethra in this region.

The **membranous urethra** is about 2 cm long. It begins just distal to the prostate gland, passes through the urogenital diaphragm, and is surrounded by the fibers of the external urethral sphincter muscle.

The **penile urethra** is about 15 cm long and passes through the corpus spongiosum of the penis, where it is surrounded by erectile tissue. This portion of the urethra terminates with the *external urethral orifice* at the tip of the penis.

---

1. Describe the events of micturition.
2. How is it possible to inhibit the micturition reflex?
3. Describe the structure of the urethra.
4. How does the urethra of a male differ from that of a female?

Clinical Terms Related to the Urinary System

anuria (ah-nu're-ah)   an absence of urine due to failure of kidney function or to an obstruction in a urinary pathway.

bacteriuria (bak-te''re-u're-ah)   bacteria in the urine.

cystectomy (sis-tek'to-me)   surgical removal of the urinary bladder.

cystitis (sis-ti'tis)   inflammation of the urinary bladder.

cystoscope (sis'to-skōp)   instrument used for visual examination of the interior of the urinary bladder.

cystotomy (sis-tot'o-me)   an incision of the wall of the urinary bladder.

diuresis (di''u-re'sis)   an increased production of urine.

dysuria (dis-u're-ah)   painful or difficult urination.

enuresis (en''u-re'sis)   uncontrolled urination.

hematuria (hem''ah-tu're-ah)   blood in the urine.

nephrectomy (ně-frek'to-me)   surgical removal of a kidney.

nephrolithiasis (nef''ro-lǐ-thi'ah-sis)   presence of a stone(s) in the kidney.

nephroptosis (nef''rop-to'sis)   a movable or displaced kidney.

oliguria (ol''ǐ-gu're-ah)   a scanty output of urine.

polyuria (pol''e-u're-ah)   an excessive output of urine.

pyelolithotomy (pi''ě-lo-lǐ-thot'o-me)   removal of a stone from the renal pelvis.

pyelonephritis (pi''e-lon-ne-fri'tis)   inflammation of the renal pelvis.

pyelotomy (pi''ě-lot'o-me)   incision into the renal pelvis.

pyuria (pi-u're-ah)   pus in the urine.

uremia (u-re'me-ah)   condition in which substances ordinarily excreted in the urine accumulate in the blood.

ureteritis (u-re''ter-i'tis)   inflammation of the ureter.

urethritis (u''re-thri'tis)   inflammation of the urethra.

Chapter Summary

Introduction
The urinary system consists of kidneys, ureters, urinary bladder, and urethra.

The Kidney
1. Location of the kidneys
   a. The kidneys are on either side of the vertebral column, high on the posterior wall of the abdominal cavity.
   b. They are positioned behind the parietal peritoneum and held in place by adipose and connective tissue.
2. Structure of a kidney
   a. A kidney contains a hollow renal sinus.
   b. The ureter expands into the renal pelvis, which in turn is divided into major and minor calyces.
   c. Renal papillae project into the renal sinus.
   d. Kidney tissue is divided into a renal medulla and a renal cortex.
3. Functions of the kidneys
   a. Kidneys remove metabolic wastes from the blood and excrete them to the outside.
   b. They also help to regulate red blood cell production, blood pressure, calcium absorption, and the volume, composition, and pH of the blood.
4. Renal blood vessels
   a. Arterial blood flows through the renal artery, interlobar arteries, arciform arteries, interlobular arteries, and afferent arterioles.
   b. Venous blood returns through a series of vessels that correspond to those of the arterial pathways.
5. The nephrons
   a. Structure of the nephron
      (1) The nephron is the functional unit of the kidney.
      (2) It consists of a renal corpuscle and a renal tubule.
         (a) The corpuscle consists of a glomerulus and a Bowman's capsule.
         (b) Portions of the renal tubule include the proximal convoluted tubule, the loop of Henle (ascending and descending limbs), the distal convoluted tubule, and the collecting duct.
      (3) The collecting duct empties into the minor calyx of the renal pelvis.
   b. Juxtaglomerular apparatus
      (1) The juxtaglomerular apparatus is located at the point of contact between the distal convoluted tubule and the afferent and efferent arterioles.
      (2) It consists of the macula densa and the juxtaglomerular cells.

c. Cortical and juxtamedullary nephrons
   (1) Cortical nephrons are most numerous and have corpuscles near the surface of the kidney.
   (2) Juxtamedullary nephrons have corpuscles near the medulla.
d. Blood supply of a nephron
   (1) The glomerular capillary receives blood from the afferent arteriole and passes it to the efferent arteriole.
   (2) The efferent arteriole gives rise to the peritubular capillary system that surrounds the renal tubule.
   (3) Capillary loops, called vasa recta, dip down into the medulla.

Urine Formation

Nephrons function to remove wastes from blood and to regulate water and electrolyte concentrations.

Urine is the end product of these functions, which involve filtration, reabsorption, and secretion of substances from renal tubules.

1. Glomerular filtration
   a. Urine formation begins when water and dissolved materials are filtered out of the glomerular capillary.
   b. The glomerular capillaries are much more permeable than the capillaries in other tissues.
2. Filtration pressure
   a. Filtration is due mainly to hydrostatic pressure inside the glomerular capillaries.
   b. Osmotic pressure of the plasma and hydrostatic pressure in the Bowman's capsule also affect filtration.
   c. Filtration pressure is the net force acting to move material out of the glomerulus and into the Bowman's capsule.
   d. The composition of the filtrate is similar to tissue fluid.
3. Filtration rate
   a. The rate of filtration varies with the filtration pressure.
   b. Filtration pressure changes with the diameters of the afferent and efferent arterioles.
   c. As osmotic pressure in the glomerulus increases, filtration decreases.
   d. Filtration rate varies with the rate of blood flow through the glomerulus.
   e. As the hydrostatic pressure in Bowman's capsule increases, the filtration rate decreases.
   f. The kidneys produce about 125 ml of glomerular fluid per minute, most of which is reabsorbed.
   g. The volume of filtrate varies with the surface area of the glomerular capillary.
4. Regulation of filtration rate
   a. When the filtration rate decreases, the macula densa causes the afferent arteriole to dilate, increasing blood flow through the glomerulus and increasing the filtration rate.

b. The macula densa also causes the juxtaglomerular cells to release renin, which triggers a series of changes leading to constriction of the efferent arteriole, increasing the glomerular hydrostatic pressure and increasing the filtration rate.
   c. Autoregulation is the ability of an organ or tissue to maintain a constant blood flow when the arterial blood pressure is changing.
5. Tubular reabsorption
   a. Substances are selectively reabsorbed from the glomerular filtrate.
   b. The peritubular capillary is adapted for reabsorption.
      (1) It carries low pressure blood.
      (2) It is very permeable.
   c. Most reabsorption occurs in the proximal tubule, where the epithelial cells possess microvilli.
   d. Various substances are reabsorbed in particular segments of the renal tubule by different modes of transport.
      (1) Glucose and amino acids are reabsorbed by active transport.
      (2) Water is reabsorbed by osmosis.
      (3) Proteins are reabsorbed by pinocytosis.
   e. Active transport mechanisms have limited transport capacities.
   f. If the concentration of a substance in the filtrate exceeds its renal plasma threshold, the excess is excreted in the urine.
   g. Substances that remain in the filtrate are concentrated as water is reabsorbed.
   h. Sodium ions are reabsorbed by active transport.
      (1) As positively charged sodium ions are transported out of the filtrate, negatively charged ions accompany them.
      (2) Water is passively reabsorbed by osmosis as sodium ions are actively reabsorbed.
6. Regulation of urine concentration and volume
   a. Most sodium is reabsorbed before urine is excreted.
   b. Sodium is concentrated in the renal medulla by the countercurrent mechanism.
      (1) Chloride ions are actively reabsorbed in the ascending limb, and sodium ions follow them passively.
      (2) Tubular fluid in the ascending limb becomes hypotonic as it loses solutes.
      (3) Water leaves the descending limb by osmosis, and NaCl enters this limb by diffusion.
      (4) Tubular fluid in the descending limb becomes hypertonic as it loses water and gains NaCl.
      (5) As NaCl repeats this circuit, its concentration in the medulla increases.
   c. The vasa recta countercurrent mechanism helps to maintain the NaCl concentration in the medulla.

d. The distal tubule and collecting duct are impermeable to water so it tends to be excreted in urine.

e. ADH from the posterior pituitary gland causes the permeability of the distal tubule and collecting duct to increase, and thus promotes the reabsorption of water.

7. Urea and uric acid excretion
   a. Urea is a by-product of amino acid metabolism.
      (1) It is reabsorbed passively by diffusion.
      (2) About 50% is excreted in urine.
      (3) A countercurrent mechanism helps in the excretion of urea.
   b. Uric acid results from the metabolism of nucleic acids.
      (1) Most is reabsorbed by active transport.
      (2) Some is secreted into the renal tubule.

8. Tubular secretion
   a. Tubular secretion is the process by which certain substances are transported from the plasma to the tubular fluid.
   b. Some substances are secreted actively.
      (1) These include various organic compounds and hydrogen ions.
      (2) Hydrogen ions are secreted by the proximal and distal segments of the renal tubule.
   c. Potassium ions are secreted passively in the distal segment and collecting duct where they are attracted by the negative charge that develops in the lumen of the tubule.

9. Composition of urine
   a. Urine is about 95% water, and it usually contains urea, uric acid, and creatinine.
   b. It may contain a trace of amino acids and varying amounts of electrolytes, depending upon the dietary intake.
   c. The volume of urine varies with the fluid intake and with certain environmental factors.

Elimination of Urine

1. The ureter
   a. The ureter is a tubular organ that extends from the kidney to the urinary bladder.
   b. Its wall has mucous, muscular, and fibrous layers.
   c. Peristaltic waves in the ureter force urine to the bladder.
   d. Obstruction in the ureter stimulates strong peristaltic waves and a reflex that causes the kidney to decrease urine production.

2. The urinary bladder
   a. The urinary bladder is a distensible organ that stores urine and forces it into the urethra.
   b. The openings for the ureters and urethra are located at the three angles of the trigone in the floor of the urinary bladder.
   c. Muscle fibers in the wall form the detrusor muscle.

3. Micturition
   a. Micturition is the process by which urine is expelled.
   b. It involves contraction of the detrusor muscle and relaxation of the external urethral sphincter.
   c. Micturition reflex
      (1) Stretch receptors in the bladder wall are stimulated by distension.
      (2) The micturition reflex center in the sacral spinal cord sends parasympathetic motor impulses to the detrusor muscle.
      (3) Urination can be controlled by means of the voluntary external urethral sphincter and nerve centers in the brain that can inhibit the micturition reflex.
      (4) When the decision to urinate is made, the external urethral sphincter is allowed to relax and nerve centers in the brain act to facilitate the micturition reflex.

4. The urethra
   a. The urethra conveys urine from the bladder to the outside.
   b. In females, it empties between the labia minor.
   c. In males, it conveys products of reproductive organs as well as urine.
      (1) There are three portions of the male urethra: prostatic, membranous, and penile.
      (2) It empties at the tip of the penis.

Application of Knowledge

1. If an infant is born with a narrowing of the renal arteries, what effect would this condition have on the volume of urine produced? Explain your answer.

2. How would you explain the observation that people with nephrotic syndrome, in which plasma proteins are lost into the urine, have increased susceptibilities to infections?

3. If a patient who has had major abdominal surgery receives intravenous fluids to the volume of blood lost during surgery, would you expect the volume of urine produced to be greater or less than normal? Why?

4. If a physician prescribed oral penicillin therapy for a patient with an infection of the urinary bladder, how would you describe for the patient the route by which the drug will reach the bladder?

Review Activities

1. Name the organs of the urinary system and list their general functions.
2. Describe the external and internal structure of a kidney.
3. List the functions of the kidneys.
4. Name the vessels through which blood passes as it travels from the renal artery to the renal vein.
5. Distinguish between a renal corpuscle and a renal tubule.
6. Name the parts through which fluid passes as it travels from the glomerulus to the collecting duct.
7. Describe the location and structure of the juxtaglomerular apparatus.
8. Distinguish between cortical and juxtamedullary nephrons.
9. Distinguish between filtration, reabsorption, and secretion as they relate to urine formation.
10. Define *filtration pressure*.
11. Compare the composition of the glomerular filtrate with that of blood plasma.
12. Explain how the diameters of the afferent and efferent arterioles affect the rate of glomerular filtration.
13. Explain how changes in the osmotic pressure of the blood plasma may affect the rate of glomerular filtration.
14. Explain how the hydrostatic pressure of Bowman's capsule affects the rate of glomerular filtration.
15. Describe two mechanisms by which the juxtaglomerular apparatus helps to regulate filtration rate.
16. Define *autoregulation*.

17. Discuss the reason tubular reabsorption is said to be a selective process.
18. Explain how the peritubular capillary is adapted for reabsorption.
19. Explain how the epithelial cells of the proximal convoluted tubule are adapted for reabsorption.
20. Explain why active transport mechanisms have limited transport capacities.
21. Define *renal plasma threshold* and explain its significance in tubular reabsorption.
22. Explain how amino acids and proteins are reabsorbed.
23. Describe the effect of sodium reabsorption on the reabsorption of negatively charged ions.
24. Explain how sodium reabsorption affects water reabsorption.
25. Explain how hypotonic tubular fluid is produced in the ascending limb of the loop of Henle.
26. Explain why fluid in the descending loop of Henle is hypertonic.
27. Describe the function of ADH.
28. Explain how urine may become concentrated as it moves through the collecting duct.
29. Compare the processes by which urea and uric acid are reabsorbed.
30. Explain how the renal tubule is adapted to carry on the secretion of hydrogen ions.
31. Explain how potassium ions may be secreted passively.
32. List the more common wastes found in urine and their sources.
33. List some factors that affect the volume of urine produced each day.
34. Describe the structure and function of a ureter.
35. Explain how the muscular wall of the ureter aids in moving urine.
36. Discuss what happens if a ureter becomes obstructed.
37. Describe the structure and location of the urinary bladder.
38. Define *detrusor muscle*.
39. Describe the micturition reflex.
40. Explain how the micturition reflex can be voluntarily controlled.
41. Compare the urethra of a female with that of a male.

21

Water and Electrolyte Balance

Cell functions and, indeed, cell survival depend upon homeostasis—the existence of a stable cellular environment. In such an environment, body cells are supplied continually with oxygen and nutrients, and the waste products resulting from their metabolic activities are carried away continually. At the same time, the concentrations of water and dissolved electrolytes in the cellular fluids and their surroundings remain relatively constant. This condition requires the maintenance of *water* and *electrolyte balance*.

Chapter Outline

Chapter Objectives

After you have studied this chapter, you should be able to

1. Explain what is meant by water and electrolyte balance and discuss the importance of this balance.

2. Describe how body fluids are distributed within compartments, how the fluid composition differs between compartments, and how fluids move from one compartment to another.

3. List the routes by which water enters and leaves the body and explain how water input and output are regulated.

4. Explain how electrolytes enter and leave the body and how the input and output of electrolytes are regulated.

5. Explain what is meant by acid-base balance.

6. Describe how hydrogen ion concentrations are expressed.

7. List the major sources of hydrogen ions in the body.

8. Distinguish between strong and weak acids and bases.

9. Explain how changing pH values of body fluids are minimized by chemical buffer systems, the respiratory center, and the kidneys.

10. Complete the review activities at the end of this chapter. Note that the items are worded in the form of specific learning objectives. You may want to refer to them before reading the chapter.

Key Terms

acid   (as'id)

acidosis   (as''i-do'sis)

alkalosis   (al''kah-lo'sis)

base   (bās)

buffer system   (buf'er sis'tem)

electrolyte balance   (e-lek'tro-līt bal'ans)

extracellular   (ek''strah-sel'u-lar)

intracellular   (in''trah-sel'u-lar)

osmoreceptor   (oz''mo-re-sep'tor)

transcellular   (trans-sel'u-lar)

water balance   (wot'er bal'ans)

Aids to Understanding Words

de-, separation from: *de*hydration—removal of water from cells or body fluids.

edem-, swelling: *edem*a—swelling due to an abnormal accumulation of extracellular fluid.

-emia, a blood condition: hypoprotein*emia*—an abnormally low concentration of blood plasma protein.

extra-, outside: *extra*cellular fluid—fluid outside of body cells.

im- (or **in-**), not: *im*balance—condition in which factors are not in equilibrium.

intra-, within: *intra*cellular fluid—fluid within body cells.

neutr-, neither one nor the other: *neutr*al—solution that is neither acidic nor basic.

-osis, a state of: acid*osis*—condition in which hydrogen ion concentration is abnormally high.

-uria, a urine condition: keto*uria*—presence of ketone bodies in the urine.

THE TERM *BALANCE* suggests a state of equilibrium, and in the case of water and electrolytes, it means the quantities entering the body are equal to the quantities leaving it. Maintaining such a balance requires mechanisms that ensure that lost water and electrolytes will be replaced and any excesses will be expelled. As a result, the quantities within the body are relatively stable at all times.

It is important to remember that water balance and electrolyte balance are interdependent, since the electrolytes are dissolved in the water of body fluids. Consequently, anything that alters the concentrations of electrolytes will necessarily alter the concentration of water by adding solutes to it or by removing solutes from it. Likewise, anything that changes the concentration of water will change the concentrations of electrolytes by making them more concentrated or more diluted.

Distribution of Body Fluids

Water and electrolytes are not evenly distributed throughout the tissues. Instead, they occur in regions or *compartments* that contain fluids of varying compositions. Movement of water and electrolytes between these compartments is regulated so that their distribution remains relatively stable.

Fluid Compartments

The body of an average adult male is about 63% water by weight. This water (about 40 liters), together with its dissolved electrolytes, is distributed into two major compartments: an intracellular compartment and an extracellular compartment (figure 21.1).

While the average adult male body is about 63% water by weight, the average adult female body is only about 52% water. This difference is related to the fact that adipose tissue has a relatively low water content, and generally there is a greater development of subcutaneous adipose tissue in females. Also, obese people of either sex contain proportionally less body water than do lean people.

The **intracellular fluid compartment** includes all the water and electrolytes enclosed by cell membranes. In other words, intracellular fluid is the fluid within cells, and in an adult it represents about 63% of the total body water.

The **extracellular fluid compartment** includes all the fluid outside cells—within the tissue spaces (interstitial fluid), the blood vessels (plasma), and the lymphatic vessels (lymph). A specialized fraction of the extracellular fluid is separated from other extracellular fluids by a layer of epithelium. It is called **transcellular**

Figure 21.1
Of the 40 liters of water in the body of an average adult male, about 63% is intracellular and 37% is extracellular.

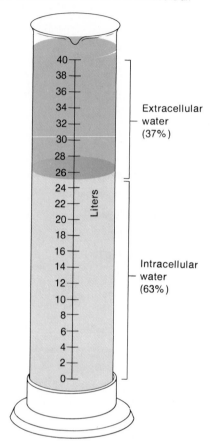

Extracellular water (37%)

Liters

Intracellular water (63%)

fluid and includes *cerebrospinal fluid* of the central nervous system, *aqueous and vitreous humors* of the eyes, *synovial fluid* of the joints, *serous fluid* within the various body cavities, and fluid *secretions* of glands. The fluids of the extracellular compartment constitute about 37% of the total body water (figure 21.2).

Composition of Body Fluids

Extracellular fluids generally have similar compositions. They are characterized by relatively high concentrations of sodium, chloride, and bicarbonate ions, and lesser concentrations of potassium, calcium, magnesium, phosphate, and sulfate ions. The plasma fraction of extracellular fluid contains considerably more protein than either interstitial fluid or lymph.

Intracellular fluid contains relatively high concentrations of potassium, phosphate, and magnesium ions. It includes a somewhat greater concentration of sulfate ions, and lesser concentrations of sodium, chloride, and bicarbonate ions than extracellular fluid. Intracellular fluid also has a greater concentration of protein than plasma. These relative concentrations are shown in figure 21.3.

Figure 21.2
Fluid within the intracellular compartment is separated from fluid
in the extracellular compartment by cell membranes. What
membranes separate the various components of the extracellular
fluid compartment?

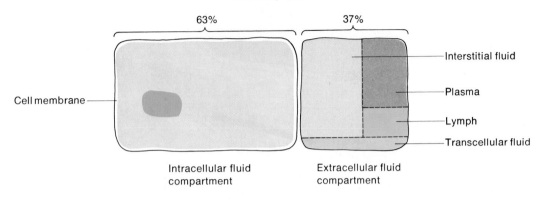

Total body water

Cell membrane

Intracellular fluid
compartment

Extracellular fluid
compartment

Interstitial fluid

Plasma

Lymph

Transcellular fluid

63%

37%

Figure 21.3
Extracellular fluid is relatively high in concentrations of sodium,
chloride, and bicarbonate ions; intracellular fluid is relatively high
in concentrations of potassium, magnesium, and phosphate ions.

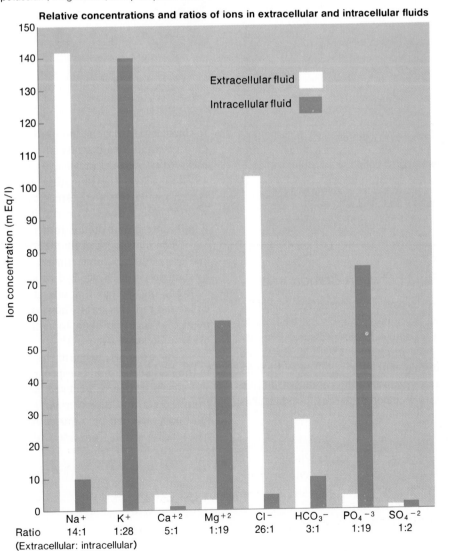

Relative concentrations and ratios of ions in extracellular and intracellular fluids

Extracellular fluid

Intracellular fluid

Ion concentration (m Eq/l)

	Na^+	K^+	Ca^{+2}	Mg^{+2}	Cl^-	HCO_3^-	PO_4^{-3}	SO_4^{-2}
Ratio (Extracellular: intracellular)	14:1	1:28	5:1	1:19	26:1	3:1	1:19	1:2

Figure 21.4
Net movements of fluids between compartments result from
differences in hydrostatic and osmotic pressures.

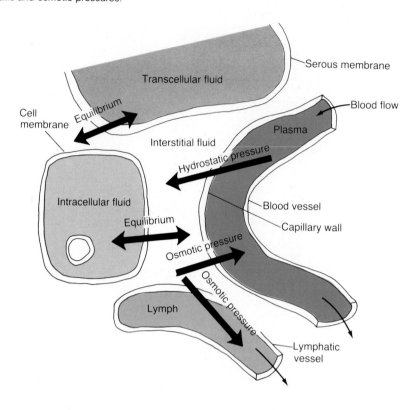

1. How are fluid balance and electrolyte balance interdependent?
2. Describe the normal distribution of water within the body.
3. What electrolytes are in higher concentrations in extracellular fluid? In intracellular fluid?
4. How does the concentration of protein vary in various body fluids?

Movement of Fluid between Compartments

The movement of water and electrolytes from one compartment to another is regulated largely by two factors: hydrostatic pressure and osmotic pressure. For example, as is explained in chapter 18, fluid leaves the plasma at the arteriole ends of capillaries and enters the interstitial spaces because of the net outward force of *hydrostatic pressure* (blood pressure). Fluid returns to the plasma from the interstitial spaces at the venule ends of capillaries because of the net inward force of *osmotic pressure*. Likewise, as mentioned in chapter 19, fluid leaves the interstitial spaces and enters the lymph capillaries due to osmotic pressure that develops within these spaces. As a result of the circulation of lymph, interstitial fluid is returned to the plasma.

The movement of fluid between the intracellular and extracellular compartments is controlled similarly by such pressures. Since hydrostatic pressure within the cells and the surrounding interstitial fluid ordinarily is equal and remains stable, however, any net fluid movement that occurs is likely to be the result of changes in osmotic pressure (figure 21.4).

For example, because the sodium concentration in the interstitial fluid is especially high, a decrease in this concentration will cause a net movement of water from the extracellular compartment into the intracellular compartment by osmosis. As a consequence, cells will tend to swell. Conversely, if the concentration of

Figure 21.5
Major sources of body water.

Average daily intake of water

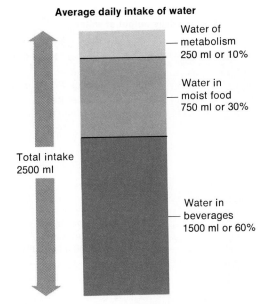

Water of
— metabolism
250 ml or 10%

Water in
— moist food
750 ml or 30%

Total intake
2500 ml

Water in
— beverages
1500 ml or 60%

sodium in the interstitial fluid increases, the net movement of water will be outward from the intracellular compartment, and the cells will shrink as they lose water.

1. What factors control movement of water and electrolytes from one fluid compartment to another?
2. How does the sodium concentration within body fluids affect the net movement of water between compartments?

Water Balance

Water balance exists when the total intake of water is equal to the total loss of water.

Water Intake

Although the volume of water gained each day varies from individual to individual, an average adult living in a moderate environment takes in about 2500 ml.

Of this amount, probably 60% (1500 ml) will be obtained from drinking water or beverages, while another 30% (750 ml) will be gained from moist foods. The remaining 10% (250 ml) will be a by-product of oxidative metabolism of various nutrients, which is called **water of metabolism** (figure 21.5).

Regulation of Water Intake

The primary regulator of water intake is thirst, and although the thirst mechanism is poorly understood, it seems to involve the osmotic pressure of extracellular fluid and a *thirst center* in the hypothalamus of the brain.

As water is lost from the body, the osmotic pressure of the extracellular fluid increases. *Osmoreceptors* in the thirst center are thought to be stimulated by such a change, and as a result, the hypothalamus causes the person to feel thirsty and seek water. The feeling of thirst usually is accompanied by dryness of the mouth, which is related to the loss of extracellular water and a consequent decreased flow of saliva.

The thirst mechanism usually is triggered whenever the total body water is decreased by 1% or 2%. As a person drinks water in response to thirst, the act of drinking and the resulting distension of the stomach wall seems to trigger nerve impulses that inhibit the mechanism. Thus, the person usually stops drinking long before the swallowed water has been absorbed. This inhibition helps to prevent the person from drinking more than is required to replace the quantity

Figure 21.6

Routes by which body water is lost. What factors influence water
loss by each of these routes?

Average daily output of water

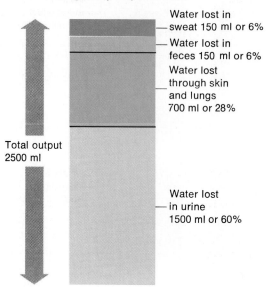

Water lost in sweat 150 ml or 6%

Water lost in feces 150 ml or 6%

Water lost through skin and lungs 700 ml or 28%

Total output 2500 ml

Water lost in urine 1500 ml or 60%

Chart 21.1   Regulation of water intake

1. The body loses 1% to 2% of its water.

2. Osmoreceptors in the thirst center are stimulated by an increase in the osmotic pressure of extracellular fluid due to water loss.

3. Activity in the hypothalamus causes the person to feel thirsty, and seek water in response.

4. The act of drinking and distension of the stomach by water stimulate nerve impulses that inhibit the thirst center.

5. Water is absorbed through the walls of the stomach and small intestine.

6. Osmotic pressure of extracellular fluid decreases.

---

1. What is meant by water balance?
2. Where is the thirst center located?
3. What mechanism stimulates fluid intake? What mechanism inhibits it?

---

lost. Thus, it avoids development of an imbalance of the opposite type. Chart 21.1 summarizes the steps in this mechanism.

The managers of bars and taverns often provide salty snacks without charge for their customers. As the patrons enjoy these snacks, their thirst mechanisms are triggered, and they are likely to order more liquid refreshments. Thus, the patrons satisfy their thirsts for fluid as well as the tavernkeeper's thirst for revenue.

Water Output

Water normally enters the body only through the mouth, but it can be lost by a variety of routes. These include obvious losses in urine, feces, and sweat (sensible perspiration), as well as less obvious losses that occur by diffusion of water through the skin (insensible perspiration) and evaporation of water from the lungs during breathing.

If an average adult takes in 2500 ml of water each day, then 2500 ml must be eliminated if water balance is to be maintained. Of this volume, perhaps 60% (1500 ml) will be lost in urine, 6% (150 ml) in feces, and 6% (150 ml) in sweat. About 28% (700 ml) will be lost by diffusion through the skin and evaporation from the lungs (figure 21.6). These percentages will, of course, vary with such environmental factors as temperature and relative humidity, and with the amount of physical exercise.

Water lost by sweating is a necessary part of the body's temperature control mechanism; water lost in feces accompanies the elimination of undigested food

Chart 21.2 Events in regulation of water balance

In case of dehydration	In case of excessive water intake
1. Extracellular fluid becomes more concentrated.	1. Extracellular fluid becomes osmotically less concentrated.
2. Osmoreceptors in the hypothalamus are stimulated by change in osmotic pressure of body fluid.	2. Osmoreceptors in the hypothalamus are stimulated by this change.
3. The hypothalamus signals the posterior pituitary gland to release ADH into the blood.	3. The posterior pituitary gland decreases the release of ADH.
4. Blood carries ADH to the kidneys.	4. Renal tubules decrease water reabsorption.
5. ADH causes distal convoluted tubules and collecting ducts to increase water reabsorption.	5. Urine output increases, and excess water is excreted.
6. Urine output decreases, and water is conserved.	

materials; and water lost by diffusion and evaporation is largely unavoidable. Therefore, the primary regulator of water output involves urine production.

Athletes who engage in endurance events may lose a relatively large volume of body water from sweating, particularly under hot and humid conditions. If more than about 2% of the body weight is lost in this manner, severe demands are placed on the person's ability to regulate body temperature (see chapter 6). Also, athletic performance is diminished, and the person may experience muscle cramps, nausea, and varying degrees of weakness.

To prevent these effects of dehydration, it is essential that an athlete replace body water as it is lost. This usually can be accomplished by drinking plain water before and during endurance events. It may be necessary, however, to drink more than is needed to satisfy thirst to completely replace the water lost.

Regulation of Water Balance

As is discussed in chapter 20, the volume of water excreted in the urine is regulated mainly by activity in the distal convoluted tubule and collecting duct of the nephron. The epithelial linings of these segments of the renal tubule remain relatively impermeable to water unless antidiuretic hormone (ADH) is present.

As is explained in chapter 13, the release of ADH involves osmoreceptors in the hypothalamus. If the plasma becomes more concentrated because of excessive water loss, these osmoreceptors tend to become dehydrated and shrink. This change triggers impulses that signal the posterior pituitary gland to release ADH. The ADH is carried by the blood to the kidneys where it causes an increase in the permeability of the distal tubule and collecting duct. Consequently, water reabsorption increases, and water is conserved. This action

resists further osmotic change in the plasma. In fact, the *osmoreceptor-ADH mechanism* can reduce a normal urine production of 1500 ml per day to about 500 ml per day at times when the body is tending to become dehydrated.

If, on the other hand, excessive amounts of water are taken into the body, the plasma becomes less concentrated, and the osmoreceptors swell as they receive extra water by osmosis. In this instance, the release of ADH is inhibited, and the distal segment and collecting duct remain impermeable to water. Consequently, less water is reabsorbed and a greater volume of urine is produced. Chart 21.2 summarizes the steps in this mechanism.

Diuretics are substances that promote the production of urine. A number of common substances such as caffeine in coffee and tea, have diuretic effects, as do a variety of drugs used to reduce the volume of body fluids.

Although diuretics produce their effects in different ways, some, such as alcohol and various narcotic drugs, promote urine formation by inhibiting the release of ADH. Others, including a group called *mercurial diuretics,* inhibit the reabsorption of sodium and other solutes in certain portions of the renal tubules. As a consequence, the osmotic pressure of the tubular fluid increases, and the reabsorption of water by osmosis is reduced.

---

1. By what routes is water lost from the body?
2. What is the primary regulator of water loss?
3. What types of water loss are unavoidable?
4. What role does the hypothalamus play in the regulation of water balance?

Water Balance Disorders
A Clinical Application

Among the more common disorders involving an imbalance in the water of body fluids are dehydration, water intoxication, and edema.

Dehydration is a deficiency condition that occurs when the output of water exceeds the intake. This condition may develop following excessive sweating, or as a result of prolonged water deprivation accompanied by continued water output. In either case, as water is lost, the extracellular fluid becomes increasingly more concentrated, and water tends to leave cells by osmosis (figure 21.7). Dehydration also may accompany illnesses in which excessive fluids are lost as a result of prolonged vomiting or diarrhea. During dehydration, the skin and mucous membranes of the mouth feel dry, and body weight is lost. Also a high fever may develop as the body temperature-regulating mechanism becomes less effective due to a lack of water needed for sweating. In severe cases, as waste products accumulate in the extracellular fluid, symptoms of cerebral disturbances including mental confusion, delirium, and coma may develop.

The kidneys of infants are less able to conserve water than those of adults. Consequently, infants are more likely to become dehydrated than adults. Elderly people also are susceptible to developing water imbalances because the sensitivity of their thirst mechanisms tend to decrease with age, and physical disabilities may make it difficult to obtain adequate fluids.

The treatment for dehydration involves the replacement of lost water and electrolytes. It is important to note that if the water alone is replaced, the extracellular fluid will become more dilute than normal, and this may produce a condition called *water intoxication.*

Water intoxication is characterized by the presence of hypotonic extracellular fluid. This condition may occur as a result of administering pure water to a dehydrated person, or it may develop in a person who drinks water for a prolonged period faster than the kidneys can excrete the excess.

In either instance, as the water is absorbed, the extracellular fluid becomes hypotonic to the cells. Then the cells tend to swell as water enters them in abnormal quantities by osmosis (figure 21.8).

The symptoms of water intoxication are related mainly to a decrease in the extracellular sodium concentration, and they include painful muscular contractions (heat cramps) and convulsions, confusion, and coma associated with swelling of brain tissues. Treatment of this condition usually involves restricting water intake and administering hypertonic salt solutions.

Edema is characterized by an abnormal accumulation of extracellular fluid within the interstitial spaces. It can be caused by a variety of factors, including a decrease in the plasma protein concentration (hypoproteinemia), obstructions in lymphatic vessels, increased venous pressure, and increased capillary permeability.

Hypoproteinemia may result from such conditions as liver disease, in which there is a failure to synthesize plasma proteins; kidney disease (glomerulonephritis), in which the glomerular capillary is damaged, allowing protein to escape into the urine; or starvation, in which the intake of amino acids is insufficient to support the synthesis of plasma proteins.

In each of these instances, the plasma protein concentration is decreased, and this is reflected in decreased plasma osmotic pressure. Consequently, the normal return of tissue fluid to the venule ends of capillaries is reduced, and tissue fluid tends to accumulate in the interstitial spaces.

As is discussed in chapter 19, *lymphatic obstructions* may result from various surgical procedures or from certain parasitic infections. In such cases, back pressure develops in the lymphatic vessels, interfering with the normal movement of tissue fluid into them. At the same time, proteins that are ordinarily removed by lymphatic circulation tend to accumulate in the interstitial spaces. The presence of these proteins causes the osmotic pressure of the interstitial fluid to rise, and this effect causes still more fluid to be attracted into the interstitial spaces.

If the outflow of blood from the liver into the vena cava is blocked, the venous pressure within the liver and portal blood vessels increases greatly. As a result, fluid with a high concentration of protein tends to be exuded from the surfaces of the liver and intestine into the peritoneal cavity. This causes a rise in the osmotic pressure of the abdominal fluid, which, in turn, attracts more water into the peritoneal cavity by osmosis. This condition, called *ascites,* is characterized by an uncomfortable distension of the abdomen.

Edema may also result from increased capillary permeability accompanying an *inflammation reaction.* As is described in chapters 5 and 19, this reaction occurs in response to tissue damage and usually involves the release of chemicals such as histamine from damaged cells. Histamine causes an increase in capillary permeability, so that excessive amounts of fluid tend to leak out of the capillary and enter the interstitial spaces.

Chart 21.3 summarizes the factors that result in edema.

Figure 21.7

If excessive amounts of extracellular fluids are lost, cells become dehydrated by osmosis. What happens to the concentration of electrolytes within the intracellular compartments at the same time?

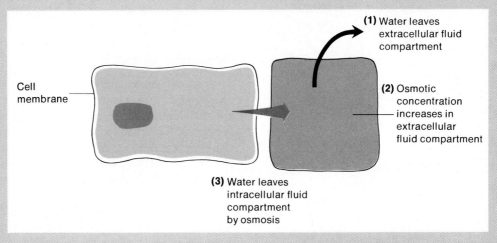

(1) Water leaves extracellular fluid compartment

(2) Osmotic concentration increases in extracellular fluid compartment

(3) Water leaves intracellular fluid compartment by osmosis

Cell membrane

Figure 21.8

If excessive water is added to the extracellular fluid compartment, cells gain water by osmosis.

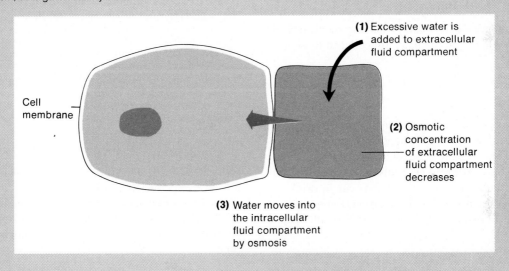

(1) Excessive water is added to extracellular fluid compartment

(2) Osmotic concentration of extracellular fluid compartment decreases

(3) Water moves into the intracellular fluid compartment by osmosis

Cell membrane

Chart 21.3 Factors associated with edema

Factor	Cause	Effect
Low plasma protein concentration	Liver disease and failure to synthesize proteins; kidney disease and loss of proteins in urine; lack of protein in diet due to starvation	Plasma osmotic pressure decreases; less fluid enters venule ends of capillaries by osmosis
Obstruction of lymph vessels	Surgical removal of portions of lymphatic pathways; parasitic infections	Back pressure in lymph vessels interferes with movement of fluid from interstitial spaces into lymph capillaries
Increased venous pressure	Venous obstructions or faulty venous valves	Back pressure in veins interferes with return of fluid from interstitial spaces into venule ends of capillaries
Inflammation	Tissue damage	Capillaries become abnormally permeable; fluid leaks from plasma into interstitial spaces

Figure 21.9
Electrolyte balance exists when the intake of electrolytes from all sources equals the output of electrolytes.

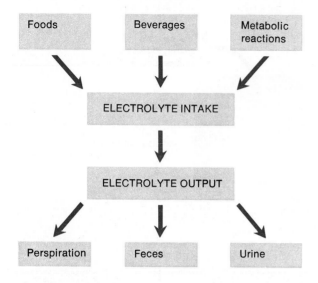

1. What electrolytes are most important to cellular functions?
2. What mechanisms ordinarily regulate electrolyte intake?
3. By what routes are electrolytes lost from the body?

Electrolyte Balance

An **electrolyte balance** exists when the quantities of electrolytes gained by the body are equal to those lost (figure 21.9).

Electrolyte Intake

The electrolytes of greatest importance to cellular functions are those that release ions of sodium, potassium, calcium, magnesium, chloride, sulfate, phosphate, and bicarbonate. These substances are obtained primarily from foods, but may also occur in drinking water and other beverages. In addition, some electrolytes occur as by-products of various metabolic reactions.

Regulation of Electrolyte Intake

Ordinarily, a person obtains sufficient electrolytes by responding to hunger and thirst. When there is a severe electrolyte deficiency, however, a person may experience a *salt craving,* which is a strong desire to eat salty foods.

Electrolyte Output

Some electrolytes are lost from the body by perspiration. The quantities leaving by this route will vary with the amount of perspiration excreted. Greater quantities are lost on warmer days and during times of strenuous exercise. Also, varying amounts of electrolytes are lost with the feces. The greatest electrolyte output, however, occurs in the urine as a result of kidney functions.

Regulation of Electrolyte Balance

The concentrations of positively charged ions, such as sodium (Na^+), potassium (K^+), and calcium (Ca^{+2}), are particularly important. Certain concentrations of these ions, for example, are necessary for the conduction of nerve impulses, the contraction of muscle fibers, and for maintaining the normal permeability of cell membranes. Thus, it is vital that their concentrations be regulated.

Sodium ions account for nearly 90% of the positively charged ions in extracellular fluid. The primary mechanism regulating these ions involves the kidneys and the hormone *aldosterone.* As is explained in chapter 13, this hormone is secreted by the adrenal cortex. Its presence causes an increase in sodium reabsorption in the collecting duct of the renal tubule.

If adrenal cortical hormones are lacking, the person develops *Addison's disease.* In this condition, because of an aldosterone deficiency, the normal concentration of extracellular sodium may be depleted within a few days. Without treatment, the person is likely to live only a short time (see chapter 13).

Aldosterone also functions in regulating potassium. In fact, the most important stimulus for aldosterone secretion is a rising potassium concentration, which seems to stimulate the cells of the adrenal cortex directly. This hormone enhances the reabsorption of sodium and causes the secretion of potassium at the same time.

As discussed in chapter 20, this action occurs because of the negative electrical charge that develops within the collecting duct as sodium ions are actively transported outward. The positively charged potassium ions (K^+) in the extracellular fluid of the kidney are attracted to this negative charge and are passively secreted into the urine. Consequently, as sodium ions are conserved, potassium ions are excreted (figure 21.10).

As mentioned in chapter 13, the concentration of *calcium ions* in extracellular fluid is regulated mainly by the parathyroid glands. Whenever the calcium concentration drops below the normal concentration, these glands are stimulated directly, and they secrete *parathyroid hormone* in response.

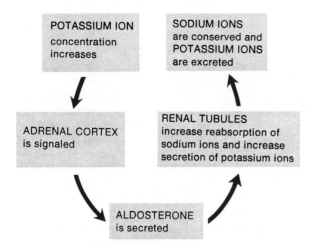

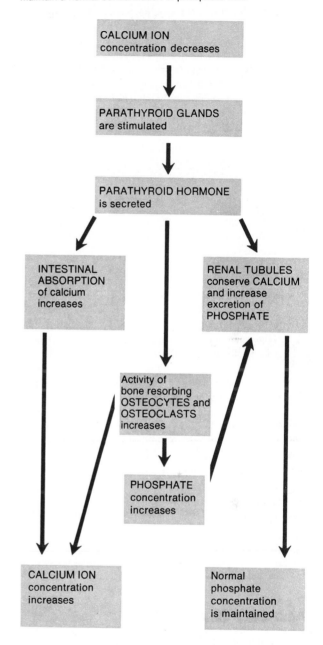

Parathyroid hormone causes increased activity in bone-reabsorbing cells (osteocytes and osteoclasts). As a result of their actions, the concentrations of calcium and phosphate in the extracellular fluid increase.

Parathyroid hormone also stimulates the absorption of calcium from the intestine indirectly (see chapter 13). Concurrently it causes the kidneys to conserve calcium and increase the excretion of phosphate. The net effect of the hormone, then, is to increase the *calcium* concentration of the extracellular fluid, but to maintain a normal *phosphate* concentration (figure 21.11).

Abnormal increases in blood calcium (hypercalcemia) sometimes result from the presence of cancers, particularly those originating in bone marrow, breasts, lungs, or prostate glands. Usually the increase in calcium occurs when ions are released from bone tissue that is being invaded and destroyed by cancerous growths. In other cases, however, the blood calcium concentration increases when cancer cells produce substances that have physiological effects similar to parathyroid hormone. Such hormonelike substances are most commonly associated with lung cancers.

The renal tubule also helps regulate the calcium concentration by conserving calcium through reabsorption when the extracellular concentration is low, and by allowing excess calcium to be excreted when the concentration is high.

Generally, the concentrations of negatively charged ions are controlled secondarily by the regulatory mechanisms that control positively charged ions. Chloride ions (Cl^-), for example (the most abundant negatively charged ions in extracellular fluid), are passively reabsorbed from the renal tubule in response to the active reabsorption of sodium ions. That is, the chloride ions are electrically attracted to the positively charged sodium ions and accompany them as they are reabsorbed.

Some negatively charged ions, such as phosphate ions (PO_4^{-3}) and sulfate ions (SO_4^{-2}), also are regulated partially by active transport mechanisms that have limited transport capacities. Thus, if the extracellular phosphate concentration is low, the phosphate ions in the renal tubules are conserved. On the other hand, if the renal plasma threshold is exceeded, the excess phosphate will be excreted in the urine.

1. What is the action of aldosterone?
2. How are the concentrations of sodium and potassium ions controlled?
3. How is calcium regulated?
4. What mechanism functions to regulate the concentrations of most negatively charged ions?

metabolic reactions, cause shifts in the distribution of other ions, or modify the actions of various hormones. Thus, acid-base balance is primarily concerned with the regulation of hydrogen ion concentrations, which are measured in pH units. (See chapter 2 for a detailed explanation of pH.)

Acid-Base Balance

As is discussed in chapter 2, electrolytes that ionize in water and release hydrogen ions are called *acids.* Substances that combine with hydrogen ions are called *bases.* The concentrations of acids and bases within body fluids must be controlled if homeostasis is to be maintained. The regulation of hydrogen ions is very important, since slight changes in hydrogen ion concentrations can alter the rates of enzyme-controlled

Sources of Hydrogen Ions

Most of the hydrogen ions in the body fluids originate as by-products of metabolic processes, although small quantities may enter in foods.

The major metabolic sources of hydrogen ions include the following:

1. **Aerobic respiration of glucose.** This process results in the production of carbon dioxide and water. Carbon dioxide diffuses out of cells and reacts with water in extracellular fluid to form *carbonic acid.*

$$CO_2 + H_2O \longrightarrow H_2CO_3$$

The resulting carbonic acid then ionizes to release hydrogen ions and bicarbonate ions.

$$H_2CO_3 \longrightarrow H^+ + HCO_3^-$$

2. **Anaerobic respiration of glucose.** When glucose is utilized anaerobically, *lactic acid* is produced and hydrogen ions are added to body fluids.
3. **Incomplete oxidation of fatty acids.** The incomplete oxidation of fatty acids results in the production of *acidic ketone bodies* that cause the hydrogen ion concentration to increase.
4. **Oxidation of amino acids containing sulfur.** The oxidation of sulfur-containing amino acids results in the production of *sulfuric acid* (H_2SO_4) that ionizes to release hydrogen ions and sulfate ions.
5. **Breakdown (hydrolysis) of phosphoproteins and nucleoproteins.** Phosphoproteins and nucleoproteins contain phosphorus, and their oxidation results in the production of *phosphoric acid* (H_3PO_4).

The acids produced by various metabolic processes vary in strength, and so their effects on the hydrogen ion concentration of body fluids vary also (figure 21.12).

---

1. Explain why the regulation of hydrogen ion concentration is so important.
2. What are the major sources of hydrogen ions in the body?

Strengths of Acids and Bases

Acids that ionize more completely are termed *strong acids,* while those that ionize less completely are termed *weak acids.* For example, the hydrochloric acid (HCl) of gastric juice is a strong acid, while the carbonic acid (H_2CO_3) produced when carbon dioxide reacts with water is weak.

The degrees to which such acids ionize can be indicated by the lengths of the arrows in chemical equations. That is, for hydrochloric acid, a long arrow indicates a considerable amount of ionization; for carbonic acid, a short arrow indicates a smaller amount of ionization.

$$HCl \rightleftharpoons H^+ + Cl^-$$

$$H_2CO_3 \rightleftharpoons H^+ + HCO_3^-$$

Figure 21.12
Some metabolic processes that serve as sources of hydrogen ions.

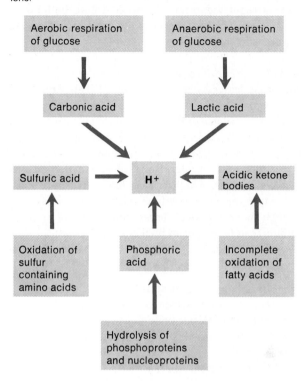

The arrows pointing to the left indicate that these chemical reactions are *reversible,* and once again the lengths of the arrows represent the degrees to which the reverse reactions tend to proceed.

Bases are substances that, like hydroxyl ions (OH^-), will combine with hydrogen ions. *Chloride ions* (Cl^-) and *bicarbonate ions* (HCO_3^-) are also bases. Furthermore, since chloride ions combine less readily with hydrogen ions (as indicated by the short arrow pointing to the left in the preceding equation), they are *weak* bases, while the bicarbonate ions, which combine more readily with hydrogen ions, are *strong bases.*

Once again, the degrees to which these substances react with hydrogen ions can be indicated by the lengths of the arrows in chemical equations, as follows:

$$Cl^- + H^+ \rightleftharpoons HCl$$

$$HCO_3^- + H^+ \rightleftharpoons H_2CO_3$$

Regulation of Hydrogen Ion Concentration

The concentration of hydrogen ions (or pH) in body fluid is regulated primarily by acid-base buffer systems, the respiratory center in the brain stem, and the nephrons in the kidneys.

Acid-Base Buffer Systems

Acid-base buffer systems occur in all body fluids and usually are composed of sets of two or more chemical substances. Such chemicals can combine with acids or bases when they occur in excess. More specifically, the substances of a buffer system can convert strong acids, which tend to release large quantities of hydrogen ions into weak acids, which release fewer hydrogen ions. Likewise, they can combine with strong bases and change them into weak bases. Such activity helps to minimize pH changes in body fluids.

The three most important acid-base buffer systems in body fluids are these:

1. **Bicarbonate buffer system.** The bicarbonate buffer system, which is present in intracellular and extracellular body fluids, consists of carbonic acid (H_2CO_3) and sodium bicarbonate ($NaHCO_3$). If a strong acid, like hydrochloric acid, is present, it reacts with the sodium bicarbonate. The products are carbonic acid, which is a weaker acid, and sodium chloride. Consequently, an increase in the hydrogen ion concentration in the body fluid is minimized.

$$HCl + NaHCO_3 \longrightarrow H_2CO_3 + NaCl$$
(strong acid) (weak acid)

If, on the other hand, a strong base like sodium hydroxide (NaOH) is present, it reacts with the carbonic acid. The products are sodium bicarbonate ($NaHCO_3$), which is a weaker base, and water. Thus, a shift toward a more basic (alkaline) state is minimized.

$$NaOH + H_2CO_3 \longrightarrow NaHCO_3 + H_2O$$
(strong base) (weak base)

2. **Phosphate buffer system.** The phosphate acid-base buffer system is also present in intracellular and extracellular body fluids. It is particularly important as a regulator of the hydrogen ion concentration in the tubular fluid of the nephrons and the urine. This buffer system consists of two phosphate compounds—sodium monohydrogen phosphate (Na_2HPO_4) and sodium dihydrogen phosphate (NaH_2PO_4).

If a strong acid is present, it reacts with the sodium monohydrogen phosphate to produce a weaker acid (sodium dihydrogen phosphate) and sodium chloride.

$$HCl + Na_2HPO_4 \longrightarrow NaH_2PO_4 + NaCl$$
(strong acid) (weak acid)

If a strong base is present, it reacts with the sodium dihydrogen phosphate, and again the products are a weak base (sodium monohydrogen phosphate) and water.

$$NaOH + NaH_2PO_4 \longrightarrow Na_2HPO_4 + H_2O$$
(strong base) (weak base)

3. **Protein buffer system.** The protein acid-base buffer system is the most important buffer system in plasma and intracellular fluid. It consists of the plasma proteins and various proteins within cells, including the hemoglobin of red blood cells.

As is described in chapter 2, proteins are composed of amino acids bound together in complex chains. Some of these amino acids have freely exposed groups of atoms, called *carboxyl groups*. Under certain conditions, a carboxyl group ($-COOH$) can become ionized, and a hydrogen ion is released.

$$-COOH \longrightarrow -COO^- + H^+$$

Some of the amino acids within a protein molecule also contain freely exposed *amino groups* ($-NH_2$). Under certain conditions, these amino groups can accept hydrogen ions.

$$-NH_2 + H^+ \longrightarrow -NH_3^+$$

Thus, protein molecules can function as acids by releasing hydrogen ions from their carboxyl groups, or as bases by accepting hydrogen ions into their amino groups. This special property allows protein molecules to operate as an acid-base buffer system. In the presence of excess hydrogen ions, the $-COO^-$ portions of the protein molecules accept hydrogen ions and become $-COOH$ groups again. This action decreases the number of free hydrogen ions in body fluid and minimizes the amount of pH change that occurs.

In the presence of excess hydroxyl ions (OH^-), the $-NH_3^+$ groups within protein molecules give up hydrogen ions and become

−NH$_2$ groups again. These hydrogen ions then combine with the hydroxyl ions to form water molecules. Once again, the pH change is reduced to a minimum, since water is a neutral substance.

$$H^+ + OH^- \longrightarrow H_2O$$

The *hemoglobin* in red blood cells provides an example of a specific protein that functions to buffer hydrogen ions. As is explained in chapter 16, carbon dioxide, which is produced by the cellular oxidation of glucose, diffuses through the capillary wall and enters the plasma and red blood cells. The red cells contain an enzyme, *carbonic anhydrase,* that speeds the reaction between carbon dioxide and water, producing carbonic acid.

$$CO_2 + H_2O \longrightarrow H_2CO_3$$

The carbonic acid quickly dissociates, releasing hydrogen ions and bicarbonate ions (HCO$_3^-$).

$$H_2CO_3 \longrightarrow H^+ + HCO_3^-$$

The hydrogen ions are accepted by the hemoglobin molecules, and thus, the pH change that would otherwise occur is reduced to a minimum.

Individual amino acids in body fluids can also function as acid-base buffers by accepting or giving up hydrogen ions, because every amino acid has an amino group (−NH$_2$) and a carboxyl group (−COOH) as part of its molecule.

To summarize, acid-base buffer systems act to take in hydrogen ions when body fluids are becoming more acidic, and give up hydrogen ions when the fluids are becoming more basic (alkaline). The buffer systems accomplish this action by converting stronger acids into weaker ones or by converting stronger bases into weaker ones, as summarized in chart 21.4.

Although buffer systems cannot prevent a pH change, they are able to reduce it to a minimum. Furthermore, the various acid-base buffer systems in body fluids act together to buffer each other. Consequently, whenever the hydrogen ion concentration begins to change, the chemical balances within all of the buffer systems change too, and the drift in pH is resisted.

Chart 21.4 Chemical acid-base buffer systems

Buffer system	Constituents	Actions
Bicarbonate system	Sodium bicarbonate (NaHCO$_3$)	Converts strong acid into weak acid
	Carbonic acid (H$_2$CO$_3$)	Converts strong base into weak base
Phosphate system	Sodium monohydrogen phosphate (Na$_2$HPO$_4$)	Converts strong acid into weak acid
	Sodium dihydrogen phosphate (NaH$_2$PO$_4$)	Converts strong base into weak base
Protein system (and amino acids)	−COO$^-$ group of a molecule	Accepts hydrogen ions in presence of excess acid
	−NH$_3^+$ group of a molecule	Releases hydrogen ions in presence of excess base

Neurons are particularly sensitive to changes in the pH of body fluids. For example, if the interstitial fluid becomes more alkaline than normal (alkalosis), neurons tend to become more excitable and the person may experience seizures. Conversely, if conditions become more acidic (acidosis), neuron activity tends to be depressed, and the person's level of consciousness may decrease.

1. What is the difference between a strong acid or base and a weak acid or base?
2. How does a chemical buffer system help regulate pH?
3. List the major buffer systems of the body.

The Respiratory Center

The **respiratory center** in the brain stem helps to regulate hydrogen ion concentrations in body fluids by controlling the rate and depth of breathing. Specifically, if the cells increase their production of carbon dioxide, as occurs during periods of physical exercise, the production of carbonic acid increases. As the carbonic acid dissociates, the concentration of hydrogen ions increases, and the pH of the fluids tends to drop (see chapter 16).

Such an increasing concentration of carbon dioxide and the consequent increase in hydrogen ion concentration in the plasma directly stimulate chemosensitive areas associated with the respiratory center.

Figure 21.13
An increase in carbon dioxide production is followed by an increase in carbon dioxide elimination.

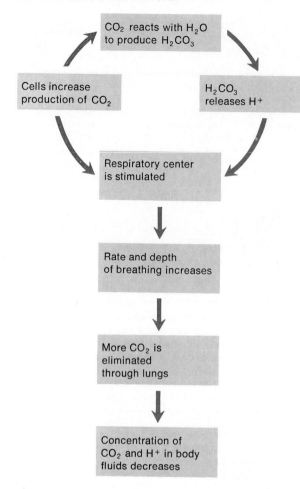

CO₂ reacts with H₂O to produce H₂CO₃

Cells increase production of CO₂

H₂CO₃ releases H⁺

Respiratory center is stimulated

Rate and depth of breathing increases

More CO₂ is eliminated through lungs

Concentration of CO₂ and H⁺ in body fluids decreases

In response, the respiratory center causes the depth and rate of breathing to increase, so that a greater amount of carbon dioxide is excreted through the lungs. This loss of carbon dioxide is accompanied by a drop in the hydrogen ion concentration in the body fluids, since the released carbon dioxide comes from carbonic acid (figure 21.13).

$$H_2CO_3 \longrightarrow CO_2 + H_2O$$

Conversely, if the body cells are less active, the concentrations of carbon dioxide and hydrogen ions in the plasma remain relatively low. As a consequence, the breathing rate and depth are decreased. In time this may result in an accumulation of carbon dioxide in the body fluids, and as the pH drops, the respiratory center is stimulated to increase the rate and depth of breathing once again.

Thus, the activity of the respiratory center is altered in response to shifts in the plasma pH, reducing these shifts to a minimum. Since most of the hydrogen ions in the body fluids originate from carbonic acid produced when carbon dioxide reacts with water, the respiratory regulation of hydrogen ion concentration is of considerable importance.

The Kidneys

The nephrons of the kidneys help to regulate hydrogen ion concentration of the body fluids by secreting hydrogen ions. As is discussed in chapter 20, these ions are secreted into the urine of the renal tubules by the epithelial cells that line the proximal and distal convoluted tubules and the collecting ducts.

This mechanism is important in balancing the quantities of sulfuric acid, phosphoric acid, and various organic acids that appear in body fluids as by-products of metabolic processes.

The metabolism of certain amino acids, for example, results in the formation of sulfuric and phosphoric acids. Consequently, a diet high in proteins (which give rise to amino acids when they are digested) will be accompanied by an excessive production of acids. The kidneys compensate for such gains in acids by altering the rate of hydrogen ion secretion into the urine. Thus, a shift in the pH of body fluids is resisted (figure 21.14).

Once hydrogen ions reach the urine, they are buffered by phosphates that were filtered into the fluid of the renal tubule. Ammonia also aids in this buffering action.

Through deamination of certain amino acids, the cells of the renal tubules produce ammonia (NH_3). It diffuses readily through the cell membranes and enters the urine. Since ammonia is a weak base, it can accept hydrogen ions, and the result is *ammonium ions* (NH_4^+).

$$H^+ + NH_3 \longrightarrow NH_4^+$$

Cell membranes are quite impermeable to ammonium ions. Consequently, these ions are trapped in the urine as they form, and they are excreted from the body with the urine.

As a result of a poorly understood mechanism, a prolonged increase in the hydrogen ion concentration of body fluids is accompanied by an increase in the production of ammonia by cells of the renal tubules. This mechanism helps to transport excessive hydrogen ions to the outside, and to control the pH of the urine.

The various regulators of hydrogen ion concentration operate at different rates. Acid-base buffers, for example, function rapidly and can convert strong acids or bases into weak acids or bases almost immediately. For this reason, these chemical buffer systems sometimes are called the body's *first line of defense* against shifts in pH.

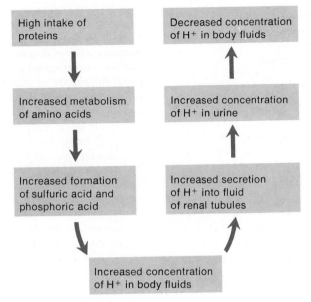

Figure 21.15
Chemical buffers act rapidly, while physiological buffers may require several minutes to begin resisting a change in pH.

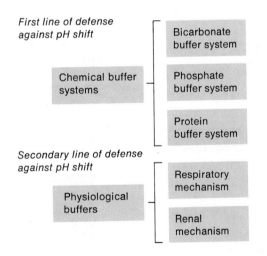

Physiological buffer systems, such as the respiratory and renal mechanisms, function more slowly, and constitute *secondary defenses*. The respiratory mechanism may require several minutes to begin resisting a change in pH, while the renal mechanism may require one to three days to control a changing hydrogen ion concentration (figure 21.15).

1. How does the respiratory system help to regulate acid-base balance?
2. How do the kidneys handle excessive concentrations of hydrogen ions?
3. What are the differences between the rates at which chemical and physiological buffer systems act?

Acid-Base Imbalances
A Clinical Application

Figure 21.16
If the pH of the arterial blood drops to 7.0 or rises to 7.8 for
more than a few hours, the person usually cannot survive.

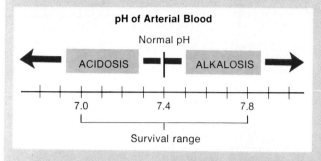

Figure 21.17
Acidosis results from an accumulation of acids or a loss of
bases; alkalosis results from a loss of acids or an accumulation
of bases.

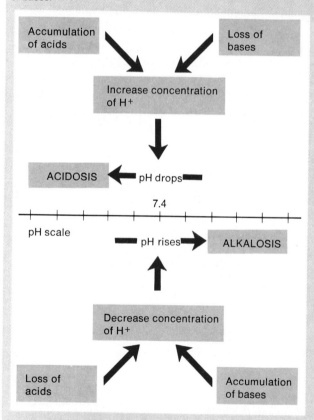

Ordinarily the hydrogen ion concentration of body fluids is maintained within very narrow pH ranges by the actions of chemical and physiological buffer systems. However, abnormal conditions may cause disturbances in the acid-base balance. For example, the pH of arterial blood is normally about 7.4, and if this value drops below 7.4, the person is said to have *acidosis*. If the pH rises above 7.4, the condition is called *alkalosis*. Such shifts in the pH of body fluids may be life-threatening, and in fact, a person usually cannot survive if the pH drops to 7.0 or rises to 7.8 for more than a few hours (figure 21.16).

Acidosis results from an accumulation of acids or a loss of bases, both of which are accompanied by abnormal increases in the hydrogen ion concentrations of body fluids. Conversely, alkalosis results from a loss of acids or an accumulation of bases accompanied by a decrease in hydrogen ion concentrations (figure 21.17).

There are two major types of acidosis: *respiratory acidosis* and *metabolic acidosis*. Respiratory acidosis is caused by factors that produce an increase in carbon dioxide, accompanied by an increase in the concentration of the respiratory acid, carbonic acid. Metabolic acidosis is due to an abnormal accumulation of any other acids in the body fluids, or to a loss of bases.

Similarly, there are two major types of alkalosis: *respiratory alkalosis* and *metabolic alkalosis*. Respiratory alkalosis is caused by an excessive loss of carbon dioxide and a consequent loss of carbonic acid. Metabolic alkalosis is due to an excessive loss of hydrogen ions or to the gain of bases.

Since **respiratory acidosis** involves an accumulation of carbon dioxide, it can be caused by factors that hinder pulmonary ventilation (figure 21.18). These include the following:

1. Injury to the respiratory center of the brain stem, followed by decreased rate and depth of breathing.
2. Obstructions in air passages that interfere with the movement of air into the alveoli.
3. Diseases that decrease gas exchanges, such as pneumonia; or those that cause a reduction in surface area of the respiratory membrane, such as emphysema.

Acid-Base Imbalances
A Clinical Application—*Continued*

As a result of such conditions, the amount of carbonic acid occurring in the body fluids increases, the concentration of hydrogen ions increases, and the pH value drops. This shift in pH may be resisted by the action of chemical buffers, such as hemoglobin. At the same time, the respiratory center may be stimulated by the increasing concentrations of carbon dioxide and hydrogen ions, so that the breathing rate and depth are increased, and the carbon dioxide concentration is thereby lowered. Also, the kidneys may begin to secrete increasing quantities of hydrogen ions.

Eventually, the pH of the body fluids may return to normal as a result of the actions of these chemical and physiological buffers. When this happens, the acidosis is said to be *compensated*.

The symptoms of respiratory acidosis include depression of activities in the central nervous system, characterized by drowsiness, disorientation, and stupor. These signs are accompanied usually by evidence of respiratory insufficiency such as labored breathing and cyanosis. In *uncompensated acidosis*, the person may become comatose and die.

Metabolic acidosis involves an accumulation of nonrespiratory acids or the loss of bases (figure 21.19). Factors that may lead to this condition include the following:

1. Kidney disease in which there is a failure to excrete acids produced by metabolic processes (uremic acidosis).
2. Prolonged vomiting in which the alkaline contents of the upper intestine are lost in addition to the stomach contents. (If only the stomach contents are lost, the result is metabolic alkalosis.)
3. Prolonged diarrhea in which there is an excessive loss of alkaline intestinal secretions (especially in infants).
4. Diabetes mellitus, in which some fatty acids are converted into acidic ketone bodies. These ketone bodies include *acetoacetic acid, beta hydroxybutyric acid,* and *acetone.* Normally these substances are produced in relatively small quantities, and they are oxidized by cells as energy sources. However, if fats are being utilized at an abnormally high rate, as may occur in diabetes mellitus, ketone bodies may accumulate faster than they can be oxidized. At such times, these compounds may be excreted in the urine (ketouria); in addition, acetone, which is volatile, may be excreted by the lungs and impart a fruity odor to the breath. More seriously, the accumulation of acetoacetic acid and beta-hydroxybutyric acid may cause a drop in pH (ketonemic acidosis).

Figure 21.18
Some factors that lead to respiratory acidosis.

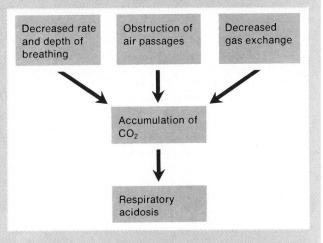

Figure 21.19
Some factors that lead to metabolic acidosis.

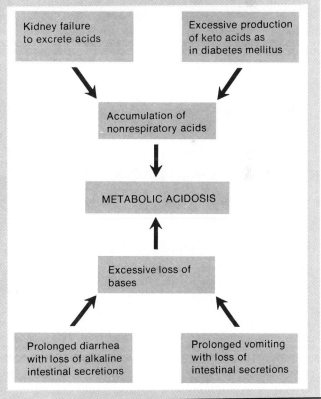

These acids also tend to combine with bicarbonate ions in the urine. As a result, bicarbonate ions are excreted excessively, and this loss interferes with the function of the bicarbonate acid-base buffer system.

In each case, the pH tends to shift toward lower values. However, this shift is resisted by the following: the actions of chemical buffer systems, which accept excessive hydrogen ions; the respiratory center, which causes the breathing rate and depth to increase; and the kidneys, which secrete increasing quantities of hydrogen ions.

Respiratory alkalosis develops as a result of *hyperventilation,* described in chapter 16. This action is accompanied by excessive loss of carbon dioxide and consequent decreases in carbonic acid and hydrogen ion concentrations (figure 21.20).

Commonly, hyperventilation occurs during periods of anxiety, although it may also accompany fever or poisoning

from salicylates such as aspirin. At high altitudes, hyperventilation may occur in response to low oxygen pressure. In any event, carbon dioxide is lost excessively by rapid, deep breathing, and the pH value of body fluids increases.

This shift in pH is resisted by chemical buffers, such as hemoglobin, that release hydrogen ions. Also, since the concentrations of carbon dioxide and hydrogen ions are lower, the respiratory center is stimulated to a lesser degree. As a result, the tendency to hyperventilate is inhibited, thus reducing further loss of carbon dioxide. At the same time, the kidneys decrease their secretion of hydrogen ions, and the urine becomes alkaline as bases are excreted.

The symptoms of respiratory alkalosis include lightheadedness, agitation, dizziness, and tingling sensations. In severe cases, impulses may be triggered spontaneously on peripheral nerves, and muscles may respond with tetanic contractions (chapter 9).

Metabolic alkalosis results from excessive loss of hydrogen ions or from a gain in bases, both of which are accompanied by a rise in pH of the body fluids (figure 21.21).

This condition may occur following gastric drainage (lavage) or prolonged vomiting in which only the stomach contents are lost. Since gastric juice is very acidic, its loss leaves the body fluids with a net increase of basic substances and a pH shift toward alkaline values. Metabolic alkalosis also may develop as a result of ingesting an excessive amount of antacids, such as sodium bicarbonate, taken to relieve the symptoms of indigestion.

Figure 21.20
Some factors that lead to respiratory alkalosis.

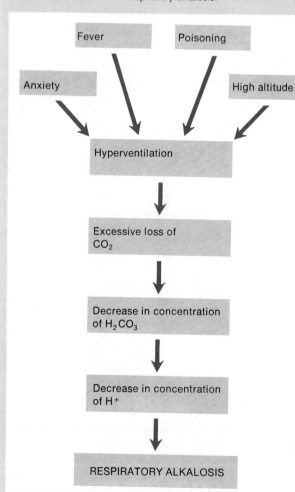

Figure 21.21
Some factors that lead to metabolic alkalosis.

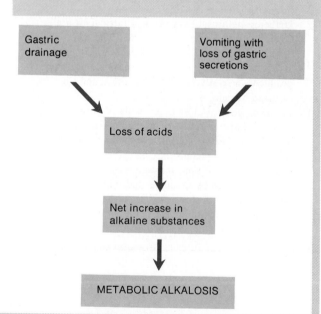

Clinical Terms Related to Water and Electrolyte Balance

acetonemia (as″ĕ-to-ne′me-ah)   the presence of abnormal amounts of acetone in the blood.

acetonuria (as″ĕ-to-nu′re-ah)   the presence of abnormal amounts of acetone in the urine.

albuminuria (al-bu″mĭ-nu′re-ah)   the presence of albumin in the urine.

antacid (ant-as′id)   a substance that neutralizes an acid.

anuria (ah-nu′re-ah)   the absence of urine excretion.

azotemia (az″o-te′me-ah)   an accumulation of nitrogenous wastes in the blood.

diuresis (di″u-re′sis)   increased production of urine.

glycosuria (gli″ko-su′re-ah)   the presence of excessive sugar in the urine.

hyperkalemia (hi″per-kah-le′me-ah)   the presence of excessive potassium in the blood.

hypernatremia (hi″per-na-tre′me-ah)   the presence of excessive sodium in the blood.

hyperuricemia (hi″per-u″rĭ-se′me-ah)   the presence of excessive uric acid in the blood.

hypoglycemia (hi″po-gli-se′me-ah)   an abnormally low level of blood sugar.

ketonuria (ke″to-nu′re-ah)   the presence of ketone bodies in the urine.

ketosis (ke″to′sis)   acidosis due to the presence of excessive ketone bodies in the body fluids.

proteinuria (pro″te-ĭ-nu′re-ah)   the presence of protein in the urine.

uremia (u-re′me-ah)   toxic condition resulting from the presence of excessive amounts of nitrogenous wastes in the blood.

Chapter Summary

Introduction
The maintenance of water and electrolyte balance requires that the quantities of these substances entering the body equal the quantities leaving it. Altering the water balance necessarily affects the electrolyte balance.

Distribution of Body Fluids
1. Fluid compartments
   a. The intracellular fluid compartment includes the fluids and electrolytes enclosed by cell membranes.
   b. The extracellular fluid compartment includes all fluids and electrolytes outside cell membranes.
      (1) Interstitial fluid within tissue spaces
      (2) Plasma within blood
      (3) Lymph within lymphatic vessels
      (4) Transcellular fluid within body cavities

2. Composition of body fluids
   a. Extracellular fluid
      (1) Extracellular fluid is characterized by high concentrations of sodium, chloride, and bicarbonate ions with lesser amounts of potassium, calcium, magnesium, phosphate, and sulfate ions.
      (2) Plasma contains more protein than either interstitial fluid or lymph.
   b. Intracellular fluid contains relatively high concentrations of potassium, phosphate, and magnesium ions; it also contains a greater concentration of sulfate ions and lesser concentrations of sodium, chloride, and bicarbonate ions than extracellular fluid.

3. Movement of fluid between compartments
   a. Movements are regulated by hydrostatic pressure and osmotic pressure.
      (1) Fluid leaves plasma because of hydrostatic pressure and returns to plasma because of osmotic pressure.
      (2) Fluid enters lymph vessels because of osmotic pressure.
      (3) Movement in and out of cells is regulated primarily by osmotic pressure.
   b. Sodium concentrations are especially important in the regulation of fluid movements.

Water Balance
1. Water intake
   a. Volume of water taken in varies from person to person.
   b. Most water is gained as liquid or in moist foods.
   c. Some water is produced by oxidative metabolism.

2. Regulation of water intake
   a. The thirst mechanism is the primary regulator of water intake.
   b. The act of drinking and the resulting distension of the stomach inhibit the thirst mechanism.

3. Water output
   a. Water is lost in a variety of ways.
      (1) It is excreted in urine, feces, and sweat.
      (2) Insensible loss occurs through evaporation from skin and lungs.
   b. The primary regulator of water output involves urine production.

4. Regulation of water balance
   a. Water balance is regulated mainly by the distal convoluted tubule and collecting duct of the nephron.
      (1) ADH from the posterior pituitary gland stimulates water reabsorption in these segments.
      (2) The mechanism involving ADH can reduce normal output of 1500 ml to 500 ml per day.
   b. If excess water is taken in, the ADH mechanism is inhibited.

Electrolyte Balance

1. Electrolyte intake
   a. The most important electrolytes in body fluids are ions of sodium, potassium, calcium, magnesium, chloride, sulfate, phosphate, and bicarbonate.
   b. These are obtained in foods and beverages or as by-products of metabolic processes.
2. Regulation of electrolyte intake
   a. Electrolytes usually are obtained in sufficient quantities in response to hunger and thirst mechanisms.
   b. In severe deficiency, a person may experience a salt craving.
3. Electrolyte output
   a. Electrolytes are lost through perspiration, feces, and urine.
   b. Quantities lost vary with temperature and physical exercise.
   c. The greatest loss is usually in urine.
4. Regulation of electrolyte balance
   a. Concentrations of sodium, potassium, and calcium ions in body fluids are particularly important.
   b. Regulation of sodium ions involves the secretion of aldosterone from the adrenal glands.
   c. Regulation of potassium ions also involves aldosterone.
   d. Regulation of calcium ions involves parathyroid hormone.
   e. In general, negatively charged ions are regulated secondarily by the mechanisms that control positively charged ions.
      (1) Chloride ions are passively reabsorbed in renal tubules as sodium ions are actively reabsorbed.
      (2) Some negatively charged ions, such as phosphate ions, are reabsorbed partially by active transport mechanisms with limited capacities.

Acid-Base Balance

Acids are electrolytes that release hydrogen ions.
Bases are substances that combine with hydrogen ions.

1. Sources of hydrogen ions
   a. Aerobic respiration of glucose
      (1) Aerobic respiration of glucose produces carbon dioxide, which reacts with water to form carbonic acid.
      (2) Carbonic acid dissociates to release hydrogen ions and bicarbonate ions.
   b. Anaerobic respiration of glucose gives rise to lactic acid.
   c. Incomplete oxidation of fatty acids gives rise to acidic ketone bodies.
   d. Oxidation of certain amino acids gives rise to sulfuric acid.
   e. Hydrolysis of phosphoproteins and nucleoproteins gives rise to phosphoric acid.
2. Strengths of acids and bases
   a. Acids vary in the degrees to which they ionize.
      (1) Strong acids, such as hydrochloric acid, ionize more completely.
      (2) Weak acids, such as carbonic acid, ionize less completely.
   b. Bases vary in strength also.
      (1) Strong bases, such as hydroxyl ions, combine readily with hydrogen ions.
      (2) Weak bases, such as chloride ions, combine with hydrogen ions less readily.
3. Regulation of hydrogen ion concentration
   a. Acid-base buffer systems
      (1) These buffer systems are composed of sets of two or more substances.
      (2) They function to convert strong acids into weaker ones, or strong bases into weaker ones.
      (3) They include the bicarbonate buffer system, phosphate buffer system, and protein buffer system.
      (4) Buffer systems cannot prevent pH changes, but they can minimize them.
   b. Respiratory center
      (1) The respiratory center is located in the brain stem.
      (2) It helps regulate pH by controlling the rate and depth of breathing.

(3) As carbon dioxide and hydrogen ion concentrations increase, chemosensitive areas associated with the respiratory center are stimulated; the breathing rate and depth increase, and the carbon dioxide concentration decreases.

(4) If the carbon dioxide and hydrogen ion concentrations are low, the respiratory center inhibits breathing.

c. Kidneys

(1) Nephrons regulate pH by secreting hydrogen ions.

(2) Hydrogen ions in urine are buffered by phosphates.

(3) Ammonia produced by renal cells helps transport hydrogen ions to the outside of the body.

d. Chemical buffers act rapidly; physiological buffers act less rapidly.

Application of Knowledge

1. An elderly, semiconscious patient is tentatively diagnosed as having acidosis. What components of the arterial blood will be most valuable in determining whether the acidosis is of metabolic or respiratory origin?

2. Some time ago a news story reported the death of several newborn infants due to an error in which sodium chloride was substituted for sugar in their formula. What symptoms would this produce? Why are infants more prone to the hazard of excess salt intake than adults?

3. Explain the threat to fluid and electrolyte balance in the following situation: a patient is being nutritionally maintained on concentrated solutions of hydrolyzed protein that are administered through a gastrostomy tube.

Review Activities

1. Explain how water balance and electrolyte balance are interdependent.

2. Name the body fluid compartments and describe their locations.

3. Explain how the fluids within these compartments differ in composition.

4. Describe how fluid movements between compartments are controlled.

5. Prepare a water budget to illustrate how the input of water equals the output of water.

6. Define *water of metabolism.*

7. Explain how water intake is regulated.

8. Explain how the nephrons function in the regulation of water output.

9. List the electrolytes of greatest importance in body fluids.

10. Explain how electrolyte intake is regulated.

11. List the routes by which electrolytes leave the body.

12. Explain how the adrenal cortex functions in the regulation of electrolyte output.

13. Describe the role of the parathyroid glands in regulating electrolyte balance.

14. Describe the role of the renal tubule in regulating electrolyte balance.

15. Distinguish between an acid and a base.

16. List five sources of hydrogen ions in body fluids and name an acid that originates from each source.

17. Distinguish between a strong acid and a weak acid, and name an example of each.

18. Distinguish between a strong base and a weak base, and name an example of each.

19. Explain how an acid-base buffer system functions.

20. Describe how the bicarbonate buffer system resists shifts in pH.

21. Explain why a protein has acidic as well as basic properties.

22. Describe how a protein functions as a buffer system.

23. Describe the function of hemoglobin as a buffer of carbonic acid.

24. Explain how the respiratory center functions in the regulation of acid-base balance.

25. Explain how the kidneys function in the regulation of acid-base balance.

26. Describe the role of ammonia in the transport of hydrogen ions to the outside of the body.

27. Distinguish between a chemical buffer system and a physiological buffer system.

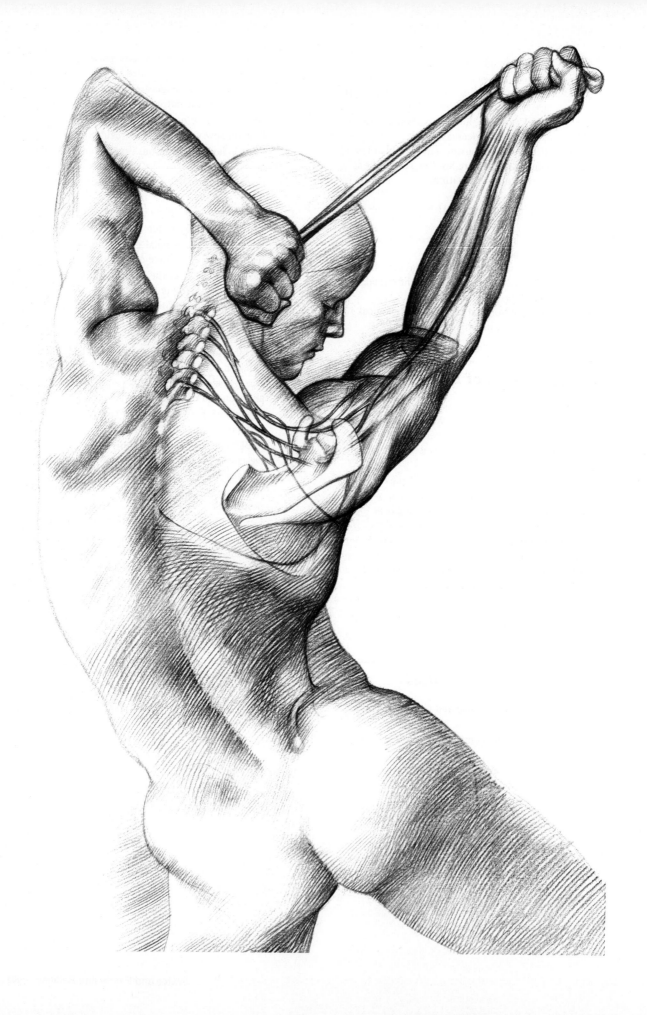

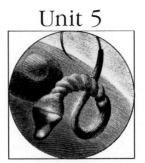

The Human Life Cycle

The chapters of unit 5 are concerned with the reproduction, growth, and development of the human organism. They describe how the organs of the male and female reproductive systems function to produce an embryo and how this offspring grows and develops, before and after birth, as it passes through the phases of its life cycle. The final chapter of this unit explains the genetic determination of individual traits.

22

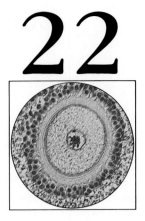

The Reproductive Systems

The male and female *reproductive systems* are specialized to produce off-spring. These systems are unique in that their functions are not necessary for the survival of each individual. Instead, their functions are vital to the continuation of the human species.

Organs of each of these systems are adapted to produce sex cells, to sustain these cells, and to transport them to a location where fertilization may occur. In addition, some of the reproductive organs secrete hormones that play vital roles in the development and maintenance of sexual characteristics and the regulation of reproductive physiology. In addition, specialized organs of the female system are adapted to support the life of an offspring that may develop from a fertilized egg and to transport this offspring to the outside of the female's body.

Chapter Outline

Chapter Objectives

After you have studied this chapter, you should be able to

1. State the general functions of the reproductive systems.

2. Name the parts of the male reproductive system and describe the general functions of each part.

3. Describe the structure of a testis and explain how sperm cells are formed.

4. Trace the path followed by sperm cells from their site of formation to the outside and explain the environmental changes that occur along this path.

5. Describe the structure of the penis and explain how its parts function to produce an erection.

6. Explain how hormones control the activities of the male reproductive organs and how they are related to the development of male secondary characteristics.

7. State the general functions of the female reproductive system.

8. Name the parts of the female reproductive system and describe the general functions of each part.

9. Describe the structure of an ovary and explain how egg cells and follicles are formed.

10. Describe how hormones control the activities of the female reproductive system and how they are related to the development of female secondary sexual characteristics.

11. Describe the major events that occur during a menstrual cycle.

12. Define *pregnancy* and describe the process of fertilization.

13. Describe the hormonal changes that occur in the maternal body during pregnancy.

14. Describe the birth process and explain the role of hormones in this process.

15. Review the structure and function of the mammary glands.

16. List several methods of birth control and describe the relative effectiveness of each method.

17. Complete the review activities at the end of this chapter. Note that the items are worded in the form of specific learning objectives. You may want to refer to them before reading the chapter.

Key Terms

androgen (an′dro-jen)

cleavage (klēv′ij)

contraception (kon″trah-sep′shun)

ejaculation (e-jak″u-la′shun)

emission (e-mish′un)

estrogen (es′tro-jen)

fertilization (fer″tĭ-lĭ-za′shun)

follicle (fol′ĭ-kl)

gonadotropin (go-nad″o-trōp′in)

implantation (im″plan-ta′shun)

menopause (men′o-pawz)

menstrual cycle (men′stroo-al si′kl)

oogenesis (o″o-jen′ĕ-sis)

ovulation (o″vu-la′shun)

placenta (plah-sen′tah)

pregnancy (preg′nan-se)

progesterone (pro-jes′tĕ-rōn)

puberty (pu′ber-te)

seminal fluid (se′men-al floo′id)

spermatogenesis (sper″mah-to-jen′ĕ-sis)

Aids to Understanding Words

andr-, man: *andr*ogens—male sex hormones.

contra-, opposed to: *contra*ception—the prevention of fertilization or of implantation.

crur-, lower part: *crur*a—diverging parts at the base of the penis by which it is attached to the pelvic arch.

ejacul-, to shoot forth: *ejacul*ation—process by which seminal fluid is expelled from the male reproductive tract.

fimb-, a fringe: *fimb*riae—irregular extensions on the margin of the infundibulum of the uterine tube.

follic-, small bag: *follic*le—ovarian structure that contains an egg cell.

genesis, origin: spermato*genesis*—process by which sperm cells are formed.

gubern-, to guide: *gubern*aculum—fibromuscular cord that guides the descent of a testis.

labi-, lip: *labi*a minora—flattened, longitudinal folds that extend along the margins of the vestibule.

mens-, month: *mens*trual cycle—monthly female reproductive cycle.

mons, mountain: *mons* pubis—rounded elevation of fatty tissue overlying the symphysis pubis in a female.

puber-, adult: *puber*ty—time when a person becomes able to function reproductively.

ORGANS OF THE male reproductive system are specialized to produce and maintain male sex cells (sperm cells), to transport these cells together with various supporting fluids to the female reproductive tract, and to produce and secrete male sex hormones.

Organs of the Male Reproductive System

The *primary sex organs* (gonads) of the male reproductive system are the two *testes* in which sperm cells and male sex hormones are formed. The other structures of the male reproductive system are termed *accessory sex organs* (secondary sex organs), and they include two groups: the internal and external reproductive organs (figure 22.1).

The Testes

The **testes** are ovoid structures, about 5 cm in length and 3 cm in diameter. Each is suspended by a spermatic cord within the cavity of the saclike *scrotum*.

Descent of the Testes

In a male fetus, the testes originate from masses of tissue located behind the parietal peritoneum, near the developing kidneys. Usually about a month or two before birth, these organs descend to regions in the lower abdominal cavity and pass through the abdominal wall into the scrotum.

The descent of the testes is stimulated by the male sex hormone, *testosterone,* which is secreted by the developing testes. The actual movement of these organs seems to be aided by a fibromuscular cord called the **gubernaculum.** This cord is attached to the developing testis and extends into the inguinal region of the abdominal cavity. It passes through the abdominal wall and is fastened to the skin on the outside. As the testis descends, apparently guided by the gubernaculum, it passes through the **inguinal canal** of the abdominal wall and enters the scrotum, where it remains anchored by the gubernaculum. Each testis carries with it a developing *vas deferens* and various blood vessels and nerves. These structures later form parts of the **spermatic cord** by which the testis is suspended in the scrotum (figure 22.2).

If the testes fail to descend into the scrotum, they will not produce sperm cells. In fact, in this condition, called *cryptorchidism,* the cells that normally produce sperm cells degenerate, and the male is sterile.

This failure to produce sperm cells is apparently caused by the unfavorable temperature of the abdominal cavity, which is a few degrees higher than the scrotal temperature.

Nurses and physicians routinely palpate the scrotal sacs of newborns in order to determine if the testes have descended. If they have not, it is sometimes possible to induce their descent by administering certain hormones, such as testosterone. This condition also may be treated by a surgical procedure in which the testes are moved into the scrotum.

---

1. What are the primary sex organs of the male reproductive system?
2. Describe the descent of the testes.
3. What happens if the testes fail to descend into the scrotum?

Structure of the Testes

Each testis is enclosed by a tough, white fibrous capsule called the *tunica albuginea*. Along its posterior border, the connective tissue thickens and extends into the organ forming a mass called the *mediastinum testis.* From this structure, thin dividers of connective tissue, called *septa*, pass into the testis and divide it into about 250 *lobules.*

A lobule contains one to four highly coiled, convoluted **seminiferous tubules,** each of which is up to 70 cm long when uncoiled. These tubules course posteriorly and become united into a complex network of channels called the *rete testis.* The rete testis is located within the mediastinum testis and gives rise to several ducts that join a tube called the **epididymis.** The epididymis, in turn, is coiled on the outer surface of the testis.

The seminiferous tubules are lined with a specialized stratified tissue called **germinal epithelium.** The cells of this tissue function to produce the male sex cells. Other specialized cells, called **interstitial cells** (cells of

Figure 22.1
(a) Sagittal view of male reproductive organs; *(b)* posterior view.

(a)

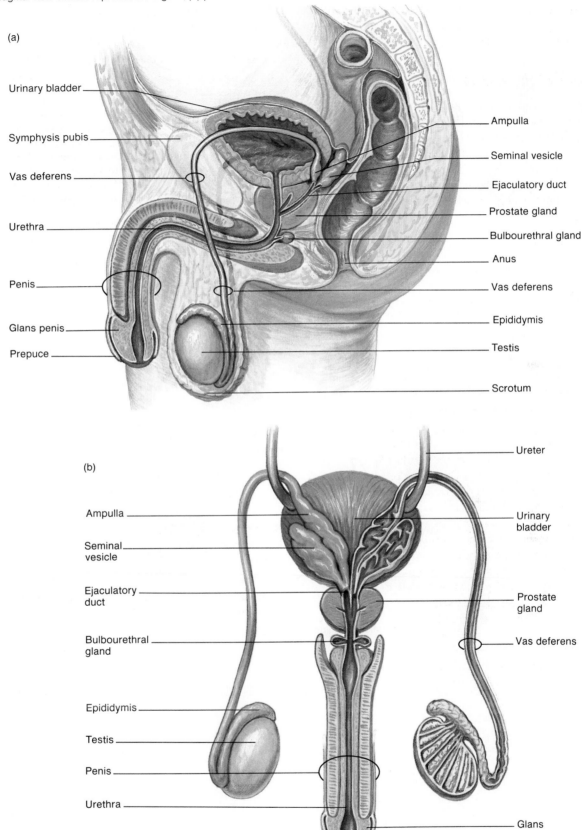

Urinary bladder

Symphysis pubis

Vas deferens

Urethra

Penis

Glans penis

Prepuce

Ampulla

Seminal vesicle

Ejaculatory duct

Prostate gland

Bulbourethral gland

Anus

Vas deferens

Epididymis

Testis

Scrotum

(b)

Ureter

Ampulla

Seminal vesicle

Ejaculatory duct

Bulbourethral gland

Epididymis

Testis

Penis

Urethra

Urinary bladder

Prostate gland

Vas deferens

Glans penis

Figure 22.2

During fetal development, each testis descends through an inguinal canal and enters the scrotum. What is the function of the gubernaculum when this occurs?

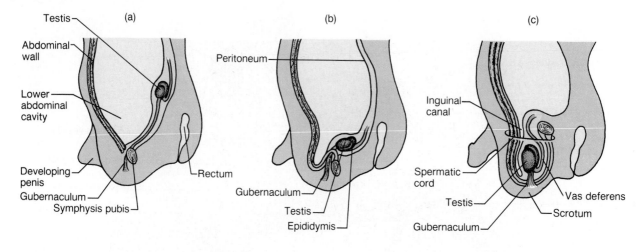

Figure 22.3

(a) Sagittal section of a testis; *(b)* cross section of a seminiferous tubule.

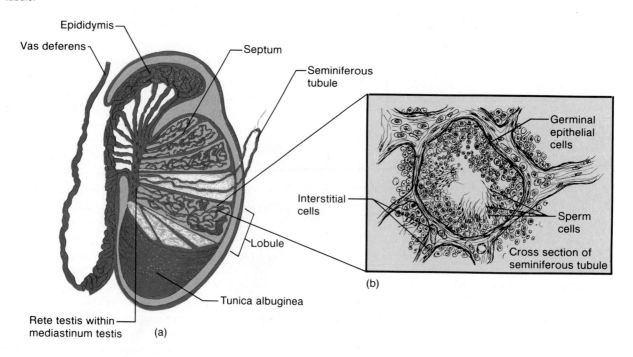

Figure 22.4
Scanning electron micrograph of cross sections of human
seminiferous tubules (about 100 X).

Leydig), are located in the spaces between the seminiferous tubules. They function in the production and
secretion of male sex hormones (figures 22.3 and 22.4).

---

1. Describe the structure of a testis.
2. Where are the sperm cells produced within the testes?
3. What cells produce male sex hormones?

Formation of Sperm Cells

The germinal epithelium consists of two types of cells:
supporting cells (Sertoli's cells) and spermatogenic
cells. The *supporting cells* are tall, columnar cells that
extend the full thickness of the epithelium from its base
to the lumen of the seminiferous tubule. Numerous, thin
processes project from these cells, filling the spaces between nearby spermatogenic cells. They function to
support, nourish, and regulate the *spermatogenic cells,*
which give rise to sperm cells (spermatozoa).

In a young male, all the spermatogenic cells are
undifferentiated and are called *spermatogonia.* Each
of these cells contains 46 chromosomes in its nucleus,
which is the usual number for human cells. (See figure
22.5.)

During early adolescence, certain hormones
stimulate spermatogonia to become active. Some of
them undergo mitosis, giving rise to new spermatogonia and providing a reserve supply of these undifferentiated cells. Others enlarge and become *primary
spermatocytes* that then divide by a special type of cell
division called **meiosis,** which is described in detail in
chapter 24.

In the course of meiosis, the primary spermatocytes each divide to form two *secondary spermatocytes.* Each of these cells, in turn, divides to form two
spermatids, which mature into sperm cells. Also during
meiosis, the number of chromosomes is reduced by one-
half. Consequently, for each primary spermatocyte that

Figure 22.5

(a) Light micrograph of seminiferous tubules; *(b)* spermatogonia give rise to primary spermatocytes by mitosis; the spermatocytes, in turn, give rise to sperm cells by meiosis.

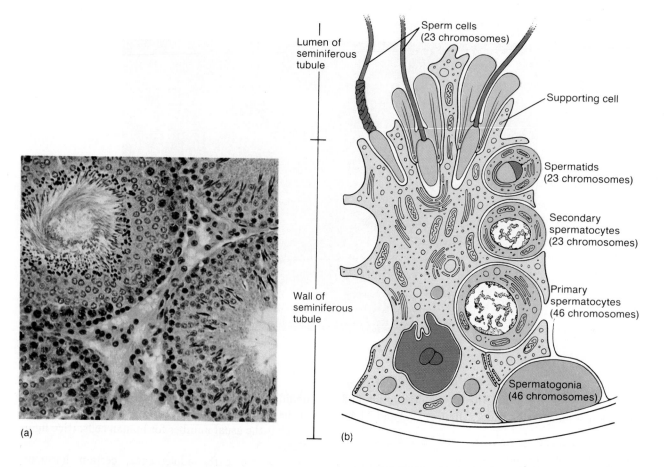

(a)

(b)

Sperm cells (23 chromosomes)

Lumen of seminiferous tubule

Wall of seminiferous tubule

Supporting cell

Spermatids (23 chromosomes)

Secondary spermatocytes (23 chromosomes)

Primary spermatocytes (46 chromosomes)

Spermatogonia (46 chromosomes)

undergoes meiosis, four sperm cells with 23 chromosomes in their nuclei are formed. This process by which sperm cells are produced is called **spermatogenesis** (figure 22.5).

The spermatogonia are located near the base of the germinal epithelium. As spermatogenesis occurs, cells in more advanced stages are pushed along the sides of supporting cells toward the lumen of the seminiferous tubule.

Near the base of the epithelium, membranous processes from adjacent supporting cells are fused by specialized junctions (occluding junctions) into complexes that divide the tissue into two layers. The spermatogonia are located on one side of this barrier, and the cells in more advanced stages are on the other side. This membranous complex seems to help maintain a favorable environment for the development of sperm cells by preventing the movement of certain large molecules from the interstitial fluid of the basal epithelium into the region of the differentiating cells.

Spermatogenesis occurs continually throughout the reproductive life of a male. The resulting sperm cells collect in the lumens of the seminiferous tubules. Then they pass through the rete testis to the epididymis, where they remain for a time and mature.

Structure of a Sperm Cell

A mature sperm cell is a tiny, tadpole-shaped structure about 0.06 mm long. It consists of a flattened head, a cylindrical body, and an elongated tail.

The *head* of a sperm cell, which is oval in outline, is composed primarily of a nucleus and contains the chromatin of 23 chromosomes in highly compacted form. It has a small part at its anterior end called the *acrosome,* which contains enzymes that aid the sperm cell in penetrating an egg cell at the time of fertilization (figure 22.6).

The *body* of the sperm cell contains a central, filamentous core and a large number of mitochondria arranged in a spiral. The *tail* consists of several

Figure 22.6

(a) The head of the sperm develops largely from the nucleus of the formative cell; (b) parts of a mature sperm cell.

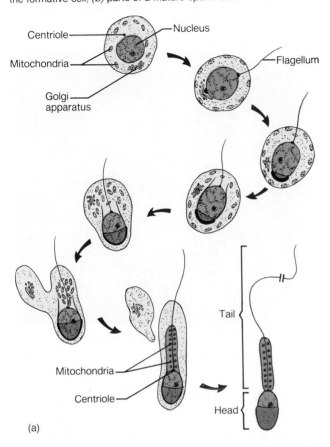

Centriole
Nucleus
Mitochondria
Golgi apparatus
Flagellum
Tail
Mitochondria
Centriole
Head

(a)

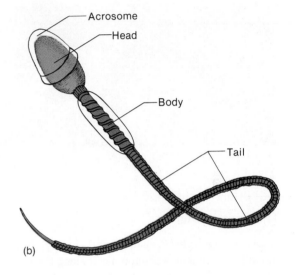

Acrosome
Head
Body
Tail

(b)

longitudinal fibrils enclosed in an extension of the cell membrane. The tail also contains a relatively large quantity of ATP, which is thought to provide the energy for the lashing movement that propels the sperm cell through fluid. The scanning electron micrograph in figure 22.7 shows a few mature sperm cells.

1. Explain the function of supporting cells in the germinal epithelium.
2. Describe the process of spermatogenesis.
3. Describe the structure of a sperm cell.

Male Internal Accessory Organs

The *internal accessory organs* of the male reproductive system include the epididymides, vasa deferentia, ejaculatory ducts, and urethra, as well as the seminal vesicles, prostate gland, and bulbourethral glands.

Figure 22.7

Scanning electron micrograph of human sperm cells.

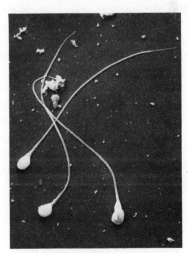

Figure 22.8
Cross section of a human epididymis.

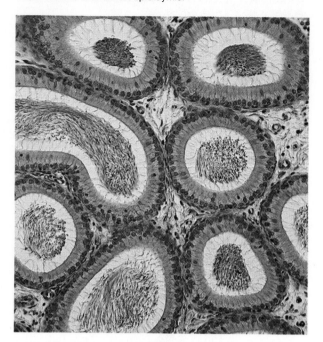

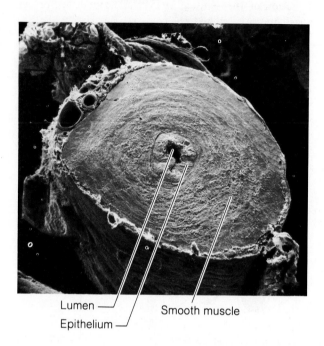

Lumen —

Epithelium —

Smooth muscle

The Epididymis

The **epididymis** (plural, epididymides) is a tightly coiled, threadlike tube that is about 6 meters long (figures 22.1 and 22.8). This tube is connected to ducts within the testis. It emerges from the top of the testis, descends along its posterior surface, and then courses upward to become the *vas deferens*.

The inner lining of the epididymis is composed of pseudostratified columnar cells that bear nonmotile cilia. These cells are thought to secrete glycogen, which helps sustain the lives of the stored sperm cells.

When immature sperm cells are moved from the ducts of the testis into the epididymis, they are completely nonmobile and are unable to fertilize an egg cell. As they travel slowly through the epididymis as a result of rhythmic peristaltic contractions, they undergo *maturation* and become capable of moving independently; however, they usually do not engage in swimming motions until after ejaculation, which is described in a subsequent section of this chapter.

The Vas Deferens

The **vas deferens** (plural, vasa deferentia; also called ductus deferens) is a muscular tube about 45 cm long. It begins at the lower end of the epididymis and passes upward along the medial side of the testis to become part of the spermatic cord. It passes through the inguinal canal, enters the abdominal cavity outside of the parietal peritoneum, and courses over the pelvic brim. From there, it extends backward and medially into the pelvic cavity, where it ends behind the urinary bladder.

Near its termination, the vas deferens becomes dilated into a portion called the *ampulla*. Just outside the prostate gland, the tube becomes slender again and unites with the duct of a seminal vesicle. The fusion of these two ducts forms an **ejaculatory duct,** which passes through the substance of the prostate gland and empties into the urethra through a slitlike opening (figures 22.1 and 22.9).

The Seminal Vesicle

A **seminal vesicle** is a convoluted, saclike structure about 5 cm long that is attached to the vas deferens near the base of the urinary bladder.

The glandular tissue lining the inner wall of the seminal vesicle secretes a slightly alkaline fluid. This fluid is thought to help regulate the pH of the tubular contents as sperm cells are conveyed to the outside. The secretion of the seminal vesicle contains a variety of nutrients and is rich in *fructose*, a monosaccharide that is thought to provide sperm cells with an energy source. It also contains *prostaglandins*, which are thought to stimulate muscular contractions within the female reproductive organs and thus aid the movement of sperm cells toward the female egg cell.

At the time of ejaculation, the contents of the seminal vesicles are emptied into the ejaculatory ducts, greatly increasing the volume of the fluid that is discharged from the vas deferens (figure 22.1).

---

1. Describe the structure of the epididymis.
2. Trace the path of the vas deferens.
3. What is the function of a seminal vesicle?

The Prostate Gland

The **prostate gland** is a chestnut-shaped structure about 4 cm across and 3 cm thick that surrounds the beginning of the urethra, just below the urinary bladder. It is enclosed by connective tissue and is composed of many branched tubular glands. These glands are separated by septa of connective tissue and smooth muscle that extend inward from the capsule. Their ducts open into the urethra (figure 22.1).

The prostate gland secretes a thin, milky fluid with an alkaline pH. It functions to neutralize the seminal fluid, which is acidic due to an accumulation of metabolic wastes produced by stored sperm cells. It also enhances the motility of the sperm cells, which remain relatively immobile in the acidic contents of the epididymis. In addition, the prostatic fluid helps neutralize the acidic secretions of the vagina, thus helping to sustain sperm cells that enter the female reproductive tract.

The prostate gland releases its secretions into the urethra as a result of smooth muscle contractions in its capsular wall. This release occurs during emission, as the contents of the vas deferens and the seminal vesicles are entering from the ejaculatory ducts, and thus the volume of the seminal fluid is increased still more.

Although the prostate gland is relatively small in male children, it begins to grow in early adolescence and reaches its adult size a few years later. As a rule, its size remains unchanged between 20 and 50 years of age. In older males the prostate gland usually enlarges. When this happens it tends to squeeze the urethra and interfere with urine excretion.

The treatment for an enlarged prostate gland is usually surgical. If the obstruction is slight, the procedure may be performed through the urethral canal and is called a *transurethral prostatic resection.*

The prostate gland is a common site of cancer in older males. Such cancers usually are stimulated to grow more rapidly by the male sex hormone, testosterone, and are inhibited by the female sex hormone, estrogen. Consequently, treatment for this type of cancer may involve removal of the testes (the main source of testosterone), the administration of drugs that block the action of testosterone, or the administration of estrogen. Although such treatment usually does not stop the cancer, it may slow its development.

The Bulbourethral Glands

The **bulbourethral glands** (Cowper's glands) are two small structures about the size of peas, which are located below the prostate gland lateral to the membranous urethra and enclosed by fibers of the external urethral sphincter muscle (figure 22.1).

These glands are composed of numerous tubes whose epithelial linings secrete a mucuslike fluid. This fluid is released in response to sexual stimulation and provides some lubrication to the end of the penis in preparation for sexual intercourse (coitus). Most of the lubricating fluid for intercourse, however, is secreted by female reproductive organs.

Seminal Fluid

The fluid conveyed by the urethra to the outside during ejaculation is called **seminal fluid** (semen). It consists of sperm cells from the testes and secretions of the seminal vesicles, prostate gland, and bulbourethral glands. It has a slightly alkaline pH (about 7.5) and a milky appearance. It contains a variety of nutrients, as well as prostaglandins, that function to enhance sperm cell survival and movement through the female reproductive tract.

The volume of seminal fluid released by emission varies from 2 to 6 ml, while the average number of sperm cells present in the fluid is about 120 million per ml.

Sperm cells remain immobile while they are in the ducts of the testis and epididymis, but become activated as they are mixed with the secretions of accessory glands during emission.

Although sperm cells are able to live for many weeks in the ducts of the male reproductive tract, they tend to survive for only a day or two after being expelled to the outside even when they are maintained at body temperature.

1. Where is the prostate gland located?
2. What is the function of its secretion?
3. What is the function of the bulbourethral glands?
4. What are the characteristics of seminal fluid?

Male External Reproductive Organs

The male external reproductive organs are the scrotum, which encloses the testes, and the penis, through which the urethra passes.

The Scrotum

The **scrotum** is a pouch of skin and subcutaneous tissue that hangs from the lower abdominal region behind the penis.

Although its subcutaneous tissue lacks fat, the scrotal wall contains a layer of smooth muscle fibers that constitute the *dartos muscle.* When these muscle fibers are contracted, the scrotal skin becomes wrinkled and is held close to the testes; when the fibers are relaxed, the scrotum hangs more loosely.

The scrotum is divided into chambers by a medial septum, and each chamber is occupied by a testis. Each chamber also contains a serous membrane that provides a covering for the front and sides of the testis and epididymis. This covering helps ensure that the testis will move smoothly within the scrotum (figure 22.1).

The Penis

The **penis** is a cylindrical organ that functions to convey urine and seminal fluid through the urethra to the outside. It is also specialized to become enlarged and stiffened by a process called *erection,* so that it can be inserted into the female vagina during sexual intercourse.

The *body,* or shaft, of the penis is composed of three columns of erectile tissue. These include a pair of dorsally located *corpora cavernosa* and a single *corpus spongiosum* below. These columns are enclosed by skin, a thin layer of subcutaneous tissue, and a layer of connective tissue. In addition, each column is surrounded by a tough capsule of white fibrous connective tissue called a *tunica albuginea* (figure 22.10).

The corpus spongiosum, through which the urethra extends, is enlarged at its distal end to form a sensitive, cone-shaped **glans penis**. This glans covers the ends of the corpora cavernosa and bears the urethral opening—the *external urethral meatus.* The skin of the glans is very thin and hairless. Also a loose fold of skin called the *prepuce* (foreskin) begins just behind the glans and extends forward to cover it as a sheath. The prepuce sometimes is removed during infancy by a surgical procedure called *circumcision.*

Figure 22.10

(a) Interior structure of the penis; (b) cross section of the penis.

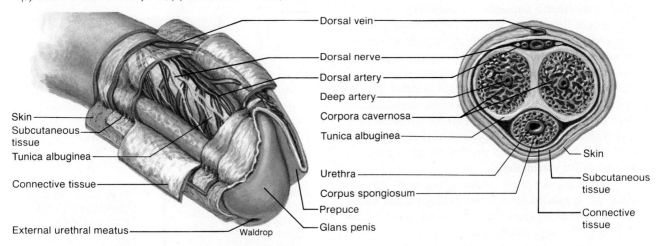

At the *root* of the penis, the columns of erectile tissue become separated. The corpora cavernosa diverge laterally in the perineum and are firmly attached to the medial surfaces of the pubic arch by connective tissue. These diverging parts form the *crura* of the penis. The single corpus spongiosum is enlarged between the crura as the *bulb* of the penis, which is attached to membranes of the perineum.

1. Describe the structure of the penis.
2. What is circumcision?
3. How is the penis attached to the perineum?

Erection, Orgasm, and Ejaculation

The masses of erectile tissue within the body of the penis contain networks of vascular spaces (venous sinusoids). These spaces are lined with endothelium and are separated from each other by bands (trabeculae) of smooth muscle and connective tissue.

Ordinarily, the vascular spaces remain small as a result of partial contractions in the smooth muscle fibers that surround them. During sexual stimulation the smooth muscles become relaxed. At the same time, *parasympathetic* nerve impulses pass from the sacral portion of the spinal cord to arteries leading into the penis, causing them to dilate. These impulses also stimulate veins leading away from the penis to constrict. As a result, arterial blood under relatively high pressure enters the vascular spaces of the erectile tissue, and the flow of venous blood away from the penis is reduced. Consequently, blood accumulates in the erectile tissues, and the penis swells, elongates, and becomes erect (figure 22.11).

Figure 22.11

Mechanism that causes erection of the penis.

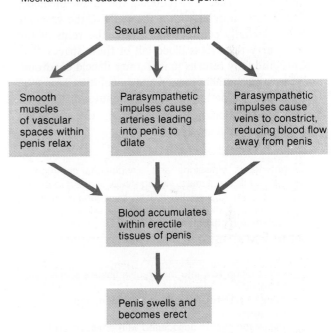

The culmination of sexual stimulation is called **orgasm** and involves a pleasurable feeling of physiological and psychological release. Also, orgasm in the male is accompanied by emission and ejaculation.

Emission is the movement of sperm cells from the testes and secretions from the prostate gland and seminal vesicles into the urethra, where they are mixed to form seminal fluid. Emission is caused by peristaltic contractions occurring in response to *sympathetic* impulses traveling from the spinal cord to smooth muscles in the walls of the testicular ducts, epididymides, vasa deferentia, and ejaculatory ducts. At the same time, other sympathetic impulses stimulate rhythmic contractions of the seminal vesicles and prostate gland.

As the urethra fills with seminal fluid, sensory impulses are stimulated and pass into the sacral portion of the spinal cord. In response, motor impulses are transmitted from the cord to certain skeletal muscles at the base of the erectile columns of the penis, causing them to contract rhythmically. This increases the pressure within the erectile tissues and aids in forcing the seminal fluid through the urethra to the outside—a process called **ejaculation.**

The sequence of events during emission and ejaculation is regulated so that the fluid from the bulbourethral glands is expelled first, followed by the release of fluid from the prostate gland, the passage of sperm cells, and finally the ejection of fluid from the seminal vesicles (figure 22.12).

Immediately after ejaculation, sympathetic impulses cause vasoconstriction of the arteries that supply the erectile tissue, thus reducing the inflow of blood. The smooth muscles within the walls of the vascular spaces partially contract again, and the veins of the penis carry the excess blood out of these spaces. The penis gradually returns to its former flaccid condition, and usually another erection and ejaculation cannot be triggered for a period of 10 to 30 minutes or longer.

The functions of the male reproductive organs are summarized in chart 22.1.

Spontaneous emissions and ejaculations commonly occur in adolescent males during sleep. These *nocturnal emissions* are apparently caused by changes in hormonal concentrations that accompany adolescent development and sexual maturation.

1. How is blood flow into the erectile tissues of the penis controlled?
2. Distinguish between orgasm, emission, and ejaculation.
3. Review the events associated with emission and ejaculation.

Hormonal Control of Male Reproductive Functions

Male reproductive functions are controlled largely by hormones secreted from the *hypothalamus,* the *anterior pituitary gland,* and the *testes.* These hormones are responsible for the initiation and maintenance of sperm cell production and for the development and maintenance of male sexual characteristics.

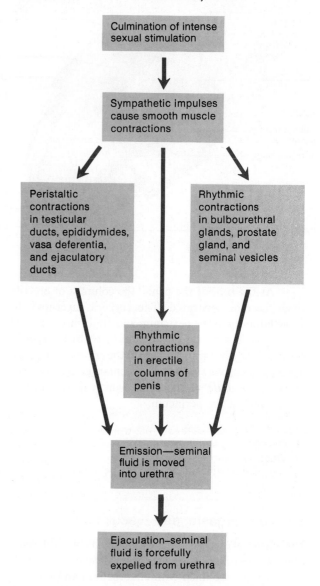

Figure 22.12
Mechanism that results in emission and ejaculation.

Hypothalamic and Pituitary Hormones

Prior to 10 years of age, the male body usually is reproductively immature. It remains childlike, and the spermatogenic cells of the testes remain undifferentiated. Then a series of changes is triggered that leads to the development of a reproductively functional adult. Although the mechanism that initiates such changes is not well understood, it involves the action of the hypothalamus.

As explained in chapter 13, the hypothalamus secretes a gonadotropin-releasing hormone (GnRH), which enters blood vessels leading to the anterior pituitary gland. In response, the anterior pituitary gland secretes the **gonadotropins** called *luteinizing hormone* (LH) and *follicle-stimulating hormone* (FSH). LH,

Chart 22.1 Functions of male reproductive organs

Organ	Function
Testes	
Seminiferous tubules	Production of sperm cells
Interstitial cells	Production and secretion of male sex hormones
Epididymis	Storage and maturation of sperm cells; conveys sperm cells to vas deferens
Vas deferens	Conveys sperm cells to ejaculatory duct
Seminal vesicle	Secretes alkaline fluid containing nutrients and prostaglandins; fluid helps neutralize acidic seminal fluid
Prostate gland	Secretes alkaline fluid that helps neutralize acidic seminal fluid and enhances motility of sperm cells
Bulbourethral gland	Secretes fluid that lubricates end of penis
Scrotum	Encloses and protects testes
Penis	Conveys urine and seminal fluid to outside of body; inserted into vagina during sexual intercourse; glans penis is richly supplied with sensory nerve endings associated with feelings of pleasure during sexual stimulation

which also is called interstitial cell-stimulating hormone (ICSH), promotes the development of the interstitial cells (cells of Leydig) of the testes, and they, in turn, secrete male sex hormones. FSH causes the supporting cells (Sertoli cells) of the germinal epithelium to become responsive to the effects of the male sex hormone, testosterone. Then, in the presence of FSH and testosterone, these supporting cells somehow stimulate the spermatogenic cells to undergo spermatogenesis, giving rise to sperm cells (figure 22.13).

Male Sex Hormones

As a group, the male sex hormones are termed **androgens,** and although most of them are produced by the interstitial cells of the testes, small amounts are synthesized in the adrenal cortex (see chapter 13).

The hormone called **testosterone** is the most abundant of the androgens, and when it is secreted it is transported in the blood loosely attached to plasma proteins. As in the case of other steroid hormones, testosterone affects its target cells by combining with receptor molecules in the cytoplasm (see chapter 13). However, in many target cells such as those in the prostate gland, seminal vesicles, and male external accessory organs, testosterone is converted to another androgen called **dihydrotestosterone,** which, in turn, stimulates the cells of these organs.

Figure 22.13
Mechanism by which the hypothalamus controls the maturation of sperm cells and the development of male secondary sexual characteristics.

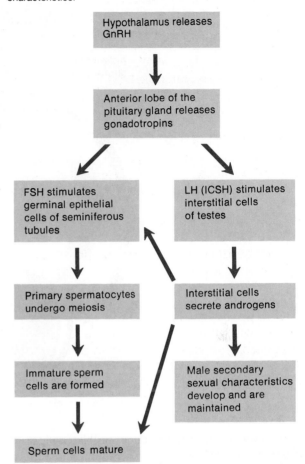

Androgen molecules that fail to become fixed in target cells usually are changed by the liver into forms that can be excreted in bile or urine.

Although the secretion of testosterone begins during fetal development and continues for a few weeks following birth, it nearly ceases during childhood. Sometime between the ages of 13 and 15, however, androgen production usually increases rapidly. This phase in development, during which a male becomes reproductively functional, is called **puberty.** After puberty, testosterone secretion continues throughout the life of the male.

Actions of Testosterone

During fetal life, testosterone is first secreted by certain embryonic cells within a structure called the *genital ridge* and later by the developing testes. It stimulates the formation of the male reproductive organs, including the penis, scrotum, prostate gland,

Figure 22.14
A negative feedback mechanism operating between the anterior
lobe of the pituitary gland and the testes controls the
concentration of testosterone.

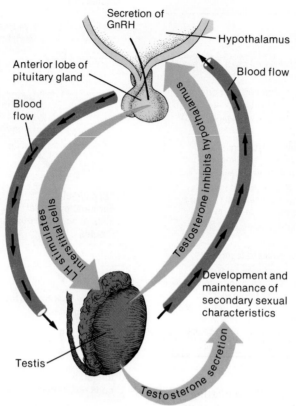

seminal vesicles, and various ducts. Still later, it causes
the testes to descend into the scrotum, as described
previously.

During puberty, testosterone stimulates enlargement of the testes and various accessory organs of the
reproductive system, and it causes the development of
the male *secondary sexual characteristics*. These secondary sexual characteristics are special features associated with the adult male body, and they include the
following:

1. Increased growth of body hair, particularly on
   the face, chest, axillary region, and pubic
   region, but sometimes accompanied by
   decreased growth of hair on the scalp.
2. Enlargement of the larynx and thickening of
   the vocal folds, accompanied by the
   development of a lower-pitched voice.
3. Thickening of the skin.
4. Increased muscular growth accompanied by the
   development of broader shoulders and a
   relatively narrow waist.
5. Thickening and strengthening of the bones.

Testosterone causes bones to thicken and strengthen
by promoting the deposition of calcium salts in osseous
tissues. Because of this, testosterone is sometimes used
to treat elderly persons suffering from *osteoporosis*, a
condition in which the bones thin due to excessive bone
resorption.

Other actions of testosterone include increasing
the rate of cellular metabolism and increasing the production of red blood cells, so the average number of red
blood cells in a cubic millimeter of blood usually is
greater in males than in females. Testosterone also
stimulates sexual activity by influencing certain portions of the brain.

Regulation of Male Sex Hormones

The degree to which male secondary sexual characteristics develop is directly related to the amount of testosterone secreted by the interstitial cells. This quantity
is regulated by a *negative feedback system* involving
the hypothalamus (figure 22.14).

As the concentration of testosterone in the blood
increases, the hypothalamus becomes inhibited, and its
stimulation of the anterior pituitary gland by GnRH is

Figure 22.15
Organs of the female reproductive system.

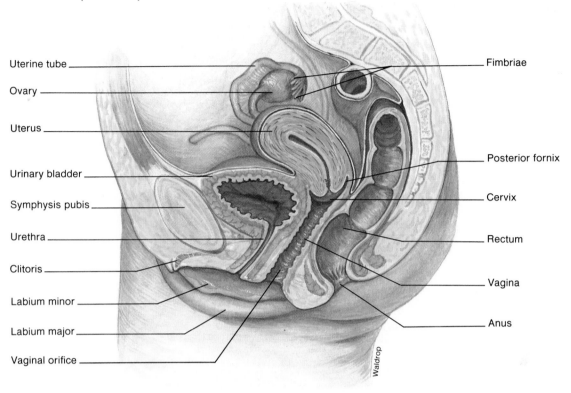

Uterine tube

Ovary

Uterus

Urinary bladder

Symphysis pubis

Urethra

Clitoris

Labium minor

Labium major

Vaginal orifice

Fimbriae

Posterior fornix

Cervix

Rectum

Vagina

Anus

Waldrop

decreased. As the pituitary's secretion of LH (ICSH) is reduced, the amount of testosterone released by the interstitial cells is reduced also.

But as the blood concentration of testosterone drops, the hypothalamus becomes less inhibited, and it once again stimulates the pituitary gland to release LH. The increasing secretion of LH causes the interstitial cells to release more testosterone, and its blood concentration rises.

Thus, the concentration of testosterone in the male body is regulated so that it remains relatively constant.

The amount of testosterone secreted by the testes usually declines gradually after about 40 years of age. Consequently, even though sexual activity may be continued into old age, males typically experience a decrease in sexual functions as they grow older. This decrease is sometimes called the *male climacteric*.

---

1. What initiates the changes associated with male sexual maturity?
2. Describe several male secondary sexual characteristics.
3. List the functions of testosterone.
4. Explain how the secretion of male sex hormones is regulated.

Organs of the Female Reproductive System

The organs of the female reproductive system are specialized to produce and maintain female sex cells, to transport these cells to the site of fertilization, to provide a favorable environment for a developing offspring, to move the offspring to the outside, and to produce female sex hormones.

The *primary sex organs* (gonads) of this system are the ovaries, which produce the female sex cells or **egg cells** and female sex hormones. The other parts of the system comprise the internal and external *accessory organs*.

The Ovaries

The **ovaries** are solid, ovoid structures measuring about 3.5 cm in length, 2 cm in width, and 1 cm in thickness. They are located, one on each side, in a shallow depression (ovarian fossa) of the lateral wall of the pelvic cavity (figure 22.15).

Figure 22.16

The ovaries are located on each side against the lateral walls of the pelvic cavity.

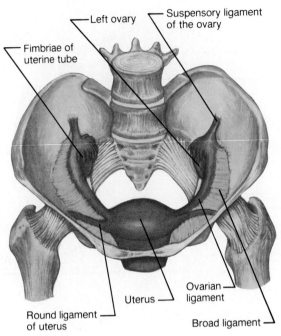

Left ovary

Suspensory ligament of the ovary

Fimbriae of uterine tube

Round ligament of uterus

Uterus

Ovarian ligament

Broad ligament

Attachments of the Ovaries

Each ovary is attached to several ligaments that help to hold it in position. The largest of these, formed by a fold of peritoneum, is called the *broad ligament*. It also is attached to the uterine tubes and uterus.

At its upper end, the ovary is held by a small fold of peritoneum, called the *suspensory ligament,* that contains the ovarian blood vessels and nerves. At its lower end, it is attached to the uterus by a rounded, cordlike thickening of the broad ligament, called the *ovarian ligament* (figure 22.16).

Descent of the Ovaries

Like the testes in a male fetus, the ovaries in a female fetus originate from masses of tissue behind the parietal peritoneum, near the developing kidneys. During development, these structures descend to locations just below the pelvic brim, where they remain attached to the lateral pelvic wall.

Structure of the Ovaries

The tissues of an ovary can be divided into two regions, an inner *medulla* and an outer *cortex*. These regions are not distinctly separated, however.

The ovarian medulla is composed largely of loose connective tissue and contains numerous blood vessels, lymphatic vessels, and nerve fibers. The ovarian cortex is composed of more compact tissue and has a somewhat granular appearance due to the presence of tiny masses of cells called *ovarian follicles*.

The free surface of the ovary is covered by a layer of cuboidal cells, called the **germinal epithelium.** Just beneath the epithelium, there is a layer of dense connective tissue called *tunica albuginea.*

---

1. What are the primary sex organs of the female reproductive system?
2. Describe the descent of the ovary.
3. Describe the structure of an ovary.

Primordial Follicles

During prenatal development (before birth), small groups of cells in the outer region of the ovarian cortex form several million **primordial follicles.** Each of these

Figure 22.17
During the process of oogenesis, a single egg cell (secondary oocyte) results from the meiosis of a primary oocyte. If the egg is fertilized, it forms a second polar body and becomes a zygote.

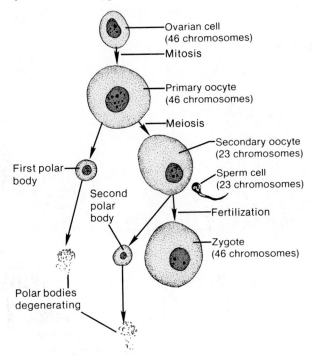

structures consists of a single, large cell called a *primary oocyte* and several epithelial cells called *follicular cells* that closely surround the oocyte.

Early in development, the primary oocytes begin to undergo meiosis, but the process is soon halted and is not continued until after puberty. Also, after the primordial follicles appear, no new ones are formed. Instead, the number of oocytes in the ovary steadily decreases throughout the life of the female, as many of these cells degenerate. Of the several million oocytes formed originally, only a million or so remain at the time of birth, and perhaps 400,000 are present at puberty. Of these, probably fewer than 1000 will be released from the ovary during the reproductive life of the person.

Oogenesis

Beginning at puberty, some of the primary oocytes are stimulated to continue meiosis, and as in the case of sperm cells, the resulting cells have one-half as many chromosomes (23) in their nuclei as the parent cells (see chapter 24).

When a primary oocyte divides, the division of the cellular cytoplasm is grossly unequal. One of the resulting cells, called a *secondary oocyte,* is quite large, and the other, called the *first polar body,* is very small.

The large secondary oocyte represents a future *egg cell* (ovum) in that it can be fertilized by uniting with a sperm cell. If this happens, the oocyte divides unequally to produce a tiny *second polar body* and a relatively large fertilized egg cell or **zygote.** (See figure 22.17.)

Thus, the result of this process, which is called **oogenesis,** is one secondary oocyte or future egg cell and a polar body. After being fertilized, the egg cell divides to produce a second polar body and a zygote, which can give rise to an embryo. Although the first polar body may divide again, the polar bodies have no further function and they soon degenerate.

1. How does the timing of egg cell production differ from that of sperm cells?
2. Describe the major events of oogenesis.

Maturation of a Follicle

The primordial follicles remain relatively unchanged throughout childhood. At puberty, however, the anterior pituitary gland secretes a greatly increased amount of FSH, and the ovaries begin to enlarge in response.

Figure 22.18
Light micrograph of the surface of a mammalian ovary.

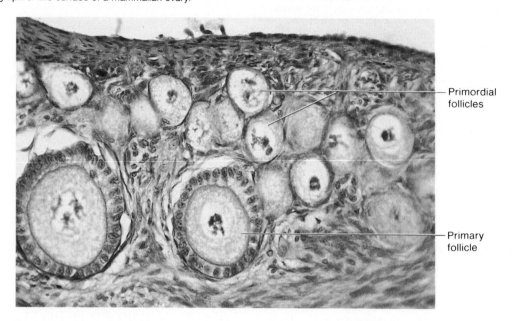

Primordial follicles

Primary follicle

At the same time, some of the primordial follicles begin to undergo maturation, becoming *primary follicles* (figure 22.18).

During maturation, the oocyte of a primary follicle grows larger, and the follicular cells surrounding it divide actively by mitosis. These follicular cells soon organize themselves into layers, and in time a cavity (antrum) appears in the cellular mass. As the cavity forms, it becomes filled with a clear *follicular fluid* that bathes the oocyte. At this stage of maturation, the structure is called a *secondary follicle* (figure 22.19).

Meanwhile, ovarian cells outside the follicle organize two cellular layers around it. One is an *inner vascular layer* (theca interna), composed largely of loose connective tissue and blood vessels; the other is an *outer fibrous layer* (theca externa), composed of tightly packed connective tissue fibers.

The fluid-filled follicular cavity continues to enlarge, and the oocyte is pressed to one side within the follicle toward the ovarian surface. In time, the follicle reaches a diameter of 10 mm or more and bulges outward on the surface of the ovary like a blister.

The oocyte within such a mature follicle (Graafian follicle) is a large, spherical cell surrounded by a thick, tough membrane (zona pellucida), and enclosed by a mantle of follicular cells (corona radiata). Processes from these follicular cells extend through the zona pellucida and are thought to supply the oocyte with nutrients.

Although as many as twenty primary follicles may begin the process of maturation at any one time, usually only one follicle reaches full development, and the others degenerate.

Ovulation

As a follicle matures, its primary oocyte undergoes oogenesis, giving rise to a secondary oocyte and a first polar body. These cells are released from the follicle by the process called ovulation.

Ovulation is stimulated by hormones from the anterior pituitary gland, which apparently cause the mature follicle to swell rapidly and the wall to weaken. Eventually the wall ruptures and the follicular fluid, accompanied by the oocyte, oozes outward from the surface of the ovary and enters the peritoneal cavity. The expulsion of a mammalian oocyte is shown in figure 22.20.

After it is expelled from the ovary, the oocyte and one or two layers of follicular cells surrounding it usually are propelled to the opening of a nearby *uterine tube*. If the oocyte is not fertilized by union with a sperm cell within a relatively short time, it will degenerate. The maturation of a follicle and the release of an oocyte are illustrated in figure 22.21.

Figure 22.19

What features can you identify in this light micrograph of a maturing follicle?

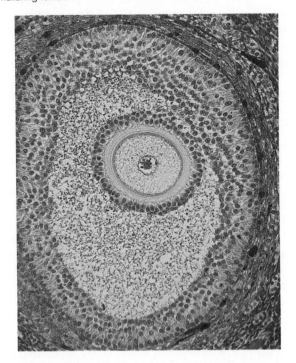

Figure 22.20

Light micrograph of a mammalian (rabbit) follicle during ovulation.

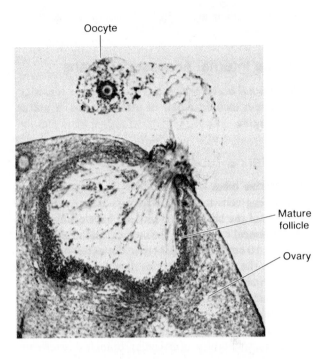

Oocyte

Mature follicle

Ovary

Figure 22.21

As a follicle matures, the egg cell enlarges and becomes surrounded by a mantle of follicular cells and fluid. Eventually, the mature follicle ruptures and the egg cell is released.

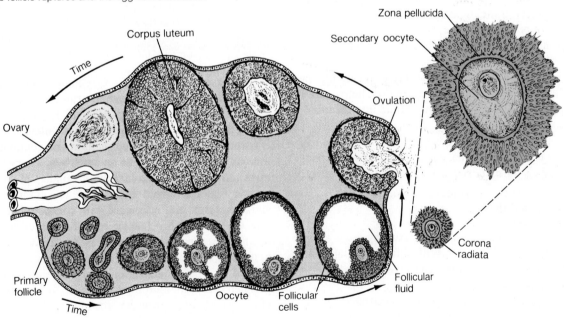

Corpus luteum

Time

Ovary

Primary follicle

Time

Oocyte

Follicular cells

Follicular fluid

Ovulation

Zona pellucida

Secondary oocyte

Corona radiata

1. What causes a primary follicle to mature?
2. What changes occur in a follicle and its oocyte during maturation?
3. What causes ovulation?
4. What happens to an egg cell following ovulation?

Female Internal Accessory Organs

The *internal accessory organs* of the female reproductive system include a pair of uterine tubes, a uterus, and a vagina.

The Uterine Tubes

The **uterine tubes** (fallopian tubes or oviducts), which convey egg cells toward the uterus, are suspended by portions of the broad ligament, and have openings into the peritoneal cavity near the ovaries. Each tube, which is about 10 cm long and 0.7 cm in diameter, passes medially to the uterus, penetrates its wall, and opens into the uterine cavity.

Near the ovary the uterine tube expands to form a funnel-shaped *infundibulum,* which partially encircles the ovary, medially. On its margin the infundibulum bears a number of irregular, branched extensions, called *fimbriae* (figure 22.22). Although the infundibulum generally does not touch the ovary, one of the larger extensions (ovarian fimbria) is connected directly to it.

The wall of a uterine tube consists of an inner mucosal layer, a middle muscular layer, and an outer covering of peritoneum. The mucosal layer is drawn into numerous longitudinal folds and is lined with simple columnar epithelial cells, some of which are *ciliated* (figure 22.23). The epithelium secretes mucus, and the cilia beat toward the uterus. These actions help to draw the egg cell and expelled follicular fluid into the infundibulum following ovulation.

Ciliary action also aids the transport of the egg cell down the uterine tube, and peristaltic contractions in the tube's muscular layer help force the egg along.

The Uterus

The **uterus** receives the embryo resulting from a fertilized egg cell and sustains its life during development. It is a hollow, muscular organ shaped somewhat like an inverted pear.

The *broad ligament,* which is also attached to the ovaries and uterine tubes, extends from the lateral walls of the uterus to the pelvic walls and floor, creating a septum across the pelvic cavity (figure 22.22). A fibrous sheet along the sides of the lower uterus and vagina form the *cardinal ligament,* which provides a deep continuation of this septum. Also, a flattened band of tissue within the broad ligament, called the *round ligament,* connects the upper end of the uterus to the pelvic wall (figures 22.16 and 22.22).

Although the size of the uterus changes greatly during pregnancy, in its adult state (before pregnancy) it is about 7 cm long, 5 cm wide (at its broadest point), and 2.5 cm in diameter. The uterus is located medially within the anterior portion of the pelvic cavity, above the vagina, and usually is bent forward over the urinary bladder.

The upper two-thirds of the uterus, or *body,* has a dome-shaped top and is joined by the uterine tubes that enter its wall at its broadest part.

The lower one-third of the uterus is the tubular **cervix,** which extends downward into the upper portion of the vagina. The cervix surrounds the opening called the *cervical orifice* (ostium uteri), through which the uterus communicates with the vagina.

Cancer developing within the tissues of the uterine cervix usually can be detected by means of a relatively simple and painless procedure called the *Pap (Papanicolaou) smear test.* This technique involves removing a tiny sample of tissue by scraping, smearing the sample on a glass slide, staining it, and examining it for the presence of abnormal cells.

Since this test can reveal certain types of cervical cancers in the early stages of development, when they may be cured completely, the American Cancer Society recommends that all women over 20 years of age have a Pap test at regular intervals.

The wall of the uterus is relatively thick and is composed of three layers: endometrium, myometrium, and perimetrium. (See figure 22.22.) The **endometrium,** the inner mucosal layer lining the uterine cavity, is covered with columnar epithelium and contains numerous tubular glands. The **myometrium,** a very thick, muscular layer, consists largely of bundles of smooth muscle fibers arranged in longitudinal, circular, and spiral patterns, and interlaced with connective tissues. During the monthly female reproductive cycles and during pregnancy, the endometrium and myometrium undergo extensive changes. (These changes are described in a subsequent section of this chapter.) The **perimetrium** is the outer serosal layer. It is composed of the peritoneal layer of the broad ligament that covers the body of the uterus and part of the cervix.

Figure 22.22

The funnel-shaped infundibulum of the uterine tube partially encircles the ovary. What factors aid the movement of an egg cell into the infundibulum following ovulation?

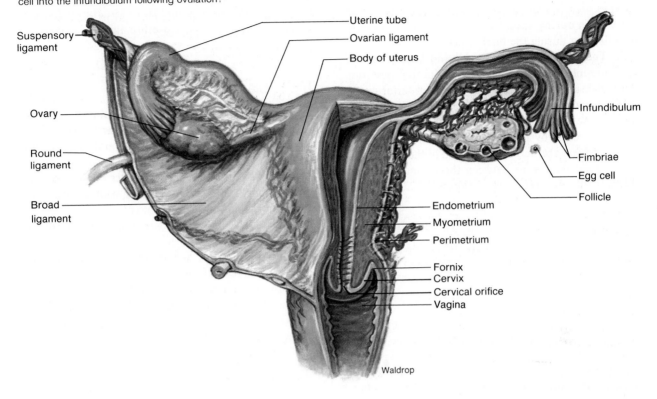

Suspensory ligament

Ovary

Round ligament

Broad ligament

Uterine tube

Ovarian ligament

Body of uterus

Infundibulum

Fimbriae

Egg cell

Follicle

Endometrium

Myometrium

Perimetrium

Fornix

Cervix

Cervical orifice

Vagina

Waldrop

Figure 22.23

Scanning electron micrograph of ciliated cells that line the uterine tube.

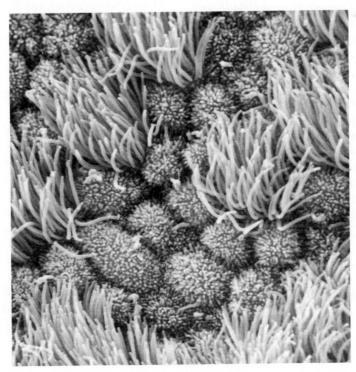

The Vagina

The **vagina** is a fibromuscular tube, about 9 cm in length, extending from the uterus to the vestibule. It conveys uterine secretions to the outside, receives the erect penis during sexual intercourse, and transports the offspring during the birth process.

The vagina extends upward and back from the vestibule into the pelvic cavity. It is located posterior to the urinary bladder and urethra and anterior to the rectum, and is attached to these parts by connective tissues. The upper one-fourth of the vagina is separated from the rectum by a pouch (rectouterine pouch). The tubular vagina also surrounds the end of the cervix, and the recesses that occur between the vaginal wall and the cervix are termed *fornices* (figure 22.22).

The fornices are clinically important since they are relatively thin-walled and allow the internal abdominal organs to be palpated during a physical examination. Also, the posterior fornix, which is somewhat longer than the others, provides a surgical access to the peritoneal cavity through the vagina.

The *vaginal orifice* opens into the vestibule as a slit and is closed partially by a thin membrane of connective tissue and stratified squamous epithelium called the **hymen.** A central opening of varying size allows uterine and vaginal secretions to pass to the outside.

The vaginal wall consists of three layers: an inner mucosa, a middle muscularis, and an outer fibrous layer. The *mucosal layer* is lined with stratified squamous epithelium and is drawn into numerous longitudinal and transverse ridges (vaginal rugae). However, this layer contains only a few glands, and much of the mucus found in the lumen of the vagina comes from glands of the cervix.

The *muscular layer* consists mainly of smooth muscle fibers arranged in longitudinal and circular patterns. At the lower end of the vagina there is a thin band of striated muscle. This band helps to close the vaginal opening, but a voluntary muscle (bulbospongiosus) is primarily responsible for closing this orifice.

The *fibrous layer* consists of dense fibrous connective tissue interlaced with elastic fibers, and it serves to attach the vagina to surrounding organs.

1. How is an egg cell moved along a uterine tube?
2. Describe the structure of the uterus.
3. What is the function of the uterus?
4. Describe the structure of the vagina.

Female External Reproductive Organs

The *external organs* of the female reproductive system include the labia major, labia minor, clitoris, and vestibular glands. As a group, these structures that surround the openings of the urethra and vagina compose the **vulva.** They are shown in figure 22.24.

The Labia Major

The **labia major** (*labium,* singular) enclose and protect the other external reproductive organs. They correspond to the scrotum of the male and are composed primarily of rounded folds of adipose tissue covered by skin. On the outside this skin includes numerous hairs, sweat glands, and sebaceous glands, while on the inside it is thinner and hairless.

The labia major lie closely together and are separated longitudinally by a cleft (pudendal cleft) that includes the urethral and vaginal openings. At their anterior ends, the labia merge to form a medial, rounded elevation of fatty tissue called the *mons pubis,* which overlies the symphysis pubis. At their posterior ends, the labia are somewhat tapered, and they merge into the perineum near the anus.

The Labia Minor

The **labia minor** are flattened longitudinal folds located within the cleft between the labia major. These folds extend along either side of the vestibule. They are composed of connective tissue that is richly supplied with blood vessels, causing a pinkish appearance. This tissue is covered with stratified squamous epithelium.

Posteriorly, the labia minor merge with the labia major, while anteriorly they converge to form a hood-like covering around the clitoris.

The Clitoris

The **clitoris** is a small projection at the anterior end of the vulva between the labia minor. Although most of it is embedded in surrounding tissues, it is usually about 2 cm long and 0.5 cm in diameter. The clitoris corresponds to the penis in the male and is somewhat similar in structure. More specifically, it is composed of two columns of erectile tissue called *corpora cavernosa.* These columns are separated by a septum and are surrounded by a covering of dense fibrous connective tissue.

At the root of the clitoris, the corpora cavernosa diverge to form *crura,* which in turn are attached to the sides of the pubic arch. At its anterior end, a small mass of erectile tissue forms a **glans,** which is richly supplied with sensory nerve fibers.

Figure 22.24
Female external reproductive organs and vestibular bulbs.

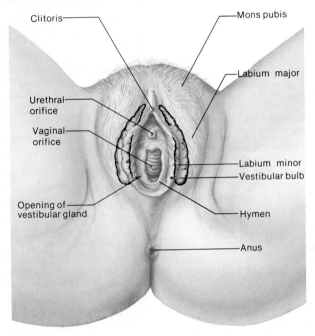

The Vestibule

The **vestibule** of the vulva is the space enclosed by the labia minora. The vagina opens into the posterior portion of the vestibule, while the urethra opens in the midline just anterior to the vagina and about 2.5 cm behind the glans of the clitoris.

A pair of **vestibular glands** (Bartholin's glands), which correspond to the bulbourethral glands of the male, lie one on either side of the vaginal opening. Their ducts open into the vestibule near the lateral margins of the vaginal orifice.

Beneath the mucosa of the vestibule on either side is a mass of vascular erectile tissue. These structures, shown in figure 22.24, are called *vestibular bulbs*. They are separated from each other by the vagina and urethra, and they extend forward from the level of the vaginal opening to the clitoris.

---

1. What is the male counterpart of the labia major? Of the clitoris?
2. What structures are located within the vestibule?

Erection, Lubrication, and Orgasm

As in the male, erectile tissues in the female located in the clitoris and around the entrance of the vagina respond to sexual stimulation. Following such stimulation, *parasympathetic nerve impulses* pass out from the sacral portion of the spinal cord causing arteries associated with the erectile tissues to dilate and veins to constrict. As a result, the inflow of blood increases, the outflow decreases, and the tissues swell. At the same time, the vagina begins to expand and elongate.

If the stimulation is sufficiently intense, parasympathetic impulses stimulate the vestibular glands to secrete mucus into the vestibule. This secretion moistens and lubricates the tissues surrounding the vestibule and the lower end of the vagina, thus facilitating the insertion of the penis into the vagina. Also, since mucus usually continues to be secreted from these glands during sexual intercourse, it helps to prevent irritation of tissues that might occur if the vagina remained dry.

The clitoris is abundantly supplied with sensory nerve fibers that are especially sensitive to local stimulation, and the culmination of such stimulation is the pleasurable sense of physiological and psychological release called *orgasm*.

Just prior to orgasm, the tissues of the outer third of the vagina become engorged with blood and swell. This action serves to increase the friction on the penis during intercourse. As orgasm is triggered, a series of reflexes involving the sacral and lumbar portions of the spinal cord are initiated.

In response to these reflexes, the muscles of the perineum contract rhythmically, and the muscular walls of the uterus and uterine tubes become active. These muscular contractions are thought to aid the transport of sperm cells through the female reproductive tract toward the upper ends of the uterine tubes, where there may be an oocyte.

Following the orgasm, the inflow of blood into the erectile tissues is reduced, and the muscles of the perineum and reproductive tract tend to relax. Consequently the organs return to a state similar to that prior to sexual stimulation.

The various functions of the female reproductive organs are summarized in chart 22.2.

---

1. What events result from parasympathetic stimulation of the female reproductive organs?
2. What changes take place in the vagina just prior to and during female orgasm?
3. What is the response of the uterus and the uterine tubes to orgasm?

Hormonal Control of Female Reproductive Functions

Female reproductive functions are controlled largely by hormones secreted by the *hypothalamus, anterior pituitary gland,* and the *ovaries.* These hormones are responsible for the development and maintenance of female secondary sexual characteristics, the maturation of egg cells, and changes that occur during the monthly reproductive cycles.

Female Sex Hormones

A female child's body remains reproductively immature until about 8 years of age. About that time, the hypothalamus begins to secrete increasing amounts of gonadotropin-releasing hormone (GnRH), which, in turn, stimulates the anterior pituitary gland to release the gonadotropins FSH and LH. These hormones play primary roles in the control of sex cell maturation and the production of female sex hormones.

Several different female sex hormones are secreted by various tissues, including the ovaries, adrenal cortices, and the placenta (during pregnancy). These hormones belong to two major groups that are called **estrogen** and **progesterone.**

The primary source of *estrogen* (in a nonpregnant female) is the ovaries. At puberty, under the influence of the anterior pituitary gland, these organs secrete increasing amounts of the hormone. The estrogen stimulates enlargement of various accessory reproductive organs, including the vagina, uterus, uterine tubes, ovaries, and the external structures. Estrogen also

Chart 22.2 Functions of the female reproductive organs

Organ	Function
Ovary	Production of egg cells and female sex hormones
Uterine tube	Conveys egg cell toward uterus; site of fertilization; conveys developing embryo to uterus
Uterus	Protects and sustains life of embryo during pregnancy
Vagina	Conveys uterine secretions to outside of body; receives erect penis during sexual intercourse; transports fetus during birth process
Labia major	Encloses and protects other external reproductive organs
Labia minor	Forms margins of vestibule; protects openings of vagina and urethra
Clitoris	Glans is richly supplied with sensory nerve endings associated with feeling of pleasure during sexual stimulation
Vestibule	Space between labia minor that includes vaginal and urethral openings
Vestibular glands	Secrete fluid that moistens and lubricates vestibule

is responsible for the development and maintenance of female *secondary sexual characteristics,* which include:

1. Development of the breasts and the ductile system of the mammary glands within the breasts.
2. Increased deposition of adipose tissue in the subcutaneous layer generally, and particularly in the breasts, thighs, and buttocks.
3. Increased vascularization of the skin (figure 22.25).

Certain other changes that occur in females at puberty seem to be related to *androgen* concentrations. For example, increased growth of hair in the pubic and axillary regions seems to be due to the presence of androgen secreted by the adrenal cortices. Conversely, the development of the female skeletal configuration, which includes narrow shoulders and broad hips, seems to be related to a lack of androgen.

The ovaries also are the primary source of *progesterone* (in a nonpregnant female). This hormone promotes changes that occur in the uterus during the rhythmic reproductive cycles. In addition, it affects the mammary glands and helps to regulate the secretion of gonadotropins from the anterior pituitary gland.

Figure 22.25
Mechanism by which female secondary sexual characteristics are stimulated to develop.

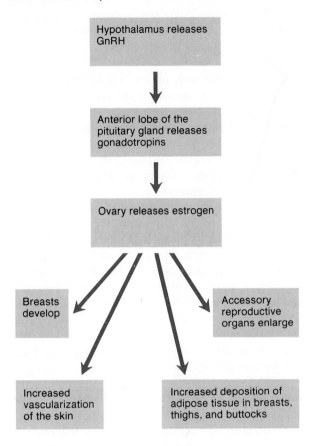

Figure 22.26
Scanning electron micrograph of a maturing human follicle. What features can you identify?

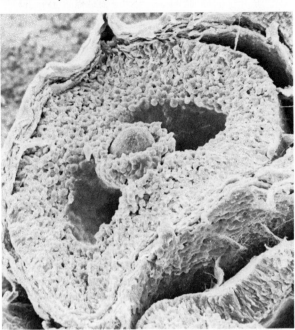

(menses). Such cycles usually begin near the thirteenth year of life and continue into middle age, after which the cycles cease.

Women athletes sometimes experience disturbances in their menstrual cycles, ranging from diminished menstrual flow (oligomenorrhea) to complete stoppage of menses (amenorrhea). The incidence of menstrual disorders generally increases with the intensity and duration of exercise periods, occurring most commonly in athletes who perform the most strenous activities and who follow the most intense training schedules.

A female's first menstrual cycle (menarche) is initiated when the hypothalamus secretes gonadotropin-releasing hormone (GnRH), which in turn stimulates the anterior pituitary gland to secrete FSH (follicle-stimulating hormone) and LH (luteinizing hormone). As its name implies, FSH acts upon the ovary to stimulate the maturation of a *follicle,* and during this development the follicular cells produce increasing amounts of estrogen. LH also plays a role in promoting this production of estrogen. (Figure 22.26 shows a maturing follicle.)

In a young female, estrogen stimulates the development of various secondary sexual characteristics. The estrogen secreted during future menstrual cycles

Female athletes who are trained for endurance events, such as the marathon, typically maintain about 6% body fat. Male endurance athletes usually have about 4% body fat. This difference of 50% in proportion of body fat is related to the actions of sex hormones. The male hormone, testosterone, tends to promote the deposition of protein throughout the body and especially in skeletal muscles; the female hormone, estrogen, causes an increased deposition of fat in the breasts, thighs, buttocks, and subcutaneous layer of the skin.

1. What factors initiate sexual maturity in a female?
2. Name the two major female sex hormones.
3. What is the function of estrogen?
4. What is the function of androgen in a female?

Female Reproductive Cycles

The female reproductive cycles or **menstrual cycles** are characterized by regular, recurring changes in the uterine lining that culminate in menstrual bleeding

Chart 22.3 Hormonal control of female secondary sexual characteristics

1. The hypothalamus releases GnRH, which stimulates the anterior pituitary gland.
2. The anterior pituitary gland secretes FSH.
3. FSH stimulates the maturation of a follicle.
4. Follicular cells produce and secrete estrogen.
5. Estrogen is responsible for the development and maintenance of most female secondary sexual characteristics.
6. Concentrations of androgen affect other secondary sexual characteristics, including skeletal growth and growth of hair.
7. Progesterone, secreted by the ovaries, affects cyclical changes in the uterus.

is responsible for continuing the development of these traits and for their maintenance. Chart 22.3 summarizes the hormonal control of female secondary sexual characteristics.

The increasing concentration of estrogen during the first few days of a cycle causes changes in the uterine lining, including thickening of the glandular endometrium (proliferative phase). This activity continues for about 14 days. Meanwhile, the developing follicle has completed its maturation, and by the fourteenth day of the cycle it appears on the surface of the ovary as a blisterlike bulge.

About this time, the anterior pituitary gland releases a relatively large quantity of LH and an increased amount of FSH. Although the mechanism responsible for these sudden increases in hormonal concentration is poorly understood, it seems to involve positive feedback. More specifically, there is a rapid rise in estrogen secretion, which causes the hypothalamus to increase its secretion of GnRH and the anterior pituitary gland to increase its secretion of LH and FSH. These hormones, in turn, stimulate the production of still more estrogen. In any case, the resulting surge in concentrations of LH and FSH, which lasts about 36 hours, seems to cause the mature follicle to swell rapidly, and the wall to weaken and rupture. Ovulation occurs as the follicular fluid, accompanied by the oocyte, leaves the follicle and enters the peritoneal cavity. The oocyte is then drawn into the uterine tube.

Following ovulation, the remnants of the follicle in the ovary undergo rapid changes. The space occupied by the follicular fluid fills with blood that soon clots, and the follicular cells enlarge greatly to form a new glandular structure within the ovary, called **corpus luteum** (figure 22.21).

Although the follicular cells secrete minute quantities of progesterone during the first part of the menstrual cycle, corpus luteum cells secrete very large quantities of progesterone and estrogen during the last half of the cycle. Consequently, as a corpus luteum becomes established the blood concentration of progesterone increases sharply.

Progesterone acts on the endometrium of the uterus, causing it to become more vascular and glandular. It also stimulates the uterine glands to secrete increasing quantities of glycogen and lipids (secretory phase). As a result, the endometrial tissues of the uterus become filled with fluids containing nutrients and electrolytes that provide a favorable environment for the development of an embryo.

Except for the time just prior to ovulation when estrogen stimulates secretion from the hypothalamus and anterior pituitary gland, high blood concentrations of estrogen and progesterone inhibit the release of GnRH from the hypothalamus and gonadotropins from the anterior pituitary gland. Consequently, no other follicles are stimulated to develop during the time the corpus luteum is active. If the oocyte that was released at ovulation is not fertilized by a sperm cell, however, the corpus luteum begins to degenerate (regress) about the twenty-fourth day of the cycle.

When the corpus luteum ceases to function, the concentrations of estrogen and progesterone decline rapidly, and in response, blood vessels in the endometrium become constricted. This action reduces the supply of oxygen and nutrients to the thickened uterine lining, and these tissues (decidua) soon disintegrate and slough away. At the same time, blood escapes from damaged capillaries, creating a flow of blood and cellular debris that passes through the vagina as the *menstrual flow* (menses). This flow usually begins about the twenty-eighth day of the cycle and continues for 3 to 5 days, while the estrogen concentration is relatively low.

The beginning of the menstrual flow marks the end of a menstrual cycle and the beginning of a new cycle. This cycle is diagrammed in figure 22.27, and summarized in chart 22.4.

Since the blood concentrations of estrogen and progesterone are low at the beginning of the cycle, the hypothalamus and pituitary gland are no longer inhibited. Consequently, the concentrations of FSH and LH soon increase, and a new follicle is stimulated to mature. As this follicle secretes estrogen, the lining of the uterus undergoes repair and the endometrium begins to thicken again.

Figure 22.27
Major events in the female ovarian and menstrual cycles.

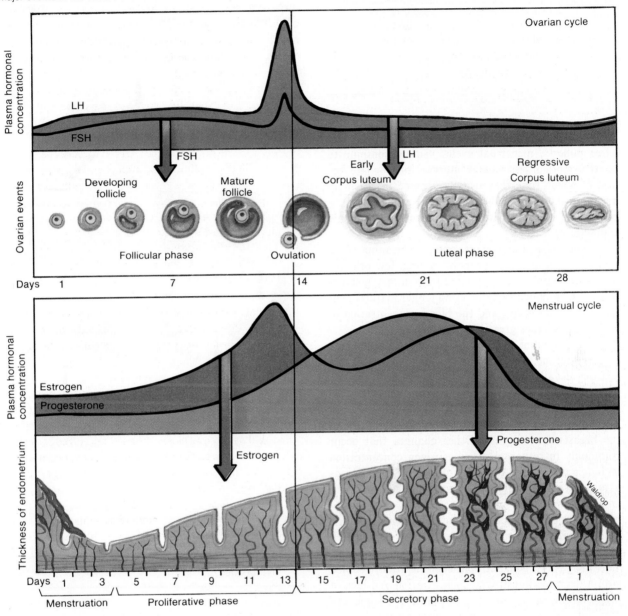

Chart 22.4   Major events in a menstrual cycle

1. The anterior pituitary gland secretes FSH.

2. FSH stimulates maturation of a follicle.

3. Follicular cells produce and secrete estrogen.

   a. Estrogen maintains secondary sexual traits.

   b. Estrogen causes uterine lining to thicken.

4. The anterior pituitary gland secretion causes a surge in the concentrations of LH and FSH, which stimulates ovulation.

5. Follicular cells become corpus luteum cells, which secrete estrogen and progesterone.

   a. Estrogen continues to stimulate uterine wall development.

   b. Progesterone stimulates uterine lining to become more glandular and vascular.

   c. Estrogen and progesterone inhibit the secretion of FSH and LH from the anterior pituitary gland.

6. If egg cell is not fertilized, corpus luteum degenerates and no longer secretes estrogen and progesterone.

7. As concentrations of estrogen and progesterone decline, blood vessels in the uterine lining constrict.

8. The uterine lining disintegrates and sloughs away, producing menstrual flow.

9. The anterior pituitary gland, which is no longer inhibited, again secretes FSH.

10. The cycle is repeated.

Some women experience a set of unpleasant changes that occur between the time of ovulation and the beginning of the menstrual flow. This disorder is called *premenstrual syndrome* (PMS), and its symptoms may include increased emotional tension, irritability, headache, fatigue, breast soreness, and swelling of hands, ankles, and abdomen. The cause of PMS is unknown, but it is believed to involve hormonal imbalance.

Menopause

After puberty, menstrual cycles usually continue to occur at more or less regular intervals into the late forties, at which time they usually become increasingly irregular and after a few months or years cease altogether. This period in life is called **menopause** (female climacteric).

The cause of menopause seems to be aging of the ovaries. After about 35 years of cycling, few primary follicles remain to be stimulated by pituitary gonadotropins. Consequently, follicles no longer mature, ovulation does not occur, and the blood concentration of estrogen decreases greatly.

As a result of low estrogen concentration and lack of progesterone, the female secondary sexual characteristics usually undergo varying degrees of change. Thus, the vagina, uterus, and uterine tubes may decrease in size, as may the external reproductive organs. The pubic and axillary hair may become thinner, and the breasts may regress. Other changes that occur commonly in response to low estrogen concentration include a decrease in thickness of the epithelial linings associated with urinary and reproductive organs, an increased loss of bone matrix (osteoporosis), and a thinning of the skin. Since estrogen and progesterone no longer inhibit the pituitary secretion of FSH and LH, these hormones are released continuously for some time. This increase in gonadotropin concentration, coupled with a decrease in estrogen concentration and lack of progesterone seems to be responsible for some unpleasant vasomotor symptoms occasionally experienced by women in menopause. For example, they may feel sensations of heat in the face and upper body, called hot flashes, which may last for 30 seconds to 5 minutes (figure 22.28).

About 50% of women reach menopause by age 48 and 85% reach it by age 52. Of these, perhaps 20% have no unusual symptoms—they simply stop menstruating. About 50% of menopausal women, however, have "hot flashes". They also may experience varying degrees of headache, backache, and fatigue.

Figure 22.28
Failure of follicles to mature results in a decreasing concentration of estrogen and may be accompanied by a regression of female secondary sexual characteristics.

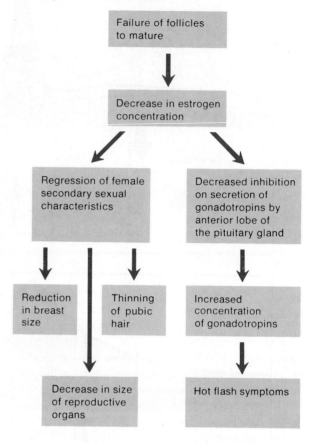

1. Trace the events of the female menstrual cycle.
2. What effect does progesterone have on the endometrium of the uterus?
3. What causes the menstrual flow?
4. What are some changes that may occur at menopause?

Pregnancy

Pregnancy is the condition characterized by the presence of a developing offspring within the uterus. It results from the union of an egg cell and a sperm cell—an event called **fertilization.**

Transport of Sex Cells

Ordinarily, before fertilization can occur, an egg cell (secondary oocyte) must be released by ovulation and be carried into a uterine tube by the action of the ciliated epithelium lining the tube.

Figure 22.29
The paths of the egg and sperm cells through the female reproductive tract. What factors aid the movements of these cells?

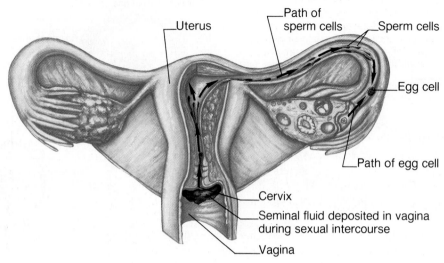

- Uterus
- Path of sperm cells
- Sperm cells
- Egg cell
- Path of egg cell
- Cervix
- Seminal fluid deposited in vagina during sexual intercourse
- Vagina

During sexual intercourse, seminal fluid containing sperm cells usually is deposited in the vagina near the cervix. To reach the egg cell, the sperm cells must then be transported upward through the uterus and uterine tube. This transport is aided by the lashing movements of the sperm's tail and by muscular contractions within the walls of the uterus and uterine tube, which are thought to be stimulated by prostaglandins in the seminal fluid. Also, under the influence of the high estrogen concentrations during the first part of the menstrual cycle, the uterus and cervix contain a thin, watery secretion that promotes sperm transport and survival. Conversely, during the latter part of the cycle, when progesterone is high in concentration, these parts secrete a viscous fluid that is unfavorable for sperm transport and survival (figure 22.29).

The sperm transport mechanism is relatively ineffective, however, because even though as many as 300 million to 500 million sperm cells may be deposited in the vagina by a single ejaculation, only a few hundred sperm cells ever reach an egg cell.

Studies indicate that an egg cell may survive for only 12 to 24 hours following ovulation, while sperm cells may live up to 72 hours within the female reproductive tract. Consequently, sexual intercourse probably must occur no more than 72 hours before ovulation or 24 hours following ovulation if fertilization is to take place.

Sperm cells are thought to reach the upper portions of the uterine tube within an hour following sexual intercourse. Although many sperm cells may reach an egg cell, only one will participate in fertilization (figure 22.30).

Figure 22.30
Although many sperm cells may reach an egg cell, only one will fertilize it.

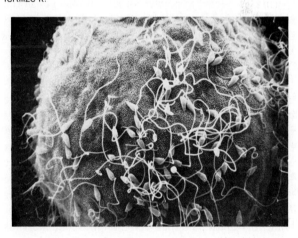

Fertilization

When a sperm cell reaches an egg cell, it moves through the follicular cells that adhere to the egg's surface (corona radiata) and penetrates the *zona pellucida* that surrounds the egg cell membrane. This penetration seems to be aided by an enzyme (hyaluronidase), released by the acrosome in the sperm head. This enzyme apparently allows the sperm cell to digest its way through the corona radiata and the zona pellucida (figure 22.21).

Figure 22.31
As a result of fertilization, a zygote is formed.

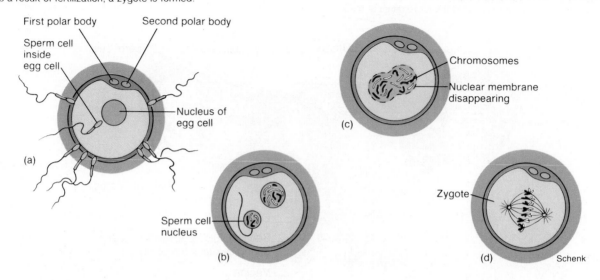

Early Embryonic Development

As the sperm cell penetrates the zona pellucida, this covering becomes impenetrable to any other sperm cells. The mechanism preventing the entrance of extra sperm cells seems to involve structural changes in the membrane that occur soon after the first sperm cell penetrates it and the release of certain enzymes that keep other sperm cells from attaching to the membrane.

Once a sperm cell reaches the egg cell membrane, it passes through the membrane and enters the cytoplasm. During this process, the sperm cell loses its tail, and its head swells to form a nucleus. As explained previously, the egg cell (secondary oocyte) then divides unequally to form a relatively large cell, and a tiny second polar body, which is expelled. The nuclei of the egg cell and that of the sperm cell then come together in the center of the larger cell. Their nuclear membranes disappear, and their chromosomes combine, thus completing the process of **fertilization.** This process is diagrammed in figure 22.31.

Since the sperm cell and the egg cell each provide 23 chromosomes, the result of fertilization is a cell with 46 chromosomes—the usual number of a human cell. This cell, called a **zygote,** is the first cell of the future offspring.

1. What factors enhance motility of sperm cells following sexual intercourse?
2. Where does fertilization normally take place?
3. List the events that occur during fertilization.

Shortly after it is formed, the zygote undergoes *mitosis,* giving rise to two daughter cells. These cells in turn divide into four cells, which divide into eight cells, and so forth. With each subsequent division, the resulting daughter cells are smaller and smaller. Consequently, this phase in development is termed **cleavage.**

Meanwhile, the tiny mass of cells is moved through the uterine tube to the cavity of the uterus. This movement is aided by the action of cilia of the tubular epithelium and by weak peristaltic contractions of smooth muscles in the tubular wall. Secretions from the epithelial lining are thought to provide the developing organism with nutrients.

The trip to the uterus takes about 3 days, and at the end of this time the structure consists of a solid ball (morula) of about sixteen cells.

Once inside the uterus, it remains free within the uterine cavity for about 3 days, during which the zona pellucida of the original egg cell degenerates, and the structure, which now consists of a hollow ball of cells (blastocyst), begins to attach itself to the uterine lining. By the end of the first week of development, it is superficially *implanted* in the endometrium (figure 22.32).

About the time of implantation, certain cells within the blastocyst organize themselves into a group (inner cell mass) that will give rise to the body of the offspring. This marks the beginning of the embryonic period of development. The offspring is termed an **embryo** until the end of the eighth week, after which it is called a **fetus.**

Eventually, the outer cells of the embryo together with cells of the maternal endometrium form a complex vascular structure called the **placenta.** This organ

Figure 22.32
Stages in early human development.

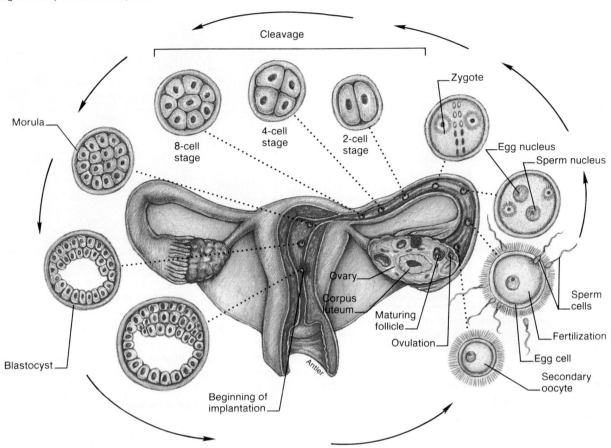

Cleavage

Zygote

Morula

8-cell stage

4-cell stage

2-cell stage

Egg nucleus

Sperm nucleus

Ovary

Corpus luteum

Maturing follicle

Ovulation

Sperm cells

Fertilization

Egg cell

Secondary oocyte

Blastocyst

Beginning of implantation

serves to attach the embryo to the uterine wall, to exchange nutrients, gases, and wastes between the maternal blood and the embryonic blood, and to secrete hormones. The placenta is described in more detail in chapter 23.

Occasionally, developing offspring may become implanted in tissues outside the uterus, including those of a uterine tube, ovary, cervix, or an organ in the abdominal cavity. The result is called an *ectopic pregnancy*. Most commonly, this condition occurs within a uterine tube and is termed a *tubal pregnancy*.

In a tubal pregnancy, the tube usually ruptures as the embryo enlarges. This is accompanied by severe pain and heavy bleeding through the vagina. The treatment involves prompt surgical removal of the embryo and repair or removal of the damaged uterine tube.

1. What is a zygote? A morula?
2. How does an embryo become implanted in the uterine wall?
3. How does a developing offspring obtain nutrients and oxygen?

Hormonal Changes during Pregnancy

During the usual menstrual cycle, the corpus luteum degenerates about 2 weeks after ovulation. Consequently, the estrogen and progesterone concentrations decline rapidly, the uterine lining is no longer maintained, and the endometrium sloughs away as menstrual flow. If this occurs following implantation, the embryo is lost (spontaneously aborted).

The mechanism that normally prevents such a termination of pregnancy involves a hormone, called HCG (human chorionic gonadotropin). This hormone is secreted by a layer of embryonic cells (trophoblast) that surrounds the developing embryo and later becomes involved with the formation of the placenta (see chapter 23). HCG has properties similar to those of LH and causes the corpus luteum to be maintained and to continue secreting relatively large amounts of estrogen and progesterone. Thus, the uterine wall continues to grow and develop. At the same time, the

Figure 22.33

Mechanism that prevents the loss of the uterine wall during the early stages of pregnancy.

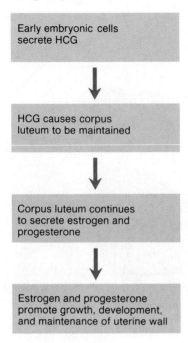

Figure 22.34

Relative concentrations of three hormones in the blood during pregnancy.

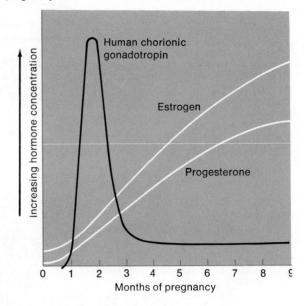

estrogen and progesterone suppress the release of FSH and LH from the pituitary gland, so that normal menstrual cycles are inhibited (figure 22.33).

The secretion of HCG continues at a high level for about two months, then declines to a relatively low level by the end of four months. Although the corpus luteum is maintained throughout the pregnancy, its function as a source of hormones becomes less important after the first three months of pregnancy (first trimester). This is due to the fact that the placenta usually is well developed by this time, and the placental tissues secrete high concentrations of estrogen and progesterone (figure 22.34).

HCG secretion by the embryonic tissues begins shortly after fertilization and increases to a peak in about 50 to 60 days. Thereafter the concentration of HCG drops to a much lower concentration that remains relatively stable throughout the pregnancy.

Since HCG is excreted in the urine, its presence in urine can be used to detect pregnancy. Such a test may indicate positive results as early as 7 days after fertilization.

For the remainder of the pregnancy, *placental estrogen* and *progesterone* maintain the uterine wall. The placenta also secretes a hormone called **placental lactogen.** This hormone is thought to stimulate breast development and preparation for milk secretion, a function that is aided by placental estrogen and progesterone. Placental progesterone and a polypeptide hormone from the corpus luteum called *relaxin* inhibit the smooth muscles in the myometrium so that uterine contractions are suppressed until it is time for the birth process to begin.

The high concentration of placental estrogen during pregnancy causes enlargement of the vagina and external reproductive organs, as well as relaxation of the ligaments holding the symphysis pubis and sacroiliac joints together. This latter action allows for greater movement at these joints and thus aids the passage of the fetus through the birth canal. Relaxation of ligaments and the softening of the cervix near the time of delivery also may be aided by relaxin.

Other hormonal changes that occur during pregnancy include increased secretion of aldosterone from the adrenal cortex and parathyroid hormone from the parathyroid glands. The aldosterone promotes renal reabsorption of sodium, leading to fluid retention, while parathyroid hormone helps to maintain a high concentration of maternal blood calcium (see chapter 13).

Chart 22.5 summarizes the hormonal changes of pregnancy.

1. What mechanism is responsible for maintaining the uterine wall during pregnancy?
2. What is the source of HCG during the first few months of pregnancy?
3. What is the source of hormones that sustain the uterine wall during pregnancy?
4. What other hormonal changes occur during pregnancy?

Other Changes during Pregnancy

A number of other changes occur in a woman's body as a result of the increased demands of a growing fetus. For example, as the fetus increases in size, the uterus enlarges greatly, and instead of being confined to its normal location in the pelvic cavity, it extends upward and may eventually reach the level of the ribs. At the same time, the abdominal organs are displaced upward and compressed against the diaphragm. Also, as the uterus enlarges, it tends to press on the urinary bladder and cause the woman to experience a need to urinate frequently.

As the placenta grows and develops, it requires more blood, and as the fetus enlarges, it needs more oxygen and produces greater amounts of wastes that must be excreted. Consequently, the mother's blood volume, cardiac output, breathing rate, and urine production all tend to increase in response to the fetal demands.

The fetal need for increasing amounts of nutrients is reflected in an increased dietary intake by the mother. Her intake must supply adequate vitamins, minerals, and proteins for herself and the fetus. The fetal tissues have a greater capacity to capture available nutrients than do the maternal tissues. Consequently, if the mother's diet is inadequate, her body usually will show symptoms of a deficiency condition before fetal growth is affected.

The Birth Process

Pregnancy usually continues for 40 weeks (280 days or 10 average menstrual cycles) or about 9 calendar months (10 lunar months), if it is measured from the beginning of the last menstrual cycle. The pregnancy terminates with the *birth process* (parturition).

Although the mechanism causing birth is not very well understood, a variety of factors seem to be involved. For example, progesterone suppresses uterine contractions during pregnancy. Estrogen, however, tends to excite such contractions. After the seventh month, the placental secretion of estrogen increases to a greater degree than the secretion of progesterone. As a result of the rising concentration of estrogen, the contractility of the uterine wall is enhanced.

These changes in estrogen and progesterone concentrations also seem to stimulate the synthesis of prostaglandins, and they, in turn, may actually initiate the birth process.

Also, as is discussed in chapter 13, the stretching of the uterine and vaginal tissues late in pregnancy is thought to initiate nerve impulses to the hypothalamus. The hypothalamus, in turn, signals the posterior pituitary gland, which responds by releasing the hormone, **oxytocin.**

Oxytocin is a powerful stimulator of uterine contractions, and its effect, combined with the greater excitability of the myometrium due to the decline in progesterone secretion, aids labor, at least in its later stages.

Labor is the term for the process in which muscular contractions force the fetus through the birth canal. Once labor starts, rhythmic contractions that begin at the top of the uterus and travel down its length force the contents of the uterus toward the cervix.

Since the fetus usually is positioned with its head downward, labor contractions force the head against the cervix. This action causes the cervix to stretch, which is thought to elicit a reflex that stimulates still stronger labor contractions. Thus, a *positive feedback system* operates in which uterine contractions result in more intense uterine contractions until a maximum effort is achieved (figure 22.35). At the same time, dilation of the cervix reflexly stimulates an increased release of oxytocin from the pituitary gland.

During childbirth the tissues of the perineum sometimes are torn by the stretching that occurs as the infant passes through the birth canal. For this reason, an incision may be made along the midline of the perineum from the vestibule to within 1.5 cm of the anus before the birth is completed. This procedure, called an *episiotomy,* ensures that the perineal tissues are cut cleanly rather than torn.

As labor continues, abdominal wall muscles are stimulated to contract by a positive feedback mechanism, and they also aid in forcing the fetus through the cervix and vagina to the outside.

Chart 22.6 summarizes some of the factors involved in promoting labor.

Following the birth of the fetus (usually within 10 to 15 minutes), the placenta, which remains inside the uterus, becomes separated from the uterine wall and is expelled by uterine contractions through the birth canal. This expulsion, which is termed the *afterbirth,* is accompanied by bleeding because vascular tissues are damaged in the process. However, the loss of blood usually is minimized by continued contraction of the uterus that constricts the bleeding vessels. This contraction is stimulated by the action of oxytocin.

Figure 22.36 illustrates the steps of the birth process.

For several weeks following childbirth, the uterus becomes smaller by a process called *involution.* Also, its endometrium sloughs off and is discharged through the vagina. This is followed by a return of an epithelial lining characteristic of a nonpregnant female.

Figure 22.35
The birth process involves this positive feedback mechanism.

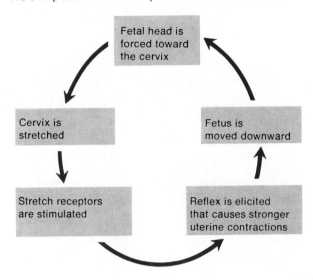

Fetal head is forced toward the cervix

Cervix is stretched

Fetus is moved downward

Stretch receptors are stimulated

Reflex is elicited that causes stronger uterine contractions

Chart 22.6 Some factors involved in the labor process

1. As the time of birth approaches, secretion of progesterone declines and its inhibiting effect on uterine contractions is lessened.
2. Changes in estrogen and progesterone concentrations stimulate synthesis of prostaglandins, which may initiate labor.
3. Stretching of uterine tissues stimulates release of oxytocin from the posterior pituitary gland.
4. Oxytocin may stimulate uterine contractions, and may aid labor in its later stages.
5. As the fetal head causes the cervix to stretch, a positive feedback mechanism results in stronger and stronger uterine contractions and a greater release of oxytocin.
6. Abdominal wall muscles are stimulated by positive feedback to contract with greater and greater force.
7. The fetus is forced through the birth canal to the outside.

Figure 22.36
Stages in the birth process.

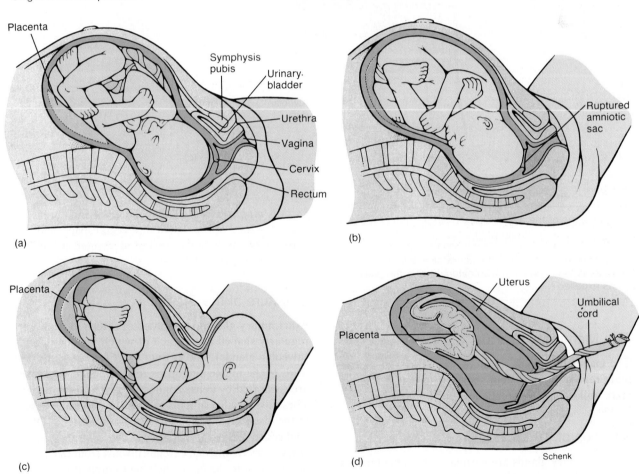

(a) Placenta, Symphysis pubis, Urinary bladder, Urethra, Vagina, Cervix, Rectum

(b) Ruptured amniotic sac

(c) Placenta

(d) Uterus, Placenta, Umbilical cord

Schenk

Figure 22.37

Structure of the breast. *(a)* Sagittal section; *(b)* anterior view.

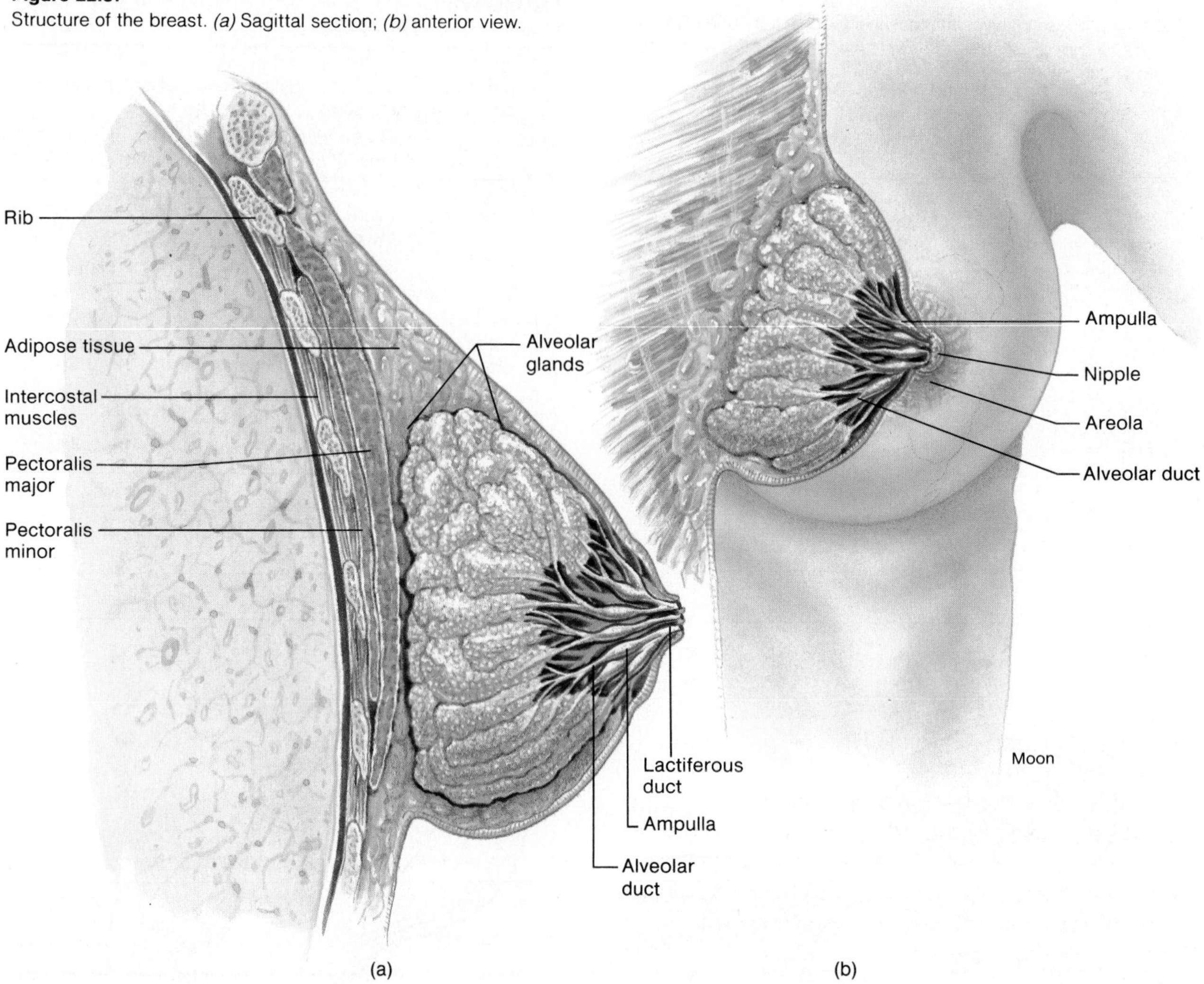

(a)

(b)

Rib

Adipose tissue

Intercostal muscles

Pectoralis major

Pectoralis minor

Alveolar glands

Lactiferous duct

Ampulla

Alveolar duct

Ampulla

Nipple

Areola

Alveolar duct

Moon

---

1. List some of the physiological changes that occur in a woman's body during pregnancy.
2. Describe the events thought to initiate labor.
3. Explain how dilation of the cervix affects labor.
4. How is bleeding controlled naturally after the placenta is expelled?

The Mammary Glands

The **mammary glands** are accessory organs of the female reproductive system that are specialized to secrete milk following pregnancy.

Location of the Glands

The mammary glands are located in the subcutaneous tissue of the anterior thorax within hemispherical elevations, called *breasts.* The breasts overlie the *pectoralis major* muscles and extend from the second to the sixth ribs, and from the sternum to the axillae.

A *nipple* is located near the tip of each breast at about the level of the fourth intercostal space, and it is surrounded by a circular area of pigmented skin called the *areola* (figure 22.37).

Structure of the Glands

A mammary gland is composed of fifteen to twenty irregularly shaped lobes, each of which includes glands (alveolar glands), and a duct (lactiferous duct) that leads to the nipple and opens to the outside. The lobes are separated by dense connective and adipose tissues. These tissues also support the glands and attach them to the fascia of the underlying pectoral muscles. Other connective tissue, which forms dense strands called *suspensory ligaments,* extends inward from the dermis of the breast to the fascia, helping to support the weight of the breast.

Breast cancer, which is one of the more common types of cancer in women, usually begins as a small, painless lump.

Since an early diagnosis of such a tumor is of prime importance in successful treatment, the American Cancer Society recommends that after age 20 women examine their breasts each month, paying particular attention to the upper, outer portions. The examination should be made just after menstruation when the breasts are usually soft, and any lump that is discovered should be checked immediately by a physician.

It also is recommended that after age 35 women have their breasts examined at regular intervals with *mammography*—a breast-cancer detection technique that makes use of relatively low dosage X ray. It has been determined that a breast cancer can be detected on a *mammogram* (an X-ray film of the breast) perhaps two years before the growing tumor can be felt.

Development of the Breasts

The mammary glands of male and female children are similar. As children reach *puberty,* the male glands fail to develop, and the female glands are stimulated to develop by ovarian hormones. As a result, the alveolar glands and ducts enlarge, and fat is deposited so that the breasts become surrounded by adipose tissue, except for the region of the areola.

During pregnancy, placental estrogen and progesterone stimulate further development of the mammary glands. Estrogen causes the ductile systems to grow and become branched and to have large quantities of fat deposited around them. Progesterone, on the other hand, stimulates the development of the alveolar glands at the ends of the ducts. These changes also are promoted by the presence of placental lactogen.

As a consequence of hormonal activity, the breasts may double in size during pregnancy. At the same time fatty tissue is largely replaced by glandular tissue, and the mammary glands become capable of secreting milk. However, no milk is produced because in the presence of the high concentrations of estrogen and progesterone that occur during pregnancy, the hypothalamus releases prolactin-inhibiting factor (PIF), which suppresses the secretion of prolactin from the anterior pituitary gland (see chapter 13).

Production and Secretion of Milk

Following childbirth and the expulsion of the placenta, the maternal blood concentrations of estrogen and progesterone decline rapidly, and the hypothalamus signals the anterior pituitary gland to release prolactin.

Figure 22.38

Mechanism that causes the ejection of milk from the breasts. What happens to milk production if the milk is not regularly removed from the breast?

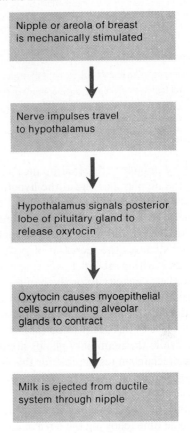

Prolactin stimulates the mammary glands to secrete large quantities of milk. This hormonal effect does not occur for 2 or 3 days following birth, and in the meantime, the glands secrete a few milliliters of a fluid called *colostrum* each day. Although colostrum contains some of the nutrients found in milk, it lacks fat.

The milk produced under the influence of prolactin does not flow readily through the ductile system of the mammary gland, but must be actively ejected by contraction of specialized *myoepithelial cells* surrounding the alveolar glands. The contraction of these cells and the consequent ejection of milk through the ducts result from a reflex action.

This reflex is elicited when the breast is sucked or the nipple or areola is otherwise mechanically stimulated. Then sensory impulses travel to the hypothalamus, which signals the posterior pituitary gland to release oxytocin. The oxytocin reaches the breast by means of the blood, and it stimulates the myoepithelial cells to contract. Consequently, milk is ejected into a suckling infant's mouth in about 30 seconds (figure 22.38).

Chart 22.7	Hormonal control of the mammary glands
Before pregnancy (beginning of puberty)	**Following childbirth**
Ovarian hormones secreted during menstrual cycles stimulate alveolar glands and ducts of mammary glands to develop. **During pregnancy** 1. Estrogen causes the ductile system to grow and branch. 2. Progesterone stimulates development of alveolar glands. 3. Placental lactogen promotes development of the breasts. 4. Secretion of the prolactin is inhibited by estrogen and progesterone, so no milk is produced.	1. Estrogen and progesterone concentrations decline, so prolactin secretion is no longer inhibited. 2. The anterior pituitary gland secretes prolactin, which stimulates milk production. 3. Mechanical stimulation of the breasts causes release of oxytocin from the posterior pituitary gland. 4. Oxytocin stimulates ejection of milk from the ducts. 5. As long as milk is removed, more prolactin is released; if milk is not removed, milk production ceases.

Sensory impulses triggered by mechanical stimulation of the nipples also signal the hypothalamus to allow a continued secretion of prolactin. Thus, prolactin is released as long as milk is removed from the breasts. However, if milk is not removed regularly, the hypothalamus causes the secretion of prolactin to be inhibited, and within about 1 week, the mammary glands lose their capacity to produce milk.

Although it is possible for a woman to become pregnant during the period that she breast-feeds her child, menstrual cycles seem to be inhibited, at least for a time, while the mammary glands are active. Although the mechanism responsible for this effect is not well understood, it is thought that prolactin may suppress the release of gonadotropins from the anterior pituitary gland. In any event, menstrual cycles may not begin for some time following the birth of an infant that is breast-fed. On the other hand, after several months of breast-feeding, FSH is usually released, and the monthly reproductive cycles are reestablished.

Chart 22.7 summarizes the hormonal effects involved in producing milk.

1. Describe the structure of a mammary gland.
2. How does pregnancy affect the mammary glands?
3. What stimulates the mammary glands to produce milk?
4. How is milk stimulated to flow into the ductile system of a mammary gland?

Birth Control

Birth control is the voluntary regulation of the number of offspring produced and of the time they will be conceived. This control usually involves some method of **contraception,** designed to avoid the fertilization of an egg cell following sexual intercourse or to prevent the implantation of an embryo.

A variety of contraceptive methods are commonly practiced, including coitus interruptus, rhythm method, mechanical barriers, chemical barriers, oral

Chart 22.8	Effectiveness of contraceptive methods*	
	Pregnancies per 100 women per year[a]	
Method	*High[b]*	*Low*
No contraceptive[c]	80	40
Coitus interruptus	23	15
Condom[d]	17	8
Chemicals (spermicides)[e]	40	9
Diaphragm and jelly	28	11
Rhythm[f]	58	14
Oral contraceptives	2	0.03
IUD	8	3
Sterilization	0.003	0

* E. Peter Volpe, *Man, Nature, and Society* (Dubuque, Ia.: Wm. C. Brown Co. Publishers, 1979). Used by permission of the publisher.

[a] Data describe the number of women per hundred who will become pregnant in a one-year period while using a given method.
[b] High and low values represent best and worst estimates from various demographic and clinical studies.
[c] In the complete absence of contraceptive practice, eight out of ten women can expect to become pregnant within one year.
[d] Effectiveness increases if spermicidal jelly or cream is used in addition.
[e] Aerosol foam is considered to be the best of the chemical barriers.
[f] Use of a clinical thermometer to record daily temperatures increases effectiveness.

contraceptives, intrauterine devices, and surgical methods. The relative effectiveness of these methods is summarized in chart 22.8.

Coitus Interruptus

Coitus interruptus involves withdrawing the penis from the vagina before ejaculation, thus preventing the entrance of sperm cells into the female reproductive tract. This method of contraception often proves unsatisfactory and may result in pregnancy, since some males find it emotionally difficult to withdraw just prior to ejaculation. Also, small quantities of seminal fluid containing sperm cells may be expelled from the penis before ejaculation occurs.

Rhythm Method

The *rhythm method,* like coitus interruptus, requires no artificial devices or chemicals. Instead, it requires abstinence from sexual intercourse a few days before and a few days after ovulation.

Since ovulation theoretically occurs on the fourteenth day of a 28-day menstrual cycle, it might seem easy to avoid intercourse near ovulation. Few women, however, have absolutely regular menstrual cycles, and the lengths of the cycles vary from time to time. Further, the variable part of the cycle occurs before ovulation, for regardless of the length of a cycle, the menstrual flow almost always begins 13 to 15 days following ovulation. Inasmuch as the length of a cycle cannot be predicted ahead of time, it is almost impossible to predict the time of ovulation accurately. Thus, it is not surprising that the rhythm method results in a relatively high rate of pregnancy.

Another disadvantage of this method is that it requires adherence to a particular pattern of behavior and restricts spontaneity in sexual activity.

The effectiveness of the rhythm method can sometimes be increased by measuring and recording the woman's body temperature when she awakes each morning for several months.

Since the body temperature typically rises about 0.6 degrees Fahrenheit immediately following ovulation, this procedure may allow a woman to more accurately predict the "unsafe times" in her reproductive cycle. On the other hand, many women apparently do not show such a change in body temperature at ovulation, and furthermore, the body temperature may vary in response to other factors such as illnesses or emotional upsets.

---

1. Why is coitus interruptus an unreliable method of contraception?
2. Describe the idea behind the rhythm method of contraception.
3. What factors make the rhythm method less reliable than some other methods of contraception?

Mechanical Barriers

Mechanical barriers sometimes are employed to prevent sperm cells from entering the female reproductive tract during sexual intercourse. One such device, for use by males, is called a *condom.* It consists of a thin rubber sheath that is placed over the erect penis before intercourse to prevent seminal fluid from entering the vagina upon ejaculation.

The condom provides a relatively inexpensive method of contraception, and it also may help protect the user against contracting venereal diseases. However, men often feel a condom decreases the sensitivity of the penis during intercourse. Also, its use has the disadvantage of interrupting the sex act.

Another rubber device, which is used by females, is the *diaphragm.* It is a cup-shaped structure with a flexible ring forming the rim. The diaphragm is inserted into the vagina so that it covers the cervix and thus prevents the entrance of sperm cells into the uterus.

To be effective, a diaphragm must be fitted for size by a physician, be inserted properly, and be used in conjunction with a chemical spermicide that is applied to the surface adjacent to the cervix and to the rim of the diaphragm. Also, the diaphragm must be left in position for several hours following sexual intercourse.

Although the diaphragm can be inserted into the vagina sometime before sexual contact is to occur, so that its use does not interrupt the sex act, some women object to it because they find the insertion of the diaphragm distasteful or uncomfortable.

Chemical Barriers

Chemical barriers for the purpose of contraception include a variety of creams, foams, and jellies with spermicidal properties. Within the vagina, such chemicals create an environment that is unfavorable for sperm cells.

Chemical barriers are fairly easy to use, but have a relatively high failure rate when used alone. They are best used along with a rubber diaphragm.

Oral Contraceptives

An *oral contraceptive* is commonly called the pill, because this method employs a series of drug-containing tablets taken by the woman.

The most common forms of oral contraceptives contain synthetic estrogen-like and progesterone-like substances.

When taken daily, these drugs seem to disrupt the normal pattern of gonadotropin secretion and to prevent the surge in LH release that triggers ovulation. They also interfere with the buildup of the uterine lining that is necessary for implantation of embryos (see chapter 23).

Successful use of oral contraceptives requires motivation and planning. However, if they are used correctly, they prevent pregnancy nearly 100% of the time. Although these hormone-like substances usually do not produce serious side effects, they commonly cause nausea, retention of body fluids, increased pigmentation of the skin, and breast tenderness. Also, some women, particularly those over 35 years of age who smoke, may experience a dangerous tendency to form

Figure 22.39

(a) Vasectomy involves removal of a portion of each vas deferens. (b) Tubal ligation involves removal of a portion of each uterine tube.

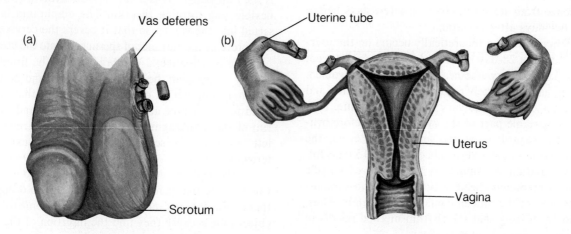

Vas deferens

(a)

Scrotum

(b)

Uterine tube

Uterus

Vagina

intravascular blood clots. Still others may develop liver disorders or high blood pressure as a result of using these substances.

A large dose of high potency estrogens sometimes is used as a "morning-after pill"—a treatment designed to prevent implantation of an embryo after sexual intercourse. Such a treatment acts by promoting powerful contractions of smooth muscles in the female reproductive tract that may cause a fertilized egg or early embryo to be expelled. However, if an embryo has become implanted already, there is danger that this treatment will injure an embryo, which then may continue to develop.

1. Describe two methods of contraception that make use of mechanical barriers.
2. How can the effectiveness of chemical contraceptives be increased?
3. What substances are contained in oral contraceptives?
4. How does an oral contraceptive prevent pregnancy?

Intrauterine Devices

An *intrauterine device,* or *IUD,* is a small, solid object in the form of a ring, coil, spiral, or loop that may be placed within the uterine cavity by a physician. Such a device is relatively effective in preventing pregnancy. It is believed that the IUD interferes with the implantation of embryos within the uterine wall, perhaps by causing inflammatory reactions in the uterine tissues.

Unfortunately, an IUD may be expelled from the uterus spontaneously, or it may produce unpleasant side effects, such as pain or excessive bleeding. It also may injure the uterus or produce other serious health problems.

Surgical Methods

Surgical methods of contraception involve sterilization of either the male or female.

In the male, a small section of each vas deferens is removed near the epididymis, and the cut ends of the ducts are tied. This procedure is called *vasectomy,* and it is a relatively simple operation that produces few side effects, although it may cause some pain for a week or two. After a vasectomy, sperm cells cannot leave the epididymis, and thus they are not included in the seminal fluid.

However, sperm cells may already be present in portions of the ducts distal to the cuts. Consequently the sperm count of the seminal fluid released by ejaculation may not reach zero for several weeks.

The corresponding procedure in the female is called *tubal ligation.* In this instance the uterine tubes are cut and tied so that sperm cells cannot reach an egg cell.

Neither a vasectomy nor a tubal ligation produces any changes in the hormonal concentrations or the sexual drives of the individuals involved. These sterilization procedures, shown in figure 22.39, provide the most reliable forms of contraception.

Although it is sometimes possible to reverse the effects of vasectomy or tubal ligation by surgically reconnecting the severed tubes, such procedures require difficult microsurgery that has a relatively low rate of success.

1. How is an IUD thought to prevent pregnancy?
2. Describe the surgical methods of contraception for a female and for a male.

Clinical Terms Related to the Reproductive Systems

abortion (ah-bor'shun)   the spontaneous or deliberate termination of pregnancy; a spontaneous abortion is commonly termed a miscarriage.

amenorrhea (a-men″o-re'ah)   the absence of menstrual flow, usually due to a disturbance in hormonal concentrations.

cesarean section (se-sa're-an sek'shun)   the delivery of a fetus through an abdominal incision.

conization (ko″ni-za'shun)   the surgical removal of a cone of tissue from the cervix for examination.

curettage (ku″re-tahzh′)   surgical procedure in which the cervix is dilated and the endometrium of the uterus is scraped (commonly called D and C).

dysmenorrhea (dis″men-o-re'ah)   painful menstruation.

eclampsia (e-klamp'se-ah)   a condition characterized by convulsions and coma that sometimes accompanies toxemia of pregnancy.

endometriosis (en″do-me″tre-o'sis)   a condition in which tissue resembling the endometrium of the uterine lining is present and growing in the abdominal cavity.

endometritis (en″do-me-tri'tis)   inflammation of the uterine lining.

epididymitis (ep″i-did″i-mi'tis)   inflammation of the epididymis.

gestation (jes-ta'shun)   the entire period of pregnancy.

hematometra (hem″ah-to-me'trah)   an accumulation of menstrual blood within the uterine cavity.

hyperemesis gravidarum (hi″per-e'me-sis grav'i-dar-um)   excessive vomiting associated with pregnancy; morning sickness.

hysterectomy (his″te-rek'to-me)   the surgical removal of the uterus.

mastitis (mas″ti'tis)   inflammation of a mammary gland.

oophorectomy (o″of-o-rek'to-me)   surgical removal of an ovary.

oophoritis (o″of-o-ri'tis)   inflammation of an ovary.

orchiectomy (or″ke-ek'to-me)   surgical removal of a testis.

orchitis (or-ki'tis)   inflammation of a testis.

prostatectomy (pros″tah-tek'to-me)   surgical removal of a portion or all of the prostate gland.

prostatitis (pros″tah-ti'tis)   inflammation of the prostate gland.

salpingectomy (sal″pin-jek'to-me)   surgical removal of a uterine tube.

toxemia of pregnancy (tok-se'me-ah)   a condition characterized by a group of metabolic disorders sometimes occurring during pregnancy.

vaginitis (vaj″i-ni'tis)   inflammation of the vaginal lining.

varicocele (var'i-ko-sel″)   distension of the veins within the spermatic cord.

Chapter Summary

Introduction
Male reproductive organs are specialized to produce and maintain sperm cells, transport these cells, and produce male sex hormones.

Organs of the Male Reproductive System
Primary sex organs are the testes, which produce sperm cells and male sex hormones. Accessory organs include internal and external reproductive organs.

The Testes
1. Descent of the testes
    a. Testes originate behind the parietal peritoneum near the level of the developing kidneys.
    b. The gubernaculum guides the descent of testes into the lower abdominal cavity and through the inguinal canal.
    c. Undescended testes fail to produce sperm cells because of the relatively high abdominal temperature.
2. Structure of the testes
    a. The testes are composed of lobules separated by connective tissue and filled with seminiferous tubules.
    b. Seminiferous tubules unite to form the rete testis that joins the epididymis.
    c. Seminiferous tubules are lined with germinal epithelium that produces sperm cells.
    d. Interstitial cells that produce male sex hormones occur between the seminiferous tubules.
3. Formation of sperm cells
    a. Germinal epithelium consists of supporting cells and spermatogenic cells.
       (1) Supporting cells support and nourish the spermatogenic cells.
       (2) Spermatogenic cells give rise to spermatogonia.
    b. Sperm cells are produced from spermatogonia by the process of spermatogenesis.
       (1) The number of chromosomes in sperm cells is reduced by one-half (46 to 23) by meiosis.
       (2) The product of spermatogenesis is four sperm cells from each primary spermatocyte.
    c. Membranous processes of adjacent supporting cells form a barrier within the germinal epithelium.
       (1) The barrier separates early and advanced stages of spermatogenesis.
       (2) It helps provide a favorable environment for differentiating cells.

4. Structure of a sperm cell
   a. Sperm head contains a nucleus with 23 chromosomes.
   b. Sperm body contains many mitochondria.
   c. Sperm tail functions to propel cell.

Male Internal Accessory Organs

1. The epididymis
   a. The epididymis is a tightly coiled tube on the outside of the testis that leads into the vas deferens.
   b. It stores immature sperm cells as they mature.
2. The vas deferens
   a. The vas deferens is a muscular tube that forms part of the spermatic cord.
   b. It passes through the inguinal canal, enters the abdominal cavity, courses medially into the pelvic cavity, and ends behind the urinary bladder.
   c. It fuses with the duct from the seminal vesicle to form the ejaculatory duct.
3. The seminal vesicle
   a. The seminal vesicle is a saclike structure attached to the vas deferens.
   b. It secretes alkaline fluid that contains nutrients and prostaglandins.
   c. This secretion is added to sperm cells during ejaculation.
4. The prostate gland
   a. This gland surrounds the urethra just below the urinary bladder.
   b. It secretes thin, milky fluid that neutralizes seminal fluid and vaginal secretions, and enhances the motility of sperm cells.
   c. It may enlarge in older males and interfere with urination.
5. The bulbourethral glands
   a. These are two small structures beneath the prostate gland.
   b. They secrete fluid that serves as a lubricant for the penis in preparation for sexual intercourse.
6. Seminal fluid
   a. Seminal fluid is composed of sperm cells and secretions of seminal vesicles, the prostate gland, and bulbourethral glands.
   b. This fluid is slightly alkaline and contains nutrients and prostaglandins.
   c. It activates sperm cells.

Male External Reproductive Organs

1. The scrotum
   a. The scrotum is a pouch of skin and subcutaneous tissue that encloses the testes.
   b. Dartos muscle in the scrotal wall causes the skin of the scrotum to be held close to the testes or to hang loosely.

2. The penis
   a. The penis functions to convey urine and seminal fluid.
   b. It is specialized to become erect for insertion into the vagina during sexual intercourse.
   c. The body of the penis is composed of three columns of erectile tissue surrounded by connective tissue.
   d. The root of the penis is attached to the pelvic arch and membranes of the perineum.
3. Erection, orgasm, and ejaculation
   a. During erection, vascular spaces within erectile tissue become engorged with blood as arteries dilate and veins constrict.
   b. Orgasm is the culmination of sexual stimulation and is accompanied by emission and ejaculation.
   c. Movement of seminal fluid occurs as result of sympathetic reflexes, involving peristaltic contraction of smooth muscles in the walls of tubular organs.
   d. Following ejaculation, the penis becomes flaccid.

Hormonal Control of Male Reproductive Functions

1. Hypothalamic and pituitary hormones
   The male body remains reproductively immature until the hypothalamus releases GnRH, which stimulates the anterior pituitary gland to release gonadotropins.
   a. FSH stimulates spermatogenesis.
   b. LH (ICSH) stimulates the interstitial cells to produce male sex hormones.
2. Male sex hormones
   a. Male sex hormones are called androgens.
   b. Testosterone is the most important androgen.
   c. Testosterone is converted into dihydrotestosterone in some organs.
   d. Androgens that fail to become fixed in tissues are metabolized in the liver and excreted.
   e. Androgen production increases rapidly at puberty.
3. Actions of testosterone
   a. Testosterone stimulates the development of the male reproductive organs and causes the testes to descend.
   b. It is responsible for the development and maintenance of male secondary sexual characteristics.
4. Regulation of male sex hormones
   a. Testosterone concentration is regulated by a negative feedback mechanism.
      (1) As its concentration rises, the hypothalamus is inhibited and pituitary secretion of gonadotropins is reduced.
      (2) As the concentration falls, the hypothalamus signals the pituitary to secrete gonadotropins.
   b. The concentration of testosterone remains relatively stable from day to day.

Organs of the Female Reproductive System

The primary sex organs of the female reproductive system are the ovaries, which produce female sex cells and sex hormones. Accessory organs include internal and external reproductive organs.

The Ovaries

1. Attachments of the ovaries
   a. The ovaries are held in position by several ligaments.
   b. Ligaments include broad, suspensory, and ovarian ligaments.
2. Descent of the ovaries
   a. The ovaries descend from behind the parietal peritoneum near the developing kidneys.
   b. They are attached to the pelvic wall just below the pelvic brim.
3. Structure of the ovaries
   a. The ovaries are divided into a medulla and a cortex.
   b. The medulla is composed of connective tissue, blood vessels, lymphatic vessels, and nerves.
   c. The cortex contains ovarian follicles and is covered by germinal epithelium.
4. Primordial follicles
   a. During development, groups of cells in the ovarian cortex form millions of primordial follicles.
   b. Each primordial follicle contains a primary oocyte and several follicular cells.
   c. The primary oocyte begins to undergo meiosis, but the process is soon halted and is not continued until puberty.
   d. The number of oocytes steadily decreases throughout the life of a female.
5. Oogenesis
   a. Beginning at puberty some oocytes are stimulated to continue meiosis.
   b. When a primary oocyte undergoes oogenesis it gives rise to a secondary oocyte in which the original chromosome number is reduced by one-half (46 to 23).
   c. A secondary oocyte represents an egg cell and can be fertilized to produce a zygote.
6. Maturation of a follicle
   a. At puberty, FSH stimulates primordial follicles to become primary follicles.
   b. During maturation, the oocyte enlarges, the follicular cells multiply, and a fluid-filled cavity appears to produce a secondary follicle.
   c. Ovarian cells surrounding the follicle form two layers.
   d. Usually only one follicle reaches full development.
7. Ovulation
   a. Oogenesis is completed as the follicle matures.
   b. The resulting oocyte is released when the follicle ruptures.
   c. After ovulation, the oocyte is drawn into the opening of the uterine tube.

Female Internal Accessory Organs

1. The uterine tubes
   a. These tubes convey egg cells toward the uterus.
   b. The end of each tube is expanded and its margin bears irregular extensions.
   c. Movement of an egg cell into the opening is aided by ciliated cells that line the tube and by peristaltic contractions in the wall of the tube.
2. The uterus
   a. The uterus receives the embryo and sustains its life during development.
   b. The cervix of the uterus is partially enclosed by the vagina.
   c. The uterine wall includes endometrium, myometrium, and perimetrium.
3. The vagina
   a. The vagina connects the uterus to the vestibule.
   b. It serves to receive the erect penis, to convey uterine secretions to the outside, and to transport the fetus during birth.
   c. The vaginal orifice is partially closed by a thin membrane, the hymen.
   d. Its wall consists of a mucosa, muscularis, and outer fibrous coat.

Female External Reproductive Organs

1. The labia major
   a. The labia major are rounded folds of fatty tissue and skin that enclose and protect the other external reproductive parts.
   b. The upper ends form a rounded, fatty elevation over the symphysis pubis.
2. The labia minor
   a. The labia minor are flattened, longitudinal folds between the labia major.
   b. They form the sides of the vestibule and anteriorly form the hoodlike covering of the clitoris.
3. The clitoris
   a. The clitoris is a small projection at the anterior end of the vulva.
   b. It is composed of two columns of erectile tissue.
   c. Its root is attached to the sides of the pubic arch.
4. The vestibule
   a. The vestibule is the space between the labia major that encloses the vaginal and urethral openings.
   b. Vestibular glands secrete mucus into the vestibule during sexual stimulation.
5. Erection, lubrication, and orgasm
   a. During periods of sexual stimulation, erectile tissues of the clitoris and vestibular bulbs become engorged with blood and swell.
   b. Vestibular glands secrete mucus into the vestibule and vagina, which lubricates these parts during sexual intercourse.
   c. During orgasm, muscles of the perineum, uterine wall, and uterine tubes contract rhythmically.

Hormonal Control of Female Reproductive Functions

Hormones from the hypothalamus, anterior pituitary gland, and ovaries play important roles in the control of sex cell maturation and the development and maintenance of female secondary sexual characteristics.

1. Female sex hormones
   a. A female body remains reproductively immature until about 8 years of age when gonadotropin secretion increases.
   b. The most important female sex hormones are estrogen and progesterone.
      (1) Estrogen from the ovaries is responsible for development and maintenance of most female secondary sexual characteristics.
      (2) Progesterone functions to cause changes in the uterus.
2. Female reproductive cycles
   a. These are called menstrual cycles; they are characterized by regularly recurring changes in the uterine lining culminating in menstrual flow.
   b. The cycle is initiated by FSH, which stimulates the maturation of a follicle.
   c. The maturing follicle secretes estrogen, which is responsible for maintaining the secondary sexual traits and causing the uterine lining to thicken.
   d. Ovulation is triggered when the anterior pituitary gland secretes a relatively large amount of LH and an increased amount of FSH.
   e. Following ovulation, follicular cells give rise to the corpus luteum.
      (1) The corpus luteum secretes progesterone, which causes the uterine lining to become more vascular and glandular.
      (2) If an egg cell is not fertilized, the corpus luteum begins to degenerate.
      (3) As concentrations of estrogen and progesterone decline, the uterine lining disintegrates, causing menstrual flow.
   f. During this cycle, estrogen and progesterone inhibit the hypothalamus and the pituitary gland; as the concentrations of estrogen and progesterone fall, the pituitary secretes FSH and LH again, stimulating a new cycle.
3. Menopause
   a. Eventually the ovaries cease responding to FSH, and cycling ceases.
   b. This results in a low estrogen concentration and a continually high concentration of FSH and LH.

Pregnancy

1. Transport of sex cells
   a. Movement of the egg cell to the uterine tube is aided by ciliary action.
   b. To move, a sperm cell lashes its tail; its movement within the female body is aided by muscular contractions in the uterus and uterine tube.
2. Fertilization
   a. A sperm cell penetrates an egg cell with the aid of an enzyme.
   b. When a sperm cell penetrates an egg cell, the entrance of any other sperm cells is prevented by structural changes that occur in the egg cell membrane and by enzyme actions.
   c. When the nuclei of a sperm and an egg cell fuse, the process of fertilization is complete.
   d. The product of fertilization is a zygote with 46 chromosomes.
3. Early embryonic development
   a. Cells undergo mitosis, giving rise to smaller and smaller cells.
   b. The developing offspring is moved down the uterine tube to the uterus, where it becomes implanted in the endometrium.
   c. The offspring is called an embryo from the second through the eighth week of development; thereafter it is a fetus.
   d. Eventually the embryonic and maternal cells together form a placenta.
4. Hormonal changes during pregnancy
   a. Embryonic cells produce HCG that causes the corpus luteum to be maintained.
   b. Placental tissue produces high concentrations of estrogen and progesterone.
      (1) Estrogen and progesterone maintain the uterine wall and inhibit the secretion of FSH and LH.
      (2) Progesterone and relaxin cause uterine contractions to be suppressed.
      (3) Estrogen causes enlargement of the vagina and relaxation of the ligaments that hold the pelvic joints together.
      (4) Relaxin also helps to soften the cervix and relax the pelvic ligaments.
   c. The placenta secretes placental lactogen that stimulates the development of the breasts.
   d. During pregnancy, increasing secretion of aldosterone promotes retention of sodium and body fluid, and increasing secretion of parathyroid hormone helps maintain a high concentration of maternal blood calcium.
5. Other changes during pregnancy
   a. The uterus enlarges greatly.
   b. The woman's blood volume, cardiac output, breathing rate, and urine production increase.
   c. The woman's dietary intake increases, but if intake is inadequate, fetal tissues have priority for use of available nutrients.

6. The birth process
   a. Pregnancy usually lasts 40 weeks.
   b. During pregnancy, estrogen excites uterine contractions and progesterone inhibits uterine contractions.
   c. A variety of factors are involved with the birth process.
      (1) Secretion of estrogen increases and secretion of progesterone decreases.
      (2) The changing concentrations of estrogen and progesterone stimulate the synthesis of prostaglandins that may initiate the birth process.
      (3) The posterior pituitary gland releases oxytocin.
      (4) Uterine muscles are stimulated to contract, and labor begins.
      (5) A positive feedback mechanism causes stronger contractions and greater release of oxytocin.
   d. Following the birth of the infant, placental tissues are expelled.

The Mammary Glands

1. Location of the glands
   a. The mammary glands are located in subcutaneous tissue of the anterior thorax within the breasts.
   b. Breasts extend between the second and sixth ribs and from sternum to axillae.
2. Structure of the glands
   a. The mammary glands are composed of lobes that contain tubular glands.
   b. Lobes are separated by dense connective and adipose tissues.
   c. The mammary glands are connected to the nipple by ducts.
3. Development of the breasts
   a. Male breasts remain nonfunctional.
   b. Estrogen stimulates female breast development.
      (1) Alveolar glands and ducts enlarge.
      (2) Fat is deposited around and within breasts.
   c. During pregnancy the breasts change.
      (1) Estrogen causes the ductile system to grow.
      (2) Progesterone causes development of alveolar glands.
      (3) No milk is produced because the hypothalamus inhibits the secretion of prolactin.
4. Production and secretion of milk
   a. Following childbirth, estrogen and progesterone concentrations decline.
      (1) The hypothalamus signals the anterior pituitary gland to release prolactin.
      (2) The mammary glands begin to secrete milk.
   b. Reflex response to mechanical stimulation of the nipple causes the posterior pituitary to release oxytocin, which causes milk to be ejected from the alveolar ducts.
   c. As long as milk is removed from glands, more milk is produced; if milk is not removed, production ceases.
   d. During the period of milk production, menstrual cycles are partially inhibited.

Birth Control

Voluntary regulation of the number of children produced and the time they are conceived is called birth control. This usually involves some method of contraception.

1. Coitus interruptus
   a. Coitus interruptus is withdrawal of the penis from the vagina before ejaculation.
   b. Some seminal fluid may be expelled from the penis before ejaculation.
2. Rhythm method
   a. Abstinence from sexual intercourse a few days before and after ovulation is the rhythm method.
   b. It is almost impossible to predict the time of ovulation accurately.
3. Mechanical barriers
   a. The condom is used by males.
   b. The diaphragm is used by females.
4. Chemical barriers
   a. Spermicidal creams, foams, and jellies are chemical barriers to conception.
   b. These provide an unfavorable environment in the vagina for sperm survival.
5. Oral contraceptives
   a. Tablets that contain synthetic estrogen- and progesterone-like substances are taken by the woman.
   b. They disrupt the normal pattern of gonadotropin secretion and prevent ovulation and the normal buildup of the uterine lining.
   c. When used correctly, this method is almost 100% effective.
   d. Some women have undesirable side effects.
6. Intrauterine devices
   a. An IUD is a solid object inserted in the uterine cavity.
   b. It is thought to prevent pregnancy by interfering with implantation.
   c. It may be expelled spontaneously or produce undesirable side effects.
7. Surgical methods
   a. These involve sterilization procedures.
      (1) Vasectomy is performed in males.
      (2) Tubal ligation is performed in females.
   b. These are the most reliable forms of contraception.

Application of Knowledge

1. What changes, if any, might be expected to occur in the secondary sexual characteristics of an adult male following removal of one testis? Following removal of both testes? Following removal of the prostate gland?
2. How would you explain the fact that new mothers sometimes experience cramps in their lower abdomens when they begin to nurse their babies?
3. If a woman who is considering having a tubal ligation asks, "Will the operation cause me to go through my change of life?" how would you answer?

Review Activities

1. List the general functions of the male reproductive system.
2. Distinguish between the primary and accessory male reproductive organs.
3. Describe the descent of the testes.
4. Define *cryptorchidism*.
5. Describe the structure of a testis.
6. Explain the function of the supporting cells in the testis.
7. List the major steps in spermatogenesis.
8. Describe a sperm cell.
9. Describe the epididymis and explain its function.
10. Trace the path of the vas deferens from the epididymis to the ejaculatory duct.
11. On a diagram, locate the seminal vesicles and describe the composition of their secretion.
12. On a diagram, locate the prostate gland and describe the composition of its secretion.
13. On a diagram, locate the bulbourethral glands and explain the function of their secretion.
14. Define *seminal fluid*.
15. Describe the structure of the scrotum.
16. Describe the structure of the penis.
17. Explain the mechanism that produces an erection of the penis.
18. Distinguish between emission and ejaculation.
19. Explain the mechanism of ejaculation.
20. Explain the role of GnRH in the control of male reproductive functions.
21. Distinguish between androgen and testosterone.
22. Define *puberty*.
23. Describe the actions of testosterone.
24. List several male secondary sexual characteristics.
25. Explain how the concentration of testosterone is regulated.
26. List the general functions of the female reproductive system.
27. Distinguish between the primary and accessory female reproductive organs.
28. Describe how the ovaries are held in position.
29. Describe the descent of the ovaries.
30. Describe the structure of an ovary.
31. Define *primordial follicle*.
32. List the major steps in oogenesis.
33. Describe how a follicle matures.
34. Define *ovulation*.
35. On a diagram, locate the uterine tubes and explain their function.
36. Describe the structure of the uterus.
37. Describe the structure of the vagina.
38. Distinguish between the labia major and the labia minor.
39. On a diagram, locate the clitoris and describe its structure.
40. Define *vestibule*.
41. Describe the process of erection in the female reproductive organs.
42. Define *orgasm*.
43. Explain the role of GnRH in regulating female reproductive functions.
44. List several female secondary sexual characteristics.
45. Define *menstrual cycle*.
46. Explain the roles of estrogen and progesterone in the menstrual cycle.
47. Summarize the major events in a menstrual cycle.
48. Define *menopause*.
49. Describe how male and female sex cells are transported within the female reproductive tract.
50. Describe the process of fertilization.
51. List the major functions of the placenta.
52. Explain the major hormonal changes that occur in the maternal body during pregnancy.
53. Describe the major nonhormonal changes that occur in the maternal body during pregnancy.
54. Discuss the events that occur during the birth process.
55. Describe the structure of a mammary gland.
56. Explain the roles of prolactin and oxytocin in milk production and secretion.
57. Define *contraception*.
58. List several methods of contraception and explain how each interferes with normal reproductive functions.

23

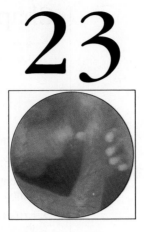

Human Growth and Development

The products of the female and male reproductive systems are egg cells and sperm cells respectively. When an egg cell and a sperm cell unite, a zygote is formed by the process of fertilization. Such a single-celled zygote is the first cell of an offspring, and it is capable of giving rise to an adult of the subsequent generation. The processes by which this is accomplished are called *growth* and *development*.

Chapter Outline

Chapter Objectives

After you have studied this chapter, you should be able to

1. Distinguish between growth and development.

2. Describe the major events that occur during the period of cleavage.

3. Explain how the primary germ layers originate and list the structures produced by each layer.

4. Describe the formation and function of the placenta.

5. Define *fetus* and describe the major events that occur during the fetal stage of development.

6. Trace the general path of blood through the fetal circulatory system.

7. Describe the major circulatory and physiological adjustments that occur in the newborn.

8. Name the stages of development that occur between the neonatal period and death, and list the general characteristics of each stage.

9. Complete the review activities at the end of this chapter. Note that the items are worded in the form of specific learning objectives. You may want to refer to them before reading the chapter.

Key Terms

amnion (am'ne-on)

chorion (ko're-on)

cleavage (klēv'ij)

embryo (em'bre-o)

fetus (fe'tus)

germ layer (jerm la'er)

neonatal (ne''o-na'tal)

placenta (plah-sen'tah)

postnatal (pōst-na'tal)

prenatal (pre-na'tal)

senescence (sĕ-nes'ens)

umbilical cord (um-bil'ĭ-kal kord)

zygote (zi'gōt)

Aids to Understanding Words

allant-, sausage-shaped: *allant*ois—tubelike structure that extends from the yolk sac into the connecting stalk of the embryo.

chorio-, skin: *chorio*n—outermost membrane that surrounds the fetus and other fetal membranes.

cleav-, to divide: *cleav*age—period of development characterized by a division of the zygote into smaller and smaller cells.

lacun-, a pool: *lacun*a—space between the chorionic villi that is filled with maternal blood.

morul-, mulberry: *morul*a—embryonic structure consisting of a solid ball of about 16 cells, thus appearing somewhat like a mulberry.

nat-, to be born: pre*nat*al—period of development before birth.

sen-, old: *sen*escence—process of growing old.

troph-, nourishment: *troph*oblast—cellular layer that surrounds the inner cell mass and helps nourish it.

umbil-, the navel: *umbil*ical cord—structure attached to the fetal navel (umbilicus) that connects the fetus to the placenta.

GROWTH REFERS TO an increase in size. In a human, growth usually reflects an increase in cell numbers as a result of *mitosis*, followed by enlargement of the newly formed cells and enlargement of the body.

Development, on the other hand, is the continuous process by which an individual changes from one life phase to another. These life phases include a **prenatal period,** which begins with the fertilization of an egg cell and ends at birth, and a **postnatal period,** which begins at birth and ends with death in old age (figure 23.1).

Prenatal Period

The prenatal period of development usually lasts for 40 weeks (ten lunar months) and can be divided into a period of cleavage, an embryonic stage, and a fetal stage.

Period of Cleavage

As is described in chapter 22, fertilization normally occurs within a uterine tube. About 24 hours after the *zygote* is formed, it undergoes mitosis, giving rise to two daughter cells. These cells, in turn, divide to form four cells, they divide into eight cells, and so forth. With each subsequent division, the resulting cells are smaller and smaller. This distribution of the zygote's contents into smaller and smaller cells is called *cleavage,* and the cells produced in this way are called *blastomeres* (figure 23.2).

It is usually not possible to determine the actual time of fertilization because reliable records concerning sexual activities are seldom available. However, the approximate time can be calculated by adding fourteen days to the date of the onset of the last menstruation. This time can then be used to calculate the fertilization age of an embryo. The expected time of birth can be estimated by adding 266 days to the fertilization date. Most fetuses are born within 10 to 15 days of this calculated time.

The mass of cells that is formed by cleavage is still enclosed in the zona pellucida of the original egg cell, and in about 3 days it consists of a solid ball of about sixteen cells, called a *morula.*

The morula enters the uterine cavity and remains unattached for about 3 days. During this time, the zona pellucida degenerates, and the morula develops a fluid-filled central cavity. Once this cavity appears, the morula becomes a hollow ball of cells called a **blastocyst** (figure 23.3).

Within the blastocyst, cells in one region group together to form an *inner cell mass* that eventually gives rise to the **embryo** proper—the body of the developing offspring. The cells forming the wall of the blastocyst make up the *trophoblast,* which will create structures that assist the embryo in its development.

Figure 23.1
(a) Growth involves an increase in size; *(b)* development refers to the process of changing from one phase of life to another.

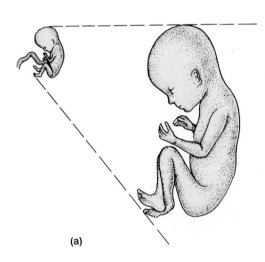

(a)

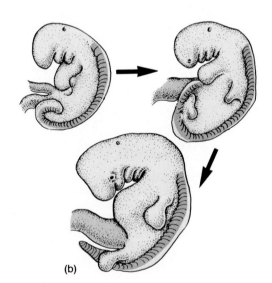

(b)

Figure 23.2
During the period of cleavage, the cells divide by mitosis and become smaller and smaller.

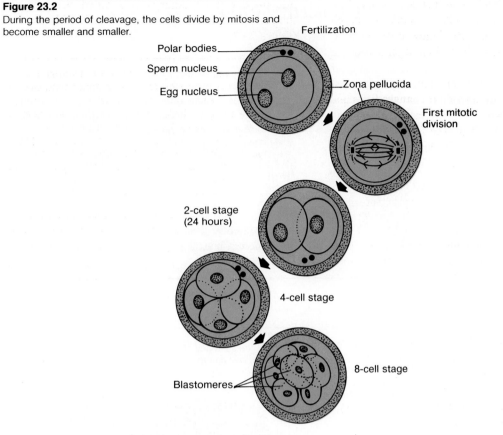

Fertilization

Polar bodies
Sperm nucleus
Egg nucleus
Zona pellucida

First mitotic division

2-cell stage
(24 hours)

4-cell stage

8-cell stage

Blastomeres

Figure 23.3
(a) A morula consists of a solid ball of cells; (b) a blastocyst has a fluid-filled cavity.

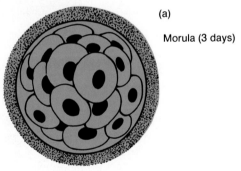

(a)

Morula (3 days)

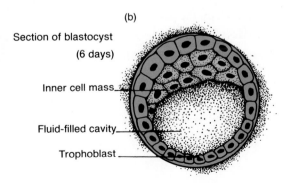

(b)

Section of blastocyst
(6 days)

Inner cell mass
Fluid-filled cavity
Trophoblast

A woman who is infertile because her uterine tubes are blocked may become pregnant as a result of *in vitro fertilization*. In one such technique, oocytes are removed from the woman's ovary and mixed with sperm in a dish to achieve fertilization (*in vitro* means "in glass" and refers to the artificial environment of the fertilization dish). Later, the resulting embryos are transferred into the woman's uterus for development.

The in vitro fertilization procedure usually begins at menses with the administration of HMG (human menopausal gonadotropin) or some other substance that will induce the development of ovarian follicles. The growth of the follicles can be monitored using ultrasonography, and when they have reached a certain diameter, the patient is given HCG (human chorionic gonadotropin) to induce ovulation.

Oocytes released from the ovary are harvested with the aid of a *laparoscope*—an optical instrument used to examine the abdominal interior. The oocytes are incubated at 37°C in a buffered medium with a pH of 7.4, and when they are mature, they are mixed in a dish with sperm that have been washed to remove various inhibitory factors.

Fertilized eggs are incubated in a special medium. After 50–60 hours, when the developing embryos have reached the 8- or 16-cell stage, normal ones are transferred through the woman's cervix and into her uterus with the aid of a specially designed catheter. She is then treated with progesterone to promote a favorable uterine environment for implantation of the embryos. As a result of this procedure, successful implantation occurs in about 20–30% of the cases.

Sometimes two ovarian follicles release egg cells simultaneously, and if both are fertilized, the resulting zygotes can develop into fraternal (dizygotic) twins. Such twins are no more alike genetically than any brothers or sisters from the same parents. In other instances, twins develop from a single fertilized egg (monozygotic twins). This can happen if two inner cell masses form within a blastocyst and each produces an embryo. Twins of this type usually share a single placenta, and they are identical genetically. Thus, they are always the same sex and are very similar in appearance.

About the sixth day, the blastocyst begins to attach itself to the uterine lining. This attachment is apparently aided by its secretion of proteolytic enzymes that digest a portion of the endometrium. The blastocyst sinks slowly into the resulting depression, becoming completely buried in the uterine lining. At the same time, the uterine lining is stimulated to thicken below the implanting blastocyst, and cells of the trophoblast begin to produce tiny, fingerlike processes (microvilli) that grow into the endometrium (figures 23.4 and 23.5).

This process of **implantation** occurs near the end of the first week of development, and completes the period of cleavage (figures 23.6 and 23.7).

Figure 23.4

(a) About the sixth day of development, the blastocyst contacts the uterine wall and *(b)* begins to become implanted within the wall.

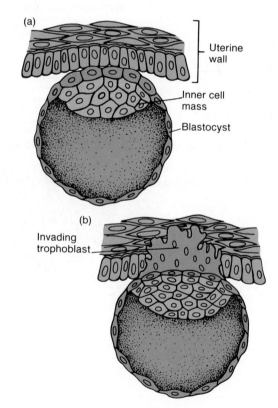

Figure 23.5

Light micrograph of a blastocyst in contact with the endothelium of the uterine wall.

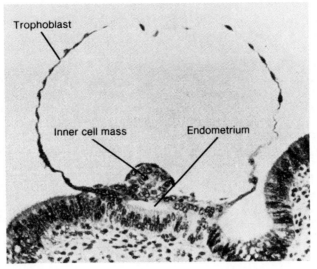

Figure 23.6

Light micrograph of a human embryo (arrow) undergoing implantation in the uterine wall.

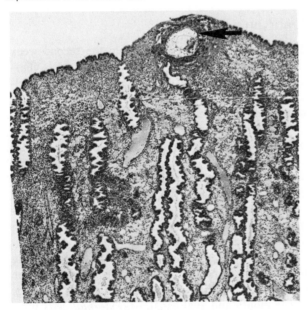

Figure 23.7

An embryo appears as a red blister bulging out from the wall of this dissected uterus.

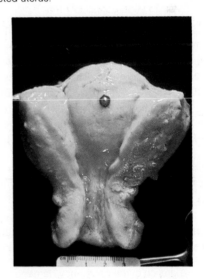

Figure 23.8

Early in the embryonic stage of development, the three primary germ layers are formed.

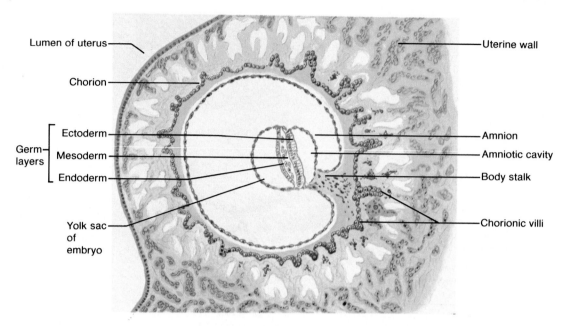

Embryonic Stage

As is discussed in chapter 22, the trophoblast of the blastocyst secretes the hormone HCG, which has properties similar to those of LH and causes corpus luteum to be maintained during the early stages of pregnancy. HCG also is thought to help protect the blastocyst against being rejected as foreign (nonself) substance by the maternal immunological system (see chapter 19) and to stimulate the synthesis of hormones from the developing placenta.

Usually the blastocyst becomes implanted in the upper posterior wall of the uterus. Sometimes, however, implantation occurs in the lower portion near the cervix, and as the placenta develops, it may partially or totally cover the opening of the cervix. This condition is called *placenta previa*, and it is likely to produce complications since any expansion of the cervix may damage placental tissues and cause bleeding.

1. Distinguish between growth and development.
2. What changes characterize the period of cleavage?
3. How does a blastocyst become attached to the endometrium of the uterus?
4. In what ways does the endometrium respond to the activities of the blastocyst?

The **embryonic stage** extends from the second week through the eighth week of development and is characterized by the formation of the placenta, the development of the main internal organs, and the appearance of the major external body structures.

Early in this stage, the cells of the inner cell mass become organized into a flattened **embryonic disk** that consists of two distinct layers: an outer *ectoderm* and an inner *endoderm.* A short time later, a third layer of cells, the *mesoderm,* forms between the ectoderm and endoderm. These three layers of cells are called the **primary germ layers,** and they are responsible for forming all of the body organs. (See figure 23.8.)

More specifically, *ectodermal cells* give rise to the nervous system, portions of special sensory organs, the epidermis, hair, nails, glands of the skin, and the linings of the mouth and anal canal. *Mesodermal cells* form all types of muscle tissue, bone tissue, bone marrow, blood, blood vessels, lymphatic vessels, various connective tissues, internal reproductive organs, kidneys, and the epithelial linings of the body cavities. *Endodermal cells* produce the epithelial linings of the digestive tract, respiratory tract, urinary bladder, and urethra (figure 23.9).

Figure 23.9

Each of the primary germ layers is responsible for the formation of a particular set of organs.

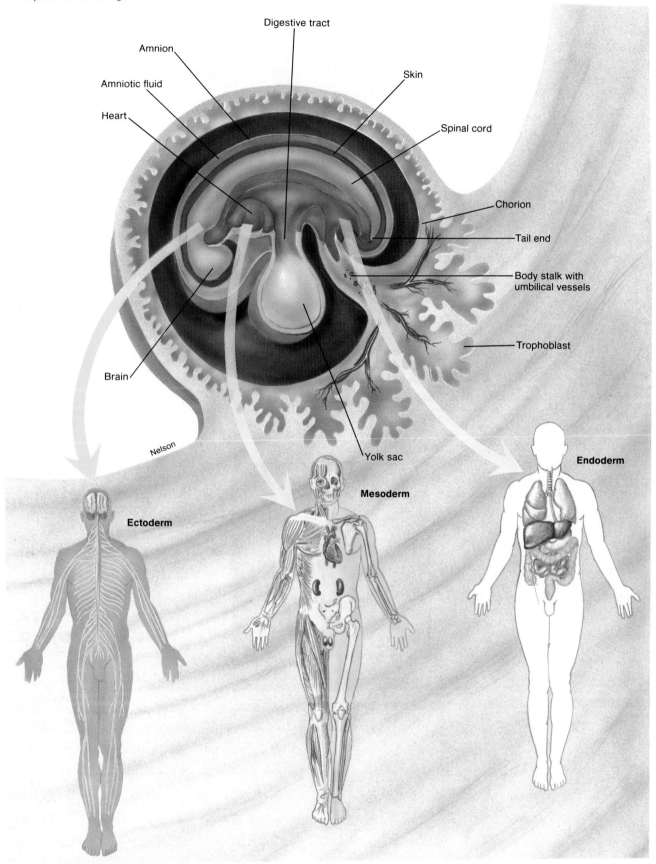

Digestive tract

Amnion

Skin

Amniotic fluid

Heart

Spinal cord

Chorion

Tail end

Body stalk with umbilical vessels

Trophoblast

Brain

Nelson

Yolk sac

Endoderm

Mesoderm

Ectoderm

Figure 23.10
During the fourth week, *(a)* the flat embryonic disk becomes *(b)* a cylindrical structure.

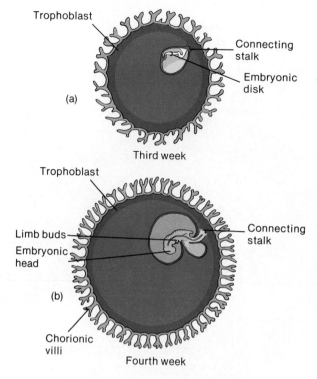

Trophoblast

Connecting stalk

Embryonic disk

(a)

Third week

Trophoblast

Limb buds

Embryonic head

Connecting stalk

(b)

Chorionic villi

Fourth week

Figure 23.11
Five-week-old embryo.

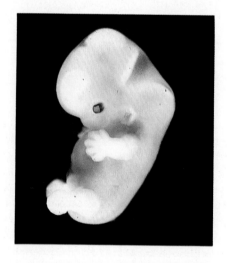

Figure 23.12
In the fifth through the seventh weeks of development, the
embryonic body and face develop a humanlike appearance.

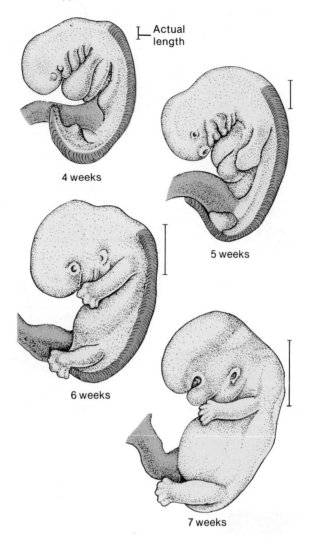

Actual
length

4 weeks

5 weeks

6 weeks

7 weeks

During the fourth week of development (figure 23.10), the flat embryonic disk is transformed into a cylindrical structure, which is attached to the developing placenta by a *connecting stalk* (body stalk). By this time, the head and jaws are appearing, the heart is beating and forcing blood through blood vessels, and tiny buds that will give rise to the arms and legs are forming (figure 23.11).

During the fifth through the seventh weeks, as shown in figure 23.12, the head grows rapidly and becomes rounded and erect. The face, which is developing eyes, nose, and mouth, becomes more humanlike. The arms and legs elongate, and fingers and toes appear (figures 23.13 and 23.14).

By the end of the seventh week, all the main internal organs have become established, and as these structures enlarge, they affect the shape of the body. Consequently, the body takes on a humanlike appearance.

Meanwhile, the embryo continues to become implanted within the uterus. As mentioned, early in this process slender projections grow out from the trophoblast into the surrounding endometrium. These extensions, which are called **chorionic villi,** become branched, and by the end of the fourth week of development they are well formed.

While the chorionic villi are developing, embryonic blood vessels appear within them, and these vessels are continuous with those passing through the connecting stalk to the body of the embryo. At the same

Figure 23.13

(a–c) Changes occurring during the fifth week of development; *(d–g)* changes during the sixth and seventh weeks.

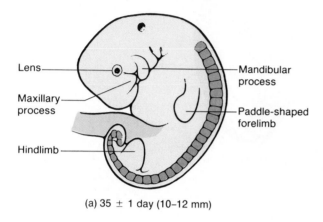

(a) 35 ± 1 day (10–12 mm)

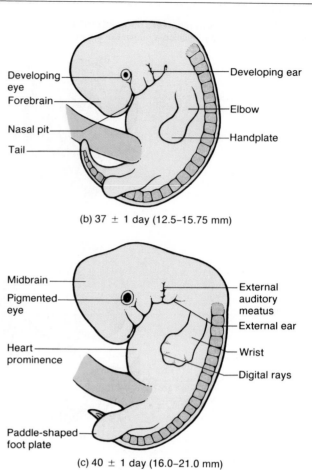

(b) 37 ± 1 day (12.5–15.75 mm)

(c) 40 ± 1 day (16.0–21.0 mm)

Figure 23.14

Human embryo after about six weeks of development.

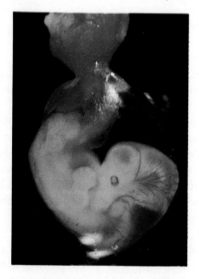

time, irregular spaces called **lacunae** are eroded around and between the villi. These spaces become filled with maternal blood that escapes from eroded endometrial blood vessels.

A thin membrane separates embryonic blood within the capillary of a chorionic villus from maternal blood in a lacuna. This membrane, called the **placental membrane,** is composed of the epithelium of the villus and the epithelium of the capillary (figure 23.15). Through this membrane, exchanges take place between maternal and embryonic blood. Oxygen and nutrients diffuse from the maternal blood into the embryonic blood, and carbon dioxide and other wastes diffuse from the embryonic blood into the maternal blood. Various substances also move through the placental membrane by active transport and pinocytosis.

Since most drugs are able to pass freely through the placental membrane, substances ingested by the mother may affect the fetus. Thus, fetal drug addiction may occur following the mother's use of drugs such as heroin.

Similarly, depressant drugs administered to the mother during labor can produce effects within the fetus and may, for example, depress the activity of its respiratory system.

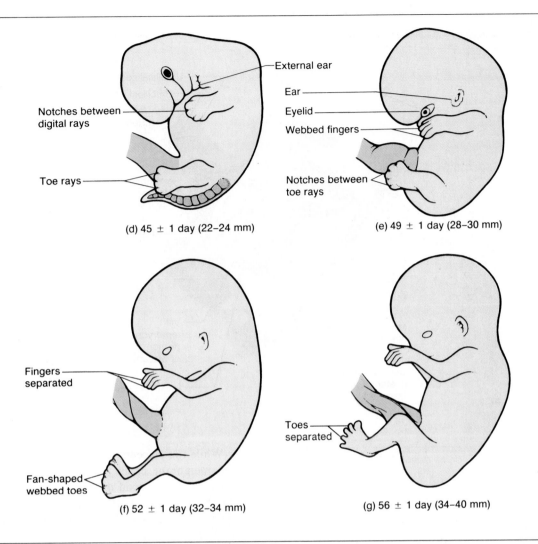

(d) 45 ± 1 day (22–24 mm)

External ear

Notches between digital rays

Toe rays

(e) 49 ± 1 day (28–30 mm)

Ear

Eyelid

Webbed fingers

Notches between toe rays

(f) 52 ± 1 day (32–34 mm)

Fingers separated

Fan-shaped webbed toes

(g) 56 ± 1 day (34–40 mm)

Toes separated

Figure 23.15
The placental membrane consists of the epithelial wall of an embryonic capillary and the epithelial wall of a chorionic villus. What is the significance of this membrane?

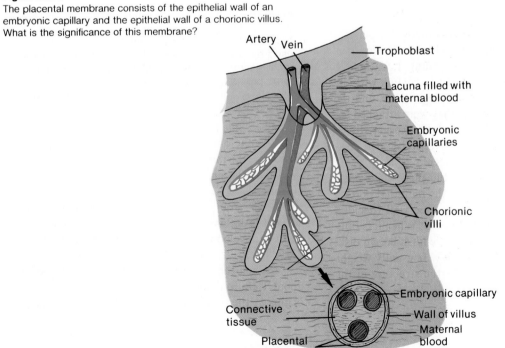

Artery

Vein

Trophoblast

Lacuna filled with maternal blood

Embryonic capillaries

Chorionic villi

Embryonic capillary

Wall of villus

Maternal blood

Connective tissue

Placental membrane

Section of villus

Figure 23.16

The placenta consists of an embryonic portion and a maternal portion.

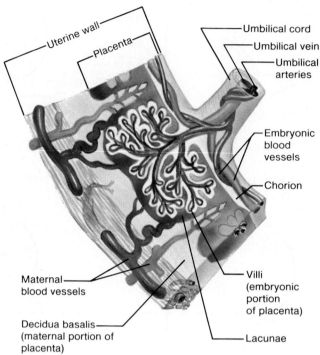

- Uterine wall
- Placenta
- Umbilical cord
- Umbilical vein
- Umbilical arteries
- Embryonic blood vessels
- Chorion
- Maternal blood vessels
- Villi (embryonic portion of placenta)
- Decidua basalis (maternal portion of placenta)
- Lacunae

1. What major events occur during the embryonic stage of development?
2. What tissues and structures develop from ectoderm? From mesoderm? From endoderm?
3. Describe the structure of a chorionic villus.
4. How are substances exchanged between the embryonic blood and the maternal blood?

Until about the end of the eighth week, the chorionic villi cover the entire surface of the former trophoblast, which is now called the **chorion.** However, as the embryo and the chorion surrounding it continue to enlarge, only those villi that remain in contact with the endometrium endure. The others degenerate, and the portions of the chorion to which they were attached become smooth. Thus, the region of the chorion still in contact with the uterine wall is restricted to a disk-shaped area that becomes the **placenta.**

The embryonic portion of the placenta is composed of the chorion and its villi; the maternal portion is composed of the area of the uterine wall (decidua basalis) to which the villi are attached. When it is fully formed, the placenta appears as a reddish brown disk, about 20 cm long and 2.5 cm thick. It usually weighs about 0.5 kg. Figure 23.16 shows the structure of the placenta.

While the placenta is forming from the chorion, another membrane, called the **amnion,** develops around the embryo. This second membrane begins to appear during the second week. Its margin is attached around the edge of the embryonic disk, and fluid, called **amniotic fluid,** fills the space between them.

As the placenta develops, it synthesizes a relatively large quantity of progesterone from cholesterol obtained from the maternal blood. This placental progesterone is then used by cells associated with the developing fetal adrenal glands to synthesize estrogens. The estrogens, in turn, promote changes in the maternal uterus and breasts, and influence the metabolism and development of various fetal organs.

As the embryo is transformed into a cylindrical structure, the margins of the amnion are folded around it so that the embryo is enclosed by the amnion and surrounded by amniotic fluid. As this process continues, the amnion envelops the tissues on the underside of the embryo, by which it is attached to the chorion and the developing placenta. In this manner, as figure 23.17 illustrates, the **umbilical cord** is formed.

When the umbilical cord is fully developed, it is about 1 cm in diameter and about 55 cm in length. It begins at the umbilicus of the embryo and is inserted

Figure 23.17

(a, b, c) As the amnion develops, it surrounds the embryo, and *(d)* the umbilical cord is formed. What structures comprise this cord?

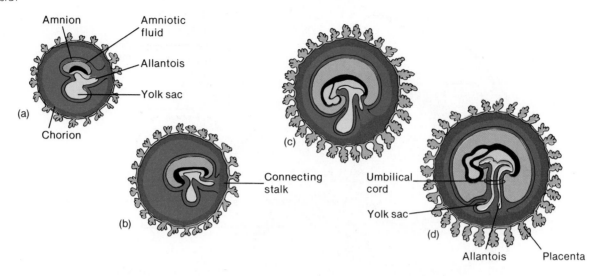

Figure 23.18

As the amniotic cavity enlarges, the amnion contacts the chorion, and the two membranes fuse to form the amniochorionic membrane.

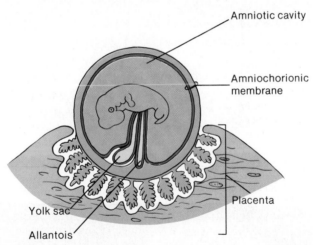

into the central region of the placenta. The cord contains three blood vessels—two *umbilical arteries* and an *umbilical vein*—through which blood passes between the embryo and the placenta (figure 23.16).

The umbilical cord also functions to suspend the embryo in the amniotic cavity, while the amniotic fluid provides a watery environment in which the embryo can grow freely without being compressed by surrounding tissues. The fluid also protects the embryo against being jarred by movements of the mother's body.

Eventually the amniotic cavity becomes so enlarged that the membrane of the amnion contacts the thicker chorion around it, and the two membranes become fused into an *amniochorionic membrane* (figures 23.18, 23.19, and 23.20).

In addition to the amnion and chorion, two other embryonic membranes appear during development. They are the yolk sac and the allantois.

The **yolk sac** appears during the second week, and it is attached to the underside of the embryonic disk (figures 23.17 and 23.18). It functions to form blood cells in the early stages of development and gives rise to the cells that later become sex cells. Portions of the

Figure 23.19

The developing placenta as it appears during the seventh week of development.

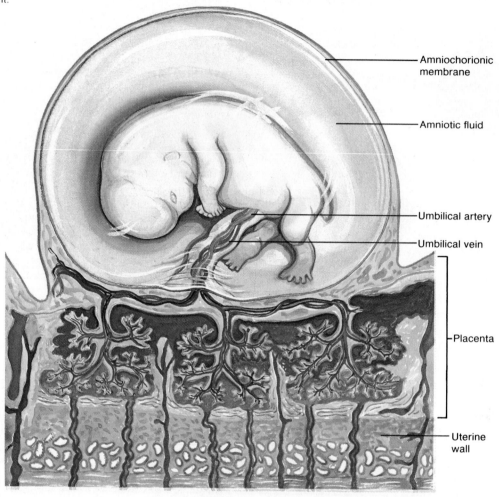

Amniochorionic membrane

Amniotic fluid

Umbilical artery

Umbilical vein

Placenta

Uterine wall

Figure 23.20

What structures can you identify in this photograph of a human embryo?

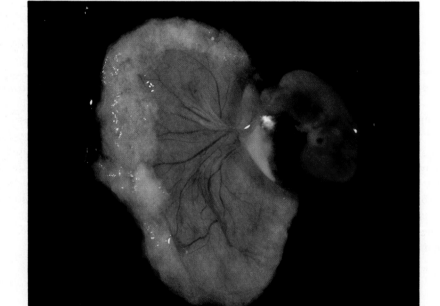

Teratogens
A Practical Application

Factors that can cause congenital malformations by affecting an embryo during its period of rapid growth and development are called *teratogens*. Such agents include various drugs and certain microorganisms.

The molecules of many drugs are able to pass freely from maternal blood to fetal blood through the placental membrane. Consequently, substances ingested by the mother-to-be may affect her embryo. For example, when a pregnant woman drinks alcohol, her embryo is exposed to an alcohol concentration equal to that of her blood, and the alcohol can produce a wide range of effects on the developing offspring.

More specifically, alcohol is able to produce a set of physical and mental abnormalities called the *fetal alcohol syndrome* (FAS). The symptoms of this syndrome include abnormal facial features, reduced head size, incomplete growth, organ malformations, and mental retardation. It is thought to result when a pregnant woman drinks as little as 3 ounces of alcohol per day or engages in a single drinking binge.

Similarly, exposure of an embryo to the drug called thalidomide during the time when the limbs are developing may lead to the *thalidomide syndrome*. In this case, the embryonic limbs may fail to form or develop as rudimentary or malformed structures.

The rubella virus that causes German measles also may cause congenital malformations. If a pregnant woman develops a rubella infection during the first four or five weeks of embryonic development, when the embryonic eyes, ears, and heart are forming, these organs may be malformed. As a result, the offspring may be blind due to cataracts, be deaf, and have various heart disorders. Exposure to the rubella virus during later stages of development may result in functional defects in the central nervous system.

---

yolk sac also enter into the formation of the embryonic digestive tube. Part of this membrane becomes incorporated into the umbilical cord, while the remainder lies in the cavity between the chorion and the amnion near the placenta.

The **allantois** forms during the third week as a tube extending from the early yolk sac into the connecting stalk of the embryo. It functions in the formation of blood cells and gives rise to the umbilical arteries and vein (figures 23.17 and 23.18).

The *embryonic stage* is completed at the end of the eighth week. It is the most critical period of development, for during this time the embryo becomes implanted within the uterine wall and all the essential external and internal body parts are formed. Any disturbances in the developmental processes occurring during the embryonic stage are likely to result in major malformations or malfunctions.

By the beginning of the eighth week, the embryo is usually 30 mm in length and weighs less than 5 grams. Although its body is quite unfinished, it is recognizable as a human being (figure 23.21).

The eighth week of development occurs two weeks after a pregnant woman has missed her second menstrual period. Since these first few weeks are critical periods in development, it is important that a woman seek health care as soon as she thinks she may be pregnant.

---

1. Describe the development of the amnion.
2. What blood vessels are found in the umbilical cord?
3. What is the function of amniotic fluid?
4. What is the significance of the yolk sac?

Figure 23.21
By the beginning of the eighth week of development, the embryonic body is recognizable as a human.

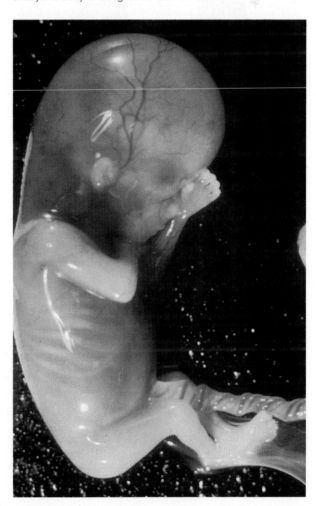

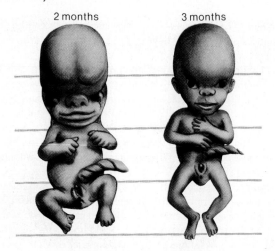

2 months   3 months

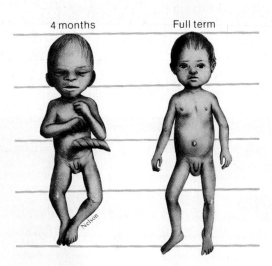

4 months   Full term

Fetal Stage

The **fetal stage** begins at the end of the eighth week of development and continues to the time of birth. During this period the offspring is called a **fetus,** and although the existing body structures continue to grow and mature, only a few new parts appear. The rate of growth is great, however, and the body proportions change considerably. For example, at the beginning of the fetal stage, the head is disproportionately large and the legs are relatively short (figure 23.22).

During the third lunar month, growth in body length is accelerated, while the growth of the head slows. (Note: a lunar month equals 28 days.) The arms achieve the relative length they will maintain throughout development, and ossification centers appear in most of the bones. By the twelfth week, the external reproductive organs are distinguishable as male or female (figure 23.23).

In the fourth lunar month, the body grows very rapidly and reaches a length of 13 to 17 cm. The legs lengthen considerably, and the skeleton continues to ossify.

In the fifth lunar month, the rate of growth decreases somewhat. The legs achieve their final relative proportions, and the skeletal muscles become active so that the mother may feel fetal movements (quickening). Some hair appears on the head, and the skin becomes covered with fine, downy hair (lanugo). The skin also is coated with a cheesy mixture of sebum from the sebaceous glands and dead epidermal cells (vernix caseosa).

During the sixth lunar month, the body gains a substantial amount of weight. The eyebrows and eyelashes appear. The skin is quite wrinkled and translucent. It is also reddish, due to the presence of dermal blood vessels.

In the seventh lunar month, the skin becomes smoother as fat is deposited in the subcutaneous tissues. The eyelids, which fused together during the third month, reopen. At the end of this month, a fetus is about 37 cm in length.

If a fetus is born prematurely, its chance of surviving increases directly with its age. One of the factors involved with the chance of survival is the development of the lungs. Thus, fetuses have increased chances of surviving if their lungs are sufficiently developed so that they have the thin respiratory membranes necessary for rapid exchange of oxygen and carbon dioxide and if the lungs produce enough surfactant to reduce the alveolar surface tension (see chapter 16).

In the eighth lunar month, the fetal skin is still reddish and somewhat wrinkled. The testes of males descend from regions near the developing kidneys into the scrotum.

During the ninth lunar month, the fetus reaches a length of about 47 cm. The skin is smooth, and the body appears chubby due to an accumulation of subcutaneous fat. The reddishness of the skin fades to pinkish or bluish pink, even in fetuses of dark-skinned parents, because melanin is not produced until the skin is exposed to light. Thus, it may be difficult to determine a newborn's race by its skin color during the first few hours or days of life.

At the end of the tenth lunar month, the fetus is said to be full term. It is about 50 cm long and weighs 6 to 8 lbs. The skin has lost its downy hair, but is still covered with sebum and dead epidermal cells. The scalp is usually covered with hair, the fingers and toes have well-developed nails, and the bones of the head are ossified. As figure 23.24 shows, the fetus is usually positioned upside down with its head toward the cervix *(vertex position)* in preparation for the birth process.

Figure 23.23
(a) Human fetus after about 10 weeks of development; (b) after 12 weeks.

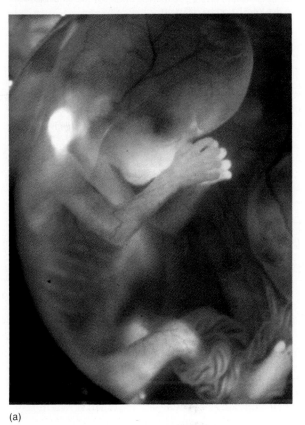

(a)

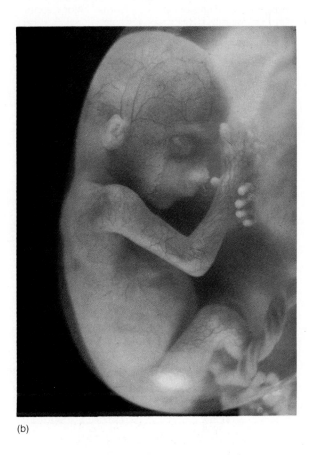

(b)

Figure 23.24
A full-term fetus usually becomes positioned with its head near the cervix.

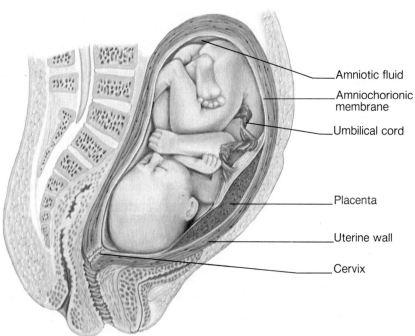

Amniotic fluid

Amniochorionic membrane

Umbilical cord

Placenta

Uterine wall

Cervix

Chart 23.1 Stages in prenatal development

Stage	Time period	Some major events
Period of cleavage	First week	Cells undergo mitosis, blastocyst forms; inner cell mass appears; blastocyst becomes implanted in uterine wall
Embryonic stage	Second through eight week	Inner cell mass becomes embryonic disk; primary germ layers form; embryo proper becomes cylindrical; main internal organs and external body structures appear; placenta and umbilical cord form; embryo proper is suspended in amniotic fluid
Fetal stage	Ninth week to birth	Existing structures continue to grow; ossification centers appear in bones; reproductive organs develop; arms and legs achieve final relative proportions; muscles become active; skin is covered with sebum and dead epidermal cells; head becomes positioned toward cervix

Figure 23.25

Oxygen and nutrients enter the fetal blood from the maternal blood by diffusion; waste substances enter the maternal blood from the fetal blood by diffusion.

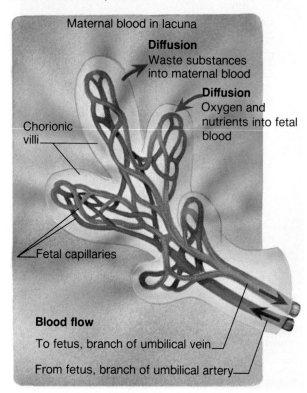

Chart 23.1 summarizes the stages of prenatal development.

The size and shape of the uterus may be used as an index of fetal growth. Thus, failure of the uterus to increase in size progressively during pregnancy suggests retarded fetal development, while a sudden increase in growth may indicate the presence of more than one fetus, the formation of excessive amniotic fluid (hydramnios) or an abnormal growth of chorionic tissue (hydatidiform mole).

---

1. What major changes characterize the fetal stage of development?
2. When can the sex of a fetus be determined?
3. How is a fetus usually positioned within the uterus at the end of the tenth lunar month?

Fetal Blood and Circulation

Throughout the fetal stage of development the maternal blood supplies the fetus with oxygen and nutrients, and carries away its wastes. These substances diffuse between the maternal and fetal blood through the placental membrane, and they are carried to and from the fetal body by means of the umbilical blood vessels (figure 23.25). Consequently, the fetal blood and vascular system must be adapted to intrauterine life in special ways, and its pattern of blood flow must differ from that of an adult. For example, the concentration of oxygen-carrying hemoglobin in the fetal blood is about 50% greater than in the maternal blood. Also, the fetal hemoglobin is chemically slightly different and has a greater attraction for oxygen than maternal hemoglobin. Thus, at a particular oxygen pressure, fetal hemoglobin can carry 20–30% more oxygen than maternal hemoglobin.

In the fetal circulatory system, the *umbilical vein* transports blood rich in oxygen and nutrients from the placenta to the fetal body. This vein enters the body through the umbilical ring and travels along the anterior abdominal wall to the liver. About half the blood passes into the liver, and the rest enters a vessel called the **ductus venosus,** which bypasses the liver.

The ductus venosus travels a short distance and joins the inferior vena cava. There the oxygenated blood from the placenta is mixed with deoxygenated blood from the lower parts of the fetal body. This blood continues through the vena cava to the right atrium.

In an adult heart, blood from the right atrium enters the right ventricle and is pumped through the pulmonary trunk and pulmonary arteries to the lungs.

Figure 23.26
The general pattern of fetal circulation. How does this pattern of
circulation differ from that of an adult?

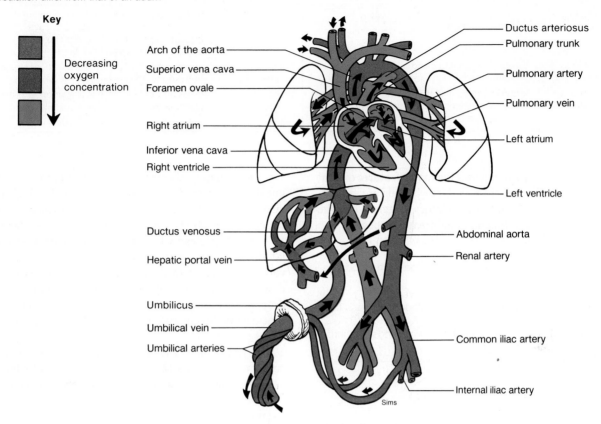

Key

Decreasing
oxygen
concentration

Arch of the aorta
Superior vena cava
Foramen ovale

Right atrium

Inferior vena cava
Right ventricle

Ductus venosus

Hepatic portal vein

Umbilicus

Umbilical vein

Umbilical arteries

Ductus arteriosus
Pulmonary trunk

Pulmonary artery

Pulmonary vein

Left atrium

Left ventricle

Abdominal aorta

Renal artery

Common iliac artery

Internal iliac artery

Sims

In the fetus, however, the lungs are nonfunctional, and the blood largely bypasses them. More specifically, as blood relatively rich in oxygen enters the right atrium of the fetal heart, a large proportion of it is shunted directly into the left atrium through an opening in the atrial septum. This opening is called the **foramen ovale,** and blood passes through it because the blood pressure in the right atrium is somewhat greater than that in the left atrium. Furthermore, a small valve (septum primum) located on the left side of the atrial septum overlies the foramen ovale and helps to prevent blood from moving from left to right.

The rest of the blood entering the right atrium, as well as a large proportion of the deoxygenated blood entering from the superior vena cava, passes into the right ventricle and out through the pulmonary trunk.

The vessels of the pulmonary circuit have a high resistance to blood flow since the lungs are collapsed, and the vessels are somewhat compressed. Enough blood reaches the lung tissues, however, to sustain them.

Most of the blood in the pulmonary trunk bypasses the lungs by entering a fetal vessel called **ductus arteriosus,** which connects the pulmonary trunk to the descending portion of the aortic arch. As a result, blood with a relatively low oxygen concentration returning to the heart in the superior vena cava bypasses the lungs.

At the same time it is prevented from entering the portion of the aorta that provides branches leading to the heart and brain.

The more highly oxygenated blood that enters the left atrium through the foramen ovale is mixed with a small amount of deoxygenated blood returning from the pulmonary veins. This mixture moves into the left ventricle and is pumped into the aorta. Some of it reaches the myocardium by means of the coronary arteries, and some reaches the tissues of the brain through the carotid arteries.

The blood carried by the descending aorta is partially oxygenated and partially deoxygenated. Some of it is carried into the branches of the aorta that lead to various parts in the lower regions of the body. The rest passes into the *umbilical arteries,* which branch from the internal iliac arteries and lead to the placenta. There the blood is reoxygenated. (See figure 23.26.)

The umbilical cord usually contains two arteries and one vein. In a small percentage of newborns, there is only one umbilical artery. Since this condition is often associated with various other cardiovascular disorders, the number of vessels within the severed cord is counted routinely following a birth.

Chart 23.2 Fetal circulatory adaptations			
Adaptation	Function	Adaptation	Function
Fetal blood	Has greater oxygen-carrying capacity than maternal blood	Foramen ovale	Conveys large proportion of blood entering the right atrium from the inferior vena cava, through the atrial septum, and into the left atrium, thus bypassing the lungs
Umbilical vein	Carries oxygenated blood from placenta to fetus	Ductus arteriosus	Conducts some blood from the pulmonary trunk to the aorta, thus bypassing the lungs
Ductus venosus	Conducts about half the blood from the umbilical vein directly to the inferior vena cava, thus bypassing the liver	Umbilical arteries	Carry blood from the internal iliac arteries to the placenta for reoxygenation

Chart 23.2 summarizes the major features of fetal circulation. At the time of birth, important adjustments must occur in the circulatory system when the placenta ceases to function and breathing begins.

---

1. Which umbilical vessel carries oxygen-rich blood to the fetus?
2. What is the function of the ductus venosus?
3. How does fetal circulation allow blood to bypass the lungs?
4. What characteristic of the fetal lungs tends to shunt blood away from them?

Postnatal Period

The postnatal period of development lasts from birth until death and can be divided into a neonatal period, infancy, childhood, adolescence, adulthood, and senescence.

Neonatal Period

The **neonatal period,** which extends from birth to the end of the first four weeks, begins very abruptly at birth (figure 23.27). At this time physiological adjustments must be made quickly, because the newborn must suddenly do for itself those things that the mother's body has been doing for it. Thus, the newborn must carry on respiration, obtain nutrients, digest nutrients, excrete wastes, regulate body temperature, and so forth. However, its most immediate need is to obtain oxygen and excrete carbon dioxide, so its first breath is critical.

The first breath must be particularly forceful, because the newborn's lungs are collapsed, and the airways are small and offer considerable resistance to air movement. Also, surface tension tends to hold the moist membranes of the lungs together.

Fortunately, the lungs of a full term fetus continuously secrete *surfactant* (see chapter 16), which reduces surface tension, and after the first powerful breath begins to expand the lungs, breathing becomes easier.

It is not clear whether the first breath is stimulated by one or several factors. Those that may be involved include an increasing concentration of carbon dioxide, a decreasing pH, low oxygen concentration, a drop in body temperature, and mechanical stimulation that occurs during and after the birth process (figure 23.28).

Prior to birth, the fetus depends primarily on glucose and fatty acids obtained from the mother's blood as energy sources. The newborn, on the other hand, is suddenly without an external source of nutrients, and its mother's milk will not be produced for 2 to 3 days. Furthermore, the newborn has a relatively high rate of metabolism, and its liver, which is not fully mature, may be unable to supply enough glucose to support the body's metabolic needs. Consequently, the newborn typically utilizes stored fat as an energy source.

---

1. Define the neonatal period of development.
2. Why must the first breath of an infant be particularly forceful?
3. What factors seem to stimulate the first breath?
4. What does a newborn use for an energy supply during the first few days after birth?

As a rule, the newborn's kidneys are unable to produce concentrated urine, so the newborn excretes a relatively dilute fluid. For this reason, the newborn may become dehydrated and develop a water and electrolyte imbalance. Also, some of the newborn's homeostatic control mechanisms may function imperfectly. The temperature regulating system, for example, may be unable to maintain a constant body temperature. As a consequence, the body temperature may be unstable during the first few days of life and may respond to slight stimuli by fluctuating above or below the normal level.

Figure 23.27
The neonatal period extends from birth to the end of the fourth
week after birth.

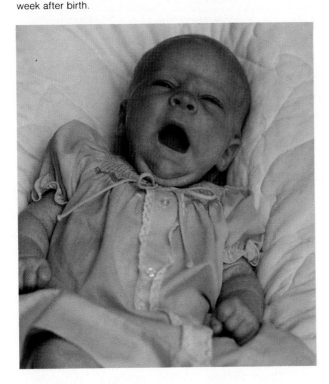

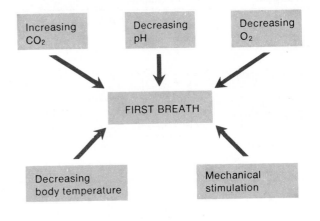

Increasing
CO_2

Decreasing
pH

Decreasing
O_2

FIRST BREATH

Decreasing
body temperature

Mechanical
stimulation

Figure 23.29
Major changes that occur in the newborn's circulatory system.

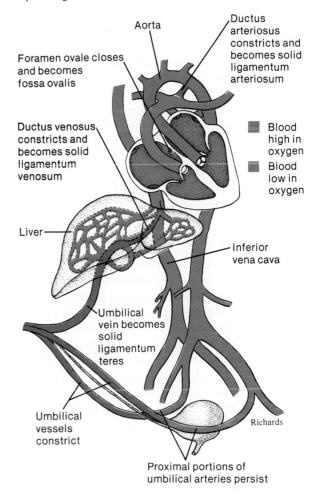

Aorta

Ductus
arteriosus
constricts and
becomes solid
ligamentum
arteriosum

Foramen ovale closes
and becomes
fossa ovalis

Ductus venosus
constricts and
becomes solid
ligamentum
venosum

Liver

Inferior
vena cava

Umbilical
vein becomes
solid
ligamentum
teres

Umbilical
vessels
constrict

Richards

Proximal portions of
umbilical arteries persist

■ Blood
high in
oxygen

■ Blood
low in
oxygen

As was mentioned, when the circulation of blood
through the placenta ceases and the lungs begin to
function at birth, adjustments must occur in the circulatory system because it is adapted to the needs of
intrauterine life (figure 23.29).

For example, following birth, the umbilical vessels constrict. The arteries close first, and if the umbilical cord is not clamped or severed for a minute or
so, blood continues to flow from the placenta to the
newborn through the umbilical vein, adding to the
newborn's blood volume.

The proximal portions of the umbilical arteries
persist in the adult as the *superior vesical arteries* that
supply blood to the urinary bladder. The more distal
portions become solid cords (lateral umbilical ligaments). The umbilical vein becomes the cordlike *ligamentum teres* that extends from the umbilicus to the
liver in an adult.

Similarly, the ductus venosus constricts shortly
after birth and is represented in the adult as a fibrous
cord (ligamentum venosum), which is superficially
embedded in the wall of the liver.

The foramen ovale closes as a result of blood
pressure changes occurring in the right and left atria
as fetal vessels constrict. More precisely, as blood ceases
to flow from the umbilical vein into the inferior vena
cava, the blood pressure in the right atrium drops. Also,

Chart 23.3   Circulatory adjustments in the newborn

Structure	Adjustment	In the adult
Umbilical vein	Becomes constricted	Becomes ligamentum teres that extends from the umbilicus to the liver
Ductus venosus	Becomes constricted	Becomes ligamentum venosum that is superficially embedded in the wall of the liver
Foramen ovale	Is closed by valvelike septum primum as blood pressure in right atrium decreases and pressure in left atrium increases	Valve fuses along margin of foramen ovale and is marked by a depression called fossa ovalis
Ductus arteriosus	Constricts in response to bradykinin released from the expanding lungs	Becomes ligamentum arteriosum that extends from the pulmonary trunk to the aorta
Umbilical arteries	Distal portions become constricted	Distal portions become lateral umbilical ligaments; proximal portions function as superior vesical arteries

as the lungs expand with the first breathing movements, the resistance to blood flow through the pulmonary circuit decreases, more blood enters the left atrium through the pulmonary veins, and the blood pressure in the left atrium increases.

As pressure in the left atrium rises and that in the right atrium falls, the valve (septum primum) on the left side of the atrial septum closes the foramen ovale. In most individuals, this valve gradually fuses with the tissues along the margin of the foramen. In an adult, the site of the previous opening is marked by a depression called the *fossa ovalis.*

In some newborns, the foramen ovale remains open (patent), and since the blood pressure in the left atrium is greater than that in the right atrium, some blood may flow from the left chamber to the right chamber. Although this condition usually produces no symptoms, it may cause a heart murmur.

The ductus arteriosus, like the other fetal vessels, constricts after birth. This constriction seems to be stimulated by a substance called *bradykinin,* which is released from the lungs during their initial expansions. Bradykinin is thought to act when the oxygen concentration of the aortic blood rises as a result of breathing. After the ductus arteriosus has closed, blood can no longer bypass the lungs by moving from the pulmonary trunk directly into the aorta. In an adult, the ductus arteriosus is represented by a cord called the *ligamentum arteriosum.*

The ductus arteriosus sometimes fails to close. Although the cause of this condition is usually unknown, it often occurs in newborns whose mothers were infected with rubella virus (German measles) during the first three months of embryonic development.

After birth, the blood pressure in the aorta is normally greater than that in the pulmonary trunk. Therefore, a patent ductus arteriosus usually results in the flow of some blood from the aorta into the pulmonary trunk.

These changes in the newborn's circulatory system do not occur very rapidly. Although the constriction of the ductus arteriosus may be functionally complete within 15 minutes, the permanent closure of the foramen ovale may take up to a year. Chart 23.3 summarizes these circulatory changes.

1. How do the kidneys of a newborn differ from those of an adult?
2. What portions of the umbilical arteries continue to function into adulthood?
3. What is the fate of the foramen ovale? The ductus arteriosus?
4. When is the closure of the ductus arteriosus functionally complete?

Infancy

The period of development extending from the end of the first four weeks to 1 year is called **infancy.** During this time, the infant grows rapidly and may triple its birth weight. Its teeth begin to erupt through the gums, and its muscular and nervous systems mature so that

coordinated muscular activities become possible. Consequently, the infant is soon able to sit, creep, and stand; to reach for objects and hold them; and to follow objects visually (figure 23.30).

Infancy also is characterized by the beginning of the ability to communicate with others: the infant learns to smile, laugh, and respond to some sounds. By the end of the first year, the infant may be able to say two or three words.

Since infancy (as well as childhood) is a period of rapid growth, the infant has particular nutritional needs. For example, in addition to an energy source, the body requires proteins to provide the amino acids necessary to form new tissues, calcium and vitamin D to promote the development and ossification of skeletal structures, iron to support blood cell formation, and vitamin C for the production of structural tissues such as cartilage and bone.

As mentioned previously, fetal hemoglobin is slightly different and has a greater affinity for oxygen than the adult type. As a rule, however, fetal hemoglobin is not produced after birth, and by the age of 4 months most of the circulating hemoglobin is the adult type.

Childhood

The period called **childhood** begins at the end of the first year and is completed at puberty (figure 23.31). During this period, growth continues at a high rate. The deciduous teeth appear and are subsequently replaced by permanent ones. The child develops a high degree of voluntary muscular control and learns to walk, run, and climb. Bladder and bowel controls are established. The child learns to communicate effectively by speaking, and later, usually learns to read, write, and reason objectively. At the same time the child is maturing emotionally.

1. Define *infancy*.
2. What developmental changes characterize this phase of life?
3. Define *childhood*.
4. What developmental changes characterize this phase of life?

Figure 23.32
Adolescence is the period between puberty and adulthood.

Figure 23.33
Adulthood extends from adolescence to old age.

Adolescence

Adolescence is the period of development between puberty and adulthood. As is discussed in chapter 22, puberty is a period of anatomical and physiological changes resulting in reproductively functional individuals. These changes are, for the most part, hormonally controlled, and they include the appearance of secondary sexual characteristics and growth spurts in the muscular and skeletal systems.

Females usually experience these changes somewhat earlier than males, so that early in adolescence, females may be taller and stronger than their male peers (figure 23.32). On the other hand, females achieve their full growth at earlier ages, and in late adolescence the average male is taller and stronger than the average female.

The periods of rapid growth in adolescence, which usually begin between the ages of 11 and 13 in females and between 13 and 15 in males, cause increased demands for certain nutrients. In addition to energy sources, foods must provide ample amounts of proteins, vitamins, and minerals to support the growth of new tissues.

Adolescence is also characterized by increasing levels of motor skill, intellectual ability, and emotional maturity.

Adulthood

Adulthood (maturity) extends from adolescence to old age. Since the anatomy and physiology of the adult has been the basis for chapters 1–22, little more needs to be said about this period of development.

Ordinarily the adult remains relatively unchanged anatomically or physiologically for many years (figure 23.33). However, sometime after the age of 30, depending on hereditary factors and the environmental factors to which the person has been exposed, degenerative changes usually begin to occur. It has been estimated that the average person in this stage of development loses about 0.8% of his or her functional capacity each year. For example, skeletal muscles tend to lose strength as more and more connective tissue appears within the muscles; the circulatory system becomes less efficient as the lumens of arterioles and arteries become narrowed due to accumulations of fatty deposits; the skin tends to become loose and wrinkled as elastic fibers in the dermis undergo changes; and the hair on the head turns gray as melanin-producing enzyme systems cease to function.

Also, the capacity of the sex cell-producing tissues decline. Females experience menopause in later adulthood, and males experience losses in sexual functions that accompany a decrease in the secretion of the male sex hormones.

Figure 23.34
The changes that characterize old age begin during adulthood
and culminate with death.

Senescence

Senescence is the process of growing old, and it culminates in the death of the individual. This process involves a continuation of the degenerative changes that begin during adulthood (figure 23.34). As a result, the body becomes less and less able to cope with the demands placed on it by the individual and by the environment.

The cause of senescence is not well understood. However, the degenerative changes that occur seem to involve a variety of factors. Some are due to the prolonged use of body parts. For example, the cartilages covering the ends of bones at joints may wear away, leaving the joints stiff and painful.

Other degenerative changes are caused by disease processes that interfere with vital functions, such as gas exchanges or blood circulation. Still others seem to be due to cellular changes, including alterations in metabolic rates and in distribution of body fluids. The rate of cell division may decline, and the ability of cells to carry on immune responses may decrease. As a result, the person becomes less able to repair damaged tissue and is more susceptible to disease.

Senescence is typically accompanied by decreasing efficiency of the central nervous system. As a result, the person may experience a loss in intellectual functions. Also, the physiological coordinating capacity of the nervous system may decrease, and homeostatic mechanisms may fail to operate effectively.

Sensory functions usually decrease with age also. As a result, perhaps one-third of older persons have significant hearing losses; they often have difficulty focusing on objects at varying distances; and they may have problems with depth perception. Typically their senses of taste and smell are diminished, and many older persons have decreased abilities to perceive environmental temperature accurately. They also may have impaired perception of touch and pressure and may have a decreased sensitivity to painful stimuli.

Death usually results, however, from mechanical disturbances in the cardiovascular system, failure of the immune system, or from disease processes that affect vital organs.

Since older people are generally less active, they need fewer calories to maintain desirable weights, but at the same time, they need to have favorable intakes of essential nutrients. Thus, if they eat simply to satisfy hunger, they may develop symptoms of malnutrition resulting from deficiencies of such nutrients as iron, magnesium, calcium, vitamin A, vitamin D, or vitamin C; or from excesses of sodium, chlorine, or phosphorus.

Aging can be defined as the structural and functional changes that occur normally in a person with the passage of time and that gradually produce functional impairments leading to death.

Although aging is considered to be part of development, there are no generally accepted explanations for it. On the other hand, both cellular and environmental factors seem to contribute to the process.

Some investigators believe that cells are programmed to age as they are programmed to undergo other developmental changes. For example, body cells seem to be able to divide a limited number of times before they lose their capacity to reproduce and finally die. Furthermore, it is known from studies of human cells cultured in the laboratory that the number of times a cell divides is related roughly to the age of its donor. Such a loss of ability to reproduce among body cells could contribute to a decline in body functions by preventing the replacement of worn or damaged cells.

Other investigators emphasize the effects of *mutations* as a cause of aging. They theorize that a decline in cellular functions may involve changes in DNA molecules, occurring as a result of exposure to certain environmental factors (see chapter 4). Then, if the mutations occur faster than the DNA repair system can correct them, molecular errors may accumulate so that, in time, cellular metabolism is disturbed. This type of damage might be particularly serious in tissues, such as skeletal muscle and nerve tissue, whose cells usually are not replaced by cell division, or in cells that provide vital control functions.

Still another possible cause of aging involves the action of highly unstable substances called *free radicals*. These substances sometimes are produced in cells as a result of metabolic processes, and they may be encountered in the environment. In either case, they can damage DNA molecules and can react with other cell structures. For example, they can cause chemical bonds to form abnormally between protein molecules. The formation of such cross-linkages may contribute to the aging changes that occur in collagen. This protein is quite flexible—a quality that is very important to maintaining the normal characteristics of many organs (see chapter 5). However, with the passage of time, collagen tends to become stiffer and less flexible, and this aging-related change may be promoted by the action of free radicals.

Other factors that are thought to contribute to the aging process include an accumulation of certain cellular wastes, such as lipofuscin in neurons, and the effects of stress acting over prolonged periods of time (see chapter 13).

Chart 23.4 summarizes the major phases of life and their characteristics, while chart 23.5 lists some of the aging-related changes.

Although the average life expectancy has been increased in relatively recent times in developed countries as a result of improved control of infectious diseases, the human life span has not changed significantly for thousands of years. This life span seems to be fixed between 90 and 100 years.

---

1. What changes characterize adolescence?
2. Define *adulthood*.
3. What changes characterize adulthood?
4. What changes characterize senescence?

Clinical Terms Related to Growth and Development

ablatio placentae (ab-la′she-o plah-cen′tā) a premature separation of the placenta from the uterine wall.

amniocentesis (am″ne-o-sen-te′sis) a technique in which a sample of amniotic fluid is withdrawn from the amniotic cavity by inserting a hollow needle through the mother's abdominal wall.

dizygotic twins (di″zi-got′ik twinz) twins resulting from the fertilization of two ova by two sperm cells.

hydatid mole (hi′dah-tid mōl) a type of uterine tumor that originates from placental tissue.

hydramnios (hi-dram′ne-os) the presence of excessive amniotic fluid.

intrauterine transfusion (in″trah-u′ter-in trans-fu′zhun) transfusion administered by injecting blood into the fetal peritoneal cavity before birth.

lochia (lo′ke-ah) vaginal discharge following childbirth.

meconium (mĕ-ko′ne-um) the first fecal discharge of a newborn.

monozygotic twins (mon″o-zi-got′ik twinz) twins resulting from the fertilization of one ovum by one sperm cell.

perinatology (per″i-na-tol′o-je) branch of medicine concerned with the fetus after 25 weeks of development and with the newborn for the first four weeks after birth.

postpartum (pōst-par′tum) occurring after birth.

teratology (ter″ah-tol′o-je) the study of abnormal development and congenital malformations.

trimester (tri-mes′ter) each third of the total period of pregnancy.

ultrasonography (ul″trah-son-og′rah-fe) technique used to visualize the size and position of fetal structures by means of ultrasonic sound waves.

Chart 23.4   Stages in postnatal development

Stage	Time period	Some major events
Neonatal period	Birth to end of fourth week	Newborn begins to carry on respiration, obtain nutrients, digest nutrients, excrete wastes, regulate body temperature, and make circulatory adjustments
Infancy	End of fourth week to one year	Growth rate is high, teeth begin to erupt, muscular and nervous systems mature so coordinated activities are possible, communication begins
Childhood	One year to puberty	Growth rate is high, deciduous teeth erupt and are replaced by permanent teeth, high degree of muscular control is achieved, bladder and bowel controls are established, intellectual abilities mature
Adolescence	Puberty to adulthood	Person becomes reproductively functional and emotionally more mature, growth spurts occur in skeletal and muscular systems, high levels of motor skills are developed, intellectual abilities increase
Adulthood	Adolescence to old age	Person remains relatively unchanged anatomically and physiologically, degenerative changes begin to occur
Senescence	Old age to death	Degenerative changes continue, body becomes less and less able to cope with the demands placed upon it, death usually results from mechanical disturbances in the cardiovascular system or from disease processes that affect vital organs

Chart 23.5   Some aging-related changes

Organ system	Aging-related changes
Integumentary system	Degenerative loss of collagenous and elastic fibers in dermis; decreased production of pigment in hair follicles; reduced activity of sweat and sebaceous glands
	Skin tends to become thinner, wrinkled, and dry; hair turns gray and then white
Skeletal system	Degenerative loss of bone matrix
	Bones become thinner, less dense, and more likely to fracture; stature may shorten due to compression of intervertebral disks and vertebrae
Muscular system	Loss of skeletal muscle fibers; degenerative changes in neuromuscular junctions
	Loss of muscular strength
Nervous system	Degenerative changes in neurons; loss of dendrites and synaptic connections; accumulation of lipofuscin in neurons; decreases in sensory sensitivities
	Decreasing efficiency in processing and recalling information; decreasing ability to communicate; diminished senses of smell and taste; loss of elasticity of lenses and consequent loss of ability to accommodate for close vision
Endocrine system	Reduced hormonal secretions
	Decrease in metabolic rate; reduced ability to cope with stress; reduced ability to maintain homeostasis
Digestive system	Decreased motility in gastrointestinal tract; reduced secretion of digestive juices
	Reduced efficiency of digestion
Respiratory system	Degenerative loss of elastic fibers in lungs; reduced number of alveoli
	Reduced vital capacity; increase in dead air space; reduced ability to clear airways by coughing
Cardiovascular system	Degenerative changes in cardiac muscle; decrease in lumen diameters of arteries and arterioles
	Decreased cardiac output; increased resistance to blood flow; increased blood pressure
Lymphatic system	Decrease in efficiency of immune system
	Increased incidence of infections and neoplastic diseases; increased incidence of autoimmune diseases
Excretory system	Degenerative changes in kidneys; reduction in number of functional nephrons
	Reduction in filtration rate, tubular secretion, and reabsorption
Reproduction systems	
Female	Degenerative changes in ovaries; decrease in secretion of sex hormones
	Menopause; regression of secondary sexual characteristics
Male	Reduced secretion of sex hormones; enlargement of prostate gland; decrease in sexual energy

Chapter Summary

Introduction

Growth refers to an increase in size; development is the process of changing from one phase of life to another.

Prenatal Period

1. Period of cleavage
   a. Fertilization occurs in a uterine tube and results in a zygote.
   b. The zygote undergoes mitosis, and the daughter cells divide.
   c. Each subsequent division produces smaller and smaller cells.
   d. A solid ball of cells (morula) is formed, and it becomes a hollow ball called a blastocyst.
   e. The inner cell mass that gives rise to the embryo proper forms within the blastocyst.
   f. The blastocyst becomes implanted in the uterine wall.
      (1) Enzymes digest the endometrium around the blastocyst.
      (2) Fingerlike processes from the blastocyst penetrate into the endometrium.
   g. The period of cleavage lasts through the first week of development.
   h. The trophoblast secretes HCG, which helps maintain corpus luteum, helps protect the blastocyst against being rejected, and stimulates the developing placenta to secrete hormones.
2. Embryonic stage
   a. The embryonic stage extends from the second through the eighth week.
   b. It is characterized by development of the placenta and the main internal and external body structures.
   c. The cells of the inner cell mass become arranged into primary germ layers.
      (1) Ectoderm gives rise to the nervous system, portions of the skin, the lining of the mouth, and the lining of the anal canal.
      (2) Mesoderm gives rise to muscles, bones, blood vessels, lymphatic vessels, reproductive organs, kidneys, and linings of body cavities.
      (3) Endoderm gives rise to linings of the digestive tract, respiratory tract, urinary bladder, and urethra.
   d. The embryonic disk becomes cylindrical and is attached to the developing placenta by the connecting stalk.
   e. The embryo develops head, face, arms, legs, and mouth, and appears more humanlike.
   f. Chorionic villi develop and are surrounded by spaces filled with maternal blood.
   g. The placental membrane consists of the epithelium of the villi and the epithelium of the capillaries inside the villi.
      (1) Oxygen and nutrients diffuse from the maternal blood through the membrane and into the fetal blood.
      (2) Carbon dioxide and other wastes diffuse from the fetal blood through the membrane and into the maternal blood.
   h. The placenta develops in the disk-shaped area where the chorion remains in contact with the uterine wall.
      (1) The embryonic portion consists of chorion and its villi.
      (2) The maternal portion consists of the uterine wall to which villi are attached.
   i. Fluid-filled amnion develops around the embryo.
   j. The umbilical cord is formed as the amnion envelops the tissues attached to the underside of the embryo.
      (1) The umbilical cord includes two arteries and a vein.
      (2) It suspends the embryo in the amniotic cavity.
   k. The chorion and amnion become fused.
   l. The yolk sac forms on the underside of the embryonic disk.
      (1) It gives rise to blood cells and cells that later form sex cells.
      (2) It helps form the digestive tube.
   m. The allantois extends from the yolk sac into the connecting stalk.
      (1) It forms blood cells.
      (2) It gives rise to the umbilical vessels.
   n. By the beginning of the eighth week, the embryo is recognizable as a human.
3. Fetal stage
   a. This stage extends from the end of the eighth week and continues until birth.
   b. Existing structures grow and mature; only a few new parts appear.
   c. The body enlarges, arms and legs achieve final relative proportions, the skin is covered with sebum and dead epidermal cells, the skeleton continues to ossify, the muscles become active, and fat is deposited in subcutaneous tissue.

d. The fetus is full term at the end of the tenth lunar month.
  (1) It is about 50 cm long and weighs 6–8 pounds.
  (2) It is positioned with its head toward the cervix.
4. Fetal blood and circulation
  a. Blood is carried between the placenta and fetus by umbilical vessels.
  b. Blood enters the fetus through the umbilical vein and partially bypasses the liver by means of the ductus venosus.
  c. Blood enters the right atrium and partially bypasses the lungs by means of the foramen ovale.
  d. Blood entering the pulmonary trunk partially bypasses the lungs by means of the ductus arteriosus.
  e. Blood enters the umbilical arteries from the internal iliac arteries.

Postnatal Period

1. Neonatal period
  a. This period extends from birth to the end of the fourth week.
  b. The newborn must begin to carry on respiration, obtain nutrients, excrete wastes, and regulate its body temperature.
  c. The first breath must be powerful in order to expand the lungs.
    (1) Surfactant reduces surface tension.
    (2) The first breath is stimulated by a variety of factors.
  d. The liver is immature and unable to supply sufficient glucose, so the newborn depends primarily on stored fat as an energy source.
  e. Immature kidneys cannot concentrate urine very well.
    (1) The newborn may become dehydrated.
    (2) Water and electrolyte imbalances may develop.
  f. Homeostatic mechanisms may function imperfectly and body temperature may be unstable.
  g. The circulatory system undergoes changes when placental circulation ceases.
    (1) Umbilical vessels constrict.
    (2) The ductus venosus constricts.
    (3) The foramen ovale is closed by a valve as blood pressure in the right atrium falls and pressure in the left atrium rises.
    (4) Bradykinin released from the expanding lungs stimulates constriction of the ductus arteriosus.

2. Infancy
  a. Infancy extends from the end of the fourth week to 1 year of age.
  b. Infancy is a period of rapid growth.
    (1) The muscular and nervous systems mature, and coordinated activities become possible.
    (2) Communication begins.
  c. Rapid growth depends on an adequate intake of proteins, vitamins, and minerals in addition to energy sources.
3. Childhood
  a. Childhood extends from the end of the first year to puberty.
  b. It is characterized by rapid growth, development of muscular control, and establishment of bladder and bowel control.
4. Adolescence
  a. Adolescence extends from puberty to adulthood.
  b. It is characterized by physiological and anatomical changes that result in a reproductively functional individual.
  c. Females may be taller and stronger than males in early adolescence, but the situation reverses in late adolescence.
  d. Adolescents develop high levels of motor skills, their intellectual abilities increase, and they continue to mature emotionally.
5. Adulthood
  a. Adulthood extends from adolescence to old age.
  b. The adult remains relatively unchanged physiologically and anatomically for many years.
  c. After age 30, degenerative changes usually begin to occur.
    (1) Skeletal muscles lose strength.
    (2) The circulatory system becomes less efficient.
    (3) The skin loses its elasticity.
    (4) The capacity to produce sex cells declines.
6. Senescence
  a. Senescence is the process of growing old.
  b. Degenerative changes continue, and the body becomes less able to cope with demands placed upon it.
  c. Changes occur because of prolonged use, effects of disease, and cellular alterations.
  d. An aging person usually experiences losses in intellectual functions, sensory functions, and physiological coordinating capacities.
  e. Death usually results from mechanical disturbances in the cardiovascular system or from disease processes that affect vital organs.

Application of Knowledge

1. How would you explain the observation that twins resulting from a single fertilized egg cell can have body parts exchanged by tissue or organ transplant procedures without the occurrence of rejection reactions?

2. One of the more common congenital cardiac disorders is a ventricular septum defect in which an opening remains between the right and left ventricles. What problem would such a defect create as blood moves through the heart?

3. If an aged relative came to live with you, what special provisions could you make in your household environment and routines that would demonstrate your understanding of the changes brought on by aging processes?

4. Why is it important for a middle-aged adult who has neglected physical activity for many years to see a physician for a physical examination before beginning an exercise program?

Review Activities

1. Define *growth* and *development*.
2. Describe the process of cleavage.
3. Distinguish between a blastomere and a blastocyst.
4. Describe the formation of the inner cell mass and explain its significance.
5. Describe the process of implantation.
6. List three functions of HCG.
7. Explain how the primary germ layers form.
8. List the major body parts derived from ectoderm.
9. List the major body parts derived from mesoderm.
10. List the major body parts derived from endoderm.
11. Describe the formation of the placenta and explain its functions.
12. Define *placental membrane*.
13. Distinguish between the chorion and the amnion.
14. Explain the function of amniotic fluid.
15. Describe the formation of the umbilical cord.
16. Explain how the yolk sac and the allantois are related and list the functions of each.
17. Explain why the embryonic period of development is so critical.
18. Define *fetus*.
19. List the major changes that occur during the fetal stage of development.
20. Describe a full term fetus.
21. Explain how the fetal circulatory system is adapted for intrauterine life.
22. Trace the pathway of blood from the placenta to the fetus and back to the placenta.
23. Distinguish between a newborn and an infant.
24. Explain why the newborn's first breath must be particularly forceful.
25. List some factors that are thought to act as stimuli for the first breath.
26. Explain why newborns tend to develop water and electrolyte imbalances.
27. Describe the circulatory changes that occur in the newborn.
28. Describe the characteristics of an infant.
29. Distinguish between a child and an adolescent.
30. Define *adulthood*.
31. List some of the degenerative changes that begin during adulthood.
32. Define *senescence*.
33. List some of the factors that seem to promote senescence.

24

Human Genetics

The complex changes an embryo undergoes as it grows and develops is controlled by mechanisms that ensure the resulting offspring will resemble its parents. At the same time, these mechanisms guarantee the offspring will differ from its parents and will become an individual with unique characteristics.

The study of the similarities and differences between parents and their offspring and the mechanisms responsible for them is called *genetics*.

Chapter Outline

Chapter Objectives

After you have studied this chapter, you should be able to

1. Explain how heredity and environment influence the development of individual characteristics.

2. Distinguish between genes and chromosomes and explain why genes and chromosomes are paired.

3. Explain why gene expression may differ with different allelic combinations.

4. Describe how environmental factors may influence gene expression.

5. Outline the process of meiosis and explain why cells resulting from this process contain different combinations of genetic information.

6. Explain the patterns by which single traits are transmitted from parents to offspring.

7. Describe how chromosomes control the inheritance of sex and how sex chromosomes act.

8. Explain the pattern by which a trait controlled by a sex-linked gene is inherited.

9. Define *nondisjunction* and explain its significance in the appearance of chromosome disorders.

10. Explain what is meant by *genetic counseling*.

11. Complete the review activities at the end of this chapter. Note that the items are worded in the form of specific learning objectives. You may want to refer to them before reading the chapter.

Key Terms

allele   (ah-lēl′)

autosome   (aw′to-sōm)

dominant   (dom′ĭ-nant)

gene   (jēn)

genetics   (jĕ-net′iks)

genotype   (je′no-tĭp)

heredity   (hĕ-red′ĭ-te)

heterozygous   (het″er-o-zi′gus)

homozygous   (ho″mo-zi′gus)

meiosis   (mi-o′sis)

multiple allele   (mul′tĭ-pl ah-lēl′)

mutation   (mu-ta′shun)

nondisjunction   (non″dis-jungk′shun)

phenotype   (fe′no-tĭp)

recessive   (re-ses′iv)

sex chromosome   (seks kro′mo-sōm)

sex-linked gene   (seks-linkt′ jēn)

Aids to Understanding Words

alb-, white: *alb*ino—condition in which normal skin pigments are lacking.

centr-, a point: *centr*omere—region at which paired chromatids are held together.

hem-, blood: *hem*ophilia—condition in which bleeding is prolonged due to a defect in the blood clotting mechanism.

hetero-, different: *hetero*zygous—condition in which the alleles of a gene pair are different.

homo-, same: *homo*logous chromosomes—a pair of chromosomes that contain similar genetic information.

pheno-, visible: *pheno*type—appearance of an individual due to the ways genes are expressed.

syn-, together: *syn*apsis—process by which homologous chromosomes become tightly intertwined.

tri-, three: *tri*somy—condition in which one kind of chromosome is triply represented.

ALTHOUGH GENETICS IS a relatively new field of study, knowledge of genetic mechanisms has expanded rapidly in recent times. It is known that individual characteristics result from the interaction of two major factors: heredity and environment.

Heredity is the transmission of genetic information from parents to offspring. This information is passed in the form of DNA molecules contained in the nuclei of egg and sperm cells, and it provides chemical instructions for human development. **Environment,** on the other hand, includes the chemical, physical, and biological factors in the surroundings of an individual that act to influence his or her characteristics.

Genes and Chromosomes

As is explained in chapter 4, genetic information within DNA molecules tells a cell how to construct specific kinds of protein molecules, which in turn function as structural materials, enzymes, and other vital substances.

The Gene

The portion of a DNA molecule containing the information for producing one kind of protein molecule is called a **gene.** The genes within the egg and sperm cells that unite to form a zygote instruct the zygote how to synthesize particular proteins. Some of these proteins act as enzymes, which in turn promote the metabolic reactions necessary for the growth and development of a unique individual.

It takes one gene to produce a particular kind of enzyme. Therefore, many sets of genes are required for the synthesis of enzymes responsible for an offspring's specific sex, eye color, hair color, skin color, blood type, and so forth.

The total number of genes in a human set is unknown. However, many kinds of enzymes, structural proteins, and proteins with special functions occur in cells. Since at least one gene is needed for the synthesis of each kind of protein, some investigators estimate that the human set includes as many as 50,000 genes.

Chromosome Numbers

Within a cell, the DNA molecules, and therefore the genes, are found within *chromosomes.* These structures, which are described in chapter 3, consist of proteins (histones) as well as DNA. They appear as rod-shaped bodies in the nucleus when a cell undergoes cell division (figure 24.1).

The number of chromosomes within the cells of organisms varies with species. For the human species *(Homo sapiens),* the number is 46. Thus, a human zygote contains 46 chromosomes, 23 of which were received from its female parent by means of an egg cell and 23 from its male parent by means of a sperm cell (figure 24.2).

Chromosome Pairs

The 23 chromosomes a zygote receives from its female parent contain a complete set of genetic information for the growth and development of an offspring, as do the 23 chromosomes it receives from its male parent. The chromosomes of these sets that contain similar genetic material are said to comprise **homologous pairs.** That is, the chromosome of the maternal set containing the information for a particular set of traits and the chromosome of the paternal set containing information for these same traits make up a homologous pair. The chromosome set shown in figure 24.3 is arranged in homologous pairs.

During interphase, the stage in a cell's life cycle between mitotic divisions described in chapter 3, the chromosomes and the genes they contain become duplicated. One member of each duplicated chromosome is transmitted to each daughter cell, so that the daughter cells receive identical sets of homologous chromosomes and therefore receive identical sets of genes.

Gene Pairs

Since the chromosomes within a cell are paired, the genes within the chromosomes are paired also. Such a pair of genes is said to consist of two **alleles,** and because the members of a gene pair are parts of homologous chromosomes, one allele of each pair originates from the female parent and the other originates from the male parent.

Figure 24.1

Chromosomes appear in a cell as it undergoes cell division.

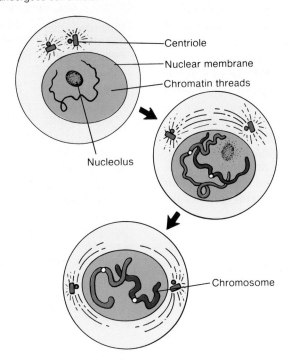

- Centriole
- Nuclear membrane
- Chromatin threads
- Nucleolus
- Chromosome

Figure 24.2

A typical human cell contains 46 chromosomes.

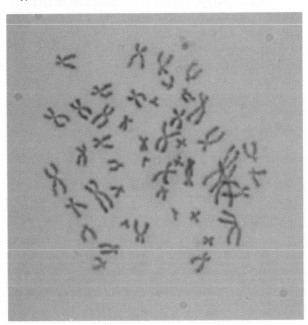

Figure 24.3

A set of human chromosomes arranged in pairs of homologous chromosomes. One chromosome of each pair originates from each parent.

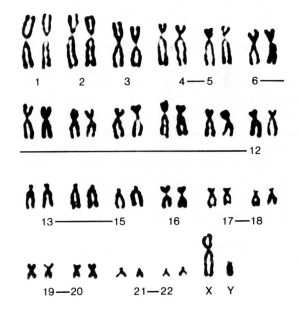

1 2 3 4—5 6——

—————————————————————————— 12

13———————15 16 17—18

19—20 21—22 X Y

Sometimes the alleles of a gene pair are alike. That is, they contain exactly the same genetic information. They may, for example, both contain information for the synthesis of a particular enzyme needed for the production of normal skin pigment (melanin). A zygote receiving such a pair of identical alleles is called **homozygous** for the genes involved, and in this case would develop into an individual with normal skin pigmentation.

In other instances, the members of a gene pair are different. One allele may contain information for the synthesis of the skin pigment enzyme, while its allele lacks information for enzyme production because of a mutation. When a zygote contains a gene pair whose alleles differ, it is said to be **heterozygous** for the genes involved.

---

1. Define *gene*.
2. What is the relationship between genes and chromosomes?
3. What is an *allele*?
4. What is meant by *homozygous?* By *heterozygous?*

Chart 24.1	Some traits determined by single pairs of dominant and recessive alleles
Phenotype due to expression of dominant gene	Phenotype due to expression of recessive gene
Full lips	Thin lips
Dark hair	Light hair
Free ear lobes	Attached ear lobes
Bridge of nose convex	Bridge of nose concave
Dark eyes (brown iris)	Light eyes (blue iris)
Farsightedness	Normal vision
Astigmatism	Normal vision
Extra fingers or toes (polydactyly)	Normal numbers of fingers and toes
Freckles	Lack of freckles
Ability to roll tongue into U-shape	Lack of this ability
Dimples in cheeks	Lack of dimples
Feet with normal arches	Flatfeet
Otosclerosis with hearing loss	Normal hearing

Gene Expression

The particular combination of genes present in a zygote and its subsequent daughter cells is said to constitute a **genotype.** The appearance of the individual that develops as a result of the ways the genes are expressed is termed the **phenotype.**

Dominant and Recessive Genes

A zygote that is *homozygous* for normal skin pigmentation will develop into an offspring with normally colored skin. Similarly, a zygote that is *heterozygous* for skin pigmentation—containing one normal gene and one mutant gene—will also develop into an offspring with normal skin pigment.

In this heterozygous condition, one gene (normal pigmentation) is expressed, but its allele (lack of pigmentation) is not. In such a case, the expressed gene is called **dominant** and the unexpressed gene is called **recessive.**

For convenience, dominant genes are symbolized with capital letters *(A),* and recessive genes are symbolized with small letters *(a).* Thus, the *genotype* of an individual who is homozygous for normal pigmentation due to a dominant gene is symbolized *AA;* a heterozygous pair is symbolized *Aa.* In both of these instances, the *phenotypes* of the individuals would be normal pigmentation, since a dominant gene is present in each genotype.

A zygote that receives two recessive genes for skin pigmentation has the genotype *aa* and will develop into an individual who lacks skin pigment. The phenotype of such an individual is called *albino* (figure 6.12).

Thus, three genotypes are possible in the instance of a gene pair involving dominant and recessive alleles. They are *homozygous dominant (AA), heterozygous (Aa),* and *homozygous recessive (aa).* However, only two phenotypes are possible, since the dominant gene is expressed whenever it is present—in both the homozygous dominant *(AA)* and the heterozygous *(Aa)* individuals. The recessive gene is expressed only in the homozygous recessive *(aa)* condition.

Chart 24.1 lists some traits due to dominant and recessive alleles.

Incomplete Dominance

Sometimes the alleles of a gene pair occur in two forms that are expressed differently, and neither is dominant to the other. In other words, when such genes are paired in a heterozygous individual, both are partially expressed. Genes of this type are said to blend or to be *incompletely dominant.*

For example, two forms of a gene needed for the synthesis of hemoglobin are *incompletely dominant.* A person with the genotype H^1H^1 develops normal hemoglobin, while one with the genotype H^2H^2 develops hemoglobin with abnormal molecules. The amino acid sequences in some of the polypeptide portions (beta chains) of these abnormal molecules differ from the normal sequence. More specifically, within the 146

Figure 24.4
Red blood cells from a person with sickle-cell anemia become elongated or sickled when the oxygen concentration is low.

amino acids in the normal sequence, the abnormal form has a valine in place of a glutamic acid, so that the abnormal form differs from the normal form in a single amino acid. As a consequence, the abnormal molecules tend to form long chains when they are exposed to low oxygen concentrations, and they cause the red blood cells containing them to become distorted or *sickle-shaped*. Consequently, a person with the genotype H^2H^2 is said to inherit *sickle-cell anemia* (figure 24.4).

A heterozygous individual, with the genotype H^1H^2, develops some normal and some abnormal hemoglobin. In this case, the red blood cells may exhibit sickling if oxygen concentrations are relatively low, but usually the red cells remain normal. This condition is called *sickle-cell trait*.

The gene for sickle-cell anemia occurs most commonly in persons with Central African ancestry. About 8% of the United States population with such ancestry possess this gene. As a result, about 1 child in 170 produced by this group develops sickle-cell anemia. This condition is characterized by symptoms associated with lack of tissue oxygen, resulting from the blockage of capillaries by sickled cells. These symptoms include severe joint and abdominal pain, skin ulcers, and chronic kidney disease.

Multiple Alleles

In the previous examples, only two forms of allelic genes were involved. In other instances, alleles occur in several forms, and such a group is called **multiple alleles.**

In multiple alleles, each person carries a single pair of genes, but since the different forms of the gene can occur in various combinations, several genotypes are possible. The inheritance of ABO *blood type,* for example, involves multiple alleles.

As is described in chapter 17, a person's ABO blood type is related to the *agglutinogens* present on the red blood cell membranes. Specifically, there are two agglutinogens involved: *agglutinogen A* and *agglutinogen B*. If the red cells have only agglutinogen A, the blood type is A; if only agglutinogen B is present, the blood type is B; if both agglutinogens A and B are present, the blood type is AB; and if neither agglutinogen A nor B is present, the blood type is O.

These blood types are determined by various combinations of three allelic genes that can be symbolized I^A, I^B, and i^O. In this instance, genes I^A and I^B are equally dominant and are said to be **codominant genes.** They are both dominant to gene i^O. Since each person inherits only two genes for ABO blood type, and there are three forms of the gene, six genotypes are possible. They are: I^AI^A, I^Ai^O, I^BI^B, I^Bi^O, I^AI^B, and i^Oi^O.

If gene I^A is present, agglutinogen A develops on the red blood cell membranes, and if gene I^B is present, agglutinogen B is produced. The presence of gene i^O causes either no agglutinogen to be produced or such a weak agglutinogen that it is usually insignificant. Thus, persons with genotypes I^AI^A or I^Ai^O have blood type A. Those with genotypes I^BI^B or I^Bi^O have blood type B. Genotype I^AI^B produces type AB blood, while genotype i^Oi^O results in type O blood.

Thus, as far as the ABO blood group is concerned, every person belongs to one of six genotypes and has one of four phenotypes.

Genotypes	Phenotypes
I^AI^A, I^Ai^O	blood type A
I^BI^B, I^Bi^O	blood type B
I^AI^B	blood type AB
i^Oi^O	blood type O

Blood and other tissue types are sometimes used as genetic evidence to prove whether a particular male is or is *not* the father of a particular child. In such cases, when the child's genotype includes a gene not included in the genotype of either the male or the mother, the male is clearly not the father.

1. Distinguish between genotype and phenotype.
2. What is meant by *dominant gene?* By *recessive gene?*
3. Define *incomplete dominance.*
4. Explain why there are six genotypes but only four phenotypes in the ABO blood group.

Figure 24.5
Variation in height is due to different combinations of several
pairs of genes.

Polygenic Inheritance

Although some traits—such as the presence or absence of skin pigment, the production of normal or abnormal hemoglobin, and the development of a certain blood type—seem to be determined by single pairs of genes, other traits are controlled by several pairs. Inheritance of this type is called *polygenic,* and examples of such inheritance include body stature and the concentration of skin pigment.

Height seems to be determined by at least four pairs of genes located in different positions within the chromosome set. Each pair causes a small effect in the development of height.

Using the symbol, *T,* for a gene causing tallness and, *t,* for a gene causing shortness, the genotype of a very tall person might be $T^1T^1\ T^2T^2\ T^3T^3\ T^4T^4$, and a very short person might be $t^1t^1\ t^2t^2\ t^3t^3\ t^4t^4$. Persons with various combinations of these genes for tallness and shortness, such as $T^1T^1\ T^2t^2\ T^3T^3\ t^4t^4$ or $T^1t^1\ T^2t^2\ t^3t^3\ t^4t^4$, would grow to intermediate heights. In fact, because many different combinations of these genes are possible, people exhibit considerable variation in height (figure 24.5).

Similarly, the amount of melanin produced in the skin seems to be determined by several pairs of genes, and as before, at least four pairs seem to be involved. Consequently, the genotype of a person with very dark skin might be symbolized $M^1M^1\ M^2M^2\ M^3M^3\ M^4M^4$, and one with very light skin might be $m^1m^1\ m^2m^2\ m^3m^3\ m^4m^4$. People with various combinations of genes for dark and light skin would develop skin of intermediate colors and would exhibit a broad range of variation (figure 24.6).

Environmental Factors

Although the traits of height and concentration of skin pigment are determined by several pairs of genes, the *development* of these and other hereditary traits is influenced by environmental factors as well. Whether or not an individual will achieve the stature that is hereditarily possible, for example, is affected by such environmental factors as diet and disease. Even if a person's genotype is for extreme tallness, the person may only grow to an intermediate height if the diet is inadequate in certain essential nutrients or if the body is affected by a disease that interferes with bone growth and development.

Figure 24.6

Variation in skin color is due in part to different combinations of several pairs of genes. What other factors influence skin color?

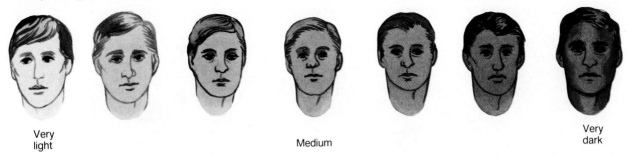

Very light   Medium   Very dark

Similarly, the amount of pigment deposited in the skin is influenced by environmental factors. If the skin is exposed to ultraviolet light from a sunlamp or from sunlight, the skin becomes more heavily pigmented. Conversely, if such exposure is avoided, the concentration of pigment is decreased (fig 24.7).

To summarize, the growth and development of a zygote is controlled by the paired genes it receives from an egg cell and a sperm cell. These genes encode genetic information for the synthesis of specific enzyme molecules, and the enzymes in turn promote the metabolic reactions that cause the development of such characteristics as sex, blood type, eye color, hair color, skin color, height, concentration of skin pigment, and so forth.

The way in which a particular gene is expressed depends on the nature of the gene and on the way in which it is combined with other genes. Gene expression also is influenced by environmental factors that may enhance the development of a trait or inhibit it.

Some genes, called *lethal genes,* cause abnormal development that ultimately results in the death of the individual who inherits them.

If, for example, a set of lethal genes causes abnormal heart development, an embryo might die about the fourth week, when its heart normally begins to function. If the lethal genes cause abnormal development of the kidneys, however, death might not occur until after birth, when the newborn must function without the aid of its mother's organs.

---

1. How does polygenic inheritance make many variations of a given trait possible?
2. How may environmental factors influence gene expression?

Figure 24.7

The gene responsible for freckles is more fully expressed when the skin is exposed to sunlight.

No sunlight   Exposed to sun

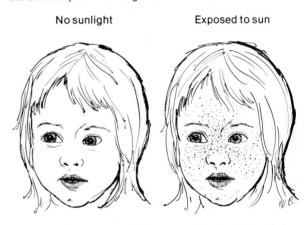

Chromosomes and Sex Cell Production

Although the offspring of a set of parents have many characteristics in common, they also differ because they originate from different egg and sperm cells and develop in different environments. Furthermore, different egg and sperm cells (from the same parents) produce zygotes with different genotypes, because the sex cells of each individual carry unique combinations of genes. This uniqueness results from the process of meiosis.

Meiosis

Meiosis occurs during *spermatogenesis* and *oogenesis,* and as a result of it, the chromosome numbers of the resulting sex cells are reduced by one-half (see chapter 22). This process involves two successive divisions: a

Figure 24.8
Spermatogenesis is a meiotic process that involves two successive divisions.

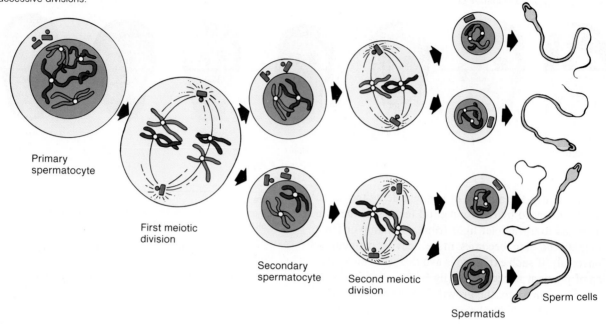

Primary spermatocyte

First meiotic division

Secondary spermatocyte

Second meiotic division

Spermatids

Sperm cells

Figure 24.9
Stages in the first meiotic division.

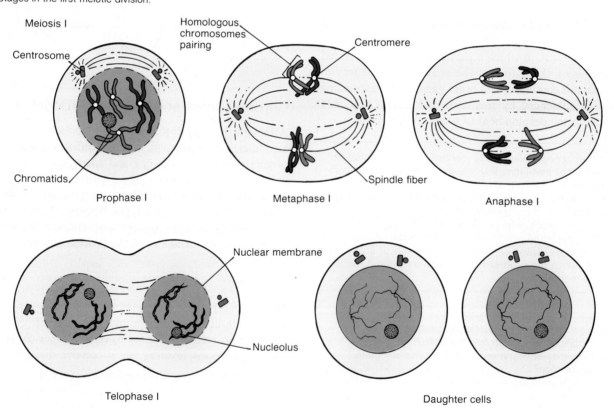

Meiosis I

Centrosome

Homologous chromosomes pairing

Centromere

Chromatids

Prophase I

Spindle fiber

Metaphase I

Anaphase I

Nuclear membrane

Nucleolus

Telophase I

Daughter cells

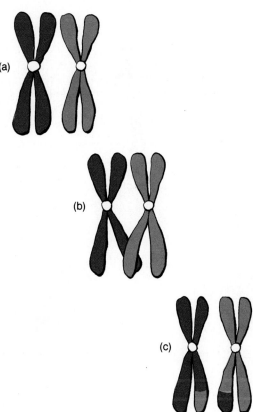

Szewczyk

first and second *meiotic division*. In the first meiotic division, the homologous chromosomes of the parent cell are separated, and the chromosome number is reduced in the daughter cells. During the second meiotic division, the chromosomes act much as they do in *mitosis*, and the result is the formation of sex cells (figure 24.8).

Like mitosis, described in chapter 3, meiosis is a continuous process without marked interruptions between steps. However, for convenience, the process can be divided into stages as follows:

1. **First meiotic prophase.** This stage (figure 24.9) resembles prophase in mitosis, although it lasts somewhat longer and is more complex. During it, the individual chromosomes appear as thin threads within the nucleus. The threads become shorter and thicker, the nucleoli disappear, the nuclear membrane fades away, and the spindle fibers become organized.

   Throughout this stage, the individual chromosomes, which became duplicated during the interphase prior to the beginning of meiosis, are composed of two identical chromatids. Each pair of chromatids is held together by a central region called the *centromere* (kinetochore).

As the prophase continues, homologous chromosomes approach each other, lie side by side, and become tightly intertwined. This process of pairing, shown in figure 24.10, is called *synapsis,* and during it the chromatids of the homologous chromosomes contact one another at various points. In fact, the chromatids may break in one or more places and exchange parts, forming chromatids with new combinations of genetic information. For example, since one chromosome of a homologous pair is of maternal origin and the other is of paternal origin, an exchange of parts between homologous chromatids results in chromatids that contain genetic information from both maternal and paternal sources. This process of exchanging genetic information is called **crossing over.**

2. **First meiotic metaphase.** During the first metaphase, shown in figure 24.9, the synaptic pairs of chromosomes become lined up about midway between the poles of the developing spindle. Each pair, consisting of two chromosomes and four chromatids, becomes associated with a spindle fiber.

Figure 24.11
Stages in the second meiotic division.

Meiosis II

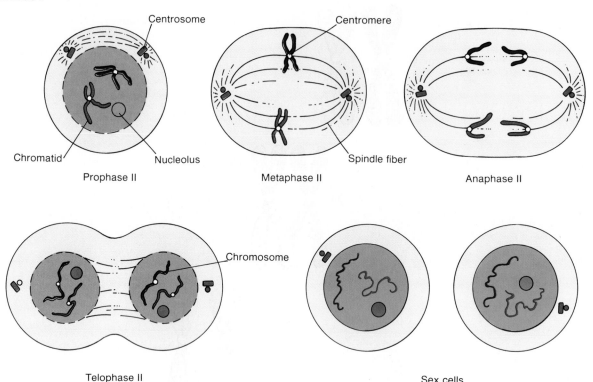

Prophase II   Metaphase II   Anaphase II

Centrosome   Centromere

Chromatid   Nucleolus   Spindle fiber

Chromosome

Telophase II   Sex cells

3. **First meiotic anaphase.** In mitosis, the centromeres that hold the chromatids of each chromosome pair together separate during anaphase, and the chromatids are free to move toward opposite ends of the spindle. However, during the first meiotic division the centromeres *do not* separate (figure 24.9), and as a result, the homologous chromosomes become separated, and the chromatids of each chromosome move together to one end of the spindle. Thus, each daughter cell receives only one member of a homologous pair of chromosomes, and the chromosome number is thereby reduced by one-half.

4. **First meiotic telophase.** Meiotic telophase is similar to mitotic telophase in that the parent cell divides into two daughter cells (figure 24.9). Also during this phase, nuclear membranes appear around the chromosome sets, the nucleoli reappear, and the spindle fibers fade away.

The first meiotic telophase is followed by a short *interphase* and the beginning of the second meiotic division. This division is essentially the same as mitosis, and it can also be divided into *prophase, metaphase, anaphase,* and *telophase.* These stages are shown in figure 24.11.

During the **second meiotic prophase,** the chromosomes reappear. They are still composed of pairs of chromatids, and the chromatids are held together by centromeres. Near the end of this phase, the chromosomes move into positions midway between the poles of the developing spindle.

In **second meiotic metaphase,** the double-stranded chromosomes become attached to spindle fibers, and during **second meiotic anaphase,** the centromeres separate so that the chromatids are free to move to opposite poles of the spindle. As a result of **second meiotic telophase,** the cell divides into two daughter cells, and new nuclei are organized around the sets of single-stranded chromosomes.

Figure 24.12

As a result of crossing over, the genetic information held within the sperm cells (and egg cells) varies from cell to cell.

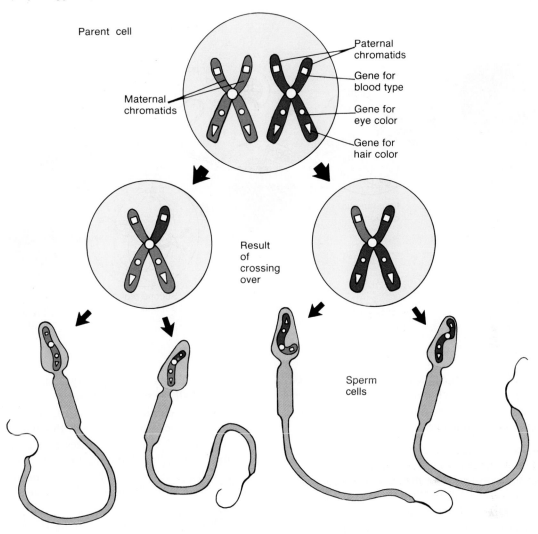

Parent cell

Paternal chromatids

Gene for blood type

Gene for eye color

Gene for hair color

Maternal chromatids

Result of crossing over

Sperm cells

Results of Meiosis

In spermatogenesis, meiosis results in the formation of *four sperm cells,* each of which contains 23 single-stranded chromosomes. These chromosomes contain a complete set of genetic information, but only one gene of each type, whereas the original parent cell (spermatogonium) contained a pair of each type of gene. Furthermore, the genetic information varies from sperm cell to sperm cell because of the crossing over that occurred during the first meiotic prophase. For example, as figure 24.12 shows, one sperm cell may contain genes for the development of eye color and hair color from an original maternal chromosome and a gene for blood type from an original paternal chromosome. Another sperm cell may contain a gene for maternal eye color and paternal hair color and blood type, while a third type contains genes for maternal hair color and blood type and a paternal gene for eye color.

In oogenesis, meiosis results in *one egg cell* (secondary oocyte) and a nonfunctional polar body. If the egg cell is fertilized, a second polar body is produced

Figure 24.13
In oogenesis, meiosis is completed following fertilization of the secondary oocyte.

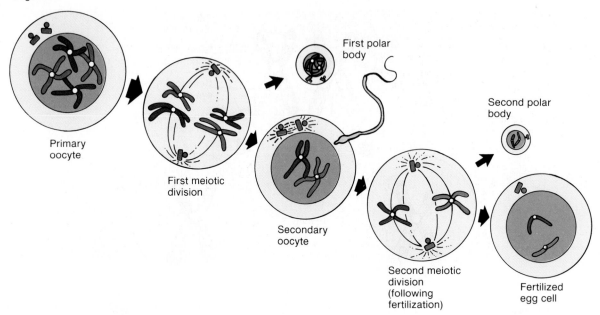

Oögenesis

Primary oocyte

First meiotic division

First polar body

Secondary oocyte

Second polar body

Second meiotic division (following fertilization)

Fertilized egg cell

(figure 24.13). As is true in sperm cells, egg cells contain 23 single-stranded chromosomes whose combinations of genetic information differ from egg cell to egg cell because of crossing over.

---

1. Distinguish between meiosis and mitosis.
2. Describe the major events that occur during meiosis.
3. What is the final result of meiosis?

Inheritance of Single Traits

Genes occur in pairs, but as a result of meiosis, a sex cell contains only one member of each gene pair. Consequently, some sex cells contain the original *maternal gene* of a gene pair, and the other sex cells contain the original *paternal gene* of such a pair.

Dominant and Recessive Inheritance

As was mentioned earlier, some genes occur in two forms—*dominant* and *recessive*. In pigmentation, for example, the dominant gene for the development of normal pigment can be symbolized as *A*, while its recessive allele can be symbolized as *a*. Thus, a person who is homozygous dominant for this trait has the genotype *AA* and will appear normal. A person who is heterozygous has the genotype *Aa* and also will appear normal, since a dominant gene for pigmentation is present. A person who is homozygous recessive, however, has the genotype *aa* and will be *albino*.

Since a person with genotype *AA* has two identical genes, all the sex cells produced by this individual will receive a dominant gene *(A)*. Likewise all of the sex cells produced by a person with genotype *aa* will receive a recessive gene *(a)*. However, a person with the genotype *Aa* will produce equal numbers of two kinds of sex cells. One kind will have a dominant gene *(A)*, and the other kind will have a recessive gene *(a)* (figure 24.14).

When information concerning the genotypes of parents is available, it often is possible to predict what types of offspring can be produced as well as the proportions in which they are likely to occur. For example, since a male parent with genotype *AA* produces only *A*-bearing sperm cells, and a female parent with genotype *aa* produces only *a*-bearing egg cells, all the zygotes resulting from the fusion of such sex cells will have the genotype *Aa:*

	Male Parent	Female Parent
Genotype:	*AA*	*aa*
Phenotype:	Normal pigment	Albino
Sex cells:	All *A*	All *a*

Egg cells

a

Possible zygotes:

Sperm cells

A | *Aa* |

Offspring genotypes: All *Aa*
Offspring phenotypes: All normal pigment

Figure 24.14

In the case of alleles *A* and *a* for skin pigmentation, there are two possible phenotypes and three possible genotypes. The sex cells produced by each person reflect the composition of his or her genotype.

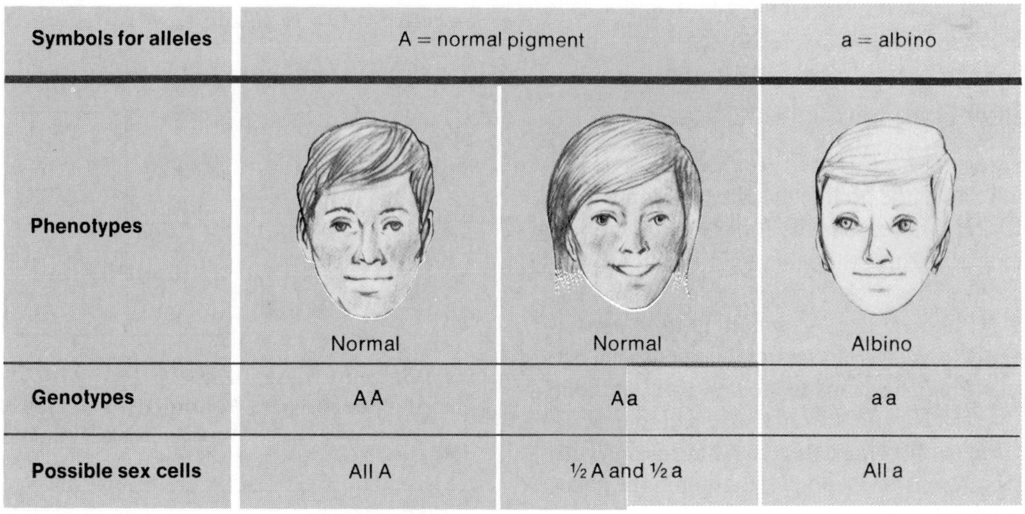

Symbols for alleles	A = normal pigment		a = albino
Phenotypes	Normal	Normal	Albino
Genotypes	A A	A a	a a
Possible sex cells	All A	½ A and ½ a	All a

If, on the other hand, the male parent has the genotype *Aa* and the female parent is *aa*, offspring of two possible genotypes can be produced—*Aa* and *aa*:

	Male Parent	Female Parent
Genotype:	*Aa*	*aa*
Phenotype:	Normal pigment	Albino
Sex cells:	½ *A* and ½ *a*	All *a*

Egg cells

a

		a
Possible zygotes:	½ *A*	½ *Aa*
	½ *a*	½ *aa*

(Sperm cells)

Offspring genotypes: ½ *Aa* and ½ *aa*
Offspring phenotypes: ½ normal pigment and ½ albino

In this case, ½ of the sperm cells carry the gene *A* and ½ carry the gene *a*. Since sperm cells have equal chances of fertilizing an egg cell, it is expected that ½ of the offspring from these parents will have the genotype *Aa* and ½ will have the genotype *aa*. Because the presence of a dominant gene *(A)* produces normal pigment, those offspring with genotype *Aa* will have normal pigmentation, and those with genotype *aa* will be albino.

A more complex situation arises if both the parents have the genotype *Aa*. In this instance, each parent produces two kinds of sex cells in equal numbers: ½ *A* and ½ *a*—and so the offspring resulting from such parents can have three possible genotypes—*AA*, *Aa*, or *aa*:

	Male Parent	Female Parent
Genotype:	*Aa*	*Aa*
Phenotype:	Normal pigment	Normal pigment
Sex cells:	½ *A* and ½ *a*	½ *A* and ½ *a*

Egg cells

		½ *A*	½ *a*
Possible zygotes:	½ *A*	¼ *AA*	¼ *Aa*
	½ *a*	¼ *Aa*	¼ *aa*

(Sperm cells)

Offspring genotypes: ¼ *AA*, ½ *Aa*, ¼ *aa*
Offspring phenotypes: ¾ normal pigment and ¼ albino

In this case, ¼ (25%) of the offspring are expected to have the genotype *AA*, ½ (50%) are expected to be *Aa*, and ¼ (25%) are expected to be *aa*. However, since those with genotypes *AA* or *Aa* will appear normal, only ¼ of the offspring from such parents are expected to be albino *(aa)*.

Since chance events, such as the fertilization of a particular kind of egg cell by a particular kind of sperm cell, occur independently, the genotype of one child has no effect upon the genotypes of other children produced by a set of parents. Thus each child conceived by parents with genotypes Aa and Aa has a 25% chance of being AA (normal), a 50% chance of being Aa (normal, but carrying the albino gene), and a 25% chance of being aa (albino).

Incompletely Dominant Inheritance

As was mentioned, there are two forms of a particular gene that function in the production of hemoglobin. One of the alleles (H^1) is responsible for the formation of normal hemoglobin, and the other (H^2) results in abnormal molecules that cause *sickling* of red blood cells. Since genes H^1 and H^2 are *incompletely dominant,* a person with genotype H^1H^1 has normal hemoglobin, one with genotype H^1H^2 develops sickle-cell trait, and one with genotype H^2H^2 suffers from sickle-cell anemia.

The kinds of offspring expected from parents with various combinations of incompletely dominant genes can be determined as follows:

1. Parent genotypes H^1H^1 and H^2H^2:

	Male Parent	Female Parent
Genotype:	H^1H^1	H^2H^2
Phenotype:	Normal	Sickle-cell anemia
Sex cells:	All H^1	All H^2

Egg cells
H^2

Possible zygotes: Sperm cells H^1 | H^1H^2

Offspring genotypes: All H^1H^2
Offspring phenotypes: All sickle-cell trait

2. Parent genotypes H^1H^2 and H^2H^2:

	Male Parent	Female Parent
Genotype:	H^1H^2	H^2H^2
Phenotype:	Sickle-cell trait	Sickle-cell anemia
Sex cells:	½ H^1 and ½ H^2	All H^2

Egg cells
H^2

Possible zygotes: Sperm cells   ½ H^1 | ½ H^1H^2
½ H^2 | ½ H^2H^2

Offspring genotypes: ½ H^1H^2 and ½ H^2H^2
Offspring phenotypes: ½ sickle-cell trait and ½ sickle-cell anemia

3. Parent genotypes H^1H^2 and H^1H^2:

	Male Parent	Female Parent
Genotype:	H^1H^2	H^1H^2
Phenotype:	Sickle-cell trait	Sickle-cell trait
Sex cells:	½ H^1 and ½ H^2	½ H^1 and ½ H^2

Egg cells
½ H^1 ½ H^2

Possible zygotes: Sperm cells   ½ H^1 | ¼ H^1H^1 | ¼ H^1H^2
½ H^2 | ¼ H^1H^2 | ¼ H^2H^2

Offspring genotypes: ¼ H^1H^1, ½ H^1H^2, ¼ H^2H^2
Offspring phenotypes: ¼ normal, ½ sickle-cell trait, ¼ sickle-cell anemia

---

1. What offspring genotypes can be produced by parents with the genotypes AA and aa? AA and Aa?
2. How many offspring genotypes can be produced by parents with genotypes Aa and Aa? How many offspring phenotypes?
3. If both parents have the sickle-cell trait, what will be the proportions of the genotypes and phenotypes of offspring that can be produced?

Sex Chromosomes

The sex of an individual is inherited in much the same manner as other traits. However, in this instance whole chromosomes are involved rather than a few paired genes.

Sex Inheritance

Within the human set of 46 chromosomes, there are 22 pairs called **autosomes** (which are chromosomes other than sex chromosomes) and one pair of **sex chromosomes.** These sex chromosomes are responsible for the development of the contrasting characteristics associated with *male* and *female.* In other words, the kinds of sex chromosomes present in a zygote determine whether the embryonic reproductive organs (which are essentially neutral in the early stages of development) will form into male or female structures.

Sex chromosomes are of two types. One type, which is relatively large, is called an **X chromosome,** and the other, which is relatively small, is called a **Y chromosome.** A female set of sex chromosomes consists

Figure 24.15

(a) A set of human male chromosomes and (b) a set of human female chromosomes arranged in pairs of homologous chromosomes. How are the sets different?

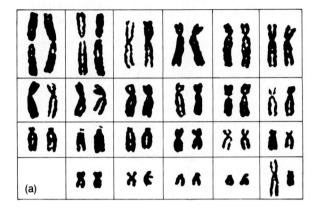

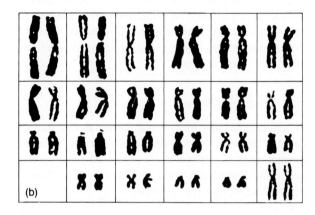

of two *X* chromosomes, and consequently all egg cells contain a single *X chromosome*. The male set consists of an *X* and a *Y* chromosome, and so equal numbers of two types of sperm cells are produced. One-half of the sperm cells possess an *X chromosome*, and ½ possess a *Y chromosome*. This difference is shown in figure 24.15.

The sex of an offspring is determined by the type of sperm cell that fertilizes the egg cell. If the sperm involved carries an *X* chromosome, it will be paired with the *X* chromosome of the egg cell, and the *XX* combination will result in the development of a female. If, on the other hand, a *Y*-bearing sperm cell fertilizes the egg, the *XY* combination results in the development of a male offspring. Since *X*-bearing and *Y*-bearing sperm cells occur in equal numbers, it is expected that equal numbers of males and females will result from the chance meeting of egg and sperm cells.

	Male Parent	Female Parent
Sex chromosomes:	*XY*	*XX*
Sex cells:	½ *X* and ½ *Y*	All *X*

Possible zygotes:

Egg cells

	X
½ *X*	½ *XX*
½ *Y*	½ *XY*

Sperm cells

Offspring sex: ½ males *(XY)* and ½ females *(XX)*

However, for reasons that are not clear, an unusually large proportion of males appear. In fact, in the United States population, the ratio of males to females at birth is 106:100.

Even though greater numbers of males than females are born, the death rate among males is higher. Consequently, the sex ratio between males and females changes as they age. This change in sex ratio is reflected in the following data:

Age	Sex Ratio (males:females)
birth	106:100
18 years	100:100
50 years	85:100
85 years	50:100
100 years	20:100

Actions of Sex Chromosomes

A zygote with an *XY* combination of sex chromosomes apparently gives rise to a male offspring because its *Y* chromosome contains information for the synthesis of a particular kind of protein molecule. This protein, which is called the *H-Y antigen,* appears on the surfaces of cells produced as the offspring develops. Its presence on the cell membranes of the undifferentiated reproductive organs somehow promotes their development into testes. At the same time, the antigen stimulates cells within the immature testes to proliferate and form spermatogonia. However, the spermatogonia are inhibited by action of the developing supporting cells (Sertoli's cells) of the testes, and sperm cell formation does not occur until puberty (see chapter 22).

Once the formation of the testes has been initiated, the function of the *Y* chromosome in determining the sex of the offspring seems to be completed. Further masculinization of the developing reproductive structures is controlled by hormones secreted by the testes as they form.

A zygote with an *XX* combination of sex chromosomes gives rise to a female offspring apparently without the aid of a substance corresponding to the H-Y antigen. Instead, in the presence of two *X* chromosomes, the embryonic reproductive organs enlarge for several weeks and then begin to differentiate into ovaries. Concurrently, several million primordial follicles are formed (see chapter 22).

Sex-linked Inheritance

In addition to their importance for sex determination, the *X* chromosomes contain genes responsible for the development of certain traits unrelated to sex. However, since these traits are determined by genes located on the *X* chromosomes, the patterns by which they are inherited differ in each sex. For example, such traits are more likely to be expressed in males *(XY)* than in females *(XX)*.

This phenomenon is related to the fact that the *X* and *Y* sex chromosomes are quite different in size, and most of the genes on the *X* chromosome do not have alleles on the relatively small *Y* chromosome. Consequently, in males the genes of the single *X* chromosome are expressed even though they may be recessive, while in females a trait determined by such a recessive gene is expressed only if it is present on both *X* chromosomes (homozygous recessive). Traits determined by recessive genes located on *X* chromosomes are called **sex linked.**

Some traits are due to genes located within a portion of the *Y* chromosome for which there is no homologous region in an *X* chromosome. These traits appear only in males and are passed from fathers to their sons. This type of genetic transmission is called *holandric inheritance. Hypertrichosis,* which is characterized by excessively hairy ears, is an example of such a trait.

Color blindness is an example of a sex-linked trait. In this instance, normal color vision is due to the presence of a dominant gene *(C)*. This gene is needed for the development of functional color receptors (cones) in the retina of the eye. The recessive allele *(c)* causes the production of defective receptors.

A male will have normal color vision if his *X* chromosome carries the dominant gene, which can be symbolized X^C. If, however, a male receives one recessive gene for color blindness (X^c), he will be color-blind, since there is no corresponding gene on the *Y* chromosome to overcome its effect.

A female, on the other hand, will have normal color vision if either of her *X* chromosomes carries the dominant gene (X^C). Thus, both the homozygous dominant $(X^C X^C)$ and the heterozygous female $(X^C X^c)$ will develop normal vision. Only a homozygous recessive female $(X^c X^c)$ will be color-blind.

A color-blind male only occurs when the female parent carries recessive genes for color blindness, since the *X* chromosome of a male offspring is always received from his mother:

	Normal Male Parent	Color-blind Female Parent
Sex chromosomes:	$X^C Y$	$X^c X^c$
Sex cells:	½ X^C and ½ Y	All X^c

Egg cells

X^c

Possible zygotes:

Sperm cells		Egg cells X^c
	½ X^C	½ $X^C X^c$
	½ Y	½ $X^c Y$

Offspring genotypes:

Females—$X^C X^c$

Males—$X^c Y$

Offspring phenotypes:

Females—normal

Males—color-blind

Although all the daughters in this case will have normal vision, they will carry the gene for color blindness.

In the case of a homozygous dominant female parent $(X^C X^C)$ and a male parent with color blindness $(X^c Y)$, all of the children will develop normal vision. However, all the daughters of a colorblind male parent will receive the recessive gene for color blindness, since all daughters receive the *X* chromosome of the father. These female offspring will be heterozygous $(X^C X^c)$ and will be carriers of the gene for color blindness.

	Color-blind Male Parent	Normal Female Parent
Sex chromosomes:	X^cY	X^CX^C
Sex cells:	½ X^c and ½ Y	All X^C

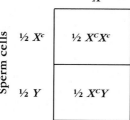

Possible zygotes:

	Egg cells
	X^C
½ X^c	½ X^CX^c
½ Y	½ X^CY

(Sperm cells)

Offspring genotypes:

        Females—X^CX^c
        Males—X^CY

Offspring phenotypes:

        Females—normal
        Males—normal

If the female parent has normal vision but carries the gene for color blindness (X^CX^c), and the male parent has normal vision (X^CY), all the female offspring will have normal vision, but ½ of them are expected to be carriers of the recessive gene. At the same time, ½ of the male offspring are expected to have normal vision (X^CY) and ½ are expected to be color-blind (X^cY).

	Normal Male Parent	Normal Female Parent
Genotypes:	X^CY	X^CX^c
Sex cells:	½ X^C and ½ Y	½ X^C and ½ X^c

Possible zygotes:

	Egg cells	
	½ X^C	½ X^c
½ X^C	¼ X^CX^C	¼ X^CX^c
½ Y	¼ X^CY	¼ X^cY

(Sperm cells)

Offspring genotypes:

        Females—½ X^CX^C, ½ X^CX^c
        Males—½ X^CY, ½ X^cY

Offspring phenotypes:

        Females—normal
        Males—½ normal, ½ color-blind

Chart 24.2	Some traits related to sex-linked recessive genes
Trait	**Characteristics**
Brown enamel	Tooth enamel appears brown rather than white
Coloboma iridis	A fissure in the iris of the eye causes the pupil to appear slitlike rather than round
Congenital night blindness	Vision is poor in dim light, but is relatively normal in bright light; however, this condition is not due to a deficiency of vitamin A
Microphthalmia	Eyes fail to develop normally and instead remain small and nonfunctional
Optic atrophy	Blindness results from degeneration of the optic nerves

Hemophilia, sometimes called the bleeder's disease, also is due to a sex-linked recessive gene. In this instance, the presence of a single recessive gene in the male (X^hY) or of two recessive genes in a female (X^hX^h) causes a defect in the blood-clotting mechanism. As a result, prolonged bleeding may accompany an otherwise minor injury.

Like color blindness, hemophilia is more common in males. The recessive gene is received by male offspring from their mothers and is passed from affected fathers to their daughters.

Actually there are two major forms of hemophilia caused by recessive genes carried on the X chromosomes. *Hemophilia A* is characterized by the lack of plasma protein called *antihemophilic globulin* (blood clotting factor VIII). *Hemophilia B* (Christmas disease) is due to a deficiency of another blood substance called *plasma thromboplastin component* (blood clotting factor IX or Christmas factor).

Chart 24.2 lists some other traits related to sex-linked genes.

An inherited disease called *pseudohemophilia* (von Willebrand's disease) is caused by the presence of a dominant gene on an autosome. In this condition, the platelets are defective and bleeding is prolonged. However, since the responsible gene is on an autosome, this disease is not sex linked.

Figure 24.16
As a result of nondisjunction, some of the sex cells formed have too many chromosomes, and others have too few.

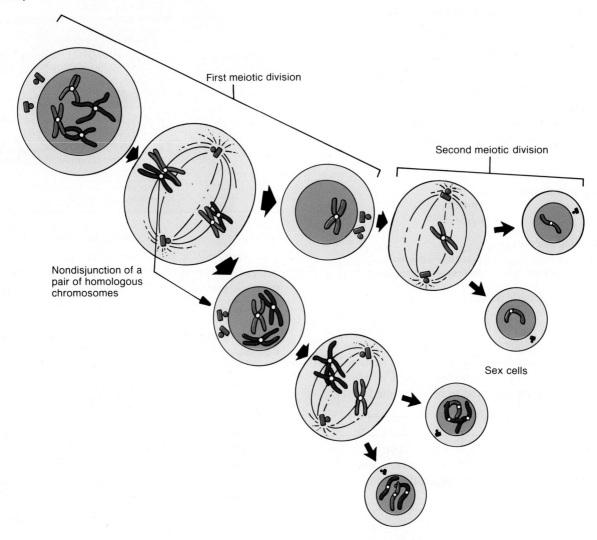

First meiotic division

Second meiotic division

Nondisjunction of a
pair of homologous
chromosomes

Sex cells

---

1. How is the sex of an offspring determined?
2. How do sex chromosomes act to promote differentiation of reproductive organs?
3. What is meant by sex-linked inheritance?
4. Why do sex-linked traits appear most commonly in males?

Baldness is a hereditary trait that is due to the expression of *sex-influenced genes*. Such genes act differently in males than in females; that is, they act as dominant genes in the hormonal environment of an adult male and as recessive genes in the hormonal environment of an adult female. Consequently, baldness occurs more frequently in males than in females.

Chromosome Disorders

Chromosome disorders are characterized by the presence of abnormal numbers of chromosomes. Since the human set usually contains 46 chromosomes, any other number is considered to be abnormal.

Cause of Chromosome Disorders

As was described previously, during meiosis, homologous chromosomes pair and undergo synapsis. Later these homologous pairs separate and move to opposite ends of the spindle. As a result, each daughter cell receives one chromosome of a homologous pair.

If, however, one pair of homologous chromosomes fails to separate during meiosis, both members of that pair will move into one daughter cell, while the other daughter cell will receive no chromosome of that type. Such a failure of homologous chromosomes to separate is called **nondisjunction.** When it occurs, some of the resulting cells have too many chromosomes in their sets, while others have too few (figure 24.16).

Sex Chromosome Disorders

If nondisjunction occurs during *oogenesis,* for example, some of the resulting egg cells may receive two *X* chromosomes *(XX),* while others receive none (figure 24.17). Both of these types of egg cells can be fertilized by sperm cells, and the zygotes formed can undergo development.

Apparently only one *X* chromosome of an *XX* set is functional, and in many female cells the other, nonfunctional *X* chromosome is found lying against the nuclear membrane, where it forms a tiny structure called a *Barr body.*

In some hospitals, cells of the amnion of each newborn are stained and observed for the presence of Barr bodies. In this way, it is possible to determine if there are abnormal numbers of *X* chromosomes in the cells.

If an *XX* egg cell is fertilized by a *Y*-bearing sperm cell, the resulting zygote will have 47 chromosomes and the genotype *XXY.* This abnormal combination of sex chromosomes causes the development of an individual with male sex organs. However, the testes usually fail to produce sperm cells, so that the person is sterile. This condition, which is called *Klinefelter's syndrome,* is also characterized by tall stature, a somewhat feminine musculature, and partial development of the breasts (figure 24.18).

An *XX* egg cell fertilized by an X-bearing sperm cell becomes a zygote with the genotype *XXX.* The individual who develops from such a zygote is a relatively normal female, and although some *XXX* females are sterile, others have produced offspring. This condition is sometimes called the *super female.*

Females with genotypes *XXXX* and *XXXXX* are known to exist, and although a consistent pattern of traits is lacking in such individuals, the chances of developing mental retardation seem to increase markedly as the number of *X* chromosomes in the genotype increases.

If an egg cell without an *X* chromosome is fertilized by an *X*-bearing sperm cell, the resulting *XO* zygote develops into an individual with *Turner's syndrome* (figure 24.19). In this instance, the person is female, but usually fails to mature reproductively. The sex organs remain juvenile, the breasts fail to develop, the stature is short, and loose folds of skin appear in the posterior region of the neck.

Figure 24.17
An egg cell containing two X chromosomes can result from nondisjunction.

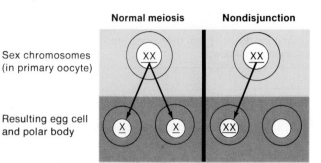

Figure 24.18
A person with Klinefelter's syndrome has the sex chromosome genotype XXY. How might such a genotype come about?

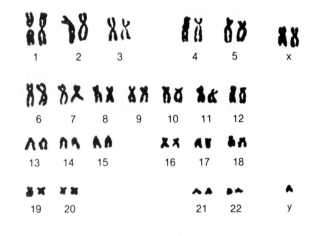

Figure 24.19
What is abnormal in this chromosome set obtained from a person with Turner's syndrome?

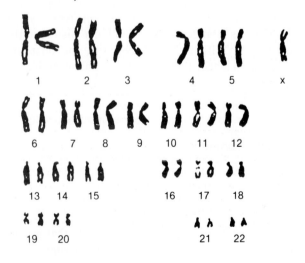

As they age, older persons seem to develop increasing numbers of body cells with 45 chromosomes. In males, this abnormal number seems to involve a lack of *Y* chromosomes, and in females, cells tend to miss an *X* chromosome. These losses apparently result from abnormal mitotic divisions, but their significance to the aging process is unknown.

Autosomal Abnormalities

Nondisjunction sometimes occurs among homologous pairs of autosomes, and as a result, a cell may receive an extra autosome. For example, if nondisjunction occurs during oogenesis, an egg cell may appear with two chromosomes of the same type. If such an egg cell is fertilized by a normal sperm cell, the zygote will have three chromosomes of one kind—a condition that is called *trisomy*.

In humans, *Down's syndrome* results from trisomy in which a zygote receives three autosomes of a particular type (chromosome 21) (figure 24.20). This condition is characterized by mental retardation, short stature, and stubby hands and feet. In addition, the reproductive organs usually remain underdeveloped, and malformations of the heart may be present.

Although the cause of nondisjunction leading to *Down's syndrome* is unknown, this condition is more likely to occur in the offspring of older mothers than in those of younger mothers. In fact, Down's syndrome occurs only about once in two-thousand live births among women of early childbearing age, but it occurs about once in fifty live births among women over 40 years of age.

Other chromosome abnormalities resulting from triply represented autosomes include *Edward's syndrome* (trisomy 18) in which the fetus develops heart and kidney defects, and *Patau's syndrome* (trisomy 13) in which the infant is usually blind and has heart abnormalities. With either abnormality, the offspring seldom lives longer than a few weeks following birth.

1. What is meant by nondisjunction?
2. What are some sex chromosome disorders due to nondisjunction?
3. Define *trisomy*.
4. What are the consequences of trisomy?

Medical Genetics

Medical genetics is the branch of genetics that is concerned with the relationship between inheritance and disease.

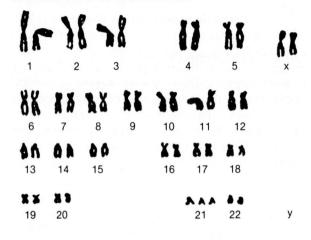

Figure 24.20
A person with Down's syndrome has three members of chromosome 21 in the chromosome set.

Heredity and Disease

Hereditary diseases comprise a diverse group of disorders including abnormalities of blood cells (such as sickle-cell anemia), defects in blood clotting mechanisms (such as hemophilia), and mental retardation (such as Down's syndrome). The causes of such diseases are *mutations*—changes in the genes or chromosomes that can be passed from parent to offspring by means of sex cells (see chapter 4). Although some of these diseases (such as hemophilia) are due to mutations of single genes, others are caused by chromosomal disorders, as in the trisomy leading to Down's syndrome.

Genetic Counseling

Genetic counseling is a service by which people who have produced children with hereditary diseases or who are members of families in which such disorders occur can obtain information about the probability of genetic disorders being transmitted to their offspring. Sometimes a genetic counselor can make rather precise predictions. For example, a counselor could tell parents who both have sickle-cell trait that their chances of having a child with sickle-cell anemia are one in four or 25%. However, if one parent is normal for hemoglobin and the other parent has sickle-cell anemia, the chance of their producing a child with sickle-cell anemia is zero, although all of their children will inherit sickle-cell trait.

The accuracy of such a prediction depends on a correct diagnosis of the genetic disorder, understanding of the pattern by which the disorder is inherited, and knowledge of the parents' genotypes. Unfortunately, some hereditary traits behave differently in different families. A trait may, for example, act as an autosomal recessive in one family and as a sex-linked recessive in another. Consequently, a genetic counselor may have to determine how a particular trait is transmitted within a family before making a prediction.

In addition, the counselor must be concerned with the psychological effects such a prediction may have upon the individuals involved, and must help them weigh the probability of a genetic defect occurring against the probability of desirable traits that might appear in their child—traits that could compensate to some degree for the possible defect. At the same time, the counselor must try to remain objective and avoid enforcing a particular set of values upon the prospective parents, since the final decision about childbearing is theirs.

---

1. What is medical genetics?
2. What is the purpose of genetic counseling?
3. What factors will affect the accuracy of predictions made by genetic counselors?

Some Hereditary Disorders of Clinical Interest

cri du chat syndrome (kre do chat sin′ drom) characterized by peculiar mewing (cat cry) vocal sounds, microcephaly, and severe mental retardation; due to a partial deletion in one autosome (probably number 5).

cystic fibrosis (sis′tik fi-bro′sis) characterized by formation of thick mucus in the pancreas and lungs, which interferes with normal digestion and breathing; an expression of autosomal recessive genes.

familial cretinism (fah-mil′e-al kre′ti-nizm) characterized by lack of thyroid secretions due to a defect in the iodine transport mechanism. Untreated children are dwarfed, sterile, and usually mentally retarded; an expression of autosomal recessive genes.

galactosemia (gah-lak″to-se′me-ah) characterized by an inability to metabolize galactose, a component of milk sugar. This results in cataract, mental retardation, and damaged liver; an expression of autosomal recessive genes.

gout (gowt) characterized by an accumulation of uric acid in the blood and tissues due to abnormal metabolism of substances called purines; an expression of an autosomal dominant gene.

hepatic porphyria (hĕ-pat′ik por-fe′re-ah) characterized by abdominal pain, gastrointestinal disorders, and neurologic disturbances due to abnormal metabolism of substances called porphyrins; an expression of an autosomal dominant gene.

hepatolenticular degeneration (Wilson's disease) (hep″ah-to-len-tik′u-lar de-jen″ĕ-ra′shun) characterized by an increase in the absorption of copper and the accumulation of copper in the brain and liver. This results in degenerative changes in the brain and cirrhosis of the liver; an expression of autosomal recessive genes.

hereditary hemochromatosis (he-red′i-ter″e he″mo-kro″mah-to′sis) characterized by an accumulation of iron in the liver, pancreas, and heart. This results in cirrhosis of the liver, diabetes, and heart failure; an expression of a sex-influenced autosomal dominant gene.

hereditary leukomelanopathy (he-red′i-ter″e lu″ko-mel″ah-nop′ah-the) characterized by decreased pigmentation in the skin, eyes, and hair, and abnormalities of the white blood cells that are accompanied by increased susceptibility to infections and early death; an expression of autosomal recessive genes.

Huntington's chorea (hunt′ing-tunz ko-re′ah) characterized by uncontrolled twitching of voluntary muscles and deterioration of mental capacities. This condition usually does not appear until after maturity, so affected persons often transmit the mutant gene to their children before the symptoms develop; an expression of an autosomal dominant gene.

Marfan's syndrome (mar-fahnz′ sin′drom) characterized by extremities of abnormally great lengths, dislocation of the lenses, and congenital defects of the cardiovascular system; an expression of an autosomal dominant gene.

nephrogenic diabetes insipidus (nef″ro-jen′ik di″ah-be′tēz in-sip′i-dus) characterized by production of large volumes of very dilute urine and consequent dehydration due to failure of the distal convoluted tubules to reabsorb water in response to ADH; an expression of a sex-linked recessive gene.

phenylketonuria (PKU) (fen″il-ke″to-nu′re-ah) characterized by an inability to normally metabolize the amino acid, phenylalanine. This results in nerve and brain damage and is accompanied by mental retardation; an expression of autosomal recessive genes.

pseudohypertrophic muscular dystrophy (su″do-hi″per-trof′ik mus′ku-lar dis′tro-fe) characterized by progressive atrophy of the muscles, which usually begins during childhood and leads to death in adolescence; an expression of a sex-linked recessive gene.

retinitis pigmentosa (ret″i-ni′tis pig″men-to′sa) characterized by progressive atrophy of the retina which causes blindness; an expression of a sex-linked recessive gene.

Tay-Sach's disease (ta saks′ di-zēz′) characterized by early blindness, deterioration of mental and physical abilities, and death; an expression of autosomal recessive genes.

thalassemia major (Cooley's anemia) (thal″ah-se′me-ah mā′jĕr) characterized by very thin and fragile red blood cells resulting in anemia and a short life expectancy; an expression of autosomal recessive genes.

Detection of Genetic Disorders
A Practical Application

Sometimes it is possible to determine the genotypes of particular individuals by performing blood tests. *Sickle-cell trait,* for example, can be detected by exposing red blood cells to low oxygen concentrations. In the test, a drop of blood is mixed with a substance that deoxygenates it, and if the red cells contain abnormal hemoglobin, they assume a sickle shape.

Similarly, blood tests can be used to reveal carriers (heterozygotes) of the recessive genes involved in the development of the following: *hemophilia,* in which bleeding is prolonged; *Tay-Sach's disease*, which leads to early blindness, deterioration of mental and physical abilities, and death; and *Cooley's anemia* (thalassemia major), which is accompanied by severe defects in the red blood cells and invariably leads to death.

If prospective parents are concerned about the genetic condition of a developing fetus (prenatal diagnosis), information can sometimes be gained by testing fetal cells, obtained by means of *amniocentesis*. In this procedure, which can be performed in a physician's office with the aid of sonography (figure 24.21), a small quantity of *amniotic fluid* is withdrawn from the amniotic cavity by passing a hollow needle through the maternal abdominal and uterine walls. When this is done between the fourteenth and sixteenth weeks of development, cells of fetal origin (fibroblasts) are usually present in the fluid, and they can be cultured in the laboratory and tested for abnormalities. (See figure 24.22.) The amniotic fluid obtained by amniocentesis also can be analyzed for the presence of a specific protein (alpha-fetal protein or AFP) that is used to detect serious defects in the developing nervous system, such as spinal bifida or anencephaly.

Another technique sometimes used to test fetal cells is called *chorionic villi biopsy*. This procedure employs sonography to guide a catheter into a pregnant woman's uterus and uses suction to obtain a sample of the hairlike projections (villi) from the embryonic membrane (chorion) surrounding a young embryo. The cells of the sample can then be analyzed for chromosomal defects. Such a biopsy must be performed between 8 and 10 weeks of pregnancy and has the advantage of providing enough tissue to examine immediately without culturing cells. The procedure, however, seems to present a greater risk to the life of the embryo than does amniocentesis.

Fetal cells also can be tested for various hemoglobin disorders, including thalassemia and sickle-cell anemia. Even though fetal fibroblasts obtained by amniocentesis do not synthesize hemoglobin, they contain a complete set of human DNA molecules. These molecules can be removed from the cells, and certain portions of their gene structures can be analyzed. This technique utilizes special enzymes (restriction enzymes) that break the DNA nucleotide sequences within the DNA molecules at particular points, producing precise patterns of fragments. The nature of these fragments reveals the presence of normal or abnormal genes for hemoglobin synthesis.

Figure 24.21
A fetus within a uterus as revealed by ultrasonography.

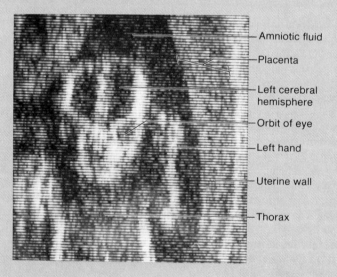

Amniotic fluid

Placenta

Left cerebral hemisphere

Orbit of eye

Left hand

Uterine wall

Thorax

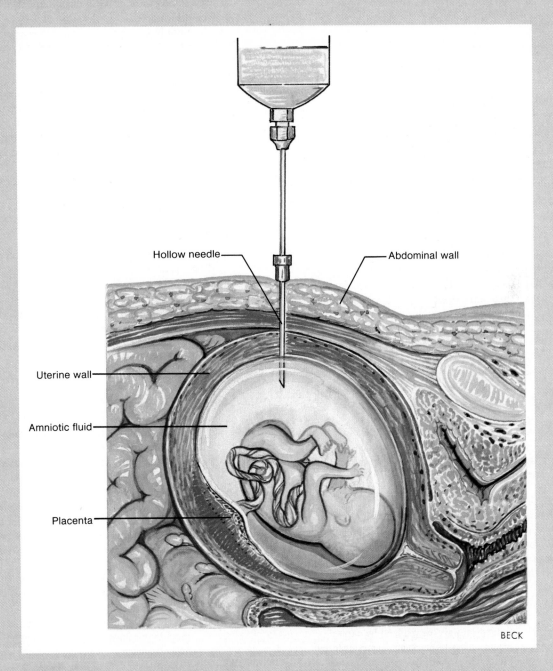

Hollow needle

Abdominal wall

Uterine wall

Amniotic fluid

Placenta

BECK

Chapter Summary

Introduction
Individual characteristics result from an interaction of heredity and environment. Heredity is the transmission of genetic information from parents to offspring; environment includes chemical, physical, and biological factors in the surroundings.

Genes and Chromosomes
1. The gene
   a. A gene is a portion of a DNA molecule that contains information for the production of one kind of protein molecule.
   b. Genes within a zygote instruct it to synthesize particular proteins, which function as enzymes that promote metabolic reactions needed for growth and development.
2. Chromosome numbers
   a. Human body cells normally contain 46 chromosomes.
   b. Twenty-three chromosomes of a zygote are received from the female parent and 23 are received from the male parent.
3. Chromosome pairs
   a. Chromosomes from male and female parents contain complete sets of genetic information.
   b. Within these sets, chromosomes that contain similar genetic information comprise homologous chromosomes.
   c. During mitosis, daughter cells receive identical sets of homologous chromosomes.
4. Gene pairs
   a. Since chromosomes are paired, the genes within the chromosomes are paired also.
   b. A pair of genes consists of two alleles.
   c. A zygote that contains a pair of identical alleles is homozygous; if the alleles of a pair are different, the zygote is heterozygous.

Gene Expression
The combination of genes present in an individual's cells constitutes a genotype; the appearance of the individual is its phenotype.
1. Dominant and recessive genes
   a. In the heterozygous condition, if one gene is expressed and its allele is unexpressed, the expressed gene is called dominant and the allele is called recessive.
   b. Dominant genes are symbolized by capital letters and recessive genes are symbolized by small letters.
   c. Three genotypes are possible in the instance of a gene pair involving dominant and recessive alleles, but only two phenotypes are possible.
2. Incomplete dominance
   a. When the alleles of a gene pair are expressed differently and neither is dominant to the other, they are called incompletely dominant.
   b. Incompletely dominant alleles result in three possible genotypes and three different phenotypes.
3. Multiple alleles
   a. When alleles occur in several forms, the group of genes is called multiple alleles.
   b. The inheritance of ABO blood types involves multiple alleles.
   c. In ABO blood type inheritance, every person belongs to one of six genotypes and has one of four phenotypes.
4. Polygenic inheritance
   a. Some traits are determined by single gene pairs, while others are determined by several pairs of genes.
   b. Body stature and concentration of skin pigment are examples of traits determined by polygenic inheritance.
   c. Polygenic inheritance results in a broad range of phenotypes, so that people vary greatly in their heights and the concentration of pigment in their skins.
5. Environmental factors
   a. Environmental factors influence the ways genes are expressed.
   b. Body stature is influenced by genes and also by such environmental factors as diet and disease.
   c. Concentration of skin pigment is influenced by genes and also by exposure to ultraviolet light.

Chromosomes and Sex Cell Production
Sex cells carry unique combinations of genes as a result of meiosis.

1. Meiosis
   a. Meiosis involves two successive divisions.
      (1) In first meiotic division, homologous chromosomes of the parent cell are separated and the chromosome number is reduced by ½.
      (2) In second meiotic division, the chromosomes act as they do in mitosis.
      (3) Meiosis results in the formation of sex cells.
   b. Meiosis is a continuous process without marked interruptions between steps, but for convenience it can be divided into stages.
      (1) Stages of the first meiotic division include prophase, metaphase, anaphase, and telophase.
      (2) Stages of second meiotic division also include prophase, metaphase, anaphase, and telophase.
2. Results of meiosis
   a. In spermatogenesis, four sperm cells are produced.
      (1) Each sperm cell contains 23 chromosomes.
      (2) The genetic information within the sperm cells varies from cell to cell as a result of crossing over.

b. In oogenesis, one egg cell (secondary oocyte) and a nonfunctional polar body are produced.
  (1) Each egg cell contains 23 chromosomes.
  (2) The genetic information within the egg cells varies from cell to cell as a result of crossing over.

Inheritance of Single Traits
1. Dominant and recessive inheritance
   a. In pigmentation, the gene for normal pigment is symbolized as *A*, and its recessive allele is symbolized as *a*.
     (1) *AA* is homozygous dominant. *Aa* is heterozygous, and *aa* is homozygous recessive.
     (2) *AA* and *Aa* individuals appear normal, while *aa* individuals are albino.
   b. When the genotypes of parents are known, it is possible to predict the types of offspring that they can produce.
2. Incompletely dominant inheritance
   a. Three genotypes and three phenotypes can result from a pair of incompletely dominant genes.
   b. When the genotypes of the parents are known, it is possible to predict the types of offspring that they can produce.

Sex Chromosomes
1. Sex inheritance
   a. A set of human chromosomes consists of 22 pairs of autosomes and one pair of sex chromosomes.
   b. The sex chromosomes are responsible for the development of characteristics associated with male and female.
     (1) A male set of sex chromosomes consists of an *X* and a *Y* chromosome.
     (2) A female set of sex chromosomes consists of two *X* chromosomes.
   c. The sex of an offspring is determined by the type of sperm cell that fertilizes the egg cell.
2. Actions of sex chromosomes
   a. The *Y* chromosome of an *XY* combination causes the synthesis of the H-Y antigen.
   b. The H-Y antigen promotes the differentiation of embryonic reproductive organs into testes and stimulates the formation of spermatogonia.
   c. The immature testes secrete hormones that control further masculinization of the reproductive organs.
   d. Embryonic reproductive organs resulting from an *XX*-bearing zygote differentiate into ovaries without the aid of a substance corresponding to the H-Y antigen.

3. Sex-linked inheritance
   a. Sex chromosomes contain genes responsible for certain traits that are unrelated to sex.
   b. The patterns in which these traits are inherited differ in each sex.
     (1) In males, genes on the single *X* chromosome are expressed even if they are recessive.
     (2) In females, such recessive genes are expressed only if they are present on both *X* chromosomes.
   c. Traits determined by recessive genes located on *X* chromosomes are called sex-linked.
     (1) Color blindness and hemophilia are examples of sex-linked traits.
     (2) Males receive sex-linked traits from their mothers.
     (3) All the daughters of a male parent expressing a sex-linked trait will receive the recessive gene for that trait.

Chromosome Disorders
Disorders are characterized by the presence of abnormal numbers of chromosomes.

1. Cause of chromosome disorders
   a. If homologous chromosomes fail to separate during meiosis, one daughter cell will receive two chromosomes of one type while the other daughter cell will receive no chromosome of that type.
   b. Such a failure of homologous chromosomes to separate is called nondisjunction.
2. Sex chromosome disorders
   a. Nondisjunction during oogenesis can result in egg cells with two *X* chromosomes and egg cells with no *X* chromosome.
   b. Such abnormal egg cells can be fertilized and the resulting zygotes may develop into offspring.
     (1) If an *XX* egg cell is fertilized by a *Y*-bearing sperm cell, the resulting *XXY* combination leads to Klinefelter's syndrome.
     (2) If an *XX* egg cell is fertilized by an *X*-bearing sperm cell, the resulting *XXX* combination produces a relatively normal female.
     (3) If an egg cell without an *X* chromosome is fertilized by an *X*-bearing sperm cell, the resulting *XO* combination leads to Turner's syndrome.
3. Autosomal abnormalities
   a. Nondisjunction may occur among autosomes and result in the presence of two chromosomes of the same type within an egg cell.
   b. If such an egg cell is fertilized by a normal sperm cell, the resulting zygote will have three chromosomes of one kind, called trisomy.
   c. Down's syndrome, Edward's syndrome, and Patau's syndrome are due to triply represented autosomes.

Medical Genetics

1. Heredity and disease
   a. Hereditary diseases include a diverse group of disorders.
   b. Such diseases are caused by mutations.
   c. Parents often seek advice concerning the chances of a hereditary disease affecting their future children.
2. Genetic counseling
   a. A service by which individuals can obtain information about the chances of genetic disorders being transmitted to their offspring is called genetic counseling.
   b. Genetic counselors can sometimes make precise predictions concerning the probability of a future child being affected by an inherited disease.
   c. The accuracy of such predictions depends upon the following:
      (1) Correct diagnosis of the genetic disorder;
      (2) Understanding of the pattern by which the disorder is inherited;
      (3) Knowledge of the parents' genotypes.

Application of Knowledge

1. Using the principles of human genetics, how could you support the stand taken by many civil and religious authorities forbidding marriages between persons who are first or second cousins?
2. If a young couple with sickle-cell trait who have a child with sickle-cell anemia said to you, "Our first child has sickle-cell anemia, but now we can have three more children before we have another one with this disease," how would you respond?
3. If a 49-year-old woman, who thought she was menopausal and thus failed to seek medical attention when she began missing her menstrual periods, finds that she is five months pregnant, what special kinds of tests are likely to be ordered by her physician? Why?

Review Activities

1. Identify two major factors that influence the development of individual characteristics.
2. Define a *gene*.
3. Discuss the origin of the 46 chromosomes in a human zygote.
4. Define *homologous chromosomes*.

5. Distinguish between homozygous and heterozygous.
6. Distinguish between genotype and phenotype.
7. Explain what is meant by a dominant gene and its recessive allele.
8. Define *incomplete dominance*.
9. Explain how ABO blood type inheritance is controlled by multiple alleles.
10. Describe how environmental factors may influence the expression of the genes that control pigmentation of the skin.
11. Outline the process of meiosis.
12. Explain the significance of synapsis during meiosis.
13. Describe the genotypes and phenotypes of the offspring expected from the following parents. (A represents the dominant gene for normal pigmentation, and a represents its recessive allele.)
    a. AA and Aa
    b. Aa and aa
    c. Aa and Aa
14. Describe the genotypes and phenotypes of the offspring expected from the following parents. (H^1 represents the normal gene for hemoglobin production, and H^2 represents the incompletely dominant allele that causes the formation of abnormal hemoglobin.)
    a. H^1H^1 and H^2H^2
    b. H^1H^1 and H^1H^2
    c. H^1H^2 and H^1H^2
15. Distinguish between autosomes and sex chromosomes.
16. Explain how sex is inherited.
17. Explain why ½ of the human zygotes are expected to develop into males and ½ into females.
18. Explain how the presence of a Y chromosome leads to the differentiation of embryonic reproductive organs into testes.
19. Define *sex-linked genes*.
20. Explain why sex-linked genes are always expressed more frequently in males than in females.
21. Describe the genotypes and phenotypes of the offspring expected from the following parents:
    a. Color-blind male and normal (homozygous dominant) female.
    b. Normal male and color-blind female.
22. Distinguish between hemophilia A and hemophilia B.
23. Explain how nondisjunction may lead to chromosome disorders.
24. Explain how an individual with a sex chromosome combination of XXY might occur.
25. Explain how an individual with a sex chromosome combination of XO might occur.
26. Define *autosomal trisomy*.

Appendixes

Appendix A
Tables

International Atomic Weights

Element	Symbol	Atomic Number	Atomic Weight
Aluminum	Al	13	26.97
Antimony	Sb	51	121.76
Arsenic	As	33	74.91
Barium	Ba	56	137.36
Beryllium	Be	4	9.013
Bismuth	Bi	83	209.00
Boron	B	5	10.82
Bromine	Br	35	79.916
Cadmium	Cd	48	112.41
Calcium	Ca	20	40.08
Carbon	C	6	12.010
Chlorine	Cl	17	35.457
Chromium	Cr	24	52.01
Cobalt	Co	27	58.94
Copper	Cu	29	63.54
Fluorine	F	9	19.00
Gold	Au	79	197.2
Hydrogen	H	1	1.0080
Iodine	I	53	126.92
Iron	Fe	26	55.85
Lead	Pb	82	207.21
Magnesium	Mg	12	24.32
Manganese	Mn	25	54.93
Mercury	Hg	80	200.61
Nickel	Ni	28	58.69
Nitrogen	N	7	14.008
Oxygen	O	8	16.0000
Palladium	Pd	46	106.7
Phosphorus	P	15	30.98
Platinum	Pt	78	195.23
Potassium	K	19	39.096
Radium	Ra	88	226.05
Selenium	Se	34	78.96
Silicon	Si	14	28.06
Silver	Ag	47	107.880
Sodium	Na	11	22.997
Strontium	Sr	38	87.63
Sulfur	S	16	32.066
Tin	Sn	50	118.70
Titanium	Ti	22	47.90
Tungsten	W	74	183.92
Uranium	U	92	238.07
Vanadium	V	23	50.95
Zinc	Zn	30	65.38
Zirconium	Zr	40	91.22

Periodic Table of Elements

1a																	0
1 H 1.008	IIa											IIIa	IVa	Va	VIa	VIIa	2 He 4.00
3 Li 6.94	4 Be 9.01											5 B 10.81	6 C 12.01	7 N 14.00	8 O 15.99	9 F 18.99	10 Ne 20.18
11 Na 22.99	12 Mg 24.31	IIIb	IVb	Vb	VIb	VIIb	⌣ VIIIb ⌣			IB	IIIB	13 Al 26.98	14 Si 28.09	15 P 30.97	16 S 32.06	17 Cl 35.45	18 Ar 39.95
19 K 39.10	20 Ca 40.08	21 Sc 44.96	22 Ti 47.90	23 V 50.94	24 Cr 51.99	25 Mn 54.94	26 Fe 55.85	27 Co 58.93	28 Ni 58.71	29 Cu 63.54	30 Zn 65.37	31 Ga 69.72	32 Ge 72.59	33 As 74.92	34 Se 78.96	35 Br 79.91	36 Kr 83.80
37 Rb 85.47	38 Sr 87.62	39 Y 88.91	40 Zr 91.22	41 Nb 92.91	42 Mo 95.94	43 Tc (99)	44 Ru 101.97	45 Rh 102.91	46 Pd 106.4	47 Ag 107.87	48 Cd 112.40	49 In 114.82	50 Sn 118.69	51 Sb 121.75	52 Te 127.60	53 I 126.90	54 Xe 131.30
55 Cs 132.91	56 Ba 137.34	57–71 see below	72 Hf 178.49	73 Ta 180.95	74 W 183.85	75 Re 186.2	76 Os 190.2	77 Ir 192.2	78 Pt 195.09	79 Au 196.97	80 Hg 200.59	81 Tl 204.37	82 Pb 207.19	83 Bi 208.98	84 Po (210)	85 At (210)	86 Rn (222)
87 Fr (223)	88 Ra (226)	89–103 see below	104 Rf (261)	105 Ha (260)	106 · 263	*newly produced											

*newly produced

57 La 138.91	58 Ce 140.12	59 Pr 140.91	60 Nd 144.24	61 Pm (147)	62 Sm 150.35	63 Eu 151.96	64 Gd 157.25	65 Tb 158.92	66 Dy 162.50	67 Ho 164.93	68 Er 167.26	69 Tm 168.93	70 Yb 173.04	71 Lu 174.97
89 Ac (227)	90 Th 232.04	91 Pa (231)	92 U 238.03	93 Np (237)	94 Pu (242)	95 Am (243)	96 Cm (247)	97 Bk (247)	98 Cf (251)	99 Es (254)	100 Fm (253)	101 Md (256)	102 No (254)	**103 Lr (257)**

Values in parentheses are approximate.

Key

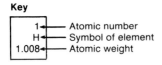

1 — Atomic number
H — Symbol of element
1.008 — Atomic weight

Appendix B
Units of Measurement and
Their Equivalents

Apothecaries' Weights and Their Metric Equivalents

1 grain (gr) =
0.05 scruple (s)
0.017 dram (dr)
0.002 ounce (oz)
0.0002 pound (lb)
0.065 gram (g)
65. milligrams (mg)

1 scruple (s) =
20. grains (gr)
0.33 dram (dr)
0.042 ounce (oz)
0.004 pound (lb)
1.3 grams (g)
1,300. milligrams (mg)

1 dram (dr) =
60. grains (gr)
3. scruples (s)
0.13 ounce (oz)
0.010 pound (lb)
3.9 grams (g)
3,900. milligrams (mg)

1 ounce (oz) =
480. grains (gr)
24. scruples (s)
8. drams (dr)
0.08 pound (lb)
31.1 grams (g)
31,100. milligrams (mg)

1 pound (lb) =
5,760. grains (gr)
288. scruples (s)
96. drams (dr)
12. ounces (oz)
373. grams (g)
373,000. milligrams (mg)

Apothecaries' Volumes and Their Metric Equivalents

1 minim (min) =
0.017 fluid dram (fl dr)
0.002 fluid ounce (fl oz)
0.0001 pint (pt)
0.06 milliliter (ml)
0.06 cubic centimeter (cc)

1 fluid dram (fl dr) =
60. minims (min)
0.13 fluid ounce (fl oz)
0.008 pint (pt)
3.70 milliliters (ml)
3.70 cubic centimeters (cc)

1 fluid ounce (fl oz) =
480. minims (min)
8. fluid drams (fl dr)
0.06 pint (pt)
29.6 milliliters (ml)
29.6 cubic centimeters (cc)

1 pint (pt) =
7,680. minims (min)
128. fluid drams (fl dr)
16. fluid ounces (fl oz)
473. milliliters (ml)
473. cubic centimeters (cc)

Metric Weights and Their Apothecaries' Equivalents

1 gram (g) =
0.001 kilogram (kg)
1,000. milligrams (mg)
1,000,000. micrograms (μg)
15.4 grains (gr)
0.032 ounce (oz)

1 kilogram (kg) =
1,000. grams (g)
1,000,000. milligrams (mg)
1,000,000,000. micrograms (μg)
32. ounces (oz)
2.7 pounds (lb)

1 milligram (mg) =
0.000001 kilogram (kg)
0.001 gram (g)
1,000. micrograms (μg)
0.0154 grains (gr)
0.000032 ounce (oz)

Metric Volumes and Their Apothecaries' Equivalents

1 liter (l) =
1,000. milliliters (ml)
1,000. cubic centimeters (cc)
2.1 pints (pt)
270. fluid drams (fl dr)
34. fluid ounces (fl oz)

1 milliliter (ml) =
0.001 liter (l)
1. cubic centimeter (cc)
16.2 minims (min)
0.27 fluid dram (fl dr)
0.034 fluid ounce (fl oz)

Approximate Equivalents of Household Measures

1 teaspoon (tsp) =
4. milliliters (ml)
4. cubic centimeters (cc)
1. fluid dram (fl dr)

1 tablespoon (tbsp) =
15. milliliters (ml)
15. cubic centimeters (cc)
0.5 fluid ounce (fl oz)
3.7 teaspoons (tsp)

1 cup (c) =
240. milliliters (ml)
240. cubic centimeters (cc)
8. fluid ounces (fl oz)
0.5 pint (pt)
16. tablespoons (tbsp)

1 quart (qt) =
960. milliliters (ml)
960. cubic centimeters (cc)
2. pints (pt)
4. cups (c)
32. fluid ounces (fl oz)

Conversion of Units from One Form to Another

Refer to the preceding equivalency lists when converting one unit to another equivalent unit.

To convert a unit shown in bold type to one of the equivalent units listed immediately below it, multiply the first number (bold type unit) by the appropriate equivalent unit listed below it.

Sample Problems:

1. Convert 320 grains into scruples (1 gr = 0.05 s).

$$320 \text{ gr} \times \frac{0.05 \text{ s}}{1 \text{ gr}} = 16.0 \text{ s}$$

2. Convert 320 grains into drams (1 gr = 0.017 dr).

$$320 \text{ gr} \times \frac{0.017 \text{ dr}}{1 \text{ gr}} = 5.44 \text{ dr}$$

3. Convert 320 grains into grams (1 gr = 0.065 g).

$$320 \text{ gr} \times \frac{0.065 \text{ g}}{1 \text{ gr}} = 20.8 \text{ g}$$

Body Temperatures in °Fahrenheit and °Celsius

°F	°C	°F	°C
95.0	35.0	100.0	37.8
95.2	35.1	100.2	37.9
95.4	35.2	100.4	38.0
95.6	35.3	100.6	38.1
95.8	35.4	100.8	38.2
96.0	35.5	101.0	38.3
96.2	35.7	101.2	38.4
96.4	35.8	101.4	38.6
96.6	35.9	101.6	38.7
96.8	36.0	101.8	38.8
97.0	36.1	102.0	38.9
97.2	36.2	102.2	39.0
97.4	36.3	102.4	39.1
97.6	36.4	102.6	39.2
97.8	36.6	102.8	39.3
98.0	36.7	103.0	39.4
98.2	36.8	103.2	39.6
98.4	36.9	103.4	39.7
98.6	37.0	103.6	39.8
98.8	37.1	103.8	39.9
99.0	37.2	104.0	40.0
99.2	37.3	104.2	40.1
99.4	37.4	104.4	40.2
99.6	37.6	104.6	40.3
99.8	37.7	104.8	40.4
		105.0	40.6

To convert °F to °C
Subtract 32 from °F and multiply by 5/9.
—— °F − 32 × 5/9 = —— °C

To convert °C to °F
Multiply °C by 9/5 and add 32.
—— °C × 9/5 + 32 = —— °F

Appendix C
Some Laboratory Tests of Clinical Importance

Common tests performed on blood

Test	Normal values* (adult)	Clinical significance
Acetone and acetoacetate (serum)	0.3–2.0 mg/100 ml	Values increase in diabetic acidosis, toxemia of pregnancy, fasting, and high-fat diet.
Albumin (serum)	3.2–5.5 gm/100 ml	Values increase in multiple myeloma and decrease with proteinuria and as a result of severe burns.
Albumin-globulin ratio or A/G ratio (serum)	1.5:1 to 2.5:1	Ratio of albumin to globulin is lowered in kidney diseases and malnutrition.
Ammonia (plasma)	50–170 μg/100 ml	Values increase in severe liver disease, pneumonia, shock, and congestive heart failure.
Amylase (serum)	80–160 Somogyi units/100 ml	Values increase in acute pancreatitis, intestinal obstructions, and mumps. They decrease in chronic pancreatitis, cirrhosis of the liver, and toxemia of pregnancy.
Bilirubin, total (serum)	0.3–1.1 mg/100 ml	Values increase in conditions causing red blood cell destruction or biliary obstruction.
Blood urea nitrogen or BUN (plasma or serum)	10–20 mg/100 ml	Values increase in various kidney disorders and decrease in liver failure and during pregnancy.
Calcium (serum)	8.5–10.5 mg/100 ml	Values increase in hyperparathyroidism, hypervitaminosis D, and respiratory conditions that cause a rise in CO_2 concentration. They decrease in hypoparathyroidism, malnutrition, and severe diarrhea.
Carbon dioxide (serum)	24–30 mEq/l	Values increase in respiratory diseases, intestinal obstruction, and vomiting. They decrease in acidosis, nephritis, and diarrhea.
Chloride (serum)	100–106 mEq/l	Values increase in nephritis, Cushing's syndrome, and hyperventilation. They decrease in diabetic acidosis, Addison's disease, diarrhea, and following severe burns.
Cholesterol, total (serum)	120–220 mg/100 ml	Values increase in diabetes mellitus and hypothyroidism. They decrease in pernicious anemia, hyperthyroidism, and acute infections.
Creatine (serum)	0.2–0.8 mg/100 ml	Values increase in muscular dystrophy, nephritis, severe damage to muscle tissue, and during pregnancy.
Creatine phosphokinase or CPK (serum)	Men: 0–20 IU/l Women: 0–14 IU/l	Values increase in myocardial infarction and skeletal muscle diseases such as muscular dystrophy.
Creatinine (serum)	0.6–1.5 mg/100 ml	Values increase in various kidney diseases.
Erythrocyte count or red cell count (whole blood)	Men: 4,600,000–6,200,000/cu mm Women: 4,200,000–5,400,000/cu mm Children: 4,500,000–5,100,000/cu mm (varies with age)	Values increase as a result of severe dehydration or diarrhea and decrease in anemia, leukemia, and following severe hemorrhage.

*These values may vary with hospital, physician, and type of equipment used to make measurements.

Common tests performed on blood—*continued*

Test	Normal values* (adult)	Clinical significance
Fatty acids, total (serum)	190–420 mg/100 ml	Values increase in diabetes mellitus, anemia, kidney disease, and hypothyroidism. They decrease in hyperthyroidism.
Ferritin (serum)	Men: 10–270 µg/100 ml Women: 5–280 µg/100 ml	Values correlate with total body iron store. They decrease with iron deficiency.
Globulin (serum)	2.5–3.5 gm/100 ml	Values increase as a result of chronic infections.
Glucose (plasma)	70–110 mg/100 ml	Values increase in diabetes mellitus, liver diseases, nephritis, hyperthyroidism, and pregnancy. They decrease in hyperinsulinism, hypothyroidism, and Addison's disease.
Hematocrit (whole blood)	Men: 40–54 ml/100 ml Women: 37–47 ml/100 ml Children: 35–49 ml/100 ml (varies with age)	Values increase in polycythemia due to dehydration or shock. They decrease in anemia and following severe hemorrhage.
Hemoglobin (whole blood)	Men: 14–18 gm/100 ml Women: 12–16 gm/100 ml Children: 11.2–16.5 gm/100 ml (varies with age)	Values increase in polycythemia, obstructive pulmonary diseases, congestive heart failure, and at high altitudes. They decrease in anemia, pregnancy, and as a result of severe hemorrhage or excessive fluid intake.
Iron (serum)	50–175 µg/100 ml	Values increase in various anemias and liver disease. They decrease in iron deficiency anemia.
Iron-binding capacity (serum)	250–410 µg/100 ml	Values increase in iron deficiency anemia and pregnancy. They decrease in pernicious anemia, liver disease, and chronic infections.
Lactic acid (whole blood)	6–18 mg/100 ml	Values increase with muscular activity and in congestive heart failure, severe hemorrhage, and shock.
Lactic dehydrogenase or LDH (serum)	90–200 milliunits/ml	Values increase in pernicious anemia, myocardial infarction, liver diseases, acute leukemia, and widespread carcinoma.
Lipids, total (serum)	450–850 mg/100 ml	Values increase in hypothyroidism, diabetes mellitus, and nephritis. They decrease in hyperthyroidism.
Oxygen saturation (whole blood)	Arterial: 94–100% Venous: 60–85%	Values increase in polycythemia and decrease in anemia and obstructive pulmonary diseases.
pH (whole blood)	7.35–7.45	Values increase due to vomiting, Cushing's syndrome, and hyperventilation. They decrease as a result of hypoventilation, severe diarrhea, Addison's disease, and diabetic acidosis.
Phosphatase, acid (serum)	1.0–5.0 King-Armstrong units/ml	Values increase in cancer of the prostate gland, hyperparathyroidism, certain liver diseases, myocardial infarction, and pulmonary embolism.
Phosphatase, alkaline (serum)	5–13 King-Armstrong units/ml	Values increase in hyperparathyroidism (and in other conditions that promote resorption of bone), liver diseases, and pregnancy.

Test	Normal values* (adult)	Clinical significance
Phospholipids (serum)	6–12 mg/100 ml as lipid phosphorus	Values increase in diabetes mellitus and nephritis.
Phosphorus (serum)	3.0–4.5 mg/100 ml	Values increase in kidney diseases, hypoparathyroidism, acromegaly, and hypervitaminosis D. They decrease in hyperparathyroidism.
Platelet count (whole blood)	150,000–350,000/cu mm	Values increase in polycythemia and certain anemias. They decrease in acute leukemia and aplastic anemia.
Potassium (serum)	3.5–5.0 mEq/l	Values increase in Addison's disease, hypoventilation, and conditions that cause severe cellular destruction. They decrease in diarrhea, vomiting, diabetic acidosis, and chronic kidney disease.
Protein-bound iodine or PBI (serum)	3.5–8.0 μg/100 ml	Values increase in hyperthyroidism and liver disease. They decrease in hypothyroidism.
Protein, total (serum)	6.0–8.4 gm/100 ml	Values increase in severe dehydration and shock. They decrease in severe malnutrition and hemorrhage.
Prothrombin time (serum)	12–14 sec (one stage)	Values increase in certain hemorrhagic diseases, liver disease, vitamin K deficiency, and following the use of various drugs.
Sedimentation rate, Westergren (whole blood)	Men: 0–15 mm/hr Women: 0–20 mm/hr	Values increase in infectious diseases, menstruation, pregnancy, and as a result of severe tissue damage.
Sodium (serum)	136–145 mEq/l	Values increase in nephritis and severe dehydration. They decrease in Addison's disease, myxedema, kidney disease, and diarrhea.
Thromboplastin time, partial (plasma)	35–45 sec	Values increase in deficiencies of blood factors VIII, IX, and X.
Thyroxine or T$_4$ (serum)	2.9–6.4 μg/100 ml	Values increase in hyperthyroidism and pregnancy. They decrease in hypothyroidism.
Transaminases or SGOT (serum)	5–40 units/ml	Values increase in myocardial infarction, liver disease, and diseases of skeletal muscles.
Uric acid (serum)	Men: 2.5–8.0 mg/100 ml Women: 1.5–6.0 mg/100 ml	Values increase in gout, leukemia, pneumonia, toxemia of pregnancy, and as a result of severe tissue damage.
White blood cell count, differential (whole blood)	Neutrophils 54–62% Eosinophils 1–3% Basophils 0–1% Lymphocytes 25–33% Monocytes 3–7%	Neutrophils increase in bacterial diseases; lymphocytes and monocytes increase in viral diseases; eosinophils increase in collagen diseases, allergies, and in the presence of intestinal parasites.
White blood cell count, total (whole blood)	5,000–10,000/cu mm	Values increase in acute infections, acute leukemia, and following menstruation. They decrease in aplastic anemia and as a result of drug toxicity.

Common tests performed on urine

Test	Normal values	Clinical significance
Acetone and acetoacetate	0	Values increase in diabetic acidosis.
Albumin, qualitative	0 to trace	Values increase in kidney disease, hypertension, and heart failure.
Ammonia	20–70 mEq/l	Values increase in diabetes mellitus and liver diseases.
Bacterial count	Under 10,000/ml	Values increase in urinary tract infection.
Bile and bilirubin	0	Values increase in melanoma and biliary tract obstruction.
Calcium	Under 300 mg/24 hr	Values increase in hyperparathyroidism and decrease in hypoparathyroidism.
Creatinine	1–2 gm/24 hr	Values increase in infections and decrease in muscular atrophy, anemia, leukemia, and kidney diseases.
Creatinine clearance	100–140 ml/min	Values increase in renal diseases.
Glucose	0	Values increase in diabetes mellitus and various pituitary gland disorders.
17-hydroxycorticosteroids	2–10 mg/24 hr	Values increase in Cushing's syndrome and decrease in Addison's disease.
Phenylpyruvic acid	0	Values increase in phenylketonuria.
Urea	25–35 gm/24 hr	Values increase as a result of excessive protein breakdown. They decrease as a result of impaired renal function.
Urea clearance	Over 40 ml blood cleared of urea/min	Values increase in renal diseases.
Uric acid	0.6–1.0 gm/24 hr as urate	Values increase in gout and decrease in various kidney diseases.
Urobilinogen	0–4 mg/24 hr	Values increase in liver diseases and hemolytic anemia. They decrease in complete biliary obstruction and severe diarrhea.

Suggestions for Additional Reading

Chapter 1 An Introduction to Human Anatomy and Physiology

Basmajian, J. V. 1980. *Grant's method of anatomy*. 10th ed. Baltimore: Williams and Wilkins.

Clemente, C. D. 1981. *Anatomy: a regional atlas of the human body*. 2d ed. Philadelphia: Lea and Febiger.

Curtis, H. 1983. *Biology*. 4th ed. New York: Worth.

Enger, E. D., et al. 1982. *Concepts in biology*. 3d ed. Dubuque, Ia.: Wm. C. Brown.

Hamilton, W. J., ed. 1976. *Textbook of human anatomy*. 2d ed. St. Louis: C. V. Mosby.

Jaffe, C. C. November–December 1982. Medical imaging. *American Scientist*.

Johnson, L. G. 1983. *Biology*. Dubuque, Ia.: Wm. C. Brown.

Lyons, A. S., and Petrucelli, R. J. 1978. *Medicine: an illustrated history*. New York: Harry N. Abrams.

Majno, G. 1975. *The healing hand: man and wound in the ancient world*. Cambridge, Mass: Harvard Univ. Press.

Mason, W. H., and Marshall, N. L. 1983. *The human side of biology*. New York: Harper and Row.

McMinn, R. M. H., and Hutchings, R. T. 1977. *Color atlas of human anatomy*. Chicago: Year Book Medical Publishers.

Pykett, I. L. May 1982. NMR imaging in medicine. *Scientific American*.

Sherman, I. W., and Sherman, V. 1983. *Biology: a human approach*. 3d ed. New York: Oxford Univ. Press.

Singer, C. A., and Underwood, E. A. 1962. *A short history of medicine*. New York: Oxford Univ. Press.

Snell, R. S. 1978. *Atlas of clinical anatomy*. Boston: Little, Brown.

Wilson, D. B., and Wilson, J. W. 1983. *Human anatomy*. 2d ed. New York: Oxford Univ. Press.

Woodburne, R. T. 1983. *Essentials of human anatomy*. 7th ed. New York: Oxford Univ. Press.

Chapter 2 The Chemical Basis of Life

Armstrong, F. B. 1983. *Biochemistry*. 2d ed. New York: Oxford Univ. Press.

Baker, J. J., and Allen, G. E. 1981. *Matter, energy, and life*. 4th ed. Palo Alto: Addison-Wesley.

Beebe, G. W. January–February 1982. Ionizing radiation and health. *American Scientist*.

Borek, E. 1981. *The atoms within us*. 2d ed. New York: Columbia Univ. Press.

Doolittle, R. F. October 1985. Proteins. *Scientific American*.

Gerard, G., and Rossi, D. R. July 1984. Nuclear magnetic resonance imaging of the brain. *Hospital Practice*.

Holum, J. R. 1983. *Elements of general and biological chemistry*. 6th ed. New York: Wiley.

Lehninger, A. 1982. *Principles of biochemistry*. New York: Worth.

Orten, J. M., and Neuhaus, O. W. 1982. *Human biochemistry*. 10th ed. St. Louis: C. V. Mosby.

Phelps, M. E., and Mazziotta, J. C. 1985. Positron emission tomography. *Science* 228:4701.

Porterfield, W. W. 1984. *Inorganic chemistry: a unified approach*. Reading, Ma.: Addison-Wesley.

Sackheim, G. I. 1974. *Introduction to chemistry for biology students*. Chicago: Educational Methods.

Upton, A. C. February 1982. The biological effects of low-level ionizing radiation. *Scientific American*.

Chapter 3 The Cell

Avers, C. J. 1982. *Cell biology*. 2d ed. New York: D. Van Nostrand Co.

Bretscher, M. S. October 1985. The molecules of the cell membrane. *Scientific American*.

Corbett, T. H. 1977. *Cancer and chemicals*. Chicago: Nelson-Hall.

Dautry-Varsat, A., and Lodish, H. May 1984. How receptors bring proteins and particles into cells. *Scientific American.*

DeDuve, C. May 1983. Microbodies in the living cell. *Scientific American.*

DeWitt, W., and Brown, E. R. 1977. *Biology of the cell.* Philadelphia: W. B. Saunders.

Dustin, P. August 1980. Microtubules. *Scientific American.*

Grivell, L. A. March 1983. Mitochondrial DNA. *Scientific American.*

Kornberg, R. D., and Klug, A. February 1981. The nucleosome. *Scientific American.*

Lake, J. A. August 1981. The ribosome. *Scientific American.*

Lodish, H. F., and Rothman, J. E. January 1979. The assembly of cell membranes. *Scientific American.*

Lyon, J. L. July 1984. Radiation exposure and cancer. *Hospital Practice.*

Reif, A. E. July–August 1981. The causes of cancer. *American Scientist.*

Rothman, J. E. 1981. The golgi apparatus. *Science* 213:1212.

Sloboda, R. D. May–June 1980. The role of microtubules in cell structure and cell division. *American Scientist.*

Staehelin, L. A., and Hull, B. E. May 1978. Junctions between living cells. *Scientific American.*

Unwin, N., and Henderson, R. February 1984. The structure of proteins in biological membranes. *Scientific American.*

Weber, K., and Osborn, M. October 1985. The molecules of the cell matrix. *Scientific American.*

Wolfe, S. L. 1981. *Biology of the cell.* 2d ed. Belmont, Calif.: Wadsworth.

Chapter 4 Cellular Metabolism

Baker, J. J., and Allen, G. E. 1981. *Matter, energy and life.* 4th ed. Palo Alto: Addison-Wesley.

Darnell, J. E. October 1983. The processing of RNA. *Scientific American.*

Devlin, T. M., ed. 1982. *Textbook of biochemistry with clinical correlations.* New York: Wiley.

Eyre, D. R. 1980. Collagen: molecular diversity in the body's protein scaffold. *Science* 207:1315.

Felsenfeld, G. October 1985. DNA. *Scientific American.*

Ganong, W. F. 1981. *Review of medical physiology.* 10th ed. Los Altos, Calif.: Lange Medical Publications.

Goldsby, R. A. 1977. *Cells and energy.* 3d ed. New York: Macmillan.

Guyton, A. C. 1986. *Textbook of medical physiology.* 7th ed. Philadelphia: W. B. Saunders.

Hinkle, P. C. March 1978. How cells make ATP. *Scientific American.*

Howard-Flanders, P. November 1981. Inducible repair of DNA. *Scientific American.*

Kazazian, H. H. February 1985. The nature of mutation. *Hospital Practice.*

Mirsky, A. E. June 1968. The discovery of DNA. *Scientific American.*

Nomura, M. January 1984. The control of ribosome synthesis. *Scientific American.*

Pederson, T. January–February 1981. Messenger RNA biosynthesis and nuclear structure. *American Scientist.*

Rich, A., and Kim, S. H. January 1978. The three-dimensional structure of transfer RNA. *Scientific American.*

Stryer, L. 1981. *Biochemistry.* 2d ed. San Francisco: W. H. Freeman.

Watson, J. D. 1968. *The double helix.* New York: Antheneum.

Weinberg, R. A. October 1985. The molecules of life. *Scientific American.*

Chapter 5 Tissues

Bevelander, G., and Ramaley, J. A. 1979. *Essentials of histology.* 8th ed. St. Louis: C. V. Mosby.

Caplan, A. I. October 1984. Cartilage. *Scientific American.*

Copenhaver, W. M., et. al. 1978. *Bailey's textbook of histology.* 17th ed. Baltimore, Md.: Williams and Wilkins.

DiFiore, M. S. 1981. *An atlas of human histology.* 5th ed. Philadelphia: Lea and Febiger.

Eyre, D. R. 1980. Collagen: molecular diversity in the body's protein scaffold. *Science* 207:1315.

Fawcett, D. W. 1986. *A textbook of histology.* 11th ed. Philadelphia: W. B. Saunders.

Leeson, T. S., and Leeson, C. R. 1981. *Histology.* 4th ed. Philadelphia: W. B. Saunders.

Ross, M. H., and Reith, E. J. 1977. *Histology: a text and atlas.* New York: Harper and Row.

Weiss, L., and Greep, R. O. 1983. *Histology.* 5th ed. New York: McGraw-Hill.

Chapter 6 The Skin and the Integumentary System

Elden, H. R., ed. 1971. *Biophysical properties of the skin.* New York: Wiley.

Marples, M. J. January 1969. Life on the human skin. *Scientific American.*

Maugh, T. H. 1978. Hair: a diagnostic tool to complement blood serum and urine. *Science* 202:1271.

Moncrief, J. A. 1973. Burns. *New Eng. Jour. Med.* 228:444.

Montagna, W. February 1965. The skin. *Scientific American.*

Ross, R. June 1969. Wound healing. *Scientific American.*

Wurtman, R. J. 1975. The effects of light on the human body. *Scientific American.*

Chapter 7 The Skeletal System

Bourne, G. W. 1976. *The biochemistry and physiology of bone.* 2d. ed. New York: Academic Press.

Chisolm, J. J. February 1971. Lead poisoning. *Scientific American.*

Frame, B., and McKenna, M. J. October 1985. Osteoporosis: postmenopausal or secondary. *Hospital Practice.*

Hall, B. K. 1970. Cellular differentiation in skeletal tissue. *Biol. Rev.* 45:455.

Harris, W. H., and Heaney, R. P. 1970. *Skeletal renewal and metabolic bone disease.* Boston: Little, Brown and Co.

Vaughan, J. M. 1981. *The physiology of bone.* 3d ed. New York: Oxford Univ. Press.

Chapter 8 Joints of the Skeletal System

Aufranc, O. E., and Turner, R. H. October 1971. Total replacement of the arthritic hip. *Hospital Practice.*

Barnett, C. H., et al. 1961. *Synovial joints: their structure and mechanics.* Springfield Ill.: Charles Thomas.

Evans, F. G. ed. 1966. *Studies on the anatomy and function of bone and joints.* New York: Springer-Verlag.

Frankel, V. H., and Nordine, M. 1980. *Basic biomechanics of the skeletal system.* Philadelphia: Lea and Febiger.

Higging, J. R. 1977. *Human movement: an integrated approach.* St. Louis: C. V. Mosby.

Koerner, M. E., and Dickinson, G. R. February 1983. Adult arthritis. *Amer. Jour. Nursing.*

Simon, W. H. 1978. *The human joint in health and disease.* Philadelphia: Univ. of Penn. Press.

Sonstegard, D. A., et al. 1978. The surgical replacement of the human knee joint. *Scientific American.*

Chapter 9 The Muscular System

Carafoli, E., and Penniston, J. T. November 1985. The calcium signal. *Scientific American.*

Carlson, F. D., and Wilkie, D. R. 1974. *Muscle physiology.* Englewood Cliffs, N.J.: Prentice-Hall.

Cheung, W. Y. June 1982. Calmodulin. *Scientific American.*

Close, R. I. 1972. Dynamic properties of mammalian skeletal muscles. *Physiological Reviews* 52:129.

Cohen, C. November 1975. The protein switch of muscle contraction. *Scientific American.*

Endo, M. 1977. Calcium release from the sarcoplasmic reticulum. *Physiological Reviews* 57:71.

Hinson, M. M. 1981. *Kinesiology.* 2d ed. Dubuque, Ia.: Wm. C. Brown.

Hoyle, G. April 1970. How a muscle is turned on and off. *Scientific American.*

Huddart, H., and Hunt, S. 1975. *Visceral muscle.* New York: Halsted Press.

Lester, H. A. February 1977. The response to acetylcholine. *Scientific American.*

Margaria, R. March 1972. The sources of muscular energy. *Scientific American.*

Marx, J. L. 1980. Calmodulin: a protein for all seasons. *Science* 208:728.

Merton, P. A. May 1972. How we control the contraction of our muscles. *Scientific American.*

Murray, J. M., and Weber, A. February 1974. The cooperative action of muscle proteins. *Scientific American.*

Oster, G. March 1984. Muscle sounds. *Scientific American.*

Peachey, L. D. 1974. *Muscle and motility.* New York: McGraw-Hill.

Sandow, A. 1970. Skeletal muscle. *Ann. Rev. Physiology* 32:479.

Chapter 10 The Nervous System I: Basic Structure and Function

Adams, R. V. 1981. *Principles of neurology.* 2d ed. New York: McGraw-Hill.

Catterall, W. A. 1984. The molecular basis of neuronal excitability. *Science* 223:653.

Dunant, Y., and Israel, M. April 1985. The release of acetycholine. *Scientific American.*

Goldberg, L. I., and Rajfer, S. I. June 1985. The role of adrenergic and dopamine receptors. *Hospital Practice.*

Kandel, E. R. September 1979. Small systems of neurons. *Scientific American.*

Keynes, R. D. September 1979. Ion channels in the nerve cell membrane. *Scientific American.*

Krieger, D. T. 1983. Brain peptides: what, where, and why. *Science* 222:975.

Kuffler, S. W., et al. 1984. *From neuron to brain: a cellular approach to function of the nervous system.* Sunderland, Ma.: Sinauer.

Lester, H. A. February 1977. The response to acetylcholine. *Scientific American.*

Levi-Montalcini, R., and Calissano, P. June 1979. The nerve growth factor. *Scientific American.*

Llinas, R. R. October 1982. Calcium in synaptic transmission. *Scientific American.*

Marx, J. L. 1980. Regeneration in the central nervous system. *Science* 209:378.

Morell, P., and Norton, W. T. 1980. Myelin. *Scientific American.*

Noback, C. T., and Demarest, R. 1980. *The human nervous system.* 3d ed. New York: McGraw-Hill.

Schwartz, J. H. April 1980. The transport of substances in nerve cells. *Scientific American.*

Shepherd, G. M. February 1978. Microcircuits in the nervous system. *Scientific American.*

Shepherd, G. M. 1983. *Neurobiology.* New York: Oxford Univ. Press.

Stevens, C. F. September 1979. The neuron. *Scientific American.*

Victor, M., and Adams, R. D. 1981. *Principles of neurology.* 2d ed. New York: McGraw-Hill.

Warman, S. G., and Ritchie, J. M. 1985. Organization of ion channels in the myelinated nerve fiber. *Science* 228:1502.

Chapter 11 The Nervous System II: Divisions of the Nervous System

Binkley, S. A., et al. 1978. The pineal gland: a biological clock in vitro. *Science* 202:1198.

Decara, L. V. January 1970. Learning in the autonomic nervous system. *Scientific American.*

Evarts, E. V. September 1979. Brain mechanisms of movement. *Scientific American.*

Galaburda, A. M., et al. 1978. Right-left asymmetries in the brain. *Science* 199:852.

Geschwind, N. September 1979. Specialization of the human brain. *Scientific American.*

Guillemin, R. 1978. Peptides in the brain. *Science* 202:390.

Hubel, D. H. September 1979. The brain. *Scientific American.*

Iversen, L. L. September 1979. The chemistry of the brain. *Scientific American.*

Loftus, E. F. May–June 1979. The malleability of human memory. *American Scientist.*

Nathanson, J. A., and Greengard, P. August 1977. Second messenger in the brain. *Scientific American.*

Nauta, J. H., and Feirtag, M. September 1979. The organization of the brain. *Scientific American.*

Nauta, J. H., and Feirtag, M. ed. 1985. *Fundamental neuroanatomy.* San Francisco: W. H. Freeman.

Norman, D. A. 1982. *Learning and memory.* San Francisco: W. H. Freeman.

Routtenberg, A. November 1978. The reward system of the brain. *Scientific American.*

Solomon, S. H. 1980. Brain peptides as neurotransmitters. *Science* 209:976.

Sperry, R. 1982. Some effects of disconnecting the cerebral hemispheres. *Science* 217:1223.

Springer, S. P., and Deutsch, G. 1981. *Left brain, right brain.* San Francisco: W. H. Freeman.

Thompson, R. F. 1985. *The brain: an introduction to neuroscience.* San Francisco: W. H. Freeman.

Willis, W. D., and Grossman, R. G. 1981. *Medical neurobiology.* 3d ed. St. Louis: C. V. Mosby.

Chapter 12 Somatic and Special Senses

Brindley, G. S. 1970. Central pathways of vision. *Ann. Rev. Physiology* 32:259.

Cain, W. S. 1979. To know with the nose: keys to odor identification. *Science* 203:467.

Daw, N. W. 1973. Neurophysiology of color vision. *Physiological Reviews* 53:571.

Durrant, J. D., and Lovrinic, J. H. 1977. *Bases of hearing science.* Baltimore: Williams and Wilkins.

Freese, A. S. 1977. *The miracle of vision.* New York: Harper and Row.

Gombrich, E. H. September 1972. The visual image. *Scientific American.*

Hudspeth, A. J. January 1983. The hair cells of the inner ear. *Scientific American.*

Loeb, G. E. February 1985. The functional replacement of the ear. *Scientific American.*

O'Brien, D. F. 1982. The chemistry of vision. *Science* 218:513.

Pettigrew, J. D. August 1972. The neurophysiology of binocular vision. *Scientific American.*

Regan, D., et al. July 1979. The visual perception of motion and depth. *Scientific American.*

Rushton, W. A. March 1975. Visual pigments and color blindness. *Scientific American.*

van Heyninger, R. December 1975. What happens to the human lens in cataract? *Scientific American.*

Werblin, F. S. January 1973. The control of sensitivity in the retina. *Scientific American.*

Wurtz, R. H., et al. June 1982. Brain mechanisms of visual attention. *Scientific American.*

Wolfe, J. M. February 1983. Hidden visual processes. *Scientific American.*

Chapter 13 The Endocrine System

Axelrod, J., and Reisine, T. D. 1984. Stress hormones: their interaction and regulation. *Science* 224:452.

Berridge, M. J. October 1985. The molecular basis of communication within cells. *Scientific American.*

Brownstein, M. J., et al. 1980. Synthesis, transport, and release of posterior pituitary hormones. *Science* 207:373.

Camunas, C. June 1983. Pheochromocytoma. *Amer. Jour. Nursing.*

Carmichael, S. W., and Winkler, H. August 1985. The adrenal chromaffin cell. *Scientific American.*

Crapo, L. 1985. *Hormones: messengers of life.* San Francisco: W. H. Freeman.

Frohman, L. A. 1975. Neurotransmitters as regulators of endocrine functions. *Hospital Practice* 10:54.

Gillie, R. B. June 1971. Endemic goiter. *Scientific American.*

Handler, J. S., and Orloff, J. 1981. Antidiuretic hormone. *Ann. Rev. Physiology* 43:611.

Johnson, G. P., and Johanson, B. C. July 1983. Beta blockers. *Amer. Jour. Nursing.*

Katzenellenbogen, B. S. 1980. Dynamics of steroid hormone receptor action. *Ann. Rev. Physiology* 42:17.

Krulich, L. 1979. Central neurotransmitters and the secretion of prolactin. *Ann. Rev. Physiology* 41:603.

Kuehl, F. A. and Egan, R. W. 1980. Prostaglandins, arachinodic acid, and inflammation. *Science* 210:978.

Martin, C. 1985. *Endocrine Physiology.* New York: Oxford Univ. Press.

Marx. J. L. 1979. Hormones and their effects in the aging body. *Science* 206:805.

Maurer, A. C. 1979. The therapy of diabetes. *American Scientist* 67:422.

Moss, R. L. 1979. Actions of hypothalamic-hypophysiotropic hormones on the brain. *Ann. Rev. Physiology* 41:617.

Notkins, A. L. November 1979. The causes of diabetes. *Scientific American.*

Oppenheimer, J. H. 1979. Thyroid hormone action at the cellular level. *Science* 203:971.

Schally, A. V. 1978. Aspects of hypothalamic regulation of the pituitary gland. *Science* 202:18.

Snyder, S. H. October 1985. The molecular basis of communication between cells. *Scientific American.*

Chapter 14 The Digestive System

Bortoff, A. 1976. Myogenic control of intestinal motility. *Physiology Review* 56:418.

Brooks, F. P., ed. 1978. *Gastrointestinal pathophysiology.* 2d ed. New York: Oxford Univ. Press.

Davenport, H. W. 1982. *Physiology of the digestive tract.* 5th ed. Chicago: Yearbook Medical Publishers.

Fredette, S. L. January 1984. When the liver fails. *Amer. Jour. Nursing.*

Grossman, M. I. 1979. Neural and hormonal regulation of gastrointestinal function. *Ann. Rev. Physiology* 41:27.

Javitt, N. B., and McSherry, C. K. July 1973. Pathogenesis of cholesterol gallstones. *Hospital Practice.*

Jones, R. S., and Myers, W. C. 1979. Regulation of hepatic biliary secretions. *Ann. Rev. Physiology* 41:67.

Kappas, A., and Alvares, A. P. June 1975. How the liver metabolizes foreign substances. *Scientific American.*

Kretchmer, N. October 1972. Lactose and lactase. *Scientific American.*

Lieber, C. S. March 1976. The metabolism of alcohol. *Scientific American.*

Marx, J. L. 1979. The HDL: the good cholesterol carrier. *Science* 205:677.

Moog, F. November 1981. The lining of the small intestine. *Scientific American.*

Neurath, R. December 1974. Protein-digesting enzymes. *Scientific American.*

Soll, A., and Walsh, J. H. 1979. Regulation of gastric acid secretion. *Ann. Rev. Physiology* 41:35.

Stroud, R. M. July 1974. A family of protein-cutting proteins. *Scientific American.*

Weems, W. A. 1981. The intestine as a fluid propelling system. *Ann. Rev. Physiology* 43:9.

Weisbrodt, N. W. 1981. Patterns of intestinal motility. *Ann. Rev. Physiology* 43:21.

Chapter 15 Nutrition and Metabolism

Ames, B. N. 1983. Dietary carcinogens and anticarcinogens. *Science* 221:1256.

Barzel, U. S. October 1984. Vitamin D deficiency and osteomalacia in the elderly. *Hospital Practice.*

Chaftz, M. E. May–June 1979. Alcohol and alcoholism. *American Scientist.*

Connell, A. M. March 1976. Dietary fiber and diverticular disease. *Hospital Practice.*

Danowski, T. S. April 1976. The management of obesity. *Hospital Practice.*

DeLuca, H. F. 1981. Recent advances in the metabolism of vitamin D. *Ann. Rev. Physiology* 43:199.

Emery, T. November–December 1982. Iron metabolism in humans and plants. *American Scientist.*

Fleck, H. 1981. *Introduction to nutrition.* 4th ed. New York: Macmillan.

Guthrie, H. A. 1979. *Introductory nutrition.* 4th ed. St. Louis: C. V. Mosby.

Maugh, T. H. 1973. Trace elements: a growing appreciation of the effects on man. *Science* 181:253.

Mayer, J. 1979. *Human nutrition.* Springfield, Ill.: Charles C. Thomas.

Page, L., and Friend, B. March 1978. The changing United States diet. *Bioscience.*

Robinson, C. H., and Weigley, E. 1978. *Fundamentals of normal nutrition.* 3d ed. New York: Macmillan.

Scrimshaw, N. S., and Young, V. R. September 1976. The requirements of human nutrition. *Scientific American.*

Williams, E. R. December 1975. Making vegetarian diets nutritious. *Amer. Jour. Nursing.*

Wurtman, R. J. April 1982. Nutrients that modify brain function. *Scientific American.*

Chapter 16 The Respiratory System

Cohen, M. I. 1981. Central determinants of respiratory rhythm. *Ann. Rev. Physiology* 43:91.

Comroe, J. H. 1974. *Physiology of respiration: an introduction.* 2d ed. Chicago: Year Book Medical Publishers.

Crandall, E. D. 1976. Pulmonary gas exchange. *Ann. Rev. Physiology* 38:69.

Culver, B. H. 1980. Mechanical influences on the pulmonary microcirculation. *Ann. Rev. Physiology* 42:187.

Cummings, G., and Semple, S. 1973. *Disorders of the respiratory system.* Philadelphia: J. B. Lippincott.

Fielding, J. E. 1985. Smoking: health effects and control. *New Eng. Jour. Medicine* 313:491.

Fraser, R. G., and Pare, J. A. 1977. *Structure and function of the lung.* 2d ed. Philadelphia: W. B. Saunders.

Guz, A. 1975. Regulation of respiration in man. *Ann. Rev. Physiology* 37:303.

Hock, R. J. February 1970. The physiology of high altitude. *Scientific American.*

Naeye, R. L. April 1980. Sudden infant death. *Scientific American.*

Simkins, R. March 1981. Croup and epiglottitis. *Amer. Jour. Nursing.*

Slonim, N. B. 1981. *Respiratory physiology.* 4th ed. St. Louis: C. V. Mosby.

West, J. B. 1984. Human physiology at extreme altitudes on Mount Everest. *Science* 223:784.

Wyman, R. J. 1977. Neural generation of the breathing rhythm. *Ann. Rev. Physiology* 39:417.

Chapter 17 The Blood

Adamson, J. W., and Finch, C. A. 1975. Hemoglobin function, oxygen affinity, and erythropoietin. *Ann. Rev. Physiology* 37:351.

Bank, A., et al. 1980. Disorders of human hemoglobin. *Science* 207:486.

Castellino, F. J. November 1983. Plasminogen activators. *Bioscience.*

Child, J., et al. 1972. Blood transfusions. *Amer. Jour. Nursing* 72:1602.

Custer, R. P. 1975. *An atlas of the blood and bone marrow.* 2d ed. Philadelphia: W. B. Saunders.

Doolittle, R. F. December 1981. Fibrinogen and fibrin. *Scientific American.*

McDonagh, A. F., et al. 1980. Blue light and bilirubin excretion. *Science* 208:145.

Perutz, M. F. December 1978. Hemoglobin structure and respiratory transport. *Scientific American.*

Platt, W. R. 1979. *Color atlas and textbook of hematology.* 2d ed. Philadelphia: J. B. Lippincott.

Rapaport, S. I. 1971. *Introduction to hematology.* New York: Harper and Row.

Seegers, W. H. 1969. Blood clotting mechanisms: three basic reactions. *Ann. Rev. Physiology* 31:269.

Sergis, E., and Hilgartner, M. W. 1972. Hemophilia. *Amer. Jour. Nursing* 72:11.

Simone, J. V. July 1974. Childhood leukemia. *Hospital Practice*.

Till, J. E. September–October 1981. Cellular diversity in the blood forming system. *American Scientist*.

Wood, W. B. February 1971. White blood cells vs bacteria. *Scientific American*.

Zucker, M. B. June 1980. The functioning of blood platelets. *Scientific American*.

Chapter 18 The Cardiovascular System

Alpert, N. R., et al. 1979. Heart muscle mechanics. *Ann. Rev. Physiology* 41:521.

Baez, S. 1977. Microcirculation. *Ann. Rev. Physiology* 39:391.

Benditt, E. P. February 1977. The origin of atherosclerosis. *Scientific American*.

Berne, R. M., and Levy, M. N. 1981. *Cardiovascular physiology*. 4th ed. St. Louis: C. V. Mosby.

Brown, M. S., and Goldstein, J. L. November 1984. How LDL receptors influence cholesterol and atherosclerosis. *Scientific American*.

Cantin, M., and Genest, J. February 1986. The heart as an endocrine gland. *Scientific American*.

Cranefield, P. F., and Wit, A. L. 1979. Cardiac arrythmias. *Ann. Rev. Physiology* 41:459.

DeBakey, M., and Gotton, A. 1977. *The living heart*. New York: David McKay Company.

Donald, D. E., and Shepherd, J. T. 1980. Autonomic regulation of the peripheral circulation. *Ann. Rev. Physiology* 42:419.

Duling, B. R., and Klitzman, B. 1980. Local control of microvascular function. *Ann. Rev. Physiology* 42:373.

Fishman, A. P. 1980. Vasomotor regulation of the pulmonary circulation. *Ann. Rev. Physiology* 42:211.

Gil, J. 1980. Organization of microcirculation in the lung. *Ann. Rev. Physiology* 42:187.

Gore, R. W., and McDonagh, P. F. 1980. Fluid exchange across single capillaries. *Ann. Rev. Physiology* 42:337.

Granger, D. N., and Kvietys, P. R. 1981. The splanchnic circulation. *Ann. Rev. Physiology* 43:409.

Hilton, S. M., and Spyer, K. M. 1980. Central nervous regulation of vascular resistance. *Ann. Rev. Physiology* 42:399.

Hurst, J. W., ed. 1979. *The heart*. New York: McGraw-Hill.

Jarvik, R. K. January 1981. The total artificial heart. *Scientific American*.

Johansen, K. July 1982. Aneurysms. *Scientific American*.

Lassen, N. A., et al. October 1978. Brain function and blood flow. *Scientific American*.

Levy, M. N., et al. 1981. Neural regulation of the heartbeat. *Ann. Rev. Physiology* 43:443.

Nuscher, R., et al. June 1984. Bone marrow transplantation. *Amer. Jour. Nursing*.

Olsson, R. A. 1981. Local factors regulating cardiac and skeletal muscle blood flow. *Ann. Rev. Physiology* 43:385.

Rushmer, R. F. 1976. *Structure and function of the cardiovascular system*. 2d ed. Philadelphia: W. B. Saunders.

Saul, L. December 1983. Heart sounds and common murmurs. *Amer. Jour. Nursing*.

Scheuer, J., and Tipton, C. M. 1977. Cardiovascular adaptation to physical training. *Ann. Rev. Physiology* 39:221.

Westfall, V. A. February 1976. Electrical and mechanical events in the cardiac cycle. *Amer. Jour. Nursing*.

Chapter 19 The Lymphatic System

Bellanti, J. A. 1978. *Immunology*. Philadelphia: W. B. Saunders.

Boxer, G. J., et al. March 1985. Polymorphonuclear leukocyte function. *Hospital Practice*.

Buisseret, P. D. August 1982. Allergy. *Scientific American*.

Capra, J. D., and Edmundson, A. B. January 1977. The antibody combining site. *Scientific American*.

Collier, R. J., and Kaplan, D. A. July 1984. Immunotoxins. *Scientific American*.

Cunningham, B. A. October 1977. The structure and functions of histocompatibility antigens. *Scientific American*.

Curren, J. W., et al. 1985. The epidemiology of AIDS. *Science* 229:1352.

Edelson, R. L., and Fink, J. M. June 1985. The immunologic function. *Scientific American*.

Eisen, H. N. 1980. *Immunology*. New York: Harper and Row.

Golub, E. S. 1981. *The cellular basis of the immune response*. 2d ed. Sunderland, Mass.: Sinauer Associates.

Hood, L. E., et al. 1978. *Immunology*. Menlo Park, Calif.: Benjamin Cummings.

Laurence, J. December 1985. The immune system in AIDS. *Scientific American*.

Lerner, R. A. February 1983. Synthetic vaccines. *Scientific American*.

Marrack, P., and Kappler, J. February 1986. The T-cell and its receptor. *Scientific American*.

McDevitt, H. O. July 1985. The HLA system and its relation to disease. *Hospital Practice*.

Milstein, C. October 1980. Monoclonal antibodies. *Scientific American*.

Nisonoff, A. 1982. *Introduction to molecular immunology*. Sunderland, Mass.: Sinauer Associates.

Old, L. J. May 1977. Cancer immunology. *Scientific American*.

Playfair, J. H. 1982. *Immunology at a glance*. 2d ed. St. Louis: C. V. Mosby.

Roitt, I. M. 1981. *Essential Immunology*. 4th ed. St. Louis: C. V. Mosby.

Rose, N. R. February 1981. Autoimmune diseases. *Scientific American*.

Rosen, F. S., et al. 1984. The primary immunodeficiencies. *New Eng. Jour. Medicine* 311:235.

Sell, S. 1980. *Immunology, immunopathology, and immunity*. 3d ed. New York: Harper and Row.

Sibley, C. H. January 1984. How do B-lymphocytes control antibody production? *Bioscience*.

Talmage, D. W. 1979. Recognition and memory in the cells of the immune system. *American Scientist* 67:174.

Tonegawa, S. October 1985. The molecules of the immune system. *Scientific American*.

Chapter 20 The Urinary System

Anderson, B. 1977. Regulation of body fluids. *Ann. Rev. Physiology* 39:185.

Baer, P. G., and McGiff, J. C. 1980. Hormonal systems and renal hemodynamics. *Ann. Rev. Physiology* 42:589.

Bauman, J. W., and Chinard, F. P. 1975. *Renal function: physiological and medical aspects.* St. Louis: C. V. Mosby.

Beeuwkes, R. 1980. The vascular organization of the kidney. *Ann. Rev. Physiology* 42:531.

Dennis, V. W., et al. 1979. Renal handling of phosphate and calcium. *Ann. Rev. Physiology* 41:257.

Harvey, R. J. 1976. *The kidneys and the internal environment.* New York: Halsted Press.

Lassiter, W. E. 1975. Kidney. *Ann. Rev. Physiology* 37:371.

Pitts, R. F. 1974. *Physiology of the kidney and body fluids.* 3d ed. Chicago: Year Book Medical Publishers.

Sullivan, L. P. 1974. *Physiology of the kidney.* Philadelphia: Lea and Febiger.

Ullrich, K. J. 1979. Sugar, amino acid, and sodium transport in the proximal tubule. *Ann. Rev. Physiology* 41:181.

Vander, A. J. 1980. *Renal physiology.* 2d ed. New York: McGraw-Hill.

Chapter 21 Water and Electrolyte Balance

Anderson, B. 1977. Regulation of body fluids. *Ann. Rev. Physiology* 39:185.

Burk, S. R. 1972. *The composition and function of body fluids.* St. Louis: C. V. Mosby.

Deetjen, P., et al. 1974. *Physiology of the kidney and water balance.* New York: Springer-Verlag.

Giebisch, F., and Stanton, B. 1979. Potassium transport in the nephron. *Ann. Rev. Physiology* 41:241.

Hills, A. G. 1973. *Acid-base balance.* Baltimore: Williams and Wilkins.

Pitts, R. F. 1974. *Physiology of the kidney and body fluids.* 3d ed. Chicago: Year Book Medical Publishers.

Share, L., et al. 1972. Regulation of body fluids. *Ann. Rev. Physiology* 34:235.

Valtin, H. 1973. *Renal function: mechanism for preserving fluid and solute balance in health.* Boston: Little, Brown.

Vander, A. J. 1980. *Renal physiology.* 2d ed. New York: McGraw-Hill.

Weldy, N. J. 1972. *Body fluids and electrolytes.* St. Louis: C. V. Mosby.

Chapter 22 The Reproductive Systems

Epel, D. November 1977. The program of fertilization. *Scientific American.*

Goldstein, B. 1976. *Introduction to human sexuality.* New York: McGraw-Hill.

Grabowski, C. T. 1983. *Human reproduction and development.* Philadelphia: W. B. Saunders.

Hatcher, R. A., and Stewart, A. K. 1980. *Contraceptive technology.* 1980–1981. 10th ed. New York: Wiley.

Jones, K. L., et al. 1985. *Dimensions of human sexuality.* Dubuque, Ia.: Wm. C. Brown.

Jones, R. E. 1984. *Human reproduction and sexual behavior.* Englewood Cliffs, N. J.: Prentice-Hall.

Katchadourian, H. A., and Lunde, D. T. 1979. *Fundamentals of human sexuality.* New York: Holt, Rinehart and Winston.

Lein, A. 1979. *The cycling female.* San Francisco: W. H. Freeman.

Marx, J. L. 1978. The mating game: what happens when sperm meets egg. *Science* 200:1256.

Parkes, A. 1976. *Patterns of sexuality and reproduction.* New York: Oxford Univ. Press.

Pengelley, E. T. 1978. *Sex and human life.* Reading, Mass.: Addison-Wesley.

Piziak, V., and Shull, B. L. February 1985. Menopausal hormone replacement. *Hospital Practice.*

Shiu, P. C., and Friesen, H. G. 1980. Mechanisms of action of prolactin in the control of mammary gland functions. *Ann. Rev. Physiology* 42:83.

Short, R. V. April 1984. Breast feeding. *Scientific American.*

Simpson, E. R., and MacDonald, P. C. 1981. Endocrine physiology of the placenta. *Ann. Rev. Physiology* 43:163.

Soloff, M. S., et al. 1979. Oxytocin receptors. *Science* 204:131.

Swanson, H. D. 1974. *Human reproduction: biology and social change.* New York: Oxford Univ. Press.

Chapter 23 Human Growth and Development

Annis, L. F. 1978. *The child before birth.* New York: Cornell Univ. Press.

Balinsky, B. I. 1981. *An introduction to embryology.* 5th ed. Philadelphia: W. B. Saunders.

Beaconsfield, P., et al. August 1980. The placenta. *Scientific American.*

Bower, T. G. 1982. *Development in infancy,* 2d ed. San Francisco: W. H. Freeman.

Diamond, M. C. January-February 1978. The aging brain. *American Scientist.*

Fischer, K., and Lazerson, A. 1984. *Human development: from conception through adolescence.* San Francisco: W. H. Freeman.

Fries, J. F., and Crapo, L. M. 1981. *Vitality and aging.* San Francisco: W. H. Freeman.

Hayflick, L. January 1980. The cell biology of human aging. *Scientific American.*

Heymann, M. A., et al. 1981. Factors affecting changes in the neonatal systemic circulation. *Ann. Rev. Physiology* 43:371.

Katchadourian, H. 1977. *The biology of adolescence.* San Francisco: W. H. Freeman.

Kolata, G. B. 1979. Sex hormones and brain development. *Science* 205:985.

Moore, K. L. 1982. *The developing human.* 3d ed. Philadelphia: W. B. Saunders.

Nilsson, L. 1973. *Behold man.* Boston: Little, Brown and Co.

Patten, B. M., and Carlson, B. M. 1974. *Foundations of embryology.* 3d ed. New York: McGraw-Hill.

Rudolph, A. M. 1979. Fetal and neonatal pulmonary circulation. *Ann. Rev. Physiology* 41:383.

Streissguth, A. P., et al. 1980. Teratogenic effects of alcohol in humans and laboratory animals. *Science* 209:353.

Tollefsbol, T. O., and Gracy, R. W. November 1983. Premature aging diseases: cellular and molecular changes. *Bioscience.*

Weisfeldt, M. L. February 1985. The aging heart. *Hospital Practice.*

Wurtman, R. J. January 1985. Alzheimer's disease. *Scientific American.*

Chapter 24 Human Genetics

Brewer, G. J., and Sing, C. F. 1983. *Genetics.* Reading, Mass.: Addison-Wesley.

Conner, J. M., and Feguson-Smith, M. A. 1984. *Essential medical genetics.* St. Louis: C. V. Mosby.

Ferguson-Smith, M. A. April 1970. Chromosomal abnormalities II: sex chromosome defects. *Hospital Practice.*

Fraser, F. C. January 1971. Genetic counseling. *Hospital Practice.*

Fuchs, F. June 1980. Genetic amniocentesis. *Scientific American.*

Gardner, E. J. 1981. *Principles of genetics.* 6th ed. New York: Wiley.

Hartl, D. L. 1983. *Human genetics.* New York: Harper and Row.

Hirschhorn, K. February 1970. Chromosomal abnormalities I: autosomal defects. *Hospital Practice.*

Jenkins, J. B. 1983. *Human genetics.* Menlo Park, Calif.: Benjamin Cummings.

Kolata, G. 1983. First trimester prenatal diagnosis. *Science* 221:1031.

Lawn, R. M., and Vehar, G. A. March 1986. The molecular genetics of hemophilia. *Scientific American.*

Lewin, B. 1983. *Genes.* New York: Wiley.

Rothwell, N. V. 1983. *Understanding genetics.* 3d ed. New York: Oxford Univ. Press.

Singer, S. 1985. *Human genetics: an introduction to the principles of heredity.* San Francisco: W. H. Freeman.

Winchester, A. M., and Mertens, T. R. 1983. *Human genetics.* 4th ed. Columbus, Ohio: C. E. Merrill.

Glossary

The words in this Glossary are followed by a phonetic guide to pronunciation.

In this guide, any unmarked vowel that ends a syllable or stands alone as a syllable is long. Thus, the word *play* would be spelled *pla*.

Any unmarked vowel that is followed by a consonant has the short sound. The word *tough,* for instance, would be spelled *tuf.*

If a long vowel does appear in the middle of a syllable (followed by a consonant), then it is marked with the macron (˘), the sign for a long vowel. For instance, the word *plate* would be phonetically spelled plāt.

Similarly, if a vowel stands alone or ends a syllable, but should have the short sound, it is marked with a breve (˘).

a

abdomen (ab-do'men)   Portion of the body between the diaphragm and the pelvis.

abduction (ab-duk'shun)   Movement of a body part away from the midline.

absorption (ab-sorp'shun)   The taking in of substances by cells or membranes.

accessory organs (ak-ses'o-re or'ganz)   Organs that supplement the functions of other organs; accessory organs of the digestive and reproductive systems.

accommodation (ah-kom''o-da'shun)   Adjustment of the lens for close vision.

acetone (as'e-tōn)   One of the ketone bodies produced as a result of the oxidation of fats.

acetylcholine (as''e-til-ko'lēn)   Substance secreted at the axon ends of many neurons that transmits a nerve impulse across a synapse.

acetyl coenzyme A (as'e-til ko-en'zīm)   An intermediate compound produced during the oxidation of carbohydrates and fats.

Achilles tendon (ah-kil'ēz ten'don)   Tendon in the back of the heel that connects muscles in the posterior lower leg to the calcaneus.

acid (as'id)   A substance that ionizes in water to release hydrogen ions.

acidosis (as''ĭ-do'sis)   Condition in which there is a relative increase in the acid content of body fluids.

ACTH   Adrenocorticotropic hormone.

actin (ak'tin)   A protein in a muscle fiber that, together with myosin, is responsible for contraction and relaxation.

action potential (ak'shun po-ten'shal)   The sequence of electrical changes occurring when a nerve cell membrane is exposed to a stimulus that exceeds its threshold.

activation energy (ak''tĭ-va'shun en'er-je)   Energy needed to initiate a chemical reaction.

active site (ak'tiv sīt)   Region of an enzyme molecule that combines temporarily with a substrate.

active transport (ak'tiv trans'port)   Process that requires an expenditure of energy to move a substance across a cell membrane; usually moved against the concentration gradient.

acupuncture (ak'u-pungk''chūr)   Procedure in which needles are inserted into various tissues to control pain sensations.

adaptation (ad''ap-ta'shun)   Adjustment to environmental conditions.

adduction (ah-duk'shun)   Movement of a body part toward the midline.

adenoids (ad'ĕ-noids)   The pharyngeal tonsils located in the nasopharynx.

adenosine diphosphate (ah-den'o-sēn di-fos'fāt)   ADP; molecule created when the terminal phosphate is lost from a molecule of adenosine triphosphate.

adenosine triphosphate (ah-den'o-sēn tri-fos'fāt)   An organic molecule that stores energy and releases energy for use in cellular processes.

adenylate cyclase (ah-den'ĭ-lāt si'klās)   An enzyme that is activated when certain hormones combine with receptors on cell membranes, causing ATP to become cyclic AMP.

ADH   Antidiuretic hormone.

adipose tissue (ad'ĭ-pōs tish'u)   Fat-storing tissue.

adolescence (ad''o-les'ens)   Period of life between puberty and adulthood.

ADP   Adenosine diphosphate.

adrenal cortex (ah-dre'nal kor'teks)   The outer portion of the adrenal gland.

adrenal glands (ah-dre'nal glandz)   Endocrine glands located on the tops of the kidneys.

adrenaline (ah-dren'ah-lin)   Epinephrine.

adrenal medulla (ah-dre'nal me-dul'ah)   The inner portion of the adrenal gland.

adrenergic fiber (ad''ren-er'jik fi'ber)   A nerve fiber that secretes norepinephrine at the terminal end of its axon.

adrenocorticotropic hormone (ah-dre''no-kor''te-ko-trōp'ik hor'mōn)   ACTH; hormone secreted by the anterior lobe of the pituitary gland that stimulates activity in the adrenal cortex.

adulthood (ah-dult'hood)   Period of life that extends from adolescence to old age.

aerobic respiration (a''er-ōb'ik res''pi-ra'shun)   Phase of cellular respiration that requires the presence of oxygen.

afferent arteriole (af′er-ent ar-te′re-ōl) Vessel that supplies blood to the glomerulus of a nephron within the kidney.

agglutination (ah-gloo′′ti-na′shun) Clumping together of blood cells in response to a reaction between an agglutinin and an agglutinogen.

agglutinin (ah-gloo′ti-nin) A substance that reacts with an agglutinogen; an antibody.

agglutinogen (ag′′loo-tin′o-jen) A substance that stimulates the formation of agglutinins; a foreign substance.

agranulocytes (a-gran′u-lo-sīt) A nongranular leukocyte.

albumin (al-bu′min) A plasma protein that helps to regulate the osmotic concentration of the blood.

aldosterone (al-dos′ter-ōn) A hormone, secreted by the adrenal cortex, that functions in regulating sodium and potassium concentrations and water balance.

alimentary canal (al′′i-men′tar-e kah-nal′) The tubular portion of the digestive tract that leads from the mouth to the anus.

alkaline (al′kah-lin) Pertaining to or having the properties of a base or alkali; basic.

alkaloid (al′kah-loid) A group of organic substances that are usually bitter in taste and have toxic effects.

alkalosis (al′′kah-lo′sis) Condition in which there is a relative increase in the alkaline content of body fluids.

allantois (ah-lan′to-is) A structure that appears during embryonic development and functions in the formation of umbilical blood vessels.

alleles (ah-lēls′) Genes that occupy corresponding positions on homologous chromosomes.

allergen (al′er-jen) A foreign substance capable of stimulating an allergic reaction.

all-or-none response (al′or-nun′ re-spons′) Phenomenon in which a muscle fiber contracts completely when it is exposed to a stimulus of threshold strength.

alpha receptor (al′fah re-sep′tor) Receptor on effector cell membrane that combines with epinephrine or norepinephrine.

alpha-tocopherol (al′′fah-to-kof′er-ol) Vitamin E.

alveolar ducts (al-ve′o-lar dukts′) Fine tubes that carry air to the air sacs of the lungs.

alveolar pores (al-ve′o-lar pōrz) Minute openings in the walls of air sacs, which permit air to pass from one alveolus to another.

alveolar process (al-ve′o-lar pros′es) Projection on the border of the jaw in which bony sockets of teeth are located.

alveolus (al-ve′o-lus) An air sac of a lung; a saclike structure.

amine (am′in) A type of nitrogen-containing organic compound, including the hormones secreted by the adrenal medulla.

amino acid (ah-me′no as′id) An organic compound of relatively small molecular size that contains an amino group ($-NH_2$) and a carboxyl group ($-COOH$); the structural unit of a protein molecule.

amniocentesis (am′′ne-o-sen-te′sis) A procedure in which a sample of amniotic fluid is removed through the abdominal wall of a pregnant woman.

amnion (am′ne-on) An embryonic membrane that encircles a developing fetus and contains amniotic fluid.

amniotic cavity (am′′ne-ot′ik kav′i-te) Fluid-filled space enclosed by the amnion.

amniotic fluid (am′′ne-ot′ik floo′id) Fluid within the amniotic cavity that surrounds the developing fetus.

ampulla (am-pul′ah) An expansion at the end of each semicircular canal that contains a crista ampullaris.

amylase (am′i-lās) An enzyme that functions to hydrolyze starch.

anabolic metabolism (an′′ah-bol′ik mě-tab′o-lizm) Metabolic process by which larger molecules are formed from smaller ones; anabolism.

anaerobic respiration (an-a′′er-ōb′ik res′′pi-ra′shun) Phase of cellular respiration that occurs in the absence of oxygen.

anal canal (a′nal kah-nal′) The last 2 or 3 inches of the large intestine that opens to the outside as the anus.

anaphase (an′ah-fāz) Stage in mitosis during which duplicate chromosomes move to opposite poles of the cell.

anaplasia (an′′ah-pla′ze-ah) A change in which mature cells become more primitive; a loss of differentiation.

anastomosis (ah-nas′′to-mo′sis) A union of nerve fibers or blood vessels to form an intercommunicating network.

anatomy (ah-nat′o-me) Branch of science dealing with the form and structure of body parts.

androgen (an′dro-jen) A male sex hormone such as testosterone.

anemia (ah-ne′me-ah) A condition characterized by a deficiency of red blood cells or of hemoglobin.

aneurysm (an′u-rizm) A sac-like expansion of a blood vessel wall.

angiotensin (an′′je-o-ten′sin) A vasoconstricting substance that is produced when blood flow to the kidneys is reduced, causing an increase in the blood pressure.

anoxia (an-ok′se-ah) Condition in which the oxygen concentration of the tissues is abnormally low.

antagonist (an-tag′o-nist) A muscle that acts in opposition to a prime mover.

antebrachium (an′′te-bra′ke-um) The forearm.

antecubital (an′′te-ku′bi-tal) The region in front of the elbow joint.

anterior (an-te′re-or) Pertaining to the front; the opposite of posterior.

anterior pituitary (an-te′re-or pi-tu′′i-tār′′e) The front lobe of the pituitary gland.

antibody (an′ti-bod′′e) A specific substance produced by cells in response to the presence of an antigen; it reacts with the antigen.

antibody-mediated immunity (an′′ti-bod′′e me′de-ātid i-mu′ni-te) Resistance to disease-causing agents resulting from the production of specific antibodies by B-lymphocytes; humoral immunity.

anticoagulant (an′′ti-ko-ag′u-lant) A substance that inhibits the action of the blood clotting mechanism.

antidiuretic hormone (an′′ti-di′′u-ret′ik hor′mōn) Hormone released from the posterior lobe of the pituitary gland that enhances the conservation of water by the kidneys; ADH.

antigen (an′ti-jen) A substance that stimulates cells to produce antibodies.

antioxidant (an′′ti-ok′si-dant) A substance that inhibits the oxidation of another substance.

antithrombin (an′′ti-throm′bin) A substance that inhibits the action of thrombin and thus inhibits the blood clotting mechanism.

anus (a′nus) Inferior outlet of the digestive tube.

aorta (a-or′tah) Major systemic artery that receives blood from the left ventricle.

aortic body (a-or′tik bod′e) A structure associated with the wall of the aorta that contains a group of chemoreceptors.

aortic semilunar valve (a-or′tik sem′′i-lu′nar valv) Flaplike structures in the wall of the aorta near its origin that prevent blood from returning to the left ventricle of the heart.

aortic sinus (a-or′tik si′nus) Swelling in the wall of the aorta that contains pressoreceptors.

apneustic area (ap-nu′stik a′re-ah) A portion of the respiratory control center located in the pons.

apocrine gland (ap′o-krin gland) A type of sweat gland that responds during periods of emotional stress.

aponeurosis (ap′′o-nu-ro′sis) A sheetlike tendon by which certain muscles are attached to other parts.

appendicular (ap′′en-dik′u-lar) Pertaining to the arms or legs.

appendix (ah-pen′diks) A small, tubular appendage that extends outward from the cecum of the large intestine.

aqueous humor (a′kwe-us hu′mor) Watery fluid that fills the anterior and posterior chambers of the eye.

arachnoid (ah-rak′noid) Delicate, weblike middle layer of the meninges; arachnoid mater.

arachnoid granulation (ah-rak′noid gran′′u-la′shun) Finger-like structures that project from the subarachnoid space of the meninges into blood-filled dural sinuses and function in the reabsorption of cerebrospinal fluid.

arbor vitae (ar'bor vi'ta) Treelike pattern of white matter seen in a section of cerebellum.

areola (ah-re'o-lah) Pigmented region surrounding the nipple of the mammary gland or breast.

arrector pili muscle (ah-rek'tor pil'i mus'l) Smooth muscle in the skin associated with a hair follicle.

arrhythmia (ah-rith'me-ah) Abnormal heart action characterized by a loss of rhythm.

arterial pathway (ar-te're-al path'wa) Course followed by blood as it travels from the heart to the body cells.

arteriole (ar-te're-ōl) A small branch of an artery that communicates with a capillary network.

arteriosclerosis (ar-te''re-o-sklě-ro'sis) Condition in which the walls of arteries thicken and lose their elasticity; hardening of the arteries.

artery (ar'ter-e) A vessel that transports blood away from the heart.

arthritis (ar-thri'tis) Condition characterized by inflammation of joints.

articular cartilage (ar-tik'u-lar kar'tĭ-lij) Hyaline cartilage that covers the ends of bones in synovial joints.

articulation (ar-tik''u-la'shun) The joining together of parts at a joint.

ascending colon (ah-send'ing ko'lon) Portion of the large intestine that passes upward on the right side of the abdomen from the cecum to the lower edge of the liver.

ascending tracts (ah-send'ing trakts) Groups of nerve fibers in the spinal cord that transmit sensory impulses upward to the brain.

ascorbic acid (as-kor'bik as'id) One of the water-soluble vitamins; vitamin C.

assimilation (ah-sim''ĭ-la'shun) The action of changing absorbed substances into forms that differ chemically from those entering.

association area (ah-so''se-a'shun a're-ah) Region of the cerebral cortex related to memory, reasoning, judgment, and emotional feelings.

astigmatism (ah-stig'mah-tizm) Visual defect due to errors in refraction caused by abnormal curvatures in the surface of the cornea or lens.

astrocyte (as'tro-sit) A type of neuroglial cell that functions to connect neurons to blood vessels.

atherosclerosis (ath''er-o-sklě-ro'sis) Condition in which fatty substances accumulate abnormally on the inner linings of arteries.

atmospheric pressure (at''mos-fer'ik presh'ur) Pressure exerted by the weight of the air; about 760 mm of mercury at sea level.

atom (at'om) Smallest particle of an element that has the properties of that element.

atomic number (ah-tom'ik num'ber) Number equal to the number of protons in an atom of an element.

atomic weight (ah-tom'ik wāt) Number approximately equal to the number of protons plus the number of neutrons in an atom of an element.

ATP Adenosine triphosphate.

ATPase Enzyme that causes ATP molecules to release the energy stored in the terminal phosphate bonds.

atrioventricular bundle (a''tre-o-ven-trik'u-lar bun'dl) Group of specialized fibers that conduct impulses from the atrioventricular node to the ventricular muscle of the heart; A-V bundle.

atrioventricular node (a''tre-o-ven-trik'u-lar nōd) Specialized mass of muscle fibers located in the interatrial septum of the heart; functions in the transmission of the cardiac impulses from the sinoatrial node to the ventricular walls; A-V node.

atrioventricular orifice (a''tre-o-ven-trik'u-lar or'i-fis) Opening between the atrium and the ventricle on one side of the heart.

atrioventricular sulcus (a''tre-o-ven-trik'u-lar sul'kus) A groove on the surface of the heart that marks the division between an atrium and a ventricle.

atrioventricular valve (a''tre-o-ven-trik'u-lar valv) Cardiac valve located between an atrium and a ventricle.

atrium (a'tre-um) A chamber of the heart that receives blood from veins.

atrophy (at'ro-fe) A wasting away or decrease in size of an organ or tissue.

audiometer (aw''de-om'ě-ter) An instrument used to measure the acuity of hearing.

auditory (aw'di-to''re) Pertaining to the ear or to the sense of hearing.

auditory ossicle (aw'di-to''re os'i-kl) A bone of the middle ear.

auricle (aw'ri-kl) An earlike structure; the portion of the heart that forms the wall of an atrium.

autoimmune disease (aw''to-i-mūn' di-zēz') Disorder characterized by an immune response directed toward a person's own tissues; autoallergy.

autonomic nervous system (aw''to-nom'ik ner'vus sis'tem) Portion of the nervous system that functions to control the actions of the visceral organs and skin.

autoregulation (aw''to-reg''u-la'shun) Phenomenon by which a mechanism within an organ or tissue maintains a constant blood flow in spite of changing arterial blood pressure.

autosome (aw'to-sōm) A chromosome other than a sex chromosome.

A-V bundle (bun'dl) A group of fibers that conduct cardiac impulses from the A-V node to the Purkinje fibers; bundle of His.

A-V node (nōd) Atrioventricular node.

axial skeleton (ak'se-al skel'ě-ton) Portion of the skeleton that supports and protects the organs of the head, neck, and trunk.

axillary (ak'si-ler''e) Pertaining to the armpit.

axon (ak'son) A nerve fiber that conducts a nerve impulse away from a neuron cell body.

b

basal ganglion (ba'sal gang'gle-on) Mass of gray matter located deep within a cerebral hemisphere of the brain.

basal metabolic rate (ba'sal met''ah-bol'ic rāt) Rate at which metabolic reactions occur when the body is at rest.

base (bās) A substance that ionizes in water to release hydroxyl ions (OH⁻) or other ions that combine with hydrogen ions.

basement membrane (bās'ment mem'brān) A layer of nonliving material that anchors epithelial tissue to underlying connective tissue.

basophil (ba'so-fil) White blood cell characterized by the presence of cytoplasmic granules that become stained by basophilic dye.

beta oxidation (ba'tah ok''si-da'shun) Chemical process by which fatty acids are converted to molecules of acetyl coenzyme A.

beta receptor (ba'tah re-sep'tor) Receptor on effector cell membrane that combines mainly with epinephrine and only slightly with norepinephrine.

bicarbonate ion (bi-kar'bon-āt i'on) HCO_3^-

bicuspid tooth (bi-kus'pid tooth) A premolar that is specialized for grinding hard particles of food.

bicuspid valve (bi-kus'pid valv) Heart valve located between the left atrium and the left ventricle; mitral valve.

bile (bīl) Fluid secreted by the liver and stored in the gallbladder.

bilirubin (bil''ĭ-roo'bin) A bile pigment produced as a result of hemoglobin breakdown.

biliverdin (bil''ĭ-ver'din) A bile pigment produced as a result of hemoglobin breakdown.

biochemistry (bi''o-kem'is-tre) Branch of science dealing with the chemistry of living organisms.

biofeedback (bi''o-fēd'bak) Procedure in which electronic equipment is used to help a person learn to consciously control certain visceral responses.

biotin (bi'o-tin) A water-soluble vitamin; a member of the vitamin B complex.

bipolar neuron (bi-po'lar nu'ron) A nerve cell whose cell body has only two processes, one serving as an axon and the other as a dendrite.

blastocyst (blas'to-sist) An early stage of embryonic development that consists of a hollow ball of cells.

B-lymphocyte (lim'fo-sit) Lymphocyte that reacts against foreign substances in the body by producing and secreting antibodies.

BMR Basal metabolic rate.

Bowman's capsule (bo'manz kap'sūl) Proximal portion of a renal tubule that encloses the glomerulus of a nephron.

brachial (bra'ke-al) Pertaining to the arm.

bradycardia (brad''e-kar'de-ah)
An abnormally slow heart rate or pulse rate.

brain stem (brān stem)   Portion of the brain that includes the midbrain, pons, and medulla oblongata.

brain wave (brān wāv)   Recording of fluctuating electrical activity occurring in the brain.

Broca's area (bro'kahz a're-ah)   Region of the frontal lobe that coordinates complex muscular actions of the mouth, tongue, and larynx, making speech possible.

bronchial tree (brong'ke-al trē)   The bronchi and their branches that function to carry air from the trachea to the alveoli of the lungs.

bronchiole (brong'ke-ōl)   A small branch of a bronchus within the lung.

bronchus (brong'kus)   A branch of the trachea that leads to a lung.

buccal (buk'al)   Pertaining to the mouth and the inner lining of the cheeks.

buffer (buf'er)   A substance that can react with a strong acid or base to form a weaker acid or base and thus resist a change in pH.

bulbourethral glands (bul''bo-u-re'thral glandz)   Glands that secrete a viscous fluid into the male urethra at times of sexual excitement.

bursa (bur'sah)   A saclike, fluid-filled structure, lined with synovial membrane, that occurs near a joint.

bursitis (bur-si'tis)   Inflammation of a bursa.

c

calcification (kal''si-fi-ka'shun)   The process by which salts of calcium are deposited within a tissue.

calcitonin (kal''si-to'nin)   Hormone secreted by the thyroid gland that helps to regulate the level of blood calcium.

calorie (kal'o-re)   A unit used in the measurement of heat energy and the energy values of foods.

calorimeter (kal'o-rim'e-ter)   A device used to measure the heat energy content of foods; bomb calorimeter.

canaliculus (kan''ah-lik'u-lus)   Microscopic canals that interconnect the lacunae of bone tissue.

cancellous bone (kan'se-lus bōn)   Bone tissue with a lattice-work structure; spongy bone.

capillary (kap'i-ler''e)   A small blood vessel that connects an arteriole and a venule.

carbaminohemoglobin (kar''bah-me'no-he''mo-glo'bin)   Compound formed by the union of carbon dioxide and hemoglobin.

carbohydrate (kar''bo-hi'drāt)   An organic compound that contains carbon, hydrogen, and oxygen, with a 2:1 ratio of hydrogen to oxygen atoms.

carbonic anhydrase (kar-bon'ik an-hi'drās)   Enzyme that promotes the reaction between carbon dioxide and water to form carbonic acid.

carbon monoxide (kar'bon mon-ok'sīd)   A toxic gas that combines readily with hemoglobin to form a relatively stable compound; CO.

carboxypeptidase (kar-bok''se-pep'ti-dās)   A protein-splitting enzyme found in pancreatic juice.

cardiac conduction system (kar'de-ak kon-duk'shun sis'tem)   System of specialized muscle fibers that conducts cardiac impulses from the S-A node into the myocardium.

cardiac cycle (kar'de-ak si'kl)   A series of myocardial contractions that constitute a complete heartbeat.

cardiac muscle (kar'de-ak mus'el)   Specialized type of muscle tissue found only in the heart.

cardiac output (kar'de-ak owt'poot)   A quantity calculated by multiplying the stroke volume by the heart rate in beats per minute.

cardiac veins (kar'de-ak vāns)   Blood vessels that return blood from the venules of the myocardium to the coronary sinus.

carina (kah-ri'nah)   A cartilaginous ridge located between the openings of the right and left bronchi.

carotene (kar'o-tēn)   A yellow, orange, or reddish pigment that occurs in plants and from which vitamin A can be synthesized.

carotid bodies (kah-rot'id bod'ēz)   Masses of chemoreceptors located in the wall of the internal carotid artery near the carotid sinus.

carpals (kar'pals)   Bones of the wrist.

carpus (kar'pus)   The wrist; the wrist bones as a group.

cartilage (kar'ti-lij)   Type of connective tissue in which cells are located within lacunae and are separated by a semi-solid matrix.

catabolic metabolism (kat''ah-bol'ik mě-tab'o-lism)   Metabolic process by which large molecules are broken down into smaller ones; catabolism.

catalase (kat'ah-lās)   An enzyme that causes the decomposition of hydrogen peroxide.

catalyst (kat'ah-list)   A substance that increases the rate of a chemical reaction, but is not permanently altered by the reaction.

cataract (kat'ah-rakt)   Condition characterized by loss of transparency of the lens of the eye.

catecholamine (kat''ě-kol'am-in)   A type of organic compound that includes epinephrine and norepinephrine.

cauda equina (kaw'da ek-win'a)   A group of spinal nerves that extends below the distal end of the spinal cord.

cecum (se'kum)   A pouchlike portion of the large intestine to which the small intestine is attached.

celiac (se'le-ak)   Pertaining to the abdomen.

cell (sel)   The structural and functional unit of an organism.

cell body (sel bod'e)   Portion of a nerve cell that includes a cytoplasmic mass and a nucleus and from which the nerve fibers extend.

cell-mediated immunity (sel me'de-ā-tid i-mu'ni-te)   Resistance to invasion by foreign cells characterized by direct attack of T-lymphocytes.

cellular respiration (sel'u-lar res''pi-ra'shun)   Process by which energy is released from organic compounds within cells.

cellulose (sel'u-lōs)   A polysaccharide that is very abundant in plant tissues, but cannot be digested by human enzymes.

cementum (se-men'tum)   Bonelike material that fastens the root of a tooth into its bony socket.

central canal (sen'tral kah-nal')   Tube within the spinal cord that is continuous with the ventricles of the brain and contains cerebrospinal fluid.

central nervous system (sen'tral ner'vus sis'tem)   Portion of the nervous system that consists of the brain and spinal cord; CNS.

centriole (sen'tre-ōl)   A cellular organelle that functions in the organization of the spindle during mitosis.

centromere (sen'tro-mēr)   Portion of a chromosome to which the spindle fiber attaches during mitosis.

centrosome (sen'tro-sōm)   Cellular organelle consisting of two centrioles.

cephalic (sě-fal'ik)   Pertaining to the head.

cerebellar cortex (ser''ě-bel'ar kor'teks)   The outer layer of the cerebellum.

cerebellum (ser''ě-bel'um)   Portion of the brain that coordinates skeletal muscle movement.

cerebral aqueduct (ser'ě-bral ak'wě-dukt'')   Tube that connects the third and fourth ventricles of the brain.

cerebral cortex (ser'ě-bral kor'teks)   Outer layer of the cerebrum.

cerebral hemisphere (ser'ě-bral hem'ī-sfēr)   One of the large, paired structures that together constitute the cerebrum of the brain.

cerebrospinal fluid (ser''ě-bro-spi'nal floo'id)   Fluid that occupies the ventricles of the brain, the subarachnoid space of the meninges, and the central canal of the spinal cord.

cerebrum (ser'ě-brum)   Portion of the brain that occupies the upper part of the cranial cavity.

cerumen (sě-roo'men)   Waxlike substance produced by cells that line the canal of the external ear.

cervical (ser'vi-kal)   Pertaining to the neck or to the cervix of the uterus.

cervix (ser'viks)   Narrow, inferior end of the uterus that leads into the vagina.

chemoreceptor (ke''mo-re-sep'tor)   A receptor that is stimulated by the presence of certain chemical substances.

chemotaxis (ke''mo-tak'sis) Phenomenon by which leukocytes are attracted by chemicals released from damaged cells.

chief cell (chēf sel) Cell of gastric gland that secretes various digestive enzymes, including pepsinogen.

childhood (chīld'hood) Phase of the human life cycle that begins at the end of the first year and ends at puberty.

chloride shift (klo'rīd shift) Movement of chloride ions from the blood plasma into red blood cells as bicarbonate ions diffuse out of the red blood cells into the plasma.

cholecystokinin (ko''le-sis''to-ki'nin) Hormone secreted by the small intestine that stimulates the release of pancreatic juice from the pancreas and bile from the gallbladder.

cholesterol (ko-les'ter-ol) A lipid produced by body cells that is used in the synthesis of steroid hormones and is excreted into the bile.

cholinergic fiber (ko''lin-er'jik fi'ber) A nerve fiber that secretes acetylcholine at the terminal end of its axon.

cholinesterase (ko''lin-es'ter-ās) An enzyme that causes the decomposition of acetylcholine.

chondrin (kon'drin) A protein that occurs in the intercellular substance of cartilage tissues.

chondrocyte (kon'dro-sīt) A cartilage cell.

chorion (ko're-on) Embryonic membrane that forms the outermost covering around a developing fetus and contributes to the formation of the placenta.

chorionic villi (ko''re-on'ik vil'i) Projections that extend from the outer surface of the chorion and help attach an embryo to the uterine wall.

choroid coat (ko'roid kōt) The vascular, pigmented middle layer of the wall of the eye.

choroid plexus (ko'roid plek'sus) Mass of specialized capillaries from which cerebrospinal fluid is secreted into a ventricle of the brain.

chromatid (kro'mah-tid) A member of a duplicate pair of chromosomes.

chromatin (kro'mah-tin) Nuclear material that gives rise to chromosomes during mitosis.

chromosome (kro'mo-sōm) Rodlike structure that appears in the nucleus of a cell during mitosis; contains the genes responsible for heredity.

chylomicron (ki''lo-mi'kron) A microscopic droplet of fat, found in the blood following the digestion of fats.

chyme (kim) Semifluid mass of food materials that passes from the stomach to the small intestine.

chymotrypsin (ki''mo-trip'sin) A protein-splitting enzyme found in pancreatic juice.

cilia (sil'e-ah) Microscopic, hairlike processes on the exposed surfaces of certain epithelial cells.

ciliary body (sil'e-er''e bod'e) Structure associated with the choroid layer of the eye that secretes aqueous humor and contains the ciliary muscle.

circadian rhythm (ser''kah-de'an rithm) A pattern of repeated behavior associated with the cycles of night and day.

circle of Willis (sir'kl uv wil'is) An arterial ring located on the ventral surface of the brain.

circular muscles (ser'ku-lar mus'lz) Muscles whose fibers are arranged in circular patterns, usually around an opening or in the wall of a tube; sphincter muscles.

circumduction (ser''kum-duk'shun) Movement of a body part, such as a limb, so that the end follows a circular path.

cisternae (sis-ter'ne) Enlarged portions of the sarcoplasmic reticulum near the actin and myosin filaments of a muscle fiber.

citric acid cycle (sit'rik as'id si'kl) A series of chemical reactions by which various molecules are oxidized and energy is released from them; Kreb's cycle.

cleavage (klēv'ij) The early successive divisions of embryonic cells into smaller and smaller cells.

clitoris (kli'to-ris) Small erectile organ located in the anterior portion of the female vulva; corresponds to the penis of the male.

clone (klōn) A group of cells that originate from a single cell and are all alike.

CNS Central nervous system.

coagulation (ko-ag''u-la'shun) The clotting of blood.

cocarboxylase (ko''kar-bok'si-lās) A coenzyme that is synthesized from thiamine and acts in the oxidation of carbohydrates.

cochlea (kok'le-ah) Portion of the inner ear that contains the receptors of hearing.

coenzyme (ko-en'zīm) A non-protein substance that is necessary to complete the structure of an enzyme molecule.

coenzyme A (ko-en'zīm) Acetyl coenzyme A.

cofactor (ko'fak-tor) A non-protein substance that must be combined with the protein portion of an enzyme before the enzyme can act.

collagen (kol'ah-jen) Protein that occurs in the white fibers of connective tissues and in the matrix of bone.

collateral (ko-lat'er-al) A branch of a nerve fiber or blood vessel.

colon (ko'lon) The large intestine.

color blindness (kul'er blīnd'nes) An inability to distinguish colors normally.

colostrum (ko-los'trum) The first secretion of the mammary glands following the birth of an infant.

common bile duct (kom'mon bil dukt) Tube that transports bile from the cystic duct to the duodenum.

complement (kom'plĕ-ment) A group of enzymes that are activated by the combination of antibody with antigen and enhance the reaction against foreign substances within the body.

complete protein (kom-plēt' pro'te-in) A protein that contains adequate amounts of the essential amino acids to maintain body tissues and to promote normal growth and development.

compound (kom'pownd) A substance composed of two or more elements joined by chemical bonds.

condom (kon'dum) A rubber sheath used to cover the penis during sexual intercourse; used as a contraceptive.

conduction (kon-duk'shun) Process by which body heat moves into the molecules of cooler objects in contact with the body surface.

condyle (kon'dīl) A rounded process of a bone, usually at the articular end.

cones (kōns) Color receptors located in the retina of the eye.

congenital (kon-jen'i-tal) Any condition that exists at the time of birth.

conjunctiva (kon''junk-ti'vah) Membranous covering on the anterior surface of the eye.

connective tissue (kŏ-nek'tiv tish'u) One of the basic types of tissue that includes bone, cartilage, and various fibrous tissues.

contraception (kon''trah-sep'shun) The prevention of fertilization of the egg cell or the development of an embryo.

contralateral (kon''trah-lat'er-al) Positioned on the opposite side of something else.

convection (kon-vek'shun) The transmission of heat from one substance to another through the circulation of heated air particles.

convergence (kon-ver'jens) The coming together of nerve impulses from different parts of the nervous system so that they reach the same neuron.

convolution (kon''vo-lu'shun) An elevation on the surface of a structure caused by an infolding of the structure upon itself.

cornea (kor'ne-ah) Transparent anterior portion of the outer layer of the eye wall.

coronary artery (kor'o-na''re ar'ter-e) An artery that supplies blood to the wall of the heart.

coronary sinus (kor'o-na''re si'nus) A large vessel on the posterior surface of the heart into which the cardiac veins drain.

corpus callosum (kor'pus kah-lo'sum) A mass of white matter within the brain, composed of nerve fibers connecting the right and left cerebral hemispheres.

corpus luteum (kor'pus lut'e-um) Structure that forms from the tissues of a ruptured ovarian follicle and functions to secrete female hormones.

corpus striatum (kor′pus stri-a′tum) Portion of the cerebrum that includes certain basal ganglia.

cortex (kor′teks) Outer layer of an organ such as the adrenal gland, cerebrum, or kidney.

cortical nephron (kor′ti-kl nef′ron) A nephron with its corpuscle located in the renal cortex.

cortisol (kor′ti-sol) A glucocorticoid secreted by the adrenal cortex.

costal (kos′tal) Pertaining to the ribs.

covalent bond (ko′va-lent bond) Chemical bond created by the sharing of electrons between atoms.

cranial (kra′ne-al) Pertaining to the cranium.

cranial nerve (kra′ne-al nerv) Nerve that arises from the brain.

creatine phosphate (kre′ah-tin fos′fāt) A substance present in muscle that acts to store energy.

crenation (kre-na′shun) Shrinkage of a cell caused by contact with a hypertonic solution.

crest (krest) A ridgelike projection of a bone.

cretinism (kre′ti-nizm) A condition resulting from a lack of thyroid secretion in an infant.

cricoid cartilage (kri′koid kar′ti-lij) A ringlike cartilage that forms the lower end of the larynx.

crista ampullaris (kris′tah am-pul′ ar-is) Sensory organ located within a semicircular canal that functions in the sense of dynamic equilibrium.

crossing over (kros′ing o′ver) The exchange of genetic material between homologous chromosomes during meiosis.

crossmatching (kros′mach″ing) A procedure used to determine whether donor and recipient blood samples will agglutinate.

cubital (ku′bi-tal) Pertaining to the forearm.

cuspid (kus′pid) A canine tooth.

cutaneous (ku-ta′ne-us) Pertaining to the skin.

cyanocobalamin (si″ah-no-ko-bal′ ah-min) Vitamin B_{12}.

cyanosis (si″ah-no′sis) A condition characterized by a bluish coloration of the skin due to a decreased blood oxygen concentration.

cyclic AMP (sik′lik) A substance produced from ATP that causes a variety of changes in cells.

cystic duct (sis′tik dukt) Tube that connects the gallbladder to the common bile duct.

cytocrine secretion (si′to-krin se-kre′shun) Process by which melanocytes transfer granules of melanin into adjacent epithelial cells.

cytoplasm (si′to-plazm) The contents of a cell surrounding its nucleus.

d

deamination (de-am″i-na′shun) Chemical process by which amino groups ($-NH_2$) are removed from amino acid molecules.

deciduous teeth (de-sid′u-us tēth) Teeth that are shed and replaced by permanent teeth.

decomposition (de-kom″po-zish′un) The breakdown of molecules into simpler compounds.

defecation (def″ē-ka′shun) The discharge of feces from the rectum through the anus.

dehydration (de″hi-dra′shun) Excessive loss of water.

dehydration synthesis (de″hi-dra′shun sin′thē-sis) Anabolic process by which molecules are joined together to form larger molecules.

dendrite (den′drit) Nerve fiber that transmits impulses toward a neuron cell body.

dental caries (den′tal kar′ēz) Process by which teeth become decalcified and decayed.

dentine (den′tēn) Bonelike substance that forms the bulk of a tooth.

deoxyhemoglobin (de-ok″si-he″mo-glo′bin) Hemoglobin that lacks oxygen.

depolarization (de-po″lar-i-za′shun) The loss of an electrical charge on the surface of a membrane.

dermatome (der′mah-tōm) An area of the body supplied by sensory nerve fibers associated with a particular dorsal root of a spinal nerve.

dermis (der′mis) The thick layer of the skin beneath the epidermis.

descending colon (de-send′ing ko′lon) Portion of the large intestine that passes downward along the left side of the abdominal cavity to the brim of the pelvis.

descending tracts (de-send′ing trakts) Groups of nerve fibers that carry nerve impulses downward from the brain through the spinal cord.

desmosome (des′mo-sōm) A specialized junction between cells, which serves as a "spot weld."

detrusor muscle (de-trūz′or mus′l) Muscular wall of the urinary bladder.

dextrose (dek′strōs) Glucose.

diabetes insipidus (di″ah-be′tēz in-sip′i-dus) Condition characterized by an abnormally great production of urine due to a deficiency of antidiuretic hormone.

diabetes mellitus (di″ah-be′tēz mel-li′tus) Condition characterized by a high blood glucose level and the appearance of glucose in the urine due to a deficiency of insulin.

dialysis (di-al′i-sis) Process by which smaller molecules are separated from larger ones in a liquid.

diapedesis (di″ah-pē-de′sis) Process by which leukocytes squeeze between the cells that make up the walls of blood vessels.

diaphragm (di′ah-fram) A sheetlike structure composed largely of muscle and connective tissue that separates the thoracic and abdominal cavities; also a caplike rubber device inserted in the vagina to be used as a contraceptive.

diaphysis (di-af′i-sis) The shaft of a long bone.

diastole (di-as′to-le) Phase of the cardiac cycle during which a heart chamber wall is relaxed.

diastolic pressure (di-a-stol′ik presh′ur) Arterial blood pressure during the diastolic phase of the cardiac cycle.

diencephalon (di″en-sef′ah-lon) Portion of the brain in the region of the third ventricle that includes the thalamus and hypothalamus.

differentiation (dif″er-en″she-a′shun) Process by which cells become structurally and functionally specialized during development.

diffusion (di-fu′zhun) Random movement of molecules from a region of higher concentration toward one of lower concentration.

digestion (di-jest′yun) The process by which larger molecules of food substances are broken down into smaller molecules that can be absorbed; hydrolysis.

dihydrotestosterone (di-hi″dro-tes-tos′ter-ōn) Hormone produced from testosterone that stimulates certain cells of the male reproductive system.

dipeptide (di-pep′tīd) A molecule composed of two amino acids bound together.

disaccharide (di-sak′ah-rīd) A sugar produced by the union of two monosaccharide molecules.

distal (dis′tal) Further from the midline or origin; opposite of proximal.

diuretic (di′u-ret′ik) A substance that causes an increased production of urine.

DNA Deoxyribonucleic acid.

dominant gene (dom′i-nant jēn) The gene of a gene pair that is expressed while its allele is not expressed.

dorsal root (dor′sal rōot) The sensory branch of a spinal nerve by which it joins the spinal cord.

dorsal root ganglion (dor′sal rōot gang′gle-on) Mass of sensory neuron cell bodies located in the dorsal root of a spinal nerve.

dorsum (dors′um) Pertaining to the back surface of a body part.

ductus arteriosus (duk′tus ar-te″re-o′sus) Blood vessel that connects the pulmonary artery and the aorta in a fetus.

ductus venosus (duk′tus ven-o′sus) Blood vessel that connects the umbilical vein and the inferior vena cava in a fetus.

duodenum (du″o-de′num)   The first portion of the small intestine that leads from the stomach to the jejunum.

dural sinus (du′ral si′nus)   Blood-filled channel formed by the splitting of the dura mater into two layers.

dura mater (du′rah ma′ter)   Tough outer layer of the meninges.

dynamic equilibrium (di-nam′ik e″kwi-lib′re-um)   The maintenance of balance when the head and body are suddenly moved or rotated.

e

eccrine gland (ek′rin gland)   Sweat gland that functions in the maintenance of body temperature.

ECG   Electrocardiogram; EKG.

ectoderm (ek′to-derm)   The outermost layer of the primary germ layers, responsible for forming certain embryonic body parts.

edema (ě-de′mah)   An excessive accumulation of fluid within the tissue spaces.

effector (ě-fek′tor)   Organ, such as a muscle or gland, that responds to stimulation.

efferent arteriole (ef′er-ent ar-te′ re-ol)   Arteriole that conducts blood away from the glomerulus of a nephron.

ejaculation (e-jak″u-la′shun)   Discharge of sperm-containing seminal fluid from the male urethra.

elastin (e-las′tin)   Protein that comprises the yellow, elastic fibers of connective tissue.

electrocardiogram (e-lek″tro-kar′de-o-gram″)   A recording of the electrical activity associated with the heartbeat; ECG or EKG.

electrolyte (e-lek′tro-līt)   A substance that ionizes in water solution.

electrolyte balance (e-lek′tro-līt bal′ans)   Condition that exists when the quantities of electrolytes entering the body equal those leaving it.

electron (e-lek′tron)   A small, negatively charged particle that revolves around the nucleus of an atom.

electrovalent bond (e-lek″tro-va′lent bond)   Chemical bond formed between two ions as a result of the transfer of electrons.

element (el′ě-ment)   A basic chemical substance.

embolus (em′bo-lus)   A substance, such as a blood clot or bubble of gas, that is carried by the blood and obstructs a blood vessel.

embryo (em′bre-o)   An organism in its earliest stages of development.

emission (e-mish′un)   The movement of sperm cells from the vas deferens into the ejaculatory duct and urethra.

emphysema (em″fĭ-se′mah)   A condition characterized by abnormal enlargement of the air sacs of the lungs.

emulsification (e-mul″sĭ-fĭ′ka′shun)   Process by which fat globules are caused to break up into smaller droplets by the action of bile salts.

enamel (e-nam′el)   Hard covering on the exposed surface of a tooth.

endocardium (en″do-kar′de-um)   Inner lining of the heart chambers.

endochondral bone (en″do-kon′dral bōn)   Bone that begins as hyaline cartilage that is subsequently replaced by bone tissue.

endocrine gland (en′do-krin gland)   A gland that secretes hormones directly into the blood or body fluids.

endocytosis (en″do-si-to′sis)   Physiological process by which substances may move through a cell membrane.

endoderm (en′do-derm)   The innermost layer of the primary germ layers responsible for forming certain embryonic body parts.

endolymph (en′do-limf)   Fluid contained within the membranous labyrinth of the inner ear.

endometrium (en″do-me′tre-um)   The inner lining of the uterus.

endomysium (en″do-mis′e-um)   The sheath of connective tissue surrounding each skeletal muscle fiber.

endoneurium (en″do-nu′re-um)   Layer of loose connective tissue that surrounds individual nerve fibers of a nerve.

endoplasmic reticulum (en-do-plaz′mic rĕ-tik′u-lum)   Cytoplasmic organelle composed of a system of interconnected membranous tubules and vesicles.

endorphin (en-dor′fin)   A neuropeptide that occurs in the pituitary gland and has a pain-suppressing action.

endothelium (en″do-the′le-um)   The layer of epithelial cells that forms the inner lining of blood vessels and heart chambers.

energy (en′er-je)   An ability to cause something to move and thus to do work.

energy balance (en′er-je bal′ans)   Condition that exists when the caloric intake of the body equals its caloric output.

enkephalin (en-kef′ah-lin)   A neuropeptide that occurs in the brain and spinal cord and inhibits pain impulses, thus relieving pain sensations.

enzyme (en′zim)   A protein that is synthesized by a cell and acts as a catalyst in a specific cellular reaction.

eosinophil (e″o-sin′o-fil)   White blood cell characterized by the presence of cytoplasmic granules that become stained by acidic dye.

ependyma (ě-pen′di-mah)   Membrane composed of neuroglial cells that lines the ventricles of the brain.

epicardium (ep″ĭ-kar′de-um)   The visceral portion of the pericardium located on the surface of the heart.

epicondyle (ep″ĭ-kon′dīl)   A projection of a bone located above a condyle.

epidermis (ep″ĭ-der′mis)   Outer epithelial layer of the skin.

epididymis (ep″ĭ-did′ĭ-mis)   Highly coiled tubule that leads from the seminiferous tubules of the testis to the vas deferens.

epidural space (ep″ĭ-du′ral spās)   The space between the dural sheath of the spinal cord and the bone of the vertebral canal.

epigastric region (ep″ĭ-gas′trik re′jun)   The upper middle portion of the abdomen.

epiglottis (ep″ĭ-glot′is)   Flaplike cartilaginous structure located at the back of the tongue near the entrance to the trachea.

epimysium (ep″ĭ-mis′e-um)   The outer sheath of connective tissue surrounding a skeletal muscle.

epinephrine (ep″ĭ-nef′rin)   A hormone secreted by the adrenal medulla during times of stress.

epineurium (ep″ĭ-nu′re-um)   Outermost layer of connective tissue surrounding a nerve.

epiphyseal disk (ep″ĭ-fiz′e-al disk)   Cartilaginous layer within the epiphysis of a long bone that functions as a growing region.

epiphysis (ě-pif′ĭ-sis)   The end of a long bone.

epithelium (ep″ĭ-the′le-um)   The type of tissue that covers all free body surfaces.

equilibrium (e″kwi-lib′re-um)   A state of balance between two opposing forces.

erythroblast (ě-rith′ro-blast)   An immature red blood cell.

erythroblastosis (ě-rith″ro-blas-to′sis)   Condition characterized by the presence of erythroblasts in the circulating blood.

erythrocyte (ě-rith′ro-sīt)   A red blood cell.

erythropoiesis (ě-rith″ro-poi-e′sis)   Red blood cell formation.

erythropoietin (ě-rith″ro-poi′ě-tin)   Substance released by the kidneys and liver that promotes red blood cell formation.

esophageal hiatus (ě-sof″ah-je′al hi-a′tus)   Opening in the diaphragm through which the esophagus passes.

esophagus (ě-sof′ah-gus)   Tubular portion of the digestive tract that leads from the pharynx to the stomach.

essential amino acid (ě-sen′shal ah-me′no as′id)   Amino acid required for health that cannot be synthesized in adequate amounts by body cells.

essential fatty acid (ě-sen′shal fat′e as′id)   Fatty acid required for health that cannot be synthesized in adequate amounts by body cells.

estrogen (es′tro-jen)   Hormone that stimulates the development of female secondary sexual characteristics.

eustachian tube (u-sta′ke-an tūb)   Tube that connects the middle ear to the pharynx; auditory tube.

evaporation (e″vap′o-ra-shun)   Process by which a liquid changes into a gas.

eversion (e-ver′zhun)   Movement in which the sole of the foot is turned outward.

exchange reaction (eks-chānj′ re-ak′shun)   A chemical reaction in which parts of two kinds of molecules trade positions.

excretion (ek-skre'shun)   Process by which metabolic wastes are eliminated.

exocrine gland (ek'so-krin gland)   A gland that secretes its products into a duct or onto a body surface.

expiration (ek''spi-ra'shun)   Process of expelling air from the lungs.

extension (ek-sten'shun)   Movement by which the angle between parts at a joint is increased.

extracellular (ek''strah-sel'u-lar)   Outside of cells.

extrapyramidal tract (ek''strah-pi-ram'i-dal trakt)   Nerve tracts, other than the corticospinal tracts, that transmit impulses from the cerebral cortex into the spinal cord.

extremity (ek-strem'ĭ-te)   A limb; an arm or leg.

f

facet (fas'et)   A small, flattened surface of a bone.

facilitated diffusion (fah-sil'ĭ-tāt''id di-fu'zhun)   Diffusion in which substances are moved through membranes from a region of higher concentration to a region of lower concentration by carrier molecules.

fallopian tube (fah-lo'pe-an tŭb)   Tube that transports an egg cell from the region of the ovary to the uterus; oviduct or uterine tube.

fascia (fash'e-ah)   A sheet of fibrous connective tissue that encloses a muscle.

fasciculus (fah-sik'u-lus)   A small bundle of muscle fibers.

fat (fat)   Adipose tissue; or an organic substance whose molecules contain glycerol and fatty acids.

fatty acid (fat'e as'id)   An organic substance that serves as a building block for a fat molecule.

feces (fe'sēz)   Material expelled from the digestive tract during defecation.

ferritin (fer'i-tin)   An iron-protein complex in which iron is stored in liver cells.

fertilization (fer''ti-li-za'shun)   The union of an egg cell and a sperm cell.

fetus (fe'tus)   A human embryo after eight weeks of development.

fibril (fi'bril)   A tiny fiber or filament.

fibrillation (fi''bri-la'shun)   Uncoordinated contraction of muscle fibers.

fibrin (fi'brin)   Insoluble, fibrous protein formed from fibrinogen during blood coagulation.

fibrinogen (fi-brin'o-jen)   Plasma protein that is converted into fibrin during blood coagulation.

fibrinolysin (fi''bri-nol'i-sin)   A protein-splitting enzyme that can digest the substance of a blood clot.

fibroblast (fi'bro-blast)   Cell that functions to produce fibers and other intercellular materials in connective tissues.

filtration (fil-tra'shun)   Movement of material through a membrane as a result of hydrostatic pressure.

fissure (fish'ūr)   A narrow cleft separating parts, such as the lobes of the cerebrum.

flaccid paralysis (flak'sid pah-ral'ĭ-sis)   A condition characterized by total loss of tone in the muscles innervated by damaged nerve fibers.

flagella (flah-jel'ah)   Relatively long motile processes that extend out from the surface of a cell.

flexion (flek'shun)   Bending at a joint so that the angle between bones is decreased.

follicle (fol'i-kl)   A pouchlike depression or cavity.

follicle-stimulating hormone (fol'i-kl stim'u-la''ting hor'mōn)   A substance secreted by the anterior pituitary gland that stimulates the development of an ovarian follicle in a female or the production of sperm cells in a male; FSH.

follicular cells (fō-lik'u-lar selz)   Ovarian cells that surround a developing egg cell and secrete female sex hormones.

fontanel (fon''tah-nel')   Membranous region located between certain cranial bones in the skull of a fetus or infant.

foramen (fo-ra'men)   An opening, usually in a bone or membrane (plural, foramina).

foramen magnum (fo-ra'men mag'num)   Opening in the occipital bone of the skull through which the spinal cord passes.

foramen ovale (fo-ra'men o-val'e)   Opening in the interatrial septum of the fetal heart.

forebrain (fōr'brān)   The anterior-most portion of the developing brain that gives rise to the cerebrum and basal ganglia.

formula (fōr'mu-lah)   A group of symbols and numbers used to express the composition of a compound.

fossa (fos'ah)   A depression in a bone or other part.

fovea (fo've-ah)   A tiny pit or depression.

fovea centralis (fo've-ah sen-tral'is)   Region of the retina, consisting of densely packed cones, which is responsible for the greatest visual acuity.

fracture (frak'tūr)   A break in a bone.

frenulum (fren'u-lum)   A fold of tissue that serves to anchor and limit the movement of a body part.

frontal (frun'tal)   Pertaining to the region of the forehead.

FSH   Follicle-stimulating hormone.

g

galactose (gah-lak'tōs)   A monosaccharide component of the disaccharide, lactose.

gallbladder (gawl'blad-er)   Saclike organ associated with the liver that stores and concentrates bile.

gamete (gam'ēt)   A sex cell; either an egg cell or a sperm cell.

ganglion (gang'gle-on)   A mass of neuron cell bodies, usually outside the central nervous system.

gastric gland (gas'trik gland)   Gland within the stomach wall that secretes gastric juice.

gastric juice (gas'trik jōōs)   Secretion of the gastric glands within the stomach.

gastric lipase (gas'trik lī'pās)   Fat-splitting enzyme that occurs in small quantities in gastric juice.

gastrin (gas'trin)   Hormone secreted by the stomach lining that stimulates the secretion of gastric juice.

gene (jēn)   Portion of a DNA molecule that contains the information needed to synthesize an enzyme.

genetic code (jĕ-net'ik kōd)   System by which information for synthesizing proteins is built into the structure of DNA molecules.

genetics (jĕ-net'iks)   The study of the mechanism by which characteristics are passed from parents to offspring.

genotype (je'no-tip)   The combination of genes present within a zygote or within the cells of an individual.

germinal epithelium (jer'mi-nal ep''i-the'le-um)   Tissue within an ovary or testis that gives rise to sex cells.

germ layers (jerm la'ers)   Layers of cells within an embryo that form the body organs during development.

globin (glo'bin)   The protein portion of a hemoglobin molecule.

globulin (glob'u-lin)   A type of protein that occurs in blood plasma.

glomerulus (glo-mer'u-lus)   A capillary tuft located within the Bowman's capsule of a nephron.

glottis (glot'is)   Slitlike opening between the true vocal folds or vocal cords.

glucagon (gloo'kah-gon)   Hormone secreted by the pancreatic islets of Langerhans that causes the release of glucose from glycogen.

glucocorticoid (gloo''ko-kor'ti-koid)   Any one of a group of hormones secreted by the adrenal cortex that influence carbohydrate, fat, and protein metabolism.

gluconeogenesis (gloo''ko-ne''o-jen'ĕ-sis)   The synthesis of glucose from noncarbohydrate materials, such as amino acid molecules.

glucose (gloo'kōs)   A monosaccharide found in the blood that serves as the primary source of cellular energy.

glucosuria (gloo''ko-su're-ah)   The presence of glucose in urine.

gluteal (gloo'te-al)   Pertaining to the buttocks.

glycerol (glis'er-ol)   An organic compound that serves as a building block for fat molecules.

glycogen (gli'ko-jen)   A polysaccharide that functions to store glucose in the liver and muscles.

glycolysis (gli-kol'i-sis)   The conversion of glucose to pyruvic acid during cellular respiration.

glycoprotein (gli''ko-pro'te-in)   A substance composed of a carbohydrate combined with a protein.

goblet cell (gob′let sel)   An epithelial cell that is specialized to secrete mucus.

goiter (goi′ter)   A condition characterized by the enlargement of the thyroid gland.

Golgi apparatus (gol′je ap′′ah-ra′tus)   A cytoplasmic organelle that functions in preparing cellular products for secretion.

Golgi tendon organ (gol′′jē ten′dun or′gan)   Sensory receptors occurring in tendons close to muscle attachments that are involved in reflexes that help to maintain posture.

gomphosis (gom-fo′sis)   Type of joint in which a cone-shaped process is fastened in a bony socket.

gonad (go′nad)   A sex-cell-producing organ; an ovary or testis.

gonadotropin (go-nad′′o-trōp′in)   A hormone that stimulates activity in the gonads.

granulocyte (gran′u-lo-sīt)   A leukocyte that contains granules in its cytoplasm.

gray matter (grā mat′er)   Region of the central nervous system that generally lacks myelin and thus appears gray.

groin (groin)   Region of the body between the abdomen and thighs.

growth (grōth)   Process by which a structure enlarges.

growth hormone (grōth hōr′mōn)   A hormone released by the anterior lobe of the pituitary gland that promotes the growth of the organism; GH or somatotropin.

h

hair follicle (hār fol′i-kl)   Tubelike depression in the skin in which a hair develops.

half-life (haf′lif)   The time it takes for one-half of the radioactivity of an isotope to be released.

hapten (hap′ten)   A small molecule that, in combination with a larger one, forms an antigenic substance and can later stimulate an immune reaction by itself.

hematocrit (he-mat′o-krit)   The volume percentage of red blood cells within a sample of whole blood.

hematoma (he′′mah-to′mah)   A mass of coagulated blood within tissues or a body cavity.

hematopoiesis (hem′′ah-to-poi-e′sis)   The production of blood and blood cells; hemopoiesis.

heme (hem)   The iron-containing portion of a hemoglobin molecule.

hemocytoblast (he′′mo-si′to-blast)   A cell that gives rise to blood cells.

hemoglobin (he′′mo-glo′bin)   Pigment of red blood cells responsible for the transport of oxygen.

hemolysis (he-mol′i-sis)   The rupture of red blood cells accompanied by the release of hemoglobin.

hemopoiesis (he′′mo-poi-e′sis)   The production of blood and blood cells; hematopoiesis.

hemorrhage (hem′ō-rij)   Loss of blood from the circulatory system; bleeding.

hemostasis (he′′mo-sta′sis)   The stoppage of bleeding.

heparin (hep′ah-rin)   A substance that interferes with the formation of a blood clot; an anticoagulant.

hepatic (hē-pat′ik)   Pertaining to the liver.

hepatic lobule (hē-pat′ik lob′ūl)   A functional unit of the liver.

hepatic sinusoid (hē-pat′ik si′nŭ-soid)   Vascular channel within the liver.

heredity (hē-red′i-te)   The transmission of genetic information from parent to offspring.

heterozygote (het′′er-o-zi′gōt)   An individual who possesses different alleles in a gene pair.

heterozygous (het′′er-o-zi′gus)   Pertaining to a heterozygote.

hindbrain (hind′brān)   Posterior-most portion of the developing brain that gives rise to the cerebellum, pons, and medulla oblongata.

histamine (his′tah-min)   A substance released from cells subjected to stressful conditions.

histology (his-tol′o-je)   The study of the structure and function of tissues.

homeostasis (ho′′me-o-sta′sis)   A state of equilibrium in which the internal environment of the body remains relatively constant.

homozygote (ho′′mo-zi′gōt)   An individual possessing identical alleles in a gene pair.

homozygous (ho′′mo-zi′gus)   Pertaining to a homozygote.

hormone (hor′mōn)   A substance secreted by an endocrine gland that is transmitted in the blood or body fluids.

humoral immunity (hu′mor-al i-mu′ni-te)   Resistance to the effects of specific disease-causing agents due to the presence of circulating antibodies; antibody-mediated immunity.

hydrolysis (hi-drol′i-sis)   The splitting of a molecule into smaller portions by the addition of a water molecule.

hydrostatic pressure (hi′′dro-stat′ik presh′ur)   Pressure exerted by fluids, such as blood pressure.

hydroxyapatite (hi-drok′′se-ap′ah-tīt)   A type of crystalline calcium phosphate found in bone matrix.

hydroxyl ion (hi-drok′sil i′on) OH^-.

hymen (hi′men)   A membranous fold of tissue that partially covers the vaginal opening.

hyperglycemia (hi′′per-gli-se′me-ah)   An excessive level of blood glucose.

hyperkalemia (hi′′per-kah-le′me-ah)   An excessive concentration of blood potassium.

hypernatremia (hi′′per-nah-tre′me-ah)   An excessive concentration of blood sodium.

hyperparathyroidism (hi′′per-par′′ah-thi′roi-dizm)   An excessive secretion of parathyroid hormone.

hyperplasia (hi′′per-pla′ze-ah)   An increased production and growth of new cells.

hyperpolarization (hi′′per-po′′lar-i-za′shun)   An increase in the negativity of the resting potential of a cell membrane.

hypertension (hi′′per-ten′shun)   Excessive blood pressure.

hyperthyroidism (hi′′per-thi′roi-dizm)   An excessive secretion of thyroid hormones.

hypertonic (hi′′per-ton′ik)   Condition in which a solution contains a greater concentration of dissolved particles than the solution with which it is compared.

hypertrophy (hi-per′tro-fe)   Enlargement of an organ or tissue.

hyperventilation (hi′′per-ven′′ti-la′shun)   Breathing that is abnormally deep and prolonged.

hypervitaminosis (hi′′per-vi′′tah-mi-no′sis)   Excessive intake of vitamins.

hypochondriac region (hi′′po-kon′dre-ak re′jun)   The portion of the abdomen on either side of the middle or epigastric region.

hypogastric region (hi′′po-gas′trik re′jun)   The lower middle portion of the abdomen.

hypoglycemia (hi′′po-gli-se′me-ah)   Abnormally low concentration of blood glucose.

hypokalemia (hi′′po-kah-le′me-ah)   A low concentration of blood potassium.

hyponatremia (hi′′po-nah-tre′me-ah)   A low concentration of blood sodium.

hypoparathyroidism (hi′′po-par′′ah-thi′roi-dizm)   An undersecretion of parathyroid hormone.

hypophysis (hi-pof′i-sis)   The pituitary gland.

hypoproteinemia (hi′′po-pro′′te-i-ne′me-ah)   A low concentration of blood proteins.

hypothalamus (hi′′po-thal′ah-mus)   A portion of the brain located below the thalamus and forming the floor of the third ventricle.

hypothyroidism (hi′′po-thi′roi-dizm)   A low secretion of thyroid hormones.

hypotonic (hi′′po-ton′ik)   Condition in which a solution contains a lesser concentration of dissolved particles than the solution to which it is compared.

hypoxia (hi-pok′se-ah)   A deficiency of oxygen in the tissues.

i

ileocecal valve (il′e-o-se′kal valv)   Sphincter valve located at the distal end of the ileum where it joins the cecum.

ileum (il′e-um)   Portion of the small intestine between the jejunum and cecum.

iliac region (il′e-ak re′jun)   Portion of the abdomen on either side of the lower middle or hypogastric region.

ilium (il′e-um)   One of the bones of a coxal bone or hipbone.

immunity (i-mu′ni-te)   Resistance to the effects of specific disease-causing agents.

immunoglobulin (im′′u-no-glob′u-lin)   Globular plasma proteins that function as antibodies of immunity.

immunosuppressive drugs
(im″u-no-sŭ-pres′iv drugz) Substances that inhibit the formation of antibodies.

implantation (im″plan-ta′shun) The embedding of an embryo in the lining of the uterus.

impulse (im′puls) A wave of depolarization conducted along a nerve fiber or muscle fiber.

incisor (in-si′zor) One of the front teeth that is adapted for cutting food.

inclusion (in-kloo′zhun) A mass of lifeless chemical substance within the cytoplasm of a cell.

incomplete protein (in″kom-plēt′ pro′te-in) A protein that lacks essential amino acids.

infancy (in′fan-se) Period of life from the end of the first 4 weeks to 1 year of age.

inferior (in-fēr′e-or) Situated below something else; pertaining to the lower surface of a part.

inflammation (in″flah-ma′shun) A tissue response to stress that is characterized by dilation of blood vessels and an accumulation of fluid in the affected region.

infrared ray (in″frah-red′ ra) A form of radiation energy, with wavelengths longer than visible light, by which heat moves from warmer surfaces to cooler surroundings.

infundibulum (in″fun-dib′u-lum) The stalk by which the pituitary gland is attached to the base of the brain.

ingestion (in-jes′chun) The taking of food or liquid into the body by way of the mouth.

inguinal (ing′gwĭ-nal) Pertaining to the groin region.

inguinal canal (ing′gwĭ-nal kah-nal′) Passage in the lower abdominal wall through which a testis descends into the scrotum.

inorganic (in″or-gan′ik) Pertaining to chemical substances that lack carbon.

insertion (in-ser′shun) The end of a muscle that is attached to a movable part.

inspiration (in″spĭ-ra′shun) Act of breathing in; inhalation.

insula (in′su-lah) A cerebral lobe located deep within the lateral sulcus.

insulin (in′su-lin) A hormone secreted by the pancreatic islets of Langerhans that functions in the control of carbohydrate metabolism.

integumentary (in-teg-u-men′tar-e) Pertaining to the skin and its accessory organs.

intercalated disk (in-ter″kah-lāt′ed disk) Membranous boundary between adjacent cardiac muscle cells.

intercellular (in″ter-sel′u-lar) Between cells.

intercellular fluid (in″ter-sel′u-lar floo′id) Tissue fluid located between cells, other than blood cells.

interferon (in″ter-fēr′on) A substance produced by cells that inhibits the multiplication of viruses.

interneuron (in″ter-nu′ron) A neuron located between a sensory neuron and a motor neuron.

interphase (in′ter-fāz) Period between two cell divisions when a cell is carrying on its normal functions.

interstitial cell (in″ter-stish′al sel) A hormone-secreting cell located between the seminiferous tubules of the testis.

interstitial fluid (in″ter-stish′al floo′id) Same as intercellular fluid.

intervertebral disk (in″ter-ver′tĕ-bral disk) A layer of fibrocartilage located between the bodies of adjacent vertebrae.

intestinal gland (in-tes′tĭ-nal gland) Tubular gland located at the base of a villus within the intestinal wall.

intestinal juice (in-tes′tĭ-nal joos) The secretion of the intestinal glands.

intracellular (in″trah-sel′u-lar) Within cells.

intracellular fluid (in″trah-sel′u-lar floo′id) Fluid within cells.

intracellular junction (in″trah-sel′u-lar jungk′shun) A connection between the membranes of adjacent cells.

intramembranous bone (in″trah-mem′brah-nus bōn) Bone that forms from membrane-like layers of primitive connective tissue.

intrauterine device (in″trah-u′ter-in de-vīs′) A solid object placed in the uterine cavity for purposes of contraception; IUD.

intrinsic factor (in-trin′sik fak′tor) A substance produced by the gastric glands that promotes the absorption of vitamin B_{12}.

inversion (in-ver′zhun) Movement in which the sole of the foot is turned inward.

involuntary (in-vol′un-tār″e) Not consciously controlled; functions automatically.

ion (i′on) An atom or a group of atoms with an electrical charge.

ionization (i″on-i-za′shun) Chemical process by which substances dissociate into ions.

ipsilateral (ip″sĭ-lat′er-al) Positioned on the same side as something else.

iris (i′ris) Colored muscular portion of the eye that surrounds the pupil and regulates its size.

irritability (ir″ĭ-tah-bil′ĭ-te) The ability of an organism to react to changes taking place in its environment.

ischemia (is-ke′me-ah) A deficiency of blood in a body part.

isometric contraction (i″so-met′rik kon-trak′shun) Muscular contraction in which the muscle fails to shorten.

isotonic contraction (i″so-ton′ik kon-trak′shun) Muscular contraction in which the muscle shortens.

isotonic solution (i″so-ton′ik so-lu′shun) A solution that has the same concentration of dissolved particles as the solution with which it is compared.

isotope (i′so-tōp) An atom that has the same number of protons as other atoms of an element but has a different number of neutrons in its nucleus.

IUD An intrauterine device.

j

jejunum (je-joo′num) Portion of the small intestine located between the duodenum and the ileum.

joint (joint) The union of two or more bones; an articulation.

juxtaglomerular apparatus (juks″tah-glo-mer′u-lar ap″ah-ra′tus) Structure located in the walls of arterioles near the glomerulus that plays an important role in regulating renal blood flow.

juxtamedullary nephron (juks″tah-med′u-lār-e nef′ron) A nephron with its corpuscle located near the renal medulla.

k

keratin (ker′ah-tin) Protein present in the epidermis, hair, and nails.

keratinization (ker″ah-tin″ĭ-za′shun) The process by which cells form fibrils of keratin and become hardened.

kernicterus (ker-nik′ter-us) Condition in which bilirubin precipitates in the brain tissues of an infant affected with erythroblastosis fetalis.

ketogenesis (ke″to-jen′ĕ-sis) The formation of ketone bodies.

ketone body (ke′tōn bod′e) Type of compound produced during fat catabolism, including acetone, acetoacetic acid, and betahydroxybutyric acid.

ketosis (ke″to′sis) A condition in which the concentration of ketone bodies in body fluids is abnormally increased.

kilocalorie (kil′o-kal″o-re) One thousand calories.

kilogram (kil′o-gram) A unit of weight equivalent to 1000 grams.

Kreb's cycle (krebz si′kl) The citric acid cycle.

Kupffer cell (koop′fer sel) Large, fixed phagocyte in the liver that removes bacterial cells from the blood.

kyphosis (ki-fo′sis) An abnormally increased convex curvature in the thoracic portion of the vertebral column.

l

labor (la′bor) The process of childbirth.

labyrinth (lab′ĭ-rinth) The system of interconnecting tubes within the inner ear, which includes the cochlea, vestibule, and semicircular canals.

lacrimal gland (lak′rĭ-mal gland) Tear-secreting gland.

lactase (lak′tās) Enzyme that converts lactose into glucose and galactose.

lactation (lak-ta′shun) The production of milk by the mammary glands.

lacteal (lak′te-al) A lymphatic vessel associated with a villus of the small intestine.

lactic acid (lak′tik as′id) An organic substance formed from pyruvic acid during anaerobic respiration.

lactose (lak′tōs) A disaccharide that occurs in milk; milk sugar.

lacuna (lah-ku′nah) A hollow cavity.

lamella (lah-mel′ah)   A layer of matrix in bone tissue.

laryngopharynx (lah-ring″go-far′ingks)   The lower portion of the pharynx near the opening to the larynx.

larynx (lar′ingks)   Structure located between the pharynx and trachea that houses the vocal cords.

latent period (la′tent pe′re-od)   Time lapse between the application of a stimulus and the beginning of a response in a muscle fiber.

lateral (lat′er-al)   Pertaining to the side.

leukocyte (lu′ko-sīt)   A white blood cell.

leukocytosis (lu″ko-si-to′sis)   An abnormally large increase in the number of white blood cells.

leukopenia (lu″ko-pe′ne-ah)   An abnormally low number of leukocytes in the blood.

lever (lev′er)   A simple mechanical device consisting of a rod, fulcrum, weight, and a source of energy that is applied to some point on the rod.

ligament (lig′ah-ment)   A cord or sheet of connective tissue by which two or more bones are bound together at a joint.

limbic system (lim′bik sis′tem)   A group of interconnected structures within the brain that function to produce various emotional feelings.

linea alba (lin′e-ah al′bah)   A narrow band of tendinous connective tissue located in the midline of the anterior abdominal wall.

lingual (ling′gwal)   Pertaining to the tongue.

lipase (li′pās)   A fat-digesting enzyme.

lipid (lip′id)   A fat, oil, or fatlike compound that usually has fatty acids in its molecular structure.

lipoprotein (lip″o-pro′te-in)   A complex of lipid and protein.

lordosis (lor-do′sis)   An abnormally increased concave curvature in the lumbar portion of the vertebral column.

lumbar (lum′bar)   Pertaining to the region of the loins.

lumen (lu′men)   Space within a tubular structure such as a blood vessel or intestine.

luteinizing hormone (lu′te-in-īz″ing hor′mōn)   A hormone secreted by the anterior pituitary gland that controls the formation of corpus luteum in females and the secretion of testosterone in males; LH (ICSH in males).

lymph (limf)   Fluid transported by the lymphatic vessels.

lymph node (limf nōd)   A mass of lymphoid tissue located along the course of a lymphatic vessel.

lymphocyte (lim′fo-sīt)   A type of white blood cell that functions to provide immunity.

lysosome (li′so-sōm)   Cytoplasmic organelle that contains digestive enzymes.

m

macrocyte (mak′ro-sīt)   A large red blood cell.

macrophage (mak′ro-fāj)   A large phagocytic cell.

macroscopic (mak″ro-skop′ik)   Large enough to be seen with the unaided eye.

macula (mak′u-lah)   A group of hair cells and supporting cells associated with an organ of static equilibrium.

macula lutea (mak′u-lah lu′te-ah)   A yellowish depression in the retina of the eye that is associated with acute vision.

malignant (mah-lig′nant)   The power to threaten life; cancerous.

malnutrition (mal″nu-trish′un)   A condition resulting from an improper diet.

maltase (mawl′tās)   An enzyme that converts maltose into glucose.

maltose (mawl′tōs)   A disaccharide composed of two glucose molecules.

mammary (mam′ar-e)   Pertaining to the breast.

marrow (mar′o)   Connective tissue that occupies the spaces within bones.

mast cell (mast sel)   A cell to which antibodies, formed in response to allergens, become attached.

mastication (mas″ti-ka′shun)   Chewing movements.

matrix (ma′triks)   The intercellular substance of connective tissue.

matter (mat′er)   Anything that has weight and occupies space.

meatus (me-a′tus)   A passageway or channel, or the external opening of such a passageway.

mechanoreceptor (mek″ah-no-re-sep′tor)   A sensory receptor that is sensitive to mechanical stimulation such as changes in pressure or tension.

medial (me′de-al)   Toward or near the midline.

mediastinum (me″de-ah-sti′num)   Tissues and organs of the thoracic cavity that form a septum between the lungs.

medulla (mĕ-dul′ah)   The inner portion of an organ.

medulla oblongata (mĕ-dul′ah ob″long-gah′tah)   Portion of the brain stem located between the pons and the spinal cord.

medullary cavity (med′u-lār″e kav′i-te)   Cavity within the diaphysis of a long bone occupied by marrow.

megakaryocyte (meg″ah-kar′e-o-sīt)   A large bone marrow cell that functions to produce blood platelets.

meiosis (mi-o′sis)   Process of cell division by which egg and sperm cells are formed.

melanin (mel′ah-nin)   Dark pigment normally found in skin and hair.

melanocyte (mel′ah-no-sīt″)   Melanin-producing cell.

melatonin (mel″ah-to′nin)   A hormone thought to be secreted by the pineal gland.

memory cell (mem′o-re sel)   B- or T-lymphocyte produced in response to a primary immune response that remains dormant and can respond rapidly if the same antigen is encountered in the future.

menarche (mĕ-nar′ke)   The first menstrual period.

meninges (mĕ-nin′jēz)   A group of three membranes that covers the brain and spinal cord (singular, meninx).

menisci (men-is′si)   Pieces of fibrocartilage that separate the articulating surfaces of bones in the knee.

menopause (men′o-pawz)   Termination of the menstrual cycles.

menstrual cycle (men′stroo-al si′kl)   The female reproductive cycle that is characterized by regularly reoccurring changes in the uterine lining.

menstruation (men″stroo-a′shun)   Loss of blood and tissue from the uterus at the end of a female reproductive cycle.

mesentery (mes′en-ter″e)   A fold of peritoneal membrane that attaches an abdominal organ to the abdominal wall.

mesoderm (mez′o-derm)   The middle layer of the primary germ layers, responsible for forming certain embryonic body parts.

messenger RNA (mes′in-jer)   Molecule of RNA that transmits information for protein synthesis from the nucleus of a cell to the cytoplasm.

metabolic rate (met″ah-bol′ic rāt)   The rate at which chemical changes occur within the body.

metabolism (mĕ-tab′o-lizm)   All of the chemical changes that occur within cells considered together.

metacarpals (met″ah-kar′pals)   Bones of the hand between the wrist and finger bones.

metaphase (met′ah-fāz)   Stage in mitosis when chromosomes become aligned in the middle of the spindle.

metastasis (mĕ-tas′tah-sis)   The spread of disease from one body region to another; a characteristic of cancer.

metatarsals (met″ah-tar′sals)   Bones of the foot between the ankle and toe bones.

microfilament (mi″kro-fil′ah-ment)   Tiny rod of protein that occurs in cytoplasm and functions in causing various cellular movements.

microglia (mi-krog′le-ah)   A type of neuroglial cell that helps support neurons and acts to carry on phagocytosis.

microscopic (mi″kro-skop′ik)   Too small to be seen with the unaided eye.

microtubule (mi″kro-tu′būl)   A minute, hollow rod found in the cytoplasm of cells.

microvilli (mi″kro-vil′i)   Tiny, cylindrical processes that extend outward from some epithelial cell membranes and increase the membrane surface area.

micturition (mik″tu-rish′un)   Urination.

midbrain (mid′brān)   A small region of the brain stem located between the diencephalon and pons.

mineralocorticoid (min″er-al-o-kor′ti-koid)   Any one of a group of hormones secreted by the adrenal cortex that influences the concentrations of electrolytes in body fluids.

mitochondrion (mi″to-kon′dre-on)   Cytoplasmic organelle that contains enzymes responsible for aerobic respiration (plural, *mitochondria*).

mitosis (mi-to′sis)   Process by which body cells divide to form two identical daughter cells.

mitral valve (mi′tral valv)   Heart valve located between the left atrium and the left ventricle; bicuspid valve.

mixed nerve (mikst nerv)   Nerve that includes both sensory and motor nerve fibers.

molar (mo′lar)   A rear tooth with a somewhat flattened surface adapted for grinding food.

molecule (mol′ĕ-kūl)   A particle composed of two or more atoms bonded together.

monocyte (mon′o-sīt)   A type of white blood cell that functions as a phagocyte.

monosaccharide (mon″o-sak′ah-rid)   A simple sugar, such as glucose or fructose, that represents the structural unit of a carbohydrate.

morula (mor′u-lah)   An early stage in embryonic development; a solid ball of cells.

motor area (mo′tor a′re-ah)   A region of the brain from which impulses to muscles or glands originate.

motor end plate (mo′tor end plāt)   Specialized portion of a muscle fiber membrane at a neuromuscular junction.

motor nerve (mo′tor nerv)   A nerve that consists of motor nerve fibers.

motor neuron (mo′tor nu′ron)   A neuron that transmits impulses from the central nervous system to an effector.

motor unit (mo′tor u′nit)   A motor neuron and the muscle fibers associated with it.

mucosa (mu-ko′sah)   The membrane that lines tubes and body cavities that open to the outside of the body; mucous membrane.

mucous cell (mu′kus sel)   Glandular cell that secretes mucus.

mucous membrane (mu′kus mem′brān)   Mucosa.

mucus (mu′kus)   Fluid secretion of the mucous cells.

multiple alleles (mul′ti-pl ah-lēls′)   Condition in which the alleles that constitute a gene pair occur in several forms.

multiple motor unit summation (mul′ti-pl mo′tor u′nit sum-mā′shun)   A sustained muscle contraction of increasing strength resulting from responses by many motor units.

multipolar neuron (mul″ti-po′lar nu′ron)   Nerve cell that has many processes arising from its cell body.

muscle spindle (mus′el spin′dul)   Modified skeletal muscle fiber that can respond to changes in muscle length.

mutagenic (mu″tah-jen′ik)   Pertaining to a factor that can cause mutations.

mutation (mu-ta′shun)   A change in the genetic information of a chromosome.

myelin (mi′ĕ-lin)   Fatty material that forms a sheathlike covering around some nerve fibers.

myocardium (mi″o-kar′de-um)   Muscle tissue of the heart.

myofibril (mi″o-fi′bril)   Contractile fibers found within muscle cells.

myoglobin (mi″o-glo′bin)   A pigmented compound found in muscle tissue that acts to store oxygen.

myogram (mi′o-gram)   A recording of a muscular contraction.

myometrium (mi″o-me′tre-um)   The layer of smooth muscle tissue within the uterine wall.

myoneural junction (mi″o-nu′ral jungk′shun)   Site of union between a motor neuron axon and a muscle fiber.

myopia (mi-o′pe-ah)   Nearsightedness.

myosin (mi′o-sin)   A protein that, together with actin, is responsible for muscular contraction and relaxation.

myxedema (mik″sĕ-de′mah)   Condition resulting from a deficiency of thyroid hormones in an adult.

n

nasal cavity (na′zal kav′i-te)   Space within the nose.

nasal concha (na′zal kong′kah)   Shell-like bone extending outward from the wall of the nasal cavity; a turbinate bone.

nasal septum (na′zal sep′tum)   A wall of bone and cartilage that separates the nasal cavity into two portions.

nasopharynx (na″zo-far′ingks)   Portion of the pharynx associated with the nasal cavity.

negative feedback (neg′ah-tiv fēd′bak)   A mechanism that is activated by an imbalance and acts to correct it.

neonatal (ne″o-na′tal)   Pertaining to the period of life from birth to the end of 4 weeks.

nephron (nef′ron)   The functional unit of a kidney, consisting of a renal corpuscle and a renal tubule.

nerve (nerv)   A bundle of nerve fibers.

neurofibrils (nu″ro-fi′brils)   Fine, cytoplasmic threads that extend from the cell body into the processes of neurons.

neuroglia (nu-rog′le-ah)   The supporting tissue within the brain and spinal cord, composed of neuroglial cells.

neurolemma (nu″ro-lem′ah)   Sheath on the outside of some nerve fibers due to the presence of Schwann cells.

neuromodulator (nu″ro-mod′u-lā-tor)   A substance that alters a neuron's response to a neurotransmitter.

neuromuscular junction (nu″ro-mus′ku-lar jungk′shun)   Myoneural junction.

neuron (nu′ron)   A nerve cell that consists of a cell body and its processes.

neuropeptide (nu″ro-pep′tid)   A peptide that occurs in the brain and seems to function as a neurotransmitter or neuromodulator.

neurotransmitter (nu″ro-trans-mit′er)   Chemical substance secreted by the terminal end of an axon that stimulates a muscle fiber contraction or an impulse in another neuron.

neutral (nu′tral)   Neither acid nor alkaline.

neutron (nu′tron)   An electrically neutral particle found in an atomic nucleus.

neutrophil (nu′tro-fil)   A type of phagocytic leukocyte.

niacin (ni′ah-sin)   A vitamin of the B-complex group; nicotinic acid.

niacinamide (ni″ah-sin-am′id)   The physiologically active form of niacin.

nicotinic acid (nik″o-tin′ik as′id)   Niacin.

Nissl bodies (nis′l bod′ēz)   Membranous sacs that occur within the cytoplasm of nerve cells and have ribosomes attached to their surfaces.

nitrogen balance (ni′tro-jen bal′ans)   Condition in which the quantity of nitrogen ingested equals the quantity excreted.

nondisjunction (non″dis-jungk′shun)   The failure of a pair of chromosomes to separate during meiosis.

nonelectrolyte (non″e-lek′tro-līt)   A substance that does not dissociate into ions when it is dissolved.

nonprotein nitrogenous substance (non-pro′te-in ni-troj′ĕ-nus sub′stans)   A substance, such as urea or uric acid, that contains nitrogen but is not a protein.

norepinephrine (nor″ep-i-nef′rin)   A neurotransmitter substance released from the axon ends of some nerve fibers.

nuclease (nu′kle-ās)   An enzyme that causes nucleic acids to decompose.

nucleic acid (nu-kle′ik as′id)   A substance composed of nucleotides bonded together; RNA or DNA.

nucleolus (nu-kle′o-lus)   A small structure that occurs within the nucleus of a cell and contains RNA.

nucleoplasm (nu′kle-o-plazm″)   The contents of the nucleus of a cell.

nucleosome (nu′kle-o-sōm)   A beadlike particle within a chromatin fiber composed of DNA and protein.

nucleotide (nu′kle-o-tid″)   A component of a nucleic acid molecule, consisting of a sugar, a nitrogenous base, and a phosphate group.

nucleus (nu′kle-us)   A body that occurs within a cell and contains relatively large quantities of DNA; the dense core of an atom that is composed of protons and neutrons.

nutrient (nu′tre-ent)   A chemical substance that must be supplied to the body from its environment.

nutrition (nu-trish′un)   The study of the sources, actions, and interactions of nutrients.

o

obesity (o-bēs'ĭ-te)   An excessive accumulation of adipose tissue; usually the condition of exceeding the desirable weight by more than 20%.

occipital (ok-sip'ĭ-tal)   Pertaining to the lower, back portion of the head.

olfactory (ol-fak'to-re)   Pertaining to the sense of smell.

olfactory nerves (ol-fak'to-re nervz)   The first pair of cranial nerves, which conduct impulses associated with the sense of smell.

oligodendrocyte (ol''ĭ-go-den'dro-sīt)   A type of neuroglial cell that functions to connect neurons to blood vessels and to form myelin.

oocyte (o'o-sīt)   An immature egg cell.

oogenesis (o''o-jen'ĕ-sis)   The process by which an egg cell forms from an oocyte.

ophthalmic (of-thal'mik)   Pertaining to the eye.

optic (op'tik)   Pertaining to the eye.

optic chiasma (op'tik ki-az'mah)   X-shaped structure on the underside of the brain created by a partial crossing over of fibers in the optic nerves.

optic disk (op'tik disk)   Region in the retina of the eye where nerve fibers leave to become part of the optic nerve.

oral (o'ral)   Pertaining to the mouth.

organ (or'gan)   A structure consisting of a group of tissues that performs a specialized function.

organelle (or''gah-nel')   A living part of a cell that performs a specialized function.

organic (or-gan'ik)   Pertaining to carbon-containing substances.

organism (or'gah-nizm)   An individual living thing.

orifice (or'ĭ-fis)   An opening.

origin (or'ĭ-jin)   End of a muscle that is attached to a relatively immovable part.

oropharynx (o''ro-far'ingks)   Portion of the pharynx in the posterior part of the oral cavity.

osmoreceptor (oz''mo-re-sep'tor)   Receptor that is sensitive to changes in the osmotic pressure of body fluids.

osmosis (oz-mo'sis)   Diffusion of water through a selectively permeable membrane due to the existence of a concentration gradient.

osmotic (oz-mot'ik)   Pertaining to osmosis.

osmotic pressure (oz-mot'ik presh'ur)   The amount of pressure needed to stop osmosis; the potential pressure of a solution due to the presence of nondiffusible solute particles in the solution.

osseous tissue (os'e-us tish'u)   Bone tissue.

ossification (os''ĭ-fĭ-ka'shun)   The formation of bone tissue.

osteoblast (os'te-o-blast'')   A bone-forming cell.

osteoclast (os'te-o-klast'')   A cell that causes the erosion of bone.

osteocyte (os'te-o-sīt)   A mature bone cell.

osteon (os'te-on)   A cylinder-shaped unit containing bone cells that surround an osteonic canal; haversian system.

osteonic canal (os'te-o-nik kan'al)   A tiny channel in bone tissue that contains a blood vessel; haversian canal.

otolith (o'to-lith)   A small particle of calcium carbonate associated with the receptors of equilibrium.

otosclerosis (o''to-sklĕ-ro'sis)   Abnormal formation of spongy bone within the ear that may interfere with the transmission of sound vibrations to hearing receptors.

oval window (o'val win'do)   Opening between the stapes and the inner ear.

ovarian (o-va're-an)   Pertaining to the ovary.

ovary (o'var-e)   The primary reproductive organ of a female; an egg-cell-producing organ.

oviduct (o'vi-dukt)   A tube that leads from the ovary to the uterus; uterine tube or fallopian tube.

ovulation (o''vu-la'shun)   The release of an egg cell from a mature ovarian follicle.

ovum (o'vum)   A mature egg cell.

oxidase (ok'sĭ-dās)   An enzyme that promotes oxidation.

oxidation (ok''sĭ-da'shun)   Process by which oxygen is combined with a chemical substance; the removal of hydrogen or the loss of electrons; the opposite of reduction.

oxygen debt (ok'sĭ-jen det)   The amount of oxygen that must be supplied following physical exercise to convert accumulated lactic acid to glucose.

oxyhemoglobin (ok''sĭ-he''mo-glo'bin)   Compound formed when oxygen combines with hemoglobin.

oxytocin (ok''sĭ-to'sin)   Hormone released by the posterior lobe of the pituitary gland that causes contraction of smooth muscles in the uterus and mammary glands.

p

pacemaker (pās'māk-er)   Mass of specialized muscle tissue that controls the rhythm of the heartbeat; the sinoatrial node.

pain receptor (pān re''sep'tor)   Sensory nerve ending associated with the feeling of pain.

palate (pal'at)   The roof of the mouth.

palatine (pal'ah-tīn)   Pertaining to the palate.

palmar (pahl'mar)   Pertaining to the palm of the hand.

pancreas (pan'kre-as)   Glandular organ in the abdominal cavity that secretes hormones and digestive enzymes.

pancreatic (pan''kre-at'ik)   Pertaining to the pancreas.

pantothenic acid (pan''to-then'ik as'id)   A vitamin of the B-complex group.

papilla (pah-pil'ah)   Tiny nipplelike projection.

papillary muscle (pap'ĭ-ler''e mus'el)   Muscle that extends inward from the ventricular walls of the heart and to which the chordae tendineae are attached.

paradoxical sleep (par''ah-dok'se-kal slēp)   Sleep in which some areas of the brain are active, producing dreams and rapid eye movements.

paralysis (pah-ral'ĭ-sis)   Loss of ability to control voluntary muscular movements, usually due to a disorder of the nervous system.

parasympathetic division (par''ah-sim''pah-thet'ik dĭ-vizh'un)   Portion of the autonomic nervous system that arises from the brain and sacral region of the spinal cord.

parathormone (par''ah-thŏr'mōn)   Hormone secreted by the parathyroid glands that helps to regulate the level of blood calcium and phosphate; parathyroid hormone or PTH.

parathyroid glands (par''ah-thi'roid glandz)   Small endocrine glands that are embedded in the posterior portion of the thyroid gland.

paravertebral ganglia (par''ah-ver'tĕ-bral gang'gle-ah)   Sympathetic ganglia that form chains along the sides of the vertebral column.

parietal (pah-ri'ĕ-tal)   Pertaining to the wall of an organ or cavity.

parietal cell (pah-ri'ĕ-tal sel)   Cell of a gastric gland that secretes hydrochloric acid and intrinsic factor.

parietal pleura (pah-ri'ĕ-tal ploo'rah)   Membrane that lines the inner wall of the thoracic cavity.

parotid glands (pah-rot'id glandz)   Large salivary glands located on the sides of the face just in front and below the ears.

partial pressure (par'shal presh'ur)   The pressure produced by one gas in a mixture of gases.

parturition (par''tu-rish'un)   The process of childbirth.

pathogen (path'o-jen)   Any disease-causing agent.

pathology (pah-thol'o-je)   The study of disease.

pectoral (pek'tor-al)   Pertaining to the chest.

pectoral girdle (pek'tor-al ger'dl)   Portion of the skeleton that provides support and attachment for the arms.

pelvic (pel'vik)   Pertaining to the pelvis.

pelvic girdle (pel'vik ger'dl)   Portion of the skeleton to which the legs are attached.

pelvis (pel'vis)   Bony ring formed by the sacrum and coxal bones.

penis (pe'nis)   External reproductive organ of the male through which the urethra passes.

pepsin (pep'sin)   Protein-splitting enzyme secreted by the gastric glands of the stomach.

pepsinogen (pep-sin'o-jen)   Inactive form of pepsin.

peptidase (pep'tĭ-dās)   An enzyme that causes the breakdown of polypeptides.

peptide (pep'tid)   Compound composed of two or more amino acid molecules joined together.

peptide bond (pep′tīd bond) Bond that forms between the carboxyl group of one amino acid and the amino group of another.

pericardial (per″ĭ-kar′de-al) Pertaining to the pericardium.

pericardium (per″ĭ-kar′de-um) Serous membrane that surrounds the heart.

perichondrium (per″ĭ-kon′dre-um) Layer of fibrous connective tissue that encloses cartilaginous structures.

perilymph (per′ĭ-limf) Fluid contained in the space between the membranous and osseous labyrinths of the inner ear.

perimetrium (per-ĭ-me′tre-um) The outer serosal layer of the uterine wall.

perimysium (per″ĭ-mis′e-um) Sheath of connective tissue that encloses a bundle of striated muscle fibers.

perineal (per″ĭ-ne′al) Pertaining to the perineum.

perineum (per″ĭ-ne′um) Body region between the scrotum or urethral opening and the anus.

perineurium (per″ĭ-nu′re-um) Layer of connective tissue that encloses a bundle of nerve fibers within a nerve.

periodontal ligament (per″e-o-don′tal lig′ah-ment) Fibrous membrane that surrounds a tooth and attaches it to the bone of the jaw.

periosteum (per″e-os′te-um) Covering of fibrous connective tissue on the surface of a bone.

peripheral (pĕ-rif′er-al) Pertaining to parts located near the surface or toward the outside.

peripheral nervous system (pĕ-rif′er-al ner′vus sis′tem) The portions of the nervous system outside the central nervous system.

peripheral resistance (pĕ-rif′er-al re-zis′tans) Resistance to blood flow due to friction between the blood and the walls of the blood vessels.

peristalsis (per″ĭ-stal′sis) Rhythmic waves of muscular contraction that occur in the walls of various tubular organs.

peritoneal (per″ĭ-to-ne′al) Pertaining to the peritoneum.

peritoneal cavity (per″ĭ-to-ne′al kav′ĭ-te) The potential space between the parietal and visceral peritoneal membranes.

peritoneum (per″ĭ-to-ne′um) A serous membrane that lines the abdominal cavity and encloses the abdominal viscera.

peritubular capillary (per″ĭ-tu′bu-lar kap′ĭ-ler″e) Capillary that surrounds a renal tubule and functions in reabsorption and secretion during urine formation.

permeable (per′me-ah-bl) Open to passage or penetration.

peroxisome (pĕ-roks′ĭ-sōm) Membranous cytoplasmic vesicle that contains enzymes responsible for the production and decomposition of hydrogen peroxide.

pH The negative logarithm of the hydrogen ion concentration used to indicate the acid or alkaline condition of a solution.

phagocytosis (fag″o-si-to′sis) Process by which a cell engulfs and digests solid substances.

phalanx (fa′langks) A bone of a finger or toe (plural, *phalanges*).

pharynx (far′ingks) Portion of the digestive tube between the mouth and esophagus.

phenotype (fe′no-tip) The appearance of an individual due to the action of a particular set of genes.

phospholipid (fos″fo-lip′id) A lipid that contains phosphorus.

photoreceptor (fo″to-re-sep′tor) A nerve ending that is sensitive to light energy.

physiology (fiz″e-ol′o-je) The branch of science dealing with the study of body functions.

pia mater (pi′ah ma′ter) Inner layer of meninges that encloses the brain and spinal cord.

pineal gland (pin′e-al gland) A small structure located in the central part of the brain.

pinocytosis (pin″o-si-to′sis) Process by which a cell engulfs droplets of fluid from its surroundings.

pituitary gland (pĭ-tu′ĭ-tār″e gland) Endocrine gland that is attached to the base of the brain and consists of anterior and posterior lobes; the hypophysis.

placenta (plah-sen′tah) Structure by which an unborn child is attached to its mother's uterine wall and through which it is nourished.

plantar (plan′tar) Pertaining to the sole of the foot.

plasma (plaz′mah) Fluid portion of circulating blood.

plasma cell (plaz′mah sel) Antibody-producing cell that is formed as a result of the proliferation of sensitized B-lymphocytes.

plasma protein (plaz′mah pro′te-in) Any of several proteins normally found dissolved in blood plasma.

platelet (plāt′let) Cytoplasmic fragment formed in the bone marrow that functions in blood coagulation.

pleural (ploo′ral) Pertaining to the pleura or membranes investing the lungs.

pleural cavity (ploo′ral kav′ĭ-te) Potential space between the pleural membranes.

pleural membranes (ploo′ral mem′brānz) Serous membranes that enclose the lungs.

plexus (plek′sus) A network of interlaced nerves or blood vessels.

pneumotaxic area (nu″mo-tax′ik a′re-ah) A portion of the respiratory control center located in the pons of the brain.

polar body (po′lar bod′e) Small, nonfunctional cell produced as a result of meiosis during egg cell formation.

polarization (po″lar-i-za′shun) The development of an electrical charge on the surface of a cell membrane due to an unequal distribution of ions on either side of the membrane.

polycythemia (pol″e-si-the′me-ah) An excessive concentration of red blood cells.

polymorphonuclear leukocyte (pol″e-mor″fo-nu′kle-ar lu′ko-sīt) A leukocyte or white blood cell with an irregularly lobed nucleus.

polypeptide (pol″e-pep′tīd) A compound formed by the union of many amino acid molecules.

polysaccharide (pol″e-sak′ah-rīd) A carbohydrate composed of many monosaccharide molecules joined together.

pons (ponz) A portion of the brain stem above the medulla oblongata and below the midbrain.

popliteal (pop″lĭ-te′al) Pertaining to the region behind the knee.

positive feedback (poz′ĭ-tiv fēd′bak) Process by which changes cause more changes of a similar type, producing unstable conditions; vicious cycle.

posterior (pos-tēr′e-or) Toward the back; opposite of anterior.

postganglionic fiber (pōst″gang-gle-on′ik fi′ber) Autonomic nerve fiber located on the distal side of a ganglion.

postnatal (pōst-na′tal) After birth.

precursor (pre-ker′sor) Substance from which another substance is formed.

preganglionic fiber (pre″gang-gle-on′ik fi′ber) Autonomic nerve fiber located on the proximal side of a ganglion.

pregnancy (preg′nan-se) The condition in which a female has a developing offspring in her uterus.

prenatal (pre-na′tal) Before birth.

presbycusis (pres″bĭ-ku′sis) Loss of hearing that accompanies old age.

presbyopia (pres″be-o′pe-ah) Condition in which the eye loses its ability to accommodate due to loss of elasticity in the lens; farsightedness of age.

pressoreceptor (pres″o-re-sep′tor) A receptor that is sensitive to changes in pressure.

primary reproductive organs (pri′ma-re re″pro-duk′tiv or′ganz) Sex-cell-producing parts; testes in males and ovaries in females.

prime mover (prim mōōv′er) Muscle that is mainly responsible for a particular body movement.

profibrinolysin (pro″fi-bri-no-li′sin) The inactive form of fibrinolysin.

progesterone (pro-jes′tĕ-rōn) A female hormone secreted by the corpus luteum of the ovary and by the placenta.

projection (pro-jek′shun)   Process by which the brain causes a sensation to seem to come from the region of the body being stimulated.

prolactin (pro-lak′tin)   Hormone secreted by the anterior pituitary gland that stimulates the production of milk from the mammary glands.

pronation (pro-na′shun)   Movement in which the palm of the hand is moved downward or backward.

prophase (pro′fāz)   Stage of mitosis during which chromosomes become visible.

proprioceptor (pro″pre-o-sep′tor)   A sensory nerve ending that is sensitive to changes in tension of a muscle or tendon.

prostaglandins (pros″tah-glan′dins)   A group of compounds that have powerful, hormonelike effects.

prostate gland (pros′tāt gland)   Gland located around the male urethra below the urinary bladder that adds its secretion to seminal fluid during ejaculation.

protein (pro′te-in)   Nitrogen-containing organic compound composed of amino acid molecules joined together.

prothrombin (pro-throm′bin)   Plasma protein that functions in the formation of blood clots.

proton (pro′ton)   A positively charged particle found in an atomic nucleus.

protraction (pro-trak′shun)   A forward movement of a body part.

proximal (prok′sĭ-mal)   Closer to the midline or origin; opposite of distal.

puberty (pu′ber-te)   Stage of development in which the reproductive organs become functional.

pulmonary (pul′mo-ner″e)   Pertaining to the lungs.

pulmonary circuit (pul′mo-ner″e ser′kit)   System of blood vessels that carries blood between the heart and lungs.

pulse (puls)   The surge of blood felt through the walls of arteries due to the contraction of the ventricles of the heart.

pupil (pu′pil)   Opening in the iris through which light enters the eye.

Purkinje fibers (pur-kin′je fi′berz)   Specialized muscle fibers that conduct the cardiac impulse from the A-V bundle into the ventricular walls.

pyloric sphincter muscle (pi-lor′ik sfingk′ter mus′l)   Sphincter muscle located between the stomach and the duodenum; pylorus.

pyramidal cell (pĭ-ram′ĭ-dal sel)   A large, pyramid-shaped neuron found within the cerebral cortex.

pyridoxine (pir″ĭ-dok′sēn)   A vitamin of the B-complex group; vitamin B_6.

pyrogen (pi′ro-jen)   A substance that causes an increase in body temperature.

pyruvic acid (pi-roo′vik as′id)   An intermediate product of carbohydrate oxidation.

r

radiation (ra″de-a′shun)   A form of energy that includes visible light, ultraviolet light, and X rays; means by which body heat is lost in the form of infrared rays.

rate-limiting enzyme (rāt lim′ĭ-ting en′zim)   An enzyme, usually present in limited amount, that controls the rate of a metabolic pathway by regulating one of its steps.

receptor (re″sep′tor)   Part located at the distal end of a sensory dendrite that is sensitive to stimulation.

recessive gene (re-ses′iv jēn)   An allele of a gene pair that is not expressed while the other allele is expressed.

recruitment (re-kroot′ment)   Increase in number of motor units activated as intensity of stimulation increases.

rectum (rek′tum)   The terminal end of the digestive tube between the sigmoid colon and the anus.

red marrow (red mar′o)   Blood-cell-forming tissue located in spaces within bones.

red muscle (red mus′el)   Slow-contracting postural muscles that contain an abundance of myoglobin.

referred pain (re-ferd′ pān)   Pain that feels as if it is originating from a part other than the site being stimulated.

reflex (re′fleks)   A rapid, automatic response to a stimulus.

reflex arc (re′fleks ark)   A nerve pathway, consisting of a sensory neuron, interneuron, and motor neuron, that forms the structural and functional bases for a reflex.

refraction (re-frak′shun)   A bending of light as it passes from one medium into another medium with a different density.

refractory period (re-frak′to-re pe′re-od)   Time period following stimulation during which a neuron or muscle fiber will not respond to a stimulus.

relaxin (re-lak′sin)   Hormone from corpus luteum that inhibits uterine contractions during pregnancy.

releasing hormone (re-le′sing hor′mōn)   A substance secreted by the hypothalamus whose target cells are in the anterior pituitary gland.

renal (re′nal)   Pertaining to the kidney.

renal corpuscle (re′nal kor′pusl)   Part of a nephron that consists of a glomerulus and a Bowman's capsule; Malpighian corpuscle.

renal cortex (re′nal kor′teks)   The outer portion of a kidney.

renal medulla (re′nal mĕ-dul′ah)   The inner portion of a kidney.

renal pelvis (re′nal pel′vis)   The hollow cavity within a kidney.

renal tubule (re′nal tu′būl)   Portion of a nephron that extends from the renal corpuscle to the collecting duct.

renin (re′nin)   Enzyme released from the kidneys that triggers a mechanism leading to a rise in blood pressure.

reproduction (re″pro-duk′shun)   The process by which an offspring is formed.

resorption (re-sorp′shun)   The process by which something is lost as a result of physiological activity.

respiration (res″pi-ra′shun)   Cellular process by which energy is released from nutrients; breathing.

respiratory center (re-spi′rah-to″re sen′ter)   Portion of the brain stem that controls the depth and rate of breathing.

respiratory membrane (re-spi′rah-to″re mem′brān)   Membrane composed of a capillary and an alveolar wall through which gases are exchanged between the blood and air.

response (re-spons′)   The action resulting from a stimulus.

resting potential (res′ting po-ten′shal)   The difference in electrical charge between the inside and outside of an undisturbed nerve cell membrane.

reticular formation (rĕ-tik′u-lar fōr-ma′shun)   A complex network of nerve fibers within the brain stem that functions in arousing the cerebrum.

reticulocyte (rĕ-tik′u-lo-sit)   A young red blood cell that has a network of fibrils in its cytoplasm.

reticuloendothelial tissue (rĕ-tik″u-lo-en″do-the′le-al tish′u)   Tissue composed of widely scattered phagocytic cells.

retina (ret′i-nah)   Inner layer of the eye wall that contains the visual receptors.

retinal (ret′i-nal)   A form of vitamin A; retinene.

retinene (ret′ĭ-nēn)   Substance used in the production of rhodopsin.

retraction (rĕ-trak′shun)   Movement of a part toward the back.

retroperitoneal (ret″ro-per″ĭ-to-ne′al)   Located behind the peritoneum.

rhodopsin (ro-dop′sin)   Light-sensitive substance that occurs in the rods of the retina; visual purple.

rhythmicity area (rith-mis′ĭ-te a′re-ah)   A portion of the respiratory control center located in the medulla.

riboflavin (ri″bo-fla′vin)   A vitamin of the B-complex group; vitamin B_2.

ribonucleic acid (ri″bo-nu-kle′ik as′id)   A nucleic acid that contains ribose sugar; RNA.

ribose (ri′bōs)   A five-carbon sugar found in RNA molecules.

ribosome (ri′bo-sōm)   Cytoplasmic organelle that consists largely of RNA and functions in the synthesis of proteins.

RNA   Ribonucleic acid.

rod (rod)   A type of light receptor that is responsible for colorless vision.

rotation (ro-ta′shun)   Movement by which a body part is turned on its longitudinal axis.

rouleaux (roo-lo′)   Groups of red blood cells stuck together like stacks of coins.

round window (rownd win′do)   A membrane-covered opening between the inner ear and the middle ear.

s

saccule (sak′ūl) A saclike cavity that makes up part of the membranous labyrinth of the inner ear.

sagittal (saj′i-tal) A plane or section that divides a structure into right and left portions.

salivary gland (sal′i-ver-e gland) A gland, associated with the mouth, that secretes saliva.

salt (sawlt) A compound produced by a reaction between an acid and a base.

saltatory conduction (sal′tah-tor-e kon-duk′shun) A type of nerve impulse conduction in which the impulse seems to jump from one node to the next.

S-A node (nōd) Sinoatrial node.

sarcolemma (sar′′ko-lem′ah) The cell membrane of a muscle fiber.

sarcomere (sar′ko-mēr) The structural and functional unit of a myofibril.

sarcoplasm (sar′ko-plazm) The cytoplasm within a muscle fiber.

sarcoplasmic reticulum (sar′′ko-plaz′mik rē-tik′u-lum) Membranous network of channels and tubules within a muscle fiber, corresponding to the endoplasmic reticulum of other cells.

saturated fatty acid (sat′u-rāt′ed fat′e as′id) Fatty acid molecule that lacks double bonds between the atoms of its carbon chain.

Schwann cell (shwahn sel) Cell that surrounds a fiber of a peripheral nerve and forms the neurolemmal sheath and myelin.

sclera (skle′rah) White fibrous outer layer of the eyeball.

scoliosis (sko′′le-o′sis) Abnormal lateral curvature of the vertebral column.

scrotum (skro′tum) A pouch of skin that encloses the testes.

sebaceous gland (se-ba′shus gland) Gland of the skin that secretes sebum.

sebum (se′bum) Oily secretion of the sebaceous glands.

secretin (se-kre′tin) Hormone secreted from the small intestine that stimulates the release of pancreatic juice from the pancreas.

semicircular canal (sem′′ī-ser′ku-lar kah-nal′) Tubular structure within the inner ear that contains the receptors responsible for the sense of dynamic equilibrium.

seminal fluid (sem′ī-nal floo′id) Fluid discharged from the male reproductive tract at ejaculation that contains sperm cells and the secretions of various glands; semen.

seminiferous tubule (sem′′ī-nif′er-us tu′būl) Tubule within the testes in which sperm cells are formed.

semipermeable (sem′′ī-per′me-ah-bl) Condition in which a membrane is permeable to some molecules and not to others; selectively permeable.

senescence (sē-nes′ens) The process of growing old.

sensation (sen-sa′shun) A feeling resulting from the interpretation of sensory nerve impulses by the brain.

sensory area (sen′so-re a′re-ah) A portion of the cerebral cortex that receives and interprets sensory nerve impulses.

sensory nerve (sen′so-re nerv) A nerve composed of sensory nerve fibers.

sensory neuron (sen′so-re nu′ron) A neuron that transmits an impulse from a receptor to the central nervous system.

serotonin (se′′ro-to′nin) A vasoconstricting substance that is released by blood platelets when blood vessels are broken and thus helps to control bleeding.

serous cell (se′rus sel) A glandular cell that secretes a watery fluid with a high enzyme content.

serous fluid (se′rus floo′id) The secretion of a serous membrane.

serous membrane (ser′us mem′brān) Membrane that lines a cavity without an opening to the outside of the body.

serum (se′rum) The fluid portion of coagulated blood.

sesamoid bone (ses′ah-moid bōn) A round bone that may occur in tendons adjacent to joints.

sex chromosome (seks kro′mo-sōm) A chromosome responsible for the development of characteristics associated with maleness or femaleness; an X or Y chromosome.

sex-linked trait (seks-linkt′ trāt) Trait determined by a recessive gene located on an X chromosome.

sigmoid colon (sig′moid ko′lon) S-shaped portion of the large intestine between the descending colon and the rectum.

simple sugar (sim′pl shoog′ar) A monosaccharide.

sinoatrial node (si′′no-a′tre-al nōd) Group of specialized tissue in the wall of the right atrium that initiates cardiac cycles; the pacemaker; S-A node.

sinus (si′nus) A cavity or hollow space in a bone or other body part.

skeletal muscle (skel′ē-tal mus′l) Type of muscle tissue found in muscles attached to skeletal parts.

smooth muscle (smooth mus′el) Type of muscle tissue found in the walls of hollow visceral organs; visceral muscle.

sodium pump (so′de-um pump) The active transport mechanism that functions to concentrate sodium ions on the outside of a cell membrane.

solute (sol′ūt) The substance that is dissolved in a solution.

solvent (sol′vent) The liquid portion of a solution in which a solute is dissolved.

somatic cell (so-mat′ik sel) Any cell of the body other than the sex cells.

somatotropin (so′′mah-to-tro′pin) Growth hormone.

spastic paralysis (spas′tik pah-ral′i-sis) A form of paralysis characterized by an increase in muscular tone without atrophy of the muscles involved.

special sense (spesh′al sens) Sense that involves receptors associated with specialized sensory organs, such as the eyes and ears.

spermatid (sper′mah-tid) An intermediate stage in the formation of sperm cells.

spermatocyte (sper-mat′o-sit) An early stage in the formation of sperm cells.

spermatogenesis (sper′′mah-to-jen′ē-sis) The production of sperm cells.

spermatogonium (sper′′mah-to-go′ne-um) Undifferentiated spermatogenic cell found in the germinal epithelium of a seminiferous tubule.

spermatozoa (sper′′mah-to-zo′ah) Male reproductive cells; sperm cells.

sphincter (sfingk′ter) A circular muscle that functions to close an opening or the lumen of a tubular structure.

sphygmomanometer (sfig′mo-mah-nom′ē-ter) Instrument used for measuring blood pressure.

spinal (spi′nal) Pertaining to the spinal cord or to the vertebral canal.

spinal cord (spi′nal kord) Portion of the central nervous system extending downward from the brain stem through the vertebral canal.

spinal nerve (spi′nal nerv) Nerve that arises from the spinal cord.

spleen (splēn) A large, glandular organ located in the upper left region of the abdomen.

spongy bone (spunj′e bōn) Bone that consists of bars and plates separated by irregular spaces; cancellous bone.

squamous (skwa′mus) Flat or platelike.

starch (starch) A polysaccharide that is common in foods of plant origin.

static equilibrium (stat′ik e′′kwī-lib′re-um) The maintenance of balance when the head and body are motionless.

sterility (stē-ril′i-te) An inability to produce offspring.

steroid (ste′roid) An organic substance whose molecules include complex rings of carbon and hydrogen atoms.

stimulus (stim′u-lus) A change in the environmental conditions that is followed by a response by an organism or cell.

stomach (stum′ak) Digestive organ located between the esophagus and the small intestine.

strabismus (strah-biz′mus) A condition characterized by lack of visual coordination; crossed eyes.

stratified (strat′i-fid) Arranged in layers.

stratum corneum (stra′tum kor′ne-um) Outer horny layer of the epidermis.

stratum germinativum (stra′tum jer′mi-na′′tiv-um) The deepest layer of the epidermis in which the cells undergo mitosis.

stress (stres) Condition produced by factors causing potentially life-threatening changes in the body's internal environment.

stressor (stres′or) A factor capable of stimulating a stress response.

stroke volume (strōk vol′ūm) The amount of blood discharged from the ventricle with each heartbeat.

structural formula (struk′cher-al fōr′mu-lah) A representation of the way atoms are united within a molecule, using the symbols for each element present and lines to indicate chemical bonds.

subarachnoid space (sub″ah-rak′noid spās) The space within the meninges between the arachnoid mater and the pia mater.

subcutaneous (sub″ku-ta′ne-us) Beneath the skin.

sublingual (sub-ling′gwal) Beneath the tongue.

submaxillary (sub-mak′si-ler″e) Below the maxilla.

submucosa (sub″mu-ko′sah) Layer of connective tissue that underlies a mucous membrane.

substrate (sub′strāt) The substance upon which an enzyme acts.

sucrase (su′krās) Digestive enzyme that acts upon sucrose.

sucrose (soo′krōs) A disaccharide; table sugar.

sulcus (sul′kus) A shallow groove, such as that between adjacent convolutions on the surface of the brain.

summation (sum-ma′shun) Phenomena in which the amount of change in membrane potential is directly related to the intensity of stimulation.

superficial (soo″per-fish′al) Near the surface.

superior (su-pe′re-or) Pertaining to a structure that is higher than another structure.

supination (soo″pi-na′shun) Rotation of the forearm so that the palm faces upward when the arm is outstretched.

suppressor cell (su-pres′or sel) A special type of T-lymphocyte that functions to interfere with the production of antibodies responsible for allergic reactions.

surface tension (ser′fas ten′shun) Force that tends to hold moist membranes together due to an attraction that water molecules have for one another.

surfactant (ser-fak′tant) Substance produced by the lungs that reduces the surface tension within the alveoli.

suture (soo′cher) An immovable joint, such as that between adjacent flat bones of the skull.

sympathetic nervous system (sim″pah-thet′ik ner′vus sis′tem) Portion of the autonomic nervous system that arises from the thoracic and lumbar regions of the spinal cord.

symphysis (sim′fi-sis) A slightly movable joint between bones separated by a pad of fibrocartilage.

synapse (sin′aps) The junction between the axon end of one neuron and the dendrite or cell body of another neuron.

synaptic knob (si-nap′tik nob) Tiny enlargement at the end of an axon that secretes a neurotransmitter substance.

synchondrosis (sin″kon-dro′sis) Type of joint in which bones are united by bands of hyaline cartilage.

syncytium (sin-sish′e-um) A mass of merging cells.

syndesmosis (sin″des-mo′sis) Type of joint in which the bones are united by relatively long fibers of connective tissue.

syndrome (sin′drōm) A group of symptoms that together characterize a disease condition.

synergist (sin′er-jist) A muscle that assists the action of a prime mover.

synovial fluid (si-no′ve-al floo′id) Fluid secreted by the synovial membrane.

synovial joint (si-no′ve-al joint) A freely movable joint.

synovial membrane (si-no′ve-al mem′ brān) Membrane that forms the inner lining of the capsule of a freely movable joint.

synthesis (sin′thě-sis) The process by which substances are united to form more complex substances.

system (sis′tem) A group of organs that act together to carry on a specialized function.

systemic circuit (sis-tem′ik ser′kit) The vessels that conduct blood between the heart and all body tissues except the lungs.

systole (sis′to-le) Phase of cardiac cycle during which a heart chamber wall is contracted.

systolic pressure (sis-tol′ik presh′ur) Arterial blood pressure during the systolic phase of the cardiac cycle.

t

tachycardia (tak″e-kar′de-ah) An abnormally rapid heartbeat.

target tissue (tar′get tish′u) Specific tissue on which a hormone acts.

tarsus (tar′sus) The bones that form the ankle.

taste bud (tāst bud) Organ containing the receptors associated with the sense of taste.

telophase (tel′o-fāz) Stage in mitosis during which daughter cells become separate structures.

tendon (ten′don) A cordlike or bandlike mass of white fibrous connective tissue that connects a muscle to a bone.

testis (tes′tis) Primary reproductive organ of a male; a sperm-cell-producing organ.

testosterone (tes-tos′tě-rōn) Male sex hormone secreted by the interstitial cells of the testes.

tetanus (tet′ah-nus) A continuous, forceful muscular contraction.

thalamus (thal′ah-mus) A mass of gray matter located at the base of the cerebrum in the wall of the third ventricle.

thermoreceptor (ther″mo-re-sep′tor) A sensory receptor that is sensitive to changes in temperature; a heat receptor.

thiamine (thi′ah-min) Vitamin B_1.

thoracic (tho-ras′ik) Pertaining to the chest.

threshold potential (thresh′old po-ten′shal) Level of potential at which an action potential or nerve impulse is produced.

threshold stimulus (thresh′old stim′u-lus) The level of stimulation that must be exceeded to elicit a nerve impulse or a muscle contraction.

thrombin (throm′bin) Blood-clotting enzyme that causes fibrinogen to become fibrin.

thrombocyte (throm′bo-sīt) A blood platelet.

thrombocytopenia (throm″bo-si″to-pe′ne-ah) A low number of platelets in the circulating blood.

thrombus (throm′bus) A blood clot in a blood vessel that remains at its site of formation.

thymosin (thi′mo-sin) A group of peptides secreted by the thymus gland that affects the production of certain types of white blood cells.

thymus (thi′mus) A two-lobed glandular organ located in the mediastinum behind the sternum and between the lungs.

thyroglobulin (thi″ro-glob′u-lin) Substance secreted by cells of the thyroid gland that serves to store thyroid hormones.

thyroid gland (thi′roid gland) Endocrine gland located just below the larynx and in front of the trachea that secretes thyroid hormones.

thyrotropin (thi″ro-trōp′in) Hormone secreted by the anterior pituitary gland that stimulates the thyroid gland to secrete hormones; TSH.

thyroxine (thi-rok′sin) A hormone secreted by the thyroid gland.

tidal volume (tīd′al vol′ūm) Amount of air that enters the lungs during a normal, quiet inspiration.

tissue (tish′u) A group of similar cells that performs a specialized function.

T-lymphocyte (lim′fo-sit) Lymphocytes that interact directly with antigen-bearing particles and are responsible for cell-mediated immunity.

trabecula (trah-bek′u-lah) Branching bony plate that separates irregular spaces within spongy bone.

trachea (tra′ke-ah) Tubular organ that leads from the larynx to the bronchi.

transcellular fluid (trans″sel′u-lar floo′id) A portion of the extracellular fluid, including the fluid within special body cavities.

transferrin (trans-fer′rin) Blood plasma protein that functions to transport iron.

transfer RNA (trans′fer) Molecule of RNA that carries an amino acid to a ribosome in the process of protein synthesis.

transverse colon (trans-vers′ ko′lon) Portion of the large intestine that extends across the abdomen from right to left below the stomach.

transverse tubule (trans-vers′ tu′būl) Membranous channel that extends inward from a muscle fiber membrane and passes through the fiber.

tricuspid valve (tri-kus′pid valv) Heart valve located between the right atrium and the right ventricle.

triglyceride (tri-glis′er-id) A lipid composed of three fatty acids combined with a glycerol molecule.

triiodothyronine (tri″i-o″do-thi′ro-nēn) One of the thyroid hormones.

trisomy (tri′so-me) Condition in which a cell contains three chromosomes of a particular type instead of two.

trochanter (tro-kan′ter) A broad process on a bone.

trochlea (trok′le-ah) A pulley-shaped structure.

trophoblast (trof′o-blast) The outer cells of a blastocyst that help to form the placenta and other embryonic membranes.

tropic hormone (trōp′ik hor′mōn) A hormone that has an endocrine gland as its target tissue.

tropomyosin (tro″po-mi′o-sin) Protein that functions in blocking muscle contraction until calcium ions are present.

troponin (tro′po-nin) Protein that functions with tropomyosin to block muscle contraction until calcium ions are present.

trypsin (trip′sin) An enzyme in pancreatic juice that acts to break down protein molecules.

trypsinogen (trip-sin′o-jen) Substance secreted by pancreatic cells that becomes trypsin as a result of enzymatic action.

tubercle (tu′ber-kl) A small, rounded process on a bone.

tuberosity (tu″bĕ-ros′i-te) An elevation or protuberance on a bone.

twitch (twich) A brief muscular contraction followed by relaxation.

tympanic membrane (tim-pan′ik mem′brān) A thin membrane that covers the auditory canal and separates the external ear from the middle ear; the eardrum.

u

umbilical cord (um-bil′i-kal kord) Cordlike structure that connects the fetus to the placenta.

umbilical region (um-bil′i-kal re′jun) The central portion of the abdomen.

umbilicus (um-bil′i-kus) Region to which the umbilical cord was attached; the navel.

unipolar neuron (u″ni-po′lar nu′ron) A neuron that has a single nerve fiber extending from its cell body.

unsaturated fatty acid (un-sat′u-rāt″ed fat′e as′id) Fatty acid molecule that has one or more double bonds between the atoms of its carbon chain.

urea (u-re′ah) A non-protein nitrogenous substance produced as a result of protein metabolism.

ureter (u-re′ter) A muscular tube that carries urine from the kidney to the urinary bladder.

urethra (u-re′thrah) Tube leading from the urinary bladder to the outside of the body.

uterine (u′ter-in) Pertaining to the uterus.

uterine tube (u′ter-in tūb) Tube that extends from the uterus on each side toward an ovary and functions to transport sex cells; fallopian tube or oviduct.

uterus (u′ter-us) Hollow muscular organ located within the female pelvis in which a fetus develops.

utricle (u′tri-kl) An enlarged portion of the membranous labyrinth of the inner ear.

uvula (u′vu-lah) A fleshy portion of the soft palate that hangs down above the root of the tongue.

v

vaccine (vak′sēn) A substance that contains antigens and is used to stimulate the production of antibodies.

vacuole (vak′u-ōl) A space or cavity within the cytoplasm of a cell.

vagina (vah-ji′nah) Tubular organ that leads from the uterus to the vestibule of the female reproductive tract.

Valsalva's maneuver (val-sal′vahz mah-noo′ver) Increasing the intrathoracic pressure by forcing air from the lungs against a closed glottis.

varicose veins (var′i-kos vānz) Abnormally swollen and enlarged veins, especially in the legs.

vasa recta (va′sah rek′tah) A branch of the peritubular capillary that receives blood from the efferent arterioles of juxtamedullary nephrons.

vascular (vas′ku-lar) Pertaining to blood vessels.

vas deferens (vas def′er-ens) Tube that leads from the epididymis to the urethra of the male reproductive tract (plural, vasa deferentia).

vasoconstriction (vas″o-kon-strik′shun) A decrease in the diameter of a blood vessel.

vasodilation (vas″o-di-la′shun) An increase in the diameter of a blood vessel.

vasopressin (vas″o-pres′in) Antidiuretic hormone.

vein (vān) A vessel that carries blood toward the heart.

vena cava (vēn′ah kāv′ah) One of two large veins that convey deoxygenated blood to the right atrium of the heart.

ventral root (ven′tral rōōt) Motor branch of a spinal nerve by which it is attached to the spinal cord.

ventricle (ven′tri-kl) A cavity, such as those of the brain that are filled with cerebrospinal fluid, or those of the heart that contain blood.

venule (ven′ūl) A vessel that carries blood from capillaries to a vein.

vermiform appendix (ver′mi-form ah-pen′diks) Appendix.

vesicle (ves′i-kal) Membranous, cytoplasmic sac formed by an action of the cell membrane.

villus (vil′us) Tiny, fingerlike projection that extends outward from the inner lining of the small intestine.

visceral (vis′er-al) Pertaining to the contents of a body cavity.

visceral peritoneum (vis′er-al per″i-to-ne′um) Membrane that covers the surfaces of organs within the abdominal cavity.

visceral pleura (vis′er-al ploo′rah) Membrane that covers the surfaces of the lungs.

viscosity (vis-kos′i-te) The tendency for a fluid to resist flowing due to the internal friction of its molecules.

vital capacity (vi′tal kah-pas′i-te) The maximum amount of air a person can exhale after taking the deepest breath possible.

vitamin (vi′tah-min) An organic substance other than a carbohydrate, lipid, or protein that is needed for normal metabolism but cannot be synthesized in adequate amounts by the body.

vitreous humor (vit′re-us hu′mor) The substance that occupies the space between the lens and retina of the eye.

vocal cords (vo′kal kordz) Folds of tissue within the larynx that create vocal sounds when they vibrate.

Volkmann's canal (fōlk′mahnz kah-nal′) A transverse channel that interconnects haversian canals within compact bone.

volt (vōlt) Unit used to measure differences in electrical potential.

voluntary (vol′un-tār″e) Capable of being consciously controlled.

vulva (vul′vah) The external reproductive parts of the female that surround the opening of the vagina.

w

water balance (wot′er bal′ans) Condition in which the quantity of water entering the body is equal to the quantity leaving it.

water intoxication (wot′er in-tok″si-ka′shun) Condition in which the extracellular body fluids become abnormally diluted due to the presence of excessive water.

water of metabolism (wot′er uv mĕ-tab′o-lizm) Water produced as a by-product of metabolic processes.

wave summation (wāv sum-ma′shun) A sustained muscle contraction that occurs when a series of twitches fuse together; summation of twitches.

white muscle (whīt mus′el) Fast-contracting skeletal muscle.

y

yellow marrow (yel′o mar′o) Fat storage tissue found in the cavities within certain bones.

z

zygote (zi′gōt) Cell produced by the fusion of an egg and sperm; a fertilized egg cell.

zymogen granule (zi-mo′jen gran′ūl) A cellular structure that stores inactive forms of protein-splitting enzymes in a pancreatic cell.

Credits

18.37a: © Edwin A. Reschke, **b:** © Victor B. Eichler, BIO-Art; **18.40(all):** © American Heart Association, Dallas, Texas.

Chapter 19
19.3: © Edwin A. Reschke; **19.5:** © Courtesy of Eastman Kodak; **19.9:** © Igaku Shoin, LTD; **19.12:** © John D. Cunningham; **19.15:** © Courtesy of Memorial Sloan-Kettering Cancer Center.

Chapter 20
20.2: © Victor B. Eichler, BIO-Art; **20.7:** from *Tissues and Organs: A Text-Atlas of Scanning Electron Microscopy*, R. G. Kessel and R. H. Kardon. © 1979 W. H. Freeman and Company; **20.9:** © Biophoto Associates/Photo Researchers, Inc.; **20.14:** © Visuals Unlimited/David M. Phillips; **20.26:** © Per H. Kjeldson, University of Michigan, Ann Arbor.

Chapter 21
No photos

Chapter 22
22.4: From *Tissues and Organs: A Text-Atlas of Scanning Electron Microscopy*, R. G. Kessel and R. H. Kardon. © 1979 W. H. Freeman and Company; **22.5:** © Biophoto Associates/Photo Researchers, Inc.; **22.7:** © David M. Phillips/The Population Council 1986/Taurus Photos, Inc.; **22.8:** © Edwin A. Reschke; **22.9:** From *Tissues and Organs: A Text-Atlas of Scanning Electron Microscopy*, R. G. Kessel and R. H. Kardon. © 1979 W. H. Freeman and Company; **22.18:** © Manfred Kage/Peter Arnold, Inc.; **22.19:** © Edwin A. Reschke/Peter Arnold, Inc.; **22.20:** © Hafez, E.S.E. and Blandaw, R. J. *The Mammalian Oviduct*, 1969, University of Chicago Press; **22.23:** © Dr. Don Fawcett/Penelope Gaddum-Rosse/Photo Researchers, Inc.; **22.26:** © P. Bagavandoss/Photo Researchers, Inc.; **22.30:** © Mia Tegner, M. J. and Epel, D: *Science* 179 (February 16, 1973), 685–88. A.A.A.S.

Chapter 23
23.5: © Roman O'Rahilly/Carnegie Laboratories of Embryology, University of California; **23.6:** © Carnegie Institute of Washington, Dept. of Embryology, Davis Division; **23.7:** © Donald Yeager/Camera M.D. Studios; **23.11:** © Landrum Shettles; **23.14:** © Donald Yeager/Camera M.D. Studios; **23.20:** © Martin M. Rother/Taurus Photos, Inc.; **23.21–23.23b:** © Donald Yeager/Camera M.D. Studios; **23.27–23.34:** © Donna Jernigan.

Chapter 24
24.2: © Runk/Schoenberger/Grant Heilman, Inc.; **24.3:** From Wisniewski, L. P., and Hirschhorn, K. "A Guide to Human Chromosome Defects," 2d ed., 1980. White Plains, N.Y. The March of Dimes Birth Defect Foundation. BD:OAS(6). **24.4:** © Bill Longcore/Photo Researchers, Inc.; **24.5:** © Donna Jernigan; **24.18–24.20:** From Wisniewski, L. P., and Hirschhorn, K. "A Guide to Human Chromosome Defects," 2d ed., 1980. White Plains, N.Y.: The March of Dimes Birth Defect Foundation. BD:OAS(6); **24.21:** © Gregory Dellore, M.D., and Steven L. Clark, M.D.

LINE ART

Chapter 1
1.4, 1.6, 1.9, 1.10, 1.11, 1.12, 1.13, 1.15, and 1.17: From Van De Graaff, Kent M. and Stuart Ira Fox, *Concepts of Human Anatomy and Physiology*. © 1986 Wm. C. Brown Publishers, Dubuque, Iowa. All Rights Reserved. Reprinted by permission.

Chapter 2
2.27: From Fox, Stuart Ira, *Human Physiology*. © 1984 Wm. C. Brown Publishers, Dubuque, Iowa. All Rights Reserved. Reprinted by permission.

Chapter 3
3.7b and 3.8b: From Mader, Sylvia S., *Inquiry into Life* 4th ed. © 1976, 1979, 1982, 1985 Wm. C. Brown Publishers, Dubuque, Iowa. All Rights Reserved. Reprinted by permission. **3.7c:** From Mader, Sylvia S., *Inquiry into Life*. © 1976 Wm. C. Brown Publishers, Dubuque, Iowa. All Rights Reserved. Reprinted by permission. **3.10b:** From Mader, Sylvia S., *Inquiry into Life* 3d ed. © 1976, 1979, 1982 Wm. C. Brown Publishers, Dubuque, Iowa. All Rights Reserved. Reprinted by permission. **3.22:** From Fox, *Human Physiology*. © 1984 Wm. C. Brown Publishers, Dubuque, Iowa. All Rights Reserved. Reprinted by permission.

Chapter 4
4.5: From Volpe E. Peter, *Biology and Human Concerns* 3d ed. © 1975, 1979, 1983 Wm. C. Brown Publishers, Dubuque, Iowa. All Rights Reserved. Reprinted by permission. **4.26 and 4.27:** From Fox, Stuart Ira, *Human Physiology*. © 1984 Wm. C. Brown Publishers, Dubuque, Iowa. All Rights Reserved. Reprinted by permission.

Chapter 5
5.1a and b, 5.8, and 5.9: From Van De Graaff, Kent M., *Human Anatomy*. © 1984 Wm. C. Brown Publishers, Dubuque, Iowa. All Rights Reserved. Reprinted by permission.

Chapter 6
None required

7.14, 7.16, 7.18, 7.19, 7.20, 7.21, 7.24, 7.27, 7.29, 7.33, 7.37, and 7.53: From Van De Graaff, Kent M., *Human Anatomy*. © 1984 Wm. C. Brown Publishers, Dubuque, Iowa. All Rights Reserved. Reprinted by permission. **7.32:** From Van De Graaff, Kent M. and Stuart Ira Fox, *Concepts of Human Anatomy and Physiology*. © 1986 Wm. C. Brown Publishers, Dubuque, Iowa. All Rights Reserved. Reprinted by permission.

Chapter 8
8.5 and 8.7: From Van De Graaff, Kent M. and Stuart Ira Fox, *Concepts of Human Anatomy and Physiology*. © 1986 Wm. C. Brown Publishers, Dubuque, Iowa. All Rights Reserved. Reprinted by permission. **8.8a, c, d, e, and f:** From Van De Graaff, Kent M., *Human Anatomy*. © 1984 Wm. C. Brown Publishers, Dubuque, Iowa. All Rights Reserved. Reprinted by permission.

Chapter 9
9.19, 9.28, and 9.36: From Van De Graaff, Kent M. and Stuart Ira Fox, *Concepts of Human Anatomy and Physiology*. © 1984 Wm. C. Brown Publishers, Dubuque, Iowa. All Rights Reserved. Reprinted by permission. **9.33, 9.37, and 9.38:** From Van De Graaff, Kent M., *Human Anatomy*. © 1984 Wm. C. Brown Publishers, Dubuque, Iowa. All Rights Reserved. Reprinted by permission.

Chapter 10
10.1: From Van De Graaff, Kent M. and Stuart Ira Fox, *Concepts of Human Anatomy and Physiology*. © 1986 Wm. C. Brown Publishers, Dubuque, Iowa. All Rights Reserved. Reprinted by permission. **10.21:** From Fox, Stuart Ira, *Human Physiology*. © 1984 Wm. C. Brown Publishers, Dubuque, Iowa. All Rights Reserved. Reprinted by permission. **10.25:** From Mader, Sylvia S., *Inquiry into Life* 4th ed. © 1976, 1979, 1982, 1985 Wm. C. Brown Publishers, Dubuque, Iowa. All Rights Reserved. Reprinted by permission.

Chapter 11
11.2a, 11.4a, and 11.17: From Van De Graaff, Kent M., *Human Anatomy Laboratory Textbook* 2d ed. © 1981, 1984 Wm. C. Brown Publishers, Dubuque, Iowa. All Rights Reserved. Reprinted by permission. **11.2b:** From Benson, Harold J., et al., *Anatomy and Physiology Laboratory Textbook, Complete Version*, 3d ed. © 1970, 1976, 1983 Wm. C. Brown Publishers, Dubuque, Iowa. All Rights Reserved. Reprinted by permission. **11.10 and 11.25:** From Mader, Sylvia S., *Inquiry into Life* 4th ed. © 1976, 1979, 1982, 1985 Wm. C. Brown Publishers, Dubuque, Iowa. All Rights Reserved. Reprinted by permission. **11.11, 11.22, 11.26, 11.27, 11.28, 11.30, and 11.31:** From Van De Graaff, Kent M., *Human Anatomy*. © 1984 Wm. C. Brown Publishers, Dubuque, Iowa. All Rights Reserved. Reprinted by permission. **11.13:** From Peele, T. L., *Neuroanatomical Basis for Clinical Neurology* 2d ed. © 1961 McGraw-Hill Book Company, New York. Reprinted by permission. **11.14:** From Van De Graaff, Kent M., and Stuart Ira Fox, *Concepts of Human Anatomy and Physiology*. © 1986 Wm. C. Brown Publishers, Dubuque, Iowa. All Rights Reserved. Reprinted by permission. **11.33, 11.35, and 11.36:** From Van De Graaff, Kent M., and Stuart Ira Fox, *Concepts of Human Anatomy and Physiology*. © 1986 Wm. C. Brown Publishers, Dubuque, Iowa. All Rights Reserved. Reprinted by permission.

Chapter 12
12.1, 12.16, and 12.17(left): From Fox, Stuart Ira, *Human Physiology*. © 1984 Wm. C. Brown Publishers, Dubuque, Iowa. All Rights Reserved. Reprinted by permission. **12.4, 12.14b, 12.29, 12.30, and 12.31:** From Van De Graaff, Kent M., and Stuart Ira Fox, *Concepts of Human Anatomy and Physiology*. © 1986 Wm. C. Brown Publishers, Dubuque, Iowa. All Rights Reserved. Reprinted by permission.

Chapter 13
13.13 and 13.31: From Van De Graaff, Kent M., and Stuart Ira Fox, *Concepts of Human Anatomy and Physiology*. © 1986 Wm. C. Brown Publishers, Dubuque, Iowa. All Rights Reserved. Reprinted by permission. **13.26:** From Van De Graaff, Kent M., *Human Anatomy*. © 1984 Wm. C. Brown Publishers, Dubuque, Iowa. All Rights Reserved. Reprinted by permission.

Chapter 14
14.1 and 14.23: From Van De Graaff, Kent M. *Human Anatomy*. © 1984 Wm. C. Brown Publishers, Dubuque, Iowa. All Rights Reserved. Reprinted by permission. **14.2, 14.27, 14.42, and 14.44:** From Van De Graaff, Kent M., and Stuart Ira Fox, *Concepts of Human Anatomy and Physiology*. © 1986 Wm. C. Brown Publishers, Dubuque, Iowa. All Rights Reserved. Reprinted by permission. **14.20:** From Enger, Eldon D., et al., *Concepts in Biology* 4th ed. © 1976, 1979, 1982, 1985 Wm. C. Brown Publishers, Dubuque, Iowa. All Rights Reserved. **14.34:** From Mader, Sylvia S., *Inquiry into Life* 4th ed. © 1976, 1979, 1982, 1985 Wm. C. Brown Publishers, Dubuque, Iowa. All Rights Reserved. Reprinted by permission.

Chapter 15
15.8: Reprinted with permission of Macmillan Publishing Company from *The Human Body*, 4th ed. by Sigmund Grollman. Copyright © 1978 by Sigmund Grollman.

Chapter 16
16.1, 16.2, and 16.12: From Van De Graaff, Kent M., *Human Anatomy*. © 1984 Wm. C. Brown Publishers, Dubuque, Iowa. All Rights Reserved. Reprinted by permission. **16.11:** From Van De Graaff, Kent M. and Stuart Ira Fox, *Concepts of Human Anatomy and Physiology*. © 1986 Wm. C. Brown Publishers, Dubuque, Iowa. All Rights Reserved. Reprinted by permission. **16.29:** From Volpe, E. Peter, *Biology and Human Concerns* 3d ed. © 1975, 1979, 1983 Wm. C. Brown Publishers, Dubuque, Iowa. All Rights Reserved. Reprinted by permission. **16.31:** From *Laboratory Studies in Biology: Observations and Their Implications* by C. A. Lawson, R. W. Lewis, M. A. Burmester and G. Hardin. Copyright © 1955 by W. H. Freeman and Company. **16.33:** From Fox, Stuart Ira, *Human Physiology*. © 1984 Wm. C. Brown Publishers, Dubuque, Iowa. All Rights Reserved. Reprinted by permission.

Chapter 17
17.6: From Van De Graaff, Kent M., and Stuart Ira Fox, *Concepts of Human Anatomy and Physiology*. © 1986 Wm. C. Brown Publishers, Dubuque, Iowa. All Rights Reserved. Reprinted by permission.

Chapter 18
18.3, 18.33, and 18.58: From Van De Graaff, Kent M. and Stuart Ira Fox, *Concepts of Human Anatomy and Physiology*. © 1986 Wm. C. Brown Publishers, Dubuque, Iowa. All Rights Reserved. Reprinted by permission. **18.4, 18.6, 18.10, 18.11, and 18.30:** From Van De Graaff, Kent M., *Human Anatomy*. © 1984 Wm. C. Brown Publishers, Dubuque, Iowa. All Rights Reserved. Reprinted by permission. **18.13:** Adapted from *Functional Human Anatomy*, James E. Crouch, 2d ed., 1972. Lea & Febiger. **18.19:** From Benson, Harold J., et al., *Anatomy and Physiology Laboratory Textbook, Complete Version* 3d ed. © 1970, 1976, 1983 Wm. C. Brown Publishers, Dubuque, Iowa. All Rights Reserved. Reprinted by permission. **18.21, 18.49, and 18.65:** From Fox, Stuart Ira, *Human Physiology*. © 1984 Wm. C. Brown Publishers, Dubuque, Iowa. All Rights Reserved. Reprinted by permission. **18.36:** From Mader, Sylvia S., *Inquiry into Life* 4th ed. © 1976, 1979, 1982, 1985 Wm. C. Brown Publishers, Dubuque, Iowa. All Rights Reserved. Reprinted by permission. **18.38:** Courtesy American Heart Association. **18.42:** From Fox, Stuart Ira, *Human Physiology Laboratory Guide* 3d ed. © 1976, 1980, 1984 Wm. C. Brown Publishers, Dubuque, Iowa. All Rights Reserved. Reprinted by permission.

Index

sphenoidal, 203, 205
venous, 431, 724
sinus headache, 576
sinusitis, 606
sinus nerve, 379
sinus rhythm, 705
skeletal muscle, 148, 149, 151, 279
contraction of, 268–73
fast, 273
fatigued, 272
interaction of, 280
and lymph flow, 720
relaxation of, 269
slow, 273
structure of, 262–68
superficial, 281, 282
use and disuse of, 277
and venous flow, 684
skeletal muscle fiber, 262–67
skeletal system, 19, 881
skeleton
appendicular, 197–98
axial, 197
organization of, 195–98
sexual differences in, 226
skeleton of heart, 653–54
skin, 158, 725
accessory organs of, 163–67
aging of, 161
cancer of, 162
color of, 169, 890, 892–93, 898–900
healing of, 170
sunlight and, 141, 162
skull, 197
age and, 240
bones of, 195, 198–209
of infant, 209, 210, 240, 241
sleep, 371
slightly movable joint, 240–42, 246
slit pore, 751
slow skeletal muscle, 273
small intestine, 36, 37, 38, 492, 518–25
absorption in, 522–24
movement of, 524–25
secretions of, 520–22
wall of, 519–20
small saphenous vein, 703, 704, 705
smell, 362, 380, 408–10
smooth endoplasmic reticulum, 75
smooth muscle, 149, 151, 276–78, 279
contraction of, 277–78
multiunit, 276
origin and insertion, 279–80
visceral, 276–77
smooth muscle fiber, 276–77
sneezing, 590, 591
sodium, 558–60
sodium atom, 47
sodium bicarbonate, 792, 793
sodium dihydrogen phosphate, 792, 793
sodium ions, 327–29, 472, 683, 757, 761, 780, 781, 788, 789
imbalances of, 790
reabsorption of, 766
sodium monohydrogen phosphate, 792, 793
sodium pump, 761
soft palate, 495, 496, 497, 574
soleus, 281, 300–2
solution, 52–53
hypertonic, 85
hypotonic, 85
isotonic, 85
somatic motor fiber, 392
somatic nervous system, 373
somatic senses, 403–8
somatostatin. See GIH
somatotrope, 458, 459
somatotropin. See GH
sour receptor, 411–12
sour taste, 411
spasm, blood vessel, 630, 632
spastic paralysis, 356
special somatic afferent fiber, 339
special visceral afferent fiber, 339
special visceral efferent fiber, 339
species resistance, 725, 727
speech, 360, 362, 363, 379, 380, 576–77, 591
spermatic artery, 689
spermatic cord, 36, 37, 808
spermatid, 811, 812, 894
spermatogenesis, 812, 893–98
spermatogenic cell, 811
spermatogonia, 811, 812
spermatozoa. See Sperm cell

sperm cell, 811–13, 893–98
maturation of, 814
transport of, 834–35
spermicide, 844, 845
sphenoidal sinus, 203, 205
sphenoid bone, 199, 201, 202–3, 576
spheroidal joint, 244
spherocytosis, 641
sphincter
anal, 294, 526, 527
precapillary, 670, 672
pyloric, 505, 506
sphincter muscle, 282
sphincter of Oddi, 516
sphincter urethrae, 295
sphygmomanometer, 678
spina bifida, 358, 908
spinal cord, 352–56
injury to, 356
spinal nerves, 339, 352, 353, 373, 381–85
injury to, 386
spinal reflex, 356
spinal shock, 356
spindle fibers, 92, 894, 896
spine, 198
spinocerebellar tract, 354
spinothalamic tract, 353, 354
spinous process, 211, 212
spiral fracture, 190
spiral lamina, 414
spirometer, 590
splanchnic nerve, 388
spleen, 37, 38, 39, 513, 724–25
splenectomy, 738
splenic artery, 688, 689, 702, 724
splenic flexure, 525, 526
splenic vein, 701, 702, 724
splenitis, 738
splenius capitis, 285–87
splenomegaly, 738
splenotomy, 738
split heart sound, 659
spondylolisthesis, 214
spondylolysis, 211
spongy bone, 183–85
spontaneous abortion, 837
spontaneous fracture, 190, 194
sprain, 255
squamosal suture, 199, 201
staircase effect, 274, 275
stapedius, 413–14, 415
stapes, 413, 415, 418, 420
starch, 55, 538, 539
Starling's law of the heart, 681
starvation, 542, 564
static equilibrium, 421–23
stem cell, 727–28
Stensen's duct, 500
stereopsis, 438
stereoscopic vision, 438, 439
sterility. See Infertility
sterilization, 844, 846
sternal notch, 580
sternal puncture, 218
sternal region, 27
sternocleidomastoid, 34, 35, 36, 281, 282, 283, 285–87, 586, 587
sternum, 36, 197, 216, 218, 587, 648
steroid hormones, 448–49, 450–51
stinger, 385
stirrup. See Stapes
stomach, 36, 37, 38, 492, 503, 504–10, 513, 518, 519
stomachache, 509
stomatitis, 529
stone, kidney, 768
storage disease, 88
strabismus, 427
stratified squamous epithelium, 136, 137
stratum basale. See Stratum germinativum
stratum corneum, 158–61
stratum germinativum, 158–61
stratum granulosum, 159–61
stratum spinosum, 159–61
stress, 478–81, 880
stressor, 478
stretch receptor, 407, 595, 596, 770
stretch reflex, 407
striated muscle, 148, 149, 264, 265
stroke, 693
stroke volume, 678–81
strong acid, 791
strong base, 791
strontium-90, 194
structural formula, 50

structural protein, 542
styloid process, 201, 203, 222
subacromial bursa, 250
subarachnoid space, 350, 351, 352, 366, 367
subcapsular bursa, 250
subclavian artery, 36, 38, 656, 688, 689, 690, 691, 693, 698
subclavian trunk, 716, 717
subclavian vein, 37, 699, 700, 704, 718
subcoracoid bursa, 250
subcutaneous fascia, 262
subcutaneous injection, 163
subcutaneous layer, 158, 162–63
subdeltoid bursa, 250
sublingual gland, 492, 499–501
subluxation, 256
submandibular gland, 492, 499–501
submaxillary gland. See Submandibular gland
submucosa, 492, 493
submucous coat of urinary bladder, 770
subscapularis, 35, 288, 289, 290
subserous fascia, 262
substance P, 334
substrate for enzyme, 106
sucrase, 106, 520, 521
sucrose, 539
sugars, 55, 56
sulcus, 358
sulfate ions, 757, 780, 781, 789
sulfur, 559, 560
sulfuric acid, 791, 794, 795
summation, 328
summation of twitches, 274
super female, 905
superficial, relative position, 23
superficial circumflex iliac artery, 696, 697
superficial partial-thickness burn, 170
superficial pudendal artery, 696, 697
superficial temporal artery, 691, 693, 698
superficial temporal vein, 704
superficial transverse perinei, 295
superior, relative position, 22
superior articulating process, 211, 212
superior concha, 575
superior constrictor muscle, 502
superior epigastric artery, 696
superior gluteal artery, 695
superior hemiazygos vein, 700, 701
superior laryngeal nerve, 379
superior meatus, 574
superior mesenteric artery, 39, 688, 689, 690, 698
superior mesenteric vein, 39, 701, 702
superior nasal conchae, 203
superior oblique, 377, 426, 427
superior ophthalmic vein, 699
superior orbital fissure, 202, 209
superior peduncle, 372, 373
superior rectus, 376, 426, 427
superior semicircular canal, 423
superior thyroid artery, 690, 691
superior vena cava, 38, 40, 651–52, 655–57, 697, 699, 700, 704, 873
superior vesical artery, 695, 875
supination, 247, 248, 290
supinator, 290, 291
supplemental air, 589
support, 19, 189
supporting cell, 811
suppression amblyopia, 427
suppressor cell, 737
suppressor factor, 729
supraorbital foramen, 198, 199, 209
supraorbital notch, 198
suprapatellar bursa, 243, 253, 254
suprarenal artery, 688, 690, 698
suprarenal gland. See Adrenal gland
suprarenal vein, 702, 749
supraspinatus, 286, 288, 289
supratrochlear lymph nodes, 722, 723
surface anatomy
of back, 310
of head, 310
of lower limb, 312, 313
of neck, 310
of thorax, 310
of torso, 311, 312
of upper limb, 310–12
surface tension, 587–88
surface-volume relationship of cells, 95, 96
surfactant, 586, 870, 874
surgical methods of sterilization, 844, 846

surgical neck of humerus, 221
suspensory ligament, 428, 429, 430, 822, 827, 842
sutural bone, 195
sutural ligament, 240, 241
suture, 195, 198, 201, 240, 241, 246
swallowing, 378–79, 380, 502–3, 577–78
sweat, 784–85, 788
sweat gland, 162, 166–67, 448
sweep circuit, 330
sweet receptor, 411–12
sweet taste, 411
swelling, 726
sympathetic division, 385, 386–89, 494
sympathetic fiber, 392
sympathetic tone, 390
sympathetic trunk, 386, 387
symphysis, 241, 242, 246
symphysis pubis, 36, 40, 225, 226, 241, 809, 821
synapse, 332–35
synapsis, 895
synaptic cleft, 267, 332, 333, 334, 335
synaptic knob, 333–34, 335
synaptic potential, 335
synaptic transmission, 332–33, 336
synaptic vesicle, 267, 333–34, 335
synarthrosis. See Immovable joint
synchondrosis, 240–41, 246
syndesmosis, 240, 246
synergist (muscle), 280
synovial cavity, 243
synovial fluid, 158, 243, 454, 780
synovial membrane, 158, 243, 251
synovitis, 254
synthesis, dehydration, 104–5
synthesis reaction, 50–51
systemic circuit, 654, 688
systemic lupus erythematosus, 739
systolic pressure, 675, 678, 680

t

tachycardia, 666, 667
tachypnea, 606
taenia coli, 525
tail, 864
tail of sperm cell, 812, 813
talus, 230
target cell, 448
tarsal gland, 424
tarsal region, 27
tarsals, 182, 198, 230, 231
tarsus, 230
taste, 362, 380, 408, 410–12
taste buds, 497
taste cell, 410–11
taste hairs, 410–11
taste nerve pathways, 412
taste pore, 410–11
taste receptor, 378, 410–11
taste sensations, 411–12
Tay-Sachs disease, 88, 907, 908
T-cell, 727–28, 731–34
tear gland, 380
tears, 425, 726
tectorial membrane, 416, 417
teeth, 240, 241, 492, 497–99, 499
decay of, 499
telencephalon, 357, 359
telophase, 93, 94, 894, 896
temperature, body, 10–11, 541
of infant, 369
maintenance of, 545
ovulation and, 845
regulation of, 167–68
temperature sense, 404
temporal artery, 677
temporal bone, 199–203
temporalis, 282, 283, 284
temporal lobe, 358, 359, 360, 363
temporal nerve, 378
temporal process, 205
temporomandibular joint syndrome, 285
tendon, 143, 262
teniae coli, 526
tennis elbow, 255
tension headache, 405
tensor fasciae latae, 35, 281, 297, 298
tensor tympani, 413, 415
tentorium cerebelli, 350, 358, 360
teratogen, 869
teratology, 880
teres major, 35, 282, 286, 288, 289
teres minor, 282, 286, 288–90
terminal bronchiole, 580–82

vestibule, 414–15, 495, 496, 829, 830
vestibulocochlear nerve, 376, 378, 380, 413, 419, 421
viral hepatitis, 514
virilism, 481
virus, 726
viscera, 14
    abdominal, 36
    thoracic, 37
visceral pain, 405–7
visceral pericardium, 16, 17, 648–50
visceral peritoneum, 16, 18, 492, 513, 519
visceral pleura, 16, 17, 584, 586
visceral smooth muscle, 276–77
viscosity, blood, 680
vision, 360, 362, 363, 380, 424–40; 550
    stereoscopic, 438, 439
visual accessory organs, 424–28
visual acuity, 434
visual cortex, 439, 440
    injury to, 440
visual nerve pathways, 438–40
visual pigments, 437–39
visual receptor, 434–38
vital capacity, 589
vital signs, 10
vitamin A, 187, 434, 437, 512, 517, 550–52
vitamin B₁. See Thiamine
vitamin B₂. See Riboflavin
vitamin B₆, 554–55, 557

vitamin B₁₂, 507, 512, 528, 619, 620. See also Cyanocobalamin
vitamin C, 187, 556, 557
vitamin D, 162, 187, 468–69, 512, 517, 551, 552, 558, 749
vitamin E, 517, 551–52
vitamin K, 517, 528, 552, 635
vitamins, 106
    fat-soluble, 550–52
    recommended daily dietary
        allowances for, 563
    water-soluble, 553–57
vitreous humor, 428, 431, 432, 780
vocal cords, 576–78
Volkmann's canal, 184, 185
volts, 328
volume of cell, 95
voluntary muscle, 148, 361
vomer bone, 199, 207, 208
vomiting, 510
vomiting center, 510
von Willebrand's disease, 641, 903
vulva, 828–29

w

wandering cell, 140
wart, 172
water, 10–11, 49, 50, 54, 55, 780, 781
    reabsorption of, 524, 760–61

water balance, 783–87
    disorders of, 786–87
water intake, 783–84
water intoxication, 786, 790
water of metabolism, 783
water output, 784–85
water-soluble vitamins, 553–57
wave summation, 274
weak acid, 791
weak base, 791
Weber test, 420
weight. See Body weight
Wharton's duct, 500
whiplash, 386
white blood cell counts, 624
white blood cells, 76, 77, 87–88, 147, 193, 480, 614, 615, 726
    development of, 622
    functions of, 624–27
    types of, 623–24
white fibers, 141
white matter, 353
white muscle, 273
white pulp, 724
white ramus, 383, 386
windpipe. See Trachea
wisdom teeth, 498
withdrawal reflex, 341
wormian bone, 195
wound healing, 170
wrist, 292–93, 864

x

X chromosome, 900–4
xiphoid process, 36, 218
X rays, 51, 126

y

yawning, 591
Y chromosome, 900–4
yellow fibers, 141
yellow marrow, 193
yolk sac, 193, 860, 867–69

z

zein, 541
zinc, 562
Z line, 264, 265
zona fasciculata, 470, 473
zona glomerulosa, 470, 472
zona pellucida, 824, 825, 835
zona reticularis, 470, 474
zygomatic arch, 205
zygomatic bone, 199, 203, 205, 208
zygomatic nerve, 378
zygomatic process, 201, 203
zygomaticus, 281, 283
zygote, 823, 836–37
zymogen granule, 511